REVIEWS IN MINERALOGY
AND GEOCHEMISTRY

Volume 60   2006

NEW VIEWS
OF THE MOON

EDITORS:

Bradley L. Jolliff   *Washington University*
St. Louis, Missouri

Mark A. Wieczorek   *Insitut de Physique de Globe de Paris*
Saint Maur, France

Charles K. Shearer   *University of New Mexico*
Albuquerque, New Mexico

Clive R. Neal   *University of Notre Dame*
Notre Dame, Indiana

FRONT COVER: The painting on the front cover depicts the Moon low on the horizon over a town in northern New Mexico, with the Moon colored in the way we might see it if our eyes were sensitive to gamma rays instead of just visible light. The scene is a reminder of the Moon's companionship with Earth and suggests a close relationship to the uniquely habitable environment of our home planet. Front cover art by *Jane Maclean.*

*Series Editor: **Jodi J. Rosso***

MINERALOGICAL SOCIᴱᵀᵧ ᴏᶠ ᴀMERICA
GEOCHEMICAl

D1158805

REVIEWS IN MINERALOGY AND GEOCHEMISTRY

(Formerly: REVIEWS IN MINERALOGY)

ISSN 1529-6466

Volume 60

New Views of the Moon

ISBN 0-939950-72-3

Additional copies of this volume as well as others in this series may be obtained at moderate cost from:

THE MINERALOGICAL SOCIETY OF AMERICA
3635 CONCORDE PARKWAY, SUITE 500
CHANTILLY, VIRGINIA, 20151-1125, U.S.A.
WWW.MINSOCAM.ORG

DEDICATION

This book is dedicated to the memory of the late Graham Ryder, who was one of the original four members of CAPTEM when the New Views of the Moon scientific initiative was first conceived. Graham, who was a perennial member of CAPTEM, owing to his broad knowledge of the lunar samples and his residence at the Lunar and Planetary Institute where CAPTEM met, was a logical choice to be one of the co-leaders of this lunar-science initiative. Not only was he an expert on every detail of the lunar samples, he also had the geological insights and knowledge to place the samples into context. He was also a stickler for the truth and for accuracy in presentation and interpretation of data. Graham's scientific skepticism, his forthright communication style, and his use of humor and wit endeared him to many in the planetary science community. For these reasons and many more, we were shocked and saddened when it became known in late 2001 that Graham was gravely ill with esophageal cancer. Even while he was very ill, he struggled to continue to work toward the goal of using the lunar samples and their geologic context to understand diverse lunar data sets. Graham died from complications associated with his cancer on January 5th, 2002. In Graham, we had a tremendous resource of knowledge about the lunar samples, and we had a dear friend and valued colleague. We hope that the synthesis of information presented in this book would find at least partial favor under Graham's keen scrutiny.

NEW VIEWS *of the* MOON

60   *Reviews in Mineralogy and Geochemistry*   **60**

FROM THE SERIES EDITOR

Unlike the Moon, this volume only took a little more than 4 years in the making, but like the Moon, it has gone through numerous changes. However, through it all, the editors of this volume have expertly coordinated the numerous authors and extensive subject matter into a comprehensive review in which the reader is now easily re-introduced to a "forgotten friend," the Moon. Their hard work and unwaivering dedication has made this volume a reality! My thanks goes out to them!

Supplemental material and errata (if any) can be found at the MSA website *www.minsocam.org*.

Jodi J. Rosso, Series Editor
West Richland, Washington
May 2006

PREFACE

Bradley L. Jolliff
Washington University

The onset of the 21st century is proving to be an exciting time for planetary exploration. A high level of activity associated with the exploration of Mars and a resurgence of activity associated with the anticipated return to the Moon have captured the imagination of much of the planetary science community and the public, alike. The present level of involvement and enthusiasm for studies of the Moon are high, perhaps as high as they have been since shortly after the end of the U.S. *Apollo* and Soviet *Luna* programs. The very successful orbital missions of the 1990s, *Clementine* and *Lunar Prospector*, have provided key mineralogical, geochemical, and geophysical data sets that will take many years to fully process and digest. These new data sets must be integrated carefully with information gained from three preceding decades of study of lunar samples and older, less complete, remotely sensed data sets. Although there have been no new lunar sample-return missions since *Apollo* and *Luna*, new samples are available in the form of meteorites, recognized to be pieces of the Moon delivered to Earth by impact and gravitational forces. These, too, play a role in improved knowledge of the Moon and in helping to couple information obtained by remote sensing with information obtained from rock and soil samples. As we stand on the edge of a new era of lunar and planetary exploration, including new missions to the Moon, Mars, and other planets and moons, we find it essential to examine in depth how the wide variety of data sets obtained during the course of lunar exploration can be used together to better understand the formation of the Moon and how it evolved to its present state. Such an understanding holds important lessons for the exploration of other planets in the Solar System and will ultimately lead to better knowledge about how our own planet Earth—with its unique environment suitable for the origin and evolution of life—originated and changed with time.

Early in the post-Apollo lunar science program, paradigms emerged for the general structure, differentiation, and thermal history of the Moon. These paradigms, which include broad acceptance of some form of an early magma ocean, a layered and magmatically

1529-6466/06/0060-0000$05.00   DOI: 10.2138/rmg.2006.60.0

differentiated mafic mantle and feldspathic crust, and a relatively well-sequenced volcanic and impact history anchored by radiometric age dates, have been detailed in numerous previous publications, notably *Basaltic Volcanism on the Terrestrial Planets* (BVSP 1981), *Planetary Science, a Lunar Perspective* (Taylor 1982), *The Geologic History of the Moon* (Wilhelms 1987), *The Lunar Sourcebook* (Heiken et al. 1991), and *Planetary Materials* (Papike 1998). These publications focused on different aspects of largely Apollo-era information and contributed a strong base toward the current paradigm of lunar evolution. Most recently, a number of contributions to *Origins of the Earth and Moon* (Canup and Righter et al. 2000) summarized key knowledge about the relationship of the Earth and Moon, and the current leading hypothesis for the origin of the Moon by a giant impact into early Earth.

Even after a decade of intense investigation that began with six landed Apollo missions and three Soviet robotic sample-return missions, and continued with vigorous investigation of lunar samples and analysis of Apollo orbital data in light of sample information, lunar scientists recognized that key data sets important to further exploration and understanding of the Moon remained to be obtained. In fact, several times during the ensuing decade of the 1980s, needs for additional data and ideas about how the information gained might impact our understanding of the Moon were documented. Such documents include *Geoscience and a Lunar Base* (Taylor and Spudis 1988), *Contributions of a Lunar Geoscience Observer (LGO) Mission to Fundamental Questions in Lunar Science* (LGO Science Workshop Members 1986), and *A Planetary Science Strategy for the Moon* (LExSWG 1992). However, no new missions to the Moon or even to orbit the Moon would follow the Apollo and Luna missions until 1990, when the *Galileo* space probe encountered the Moon during gravity-assist maneuvers and again in 1992 enroute to Jupiter. *Clementine*, a joint DoD-NASA mission, designed to investigate both the Moon and an asteroid by remote sensing, flew in 1994. Clementine spent four months in mapping orbit around the Moon and obtained multispectral, gravity, laser-ranging, and radar data (Nozette et al. 1994). Then, in 1998, *Lunar Prospector*, a NASA Discovery-class mission with the Moon as its *primary* target, gathered long-sought orbital data sets including gamma-ray, magnetic, gravity, and neutron data (Binder 1998).

In large measure, data sets that had been identified as key to improved understanding of the Moon have now been obtained. In light of the wealth of data thus obtained, the time is at hand to revisit the post-Apollo lunar paradigms. Some of the hypotheses and models that were advanced during the 1970s and 1980s will no doubt persevere, and the new data will simply add to the level of detail of our understanding. Other concepts, however, will undergo significant changes, and a few may have to be completely revised. The purpose of this book is to assess the current state of knowledge of lunar geoscience given the new data sets provided by missions of the 1990s and the continuing investigations of samples (including lunar meteorites), to assess the remaining key questions that will guide future lunar exploration. It is also our intent to document how a planet or moon other than the world on which we live can be studied and understood using integrated suites of specific kinds of information. The Moon is the only body other than Earth for which we have material samples of known geologic context for study and for which in-situ geophysical experiments have been done. We seek in this book to show how the different kinds of information gained about the Moon are related to each other and thus to aid in planning for the exploration of other worlds.

New Views of the Moon lunar science initiative

In the fall of 1997, with the *Lunar Prospector* mission gearing up to launch, NASA's Curation and Analysis Planning Team for Extraterrestrial Materials (CAPTEM), began planning a lunar science initiative as an integrative project. It was evident then that a successful *Lunar Prospector* mission, coupled with the results of the mapping phase of the *Clementine* mission, could yield new insights through global data sets of the Moon. CAPTEM and others realized that the maximum return on these data required synthesis and integration with other data sets, including knowledge gained from the lunar samples. A primary purpose of this lunar science initiative was to foster interdisciplinary science approaches to address problems of the origin and evolution of the Moon, and to pursue, in the most efficient manner possible, the new ideas emerging from examination of the new global data. We also sought to optimize the usefulness of existing data through this integrative approach. One logical outcome of this effort would be to place the lunar science and exploration community in a better position to contribute to planning future missions to the Moon. Focused workshops and special sessions at national meetings were held to promote interaction and continuity of effort, and to culminate with this publication showing how integration creates a better understanding of the Moon. The lessons learned from lunar exploration, although not a single, integrated and coordinated program, could then be used in consideration of future exploration strategies for other planets, including Mars.

A vast amount of information has been obtained from the study of rocks and soils from the *Apollo* and *Luna* sample collections (1969–1976) and from study of several dozen (and counting) lunar meteorites collected on Earth (1981 to present). From sample studies, many constraints have been established for the age, early differentiation, crust and mantle structure, and contemporaneous and subsequent impact modification of the Moon. These sample studies continue with the analysis of more samples and subsamples, and with the application of new analytical techniques and methodologies, which produce new kinds of data and inferences. Observations made by Apollo astronauts and geophysical experiments performed on the surface provided important data and constraints that must be kept in mind. Remote sensing from lunar orbit and Earth-based telescopic studies have provided valuable additional data that constrain the composition and geology of the Moon's surface and internal structure, but prior to the 1990's, these sources did not produce full, global data sets.

In 1990 and 1992, the *Galileo* spacecraft encountered the Earth-Moon system and provided multispectral views of the Moon and a new perspective, which added significantly to the body of knowledge (e.g., Pieters et al. 1993). In 1994, the *Clementine* mission provided the first global or near-global data sets for lunar gravity, topography, and multispectral imaging. Scientific results from the *Clementine* data exceeded expectations, and refinements to those data sets continue to provide a useful base for further investigations (Eliason et al. 1999; Robinson et al. 1999; Gaddis et al. 2000; Lucey et al. 2000a,b). Results from the *Lunar Prospector* mission exceeded expectations in terms of the compositional, gravity, and magnetic-field data sets (Feldman et al. 2000; Elphic et al. 2000; Hood and Zuber 2000; Lawrence et al. 1999, 2000; Maurice et al. 2000; and many workshop and Lunar and Planetary Science Conference abstracts).

Have we really "been there and done that?"

Some may consider the Moon to be a relatively simple geologic body. However, the *Lunar Prospector* and *Clementine* missions, as well as recent sample-based geochemical, isotopic, experimental, and theoretical studies, have provided important evidence of the real geological complexity of the Moon, and have shown us that we still do not yet adequately understand its early geologic history. The *Galileo, Clementine,* and *Lunar Prospector* missions provided global information viewed through new kinds of windows—and a changed context for models of lunar origin, evolution, and resources, with the impetus for new questions and new hypotheses. A simple example of the usefulness of combined studies is that of the distribution of KREEP (an acronym for lunar material rich in potassium, rare-earth elements, and phosphorus) and mare basalts on the Moon. The remotely sensed data alone would probably combine KREEP and mare basalts as a single entity, because they occupy to a large extent the same areas. There is at least a rough correlation of Th and Fe in the remote global chemical data. However, sample studies show the critical distinction of mare basalts and KREEP in chemistry, origin, and age. However, the remotely sensed data can then utilize this distinction to thoroughly map the distributions of these two basic lunar rock types, and to map them as distinct from the more feldspathic highlands, providing important constraints on the development of the lunar crust and the melting of the mantle. The possible detection and characterization of water-ice at the poles, and the high-resolution gravity modeling enabled by the global missions, are other examples of the kinds of new results that must be integrated with the extant body of knowledge from sample studies, *in situ* experiments, and prior remote-sensing missions. The data have greatly benefited from the Lucey et al. (1995) method of deriving iron contents from the multispectral data, but this derivation required a contextual understanding of the collected lunar samples.

Numerous research groups continue to capitalize on the *Clementine* and *Lunar Prospector* data sets and to integrate the data with results from studies of rocks and soils collected during the *Apollo* missions (e.g., Keller et al. 1999; Pieters et al. 2000; Taylor et al. 2001), from studies of the geology of the landing sites (e.g., Blewett et al. 1997; Jolliff 1999; Lucey et al. 1998, 2000a; Robinson and Jolliff 2002), and from studies of the results of *Apollo* geophysical experiments (e.g., Neumann et al. 1996; Wieczorek and Phillips 1998, 1999; Hood and Zuber 2000; Khan et al. 2000; Khan and Mosegaard 2002). Research groups that focus primarily on the lunar samples recognize the importance and utility of remotely sensed data and the need to couple our under-standing of the samples with the information contained in these new data sets. Examples include the method of extracting compositional information from multispectral data developed by Lu-cey and coworkers, and the quantification of the method using samples and landing-site data; the coupling of information on different soil grain-size fractions to remote spectral characteristics by the Lunar Soil Consortium (e.g., Keller et al. 1999; Pieters et al. 2000; Taylor et al. 2001); and other collaborative efforts to tie sample and landing-site information to *Clementine* and *Lunar Prospector* compositional information. Extending our understanding of lunar geology and resources from the landing-site scale to a global scale provides the foundation for new para-digms of the Moon's geologic evolution as well as a foundation for studies that pave the way to future resource utilization, on-surface experiments, and manned lunar-outpost missions.

Why the emphasis on the integration of diverse data sets?

Data sets obtained by the 1994 Clementine mission have reached a state of maturity. Gravity data derived from the Clementine mission have fueled a new set of studies and interpretations regarding the internal structure of the Moon (e.g., Zuber et al. 1994; Neumann et al. 1996; Wieczorek and Phillips 1998; Konopliv et al. 1998). Established and tested calibration procedures have been developed and are in use for the UVVIS and NIR data. The *Clementine* multispectral data are providing the means to do local and regional studies of lunar surface materials and major mineralogy almost anywhere on the Moon, as well as global studies of surface composition. *Lunar Prospector* neutron and gamma-ray data (Binder et al. 1998; Feldman et al. 1998; Lawrence et al. 1998, 2003) provide a synergistic cross correlation with the *Clementine* data for information that can be determined using results from both missions, such as FeO concentration of surface soils (e.g., Munoz et al. 1998). The high spatial resolution of the *Clementine* multispectral data and the extended set of major-element compositions determined from the *Lunar Prospector* gamma-ray data will each continue to enhance the value of the other data set. Such cross-correlated data enable direct tests of specific hypotheses regarding the global elemental distribution that are important to interpretations of lunar crustal genesis and present-day distribution of materials. For example, using *Clementine* data, Lucey et al. (1995, 1998) developed procedures for estimating Fe and Ti concentrations; these have also been extracted from the *Lunar Prospector* gamma-ray data, and both data sets can be cross correlated to data taken directly on soils at the Apollo and Luna landing sites (Gillis et al. 2003). As the spatial resolution of the gamma-ray experiment is low (~50 km, Feldman et al. 1996), *Clementine* data are crucial to provide a link enabling this three-way cross correlation.

The Moon is the keystone to our understanding of the silicate bodies of the Solar System and a stepping stone to exploration of the Solar System. The Moon is the only other object for which we have samples of known spatial context. Studies of the Moon and the lunar samples established the concept of primary differentiation of planetary crusts, the relative time scale based on cratering statistics, and a record of early solar system exogenic processes (LGO 1986). Understanding the Moon's origin is key to understanding the Earth's early history, and the Moon, because of its relatively simple silicate differentiation and lack of surface chemical weathering, serves as a baseline to study more complex planetary processes. NASA is now actively involved in programs to study other solar system objects using remote sensing and, eventually, sample returns. With the Moon, we have the opportunity *and the obligation* not only to understand its origin and history, but to determine the best ways to integrate diverse types of data for the maximum scientific return. What kinds of samples and sampling strategies will be the most useful for interpretation of remotely sensed data? What laboratory measurements can be made to further our understanding of the remotely sensed data? What are the common questions that the different disciplines seek to answer and how can we best work together toward that end?

Approach and methodology: the role of integration in fundamental problems of lunar geoscience

The 1986 Report *Contributions of a Lunar Geoscience Observer (LGO) Mission to Fundamental Questions in Lunar Science* laid out in detail many of the important questions remaining in lunar geoscience that could be addressed by a mission involving global multispectral imaging, X-ray and gamma-ray mapping, radar altimetry and Doppler tracking, and magnetometer and electron reflectometer experiments. These questions remain just as relevant today as they were then. The fundamental problems as listed in the LGO report are not repeated here, but to provide an idea of the scope of what is covered in this book, we summarize in Table 1 and below some of the overarching questions for which multidisciplinary approaches can potentially make significant advances.

Table 1. Key current questions of lunar science.

What elemental and molecular species make up the lunar exosphere and how has it changed with time?
What was the initial thermal state of the Moon?
What was the cause of global-scale asymmetry?
What are the characteristics of the lunar core (size, composition), and did the Moon ever support a dynamo-driven magnetic field?
Was there a significant late veneer of accretion (post-core formation/early differentiation)?

Is there an undifferentiated lower mantle (limited or no involvement in magma-ocean melting); if so, what was its role in lunar magmatism?
   • Did the volcanic glasses come from a deep, garnet-bearing region beneath the cumulate mantle?
   • What volatiles are (were) present in the deep lunar interior and what was their role in magmatic processes and eruptive styles?
   • What was the depth of lunar magma-ocean differentiation?

What were the sources and magnitude of heating to drive secondary magmatism?
   • How was heat transferred from Th, U, K-rich crustal reservoirs to the mantle?
   • What was the role of large-scale crustal insulation?
   • How are the different suites of plutonic rocks related to specific or localized geologic terranes and to the global geochemical asymmetry?

What and where are the most concentrated, extensive, and readily extractable deposits of H and ^{3}He?
   • What is the origin and mineralogical or physical form, thickness, and concentration of H or H_2O ice deposits in permanently shadowed craters at the poles?
   • Is H at the poles a viable resource?
   • Where are the best sites for such facilities located? (H, ^{3}He, protection from radiation, communications, transportation)

1) ***What is the vertical and lateral structure of the lunar crust and how did the crust evolve?*** Geophysical crustal thickness models based on seismic, gravity, and topography data currently assume a single or dual layered crust (e.g., Wieczorek and Phillips 1998). Multispectral studies of central peaks of craters (e.g., Tompkins and Pieters 1999) and basin uplift structures may be able to constrain or improve upon these simple models. By carefully modeling of the composition of ejecta deposits from basins, it is possible to infer the structure beneath the largest basins. Apollo samples provide real constraints for the geophysical models.

2) ***What is the composition and structure of the lunar mantle?*** Seismic velocity models provide geophysical constraints on lunar mantle structure and for changes in mineralogical phases and density. Constraints on composition are provided by petrologic studies of lunar basalts and pyroclastic materials thought to have erupted from different depths within the upper mantle (e.g., BVSP 1981; Neal 2001). Other constraints are derived from consideration of bulk lunar composition and the proportion of the Moon that must have undergone differentiation to produce the observed crust. Thus, improved models of crustal structure contribute to better models for the mantle. Limits on the size of a metallic core from electromagnetic sounding will also help in the evaluation of mantle density models through moment-of-inertia and mean density constraints (LGO 1986).

3) ***What was the extent of a lunar magma ocean?*** The concept of the lunar magma ocean depends almost entirely on an understanding of the composition and structure of the Moon's crust and the bulk composition of the Moon. One of the keys to that understanding is knowing the distribution of plagioclase, corresponding (cogenetic) mafic minerals, and incompatible elements (KREEP). Another key is knowing the extent of variability and whether those materials are related to global differentiation layers or bodies, serial or isolated intrusive rocks, or differentiates of thick basin impact melt. Where do large bodies of anorthosite crop out on the Moon (e.g., Hawke et al. 1992, 2003) and how large must the system have been that produced them? Global, high-resolution multispectral data coupled with global gamma-ray data will help to place limits on how these materials vary within the crust. The Apollo samples are the key to these interpretations because they provide direct knowledge or "ground-truth" of rock types, lithologic associations, detailed chemical compositions, age dates, and depth constraints.

4) ***How is the surface expression of lunar materials related to the Moon's internal structure and evolution? (or Where exactly do the different rock types come from?)*** The major lunar highland rock types are known from the Apollo samples. Little is known, however, about the exact place of origin of the igneous rocks because none were sampled in place. Uplift structures associated with large impact craters and basins may expose crustal igneous rocks, and mineralogical remote sensing of these structures can provide important clues to the lateral distribution of different types of igneous rocks and their pre-impact depths of formation (e.g., Tompkins and Pieters 1998). The composition of regions of megaregolith, determined by remote geochemical analysis, can provide information about the types of rocks exhumed from basin impacts, especially for those formations that can be associated with a specific basin of origin (e.g., Haskin 1998). From these clues, it should be possible to infer the distribution and abundance of important rock types such as anorthosite (assuming large bodies of anorthosite to be of the ferroan variety and original products of a magma ocean), and the magnesian suite of plutonic rocks including norite, troctolite, and gabbro. Geophysical models will provide critical tests of whether the vertical and

lateral distributions of rock types inferred from surface data are consistent with gravity and topography data.

5) *What is the nature of the Moon's asymmetry, what caused it, and what are the implications for the Moon's internal evolution and present-day distribution of materials?* What is the nature of the lunar center-of-mass/center-of-figure offset? Geophysical models suggest either an increase in crustal density for the lunar nearside, or a thickened farside crust (Neumann et al. 1996; Wieczorek and Phillips 1998). From remote compositional analysis of the regolith, it may be possible to infer/confirm lateral density variations, and the presence or absence of ejected mantle material from large basins may help constrain the thickness of the crust far from the Apollo seismic stations. Knowledge of the distribution of surface materials using multispectral and geochemical analysis on a global scale may show a global compositional asymmetry as well. If that can be shown to predate the formation of the major nearside basins, then it may be shown that the early lunar crustal differentiation was heterogeneous on a global scale with important consequences such as the concentration of KREEP residua under the Procellarum region, as suggested by Haskin (1998). If such asymmetry existed early in the Moon's history, subsequent thermal evolution may have been driven in large part by the non-uniform concentration of radioactive elements. Key to answering this question is translating from the Apollo samples of known composition and rock type to the successful identification of rock types and compositions by remote techniques.

6) *What is the origin, evolution, and distribution of mare volcanism?* Mare basalts, although volumetrically minor, formed by partial melting of the lunar mantle and thus record compositions, mineralogy, and processes from as deep as 200 to 400 km (Neal and Taylor 1992; Shearer and Papike 1999). Basalts sampled by the Apollo and Luna missions range in age from 3.9 to 3.1 Ga; however, significantly younger volcanism is indicated by crater densities (Hiesinger et al. 2000, 2003) and volcanism as old as 4.2 Ga is recorded by basalt clasts in impact breccias (Dasch et al. 1987). The sampled basalts and related pyroclastic glasses cover a broad range in composition, for example, from <1 to 16 wt% TiO_2. However, spectroscopic data suggest that perhaps less than half of the mare basalt types on the Moon have been sampled and little is known about the smaller farside maria. Systematic relationships between mare basalt age and chemistry are proving to be more elusive than thought in early studies of Apollo basalts. Volatile-rich pyroclastic eruptions were an important part of lunar mare volcanism (e.g., Delano and Livi 1981; Shearer and Papike 1993); these deposits need to be mapped, dated, and understood in terms of eruptive volume and duration.

7) *What were the timing and effects of the major basin-forming impacts on lunar crustal stratigraphy? What is the nature of the South Pole-Aitken Basin and how did it affect early lunar crustal evolution?* Impact is perhaps the most important process in the assembly and early history of the terrestrial planets, and basin impacts are the most important events in shaping the large-scale features of the Moon's present-day surface. Owing to its relative tectonic simplicity, the Moon preserves a record of its early bombardment. Much work remains to be done to sort out the detailed stratigraphy of the lunar basin deposits. In the absence of direct, in-place samples of impact-melt sheets for geochronology, it will be the task of mineralogical and geochemical remote sensing coupled with photogeology to improve our knowledge of the timing and the effects of basin formation. Understanding when the major basin impacts occurred with respect to the likely timing of the Moon's thermal evolution may provide explanations for the nature of the geophysical (mascon) anomalies, for example, where large basins such

as Imbrium and South Pole-Aitken appear not to have excavated as deeply as might be expected on the basis of their diameters (Wieczorek and Phillips 1998). The implications of the enormous South Pole-Aitken basin for early lunar evolution are currently a topic of much investigation and will continue to be a target of intensive study.

8) **What are the origins of lunar paleomagnetism?** The magnetization of surface materials, if related to a core dynamo, imply the presence of a metallic core and thus have great significance for the early differentiation and thermal evolution of the Moon. Strong localized magnetic anomalies have been detected from orbit, but their origins have not been determined. A leading hypothesis relates to magnetization of materials located in regions antipodal to major and relatively young impact basins (e.g., Lin et al. 1988; Hood and Williams 1989). Experiments involving the magnetometer and electron reflectometer (*Lunar Prospector*: Hood et al. 1999) have yielded a global map of magnetic anomalies, and correlation to global mineralogical and geochemical data sets, coupled with known magnetization properties of the Apollo samples, may go a long way toward resolving the causes of lunar paleomagnetism.

9) **What are the Moon's important resources, where are they concentrated, and how can they be accessed?** The Moon, as a stepping stone, will figure prominently in future space exploration as a place where human beings will learn how to survive on the inhospitable surface of another planet. The Moon has abundant resources of oxygen, hydrogen, and other solar-wind gases trapped in its regolith. And, based on the results of the *Clementine* bi-static radar experiment and the *Lunar Prospector* neutron spectrometer, there may be significant quantities of water-ice in the regolith in permanently shadowed craters at both poles (Feldman et al. 2001). Some soils have high concentrations of iron and titanium, which could be recovered during the processing of regolith for its gases, which are stored in Fe-Ti minerals. Understanding the siting of such resources, from the perspectives of mineralogy, lithology, and regional and local geology, is prerequisite to a human presence on the Moon. Chapter 6 of this book provides ample discussion and linkages between the scientific understanding of the Moon and distribution of materials on its surface, and the eventual utilization of the Moon and its resources for human activities.

In addition to (and in many cases, prerequisite to) addressing the fundamental science questions, we address issues related to usage and constraints of the remote-sensing, geophysical, and sample data sets. Underlying each of the questions above is the need to calibrate and interpret correctly the remotely sensed data. This task is made difficult by the effects of solar and cosmic radiation, and the nearly ubiquitous regolith, the composition of which is affected by impact mixing of surface materials. The advantage of the Moon is that rock samples exist for known landing sites, and the composition of the regolith at those sites is also known. Thus, there exists a natural means for calibrating the remotely sensed data. With this book, we seek to provide essential data sets that are not found elsewhere such as a comprehensive table of soil compositions that represent the landing-site sample stations and a table of compositions of lunar meteorites known at present.

The *New Views of the Moon* lunar science initiative has been a sparkplug to catalyze activities that promote cooperation between the remote-sensing, geophysical, and sample-analysis communities. Activities were organized specifically with the intent to foster the kinds of interest and collaborations necessary to integrate the different data sets with one another and with the sample data base. These activities were focused on the topics of planetary structure, crustal evolution, planetary volcanism, and regolith formation. In addition, this integration highlights the fundamental scientific questions that remain unanswered regarding the origin and evolution of the Moon and points to the specifics kinds of data required to answer them. We emphasize the need to connect lunar science to broader topics of inner solar-system evolution and human exploration and habitation. Now is a crucial time for a focused assessment of the state of Lunar science at the beginning of the 21st century, with the specific goals of showing how the diverse data sets can be integrated and how the resulting synthesis of information applies not only to the Moon, but to our understanding of Earth and the Earth-Moon system, the Sun, and the inner Solar System, as well.

Summary

The purpose of this book is to assess the current state of knowledge of lunar geoscience, given the data sets provided by missions of the 1990's, to assess the remaining key questions and identify new ones for future exploration to address. It is also our intention to document how a planet or moon other than the world on which we live can be studied and understood in light of integrated suites of specific kinds of information. The Moon is the only body other than Earth for which we have material samples of known geologic context for study. We seek in this book to show how the different kinds of information gained about the Moon are related to each other and thus to aide in planning for the exploration of other worlds. We must get it "right" for the Moon if we are going to optimize our study of other planets, such as Mars.

Acknowledgments

We are grateful to several organizations for supporting this endeavor. The NASA Cosmochemistry program provided partial support through NAG5-9417. The Lunar and Planetary Institute provided expert support for several of the workshops and graphics support for several of the chapters. We are especially grateful to Renee Dotson, Leanne Woolley, Stephen Tellier, and LPI Director Stephen Mackwell for their assistance and support. We also appreciate Jeff Gillis for his work on the images for color plates and Ryan Zeigler for his work on reference lists. Finally, we are deeply grateful to the Mineralogical Society of America for agreeing to publish this work as part of the RiMG series and for the patience, encouragement, and excellent work of series editor Jodi Rosso.

References

Binder A (1998) Lunar Prospector: Overview. Science 281:1475–1476

Binder AB, Feldman WC, Lawrence DJ, Maurice S, Barraclough BL, Elphic RC (1998) First results from the Lunar Prospector Gamma Ray Spectrometer: The KREEP distribution on the Moon as delineated by thorium and potassium. Eos Trans AGU 79(17), Spring Meet Suppl, S189

Blewett DT, Lucey PG, Hawke BR, Jolliff BL (1997) Clementine images of the lunar sample-return stations: Refinement of FeO and TiO_2 mapping techniques. J Geophys Res 102:16,319–16,325

BVSP (Basaltic Volcanism Study Project) (1981) Basaltic Volcanism on the Terrestrial Planets. Pergamon

Canup RM, Righter K (eds) (2000) Origin of the Earth and Moon. Univ Arizona Press

Dasch EJ, Shih CY, Bansal BM, Wiesmann H, Nyquist LE (1987) Isotopic analysis of basaltic fragments from lunar breccia 14321:Chronology and petrogenesis of pre-Imbrium mare volcanism. Geochim Cosmochim Acta 51: 3241-3254

Delano JW, Livi K (1981) Lunar volcanic glasses and their constraints on mare petrogenesis. Geochim Cosmochim Acta 45:2137–2149

Eliason EM, McEwen AS, Robinson MS, Lee EM, Becker T, Gaddis L, Weller LA, Isbell CE, Shinaman JR, Duxbury T, Malaret E (1999) Digital processing for a global multispectral map of the Moon from the Clementine UVVIS imaging instrument. Lunar Planet Sci 30:1933

Elphic R, Lawrence D, Feldman W, Barraclough B, Maurice S, Binder A, Lucey P (2000) Lunar rare earth element distribution and ramifications for FeO and TiO_2: Lunar Prospector neutron spectrometer observations. J Geophys Res 105:20,333-20,345

Feldman WC, Binder AB, Hubbard GS, McMurray RE Jr, Miller MC, Prettyman TH (1996) The Lunar Prospector gamma-ray spectrometer. Lunar Planet Sci 27:355-356

Feldman WC, Binder AB, Maurice S, Lawrence DJ, Barraclough BL, Elphic RC (1998) First positive identification of water ice at the lunar poles. Eos Trans AGU, 79(17), Spring Meet Suppl, S190

Feldman WC, Lawrence DJ, Elphic RC, Vaniman DT, Thomsen DR, Barraclough BL (2000) The chemical information content of Lunar thermal and epithermal neutrons. J Geophys Res 105:20,347-20,363

Feldman W, Maurice S, Lawrence DJ, Little RC, Lawson SL, Gasnault O, Wiens RC, Barraclough BL, Elphic RC, Prettyman TH, Steinberg JT, Binder AB (2001) Evidence for water ice near the lunar poles. J Geophys Res 106: 23,231-23,251

Gaddis LR, Hawke BR, Robinson MR, Coombs C (2000), Compositional analyses of small lunar pyroclastic deposits using Clementine multispectral data. J Geophys Res 105:4245-4262

Gillis JJ, Jolliff BL, Elphic RC (2003) A revised algorithm for calculating TiO_2 from Clementine UVVIS data: A synthesis of rock, soil, and remotely sensed TiO_2 concentrations. J Geophys Res 108(E2):10.1029/2001JE001515

Haskin LA (1998) The Imbrium impact event and the thorium distribution at the lunar highlands surface. J Geophys Res 103:1679-1689

Hawke BR, Lucey PG, Taylor GJ (1992) The distribution of anorthosite on the nearside of the Moon. *In:* Workshop on the Physics and Chemistry of Magma Oceans from 1 bar to 4 mbar. Lunar and Planetary Institute, p 20–21

Hawke BR, Peterson CA, Blewett DT, Bussey DBJ, Lucey PG, Taylor GJ, Spudis PD (2003) Distribution and modes of occurrence of lunar anorthosite. J Geophys Res 108:10.1029/2002JE001890

Heiken GH, Vaniman DT, French BM (1991) Lunar Sourcebook: A User's Guide to the Moon. Cambridge Univ Press

Hiesinger H, Jaumann R, Neukum G, Head JW III (2000) Ages of mare basalts on the lunar nearside. J Geophys Res 105:29239-29175

Hiesinger H, Head JW III, Wolf U, Jaumann R, Neukum G (2003) Ages and stratigraphy of mare basalts in Oceanus Procellarum, Mare Nubium, Mare Cognitum, and Mare Insularum. J Geophys Res 108, doi:10.1029/2002JE001985

Hood LL, Williams CR (1989) The lunar swirls: Distribution and possible origins. Proc Lunar Planet Sci Conf 19: 99-113

Hood LL, Mitchell DL, Lin RP, Acuña M, Binder A (1999) Initial measurements of the lunar induced magnetic dipole moment using Lunar Prospector magnetometer data. Geophys Res Lett 26:2327-2330

Hood LL, Zuber MT (2000), Recent refinements in geophysical constraints on lunar origin and evolution. *In:* The Origin of the Earth and Moon. Canup R, Righter K (eds) Univ Ariz Press, p 397-409

Jolliff BL (1999) Clementine UVVIS multispectral data and the Apollo 17 landing site: What can we tell and how well? J Geophys Res 104:14,123-14,148

Khan A, Mosegaard K (2002) An inquiry into the lunar interior: A nonlinear inversion of the Apollo lunar seismic data. J Geophys Res 107, doi:10.1029/2001JE001658

Khan A, Mosegaard K, Rasmussen K (2000) A new seismic velocity model for the Moon from a Monte Carlo inversion of the Apollo lunar seismic data. Geophys Res Lett 27:1591

Keller L, Wentworth S, McKay D, Taylor L, Pieters C, Morris R (1999) Space weathering in the fine size fraction of Lunar soils: Soil maturity effects. *In:* New Views of the Moon II. Lunar and Planetary Institute 980:32-33

Konopliv AS, Kucinskas AB, Sjogren WL (1998) Gravity results from Lunar Prospector. Eos Trans AGU, 79(17), Spring Meet Suppl, S190

Lawrence DJ, Feldman WC, Barraclough BL, Binder AB, Elphic RC, Maurice S, Thomsen DR (1998) Global elemental maps of the Moon: The Lunar Prospector gamma-ray spectrometer. Science 281:1484-1489

Lawrence DJ, Feldman WC, Barraclough BL, Binder AB, Elphic RC, Maurice S, Miller MC, Prettyman TH (1999) High resolution measurements of absolute thorium abundances on the lunar surface. Geophys Res Lett 26: 2681-2684

Lawrence DJ, Feldman WC, Barraclough BL, Binder AB, Elphic RC, Maurice S, Miller MC, Prettyman TH (2000) Thorium abundances on the lunar surface. J Geophys Res 105:20,307-20,331

Lawrence DJ, Elphic RC, Feldman WC, Prettyman T, Gasnault O, Maurice S (2003) Small-area thorium features on the lunar surface. J Geophys Res 108:doi:10.1029/2003/JE002050

LGO Science Workshop Members (1986) Contributions of a Lunar Geoscience Observer (LGO) Mission to Fundamental Questions in Lunar Science. Southern Methodist University

Lin RP, Anderson KA, Hood LL (1988) Lunar surface magnetic field concentrations antipodal to young large impact basins. Icarus 74:529-541

Lucey PG, Taylor GJ, Malaret E (1995) Abundance and distribution of iron on the Moon. Science 268:1150-1153

Lucey PG, Blewett DT, Hawke BR (1998) Mapping the FeO and TiO_2 content of the lunar surface with multispectral imagery. J Geophys Res 103:3679-3699

Lucey PG, Blewett DT, Jolliff BL (2000a) Lunar iron and titanium abundance algorithms based on final processing Clementine UVVIS images. J Geophys Res 105:20,297-20,305

Lucey PG, Blewett DT, Taylor GJ, Hawke BR (2000b) Imaging of lunar surface maturity. J Geophys Res 105:20,377-20,386

Lunar Exploration Working Group (LExSWG) (1992) A Planetary Science Strategy for the Moon. NASA, Solar System Exploration Division, Lyndon B Johnson Space Center publication JSC-25920

Maurice S, Feldman WC, Lawrence DJ, Elphic RC, Gasnault O, d'Uston C, Genetay I, Lucey PG (2000) High-energy neutrons from the Moon. J Geophys Res 105:20,365-20,375

McEwen AS, Robinson MS (1997) Mapping of the Moon by Clementine. Adv Space Res 19:1523-1533

Munoz ES, Elphic RC, Maurice S, Lawrence DJ, Feldman WC, Barraclough BL, Binder AB, Lucey PG (1998) Lunar Prospector measurements of lunar Fe and Ti abundance: Comparison with spectroscopic determinations. Eos Trans AGU, 79(17), Spring Meet Suppl, S190

Neal CR (2001) Interior of the Moon: The presence of garnet in the primitive deep lunar mantle. J Geophys Res 106: 27865-27885

Neal CR, LA Taylor (1992) Petrogenesis of mare basalts: A record of lunar volcanism. Geochim Cosmochim Acta 56, 2177-2211

Neumann GA, Zuber MT, Smith DE, Lemoine FG (1996) The lunar crust: Global structure and signature of major basins. J Geophys Res 101:16,841-16,863

Nozette S, and The Clementine Team (1994) The Clementine Mission to the Moon: Scientific overview. Science 266: 1835-1839

Papike JJ (ed) (1998) Planetary Materials. Rev Mineral 36. Mineralogical Society of America,

Pieters CM, Head JW, Sunshine JM, Fischer EM, Murchie SL, Belton MJS, McEwen A, Gaddis L, Greeley R, Neukum G, Jaumann R, Hauffmann H (1993) Crustal diversity of the Moon: Compositional analyses of Galileo SSI data. J Geophys Res 98:17,127-17,148

Pieters C, Taylor L, Noble S, Keller L, Hapke B, Morris R, Allen C, McKay D, Wentworth S (2000) Space weathering on asteroids: Resolving a mystery with Lunar samples. Meteorit Planet Sci 35:1101-1107

Robinson MS, Jolliff BL (2002) Apollo 17 landing site: Topography, photometric corrections, and heterogeneity of the surrounding highland massifs. J Geophys Res 107(E11), doi:10.1029/2001JE001614

Robinson MS, McEwen AS, Eliason E, Lee EM, Malaret E, Lucey PG (1999) Clementine UVVIS global mosaic: A new tool for understanding the lunar crust. Lunar Planet Sci 30:1931

Shearer CK, Papike JJ (1993) Basaltic magmatism on the Moon: A perspective from volcanic picritic glass beads. Geochim Cosmochim Acta 57:4785-4812

Shearer CK, Papike JJ (1999) Magmatic evolution of the Moon. Am Mineral 84:1469–1494

Taylor LA, Pieters C, Keller LP, Morris RV, McKay DS, Patchen A, Wentworth S (2001) The effects of space weathering on Apollo 17 mare soils: Petrographic and chemical characterization. Meteorit Planet Sci:285-299

Taylor SR (1982) Planetary Science, A Lunar Perspective, Lunar and Planetary Inst

Tompkins S, Pieters CM (1999) Mineralogy of the lunar crust: Results from Clementine. Meteorit Planet Sci 34: 25-41

Wieczorek MA, Phillips RJ (1998) Potential anomalies on a sphere: Application to the thickness of the lunar crust. J Geophys Res 103:1715-1724

Wieczorek MA, Phillips RJ (1999) Lunar multiring basins and the cratering process. Icarus 139:246-259

Wilhelms DE (1987) The Geologic History of the Moon. US Geol Surv Prof Pap 1348, 302 pp

Yingst RA, Head JW (1997) Multispectral analysis of mare deposits in South Pole/Aitken basin. Lunar Planet Sci 28: 1609-1610

Zuber MT, Smith DE, Lemoine FG, Neumann GA (1994) The shape and internal structure of the Moon from the Clementine mission. Science 266:1839-1843

TABLE OF CONTENTS

1   New Views of Lunar Geoscience: An Introduction and Overview

Harald Hiesinger, James W. Head III

2   Understanding the Lunar Surface and Space-Moon Interactions

Paul Lucey et al.

3   The Constitution and Structure of the Lunar Interior

Mark A. Wieczorek et al.

4   Thermal and Magmatic Evolution of the Moon

Charles K. Shearer et al.

5   Cratering History and Lunar Chronology

Dieter Stöffler et al.

6   Development of the Moon

Michael B. Duke et al.

9   Earth-Moon System, Planetary Science, and Lessons Learned

S. Ross Taylor et al.

Additional Volume Contents

Reviews in Mineralogy & Geochemistry
Vol. 60, pp. 1-81, 2006
Copyright © Mineralogical Society of America

1

New Views of Lunar Geoscience: An Introduction and Overview

Harald Hiesinger and James W. Head III

Department of Geological Sciences
Brown University
Box 1846
Providence, Rhode Island, 02912, U.S.A.
Harald_Hiesinger@Brown.edu   James_Head_III@Brown.edu

1. INTRODUCTION

Beyond the Earth, the Moon is the only planetary body for which we have samples from known locations. The analysis of these samples gives us "ground-truth" for numerous remote sensing studies of the physical and chemical properties of the Moon and they are invaluable for our fundamental understanding of lunar origin and evolution. Prior to the return of the Apollo 11 samples, the Moon was thought by many to be a primitive undifferentiated body (e.g., Urey 1966), a concept shattered by the data returned from the Apollo and Luna missions. Ever since, new data have helped to address some of our questions, but of course, they also produced new questions. In this chapter we provide a summary of knowledge about lunar geologic processes and we describe major scientific advancements of the last decade that are mainly related to the most recent lunar missions such as Galileo, Clementine, and Lunar Prospector.

1.1. The Moon in the planetary context

Compared to terrestrial planets, the Moon is unique in terms of its bulk density, its size, and its origin (Fig. 1.1a-c), all of which have profound effects on its thermal evolution and the formation of a secondary crust (Fig. 1.1d). Numerous planetary scientists considered the Moon as an endmember among the planetary bodies in our solar system because its lithosphere has been relatively cool, rigid, and intact throughout most of geological time (a "one-plate" planet), and its surface has not been affected by plate recycling, an atmosphere, water, or life. Therefore the Moon recorded and preserved evidence for geologic processes that were active over the last 4–4.5 b.y. and offers us the unique opportunity to look back into geologic times for which evidence on Earth has long been erased (Fig. 1.1c,d). Impact cratering, an exterior process, is considered the most important surface process on the Moon. Internal processes, such as volcanism and tectonism, also have played an important role.

The Moon represents a keystone in the understanding of the terrestrial planets. For the Moon we have a data set for geology, geochemistry, mineralogy, petrology, chronology, and internal structure that is unequaled for any planetary body other than the Earth. These data are fundamental to understanding planetary surface processes and the geologic evolution of a planet, and are essential to linking these processes with the internal and thermal evolution. The Moon thus provides a planetary process and evolutionary perspective.

Specifically, for important planetary processes such as impact cratering, the Moon records and preserves information about depths of excavation, the role of oblique impact, modification stages, composition and production of impact melt, ejecta emplacement dynamics, and the role of volatile-element addition. By virtue of the lunar samples returned from known geological units, the Moon also provides the foundation of crater size-frequency distribution chronologies

    DOI: 10.2138/rmg.2006.60.1

for the solar system. These data are key to further understanding the importance of this fundamental process in shaping planetary crusts, particularly early in solar-system history. For example, from crater counts it is apparent that the impactor flux was much higher in the early history of the Moon, the period of the "heavy bombardment," which lasted until ~3.8 b.y. ago (e.g., Melosh 1989; Neukum et al. 2001).

The Moon also provides key information about planetary magmatic activity. We have a general picture of many aspects of plutonism (intrusion) and volcanism (extrusion), and can assess the role of magmatism as a major crust building and resurfacing process throughout history. The ages, distribution, and volumes of volcanic materials provide a record of the distribution of mantle melting processes in space and time. Furthermore, the detailed record coupled with the samples permit an assessment of the processes in a manner that can be used to infer similar processes on other planets. These data have provided a picture of the role of magmatic activity during the heavy bombardment (intrusion, extrusion, cryptomaria), and more recently in lunar history, the mare stratigraphic record, the distribution of basalt types, and the implied spatial and temporal distribution of melting. Stratigraphic information and crater ages are also providing an emerging picture of volcanic volumes and fluxes. In addition, the Moon allows us to assess a wide range of eruption styles, including pyroclastics and their petrogenetic significance.

The Moon provides a type locality for tectonic activity on a one-plate planet. Tectonic processes and tectonic activity can be understood in the context of the complete lunar data set, including internal structure and thermal evolution

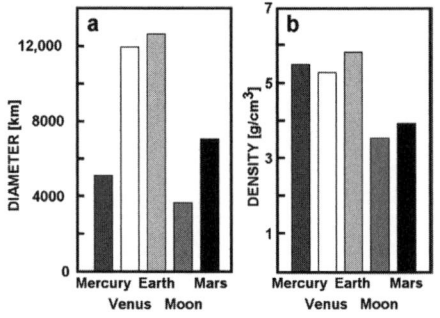

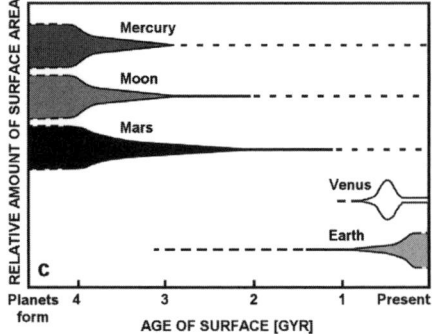

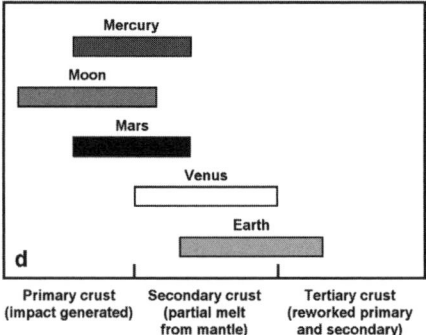

Figure 1.1. Comparison of the Moon with terrestrial planets in terms of (a) diameter, (b) bulk density, (c) surface age, and (d) types of crust (Used by permission of Sky Publishing, after Head 1999, *The New Solar System*, 4th ed., Fig. 1, p. 157, Fig. 18, p. 171, Fig. 20, p. 173).

data. Graben illustrate deformation associated with mascon loading, and wrinkle ridges appear to document the change in the net state of stress in the lithosphere from initially extensional to contractional in early lunar history. Finally we are able to link individual tectonic features to the lunar thermal evolution.

1.2. Lunar missions of the last decade

Several spacecraft visited the Moon during the 1990s and returned new data that have significantly expanded our knowledge about lunar topography, composition, internal structure, magnetic field, and the impact flux in the inner solar system. This is only a partial list of specific fields for which a better understanding exists now compared to a decade ago. The results of these missions will be highlighted throughout various chapters of this volume.

In 1990 and 1992, the Galileo spacecraft used the Earth/Moon system for gravity assist maneuvers to gain enough momentum for its travel to Jupiter. During the first flyby, Galileo imaged the western nearside and parts of the farside that were not illuminated during the Apollo missions, thus becoming the first spacecraft to obtain multispectral images of the Moon since Mariner 10 days (Head et al. 1993). The images from the SSI camera (e.g., Belton et al. 1992) led to investigations of the crustal diversity of the western hemisphere, the geology of several lunar impact basins such as the Orientale basin and South Pole-Aitken basin, the western maria and their related deposits, and the post-Imbrium impact craters (e.g., Greeley et al. 1993; Head et al. 1993; McEwen et al. 1993; Pieters et al. 1993a). During the second flyby, Galileo took multispectral images from the north-central nearside (e.g., Belton et al. 1994) that allowed detailed studies, for example, of the Humboldtianum basin, a large impact structure only partially visible from Earth.

In 1994, Clementine spent two months in lunar orbit and acquired a global data set of just under two million digital images at visible and near-infrared wavelengths, which allowed detailed mapping of the lunar major mineralogy (i.e., mafic silicates and ilmenite) and rock types (Nozette et al. 1994). Clementine, which was originally intended to observe asteroid Geographos, had four cameras on board and mapped the Moon in 11 colors at an average surface resolution of ~200 m. Most widely used and best calibrated are data from the UV/VIS camera, which provided information in 5 narrow-band filters and 1 broad-band filter that have been used to derive the major mineralogy and global composition (FeO, TiO_2) of the lunar surface (e.g., Lucey et al. 1994, 1995, 1996, 1997, 1998, 2000; McEwen et al. 1994; Pieters et al. 1994; Shoemaker et al. 1994; Blewett et al. 1997; Giguere et al. 2000; Gillis et al. 2003). UV/VIS data have also been used to study specific lunar surface features in great detail (e.g., Staid et al. 1996; Hawke et al. 1999; Li and Mustard 2000; Staid and Pieters 2001), identify hidden mare deposits (e.g., Antonenko and Yingst 2002), estimate the thickness of mare basalts (e.g., Budney and Lucey 1998), and to investigate the structure and composition of the lunar crust (e.g., Tompkins et al. 1994, 2000; Neumann et al. 1996; Wieczorek and Phillips 1998, 2000; Jolliff et al. 2000). In addition, Clementine carried a laser altimeter, which provided the first global view of lunar topography (Zuber et al. 1994) and although the resolution of these data was not optimal, several previously unmapped impact basins were revealed (Spudis et al. 1994). Clementine also gave us the first look, albeit preliminary, at the global lunar gravity field (Zuber et al. 1994; Lemoine et al. 1997), which provided new insights into the internal structure and thermal evolution of the Moon. The new Clementine data gave us the first total view of the South Pole-Aitken basin, the oldest discernible impact structure on the Moon, which is ~2500 km in diameter and about 13 km deep (Spudis et al. 1994; Pieters et al. 2001; Petro and Pieters 2004). This basin is larger and deeper than the Hellas basin on Mars (Spudis 1993; Smith et al. 1999) and is the largest and deepest impact crater yet discovered in the solar system.

Lunar Prospector was launched in 1998 and was the first NASA-supported lunar mission in 25 years (Binder 1998). The main goal of the Lunar Prospector mission was to map the surface abundances of a series of key elements such as H, U, Th, K, O, Si, Mg, Fe, Ti, Al, and Ca with special emphasis on the detection of polar water-ice deposits. For this purpose, Lunar Prospector had several spectrometers on board, including a gamma-ray spectrometer, a neutron spectrometer, an alpha-particle spectrometer, a magnetometer/electron reflectometer, and a Doppler gravity experiment. The interpretation of high radar reflectivities and high polarization ratios associated with permanently shaded craters in the polar areas of Mercury (e.g., Slade et al. 1992) as water-ice deposits suggested that similar deposits might also exist on the Moon (Arnold 1979). Although an initial analysis of the Clementine bistatic radar data was consistent with an occurrence of water-ice near the poles (e.g., Nozette et al. 1996), more detailed studies of the same data suggest that this interpretation is not unique (e.g., Simpson and Tyler 1999, Nozette et al. 2001, Vondrak and Crider 2003). Feldman et al. (1998, 2001) concluded that Lunar Prospector data are consistent with deposits of hydrogen in the form of water ice in the permanently shaded craters of the lunar poles. Lunar Prospector provided for the first time an entire suite of global elemental abundance maps, although at various resolutions. These maps are described in detail elsewhere (e.g., Lawrence et al. 1998, 2000, 2001, 2002; Feldman et al. 1999, 2001; Elphic et al. 2000; Maurice et al. 2001; Lawson et al. 2002; Prettyman et al. 2002) and in Chapter 2. Another important Lunar Prospector contribution is to completely outline a large area on the lunar nearside with high thorium concentrations (e.g., Lawrence et al. 1998) and to show the uniqueness of this terrane, the formation of which is enigmatic (see Chapters 2, 3, and 4). Lunar Prospector improved our knowledge of the global gravity field of the Moon and detected several new areas with large mass concentrations, so-called "mascons" (e.g., Konopliv et al. 1998, 2001). Finally, Lunar Prospector measured the lunar crustal magnetic field and provided further evidence that basin-forming impacts magnetize the lunar crust at their antipodes (e.g., Lin et al. 1988; Halekas et al. 2001; Hood et al. 2001).

1.3. Origin and evolution of the Moon

Numerous models, summarized in Hartmann et al. (1986), have been proposed for the origin of the Moon, including fission from Earth (e.g., Darwin 1879; Binder 1980), formation along with Earth as a sister planet (e.g., Schmidt 1959), and gravitational capture of a body formed elsewhere in the solar system (e.g., Gerstenkorn 1955). Today it is widely accepted that the Moon formed early in solar-system history when a Mars-sized object collided with the proto-Earth (Fig. 1.2a), ejecting crust and upper mantle material (Fig. 1.2b), which re-accreted in Earth orbit (Fig. 1.2c,d) (e.g., Hartmann and Davis 1975; Cameron and Ward 1976; Hartmann et al. 1986; Kipp and Melosh 1986; see Chapter 4). In order to create a Moon with the observed geochemical characteristics (see Chapter 7), the impactor's iron and siderophile elements must have been concentrated into a core before the collision. While this core became incorporated into the Earth's mantle, the outer portions of the impactor and the ejected terrestrial material accreted to the Moon.

Over the last decade significant improvements in numerical modelling of the origin and accretion of the Moon have been made and additional information can be found in Canup (2004), Canup et al. (2001), Levison et al. (2001), Agnor et al. (1999), Cameron and Canup (1998), Cameron (1997), Canup and Esposito (1996), Tonks and Melosh (1992, 1993), Cameron and Benz (1991), and Benz et al. (1986, 1987). Energy release associated with the large impact and accretion produced large-scale melting, that is a magma ocean, accompanied by density segregation of crystals from the melt and formation of a low-density, plagioclase-rich crust (Fig. 1.3). Chapter 4 regards the initial differentiation of the Moon in detail, including the magmaocean concept and alternative models. Figure 1.4 is an interpretative diagram of the thermal evolution of the Moon, which is discussed in detail in Chapter 4.

Seismic and remotely sensed data, as well as the lunar samples, suggest that the Moon has been differentiated into a crust, mantle, and possibly a small core, although higher resolution seismic data that come from a global (rather than just the nearside as in the Apollo Seismic Experiement) seismometer network are required for definitive conclusions. The formation of a globally continuous low-density crust is thought to be responsible for the lack of plate tectonics on the Moon, leading to dominantly conductive cooling through this layer and producing a globally continuous lithosphere (i.e., a one-plate planet) instead of multiple laterally moving and subducting plates as on Earth. It is currently thought that during and several hundred million years after the solidification of the magma ocean, a massive influx of projectiles, termed the "heavy bombardment," impacted on the Moon (e.g., Ryder et al. 2000; Ryder 2002; see Chapter 5). The exact timing of the heavy bombardment remains an open question. This

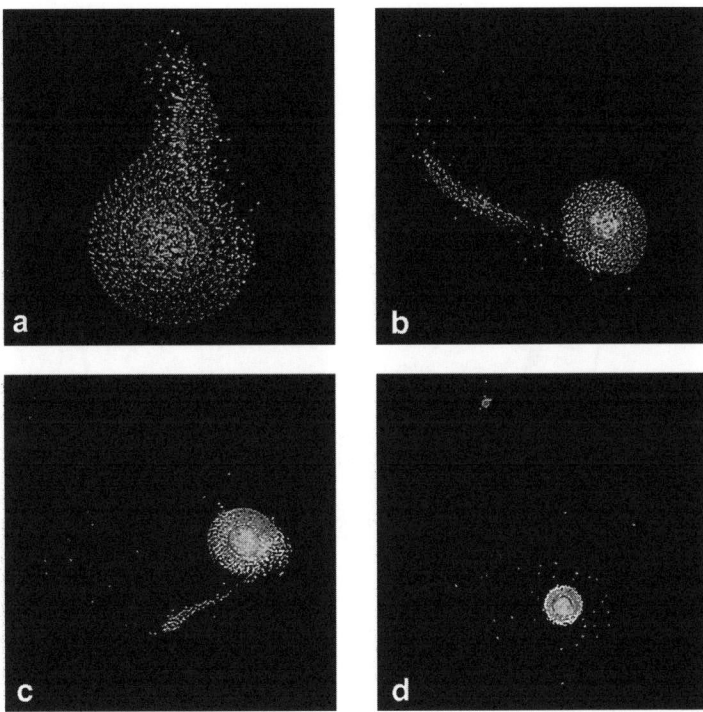

Figure 1.2. Computer simulation of the formation of the Moon. A Mars-size body impacts on proto-Earth (a), ejects crustal and mantle material (b-c), which later accretes to the Moon in Earth orbit (d). (Used by permission of Sky Publishing, in Spudis 1990, *The New Solar System*, 3rd ed., Fig. 14, p. 51).

heavy bombardment formed numerous craters and basins and obscured most evidence for early volcanism. Most evidence for volcanic flooding is only preserved since the waning stages of the heavy bombardment (3.8–3.7 Ga) when lavas could retain observable morphologies and were no longer covered and modified by regional impact-ejecta deposits. By about 2.0–1.5 Ga, basaltic lavas covered the surface to the presently observed extent (~17%) and made up ~1% of the volume of the lunar crust (Head 1976). Virtually no major internal geologic activity has occurred for the last 1.5 Ga.

1.4. Internal structure of the Moon

Since the publication of the Lunar Source Book (Heiken et al. 1991), new Clementine and Lunar Prospector data have provided improved models of the topography and internal structure of the Moon (e.g., Zuber et al. 1994; Smith et al. 1997; Konopliv et al. 1998, 2001; Wieczorek and Phillips 1998; Arkani-Hamed et al. 1999), which are discussed in Chapter 3.

A decade ago, topographic data were only available for limited areas of the Moon, that is, the areas covered by the Apollo laser altimeter and stereo imagery as well as by earth-based radar interferometry (e.g., Zisk 1978; Wu and Moore 1980). The Clementine topographic map of Zuber et al. (1994) was the first reliable global characterization of surface elevations on the Moon. Compared to data derived during the Lunar Orbiter missions (e.g., Müller and Sjogren 1968, 1969), Clementine also significantly improved knowledge of the lunar gravity field (e.g., Zuber et al. 1994), although for a variety of reasons these data are still subject to relatively large errors. From an assessment of the combined Clementine data, the lunar highlands appear to be in a state of near-isostatic compensation; however, basin structures show a wide range

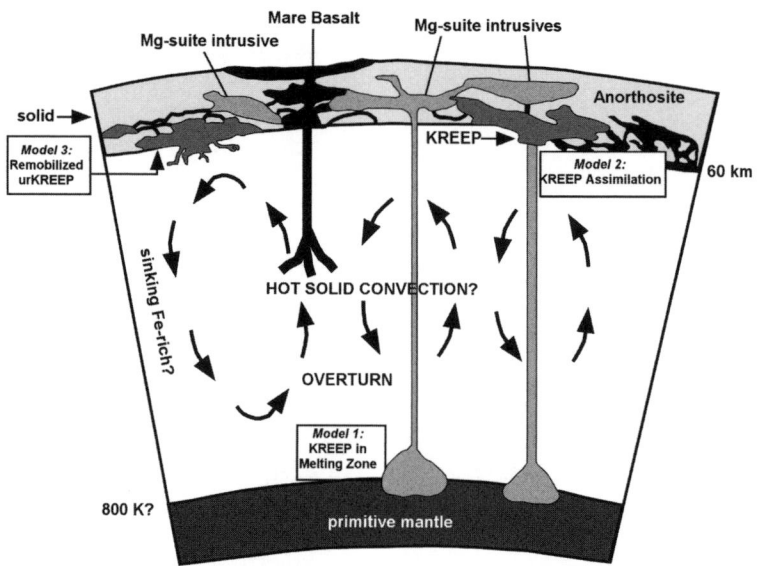

Figure 1.3. Pre-Lunar Prospector schematic cross section through the magma ocean down to ~800 km depth. An ~60 km thick crust of anorthosite forms from buoyant plagioclase cumulates. Also shown are three models for the origin of the younger Mg-suite rocks. In model 1 KREEP is delivered to depth by convective overturn and mixing with Mg-rich ultramafic cumulates. Model 2 is the same as model 1 with the exception that KREEP is assimilated at the base of the crust. Model 3 involves mobilization of urKREEP caused by decompressional melting triggered by basin-forming impact events (After Papike et al. 1998, *Planetary Materials*, Fig. 4, p. 5-5).

of compensation states that do not correlate with basin size or age (Zuber et al. 1994). Crustal thinning was observed beneath all resolvable basin structures and it was concluded that the structure and thermal history of the Moon are more complex than was appreciated ten years ago. Using Clementine data, the crustal thickness was modeled as ranging between ~20 and ~120 km, averaging at ~61 km (e.g., Neumann et al. 1996; Arkani-Hamed 1998; Khan et al. 2000; Wieczorek et al. 2001; Logonné et al. 2003; Wieczorek 2003). The average farside crust was estimated to be ~12 km thicker than on the nearside. Using Clementine data, von Frese et al. (1997) estimated a minimum crustal thickness of ~17 km beneath the Orientale basin, and Arkani-Hamed (1998) derived crustal thicknesses of 30–40 km beneath the mascon basins except for Crisium and Orientale where they estimated a thickness of ~20 km. Recently, from reinvestigation of Apollo seismic data, crustal thicknesses as low as 30±2.5 km have been inferred for some regions of the nearside crust, and the region of thinnest crust is found to be located beneath Crisium (e.g., Khan et al. 2000; Wieczorek et al. 2001; Logonné et al. 2003; Wieczorek 2003). Using Lunar Prospector data, Konopliv et al. (1998) detected three new large mass concentrations (mascons) beneath Mare Humboldtianum, Mendel-Rydberg, and Schiller-Zucchinus, and possibly several others on the lunar farside (Konopliv et al. 2001).

Seismic data collected by seismometers emplaced by the Apollo astronauts revealed apparent divisions within the mantle at ~270 and ~500 km (e.g., Goins et al. 1978; Nakamura 1983). Below ~1000 km P- and S-waves are attenuated, suggesting minor partial melting. Because only P-waves are able to move through the core, it is reasonable to assume that the core is at least partially molten (Hood and Zuber 2000). Work by Khan et al. (2000), reanalyzing Apollo seismic data, resulted in a more detailed lunar velocity structure than was previously obtainable. On the basis of their reinterpretation, they found that the velocity increases from the surface to the base

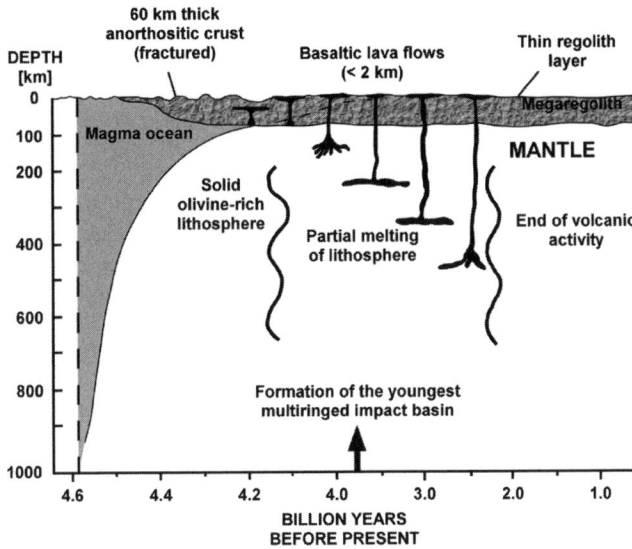

Figure 1.4. Evolution of lunar magma ocean, crust and mantle as a function of time. Simplified geological cross section of the outer 1000 km of the Moon showing (1) the evolution of the magma ocean by fractional crystallization (formation of olivine-rich residual mantle and plagioclase-rich flotation crust), (2) the simultaneous impact brecciation of the crust (megaregolith and regolith) and (3) the later partial melting of the mantle and the formation of chemically distinct basaltic lavas due to mineralogical zoning of the mantle established about 4.3 Ga. Timescale not linear. (Used by permission of Kluwer Academic Publishers, after Stöffler and Ryder 2001, *Chronology and Evolution of Mars*, Fig. 2, p. 11).

of the crust at 45±5 km. Velocities remain constant throughout the upper mantle extending to 560±15 km depth but are about 1 km/s higher in the middle mantle (Fig. 1.5). This work, coupled with that of Nakamura (2003, 2005), also resulted in an improvement in locating the origin of moonquakes. Shallow moonquakes were found to originate at ~50–220 km depths, deep moonquakes are located in 850–1000 km depth with a rather sharp lower boundary. Finally, Lognonné et al. (2003) presented a new seismic model for the Moon based upon a complete reprocessing of the Apollo lunar seismic data that suggested that the crust-mantle boundary is at ~30 km depth, with a pyroxenitic mantle. This model utilized electrical conductivity observations to suggest a liquid Fe core was not possible, but that their model was compatible with the Moon having a Fe–S liquid core. However, with the Apollo seismometers only located over a relatively small area on the lunar nearside, definitive conclusions about the global interior structure of the Moon based upon seismic data, including core size and composition, remain illusive.

The improved normalized polar moment of inertia provides an upper bound of the core radius, with core-free scenarios possible. The polar moment of inertia data were interpreted by Konopliv et al. (1998) to be consistent with an iron core of 220–450 km radius. However, moment of inertia data are also consistent with a core-free Moon (e.g., Kuskov and Konrod 2000). Support for the existence of a core comes from Lunar Prospector magnetometer data that indicate a core size of 340 km (Hood et al. 1999; Khan et al. 2004) and from analysis of laser ranging data, which indicates a liquid core having a radius less than 374 km (Williams et al. 2001a). The internal structure of the Moon is considered in more detail in Chapter 3.

1.5. Diversity of lunar rocks

The 6 Apollo missions returned 381.7 kg of lunar rocks or 2196 individual samples, and the Luna missions brought 276 g of lunar regolith to Earth (Table 1.1). Figure 1.6a is a view of the

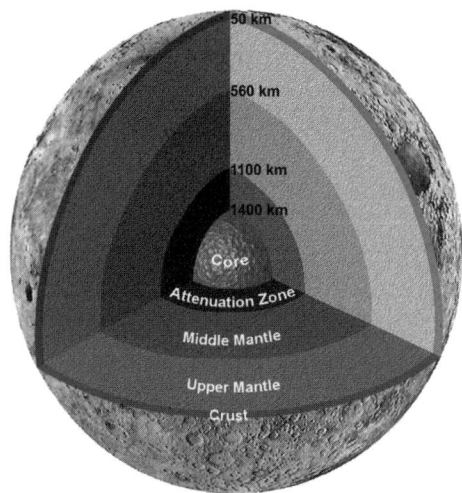

Figure 1.5. Model of the internal structure of the Moon. The thickness of the crust and mantle are based on work by Khan et al. (2000), the size of the core is based on work by Konopliv et al. (1998) and Hood et al. (1999).

lunar nearside and shows the location of the Apollo and Luna landing sites from which the samples originated, as well as the names and locations of major impact structures and maria. Figure 1.6b shows the lunar farside with several impact basins, including the largest basin, South Pole-Aitken. The importance of the returned samples, which are the only extraterrestrial samples that were specifically selected by humans, cannot be overemphasized. The returned samples have allowed scientists to study in great detail the age, mineralogy, chemistry, and petrology of lunar rocks as well as their physical properties (Fig. 1.7) and, maybe most importantly, the samples have allowed remotely sensed data to be properly calibrated, interpreted, and expanded to areas where no samples have been returned. Having lunar samples on Earth makes the Moon a true and unique keystone in our understanding of the entire solar system. The lunar experience has shown the utility of samples from other planets for comparative planetology. For example, samples from Mars will provide a record of the history of water on the surface of that planet, samples from Mercury will record whether there was active basaltic volcanism in its past, samples from Venus will be key to understanding its evolution and its heat- and volatile-loss mechanisms, and additional samples of the Moon, for example from the South Pole-Aitken basin, will lead to a better understanding of its internal structure, thermal evolution, and geologic history (see Chapters 3, 5 and 7). Here we provide only a very brief summary of the returned samples and point the reader to extensive summaries (and references therein) in BVSP (1981), Taylor et al. (1991), Haskin and Warren (1991), Papike et al. (1998), and in the following chapters.

Lunar materials can be classified on the basis of texture and composition into four distinct groups: (1) pristine highland rocks that are primordial igneous rocks, uncontaminated by impact mixing; (2) pristine basaltic volcanic rocks, including lava flows and pyroclastic deposits; (3) polymict clastic breccias, impact melt rocks, and thermally metamorphosed granulitic breccias; and (4) the lunar regolith. Details on the classification are given in Tables 1.1 and 1.2. A detailed discussion of lunar rock types is given in Chapters 2 and 3.

Pristine highland rocks can be subdivided into two major chemical groups based on their molar Na/(Na+Ca) content versus the molar Mg/(Mg+Fe) content of their bulk rock compositions (e.g., Warner et al. 1976; Papike et al. 1998). The ferroan anorthosites yield ages in the range 4.5–4.3 Ga; the magnesian-suite rocks (high Mg/Fe) are, as a group, somewhat younger (4.43–4.17 Ga) (Taylor et al. 1993) and contain dunites, troctolites, norites, and gabbronorites. A less abundant alkali suite contains similar rock types, but enriched in alkali and other trace elements, and extending to somewhat younger ages. This rather simple picture has been complicated by more recent work (e.g., see Shearer and Newsom 2000). The magnesian suite may mark the transition between magmatism associated with the magma ocean and serial magmatism. This transition period may have occurred as early as 30 Ma (Shearer and Newsom 2000) or as late as 200 Ma after the formation of the magma ocean (Solomon and Longhi

1977; Longhi 1980). The duration of magnesian-suite emplacement into the crust is also an open issue. The absence of magnesian-suite rocks with ages similar to products of the younger episodes of KREEP basaltic magmatism can be explained by a lack of deep sampling after 3.9 Ga due to a decrease in impact flux. "KREEPy" rocks are enriched in potassium (K), rare-earth elements (REE), and phosphorus (P), and are recognized in many breccias owing to the unique chemical signature. Recent Lunar Prospector data indicate that KREEP-rich rocks are most abundant around the Imbrium basin. For a more detailed description of these rock types

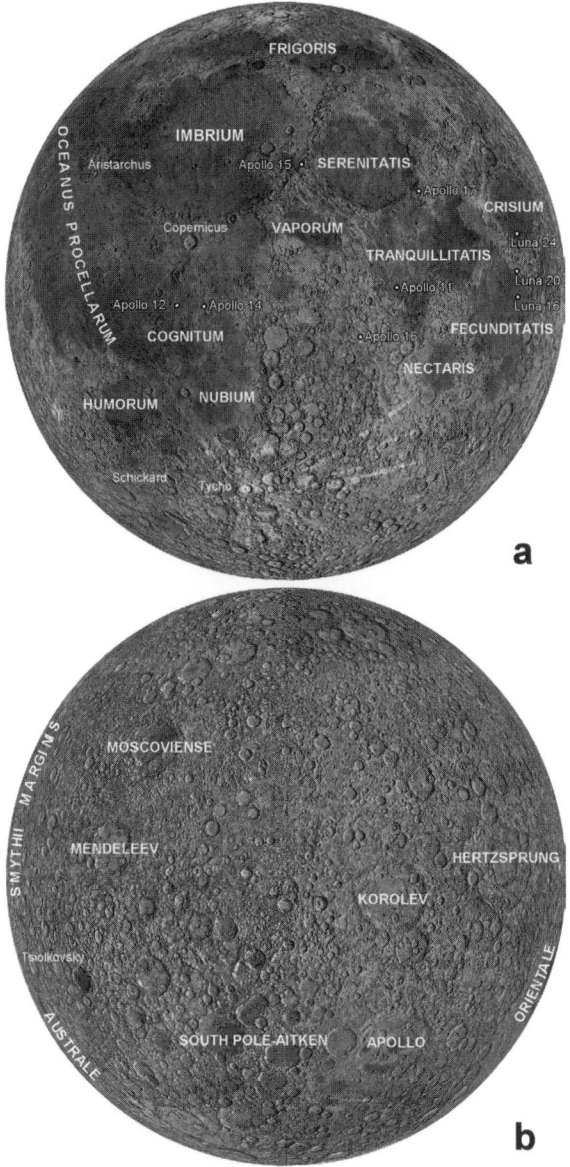

Figure 1.6. Global views of the Moon (USGS shaded relief map) which show the location of the Apollo and Luna landing sites, basins, and other prominent morphologic features; (a) nearside, (b) farside.

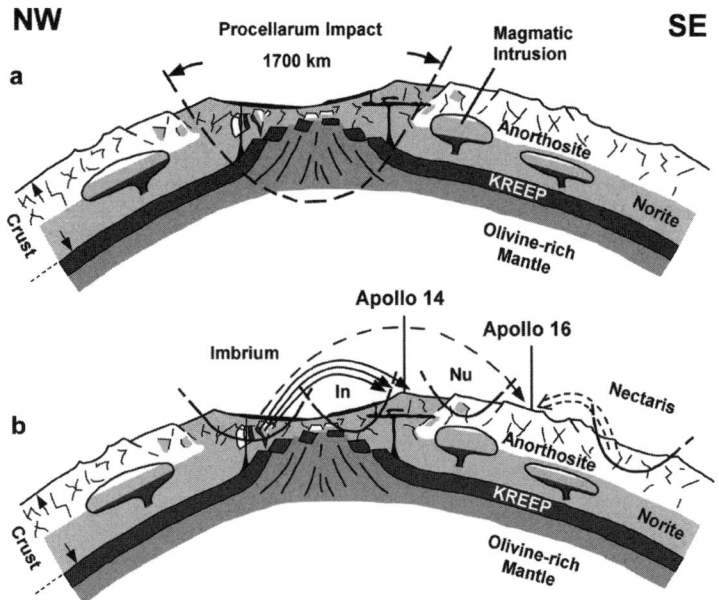

Figure 1.7. Conceptual cross sections of lunar crust and upper mantle at two different early stages. (a) Simplified cross section of lunar crust and upper mantle ~4.2 Ga shortly after the formation of a hypothetical Procellarum impact basin, which would have uplifted lower crust and upper mantle, bringing material from the layer of residual melt (ur-KREEP) and from some layered intrusive plutons containing Mg-suite igneous rocks near the surface. (b) Simplified cross section as in (a) but at ~3.8 Ga, shortly after the formation of the Imbrium impact basin, explaining the widespread occurrence of KREEP-rich lithologies around this basin; note that the Apollo 14 and 16 landing sites collected mainly Imbrium ejecta and Nectaris and Imbrium ejecta, respectively; In = Insularum basin, Nu = Nubium basin; Figures adapted from Stöffler (1990).

see Chapters 2 and 3. Most highland rocks were formed during the early differentiation of the Moon when upward separation of buoyant plagioclase within a magma ocean produced a thick anorthositic crust (see Chapter 4). Taylor et al. (1993) estimated that the magnesian-suite rocks make up ~20% of the uppermost 60 km of the crust; the rest is composed of ferroan anorthosite. Studies conducted since Lunar Prospector, however, suggest a more restricted provenance for the magnesian and alkali suite rocks (e.g., Jolliff et al. 2000; Korotev 2000; see Chapter 3). Non-volcanic rocks and the breccias and regolith derived from them cover about 83% of the lunar surface and are likely to represent over 90% of the volume of the crust (Head 1976).

Mare basalts are variably enriched in FeO and TiO_2, depleted in Al_2O_3, and have higher CaO/Al_2O_3 ratios than highland rocks (Taylor et al. 1991). Mineralogically, mare basalts are enriched in olivine and/or pyroxene, especially clinopyroxene, and depleted in plagioclase compared to highland rocks. Mare basalts originate from remelting of mantle cumulates produced in the early differentiation of the Moon. KREEP basalts, with enriched concentrations of incompatible elements, differ fundamentally from mare basalts and are thought to have formed by remelting or assimilation by mantle melts of late-stage magma-ocean residua, so-called urKREEP (Warren and Wasson 1979; see Chapter 4). Several classification schemes exist for lunar mare basalts based on their petrography, mineralogy, and chemistry (e.g., Neal and Taylor 1992; Papike et al. 1998). For example, using TiO_2 contents, mare basalts have been classified into three groups: high-Ti basalts (>9 wt% TiO_2), low-Ti basalts (1.5–9 wt% TiO_2), and very low-Ti (VLT) basalts (<1.5 wt% TiO_2) (Taylor et al. 1991).

Table 1.1. Classification of lunar rocks and statistics of samples at the lunar landing sites.

Sample Summary	A 11	A 12	A 14	A 15	A 16	A 17	L 16	L 20	L 24
Number of samples	58	69	227	370	731	741	1 core	1 core	1 core
Total mass (kg)	21.6	34.3	42.3	77.3	95.7	110.5	0.101	~0.05	0.17
Wt% of rocks >10 mm	44.9	80.6	67.3	74.7	72.3	65.9	-	-	-
plutonic	2	<1	<1	<1	12	5			
volcanic (basalt)	20	79	9	38	<1	29			
impactite	23	2	58	36	62	32			
Wt% soil fines < 10 mm	54.6	16.8	30.6	17	19.3	26.7	*100*	*100*	*100*
Wt% drill cores	0.4	1.2	0.9	6	7.4	6.6	*100*	*100*	*100*

IGNEOUS ROCKS (FIRST GENERATION)

Plutonic rocks   *Volcanic rocks*

Anorthosite   Granite   Basalts
Gabbronorite   Quartz monzodiorite   Aluminous basalt
Troctolite   Monzogabbro   KREEP-basalt
Dunite   Mare basalt
Norite   Basaltic glass

METAMORPHIC ROCKS (SECOND GENERATION)

Recrystallized rocks/breccias   *Polymict (impact) breccias*
Granulites   Impact melt rocks, clast-bearing, feldspathic
Granulitic breccias   Impact melt rocks, clast bearing, mafic
Monomict (impact) breccias   Impact melt rocks, clast-free
Cataclastic plutonic rocks   Fragmental (lithic) breccias
Cataclastic metamorphic rocks   Impact glass

POLYMICT IMPACT BRECCIAS (THIRD GENERATION)

Fragmental (lithic) breccias
Regolith breccias
Impact glass

Data from Heiken et al. (1991); weight percentages of the total weight of all samples of each mission given in italics; *Wt.% of total of rocks > 10 mm; from Stöffler and Ryder (2001).

The TiO_2 content, which is the most useful discriminator to classify lunar mare basalts, can be derived from remotely sensed data. The TiO_2 concentrations of the lunar samples have been studied extensively (e.g., Papike et al. 1976; Papike and Vaniman 1978; Neal and Taylor 1992; Papike et al. 1998) and numerous attempts have been made to expand this knowledge with remote-sensing techniques in order to derive global maps of the major mineralogy/chemistry of the Moon (e.g., Charette et al. 1974; Johnson et al. 1991; Melendrez et al. 1994; Shkuratov et al. 1999; Lucey et al. 2000; Giguere et al. 2000). From the Apollo and Luna samples, an early reading of the data suggested that Ti-poor basalts were generally younger than Ti-rich basalts, and models were proposed in which lunar mare volcanism began with high-TiO_2 content but decreased with time, and that this was coupled to depth of melting (e.g., Taylor 1982). However remote-sensing data indicate that young basalts exist with high TiO_2 concentrations (e.g., Pieters et al. 1980) and lunar basaltic meteorites have been found that are very low in Ti content and are old (mostly >3 Ga) (Cohen et al. 2000a). Based on Clementine data, Lucey et al. (2000) produced maps of the iron and titanium concentrations of the lunar surface. Combining these maps with crater size-frequency distribution ages, Hiesinger et al. (2001) found no distinct correlation between the deposit age and the composition. Instead, FeO and TiO_2 concentrations appear to vary independently with time, and generally eruptions of TiO_2-rich and TiO_2-poor basalts have occurred contemporaneously, as is the case for basalts with varying FeO contents.

Lunar impact breccias are produced by single or multiple impacts and are a mixture of materials derived from different locations and different kinds of bedrock (Table 1.2). They contain various proportions of clastic rock fragments and impact melts, and show a wide variety of textures, grain sizes, and chemical compositions. A widely adopted classification of breccias was presented by Stöffler et al. (1980), who discriminated between fragmental, glassy melt, crystalline melt, clast-poor impact melt, granulitic, dimict, and regolith breccias. A detailed discussion of these breccias is provided in Taylor et al. (1991), Papike et al. (1998), and in subsequent chapters of this volume.

"Lunar regolith" usually refers to the fine-grained fraction (mostly <1 cm) of unconsolidated surface material. The term "soil" has often been misused in the literature to describe this material. Lunar regolith ranges in composition from basaltic to anorthositic (with a small meteoritic component usually <2%), has an average grain size of ~60–80 μm, and consists of mainly five particle types: mineral fragments, pristine crystalline rock fragments, breccia fragments, glasses, and agglutinates (McKay et al. 1991). Agglutinates are aggregates of smaller particles welded together by glasses produced in micrometeorite impacts, which also produces an auto-reduction of FeO to metallic Fe, increasing the abundance in nanophase iron in the more "mature" regolith (see Chapter 2). Lunar regolith exhibits variations in maturity, a quality which is roughly proportional to the time the soil is exposed to micrometeorite bombardment and the agglutinate abundances. With the exception of a few small-scale exposures of bedrock, lunar regolith covers more or less the entire lunar surface. Thus, understanding the optical properties, for example the effects of grain size and maturity of the lunar regolith, is extremely important for optical remote-sensing techniques (see Chapter 2) such as the mapping of TiO_2 and FeO abundances (e.g., Pieters 1993b; Lucey et al. 2000; Taylor 2002).

1.6. Lunar meteorites

Since about 1980, more than 36 meteorites have been recovered, mostly from Antarctica and the Sahara desert, that have proven to be pieces of the Moon delivered to Earth as debris from large impacts. Extensive work has been done on the lunar meteorites (e.g., see Warren 1994; Korotev et al. 1996, 2003a,b). In contrast to martian meteorites, which are mostly igneous rocks from terrains younger than 1.3 Ga (McSween 1999), the lunar meteorites are mostly samples of the very old highlands, though some basaltic lavas from the maria are present in the collection. Ages derived by ^{40}Ar-^{39}Ar techniques of 31 impact-melt clasts in lunar meteorites range from ~2.43–4.12 Ga with most of the meteorites being older than 3 Ga (Cohen et al.

Table 1.2. Classification of impactites according to a provisional proposal by the IUGS Subcommission of the Systematics of Metamorphic Rocks (SCMR), Subgroup "Impactites" (Stöffler and Grieve 1994, 1996).

I. CLASSIFICATION OF IMPACTITES FROM SINGLE IMPACTS

Proximal impactites

Shocked rocks[a]	Impact melt rocks[b]	Impact breccias
4–6 stages of progressively increasing shock metamorphism[d]	Clast-rich Clast-poor Clast-free	Cataclastic (monomict) breccia Lithic breccia (without melt particles)[c] Suevite (with melt particles)[c]

Distal impactites

Consolidated	Unconsolidated
Tektite* Microtektite* Microkrystite*	Air fall bed

II. CLASSIFICATION OF IMPACTITES FROM MULTIPLE IMPACTS[1]

Impact detritus (unconsolidated impactoclastic debris)	Shock lithified impact detritus (consolidated impactoclastic debris)	
Regolith[2]	Regolith breccias[2] (breccias with matrix melt and melt particles)	Lithic ("fragmental" [3]) breccias (breccias without matrix melt and melt particles

[a] either as clasts in polymict breccias and impact melt rocks or as zones in the crater basement
[b] may be subclassified into glassy, hypocrystalline, and holocrystalline varieties
[c] generally polymict but can be monomict in a single lithology target
[d] depending on type of rock (see Table 1.4)
* impact melt with very minor or minor admixed shocked and unshocked clasts
[1] best known from lunar rocks (Apollo and Luna) and from lunar and asteroidal meteorites
[2] contain solar wind gases
[3] previously used for lunar rocks and meteorites

2000a). These ages have been attributed to seven different impact events ranging in age from 2.76 to 3.92 Ga (Cohen et al. 2000a). Most of the lunar meteorites are feldspathic regolith breccias or impact-melt breccias, and there is evidence that some of the lunar meteorites might share the same source crater. For example Warren (1994) proposed that Asuka-881757 and Y-793169 (most probable) and Y-793274 and EET875721 were derived from the same source crater, respectively. Similarly, Arai and Warren (1999) found that QUE94281 is remarkably similar to Y-793274 and interpreted this as evidence for a shared launch. Warren (1994) also suggested that some lunar meteorites were launched from <3.2 m depth within the last 1 million years. In addition to the Apollo and Luna sample collection lunar meteorites are "new" samples, which provide significant new information about the Moon because they potentially represent areas not sampled by the Apollo and Luna missions. Unlike the returned samples, however, the lunar meteorites do contain a terrestrial weathering component.

1.7. The stratigraphic system of the Moon

Numerous attempts have been made to subdivide the history of the Moon into time-stratigraphic systems (e.g., Shoemaker and Hackman 1962; Shoemaker 1964; McCauley 1967,

Wilhelms 1970, Stuart-Alexander and Wilhelms 1975; Wilhelms 1987). The basic idea is to use the ejecta blankets of large basins and craters as global marker horizons, similar to time-strati-graphic units on Earth, in order to establish a relative stratigraphy of the Moon. The most widely accepted stratigraphy is based on work by Wilhelms (1987), who divided the lunar history into five time intervals, the pre-Nectarian (oldest), the Nectarian, the Imbrian, the Eratosthenian, and the Copernican System (Fig. 1.8). Assigning absolute ages to these stratigraphic periods is controversial and depends mostly on an interpretation of which sample represents the "true" age of a specific impact event, i.e., the Imbrium or Nectaris event (Fig. 1.9) (e.g., Baldwin 1974, 1987; Jessberger et al. 1974; Nunes et al. 1974; Schaefer and Husain 1974; Maurer et al. 1978; Neukum 1983; Wilhelms 1987; Stöffler and Ryder 2001). Detailed discussions of this issue occur elsewhere (e.g., Spudis 1996; Hiesinger et al. 2000; and in Chapter 5). The chronostrati-graphic systems such as *Imbrian*, *Eratosthenian*, and *Copernican* were defined in different ways that couple measured ages of samples to specific key lunar impact events (e.g., Wilhelms 1987; Neukum and Ivanov 1994; Stöffler and Ryder 2001). For example, the beginning of the Eratosthenian system, 3.2 Ga, is based on the measured ages of basalts onto which ejecta from the crater Eratosthenes (which has no rays) are superposed. However, the stratigraphies vary

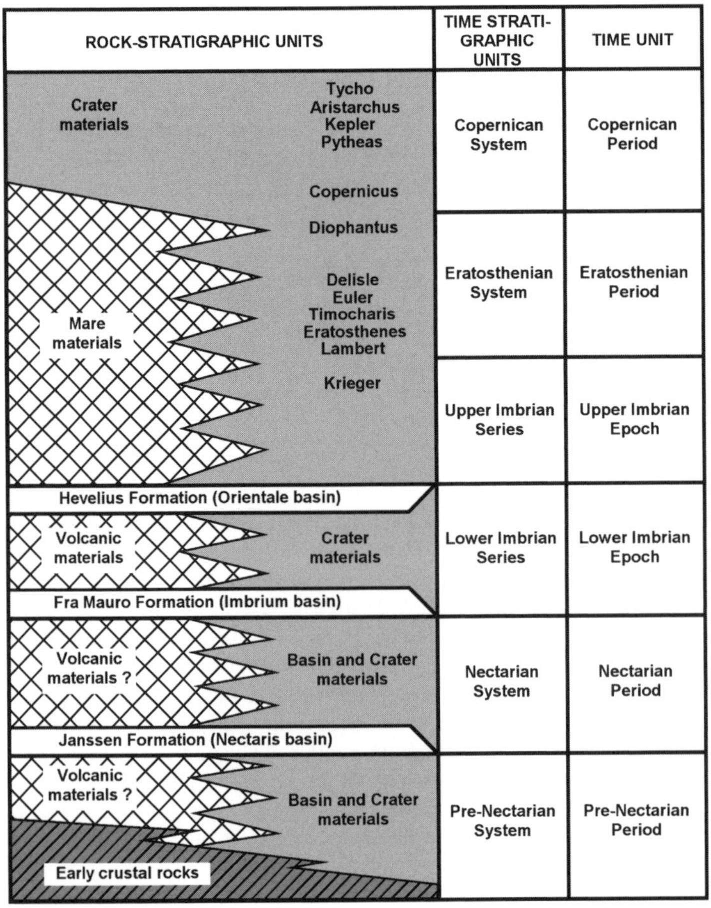

Figure 1.8. The stratigraphic system of the Moon. (Courtesy of the U. S. Geological Survey, after Wilhelms 1987, *The Geologic History of the Moon*, Fig. 7.1, p. 121).

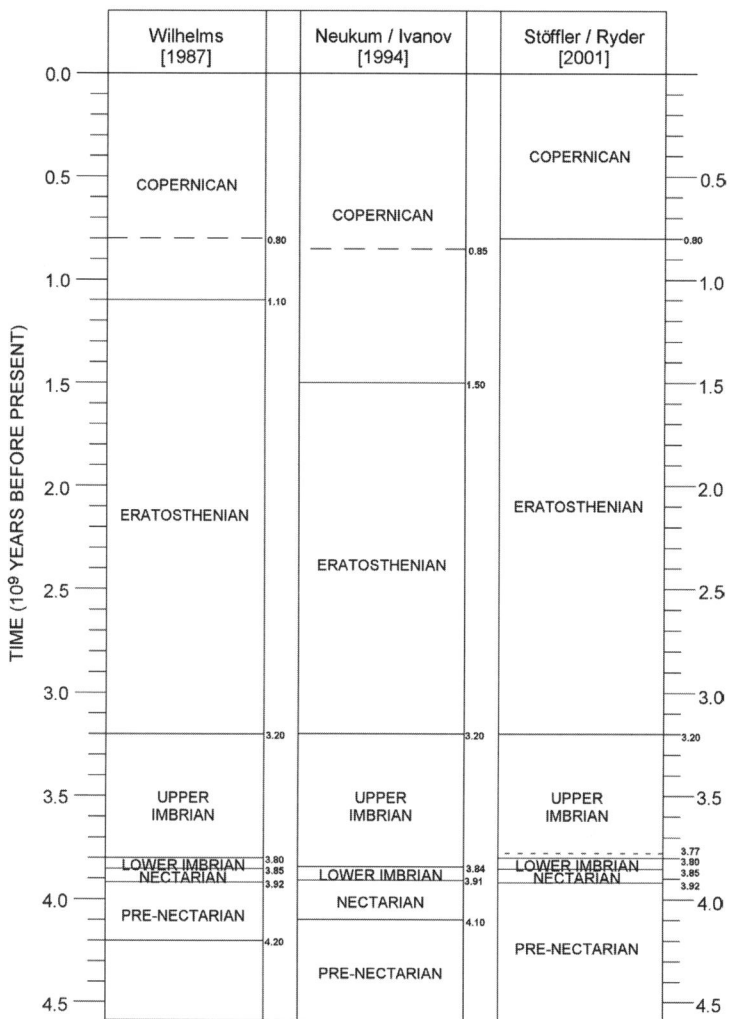

Figure 1.9. Comparison of stratigraphies of Wilhelms (1987), Neukum and Ivanov (1994, and Stöffler and Ryder (2001). Dashed lines in the stratigraphies of Wilhelms (1987) and Neukum and Ivanov (1994) indicate radiometric ages, which these authors attribute to the formation of the crater Copernicus. In Stöffler and Ryder (2001) two formation ages for the Imbrium basin have been proposed, that is, 3.85 Ga and 3.77 Ga (dashed line).

substantially in their definition of the beginning of the Copernican system i.e., 0.8–1.5 b.y., based on different interpretations of the age of Copernicus. A detailed and updated review of the available sample data and their stratigraphic interpretations is given in Chapter 5.

On the basis of Wilhelms' (1987) work, the pre-Nectarian Period spans from the time of accretion (~4.5 b.y.) to ~3.92 b.y., the time of the formation of the Nectaris basin, which is based on the ages of impact-melt breccia from Apollo 16. The pre-Nectarian Period is characterized by an intense impact bombardment that melted and mixed the lunar surface to significant depth. This bombardment is accompanied by the differentiation of a feldspathic crust and a more mafic mantle from a magma ocean prior to about 4.4–4.3 b.y. ago. The age of the Nectaris basin of ~3.92 b.y. was tentatively inferred from samples of the Apollo 16 landing site, but recently this has been challenged by Haskin et al. (1998, 2003), who argued that these samples may represent Imbrium ejecta. Accepting the age of 3.92, then the Nectarian Period lasted from ~3.92 b.y. until ~3.85 b.y. when the Imbrium impact occurred. However, Stöffler and Ryder (2001) presented arguments for two distinctly different ages of the Imbrium event, one 3.77 (Stöffler), the other 3.85 b.y. (Ryder). During the Nectarian Period, continued heavy impact bombardment resulted in the formation of a number of large impact basins, of which two were sampled (Serenitatis, Crisium). There is also evidence for early volcanism during the Nectarian but the extent remains undetermined.

Wilhelms (1987) subdivided the Imbrian Period into Early (~3.85–3.8 b.y) and Late Imbrian Epochs (~3.8–3.2 b.y.). During the Early Imbrian, two major impact basins (Imbrium, Orientale) were formed that had profound effects on the subsequent and present day appearance of the Moon. Cessation of giant impacts during the Late Imbrian enabled mare basalts that were continuously extruded, finally to remain preserved on the surface (Fig. 1.10). Wilhelms (1987) argued that Imbrian mare basalts probably lie beneath the entire area now covered with mare basalts.

The Eratosthenian Period, which began at ~3.2 b.y. ago and lasted for ~2.1 b.y., is characterized by continued volcanic eruptions and interfingering of volcanic deposits with crater materials. However, both impact rates and volcanic activity declined drastically during the Eratosthenian Period.

The beginning of the Copernican Period has not been accurately dated on either an absolute or a relative time scale. Apollo 12 samples, thought to represent ray material from Copernicus,

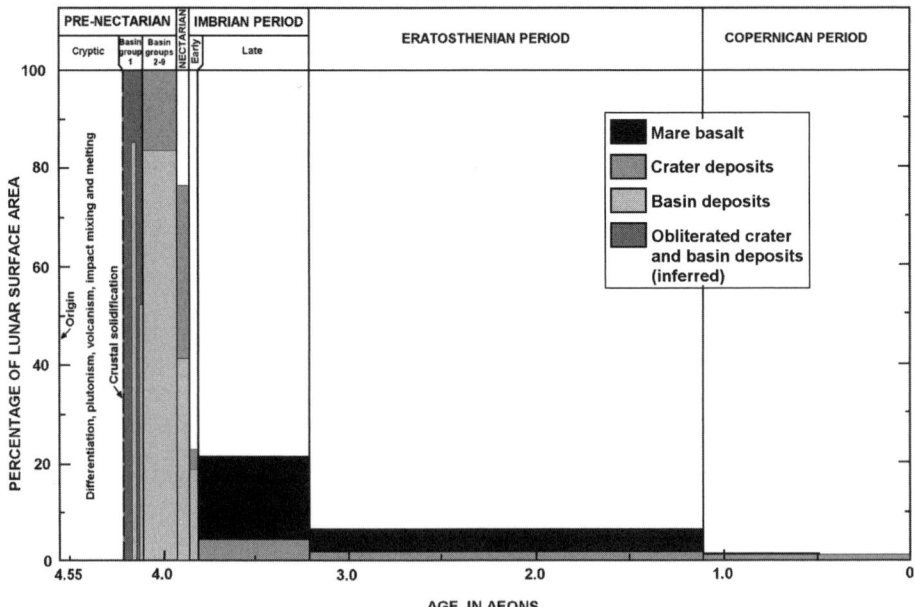

Figure 1.10. Stratigraphy of the Moon (after Wilhelms, 1987). The vertical axis indicates the surface area inferred to have been originally covered by each group of deposits. Basin and crater deposits are assumed to extend one diameter from the crater (i.e., the continuous ejecta), mare deposits of each age are assumed to extend beneath the entire area of the younger maria. Procellarum and South Pole-Aitken basin (pre-Nectarian group 1) are shown diagrammatically as narrow vertical bars. (Courtesy of the U. S. Geological Survey, after Wilhelms 1987, *The Geologic History of the Moon*, Fig. 14.3, p. 277).

yield an uncertain age of 0.81 b.y.; crater frequencies suggest an age of 1.1 b.y. This age is extremely important to know accurately because it provides a guideline for the interpretation of rayed craters in general and of the length required for impact gardening to homogenize ray material and substrate, thus erasing the ray signature. Volcanism finally stopped during the Copernican Period, and mare units younger than ~1.5 b.y. are known only from a few regions (e.g., Schultz and Spudis 1983; Hiesinger et al. 2000, 2003). Impacts, mostly of small scale, still occur on the Moon and are responsible for the formation of the uppermost layer of comminuted rock and glass particles, termed the regolith. Figure 1.11 shows conceptually the various evolutionary stages of the Moon, depicting how the Moon might have looked (a) 3.8–3.9 b.y. ago, (b) 3.0 b.y. ago, and (c) today.

2. GEOLOGIC PROCESSES

The next sections are intended as brief descriptions and syntheses of lunar processes and surface features that are crucial for understanding the geologic history and evolution of the Moon. An introduction to impact cratering, followed by introductions to volcanism and tectonism summarize important basic concepts, provide background knowledge, and set the stage for the following chapters.

2.1. Impact processes

2.1.1. Origin of craters. Although craters on the Moon were once thought to be mostly volcanic in origin (e.g., Dana 1846; Spurr 1944, 1945, 1948, 1949), we know today that most of these craters were formed by impact processes. The Moon has been struck by 15–20 km/sec projectiles that range over 35 orders of magnitude in mass from microscopic dust particles of 10^{-15} g to asteroids as massive as 10^{20} g. The associated kinetic energies vary from a fraction of an erg to ~10^{32} erg per individual impact and can exceed the total internal energy of 10^{26}–10^{27} erg that is released by the Earth during one year (Lammlein et al. 1974). A non-trivial amount of meteoritic material has been added since the solidification of the Moon's crust. For example, Chyba (1991) estimated addition of ~10^{20} kg of material from impacts, with perhaps half of it retained (compared to a lunar mass: 7.35×10^{22} kg). An extensive review

Figure 1.11. Artist conception of the evolution of the nearside of the Moon. (a) Surface of the Moon as it probably appeared after the formation of most of the lunar impact basins but before the formation of the Imbrium basin 3.8-3.9 Ga; (b) features as they probably appeared after the emplacement of most of the extensive mare lava floods 3.0 Ga, (c) the present appearance of the Moon. [After Fig. 2.7, p. 20-21 in *Lunar Source Book*, edited by Heiken et al. (1991). Artist: Don Davis].

of impact cratering as a geologic process was provided by Melosh (1989), including discussions on the influence of an atmosphere and different gravities, impact angles, and velocities.

2.1.2. Morphology of craters. From examination of the morphology of lunar impact craters in different size ranges, it is apparent that with increasing diameter, craters become proportionally shallower (e.g., Pike 1980) and develop more complex floor and rim morphologies, including the formation of multiple rings and central peaks (Fig. 1.12a–h). Impact crater morphologies can be characterized as (1) simple craters, (2) complex craters, and (3) basins. Simple craters are usually bowl-shaped, with rounded or small flat floors, smooth rims, and no wall terraces. With increasing diameter, simple craters transition into complex craters showing terraces and crenulated rims, zones of broad-scale inward slumping, and uplifted central peaks protruding through a wide, flat floor. On the Moon this transition from simple to complex craters occurs at about 15–20 km (Pike 1977). The diameter of this transition scales to 1/g for planetary surfaces when strength is not strain-rate dependent (O'Keefe and Ahrens 1993, 1994). At diameters in excess of 100 km, peak rings occur on the crater floor, defining the transition from craters to basins (Hartmann and Kuiper 1962). Central-peak basins with a fragmentary ring of peaks, such as in crater Compton, are found in the 140–175 km diameter range and are transitional to peak-ring basins, which have a well-developed ring but lack a central peak. Peak-ring basins like the Schrödinger basin are generally 175–450 km in diameter. Finally, multiring basins on the Moon, such as the Orientale basin, are larger than 400 km and have up to six concentric rings. Spudis (1993) reviewed the geology of lunar multiring impact basins; for a more detailed discussion of morphologic parameters of lunar craters and basins such as crater diameter/depth ratios etc., see for example Pike (1977, 1980) and Williams and Zuber (1998). A discussion of the influence of a planet's curvature on the shape of impact craters is provided, for example, by Fujiwara et al. (1993). The effects of target porosities are discussed by Love et al. (1993) and the influence of impact angle on crater shape are discussed in Burchell and Mackay (1998).

On the Moon, an airless body with most "erosion" related to impact processes, we can observe a large number of craters over a wide range of sizes, degradation stages, and morphologies. The large number of lunar impact craters and their characteristics provide excellent statistics for detailed quantitative studies of crater morphologies (e.g., Pike 1980), which can be used for comparison with other planetary bodies, such as Mercury, Mars, Venus, the asteroids, or the moons of the outer planets. Studies of terrestrial impact craters complement the investigations of lunar craters. On Earth, erosional processes can remove most of the surface features of impact craters (e.g., ejecta blankets, crater rims) and expose deeper levels within or beneath the original crater. This allows a three-dimensional study of large-scale impact craters not possible on other planets or in the laboratory. As a result, many fundamental concepts of cratering mechanics have been established on terrestrial impact structures and then applied to craters elsewhere in the solar system.

2.1.3. Cratering mechanics. An understanding of the observed variety of lunar impact crater morphologies requires a basic knowledge of cratering mechanics. On the basis of observations and modelling of lunar impact crater morphologies, we now know that the morphology of a crater is strongly dependent on the interaction of stress waves with free surfaces as well as upon the thermodynamics of the stress wave itself (e.g., Melosh 1989). Rocks that were formed during the impact process, i.e., the breccias returned by the Apollo missions, provide key information on the cratering process. For example, from these samples we can estimate the temperature and pressure regimes to which the breccias were exposed during their formation.

The mechanics of impacts that produce small, simple craters have been extensively studied and are relatively well understood from observations of experimental craters. During the very first stages of an impact, the projectile's kinetic energy is transformed into shock waves that travel forward into the target surface and backward into the projectile. Within the target, the

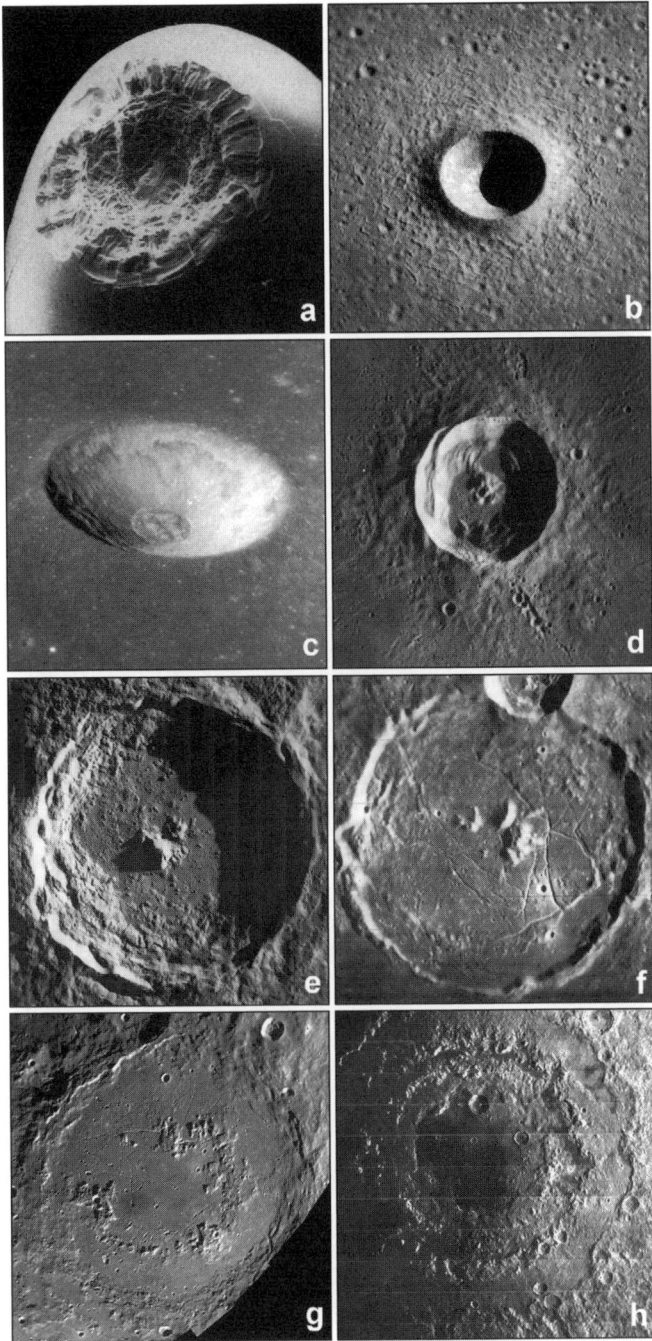

Figure 1.12. Morphologies of craters and basins; (a) micro crater on a lunar glass spherule (<1 mm), (b) bowl-shaped crater (Linné, 2.5 km, Apollo 15 frame P-9353), (c) transitional crater (Taruntius, 8.5 km, Apollo 10 frame H-4253), (d) complex crater (Euler, 28 km, Apollo 17 frame M-2923), (e) central peak crater (Tycho, 85 km, LO V M-125), (f) floor-fractured crater (Gassendi, 110 km, LO IV 143H2), (g) central peak ring basin (Schrödinger, 320 km, Clementine mosaic), (h) multi-ringed basin (Orientale, 930 km, LO IV M-187).

shock wave causes particle motion, which accelerates the impacted material radially downward and outward. At the same time the return shock wave decelerates the projectile. During this phase, the specific energies of the highly compressed target and projectile are increased, and after the passage of the shock wave, adiabatic decompression causes the release of this energy in the form of heat. Because pressure release associated with hypervelocity impacts can be up to several hundred GPa, large volumes of the target and virtually the entire projectile are melted and vaporized. Unloading from the high-pressure levels is initiated by the rarefaction waves that form as the shock wave reaches free surfaces such as the backside of the projectile or the ground surface at some distance from the impact point. Rarefaction waves modify the initial shock particle motions and ultimately set up a flow field that initiates and eventually completes the actual crater excavation (e.g., Croft 1980).

At the moment the shock-induced particle motion ceases, the total excavated and temporarily displaced target materials form the "transient cavity." The transient cavity is significantly deeper than the final excavation cavity even if their diameters are similar (Fig. 1.13). Later, during the modification stage, the transient cavity is modified by rim collapse and floor rebound. Both collapse of the rim of the transient cavity with downward and inward slumping material and the unloading of the compressed materials leading to upward motion in the crater center are responsible for the shallower apparent crater compared to its transient cavity. A more detailed discussion of cratering mechanics can be found in Melosh (1989) and in Chapter 5.

2.1.4. Nature of ejecta. The nature of impact ejecta varies with radial distance from the crater (Fig. 1.13). Generally, ejecta deposits are highly centrosymmetric; non-centrosymmetric ejecta deposits and elongated crater cavities are observed only for craters formed by oblique impacts (e.g., Gault and Wedekind 1978; Melosh 1989; Schultz and Gault 1990). Although individual ejecta fragments follow ballistic trajectories, they cumulatively form a so-called ejecta curtain that gradually extends outward and thins. Continuous ejecta deposits are located closest to the crater, discontinuous ejecta deposits are farther away, and ejecta rays form the most distal parts of the ejecta deposits. Continuous ejecta deposits completely cover or disrupt the preexist-

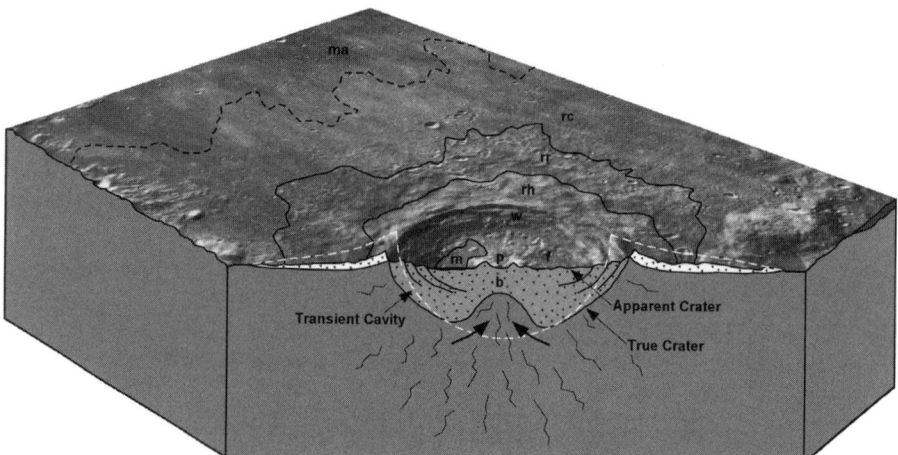

Figure 1.13. Block diagram of an idealized typical large impact crater. Crater subunits are the rim (rh), floor (f), central peak (p), wall (w), and impact melt pond (m). Ejecta deposits consist of hummocky deposits (rh), radial deposits (rr), and deposits cratered by secondary craters (rc); mare regions which are not influenced by the crater ejecta (ma). Dashed white line shows the transient cavity and its relationship to the apparent crater and the true crater. Collapsed wall material and brecciated material (b) filled the true crater to form the apparent crater.

ing surface in contrast to discontinuous ejecta deposits, which are patchy and characterized by localized deposits and shallow elongated secondary craters that frequently occur in clusters. Ejecta deposits form by ballistic sedimentation and thus incorporate local substrate, with the proportion of local material increasing with distance from the crater (e.g., Oberbeck et al. 1974; Oberbeck 1975; Haskin et al. 2003; Li and Mustard 2003). Ejecta rays occurring beyond the discontinuous ejecta deposits are relatively thin, long streaks that are oriented radially to the crater.

Recent experiments using two and three-dimensional particle image velocimetry (PIV) techniques (e.g., Cintala et al. 1999; Anderson et al. 2000, 2001, 2003) allow the study of individual particle motions within the ejecta curtain (Fig. 1.14a,b). These studies indicate that the angle of ejection of particles is not constant throughout an impact but decreases gradually with increasing radial distance from the impact point. Within an ejecta curtain, finer material that has been ejected at higher velocities and higher ejection angles during the early phases of the cratering process is found high in the ejecta curtain and deposits farthest away from the crater. Coarser material that originates from deeper target levels is ejected with lower velocities and at lower angles, and is deposited closer to the crater. This sequence produces an inverted stratigraphy of the target in the ejecta deposits with deepest target strata close to the crater and shallow target strata deposited farthest away from the crater.

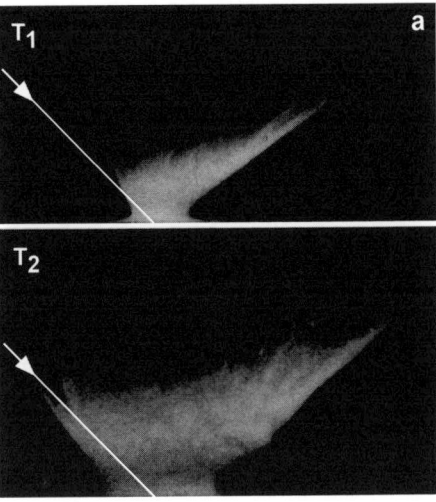

The Moon is easily observed and thus allows detailed investigation of craters as products of impact-cratering processes in the inner solar system. By understanding the lunar cratering processes, we have been able to infer processes on the other terrestrial planets and their moons (e.g., Sharpton 1994; Pike 1980, 1988; Ivanov 1992) (Fig. 1.15). On Mars, for example, Mariner and Viking images showed numerous craters with multiple lobate ejecta deposits and pedestal craters. These craters have been interpreted as having formed by incorporation of ground water or ground ice into the ejecta (e.g., Carr et al. 1977; Kuzmin et al. 1988) or by atmospheric effects, that is, the entrainment and interaction of the atmosphere with the ejecta curtain (e.g., Schultz and Gault 1979; Barnouin-Jah and Schultz 1996, 1998). Similarly, flow lobes of craters on Venus have been interpreted to form by entrainment of thick atmosphere into the ejecta curtain (e.g., Barnouin-Jah and Schultz 1996). Modeling and extrapolating lunar cratering processes to conditions on Mercury revealed that the impact rate there

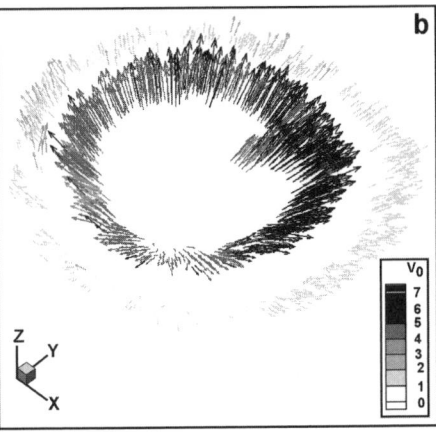

Figure 1.14. Two time steps (a) of an oblique impact experiment at 45° impact angle (Schultz and Anderson 1996) and 3D PIV velocity vectors (b) for a 30° oblique impact (Anderson et al. 2000).

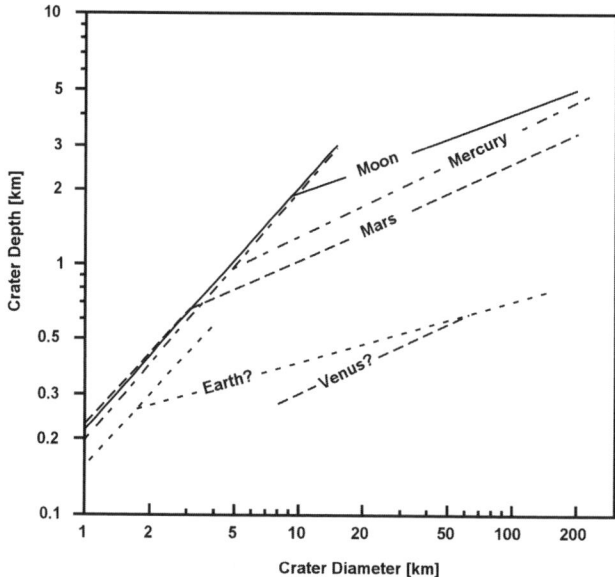

Figure 1.15. Depth/Diameter ratios of impact craters of the terrestrial planets and the Moon. (Used by permission of The Geological Society of America, after Sharpton 1994, *Large Meteorite Impacts and Planetary Evolution*, Fig. 1, p. 20).

is 5.5× higher than on the Moon, impact velocities are 60% higher, impact-melt production is 14× greater, and vapor production is 20× greater than on the Moon (Cintala 1992). Apparent crater morphologies on the icy satellites of Jupiter, especially Europa, can vary significantly from those on the Moon. Shallow depressions seen along the terminator zone have irregular shapes characteristic of endogenic origin (Chapman et al. 1997), others have slightly raised rims suggesting an impact origin (Greeley et al. 1998).

2.1.5. Shock metamorphism. In addition to ejection from the crater, target materials are also highly modified by the shock wave of the impact with pressures that exceed pressures known from internally-driven processes, especially within planetary crusts (Fig. 1.16). In addition, temperatures and strain rates associated with impacts can be orders of magnitude higher than those of internal processes and can lead to shock-metamorphism (see Chapter 5). At high shock pressures large amounts of energy are deposited in the target material, raising the temperature above the melting or vaporization points. The degree of shock metamorphism is heavily dependent on the material behavior at ultrahigh temperatures and pressures. At pressures of 5–12 GPa, minerals and rocks behave plastically instead of elastically. Between 40 and 100 GPa thermal effects become more important and melting begins. Vaporization occurs at pressures in excess of about 150 GPa, and finally ionization occurs at pressures of a few hundred GPa (e.g., Hörz et al. 1991). Table 1.3 summarizes shock effects in rocks and minerals calibrated for the shock pressure by recovery experiments.

All types of lunar rocks returned to Earth have, to some degree, been affected by impact processes. The last shock event, which may have only displaced the rock fragment at very low shock pressure or may have strongly shock metamorphosed it in this "last" impact event, can be diagnosed by specific shock effects in the constituent minerals (Chao 1968, Stöffler 1972, 1974, 1984; Bischoff and Stöffler 1992; see also compilation by French 1998). Progressive stages of shock metamorphism are defined by an increasing intensity of shock, which is inversely proportional to the original distance of the shocked rock unit from the impact center. The

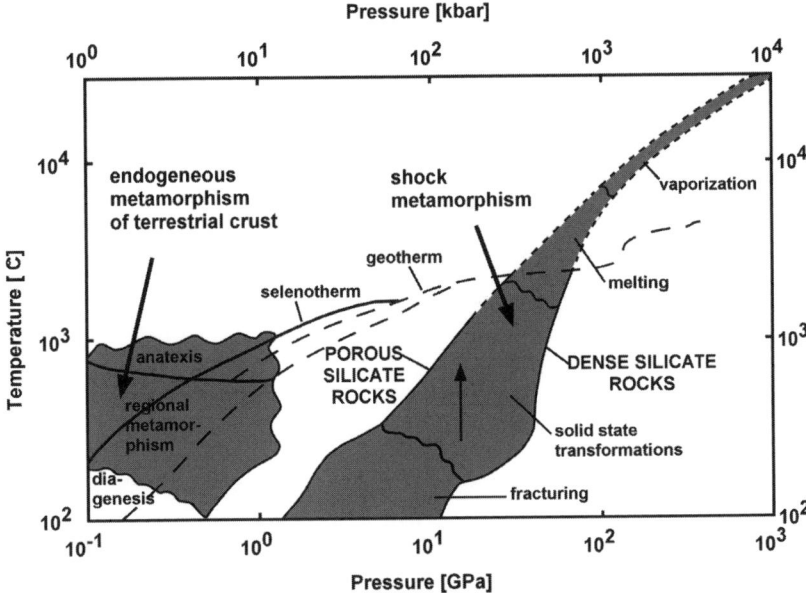

Figure 1.16. Pressure temperature fields for progressive shock metamorphism of lunar rocks in comparison to terrestrial endogenic metamorphism; upper curve of P-T-field of shock metamorphism holds for porous regolith and regolith breccias (data sources see Tables 1.3 and 1.4) and lower curve holds for dense, non porous mafic rocks of basaltic to peridotitic to anorthositic composition (data sources see Tables 1.3 and 1.4); dashed parts of boundaries of shock metamorphism is based on Davies (1972); geotherms from Anderson (1981) (upper curve = oceanic, lower curve = continental).

complete system of progressive shock metamorphism is given in Table 1.4 for the main types of lunar rocks (see also Chapter 5).

2.1.6. Thermal metamorphism. The collections of returned lunar samples and of meteorites contain a group of rocks called "feldspathic granulitic impactites." Previously, some of these rocks were interpreted to be products of igneous plutons or thermal metamorphism (Heiken et al. 1991). Compositionally, these samples are exclusively derived from the lunar highlands (70–80% modal plagioclase; Cushing et al. 1999) and appear to represent an essential fraction of the lunar crust (Warner et al. 1977). They occur in all sample suites that sampled lunar highlands materials (Ap 14, 15, 16, 17; Luna 20) and are ubiquitous in lunar meteorites of highland provenance (Korotev et al. 2003b). Recently, three different textural types of granulitic breccias were identified: poikilitic, granoblastic, and poikilitic-granoblastic breccias (Cushing et al. 1999). The equilibrium temperatures of granulitic breccias are near 1000–1100 °C as deduced from pyroxene thermometry and other observations (Warner et al. 1977; Ostertag et al. 1987; Cushing et al. 1999). Cushing et al. (1999) suggested that some granulitic breccias cooled relatively rapidly (0.5–50 °C/year) at shallow depths of <200 m, and argued that they were formed in craters of 30–90 km in diameter, physically associated with impact-melt breccias or fine-grained fragmental precursor lithologies. Others (see, for example, Korotev and Jolliff 2001) think it more likely that granulitic rocks were assembled by very large impacts (basins) that penetrated to mid-crustal levels, and that later impacts re-excavated these rocks and brought them to the surface. Gibson et al. (2001) proposed that granulites do not need to have formed from contact metamorphism of older breccias in or close to impact-melt sheets, because brecciation, melting, and high-temperature metamorphism can occur at a variety of crustal depths in the core of a central uplift during an impact. Granulitic lithologies have radiometric

Table 1.3. Shock wave barometry for lunar rock-forming minerals
and for whole rock melting.

Mineral/Rock	Shock Effects	Shock Pressure (GPa)
Olivine	undulatory extinction	4–5 to 10–15
	mosaicism	10–15 to 60–65
	planar fractures	15–20 to 60–65
	planar deformation features	35–40 to 60–65
	melting and recrystallization	> 60–65
Plagioclase	undulatory extinction	5–10 to 10–12
	mosaicism	10–12 to 28/34*
	diaplectic glass	28/34* to 45
	melting	> 45
Pyroxene	undulatory extinction	5–10 to 20–30
	mechanical twinning	> 5
	mosaicism	20–30 to 75–80
	planar deform. features	30–35 to 75–80
	incipient melting	> 75–80
Quartz	planar fractures (0001) and {1011}	> 5–10
(minor phase in lunar rocks)	planar deformation features {1013}	>10
	planar deformation features {1012}	>20
	diaplectic quartz glass	34 to 50
	stishovite	12 to 45
	melting	> 50
Basalt/gabbro	whole rock melting	> 75–80
Dunite	whole rock melting	> 60–70
Anorthosite	whole rock melting	> 45–50
Regolith	whole rock melting	> 40

* noritic anorthosite composition
Compilation by Stöffler 2003, unpublished; data from Müller and Hornemann 1969; Hornemann and Müller 1971; Stöffler and Hornemann 1972; Stöffler 1972, 1974; Snee and Ahrens 1975; Kieffer et al. 1976; Schaal and Hörz 1977; Stöffler and Reimold 1978; Schaal et al. 1979; Bauer 1979; Ostertag 1983; Stöffler et al. 1986, 1991; Bischoff and Stöffler 1992; Stöffler and Langenhorst 1994; Schmitt 2000).

ages ranging from 3.75 to >4.2 b.y. (Bickel 1977; Ostertag et al. 1987; Stadermann et al. 1991) and therefore mostly predate the postulated "terminal cataclysm," if it existed. The source of heat and setting of ancient thermal metamorphism on the Moon remains enigmatic.

2.1.7. Impact melts. In the Apollo collection, impact-melt rocks, breccias, and glasses constitute some 30–50% of all hand-sized samples returned from the highland landing sites and about 50% of all soil materials, including the mare soils. Figure 1.17a,b shows a numerical model of impact-melt generation and an example of a lunar impact-melt pond associated with King crater. Impact melts are a homogenized mixture of the target lithologies and several compositional and textural characteristics distinguish them from conventional igneous rocks. One is the occurrence of clasts of target materials and schlieren within melted matrices. Lunar impact-melts can be identified by remnants of the projectile itself, particularly by the detection of high concentrations in Ni, Co, Ir, Au, and other highly siderophile elements (e.g., Korotev 1987, 1990; Warren 1993; Warren et al. 1997).

Identification and proper assignment of impact melts to specific basin-forming events is crucial for dating these events in order to derive a coherent, absolute lunar stratigraphy (e.g., Stöffler and Ryder 2001). Cintala (1992) compared the impact-melt production of lunar and

Table 1.4. Progressive shock metamorphism of most common lunar non-porous crystalline rocks and of lunar regolith (Stöffler 2003, unpublished); Shock pressure corresponds to the final equilibration peak shock pressure for polycrystalline rocks (in GPa); post-shock temperature is the temperature increase (in °C) after pressure release relative to any ambient pre-shock temperature; PDF's = planar deformations features = isotropic lamellae of distinct crystallographic orientation (low Miller indices planes); Table based on data as referenced in Stöffler (1972, 1974, 1984), Stöffler et al. (1980, 1986, 1988), Bischoff and Stöffler (1992), and Schmitt (2000).

Anorthosite (< 10 vol. % of mafics)				Basalt, gabbro, norite, granulitic rocks and breccias*				Regolith (impactoclastic detritus)			
Shock stage	Shock P	Post-shock T	Shock effects	Shock stage	Shock P	Post-shock T	Shock effects	Shock stage	Shock P	Post-shock T	Shock effects
0			sharp opt. extinct. of minerals; fract.	0			fractures, sharp optical extinction of minerals	0			unshocked and relic porosity
1a	< 5	50	plag with undulat. extinct.; px with mechanic. twinn.	1a	< 5	50	fractured minerals with undulatory optical extinction	1	~ 3	~ 250	compaction of regolith (zero porosity)
1b	26	220	plag partially isotropic and PDF's; mosaic. in px	1b	20-22	200	plag with PDF's and und. ext.; mechanic. twinn. px and ilm	2	~ 5-6	350	incipient lithification by submicr. intergr. melt (glass) cement
2	28	300	diaplectic plag glass; mosaicism and PDF's in px	2	28	300	diaplectic plag glass; px with mosaicism	3	10	700	lithification by intergranular melting and cementation
3	40-45	800-900	plag with flow struct. and vesicul.; strong mosaic. and PDF's in px; mixed melt of plag and px	3	40-45	~ 900	plag with incipient flow structure, mafics as in stage 2 but strong mosaic.	4	20	1500	strong intergranular melting and cementation, incipient formation of vesiculated melt
4	60	1850	complete whole rock melting and fine-grained crystallization of quenched melt	4	~ 60	~ 1400	normal plag glass with flow structures and vesicles; px with strong mosaic. and PDF's; incipient edge melting of mafics	5	> 40	> 2000	large scale formation of vesiculated "mixed" melt and whole rock melting
5	> 100	> 3000	whole rock vaporization	5	~ 80	~ 1600	vesicul. plag glass increasingly mixed with melt products of coexisting minerals				
				6	> 100	> 1800	complete whole rock melting and incr. vaporization				

Abbreviations: ol = olivine, plag = plagioclase, px = pyroxene, ilm = ilmenite; mosaic. = mosaicism, ext. or extinct. = extinction, incr. = increased, fract. or fr. = fractures, mechanic. = mechanical, twinn. twinning, pl. or plan. = planar, opt. = optical, undulat.or und. = undulatory, struct. = structure, vesicul. = vesiculation; submicr. = submicroscopic, intergr. = intergranular; diapl. = diaplectic, gl. = glass; note that pressure calibration is adjusted to plag of An>90 for anorthosite and related rocks and for plag of An>80 for basaltic/gabroic rocks (Ostertag 1983; Stöffler et al. 1986).

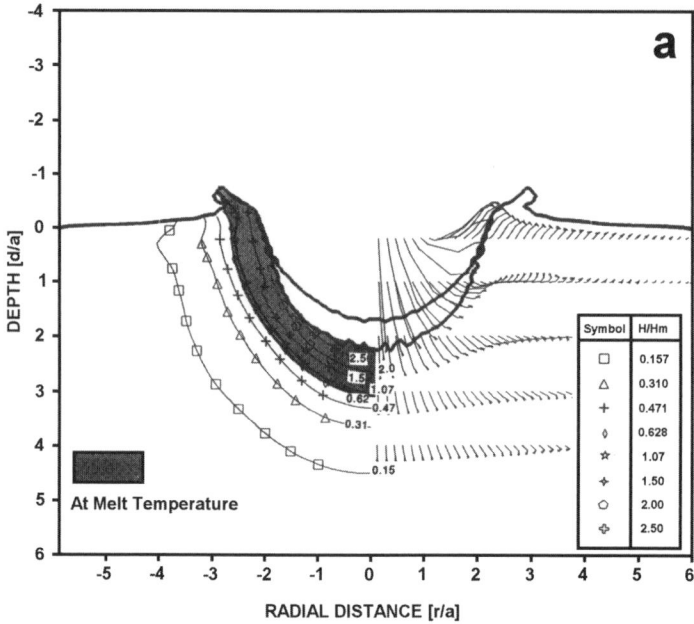

Figure 1.17. Simulation of impact-melt generation (O'Keefe and Ahrens 1994) and example of impact-melt ponds associated with King crater (Ap 16 frame H-19580). (Used by permission of The Geological Society of America, after O'Keefe and Ahrens 1994, *Large Meteorite Impacts and Planetary Evolution*, Fig. 3, p. 107).

mercurian craters. From his model, Cintala (1992) concluded that the surface temperature is important but that the impact velocity is the dominant factor controlling the amount of impact-melt produced by an impact. Cintala and Grieve (1994) pointed out that craters on the Moon should contain proportionately less impact melt than terrestrial craters and that the clast content in lunar impact melt should be higher.

  2.1.8. Crater frequency and bombardment history. Statistical investigations of impact-crater populations are providing key data for understanding the history and evolution of the Moon and the solar system. Because the distribution of impact-crater diameters reflects the number and mass distribution of the incoming projectiles, it provides fundamental information about the collision dynamics in the solar system, such as the mass frequency and the time-integrated flux of impacting objects.

  The exact shape of the size frequency distribution of craters created on a fresh surface has long been debated (e.g., Shoemaker 1970; Hartmann 1971; Baldwin 1971; Hartmann et al. 1981; Neukum 1983; Neukum et al. 2001).Part of the problem is the difficulty of finding large enough contiguous units that are much younger than the mare areas. Since the publication of the Lunar Source Book (Heiken et al. 1991), Galileo and Clementine data allowed McEwen et al. (1993, 1997) to estimate the size frequency distribution of rayed (Copernican) craters. Neukum et al. (2001) showed that a non-power-law shape of the production function gives a good fit to the new data (Fig. 1.18). With this production function at hand, crater statistics can be used to date planetary surfaces because older surfaces accumulate more craters than younger ones. The density of craters in any given region therefore yields the relative age of a geologic unit. Correlating crater statistics of a region with radiometric ages from returned samples not only allows calculation of absolute crater production rates and absolute projectile fluxes, but also provides absolute model ages for unsampled lunar regions (Fig. 1.19a,b). Neukum et al. (2001) concluded that for the last 4 b.y., the shape of the production function did not change within the limits of observational accuracy. The size-frequency distribution of the crater-forming bodies on the Moon is similar to the size-frequency distribution of asteroids in the Main Belt; from this, it appears that asteroids from the Main Belt and Near Earth Asteroids (NEAs) are the main source for lunar impact craters, and comets play only a minor role.

  Most non-volcanic rocks in the Apollo sample collection have ages rang-

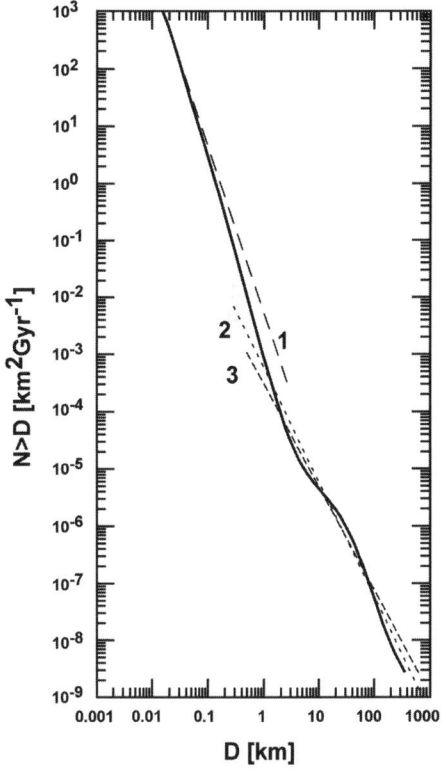

Figure 1.18. Comparison of the lunar production function (Neukum et al. 2001) with power-law distributions from (1) Shoemaker (1970), (2) Hartmann (1971), and (3) Baldwin (1971), and Hartmann et al. (1981). (Used by permission of Kluwer Academic Publishers, after Neukum et al. 2001, *Chronology and Evolution of Mars*, Fig. 2, p. 61).

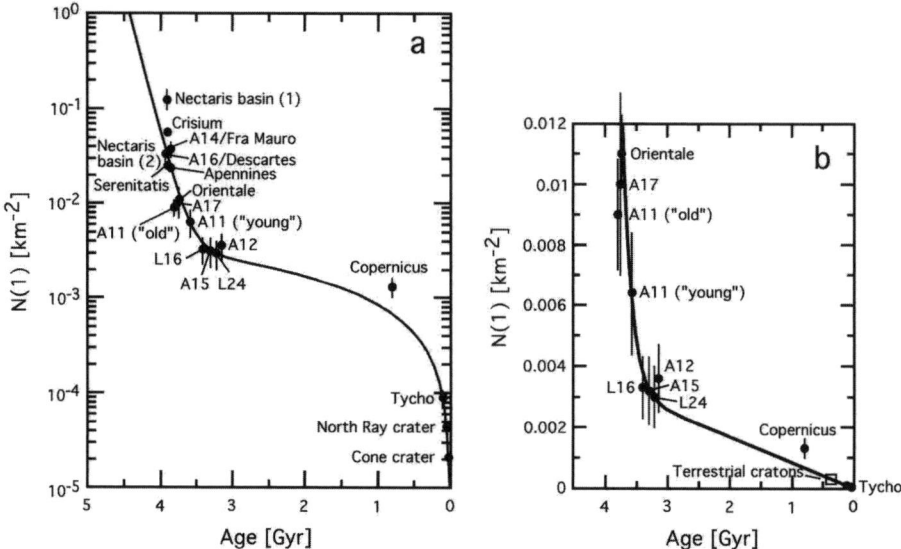

Figure 1.19. Lunar cratering chronology in (a) log form and (b) linear form (after Neukum et al. 2001; see also Stöffler and Ryder 2001, discussion in chapter 5, Fig. 5.33).

ing from 4.5 to 3.8 b.y. and are thus only 10–30% older than most mare basalts, yet the high-lands surfaces display 10–50× more craters >15 km than do the mare-basalt surfaces. Thus, it has been concluded that the crater production was significantly higher between 4.5 and 3.8 b.y. compared to the last 3.8 b.y. Interpretations differ on this finding. One group of workers sees this difference as the tail of the original planetary accretion process (e.g., Hartmann 1975; Neukum and Wise 1976; Neukum et al. 2001), whereas others propose that a cataclysmic increase in the infalling projectiles occurred 3.8–4.0 b.y. ago (e.g., Tera et al. 1974; Ryder 1990; Cohen et al. 2000; Ryder 2002; Kring and Cohen 2002), as evidenced by the narrow range of ages of the sampled impact-melt breccias (mostly 3.8–4.0 b.y.). The issue of a terminal lunar cataclysm is not resolved and remains one of the key objectives in crater-chronology research.

2.1.9. Impact erosion. Meteorite impacts are a major process in transporting material verti-cally and laterally, and they are the major erosional process on the Moon. Very small impacts have an abrasive effect similar to sandblasting, whereas larger projectiles up to centimeters in size can completely shatter a target by "collisional fragmentation" (Gault and Wedekind 1969). Erosion rates by abrasion have been estimated to be about 1 mm/10^6 years for kilogram-sized lunar rocks, with faster erosion rates for larger rocks (Ashworth 1977). Collisional fragmenta-tion is a faster and more effective process than abrasion and it is estimated that on average, a rock of 1 kg survives at the lunar surface for about 10 Ma before becoming involved in another impact event. Craddock and Howard (2000) recently compared the topography of simulated craters with the topography of degraded craters on the Moon and estimated that the average erosion rate since the Imbrian Period has been ~2.0±0.1 × 10^{-1} mm/million years.

2.1.10. Regolith formation. Impacts are the most important metamorphic process on the Moon, altering the textures of rocks and generating new ones, such as glasses, impact melts and fragmental rocks (breccias). Repetitive and frequent impacts cause shattering, burial, exhumation, and transportation of individual particles in a random fashion. The numerous large impacts that occurred during the period of heavy bombardment shattered and fragmented the lunar crust down to several kilometers and produced a global layer of chaotically mixed impact debris, termed "megaregolith" (Fig. 1.20). Seismic evidence and estimates on the effects of large-scale bombardment of the highlands during the epoch of intense bombardment suggested that the cumulative ejecta thickness on the lunar highlands is 2.5–10 km, possibly tens of kilometers (e.g., Short and Foreman 1972; Toksöz et al. 1973; Hörz et al. 1977; Cashore and Woronow 1985). The average depth to which the lunar crust is mechanically disturbed by impacts is not well known but conservative estimates indicate a thickness of the ejecta blankets of at least 2–3 km (megaregolith), a structural disturbance to depths of more than 10 km, and fracturing of the *in-situ* crust reaching down to about 25 km (Fig. 1.20).

In terms of moving large masses of material and reshaping the Moon's crust, basin-forming impacts are by far the dominant force. Large basin-forming events can distribute pervasive ejecta deposits to about 6 final crater radii, and can make significant contributions to the megaregolith at any location on the lunar surface (Haskin 2000). For example, at the Apollo 16 landing site ejecta deposits of Nectaris, Serenitatis, and Imbrium each are expected to be, on average, in excess of 500 m thick (Haskin 2000) with younger basin deposits eroding into older, underlying deposits.

Over lunar geologic time, bombardment by large impactors decreased and smaller impacts became relatively more important. The result of these smaller impacts is an accumulated fine-grained powdery layer on the lunar surface above the megaregolith. This uppermost part of the megaregolith layer is called "regolith" and is continuously "gardened" or turned over by

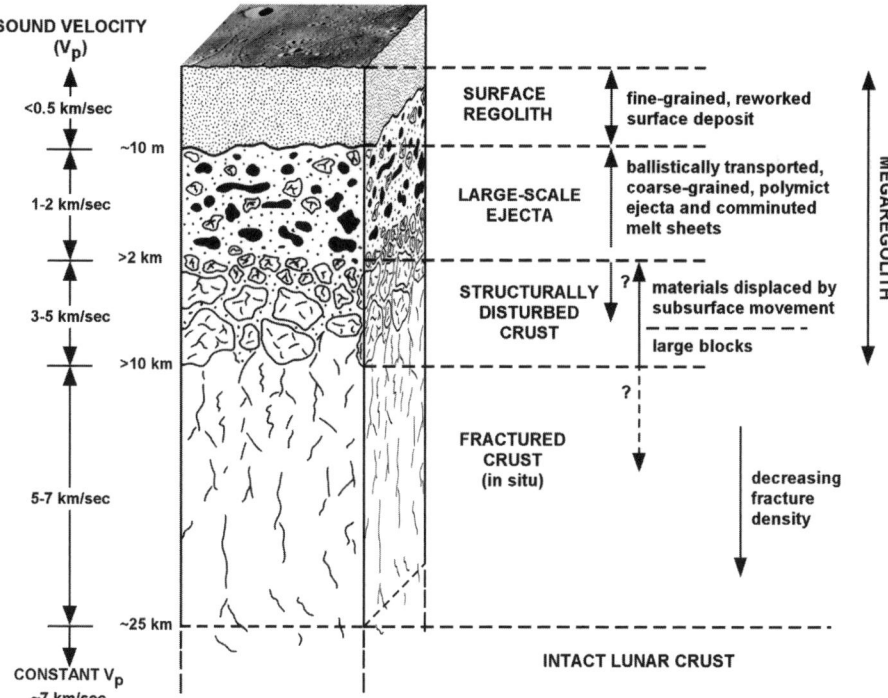

Figure 1.20. Highly idealized cross-section through the internal structure of the megaregolith illustrating the effects of large-scale cratering on the structure of the upper crust. The depth scale of this figure is highly uncertain because regional variations are expected depending on the degree to which a region has been influenced by basin-sized impacts. (Used by permission of Cambridge University Press, after Hörz et al. 1991, *Lunar Sourcebook*, Fig. 4.22, p. 93).

impacts (Fig. 1.20). In the development and evolution of the regolith, rare large craters play an important role because they can create a layered regolith column despite the tendency of the smaller craters to homogenize the upper parts of the regolith. In general, the composition of the regolith is largely controlled by the underlying local "bedrock," which may itself be megaregolith, and regolith samples from any given Apollo site usually show a relatively narrow compositional variation (see Chapter 2).

Although much of the regolith derives from local sources (e.g., Rhodes 1977; Hörz 1978), remotely sensed data of highland/mare boundary regions indicate clear evidence that lateral transport of material owing to meteorite impacts occurred on the Moon (e.g., Mustard and Head 1996; Li and Mustard 2000). Projectiles that are responsible for the macroscopic evolution of the regolith are typically 10–1000 cm in diameter. Based on the number of such impact events, the regolith thicknesses are typically only a few meters for mare areas and >10 m for the highlands. Shkuratov et al. (2001), on the basis of new radar data, estimated that the average thickness of the regolith is ~5–12 m. Thin regoliths (~4 m) were derived for Mare Serenitatis, Mare Tranquillitatis, and Mare Humorum, and a thick regolith was derived for Mare Nectaris (~9 m). Shkuratov et al. (2001) argued that the regolith thickness correlates with age and that a higher regolith production rate existed prior to ~3.5 b.y. ago. The overall growth rate of regolith decreases with time because the regolith itself acts as a shield that protects the bedrock from incorporation into the regolith. Only rare large craters can penetrate the regolith to incorporate "fresh" bedrock; smaller impacts only redistribute the existing regolith.

2.1.11. Bright ray craters. Bright rays and topographic freshness were used as the main criteria to assign a Copernican age to craters (e.g., Wilhelms 1987). However, using recent Clementine data, Hawke et al. (1999) found that bright rays of fresh craters can be due to differences in maturity (e.g., Messier), composition (e.g., Lichtenberg), or a combination of both (e.g., Tycho, Olbers A). Grier et al. (1999) used bright ray craters to study the cratering rate in the last few hundred million years. On the basis of maturity maps derived from Clementine images, Grier et al. (1999) determined relative ages of rayed craters and concluded that there is no evidence for a change in the cratering rate since Tycho (~109 Ma; Grier et al. 1999) compared to the cratering rate since Copernicus (~810 Ma; Grier et al. 1999).

2.1.12. Lunar swirls. The origin of light and dark colored "swirl-like" markings, up to hundreds of kilometers across, (e.g., Reiner Gamma) is still under debate and a variety of models including cometary impacts, magnetic storms, volcanism, or alteration by gases from the lunar interior have been suggested for their origin (e.g., El-Baz 1972; Schultz and Srnka 1980). However, recent magnetic data suggest a link to impact-related magnetism because the swirls in Mare Ingenii and near Gerasimovich crater are associated with relatively strong magnetic anomalies antipodal to some of the young impact basins such as Imbrium and Crisium (Hood 1987; Lin et al. 1988; Hood and Williams 1989; Halekas et al. 2001; Hood et al. 2001; Richmond et al. 2005). Because the Reiner Gamma and Rima Sirialis magnetic anomalies are oriented radially to the Imbrium basin, Hood et al. (2001) proposed that the most likely sources are Imbrium ejecta beneath a thin veneer of mare basalts. If so, the albedo markings associated with Reiner Gamma may be consistent with a model involving magnetic shielding of freshly exposed mare materials from the solar wind ion bombardment, resulting in a reduced rate of surface darkening relative to the surrounding basalts (Hood et al. 2001). In an independent study, Halekas et al. (2001) argued that a cometary origin of the swirls appears to be unlikely because some of them are found in locations antipodal to large farside impact basins. They favor an origin as buried, magnetized Imbrium ejecta similar to Hood et al. (2001).

2.2. Volcanic processes

2.2.1. Nature of lunar volcanism. The presence, timing and evolution of volcanic activity on the surface of a planet are indicative of its thermal evolution. On the Moon, a one-plate planetary body, volcanism apparently resulted from melting of mantle rocks that was unlikely to be contaminated by recycled crust. Such melting was caused by the decay of naturally radioactive elements and resulted in the production of partial melts, most commonly of basaltic composition (45–55% SiO_2 and relatively high MgO and FeO contents). In order to produce basaltic melts on the Moon, temperatures of >1100 °C and depths of >150–200 km would have been required. Radiometric ages of returned lunar samples indicate that most volcanism on the Moon ceased approximately 3 b.y. ago, implying that the mantle long ago cooled below the temperature necessary to produce partial melts. However, crater counts on mare basalt surfaces suggest that some basalts erupted as late as ~1–2 b.y. ago (Hiesinger et al. 2003). Recent two- and three-dimensional modeling of the thermal evolution of the Moon suggests the growth of a 700–800 km thick lithosphere while the lower mantle and core only cooled 100–200 °C (Spohn et al. 2001). According to such models, the zone of partial melting necessary for the production of basaltic magmas would migrate to depths too great for melts to reach the surface at ~3.4–2.2 b.y. ago. However, a non-uniform distribution of heat-producing elements in the mantle as indicated by Lunar Prospector data could extend the potential for melting to more recent times.

Basaltic lavas cover about 17% or 7×10^6 km^2 of the lunar surface and make up ~1% of the volume of the lunar crust (Head 1976) (Fig. 1.21). These basalts are almost exclusively exposed within the nearside basins, are rare on the lunar farside, and form the relatively smooth dark areas visible on the lunar surface. They appear to be correlated with the distribution of KREEP (especially Oceanus Procellarum) suggesting a possible genetic relationship between the two phenomena (Haskin et al. 2000a,b; Wieczorek and Phillips 2000; see also Chapters 3 and 4). Besides these lava flows, other products of basaltic volcanism occur on the Moon. Fire fountaining, driven by gas exsolution from erupting lava, dispersed melt as fine droplets in the form of pyroclastic deposits or volcanic ash (Chou et al. 1975; Delano and Livi 1981; Delano 1986). On the basis of either their glass-rich nature or their abundant titanium-rich black spheres, pyroclastic deposits can be clearly distinguished from lava flow products with

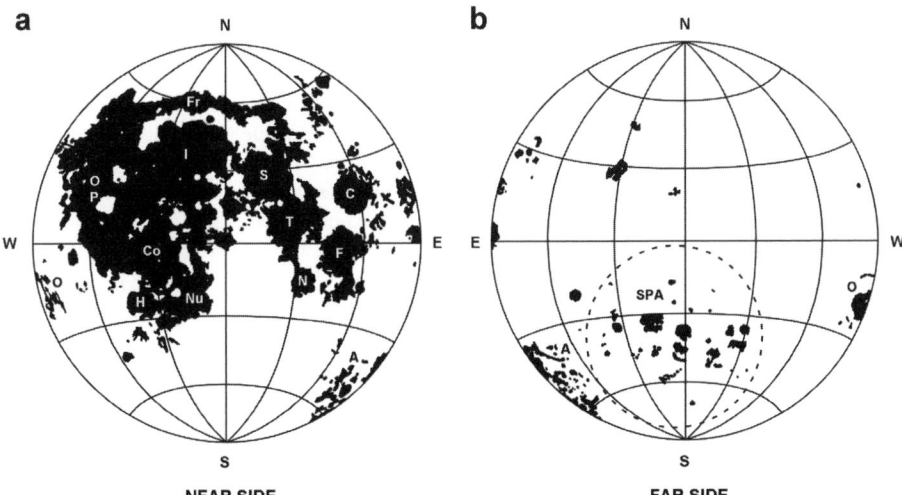

Figure 1.21. Map of the distribution of mare basalts on the lunar nearside (a) and the farside (b) (Head 1976). Note the highly asymmetric distribution of mare basalts on the nearside and the farside. Australe (A), Crisium (C), Cognitum (Co), Fecunditatis (F), Frigoris (Fr), Humorum (H), Imbrium (I), Nectaris (N), Nubium (Nu), Orientale (O), Oceanus Procellarum (OP), Serenitatis (S), South Pole-Aitken (SPA), Tranquillitatis (T).

multispectral remote-sensing techniques. However, identifying pyroclastic deposits of low-Ti content with current remote sensing techniques is problematic.

2.2.2. Lava flows. Very early, Galileo Galilei interpreted the smooth dark mare plains as oceans, thus the Latin name "mare," and other interpretations included emplacement as pyroclastic flows (Mackin 1964) and deep dust mass-wasted from the highlands (Gold 1966). However, samples returned from the Apollo and Luna missions provided firm evidence that the mare areas formed by large-volume eruptions of low-viscosity basaltic lavas, similar in style to the Columbia River flood basalts on Earth. Additional support for such an origin comes from telescopic and spacecraft images, which show lava flows several tens of meters thick that extend for hundreds to thousands of kilometers across the surface. In Mare Imbrium, the youngest lavas flowed in three successive events for 1200, 600, and 400 km across slopes of 1:1000 (Fig. 1.22a) (Schaber 1973; Schaber et al. 1976). Lobate flow fronts bounding these flows were found to be 10–63 m high, averaging ~35 m (Gifford and El Baz 1978).

Measurements of molten lunar basalts revealed that their viscosity is unusually low, only a few tens of poise at 1200 °C, allowing them to flow for long distances across the surface before solidifying (Hörz et al. 1991). Thin lava flows such as the ones in Mare Imbrium are not unusual and have been observed elsewhere on the Moon, for example, within the Hadley Rille at the Apollo 15 landing site (Fig. 1.22b). In Hadley Rille at least three flows are exposed in the upper 60 m below the surface (Howard et al. 1973). Further support for thin lunar lava flows stems from investigations of the chemical/kinetic aspects of lava emplacement and cooling. Basalt samples of the Apollo 11, 12, and 15 sites indicate that individual flow units are no thicker than 8–10 m (Brett 1975). Many lunar lava flows lack distinctive scarps and this has been interpreted as indicating even lower viscosities, high eruption rates, ponding of lava in shallow depressions, subsequent destruction by impact processes, or burial by younger flows.

Distinctive kinks in the cumulative distributions of crater size-frequencies derived from Lunar Orbiter images have been used to estimate the thickness of lava flows (e.g., Hiesinger et al. 2002). This technique expands considerably the ability to assess lava flow unit thicknesses and allows one to obtain thicknesses and volumes for additional flow units that have not been detected in low-sun angle images. Using this method, Hiesinger et al. (2002) found lava flow units to average 30–60 m in thickness. These thicknesses are commonly greater than those typical of terrestrial basaltic lava flows and more comparable to those of terrestrial flood basalts; a correlation consistent with evidence for high effusion rates and volumes for basalt eruptions in the lunar environment (Head and Wilson 1992). An alternative technique to estimate the thickness of lava flows is to look at distinctive compositions of the ejecta deposits of small-scale craters in high-resolution Clementine color data. Using the depth/diameter relationship of lunar craters, one can derive the depth to the compositionally distinctive underlying material, which is equivalent to the thickness of the uppermost layer (e.g., Hawke et al. 1998; Antonenko et al. 1995).

2.2.3. Sinuous rilles. Besides the prominent lava flows of the lunar maria, other volcanic landforms have been observed on the Moon, such as sinuous rilles, domes, lava terraces, cinder cones, and pyroclastic deposits. Sinuous rilles are meandering channels that commonly start at a crater-like depression and end by fading downslope into the smooth mare surface (Greeley 1971) (Fig. 1.23a). Most sinuous rilles appear to originate along the margins of the basins and to flow towards the basin center. Schubert et al. (1970) reported that sinuous rilles vary from a few tens of meters to 3 km in width and from a few kilometers to 300 km in length. The mean depth of these rilles is ~100 m. The Apollo 15 landing site was specifically selected to study the formation process of sinuous rilles. No evidence for water or pyroclastic flows was found and it was concluded that sinuous rilles formed by widening and deepening of channeled lava flows due to melting the underlying rock by very hot lavas (Hulme 1973; Coombs et al. 1987). On the basis of geochemical studies of a lava tube in the Cave Basalt flows, Mount St. Helens, Williams et al. (2001) concluded that mechanical and thermal erosion operated simultaneously

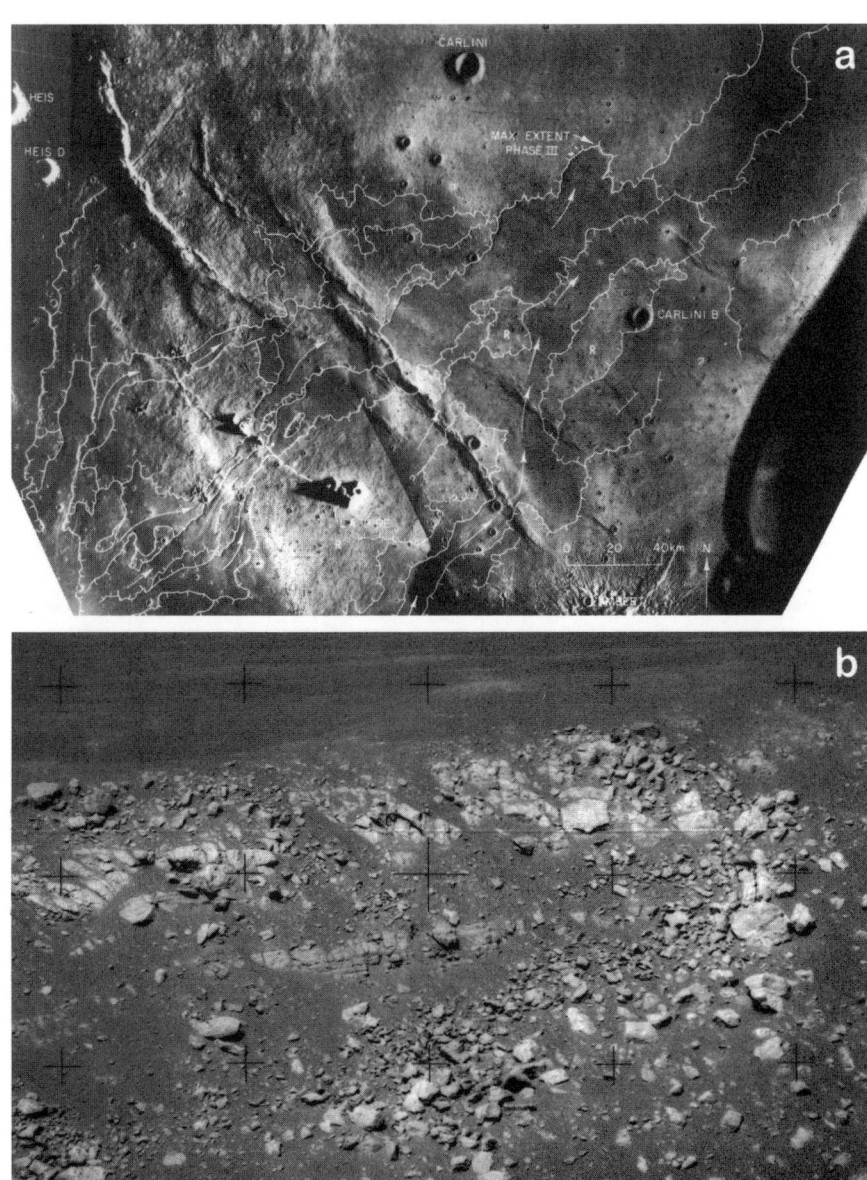

Figure 1.22. Examples of volcanic landforms; (a) mare flows in Mare Imbrium (Ap 15 frame M-1556) (Schaber 1973; Schaber et al. 1976) and (b) flows exposed within the wall of Hadley Rille (Ap 15 frame H-12115).

to form the lava tube. The much larger sizes of lunar sinuous rilles compared to terrestrial lava tubes are thought to result from lower gravity, high melt temperatures, low viscosities, and high extrusion rates. Since the publication of the Lunar Source Book (1991), new work by Bussey et al. (1997) showed that the process of thermal erosion is very sensitive to the physical conditions in the boundary layer between lava and solid substrate. They found that thermal erosion rates

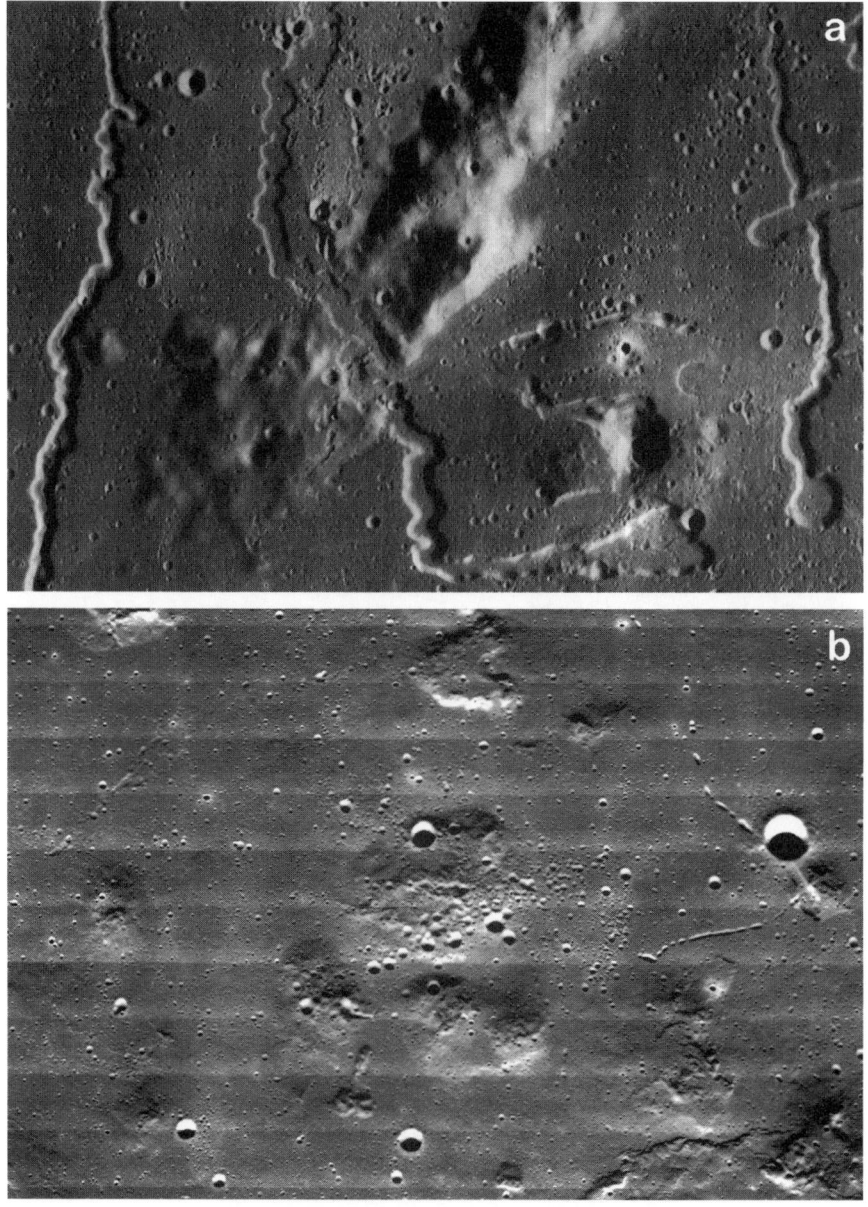

Figure 1.23. Examples of volcanic landforms; (a) sinuous rilles of Rimae Prinz (LO V 191H1) and (b) mare domes of the Marius Hills (LO V M-210).

depend on the slopes, effusion rates, and thermal conductivities of the liquid substrate boundary layer. In their model, the highest thermal erosion rates occur when thermal conductivities are high for the liquid substrate and low for the solid substrate.

2.2.4. Cryptomaria. Cryptomaria are defined as mare-like volcanic deposits that have been obscured from view by subsequent emplacement of lighter material, commonly ejecta from craters and basins (Head and Wilson 1992). These deposits have been studied through investigation of dark halo craters (e.g., Schultz and Spudis 1979, 1983; Hawke and Bell 1981), through multispectral images (e.g., Head et al. 1993; Greeley et al. 1993; Blewett et al. 1995; Mustard and Head 1996), and Apollo orbital geochemistry (e.g., Hawke and Spudis 1980; Hawke et al. 1985). Significant progress has been made in the last ten years and the new data show that if cryptomaria are included, the total area covered by mare deposits exceeds 20% of the lunar surface, compared to ~17% of typical mare deposits alone (Head 1976; Antonenko et al. 1995). Recent studies using Clementine color data further expand the extent of cryptomare, especially in the Schickard region (e.g., Hawke et al. 1998; Antonenko and Yingst 2002). Similarly, Clark and Hawke (1991), on the basis of Apollo X-ray fluorescence data, found a mafic enrichment of areas south of crater Pasteur apparently associated with buried basalts. The data not only show a wider spatial distribution of products of early volcanism, but also indicate that mare volcanism was already active prior to the emplacement of Orientale ejecta. Even older cryptomare may exist but evidence for such ancient cryptomare is likely to be even more obscured by superposed ejecta deposits. Using Clementine multispectral images and Lunar Prospector gamma-ray spectrometer data, Giguerre et al. (2003) and Hawke et al. (2005) report that the buried basalts in the Lomonosov-Fleming and the Balmer-Kapteyn regions are very low to intermediate titanium basalts. Sampling and subsequent analysis of such deposits is required to understand the true nature of cryptomaria. Until then, careful study of crater ejecta using, for example, high-resolution reflectance spectra, should lead to a better understanding of the extent and importance of the very earliest volcanism on the Moon.

2.2.5. Volcanic centers. Several areas of the Moon are characterized by an anomalously high concentration of volcanic features. Examples include the Marius Hills (Fig. 1.23b) and the Aristarchus Plateau/Rima Prinz region in Oceanus Procellarum. The Marius Hills (~35,000 km^2) consist of more than 100 domes and cones and 20 sinuous rilles (Weitz and Head 1998), and the Aristarchus Plateau/Rima Prinz region (~40,000 km^2) shows at least 36 sinuous rilles (Guest and Murray 1976; Whitford-Stark and Head 1977). The large number of sinuous rilles suggests that these two regions are loci of multiple high-effusion-rate, high-volume eruptions, which may be the source for much of the lavas exposed within Oceanus Procellarum (Whitford-Stark and Head 1980). Work by McEwen et al. (1994), using Clementine color data, indicates the presence of mare basalts that underlie a 10–30 m thick layer of pyroclastic material and anorthosite within the Aristarchus crater. Clementine altimetry profiles show the Aristarchus Plateau sloping ~1° to the north-northwest and rising about 2 km above the lavas of the surrounding Oceanus Procellarum (McEwen et al. 1994).

2.2.6. Domes, sills, and shields. Lunar mare domes are generally broad, convex, semi-circular landforms with relatively low topographic relief (Fig. 1.23b). Guest and Murray (1976) mapped 80 low domes with diameters of 2.5–24 km, 100–250 m heights, and 2–3° slopes with a high concentration in the Marius Hills complex. Some of the domes of the Marius Hills have steeper slopes (7–20°) and some domes have summit craters or fissures. The formation of mare domes is thought to be related to eruptions of more viscous (more silicic) lavas, intrusions of shallow laccoliths, or mantling of large blocks of older rocks with younger lavas (e.g., Heather et al. 2003; Lawrence et al. 2005). Importantly, no shield volcanoes larger than ~20 km have been identified on the Moon (Guest and Murray 1976).

Large shield volcanoes (>50 km) are the most prominent volcanic features on Earth, Venus, and Mars. They are built by a large number of small flows derived from a shallow magma reservoir where the magma reaches a neutral buoyancy zone (e.g. Ryan 1987; Wilson and Head 1990). The presence of shield volcanoes and calderas indicates shallow buoyancy zones, stalling and evolution of magma there, production of numerous eruptions of small volumes and durations, and shallow magma migration to cause caldera collapse. The general absence of such shield volcanoes implies that shallow buoyancy zones do not occur on the Moon, and that lavas did not extrude in continuing sequences of short-duration, low-volume eruptions from shallow reservoirs. This is a fundamental observation and it makes the Moon an unique endmember compared to other terrestrial planets. However, in a few cases magma may have stalled near the surface and formed shallow sills or laccoliths, as possibly indicated by the formation of floor-fractured craters (Schultz 1976). This interpretation is supported by studies of Wichman and Schultz (1995, 1996) and by modeling of floor-fractured craters, which led Dombard, and Gillis (2001) to conclude that, compared to topographic relaxation, laccolith emplacement is the more viable formation process. Wichman and Schultz (1996) presented a model that allowed them to estimate the minimum depth of a 30 km wide and 1900 m thick intrusion beneath crater Tauruntius to be on the order of 1–5 km. The magma excess pressure was modeled to be on the order of 9 MPa.

2.2.7. Cones. Cinder cones, the most common terrestrial subaerial volcanic landform, are frequently associated with linear rilles on the Moon, for example in crater Alphonsus (Head and Wilson 1979). Lunar cones are less than 100 m high, are 2–3 km wide, have summit craters of less than 1 km, and have very low albedos (Guest and Murray 1976).

2.2.8. Lava terraces. In some craters small lava terraces have been observed and inter-preted as remnants of lava that drained back into a vent or by flow into a lower basin (Spudis and Ryder 1986). As for all other small-scale, low-relief features (e.g., cones, flow fronts), lava terraces are difficult to recognize and require special image qualities such as high spatial image resolution combined with favorable low-sun angle illumination to enhance their morphology.

2.2.9. Pyroclastic deposits. Regional pyroclastic deposits are extensive (>2500 km^2) and are located on the uplands adjacent to younger maria (e.g., Gaddis et al. 1985; Weitz et al. 1998). In contrast, localized pyroclastic deposits are smaller in extent and are more widely dispersed across the lunar surface (Head 1976; Hawke et al. 1989; Coombs et al. 1990). Pyroclastic glass beads and fragments are especially abundant at the Apollo 17 site (e.g., Figure 1.24a). The Apollo 17 orange glasses and black vitrophyric beads formed during lava fountaining of gas-rich, low-viscosity, Fe-Ti-rich basaltic magmas (Heiken et al. 1974). Experiments by Arndt and von Engelhardt (1987) indicated that crystallized black beads from the Apollo 17 landing site had cooling rates of 100 °C/s, which is much slower than expected from blackbody cooling in a vacuum. Apollo 15 green glasses are also volcanic in origin as are other pyroclastic glasses found in the regolith of other landing sites (Delano 1986). Although trivial in volume, these glasses are important because they indicate that lava fountaining occurred on the Moon and because, as melts generally unmodified by crystal fractionation, they may represent the best samples for studying the lunar mantle. Regional dark-mantle deposits, which cover significant areas of the lunar surface, can be detected in remotely sensed data (e.g., Hawke et al. 1979, 1989; Head and Wilson 1980; Gaddis et al. 1985; Coombs et al. 1990; Greeley et al. 1993; Weitz et al. 1998; Weitz and Head 1999; Head et al. 2002). These deposits tend to occur along the margins of impact basins and in association with large vents and sinuous rilles, implying that they were formed by large-volume sustained eruptions.

Regional dark-mantle deposits are interpreted to result from eruptions in which continuous gas exsolution in the lunar environment caused Hawaiian-style fountaining that distributed pyroclastic material over tens to hundreds of kilometers (Wilson and Head 1981, 1983). Figure 1.24b is an example for near-surface propagation of a dike, which resulted in an extensional surface stress field forming a graben (i.e., Hyginus Rille). Gas exsolution from the dike formed numerous pit craters along the graben floor (e.g., Head et al. 1998). Significant progress in understanding the ascent and eruption conditions of pyroclastic deposits has been made over the last decade and we point the reader, for example, to Head et al. (2002), references therein, and to Chapter 4 of this volume. There is also strong evidence that at least some of the observed dark-halo craters are volcanic (i.e., pyroclastic) in origin, whereas others are impact craters

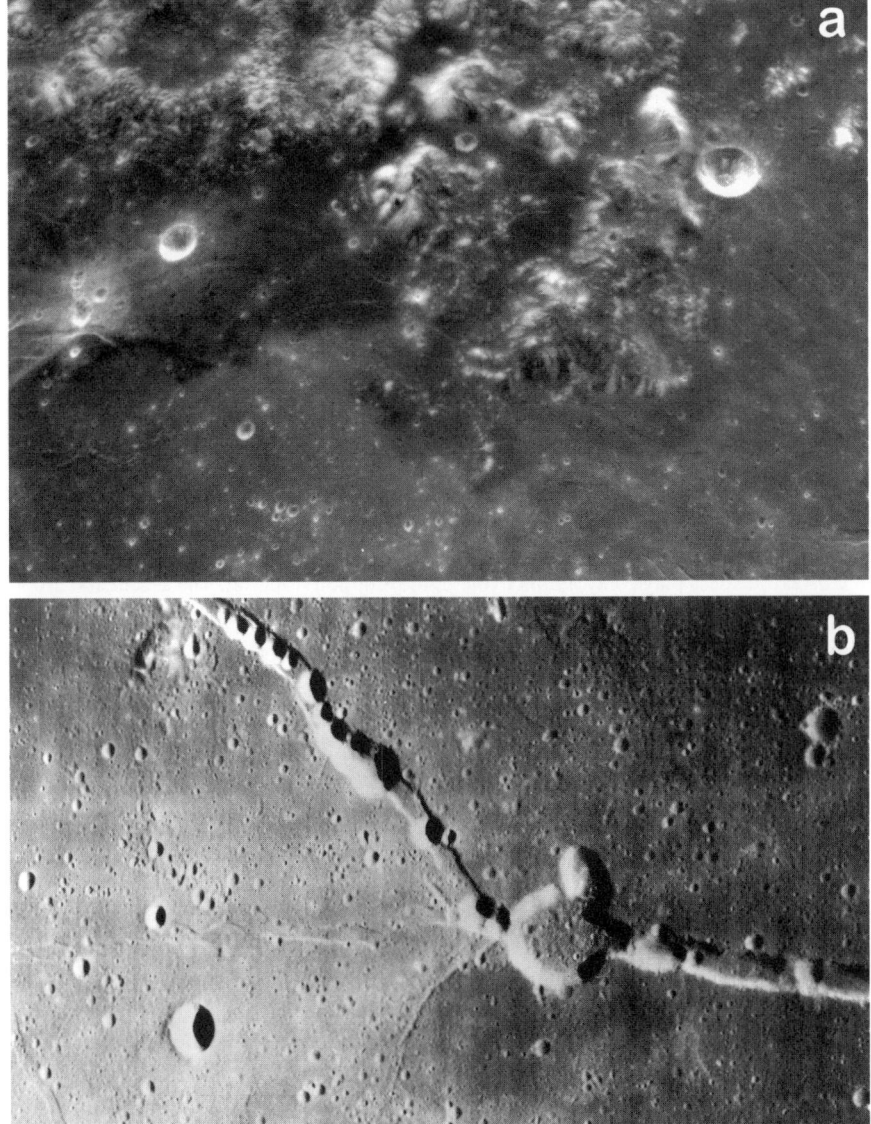

Figure 1.24. Examples of volcanic landforms; (a) pyroclastic deposits close to the Apollo 17 landing site (Ap 15 frame M-1404) and (b) pit craters of Rima Hyginus (LO V M-97).

that excavated darker material from subsurface levels. Dark-halo craters that are volcanic in origin are commonly located along fractures and on the floors of larger craters and do not have the same morphologies as impact craters. From studies of dark-halo craters within crater Alphonsus, it appears likely that they formed from vulcanian-style eruptions (Head and Wilson 1979; Coombs et al. 1990).

2.2.10. Filling of the basins. Studies of impact-basin formation indicate that immediately after the impact excavation, the transient cavity collapses, leading to large-scale fracturing of the crust and providing structures along which magma could ascend to the surface (also see the discussion in Chapter 4 and in section 1.2.2.13 of this chapter). It appears that physical uplift of mantle material during basin formation occurred in association with a decrease of lithostatic pressure caused by sudden removal of overburden (Brett 1976). If uplifted mantle materials were already near their melting temperatures, decreasing confining pressure might have led to melting and magma generation below the basin. However, from the absolute age dating of specific rock samples, it would appear that significant periods of time elapsed between basin formation and extrusion of mare basalts into the basin. For example, the Imbrium basin is thought to be 3.85 b.y. old, but the lavas sampled at the Apollo 15 site are about 3.3 b.y. old. Nonetheless, deeper, buried basalts at the Apollo 15 site and elsewhere in Mare Imbrium could be much older. Volcanism in some of the lunar nearside basins lasted several billion years after the basin impact events (Hiesinger et al. 2000, 2003), and it appears unlikely that magma would ascent along impact-induced fractures for such extended periods of time. Also, recent thermal modeling suggests that craters of 300–350 km in diameter likely did not raise uplifted mantle material above its solidus in order to produce mare basalts (Ivanov and Melosh 2003).

Differences also exist in the distribution and amount of lava filling lunar basins. Nearside basins were flooded with basalts to a significantly greater degree than farside basins, seemingly corresponding to differences in crustal thickness. However, Wieczorek et al. (2001) have argued that crustal thickness plays only a minor role in the eruption of basalt. The ascent and eruption of basalt remains a complex issue and is discussed in more detail in Chapters 3 and 4.

2.2.11. Volumes of basalts. The total volume of volcanic products can be used to estimate the amount of partial melting and how much of the lunar mantle was involved in the production of the mare basalts. Establishing the volume of mare basalts emplaced on the surface as a function of time (the flux) sets an important constraint on the petrogenesis of mare basalts and their relation to the thermal evolution of the Moon (Head and Wilson 1992). A variety of techniques have been used to estimate the thicknesses of basalts within the large lunar impact basins (summarized by Head 1982). Crater-geometry techniques using pristine crater morphometric relationships and the diameter of partially to almost entirely flooded craters yield values of basalt thickness up to 2 km, ranging 200–400 m on average (DeHon and Waskom 1976). Hörz (1978) reviewed the assumptions underlying the thickness estimates of DeHon and Waskom (1976) and concluded these values were overestimates.

More recently, thickness estimates have been updated using new Clementine spectral data by Budney and Lucey (1998), who investigated craters that penetrated the maria and excavated underlying nonmare material to derive the thickness of basalts in Mare Humorum. Their results agree better with results from Hörz (1978) than with results from DeHon (1975, 1977, 1979) and Head (1982). Figure 1.25 is a map of the DeHon and Waskom (1976) data; insets compare thickness estimates of (a) Budney and Lucey (1998) with those of (b) DeHon and Waskom (1976). Studies of mascons, that is, positive gravity anomalies observed for many of the mare basins, indicate basalt thicknesses of 1–2 km and volumes of 6×10^6 km^3 (e.g., Head 1975a, Thurber and Solomon 1978). Because even the largest estimates are still <1 % of the volume of the lunar mantle, it appears that high degrees of partial melting involving large regions of the mantle did not occur within the Moon.

Equally important to estimates of the total erupted basalt volumes are estimates of volumes for individual basalt flow units. Since the publication of the Lunar Source Book (1991), several studies on this aspect of lunar volcanism have been conducted (e.g., Yingst and Head 1997, 1998; Hiesinger et al. 2002). Yingst and Head (1998) investigated individual isolated lava ponds in Mare Smythii and Mare Marginis and, from morphologic evidence, concluded that these lavas were emplaced during a single eruptive phase. They found that pond volumes in both

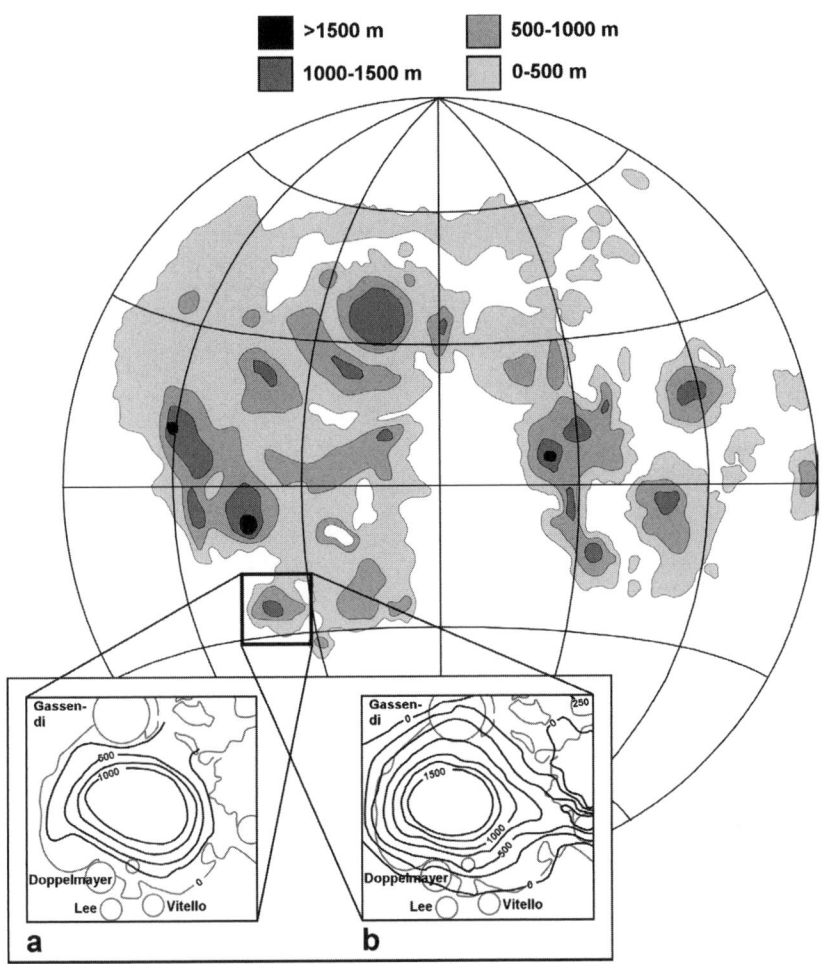

Figure 1.25. Thickness of basalts on the lunar nearside (DeHon and Waskom 1976). Inserts compare estimates for Mare Humorum of (a) Budney and Lucey (1998) and (b) DeHon and Waskom (1976).

maria range from 15 km³ to 1045 km³. The mean pond volume in Mare Smythii is ~190 km³ and about ~270 km³ in Mare Marginis. Ponds in the South Pole-Aitken basin that were also interpreted to be single eruptive phases have volumes ranging from 35 to 8745 km³ and average 860 km³ (Yingst and Head 1997). Flow units in the Mare Orientale/Mendel-Rydberg basins have volumes of 10 to 1280 km³ and a mean volume of 240 km³ (Yingst and Head 1997). From a study of 58 mare flow units in the major nearside basins, Hiesinger et al. (2002) found that the range of volumes is 30–7700 km³. The minimum average volume of all investigated flow units was estimated to be ~590 km³ and the maximum average volume, ~940 km³.

2.2.12. Ages of lunar mare basalts. The onset and extent of mare volcanism are not very well understood (summarized by Nyquist et al. 2001). The returned samples revealed that mare volcanism was active at least between ~3.9 and 3.1 Ga (Head 1976; Nyquist and Shih 1992). Ages of some basaltic clasts in older breccias point to an onset of mare volcanism prior to 3.9 b.y. (Ryder and Spudis 1980), perhaps as early as 4.2–4.3 b.y. in the Apollo 14 region (Taylor et al. 1983; Dasch et al. 1987; Nyquist et al. 2001). Early volcanism is also supported by remote-sensing data. For example, Schultz and Spudis (1979), Hawke and Bell (1981), and Antonenko et al. (1995) interpreted dark halo craters as impacts into basaltic deposits that are now buried underneath a veneer of basin ejecta. These underlying basalts might be among the oldest basalts on the Moon, implying that volcanism was active prior to ~3.9 b.y. ago.

Remote-sensing data suggest that the returned samples represent only a small number of basalt types from a few limited locations and that the majority has still not been sampled (Pieters 1978; Giguere et al. 2000). On the basis of crater degradation stages, Boyce (1976) and Boyce and Johnson (1978) derived absolute model ages that indicate volcanism might have lasted from 3.85 ± 0.05 b.y. until 2.5 ± 0.5 b.y. ago. Support for such young basalt ages comes from a recently collected lunar meteorite, Northwest Africa 032, which shows a Ar-Ar whole rock age of ~2.8 b.y. (Fagan et al. 2002). Schultz and Spudis (1983) made crater size-frequency distribution measurements for basalts embaying the Copernican crater Lichtenberg and concluded that these basalts might be less than 1 b.y. old.

Over the last five years, a large effort has been made to systematically determine model ages of all basalts on the lunar surface using crater counts (e.g., Hiesinger et al. 2000, 2001, 2003). At the time of writing ~220 basalt units have been dated (Fig. 1.26). Hiesinger et al. (2000, 2001, 2003) applied a new approach based on color-ratio composites to ensure the definition of spectrally homogeneous lithologic units for which they performed crater counts. The model ages so derived indicate that mare volcanism lasted from ~4.0 to 1.2 b.y. Their results also indicate that lunar volcanism was not evenly distributed throughout time but peaked in the Late Imbrian and drastically declined during the Eratosthenian Period. The ages of lunar mare basalts are discussed in more detail in Chapter 4.

2.2.13. Magma generation and eruption. Petrologic models are discussed in Chapter 4; here we convey only some basic concepts about the ascent and eruption styles of lunar magmas.

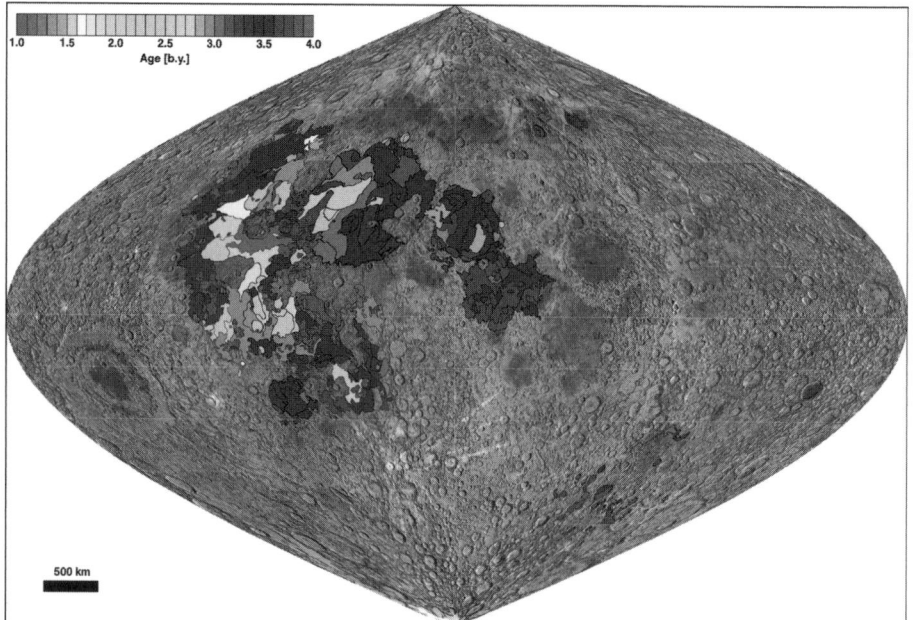

Figure 1.26. Absolute model ages of mare basalts based on crater counts on spectrally and morphologically defined mare units (Hiesinger et al. 2000, 2001, 2003). Base map is a shaded relief in sinusoidal projection. *See Plate 1.1 for a color version of this figure.*

Most petrologic models of lunar basaltic magmas suggest an origin by partial melting at 200–400 km depth (e.g., Ringwood and Essene 1970; Hubbard and Minear 1975; Ringwood and Kesson 1976; BVSP 1981; Taylor 1982). Interestingly, experimental data suggest the volcanic (pyroclastic) glasses were derived from greater depths within the lunar mantle (360–520 km) than the crystalline mare basalts (e.g., Green et al. 1975; Longhi 1992a,b, 1993). It was thought early on that magma ascent was related to impact-induced fracture and fault zones, with the location of conduits having been affected by impact structures (Solomon and Head 1980). However, it is unlikely that basin-induced fractures would have remained open for the extended periods of time (several hundred million years) that are indicated for extrusion of lavas by radiometrically dated samples and crater counts (Hiesinger et al. 1999; Melosh 2001). Dike propagation models further suggest that fractures played little to no role in controlling the paths of dikes. In addition, petrologic and geochemical evidence suggest that the basaltic melts are less dense than the lunar mantle but denser than the overlying crust. Hence, basaltic diapirs would rise buoyantly through the lunar mantle, but would stall at the base of the crust and would not erupt onto the surface if they simply followed fracture zones in the crust (Head and Wilson 1992).

Head and Wilson (1992) suggested instead that overpressurization of basaltic diapirs was the key to upward dike propagation into overlying rocks (Fig. 1.27). The excess pressure required to propagate a dike to the surface is on the order of ~15 MPa for a ~64 km thick

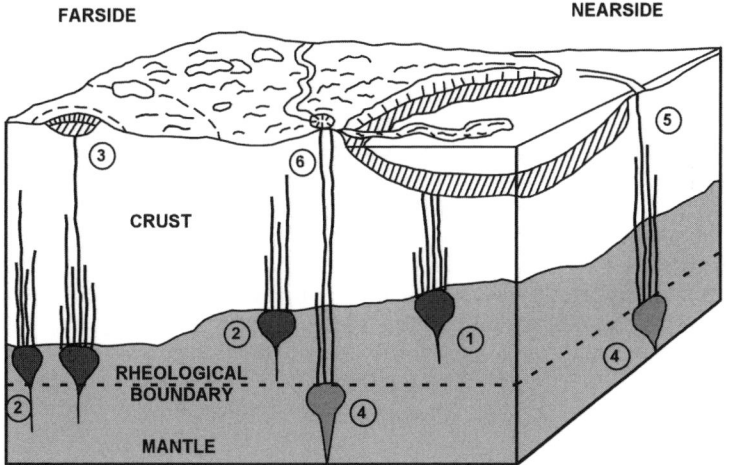

Figure 1.27. Magma transport from neutral buoyancy zones (Head and Wilson 1992). Diagrammatic representation of the emplacement of secondary crust on the Moon. At (1), early basaltic magmas rise diapirically to the density trap at the base of the crust. Those below topographic lows (thin crust) associated with lunar impact basins are in a favorable environment for dike propagation and extrusion of lavas to fill the basin interior. Diapirs reaching the base of the thicker crust on the farside and parts of the nearside (2) at the same time stall and propagate dikes into the crust, most of which solidify in the crust and do not reach the surface. Variations in regional and local compensation produce a favorable setting for emplacement of some lavas in craters and the largest basins on the farside (3). With time, the lithosphere thickens and ascending diapirs stall at a rheological boundary, (4) build up of excess pressure to propagate dikes toward the surface. At the same time, loading and flexure of the earlier mare deposits creates a stress environment that favors extrusion at the basin edge (5); lavas preferentially emerge at the basin edge, flowing into the subsiding basin interior. The latest eruptions are deepest and require high stress buildups and large volumes in order to reach the surface; thus, these tend to be characterized by high-volume flows and sinuous rilles (6). Deepening of source regions over time and cooling of the Moon causes activity to diminish and eventually cease.

nearside crust and about ~20 MPa for a ~86 km thick farside crust (Head and Wilson 1992). To erupt magma onto the surface, slightly higher pressures of ~21 and ~28 MPa, respectively, are required. These conditions correspond to dike widths of a few to several hundred meters (Head and Wilson 1992). This model predicts that volcanic material should preferentially occur on the lunar nearside where the crust is thinner and more dikes can reach the surface. On the farside more dikes would stall within the crust, extruding only in the deepest basins. This argument, however, does not explain the paucity of basalts in the South Pole-Aitken basin. From recent crustal thickness models and the presumption of a mafic lower crust, Wieczorek et al. (2001) concluded that all basaltic magmas would be less dense than the lower crust; thus, when a basin-forming impact excavated the upper portion of the crust, any basaltic magma could reach the surface. A more comprehensive knowledge of crustal thickness, crustal composition with depth, and basalt compositions around the globe are needed to improve our understanding of eruption of basalts on the Moon.

The common occurrence of volcanic glasses among the lunar samples and as observed with remote sensing indicates that some lavas rose through the crust rapidly in order to reach the surface with little or no crystallization. Head and Wilson (1979) calculated that the rising velocity of magma would have to be >0.5–1.0 m/s to maintain lava fountains even if no explosive lava fountaining occurred. Vesicles indicate that volatile phases, most likely CO, were present when the rocks were molten and only 250–750 ppm of CO are required to disrupt magmas at depths of 15–40 km, causing explosive eruptions. In a study of a 180 km wide dark pyroclastic ring south of the Orientale basin, Weitz et al. (1998) estimated that the plume height was on the order of 20 km and that ejection velocities were ~320 m/s. In a recent study, Head et al. (2002) derived similar velocities of ~350–420 m/s.

By coupling the estimated eruptive volumes (discussed above) with basalt ages (absolute and relative), estimates of the flux of mare volcanism on the Moon are possible (e.g., Hartmann et al. 1981; Wilhelms 1987; Head and Wilson 1992). Head and Wilson (1992) pointed out that during the Upper Imbrian ~9.3 × 10^6 km^3 of mare basalts were emplaced. Assuming steady emplacement throughout the Upper Imbrian yields an average eruption rate of ~0.015 km^3/yr. Applying the same approach to the Eratosthenian and Copernican Period indicates an average eruption rate of ~1.3 × 10^{-4} km^3/yr and ~2.4 × 10^{-6} km^3/yr respectively. Head and Wilson (1992) reported that at its peak average flux, lunar global eruption rates are ~10^{-2} km^3/yr, an amount comparable to eruption rates of a single terrestrial volcano such as Vesuvius or Kilauea, Hawaii. However, some lunar sinuous rilles have been estimated to emplace ~1000 km^3/yr (Hulme 1973), hence representing about 70,000 years of the average flux. On the basis of their investigation of the dark Orientale ring deposit, Head et al. (2002) estimated the total erupted mass to be on the order of 7–17 × 10^{13} kg, an eruption rate of ~2 × 10^8 kgs^{-1}, and a total duration of the eruption of ~4–11 days.

2.2.14. Non-mare domes. Several presumably volcanic features on the Moon have albedos, spectral characteristics, and morphologies that differ from typical mare volcanic features. Examples of such non-mare volcanism are the Gruithuisen domes (~36°N, ~40°W; Fig. 1.28), the Mairan domes and cones (~40°N, ~46°W), as well as Hansteen Alpha (~12°S, ~50°W) and Helmet (~17°S, ~32°W) (see Hawke et al. 2003). These domes are up to 20 km in diameter with a topographic relief of more than 1000 m and their shapes are consistent with extrusion of viscous lava (Gruithuisen domes) and explosive volcanism (Mairan cones) (Head and McCord 1978; Chevrel et al 1998; Wilson and Head 2003b). Spectrally they are characterized by a downturn in the ultraviolet, similar to the "red spots" mapped elsewhere (Wood and Head 1975; Bruno et al. 1991; Hawke et al. 2002a). No samples collected and returned from the landing sites match their spectral characteristics. Due to the relatively small size and steep slopes of these features their age is difficult to determine. From recent crater counts, the Gruithuisen domes appear to be contemporaneous with the emplacement of the surrounding mare basalts but postdate the

Figure 1.28. Oblique view of the non-mare Gruithuisen domes.

formation of post-Imbrium crater Iridum (Wagner et al. 1996, 2002). Contemporaneity with the maria has been interpreted to indicate petrogenetic linkages; one possibility is that mare diapirs stalled at the base of the crust and partially remelted the crust, which produced more silicic viscous magmas (e.g., Head et al. 2000). Malin (1974) argued that lunar "red spots" are the surface manifestation of more radioactive pre-mare basalts such as occur among the Apollo 12 and 14 KREEP-rich material). Hawke et al. (2003), however, showed that one of these, Hansteen Alpha, although occurring in an area of elevated Th concentrations, is not enriched in Th relative to surrounding mare basalts. However, more recent work by Lawrence et al. (2005), who re-modeled Lunar Prospector Th data for this area, showed Hansteen Alpha contains up to 25 ppm Th. Head and Wilson (1999) reported that the yield strength ($\sim 10^5$ Pa), plastic viscosity ($\sim 10^9$ Pa s), and effusion rates (~ 50 m^3/s) of the Gruithuisen domes are similar to those of terrestrial rhyolites, dacites, and basaltic andesites. The viscosity of lava that formed the Gruithuisen Domes is orders of magnitudes larger than for typical mare basalts (0.45–1 Pa s; Murase and McBirney 1970; Wilson and Head 2003). On the basis of near-IR spectra, Hawke et al. (2002a) concluded that Helmet is noritic in composition, implying on average slightly higher SiO$_2$ compared to mare basalts. For the most recent discussion of the composition of non-mare domes, we defer to Hawke et al. (2001, 2002a,b, 2003) and Lawrence et al. (2005).

    2.2.15. Light plains. The origin of lunar "light plains" is still not well understood and is a controversial topic. Prior to the Apollo 16 mission, light plains (in this case, the Cayley Formation) were thought to be products of some sort of highland volcanism because of their smooth texture and their filling of highland craters. Crater counts suggested that these plains formed between the latest basin-scale impact event and the emplacement of the maria. Thus it was thought that significant amounts of highland volcanism occurred during the Early Imbrian. However, samples from the Apollo 16 site showed that light plains mainly consist of impact breccias and not volcanic material. Fieldwork and laboratory studies indicate that light plains can form by ballistic erosion and sedimentation processes (Oberbeck et al. 1974; Oberbeck 1975). Light plains appear not to represent a single ejecta blanket (e.g., from Orientale or Imbrium), but consist of deposits mixed dynamically from primary ejecta and local sources. New crater counts of light plains deposits, however, show a wide variety of ages that cannot be exclusively attributed to either the Orientale or Imbrium events, suggesting that at least some of the light plains might be of volcanic origin (Neukum 1977; Köhler et al. 1999, 2000). Spudis (1978) and Hawke and Head (1978) suggested that some light plains might be related to KREEP volcanism. Hawke and Bell (1981), Antonenko et al. (1995), and Robinson and Jolliff (2002) presented evidence that ancient lava plains (cryptomare) might be present beneath some light plains. In summary, these studies suggest that light plains may have formed by a variety of processes and that the interpretation of their origin should be addressed on a case-by-case basis.

2.3. Tectonic processes

2.3.1. A one-plate planetary body. Tectonic features suggest that the Moon underwent global expansion before about 3.6 b.y. ago, followed by net global contraction from that point until today (e.g., Solomon and Head 1980). However, compared to Earth, there is no evidence for plate tectonics on the Moon. This raises several questions. Why are these planetary bodies so different and how do they compare to other planets? What is the driving force in their evolution? How do tectonic styles and heat-loss dynamics change with time?

Crystallization of a globally continuous, low-density crust from a magma ocean may have precluded the development of plate tectonics early in lunar history. Following the establishment of such a nearly continuous low-density crust or stagnant lid, conductive cooling dominated the heat transfer on the Moon, which resulted in the production of a globally continuous lithosphere rather than multiple, moving, and subducting plates as on Earth. Heat flow experiments carried out during the Apollo missions revealed values much less than those on Earth (e.g., Langseth et al. 1976) and are consistent with heat loss predominantly by conduction. Furthermore, nearside seismic data indicate the presence of an outer relative rigid, 800–1000 km thick lithosphere. The large ratio of surface area to volume of the Moon is very effective in cooling the planet by conduction, i.e. radiating heat into space. Thus the lithosphere of the Moon thickened rapidly and the Moon quickly became a one-plate planet, which lost most of its heat through conduction (Solomon 1978). In this respect the Moon is more similar in its lithospheric heat transfer mechanism, that is conduction, to Mercury and Mars, and even Venus, than it is to Earth (Solomon and Head 1982). Venus is presently characterized by conductive heat loss, but there is evidence that the heat loss on Venus may have been episodic. On the other hand, Earth is dominated by plate recycling and radioactive decay of elements sequestered in the continental crust, while yet another heat loss mechanism, advective cooling through constant volcanic eruptions, characterizes the innermost Galilean satellite, Io.

2.3.2. Moonquakes. Most information about lunar tectonic processes has been gathered from the interpretation of surface images, the Apollo seismometers, and from modeling of the thermal state of the crust and the mantle. Compared to Earth, the internal tectonic activity on the Moon, i.e., moonquakes, is a minor process, probably releasing less than 10^{-12} of the energy of terrestrial seismic activity (e.g., Goins et al. 1981). The crust of the Moon appears to be thick, rigid, immobile, and cool, inhibiting large-scale motion. The lack of crustal deformation is consistent with the returned samples that show virtually no textures typical of plastic deformation. There are two types of moonquakes, which differ in the location of their source region. Based on a re-evaluation of Apollo seismic data, the source depths of moonquakes have recently been re-investigated (e.g., Khan et al. 2000; Nakamura 2003, 2005). Shallow moonquakes are now thought to originate at depths of ~50–220 km and deep moonquakes, at depths of ~850–1100 km (Khan et al. 2000). Deep-seated moonquakes, which are correlated with the Earth's tides, originate from so-called nests that repeatedly release seismic energy (e.g., Nakamura 1983). All but one of these nests are located on the lunar nearside, a phenomenon not well understood (e.g., Oberst et al. 2002). One explanation could be that if the lunar mantle is partially molten at depths of ~1000 km below the surface, one would not be able to measure farside S-waves, which is necessary for the location of moonquakes on the lunar farside (Nakamura et al. 1973). However, in a re-evaluation of the Apollo seismic data, Nakamura (2005) identified ~30 nests of deep moonquakes that are likely to be on the lunar farside, although he noted that only a few of them are locatable with the currently available Apollo seismic data.

2.3.3. Impact-induced tectonism. Impacts extensively shatter the subsurface and cause lateral and vertical movements of the crust. Ahrens and Rubin (1993) pointed out that damage caused by impact-induced rock failure decreases as a function of $\sim r^{-1.5}$ from the crater, indicating a dependence on the magnitude and duration of the tensile pulse. In order to adjust to the post-impact stress field, subsequent movements likely occurred along impact-induced faults, which may be kept active for long periods of time or may be reactivated by seismic energy from subsequent impacts (e.g., Schultz and Gault 1975). Post-impact isostatic adjustments of large impact structures might have occurred; however, the rigidity of the lunar crust inhibits this process, resulting in numerous basins that remain in isostatic disequilibrium even several billion years after the impact. Although the existence of the Procellarum basin is still debated (e.g., Wilhelms 1987; Spudis 1993; Neumann et al. 1996; Schmitt 2000a, 2001), Cooper et al. (1994) proposed that the Procellarum basin is the result of faulting associated with the Imbrium event. A similar argument has been made for Frigoris (see review in Spudis 1993).

2.3.4. Loading-induced tectonism. Loading of large basin structures by infilling with basalts produced extensional stress fields that may have led to the formation of concentric graben at the edges of the basins (Fig. 1.29a). The loading also produced down-warping of the basin center accompanied by compressional stresses that caused the formation of wrinkle ridges (Fig. 1.29b). Graben in the Humorum area that are concentric to the basin structure are a few hundred kilometers long and are filled with basaltic lavas, indicating that they were relatively early extensional features. These graben extend virtually unobstructed from the mare into the adjacent highlands and cut across preexisting craters, which suggests that they were formed by a substantial, possibly deep-seated, basin-wide stress field. Schultz and Zuber (1994) investigated faulting caused by axisymmetric surface loading and pointed out that geophysical models of flexural stresses in an elastic lithosphere due to loading typically predict a transition with increased distance from the center from radial thrust faults to strike-slip faults to concentric normal faults. These models (e.g., Melosh 1978; Golombek 1985) are inconsistent with the absence of annular zones of strike-slip faults around the lunar maria. Schultz and Zuber (1994) suggested that this paradox is caused by difficulties in relating failure criteria for brittle rocks to stress models. Their findings apply not only to basin loading with lunar maria but also to loading of the venusian and martian lithospheres with large volcanoes or, for example, the Tharsis rise. Freed et al. (2001) showed that lunar curvature, certain initial stress distributions, and certain failure criteria reduce and perhaps eliminate the zone of strike-slip faulting.

2.3.5. Tidal forces. Dynamical considerations indicate that during its early history the Moon was closer to Earth than it is today (e.g., Thompson and Stevenson 1983; Kokubo et al. 2000; Touma 2000). If this is true or if the Moon underwent any reorientation (e.g., Melosh 1975), tidal and synchronous rotational stresses could have built up to 100 kbar (at Earth-Moon distances of <8 R_e) within the early lunar lithosphere, an amount large enough to produce major tectonic deformation. However, this deformation would have occurred during the times of intense impact bombardment, which would have destroyed the evidence for such deformation. Today, tidal forces deform the Moon into a triaxial ellipsoid and tidal stresses of ~0.2 bar at depths of 700–1100 km appear sufficiently large to trigger deep moonquakes. Recently, detailed maps of lunar lineaments have been compared to predicted stress patterns caused by tidal forces. Chabot et al. (2000) found that patterns on the near and farside are similar. However, because patterns of the sub-Earth and anti-Earth regions are indistinguishable from patterns in

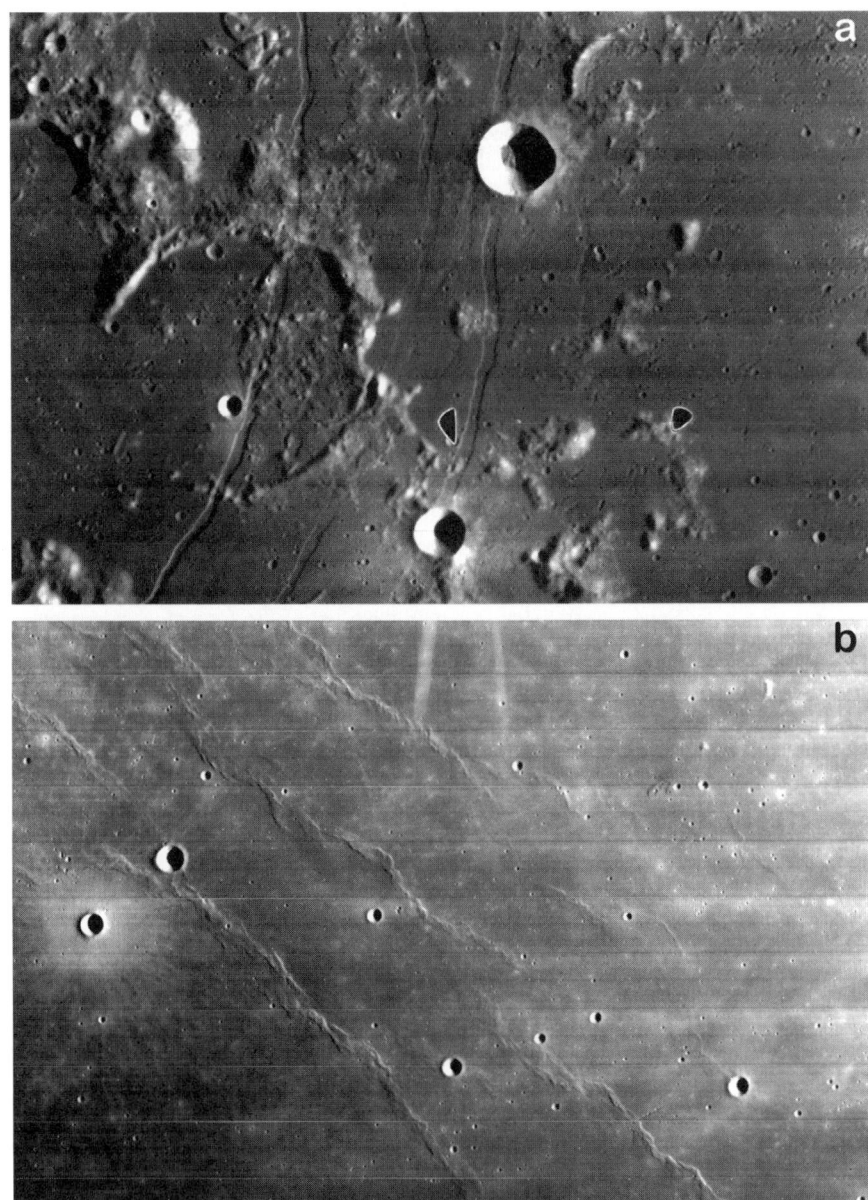

Figure 1.29. Examples of tectonic landforms; (a) linear rilles (LO IV 123H1)
and (b) wrinkle ridges (LO IV 163H1).

other areas, they concluded that the lunar lineament patterns do not support the predictions of a
global tectonic pattern due to the collapse of a once larger tidal bulge on the Moon.

2.3.6. Thermal effects. Models of the thermal evolution of the Moon indicate that during
the first billion years, thermal expansion produced extensional stresses in the lunar crust and that
during the next 3.5 b.y. until the present, cooling and contraction have caused compressional

stresses (e.g., Solomon and Chaiken 1976). Pritchard and Stevenson (2000) point out that the thermal models have many adjustable and unconstrained parameters that influence the evolution of the planetary radius and the resulting stress field. The amount of these stresses are still not well understood, but early lunar tectonic activity may have been dominated by tidal stresses and internally generated stresses may have been insignificant. However, graben that are most likely related to loading of basin centers with basalts could not have formed without the presence of global, mildly tensile stresses (e.g., Solomon and Head 1979). After about 3.6 b.y., the lunar stress field became compressional and internally driven tectonic activity may have ceased for 2.5–3 b.y. until sufficient stress (>1 kbar) had accumulated to produce small-scale thrust faults. Pritchard and Stevenson (2000) argued that a cessation of graben formation at 3.6 b.y. is a nonunique constraint because local effects including flexure and magmatic pressures could mask the signal from global stress.

2.3.7. Formation of ridges. The en-echelon offsets of wrinkle ridges in Mare Serenitatis indicate a formation due to compressional stresses. The ridges in Mare Serenitatis are concentric to the basin, and Muehlberger (1974) and Maxwell (1978) estimated a centrosymmetric foreshortening of ~0.5–0.8% in order to produce the ridges. Ground-penetrating radar data from the Apollo Lunar Sounding Experiment (ALSE) revealed significant upwarping and possibly folding and faulting of the basaltic surface down to ~2 km below the wrinkle ridges. Lucchitta (1976) suggested that wrinkle ridges are associated with thrust faulting and folding, a concept recently supported by Golombek et al. (1999, 2000) on the basis of new MOLA topographic data of martian wrinkle ridges.

Although most ridges on the Moon are interpreted to have formed by compressive stresses, some may result from subsurface magma emplacement (Strom 1964). Similarly, Schultz (1976) suggested that some floor-fractured craters formed by uplift and expansion due to the emplacement of sills below the crater floor.

3. GEOLOGIC SETTING AND SIGNIFICANCE OF THE APOLLO AND LUNA LANDING SITES

3.1. Overview

Six manned American Apollo missions (1969–1972) and three automated soviet Luna missions (1970–1976) returned samples from the nearside lunar surface. Each of these landing sites is, in its unique way, useful for understanding and interpreting the new global data sets from Clementine and Lunar Prospector because each provides ground truth for numerous geologic investigations and experiments, the calibration of data, and the development of new techniques of data reduction. Here we briefly review the geologic setting of each landing site in order to provide a framework for the following chapters. In addition, geologic cross-sections and the traverses of extravehicular activities (EVA) are intended to put the returned samples into geologic context and to familiarize the reader with the local topography, morphology, geology, and the sites where scientific experiments were carried out.

3.2. Apollo 11 (July 1969)

The Apollo 11 mission returned samples of basalt from Mare Tranquillitatis, confirming the hypothesis that the dark, circular basins contained extruded lavas formed by partial melting of the lunar mantle (e.g., Smith et al. 1970). Unlike terrestrial basalt, however, these were marked by extraordinarily high TiO_2 concentrations, and the high proportions of ilmenite were found to be consistent with their spectrally "blue" character. Their ages were determined to be 3.57–3.88 b.y. old (BVSP 1981 and references therein), which, coupled with crater densities, provided an early calibration point in the lunar chronology and showed that volcanism on the Moon was indeed ancient by terrestrial standards. Also found among the samples were pieces of non-volcanic material, including fragments of plagioclase-rich anorthosites, which were interpreted to be from the adjacent light-colored highlands. From these pieces of rock, the general feldspathic character of the lunar crust and the first concepts of lunar differentiation were correctly inferred (e.g., Wood et al. 1970; Smith et al. 1970). The Apollo 11 landing site, which is about 40–50 km from the nearest mare/highland boundary within Mare Tranquillitatis (0.7°N, 24.3°E), was chosen primarily for safety reasons (Fig. 1.30a,b). During their 2.5 hours EVA, the astronauts collected 21.6 kg of lunar samples. Samples were collected about 400 m west of West crater, a sharp-rimmed, rayed crater ~180 m in diameter and ~30 m deep. The regolith there is about 3–6 m thick, and Beaty and Albee (1978) suggested that most of the collected samples were ejected from West crater. Among

Figure 1.30. The geologic setting of the Apollo 11 landing site;
(a) LO IV 85H1 context image. *(b) is on facing page.*

these samples, five geochemically distinctive groups of mare basalts have been identified (see Neal and Taylor 1992, for a summary). In addition, some of the returned regolith samples consist of feldspathic lithic and mineral fragments that were derived from highland regions (Wood et al. 1970; Smith et al. 1970). Despite its location in a mare region, Apollo 11 regolith contains up to ~28% of nonmare material, which is similar in composition to the regolith of the Cayley Plains at the Apollo 16 landing site (Korotev and Gillis 2001). Korotev and Gillis (2001) suggested that the nonmare material is most likely Imbrium ejecta deposits excavated from beneath the basalt flows by impact craters. For further information about the geology of the Apollo 11 landing site see Chapter 2 and, for example, Shoemaker et al. (1970a) and Beaty and Albee (1978, 1980).

3.3. Apollo 12 (November, 1969)

In order to demonstrate the capability of pinpoint landings, Apollo 12 was sent to the Surveyor 3 site, a flat mare site with only a few large boulders. Apollo 12 touched down within 200 m of the Surveyor 3 landing site in southeastern Oceanus Procellarum (3.2°N, 23.4°W), and was the first spacecraft to demonstrate the capability of pinpoint landings on another planetary body (Fig. 1.31a,b). Compared to the Apollo 11 site, this site is less cratered, hence younger, and sampled mare basalts differ in their spectral characteristics and composition from the Apollo 11 basalts. Differences in age and chemistry between basalts of the Apollo 11 and Apollo 12 landing sites demonstrated variability in mantle sources and basalt production processes on the Moon. Exposures of non-mare materials near the landing site, mostly part of the Fra Mauro Formation, indicate that the basalts are relatively thin (Head 1975b). Exposures of non-basaltic formations and mare basalts form a complex topography and crater counts and D_L values (defined as the diameter of craters with wall slopes of 1°) reveal that the landing site basalts are older than basalts 1 km away to the east and west (Wilhelms 1987). Despite the fact that the landing site is dominated by the ejecta of several craters larger than ~100 m, the regolith is only half the thickness of the Apollo 11 regolith and craters only 3 m deep penetrate into basaltic bedrock. Non-volcanic materials are abundant, as anticipated because of the crossing of the site by a prominent ray from crater Copernicus. Although there are probably multiple sources of

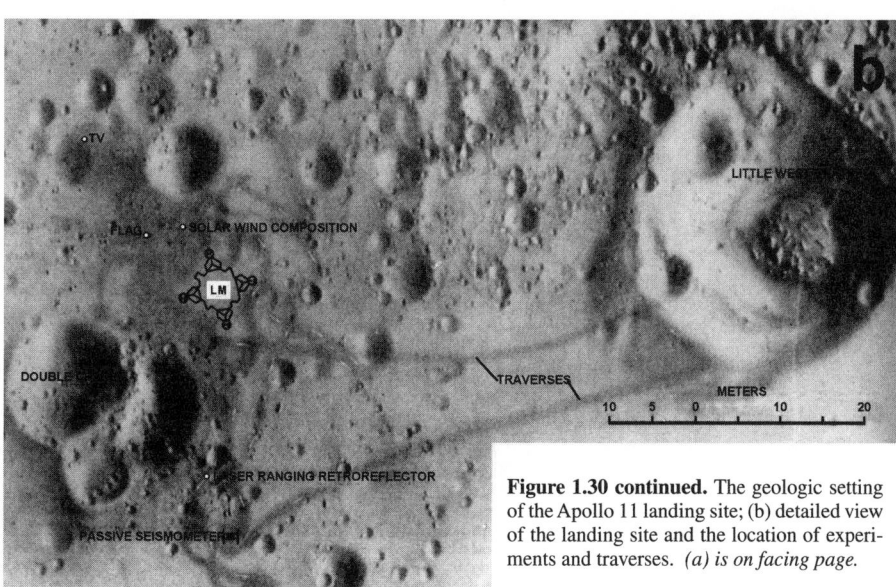

Figure 1.30 continued. The geologic setting of the Apollo 11 landing site; (b) detailed view of the landing site and the location of experiments and traverses. *(a) is on facing page.*

the non-volcanic materials at the Apollo 12 site, including material mixed into the surface from impact-melt formations that underlie the basalts, analyses of non-volcanic materials record a major disturbance at about 800–900 m.y. ago, and this has been inferred to be the age of crater Copernicus, although no sample has been unambiguously related to this crater.

The astronauts performed two EVAs and collected 34.3 kg of samples, mostly basalts. Radiometric dating showed the basalts to range in age between 3.29 and 3.08 b.y. Based on the large rock samples, at least three chemically distinctive groups of mare basalts were identified (e.g., James and Wright 1972; Neal et al. 1994 a,b). The mare basalt flows (20–21% FeO) cover the older Fra Mauro Formation, which contains about 10% FeO (Wilhelms 1987; Jolliff et al. 1991, 2000). Apollo 12 was the first mission to bring back KREEP material with a very specific and unique geochemical signature (enriched in potassium, rare-earth elements, phosphorus, and other incompatible elements) that requires substantial magmatic differentiation, consistent with

Figure 1.31. The geologic setting of the Apollo 12 landing site;
(a) LO IV 125H3 context image, *(b) is on facing page.*

a strongly differentiated Moon that differs significantly from all known classes of meteorites. This KREEP material occurs in form of dark, ropy glasses similar in composition to Apollo 14 soils and a subset of Apollo 14 impact glasses (Wentworth et al. 1994). Because these glasses contain lithic fragments including anorthosite, KREEP basalt and felsite (Wentworth et al. 1994), they appear to represent impacted surface material similar to that exposed in the Fra Mauro Formation (Jolliff et al. 2000). The KREEP material is derived mostly from the Procellarum KREEP Terrane and was probably incorporated into the regolith of the landing site by a combination of lateral transport and vertical mixing (Jolliff et al. 2000). For additional discussions of the Apollo 12 landing site, the reader is referred to Shoemaker et al. (1970b), Warner (1970), Rhodes et al. (1977), and Wilhelms (1984).

3.4. Apollo 14 (January-February, 1971)

Apollo 14 landed on a hilly terrain north of Fra Mauro crater (3.7°S, 17.5°W) and became the first mission to sample the lunar "highlands" (Fig. 1.32a,b). The landing site is ~550 km south of the Imbrium basin and was chosen primarily to collect samples from this basin in order to characterize its ejecta, which was thought to be derived from deep crustal levels, and to date the Imbrium event, which serves as a major stratigraphic division of the lunar history. During two EVAs 42.3 kg of samples were collected, including complex fragmental breccias, impact-melt breccias, and clast-poor impact melts with generally basaltic and KREEP-rich compositions. These breccias were formed about 3.9–3.8 b.y ago but whether they formed in the Imbrium event itself, or through later, smaller impacts into Imbrium ejecta is still contended,

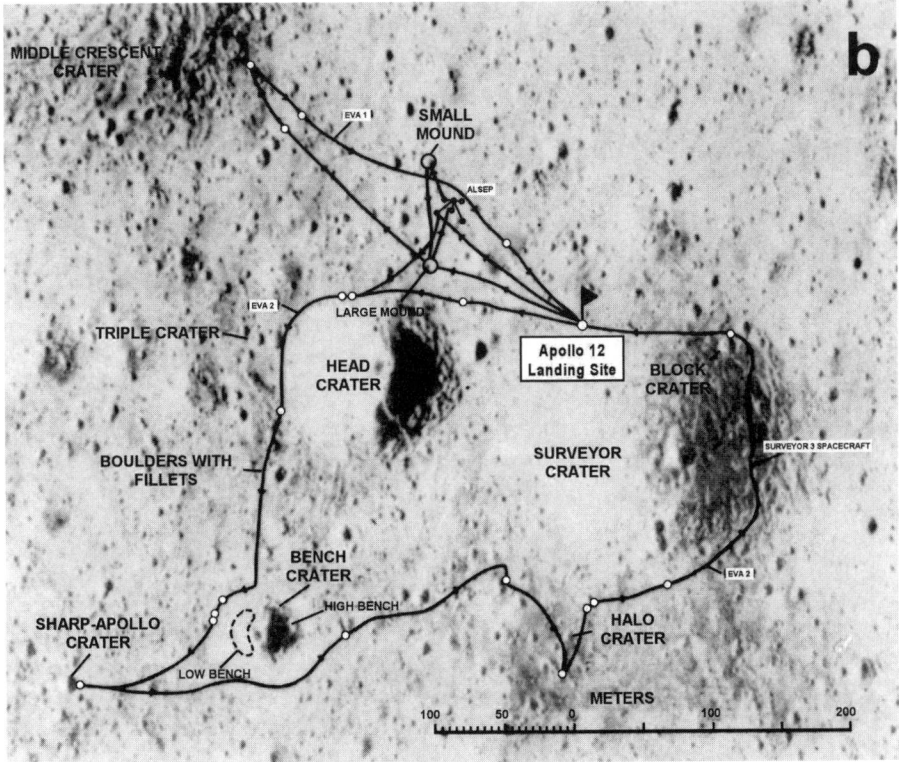

Figure 1.31 continued. The geologic setting of the Apollo 12 landing site; (b) detailed view of the landing site and the location of experiments and traverses. *(a) is on facing page.*

and it is still unclear which Apollo 14 samples represent true Imbrium ejecta or locally derived materials. The Fra Mauro Formation has been interpreted as primary Imbrium ejecta (Wilhelms 1987). Haskin et al. (2002) argued that it consists of about 58% Imbrium ejecta and Morrison and Oberbeck (1975) proposed that the Fra Mauro Formation is mostly dominated by local material with intermixed 15–20% Imbrium ejecta.

The Apollo 14 landing site is usually considered as a "highland" landing site, but from a geochemical point of view it is neither mare nor feldspathic highlands. Rather, the Apollo 14 site is located within the Procellarum KREEP terrane, which recent Lunar Prospector gamma-ray data indicated it contained exceptionally high thorium concentrations. However, this anomaly was not known at the time of the site selection. The Apollo 14 landing site is also characterized by relatively aluminous basalts, a subset of which are also enriched in K. These

Figure 1.32. The geologic setting of the Apollo 14 landing site;
(a) LO IV 120H3 context image, *(b) is on facing page.*

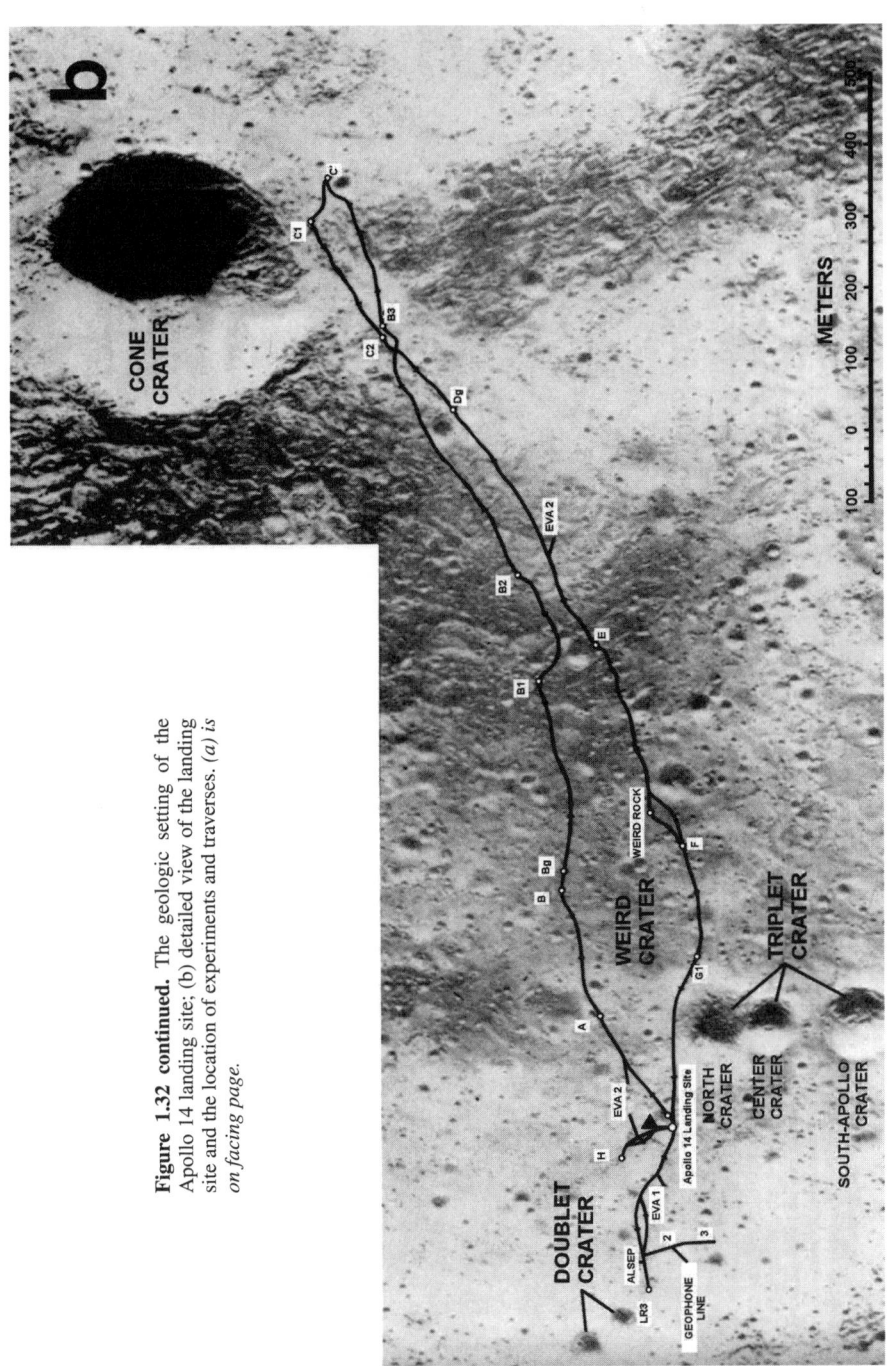

Figure 1.32 continued. The geologic setting of the Apollo 14 landing site; (b) detailed view of the landing site and the location of experiments and traverses. *(a) is on facing page.*

aluminous and high-K basalts are unknown from other Apollo mare landing sites (e.g., Shervais et al. 1985a,b), although aluminous basalts were returned by the Soviet Luna missions (see below). A "model" mixture of the regolith at the Apollo 14 site consists of ~60% impact-melt breccias, ~20–30% noritic lithologies, ~5–10% each mare basalts and troctolitic anorthosites, and minor amounts of meteoritic components (Jolliff et al. 1991). Compared to the regolith of the Apollo 11 landing site, the Apollo 14 regolith shows little variation in composition, both vertically (tens of centimeters) and laterally (kilometers around the landing site). The landing site is located ~1100 m west of Cone crater, a crater ~340 m in diameter and 75 m deep that ejected blocks of up to ~15 m across. The landing site shows numerous subdued craters up to several hundreds of meters across and the regolith was estimated to be 10–20 m thick (Swann et al. 1971). The geology of the Fra Mauro area is discussed in detail in Chapter 2 and by Chao (1973), Swann et al. (1977), Hawke and Head (1977), and Simonds et al. (1977).

3.5. Apollo 15 (July-August, 1971)

The Apollo 15 mission was the first advanced ("J") mission that carried the Lunar Roving Vehicle (LRV) and was sent to a complex multi-objective landing site in the Hadley-Apennine

Figure 1.33. The geologic setting of the Apollo 15 landing site;
(a) LO IV 102H3 context image, *(b) is on facing page.*

region (26.1°N, 3.7°E) (Fig. 1.33a,b). The purpose of this mission was to sample and study the massifs and highlands of the Imbrium rim, and the mare lavas and landforms of Palus Putredinis (e.g., Hadley Rille). Extensive lava plains are exposed west of the landing site and the main Imbrium ring rises ~3.5 km above the plains at Hadley Delta, only 4 km south of the Apollo 15 landing site. Exposed inside the Imbrium ring and within a few kilometers from the landing site is the Apennine Bench Formation, a light plains unit that probably underlies the Upper Imbrian basalts at the Apollo 15 landing site. Rays from crater Autolycus and Aristillus cross the landing site and the regolith thickness varies widely depending on the local terrain. The regolith, which is only ~5 m thick at the landing site, is absent at the Hadley Rille. Photographs show the wall of Hadley Rille to be a layered basalt-flow sequence. In total, the astronauts performed three

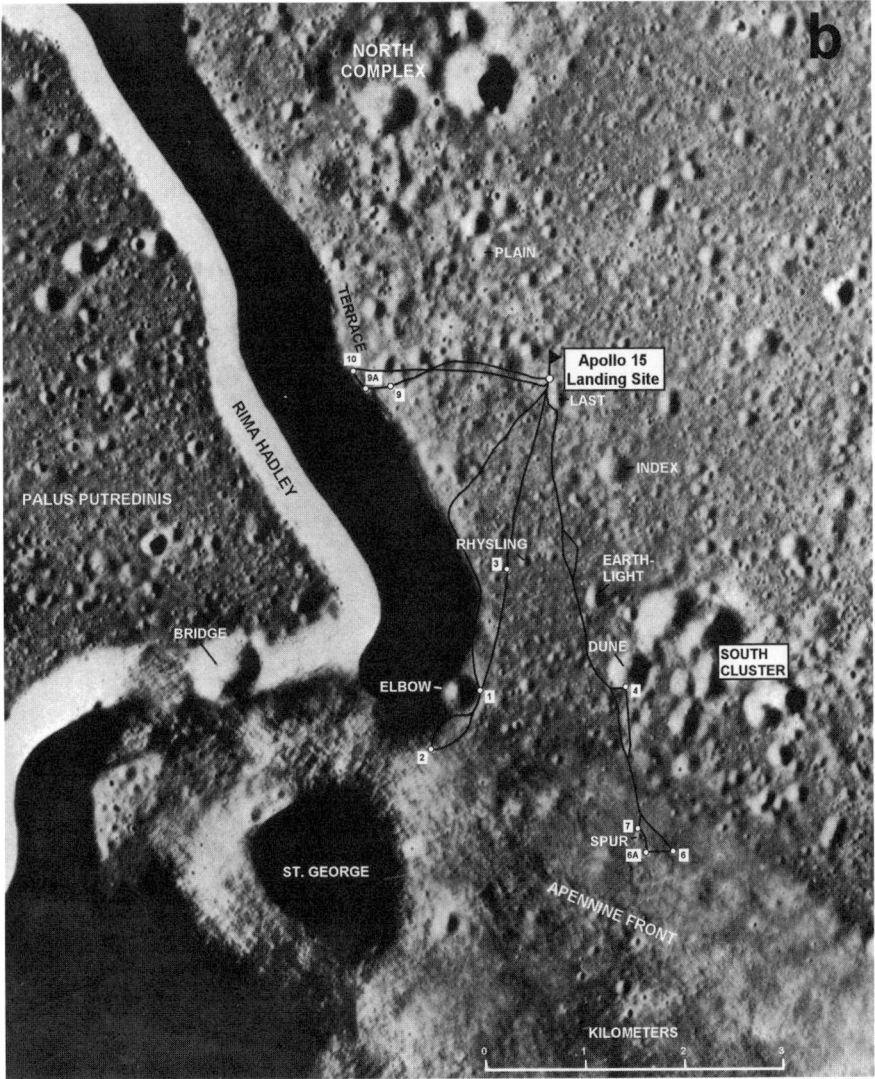

Figure 1.33 continued. The geologic setting of the Apollo 15 landing site; (b) detailed view of the landing site and the location of experiments and traverses. *(a) is on facing page.*

EVAs with the LRV collecting 77.3 kg of samples, both mare and nonmare rocks. Two types of lavas were collected at this landing site (quartz normative and olivine normative; e.g., Rhodes and Hubbard 1973; Chappell and Green 1973; Dowty et al. 1973), both basically identical in age (3.3 b.y.). Non-volcanic samples consist of anorthosites, magnesian-suite plutonic rocks, impact melts, and granulites, many of which occur as individual clasts in regolith breccias. In addition, KREEP-rich non-mare basalts and green ultramafic volcanic glasses were collected at this site. Based on evidence from the Apollo 14 and Apollo 15 sites, the Imbrium basin is interpreted to be 3.85 b.y. old. Swann et al. (1972), Spudis and Ryder (1985, 1986), Spudis et al. (1988), and Spudis and Pieters (1991) provide a further discussion of the geology of the Apollo 15 landing site.

3.6. Apollo 16 (May, 1972)

This mission was to explore the lunar highlands, especially the smooth Cayley Formation and the hilly and furrowed Descartes Formation, both of which were thought to be volcanic in origin (e.g., Hinners 1972) (Fig. 1.34a,b). Other goals were to sample highland material far from any mare region and to investigate two nearby, fresh 1–2 km craters. Apollo 16 landed at 9°S latitude and 15.5°E longitude near Descartes crater. Three long EVAs were performed with the

Figure 1.34. The geologic setting of the Apollo 16 landing site;
(a) LO IV 89H3 context image, *(b) is on facing page.*

LRV, and 95.7 kg of samples were returned. The Apollo 16 landing site is considered the only true highland landing site of the Apollo program and is sometimes taken to be representative of typical lunar highlands. However, surface morphology, vicinity to the Imbrium basin distributing material of the Procellarum KREEP terrane across the site, composition of the regolith, and lithologic components of the regolith argue against Apollo 16 having landed on "typical" highlands. The landing site is characterized by numerous overlapping subdued craters of ~500 m size, two young fresh craters, North Ray (1 km wide, 230 m deep) and South Ray (680 m wide, 135 m deep), as well as Stone Mountain (Fig. 1.34b). Freeman (1981) reported that the regolith thickness on both the Cayley and the Descartes Formations ranges from 3–15 m and averages about 6–10 m. Contrary to pre-mission expectations, the returned samples are

Figure 1.34 continued. The geologic setting of the Apollo 16 landing site; ((b) detailed view of the landing site and the location of experiments and traverses. *(a) is on facing page.*

all impact products, most of them impact melt or fragmental breccias, and some anorthositic rocks. On the basis of the returned samples, neither the Cayley Formation nor the Descartes Formation are volcanic in origin but instead are related to ejecta deposition of the Imbrium, Serenitatis, and Nectaris basins, and the relative amounts of ejecta contributed to the landing site by each are still open questions (e.g., Head 1974; Muehlberger et al. 1980; Spudis 1984; Haskin et al. 2002). Some impact melts of the Apollo 16 site have been interpreted as products of the Nectaris event, indicating a formation of the Nectaris basin 3.92 b.y. ago (e.g., Wilhelms 1987; Stöffler and Ryder 2001). For more information about the geology of the Apollo 16 landing site see Muehlberger et al. (1980), Ulrich et al. (1981), James and Hörz (1981), James (1981), Spudis (1984), Stöffler et al. (1985), and Spudis et al. (1989).

3.7. Apollo 17 (December, 1972)

Apollo 17 marked the end of the Apollo program and was sent to the highland/mare boundary near the southeastern rim of the Serenitatis basin, the Taurus-Littrow Valley (20.2°N, 30.8°E) (Fig. 1.35a,b). This site was selected in order to examine two highland massifs where rocks from deep crustal levels could be collected, to investigate the presumably basaltic valley floor, and to study the low-albedo deposit that mantles both highlands and mare at the site. Thus, the Apollo 17 and 15 sites are the most geologically complex of the landing sites. Using the

Figure 1.35. The geologic setting of the Apollo 17 landing site;
(a) LO IV 78H3 context image, *(b) is on facing page.*

LRV the astronauts performed three EVAs and collected 110.5 kg of samples, bringing the total amount of returned samples to 381.7 kg. Samples from this site confirmed a Ti-rich basaltic composition of the valley floor and several chemical subgroups of basalts were identified, ranging in age from 3.8 to 3.7 b.y.

The Apollo 17 impact melts were collected from thick impact ejecta deposits on the rim of the Serenitatis basin and are likely of Serenitatis origin. A small subset of the impact-melt breccias have trace element signatures similar to some groups from Apollo 15 and Apollo 14, suggesting a minor component from the later Imbrium basin. Deep-seated magnesian-suite rocks and granulitic breccias are also found in the highland massif regolith (e.g., Warner et al. 1977; Warner et al. 1978; Lindstrom and Lindstrom 1986).

The regolith at the Apollo 17 landing site was found to be on the order of 15 m thick. The highland regions along this part of the Serenitatis rim are relatively FeO-rich because of

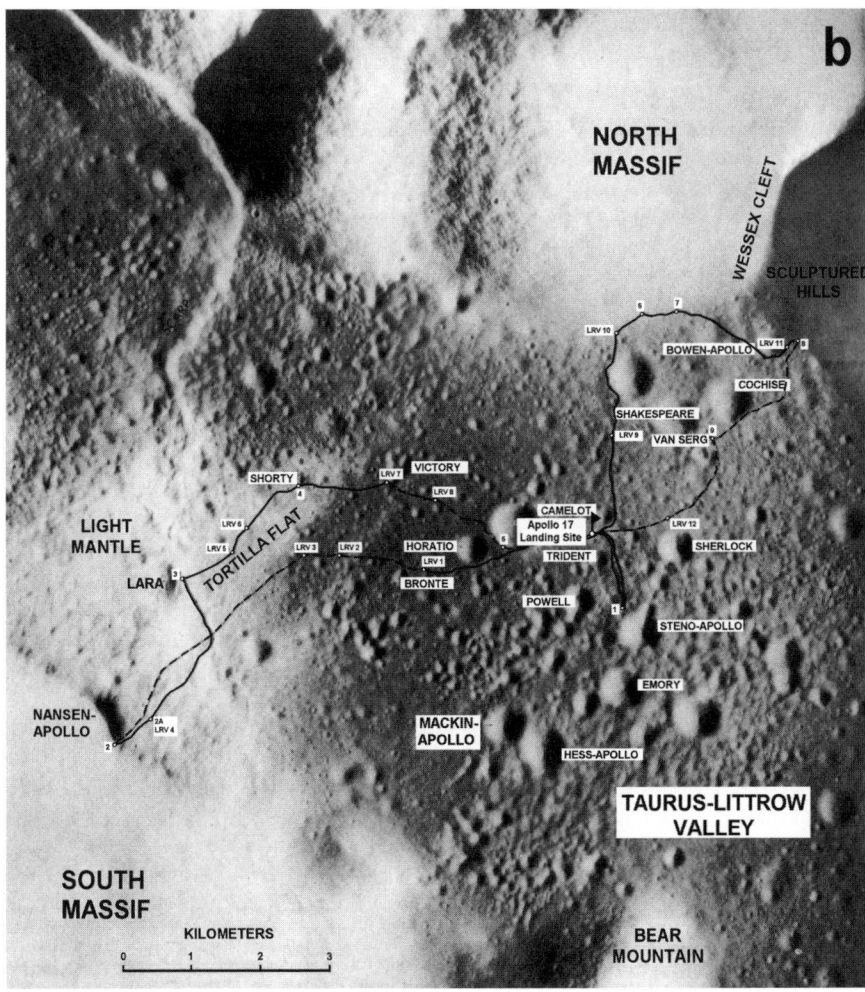

Figure 1.35 continued. The geologic setting of the Apollo 17 landing site; (b) detailed view of the landing site and the location of experiments and traverses. *(a) is on facing page.*

abundant mafic impact melt breccias and additions of basaltic material as impact debris and pyroclastic deposits. Recent work coupling Clementine data to landing site and sample data include Weitz et al. (1998) and Jolliff (1999). On the basis of topographic-photometrically corrected Clementine data, massif compositions are consistent with mixtures of noritic impact melt and feldspathic granulitic material, plus variable amounts of high-Ti basalts on the flanks at low elevations and pyroclastic deposits at high elevations (Robinson and Jolliff 2002). Hence, the highlands at this site consist of complex impact-melt breccias and plutonic rocks of the Mg-suite. Impact-melt breccias have been dated to be 3.87 b.y. old and have been interpreted to reflect the formation of the Serenitatis basin. The dark mantling deposits consist of orange and black volcanic glass beads of high-Ti basaltic composition, approximately 3.64 b.y. old. Photogeology suggests that a light colored mantle unit results from an avalanche triggered by secondary impacts from Tycho crater (e.g., Lucchitta 1977). Cosmic-ray exposure ages were used to derive an age of ~100 m.y. for the avalanche materials, and it has been argued that this age represents the age of the Tycho event. A detailed discussion of the geology of the Taurus-Littrow Valley and the Apollo 17 landing site may be found in Schmitt (1973), Wolfe et al. (1981), Spudis and Ryder (1981), Spudis and Pieters (1991), and Ryder et al. (1992).

3.8. Luna 16 (September, 1970)

Luna 16 was the first successful automated Soviet sample return mission and landed in northern Mare Fecunditatis at 0.7°S and 56.3°E (Fig. 1.36). Work by DeHon and Waskom (1976) indicated that Luna 16 landed on a sequence of relatively thin basalt flows ~300 m in thickness. The landing site is influenced by ejecta and ray material from Eratosthenian crater Langrenus (132 km), Copernican crater Tauruntius (56 km), as well as the more distant craters Theophilus and Tycho (McCauley and Scott 1972). Like all Luna samples, the Luna 16 samples are derived from shallow regolith drill cores. The drill core reached a depth of 35 cm and provided 101 g of dark gray regolith with preserved stratigraphy (Vinogradov 1971). No visible layering was observed within the core but 5 zones of increasing grain sizes with depth were recognized. Luna 16 samples are from a mare basalt regolith, which consists of moderately-high-Ti, high-Al basalt fragments, approximately 3.41 b.y. old. Mare basalts from this landing site are among the most Fe- and Mg-poor basalt samples returned from the Moon and, with ~13.5% Al_2O_3, are the most Al-rich basalts yet sampled (e.g., Ma et al. 1979). More information about the Luna 16 site can be found in McCauley and Scott (1972).

3.9. Luna 20 (February, 1972)

This mission landed in the highlands south of Mare Crisium (3.5°N, 56.5°E) and returned samples from an anorthositic highland regolith that contains lithic fragments of granulites, anorthosites, impact melts, and polymict breccias (Fig. 1.36). The Luna 20 landing site sampled the rim of the Nectarian Crisium basin and the landing site is characterized by smooth rounded hills and shallow linear valleys that give the region a hummocky appearance. The landing site is topographically ~1 km above the basaltic surface of Mare Fecunditatis and is influenced by Apollonius C, a 10 k.y. fresh Copernican crater located only a few kilometers to the east. Luna 20 returned a drill core that contained ~50 g of fine-grained light-gray regolith. No stratification has been observed within the core but mixing during transport might have occurred. The majority of lithic regolith fragments are breccias of anorthositic-noritic-troctolitic composition and impact-melt rocks of noritic-basaltic composition. Compared to the regolith of the Apollo 16 landing site, the Luna 20 regolith contains less K and P, fragments of anorthosites are rare, and MgO-rich spinel troctolites are common (e.g., Brett et al. 1973; Taylor et al. 1973; Papike et al. 1998). The higher concentrations of MgO in the Luna 20 samples are important to note because they may reflect the addition of an unknown mafic highland rock material, that is present at the Luna 20 site itself but is absent at the Apollo 16 landing site (Papike et al. 1998). Because K and P concentrations are low it has been argued that the samples of the Luna 20 landing site have not been contaminated with KREEP-rich Imbrium ejecta but instead represent

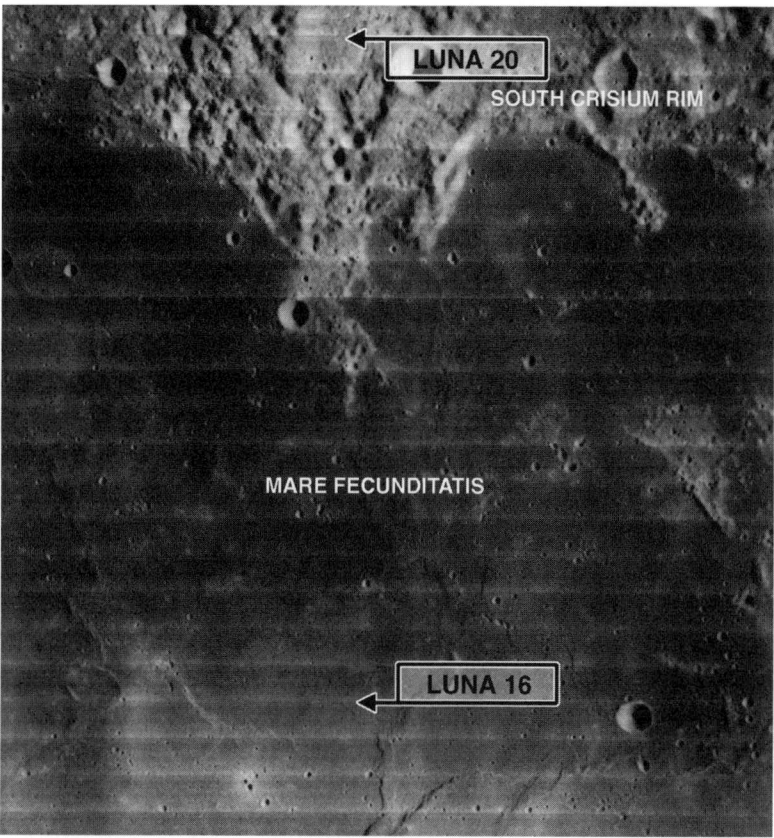

Figure 1.36. The geologic settings of the Luna 16 and Luna 20 landing sites (LO I M-33).

middle to lower crustal material ejected by the Crisium impact. Heiken and McEwen (1972) and Vinogradov (1973) discuss the Luna 20 landing site in more detail.

3.10. Luna 24 (August, 1976)

The last and most successful Luna mission was Luna 24, which landed in southern Mare Crisium (12.7°N, 62.2°E) (Fig. 1.37). The landing site is located within the inner ring of Crisium, which is marked by wrinkle ridges, and about 40 km north of the main basin ring, which is ~3.5–4.0 km higher than the landing site (Florenskiy et al. 1977; Butler and Morrison 1977). Head et al. (1978a,b), using remote-sensing techniques, identified several distinctive basalt types in Mare Crisium, the thickness of which was estimated to be ~1–2 km. Several bright patches and rays indicate that nonmare material has been dispersed across the basin by impacts such as Giordano Bruno and Proclus (e.g., Maxwell and El-Baz 1978; Florenskiy et al. 1977). Luna 24 returned 170 g of mostly fine-grained mare regolith in a 1.6 m long drill core. Stratification was well preserved in the core sample and, on the basis of color and grain size differences, four layers were identified (Florenskiy et al. 1977; Barsukov 1977). Basaltic fragments from this core are very low in TiO_2, low in MgO, high in Al_2O_3 and FeO, and are 3.6–3.4 b.y. old. Concentrations of FeO in the mare basalts and soils of Luna 24 are very similar, suggesting that only minor amounts of nonmare material are present at the landing site. The local and regional geology of the Luna 24 landing site is further discussed by Head et al.

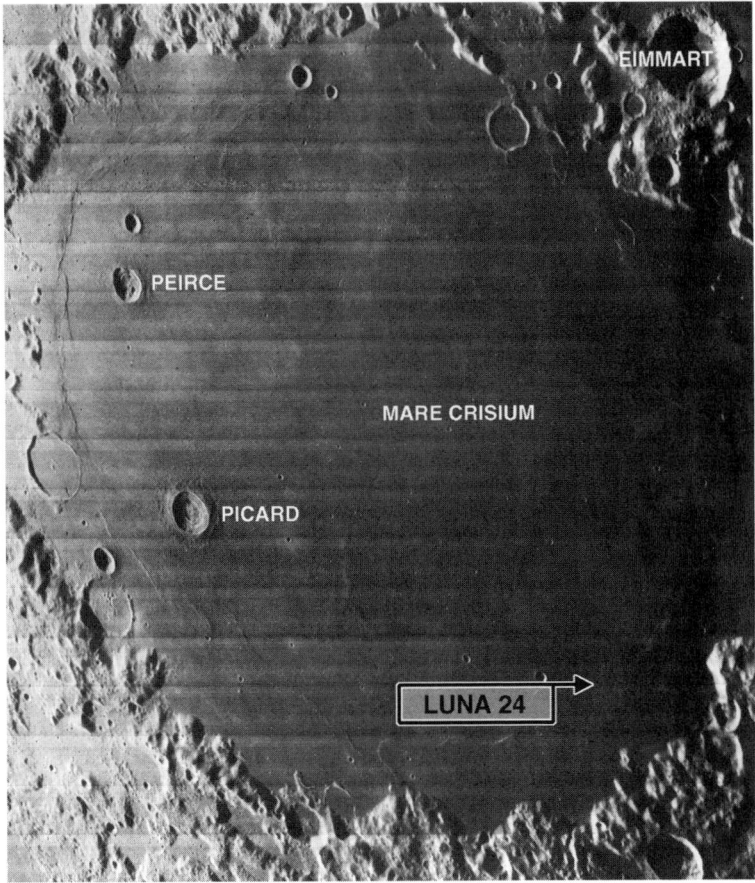

Figure 1.37. The geologic setting of the Luna 24 landing site (LO IV 191H3).

(1978a,b). A comprehensive summary of numerous contributions about the Luna 24 landing site and Mare Crisium can be found in Merrill and Papike (1978).

3.11. Significance of landing sites for the interpretation of global data sets

Our geologic understanding of the Moon relies heavily on the information derived from the lunar landing sites. A key advantage of landing-site exploration and sample return is the strength and synergy of having samples with known geologic context. Samples returned from these sites can be analyzed in great detail in laboratories with sophisticated, state-of-the-art techniques. Portions of samples set aside for future analysis with new methods offer many advantages over *in-situ* robotic measurements, which are restricted to techniques available at the time of the mission. Equally important, numerous geophysical experiments done by astronauts while on the surface improved our knowledge of the physical properties and the internal structure of the Moon. Again, these data can be used in new investigations, and on the basis of these data, new models have been and will be developed and tested long after the end of the Apollo missions. The landing sites also provide important ground-truth for all remote-sensing experiments. Calibration of remotely-sensed data to the landing sites and the lunar samples allows derivation of information about areas not yet sampled.

We consider briefly examples of how each individual landing site contributes to a comprehensive understanding of impact-basin structures and materials, because the impact-basin process is largely responsible for shaping the first-order features of the lunar surface. Figure 1.38a provides a schematic cross-section that illustrates the geologic setting of the Apollo landing sites relative to a multi-ringed impact structure. Figure 1.38b shows the Apollo landing sites in relation to morphologic features of the youngest lunar impact basin, Orientale. As shown in the figure, each landing site sampled very specific terrain types associated with large lunar impact basins. Apollo 16 sampled old pre-basin highland terrain, Apollo 14 landed on radially textured

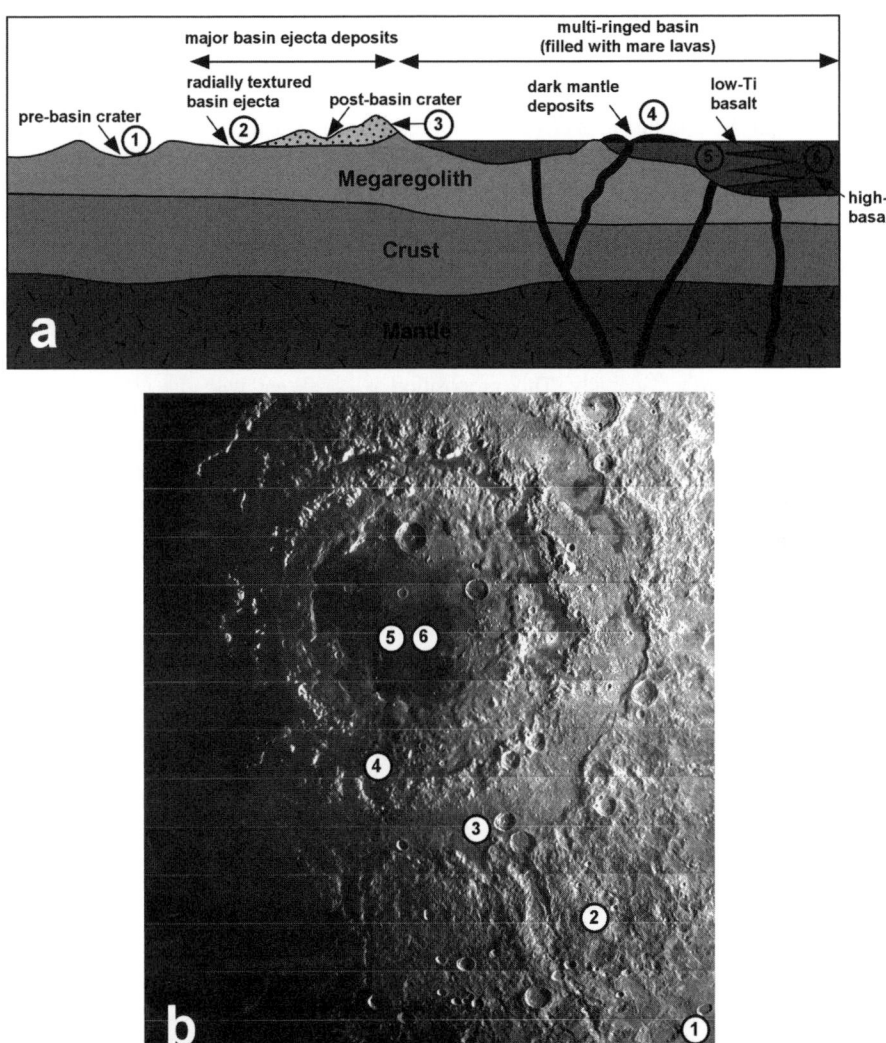

Figure 1.38. The Apollo and Luna landing sites; (a) schematic cross-section showing the Apollo and Luna landing sites in the context of a multi-ringed impact basin, (b) plan-view of the Orientale basin with superposed locations similar to the Apollo and Luna landing sites; (1) similar to Apollo 16 and Luna 20, (2) similar to Apollo 14, (3) similar to Apollo 15, (4) similar to Apollo 17, (5) similar to Apollo 12 and Luna 24, (6) similar to Apollo 11 and Luna 16.

basin-ejecta deposits, and Apollo 15 samples contain polymict breccias, which were formed by a post-basin crater. Besides other investigations, Apollo 17 examined dark mantle deposits in detail, and Apollo 11 and 12 investigated different types of basalts. All sample return sites, except Apollo 16 and Luna 20, returned basaltic samples that have provided our best window into the lunar mantle as no mantle xenoliths have been found in the returned basalt samples.

One of the major findings of Lunar Prospector was the identification of a large area with elevated thorium concentrations on the lunar nearside (e.g., Lawrence et al. 1998, 2000; Elphic et al. 2000). The origin of this anomaly is still debated and is discussed elsewhere in this volume. Several models have been put forward such as accumulation of KREEP-rich magma beneath a region of thinned crust presently corresponding to the region of Oceanus Procellarum or accumulation above a degree-1 downwelling of dense ilmenite cumulates (e.g., Warren 2001; Parmentier et al. 2000). The precise carrier of the Th-rich compositional signature in broad areas of the Procellarum region is not clearly known from remote sensing alone. Yet the question of what carries the Th enrichment is of great importance for the magmatic and thermal evolution of the Moon due to radiogenic heat production (e.g., Wieczorek and Phillips 2000). As it turned out, several Apollo landing sites, i.e., Apollo 12, Apollo 14, and Apollo 15, were located within the Procellarum KREEP terrane and thus allow study of the KREEP signature from returned samples. Investigation of samples from the Apollo 12 and 14 landing sites indicate that impact-melt breccias and derivative breccias, such as regolith breccias and agglutinates, are the main carriers of KREEP. However, at the Apollo 15 landing site, internally generated basalt is an important carrier of KREEP. Haskin (1998) and Haskin et al. (1998) proposed that the Imbrium basin impacted into the Procellarum KREEP terrane and distributed KREEP-rich material across the lunar surface globally. This proposition is consistent with the observation that abundant KREEP-rich material occurs in impact-melt breccias of landing sites located outside the Procellarum KREEP terrane, such as Apollo 11, 16, 17, but less KREEP–rich material in soils from the Luna 16, 20, and 24 landing sites, which are located at greater distances from the Procellarum KREEP terrane (see subsequent chapters).

Clementine was the first spacecraft in orbit around the Moon that provided high-resolution multispectral reflectance data. The Clementine mission permitted, for the first time, location of individual sampling stations at the Apollo 15, 16, and 17 landing sites (Blewett et al. 1997; Lucey et al. 1998) because these missions included rovers that allowed the astronauts to travel distances that can be resolved in the high-resolution Clementine data. Comparing spectra from 39 sampling stations with their average soil composition, Blewett et al. (1997) found a strong correlation between spectral variations and compositions. On the basis of this correlation, Lucey et al. (1998) developed an algorithm that estimates the FeO and TiO_2 concentrations on the lunar surface globally to about 1–2 wt% accuracy. Knowledge of the variability and spatial distribution of basalts relative to these two compositional parameters is important for understanding the petrogenesis of mare basalts and the history and evolution of magmatic processes on the Moon.

The lunar landing sites and their returned samples are extremely important for understanding the absolute chronology in order to anchor the Moon's geologic record. One of the major geologic goals of the Apollo and Luna missions was to return lunar samples. These returned samples could be dated radiometrically (e.g., U-Pb, Rb-Sr, Sm-Nd, and ^{40}Ar-^{39}Ar), but this provides highly accurate absolute ages for only a few small areas on the Moon (i.e., the landing sites). Obtaining ages for the vast unsampled regions relies on remote-sensing techniques such as crater counts. In order to derive absolute model ages from crater counts for any unsampled lunar region, we first have to correlate the radiometric ages from returned samples with the crater counts from the landing sites to establish the lunar cratering chronology. For this purpose, crater counts for the Apollo 11, 12, 14, 15, 16, and 17 and the Luna 16 and 24 landing sites were performed and correlated with the corresponding radiometric ages of these sites (e.g., BVSP 1981; Neukum 1983; Strom and Neukum 1988; Neukum and Ivanov

1994; Stöffler and Ryder 2001). A variety of interpretations as to which radiometric sample ages are most representative for a given landing site has resulted in several empirically derived lunar impact chronologies. These are discussed in Chapter 5 of this volume. We can use the lunar chronology to derive ages for any area for which crater counts have been conducted. It is important to note here that for the accurate derivation of the lunar chronology it is important to have data points over the entire period of time from lunar formation to present. Rocks from the sample collection yield ages from about 4.5 b.y. (ferroan anorthosite) to 2 m.y. (exposure age of Cone crater), allowing accurate modeling of the impact flux on the Moon.

Although this section highlights the importance of samples and surface exploration as ground-truth for remote sensing, information derived from orbital data feeds back into our understanding of the planet and as context for the landing sites. As an example, global compositional data, as recorded by both the Clementine and Lunar Prospector missions, show that a vast area of the Moon is not particularly well represented by the returned samples (Feldspathic Highlands terrane) or not at all (South Pole-Aitken region). This statement is supported by the lunar meteorites, which now include enough specimens to approach a statistically significant and more-or-less random sampling of the Moon. These rocks are dominated by feldspathic breccias and by low- to intermediate-Ti basalts. The feldspathic breccias appear to represent the Feldspathic Highlands terrane better than Apollo 16. The distribution of basalt compositions also appears to be more like what is anticipated from remote sensing, unlike the Apollo basalts, which disproportionately sampled high-Ti regions (e.g., Giguere et al. 2000). An optimal understanding of the planet, however, comes from integration of all of these datasets. This theme is elaborated in subsequent chapters of this volume.

4. SUMMARY AND OUTSTANDING QUESTIONS

The exploration of the Moon has revealed a geologically complex planetary body. Although at the beginning of modern lunar exploration most scientists would have agreed on the Moon being a rather simple object, exploration readily showed that the Moon is in fact more complicated than initially thought. Crew observations, lunar samples, geophysical measurements made on the lunar surface, the wealth of Galileo and Clementine multispectral reflectance data, and the Lunar Prospector gamma-ray and neutron spectrometer data, as well as magnetic and gravity data, all painted a picture vastly different from the one we had when we began the journey more than 35 years ago. The Moon is differentiated, has a complex history of volcanic and tectonic events, has weak crustal magnetic anomalies, shows crustal gravity anomalies, and is covered with craters of all sizes that were formed by impact rather than volcanic processes. In the planetary context the Moon is especially important because it provides an excellent record of primordial crust formation and complete records of the cratering history and the formation and evolution of its secondary crust.

As a result of many detailed investigations we now have a much better concept of the origin and evolution of the Moon, its structure, composition, stratigraphy, and physical properties, to name only a few fields in which enormous progress has been made over the past few decades. However, we are far from adequately "understanding" the Moon. Numerous questions remain unanswered and below is by no means an exhaustive list of such questions.

- What is the cause of the global asymmetry of the Moon? Is it related to convection and density inversion dynamics, early giant impacts, asymmetric crystallization of the magma ocean, or Earth-Moon tidal effects?

- How does a magma ocean work? Did this stage involve whole or partial melting of the Moon?

- What was the early thermal evolution of the Moon?

- What is the vertical and lateral structure of the lunar crust and how did it develop?

- What is the composition and structure of the lunar mantle?

- What is the size and composition of the postulated lunar core?

- What role did early (i.e., >4 Ga) volcanism play?

- What were the timing and the effects of major basin-forming events on the lunar stratigraphy?

- What is the nature of the South Pole-Aitken basin? Did it penetrate into the mantle, and how did it affect early crustal evolution?

- What was the impactor flux in the inner Solar System and how has this varied over time? Was there a terminal cataclysm at ~3.9 Ga?

- How and why is the Moon different from other planets and how do planets work in terms of, for example, surface processes, heat transfer, and geologic evolution?

- Are the Apollo geophysical measurements representative of the Moon, or are they only valid for the small regions encompassed by the Apollo landing sites?

- How does the formation and evolution of this small body and its well preserved record of impact modification and solar activity relate to Earth's ancient history and its development of habitability?

With new global data, e.g., from Clementine and Lunar Prospector in hand, we can revisit the post-Apollo paradigms. By integrating old and new data sets, and sample and remotely sensed information collected in the past decades, we have been able to test old ideas, put forward new ones, and so build a solid foundation for future lunar exploration. As a result, some of the post-Apollo models and hypotheses persevered and will no doubt continue to do so, while others underwent or will undergo substantial changes, and yet others have been or will be dismissed entirely. This is only possible because the lunar science community realized that it is beneficial to foster interdisciplinary approaches to address the major scientific questions about the origin and evolution of the Moon. Even with this integrated approach, we might not be able to answer all questions, but it will greatly advance our geologic understanding, will enable us to ask better questions in exploring the other planets, and will put us in a position to develop much better future planetary missions.

To optimize the Moon as the foundation and cornerstone of our understanding of planetary processes and evolution, future exploration should concentrate on some or all of the following goals: deconvolution of the complex record of early lunar crustal formation and evolution; determination of the nature of basic processes and their role in the evolution of the Moon; investigation of the relationships of these processes to the thermal evolution of the Moon and other one-plate planets; establishment of planetary perspectives on the first half of solar-system history; and extrapolation to the nature and evolution of terrestrial planetary bodies including Earth.

As we stand on the edge of a new era of planetary exploration, including new missions to the Moon, Mars, Mercury, Saturn and some asteroids and comets, we have the opportunity and obligation not only to understand the Moon's origin and evolution but also to use the lessons learned from lunar exploration to define the best ways of exploring the rest of our Solar System. In this sense the Moon is a keystone to our understanding of the terrestrial planets and a stepping-stone for future exploration of our Solar System. How do we explore planets? What types of measurements do we need to make? What strategies are most useful? Orbiting spacecraft or sample return missions, or both? Human exploration? From the experience with lunar exploration we know that each exploration strategy has its own benefits, but that it is the integration and simultaneous interpretation of various data sets that are most powerful to further our knowledge. This is a very important lesson that needs to be considered in plans for the exploration of other planets.

The Moon is a geologically complex planetary body. And it is an excellent natural, easy-to-reach laboratory to test equipment and ideas on our way to other planetary bodies. In this sense the Moon might well prove to be the Rosetta stone for planetary exploration.

5. REFERENCES

Agnor CB, Canup RM, Levison HF (1999) On the character and consequences of large impacts in the late stage of terrestrial planet formation. Icarus 142:219-237

Ahrens TJ, Rubin AM (1993) Impact-induced tensional failure in rock. J Geophys Res 98(E1):1185-1203

Anderson OL (1981) A decade of progress in Earth's internal properties and processes. Science 213:76-82

Anderson JLB, Schultz PH, Heineck JT (2000) A new view of ejecta curtains during oblique impacts using 3D particle imaging velocimetry. Lunar Planet Sci XXXI:#1749

Anderson JLB, Schultz PH, Heineck JT (2001) Oblique impact ejecta flow fields: An application of Maxwell's Z model. Lunar Planet Sci XXXII:#1352

Anderson JLB, Schultz PH, Heineck JT (2003) Asymmetry of ejecta flow during oblique impacts using three-dimensional particle image velocimetry. J Geophys Res 108, doi 10.1029/2003JE002075

Antonenko I, Yingst RA (2002) Mare and cryptomare deposits in the Schickard region of the Moon: New measurements using Clementine FeO data. Lunar Planet Sci XXXIII:#1438

Antonenko I, Head JW, Mustard JF, Hawke BR (1995) Criteria for the detection of lunar cryptomaria. Earth, Moon and Planets 69:141-172

Arai T, Warren PH (1999) Lunar meteorite QUE 94281: Glass compositions and other evidence for launch pairing with Yamato-793274. Meteorit Planet Sci 34:209-234

Arkani-Hamed J (1998) The lunar mascons revisited. J Geophys Res 103(E2):3709-3739

Arkani-Hamed J, Konopliv AS, Sjogren WL (1999) On the equipotential surface hypothesis of lunar maria floors, J Geophys Res 104(E3):5921-5932

Arndt J, von Engelhardt W (1987) Formation of Apollo 17 orange and black glass beads. J Geophys Res 92:372-376

Arnold JR (1979) Ice in the lunar polar regions. J Geophys Res 84(B10):5659-5668

Ashworth DG (1977) Lunar and planetary impact erosion. *In:* Cosmic Dust. McDonnell JAM (ed) Wiley, p 427-526

Baldwin RB (1971) On the history of lunar impact cratering: The absolute time scale and the origin of planetesimals. Icarus 14:36-52

Baldwin RB (1974) On the origin of mare basins. Proc Lunar Sci Conf 5:1-10

Baldwin RB (1987) On the relative and absolute ages of seven lunar front face basins. II - From crater counts. Icarus 71:19-29

Barnouin-Jha OS, Schultz PH (1996) Ejecta entrainment by impact-generated ring vortices: Theory and experiments. J Geophys Res 101(E9):21099-21116

Barnouin-Jha OS, Schultz PH (1998) Lobateness of impact ejecta deposits from atmospheric interactions. J Geophys Res 103(E11):25739-25756

Barsukov VL (1977) Preliminary data for the regolith core brought to Earth by automatic lunar station Luna 24. Proc Lunar Sci Conf 8:3303-3318

BVSP (Basaltic Volcanism Study Project) (1981) Basaltic Volcanism on the Terrestrial Planets. Pergamon Press

Bauer JF (1979) Experimental shock metamorphism of mono- and polycrystalline olivine - A comparative study. Proc Lunar Planet Sci Conf 10:2573-2596

Beaty DW, Albee AL (1978) Comparative petrology and possible genetic relations among the Apollo 11 basalts. Proc Lunar Planet Sci Conf 9:359-463

Beaty DW, Albee AL (1980) The geology and petrology of the Apollo 11 landing site. Proc Lunar Planet Sci Conf 11:23-35

Belton MJS, Head JW, Pieters CM, Greeley R, McEwen AS, Neukum G, Klaasen KP, Anger CD, Carr MH, Chapman CR (1992) Lunar impact basins and crustal heterogeneity - New western limb and far side data from Galileo. Science 255:570-576

Belton MJS, Greeley R, Greenberg R, McEwen A, Klaasen KP, Head JW, Pieters C, Neukum G, Chapman CR, Geissler P, Heffernan C, Breneman H, Anger C, Carr MH, Davies ME, Fanale FP, Gierasch PJ, Ingersoll AP, Johnson TV, Pilcher CB, Thompson WR, Veverka J, Sagan C (1994) Galileo multispectral imaging of the north polar and eastern limb regions of the Moon. Science 264:1112-1115

Benz W, Slattery WL, Cameron AGW (1986) The origin of the moon and the single-impact hypothesis I. Icarus 66:515-535

Benz W, Slattery WL, Cameron AGW (1987) The origin of the moon and the single-impact hypothesis II. Icarus 71:30-45

Bickel CE (1977) Petrology of 78155 - An early, thermally metamorphosed polymict breccia. Proc Lunar Sci Conf 8: 2007-2027

Binder AB (1980) On the internal structure of a moon of fission origin. J Geophys Res 85:4872-4880

Binder AB (1998) Lunar Prospector: Overview. Science 281:1475-1476

Bischoff A, Stöffler D (1992) Shock metamorphism as a fundamental process in the evolution of planetary bodies: Information from meteorites. Europ J Mineral 4:707-755

Blewett DT, Hawke BR, Lucey PG, Taylor GJ, Jaumann R, Spudis PD (1995) Remote sensing and geologic studies of the Schiller-Schickard region of the Moon. J Geophys Res 100:16959-16978

Blewett DT, Lucey PG, Hawke BR, Jolliff BL (1997) Clementine images of the lunar sample-return stations: Refinement of FeO and TiO$_2$ mapping techniques. J Geophys Res 102:16319-16325

Boyce JM (1976) Ages of flow units in the lunar nearside maria based on Lunar Orbiter IV photographs. Proc Lunar Sci Conf 7:2717-2728

Boyce JM, Johnson DA (1978) Ages of flow units in the far eastern maria and implications for basin-filling history. Proc Lunar Planet Sci Conf 9:3275-3283

Brett R (1975) Thicknesses of some lunar mare basalt flows and ejecta blankets based on chemical kinetic data. Geochim Cosmochim Acta 39:1135-1143

Brett R (1976) Reduction of mare basalts by sulfur loss. Geochim Cosmochim Acta 40:997-1004

Brett R, Gooley RC, Dowty E, Prinz M, Keil K (1973) Oxide minerals in lithic fragments from Luna 20 fines. Geochim Cosmochim Acta 37:761-773

Bruno BC, Lucey PG, Hawke BR (1991) High-resolution UV-visible spectroscopy of lunar red spots. Proc Lunar Planet Sci Conf 21:405-415

Budney CJ, Lucey PG (1998) Basalt thickness in Mare Humorum: The crater excavation method. J Geophys Res 103: 16855-16870

Burchell and Mackay (1998) Crater ellipticity in hypervelocity impacts on metals, J Geophys Res 103:22761-22774

Bussey DBJ, Guest JE, Sørensen S-A (1997) On the role of thermal conductivity on thermal erosion by lava. J Geophys Res 102:10905-1090

Butler P, Morrison DA (1977) Geology of the Luna 24 landing site. Proc Lunar Sci Conf 8:3281-3301

Cameron AGW (1997) The origin of the Moon and the single impact hypothesis V. Icarus 126:126-137

Cameron AGW, Ward WR (1976) The origin of the Moon. Lunar Planet Sci XII:120

Cameron AGW, Benz W (1991) The origin of the Moon and the single impact hypothesis IV. Icarus 92:204-216

Cameron AGW, Canup RM (1998) The giant impact and the formation of the Moon. Proceedings of Origin of the Earth and Moon, LPI Contribution No. 957, Lunar and Planetary Institute

Canup RM, (2004) Simulations of a late lunar-forming impact. Icarus 168:433-456

Canup RM, Esposito LW (1996) Accretion of the Moon from an impact-generated disk. Icarus 119:427-446

Canup RM, Ward WR, Cameron AGW (2001) A scaling relationship for satellite-forming impacts. Icarus 150:288-296

Carr MH, Crumpler LS, Cutts JA Greeley R, Guest JE, Masursky H (1977) Martian impact craters and emplacement of ejecta by surface flow. J Geophys Res 82:4055-4065

Cashore J, Woronow A (1985) A new Monte Carlo model of lunar megaregolith development. J Geophys Res 90:811-816

Chabot NL, Hoppa GV, Strom RG (2000) Analysis of lunar lineaments: Far side and polar mapping. Icarus 147:301-308

Chao ECT (1968) Pressure and temperature histories of impact metamorphosed rocks – based on petrographic observations. *In:* Shock Metamorphism of Natural Materials. French BM, Short NM (eds) Mono Book Corp., 135-158

Chao ECT (1973) Geologic implications of the Apollo 14 Fra Mauro breccias and comparison with ejecta from the Ries crater, Germany. J Res US Geol Surv 1:1-18

Chapman CR, Merline WJ, Bierhaus B, Keller J, Brooks S, McEwen A, Tufts BR, Moore J, Carr M, Greeley R, Bender KC, Sullivan R, Head JW, Pappalardo RT, Belton MJS, Neukum G, Wagner R, Pilcher C, and the Galileo Imaging Team (1997) Populations of small craters on Europa, Ganymede, and Callisto: Initial Galileo imaging results. Lunar Planet Sci XXVIII:217-218

Chappell BW, Green DH (1973) Chemical composition and petrogenetic relationships in Apollo 15 mare basalts. Earth Planet Sci Lett 18:237-246

Charette MP, McCord TB, Pieters C, Adams JB (1974) Application of remote spectral reflectance measurements to lunar geology classification and determination of titanium content of lunar soils. J Geophys Res 79:1605-1613

Chevrel SD, Rosemberg C, Pinet PC, Shevchenko VV, Daydou Y (1998) Lunar swirl-like terrains exploration: The case of Mare Ingenii. Lunar Planet Sci XXIX:1660

Chou C-L, Boynton WV, Sundberg LL, Wasson JT (1975) Volatiles on the surface of Apollo 15 green glass and trace element distributions among Apollo 15 soils. Proc Lunar Sci Conf 6:1701-1727

Chyba CF (1991) Terrestrial mantle siderophiles and the lunar impact record. Icarus 92:217-233

Cintala MJ (1992) Impact-induced thermal effects in the lunar and mercurian regoliths. J Geophys Res 97:947-973

Cintala MJ, Grieve RAF (1994) The effects of differential scaling of impact melt and crater dimensions on lunar and terrestrial craters: Some brief examples. *In:* Large Meteorite Impacts and Planetary Evolution. BO Dressler, RAF Grieve, VL Sharpton (eds), The Geological Society of America, Special Paper 293, p 51-59

Cintala MJ, Berthoud L, Hörz F (1999) Ejection-velocity distributions from impacts into coarse-grained sand. Meteorit Planet Sci 34:605-623

Clark PE, Hawke BR (1991) The lunar farside - The nature of highlands east of Mare Smythii. Earth, Moon, and Planets 53:93-107

Cohen BA, Swindle TD, Kring DA (2000) Support for the lunar cataclysm hypothesis from lunar meteorite impact melt ages. Science 290:1754-1756

Cohen BA, Swindle TD, Kring DA (2000a) Argon-40-Argon-39 geochronology of lunar meteorite impact melt clasts. Meteorit Planet Sci 35, Supplement:A43

Coombs CR, Hawke BR, Gaddis LR (1987) Explosive volcanism on the Moon. Lunar Planet Sci XVIII:197-198

Coombs CR, Hawke BR, Peterson CA, Zisk SH (1990) Regional pyroclastic deposits in the north-central portion of the lunar nearside. Lunar Planet Sci XXI:228-229

Cooper BL, Carter JL, Sapp CA (1994) New evidence for graben origin of Oceanus Procellarum from lunar sounder optical imagery. J Geophys Res 99:3799-3812

Craddock RA, Howard AD (2000) Simulated degradation of lunar impact craters and a new method for age dating farside mare deposits. J Geophys Res 105:20387-20401

Croft SK (1980) Cratering flow fields: Implications for the excavation and transient expansion stages of crater formation. Proc Lunar Planet Sci Conf 11:2347-2378

Cushing JA, Taylor GF, Norman MD, Keil K (1999) The granulitic impactite suite: Impact melts and metamorphic breccias of the early lunar crust. Meteorit Planet Sci 34:185-195

Dana J. D. (1846) On the volcanoes of the Moon. Am J Sci 2:335-353

Darwin G (1879) On the procession of a viscous spheroid and on the remote history of Earth. Philos Trans R Soc London 170:447-538

Dasch EJ, Shih C-Y, Bansal BM, Wiesmann H, Nyquist LE (1987) Isotopic analysis of basaltic fragments from lunar breccia 14321: Chronology and petrogenesis of pre-Imbrium mare volcanism. Geochim Cosmochim Acta 51: 3241-3254

Davies GF (1972) Equations of state and phase equilibria of stishovite and a coesite-like phase from shock-wave and other data. J Geophys Res 77:4920-4933

DeHon RA (1975) Mare Spumans and Mare Undarum - Mare thickness and basin floor. Proc Lunar Sci Conf 6:2553-2561

DeHon RA (1977) Mare Humorum and Mare Nubium - Basalt thickness and basin-forming history. Lunar Planet Sci Conf 8:633-641

DeHon RA (1979) Thickness of the western mare basalts. Proc Lunar Planet Sci Conf 10:2935-2955

DeHon RA, Waskom JD (1976) Geologic structure of the eastern mare basins. Proc Lunar Sci Conf 7:2729-2746

Delano JW (1986) Pristine lunar glasses. J Geophys Res 91:D201–D215

Delano JW, Livi K (1981) Lunar volcanic glasses and their constraints on mare petrogenesis. Geochim Cosmochim Acta 45:2137-2149

Dombard AJ, Gillis JJ (2001) Testing the viability of topographic relaxation as a mechanism for the formation of lunar floor-fractured craters. J Geophys Res 106:27901-27910

Dowty E, Prinz M, Keil K (1973) Composition, mineralogy, and petrology of 28 mare basalts from Apollo 15 rake samples. Proc Lunar Planet Sci Conf 4:423-444

El-Baz F (1972) The Alhazen to Abul Wafa swirl belt: An extensive field of light-colored, sinuous markings. Apollo 16 Preliminary Science Report, NASA SP-315:29-93 – 29-97

Elphic RC, Lawrence DJ, Feldman WC, Barraclough BL, Maurice S, Binder AB, Lucey PG (2000) Lunar rare earth element distribution and ramifications for FeO and TiO$_2$: Lunar Prospector neutron spectrometer observations. J Geophys Res 105:20,333-20,345

Fagan TJ, Taylor GJ, Keil K, Bunch TE, Wittke JH, Korotev RL, Jolliff BL, Gillis JJ, Haskin LA, Jarosewich E, Clayton RN, Mayeda TK, Fernandes VA, Burgess R, Turner G, Eugster O, Lorenzetti S (2002) Northwest Africa 032: Product of lunar volcanism. Meteorit Planet Sci 37:371-394

Feldman WC, Binder AB, Barraclough BL, Belian RD (1998) First results from the Lunar Prospector Spectrometers. Lunar Planet Sci XXIX:1936

Feldman WC, Lawrence DJ, Maurice S, Elphic RC, Barraclough BL, Binder AB, Lucey PG (1999) Classification of lunar terranes using neutron and thorium gamma-ray data. Lunar Planet Sci XXX:2056

Feldman WC, Maurice S, Lawrence DJ, Little RC, Lawson SL, Gasnault O, Wiens RC, Barraclough BL, Elphic RC, Prettyman TH, Steinberg JT, Binder AB (2001) Evidence for water ice near the lunar poles. J Geophys Res 106, E10:23,231–23,251

Florenskiy KP, Bazilevskiy AT, Ivanov AV (1977) The role of exogenic factors in the formation of the lunar surface. *In:* The Soviet-American Conference on Cosmochemistry of the Moon and Planets. Moscow USSR, June 4-8, 1974. Pomeroy H, Hubbard NJ (eds) p 571-584

Freed AM, Melosh HJ, Solomon SC (2001) Tectonics of mascon loading: Resolution of the strike-slip faulting paradox. J Geophys Res 106:20,603-20,620

Freeman VF (1981) Regolith of the Apollo 16 site. *In:* Geology of the Apollo 16 area, Central Highlands. Ulrich GE, Hodges CA, Muehlberger WR (eds) USGS Prof Paper 1048:147-159

French BM (1998) Traces of Catastrophes: A Handbook of Shock-Metamorphic Effects in Terrestrial Meteorite Impact Structures. LPI Contribution 954, Lunar and Planetary Institute

Fujiwara A, Kadono T, Nakamura A (1993) Cratering experiments into curved surfaces and their implication for craters on small satellites. Icarus 105:345-350

Gaddis LR, Pieters CM, Hawke BR (1985) Remote sensing of lunar pyroclastic mantling deposits. Icarus 61:461-489

Gault DE, Wedekind JA (1969) The destruction of tektites by micrometeoroid impact. J Geophys Res 74:6780-6794

Gault DE, Wedekind JA (1978) Experimental studies of oblique impact. Lunar Planet Sci IX:374-376

Gerstenkorn H (1955) Über Gezeitenreibung beim Zweikörper-Problem. Zeit Astrophys 36:245-274

Gibson RL, Reimold WU, Ashley AJ, Koeberl C (2001) Granulitic melt breccias in the Vredefort impact structure, South Africa - A terrestrial analog for lunar granulites. Lunar Planet Sci XXXII:1013

Gifford AW, El-Baz F (1978) Thickness of mare flow fronts. Lunar Planet Sci IX:382-384

Giguere TA, Taylor GJ, Hawke BR, Lucey PG (2000) The titanium content of lunar mare basalts. Meteorit Planet Sci 35:193-200

Giguere TA, Hawke BR, Blewett DT, Bussey DBJ, Lucey PG, Smith GA, Spudis PD, Taylor GJ (2003) Remote sensing studies of the Lomonosov-Fleming region of the Moon. J Geophys Res 108, doi 10.1029/2003JE002069

Gillis JJ, Jolliff BL, Elphic RC (2003) A revised algorithm for calculating TiO_2 from Clementine UVVIS data: A synthesis of rock, soil, and remotely sensed TiO_2 concentrations. J Geophys Res 108, doi 10.1029/2001JE001515

Goins NR, Toksöz MN, Dainty AM (1978) Seismic structure of the lunar mantle - An overview. Proc Lunar Planet Sci Conf 9:3575-3588

Goins NR, Dainty AM, Toksöz MN (1981) Seismic energy release of the Moon. J Geophys Res 86:378-388

Gold T (1966) The Moon's surface. *In:* The Nature of the Lunar Surface, Proc 1965 IAU-NASA Symp (WN Hess, DH Menzel, JA O'Keefe, eds.), 107-124. Johns Hopkins, Baltimore

Golombek MP (1985) Fault type predictions from stress distributions on planetary surfaces - Importance of fault initiation depth. J Geophys Res 90:3065-3074

Golombek MP, Anderson FS, Zuber MT (1999) Topographic profiles across wrinkle ridges indicate subsurface faults. EOS Trans, Am Geophys Union, 80:610

Golombek MP, Anderson FS, Zuber MT (2000) Martian wrinkle ridge topography: Evidence for subsurface faults from MOLA. Lunar Planet Sci XXXI:1294

Greeley R (1971) Lava tubes and channels in the lunar Marius Hills. Earth, Moon, and Planets 3:289

Greeley R, Kadel SD, Williams DA, Gaddis LR, Head JW, McEwen AS, Murchie SL, Nagel E, Neukum G, Pieters CM, Sunshine JM, Wagner R, Belton MJS (1993) Galileo imaging observations of lunar maria and related deposits. J Geophys Res 98:17183-17206

Greeley R, Sullivan R, Klemaszewski J, Homan K, Head JW, Pappalardo RT, Veverka J, Clark B, Johnson TV, Klaasen KP, Belton M, Moore J, Asphaug E, Carr MH, Neukum G, Denk T., Chapman CR, Pilcher CB, Geissler PE, Greenberg R, Tufts R (1998) Europa: Initial Galileo geological observations. Icarus 135:4-24

Green DH, Ringwood AE, Hibberson WO, Ware NG (1975) Experimental petrology of Apollo 17 mare basalts. Proc Lunar Sci Conf 6:871-893

Grier JA, McEwen AS, Lucey PG, Milazzo M, Strom RG (1999) The optical maturity of ejecta from large rayed craters: Preliminary results and implications. *In:* Workshop on New Views of the Moon II: Understanding the Moon Through the Integration of Diverse Datasets. LPI Contribution No. 980, 19

Guest JE, Murray JB (1976) Volcanic features of the nearside equatorial lunar maria. J Geol Soc Lond 132:251-258

Halekas JS, Mitchell DL, Lin RP, Frey S, Hood LL, Acuna MH, Binder AB (2001) Mapping of crustal magnetic anomalies on the lunar near side by the Lunar Prospector electron reflectometer. J Geophys Res 106, E11:27,841-27,852

Hartmann WK (1971) Martian Cratering III: Theory of crater obliteration. Icarus 15:410-428

Hartmann WK (1975) Lunar 'cataclysm' - A misconception. Icarus 24:81-187

Hartmann WK, Kuiper GP (1962) Concentric structures surrounding lunar basins. Commun Lunar Planet Lab 1:51-66

Hartmann WK, Davis DR (1975) Satellite-sized planetesimals and lunar origin. Icarus 24:504-515

Hartmann WK, Strom RG, Weidenschilling SJ, Blasius KR, Woronow A, Dence MR, Grieve RAF, Diaz J, Chapman CR, Shoemaker EM, Jones KL (1981) Chronology of planetary volcanism by comparative studies of planetary craters. *In:* Basaltic Volcanism on the Terrestrial Planets. Pergamon Press, p 1050-1127

Hartmann WK, Phillips RJ, Taylor GJ (1986) Origin of the Moon. Lunar and Planetary Institute

Haskin LA (1998) The Imbrium impact event and the thorium distribution at the lunar highland surface. J Geophys Res 103(E1):1679-1689

Haskin LA (2000) Basin impacts, especially large and late Imbrium. *In:* Workshop on New Views of the Moon III: Synthesis of sample analysis, on-surface investigation, and remote sensing information. Lunar and Planetary Institute, p 12

Haskin LA, Warren P (1991) Lunar chemistry. *In:* The Lunar Sourcebook: A user's guide to the Moon. Heiken G, Vaniman D, French BM (eds) Lunar and Planetary Institute and Cambridge Univ Press 357-474

Haskin LA, Jolliff BL (1998) On estimating provenances of lunar highland materials. *In:* Workshop on New Views of the Moon: Integrated Remotely Sensed, Geophysical, and Sample Datasets, 35

Haskin LA, Korotev RL, Rockow KM, Jolliff BL (1998) The case for an Imbrium origin of the Apollo Th-rich impact-melt breccias. Meteorit Planet Sci 33:959-975

Haskin LA, Gillis JJ, Korotev RL, Jolliff BL (2000a) The nature of mare basalts in the Procellarum KREEP Terrane. Lunar Planet Sci XXXI:#1661

Haskin LA, Gillis JJ, Korotev RL, Jolliff BL (2000b) The materials of the lunar Procellarum KREEP Terrane: A synthesis of data from geomorphological mapping, remote sensing and sample analyses. J Geophys Res 105: 20,403-20,415

Haskin LA, Korotev RL, Gillis JJ, Jolliff BL (2002) Stratigraphies of Apollo and Luna highland landing sites and provenances of materials from the perspective of basin impact ejecta modeling. Lunar Planet Sci XXXIII:#1364

Haskin LA, Moss BE, McKinnon WB (2003) On estimating contributions of basin ejecta to regolith deposits at lunar sites. Meteorit Planet Sci 38(Nr 1):13-33

Hawke BR, Head JW (1977) Pre-Imbrian history of the Fra Mauro region and Apollo 14 sample provenance. Proc Lunar Sci Conf 8: 2741-2761

Hawke BR, Head JW (1978) Lunar KREEP volcanism - Geologic evidence for history and mode of emplacement. Lunar Planet Sci IX:3285-3309

Hawke BR, Spudis PD (1980) Geochemical anomalies on the eastern limb and farside of the moon. Conference on the Lunar Highlands Crust, Houston, Tex., November 14-16, 1979, Proceedings. (A81-26201 10-91) Pergamon Press, 467-481

Hawke BR, Bell JF (1981) Remote sensing studies of lunar dark-halo craters. Bull Am Astron Soc 13:712

Hawke BR, Mac Laskey D, McCord TB, Adams JB, Head JW, Pieters CM, Zisk S (1979) Multispectral mapping of lunar pyroclastic deposits. Bull Am Astron Soc 12:582

Hawke BR, Spudis PD, Clark PE (1985) The origin of selected lunar geochemical anomalies: Implications for early volcanism and the formation of light plains. Earth, Moon and Planets 32:257-273

Hawke BR, Coombs CR, Gaddis LR, Lucey PG, Owensby PD (1989) Remote sensing and geologic studies of localized dark mantle deposits on the Moon. Proc Lunar Planet Sci Conf 19:255-268

Hawke BR, Giguere TA, Lucey PG, Peterson CA, Taylor GJ, Spudis, PD (1998) Multidisciplinary studies of ancient mare basalt deposits. Workshop on New Views of the Moon, Lunar and Planetary Institute, 37-38

Hawke BR, Blewett DT, Lucey PG, Peterson CA, Bell JF, Campbell BA, Robinson MS (1999) The composition and origin of selected lunar crater rays. Workshop on New Views of the Moon II: Understanding the Moon Through the Integration of Diverse Datasets, #8035

Hawke BR, Blewett DT, Lucey PG, Smith GA, Taylor GJ, Lawrence DJ, Spudis PD (2001) Remote sensing studies of selected spectral anomalies on the Moon. Lunar Planet Sci XXXII:#1241,

Hawke BR, Lawrence DJ, Blewett, DT, Lucey, PG, Smith GA, Taylor GJ, Spudis PD (2002a) Remote sensing studies of geochemical and spectral anomalies on the nearside of the Moon. Lunar Planet Sci XXXIII:#1598

Hawke BR, Lawrence DJ, Blewett, DT, Lucey, PG, Smith GA, Taylor GJ, Spudis PD (2002b) Lunar highlands volcanism: The view from a millennium. *In:* The Moon Beyond 2002: Next Steps in Lunar Science and Exploration. LPI Contribution No. 1128, Lunar and Planetary Institute, p 22

Hawke BR, Lawrence DJ, Blewett DT, Lucey PG, Smith GA, Spudis PD, Taylor GJ (2003) Hansteen Alpha: A volcanic construct in the lunar highlands. J Geophys Res 108(E7), doi:10.1029/2002 JE002013

Hawke BR, Gillis JJ, Giguere TA, Blewett DT, Lawrence DJ, Lucey PG, Smith GA, Spudis PD, Taylor GJ (2005) Remote sensing and geologic studies of the Balmer-Kapteyn region of the Moon. J Geophys Res 110, doi 10.1029/2004JE002383

Head JW (1974) Lunar dark-mantle deposits: Possible clues to the distribution of early mare deposits. Proc Lunar Sci Conf 5:207-222

Head JW (1975a) Lunar mare deposits: Areas, volumes, sequence, and implication for melting in source areas. *In:* Origins of Mare Basalts and their Implications for Lunar Evolution. Lunar and Planetary Institute, p 66-69

Head JW (1975b) Some geologic observations concerning lunar geophysical models. Proc Soviet-American Conference on the Cosmochemistry of the Moon and Planets, Moscow 407-416

Head JW (1976) Lunar volcanism in space and time. Rev Geophys Space Phys 14:265-300

Head JW (1982) Lava flooding of ancient planetary crusts: Geometry, thickness, and volumes of flooded lunar impact basins. Moon and Planets 26:61-88

Head JW (1999) Surfaces and interiors of the terrestrial planets. *In:* The New Solar System. Beatty JK, Petersen CC, Chaikin A (eds) Cambridge University Press, p 157-173

Head JW, McCord TB (1978) Imbrian-age highland volcanism on the moon - The Gruithuisen and Mairan domes. Science 199:1433-1436

Head JW, Wilson L (1979) Alphonsus-type dark-halo craters: Morphology, morphometry and eruption conditions. Proc Lunar Planet Sci Conf 10:2861-2897

Head JW, Wilson L 1980) The formation of eroded depressions around the sources of lunar sinuous rilles: Observations. Lunar Planet Sci XI:426-428

Head JW, Wilson L (1992) Lunar mare volcanism: Stratigraphy, eruption conditions, and the evolution of secondary crusts. Geochim Cosmochim Acta 56:2155-2175

Head JW Wilson L (1999) Lunar Gruithuisen and Mairan Domes: Rheology and mode of emplacement. *In:* Workshop on New Views of the Moon II: Understanding the Moon Through the Integration of Diverse Datasets. LPI Contribution 980, Lunar and Planetary Institute, p 23

Head JW, Pieters C, McCord TB, Adams J, Zisk SH (1978a) Definition and detailed characterization of lunar surface units using remote observations. Icarus 33:145-172

Head JW, Adams JB, McCord TB, Pieters C, Zisk A (1978b) Regional stratigraphy and geologic history of Mare Crisium. *In:* Mare Crisium: The view from Luna 24. Merrill RB, Papike JJ (eds) Geochim Cosmochim. Acta Suppl. 9. Pergamon Press, New York, 727

Head JW, Murchie S, Mustard JF, Pieters CM, Neukum G, McEwen A, Greeley R, Nagel E, Belton MJS (1993) Lunar impact basins: New data for the western limb and far side (Orientale and South Pole-Aitken basins) from the first Galileo flyby. J Geophys Res 98:17149-17182

Head JW, Reed JS, Weitz C (1998) Lunar Rima Parry IV: Dike emplacement processes and consequent volcanism. Lunar Planet Sci XXIX:#1914

Head JW, Wilson L, Robinson M, Hiesinger H, Weitz C, Yingst A (2000) Moon and Mercury: Volcanism in Early Planetary History. *In:* Environmental Effects on Volcanic Eruptions: From Deep Oceans to deep Space. Zimbelman JR, Gregg TKP (eds) Kluwer Academic Press, p 143-178

Head JW, Wilson L, Weitz CM (2002) Dark ring in southwestern Orientale Basin: Origin as a single pyroclastic eruption. J Geophys Res 107(E1), doi:10.1029/2000JE001438

Heather DJ, Dunkin SK, Wilson L (2003) Volcanism on the Marius Hills plateau: Observational analyses using Clementine multispectral data. J Geophys Res 108, doi:10.1029/2002JE001938

Heiken G, McEwen MC (1972) The geologic setting of the Luna 20 site. Earth Planet Sci Lett 17:3-6

Heiken G, McKay DS, Brown RW (1974) Lunar deposits of possible pyroclastic origin. Geochim Cosmochim Acta 38:1703-1718

Heiken G, Vaniman D, French BM (eds) (1991) The Lunar Sourcebook: A User´s Guide to the Moon. Lunar and Planetary Institute and Cambridge Univ Press

Hiesinger H, Head JW, Jaumann R, Neukum G (1999) Lunar mare volcanism. Lunar Planet Sci XXX:#1199

Hiesinger H, Jaumann R, Neukum G, Head JW, Wolf U (2000) Ages and stratigraphy of mare basalts on the lunar nearside. J Geophys Res 105:29,239-29,275

Hiesinger H, Head JW, Wolf U, Neukum G (2001) New age determinations of lunar mare basalts in Mare Cognitum, Mare Nubium, Oceanus Procellarum, and other nearside mare. Lunar Planet Sci XXXII:#1815

Hiesinger H, Head JW, Wolf U, Jaumann R, Neukum G (2002) Lunar mare basalt flow units: Thicknesses determined from crater size-frequency distributions. Geophys Res Lett 29, doi:10.1029/2002GL014847

Hiesinger H, Head JW, Wolf U, Jaumann R, Neukum G (2003) Ages and stratigraphy of mare basalts in Oceanus Procellarum, Mare Nubium, Mare Cognitum, and Mare Insularum. J Geophys Res 108, doi 10.1029/2002JE001985

Hinners NW (1972) Apollo 16 site selection. Apollo 16 Preliminary Science Report, NASA-SP315: 1-1 – 1-3

Hood LL (1987) Magnetic field and remanent magnetization effects of basin-forming impacts on the moon. Geophys Res Lett 14:844-847

Hood LL, Williams CR (1989) The lunar swirls - Distribution and possible origins. Lunar Planet Sci XIX:99-113

Hood LL, Zuber MT (2000) Recent refinement in geophysical constraints on lunar origin and evolution. *In:* Origin of the Earth and Moon Canup. RM, Righter K (eds) University of Arizona Press, p 397-409

Hood LL, Mitchell DL, Lin RP, Acuna MH, Binder AB (1999) Initial measurements of the lunar induced magnetic dipole moment using Lunar Prospector magnetometer data. Geophys Res Lett 26:2327-2330

Hood LL, Zakharian A, Halekas J, Mitchell DL, Lin RP, Acuna MH, Binder AB (2001) Initial mapping and interpretation of lunar crustal magnetic anomalies using Lunar Prospector magnetometer data. J Geophys Res 106, E11:27,825-27,839

Hornemann U, Müller WF (1971) Shock-induced deformation twins in clinopyroxene. Neues Jb Min, Mh 6: 247-256

Hörz F (1978) How thick are lunar mare basalts? Proc Lunar Planet Sci Conf 9:3311-3331

Hörz F, Morrison DA, Gault DE, Oberbeck VR, Quaide WL, Vedder JF, Brownlee DE, Hartung JB (1977) The micrometeoroid complex and evolution of the lunar regolith. *In:* The Soviet-American Conference on Cosmochemistry of the Moon and Planets, 605-635

Hörz F, Grieve R, Heiken G, Spudis P, Binder A (1991) Lunar surface processes. *In:* Lunar Sourcebook - A User Guide to the Moon .Heiken G, Vaniman D, French B (eds) Cambridge University Press, p 61-120

Howard KA, Head JW, Swann GA (1973) Geology of Hadley Rille. Proc Lunar Sci Conf 3:1-14

Hubbard NJ, Minear JW (1975) A chemical and physical model for the genesis of lunar rocks: Part II mare basalts. *In:* Lunar Planet Sci XI:405

Hulme G (1973) Turbulent lava flows and the formation of lunar sinuous rilles. Mod Geol 4:107-117

Ivanov BA (1992) Impact craters. *In:* Venus Geology, Geochemistry, and Geophysics - Research results from the USSR. Barsukov V, Basilevsky A, Volkov V, Zharkov V (eds) Univ. Arizona Press, p 113-128

Ivanov BA, Melosh HJ (2003) Impacts do not initiate volcanic eruptions: Eruptions close to the crater. Geology 31: 869–872

James OB (1981) Petrologic and age relations in Apollo 16 rocks: Implications for subsurface geology and the age of the Nectaris basin. Proc Lunar Planet Sci Conf 12B:209-233

James OB, Wright TL (1972) Apollo 11 and 12 mare basalts and gabbros: Classification, compositional variations, and possible petrogenetic relations. Bull Geol Soc Am 83:2357-2382

James OB, Hörz F (eds) (1981) Workshop on Apollo 16. LPI Technical Report 81-01, Lunar and Planetary Institute

Jessberger EK, Huneke JC, Podosek FA, Wasserburg GJ (1974) High-resolution argon analysis of neutron-irradiated Apollo 16 rocks and separated minerals. Proc Lunar Sci Conf 5:1419-1449

Johnson JR, Larson SM, Singer RB (1991) Remote sensing of potential lunar resources. I - Near-side compositional properties. J Geophys Res 96:18,861-18,882

Jolliff BL (1999) Clementine UVVIS multispectral data and the Apollo 17 landing site: What can we tell and how well? J Geophys Res 104:14123-14148

Jolliff BL, Korotev RL, Haskin LA (1991) Geochemistry of 2-4-mm particles from Apollo 14 soil (14161) and implications regarding igneous components and soil-forming processes. Proc Lunar Planet Sci Conf 21:193-219

Jolliff BL, Gillis JJ, Haskin LA, Korotev RL, Wieczorek MA (2000) Major lunar crustal terrains: Surface expression and crust-mantle origins. J Geophys Res 105(E2):4197-4216

Khan A, Mosegaard K, Rasmussen KL (2000) A new seismic velocity model for the Moon from a Monte Carlo inversion of the Apollo lunar seismic data. Geophys Res Lett 27:1591-1594

Khan A, Mosegaard K, Williams JG, Logonné P (2004) Does the Moon possess a molten core? Probing the deep lunar interior using results from LLR and Lunar Prospector. J Geophys Res 109, doi 10.1029/2004JE002294

Kieffer SW, Schaal RB, Gibbons RV, Hörz F, Milton DJ, Dube A (1976) Shocked basalt from Lonar impact crater, India, and experimental analogues. Proc Lunar Science Conference 7:1391-1412

Kipp ME, Melosh HJ (1986) Origin of the Moon: A preliminary numerical study of colliding planets. Lunar Planet Sci XVII:420-421

Köhler U, Head JW, Neukum G, Wolf U (1999) North-polar lunar light plains: Ages and compositional observations. *In:* Workshop on New Views of the Moon II: Understanding the Moon Through the Integration of Diverse Datasets, 34

Köhler U, Head JW, Neukum G, Wolf U (2000) Lunar light plains in the northern nearside latitudes: Latest results on age distributions, surface composition, nature, and possible origin. Lunar Planet Sci XXXI:#1822

Kokubo E, Canup RM, Ida S (2000) Lunar accretion from an impact-generated disk. *In:* Origin of the Earth and Moon (Canup RM, Righter K, eds.) Univ. Arizona Press, 145-163

Konopliv AS, Binder AB, Hood LL, Kucinskas AB, Sjogren WL, Williams JC (1998) Improved gravity field of the Moon from Lunar Prospector. Science 281:1476-1480

Konopliv AS, Asmar SW, Yuan DN (2001) Recent gravity models as a result of the Lunar Prospector mission. Icarus 150:1-18

Korotev RL (1987) The nature of the meteoritic components of Apollo 16 soil as inferred from correlations of iron, cobalt, iridium, and gold with nickel. Proc Lunar Planet Sci Conf 17:447-461

Korotev RL (1990) Cobalt and nickel concentrations in the 'komatiite component' of Apollo 16 polymict samples. Earth Planet Sci Lett 96:481-489.

Korotev RL (2000) The great lunar hot spot and the composition and origin of the Apollo mafic ("LKFM") impact-melt breccias. J Geophys Res 105:4317-4345

Korotev RL, Jolliff BL (2001) The curious case of the lunar magnesian granulitic breccias. Lunar Planet Sci XXXII: #1455

Korotev RL, Gillis JJ (2001) A new look at the Apollo 11 regolith and KREEP. J Geophys Res 106:12339-12353

Korotev RL, Jolliff BL, Rockow KM (1996) Lunar meteorite Queen Alexandra Range 93069 and the iron concentration of the lunar highlands surface. Meteorit Planet Sci 31:909-924

Korotev RL, Jolliff BL, Zeigler RA, Haskin LA (2003a) Compositional constraints on the launch pairing of three brecciated lunar meteorites of basaltic composition. Antarctic Met Res 16:152-175

Korotev RL, Jolliff BL, Zeigler RA, Gillis JJ, Haskin LA (2003b) Feldspathic lunar meteorites and their implications for compositional remote sensing of the lunar surface and the composition of the lunar crust. Geochim Cosmochim Acta 67(24):4895-4923

Kring DA, Cohen BA (2002) Cataclysmic bombardment throughout the inner solar system 3.9-4.0 Ga. J Geophys Res 107, doi 10.1029/2001JE001529

Kuskov OL, Konrod VA (2000) Resemblance and difference between the constitution of the Moon and Io. Planet Space Sci 48:717-726

Kuzmin RO, Bobina NN, Zabalueva EV, Shashkina VP (1988) The Structure of the martian cryolithosphere upper levels. *In:* Workshop on Mars Sample Return Science. LPI Technical Report 88-07, Lunar and Planetary Institute, p 108

Lammlein DR, Latham GV, Dorman J, Nakamura Y, Ewing M (1974) Lunar seismicity, structure and tectonics. Rev Geophys Space Phys 12:1-21

Langseth MG, Keihm SJ, Peters K (1976) Revised lunar heat-flow values. Proc Lunar Sci Conf 7:3143-3171

Lawrence DJ, Feldman WC, Barraclough BL, Binder AB, Elphic RC, Maurice S, Thomsen DR (1998) Global elemental maps of the Moon: The Lunar Prospector gamma-ray spectrometer. Science 281:1484-1489

Lawrence DJ, Feldman WC, Barraclough BL, Binder AB, Elphic RC, Maurice S, Miller MC, Prettyman TH (2000) Thorium abundances on the lunar surface. J Geophys Res 105(E8):20,307-20,331

Lawrence DJ, Feldman WC, Blewett DT, Elphic RC, Lucey PG, Maurice S, Prettyman TH, Binder AB (2001) Iron abundances on the lunar surface as measured by the Lunar Prospector gamma-ray spectrometer. Lunar Planet Sci XXXII:#1830

Lawrence DJ, Feldman WC, Elphic RC, Little RC, Prettyman TH, Maurice S, Lucey PG, Binder AB (2002) Iron abundances on the lunar surface as measured by the Lunar Prospector gamma-ray and neutron spectrometers. J Geophys Res 107, doi:10.1029/2001JE001530

Lawrence DJ, Hawke BR, Hagerty JJ, Elphic RC, Feldman WC, Prettyman TH, Vaniman DT (2005) Evidence for a high-Th, evolved lithology on the Moon at Hansteen Alpha. Geophys Res Lett 32, doi 10.1029/2004GL022022

Lawson SL, Feldman WC, Lawrence DJ, Moore KR, Maurice S, Belian RD, Binder AB (2002) Maps of Lunar Radon-222 and Polonium-210. Lunar Planet Sci XXXIII:#1835

Lemoine FG, Smith DE, Zuber MT, Neumann GA, Rowlands DD (1997) A 70th degree lunar gravity model (GLGM-2) from Clementine and other tracking data. J Geophys Res 102:16339-16359

Levison HF, Dones L, Chapman CR, Stern SA, Duncan MJ, Zahnle K (2001) Could the lunar "Late Heavy Bombardment" have been triggered by the formation of Uranus and Neptune? Icarus 151:286-306

Li L, Mustard JF (2000) The compositional gradients and lateral transport by dark-halo and dark-ray craters. Lunar Planet Sci XXXI:#2007

Li L, Mustard JF (2003) Highland contamination in lunar mare soils: Improved mapping with multiple end-member spectral mixture analysis (MESMA). J Geophys Res 108, doi 10.1029/2002JE001917

Lin RP, Anderson KA, Hood LL (1988) Lunar surface magnetic field concentrations antipodal to young large impact basins. Icarus 74:529-541

Lindstrom MM, Lindstrom DJ (1986) Lunar granulites and their precursor anorthositic norites of the early lunar crust. J Geophys Res 91:D263-D276

Logonné P, Gagnepain-Beyneix J, Chenet H (2003) A new seismic model of the Moon: Implications for structure, thermal evolution and formation of the Moon. Earth Planet Sci Lett 211:27-44

Longhi J (1980) A model of early lunar differentiation. Proc Lunar Planet Sci Conf 11:289-315

Longhi J (1992a) Origin of picritic green glass magmas by polybaric fractional fusion. Proc Lunar Planet Sci Conf 22: 343-353

Longhi J (1992b) Experimental petrology and petrogenesis of mare volcanics. Geochim Cosmochim Acta 56:2235-2252

Longhi J (1993) Liquidus equilibria of lunar analogs at high pressure. Lunar Planet Sci XXIV:895-896

Love SG, Hörz F, Brownlee DE (1993) Target porosity effects in impact cratering and collisional disruption. Icarus 105:216-224

Lucchitta BK (1976) Mare ridges and related highland scarps - Result of vertical tectonism. Proc Lunar Sci Conf 7: 2761-2782

Lucchitta BK (1977) Crater clusters and light mantle at the Apollo 17 site - A result of secondary impact from Tycho. Icarus 30:80-96

Lucey PG, Spudis PD, Zuber M, Smith D, Malaret E (1994) Topographic compositional units on the Moon and the early evolution of the lunar crust. Science 266:1855

Lucey PG, Taylor GJ, Malaret E (1995) Abundance and distribution of iron on the Moon. Science 268:1150-1153

Lucey PG, Blewett DT, Johnson JL, Taylor GJ, Hawke BR (1996) Lunar titanium content from UV-VIS measurements. Lunar Planet Sci XXVII:781-782

Lucey PG, Taylor GJ, Malaret E (1997) Global abundance of FeO on the Moon - Improved estimates from multispectral imaging and comparisons with the lunar meteorites. Lunar Planet Sci XXVIII:849-850

Lucey PG, Blewett DT, Hawke BR (1998) Mapping the FeO and TiO_2 content of the lunar surface with multispectral imagery. J Geophys Res 103:3679-3699

Lucey PG, Blewett DT, Jolliff BL (2000) Lunar iron and titanium abundance algorithms based on final processing of Clementine ultraviolet-visible images. J Geophys Res 105:20297–20305

Ma M-S, Schmitt RA, Nielsen RL, Taylor GJ, Warner RD, Keil K. (1979) Petrogenesis of Luna 16 aluminous mare basalts. Geophys Res Lett 6:909–912

Mackin JH (1964) Origin of lunar maria. Geol Soc Am Bull 80:735-747

Malin MC (1974) Lunar red spots: Possible pre-mare materials. Earth Planet Sci Lett 21:331-341

Maurer P, Eberhardt P, Geiss J, Grogler N, Stettler A, Brown GM, Peckett A, Krähenbühl U (1978) Pre-Imbrian craters and basins - Ages, compositions and excavation depths of Apollo 16 breccias. Geochim Cosmochim Acta 42: 1687-1720

Maurice S, Feldman WC, Lawrence DJ, Elphic RC, Johnson JR, Chevrel S, Genetay I, Binder AB (2001) A maturity parameter of the lunar regolith from neutron data. Lunar Planet Sci XXXII:#2033

Maxwell TA (1978) Origin of multi-ring basin ridge systems: An upper limit to elastic deformation based on a finite-element model. Proc Lunar Planet Sci Conf 9:3541-3559

Maxwell TA, El-Baz F (1978) The nature of rays and sources of highland material in Mare Crisium. *In:* Mare Crisium: The View from Luna 24. Geochim Cosmochim Acta Suppl. 9, p 89-103

McCauley JH (1967) Geologic map of the Hevelius Region of the Moon. USGS Misc. Inv. Map I-491

McCauley JF, Scott DH (1972) The geologic setting of the Luna 16 landing site. Earth Planet Sci Lett 13:225-232

McEwen AS, Gaddis LR, Neukum G, Hoffmann H, Pieters CM, Head JW (1993) Galileo observations of post-Imbrium lunar craters during the first Earth-Moon flyby. J Geophys Res 98:17207-17234

McEwen AS, Robinson MS, Eliason EM, Lucey PG, Duxbury TC, Spudis PD (1994) Clementine observations of the Aristarchus region of the Moon. Science 266:1858-1862

McEwen AS, Moore JM, Shoemaker EM (1997) The Phanerozoic impact cratering rate: Evidence from the farside of the Moon. J Geophys Res 102:9231-9242

McKay DS, Heiken G, Basu A, Blanford G, Simon S, Reedy R, French B, Papike J (1991) The lunar regolith. *In:* The Lunar Source Book: A user's guide to the Moon. Heiken G, Vaniman D, French B (eds) Cambridge Univ. Press, p 285-356

McSween HY (1999) Meteorites and Their Parent Planets. Cambridge University Press

Melendrez DE, Johnson JR, Larson SM, Singer RB (1994) Remote sensing of potential lunar resources. 2: High spatial resolution mapping of spectral reflectance ratios and implications for nearside mare TiO_2 content. J Geophys Res 99:5601-5619

Melosh HJ (1975) Large impact craters and the moon's orientation. Earth Planet Sci Lett 26:353-360

Melosh HJ (1978) The tectonics of mascon loading. Proc Lunar Planet Sci Conf 9:3513-3525

Melosh HJ (1989) Impact Cratering: A Geological Process. Oxford University Press

Melosh HJ (2001) Can impacts induce volcanic eruptions? *In:* International Conference on Catastrophic Events and Mass Extinctions: Impacts and Beyond, Vienna, 3144

Merrill RB, Papike JJ (eds) (1978) Mare Crisium: The View from Luna 24. Geochim Cosmochim Acta Suppl. 9. Pergamon Press

Morrison RH, Oberbeck VR (1975) Geomorphology of crater and basin deposits–emplacement of the Fra Mauro formation. Proc Lunar Sci Conf 6:2503–2530

Muehlberger WR (1974) Structural history of southeastern Mare Serenitatis and adjacent highlands. Proc Lunar Sci Conf 5:101-110

Muehlberger WR, Hörz F, Sevier JR, Ulrich GE (1980) Mission objectives for geological exploration of the Apollo 16 landing site. *In:* Proc Conf Lunar Highlands Crust. JJ Papike, RB Merrill (eds) Pergamon Press, p 1–49

Müller PM, Sjogren WL (1968) Mascons: Lunar mass concentrations. Science 161:680-684

Müller PM, Sjogren WL (1969) Lunar gravimetry and mascons. Appl Mech Rev 22:955-959

Müller WF, Hornemann U (1969) Shock-induced planar deformation structures in experimentally shock-loaded olivines and in olivines from chondritic meteorites. Earth Planet Sci Lett 7:251-264

Murase T, McBirney AR (1970) Viscosity of lunar lavas. Science 167:1491-1493

Mustard JF, Head JW (1996) Buried stratigraphic relationships along the southwestern shores of Oceanus Procellarum: Implications for early lunar volcanism. J Geophys Res 101:18913-18926

Nakamura Y (1983) Seismic velocity structure of the lunar mantle. J Geophys Res 88:677-686

Nakamura Y (2003) New identification of deep moonquakes in the Apollo lunar seismic data. Physics Earth Planet Int 139:197-205

Nakamura Y (2005) Farside deep moonquakes and deep interior of the Moon. J Geophys Res 110, doi 10.1029/2004JE002332

Nakamura Y, Lammlein D, Latham G, Ewing M, Dorman J, Press F, Toksöz N (1973) New seismic data on the state of the deep lunar interior. Science 181:49-51

Neal CR, Taylor LA (1992) Petrogenesis of mare basalts - A record of lunar volcanism. Geochim Cosmochim Acta 56: 2177-2211

Neal CR, Hacker MD, Snyder GA, Taylor LA, Liu Y-G, Schmitt RA (1994a) Basalt generation at the Apollo 12 site, Part 1: New data, classification, and re-evaluation. Meteoritics, 29:334-348

Neal CR, Hacker MD, Snyder GA, Taylor LA, Liu Y-G, Schmitt RA (1994b) Basalt generation at the Apollo 12 site, Part 2: Source heterogeneity, multiple melts, and crustal contamination. Meteoritics 29:349-361

Neukum G (1977) Different ages of lunar light plains. The Moon 17:383-393

Neukum G (1983) Meteoritenbombardement und Datierung planetarer Oberflächen. Habilitationsschrift, Univ. München, Munich, Germany

Neukum G, Wise DU (1976) Mars - A standard crater curve and possible new time scale. Science 194:1381-1387

Neukum G, Ivanov BA (1994) Crater size distributions and impact probabilities on Earth from lunar, terrestrial-planet, and asteroid cratering data. *In:* Hazard Due to Comets and Asteroids. Gehrels T (ed) Univ. of Ariz. Press, p 359-416

Neukum G, Ivanov BA, Hartmann WK (2001) Cratering records in the inner solar system in relation to the lunar reference system. Space Sci Rev 96:55-86

Neumann GA, Zuber MT, Smith DE, Lemoine FG (1996) The lunar crust: Global structure and signature of major basins. J Geophys Res 101:16841-16843

Nozette S, and the Clementine team (1994) The Clementine mission to the Moon: Scientific overview. Science 266: 1835-1839

Nozette S, Lichtenberg CL, Spudis P, Bonner R, Ortr W, Malaret E, Robinson M, Shoemaker EM (1996) The Clementine bistatic radar experiment. Science 274:1495-1498

Nozette S, Spudis PD, Robinson MS, Bussey DBJ, Lichtenberg C, Bonner R (2001) Integration of lunar polar remote-sensing data sets: Evidence for ice at the lunar south pole. J Geophys Res 106:23,253-23,266

Nunes PD, Tatsumoto M, Unruh DM (1974) U-Th-Pb and Rb-Sr systematics of Apollo 17 boulder 7 from the North Massif of the Taurus-Littrow valley. Earth Planet Sci Lett 23:445-452

Nyquist LE, Shih C-Y (1992) The isotopic record of lunar volcanism. Geochim Cosmochim Acta 56:2213-2234

Nyquist LE, Bogard DD, Shih C-Y (2001) Radiometric chronology of the Moon and Mars. *In:* The Century of Space Science. Bleeker JA, Geiss J, Huber M (eds) Kluwer Academic Publishers, p 1325-1376

Oberst J, Mizutani H (2002) A new inventory of deep moonquake nests visible in the Apollo 12 area. Lunar Planet Sci XXXIII:#1704

O'Keefe JD, Ahrens TJ (1993) Planetary cratering mechanics, J Geophys Res 98:17,011-17,028

O'Keefe JD, Ahrens TJ (1994) Impact-induced melting of planetary surfaces. *In:* Large Meteorite Impacts and Planetary Evolution. Dressler BO, Grieve RAF, Sharpton VL (eds), Special Paper 293, The Geological Society of America, Boulder, p 103-109

Ostertag R (1983) Shock experiments on feldspar crystals. J Geophys Res (Suppl.) 88:B364-B376

Ostertag R, Stöffler D, Borchardt R, Palme H, Spettel B, Wänke H (1987) Precursor lithologies and metamorphic history of granulitic breccias from North Ray crater, Station 11, Apollo 16. Geochim Cosmochim Acta 51:131-142

Oberbeck VR (1975) The role of ballistic erosion and sedimentation in lunar stratigraphy. Rev Geophys Space Phys 13:337-362

Oberbeck VR, Quaide WL, Gault DE, Morrison RH, Hörz F (1974) Smooth plains and continuous deposits of craters and basins. Proc Lunar Sci Conf 5:111-136

Papike JJ, Vaniman DT (1978) The lunar mare basalt suite. Geophys Res Lett 5:433-436

Papike JJ, Hodges FN, Bence AE, Cameron M, Rhodes JM (1976) Mare basalts: Crystal chemistry, mineralogy, and petrology. Rev Geophys Space Phys 14:475-540

Papike JJ, Ryder G, Shearer CK (1998) Lunar samples. Rev Mineral 36:5-1–5-234

Parmentier EM, Zhong S, Zuber MT (2000) On the relationship between chemical differentiation and the origin of lunar asymmetries. Lunar Planet Sci XXXI:#1614

Petro NE, Pieters CM (2004) Surviving the heavy bombardment: Ancient material at the surface of South Pole-Aitken Basin. J Geophys Res 109, doi 10.1029/2003JE002182

Pieters CM (1978) Mare basalt types on the front side of the Moon. Proc Lunar Planet Sci Conf 9:2825-2849

Pieters CM (1993b) Compositional diversity and stratigraphy of the lunar crust derived from reflectance spectroscopy. *In:* Topics in Remote Sensing 4 - Remote geochemical analysis: elemental and mineralogical composition. Pieters CM, Englert PAJ (eds) Cambridge Univ. Press, p 309-339

Pieters CM, Head JW, Adams JB, McCord TB, Zisk SH, Whitford-Stark JL (1980) Late high-titanium basalts of the western maria: Geology of the Flamsteed Region of Oceanus Procellarum. J Geophys Res 85:3919-3938

Pieters CM, Sunshine JM, Fischer EM, Murchie SL, Belton M, McEwen A, Gaddis L, Greeley R, Neukum G, Jaumann R, Hoffmann H (1993a) Crustal diversity of the Moon: Compositional analyses of Galileo solid state imaging data. J Geophys Res 98:17,127-17,148

Pieters CM, Staid MI, Fischer EM, Tompkins S, He G (1994) A sharper view of impact craters from Clementine data. Science 266:1844

Pieters CM, Head JW, Gaddis L, Jolliff B, Duke M (2001) Rock types of South Pole-Aitken basin and extent of basaltic volcanism. J Geophys Res 106(E11):28,001-28,022

Pike RJ (1977) Apparent depth/apparent diameter relation for lunar craters. Proc Lunar Planet Sci 8:3427-3436

Pike RJ (1980) Geometric interpretation of lunar craters. U.S. Geol. Survey Prof. Paper, 1046-C

Pike RJ (1988) Geomorphology of impact craters on Mercury. *In:* Mercury. Vilas F, Chapman CR, Matthews MS (eds), Univ. Arizona Press, p 165-273

Prettyman TH, Feldman WC, Lawrence DJ, McKinney GW, Binder AB, Elphic RC, Gasnault OM, Maurice S, Moore KR (2002) Library least squares analysis of lunar Prospector gamma ray spectra. Lunar Planet Sci XXXIII: #2012

Pritchard ME, Stevenson DJ (2000) The thermochemical history of the Moon: Constraints and major questions. Lunar Planet Sci XXXI:#1878

Rhodes JM (1977) Some compositional aspects of lunar regolith evolution. Phil Trans A 285, 1327:293-301

Rhodes JM, Hubbard NJ (1973) Chemistry, classification, and petrogenesis of Apollo 15 mare basalts. Proc Lunar Sci Conf 4:1127-1148

Rhodes JM, Brannon JC, Rodgers KV, Blanchard DP, Dungan MA (1977) Chemistry of Apollo 12 mare basalts - Magma types and fractionation processes. Proc Lunar Sci Conf 8:1305-1338

Richmond NC, Hood LL, Mitchell DL, Lin RP, Acuña MH, Binder AB (2005) Correlations between magnetic anomalies and surface geology antipodal to lunar impact basins. J Geophys Res 110, doi 10.1029/2005JE002405

Ringwood AF, Essene E (1970) Petrogenesis of Apollo 11 basalts, internal constitution, and origin of the Moon. Proc Apollo 11 Lunar Sci Conf:769-799

Ringwood AF, Kesson SE (1976) A dynamic model for mare basalt petrogenesis, Proc Lunar Sci Conf 7:1697-1722

Robinson MS, Jolliff BL (2002) Apollo 17 landing site: Topography, photometric corrections, and heterogeneity of the surrounding highland massifs. J Geophys Res 107, doi 10.1029/2001JE001614

Ryan MR (1987) Elasticity and contractancy of Hawaiian olivine tholeiite and its role in the stability and structural evolution of subcaldera magma reservoirs and rift systems. *In:* Volcanism in Hawaii I. Decker RW, Wright TL, Stauffer PH (eds) USGS Prof. Paper 1350:1395-1447

Ryder G (1990) Accretion and bombardment in the early Earth-Moon system: The lunar record. LPI Contribution 746: 42, Lunar and Planetary Institute

Ryder G (2002) Mass flux in the ancient Earth-Moon system and benign implications for the origin of life on Earth. J Geophys Res 107, doi 10.1029/2001JE001583

Ryder G, Spudis PD (1980) Volcanic rocks in the lunar highlands. *In:* Proc of the Conference on the Lunar Highlands Crust. Pergamon Press, p 353-375

Ryder G, Schmitt HH, Spudis PD (eds) (1992) Workshop on Geology of the Apollo 17 Landing Site, LPI Tech. Rpt. 92-09, Part 1, Lunar and Planetary Institute

Ryder G, Koeberl C, Mojzsis SJ (2000) Heavy bombardment of the Earth at ~3.85 Ga: The search for petrographic and geochemical evidence. *In:* Origin of the Earth and Moon. Canup RM, Righter K (eds) Univ. of Arizona Press, p 475-492

Schaal RB, Hörz F (1977) Shock metamorphism of lunar and terrestrial basalts. Proc Lunar Science Conference 8: 1697-1729

Schaal RB, Hörz F, Thompson TD, Bauer JF (1979) Shock metamorphism of granulated lunar basalt. Proc Lunar Planet Sci Conf 10:2547-2571

Schaber GG (1973) Lava flows in Mare Imbrium: Geologic evaluation from Apollo orbital photography. Proc Lunar Planet Sci Conf 4:73-92

Schaber GG, Boyce JM, Moore HJ (1976) The scarcity of mapable flow lobes on the lunar maria: Unique morphology of the Imbrium flows. Proc Lunar Sci Conf 7:2783-2800

Schaefer OA, Husain L (1974) Chronology of lunar basin formation and ages of lunar anorthositic rocks. Lunar Planet Sci V:663-665

Schmidt OY (1959) A theory of the origin of the Earth. Lawrence and Wishart

Schmitt HH (1973) Apollo 17 report on the valley of Taurus-Littrow. Science 182:681-690

Schmitt HH (2000a) Source and implications of large lunar basin-forming objects. Lunar Planet Sci XXXI:#1821

Schmitt HH (2001) Lunar cataclysm? Depends on what "cataclysm" means. Lunar Planet Sci XXXII:#1133

Schmitt RT (2000) Shock experiments with the H6 chondrite Kernouvé: Pressure calibration of microscopic shock effects. Meteoritics 35:545-560

Schubert G, Lingenfelter RE, Peale SJ (1970) The morphology, distribution, and origin of lunar sinuous rilles. Rev Geophys Space Phys 8:199-224

Schultz PH (1976) Floor-fractured lunar craters. The Moon 15:241-273

Schultz PH, Gault DE (1975) Seismic effects from major basin formations on the Moon and Mercury. The Moon 12: 159-177

Schultz PH, Gault DE (1979) Atmospheric effects on martian ejecta emplacement. J Geophys Res 84:7669-7687

Schultz PH, Spudis PD (1979) Evidence for ancient mare volcanism. Proc Lunar Planet Sci Conf 10:2899-2918

Schultz PH, Srnka LJ (1980) Cometary collisions on the Moon and Mercury. Nature 284:22-26

Schultz PH, Spudis PD (1983) Beginning and end of lunar mare volcanism. Nature 302:233-236

Schultz PH, Gault DE (1990) Prolonged global catastrophes from oblique impacts. Geol Soc Am Spec Paper 247: 239-261

Schultz PH, Anderson RR (1996) Asymmetry of the Manson impact structure: Evidence for impact angle and direction. Geol Soc Am Special Paper 302:397-417

Schultz RA, Zuber MT (1994) Observations, models, and mechanisms of failure of surface rocks surrounding planetary surface loads. J Geophys Res 99:14,691-14,702

Sharpton VL (1994) Evidence from Magellan for unexpectedly deep complex craters on Venus. *In:* Large Meteorite Impacts and Planetary Evolution. Dressler BO, Grieve RAF, Sharpton VL (eds), Special Paper 293, The Geological Society of America, Boulder, p 19-27

Shearer CK, Newsom HE (2000) W-Hf abundances and the early origin and evolution of the Earth-Moon system. Geochim Cosmochim Acta 64:3599-3613

Shervais JW, Taylor LA, Lindstrom MM (1985a) Apollo 14 mare basalts: Petrology and geochemistry of clasts from consortium breccia 14321. J Geophys Res 90:C375-C395

Shervais JW, Taylor LA, Laul JC, Shih C-Y, Nyquist LE (1985b) Very high potassium (VHK) basalt: Complications in mare basalt petrogenesis. Proc Lunar Planet Sci Conf 16:D3-D18

Shkuratov YG, Kaydash VG, Opanasenko NV (1999) Iron and titanium abundance and maturity degree distribution on the lunar nearside. Icarus 137:222-234

Shkuratov YG, Bondarenko NV (2001) Regolith layer thickness mapping of the Moon by radar and optical data. Icarus 149:329-338

Shoemaker EM (1964) The geology of the Moon. Scientific American 211:38-47

Shoemaker EM (1970) Origin of fragmental debris on the lunar surface and history of bombardment of the Moon. Presentation at I Seminario de Geologia Lunar, Univ. of Barcelona (Rev. January 1971)

Shoemaker EM, Hackman R (1962) Stratigraphic basis for a lunar time scale. *In:* The Moon: Symposium 14 of the International Astronomical Union. Kopal Z, Mikhailov ZK (eds), Academic, p 289-300

Shoemaker EM, Bailey NG, Batson RM, Dahlem DH, Foss TH, Grolier MJ, Goddard EM, Hait MH, Holt HE, Larson KB, Rennilson JJ, Schaber GG, Schleicher DL, Schmitt HH, Sutton RL, Swann GA, Waters AC, West MN (1970a) Geologic setting of the lunar samples returned by the Apollo 11 mission. *In:* Apollo 11 Preliminary Science Report 41-84, NASA SP-214

Shoemaker EM, Batson RM, Bean AL, Conrad C Jr., Dahlem D, Goddard EN, Hait MT, Larson KB, Schaber GG, Schleicher DL, Sutton RL, Swann GA, Waters AC (1970b) Preliminary geologic investigation of the Apollo 12 landing site. Part A Geology of the Apollo 12 landing site. *In:* Apollo 12 Preliminary Science Report 113-182, NASA SP-235

Shoemaker EM, Robinson MS, Eliason EM (1994) The south pole region of the Moon as seen by Clementine. Science 266:1851-1854

Short NM, Foreman ML (1972) Thickness of impact crater ejecta on the lunar surface. Mod Geol 3:69-91

Simonds CH, Phinney WC, Warner JL, McGee PE, Geeslin J, Brown RW, Rhodes JM (1977) Apollo 14 revisited, or breccias aren't so bad after all. Proc Lunar Sci Conf 8:1869-1893

Simpson RA, Tyler GL (1999) Reanalysis of Clementine bistatic radar data from the lunar south pole. J Geophys Res 104:3845–3862

Slade MA, Butler BJ, Muhleman DO (1992) Mercury radar imaging - Evidence for polar ice. Science 258:635-640

Smith DE, Zuber MT, Neumann GA, Lemoine FG (1997) Topography of the Moon from the Clementine LIDAR. J Geophys Res 102:1591–1611

Smith DE, Zuber MT, Solomon SC, Phillips RJ, Head JW, Garvin JB, Banerdt WB, Muhleman DO, Pettengill GH, Neumann GA, Lemoine FG, Abshire JB, Aharonson O, Brown DC, Hauck SA, Ivanov AB, McGovern PJ, Zwally HJ, Duxbury TC (1999) The global topography of Mars and implications for surface evolution. Science 284: 1495-1503

Smith JV, Anderson AT, Newton RC, Olson EJ, Wyllie PJ, Crewe AV, Isaacson MS, Johnson D (1970) Petrologic history of the Moon inferred from petrography, mineralogy, and petrogenesis of Apollo 11 rocks. Proc Apollo 11 Lunar Sci Conf:897-925.

Snee LW, Ahrens TJ (1975) Shock-induced deformation features in terrestrial peridot and lunar dunite. Proc Lunar Sci Conf 6:833-842

Solomon SC (1978) The nature of isostasy on the Moon: How big of a Pratt-fall for Airy methods. Proc Lunar Planet Sci Conf 9:3499-3511

Solomon SC, Chaiken J (1976) Thermal expansion and thermal stress in the Moon and terrestrial planets: Clues to early thermal history. Proc Lunar Sci Conf 7:3229-3243

Solomon SC, Longhi J (1977) Magma oceanography 1: Thermal evolution. Proc Lunar Sci Conf 8:583-599

Solomon SC, Head JW (1979) Vertical movement in mare basins: Relation to mare emplacement, basin tectonics, and lunar thermal history. J Geophys Res 84:1667-1682

Solomon SC, Head JW (1980) Lunar mascon basins: Lava filling, tectonics, and evolution of the lithosphere. Rev Geophys Space Phys 18:107-141

Solomon SC, Head JW (1982) Mechanisms for lithospheric heat transport on Venus: Implications for tectonic style and volcanism. J Geophys Res 87:9236-9246

Spohn T, Konrad W, Breuer D, Ziethe R (2001) The longevity of lunar volcanism: Implications of thermal evolution calculations with 2D and 3D mantle convection models. Icarus 149:54-65

Spudis PD (1978) Composition and origin of the Apennine Bench Formation. Lunar Planet Sci IX:1086-1088

Spudis PD (1984) Apollo 16 site geology and impact melts: Implications for the geologic history of the lunar highlands. Proc Lunar Planet Sci Conf 15:C95–C107

Spudis PD (1993) The Geology of Multi-ring Impact Basins. Cambridge Planet. Sci Ser. 8, Cambridge Univ. Press

Spudis PD (1996) The Once and Future Moon. Smithsonian Institution Press

Spudis PD, Ryder G (1981) Apollo 17 impact melts and their relation to the Serenitatis basin. *In:* Multi-Ring Basins, Proc Lunar Planet Sci Conf 12, Part A. Schultz PH, Merrill RB (eds) Pergamon, p 133–148

Spudis PD, Ryder G (1985) Geology and petrology of the Apollo 15 landing site - Past, present, and future understanding. EOS 66:721, 724-726

Spudis PD, Ryder G (eds) (1986) Geology and petrology of the Apollo 15 landing site, LPI Tech. Rpt. 86-03, Lunar and Planetary Institute

Spudis PD, Pieters CM (1991) Global and regional data about the Moon. *In:* Lunar Sourcebook. Heiken G, Vaniman D, French B (eds) Cambridge University Press, p 595-632

Spudis PD, Swann GA, Greeley R (1988) The formation of Hadley Rille and implications for the geology of the Apollo 15 region. Proc Lunar Planet Sci Conf 18:243-254

Spudis PD, Hawke BR, Lucey PG (1989) Geology and deposits of the lunar Nectaris basin. Proc Lunar Planet Sci Conf 19:51-59

Spudis PD, Reisse RA, Gillis JJ (1994) Ancient multiring basins on the Moon revealed by Clementine laser altimetry. Science 266:1848-1851

Spurr JE (1944) Geology applied to selenology, Vol. I The Imbrian plain region of the Moon. Science Press

Spurr JE (1945) Geology applied to selenology, Vol. II The features of the Moon. Science Press

Spurr JE (1948) Geology applied to selenology, Vol. III Lunar catastrophic history. Rumford

Spurr JE (1949) Geology applied to selenology, Vol. IV The shrunken Moon. Rumford

Stadermann FJ, Heusser E, Jessberger EK, Lingner S, Stöffler D (1991) The case for a younger Imbrium basin - New Ar-40 - Ar-39 ages of Apollo 14 rocks. Geochim Cosmochim Acta 55:2339-2349

Staid MI, Pieters C (2001) Mineralogy of the last lunar basalts: Results from Clementine, J Geophys Res 106:27887-27900

Staid MI, Pieters C, Head JW (1996) Mare Tranquillitatis: Basalt emplacement history and relation to lunar samples. J Geophys Res 101:23213–23228

Stöffler D (1972) Deformation and transformation of rock-forming minerals by natural and experimental shock processes: I. Behavior of minerals under shock compression. Fortschr Miner 49:50-113

Stöffler D (1974) Deformation and transformation of rock-forming minerals by natural and experimental shock processes: II. Physical properties of shocked minerals. Fortschr Miner 51:256-289

Stöffler D (1984) Glasses formed by hypervelocity impact. J Non-Crystalline Solids 67:465-502

Stöffler D (1990) Die Bedeutung des Rieskraters für die Planeten- und Erdwissenschaften, *In:* Rieskrater-Museum Nördlingen, Hrsg. von der Stadt Nördlingen, Verlag F. Steinmeier, Nördlingen, 2. Auflage 1991, 97-114

Stöffler D, Hornemann U (1972) Quartz and feldspar glasses produced by natural and experimental shock. Meteoritics 7:371-394

Stöffler D, Reimold (1978) Experimental shock metamorphism of dunite. Proc Lunar Planet Sci Conf 9:2805-2824

Stöffler D, Grieve RAF (1994) Classification and nomenclature of impact metamorphic rocks: A proposal to the IUGS Subcommission on the Systematics of Metamorphic Rocks. Lunar Planet Sci XXV:1347-1348. Lunar Planet Sci Institute

Stöffler D, Langenhorst F (1994) Shock metamorphism of quartz in nature and experiment: I. Basic observation and theory. Meteoritics 29:155-181

Stöffler D, Grieve RAF (1996) IUGS classification and nomenclature of impact metamorphic rocks: Towards a final proposal. International Symposium on the Role of Impact Processes in the Geological and Biological Evolution of Planet Earth, Postojna, Slovenia, 27.9.-2.10.1996, Abstract

Stöffler D, Ryder G (2001) Stratigraphy and isotope ages of lunar geologic units: Chronological standard for the inner solar system. *In:* Chronology and Evolution of Mars. Kallenbach R, Geiss J, Hartmann WK (eds), Space Science Series of ISSI, Kluwer Academic Publishers, Dordrecht, Space Science Rev 96:9-54

Stöffler D, Knöll HD, Marvin UB, Simonds CH, Warren PH (1980) Recommended classification and nomenclature of lunar highland rocks, *In:* Proc Conf Lunar Highland Crust. Papike JJ, Merrill RB (eds) Pergamon Press, 51-70

Stöffler D, Bischoff A, Borchardt R, Burghele A, Deutsch A, Jessberger EK, Ostertag R, Palme H, Spettel B, Reimold WU (1985) Composition and evolution of the lunar crust in the Descartes highlands, Apollo 16. J Geophys Res 90:C449-C506

Stöffler D, Ostertag R, Jammes C, Pfannschmidt G, Sen Gupta PR, Simon SB, Papike JJ, Beauchamp RM (1986) Shock metamorphism and petrography of the Shergotty achondrite. Geochim Cosmochim Acta 50:889-903

Stöffler D, Bischoff A, Buchwald U, Rubin AE (1988) Shock effects in meteorites. *In:* Meteorites and the Early Solar System. Kerridge JF, Matthews MS (eds) University of Arizona Press, p 165-205

Stöffler D, Keil K, Scott ERD (1991) Shock metamorphism of ordinary chondrites. Geochim Cosmochim Acta 55:3845-3867

Strom RG (1964) Analysis of lunar lineaments, I: Tectonic maps of the Moon. Univ. of Arizona Lunar Planet Lab Comm 2:205-216

Strom RG, Neukum G (1988) The cratering record on Mercury and the origin of impacting objects. *In:* Mercury. Vilas F, Chapman CR, Matthews MS (eds), Univ. Arizona Press, p 336-373

Stuart-Alexander DE, Wilhelms DE (1975) The Nectarian system, a new time-stratigraphic unit. USGS Journal of Research 3:53-58

Swann GA, Trask NJ, Hait MH, Sutton RL (1971) Geologic setting of the Apollo 14 samples. Science 173:716-719

Swann GA, Bailey NG, Batson RM, Freeman VL, Hait MH, Head JW, Holt HE, Howard KA, Irwin JB, Larson KB, Muehlberger WR, Reed VS, Rennilson JJ, Schaber GG, Scott DR, Silver LT, Sutton RL, Ulrich GE, Wilshire HG, Wolfe EW (1972) Preliminary geologic investigation of the Apollo 15 landing site. *In:* Apollo 15 Preliminary Science Report 5-1 – 5-112, NASA-SP289

Swann GA, Bailey NG, Batson RM, Eggleton RE, Hait MH, Holt HE, Larson KB, Reed VS, Schaber GG, Sutton RL, Trask NI, Ulrich GE, Wilshire HG (1977) Geology of the Apollo 14 landing site in the Fra Mauro highlands. USGS Prof. Paper 800:103

Taylor GJ, Drake MJ, Wood JW, Marvin UB (1973) The Luna 20 lithic fragments, and the composition and origin of the lunar highlands. Geochim Cosmochim Acta 37:1087–1106

Taylor GJ, Warren P, Ryder G, Delano J, Pieters C, Lofgren G (1991) Lunar rocks. *In:* The Lunar Source Book: A User's Guide to the Moon. Heiken G, Vaniman D, French B (eds) Cambridge Univ. Press, 183-284

Taylor LA (2002) Origin of nanophase Fe^0 in agglutinates: A radical new concept. *In:* The Moon Beyond 2002: Next Steps in Lunar Science and Exploration, LPI Contribution No. 1128, Lunar and Planetary Institute, p 62

Taylor LA, Shervais JW, Hunter RH, Shih C-Y, Bansal BM, Wooden J, Nyquist LE, Laul LC (1983) Pre-4.2 AE mare-basalt volcanism in the lunar highlands. Earth Planet Sci Lett 66:33-47

Taylor SR (1982) Planetary Science. Lunar and Planetary Institute

Taylor SR, Norman MD, Esat T (1993) The lunar highland crust: The origin of the Mg suite. Meteoritics 28:448

Tera F, Papanastassiou DA, Wasserburg GJ (1974) Isotopic evidence for a terminal lunar cataclysm. Earth Planet Sci Lett 22:1-21

Thompson AC, Stevenson DJ (1983) Two-phase gravitational instabilities in thin disks with application to the origin of the Moon. Lunar Planet Sci XIV:787-78

Thurber CH, Solomon SC (1978) An assessment of crustal thickness variations on the lunar near side: Models, uncertainties, and implications for crustal differentiation. Proc Lunar Planet Sci Conf 9:3481-3497

Toksöz MN, Dainty AM, Solomon SC, Anderson KR (1973) Velocity structure and evolution of the Moon. Proc Lunar Sci Conf 4:2529

Tompkins S, Pieters CM, Mustard JF, Pinet P, Chevrel S (1994) Distribution of materials excavated by the lunar crater Bullialdus and implications of the geologic history of the Nubium Region. Icarus 110:261-274

Tompkins S, Margot JL, Pieters CM (2000) Effects of topography on interpreting the composition of materials within the crater Tycho. Lunar Planet Sci XXXI:#1401

Tonks WB, Melosh HJ (1992) Core formation by giant impacts. Icarus 100:326-346

Tonks WB, Melosh HJ (1993) Magma ocean formation due to giant impacts. J Geophys Res 98:5319-5333

Touma J (2000) The phase space adventure of the Earth and Moon. *In:* Origin of the Earth and Moon. Canup RM, Righter K (eds) University of Arizona Press, p 165-178

Ulrich GE, Hodges CA, Muehlberger WR (1981) Geology of the Apollo 16 area, central lunar highlands. USGS Prof. Paper 1048:539

Urey HC (1966) The capture hypothesis of the origin of the Moon. *In:* The Earth-Moon system. Marsden BG, Cameron AGW (eds) Plenum Press, p 210-212

Vinogradov AP (1971) Preliminary data on lunar ground brought to Earth by automatic probe "Luna-16". Proc Lunar Sci Conf 2:1-16

Vinogradov AP (1973) Preliminary data on lunar soil collected by the Luna 20 unmanned spacecraft. Geochim Cosmochim Acta 37:721-729

von Frese RRB, Tan L, Potts LV, Kim JW, Merry CJ, Bossler JD (1997) Lunar crustal analysis of Mare Orientale from topographic and gravity correlations. J Geophys Res 102:25,657-25,676

Vondrak RR and Crider DH (2003) Ice at the lunar poles. American Scientist 91(4):322-329

Wagner RJ, Head JW, Wolf U, Neukum G (1996) Age relations of geologic units in the Gruithuisen region of the Moon based on crater size-frequency measurements. Lunar Planet Sci XXVII:1367-1368

Wagner R, Head JW, Wolf U, Neukum G (2002) Stratigraphic sequence and ages of volcanic units in the Gruithuisen region of the Moon. J Geophys Res 107, doi 10.1029/2002JE001844

Warner JL (1970) Apollo 12 Lunar sample information. NASA TR-R-353:391

Warner JL, Simonds CH, Phinney WC (1976) Genetic distinction between anorthosites and Mg-rich plutonic rocks: New data from 76255. Lunar Planet Sci VII:915-917

Warner JL, Phinney WC, Bickel CE, Simonds CH (1977) Feldspathic granulitic impactites and pre-final bombardment lunar evolution. Proc Lunar Sci Conf 8:2051-2066

Warner RD, Taylor GJ, Mansker WL, Keil K (1978) Clast assemblages of possible deep-seated (77517) and immiscible-melt (77538) origins in Apollo 17 breccias. Proc Lunar Planet Sci Conf 9:941-958

Warren PH (1993) A concise compilation of petrologic information on possibly pristine nonmare moon rocks. Am Mineral 78:360-376

Warren PH (1994) Lunar and martian meteorite delivery services. Icarus 111:338-363

Warren PH (2001) Early lunar crustal genesis: The ferroan anorthosite epsilon-neodymium paradox as a possible result of crustal overturn. Meteoritics 36:219

Warren PH, Wasson JT (1979) The origin of KREEP. Rev Geophys Space Phys 17:73-88

Warren PH, Kallemeyn GW, Kyte FT (1997) Siderophile element evidence indicates that Apollo 14 high-Al mare basalts are not impact melts. Lunar Planet Sci XXVIII:1501-1502

Wentworth SJ, McKay DS, Lindstrom DJ, Basu A, Martinez RR, Bogard DD, Garrison DH (1994) Apollo 12 ropy glasses revisited. Meteoritics 29:323-333

Weitz C, Head JW (1998) Diversity of lunar volcanic eruptions at the Marius Hills complex. Lunar Planet Sci XXIX: #1229

Weitz C, Head JW (1999) Spectral properties of the Marius Hills volcanic complex and implications for the formation of lunar domes and cones. J Geophys Res 104:18933-18956

Weitz CM, Head JW, Pieters CM (1998) Lunar regional dark mantle deposits: Geologic, multispectral, and modeling studies. J Geophys Res 103:22725-22760

Whitford-Stark JL, Head JW (1977) The Procellarum volcanic complexes: Contrasting styles of volcanism. Proc Lunar Sci Conf 8:2705-2724

Whitford-Stark JL, Head JW (1980) Stratigraphy of Oceanus Procellarum basalts: Sources and styles of emplacement. J Geophys Res 85:6579-6609

Wichman RW, Schultz PH (1995) Floor-fractured impact craters on Venus: Implications for igneous crater modification and local mechanism. J Geophys Res 100:3233-3244

Wichman RW, Schultz PH (1996) Crater-centered laccoliths on the Moon: Modeling intrusion depth and magmatic pressure at the crater Taruntius. Icarus 122:193-199

Wieczorek MA (2003) The thickness of the lunar crust: How low can you go? Lunar Planet Sci XXXIV:#1330

Wieczorek MA, Phillips RJ (1998) Potential anomalies on a sphere: Applications to the thickness of the lunar crust. J Geophys Res 103:1715-1724

Wieczorek MA, Phillips RJ (2000) The Procellarum KREEP Terrane: Implications for mare volcanism and lunar evolution. J Geophys Res 105:20,417-20,430

Wieczorek MA, Zuber MT, Phillips RJ (2001) The role of magma buoyancy on the eruption of lunar mare basalts. Earth Planet Sci Lett 185:71-83

Wilhelms DE (1970) Summary of lunar stratigraphy – Telescopic observations. USGS Prof. Paper 599:F1-F47

Wilhelms DE (1984) Moon. *In:* The Geology of the Terrestrial Planets 107-206, NASA SP-469

Wilhelms DE (1987) The geologic history of the Moon. U. S. Geol. Survey Prof. Paper 1348

Williams DA, Kadel SD, Greeley R, Lesher CM (2001) Erosion by flowing lava: Geochemical evidence in the Cave Basalt, Mount St. Helens, Washington. American Geophysical Union abstract #V22B-1046

Williams JG, Boggs DH, Yoder CF, Ratcliff JT, Dickey JO (2001a) Lunar rotational dissipation in solid body and molten core. J Geophys Res 106:27,933-27,968

Williams KK, Zuber MT (1998) Measurement and analysis of lunar basin deposits from Clementine altimetry. Icarus 131:107-122

Wilson L, Head JW (1981) Ascent and eruption of basaltic magma on the Earth and Moon. J Geophys Res 86:2971-3001

Wilson L, Head JW (1983) A comparison of volcanic eruption processes on Earth, Moon, Mars, Io, and Venus. Nature 302:663-669

Wilson L, Head JW (1990) Factors controlling the structures of magma chambers in basaltic volcanoes. Lunar Planet Sci XXI:1343-1344

Wilson L, Head JW (2003) Deep generation of magmatic gas on the Moon and implications for pyroclastic eruptions. Geophys Res Lett 30, doi:10.1029/2002GL016082

Wilson L, Head JW (2003b) Lunar Gruithuisen and Mairan domes: Rheology and mode of emplacement. J Geophys Res 108, doi 10.1029/2002JE001909

Wolfe EW, Bailey NG, Lucchitta BK, Muehlberger WR, Scott DH, Sutton RL, Wilshire HG (1981) The geologic investigation of the Taurus-Littrow valley: Apollo 17 landing site. USGS Prof. Paper 1080:280

Wood CA, Head JW (1975) Geologic setting and provenance of spectrally distinct pre-mare material of possible volcanic origin. *In:* Conference on Origins of Mare Basalts and their Implications for Lunar Evolution. LPI Contribution number 234:189

Wood JA, Dickey JS, Marvin UB, Powell BN (1970) Lunar anorthosites and a geophysical model of the Moon. Proc Apollo 11 Science Conference 965-988

Wu SSC, Moore HJ (1980) Experimental photogrammetry of lunar images – Apollo 15-17 orbital investigations. USGS Prof. Paper 1046:D1-D23

Yingst RA, Head JW (1997) Volumes of lunar lava ponds in South Pole-Aitken and Orientale Basins: Implications for eruption conditions, transport mechanisms, and magma source regions. J Geophys Res 102(E5):10909-10932

Yingst RA, Head JW (1998) Characteristics of lunar mare deposits in Smythii and Marginis basins: Implications for magma transport mechanisms. J Geophys Res 103(E5):11,135-11,158

Zisk SH (1978) Mare Crisium area topography: A comparison of Earth-based radar and Apollo mapping camera results. *In:* Mare Crisium: The View from Luna 24. Geochim Cosmochim Acta Suppl. 9, Lunar and Planetary Institute. Pergamon Press, p 75-80

Zuber MT, Smith DE, Lemoine FG, Neumann GA (1994) The shape and internal structure of the Moon from the Clementine mission. Science 266:1839-1843

Reviews in Mineralogy & Geochemistry
Vol. 60, pp. 83-219, 2006
Copyright © Mineralogical Society of America

2

Understanding the Lunar Surface and Space-Moon Interactions

Paul Lucey[1], Randy L. Korotev[2], Jeffrey J. Gillis[1], Larry A. Taylor[3], David Lawrence[4], Bruce A. Campbell[5], Rick Elphic[4], Bill Feldman[4], Lon L. Hood[6], Donald Hunten[7], Michael Mendillo[8], Sarah Noble[9], James J. Papike[10], Robert C. Reedy[10], Stefanie Lawson[11], Tom Prettyman[4], Olivier Gasnault[12], Sylvestre Maurice[12]

[1]*University of Hawaii at Manoa, Honolulu, Hawaii, U.S.A.*
[2]*Washington University, St. Louis, Missouri, U.S.A.*
[3]*University of Tennessee, Knoxville, Tennessee, U.S.A.*
[4]*Los Alamos National Laboratory, Los Alamos, New Mexico, U.S.A.*
[5]*Smithsonian Institution, Washington D.C., U.S.A.*
[6]*Lunar and Planetary Laboratory, Univ. of Arizona, Tucson, Arizona, U.S.A.*
[7]*University of Arizona, Tucson, Arizona, U.S.A.*
[8]*Boston University, Cambridge, Massachusetts, U.S.A.*
[9]*Brown University, Providence, Rhode Island, U.S.A.*
[10]*University of New Mexico, Albuquerque, New Mexico, U.S.A.*
[11]*Northrop Grumman, Van Nuys, California, U.S.A.*
[12]*Centre d'Etude Spatiale des Rayonnements, Toulouse, France*

Corresponding authors e-mail:
Paul Lucey <lucey@higp.hawaii.edu>   Randy Korotev <korotev@wustl.edu>

1. INTRODUCTION

The surface of the Moon is a critical boundary that shapes our understanding of the Moon as a whole. All geologic mapping and remote sensing techniques utilize only the outermost portion of the Moon. Before leaving the Moon for study in our laboratories, all lunar samples that have been studied existed at or very near the surface. With the exception of the deeply probing geophysical techniques, our understanding of the interior of the Moon is derived from surficial, but not superficial, information, coupled with geologic reasoning. While the surface is the upper boundary of the lunar crust, it is the lower boundary layer of the tenuous lunar atmosphere and constitutes both a source and a sink for atmospheric gases. The surface is also where the Moon interacts with the space environment, causing changes in the physical nature of lunar materials, and provides a laboratory for the study of processes that occur on all airless bodies.

The data obtained remotely by the Galileo, Clementine, and Lunar Prospector missions, as well as data derived from lunar meteorites, have resulted in major changes to our understanding of global distributions of chemistry and rocks. This chapter summarizes the current understanding of this critical interface, the surface of the Moon, in its role as the lower boundary of the lunar atmosphere, the upper boundary of the crust, and the window through which we view, through remote sensing, the composition of the crust and the history of the Moon. In this post-Lunar Prospector time, the view of the Moon has changed, lending new perspectives to lunar samples and lunar processes. But the New View will likely remain in flux as we continue to digest the results from these recent space missions and move forward to a new era of lunar exploration.

1529-6466/06/0060-02$15.00   DOI: 10.2138/rmg.2006.60.2

Despite the freshness of our perspective, this is an important moment to capture, before the next generation of lunar scientists are forced to relearn old lessons. We can examine the new data through the lens of the Apollo era with its hard won lessons and years of anticipation of new data such as is now in our hands. We can attack these new data with tools forged for other missions that did not aim at the Moon. In addition to capturing a snapshot of rapidly evolving new scientific perspectives on the Moon, this Chapter will present an overview of the materials that constitute the surface of the Moon and its interior—its rocks and minerals and the regolith that constitutes the entire outer surface of the Moon. The new discoveries in the space weathering of the lunar soil aids the further understanding and quantification of the remotely sensed observations, and provide their ground-truth.

Of necessity, this chapter is a snapshot of a field in flux, but not a new field, one lacking introspection. An integration of data obtained during the last 30+ years is revealed herein. It will become apparent to the reader that new views of the Moon have been accrued by the coalition of several planetary science disciplines focused on a common cause, the science of the Moon.

Excellent and detailed reviews of lunar rocks, minerals, and soils have been presented in the *Lunar Sourcebook* (1991) and in *Planetary Materials* (1998). Readers seeking a more thorough treatment of these topics are referred to these monumental books.

2. THE LUNAR REGOLITH

Regolith is the term for the layer or mantle of fragmental and unconsolidated rock material, whether residual or transported and of highly varied character, that nearly everywhere forms the surface. This definition applies to the surface of all heavenly bodies, including the Moon and Earth. The entire lunar surface consists of a layer of regolith that completely covers the underlying bedrock, except perhaps on some very steep-sided crater walls and lava channels, where bedrock may be exposed. All samples collected by the Apollo and Luna Missions, as well as the lunar meteorites, are from the regolith; no "bedrock" was sampled directly in place. The lunar regolith is the product of the more than 4 billion years of impacts of meteoroids, big and small, into the Moon.

The lunar regolith is the boundary layer between the solid Moon and the matter and energy that fill the solar system. It contains information about both of these regions, and the complexities of studying the regolith are exceeded only by its importance to understanding the Moon and the space environment around it. The regolith is the source of virtually all our information about the Moon. All direct measurements of physical and chemical properties of lunar material have been made on samples collected from the regolith. Remotely sensed X-ray fluorescence, optical and infrared spectra, and gamma-ray signals come from the very top of the lunar regolith; in fact, from depths of no more than 20 μm, 1 mm, and 10–20 cm respectively (Adler et al. 1973; Metzger et al. 1973; Pieters 1983; Morris 1985). In addition, because of its surficial, unconsolidated, and fine-grained nature, it is likely that the regolith will be the raw material used for construction, mining, road building, and resource extraction when permanent lunar bases are established. As a resource, the lunar regolith is far more useful and accessible than the underlying "bedrock."

The lunar regolith also preserves information from beyond the Moon. Trapped in the solid fragments that make up the regolith are atoms from the Sun and cosmic-ray particles from beyond the solar system. In the regolith, data about the nature and evolution of the Moon are mixed with other records. These records include the composition and early history of the sun, and the nature and history of cosmic rays. The regolith also contains information about the rate at which meteoroids and cosmic dust have bombarded the Moon and, by inference, the Earth. Unscrambling these intertwined histories is a major challenge—and a major reward—of lunar research.

2.1. Some properties of lunar regolith and soil

We review here a few first-order properties of lunar regolith that are important for subsequent discussion. Although *soil* and *regolith* are terms that are sometimes used interchangeably, lunar *soil* usually refers to the fine portion of the regolith, operationally, the <1-mm grain-size fraction. Lunar soil is somewhat cohesive and dark gray to light gray in color. The mean grain size of lunar soils ranges from 40 to 800 μm and averages between 60 and 80 μm. Using conventional terrestrial descriptions, most lunar regolith samples would correspond to pebble- or cobble-bearing silty sands. Sorting values (the standard deviation of Folk 1968) range from 1.99 to 3.73 ϕ. In other words, lunar soils are poorly to very poorly sorted because mechanisms such as wind and water that sort terrestrial sediments do not occur on the Moon. There is also an inverse correlation between mean grain size and sorting (standard deviation); the coarsest samples are the most poorly sorted. Soils from all the Apollo sites are nearly symmetrically to coarsely skewed (skewness = 0–0.3). Exceptions to these generalizations are soils dominated by pyroclastic (volcanic) ash, which are finer-grained (mean grain sizes of ~40 μm) and better sorted (σ = 1.6 to 1.7 ϕ) than impact-produced soils.

The lunar regolith consists of fragments of igneous intrusive and extrusive rocks, crystalline impact-melt rocks that texturally resemble igneous rocks, various types of crystalline and glassy breccias produced by meteoroid impacts, mineral grains derived from rocks and breccias as a result of impact processes, glassy and crystallized spherules of volcanic origin, impact glass (spherules, ropy glass, glass coatings, glass fragments), meteorites, meteoritic metal, and agglutinates.

Agglutinates are a special and common lithic component of the lunar regolith, one that was unanticipated and that has no terrestrial analog, a direct result of the lack of an atmosphere on the Moon, which does not slow impinging micrometeorites. (Fig. 2.1). Each agglutinate particle is produced during the collision between a micrometeoroid traveling at 15–30 km/s and lunar regolith. Agglutinates, sometimes called glass-welded aggregates in early literature, consist of lithic and mineral fragments welded together by the glass formed as the small volume of resulting impact melt quenches. Agglutinates are mostly glassy, with complex morphologies and high vesicularity. The vesicles form from release of solar-wind implanted gases (Section 8) during the nearly instantaneous heating and melting of fined-grained surface regolith. Agglutinates make up a high proportion of the lithic fragments in lunar soils, about 25–30 vol% on average, although their abundances may range from essentially zero on the ejecta of a fresh impact crater to about 65 vol% in a mature soil (Section 9).

Figure 2.1. Photomicrograph of a thin section of an agglutinate from the Apollo 17 regolith (76503,7020; Jolliff et al. 1996). The particle is 3 mm in longest dimension. Most agglutinates are much smaller.

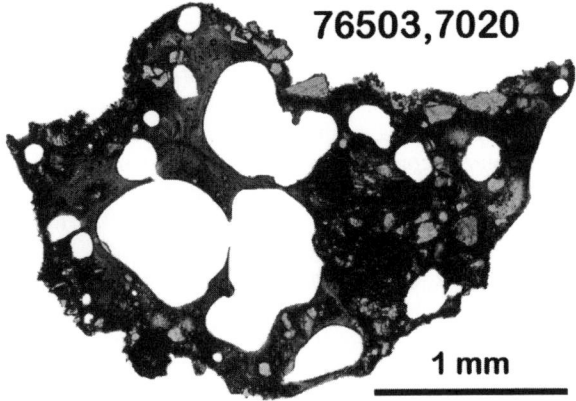

3. LUNAR SAMPLES

Nearly all of the direct information about the Moon's minerals, chemical composition, and age have been derived from samples. There are three classes of lunar samples available for study: (1) rocks and (2) soils collected on the six U.S. Apollo and three Russian Luna missions, and (3) lunar meteorites. A total of 382 kg of samples were collected on the Apollo missions (July, 1969, to December, 1972) and 0.3 kg on the Luna missions (September, 1970, to August, 1976). About 70% of the mass of the Apollo collection consists of rocks >1 cm in diameter; the rest is in the form of *fines* (<1 mm) and *coarse fines* (1–10 mm, but see Section 3.3). About 27 kg of lunar meteorites have been found to date (Section 3.2).

3.1. Apollo rock samples

In the study of terrestrial rocks, geologists typically collect samples by chipping, cutting, or drilling rocks that are several kilograms in mass from outcrops. As noted above (and perhaps with one exception, an Apollo 15 basalt), all lunar rock samples are isolated, loose rocks that were displaced by impacts from the locations where they originally formed. Most compositional and petrographic data that exist for lunar rocks are based on rocks that are small by terrestrial standards. Of the approximately 1500 numbered rocks collected on the Apollo missions (i.e., rocks >1 cm in diameter), 66 exceeded 1 kg in mass; the largest was 11.7 kg. About half (180 kg) of the total mass of material (382 kg) collected on the Apollo missions consists of >1-kg rocks. Of the numbered rock samples, the mean mass is about a quarter of a kilogram. Many of the numbered rocks, however, are less than a gram in mass. Tens of thousands of small lithic fragments (<1 cm) occur in the regolith samples and these have provided important information on the relative abundance of rock types in the regolith as well as a few unique types of rocks. The largest rock from the Luna missions is only about a gram in mass. Only 5 of ~37 lunar meteorites exceed 1 kg (some as more than one stone). Some rock types discussed here (e.g., alkali anorthosite and granite/felsite) are known only from small clasts in breccias or lithic fragments in the regolith. Before the discovery of the lunar meteorites, very low-Ti basalt was known only as small rock fragments.

For geochemical studies of terrestrial rocks, samples of several kilograms are pulverized and small subsamples of the resulting powder are analyzed to obtain data that are representative of the whole rock. In stark contrast, NASA has not pulverized any Apollo rocks and only small subsamples have been allocated to investigators for chemical analysis because of the rarity of lunar rocks. Most compositional data for lunar rocks have been obtained from fragments of a few tens of milligrams (sometimes smaller, e.g., Luna rocks) to, at most, a few grams (only for a few large Apollo rocks). Similarly, petrographic thin sections of all but the largest Apollo rocks are typically <1 cm^2 in area. For these reasons, compositional or petrographic results for different samples of a given "rock" can be expected to differ. The only lunar rock that is generally available in pulverized form is a portion of the MAC 88105 lunar meteorite (Lindstrom et al. 1991).

Each Apollo rock was given a 5-digit sample number. As an illustration, examples of some well known Apollo rock samples from each of the six missions are 10017 (Apollo 11 basalt), 12013 (Apollo 12 complex breccia), 14310 (Apollo 14 impact-melt rock), 15415 (Apollo 15 anorthosite), 65015 (Apollo 16 impact-melt breccia), and 72415 (Apollo 17 dunite). For the first four missions, the first two digits provide the key to the mission number; for the latter two missions, only the first digit provides the key. For Apollo 16 and 17, the second number represented the station at which the sample was collected. The last of the five numbers signifies the nature of the sample. Numbers of '5' and above represent rocks >1 cm; the number '0' refers to a soil, with '1' indicating the <1 mm sieved fraction, '2' the 1–2 mm fraction, '3' the 2–4 fraction, and '4' the 4–10 "rocklets." Luna samples provided to NASA for distribution use the same scheme, with numbers of the form 21xxx, 22xxx, and 24xxx representing the Luna 16, 20, and 24 missions, respectively.

3.2. Lunar meteorites

Since the end of the Apollo and Luna missions, numerous rocks have been found on Earth that are pieces of the Moon. All of the lunar meteorites known at this time (Table 2.1) are rocks that experienced a sequence of improbable events (Korotev et al. 2003a): They were blasted off the Moon by the impact of meteoroids, they eventually fell to Earth at remote locations (some up to several million years after they left the Moon), they were found by humans, and they were recognized to be of lunar origin. The first lunar meteorite to be so recognized was ALHA 81005 in 1982. Its identification as a piece of the Moon was uncontroversial (Marvin 1983 and other papers therein) because of our acquired experience with the Apollo and Luna samples. It is not known where on the Moon any of the lunar meteorites originated and it is not known with certainty how many impacts on the Moon launched the 37 currently known lunar meteorites (Table 2.1). The number of source craters is almost certainly a high fraction of the number of meteorites, however (Korotev et al. 2003b).

3.3. Lunar regolith and soil samples

During the Apollo and Soviet Luna missions, regolith samples were obtained with scoops, drive tubes, and rotary drill cores (Allton 1989). Apollo scoop samples were taken from near the surface (<10 cm depth) as well as from trenches as deep as 30 cm. Scoop and trench samples range in mass from tens of grams to several kilograms, although most are several hundred grams in mass. The number of distinct samples of surface and trench soils range from 2 at Apollo 11 to 68 at Apollo 17. With each successive Apollo mission, the distance separating the most distant collection points increased, from 10–30 m for Apollo 11 to 11 km for Apollo 17. A total of 24 drive tubes and cores (10 cm to 3 m in length) were taken on six Apollo missions and one on each of the three Luna missions. The compositional database for surface and trench soils is poorest for Apollo 12 and Apollo 15 samples.

As with the rocks, NASA assigned each scoop and trench sample and each core or drive tube section a unique 5-digit number. Some nominal samples of "soil" are fines derived largely from a disaggregation of a single rock or abrasion of rocks after collection (e.g., samples 10011, 12057, 67700, 73130, 76320), not true regolith. Most Apollo scoop and trench samples were passed through sieves of 10 mm, 4 mm, 2 mm, and 1 mm mesh size in the curatorial facility at the NASA Johnson Space Center. Each of the sub-cm grain-size fractions was assigned a new sample number and any fragment that was held by the 1-cm sieve was treated as a "rock" and given a unique sample number. In this section, "sample" is usually reserved for a mass of regolith with a distinct NASA 5-digit sample number. "Subsample" refers to that small portion of a sample allocated to an investigator or analyzed by some technique. Because of concerns about contamination and sample rarity, NASA did not pulverize or split by conventional sample splitting techniques masses of soil prior to allocation.

Soil from some regolith cores and drive tubes was also passed through a 0.25-mm sieve, and some investigators sieved or otherwise sized the <1-mm material in their own laboratories to obtain sub-millimeter grain-size fractions (e.g., McKay et al. 1974; Laul et al. 1988). Most chemical and physical measurements on lunar soil have been made only on subsamples of material that passed through a 1-mm sieve, i.e., the *<1-mm fines* or *fine fines*. Unless otherwise stated, nearly all available compositional data are based on subsamples of <1-mm fines. For the Luna samples and Apollo core samples, data are usually for <0.25-mm fines, however. Most petrographic data are only for grain-size fractions of even an narrower range, e.g., 90–150 μm (e.g., Heiken et al. 1973; Heiken and McKay 1974).

Coarse fines is a term that has been used to refer to both 1–10-mm and 4–10-mm material. With few exceptions (e.g., Jolliff et al. 1991a, 1996), data for bulk composition or other bulk properties of material in the 1–2 mm, 2–4 mm, and 4–10-mm grain-size fractions are nonexistent because of the problem of obtaining small, representative samples of coarse-

Table 2.1. List of lunar meteorites and some of their properties, in approximate order of increasing Al_2O_3 concentration, as of October 2005 (after Korotev et al. 2003b).

N	name	lunar rock type	mass (g)	TiO_2 (%)	Al_2O_3 (%)	FeO (%)	MgO (%)	Mg' (%)	Th µg/g
mare basalts									
1	Dhofar 287	mare basalt &	154	2.8	8.4	22.	13.	52.	0.9
		regolith breccia		n.a.	n.a.	n.a.	n.a.	n.a.	n.a.
2	Northwest Africa 032/479	mare basalt	~456	3.1	9.2	22.	7.8	39.	2.0
3	Asuka 881757	mare basalt	442	2.4	9.8	23.	6.3	33.	0.4
4	LAP/02205/02224/02226/ 02436/03632	mare basalt	1875	3.2	9.9	22.	6.3	34.	2.0
5	Yamato 793169	mare basalt	6	2.2	10.8	22.	5.9	33.	0.7
KREEP impact-melt breccia									
6	Sayh al Uhaymir 169	impact-melt breccia &	206	2.2	15.9	10.7	11.1	65	32.7
		regolith breccia		2.5	17.4	11.1	7.9	56	8.4
"mixed"(feldspathic-basaltic) breccias									
7	Northwest Africa 773	olivine gabbro &	633	0.3	4.7	19.4	26.3	71.	1.3
		regolith breccia		0.9	9.0	19.0	13.6	56.	2.1
8	Kalahari 008/009	fragmental breccia &	13,500	n.a.	13.	16.	8.5	49.	n.a.
		regolith breccia	598	n.a.	28.	4.5	4.4	64.	
9	EET 87521/96008	fragmental breccia	84	0.8	14.	18.	8.	43.	0.9
10	Northwest Africa 3136	regolith breccia	95	1.2	14.	15.	10.	55.	1.3
11	QUE 94281	regolith breccia	23	0.7	16.	14.	8.3	52.	0.9
12	MET 01210	regolith breccia	23	1.6	17.	16.	6.1	40.	0.9
13	Yamato 793274/981031	regolith breccia	195	0.7	18.	12.	8.9	56.	1.1
14	Calcalong Creek	regolith breccia	19	0.8	21.	9.	8.	60.	4.
15	Yamato 983885	regolith breccia	289	0.5	22.	9.	8.	62.	2.
feldspathic breccias									
16	ALHA 81005	regolith breccia	31	0.27	25.9	5.5	8.2	73.	0.31
17	Yamato 791197	regolith breccia	52	0.34	26.2	6.2	6.1	64.	0.34
18	Northeast Africa 001	regolith breccia	262	0.27	26.4	5.6	5.6	64.	0.24
19	PCA 02007	regolith breccia	22	0.29	26.5	6.2	6.9	66.	0.41
20	Dhofar 025/301/304/308	regolith breccia	772	0.30	27.	4.9	6.6	71.	0.6
21	"specimen 1153"	regolith breccia	?	0.18	27.	5.2	3.9	57.	n.a.
22	Dar al Gani 262	regolith breccia	513	0.21	27.9	4.5	5.5	68.	0.39
23	QUE 93069/94269	regolith breccia	25	0.25	28.3	4.4	4.6	65.	0.52
24	Yamato 82192/82193/86032	fragm. or reg. breccia	712	0.19	28.5	4.4	5.2	68.	0.20
25	Dar al Gani 400	regolith breccia	1425	0.18	28.5	3.6	4.8	70.	0.34
26	MAC 88104/88105	regolith breccia	724	0.24	28.7	4.3	4.1	63.	0.39
27	Dhofar 026/457–468	granulitic breccia	709	0.2	28.8	4.3	4.8	67	0.39
28	Northwest Africa 482	impact-melt breccia	1015	0.17	29.1	3.8	4.2	66.	0.23
29	Dhofar 302/303/305/306/307/ 309/310/311/489/730/731/ 908/909/911/950/1085	impact-melt breccia	1041	0.14	29.	3.	5.	75.	0.06
30	Dhofar 081/280/910/1224	fragmental breccia	572	0.12	32.	3.1	2.6	60.	0.2
31	Dar al Gani 996	fragmental ? breccia	12	n.a.	n.a.	n.a.	n.a.	n.a.	n.a.
32	Dhofar 490/1084	fragmental breccia	124	n.a.	n.a.	n.a.	n.a.	n.a.	n.a.
33	Dhofar 733	granulitic? breccia	459	n.a.	n.a.	n.a.	n.a.	n.a.	n.a.
34	Dhofar 925/960/961	impact melt ? breccia	106	n.a.	n.a.	n.a.	n.a.	n.a.	n.a.

ALHA = Allen Hills, EET = Elephant Moraine, LAP = LaPaz Icefield, MAC = MacAlpine Hills, MET = Meteorite Hills; PCA = Pecora Escarpment, QUE = Queen Alexandra Range. Multiple meteorites listed on the same line are known or suspected to be terrestrially paired (different fragments of a single meteorite fall); other unrecognized pairings might exist. In particular, pairing relationships among the various Dhofar meteorites are not well established. Concentration values in italics are uncertain. Mg' = bulk mole % Mg/(Mg+Fe). Data from many literature sources and unpublished data (R. L. Korotev).

grained material. Many properties of lunar soil, e.g., composition and spectral reflectance (Sections 9.5, 10.4.4), are a function of grain size. Thus for some purposes, data derived from the <1-mm *fines* may not be comparable to data for other grain-size fractions or data obtained remotely on exposed surface material.

3.4. The mare-highlands dichotomy and KREEP

3.4.1. The dichotomy. One of the best-known concepts in lunar geoscience is that the Moon has two types of surfaces, the highlands or terra (land) and the maria (seas; singular: mare). The mare-terra dichotomy originates from early astronomical observations that the light-colored areas of the Moon are rugged and mountainous and the dark areas are smooth like water, circular, and occur at lower elevation. Largely as a result of the Apollo, Clementine, and Lunar Prospector missions, it is now known that the highlands consist largely of feldspathic rocks termed *anorthosite*. The highlands are rugged because they are old and have experienced impacts by countless meteoroids. The maria are smooth because volcanic magmas were extruded into the low spots resulting from the largest and deepest impacts after the most intense period of meteoroid bombardment had ended. The maria are darker than the highlands because the basalts of the maria are richer in iron than the anorthosites of the highlands.

Despite the mixed etymology, lunar scientists usually use the terms *mare* and *highlands* to refer to the two types of surfaces and samples derived from them. For various reasons, the distinction, which appears to be conveniently black and white or low and high, is not always easily or accurately applicable to all situations. For example, there is not the one-to-one correspondence among elevation, landform morphology, composition, and geologic history that one might expect from a simple dichotomy in topography or albedo. Some areas of morphologic highlands are actually at low elevation, some rocks that are not mare basalts are also not feldspathic, and many places on the lunar surface the regolith consists of material of both highlands and mare origin. With regard to samples, the term *nonmare* is commonly used refer to rocks that are clearly not of mare volcanic origin but for which the relationship to the feldspathic highlands is not clear or firmly established (e.g., Warren and Wasson 1977).

Geochemically, the mare-highlands dichotomy is inadequate because three general compositional groups of common rocks occur (Fig. 2.2). Basalts, demonstrably from the maria, are rich in Fe. Rocks and breccias from the feldspathic highlands are rich in aluminum. However, a great many rocks and breccias in the Apollo collection have intermediate concentrations of Fe and Al, but are much richer in incompatible elements than mare basalts or feldspathic breccias. Such rocks are not simple mixtures of materials of the feldspathic highlands and the maria but are a third type of material. Subsequent discussion requires a brief introduction to one of the most crucial, but confusing, concepts in lunar geoscience: KREEP.

3.4.2. KREEP. Incompatible elements are those lithophile (rock-loving) elements that do not fit into the crystal structure of the major rock-forming minerals. As a consequence, incompatible elements tend to become increasingly concentrated in the residual liquid phase as a magma solidifies. Many types of terrestrial igneous rocks and many samples from the Apollo collection have high to very high concentrations of incompatible elements compared to meteorites. Large regions of the near-side lunar surface are enriched in Th, a quintessential incompatible element, by factors of 100–300 over chondritic meteorites (Plate 2.1). These simple observations indicate that both the Earth and Moon differentiated to a much greater degree than did the asteroids.

A distinguishing characteristic of the Moon, however, is that in nearly all lunar samples that have moderate to high concentrations of incompatible elements, the ratio of the concentrations of any two incompatible elements is nearly always the same. It is as though there is but one carrier of incompatible elements, concentrations of incompatible elements in that material are very high, and the samples contain variable proportions of that material. This simple characteristic is one of the important geochemical arguments used to support the hypothesis of a global magma

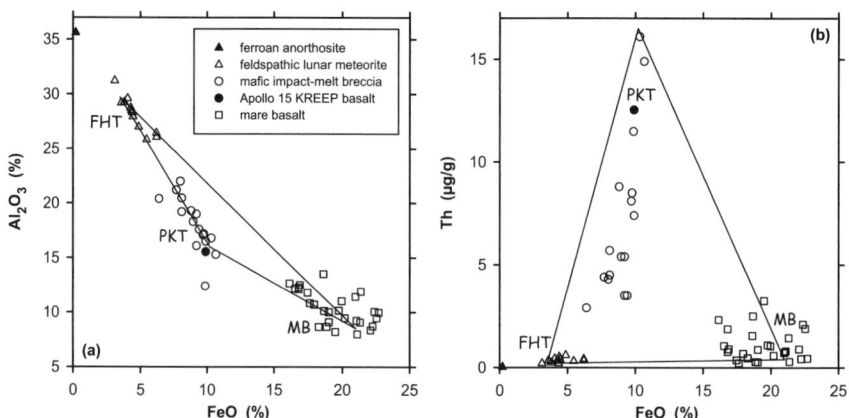

Figure 2.2. Geochemical variations among lunar samples corresponding to three "terrane" types that represent compositional extremes among common rocks of the Apollo missions: (1) Feldspathic Highlands Terrane (FHT), (2) Procellarum KREEP Terrane (PKT) (see Chapter 3), and (3) mare basalts (MB). The FHT is represented here by the feldspathic lunar meteorites (Korotev et al. 2003b) and highly feldspathic ferroan anorthosite (sample 15415), which have very low Th concentrations. The PKT is represented by Apollo 15 KREEP basalt and mafic impact-melt breccias, which are dominated by material (KREEP) with a noritic composition and high concentrations of incompatible elements such as Th. Each point represents one the 12 compositional groups defined by Korotev (2000) among samples from Apollo 14, 15, 16, and 17. The most Fe- and Th-rich ones are from Apollo 12 and 14 (Korotev 2000). The MB points each represent the mean of one of the major lunar basalt types. The triangles shown on these plots, are the same as those of Figures 2.5, 2.11, 2.12, 2.13, and 2.14.

ocean (e.g., Warren 1985). In this model the incompatible-element-rich material is, or derives from, the last liquid to solidify starting with a mostly or entirely molten Moon. Because these observations about lunar samples that are rich in incompatible elements (1) had no terrestrial analog, (2) were made very early in the study of lunar samples when lunar geochemistry was not well understood, (3) and the first materials to be recognized with these properties were glasses and breccias, not igneous rocks, the samples and the implied chemical component were given a non-standard designation: KREEP (Hubbard et al. 1971). The word is an acronym for K (potassium), REE (rare-earth elements), and P (phosphorous) which, along with Th and U, are among the many incompatible elements that are characteristic of materials identified as KREEP.

As a noun, KREEP usually refers to a rock, residual liquid, or chemical component (real or imaginary) that has concentrations of incompatible elements like those of samples that are rich in incompatible elements from Apollo 12 and Apollo 14. However, the term is not well defined or consistently used, and even experienced lunar geoscientists disagree on a definition and what the term means. Because the term KREEP has been applied to any and all materials with high concentrations of incompatible elements, largely regardless of major-element composition or whether the materials are igneous rocks, polymict samples (breccias, impact glass, regolith), or regions of the surface observed form orbit, its use sometimes obscures petrogenetic and chemical differences that, in fact, do occur among such materials.

As discussed in more detail below, samples identified as KREEP or as containing a large component of KREEP are usually noritic or basaltic in composition (Fig. 2.2, 2.3), although some are troctolitic (Shervais et al. 1988). The lithologic carriers of the chemical signature of KREEP in most Apollo regoliths are Fe- and Mg-rich impact-melt breccias and glasses. KREEP also occurs as a feldspathic basalt (Dymek 1986) with igneous texture at the Apollo 15 site, and some rocks of Apollo 17 have also been identified as KREEP basalt (Ryder et al.

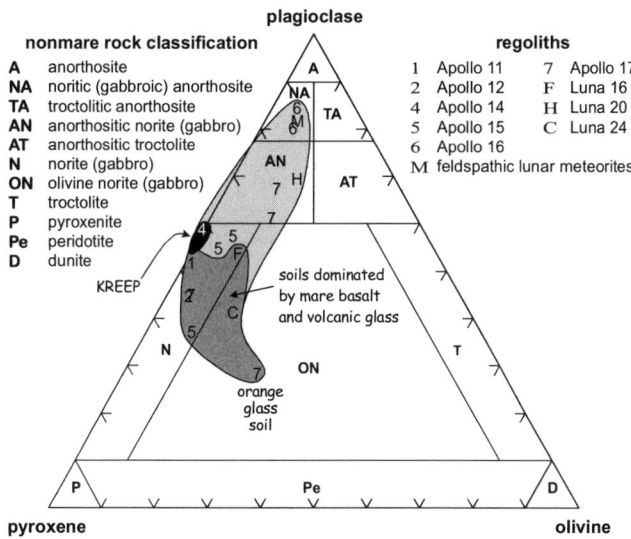

Figure 2.3. Normative mineralogy (idealized mass fraction based on chemical composition) of the three major minerals of lunar regolith converted to estimated modal mineralogy (volume fraction). Although lunar soil is composed mainly of plagioclase, pyroxenes, and olivine (most lunar regolith contain <5% ilmenite and other minerals), known soils occupy only a small portion (shaded region) of the triangle. For Apollo 11 (1) and 14 (4) and the three Luna missions (F, H, and C), a single point is plotted representing the mean composition of soil from each of those sites. For Apollo 16 (6), the two points represent the means of mature soils from the Cayley Plains and soils from North Ray crater (closer to plagioclase apex). For Apollo 15 (5) and 17 (7), the three points represent the compositional extremes; all other soils from either of these sites would plot within the triangle formed by these three points. The point for feldspathic lunar meteorites (M) is based on the mean composition of Table 2.1. The field for KREEP is based on Apollo 15 KREEP basalt and the Th-rich impact-melt breccias of Apollo 12, 14, and 15. The classification of Stöffler et al. (1980) for lunar highlands rocks is shown by the various fields (anorthositic norite, etc.). The classification does not strictly apply to soils dominated by mare basalt and volcanic glass.

1977; Salpas et al. 1987). The issue of "what is KREEP?" has been confused because the most common types of KREEP rocks are polymict breccias that vary substantially in major-element composition and some of which have greater concentrations of incompatible elements than their apparently igneous counterparts (Korotev 2000; Section 5.2.2). Many samples that are said to "contain KREEP" are polymict breccias or regoliths that contain clasts of impact-melt breccia of KREEP composition. However, even some mare basalts are described as containing KREEP as an assimilant (Dickinson et al. 1985; Neal et al. 1988; Anand et al. 2003).

More extensive discussions and summaries of the petrogenesis and significance of KREEP can be found elsewhere in this volume (Chapters 3 and 4) and in Schonfeld (1974), McKay and Weill (1977), Palme (1977), Hawke and Head (1978), Warren and Wasson (1979), Taylor (1982), Warren (1985), Dymek (1986), Haskin and Warren (1991), Ryder (1994), Warren (1998), Jolliff (1998), Wieczorek and Phillips (2000), Korotev (2000), and Korotev and Gillis (2001). The summary of McKay et al. (1986) is particularly enlightening.

4. LUNAR MINERALOGY

This section provides a summary of lunar mineralogy derived from direct examination of lunar samples. Even though the lunar samples were collected from a spatially limited region of the Moon, impacts have redistributed material over the lunar surface so there is a finite probability at any location on the surface of the Moon of obtaining material that formed at any other location. Also the Apollo samples provide a basis for understanding lunar mineralogy within the context of parent-body history and processes. For these reasons, the roster of lunar minerals is not expected to differ significantly in the unsampled regions except perhaps at the lunar poles for reasons described below. The lunar meteorites, which are random samples of the lunar surface that were "redistributed" by impacts to the extreme, confirm this expectation. Thus far no common minerals have been found in lunar meteorites that are not present in the Apollo collection (discounting terrestrial alteration products). With few exceptions, the minerals found on the Moon and their chemistry constitute a subset of known minerals that occur on Earth and that are consistent with the specific settings, history, and origin of the Moon. For a detailed coverage of mineralogy based on lunar samples, see Smith and Steele (1976), Papike et. al. (1991, 1998), and Papike (1998).

Minerals have provided the keys to understanding lunar rocks because their compositions and atomic structures reflect the physical and chemical conditions under which the rocks formed. Analyses of lunar minerals, combined with the results of laboratory experiments, have enabled scientists to determine key parameters—temperature, pressure, cooling rate, and the partial pressures of such gases as oxygen, sulfur, and carbon monoxide—that existed during formation of the lunar rocks. The array of minerals found in the lunar samples demonstrates strong differences between the Earth and Moon, at least since formation. The lack of water-bearing minerals shows the Moon is extremely dry; the samples are depleted in moderately volatile elements such as Na; the presence of metallic iron and details of the oxidation state of various elements in lunar minerals shows the Moon has very reducing conditions with oxygen fugacity near and mostly below the Fe-FeO buffer. High pressure phases unrelated to shock are not found at the Moon's surface.

As on the Earth, we recognize that minerals that occur on the surface of the Moon may not represent its deep interior because few delivery mechanisms exist to bring deep-seated solid materials to the lunar surface. On Earth, mantle materials are borne by kimberlites or diamond-bearing pipes that have sampled the Earth's mantle, but these occurrences are extremely rare. Similar features either do not occur or have not been recognized on the Moon. Asteroidal impacts may have excavated deep seated material in some areas, but no such material has been recognized in the sample collection. However, there is indirect evidence that garnet, a mineral not found in the samples, may occur in the Moon's mantle (see Chapter 3).

One area where there may yet be significant additions to the lunar mineral suite is in regions where hydrogen has been concentrated in the regolith, preserved by burial, and subjected to enough heat to form H-bearing minerals. Remote sensing has detected H at the lunar poles that may have been added to the Moon by many plausible sources, including cometary impacts or by solar wind implantation. Other volatile elements could have been added in the same events. The rare occurrences of volatile-element-bearing phases such as iron oxyhydroxides and carbonates in lunar samples (if indeed they formed on the Moon) may give evidence to these processes. The H deposits are most significant near the lunar poles and are at least partly associated with permanently shadowed locations in impact craters which may have acted as cold-traps (Watson et al. 1961; Feldman et al. 2001).

The most common minerals found in the lunar samples are the silicates plagioclase feldspar $(Ca,Na)(Al,Si)_4O_8$, pyroxene, $(Ca,Fe,Mg)_2Si_2O_6$, and olivine $(Mg,Fe)_2SiO_4$. Potassium feldspar $(KAlSi_3O_8)$ and the silica (SiO_2) minerals (e.g., quartz), although abundant on Earth, are notably rare on the Moon. Minerals containing ferric iron (Fe^{3+}) and carbonate (CO_3^{2-}) are absent on the Moon. The most striking aspect of lunar mineralogy, however, is the lack of minerals that contain water, such as phyllosilicate clays, micas, amphiboles, and oxyhydroxides. These minerals may yet be found near the lunar poles where increased H concentrations occur, but these minerals would not be indicative of Moon-wide processes or compositions.

After the silicate minerals, oxide minerals are next in abundance. They are particularly concentrated in the mare basalts, and they can make up more than 20% by volume of these rocks. The most abundant oxide mineral is ilmenite, $(Fe,Mg)TiO_3$, a black, opaque mineral that is the carrier of high TiO_2 concentrations of many mare basalts. The second most abundant oxide mineral, spinel, has a widely varying composition and actually consists of a complex series of solid solutions. Members of this series include: chromite, $FeCr_2O_4$; ulvöspinel, Fe_2TiO_4; hercynite, $FeAl_2O_4$; and spinel (*sensu stricto*), $MgAl_2O_4$ (Table 2.2). Another oxide phase, which is only abundant in Ti-rich lunar basalts, is a mineral first described from the Moon, armalcolite $(Fe,Mg)Ti_2O_5$.

Two additional minerals are noteworthy because, although they occur only in small amounts, they are important indicators of a highly reducing, low-oxygen environment under which the lunar rocks formed. Native iron (Fe) is ubiquitous in lunar rocks, and commonly contains small amounts of Ni and Co. Troilite, relatively pure FeS, is a common minor component; it holds most of the sulfur in lunar rocks. As a result of selective volatilization of silica by meteoroid bombardment, some high-alumina and silica-poor glasses occur in the regolith (Naney et al. 1976), as well as new minerals resulting from this process, yoshiokaite, a Ca-Al silicate (Vaniman and Bish 1990) and hapkeite, an Fe_2Si phase (Anand et al. 2004).

4.1. Silicate minerals

Silicate minerals are the most abundant minerals in the lunar crust. Modal data for mineral grains in regoliths show the predominance of silicate minerals, especially pyroxene, plagioclase feldspar, and olivine (Table 2.2). It is expected that in some regions of the Moon the range of abundance is likely to be greater, especially in anorthosite terrains where the abundance of feldspar could locally approach 100%.

4.1.1. Pyroxene. Pyroxenes are the most chemically complex of the major silicate phases in lunar rocks and their chemistry records conditions of formation and evolutionary history of rocks in which they occur. Pyroxenes are compositionally variable solid solutions and they contain most of the major elements present in the host rocks. Lunar pyroxenes are primarily "quadrilateral" in that their compositions are defined within the compositional space bounded by enstatite-ferrosilite-diopside-hedenbergite. The most common varieties are hypersthene, augite, and pigeonite.

For a review of pyroxene crystal chemistry, see Papike (1987). Briefly, in the pyroxene structure, the M1 and M2 sites provide a range of atomic environments; as a result, pyroxenes can accommodate a wide variety of cations, and these cations reflect much of the chemistry and crystallization history of the rocks in which they occur. Ca, Na, Mn, Mg, and Fe^{2+} are accommodated in the large, distorted six- to eight-coordinated M2 site; Mn, Fe^{2+}, Mg, Cr^{3+}, Cr^{2+}, Ti^{4+}, Ti^{3+}, and Al occur in the six-coordinated M1 site; and Al and Si occupy the small, tetrahedral site. Potassium is too large to be accommodated in any of the pyroxene crystallographic sites at low pressure.

Although the major elements that define the pyroxene quadrilateral plot (Ca, Mg, and Fe) show important variations, the other, less abundant elements also show important trends. In lunar pyroxenes, these other elements include Al, Ti, and Cr. Reduced valence states for some Ti cations (including some Ti^{3+}) and Cr cations (including some Cr^{2+}) are important indications of very low oxygen fugacity during the crystallization of lunar rocks.

Lunar pyroxenes also show evidence of substantial *subsolidus* reactions (i.e., recrystallization and other changes that take place below melting temperatures). Considerable work has been done to interpret the resulting features. It was discovered soon after the return of the Apollo 11 samples that subsolidus exsolution of two distinct pyroxenes, augite and pigeonite, had taken place within originally uniform pyroxene crystals (e.g., Ross et al. 1970). This process produced distinctive microscopic and submicroscopic exsolution lamellae, i.e.,

Table 2.2. Lunar mineralogy summary.

Major minerals
   Pyroxene
      orthopyroxene $(Mg,Fe)_2Si_2O_6$
      clinopyroxene $(Ca,Mg,Fe)_2Si_2O_6$
   Plagioclase $(Ca,Na,K)Al_2Si_2O_8$
   Olivine $(Mg,Fe)_2SiO_4$
   Ilmenite $FeTiO_3$

Minor/accessory minerals
   K-feldspar $(K,Ba)AlSi_3O_8$
   Phosphates
      apatite $Ca_5(PO_4)_3(F,Cl)$
      whitlockite (RE-merrillite) $Ca_{18-x}(Mg,Fe)_2(Y,REE)_xNa_{2-x}(P,Si)_{14}O_{56}$
      monazite $(La,Ce,Nd)PO_4$
      farringtonite $(Mg,Fe)_3(PO_4)_2$ (tentative—65785 Sp. Troct.)
      graftonite $(Fe,Mn,Ca)_3(PO_4)_2$ (not verified)
   Pyroxenoid
      pyroxferroite $Ca_{1/7}Fe_{6/7}SiO_3$
   Silica (SiO_2) polymorphs: cristobalite, quartz, tridymite
   Spinels[1]
      pleonaste $(Mg,Fe)(Al,Cr)_2O_4$ Mg>Fe, Al>>Cr ("normal" spinel)
      chromite $(Fe,Mg)(Cr,Al)_2O_4$ ideally $Fe^{2+}Cr_2O_4$ ("normal" spinel)
      ulvöspinel Fe_2TiO_4 inverse ($TiFe_2O_4$)
   Sulfides
      troilite FeS
      sphalerite ZnS
      wurtzite ZnS see Bogatikov et al. (2001)
   Ti-bearing minerals other than ilmenite
      rutile TiO_2 (typically niobian)
      armalcolite $(Fe,Mg)Ti_2O_5$
      perovskite $CaTiO_3$
      sphene $(Ca,Cr)(Ti,Zr)SiO_5$
   Zr-bearing minerals
      zircon $(Zr,Hf)SiO_4$
      baddeleyite $(Zr,Hf)O_2$
      zirconolite $CaZrTi_2O_7$
      zirkelite $(Ca,Fe)(Zr,Ti)_2O_5$
      tranquillityite $Fe_8(Zr,Y)_2Ti_3Si_3O_{24}$
   Others
      cohenite Fe_3C
      calcite/arag $CaCO_3$ (tentative, contaminant(?) in 10058)
      cordierite $Al_3(Mg,Fe)_2[Si,AlO_{18}]$
      eskolaite Cr_2O_3 see Bogatikov et al. (2001)
      lawrencite $FeCl_2$
      schreibersite $(Fe,Ni,Co)_3P$
      yoshiokaite $Ca_{8-(x/2)}Al_{16-x}Si_xO_{32}$ see Vaniman & Bish (1990)
      yttrobetafite $(Ca,Y,Fe,REE,Th,U)_2(Ti,Nb,Ta)_2O_7$

[1]Spinel explanation: general formula is $(A^{IV})(B^{VI})_2O_4$; normal spinels have $8R^{2+}$ in A, $16R^{3+}$ in B; inverse spinels have $8R^{3+}$ in A, $8R^{2+}$ in B, $8R^{3+}$ in B (e.g., magnetite is inverse).

thin layers of pigeonite in augite, or vice versa. Papike et al. (1971) attempted to relate these exsolution features to the relative cooling histories of mare basalts. They pointed out that certain parameters of the pyroxene crystal unit cell (the length b and the angle β) could also be used to estimate the composition of the intergrown augite and pigeonite. They also suggested that $\Delta\beta$ (β-pigeonite – β-augite) could be used to indicate the degree of subsolidus exsolution and thus the relative annealing temperatures of the exsolved pyroxenes. Takeda et al. (1975) summarized similar exsolution data for pyroxene grains from Apollo 12 and 15 basalts. They compared the relative cooling rates (determined from exsolution studies) with absolute cooling rates determined from experimental studies. Ross et al. (1973) experimentally determined the 1-atmosphere augite-pigeonite stability relations for pyroxene grains from mare basalt 12021. Grove (1982) used exsolution lamellae in lunar clinopyroxenes as cooling rate indicators, and his results were calibrated experimentally. These studies all indicate that the cooling and subsolidus equilibration of igneous and metamorphic pyroxenes is a slow process; estimated cooling rates range from 1.5–0.2 °C/hr for lava flows 6 m thick for Apollo 15 mare lavas (Takeda et al. 1975). McCallum and O'Brien (1996) studied the width of exsolution lamellae and chemistry of exsolved pyroxenes in lunar highlands rocks to infer the stratigraphy of the highland crust, and the method has been applied to other rock types (e.g., Jolliff et al. 1999).

Shock lamellae can be produced in pyroxenes by the shock waves due to meteoroid impact. However, these features are rarely observed, and they are much less well characterized than the analogous shock lamellae in plagioclase (Schaal and Hörz 1977).

4.1.2. Plagioclase feldspar. The silicate mineral feldspar is the most common mineral in the lunar crust. It has a framework structure of three-dimensionally linked SiO_4 and AlO_4 tetrahedra (reviewed by Papike 1988). The Si:Al ratio varies between 3:1 and 1:1. Ordering of Si and Al in specific tetrahedral sites can lead to complexities such as discontinuities in the crystal structure. Within this three-dimensional framework of tetrahedra containing Si and Al, much larger sites with 8 to 12 coordination occur that accommodate large cations (Ca, Na, K, Fe, Mg, Ba).

Aside from rare K- and Ba-enriched feldspars, most lunar feldspars belong to the plagioclase series, which consists of solid solutions between albite ($NaAlSi_3O_8$) and anorthite ($CaAl_2Si_2O_8$). Lunar plagioclases are also depleted in Na (the albite component) relative to terrestrial plagioclases indicating the alkali depleted nature of the Moon. The maximum chemical variation involves solid solution between albite and anorthite, a variability that can also be described as the coupled substitution between NaSi and CaAl, in which the CaAl component represents anorthite. The Ca abundance in the plagioclase, and therefore mol% anorthite correlate positively with the Ca/Na ratio in the host basalts (e.g., Papike et al. 1976; BVSP 1981).

Plagioclase from impact-melt breccias that are rich in incompatible elements (Section 5.2) has more Na-rich compositions than those in highland plutonic rocks (i.e., lower anorthite content). Plagioclase from coarsely crystalline igneous rocks has more restricted compositions; however, there is a positive correlation between the alkali content of the host rock and that of the plagioclase.

4.1.3. Olivine. The crystal structure of olivine, $(Mg,Fe)_2SiO_4$, consists of serrated chains formed of edge-sharing octahedra, which parallel the crystallographic c-axis. The octahedral chains are cross-linked by isolated SiO_4 tetrahedra. The major cations in the octahedral sites, Fe^{2+} and Mg^{2+}, are distributed randomly over both the M1 and M2 octahedral sites; however, the small amounts of Ca that may occur in olivine occupy only the M2 site (see Papike 1987).

The major compositional variation within olivines is caused by exchange of Fe and Mg; this exchange, and the resulting variations in composition, are represented by the ratio Fe/(Fe+Mg). The Fe end member, Fe_2SiO_4, is fayalite, and the Mg end member, Mg_2SiO_4, is forsterite. The most magnesian mare basalt olivine grains contain only 20 mol% fayalite (Fa), represented by the notation Fa_{20}. Most mare basalt olivines have compositions in the range Fo_{80}-Fo_{30}; however, a number of mare basalts contain very Fe-rich olivine (Fa_{90}-Fa_{100}). These olivines are part of an equilibrium three-phase assemblage (Ca,Fe-pyroxene, Fe-olivine, silica) that crystallized stably from late-stage, Fe-enriched basaltic melts, typically occurring as mesostasis. Some mare basalts that cooled quickly during the late stages of crystallization instead contain either an Fe-rich pyroxene that crystallized metastably relative to the normal three-phase assemblage, or as Fe-rich pyroxenoid, pyroxferroite. The formation of extremely Fe-rich pyroxene violates a so-called "forbidden region" at the Fe-apex of the "pyroxene quadrilateral."

Other significant elements in lunar olivines are Ca, Mn, Cr, Ni, and Al. Calcium varies directly with the Fe content, and it may be an indicator of the cooling rate (Smith 1974). The experimental data of Donaldson et al. (1975) supported this contention. Olivine in mare basalt is significantly enriched in Cr relative to olivine in terrestrial basalts. Cr_2O_3 values, which are commonly below detection limits (~0.1 wt%) in terrestrial olivine (Smith 1974) range up to 0.6 wt% in lunar olivine. Much or all of this Cr may be in the reduced Cr^{2+} valence state, and Haggerty et al. (1970) identified significant Cr^{2+} in lunar olivine using optical absorption techniques. Cr^{2+} is more readily accommodated in the olivine structure than is Cr^{3+}, which is the normal valence state for Cr in terrestrial olivine. The presence of Cr^{2+} is another strong indicator of the low oxygen fugacities that existed during mare-basalt crystallization. Similarly, Cr^{2+} is much more abundant in lunar pyroxenes than in terrestrial pyroxenes (BVSP 1981).

4.1.4. Silica minerals: quartz, cristobalite, and tridymite. Silica minerals include several structurally different minerals, all of which have the simple formula SiO_2. These minerals are generally rare on the Moon. This rarity is one of the major mineralogic differences between the Moon and the Earth, where silica minerals are abundant in such common rocks as granite, sandstone, and chert.

The silica minerals found on the Moon are cristobalite, quartz, and tridymite. In spite of the intense impact cratering of the Moon, it is interesting that the high-pressure polymorphs of SiO_2, coesite and stishovite, which are known from young terrestrial impact craters, have not been found on the Moon. Explanations for their absence include the rarity of silica grains in the original target rocks and volatilization of silica during impact events in the high vacuum at the lunar surface (Papike et al. 1997).

The crystal structures of the lunar silica minerals are distinctly different from each other, but they all consist of frameworks of SiO_4 tetrahedra in which each tetrahedral corner is shared with another tetrahedron. A comparison of silica mineral structures, along with structure diagrams, can be found in Papike and Cameron (1976). All of the silica mineral structures contain little or no room for cations larger that Si^{4+}, hence the relatively pure SiO_2 composition of these minerals. The structures become more open in going from quartz to tridymite and cristobalite, and the general abundance of impurities increases accordingly.

Quartz occurs in a few granite-like (felsite) clasts as needle-shaped crystals that probably represent structural transformation (inversion) of original tridymite (Quick et al. 1981). Some tridymite is preserved in these felsite clasts. The other rock type in which quartz is abundant is coarse-grained lunar granites occurring in rare fragments. The largest lunar granite clast yet found, from Apollo 14 breccia 14321, weighs 1.8 g and contains 40 vol% quartz (Warren et al. 1983). A smaller granite clast, from sample 14303, was estimated to have 23 vol% quartz (Warren et al. 1983). Based on their isotopic work on the large clast, Shih et al. (1985, 1993) suggested that the sample crystallized in a deep-seated plutonic environment about 4.1 Ga. Consistent with the general absence of hydrous minerals on the Moon, the lunar granites do not contain mica or amphibole, as do granites on Earth.

The most common silica mineral in lunar basalts is not quartz but cristobalite, which can constitute up to 7 vol% of some basalts. This situation contrasts with the general absence of all silica minerals in terrestrial basalts. Lunar cristobalite commonly has twinning and curved fractures, indicating that it has inverted from a high-temperature to a low-temperature crystal structure during cooling of the lavas (Dence et al. 1970, Champness et al. 1971). Other mare basalts contain crystals of the silica mineral tridymite that have incompletely inverted from cristobalite, producing rocks that contain both tridymite and cristobalite. Cristobalite tends to occur as irregular grains wedged between other crystals, while tridymite forms lathlike crystals. Klein et al. (1971) observed tridymite laths enclosed by pyroxene and plagioclase and suggested that tridymite was an early crystallizing phase. In a study of Apollo 12 basalts, Sippel (1971) found that the coarser-grained samples contained cristobalite and quartz. Unfortunately, these mineral pairs are stable over fairly large temperature ranges and can also form metastably, outside of their equilibrium stability fields, so they are not useful for inferring the temperatures of lava crystallization.

The paucity of silica minerals in the lunar samples and presumably the Moon as a whole has several implications. For one, the Moon has apparently not evolved chemically beyond the formation of a low-silica, high-alumina anorthositic crust, so that high-silica granitic rocks are rare. For another, the Moon lacks hydrous and hydrothermal systems like those that can crystallize silica on Earth. Despite the scarcity of lunar silica minerals, some lunar geologic features exhibit a morphology that suggests they may have formed from viscous lavas, presumably silica-rich. Some of these locations are spectrally anomalous in their UV-Visible ratio giving them (and other color anomalies) the nickname "red spots." There is no direct evidence that these locations contain silica minerals (or other minerals indicative of a viscous magma composition) but such evidence could be derived from thermal-infrared spectroscopy.

Despite their rarity in the samples, the silica minerals are nevertheless important in classifying and unraveling the origin of some lunar rocks. Furthermore, lunar crustal rocks that contain silica minerals may be more abundant than their meager representation among the returned Apollo and Luna samples suggests. The silica minerals tend to concentrate along with incompatible elements. For these reasons, the lunar silica minerals deserve greater consideration than their rarity would otherwise warrant.

4.2. Oxide minerals

The oxide minerals, although less abundant than silicates in lunar rocks, are of great significance because they retain signatures of critical conditions of formation (e.g., limited availability of oxygen) of the rocks in which they occur. Whereas most of the silicate minerals differ little from those on Earth, the opaque oxide phases indicate the reducing, anhydrous conditions that prevailed during their formation. By combining analyses of lunar oxide minerals with the results of laboratory experiments on their synthetic equivalents, the temperature and oxygen pressure conditions during formation of lunar rocks can be estimated (see Sato et al. 1973; Usselman and Lofgren 1976). Several oxide minerals are important constituents of lunar samples: ilmenite, $FeTiO_3$; spinels with extensive chemical variations: $(Fe,Mg)(Cr,Al,Fe,Ti)_2O_4$; and armalcolite, $(Fe,Mg)Ti_2O_5$. The less abundant lunar oxide minerals include rutile, TiO_2, baddeleyite, ZrO_2, and zirconolite, $(Ca,Fe)(Zr,REE)(Ti,Nb)_2O_7$ (Table 2.2).

Because their oxygen is more weakly bonded than that in silicate minerals, oxide minerals are obvious and important potential feedstocks for future production of lunar oxygen and metals. On Earth, similar oxide minerals commonly occur in economically recoverable quantities called ore deposits. However, most of these deposits have formed from hydrothermal waters (100–300 °C or hotter). The Moon has little, if any, water, and the presence of similar hydrothermal ore deposits on the Moon is improbable. However, there are other means of concentrating oxide minerals into exploitable ores. Crystal settling of dense minerals (e.g., chromite, ilmenite, and minerals containing the platinum-group elements) is possible within silicate magmas, if the magma remains liquid for a long enough time. On Earth, such accumulations are normally found in layered intrusions. These bodies form from large masses of magma that have been emplaced into deep crustal rocks without reaching the Earth's surface. Under such conditions, cooling is slower, and physical separation processes have time to act and include convection as well as settling. Well-known examples of ore deposits resulting from these processes occur in the Stillwater Anorthosite Complex (Montana) and the Bushveld Igneous Complex (South Africa).

The major differences between the oxide minerals in lunar and terrestrial rocks arise from fundamental differences between both the surfaces and the interiors of these two planets. On the Moon, meteoroid impact and shock-metamorphic processes play a major role in altering rocks. These effects are not the same for all minerals. Shock damage and the formation of shock glasses from minerals, e.g., maskelynite from plagioclase feldspar, are observed chiefly in silicate minerals. Oxide minerals also record shock damage, but another effect of impact on oxide (and sulfide) minerals is to produce small amounts of chemical reduction.

4.2.1. Ilmenite. With the ideal formula $FeTiO_3$, ilmenite is the most abundant oxide mineral in lunar rocks. The amount of ilmenite in a rock is a function of the bulk composition of the magma from which the rock crystallized (Campbell et al. 1978; Norman and Ryder 1980; Rutherford et al. 1980); the higher the TiO_2 content of the original magma, the higher the ilmenite content of the rock. Ilmenite forms as much as 15 to 24 vol% of many Apollo 11 and 17 mare basalts (McKay and Williams 1979). However, the volume percentages of ilmenite in mare basalts vary widely across the Moon, as indicated by the range of TiO_2 contents in samples from different lunar missions.

The ilmenite crystal structure is hexagonal and consists of alternating layers of Ti- and Fe-containing octahedra. Most lunar ilmenite contains some Mg substituting for Fe, which arises from the solid solution that exists between ilmenite ($FeTiO_3$) and $MgTiO_3$, the mineral geikielite. Other elements are present only in minor to trace amounts (i.e., <1 wt%); these include Cr, Mn, Al, and V. In addition, ZrO_2 contents of up to 0.6 wt% have been reported from ilmenite in Apollo 14 and 15 basalts (El Goresy et al. 1971a,b; Taylor et al. 1973). In fact, the partitioning of ZrO_2 between ilmenite and coexisting ulvöspinel (Fe_2TiO_4) has been experimentally determined (Taylor and McCallister 1972) and has been used as both a geothermometer (to deduce temperatures during crystallization) and as a cooling-rate indicator (Taylor et al. 1975, 1978; Uhlmann et al. 1979). Although terrestrial ilmenite almost always contains some Fe^{3+}, lunar ilmenite contains none.

Ilmenite commonly occurs in mare basalts as bladed crystals up to a few millimeters long. It typically forms near the middle of the crystallization sequence, where it is closely associated with pyroxene. It also forms later in the sequence and at lower temperatures, where it is associated with native Fe and troilite. In Apollo 17 rocks, ilmenite is frequently associated with armalcolite and occurs as mantles on armalcolite crystals (e.g., Haggerty 1973a; Williams and Taylor 1974). In these instances, ilmenite has possibly formed by the reaction of earlier armalcolite with the melt during crystallization.

The composition of lunar ilmenite plots along the $FeTiO_3$-$MgTiO_3$ join; variation from $FeTiO_3$ is often expressed in wt% of MgO. In general, the ilmenite with the highest Mg contents tends to come from relatively high-Mg rocks; ilmenite composition correlates with the bulk composition of the rock and therefore reflects magmatic chemistry rather than pressure. In detail, the distribution of Mg between ilmenite and coexisting silicate minerals in a magma is related to the timing of ilmenite crystallization relative to the crystallization of the other minerals. This crystallization sequence is itself a function of cooling rate and other factors, most notably the oxygen fugacity (Usselman et al. 1975). However, it is doubtful that the Mg contents in ilmenite all represent equilibrium conditions, because ilmenite compositions can vary significantly, even within distances of a few millimeters, within a single rock.

The stability curve of pure ilmenite as a function of temperature and fO_2 is significantly different from that of ulvöspinel (Taylor et al. 1973), the spinel phase with which it is commonly associated, implying that the two minerals did not crystallize together. The data suggest that, in these mineral assemblages, ilmenite has formed by solid-state reduction of this high-Ti spinel at temperatures below their melting points. Rare grains of ilmenite also contain evidence for subsolidus reduction of ilmenite to rutile (TiO_2) + native Fe; other grains show reduction to chromite ($FeCr_2O_4$) + rutile + native Fe.

4.2.2. Spinels. Spinel is the name for a group of oxide minerals, all with cubic crystal symmetry, that have extensive solid solution within the group. Spinels are the second most abundant opaque mineral on the Moon, second only to ilmenite, and they can make up as much as 10 vol% of certain basalts, most notably those from the Apollo 12 and 15 sites. The general structural formula for these minerals is $^{IV}A^{VI}B_2O_4$, where IV and VI refer to cations with tetrahedral and octahedral coordination, respectively.

The basic spinel structure is a cubic array of oxygen atoms. Within the array, the tetrahedral A-sites are occupied by one-third of the cations, and the octahedral B-sites are occupied by the remaining two-thirds of the cations. In a normal spinel structure, the divalent cation, such as Fe^{2+}, occupies only the tetrahedral sites, and the two different sites each contain only one type of cation (e.g., $FeCr_2O_4$). If the divalent cation occurs in one-half of the B-sites, the mineral is referred to as an inverse spinel [e.g., $Fe(Fe,Ti)_2O_4$]. In lunar spinels, the divalent cations (usually Fe^{2+} or Mg^{2+}) occupy either the A- or both A- and B-sites (i.e., there are both normal and inverse lunar spinels), and higher-charge cations (such as Cr^{3+}, Al^{3+}, Ti^{4+}) are restricted to the B-sites.

The relations of the various members of the spinel group can be displayed in a diagram known as the Johnston compositional prism (e.g., Haggerty 1978a). The end-members represented include chromite, $FeCr_2O_4$; ulvöspinel, $FeFeTiO_4$ (commonly written as Fe_2TiO_4; but this is an inverse spinel with Fe^{2+} in both A- and B-sites); hercynite, $FeAl_2O_4$; and spinel (*sensu stricto*), $MgAl_2O_4$. Intermediate compositions among these end-members are designated by using appropriate modifiers (e.g., chromian ulvöpsinel or titanian chromite).

Most lunar spinels have compositions generally represented within the three-component system: $FeCr_2O_4$ - $FeFeTiO_4$ - $FeAl_2O_4$, and their compositions can be represented on a simple triangular plot. The addition of Mg as another major component provides a third dimension to this system; the compositions are then represented as points within a limited Johnston compositional prism in which the Mg-rich half (Mg>Fe) is deleted because most lunar spinels are Fe-rich (e.g., Agrell et al. 1970; El Goresy et al. 1971b, 1976; Haggerty 1971a, 1972b,c, 1973b, 1978a; Taylor et al. 1971; Busche et al. 1972; Dalton et al. 1974; Nehru et al. 1974, 1976). Most lunar spinel compositions fall between chromite and ulvöspinel. The principal cation substitutions in these lunar spinels can be represented by $Fe^{2+} + Ti^{4+} = 2(Cr,Al)^{3+}$. Other cations commonly present include V, Mn, and Zr.

Spinels are ubiquitous in mare basalts, where they occur in various textures and associations. The spinels are invariably zoned chemically. Such zoning occurs particularly in Apollo 12 and 15 rocks, in which chromite is typically the first formed mineral. As the chromite crystals grow, their TiO_2 and FeO contents increase and Al_2O_3, MgO, and Cr_2O_3 contents decrease, with the overall composition moving toward ulvöspinel. In most of the basalts that contain both titanian chromite and chromian ulvöspinel, the ulvöspinel occurs as overgrowths and rims surrounding the chromite crystals. Some individual ulvöspinel grains also occur as intermediate to late-stage crystallization products.

In reflected light, the ulvöspinel in these composite crystals appears as tan to brown rims around the bluish chromite. The contact between the two is commonly sharp, indicating a discontinuity in the compositional trend from core to rim. This break probably records a cessation in growth, followed later by renewed crystallization in which the early chromite grains acted as nuclei for continued growth of ulvöspinel (Cameron 1971). Some rocks (e.g., basalt 12018) contain spinel grains with diffuse contacts that reflect gradational changes in the composition of the solid solution. These textures could result from continuous crystallization of the spinel or from later solid-state diffusion within the crystal (Taylor et al. 1971).

Although most abundant in mare basalts, spinels also occur in highland rocks such as anorthosites, anorthositic gabbros, troctolites, and impact mixtures of these rock types (e.g., Haselton and Nash 1975). The spinels in anorthositic (plagioclase-rich) highland rocks tend to be chromite with lesser amounts of MgO, Al_2O_3, and TiO_2. However, certain highland rocks, notably the olivine-feldspar types (troctolites), contain pleonaste spinel. The composition of this spinel is slightly more Fe- and Cr-rich than an ideal composition precisely between the end members $MgAl_2O_4$ and $FeAl_2O_4$. This spinel is not opaque; viewed with a petrographic microscope, it stands out because of its pink color, high index of refraction, and isotropic character in cross-polarized light.

4.3. Phosphate minerals

Phosphates are found as accessory minerals in most lunar rocks, typically as apatite $[Ca_5(PO_4)_3(F,Cl)]$ or RE (rare earth)-merrillite $[Ca_{16}(Mg,Fe)_2(REE)_2(PO_4)_{14}]$, and commonly these two are found together. Here we use the mineral name RE-merrillite instead of whitlockite because structural and chemical studies indicate a closer affiliation with meteoritic merrillite than with terrestrial whitlockite (Dowty 1977; Jolliff et al. 1993; Rubin 1997). Lunar merrillite is invariably rich in rare-earth elements (REE) because of the lack of structural H, which sets it apart from terrestrial whitlockite (Dowty 1977). In fact, RE-merrillite is the main mineral host for the REE, containing up to 10% or more RE_2O_3, including Y (Jolliff et al. 1993). In crystalline rocks, the phosphates occur in late-stage mesostasis, commonly with K-rich glass, K-feldspar, ilmenite, zircon, fayalite, and cristobalite. The phosphate mineral monazite $[(La,Ce,Nd)PO_4]$, which is rich in light REEs (La, Ce, Nd), has been found in lunar rocks, but it is extremely rare and occurs in association with RE-merrillite (Lovering et al. 1974; Jolliff 1993). The Mg-Fe phosphate, farringtonite $[(Mg,Fe)_3(PO_4)_2]$, was tentatively identified by Dowty et al. (1974). Phosphate hydrates, which are extremely diverse and numerous on Earth, have not been found in lunar samples, consistent with the lack of other water or hydroxylated minerals; the halogen site of apatite contains essentially stoichiometric F and Cl.

4.4. Iron metal

Native iron metal occurs commonly in lunar rocks and regolith. In pristine igneous rocks, Fe metal forms in trace to minor amounts by indigenous igneous processes (Reid et al. 1970; Goldstein et al. 1974; Ryder et al. 1980; Warren et al. 1987). Such metal typically exists as grains with dimensions of 1–100 μm. Greater concentrations of Fe metal occur in lunar regolith, where it forms by reduction of Fe^{2+} and by addition of metal from impacting meteoroids. Iron contents and sources in lunar materials are summarized in Papike et al. (1991) and Papike et al. (1998). In this section, we focus on metal found in regolith.

Metal grains of meteoritic composition are found in virtually all lunar breccias and throughout the regolith. Because the Moon has no atmosphere, most metal of meteoritic origin in the regolith has either (1) melted and re-solidified as part of an impact-melt breccia or agglutinate or (2) vaporized and been deposited on the surface of regolith grains. Some types of impact-melt breccias contain as much as 2% FeNi metal, on average (Korotev 1994). Metal grains up to several millimeters in size that have solidified from melts of lunar silicates are found in the regolith (Korotev 1987a, 1991, 1997; Korotev and Jolliff 2000). Iron metal also occurs in the regolith from reduction of Fe^{2+} in silicates and oxides during agglutinate formation and vaporization and deposition. A more detailed discussion of metallic iron in the lunar regolith is presented in Section 6.2.1.

4.5. Importance of the ratio of magnesium to iron

When lunar magmas cooled and solidified, the first-formed mafic (Mg- and Fe-bearing) minerals to crystallize, olivine or pyroxene, had a greater Mg/Fe ratio than the magma. By the time the last-formed olivine and pyroxene crystallized, magmas were depleted in Mg, so the late-formed minerals have lower Mg/Fe ratio than the original magma. The first-formed crystals are said to be more *magnesian* (greater Mg/Fe) than the original liquid, and the last-formed crystals are more *ferroan* (lesser Mg/Fe). Similarly, the first-formed plagioclase has a greater Ca/(Na+K) ratio than the original magma and as crystallization proceeds the Ca/(Na+K) ratio of the crystallizing plagioclase decreases. In other words, early formed plagioclase is at the anorthitic end of the range and later formed plagioclase is at the more albitic end. There are important exceptions to these generalizations, but rocks with high Mg/Fe and high Ca/Na are usually older (more primitive) than geologically related rocks that are more ferroan and more albitic. This simple concept is the heart of why the Mg/Fe ratio of lunar rocks is one of the most fundamentally important petrologic parameters.

For mafic minerals such as olivine and pyroxene, the molar or atom ratio of Mg/(Mg+Fe) is a more useful quantity than the simple mass or molar ratio of Mg/Fe or MgO/FeO. To state that olivine has a composition of Fo_{81} is a shorthand way of saying that the forsterite (Mg_2SiO_4) to fayalite (Fe_2SiO_4) ratio of the olivine solid solution series is 81:19 or that the mole percent of Mg/(Mg+Fe) is 81. When discussing the Mg/Fe ratio of rocks it is similarly convenient to use the "magnesium number," which is simply the whole-rock bulk mole percent MgO/(MgO+FeO). Here, we symbolize the whole rock magnesium number as *Mg´* but the quantity is sometimes symbolized as "Mg#." For lunar rocks, if the abundance of ilmenite and metallic iron is low, *Mg´* approximates the mass-weighted average Fo (forsterite) and En (enstatite) proportions of the olivine and pyroxene.

5. LUNAR ROCKS

In this section, we discuss aspects of the major rock types that are important to understanding the Moon's surface mineralogy and geochemistry. In the following paragraphs, each of the major rock types is described briefly. Additional descriptions and details of petrogenesis are in Chapters 3 and 4.

As on Earth, lunar rocks come in igneous, metamorphic, and sedimentary varieties, although the metamorphic and sedimentary varieties are almost entirely impact-related. Mare basalts are the most common igneous rocks in the sample collection. In the more ancient highlands, unbrecciated igneous rocks are rare, at least among the samples, because the original lunar crust has been thoroughly reworked by meteoroid impacts. Much effort has gone into seeking, within breccias and the regolith, fragments of igneous and plutonic rocks of the original lunar crust that are largely unscathed with respect to composition, texture, and mineralogy. Such rocks are commonly called *pristine* rocks (Warren and Wasson 1977).

5.1. Pristine nonmare rocks

Pristine rocks are identified by a combination of texture and composition (Warren and Wasson 1977, 1978, 1979). Distinguishing characteristics include coarse-grained plutonic textures, absence of petrographic or chemical evidence that the rock is a mixture of more than one rock type, and absence of contamination by siderophile elements of meteoritic origin. Warren (1993) compiled a list of 260 "possibly pristine" rocks, of which 89 were confidently pristine. Less than half of these were over 1 gram in mass, and many were small clasts from breccias.

Until recently, petrologists and geochemists commonly distinguished four suites of pristine nonmare igneous rocks: (1) ferroan anorthosite (FAN) or ferroan-anorthositic-suite (FAS), (2) magnesian suite (aka. magnesium suite, Mg-suite, Mg-rich), (3) alkali suite (alkalic suite), and (4) KREEP basalt, including possibly related rocks such as felsite (granite) and quartz monzogabbro. More recently, (2)-(4) have been shown to be related (discussed at length in chapters 3 and 4). Pristine rocks are rarely found with preserved, original plutonic-rock texture (coarsely crystalline); most occur as monomict breccias, i.e., breccias or aggregates of rock clasts, all of the same original rock type, or as cataclastic rocks with original textures thoroughly disrupted by impact processes. Volumetrically, among the Apollo and Luna samples, alkali-suite pristine rocks and KREEP rocks are far less abundant than are FAS and Mg-suite rocks. In early lunar literature the acronym *ANT* was used to refer collectively to anorthosites, norites, and troctolites (e.g., Keil et al. 1972), but subsequent work has shown that the *A* is not closely related to the *N* and *T*. Lunar anorthosites tend to be ferroan, that is, they have low Mg/Fe ratios, whereas the norites and, particularly, troctolites are magnesian – they have high Mg/Fe ratios. More importantly, mineral compositions differ. The ferroan-anorthosite suite is distinct from the magnesian and alkali suites on plots of *Mg´* of mafic silicates against *AN* [mol% Ca/(Ca+Na)] in plagioclase (Fig. 2.4). Other mineralogical differences occur (Papike et al. 1998), and some trace element ratios also tend differ (Norman and Ryder 1980).

5.1.1. Ferroan-anorthositic suite.
5.1.1. Ferroan-anorthositic suite. Two early surprises in the study of Apollo samples were that (1) anorthosite was common in the lunar highlands and (2) lunar anorthosite is highly anorthitic. Plagioclase in rocks from the lunar highlands is typically An_{96} in composition. This is a much higher anorthite content than that typically found in plagioclase in terrestrial rocks and ultimately reflects the Moon's depletion in volatile elements like sodium. For example, the composition of plagioclase in terrestrial massif anorthosites is typically in the An_{35} to An_{65} range. A third surprise was that by terrestrial standards, the *Mg´* of pyroxene and olivine in lunar anorthosite was much more ferroan than in terrestrial rocks of such high Ca/Na ratio and any other nonmare lunar rocks, such as impact-melt breccias and troctolites (range: En_{44-76}, mean: $En_{62\pm3}$; Fig. 2.4). For these reasons, since the time of Dowty et al. (1974), lunar anorthosite with a plutonic or relic plutonic textures has been called *ferroan anorthosite*.

The term ferroan-anorthositic suite or *FAS* (Warren 1993) encompasses ferroan anorthosites (>90% plagioclase) as well as their more mafic but less common variants, ferroan noritic anorthosite and ferroan anorthositic norite (Fig. 2.5). Some ferroan anorthosites contain olivine, but pyroxene usually predominates. Lunar ferroan anorthosites formed as coarse-grained igneous rocks. The coarse grain size indicates that they are intrusive rocks, formed during slow cooling at some depth below the surface. The high concentration of plagioclase feldspar suggests that they are also cumulate rocks, produced by the separation and accumulation of just-formed crystals (in this case, plagioclase feldspar) from the remaining melt. Although the vast majority of ferroan anorthosites are severely brecciated, the few samples that show vestiges of their former igneous texture tend to be coarse-grained, with large subhedral to euhedral plagioclase crystals surrounded by smaller anhedral pyroxene or olivine. Geochemically, ferroan anorthosites are distinct in having very low concentrations of both FeO and incompatible trace elements, as exemplified by Th, compared to other lunar rocks (Fig. 2.5).

Ferroan anorthosite is a fascinating rock type in that several large Apollo rocks are nearly monomineralic plagioclase (e.g., Warren 1990) and outcrops large enough to be seen telescopically from Earth occur (Hawke et al. 2003). However, its perceived importance to lunar crustal formation may have been enhanced somewhat by historical accidents. One large sample dubbed the "Genesis Rock" (sample 15415, e.g., Ryder 1985) was found at the Apollo 15 site, and ferroan anorthosite is by far the most common type of pristine rock at the Apollo 16 site, although it is uncommon to rare at other sites. Because (1) the Apollo 16 mission was the only Apollo mission to sample the feldspathic highlands distant from a mare basin, (2) highly feldspathic (typically >98% plagioclase) ferroan anorthosite was common at the site, and (3) the first feldspathic lunar meteorite was not discovered until 10 years after the Apollo 16 mission, the notion persists that the lunar highlands, in general, are dominated by ferroan anorthosite and

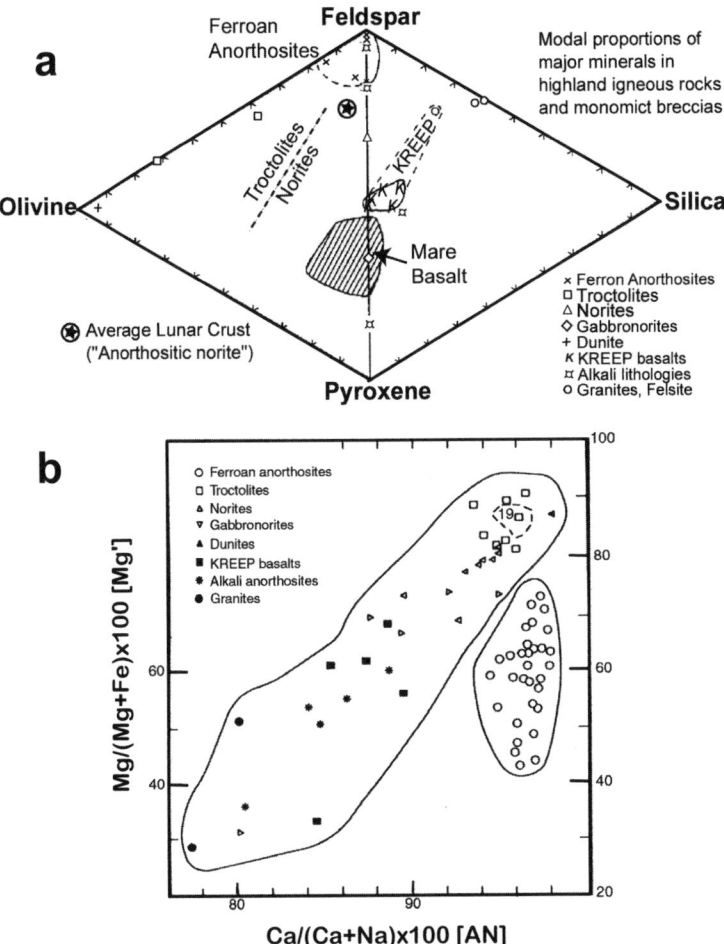

Figure 2.4. (a) Modal (volume) proportions of four lunar minerals in highland pristine igneous rocks and monomict breccias. The KREEP basalts, in general, have higher abundances of feldspar and silica minerals. Part (a) modified after Taylor et al. (1991). (b) Mineral compositions of highland pristine rocks plotted in terms of orthopyroxene Mg/(Mg+Fe) versus plagioclase anorthite content [AN]. The envelope to the right at high AN is the Ferroan Anorthosite (FAN) field. The larger field is for the magnesian-suite rocks.

that the FAS component of the crust is highly feldspathic. However, evidence and arguments based on the feldspathic lunar meteorites, orbital geochemistry, as well as regolith samples from Apollo 16 and Luna 20 suggest a different view. Highly feldspathic ferroan anorthosite is not necessarily typical of the highlands surface, at least, not on the Moon's near side. Most feldspathic lunar meteorites, which are all but devoid of mare volcanic material, are more mafic than the feldspathic material of the Apollo 16 regolith (Korotev 1996, 1997; Korotev et al. 2003). Also, a significant fraction of the plagioclase in the lunar highlands must derive from magnesian feldspathic rocks in order to account for the high *Mg'* of the upper feldspathic crust (70 ± 3), which at the surface is at the high end of the range for FAS rocks (Fig. 2.4; Section 7.1).

5.1.2. Magnesian suite. The precise compositional range of the magnesian suite of pristine nonmare rocks is not well defined (Figs. 2.4, 2.6). Rocks identified as magnesian-suite rocks

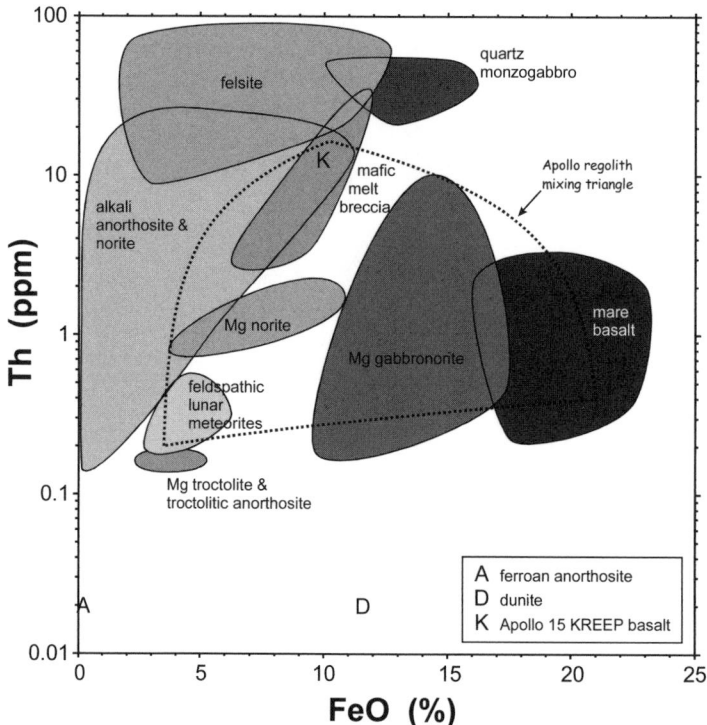

Figure 2.5. Lunar rocks in FeO-Th space. The dotted "triangle" is the Apollo regolith mixing triangle of Figure 2.2, distorted by the logarithmic Th scale. Nearly all regolith compositions from the Apollo and Luna missions plot within the triangle, which is defined by the fields for the feldspathic lunar meteorites, mare basalt, and mafic, KREEP-bearing impact-melt breccias of the Apollo missions. Apollo 15 KREEP basalt (K) plots in the field for the mafic melt breccias and is compositionally similar. Norites and gabbronorites of the magnesian suite mostly plot within the regolith mixing triangle, although they are not mixtures of the rocks represented by the triangle apices. Felsite ("granite") and quartz monzogabbro (alias, quartz monzodiorite) have the greatest concentrations of Th among lunar rocks. Alkali-suite rocks vary greatly in composition. The "A" point represents highly feldspathic ferroan anorthosite sample 15418. Other ferroan anorthosites would plot between "A" and the field for feldspathic lunar meteorites. The "D" point represents the Apollo 17 dunite, sample 72415–7.

are usually substantially more mafic than FAS rocks, although a few plagioclase-rich rocks (magnesian anorthosite) are included. Most samples are norites ($Mg' = 78 \pm 8$), troctolites ($Mg' = 87 \pm 5$, including anorthositic troctolite and troctolitic anorthosite), or ultramafic rocks ($Mg' = 88 \pm 2$). There are also some gabbros and gabbronorites, i.e., rocks with a significant proportion of Ca-rich clinopyroxene. Some of the gabbronorites are not actually magnesian in that they have moderate to low Mg' (Fig. 2.4). As originally applied, the term "Mg-rich" rocks covered all coarse-grained nonmare rocks that could be distinguished from the ferroan anorthosites by such chemical indicators as higher Mg' or lower AN contents (Warner et al. 1976a). Subsequently, many petrologists have come to suspect that there are fundamental genetic distinctions and that different magnesian-suite rocks were produced by separate and dissimilar magmas (e.g., James and Flohr 1983).

Most magnesian-suite rocks were collected at the Apollo 15 and 17 sites. Geochemically, these rocks (norites, troctolites, dunite) mostly appear to have crystallized from magmas rich in incompatible elements. Magnesian-suite rocks among the Apollo 16 collection are restricted

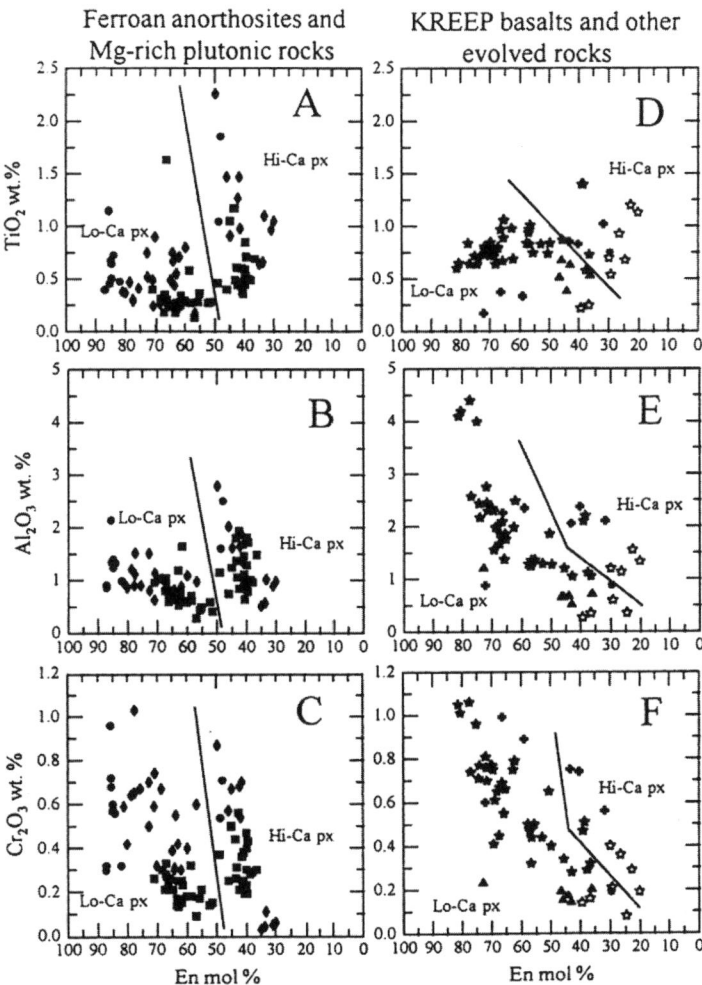

Figure 2.6. Chemistry of pyroxenes in pristine igneous highland rocks. Symbols: = ferroan anorthosites; = magnesian plutonic olivine-poor; = magnesian plutonic olivine-rich; = A15 KREEP Basalts; = quartz monzodiorites. Figure after Papike et al. (1998).

to a few troctolites and gabbronorites. Mafic, magnesian norites and troctolites are also rare to absent in the feldspathic lunar meteorites. These observations all suggest that many to most of the mafic, magnesian-suite rocks of the Apollo collection are not characteristic of the highlands globally, but instead may derive mainly from parts of the crust where incompatible-element rich materials are inferred to occur on the basis of surface compositions, primarily the Procellarum KREEP Terrane (Korotev 2000; Korotev et al. 2003; see also Chapter 3 and Plate 3.10). Magnesian, but not pristine, rocks occur in the feldspathic highlands (e.g., granulitic breccias, Section 5.2.4), but most are feldspathic and their relationship to the mafic magnesian-suite rocks of the Apollo collection has not been established.

5.1.3. Alkali Suite, KREEP basalt, felsite, and quartz monzogabbro. Alkali-suite rocks range in composition from anorthosite to norite and gabbronorite. They are distinguished by plagioclase that is more sodic ($An_{82 \pm 8}$) than that of ferroan anorthosite (An_{96}), e.g., Na concentrations are about a factor of 4 greater. Alkali-suite rocks are also tend to be enriched in incompatible elements. Most alkali-suite rocks occur in the Apollo 12 and 14 collection (e.g., Warren and Wasson 1980, 1981; Warren 1993), suggesting that they are petrogenetically related to KREEP (Shervais and McGee 1998). Because alkali-suite rocks are all but nonexistent in the feldspathic lunar meteorites, they are almost certainly a nonmare but also non-highlands lithology. The petrogenesis of alkali-suite rocks is discussed in more detail in Chapters 3 and 4.

Few rocks of KREEP composition occur in pristine form. Nearly all pristine KREEP rocks are basalts from Apollo 15, and the term KREEP basalt usually refers to these rocks (e.g., samples 15382 and 15386). Some workers have advocated that KREEP basalt is crystallized impact melt whereas others regard it as an extruded magma (Taylor 1982; McKay et al. 1986; Dymek 1986; Spudis and Hawke 1986). The Apennine Bench Formation in Mare Imbrium is a candidate site of origin for the Apollo 15 KREEP basalts (Spudis and Ryder 1985; Spudis and Hawke 1986). The main arguments against the impact-melt hypothesis is that KREEP basalt has low concentrations of siderophile elements and essentially no clasts compared impact-melt breccias. However, most rocks of KREEP composition are impact-melt breccias, and the possibility that KREEP basalts are ponded impact melt that extruded onto the surface has not been ruled out. Virtually all KREEP impact-melt breccias in the Apollo collection were ejected from a basin by the impact that formed them whereas KREEP basalts likely formed in a basin. Ejected melt is more likely to be clast rich than ponded melt and iron metal containing siderophile elements may have settled from ponded impact melt (e.g., Delano and Ringwood 1978). The origin of KREEP basalt is an important unanswered question in lunar science.

KREEP basalts typically contain 40–50% plagioclase, 30–40% low-Ca pyroxene, and minor cristobalite, ilmenite, apatite, merrillite, and Si- and K-rich glass (Spudis and Hawke 1986). Concentrations of highly incompatible elements in Apollo 15 KREEP basalts are typically 90 (Lu) to 310 (Th, U) times those of ordinary chondrites. The bulk $Mg´$ is 60 ± 6. KREEP basalt presents a petrogenetic enigma: the $Mg´$ is much too primitive (high) for such an evolved rock (high concentrations of silica and incompatible elements). The enigma is discussed in more detail in Chapter 4 and in the next section on breccias.

Some rare rocks in the Apollo collection are probably related to KREEP in that they appear have differentiated from magmas of KREEP composition. These include felsite (granite), quartz monzogabbro (quartz monzodiorite), and other rocks that occur as clasts in Apollo 14 breccias (Neal et al. 1989; Jolliff 1991; Marvin et al. 1991; Snyder et al. 1992, 1994; Jolliff et al. 1993).

5.2. Breccias

Lunar breccias are rocks composed of materials from older rocks that were disaggregated or melted by meteoroid impacts. Those materials can exist as (a) mineral and lithic (rock) fragments, (b) crystallized impact melt, or (c) glassy impact melt. Large fragments are called *clasts*; the material binding the clasts is called *matrix*. Breccias are lithified (converted to a solid rock) by the heat and shock associated with meteoroid impacts. Most breccias are *polymict*—they contain fragments from many different older rocks. Some are *dimict*—they contain material from only two sources. *Monomict* breccias are rocks that are re-lithified fragments from a single precursor igneous rock. Idealized locations showing where different kinds of breccia would occur in an impact structure are shown in Figure 2.7.

At the time of the Apollo missions, little work had been done on terrestrial impact craters and no generally accepted classification system existed for impact-produced rocks. As a result, the nomenclature for lunar breccias developed and evolved during the Apollo missions on a piecemeal basis. Terms used in most old, and some new, literature is confusing and duplicative. For example, a common breccia that occurs at the Apollo 17 site has been identified by terms emphasizing, variously, mineralogy (*noritic* breccia; Rhodes et al. 1974),

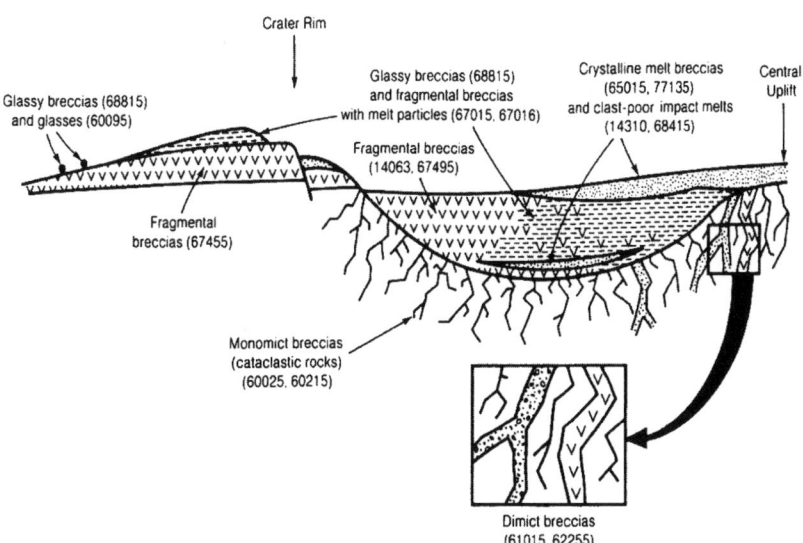

Figure 2.7. Cross-section of an ideal lunar crater showing the relations of different polymict breccia types and possible monomict breccias to the geological environment of the crater (adapted from Stöffler 1981). Numbers indicate actual lunar samples that are examples of the different breccia types. Many lunar polymict breccias may be the result of a series of impacts, such that the target for each new breccia-producing impact is already a polymict breccia.

a chemical composition that is rich, but not too rich, in incompatible elements (*low-K* KREEP; Laul and Papike 1980), texture (*poikilitic breccia*; Spudis and Ryder 1981), implied components (*KREEP-bearing* breccia; Korotev et al. 2003), geographic location of the type specimen (low-K *Fra Mauro*; Vaniman and Papike 1980), and mode of formation (*impact melt* breccia; Stöffler et al. 1980). Such inconsistency provides a hurdle to the newcomer and still confuses the old-timers. Stöffler et al. (1980) produced a consistent, unified classification and nomenclature for lunar highland rocks, including breccias. This is the classification that is followed by most subsequent workers and is given in Table 2.3.

5.2.1. Glassy melt breccias and impact glasses. Glassy materials produced by meteoroid impacts vary widely in texture, depending on the amount of glass present. These materials range from extremely clast-rich breccias with glassy matrices (glassy melt breccias) to glass-rich bodies that are clast-poor or clast-free (impact glasses). Shapes of samples of glassy melt breccias are irregular. Many glassy breccias are ovoid, some are ropy or hollow, and others occur as splashes on rocks (Fruland et al. 1977; Morris et al. 1983; Borchardt et al. 1986; Wentworth et al. 1994). Glassy melt breccias are polymict; they show a wide range of fragment types and there may be more than one generation of glass involved in their formation and lithification. Impact glasses appear to have formed and cooled in single impacts. A distinctive feature of most impact glasses is the presence of a free melt surface, in contrast with most breccias, including glassy melt breccias, which have mechanically broken surfaces. Small impacts lead to glassy melt breccias, and most glassy breccias are younger in solidification age than crystalline melt breccias (Borchardt et al. 1986).

5.2.2. Crystalline melt breccias and impact-melt breccias. A large impact produces a large volume of melt that cools more slowly than melt formed in a small crater. Thus, large impacts produce crystalline melt breccias. Crystalline melt breccias can be formed from impact melt pooled in craters as small as a few kilometers in diameter (Deutsch and

Table 2.3. Classification, characteristics, and examples of nonmare polymict breccias (adapted from Stöffler et al. 1980)

Class	Main Characteristics	Examples
Fragmental Breccia	Angular clasts in porous, clastic matrix of rock, mineral, and rare glass debris. Some melt clasts may be cogenetic with assembly (suevite). Most are friable.	14063 67015 67455
Glassy Melt Breccia	Coherent glassy or devitrified glass matrix with clasts (melt breccia) or without clasts (impact glass).	60115 68815 60095 79175
Crystalline Melt Breccia (Impact-Melt Breccia)	Rock or mineral clasts or both in an igneous (extrusive)-textured matrix (ophitic, subophitic, poikilitic, dendritic, etc.) May be fine-grained or coarse-grained, and clast-poor to clast-rich.	15455 62295 76015
Clast-Poor Impact Melt	Igneous (extrusive)-textured rock containing meteoritic siderophile-element contamination. Textures can be slightly more heterogeneous than igneous rocks, and rare clasts may be present. Compositions can be unlike those generated by igneous processes.	14310 68415
Granulitic Breccia and Granulite	Rock and/or mineral clasts in an equilibrated granoblastic to poikiloblastic matrix. Clasts may not be obvious, and poikilitic textures may mimic those of, or be transitional to, plutonic rocks. Compositions reflect siderophile-element contamination.	79215 77017 67955 78155
Dimict Breccia	Veined texture of intrusive, dark, fine-grained crystalline melt breccia with coarser-grained light-colored breccia consisting of plutonic or metamorphic fragments or both. In some cases, the dark-light relationship appears mutually intrusive.	61015 62255 64475
Regolith (Soil) Breccia	Lithified regolith. Regolith fragments including impact glass and volcanic glass and volcanic debris with a glassy matrix. Commonly retains some solar wind gases through the lithification process. See also the lunar meteorite regolith breccias listed in Table 2.1.	14318 15299 60019

Stöffler 1987). Crystalline impact-melt breccias resemble rocks crystallized from a magma, with clast proportions ranging from a few percent to more than half the rock. As the amount of clastic debris increases, the matrix tends to be more glassy, thus crystalline melt breccias grade into the glassy melt breccia group. Many of the clasts show evidence of shock metamorphism or resorption by the melt. Melt breccias in the Apollo collection typically have high concentrations of siderophile elements resulting from contamination by the impacting meteoroid. The siderophile elements are carried in FeNi metal from the impactor, which has melted and re-crystallized along with the silicate and oxide minerals. Vesicles are abundant in some samples, and there is evidence that vesicle abundance correlates with clast load, i.e., vesicles are more likely to be frozen into melt that cooled rapidly as a result of a high clast/melt (Korotev 1994).

Crystalline melt breccias from the Apollo missions are composed of nonmare material. Crystalline impact-melt breccias composed mainly of mare material are not known; most mare breccias are glassy. Crystalline impact-melt breccias composed mainly of mare material likely exist on the Moon, but for impacts into the maria large enough to produce a significant volume of crystalline melt, the zone of impact melting lies mainly beneath the surficial layer of mare basalt. Crystalline melt breccias are common at all the Apollo highland sites and they also occur in the regolith of the mare-basalt-dominated Apollo 11 and 12 sites (e.g., Simon et al. 1983; Simon and Papike 1985).

The chemical compositions of crystalline melt breccias span nearly the entire range of compositions observed in nonmare samples, from 2% to 11% FeO (33–14% Al_2O_3; Fig. 2.2). Breccias at the Fe-poor end of the range were clearly formed by impacts into the feldspathic highlands. It is the melt breccias on the mafic (high-Fe) end of the trend that are enigmatic. Virtually all mafic impact-melt breccias are rich in incompatible elements and those from Apollo 12 and 14 are essentially identified with KREEP. Early studies reasonably concluded that (1) given that mafic impact-melt breccias are more mafic and richer in KREEP than the surface of the feldspathic crust (FHT of Fig. 2.2) and (2) the breccias are found in basin ejecta deposits, the crust must become more mafic with depth and KREEP is concentrated beneath the feldspathic crust (Ryder and Wood 1977). However, one of the most significant new views provided by Lunar Prospector is that KREEP is concentrated at the surface in and only in the Procellarum KREEP Terrane and that, except perhaps for South Pole-Aitken, basins distant from the Procellarum KREEP Terrane show no hint of having exhumed KREEP. These observations lead to an alternate hypothesis, that all mafic, KREEP-rich impact-melt breccias were formed by impacts into the Procellarum KREEP Terrane, most likely the impacts forming the Imbrium and, perhaps, Serenitatis basins (Haskin et al. 1998).

To a first approximation, the composition of all of the mafic impact-melt breccias of the Apollo collection can be modeled as mixtures of (1) a KREEP-norite component similar to KREEP basalt of the Apollo 15 site, (2) an olivine or dunite (Fo_{90}) component (possibly a troctolite), and (3) material of the feldspathic highlands (Fig. 2.8). The KREEP component is the dominant component of the mixture (40–95%). Proportions of the three component vary among mafic melt breccias found at a given site and among different Apollo sites. The feldspathic component of the breccias (4–46%) is mainly clastic and is more abundant in breccias from sites distant from the Procellarum KREEP Terrane (Apollo 16 and 17; Fig. 2.8). Melt breccias with greater concentrations of incompatible elements tend to be less diluted with the olivine and feldspathic components. Overall, the compositional variation among mafic impact-melt breccias is consistent with mixing between the Procellarum KREEP Terrane and Feldspathic Highlands Terrane.

The mafic impact-melt breccias of the Apollo missions tell us that the melt zone of the impactor(s) that formed them was dominated by KREEP, in some form, but that there was also a substantial amount of non-uniformly distributed, highly magnesian olivine in the target area (0–30%; Fig. 2.8). Mass balance allows that the actual KREEP component of the breccias can have a more primitive composition, one corresponding to Apollo 15 KREEP basalt minus a normative component of Fo_{90} olivine (Fig. 2.8). This observation may account in part for the KREEP enigma (Section 3.4.2)—perhaps all KREEP sampled by the Apollo missions is an impact mixture of a low-Mg' magma ocean residuum, the urKREEP of Warren and Wasson (1979), and a mantle-derived, olivine-rich component (Korotev 2000).

Most of the crystallization ages of about 3.9 Gy that have been obtained on lunar rocks have been obtained on mafic, KREEP-rich impact-melt breccias. Compared to crystalline breccias formed from feldspathic highlands material, mafic breccias have high concentrations of all of the radioactive elements used in geochronology. Among the mafic breccias discussed here, only the group-A breccias of Apollo 15 (on Fig. 2.8, represented by the point lying closest to the KREEP apex) may to have a substantially different age, 1–1.5 Gy [see discussion of Ryder (1985) for sample 15405].

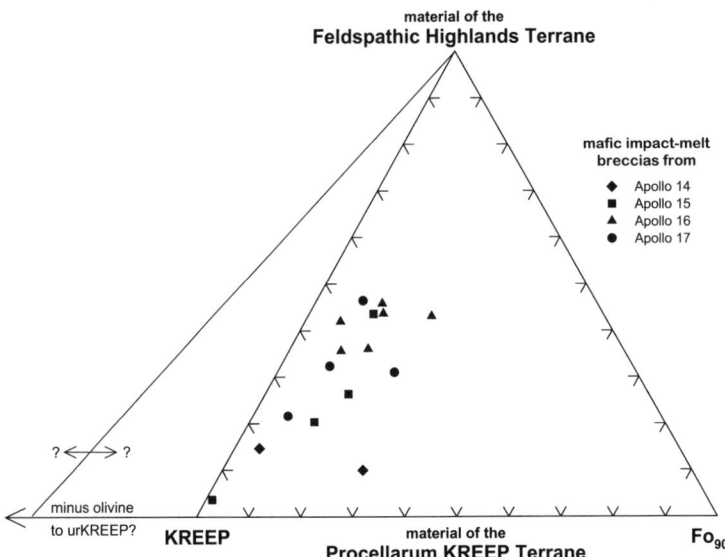

Figure 2.8. Mafic impact-melt breccias from the Apollo missions vary in composition, but compositional groupings occur (e.g., Apollo 15 type D), each represented by a point. The bulk composition of each breccia group corresponds to a mixture of a (1) KREEP component generally similar to Apollo 15 KREEP basalt, (2) a dunite or olivine component of Fo_{90} composition, and (3) feldspathic material of the highlands (Korotev 2000). Variation in the proportions of the three components accounts for the compositional variation among the breccias (e.g., Al/Fe, Mg/Fe, and Th/Fe ratios).

5.2.3. Clast-poor impact melts. Crystalline melt breccias that have so few clasts that they have textures similar to igneous rocks are *clast-poor impact melt breccias* or, in the total absence of clasts, *impact-melt rocks*. In many cases it is difficult to distinguish between such impact-produced rocks and pristine lunar igneous rocks, particularly on the basis of small thin sections typical of lunar samples. Much research and discussion has been directed to distinguishing between clast-poor impact melts and extrusive igneous rocks. High concentrations of siderophile elements in proportions characteristic of meteorites are typically used as a criterion of impact-melt origin. Another test is whether a rock has a bulk composition that is representative of a melt. On this basis, KREEP-like rocks from Apollo 14, which have more plagioclase than a corresponding low-pressure lunar lava (constrained to the pyroxene-plagioclase cotectic) have been classified as clast-poor or clast-free impact melts, as opposed to volcanic KREEP basalts (as found at Apollo 15).

5.2.4. Granulitic breccias and granulites. Granulitic breccias were formed during the alteration and recrystallization of older breccias that have been subjected to high temperatures (~1000°C) since they were formed. Under these conditions the original textures have been obliterated during recrystallization and replaced by an even-grained granulitic texture. A completely recrystallized rock that retains none of its original texture is a granulite. The recrystallization textures are characterized by rounded or *equant* polyhedral grain shapes with triple-junction grain boundaries common (i.e., granoblastic texture) or the presence of small bead-like crystals enclosed within larger crystals (poikiloblastic texture), and by an absence of typical volcanic textures (Bickel et al. 1977; James and Hammarstrom 1977; Warner et al. 1977; Stöffler et al. 1980; Cushing et al. 1999).

Granulitic breccias contain mineral or lithic clasts or relicts of clasts, whereas granulites do not. Granulites are less abundant than granulitic breccias. Both rock types commonly consist of a mosaic of grains whose boundaries meet at angles of about 120°, a texture produced by thorough recrystallization (Stewart 1975). Mineral compositions are also uniform and equilibrated. All samples are aluminous (25–29% Al_2O_3) with wide a range of Mg' (56–77; Lindstrom and Lindstrom 1986) and low abundances of the incompatible elements. All granulitic breccias and granulites are contaminated with siderophile elements of meteoritic origin, indicating that they were produced in meteoroid impact melts before they were recrystallized. Ferroan granulitic breccias have compositions consistent with derivation from igneous rocks of the ferroan-anorthositic suite. However, the composition of magnesian granulitic breccias (Mg' > 70) are not consistent with any mixture of known pristine rocks (Korotev and Jolliff 2001; Korotev et al. 2003). The magnesian precursor of magnesian granulitic breccias appears to be a feldspathic lithology largely unrecognized in pristine form in the Apollo collection. Most granulitic breccias were found in the Apollo 16 and 17 collection, but they are also common as clasts in the feldspathic lunar meteorites (Korotev et al. 2003b).

5.2.5. Dimict breccias. Dimict breccias are relatively rare among the lunar samples. They consist of two distinct lithologies combined into a single rock. In most examples, the two lithologies are mutually intrusive, i.e., neither one can be defined as the host rock. The samples generally resemble complex veins or dikes of dark and light components. The dark material is a fine-grained crystalline melt breccia or a nearly clast-free impact melt rock. The light-colored material is, in most cases, an anorthositic breccia whose main constituents are crushed and shattered (cataclastic) anorthosite fragments. Dimict breccias cannot be recognized in samples less than a few centimeters across, because the scale of mutual intrusion is on that order.

All of the few recognized dimict breccias are from the Apollo 16 site. (In an older classification as "black-and-white rocks," some similar Apollo 15 samples were included with them.) The dark crystalline melt breccia is similar in all samples. It has about 21% Al_2O_3, it is moderately enriched in incompatible elements (James et al. 1984; McKinley et al. 1984), and it has among the highest concentrations of meteoritic siderophile elements of any lunar rock type (Korotev 1994, 1997). The impact melt phase of the Apollo 16 dimict breccias is one of the least mafic of mafic impact-melt breccias of Figure 2.8. The white portion of the breccias varies from nearly pure cataclastic ferroan anorthosite to less feldspathic breccias (but nearly all containing more than 30% Al_2O_3; James et al. 1984; McKinley et al. 1984). Dimict breccias are believed to have been produced in the shattered rocks injected by and mixed with impact melt below an impact crater (Stöffler et al. 1980; James et al. 1984). In large impact events the melt could have been created and the older anorthositic breccia could have been shattered and remobilized at the same time.

5.2.6. Fragmental breccias. Fragmental breccias are composed of fragments of rocks, commonly including earlier-formed breccias, that have been lithified by shock compaction during a meteoroid impact. The matrix consists of finer-grained fragments of the same material; there is no chemical cementing agent. Brecciation and lithification occurs only by impact processes, thus the degree of coherency is a function of shock pressure, which in turn is a function of location within the impact target region and amount of intergranular melt produced, if any. Fragmental breccias can be composed of any rock' type, and some consist entirely of feldspathic highlands material and others consist mainly of mare material. Several of the lunar meteorites are fragmental breccias.

5.2.7. Regolith breccias. Regolith breccias are rocks composed of regolith that was lithified by shock compaction or heating. Regolith breccias differ from fragmental breccias in containing lithologies such as glass spherules and agglutinates that can only be produced or acquired at or above the lunar surface. The distinction between regolith breccias and fragmental breccias is not always clear because regolith lithologies may be rare or not present in every thin section of a regolith breccia. Another characteristic of regolith breccias is that they are rich in solar-wind implanted gases compared to other types of breccias because they contain a high proportion of fine-grained material (high surface to volume ratio) that was exposed at the lunar surface (Section 8). Some regolith breccia are glassy, but many are not. Regolith breccias range from friable clods to highly coherent rocks. For example, some Apollo regolith breccias largely disaggregated between collection on the Moon and arrival at the NASA curatorial facility in Houston (e.g., sample 73131; Korotev and Kremser 1992). In contrast, many of the lunar meteorites are regolith breccias, and these rocks survived being blasted off the Moon, the stress of passing through Earth's atmosphere, and a hard landing on Earth (Warren 1994, 2001). Vesicles in glassy regolith breccias are abundant and can result from escaping solar wind-volatiles; in fact, highly vesicular fusion crusts on regolith-breccia lunar meteorites likely form in this way. Regolith breccias have a wide range of compositions because they can be formed in any regolith.

Regolith breccias are important for at least three reasons. First, they are fossil regoliths that at some point became closed to further input of material. Consequently, some regolith breccias represent very ancient regoliths and thus provide information about conditions in the past (McKay et al. 1986). For example, the Apollo 16 regolith developed upon the ejecta deposit of the Imbrium basin. The ancient regolith breccias of Apollo 16 contain virtually no mare volcanic material whereas the present Apollo 16 regolith does (McKay et al. 1986; Korotev 1996). This difference implies that the mare component of the Apollo 16 regolith results from post-Imbrium redistribution of material. Second, some regolith breccias represent regoliths from places that are distant from where the breccias were collected. For example, the Apollo 12 and 14 sites are in the Procellarum KREEP Terrane and their regoliths are dominated by mafic, KREEP-rich impact-melt breccias and mare basalt (Apollo 12). Yet, a few feldspathic and KREEP-poor regolith breccias that surely formed in the Feldspathic Highlands Terrane have been found among samples collected at both sites (Simon et al. 1985; Jerde et al. 1990). These breccias were delivered to the Apollo 12 and 14 sites by impacts in the feldspathic highlands. The ultimate examples, of course, are the numerous lunar meteorites found on Earth that are regolith breccias from the Moon. Some regolith breccias are dissimilar in detail to any sample of the Apollo or Luna regolith fines (e.g., Jerde et al. 1987; Jolliff et al. 2003), and these breccias provide information about otherwise unsampled places on the Moon. Third, because a regolith breccia is a thoroughly polymict rock consisting of soil, its composition is more likely to represent the average composition of the surface of the area at which it formed than is any more primary rock type such as mare basalt or anorthosite. This aspect is particularly important for the lunar meteorites because, although we do not know where on the Moon any of them originate, we can be reasonably certain their compositions represent the regolith at their points of origin (e.g., Section 7).

5.3. Mare volcanics

The mare basaltic lavas formed by melting of the solid interior (mantle) of the Moon, probably at depths of 100-500 km, followed by the buoyant rise of molten rock through the mantle and eruption to the lunar surface as a result of overpressurization. Two types of volcanic materials have been erupted at the lunar surface:

(1) Basaltic lavas/flows. Lunar lava flows erupted from fissures in the lunar surface. Because lunar basalts contain more Fe and less Si and Al than terrestrial basalts and were much hotter because of a lack of water, the lavas possessed low viscosities and readily formed thin, widespread flows. The lavas cooled at different rates, dependent on the thickness of individual flows, to produce a variety of mineral textures. Many lava flows accumulated in thick stacks that partly filled many of the mare basins.

(2) Pyroclastic deposits. Gases, probably CO, contained within rising magmas, were explosively released when the magmas approached the surface of the Moon (e.g., Nicholis and Rutherford 2005). As lavas poured from a fissure vent, gases were released that effectively expelled a fountain of molten droplets. Such a "fire fountain" produces volcanic droplets,

subsequently found as glass beads of various compositions and colors (e.g., the well known Apollo 17 'orange' soil).

On the Moon, the lava fountains associated with eruptions of basalt formed small, sub-millimeter glassy beads. Deposits of these beads are widely dispersed around lunar volcanic vents because of low gravity and eruption into a vacuum. Such pyroclastic deposits are similar to the volcanic ash deposited around lava fountains on Earth. Two striking examples of lunar pyroclastic rocks are the orange soils (although they really are not soils, but rather volcanic ash deposits) from the Apollo 17 landing site and the green glass from the Apollo 15 landing site. Green-glass "clod" sample 15426 contains variable proportions of glass beads in different subsamples, but a thin section of green glass-rich sample 15427 regolith breccia consists of 90% green glass shards and spherules (Ryder 1985). The "orange soil" from Shorty crater – 74220 – is >90% glass beads with an average size of 44 μm. The "black soil" just below the orange soil contains the same proportion of glass beads, but they appear black because they cooled a bit slower and contain skeletal crystals of olivine and ilmenite. In the 74001/2 drill core, the deposit is >99% pyroclastic material. Orbital photographs of the lunar surface reveal dark-mantle deposits, many surrounding apparent volcanic vents, which have been interpreted as representing such volcanic ash and glass of pyroclastic origin.

Lunar basalts have been classified, to a first approximation, by their TiO_2 contents (Fig. 2.9), with additional divisions based upon Al_2O_3, and further subdivided by K_2O contents (Neal and Taylor 1992). Six of the nine missions to the Moon that retrieved samples returned mainly mare basalts and fragmented, consolidated basaltic breccias. A useful plot of the various basalt and volcanic glass compositional data is shown in Figure 2.10, which addresses the subdivisions of these extrusive materials within the context of elements that can be detected by remote sensing (Fe, Ti, K). On this plot, samples of basalt types have compositions that form clusters, making this a useful classification scheme.

5.3.1. Mare basalts.
Mare basalts are exposed over 17% of the lunar surface area and are thought to make up less than 1% of the lunar crustal volume (Head 1976, Head and Wilson 1992). They primarily fill multi-ringed basins and irregular depressions on the Earth-facing hemisphere of the Moon. On the lunar far side, limited patches of mare basalts occur in younger craters (e.g., Tsiolkovsky) and in some basins, such as Ingenii and Apollo (see Fig. 1.21, Chapter 1). This preponderance of mare basalts on the Earth-facing hemisphere of the Moon has been attributed to the thicker highland crust on the far side of the Moon and the gravitational attraction of the Earth (Kaula et al. 1972, 1974; Toksoz et al. 1974). More recently, Haskin et al. (2000) suggested the possibility that this distribution is broadly related to non-uniform distribution of heat-producing residua relating to early lunar differentiation and thermal evolution (see Chapter 4). The general style of volcanic activity was the eruption of large volumes of magma from relatively deep sources (not shallow crustal reservoirs) with high effusion rates (Head and Wilson 1992). The mare basalts and pyroclastic glasses vary considerably in terms of their compositions, textures, and distribution across the Moon. A wide range of variation in terms of TiO_2 concentrations is seen both in the samples and in remote sensing. A key question, to be addressed in a later section, is this: do the samples fairly represent the diversity of basalt and pyroclastic glasses on the Moon? In some ways, the diversity in the samples far exceeds what can be discerned with remote sensing, and certainly, the diversity observed among the samples provides a key to interpreting variations in the remotely sensed data. This is one of the fundamental lessons to be learned with regard to the exploration of Mars and other planetary objects.

Below, an overview is given of the major basalt types. Interested readers are referred to more thorough treatments in several review articles (e.g., Neal and Taylor 1992; Papike et al. 1998). A tabulation of some 500 chemical analyses of the basalts and volcanic glasses is given on the electronic archive associated with this book.

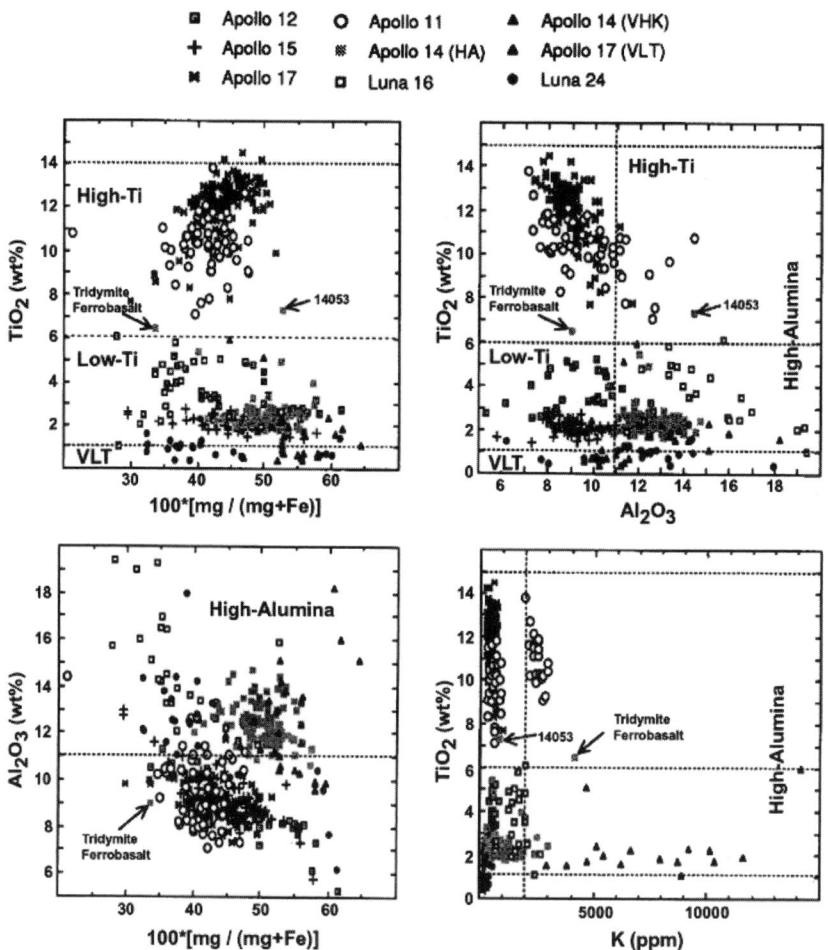

Figure 2.9. Plot of 500 basalt compositions illustrating the lunar classification (data and figures after Neal and Taylor 1992). Notice the distinctions between the different classes of basalt as based initially on TiO_2, secondarily on Al_2O_3, and lastly on K_2O.

High-Ti mare basalts. Basalts collected from the Apollo 11 and 17 sites are mainly high-Ti mare basalts (Figs. 2.8, 2.9). The Apollo 11 high-Ti basalts have been subdivided into two main groups, low-K and high-K. The Apollo 17 high-Ti basalts include one group that is much like Apollo 11 low-K basalts, as well as another high-Ti group, which is more abundant, that contains even more TiO_2 and more of the opaque mineral ilmenite (Table 2.4). Even within these groups, the compositions of high-Ti basalts vary significantly from sample to sample. Explanations for these differences are complicated by the fact that all mare basalt samples were collected as loose blocks in the regolith and no identifiable lava flows were sampled directly on the lunar surface. Nevertheless, studies of terrestrial lava flows suggest that the observed variations from sample to sample reflect variation both within and between individual lava flows.

Low-Ti mare basalts. A wide variety of low-Ti mare basalts were collected at the Apollo 12, Apollo 15, and Luna 16 landing sites on various maria, as well as at the Apollo 14 landing site (Figs. 2.9, 2.10). The Apollo 12 and 15 varieties each consist of two main basalt types,

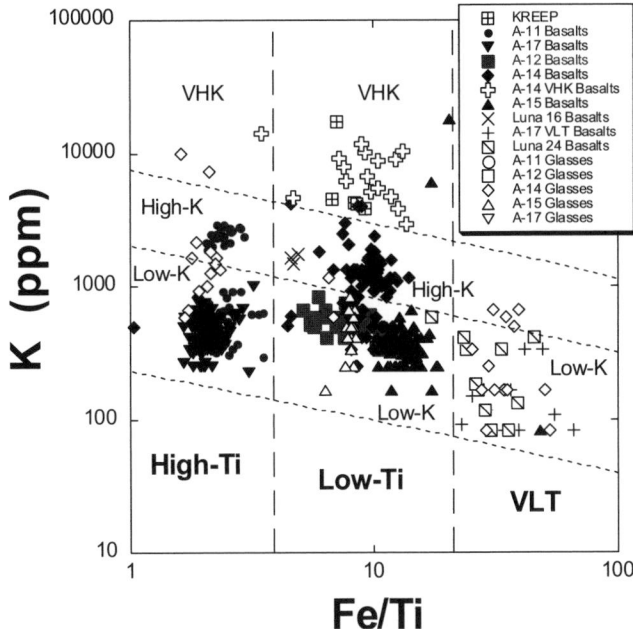

Figure 2.10. Plot of K vs. Fe/Ti shows the major types of basalts and volcanic glasses is terms of chemical components that can be determined remotely (database from Neal and Taylor 1992). A similar pattern exists for Fe/Ti vs. Th, but the database is much smaller.

olivine basalts and pigeonite (low-Ca pyroxene) basalts. In general, the olivine basalts have more MgO and less CaO, Al_2O_3, and TiO_2, than do the pigeonite basalts at the same site. Like the high-Ti mare basalts, none of these low-Ti basalts crystallize plagioclase feldspar early as they cool. Both types show clear evidence for the early separation of olivine or pigeonite, but not of ilmenite. The resulting trends of chemical composition show increasing TiO_2 and Al_2O_3 with decreasing MgO (Fig. 2.9). In addition to these basalts, some of the low-Ti basalts collected at the Apollo 12 site are noticeably richer in ilmenite than the others; these have been designated ilmenite basalts. In general, the REE abundances in low-Ti basalts are lower than in the high-Ti basalts, although both groups have the same bow-shaped pattern and both have negative Eu anomalies. Concentrations of other incompatible elements (i.e., those elements that tend to stay in the melt rather than in the early crystallizing minerals), such as Sr and U, are also lower in the low-Ti basalts than in the high-Ti basalts.

Aluminous, low-Ti mare basalts contain more Al_2O_3 than do those from the Apollo 12 and 15 sites (Fig. 2.9). Three major categories of aluminous mare basalts can be distinguished on the basis of their location and chemistry: those from the Luna 16 site and two types from the Apollo 14 site. Although the petrologic record has been obscured by the early catastrophic impact history of the Moon, a record of pre-mare volcanics retained in highland high-Al soils and breccias as clasts (e.g., Taylor et al. 1983; Dickinson et al. 1985, Neal et al. 1988). In addition, there are very-high-K, low-Ti basalts associated with the high-alumina basalts at the Apollo 14 site (Fig. 2.10). Although this 2000-12,000 ppm K_2O range is "very-high" for the Moon, it is not to be confused with values from terrestrial rocks many times higher. The Luna 16 mare basalts contain the most TiO_2 of any of the aluminous low-Ti basalts, averaging about 5 wt%. They also have distinctive REE patterns that are characterized by higher concentrations of the light REE (La through Sm) than the heavier REE. The Luna 16 basalts do, however, display the

Table 2.4. Summary of average modal data for the mare basalts
(modified after Papike and Vaniman 1978)

Basalt type	Opaques	Pyroxene	Feldspar	Olivine
A-17 high-Ti	24.4	47.7	23.4	4.6
A-11 high-K/high-Ti	20.6	57.5	21.7	0.1
A-17 low-K/high Ti	15.1	51.6	33.3	--
A-11 low-K/high Ti	14.6	50.9	32.2	2.3
A-12 ilmenite	9.3	61.1	25.9	3.6
A-12 pigeonite	9.1	68.4	21.1	1.4
A-12 olivine	7.1	53.5	19.2	20.2
A-15 olivine-normative	5.5	63.3	24.1	7.0
A-15 quartz-normative	3.7	62.5	33.8	--
Luna 16 high-Al	7.1	51.5	41.2	0.1
A-14 high-Al	9.0	37.1	43.9	10.0
A-14 VHK/high-Al	8.0	33.2	46.8	12.0
Luna 24 "VLT"	1.8	48.6	39.1	10.4
A-17 "VLT"	1.0	61.7	31.9	5.4

Apollo 14 data from Neal et al. (1988a, 1989). The opaque minerals do not transmit light as viewed with a microscope. The major opaque mineral in lunar basalts is ilmenite (up to 25 vol%), with lesser abundances of spinel phases (chromite, ulvöspinel), troilite, and native FeNi metal.

same typical bow-shaped pattern, with a negative Eu anomaly, that is characteristic of almost all mare basalts.

Very-low-Ti (VLT) mare basalts. VLT basalts were first discovered as tiny fragments in the Apollo 17 drill core (Vaniman and Papike 1977). The basalt particles collected by Luna 24 from Mare Crisium were also VLT mare basalts (Figs. 2.9, 2.10). Besides containing lower concentrations of TiO_2 (<1.5) than other mare basalts, these basalts are distinctive in having low concentrations of REE and other incompatible elements; they also have REE patterns that are not bow-shaped. Instead, the REE contents rise continuously from La to Lu. All but one fragment of the Luna 24 basalts have small negative Eu anomalies. A similar basaltic composition, the Apollo 15 green-glass pyroclastic material, has comparable chemical characteristics.

5.3.2. Pyroclastic deposits. Glass beads and fragments are a common component in the lunar soil; they are found at all the Apollo sites, in several of the Luna samples, and in lunar meteorites. The glasses have a broad range in both composition and texture. Lunar glasses have been produced by two very different processes, meteoroid impacts (impact melting) and volcanic eruptions. Glasses formed by impact melting can generally be recognized by their heterogeneous textures (e.g., swirls and schlieren, and incompletely melted clasts) and compositions. The volcanic glasses formed in a fundamentally different way, in gas-driven "fire-fountain" eruptions that leave deposits made up of glass droplets that have chilled from the spray of molten lava. The volcanic glasses have volatile coatings produced during fire-fountaining (Meyer et al. 1975). Except for Apollo 17 orange glass, which was most likely collected in place, i.e., where the glass droplets landed, no field or stratigraphic evidence exists to show that any other sample of lunar glass is volcanic, but investigators have developed other criteria for distinguishing between these two types of glasses. These criteria are discussed in detail by Delano (1986).

Evidence of lunar fire-fountaining is preserved not only as spherical glass beads, but also as dark mantling deposits (Wilhelms and McCauley 1971, Head 1974, 1976; Head and Wilson 1992). Remote-sensing data indicate that deposits of volcanic glasses are fairly abundant on the Moon. The dark-mantle deposits include the one sampled in the form of orange and black glass by the Apollo 17 mission. Such pyroclastic deposits often occur on the edges of the maria and overlap onto the adjacent highland regions. As their name implies, these deposits have low albedo (they are poor reflectors of visible light) and so appear dark. Gaddis et al. (1985) have shown that not all pyroclastic deposits are necessarily dark. Some thick dark-mantle deposits exhibit low intensities on depolarized 3.8-cm radar maps, indicating the absence of surface scatterers. Gaddis et al. (1985) suggest that numerous other areas are likely to contain pyroclastic deposits. Some of these other pyroclastic units may be as large as the observed dark mantle deposits. If the dark mantle deposits and similar units are indeed pyroclastic, then they should be composed of relatively homogeneous glasses with concentrations of volatile elements on their surfaces, similar to Apollo 15 green glass and the Apollo 17 orange glass.

In these dark-mantle deposits, there is abundant evidence of basaltic volcanism that pre-dated formation of the younger basins (i.e., pre-Nectarian). These have been identified through remote sensing as "cryptomaria" (e.g., Hawke et al. 1990; Head and Wilson 1992; Head et al. 1997; Antonenko et al. 1995). However, numerous studies imply that large volumes of these basaltic magmas ("cryptomaria") were erupted (Metzger and Parker 1979; Davis and Spudis 1985, 1987; Head and Wilson 1992; Antonenko et al. 1995). Head and Wilson (1992) have suggested that perhaps up to a third of the erupted basalts at the lunar surface were such cryptomaria, now buried by later-formed impact-ejecta deposits.

6. REGOLITH COMPOSITION

This section reviews aspects of regolith composition, as it is known from samples, that are relevant to interpreting data obtained remotely from the lunar surface. The discussion focuses on chemical elements for which the surface concentrations can be determined directly or indirectly from orbit.

6.1. Chemical composition data for regolith samples

Data on the chemical composition of lunar regolith are abundant. Most of these data are available in the annual *Proceedings of the Lunar [and Planetary] Science Conference* published from 1970 through 1992. Although compilations of averages are also available (McKay et al. 1991; Haskin and Warren 1991; Korotev and Gillis 2001 [Apollo 11]; Korotev 1996, 1997 [Apollo 16]; Korotev and Kremser 1992 [Apollo 17]), there are many problems and pitfalls associated with obtaining "the composition" of a lunar soil sample (and some of these problems are even worse for rocks). Problems include: (1) There is no Apollo or Luna soil sample for which data are available for all chemical elements. For many samples, no data are available for some important elements, e.g., MgO. (2) Most laboratories that have analyzed lunar samples have used only one or two analytical techniques and, consequently, have determined the concentrations of only some elements; some labs determined only a few. Some important element pairs (e.g., Ti and Fe, Fe and Mg, K and Th, Gd and Sm) were not always precisely measured together on a single subsample. (3) For any given element, some techniques are more precise and reliable than others and some laboratories consistently provided better data than others. Even for data from a given laboratory, some elements were measured more precisely than others. It is not always easy to identify the precise and imprecise data. (4) Some available data are systematically erroneous. (5) A common test of the veracity of data reported for the major elements is whether the sums-of-oxides is close to 100%. SiO_2 accounts for almost half the mass of any sample, thus SiO_2 data are essential for checking the oxide sum. However, SiO_2 data are often not available or are imprecise. (6) Analyzed subsample masses are highly variable and often not reported. (7) Finally, data for lunar samples suffer from sampling error, which acknowledges that no two subsamples of any given sample have exactly the same composition. Sampling error occurs because lunar samples are not homogenized prior to allocation, analyzed subsample masses are small, typically 10–500 mg for lunar soils, and any given subsample contains only a finite number of particles or mineral grains, some of which are large or have unusual compositions. Elements typically affected by sampling error include Ti (in ilmenite), Zr, Hf, Th, and U (in zircon), REEs (rare earth elements), Th, and U (in RE-merrillite), K, Rb, Cs, Ba, Th, U (in minerals associated with felsite), Na, Sr, and Eu (in alkali feldspar), and Fe, Ni, Co, Ir, Au, etc. (in metal of meteoritic origin). Sampling error is common in the lunar-sample literature and cannot be overlooked when compiling data.

For these various reasons, calculation of simple means from all available data can lead to systematic errors that are unacceptable for some purposes. For example, if in a given source of data the reported concentration of FeO or La is anomalously high because of sampling error (an excess of olivine or RE-merrillite), other elements will be affected because of the various mineral-controlled interelement correlations discussed in Section 6.2. If the report does not also include data for MgO or Sm, inclusion of the FeO or La data in the average will lead to systematic errors in calculated mean ratios of correlated elements (e.g., FeO/MgO or La/Sm).

Sparse data for a sample can lead to other types of errors as well. For example, the reported I_s/FeO (Section 9.2.2) value of sample 64501 is 61 (Morris 1976). The value is based on an FeO concentration of 5.2%, which is the mean of pre-1976 data from three sources. However, the value of one of those sources is highly anomalous. Assuming the anomalous datum is the result of analytical or computational error, an assumption confirmed by post-1976 data, the FeO concentration of 64501 is more nearly 4.45%, which leads to I_s/FeO of 71, not 61. This difference is a significant error associated with a erroneous data from a single laboratory.

With these various caveats in mind, we present selected averages of concentrations of a number chemical elements in surface and trench soils from the Apollo and Luna missions in Table 2.5. For the simple missions in which only one or two regolith samples were taken (e.g., Apollo 11 and all Luna missions), the values presented are based on all available data. For the more complex missions (the later Apollo missions), the compositional extremes are listed.

6.2. Mineral and rock control of regolith composition

6.2.1. FeO and metallic iron. Most analytical techniques that have been used to determine the bulk concentration of iron in lunar samples (mainly X-ray fluorescence and neutron activation) provide the total concentration of iron regardless of its oxidation state. By geochemical convention, this concentration (and that of other major elements) is usually expressed as the percent oxide (e.g., Table 2.5). Because essentially no Fe on the Moon exists as Fe^{3+}, iron concentrations in lunar samples are commonly reported as percent FeO. In lunar breccias and soil, however, particularly those from the feldspathic highlands, a significant fraction of the Fe exists as metal (Fe^0). Some studies have measured the concentration of metallic Fe in the regolith directly by magnetic techniques. These studies conclude that the lunar regolith contains, on average, 5.4 ± 1.8 mg/g of Fe metal ("~0.5 equivalent wt%" as Fe^0; Morris 1980). This mean and standard deviation corresponds to 0.69 ± 0.23% as FeO. Because most of the metallic iron in feldspathic soils derives from meteorites, a greater fraction of the total iron exists in the metallic state for feldspathic soils than for mare soils. For example, approximately 10% of the (total) iron in the feldspathic lunar meteorites, which represent typical surface material of the highlands (Section 7), is of extralunar origin (asteroidal meteorites; Korotev et al. 2003b) and most of this iron likely occurs as metal. For example, in mature Apollo 16 soil, which contains a high proportion of metal-rich, mafic impact-melt breccia, about 15% of the iron is metal derived from meteorites (Korotev 1997, 2000) and another 3% is nanophase iron (Morris 1980).

Although no quantitative data are available, it is likely that only a small fraction of the metallic iron in the lunar regolith is indigenous to igneous and plutonic rocks. Most metallic iron in the regolith derives from four other sources.

Table 2.5. Compositions of soils from the Apollo (A) and Luna (L) missions

mission	type	location or station no.	note	SiO$_2$ %	TiO$_2$ %	Al$_2$O$_3$ %	FeO %	MnO %	MgO %	CaO %	Na$_2$O %	K$_2$O %	Cr$_2$O$_3$ %	P$_2$O$_5$ %	S %	Σ %	Mg' %	Li ppm	Na ppm	P ppm
A11	only sample	lunar module	1	42.0	7.5	13.5	15.8	0.21	7.9	12.0	0.44	0.14	0.30	0.10	0.11	100.1	47	12.	3200	400
A12	typical	most of site	2	46.1	2.7	12.6	16.5	0.21	10.2	10.3	0.46	0.24	0.38	0.30	0.08	100.0	52	17.	3400	1300
A12	atypical	Head Crater	3	47.0	2.5	14.3	14.6	0.20	9.1	10.6	0.64	0.39	0.33	0.39	0.07	100.1	53	23.	4800	1700
A14	typical	most of site	4	47.7	1.7	17.4	10.5	0.14	9.4	10.9	0.70	0.52	0.20	0.49	0.10	99.8	61	27.	5200	2100
A14	atypical	Cone Crater	5	n.a.	1.6	16.8	10.0	0.13	9.8	10.6	0.81	0.64	0.20	0.51	0.08	–	64	n.a.	6000	2200
A15	highest Fe	9a, edge of rille	6	46.2	2.0	10.4	19.8	0.25	11.1	9.6	0.30	0.094	0.53	0.11	0.06	100.4	50	10.	2200	500
A15	highest Mg	7, Apennine Front	7	46.3	1.3	15.6	13.0	0.17	11.9	10.4	0.41	0.16	0.36	0.16	0.07	99.8	62	n.a.	3000	700
A15	highest Th	6, Apennine Front	8	46.7	1.5	16.4	12.2	0.16	10.4	11.2	0.46	0.21	0.34	0.22	0.08	99.8	60	9.	3400	1000
A15	lowest Fe	2, Apennine Front	9	46.5	1.3	17.6	11.6	0.16	10.4	11.5	0.42	0.18	0.31	0.16	0.07	100.1	62	10.	3100	700
A16	typical mature	all Cayley	10	44.9	0.59	26.7	5.44	0.07	6.0	15.3	0.46	0.121	0.111	0.12	0.07	99.9	66	8.	3400	540
A16	highest Fe	5 & 6	11	45.0	0.66	26.2	5.85	0.07	6.3	15.1	0.46	0.131	0.119	0.14	0.08	100.1	66	8.	3400	600
A16	lowest Fe	1	12	45.3	0.50	28.3	4.36	0.06	5.0	15.9	0.48	0.083	0.084	0.10	n.a.	100.1	67	7.	3600	420
A16	lowest Fe	4	13	45.3	0.47	28.0	4.45	0.06	5.4	16.4	0.46	0.107	0.086	0.11	0.05	100.8	68	7.	3400	480
A16	atypical	11, North Ray Crater	14	44.7	0.39	28.7	4.14	0.05	4.6	16.3	0.53	0.084	0.080	0.08	0.03	99.6	66	6.	3900	300
A17	orange glass	4, Shorty Crater	15	38.7	8.8	6.5	22.3	0.29	14.5	7.5	0.37	0.077	0.70	0.06	0.05	99.8	54	11.	2800	300
A17	highest-Fe	1 & 5, valley floor	16	39.9	9.6	10.9	17.7	0.24	9.5	10.7	0.38	0.078	0.46	0.07	0.12	99.8	49	9.	2800	300
A17	lowest-Fe	6, North Massif	17	43.5	3.3	18.2	10.7	0.15	10.8	12.2	0.40	0.116	0.28	0.09	0.08	99.8	64	10.	3000	400
A17	lowest-Fe	2A, South Massif	18	45.1	1.3	21.3	8.3	0.11	9.8	12.9	0.43	0.144	0.22	0.13	0.06	99.9	68	10.	3200	600
L16	only sample	lander	19	44.3	3.4	15.6	16.3	0.21	8.4	11.9	0.39	0.11	0.30	0.05	0.21	101.2	48	7.	2900	200
L20	only sample	lander	20	45.2	0.49	22.8	7.3	0.11	9.5	14.4	0.35	0.07	0.19	0.12		100.5	70	6.	2600	500
L24	only sample	lander	21	44.8	1.1	11.8	19.7	0.26	9.7	11.3	0.28	0.03	0.46	0.04	0.14	99.6	47	6.	2100	200

table continued on following page

Table 2.5. continued

mission	sample type or location	K ppm	Sc ppm	V ppm	Cr ppm	Mn ppm	Co ppm	Ni ppm	Zn ppm	Ga ppm	Rb ppm	Sr ppm	Y ppm	Zr ppm	Nb ppm	Cs ppm	Ba ppm	La ppm	Ce ppm	Pr ppm
A11	only	1100	63.	65.	2100	1600	29.	200	24.	4.5	2.8	163	115	300	19.	0.11	170	16.	47.	7.
A12	typical	2000	39.	110.	2600	1700	43.	200	7.	4.2	6.2	137	131	480	29.	0.30	390	34.	89.	12.
A12	atypical	3200	35.	100.	2300	1500	32.	110	7.	4.7	9.1	164	180	720	42.	0.41	560	49.	128.	17.
A14	typical	4300	22.	55.	1400	1100	35.	370	25.	6.0	15.	179	242	880	55.	0.65	800	67.	176.	24.
A14	atypical	5300	22.	45.	1300	1000	27.	250			20.	182	248	860	n.a.	0.70	910	69.	186.	25.
A15	highest Fe	800	38.	190.	3600	1900	50.	180	~6.	3.3	2.7	114	46	240	13.	0.22	150	11.	31.	4.
A15	highest Mg	1300	25.	100.	2500	1300	46.	210	29.	4.4	3.9	117	69	270	17.	0.20	210	19.	49.	7.
A15	highest Th	1700	24.	85.	2300	1300	41.	260	25.	4.2	5.8	140	90	380	23.	0.26	280	25.	68.	9.
A15	lowest Fe	1500	21.	85.	2100	1200	38.	280	19.	3.1	5.1	157	75	290	16.	0.22	280	22.	54.	7.
A16	typical mature	1010	9.5	25.	760	540	31.	440	26.	3.6	2.9	176	46	180	13.	0.14	140	13.0	34.	4.5
A16	highest Fe, station 5	1090	10.3	26.	820	580	33.	480	23.	3.5	3.2	176	51	190	14.	0.15	150	14.4	38.	5.
A16	lowest Fe, station 1	690	7.3	20.	570	460	17.	220	22.	4.6	2.0	177	31	120	9.	0.10	100	8.7	23.	3.
A16	lowest Fe, station 4	890	7.5	20.	580	470	32.	440	20.	4.5	2.4	173	40	170	11.	0.12	120	11.1	29.	4.
A16	atypical, NRC	700	7.2	20.	550	410	15.	170	13.	4.1	2.0	194	25	90	6.	0.09	85	6.6	18.	2.3
A17	orange glass, station 4	640	49.	130.	4800	2200	62.	90	250.		1.1	209	49	190	15.	0.12	80	6.4	19.	3.
A17	highest-Fe, stations 1 & 5	640	67.	91.	3200	1900	32.	120	36.	10.	1.2	151	78	230	22.	0.12	100	7.4	23.	4.
A17	lowest-Fe, NM station 6	1000	29.	55.	1900	1100	32.	210	22.	5.1	2.7	159	52	190	13.	0.13	130	10.	28.	4.
A17	lowest-Fe, SM station 2A	1200	17.	40.	1500	900	32.	270	15.	2.8	3.4	152	58	220	13.	0.18	170	15.	39.	5.
L16	only	900	51.	80.	2000	1600	31.	180	23.	n.a.	1.8	276	62	240	n.a.	0.07	170	11.	35.	5.
L20	only	600	16.	45.	1300	900	30.	230	~50.	n.a.	1.6	139	26	110	n.a.	0.09	90	6.9	17.	2.4
L24	only	200	43.	145.	3100	2000	50.	140	15.	1.1	0.5	91	18	50	n.a.	0.06	38	3.0	8.	1.2

table continued on following page

Table 2.5. continued

mission	sample type or location	Nd ppm	Sm ppm	Eu ppm	Gd ppm	Tb ppm	Dy ppm	Ho ppm	Er ppm	Tm ppm	Yb ppm	Lu ppm	Hf ppm	Ta ppm	W ppm	Ir ppb	Au ppb	Th ppm	U ppm	I_s/FeO
A11	only	38.	13.	1.77	17.	2.9	19.	4.6	11.	1.7	10.6	1.54	10.	1.3	0.2	7.	2.9	2.0	0.51	78
A12	typical	55.	16.	1.75	20.	3.5	22.	4.8	13.	1.9	12.3	1.74	13.	1.5	0.7	7.	3.5	5.8	1.5	55
A12	atypical	82.	22.	2.3	27.	4.6	30.	6.6	18.	2.7	17.	2.4	18.	2.2	n.a.	3.	2.2	9.2	2.6	8
A14	typical	105.	30.	2.5	35.	6.2	39.	8.7	23.	3.5	22.	3.1	23.	3.0	1.7	13.	7.	13.	3.5	65
A14	atypical	106.	31.	2.6	36.	6.5	42.	9.1	25.	3.6	23.	3.2	24.	3.1	1.8	6.	7.	14.	3.8	6
A15	highest Fe	22.	5.8	0.98	8.	1.3	8.	1.7	5.	0.7	4.3	0.66	4.	0.6	0.5	4.	1.6	1.5	0.45	28
A15	highest Mg	30.	8.8	1.15	10.	1.7	11.	2.5	7.	1.0	6.5	1.01	7.	0.9	n.a.	4.	3.	3.4	0.9	43
A15	highest Th	39.	12.	1.45	13.	2.3	15.	3.4	9.	1.3	8.5	1.24	10.	1.1	1.2	7.	4.	4.2	1.2	65
A15	lowest Fe	34.	10.	1.28	12.	2.0	13.	2.9	8.	1.1	6.9	0.96	8.	0.9	n.a.	8.	2.2	3.4	0.90	64
A16	typical mature	21.	6.0	1.20	8.	1.2	8.	1.7	5.	0.7	4.3	0.61	4.5	0.55	<0.5	15.	10.	2.2	0.62	82
A16	highest Fe, station 5	23.	6.7	1.22	8.	1.3	9.	1.9	5.	0.7	4.8	0.68	5.0	0.63	n.a.	16.	11.	2.4	0.68	83
A16	lowest Fe, station 1	14.	4.2	1.19	5.	0.9	6.	1.2	3.	0.5	3.0	0.42	2.9	0.38	n.a.	6.	4.	1.6	0.35	9
A16	lowest Fe, station 4	19.	5.1	1.11	6.	1.0	7.	1.5	4.	0.6	3.6	0.50	3.9	0.46	n.a.	15.	19.	1.9	0.50	61
A16	atypical, NRC	11.	3.1	1.24	4.	0.63	4.	0.9	2.	0.4	2.3	0.33	2.2	0.30	<0.3	6.	3.	1.1	0.34	25
A17	orange glass, station 4	18.	6.6	1.84	9.	1.5	9.	1.8	5.	0.7	4.2	0.61	5.	1.0	n.a.	0.8	0.3	0.5	0.16	1
A17	highest-Fe, stations 1 & 5	22.	8.3	1.71	11.	2.1	13.	2.9	8.	1.2	7.6	1.06	7.	1.3	<0.3	6.	2.	0.7	0.21	31
A17	lowest-Fe, NM station 6	20.	6.2	1.33	8.	1.4	9.	1.9	5.	0.8	5.0	0.70	5.	0.8	n.a.	8.	3.	1.6	0.45	63
A17	lowest-Fe, SM station 2A	26.	7.3	1.24	9.	1.5	10.	2.1	6.	0.9	5.5	0.78	6.	0.7	n.a.	12.	5.	2.6	0.73	65
L16	only	26.	8.2	2.3	11.	1.6	11.	2.3	6.	0.9	5.6	0.82	7.	0.6	45.	10.	2.6	1.3	0.31	n.a.
L20	only	11.	3.1	0.91	4.	0.64	4.2	0.9	2.7	0.4	2.5	0.38	2.5	0.3	64.	10.	3.4	1.3	0.33	n.a.
L24	only	6.	1.9	0.65	2.6	0.46	3.0	0.7	1.9	0.3	1.8	0.27	1.5	0.2	n.a.	5.	7.	0.4	0.11	n.a.

table continued on following page

Table 2.5 continued:

Concentration values in the table are selected means based on literature data too numerous to cite and some unpublished data from Washington University (Korotev). The table presents typical compositions and extreme compositions. The Apollo data are based only on samples of surface and trench soils. Some soils in cores and drive tubes are more extreme in composition, but less well characterized. The Luna samples were taken with coring equipment and are contaminated with W, presumably from tungsten carbide. Concentrations of some elements for which no data exist, e.g., Y, Pr, Gd and Er, have been estimated on the basis of interelement correlations. Oxide concentrations represent the total concentration of the element expressed as the oxide. All units are mass ratios, e.g., % = g oxide per g of sample and ppm = μg element per g of sample. I/FeO values are values or means of values from Morris et al. (1983). Mg' = mole percent MgO/(MgO+FeO). Compositionally atypical samples tend to be less mature (lower I_s/FeO).

Notes:

1 Apollo 11: Based on many data for sample 10084 and a few data for samples 10085 and 10010, which are both essentially identical to 10084 (Korotev and Gillis, 2001).

2 Apollo 12: Based on data for typical samples 12001, 12023, 12029, 12041, 12042, 12044, 12057, and 12070, which are all very similar in composition (e.g., FeO range: 15.9% to 17.1%; Th range: 5.3 ppm to 6.3 ppm). Some Apollo 12 samples (12003, 12030, 12037, 12060) are contaminated with fines abraded from mare basalts carried in the same container; data for these samples are not included.

3 Apollo 12: Based on data for atypical samples 12032 and 12033 from Head Crater.

4 Apollo 14: Based on data for all samples except 14141. Except for 14141, Apollo 14 soil samples are virtually identical in composition (e.g., FeO range: 10.1% to 10.9%; Th range: 12.3 ppm to 13.6 ppm).

5 Apollo 14: Data for sample 14141 from Cone Crater, which appears to have a greater proportion of felsite (K, Ba, Th) and alkali anorthosite (Na) than the typical soils.

6 Apollo 15: Based on data for samples 15531 & 15601 from station 9a at the edge of Hadley Rille. These samples contain the greatest proportion of mare volcanic material and are consequently richest in FeO.

7 Apollo 15: Based on data for all samples from station 7 (N=4) on the Apennine Front. Most data are from samples 15301 and 15411. The station 7 regolith contains the greatest proportion of 15426-type green picritic glass and is consequently the richest in MgO.

8 Apollo 15: Based on data for all samples from station 6 (N=6). The station 6 regolith contains the greatest proportion of KREEP-rich lithologies and consequently has the greatest concentrations of incompatible elements.

9 Apollo 15: Based on data for all samples (N=6) from station 2. The station 2 regolith contains the greatest proportion of nonmare lithologies and consequently has the lowest concentration of FeO. All other surface and trench soils from Apollo 15 are intermediate in composition to the four extremes listed here.

10 Apollo 16: Mean of data for all 24 samples of mature soil from Apollo 16. This composition probably best represents the surface of the Cayley Plains (Korotev, 1997). The compositional range of mature soils is moderate (e.g., FeO: 4.4% to 6.0 %; Th: 1.6 ppm to 2.5 ppm) and mainly represents variation in the relative abundance of anorthosite.

11 Apollo 16: Based on data for all samples (N=7) from stations 5 and 6. The soils of these stations are all similar in composition and are at the high-Fe extreme of the range of Apollo 16 soils because they contain the lowest proportion of anorthosite.

12 Apollo 16: Based on data for sample 61121 from station 1 on the Cayley Plains. This sample is the most feldspathic (lowest FeO) of the surface and trench soils from the central part of the site (stations LM, 1, & 2). Some soils at depth in the cores of the central area are more feldspathic, however (Korotev and Morris, 1993).

13 Apollo 16: Based on data for sample 64601 from station 4 on Stone Mountain. This sample is the most feldspathic (lowest FeO) of the surface and trench soils from the southern part of the site (stations 4, 5, 6, 8, & 9).

14 Apollo 16: Based on data for all true soil samples (N=8) from station 11 at North Ray Crater (some 67xx0 and 67xx1 samples are not true soils; see Korotev, 1996). This is the only set of means presented here for which the samples upon which they are based vary considerably in composition (e.g., FeO range: 2.9% to 4.6%; Th range: 0.5 ppm to 2.0 ppm) and Mg' (range: 62 to 70). Soils from station 13 (63xx1) about a kilometer from North Ray Crater are intermediate in composition between the station-11 soils and the mature soils.

15 Apollo 17: Based on many data for the "orange glass" soil, sample 74220, taken near Shorty Crater at station 4. This soil is the most anomalous in the Apollo collection because it consists nearly entirely of pyroclastic glass beads, with some minor contamination by "normal" soil. (Soil from the double drive tube 74001/2 taken at the same location is less contaminated, but fewer data are available.)

16 Apollo 17: Based on data for all samples (N=6) from stations 1, 1A, and 5 on the valley floor. These samples are all similar in composition and represent the most Fe-rich extreme (excluding 74220) of soils found at the Apollo 17 site in that they contain the highest proportion of mare basalt.

17 Apollo 17: Based on data for all samples (N=6) from station 6 at the base of the North Massif. The station-6 soils are the least contaminated with mare material of nominally nonmare soils collected from the north side of the site, although the degree of contamination is still substantial (about one 27%, Korotev and Kremser, 1992).

18 Apollo 17: Based on data for all samples (N=3) from station 2A (LRV stop 4) on the light mantle deposit at the base of the South Massif. The station-2A soils are the least contaminated with mare material of any soils collected at the Apollo 17 site (about 6%, Korotev and Kremser, 1992). Other soils from Apollo 17 are intermediate in composition to the four extremes listed here.

19-21 Based on all available data for Luna soils.

(1) Ancient impact-melt breccias. A peculiar and fascinating feature of nearly all of the Th-rich, mafic impact-melt breccias of the Apollo collection (Section 5.2.2) is that they contain FeNi metal of meteoritic origin in moderately high abundances, 0.2–1.7% Fe^0 by mass, depending upon the type of breccia (Korotev 2000). The composition of the metal (typically $6 \pm 2\%$ Ni and Ni/Co $\approx$ 15) is not like that in ordinary chondrites but resembles that in some iron meteorites (Hewins and Goldstein 1975; Korotev 1987a,b; James 1995, 1996). Nearly all of the breccias are ancient, ~3.9 Gy, and all or most are products of impacts that formed basins or large craters. Like the metal in the igneous and plutonic rocks, this metal is very old. In the melt breccias, the metal grains range in size from micrometers or less to a few millimeters. Because some regoliths contain a high proportion of such breccias, this ancient metal is a major source of metallic Fe in some Apollo soils, most notably those of Apollo 14 and 16. Grains of metal that exceeding 1 mm in size have been found as fragments in the Apollo 16 soil samples (Goldstein et al. 1972; Reed and Taylor 1974; Ali and Ehmann 1977; Korotev and Jolliff 2000). Most large metal grains in lunar soil appear to derive mainly from disaggregation of melt breccias, not directly as meteorites (Korotev 1987a; Korotev 1991, 1997; Korotev and Jolliff 2000). Metal derived from ancient melt breccias is identifiable because it is compositionally distinct in having high concentrations of W, compared to metal in meteorites, as a result of having solidified from W-rich, mainly-silicate impact melt (Wlotzka et al. 1973; Korotev et al. 2003a). These large metal nuggets are a common source of anomalously high Fe concentrations in small subsamples of lunar soils (Korotev 1987a).

In lunar soils, metal contained in igneous rocks and ancient impact-melt breccias together correspond to the $Fe^0{}_{SM}$ (source material) component of Morris (1980), that is, the metal component which is uncorrelated with the magnitude of surface exposure. On average, this component accounts for a few tenths of a percent of the mass of Apollo soils (0.17 ± 0.08 "equivalent wt %" as Fe^0, mean and standard deviation; Morris 1980). For some regoliths it is higher. Based on siderophile-element abundances, the soils of the Apollo 14 and 16 sites each contain 0.5% "FeO" as metal derived from ancient impact-melt breccias. These concentrations correspond, respectively, to 5% and 9% of the total iron in the regoliths and a large fraction of the metallic iron.

(2) Macrometeoroids. Most of the craters produced since the time of basin formation have been caused by impacts of ordinary chondrites and iron meteorites, all of which contain FeNi metal. The amount of metal in the regolith derived from post-basin macrometeorites is not known. At the Apollo 16 site, glass from the South Ray crater impact 2 Ma ago is commonly found as spherules in the regolith and as glass coatings and splashes on rocks (Morris et al. 1986; Eugster 1999). The glass is rich in siderophile elements (mean Ni: 800 µg/g; Morris et al. 1986) and Ni/Co and Ir/Au ratios of the glass are consistent with a chondritic impactor. The glass contains "minute" metal spherules (James et al. 1984). Metal spherules in glass "bombs" found in the North Ray Crater ejecta range up to at least 50 µm in diameter (Fig. 3 of Borchardt et al. 1986).

(3) Micrometeoroids. Morris (1980) noted that the concentration of some Fe metal that exists in the regolith as grains >33 nm in diameter increases with regolith maturity and concluded that this maturity-correlated, multiple-domain metal component, designated $Fe^0{}_{MM}$, derived from micrometeorites. This conclusion is consistent with the observation that FeNi metal occurs in the glass of 0.1–1 mm craters formed by impacts of interplanetary dust particles (Brownlee et al. 1991, 1992). It is also likely that a portion of the nominal micrometeorite component is actually metal from macrometeorites because metal-bearing glass spherules, such as those produced when South Ray Crater was formed, accumulate at the surface and thus increase in concentration with time.

Morris (1980) concluded that on average, the regolith contains 0.17 ± 0.08 "equivalent wt%" Fe^0 as the micrometeorite component. In samples of submature and mature soils, Ni concentrations in excess of that portion likely carried by source rocks is typically 120–200 μg/g, which corresponds to 0.21–0.35% Fe or 0.27–0.45% FeO, using CI chondrite abundance ratios (Anders and Grevesse 1989). Thus, half to most of the Fe that is contributed by meteorites associated with surface exposure occurs in metallic form in the regolith. In the Apollo 16 regolith (5.5% Fe as FeO), 13–17% of the Fe is carried by metal and another 4–9% is carried by ilmenite, leaving only 73–83% in silicate minerals and glass. Clearly, not all the Fe contained in lunar soil leads to UV-VIS spectral absorption bands.

(4) Nanophase iron. "Nanophase" or "single-domain" iron occurs as very small (submicrometer) grains, usually in agglutinates and coatings on regolith grains. Some nanophase Fe is pure iron that is produced by reduction of Fe^{2+} in lunar minerals as a consequence of exposure of the mineral grains at the lunar surface (Morris 1980). Other nanophase Fe is produced when vapors formed during micrometeorite impacts condense on the surface of regolith materials (Keller and McKay 1997). Because micrometeorites totally vaporize upon impact, some nanophase iron must be of meteoritic origin. There are no studies, however, that quantify the relative importance of the two possible sources of nanophase iron, *in situ* reduction, and vapor deposition. The relative abundance of nanophase iron correlates both with the iron concentration of the regolith and with regolith maturity, as determined by other parameters. Morris (1980) concluded that, on average, 40% of the metallic Fe in the lunar regolith occurs as nanophase iron (0.20 ± 0.10 "equivalent wt%" as Fe^0). Nanophase iron is discussed in more detail below (Section 9), because it is main cause of spectral reddening of lunar regolith with increased surface exposure.

6.2.2. The regolith as a mixture. Despite the wide variety of materials in the lunar regolith, the compositions of soils from the Apollo and Luna landing sites correspond to mixtures of only three compositional classes of material: (1) rocks and breccias of the feldspathic highlands ("anorthosites," including rocks with compositions corresponding to noritic and troctolitic anorthosite and anorthositic norite and troctolite; Fig. 2.3), (2) some form of KREEP, usually impact-melt breccia of basaltic, noritic, or troctolitic composition or (rarely) KREEP basalt, and (3) mare basalt and volcanic glass (Fig. 2.2). Thus on many 2-element plots of regolith composition, regolith samples define a triangular array (Figs. 2.11, 2.12d, 2.13). The lithologies representing the three classes typically differ in composition from site to site (e.g., different types of melt breccia or mare basalt) and several different compositionally distinct lithologies may represent a given class at a given landing site (e.g., several types of mare basalt at Apollo 11 or several kinds of impact-melt breccia at Apollo 17). Thus, the apices of the triangle are not discrete points but are "fuzzy" (e.g., Garrison and Taylor 1980).

Regolith samples collected on the Apollo and Luna missions cover much of the range of the mixing triangle. Certain possible mixtures, however, do not occur among the Apollo and Luna samples. For example, Apollo collection does not contain soils composed subequally of mare basalt and feldspathic material with negligible KREEP but such soils might be expected to occur elsewhere. The QUE 94281 and Yamato-793274 lunar meteorites apparently derive from such a regolith (Jolliff et al. 1998a; Arai and Warren 1999). At points distant from the Apollo landing sites, e.g., the South Pole-Aitken area, other classes of material may exist, although there is no strong reason to suspect that anomalous regoliths occur elsewhere. At the Apollo and Luna sites, soils that consist mainly (>90%) of a single lithology or of rocks of a single compositional class are rare. The orange-glass soil at station 4 of Apollo 17 would be an example of such a rare regolith. Most soil samples contain at least a few percent of material from all three compositional classes. For example, all mare soils contain nonmare material (Luna 24 soil appears to be the "cleanest" in this regard) and mare material is rare but present to varying degrees even in the feldspathic lunar meteorites.

At Apollo and Luna landing sites that are geologically simple or where only one or a few samples were collected, soil compositions tend to occupy only a small portion of the mixing

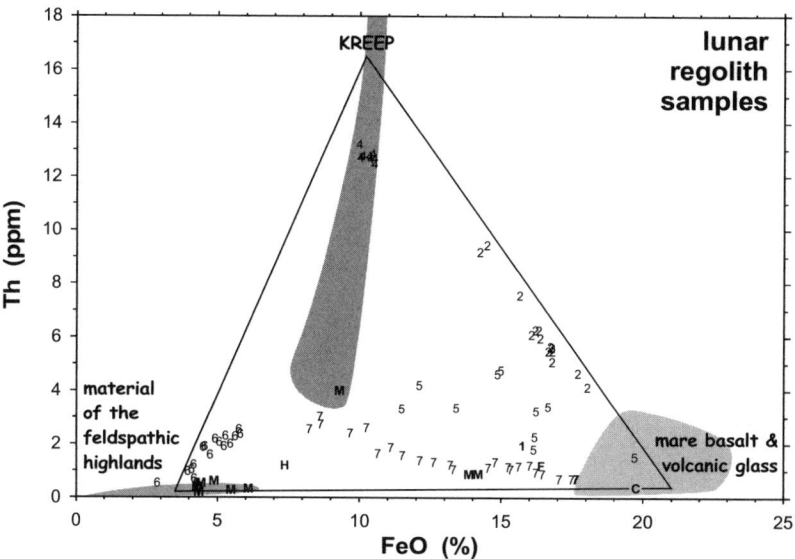

Figure 2.11. The Apollo regolith mixing triangle in terms of FeO (total Fe as FeO) and Th concentrations (adopted from Jolliff et al. 2000). Regolith samples from the Apollo missions are mainly mixtures of the three classes of material represented by the apices of the triangle. Mean FeO and Th concentrations of the actual rocks representing the three classes differ from site to site, but for any given site the apices lie within the shaded fields. For example, in Apollo soils, KREEP is carried by mafic impact-melt breccias; those from the Apollo 16 and 17 sites have low Th concentrations (4–8 ppm) whereas those from the Apollo 12 and 14 sites have high concentrations (16–22 ppm). Symbols are as in Figure 2.3, except that all lunar meteorites that are plotted (M). For Apollo 12 (2) and 14 (4), each point represents a numbered surface or trench soil sample (e.g., 12070 and 14163). For Apollo 15 (5), 16 (6), and 17 (7), each point represents the mean of all surface and trench soils from a sampling station (e.g., station 2), except that all samples from station 11 (North Ray Crater) at Apollo 16 are plotted.

triangle. In contrast, soils from the Apollo 15 and 17 missions cover a large range because the landing sites were at highlands-mare interfaces, the ratio of mare to nonmare material in the regolith varies from place to place, and the astronauts collected samples over a wide area. Although material of all three compositional classes occurs in the regoliths of all sites (Fig. 2.14), soil compositions from a given site do not scatter all over the mixing triangle. This simple observation indicates that the material of the regolith is so well mixed that even the tiny samples used for chemical analysis are remarkably uniform in composition from sample to sample. We might expect that for larger samples sizes, e.g., the area of a Clementine pixel, the degree of sample-to-sample (pixel-to-pixel) uniformity would be even greater.

At landing sites where regolith compositions are not uniform, the variation is systematic, typically leading to linear trends on plots of element Y vs. element X. For example, Apollo 12 soils are largely binary mixtures of mare basalt and KREEP impact-melt breccias. The basalt/KREEP ratio varies among different regolith samples, leading to a linear trend on 2-element plots (Fig. 2.11). Such trends can be used to infer information about local geology. For example, a plot of the concentration of any lithophile element against that of FeO for Apollo 17 soils yields a linear trend such as that of Figure 2.11. Extrapolation of the trend to high FeO concentration provides the mean composition of the mare components, essentially the Apollo 17 high-Ti basalts and orange volcanic glass. Extrapolation to low FeO provides the mean composition of the nonmare components. That composition does not, in fact, correspond to any single rock type that occurs at the Apollo 17 site but to a mixture of various nonmare li-

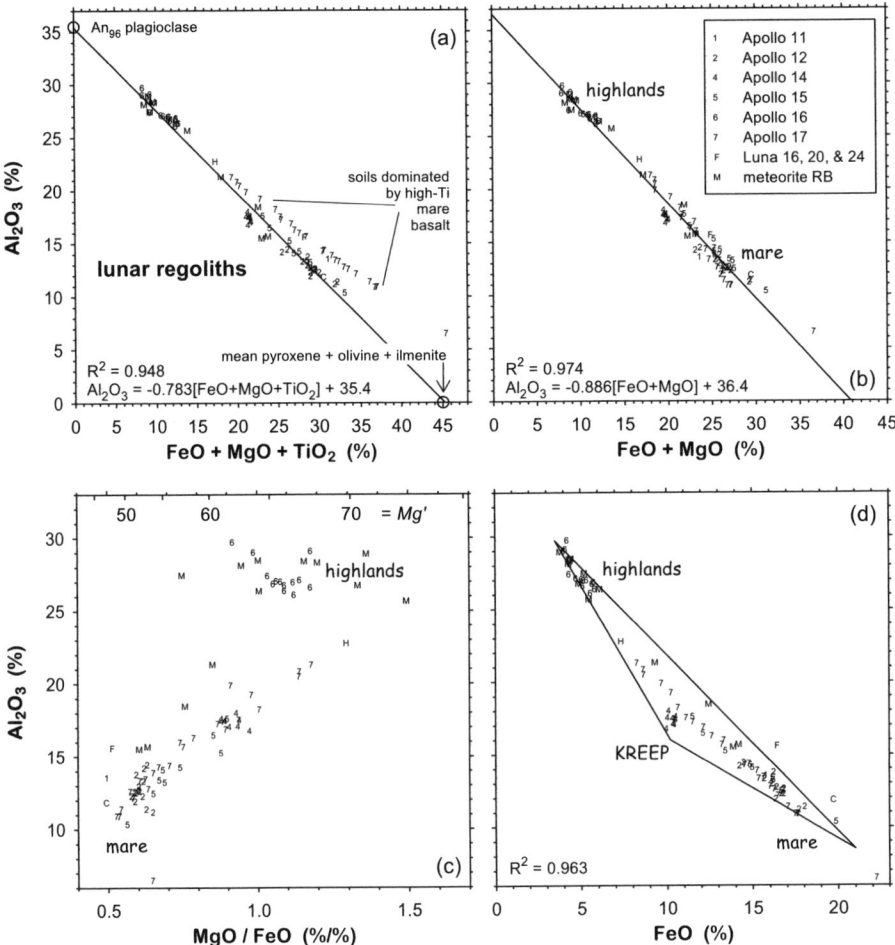

Figure 2.12. In the lunar crust, nearly all Al is carried by plagioclase whereas Fe, Mg, and Ti are carried by pyroxene, olivine, and ilmenite. Thus, to a first approximation, all regolith samples plot on or about a mixing line between the composition of plagioclase and a point corresponding to the mean composition of the three Fe-bearing minerals, which varies from site to site. Soils from the feldspathic highlands plot at the high-Al_2O_3 end of the trend, those dominated by mare basalt plot at the low-Al_2O_3 end, and soils dominated by KREEP (Apollo 14) plot in the middle (Fig. 2.11). Symbols are as in Figure 2.3. (a) Soils from Apollo 11 and 17, which are dominated by high-Ti mare basalt, deviate from the trend defined by the low-Ti samples because of the excess of Ti. The equation is that for the line, which excludes the Apollo 11 and 17 data; R^2 is for all data. (b) The correlation is best between Al_2O_3 and FeO+MgO. (c) MgO/FeO ratios for highland soils have a wide range while those for mare soils are uniformly low. Mg' = bulk mole percent MgO/(MgO+FeO) (d) The nonlinearity of the trend occurs because the Apollo 14 soils in the middle of the array are not mixtures of feldspathic highlands material and mare basalt (Fig. 2.11) and because mare basalts have higher FeO/MgO ratios than the nonmare materials (c). The triangle represents the mixing triangle of Figure 2.11; in (a) and (b) the mixing triangle is compressed to a line. See also Figure 8 of Fischer and Pieters (1995).

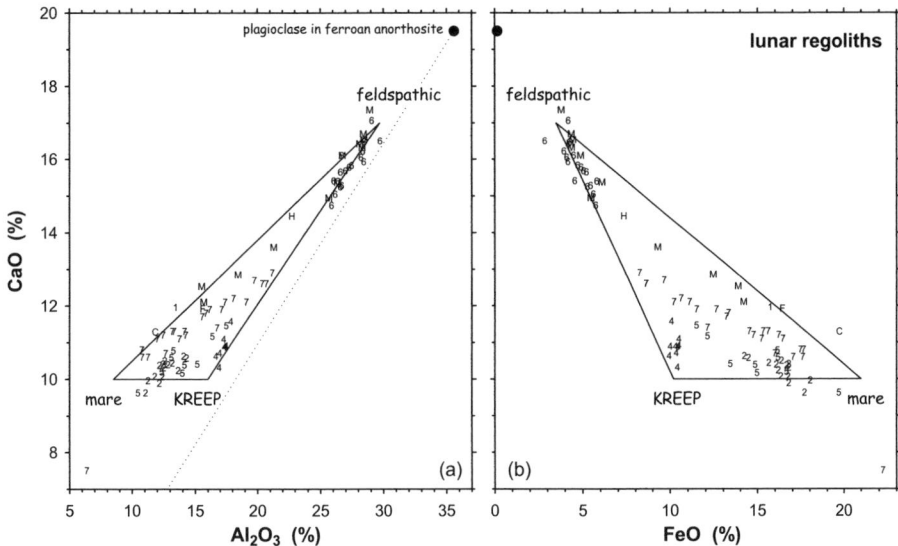

Figure 2.13. Variation of CaO with Al$_2$O$_3$ and FeO in lunar regolith (total Fe as FeO). The triangles are the Apollo regolith mixing triangle of Figure 2.11. In (a), the CaO/Al$_2$O$_3$ ratio of plagioclase in ferroan anorthosite () is shown by the dotted line. Mare basalts have high CaO/Al$_2$O$_3$ ratios compared to plagioclase in ferroan anorthosite and KREEP because in mare basalts some CaO is carried by Ca-rich pyroxene (clinopyroxene) whereas the pyroxenes of feldspathic and KREEP-rich samples are low-Ca pyroxenes.

thologies. This observation indicates that the nonmare lithologies were moderately well mixed before the admixture of the mare component. We can expect similar trends on a regional basis in data taken from orbit.

The mixing triangle of Figure 2.11 does not contain apices for magnesian-suite or alkali-suite nonmare rocks (e.g., Spudis and Davis 1986) because, with two possible exceptions, in Apollo and Luna regoliths such rocks are volumetrically minor subcomponents of the KREEP or, possibly, feldspathic highlands components. The exceptions are that soils from the North Massif at Apollo 17 appear to contain as much as a few tens of percent troctolite or troctolitic anorthosite for which the abundance is not correlated with the abundance of the other nonmare components (Korotev and Kremser 1992; Jolliff et al. 1996). Much of the Luna 20 regolith also contains anorthositic norite, troctolitic anorthosite, and spinel troctolite with moderately high Mg concentrations. Some of these compositions would plot outside the field of "material of the feldspathic highlands" on Figure 2.11.

One implication of Figure 2.11 requires some words of clarification and caution. In terms of the KREEP concept, the figure implies, for example, that the Luna 20 regolith contains 5–10% of a high-Th KREEP lithology that was mixed into the regolith. This inference is probably incorrect or, at least, misleading for Luna 20 and perhaps for other regoliths of the figure as well. The Luna 20 regolith, which is not well understood, may instead derive ultimately from early crustal rocks that never differentiated to the point where they developed high concentrations of incompatible elements (Korotev et al. 2003b). The components of the Luna 20 regolith are moderately low-Th rocks that happen to have concentrations of incompatible elements that occur in the same ratios as KREEP because early crustal rocks formed from a global magma ocean and they contain some trapped residual liquid of KREEP-like composition (Jolliff and Haskin 1995; Jolliff 1998). This example highlights one the ambiguities associated with the

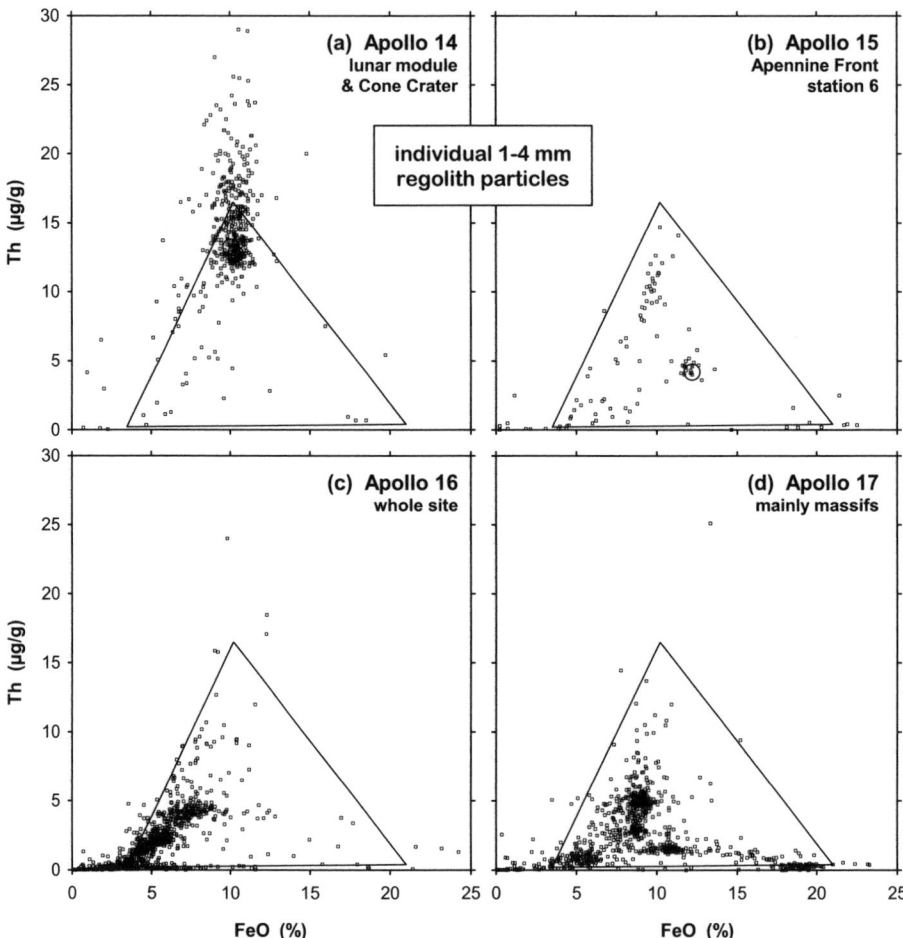

Figure 2.14. Concentrations of FeO (total Fe as FeO) and Th in individual rock particles from the 1–2 mm and 2–4 mm fractions of lunar soils shown with the Apollo regolith mixing triangle of Figure 2.11. (a) Data are mainly for 14161, a surface scoop sample taken near the lunar module; the rest are for 14142 and 14143 taken near Cone Crater (Jolliff et al. 1991, and unpublished data). Large circles obscured by the main cluster of points represent the composition of the <1-mm grain-size fractions of the two soils (14163 and 14141). Apollo 14 soils are dominated by impact-melt, fragmental, and regolith breccias of KREEP composition (i.e., most points with >10 µg/g Th). Feldspathic lithologies and mare basalt are rare. Six particles in this data set contain >30 µg/g Th. (b) Data for 15272 and 15273 from station 6 on the Apennine Front (Korotev 1987c, and unpublished data). The station-6 soils of Apollo 15 consist of subequal amounts of rock types representing all three apices of the mixing triangle. The large circle represents the composition of the <1-mm grain-size fraction of 15271; the cluster of points near the circle are for regolith breccias of bulk-soil composition. Points with >8 µg/g Th mainly represent KREEP basalts, with some impact-melt breccias. (c) Data for particles from all over the Apollo 16 site (Korotev 1983a,b; Jolliff and Haskin 1995; Korotev et al. 1997b). Some of the points with >18% FeO represent gabbronorites (James and Flohr 1983), not mare basalts. (d) Most of the points are for 2–4 mm particles from the North and South Massifs of Apollo 17, so mare basalts are underrepresented (Jolliff et al. 1996, 1998). Nearly every point with >3 µg/g Th in (c) and (d) represents a KREEP-bearing impact-melt breccia.

KREEP concept. The term is used to refer to a discrete high-K, high-Th lithologic component of polymict materials such as regoliths and breccias (e.g., Simon et al. 1986), as a compositional component that may or may not represent a lithologic component (e.g., Schonfeld 1974), and as the residual liquid phase of a crystallizing igneous system (Warren and Wasson 1979) that may or may not have been separated from rock in which it occurs.

6.2.3. Compositional variation with depth in the regolith. A several locations where regolith core samples were taken, the range of compositional variation with depth over tens of centimeters is equivalent to the range among surface samples taken over a few kilometers of lateral distance (Fig. 2.15). Compositional variation with depth reflects less-than-ideal vertical mixing of the components of which the soil is composed. Regolith stratigraphy results from overlapping ejecta deposits from many craters, big and small. Compositionally anomalous layers almost always contain an excess of some specific lithology of which the soil is composed compared to the average for the whole core (Fig. 2.16). Such layers probably have only limited lateral extent. Apollo 16 cores 60001–7, 60009/10, and 60013/14 were taken at apices of a triangle 40–50 m apart (Fig. 2.15). Each of the cores contains a layer a few centimeters thick that is richer in anorthosite that surface soils, but there is no way to know whether these layers are part of a continuous deposit. Core 60009/10 shows the widest variation in composition with depth of any lunar core (Figs. 2.15, 2.16).

It is not uncommon for soil in the upper 0.5–1 cm to have concentrations of several elements that are 10 or more percent different from the average for the upper half meter or meter. For example, the FeO concentration of the upper centimeter of the Apollo 17 deep drill core is 9% greater than the mean for the whole 3 m core (Fig. 2.15). However any such effects probably cancel over a large areal extent unless there is some large scale stratigraphy. In the absence of such stratigraphy there is no reason to believe that the upper millimeter, centimeter, or meter, that is, the portion of the regolith observed by orbiting spectrometers, is a systematic biased sample of the material of the upper 10 m with respect to concentrations of lithophile elements. Siderophile (iron-metal loving) and volatile elements, on the other hand, are concentrated toward the surface because they derive mainly from micrometeorites.

6.2.4. Correlations among major elements resulting from mineralogy. More than 98% of the crystalline material of the lunar crust consists of only four minerals: plagioclase feldspar, pyroxenes, olivine, and ilmenite. Thus, seven chemical elements, O, Mg, Al, Si, Ca, Ti, and Fe, account for >98% of the mass of the lunar crust. Ilmenite is minor, <5% in most rocks, except for mare basalts which may contain up to 25% by mass ("high-Ti basalts") of ilmenite plus other Fe-Ti oxides. In soils, the three remaining minerals, all silicates, each contain average concentrations of Si and O that are similar to each other, thus there is little variation in these two elements among regolith samples (Fig. 2.17). For example, total Si as SiO_2 ranges from 42% in the ilmenite-rich Apollo 11 and 17 soils to 48% in KREEP-rich Apollo 14 soils. This range narrow range compares with essentially 0% to 100% in terrestrial sediments (e.g., carbonates to sandstones). Because of the narrow range of Si variation in lunar soils, Al/Si and Mg/Si ratios obtained by the Apollo 15 and 16 orbiting X-ray spectrometers (Section 10.3) provided accurate estimates of absolute Al and Mg concentrations. Total oxygen ranges from 42% in the ilmenite-rich Apollo 11 and 17 soils (40.6% in the anomalous Apollo 17 orange-glass soil) to 45% in plagioclase-rich highlands regoliths (Fig. 2.18). This leaves only five major elements that vary significantly in concentration: Mg, Al, Ca, Ti, and Fe.

In terms of the Apollo mixing triangle, feldspathic highlands materials are rich in plagioclase (70–99%) and consequently poor in pyroxene, olivine, and ilmenite. Thus highlands materials are rich in Al and Ca and poor in Fe, Mg, and Ti (Figs. 2.13, 2.19). At the other extreme, mare basalts are rich in pyroxene, olivine, and sometimes ilmenite ($\Sigma \approx 80 \pm 10\%$ by mass, leaving $20 \pm 10\%$ plagioclase) and are consequently rich in Fe, Mg, and Ti. KREEP-rich impact-melt breccias and basalt are intermediate (40–70% plagioclase by mass), which leads

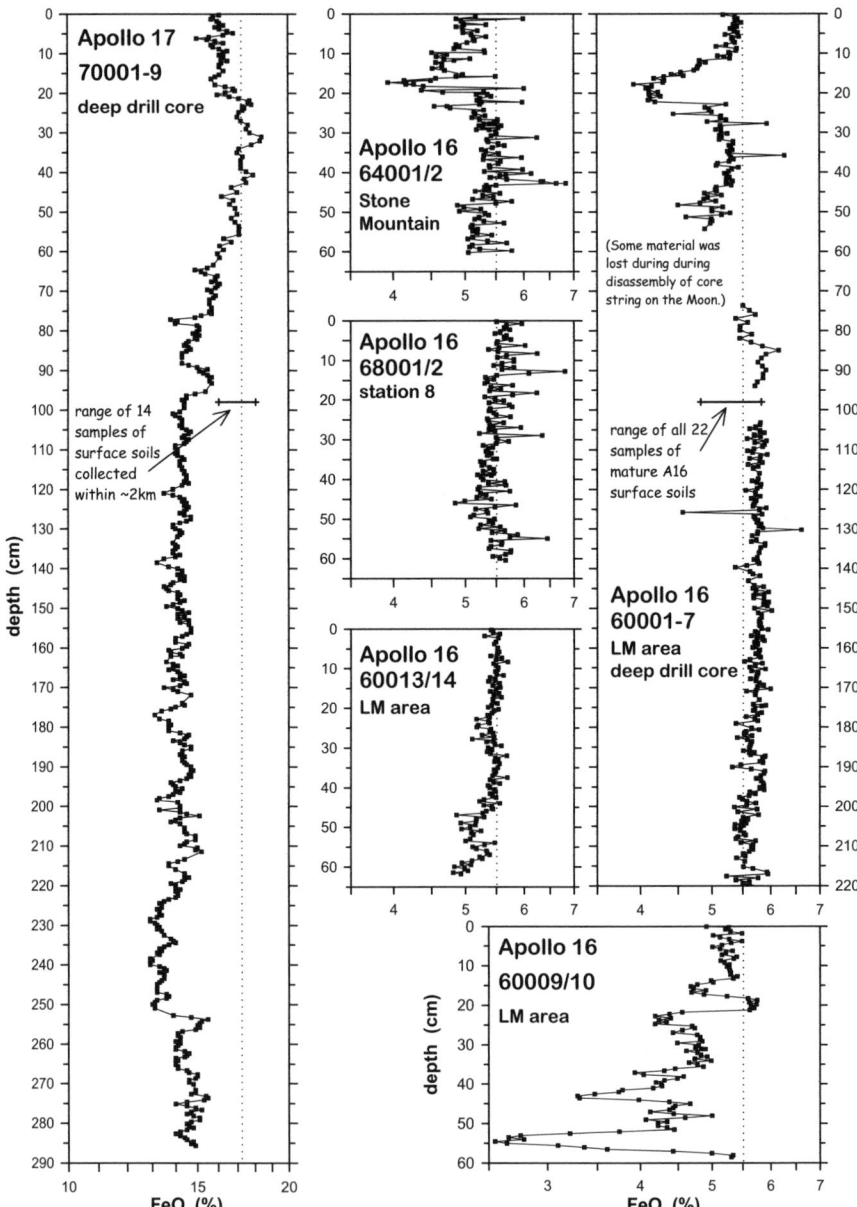

Figure 2.15. Variation in FeO concentrations with depth in the Apollo 16 and 17 deep drill cores and four Apollo 16 double drive tubes. For Apollo 17, the vertical dotted line represents the mean concentration of 14 samples of surface and trench soils collected within 2 km of the core location; the range is also shown (the soil from LRV stop 9, which has 14.6% FeO is excluded). For Apollo 16, the vertical dotted line is the mean FeO concentration of all 22 mature surface and trench soils from the site; the range is shown in the 60001-7 plot. The FeO axes are logarithmic and, e.g., a 10% change is represented by the same horizontal distance on all plots. Data sources: Morris et al. (1979), Korotev et al. (1984), Korotev (1991), Korotev and Morris (1993), and Korotev et al. (1997a).

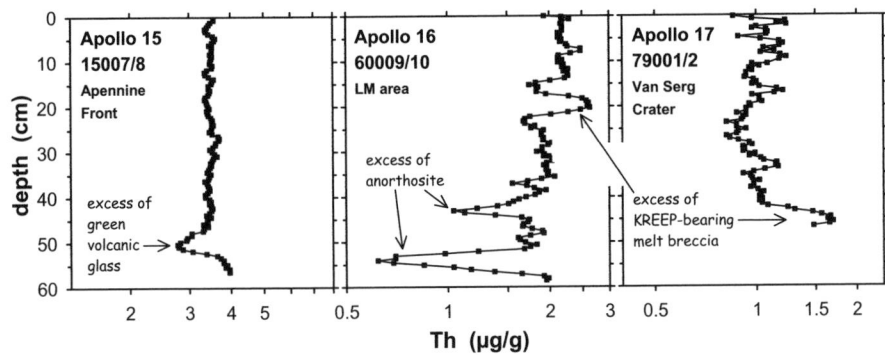

Figure 2.16. Variation in Th concentration with depth in three Apollo double drive tubes. Compositionally anomalous layers usually contain an excess of one of the lithologies of which the regolith is composed. Each of the lithologies causing the anomalies illustrated here corresponds to one of the apices of the mixing triangle of Figure 2.11. Data are from Korotev (1995), Korotev (1991), and Morris et al. (1989).

to FeO concentrations (~10%) that are about half those of mare basalt (~20%) yet considerably greater than those of rocks of the feldspathic highlands (<6%; Figs. 2.11, 2.19). Thus although soils typically contain material of all three compositional classes, most compositional variation ultimately involves variation in the ratio of plagioclase to the three Fe-bearing minerals (Fig. 2.10). Closure leads to a tight anticorrelation between Al or Ca with Fe+Mg+Ti (Fig. 2.12, Fig. 2.13) and many regions of compositional space are disallowed, e.g., high Al coupled with high Fe (Fig. 2.12) or high Ti coupled with low Fe (Fig. 2.19a). Most correlations among major, minor, and trace elements in lunar materials ultimately result from two factors: (1) the simplicity of lunar mineralogy and (2) the fact that there is only one apparent significant carrier of incompatible elements, KREEP, which in addition to having much higher concentrations of incompatible elements is more mafic (i.e., richer in olivine, pyroxene, and ilmenite) than feldspathic highlands rocks and less mafic than mare basalts.

There are at least three second-order effects that lead to imperfect correlations among elements: (1) the high abundance of ilmenite in some mare basalts but not others (Figs. 2.12a, 2.19), (2) variation in the MgO/FeO ratios of pyroxenes and olivine, with low ratios characteristic of mare basalts and volcanic glass and high ratios characteristic of nonmare rocks (Figs. 2.12c, 2.19), and (3) the presence of significant Ca in the clinopyroxene of mare basalts and its low abundance in the pyroxenes of most nonmare samples (Fig. 2.13). Nonmare regoliths and breccias vary considerably in MgO/FeO, from ferroan (low MgO/FeO) to magnesian (high MgO/FeO) extremes (Fig. 2.12c).

6.2.4. Correlations among incompatible elements. To a first approximation, ratios of any two incompatible elements are constant among Apollo regolith samples (Taylor 1975; Haskin and Warren 1991) because KREEP lithologies (Section 3.4.2) such as impact-melt breccias are the major carriers of incompatible elements. To a second approximation, important differences exist and the reasons are well understood. There are two causes.

(1) Among rocks and breccias that do not contain KREEP or that contain only a small proportion of KREEP, ratios of incompatible elements vary because of differences in compositions of magmas from which the rocks ultimately derived and the crystallization history of the magmas. For example, Gd/Sm and Th/Sm ratios are quite variable among different types of mare basalt (next sections) and ratios of light to heavy REE vary among the various feldspathic lunar meteorites (Korotev et al. 2003b). Although these differences can be measured in the laboratory where high analytical precision can be achieved, they are difficult to detect

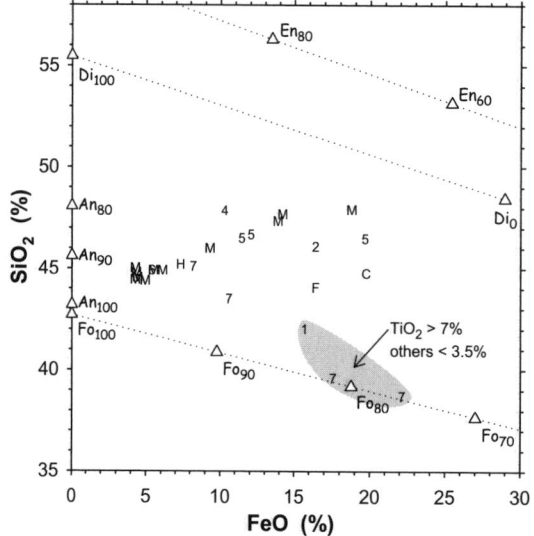

Figure 2.17. Variation of SiO_2 with FeO (total Fe as FeO) in lunar regolith (symbols as in Fig. 2.3) and comparison to stoichiometric concentrations in some minerals (). High TiO_2 soils (Apollo 11 and 17) have low SiO_2 concentrations because the main carrier of Ti, ilmenite, contains essentially zero SiO_2. SiO_2 concentrations are nearly constant in Ti-poor soils, but increase somewhat with FeO because (1) FeO is carried mainly by pyroxenes in low-Ti soils (Fig. 2.3) and pyroxenes have a greater SiO_2 concentrations than the plagioclase ($\sim An_{96}$) typical of Fe-poor soils and (2) average anorthite (An) content of plagioclase from Fe-rich rocks tends to be less than that of Fe-poor rocks and SiO_2 increases with decreasing An content of plagioclase. Mineral abbreviations: An, anorthite; Fo, forsterite; En, enstatite; and Di, diopside.

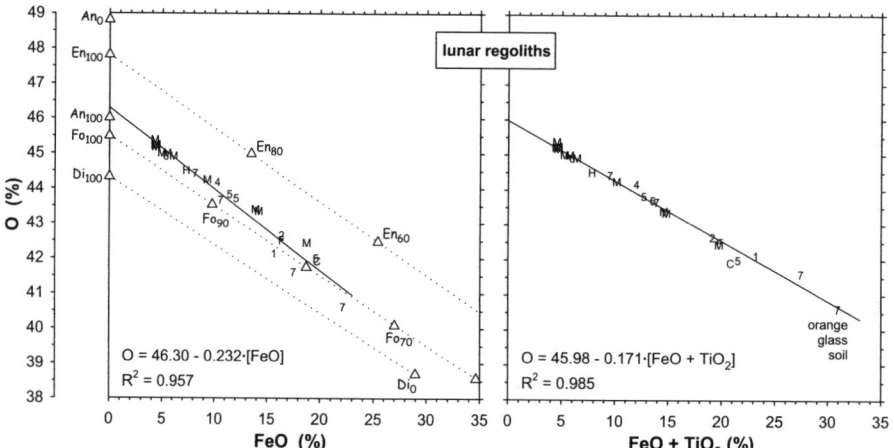

Figure 2.18. Variation of oxygen (O) with FeO and FeO plus TiO_2 (total Fe as FeO) in lunar regolith and comparison to stoichiometric concentrations in some minerals (see Fig. 2.17).

from orbit by remote techniques because in low-KREEP samples the absolute concentrations of incompatible elements are usually low and the relative precision is consequently poor.

(2) Although largely excluded from the major rock-forming minerals, when the concentrations of incompatible elements become sufficiently high in a cooling magma, they are incorporated into specific minor or accessory minerals, e.g., K and Ba into alkali feldspar, REE into RE-merrillite, and Th into zircon. In an igneous system where concentrations of incompatible elements are high, such as when crystallization of a liquid is nearly complete, minor minerals form and magmatic processes can separate one type of mineral from another. Feldspar may rise buoyantly in a magma chamber, concentrating alkali elements like Na and K toward the top. (This is a hypothetical example, one not actually demonstrated to be the cause

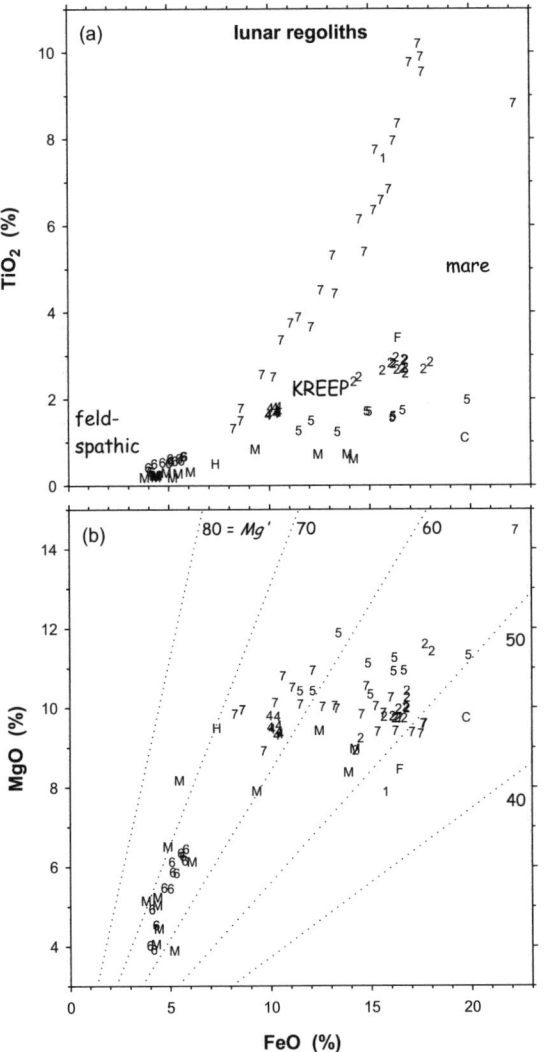

Figure 2.19. Variation of TiO_2 and MgO with FeO (total Fe as FeO) in lunar regolith samples. $Mg' =$ mole percent Mg/[Mg+Fe]. (a) TiO_2 is uncorrelated with FeO in mare basalts, so high-FeO soils have highly variable TiO_2 abundances. TiO_2 concentrations in feldspathic regolith are consistently low. (b) MgO is only poorly correlated with FeO in lunar soils. Nonmare soils tend to have greater MgO/FeO (and Mg') than mare soils (see also Fig. 2.12c).

of formation of alkali anorthosites.) In highly differentiated lunar magmas, two immiscible silicate liquids can form, one rich in Fe, Ti, P, and REE and the other rich in K and Si. This mechanism has been suggested to be the one leading to formation of lunar granite or felsite (Quick et al. 1977; Taylor et al. 1980; Neal and Taylor 1989; Jolliff 1991, 1998; Jolliff et al. 1999). If separation, for example, of K feldspar from RE-rich phosphate minerals occurs, the resulting rocks will not have KREEP-like K/REE ratios. Although on the basis of the Apollo and Luna samples, such separations are known to occur, all incompatible-element-rich rocks with non-KREEP-like ratios of incompatible elements are very small rocks, usually less than a gram. A small felsite clast in a breccia or a small felsite fragment in the regolith may represent a segregation and separation on the order of centimeters or kilometers. That such segregations occur on a small scale is clear; however, it is not known if they occur in lunar magmatic systems on a large scale. In principle, the chemical effects of large-scale separations might be observed with detectors on orbiting spacecraft. Any such anomalies in ratios of incompatible elements are of special importance because they ultimately relate to whether there was a global magma ocean and, if so, how it evolved.

Samarium and Gadolinium. Two incompatible elements of special interest are the rare earth elements Sm and Gd because, like Th, their concentrations provide a measure of the relative abundance of KREEP (Section 1.3.2) and the sum of their concentrations can be estimated from the Lunar Prospector neutron spectrometer (LP-NS) data (e.g., Elphic et al. 2000). Because the geochemical behavior of the two elements is very similar, Gd/Sm ratios in lunar rocks span a narrow range, about a factor of 2, and soils show an even narrower range, less than a factor of 1.4 (Figs. 2.20, 2.21). Samples with high concentrations of REEs, that is, samples with a large component of KREEP, consistently have a Gd/Sm ratios of about 1.18. Gd/Sm ratios in mare basalts are greater than in KREEP and they cover a broader range, 1.3–1.6. The most extreme Gd/Sm ratios occur in samples with very low concentrations of REE.

For the purpose of comparing data obtained remotely for Gd and Sm with samples, one problem is the lack of data for samples. Concentrations of Sm in lunar samples are widely available because Sm concentrations have been determined by INAA (instrumental neutron activation analysis), mass spectrometry, and other techniques. The most precise data for Gd and Th in lunar samples have been obtained by mass spectrometry, a technique that has been rarely used for these elements since the 1970's. Most data for the REEs and Th have been

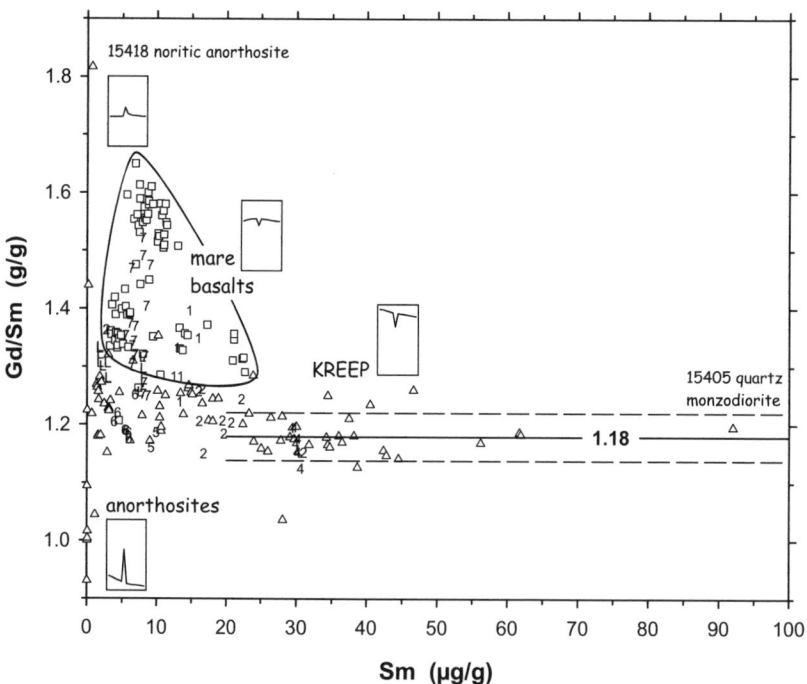

Figure 2.20. Data for Sm and Gd obtained by mass-spectrometric isotope dilution in lunar samples. Letters and numbers represent soils (e.g., Fig. 2.11), squares represent mare basalts, and triangles represent nonmare rocks. Mare basalts have Gd/Sm ratios that range from 1.3–1.6. For nonmare samples with high concentrations of rare earth elements (Sm >20 μg/g), that is, those with a large proportion of KREEP component, Gd/Sm ratios are essentially constant at 1.18 ± 0.04 (mean ± standard deviation). For nonmare samples with very low concentrations of Sm, Gd/Sm ratios are highly variable. Idealized chondrite-normalized REE patterns (e.g., Fig. 2.21) are shown for different portions of the plot.

obtained by INAA, a technique that determines Sm with high precision and Th with poor precision at low concentration and moderate precision at high concentration but which does not determine Gd accurately. As a consequence, there are no Gd data for most lunar samples. However, for any given sample or rock type, the Gd concentrations can be estimated from other trivalent REEs with high accuracy because of the strong interelement correlations among the trivalent REE. For nonmare soils and breccias with moderate to high concentrations of Sm, the Gd concentration can be estimated well by multiplying the Sm concentration by 1.18 (Fig. 2.20). Gd concentrations can also be estimated graphically in any sample using plots of chondrite-normalized concentrations of other REE against atomic number (e.g., Fig. 2.21). From INAA data for Ce, Sm, Tb, and Yb, the concentration of Gd can be estimated well in any

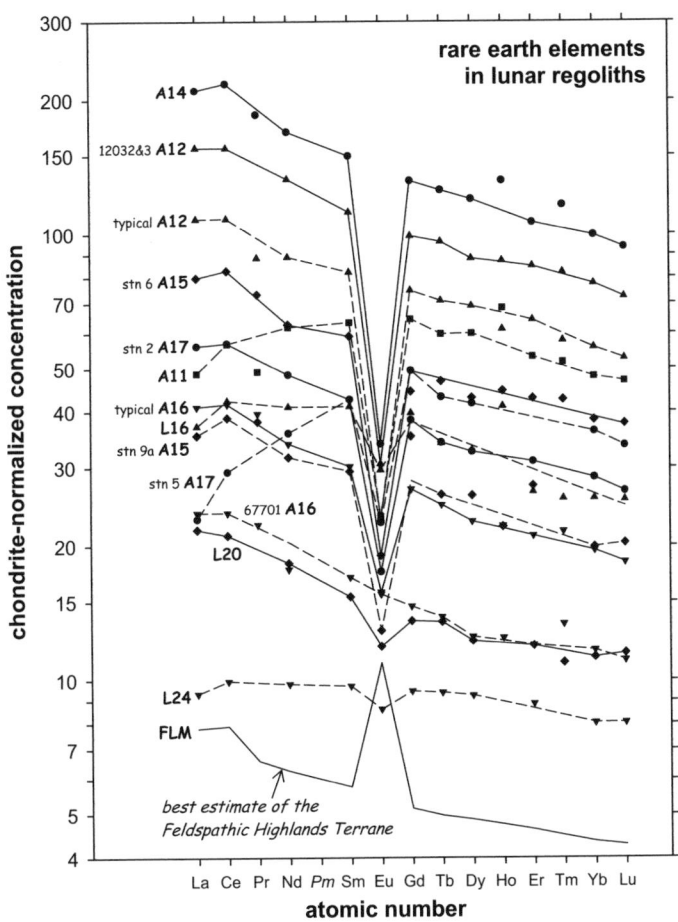

Figure 2.21. Chondrite-normalized concentrations of rare earth elements in lunar regolith as a function of REE atomic number (e.g., Haskin, 1985). Each regolith pattern represents averages of all data for that regolith. Data for Pr, Ho, and Tm are rare and imprecise. Dashed lines represent soils dominated by mare basalt. Samples 12032 and 12033 are the two most Th rich soil samples from Apollo 12 (Fig. 2.11). At Apollo 15, the soil from station 9a is the richest in Fe and that from station 6 is poor in FeO (Fig. 2.11). Sample 67701 is a soil from North Ray Crater at Apollo 16. At Apollo 17, soil from station 5 is the most Fe rich and that from station 2 is among the most Fe poor (Fig. 2.11). The feldspathic lunar meteorites (FLM), provide our best estimate of the surface composition of the Feldspathic Highlands Terrane (Table 2.6). Normalization values: 1.36C, where C represents the "Mean C1 Chondr." values of Table 1 of Anders and Grevesse (1989).

lunar sample by formulas based on logarithmic plots of chondrite-normalized concentrations using the CI chondrite values of Anders and Grevesse (1989):

$$Gd = 0.267 \ln[(A+B+C)/3] \qquad \text{(mean)}$$

$$A = \exp[3 \ln(Sm/0.2) - 0.5 \ln(Ce/0.82)] \qquad \text{(Ce-Sm extrapolation)}$$

$$B = \exp[(\ln(Sm/0.2) + 2 \ln(Tb/0.0493))/3] \qquad \text{(Sm-Tb interpolation)}$$

$$C = \exp[(6 \ln(Tb/0.047) - \ln(Yb/0.221))/5] \qquad \text{(Tb-Yb extrapolation)}$$

where *Gd*, *Ce*, *Sm*, *Tb*, and *Yb* are element concentrations in µg/g (ppm). The accuracy of the estimate depends strongly on the accuracy of the Tb concentration, an element that is determined with poor to moderate precision by INAA and for which systematic errors occur in some data sets. (The Ce-Sm extrapolation could be formulated with Nd instead of Ce, but Nd is determined with lower precision by INAA.)

Correlations with Thorium. In all lunar regolith, Th and U are highly correlated, with a mean U/Th ratio of 0.28 (Fig. 2.22). The correlation is strong because both elements are carried by the same few minerals, mainly zircon and RE-merrillite, the latter also being the main carrier of the REEs.

In nonmare regoliths, REEs such as Sm correlate well with Th, and Sm/Th ratios are typically about 2.8 (Fig. 2.23). Because the Apollo 14 rocks and regolith contain felsite (granite), a rare evolved lithology that has high concentrations of Th and low Sm/Th ratios (highly variable, but 0.9 ± 0.3 is typical) compared to KREEP (Jolliff 1998), the Sm/Th ratio of the Apollo 14 regolith is lower, 2.36. This difference is consistent with the observation that the ratio of RE-merrillite to zircon is less in felsite than in generic KREEP and zircon is not an important carrier of Sm. In contrast, the RE-merrillite to zircon ratio is high in alkali anorthosite compared to KREEP, leading to high Sm/Th ratios (6 ± 3). Although the Apollo 14 regolith also contains a high proportion of alkali anorthosite compared to other nonmare regoliths, Sm concentrations are not usually so high in alkali anorthosites that their presence modifies the Sm/Th ratio. Regoliths from mare sites also have high Sm/Th ratios, particularly

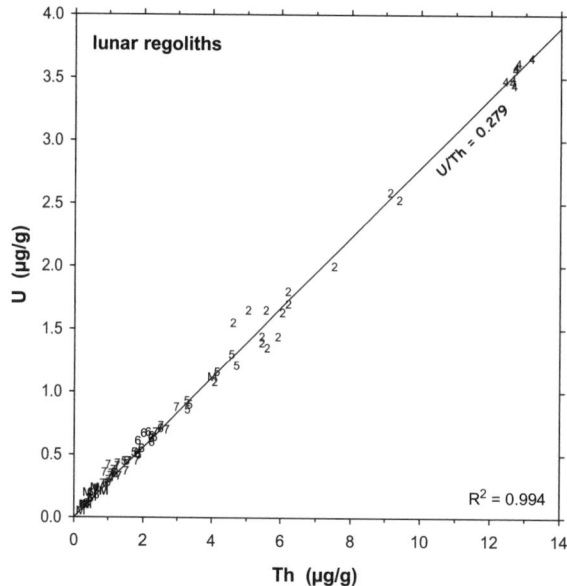

Figure 2.22. Th and U are highly correlated in lunar regolith. The line is defined by the origin and the mean Th and U concentration of the plotted points. Most of the scatter results from analytical uncertainty.

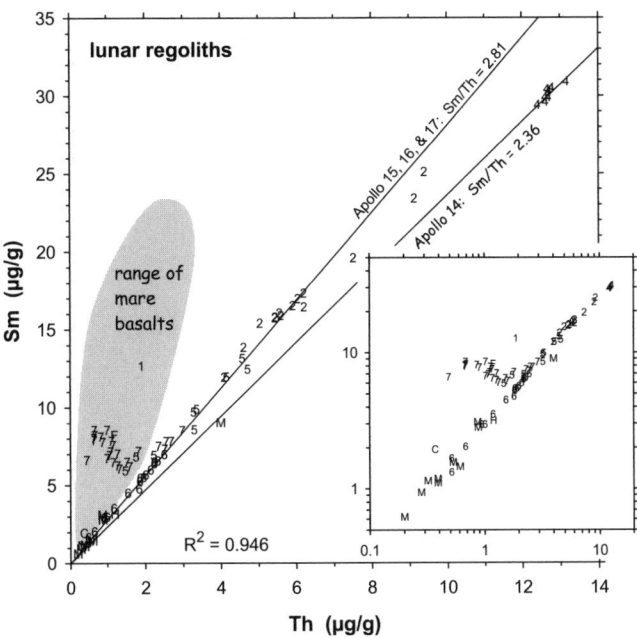

Figure 2.23. Variation of Sm with Th in lunar regolith (same symbol key as Fig. 2.12). Soils with high concentrations of Th are dominated by KREEP (Fig. 2.11). All KREEP-rich soils have similar Sm/Th ratios, although breccias and soils from Apollo 14 have somewhat lower ratios (2.36 ± 0.04) than those from other sites (2.81 ± 0.10; see text). Mare basalts have greater Sm/Th ratios than KREEP, thus soils dominated by mare basalt (Apollo 11, 15, and 17) plot off the KREEP mixing lines. The inset shows the same data on logarithmic scales.

those formed from high-Ti mare basalts (Figs. 2.23, 2.24). For example, the Sm/Th ratio of the Apollo 11 regolith is 6.5 and for the mare soils of Apollo 17 the Sm/Th ratio is 12.

Although also an elemental characteristic of KREEP, K does not correlate perfectly with Th in lunar regolith (Fig. 2.25). K/Th ratios in KREEP-rich samples are variable because K and Th are carried by different phases (K is in potassium feldspar) that are not exactly in the same proportions in all rocks of nominal KREEP composition. Mare basalts and KREEP-poor nonmare samples also have variable K/Th ratios. Anomalous regolith samples typically have greater K/Th ratios than Apollo 14-type KREEP.

7. FELDSPATHIC LUNAR METEORITES AS REGOLITH

Most of the information provided in this section is summarized from Korotev et al. (2003b) without further citation.

7.1. Composition of the feldspathic upper crust

Remarkably, a large fraction of the lunar meteorites are either regolith or fragmental breccias that are composed of fine-grained material that occurred within a few meters of the lunar surface (Warren 1994, 2001). Most lunar meteorites have compositions that are consistent only with an origin in the feldspathic highlands (Fig. 2.26). All of the feldspathic lunar meteorites have lower concentrations of the elements associated with KREEP (e.g., 0.2–0.5 µg/g Th; Table 2.1 and Fig. 2.27) than does typical regolith of the Apollo 16 site

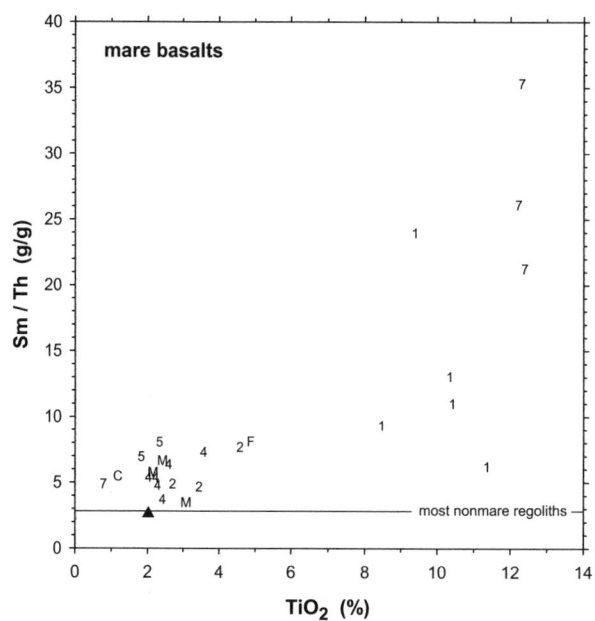

Figure 2.24. Among mare basalts there is a tendency for high-Ti basalts (Apollo 11 and 17) to have greater Sm/Th ratios than low-Ti basalts (2 = Apollo 12, etc., F = Luna 16, C = Luna 24, and M = lunar meteorite). Sm/Th ratios are greater in mare basalts than in KREEP basalt (triangle) and nonmare regolith (Fig. 2.23). Each symbol represents the mean composition of a type of mare basalt or a mare basalt meteorite.

(1.9–2.5 µg/g Th; Fig. 2.27), suggesting that the meteorites all originate from points more distant from the Procellarum KREEP Terrane than the Apollo 16 site. There are no reasons to suspect that the source craters of the meteorites are not randomly distributed about the Moon. Thus, ironically, the feldspathic lunar meteorites provide the best estimate of the composition of the surface of the Feldspathic Highlands Terrane (Table 2.6) even though their point of origin on the lunar surface is not known. Among the brecciated lunar meteorites, only Sayh al Uhaymir 169 and, possibly, Calcalong Creek have concentrations of incompatible elements great enough to have an origin in the Procellarum KREEP Terrane.

On the basis of the feldspathic lunar meteorites, the surface of the Feldspathic Highlands Terrane has the following composition: $0.22 \pm 0.04\%$ TiO_2, $28.2 \pm 1.0\%$ Al_2O_3, $4.4 \pm 0.5\%$ FeO (total Fe as FeO), $5.4 \pm 1.4\%$ MgO, and 0.37 ± 0.11 µg/g Th (Table 2.6). These estimates apply strictly to the lunar surface because the meteorites are surface samples. Consequently they include about 1.6% meteoritic material as CI chondrite, based on the mean concentrations of Ni, Ir, and Au in the meteorites. If we estimate the mean composition of, say, the upper kilometer of regolith by removal of the meteoritic component (Sections 2.1 and 6.2.1), concentrations of FeO and MgO are distinctly lower (4.0% and 5.3%) and that of Al_2O_3 is a bit greater (28.5%). This composition corresponds to 78% plagioclase by mass or about 82% by volume and is equivalent to that of a noritic anorthosite (Fig. 2.3).

7.2. Lunar meteorites as ground truth

Lunar meteorites are important to remote sensing because they provide a form of ground truth. For example, the distribution of FeO concentrations on the lunar surface as derived from Clementine spectral reflectance data compares well with the distribution of FeO among the lunar meteorites (Fig. 2.26). This concurrence provides support that the algorithm used to derive FeO concentrations from spectral reflectance parameters (Lucey et al. 2000) is accurate to first order. Conversely, it provides support to the assumption that estimates of the composition of the upper crust of the Feldspathic Highlands Terrane as derived from lunar meteorites are reasonable.

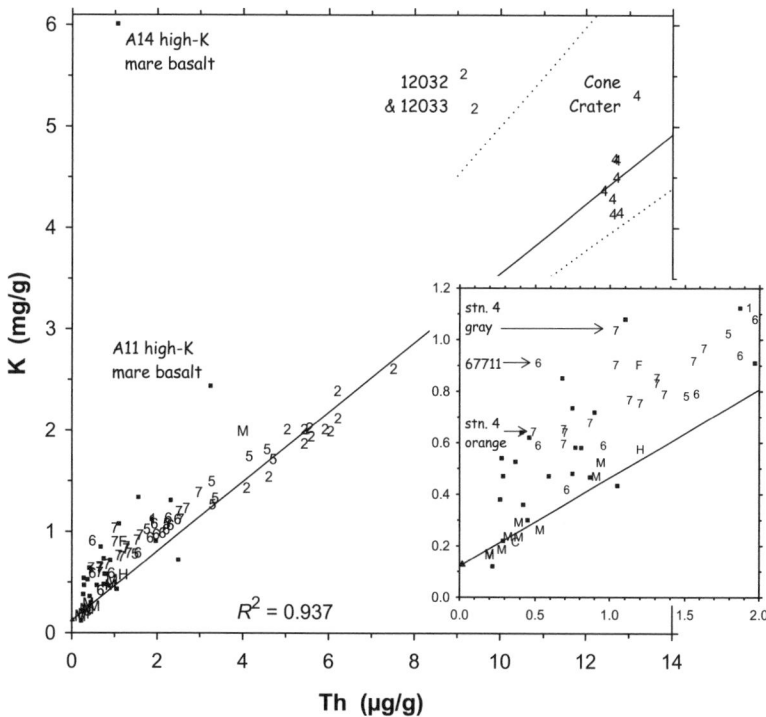

Figure 2.25. Variation of K with Th in lunar regolith (symbols as in Fig. 2.11) and all known types of mare basalt (small squares). The solid diagonal line (K = 120 + 343•Th) is defined by highly feldspathic ferroan anorthosite (sample 15415, triangle) and Apollo 14 KREEP-rich impact-melt breccias (off scale at 16.1 µg/g Th and K/Th = 350; Korotev 2000). R^2 refers to the regolith points only. The dotted lines represent the range of K/Th ratios in KREEP basalt and KREEP-rich impact-melt breccias from the Apollo sites (310 < K/Th < 500). Most mare basalts have K/Th ratios greater than that of KREEP and the feldspathic lunar meteorites (M's at low Th), with the high-K basalts of Apollo 11 and 14 being the most anomalous. Some regolith samples have anomalously high K/Th ratios compared other regolith samples collected at the same site. These include the ropy glass soils of Apollo 12 (samples 12032 and 12032, Korotev et al. 2000) and Apollo 17 (gray soils of station 4; Korotev and Kremser 1992), the orange glass soil of Apollo 17 (74220), the soil from Cone Crater at Apollo 14 (14141), and the alkali-rich immature soil, 67711, of North Ray Crater at Apollo 16 (Korotev 1983b, 1996).

A similar comparison of Th between the feldspathic lunar meteorites and data from the Lunar Prospector GRS suggests that at low Th concentration, the LP-derived concentration values (Lawrence et al. 2000) are too high (Fig. 2.27; Gillis et al. 2004; Warren 2005). Similarly, all feldspathic lunar meteorites have TiO_2 concentrations in the range of 0.2–0.3% (Table 2.1), much lower than the average of 1.5% TiO_2 derived for farside highlands from the Apollo GRS (Metzger and Parker 1979; Davis 1980). This discrepancy confirms the existence of a systematic error in the Apollo-derived data. Within the large uncertainties, Clementine-derived TiO_2 concentrations (Lucey et al. 2000; Section 10.4.6) are consistent with the feldspathic lunar meteorites.

7.3. The magnesium to iron ratio of the feldspathic crust

The numerous feldspathic lunar meteorites have the following interesting combination of properties: (1) With 72–86 wt% plagioclase (approximately 76–89 vol%, they are highly feldspathic, but not as feldspathic as ferroan anorthosites of the Apollo collection,

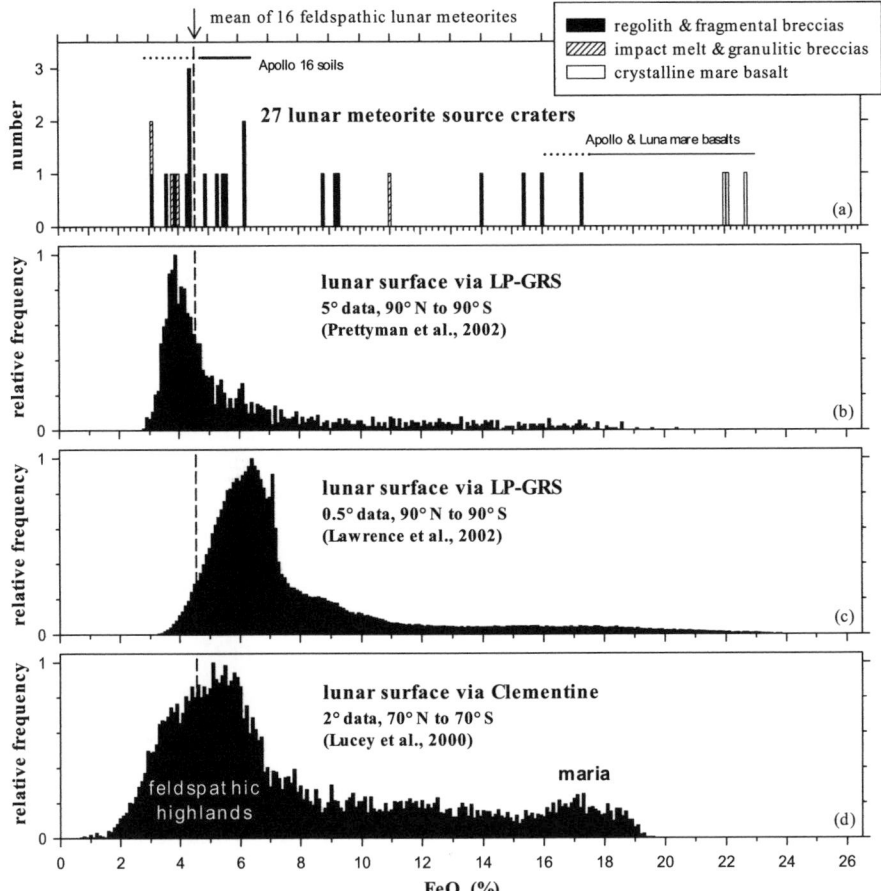

Figure 2.26. Comparison of estimates and measurements of the distribution of FeO on the lunar surface (updated from Korotev et al. 2003b). (a) Distribution of FeO concentrations (total iron as FeO) in source regions of lunar meteorites. The range of compositions of mature surface and trench soils from Apollo 16 is shown by the solid horizontal line at low FeO; the dotted line represents the range of immature and submature soils from North Ray Crater (Korotev 1996, 1997). The solid horizontal line at high FeO represents the range of most mare basalts from the Apollo and Luna missions; the dotted portion corresponds to the aluminous basalts of Apollo 14 (Dickinson et al. 1985). (b–d) Concentrations derived from data acquired from orbit. The LP (Lunar Prospector) GRS (gamma-ray spectrometer) results are those of Prettyman et al. (2002a; June 2002 data) and Lawrence et al. (2002; January 2002 data). For the Clementine results of Lucey et al. (2000), 94% of the lunar surface lies between 70° S and 70° N. The peak at low-FeO concentration corresponds mainly to farside, northern highlands.

which average 92 wt% plagioclase, (8.1 ± 8.6 wt% mafic silicates; Warren 1990). (2) To a first approximation the meteorites are all similar in composition in that they all have low concentrations of incompatible elements and a moderately narrow range of Al_2O_3 concentrations (25–31%) but they differ in having a wide range of MgO to FeO ratios (Fig. 2.12). The MgO/FeO variation leads to a wide range for Mg' of 57 to 77 (Fig. 2.19). (If only those meteorites for which compositional data have been published in peer-reviewed literature are considered, the range is reduced to 60–73.) (3) The MgO/FeO ratio of feldspathic lunar meteorites correlates with the proportion of normative olivine and high Mg' meteorites are rich

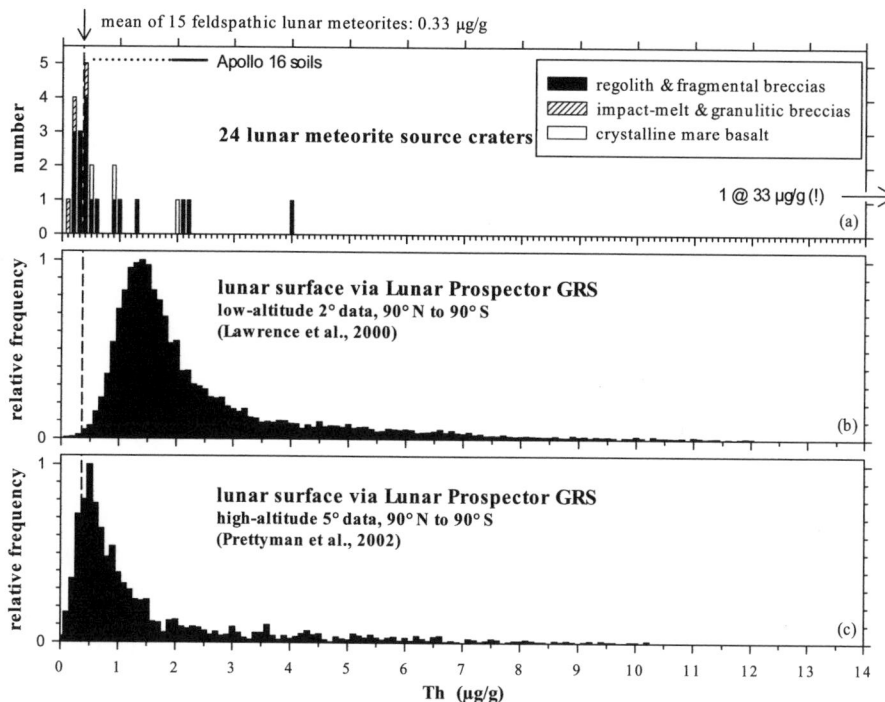

Figure 2.27. Like Figure 2.26, but for Th (updated from Korotev et al. 2003b). All feldspathic lunar meteorites have ≤0.6 µg/g Th. (a) One lunar meteorite, Sayh al Uhaymir 169 with 33 µg/g Th (Gnos et al. 2004), plots far off scale. No Th data are yet available for two meteorites of Figure 2.26. (b) Th concentrations derived from LP-GRS (gamma-ray spectrometer) data by Lawrence et al. (2000; with June 2001 update of data) using a simple peak-over-background technique are systematically high, at least at low Th concentration, compared the lunar meteorites. The mode (1.4 µg/g) exceeds the mean of the feldspathic lunar meteorites (0.36 µg/g) by a factor of 3.6. (c) The distribution of Th concentrations derived by Prettyman et al. (2002a; data of June 2002), using a library least-squares technique, more closely match the distribution of the brecciated lunar meteorites and yields a highlands mode (0.5 µg/g) only a bit greater than the mean of the feldspathic lunar meteorites.

in modal forsterite. (4) The mean Mg' of the feldspathic lunar meteorites (~69) is greater than the mean of ferroan anorthosite (~60) and the most magnesian lunar meteorites are outside the range for ferroan anorthosite (Fig. 2.4).

These observations have important implications if one accepts the premise that the meteorites represent the upper crust of the Moon, at least with respect to composition, and that they derive from numerous random locations. For example, they support the hypothesis that a significant fraction of the Mg and Fe in the feldspathic lunar crust does not derive from the ferroan-anorthositic suite of lunar plutonic rocks (Korotev and Haskin 1988). Qualitatively, the various observations are consistent with models that the early crust of the Moon was a flotation cumulate of ferroan anorthosite into which magnesian-suite magmas intruded. In terms of such models, the feldspathic lunar meteorites are mainly mixtures of FAS rocks and Mg-suite rocks such as those at the high-Mg' end of the trend of Figure 2.4 because most of the high-Mg' rocks of the figure are olivine-rich, i.e., troctolite and dunite. In detail, however, this simple model does not account for the feldspathic lunar meteorites and certain other observations. Problems include the following. (1) In feldspathic lunar meteorites, mafic, high-Mg' rocks such as troctolite and dunite are all but absent as clasts. (2) The actual lithologic

Table 2.6. Estimates of the concentrations some mostly lithophile elements in the upper few meters (Surface) and upper few kilometers (FUpCr) of typical feldspathic crust based on feldspathic lunar meteorites (from Korotev et al. 2003b)

	Unit	Surface	±	FUpCr		Unit	Surface	±	FUpCr
SiO_2	%	44.7	0.3	44.9	Zr	ppm	35	11	35
TiO_2	%	0.22	0.04	0.22	Ba	ppm	33	10	33
Al_2O_3	%	28.2	1.0	28.5	La	ppm	2.3	0.6	2.4
Cr_2O_3	%	0.096	0.014	0.092	Ce	ppm	6.0	1.6	6.0
FeO	%	4.4	0.5	4.0	Pr	ppm	0.8	0.2	0.8
MnO	%	0.063	0.004	0.060	Nd	ppm	3.6	0.9	3.7
MgO	%	5.4	1.4	5.3	Sm	ppm	1.1	0.3	1.1
CaO	%	16.3	0.9	16.4	Eu	ppm	0.78	0.05	0.79
Na_2O	%	0.35	0.03	0.34	Gd	ppm	1.3	0.3	1.3
K_2O	%	0.027	0.008	0.026	Tb	ppm	0.23	0.05	0.23
P_2O_5	%	0.027	0.009	0.023	Dy	ppm	1.5	0.4	1.5
Σ	%	99.8		99.9	Ho	ppm	0.33	0.08	0.33
Mg´	%	69	3	70	Er	ppm	0.9	0.2	0.9
K	ppm	220	60	210	Tm	ppm	0.14	0.03	0.14
Sc	ppm	8.0	1.0	8.0	Yb	ppm	0.89	0.2	0.90
Cr	ppm	660	90	630	Lu	ppm	0.13	0.03	0.13
Mn	ppm	490	40	460	Hf	ppm	0.8	0.2	0.8
Co	ppm	17	3	10	Ta	ppm	0.11	0.02	0.11
Ni	ppm	185	45	~16	Ir	ppb	7.5	2.8	=0
Rb	ppm	0.7	0.3	0.7	Au	ppb	2.8	1.0	~0.6
Sr	ppm	150	12	151	Th	ppm	0.37	0.11	0.38
Y	ppm	9	2	9	U	ppm	0.16	0.10	0.16

"Surface" is essentially the mean composition of lunar meteorites ALHA 81005, MAC 88105, QUE 93069, Yamato 86032, Dar al Gani 262, Dar al Gani 400, Dhofar 025, and NWA 482 based all literature data (Table 1), except that estimated concentration values were used for data suspected to be compromised by terrestrial contamination (K, Ca, Rb, Sr, Ba, and Au). Values for Y, Pr, Gd, Dy, Ho, Er, and Tm were estimated from the other REE. The uncertainty (±) is the 95% confidence limit on the mean values. "FUpCr" (feldspathic upper crust) is an estimate of the composition of the upper few kilometers of the crust and differs from the surface composition only in the absence of a component of CI chondrite (1.58%; values of Anders and Grevesse, 1989, or 1.16% on a volatile-free basis).

carriers of Mg are feldspathic lithologies, mainly magnesian granulitic breccias (Section 5.2.4). (3) Compositions of the magnesian granulitic breccias are not consistent with mixtures of ferroan anorthosite and mafic Mg-suite rocks such as the norites, troctolites, and dunites of the Apollo collection. (4) Among the Apollo samples, mafic Mg-suite rocks were found mainly at the Apollo 15 and 17 sites, which are in or near the Procellarum KREEP Terrane; they were not found at the Apollo 16 site, which is in the Feldspathic Highlands Terrane.

Compositions of the feldspathic lunar meteorites suggest that *Mg´* varies considerably from place to place at the surface of the feldspathic highlands, even in regions largely devoid if Imbrium ejecta. There is little information on the nature of scale and distribution of the variation because the Mg concentration of the feldspathic highlands is low (3–8% MgO) and has, as of this writing, not been determined precisely by orbital spacecraft. Obtaining such information is a high priority goal of future orbiting missions because the data would provide an important constraint for models of formation of the lunar crust. For example, the Luna 20 regolith (23% Al_2O_3) is more mafic than the feldspathic lunar meteorites and has a moderately high Mg' (70). Does the unusual composition reflect that the regolith consists mostly of ejecta from Crisium, a basin that penetrated deeper into the crust than most? Are the ejecta of other basins in the Feldspathic Highlands Terrane more magnesian than regions distant from basins? Is the lower crust more magnesian than the upper crust?

In order to constrain Mg' to the ± 2 units (e.g., 69 ± 2), which would provide truly useful information, Fe and Mg concentrations need be determined to within 2–3% of their values, e.g., 4.4 ± 0.11% FeO and 5.4 ± 0.14% MgO for typical feldspathic highlands (Table 2.6). Alternatively, the Mg/Fe ratio would need to be determined to within 4%. Such precision would be a challenge to any presently available technology. If the uncertainty in Fe and Mg concentrations are each 10% and not correlated with each other, then the uncertainty in Mg' is 8 units (e.g., 69 ± 8) for typical highlands compositions. For certain types of questions, for example, "Is proximal basin ejecta more magnesian than distal ejecta?," this level of precision would be adequate if the spatial resolution were small enough such that numerous measurements could be obtained on the proximal deposit.

8. REGOLITH EVOLUTION: MICROMETEORITES AND ENERGETIC PARTICLE INTERACTIONS

The Moon's surface is a hostile environment. Unprotected by a planetary magnetic field and atmosphere, it is assailed by meteoroids and energetic particles. These influences constitute the lunar weathering agents. Unlike on Earth where water and wind are major weathering forces, the forces of weathering on the Moon originate external to the planet. Thus the term *space weathering* is often used to denote the *in situ* modification of lunar surface material by external forces. The result of space weathering of lunar soil is *maturation*, the accumulated effects of space weathering. Space weathering denotes processes at the finest scales, of grains or a few grains. The process that operates at larger scales, impacts of objects larger than dust particles, produce the lunar soil upon which space weathering operates. These effects include disaggregation, lithification, melting, and vaporization of material from meteoroid impacts, implantation of ions from solar and galactic cosmic rays and the solar wind, sputtering of surface atoms from interaction of cosmic rays, condensation of sputtered and impact-vaporized material, accumulation of meteoritic material and agglutinates. Micrometeorite impact also produces "nanophase," "single-domain," or "submicroscopic" iron grains not of meteoritic origin that are very small, typically 4–33 nm in diameter, thought to consist of essentially pure iron that is produced by reduction of Fe^{2+}.

Many of the effects of maturation can be quantified, e.g., the concentrations of solar-wind derived gases or the relative abundance of agglutinates, and these quantities are used as a measure of maturity. This section discusses progress in understanding regolith evolution and maturation from a sample perspective. Space weathering products, principally the production of nanophase Fe, have major effects on the optical properties of lunar soil. Optical effects of space weathering are discussed in Section 10.4.3.

8.1. The Moon's micrometeorite and energetic-particle environment

The scientific significance of the energetic particles in lunar science is two-fold. First, these particles alter the composition of the lunar surface physically and chemically and these influences are used to make inferences about the history of the lunar surface. Second, the immediate effects of these particles enable gamma-ray and neutron spectroscopic remote sensing (Sections 10.1, 10.2). Micrometeorite impact causes effects not found in larger impacts owing to very short timescales. These effects are central to many space-weathering effects.

There are a variety of energetic particles in the Moon's environment. Because the Moon has essentially no atmosphere and no or very weak magnetic fields, these particles interact with the lunar surface and often leave a characteristic and measurable record of their interactions. Some interaction products, such as energy-rich defects and stable cosmogenic nuclides, may be preserved over the entire history of the Moon and provide windows into the distant past of the Moon and of these particles. Other products, such as cosmogenic radionuclides, can be used to study specific time periods.

The most numerous energetic particles in the lunar environment are those in the solar wind, in the solar energetic particles, and in the galactic cosmic rays (for details on these particles and their lunar interactions, see, e.g., Reedy et al. 1983; Sonett et al. 1991; Vaniman et al. 1991). These particles are either implanted or induce nuclear reactions in lunar materials.

The solar wind is a plasma continuously emitted from the Sun and contains an equal number of ions and electrons. At 1 A.U. from the Sun, the solar wind ions are moving away from the Sun at velocities of ~300-800 km/s and have energies of ~1 keV/nucleon. Solar wind ions penetrate only a few tens of nanometers into lunar material before stopping. Many solar wind-implanted species diffuse out of the surface and escape back into space. Some, however, remain implanted indefinitely.

Solar energetic particles, sometimes called solar cosmic rays, are particles emitted from the Sun that have been accelerated to energies of ~1-100 MeV/nucleon either at the Sun (small impulsive events) or by shock processes in the interplanetary medium (large gradual events). Occasionally, solar particles can be accelerated to GeV and higher energies. Solar energetic particles occur very irregularly and for most of the time are not present in the solar system. Solar particle events last a few days to a week or so. Most solar-particle events occur two or more years away from the time of minimum solar activity. Modern solar energetic particles have been studied since 1956, and are usually about 98% protons with a proton/alpha-particle ratio of ~50. There is also a small fraction of heavier ions. Solar protons penetrate up to ~1 cm in lunar material. Lunar surface materials contain noble-gas isotopes implanted deeper into grains than those isotopes in the solar wind. Although these more-energetic implanted particles are referred to as solar energetic particles, they appear to be different from those that make up modern solar energetic particles (Wieler 1998).

The galactic-cosmic-ray particles come from outside the solar system and typically have energies of ~0.1-10 GeV/nucleon. They are about 87% protons, 12% alpha particles, and 1% heavier ions. The flux of galactic cosmic-ray particles vary by factors of ~2 over a typical 11-year solar-activity cycle. Galactic cosmic-ray particles can penetrate meters into the Moon and produce a large cascade of secondary energetic particles, including many neutrons.

Energetic charged particles are slowed down in material by ionization energy losses. While ionizing matter, they can induce radiation damage that can be observed as tracks or as thermoluminescence. Most solar energetic particles and galactic cosmic-ray particles are energetic enough to induce nuclear reactions. The nuclear reactions that are produced by these primary and secondary energetic particles make a variety of stable and radioactive nuclides. Some of these product nuclides can be identified as having been made by such nuclear reactions, including rare noble-gas isotopes such as ^{21}Ne and radionuclides such as ^{26}Al. Recent reviews on cosmogenic nuclides and their applications include Vogt et al. (1990), Herzog (1994), Tuniz et al. (1998), and Michel (1999).

The impact of micrometeoroids, essentially interplanetary material ("dust") of less than a millimeter in diameter, has strongly affected the lunar surface. On the basis of estimates for the Earth (Love and Brownlee 1993), approximately 80 g/km^2 of micrometeoroids accrete to the Moon each year. Assuming, for convenience, that they each have the typical diameter of 0.22 mm (Love and Brownlee 1993), a typical square meter of lunar surface will be hit 5 to 10 times per year by a micrometeoroid. Each impact delivers on the order of a Joule of energy. The micrometeoroid vaporizes, but the lunar regolith both melts and vaporizes. This impact flux also causes erosion of rocks. The rates that micrometeoroid impacts (and sputtering reactions) erode the surfaces of lunar rocks have been inferred from profiles for solar cosmic-ray-produced nuclides. The erosion rates of ~1–2 mm/Ma inferred by recent measurements generally agree with previous erosion rates. The erosion rate determined for the hard layer on the surface of rock 64455 (Nishiizumi et al. 1995) was less than that for 74275 (Fink et al. 1998).

8.2. Recent developments in the detection and studies of these interaction products

The products from the interactions of energetic particles with the Moon are measured by counting the atoms themselves or, for some radionuclides, by counting their decay radiations. New and improved techniques have been developed over the last decade that have improved the ability to detect and study these products in lunar samples. For long-lived radionuclides, the use of accelerator mass spectrometry (AMS) has greatly improved detection sensitivities, accuracies, and precision (e.g., Tuniz et al. 1998). The measurements of atoms in gaseous and solid samples have improved, such as better methods for making and extracting ions into mass spectrometers. New techniques have also been developed to get these ions from a sample, such as using closed-system stepped etching (Wieler and Baur 1995). Some measurements of implanted isotopes are now being done with ion microprobes (e.g., Hashizume et al. 2000).

New auxiliary data that are critical for interpreting the measurements have become available, especially for cosmogenic nuclides. Many cross sections have recently been measured for nuclear reactions that make cosmogenic nuclides (e.g., Michel et al. 1997; Sisterson et al. 1997; Jull et al. 1998). Much better computer codes are now available to numerically simulate the propagation of cosmic-ray particles in the Moon and the attendant production of cosmogenic nuclides (e.g., Reedy 2000; Leya et al. 2001; Masarik et al. 2001). Experimental simulations have been performed to test these models (Leya et al. 2000a). The calculated production rates for making cosmogenic nuclides are now much better than they were previously and have been extended to unusual compositions and irradiation geometries (such as very deep in the Moon).

8.3. Implanted solar-wind gases

Mature lunar soils are rich in gas, and the abundance of this gas correlates well with other maturity indices. Many of the noble-gas isotopes found in lunar samples were implanted. Such isotopes can be used to study the history of the lunar sample in which they are found. Samples now buried up to several meters deep in the regolith and some grains inside breccias record such implantations, indicating that they once were on the very surface of the Moon.

Some isotopes have been identified that escaped from the Moon, became ionized, were accelerated in the interplanetary medium, and were then implanted in the Moon. Many atoms of ^{40}Ar made by the decay of ^{40}K escape and are later implanted in the Moon. The amount of this "parentless" ^{40}Ar relative to solar wind-implanted ^{36}Ar changed with time and has been used to study the antiquity of the samples into which they were implanted (Eugster et al. 2001). Some lunar samples had their parentless ^{40}Ar implanted at about 4 Ga.

The record of noble-gas isotopes implanted in the Moon has been well studied. By using samples with known dates for their surface exposures, it has been shown that the relative abundances of noble-gas isotopes from the Sun have changed little during the last 4 Ga (Wieler and Baur 1995; Wieler 1998). Elemental ratios of heavy noble gases have also provided some evidence on how atoms in the solar wind are ionized at the Sun (Wieler 1998).

Some of the nitrogen observed in lunar samples, especially near the very surface of lunar grains, is implanted (e.g., Hashizume et al. 2000). The isotope ratio for this implanted nitrogen in the past has varied by ~30%. The source of this variation is still in question (e.g., Kerridge 2001; Hashizume et al. 2001; Mathew and Marti 2001).

Excesses of 5730-year ^{14}C and 1.5-Ma [10]Be have been observed in the surface of lunar samples (e.g., Jull et al. 2000; Nishiizumi and Caffee 2001). These radionuclides are believed to have been made at the Sun and can serve as tracers as nuclear processes occurring in the outer regions of the Sun and for the escape of isotopes from the Sun into the solar wind.

8.4. Solar cosmic rays

8.4.1. Recent measurements.

Several depth-versus-concentration profiles have been measured for nuclides made by solar energetic particles. Thin layers were ground from rock 64455 to give a detailed profile near the very surface where most SCR nuclides are produced (Nishiizumi et al. 1995). Better profiles have been measured using AMS for [10]Be, ^{14}C, and [26]Al in several lunar rocks (e.g., Nishiizumi et al. 1995; Fink et al. 1998; Jull et al. 1998). Because [10]Be is mainly made by higher-energy solar protons, its measurement with other SCR-produced radionuclides such as [26]Al helps to constrain the spectral shape of the solar protons. Profiles have also been measured by AMS for radionuclides that previously were very hard to measure, [36]Cl, [41]Ca, and [59]Ni (Fink et al. 1998; Reedy and Nishiizumi 1998; Schnabel et al. 2000). Near-surface depth profiles for cosmogenic [3]He and Ne isotopes were also measured and used to study the records of the solar cosmic rays over the last 2 Ma (Rao et al. 1994).

8.4.2. SCR fluxes in the past.

These new and some older profiles of cosmogenic nuclides in the top ~1 cm of lunar samples have been used to determined average fluxes of solar protons during the past ~10 Ma. The profile for [59]Ni was used to study the fluxes of solar alpha particles averaged over the last ~0.1 Ma (Schnabel et al. 2000). Measured profiles for solar-proton-produced nuclides have been interpreted by using newly measured cross sections and improved models for GCR production of these nuclides. These new and better average fluxes of solar protons during the last few Ma appear to be less than those for more recent periods (~10-500 kyr) and for SEP measurements since 1956 (Reedy 1998).

8.5. Products of galactic-cosmic-ray interactions

8.5.1. Neutron-capture reactions.

Several profiles for neutron-capture reactions as a function of depth have recently been reported in Apollo deep-drill cores. The measurements done by AMS for the 0.1-kyr radionuclide [41]Ca agreed well with theoretical calculations (Nishiizumi et al. 1997). Regular mass spectrometry was used to study Sm and Gd isotopic anomalies induced by thermal-neutron-capture reactions as a function of depth (Hidaka et al. 2000). The depth profile for such neutron-capture reactions is quite different from those for most cosmogenic nuclides, which are made by high-energy spallation reactions. Having cosmogenic nuclides with very different depth profiles helps in determining the depth at which a sample, such as a lunar meteorite, was exposed. Other isotopic anomalies have been observed due to reactions on Eu (Hidaka et al. 2000) and cadmium (Sands et al. 2001). Leya et al. (2000b) reported that neutron-capture reactions on [181]Ta can affect the isotopic ratios of tungsten isotopes and could affect the use of W in the Hf-W cosmochronometer.

8.5.2. Spallation reactions.

Cosmogenic ^{15}N was measured by Mathew and Marti (2001) in nitrogen released at high temperatures from lunar samples. These data were used to infer its production rate near the lunar surface, which was then compared with previously inferred or calculated production rates.

New numerical simulations of the production of spallogenic nuclides have been used to calculate production rates for many cosmogenic nuclides in the Moon. These calculated production rates have been used to unfold the irradiation histories of lunar meteorites and to predict effects not yet seen. Leya et al. (2001) reported depth-versus-production rate profiles for several spallogenic radionuclides and for [21]Ne and [22]Ne. Masarik et al. (2001) reported calculated production rates as a function of lunar depth for [3]He, [21]Ne, and [22]Ne and noted that the [22]Ne/[21]Ne ratio starts to increase at great depths after decreasing with depth, an effect not yet seen experimentally in lunar samples.

8.5.3. Regolith gardening. SCR-produced radionuclides can be used to study the top ~1 cm of lunar cores. Rates for the mixing of grains due to micrometeoroid impacts, called "gardening," are of the order of a few centimeters per million years (e.g., Reedy et al. 1983). Jull et al. (1996) reported that the top few centimeters of core 76001 was disturbed very recently, probably less than 1 kyr ago by a nearby cratering event. Stable cosmogenic nuclides can be used to estimate gardening effects to lunar depths of meters on time scales of billions of years. The measured profiles of Hidaka et al. (2000) for Sm and Gd isotopic anomalies confirm earlier results that the Apollo 15 deep-drill core has been relatively undisturbed over a very long part of the Moon's history.

8.5.4. Rock erosion. The rates that micrometeoroid impacts and sputtering reactions erode the surfaces of lunar rocks have been inferred from profiles for SCR-produced nuclides. The erosion rates of ~1-2 mm/Ma inferred by recent measurements generally agree with previous erosion rates. The erosion rate determined for the hard layer on the surface of rock 64455 (Nishiizumi et al. 1995) was less than that for 74275 (Fink et al. 1998).

8.5.5. Crater ages. Cosmogenic nuclides have been used to determine when many craters were formed. New measurements by Eugster (1999) confirm the previously reported 2 Ma age for the Apollo 16 South Ray crater.

8.6. Exposure records of lunar meteorites

8.6.1. Ages. By measuring several cosmogenic nuclides with a range of half-lives, the various ages of lunar meteorites can be determined (e.g., Nishiizumi et al. 1996; Polnau and Eugster 1998). Lunar meteorites are finds. The length of time that they were on the Earth's surface and shielded from cosmic rays can be determined from the decay of the shorter-lived radionuclides such as ^{14}C. Most lunar meteorites have terrestrial ages of less than 100 kyr. The length of time that the meteorite was exposed to the omnidirectional (4π) cosmic rays in space while in transit from the Moon to the Earth can usually be distinguished from any cosmogenic-nuclide production while the meteorite was in the lunar surface (a 2π irradiation). The sum of the meteorite's terrestrial age and its 4π-exposure age in space gives the time that the meteorite was ejected from the Moon. Many cratering events on the Moon ejected lunar meteorites.

8.6.2. Lunar exposure depths. Not only can the exposure ages for lunar meteorites in the Moon's surface be determined, but also the depth at which that exposure took place usually can be determined. Measurements for several cosmogenic nuclides are needed to determine this lunar exposure depth and the various ages for a lunar meteorite (e.g., Nishiizumi et al. 1999). The exposure depths for eight lunar meteorites are summarized in Warren (1994), with six of them having been exposed at depths in the Moon of less than about 600 g/cm^2 (less than ~ 3 meters). Warren (1994) noted that none of the 10 martian meteorites then studied had a record of a 2π exposure in their parent body.

9. REGOLITH EVOLUTION: MIXING, SPACE WEATHERING, AND MATURITY

9.1. Reworking and mixing

In addition to energetic particle interactions at the very top surface, major physical changes are induced by the interaction of micrometeorite impacts. Two major physical changes that occur with regolith maturation are (1) comminution, the disaggregation or breaking of rocks and minerals into smaller fragments, and (2) agglutination, the formation of agglutinates (Section 2.1). The two processes compete to decrease and increase, respectively, the mean grain size of soil particles at the very surface. Beginning with a fresh deposit of mare basalt or impact ejecta from a crater, maturation leads to a decrease in the mean grain size of the surface material with time because comminution by micrometeorite impacts dominates at first. As the soil matures *in situ* and more fine-grained material is produced, agglutination converts fine material to coarser-grained material. This process was termed *soil evolution path 1* ("reworking dominates mixing") by McKay et al. (1974). In principle, in an undisturbed area a steady state might be reached where the agglutination rate equals the comminution rate. However, the mean grain size of lunar regolith increases with distance below the surface, thus on average large impacts deposit material at the surface that has a coarser mean grain size than the material that was there before the impact. The mixing of rock fragments with soil, or soil with soil, was called *soil evolution path 2* ("mixing dominates reworking") by McKay et al. (1974). In effect, the grain size distribution of a lunar soil is controlled by three processes: comminution, agglutination, and mixing. On average, mature soils, that is, soils with a high percentage of agglutinates, a high density of energetic particle tracks, and high concentrations of solar-wind-implanted ions, are finer grained than immature soils (McKay et al. 1974; Morris et al. 1978b; Fig. 2.28).

Although maturation effects occur only at the very surface, the turnover of the regolith by small impacts, a process termed *in situ reworking*, mixes and buries mature soil, thus units of mature soil exist to depths of at least a few meters (Morris 1978a). Data for Apollo cores show that any of a number of parameters, such as maturity or chemical composition, vary unsystematically with depth at any given location. The cores do not represent continuous slow-deposition, as in a terrestrial ocean environment, but sporadic impact events intermixed with periods of quiescent soil development. Stratigraphic units in the cores (Figs. 2.15, 2.16) represent discontinuous pods, rather than layers with horizontal extent and time significance (e.g., Korotev and Morris 1993). On average, however, maturity decreases by a factor of two within the upper half meter (Fig. 2.29; see also Plate 2.2).

9.2. Space weathering products

9.2.1. Nanophase iron and agglutinates.
An important observation made on the first Apollo samples to be studied was that the intensity of the ferromagnetic resonance (FMR) signal from the lunar soil samples was almost an order of magnitude greater than that from rock samples (e.g., Manatt et al. 1970; Tsay et al. 1971a,b). Tsay et al. (1971a,b) demonstrated that the FMR signal was due to small particles of metallic Fe. The FMR signal is dominated

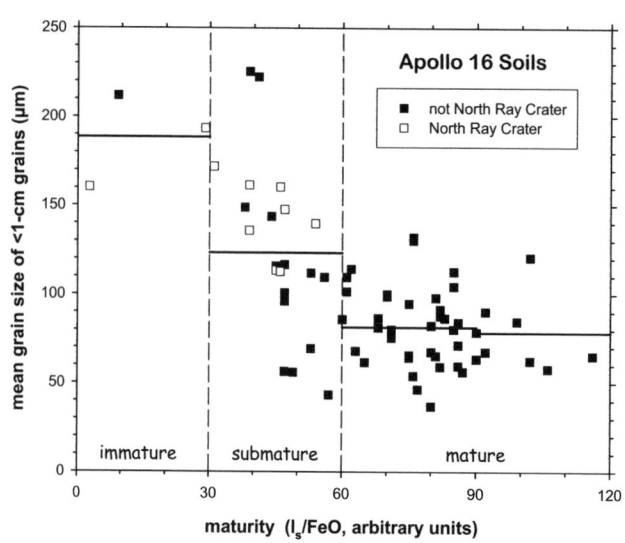

Figure 2.28. Variation in the mean grain size of the <1-cm grain-size fraction of Apollo 16 regolith samples (Graf 1993) with regolith maturity as measured by I_s/FeO (Morris 1978b). Open squares represent surface and trench soils developed on the ejecta deposit of North Ray crater, a fresh crater formed 50 Ma ago; closed squares represent all other surface and trench soils. The immature, submature, and mature divisions are those of Morris (1976, 1978b). The horizontal lines represent the mean concentration of all points within the range of I_s/FeO that they span. The plot demonstrates that mature soils tend to be finer grained than immature soils (McKay et al. 1974).

by single-domain grains of Fe^0 in the size range of 4 to 33 nm, grains that today are usually called nanophase Fe. Housley et al. (1974) and Cirlin et al. (1974, 1975) showed that for soil particles, the characteristic resonance is predominantly associated with agglutinates and regolith breccias. Agglutinates are the carriers of much of the nanophase iron in lunar soil and that most of the iron grains are in the 10–20 nm range. FMR Curie-point measurements show that these grains are essentially pure metallic Fe in composition (Morris et al. 1975). These tiny blebs of Fe^0 are found as a surface-correlated feature in the rims of grains, and also incorporated throughout agglutinates (Keller and McKay 1997).

Recent discoveries have been made of another major reservoir of nanophase iron in lunar soils (Keller and McKay 1993, 1997). This Fe^0 is present in thin patinas (~0.1 μm thick) on the surfaces of most soil particles (Wentworth et al. 1999). Several high-resolution TEM and EMP studies have described and documented the significant contributions from vapor-deposited, nanophase iron present in the patinas of most lunar soil grains (e.g., Keller and McKay 1997; Keller et al. 1999, 2000; Pieters et al. 2000; Taylor et al. 2000a,b, 2001a,b). Figure 2.30 shows BSE and X-ray images produced from both mature and immature mare soils, where even plagioclase grains contain thin rims with appreciable Fe contents, and silica-rich patinas are developed on ilmenite and olivine. These last two minerals also contain Fe-rich rims, as confirmed by high resolution transmission-electron-microscopic examination. Figure 2.30 also shows TEM

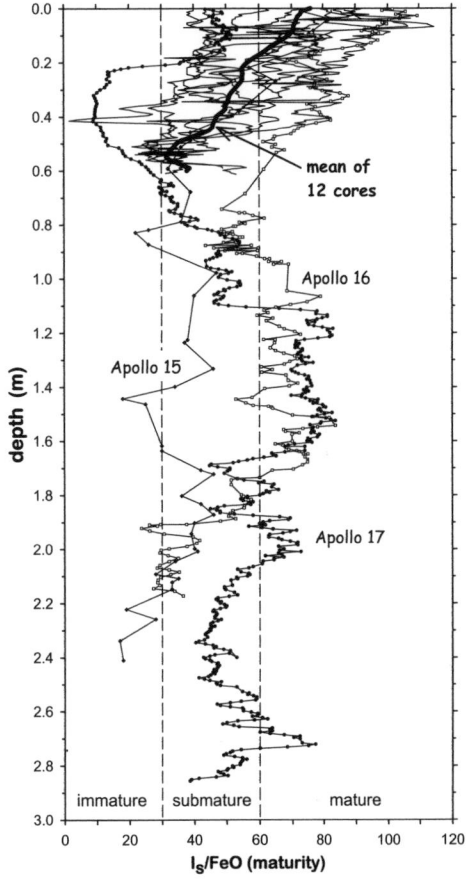

Figure 2.29. Variation in the regolith maturity parameter I_s/FeO with depth in 12 Apollo regolith cores. "Deep drill cores," which penetrated to 2–3 m, were collected on the last three Apollo missions. Other data derive from drive tubes of ~30 or ~60 cm length. It takes about 500 m.y. to garden to a depth of ~50 cm and ~100 m.y. integrated surface time to produce a mature regolith on the surface (Morris 1978a). All data are from R.V. Morris (Heiken et al. 1976; Morris and Gose 1976; Gose et al. 1977; Morris et al. 1979, 1989; Bogard et al. 1980, 1982; Korotev and Morris 1993; Korotev et al. 1997a; and unpublished data). See Plate 2.2 for color version.

photos from Keller et al. (2000) where the thin patinas on plagioclase and agglutinate grains show numerous nanophase iron particles. The spheres of Fe^0 found in agglutinates are roughly twice as large as those found in amorphous rims, averaging ~7 nm in diameter vs. ~3 nm for the rims (Fig. 2.30). The image shows layers of patinas on the agglutinate as well as the finer-grain size of the nanophase iron relative to that in the interior of the agglutinate. The major portion of this nanophase iron formed by deposition of vapor produced by abundant micrometeorite impacts, as documented by the presence of multiple and overlapping patinas. A smaller portion seems to have formed by radiation sputtering (Bernatowicz et al. 1994). The exact mechanism of formation of this surface-correlated Fe^0 is not well understood (Keller and McKay 1997), but

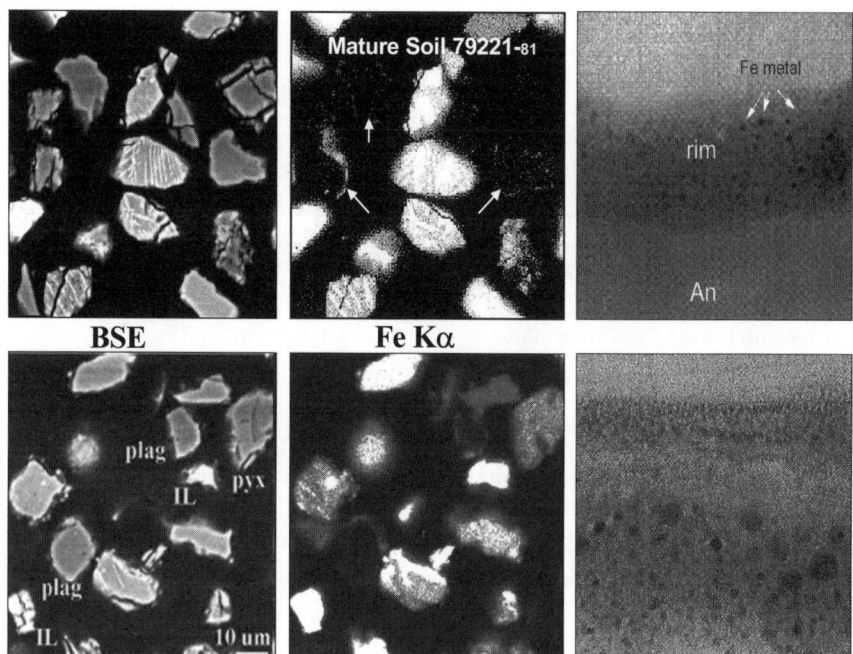

Figure 2.30. Plagioclase grains in lunar soil. Note the plagioclase grains ($CaAl_2Si_2O_8$) in the Back-Scattered Electron (BSE) image (left). In the Fe X-ray map (center), a thin Fe rim is present on these plagioclase grains, giving them a significant bulk magnetic susceptibility. Both mature (79221; top row) and immature (71061; bottom row) soils have vapor-deposited coatings on rims of most grains. The TEM images (right) show the fine-grained nature of the nanophase Fe^0 on the plagioclase (An) grains. Modified after Keller et al. (2000), Pieters et al. (2000), and Taylor et al. (2001a).

its presence was predicted by Hapke et al. (1975), but largely overlooked for over 20 years. Recent work (e.g., Basu et al. 2001) suggests that the nanophase iron in agglutinates was originally formed in vapor-deposited rims. The grain size of the metal in agglutinates is typically larger than in the rims, consistent with aggregation of the metal particles in the melt phase.

The relative importance of patinas vs. agglutinates as carriers of nanophase iron was addressed by Taylor et al. (2001b), who presented a comparison of the FMR values for each size fraction of nine mare soils versus agglutinitic glass abundances for these same soil splits, as shown in Figure 2.31. The numbers above each data point represent the percentage increase in the value with reference to the next coarsest size. For example for 79221, the FMR value for the <10 μm fraction is 117% greater than that for the 10-20 μm size, while the agglutinitic glass content has only increased by 15%. The large increases in the FMR values with decreasing grain size contrast with the small increases in the agglutinitic glass abundances for the same grain-size change.

Assuming spherical particles, for the same masses of two soil-size fractions, the surface area of the size fraction increases by a factor of 4, as the average grain size decreases by only 50%. If the increase in FMR signal that is attributable to the increase in agglutinitic glass is accounted for in each change in grain size, the "residual" is the surface-correlated FMR contribution. On average, there is about a 100% increase in FMR signal value between the finest two size fractions (Fig. 2.31). Therefore to a first approximation, the increase of 2× in FMR correlates well with the predicted 4 times increase in particle surface area. The contribution to the

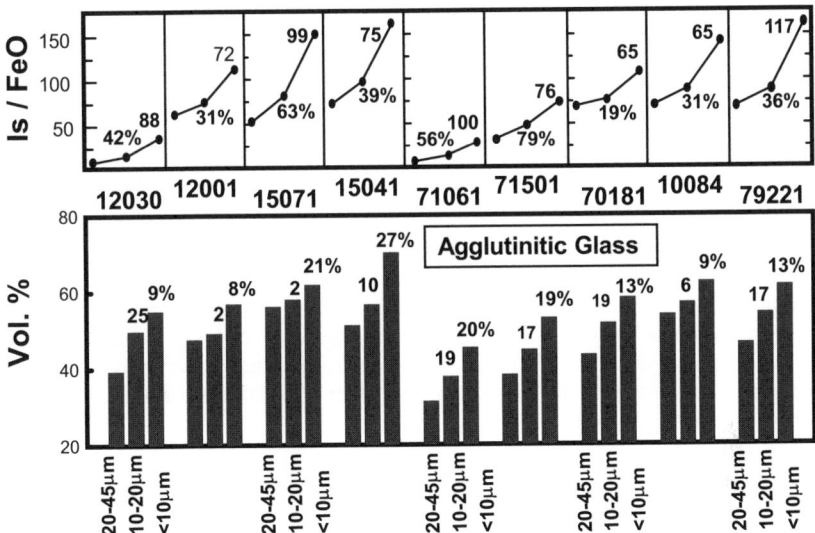

Figure 2.31. Comparisons of the I_s/FeO values for the finest fractions of mare soils with changes in agglutinitic-glass abundances between grain-size fractions. The numbers above each data point represent the stepwise percentage increase in the value on going from each coarser to finer grain size.

FMR from surface-correlated nanophase iron in the <10 μm fraction may be even greater than that made by the nanophase iron in the agglutinitic glass (Taylor et al. 2001b). This surface-correlated nanophase iron can account for the additional contribution to the FMR values of the finest fractions of lunar soils. Taylor et al. (2005) showed that this greatly increased nanophase Fe in the finest fractions yields an increase in magnetic susceptibility that virtually all grains <20 μm can be picked up by a simple hand-held magnet. This magnetic property has direct applications to dust mitigation during future lunar missions.

9.2.2. Nanophase iron and the maturity index I_s/FeO. Cirlin et al. (1974), Housley et al. (1975), and Pearce et al. (1974) suggested that the FMR signal intensity, I_s, normalized to the total concentration of iron expressed as percent FeO, is a measure of the relative surface exposure ages of lunar soils. The normalization to FeO concentration was proposed because more Fe^{2+} is available for reduction in a high-FeO soil than a low-FeO soil. The parameter I_s/FeO, usually taken to be dimensionless, conveniently ranges from essentially zero in rocks and soils with no surface exposure to about 100 in highly mature lunar regolith. I_s/FeO correlates reasonably well with other maturity indices such as the modal abundance of agglutinates in the 90–150 grain-size fraction and the concentration of solar-wind-implanted gasses (Pillinger et al. 1974; Morris 1976; Morris et al. 1989). I_s/FeO has become the standard measure of regolith maturity because it is much easier and quicker to measure than other indices of maturity, and data for nearly all Apollo soil samples have been generated (Cirlin et al. 1975; Housley et al. 1976; Heiken et al. 1976; Morris 1976, 1977, 1978a,b, 1979, 1980, 1989; Morris and Gose 1976, 1977; Gose and Morris 1977; McKay et al. 1977, 1978a,b, 1979, 1980; Morris et al. 1979; Morris and Lauer 1979, 1980; Korotev et al. 1984, 1997a; Korotev and Morris 1993; Taylor et al. 2001a,b). From definitions originally established on the basis of agglutinate abundance (McKay et al. 1974), the categories *immature, submature,* and *mature* now correspond to I_s/FeO ranges of <30, 30–60, and >60 (e.g., Fig. 2.28).

Although I_s/FeO is a convenient and common measure of soil maturity, there are several factors that should be considered with respect to this measure.

Sampling. Most I_s/FeO data have been obtained on very small subsamples of material with a nominal grain size of <0.25 mm or, in some cases, 90–150 μm (Morris 1976, 1978b). More precisely, the ferromagnetic resonance intensity, I_s, was obtained from ~2 mg of material from the <1-mm fines that fit into a tube with a nominal inside diameter of 0.5 mm, whereas in most cases the FeO concentration was obtained from literature data for <1-mm fines. In other words, the numerator and denominator were obtained from different subsamples of different mean grain size. For submature and mature soils, 84 ± 8% of the mass of the <1-mm grain-size fraction also passes through a 0.25-mm sieve (Graf 1993), so this difference is a likely source of error only for samples for which there is a strong variation in composition with grain size (Section 9.7).

Analysis. For most samples, published data for I_s/FeO are based on a single FMR determination. For some core samples, the denominator, the total FeO concentration, is obtained by imprecise magnetic methods, not by chemical analysis. Even when the FeO concentration is determined by chemical analysis, the value determined may not necessarily be representative of the soil. For this reason, published I_s/FeO data for some samples may be in error by as many as 10 units.

Grain Size. Some maturation effects are not strongly related to grain size. For example, neutron capture and galactic-cosmic-ray effects are volume correlated, not surface correlated. Many maturation effects, however, occur only at a surface of a soil grain, e.g., condensation of volatiles or implantation of solar wind ions. As a consequence, some space-weathering effects are a strong function of grain size (e.g., Fig 2.28) and others are not. Even among those effects that are surface correlated, the depth of each effect within a given grain differs. The process of agglutinate formation, which converts finer-grained material to coarser-grained material, may release or erase some effects (solar-wind-derived gases, energetic particle tracks) and concentrate others into coarser particles (nanophase iron). Thus, there is no particular reason to expect that the pattern of variation of any given maturity parameter with grain size will match that of another.

Resetting. There is one known case where a layer of soil in a core is submature on the basis of agglutinate abundance and siderophile-element concentrations yet I_s/FeO is very low (i.e., <<30; Korotev et al. 1997a). The FMR signal was apparently reset by a heating event that agglomerated nanophase iron particles into larger, multi-domain particles that gave no single-domain range FMR signal.

Vector paths. A path-1 soil (*in situ* reworking) will have a different exposure history than a path-2 soil (mixing), yet the two soils can have identical values of I_s/FeO or any other maturity parameter. In particular, a path-2 soil that is a mixture of regoliths with differing I_s/FeO values and different FeO concentrations can have misleading and non-intuitive value of I_s/FeO, one that does not reflect its history if it is interpreted as a path-1 soil (Fig. 2.32).

Despite the various problems discussed here, maturity parameters such as nitrogen abundance and agglutinate abundance correlate reasonably well with I_s/FeO among soil samples of varying maturity (e.g., Morris et al. 1989; Fig. 2 in Lucey et al. 2000). Imperfect intercorrelations among maturity parameters ultimately reflects that every grain in the regolith has a different exposure history, no two maturation parameters record that history in exactly the same way, and the concept of "the maturity" of a soil or surface is an average characteristic.

9.3. Compositional variation with maturity in path-1 soils

Two types of changes in bulk composition occur in regolith that matures *in situ*, that is, regolith that follows soil evolution path 1 of McKay et al. (1974). First, concentrations of those solar-wind elements that are all but nonexistent in lunar igneous rocks (H, He, C, N, and other light elements), as well as isotopes produced by cosmic-ray interactions, increase dramatically (McKay et al. 1991). Second, impacts of micrometeorites cause concentrations of siderophile elements (Ni, Ir, Au) and some volatile elements (Zn, Ga) to increase because these

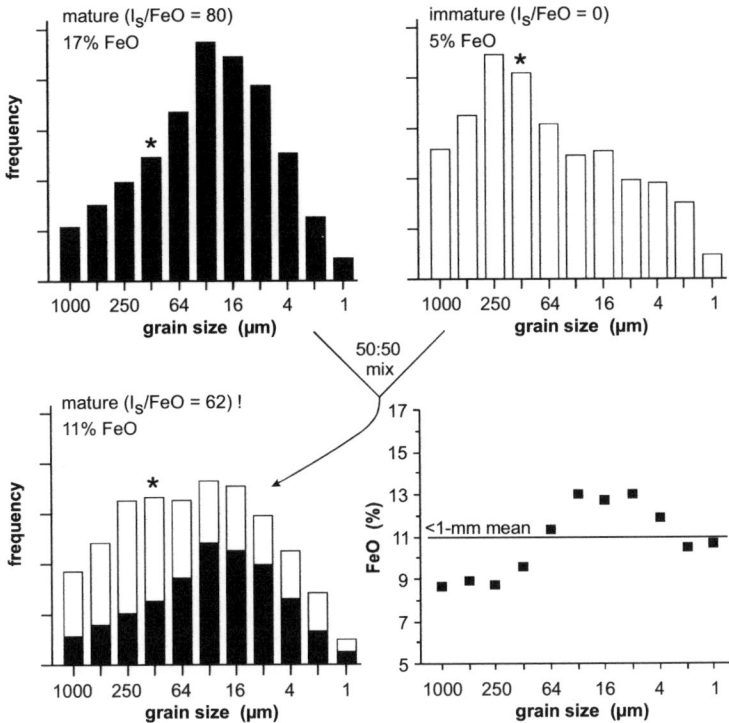

Figure 2.32. Schematic diagram showing how mixing of two compositionally distinct regoliths with different grain-size distributions leads to variation in composition with grain size (after Fig. 6 of McKay et al. 1974) as well as a non-intuitive final value of I_s/FeO. Although exaggerated, the example corresponds to mixing mature, fine-grained mare regolith with a totally immature, coarse-grained highlands regolith. Note that although half the material in the mixture has never had any surface exposure, the mixed soil would still be classified as mature on the basis of the I_s/FeO scale of Morris (1976, 1978b) ($I_s/FeO = [17(80)0.5 + 5(0)0.5]/11 = 62 > 60$). Conversely, mixing a highly mature feldspathic soil with freshly disaggregated mare basalt will yield an immature I_s/FeO even though other measures of maturity may still be high. That other maturity indices (e.g., nitrogen and agglutinate abundances) correlate well with I_s/FeO (e.g., Fig. 2 of Morris et al. 1989) for many soils indicates that these extreme examples are rare. The asterisk represents the grain-size range for which most modal petrography data are available.

elements are more highly concentrated in meteorites than in lunar igneous rocks (Wasson et al. 1975; Haskin and Warren 1991; Korotev and Morris 1993). In highly feldspathic soils, maturation can lead to nontrivial increases in concentrations of some lithophile elements by the same mechanism because chondritic meteorites have high concentrations of certain elements compared to feldspathic lunar rocks. For example, lunar meteorite QUE 93069 is a feldspathic regolith breccia (4.4% FeO) formed from a mature regolith (Thalmann et al. 1996). Siderophile-element concentrations (Spettel et al. 1995; Korotev et al. 1996; Lindstrom et al. 1996) are high and equivalent to 2.9% CI chondrite. Thus, about 14% of the Fe and 10% of the Mg and Cr in the regolith breccia derives from meteorites. Much of the meteoritic material was delivered as micrometeoroids associated with the soil maturation process. This component probably exists in agglutinates or as vapor-deposited coatings on soil grains. The rest is carried in glassy and crystalline breccias of which the regolith is in part composed. The latter were formed during crater-forming impacts that were not related to the maturation process. Some to most of the meteoritic Fe occurs as metal (Sections 4.4 and 6.2.1).

9.4. Compositional variation with maturity in path-2 soils

Impacts that form craters with diameters in the decameter to kilometer size range will mix less mature regolith with more mature surface regolith. Soils formed in this manner, that is, regolith that follows soil evolution path 2 of McKay et al. (1974), will typically have intermediate or low values for soil maturity parameters such as I_s/FeO. Freshly exhumed rock debris or buried regoliths with little surface exposure are not likely to have the same composition as mature surface soils because they have not been as thoroughly mixed with other local materials. Thus, at the Apollo sites, soils of low maturity (e.g., samples 12032, 15531, 61221, 67511) are commonly compositionally distinct from mature soils collected nearby, and all mature soils at some sites (Apollo 12 and 16) tend to be compositionally similar to each other.

Mixing by impacts can lead to correlations between a regolith maturity parameter such as I_s/FeO and some other property such as bulk composition. These correlations, which result from mixing (soil evolution path 2), can be erroneously attributed to effects of the *in situ* maturation process (path 1). For example, Apollo 16 soils are mixtures dominated by feldspathic lithologies but also contain mafic, KREEP-rich impact-melt breccias (Fig. 2.11). Prior to the Imbrium impact, the site was likely entirely dominated by KREEP-poor, feldspathic lithologies. The Imbrium impact emplaced an ejecta deposit at the site that was a mixture of KREEP-rich breccias from the Imbrium area and KREEP-poor anorthositic rocks of local origin. Immature soils of Apollo 16 consistently contain a higher proportion of feldspathic material having little surface exposure than do the mature soils. As a result, Apollo 16 soil compositions vary systematically with maturity (Fig. 2.33). The variation is not caused by the maturation process because there is no mechanism by which the soils can become richer in Fe and Th by up to a factor of 2 simply as a result of *in situ* maturation. As there are no outcrops of feldspathic rocks at the site, the feldspathic material must derive from beneath the surface. Impacts that are sufficiently large to penetrate the regolith, such as that which formed North Ray crater, encounter rocks at depth that are more feldspathic (lower FeO, Fig. 2.33a) than the typical mature surface soil (Stöffler et al. 1985; Korotev 1991, 1997). None of the soils collected near North Ray crater are mature because the time elapsed since the crater was formed (50 m.y.) is not long enough for the crater ejecta to have become mature (Morris 1978a). Soils collected at station 13, about one crater-diameter from the crater (Fig. 2.33b), are intermediate in both composition and maturity (submature, as measured by I_s/FeO) to soils collected on the rim of the crater and to the soils collected more distant because they are mixtures of the two types of material. However, submature soils with Th concentrations greater than that of the main trend of Figure 2.33b (all from station 5 and from 18–21 cm depth in the 60009/10 core) cannot be simple mixtures of the two end-member regolith types; they may instead be products of *in situ* maturation.

9.5. Compositional variation with grain size

Many early studies of lunar soil showed that the composition of the regolith changed with grain size. Because optical and infrared remote sensing is sensitive to particle size, such changes are important for interpreting data from these techniques. One consistent finding is that the finest grain-size fractions of soils developed on the maria are consistently richer in elements associated with plagioclase (Al, Na) and late-crystallizing minerals (incompatible elements) and depleted in elements associated with ferromagnesian minerals (Fe, Mg, Ti, Sc, V, Cr, etc.) compared to coarser grain-size fractions and the bulk (<1-mm) soil (Evensen et al. 1974; Korotev 1976; Laul et al. 1978, 1979, 1980, 1988; Papike et al. 1981, 1982; Taylor et al. 2001a,b). The effect is not often apparent among grain-size fractions of >45-μm material but becomes pronounced with decreasing grain size for <45 μm grain-size fractions (e.g., Laul et al. 1988; Taylor et al. 2001a,b). For example, in soil from the Apollo 17 deep drill core, which is composed of about 50 ± 20% high-Ti mare basalt, Fe and Ti are 7–8% lower, and Al is 12% higher in the <20 μm grain-size fractions compared to the <1-mm fractions; incompatible elements Sm and Th are enriched by 6% and 36% in the <20 μm fractions (Table 2.7). There

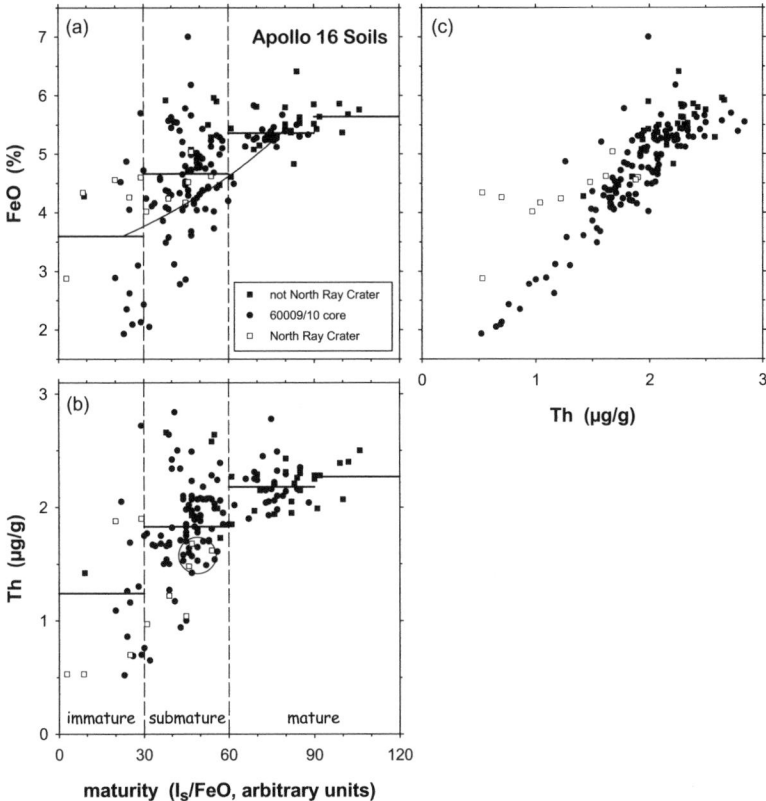

Figure 2.33. (a,b) Concentrations of FeO and Th as a function of the maturity parameter I_s/FeO in Apollo 16 soils. Open squares represent surface and trench soils from station 11 on the rim of North Ray crater and station 13 about a kilometer from the crater, closed squares represent all other surface and trench soils, and circles represent samples from the 60009/10 double drive tube. The immature, submature, and mature divisions are those of Morris (1976, 1978b). The horizontal lines represent the mean concentration of all points within the range of I_s/FeO that they span. (c) The correlations between composition and maturity of (a) and (b) are not caused by maturation of regolith, they are mainly the result of mixing mature surface regolith that is rich in FeO and Th with immature subsurface material that is poorer in FeO and Th. For example, many submature soils, such as those from station 13 [represented by the three North Ray Crater points inside the circle in (b)], are mixtures of mature surface soil and immature North Ray crater ejecta. The curved line in (a) is the mixing line between mean immature and mean mature soil. In (a), most points with I_s/FeO >30 and which have anomalously high FeO also have anomalously high concentrations of Ni and other siderophile elements because they contain nuggets of Fe-Ni metal (Korotev 1987a, 1997). FeO and Th data are from Korotev (1991, 1996, 1997); I_s/FeO data are from Morris (1978b) and Morris and Gose (1976).

is some evidence that the degree of enrichment or depletion may be somewhat greater for immature soils, but the effect is still strong in mature soils.

What causes such variations? Four processes have been proposed for why the composition of the lunar regolith varies with grain size (Korotev 1976).

(1) Surface processes. As the grain-size decreases, the concentrations of highly volatile elements such as Cd, Zn, In, and Ga increase. Some of the surface-correlated volatiles derive from volcanic-gas condensates, whereas others derive mainly from vaporization of micrometeorites (Boynton et al. 1976). Similarly, concentrations of solar-wind implanted gases

Table 2.7. Mean concentration of some elements in two grain-size fractions of 16 samples from the Apollo 17 deep drill core.

	unit	<1000 μm mean	<20 μm mean	<20/<1000 mean	±
TiO$_2$	%	6.32	5.79	0.92	0.03
Al$_2$O$_3$	%	13.41	14.97	1.118	0.015
FeO	%	16.27	15.14	0.931	0.017
Sm	μg/g	7.78	8.24	1.06	0.06
Th	μg/g	1.40	1.83	1.36	0.11

The uncertainties (±) on the mean ratios are 95% confidence limits. Data are from Laul et al. (1978, 1979) and Laul and Papike (1980).

(Eberhardt et al. 1970; Bernatowicz et al. 1979) and nanophase iron (I$_s$/FeO) increase with decreasing grain-size (Taylor et al. 2001a,b; Fig. 2.34) because of the increase in surface-to-volume ratio. Among mainly-lithophile elements, surface processes are probably not the cause of any detectable compositional variations with grain size, except perhaps in part for Na.

(2) Mixing processes. As noted in the previous section, some soils are mixtures of fine-grained, mature soil of one composition with coarse-grained, immature material of another. The mixing of compositionally distinct materials with different grain-size distributions is one important cause of variation in regolith composition with grain size (e.g., Fig. 2.32). One extreme example is the orange and black volcanic glass spherules of the Apollo 17 site. All samples of soil collected at the Apollo 17 site contain volcanic glass spherules. The mean grain size of the spherules (~44 m for 74220 "orange soil") is less than that of nearly any mature soil (Fig. 2.28) and the spherules have higher concentrations of Fe and Cr than any other component of the soil. Thus, among the various submillimeter grain-size fractions of Apollo 17 soils analyzed by Laul and coworkers, the 20–90-μm fractions consistently had the greatest concentrations of Fe and Cr (Laul and Papike 1980; Laul et al. 1978, 1979, 1981) because they contained the greatest relative abundances of volcanic spherules. This effect is particularly evident in the nominally nonmare soils, those of the South and North Massifs.

As noted above, in mare soils, the finer grain-size fractions tend to be richer in Al and incompatible elements than the coarser fractions or the bulk soil. Given that all nominally mare soils from the Apollo sites contain considerable nonmare material, which consists of high-Al feldspathic lithologies or high-Th impact-melt breccias (Fig. 2.11), a reasonable explanation for the data is that the proportion of nonmare material increases with decreasing grain size (Korotev 1976; Laul et al. 1979; Pieters and Taylor 2003). This explanation is particularly appealing at the Apollo 15 and 17 sites, which are on the boundary between maria and Al- and Th-rich highlands and where Th-rich lithologic components are local, not exotic (Papike et al. 1981).

(3) Mineral property processes. An hypothesis that also accounts for the observations about mare soils is that a greater proportion of plagioclase and mesostasis from disaggregated mare basalt is concentrated in the finer grain-size fraction (Korotev 1976; Papike et al. 1982; Laul et al. 1987). Mesostasis is the last material to crystallize and commonly occurs between grains of the major mineral phases in igneous rocks. It consists of accessory minerals and glass with high concentrations of incompatible elements. In laboratory studies, during the disaggregation of basaltic rocks by impacts, there is preferential enrichment of plagioclase in the finest grain-size fractions and mesostasis phases (Haskin and Korotev 1977; Hörz et al. 1984; Cintala and Hörz 1992). For mesostasis phases, the enrichment occurs because the

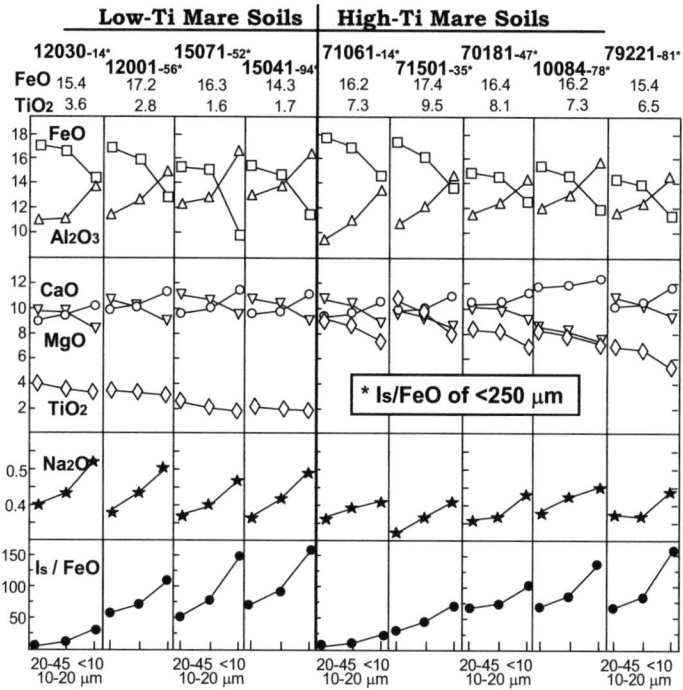

Figure 2.34. Chemistry of the finest size fractions of lunar mare soils, from Taylor et al. (2001b). The soils are divided into high- and low-Ti soils for this figure and further arranged in left to right order from least to most mature. The small number after the sample number (e.g., 12030-14) is the I_s/FeO value of the <250 μm fraction (Morris 1978). The data for a given soil are presented from left to right for decreasing grain size of the soil fractions. With decreasing grain size, there is a systematic increase in plagioclase components (CaO, Al$_2$O$_3$, Na$_2$O) and decrease in ferromagnesian-mineral components (FeO, MgO, TiO$_2$). Also, there is a increase in I_s/FeO with decreasing grain size.

intrinsic grain size of the minerals is small. For plagioclase, the enrichment occurs because the mineral shatters to a finer grain size with shock than does pyroxene (Hörz et al. 1984).

X-ray imaging of soils has been used to better understand relationships between mineralogy of soil grain-size fractions and chemical composition, and on variations as a function of grain size. As part of the characterization of lunar soils for remote-sensing purposes, detailed chemical and modal analyses were made of the fine-size fractions of a set of soils of contrasting chemistry and maturity (9 mare and 10 nonmare) (Taylor et al. 2001b). The modal percentages of mineral and glass components for all the 45-20 μm, 20-10 μm, and <10 μm size fractions of the nine mare soils are shown graphically in Figure 2.35. The bulk soil maturities, expressed as I_s/FeO values for the <250 μm portion of each soil (Morris 1978), are given for comparison, in addition to bulk-soil FeO and TiO$_2$ contents.

Examining the properties of the size fractions for individual soils reveals systematic changes. Impact-produced agglutinitic-glass content increases with decreasing grain size. The modal abundances of crystalline plagioclase are relatively constant to slightly increasing, with decreasing grain size of the fractions. However, the abundances of all other components (pyroxene, oxides, volcanic glass, and olivine) decrease with particle size. This decrease for pyroxene is pronounced and very significant since pyroxene is probably the most optically active of the lunar minerals.

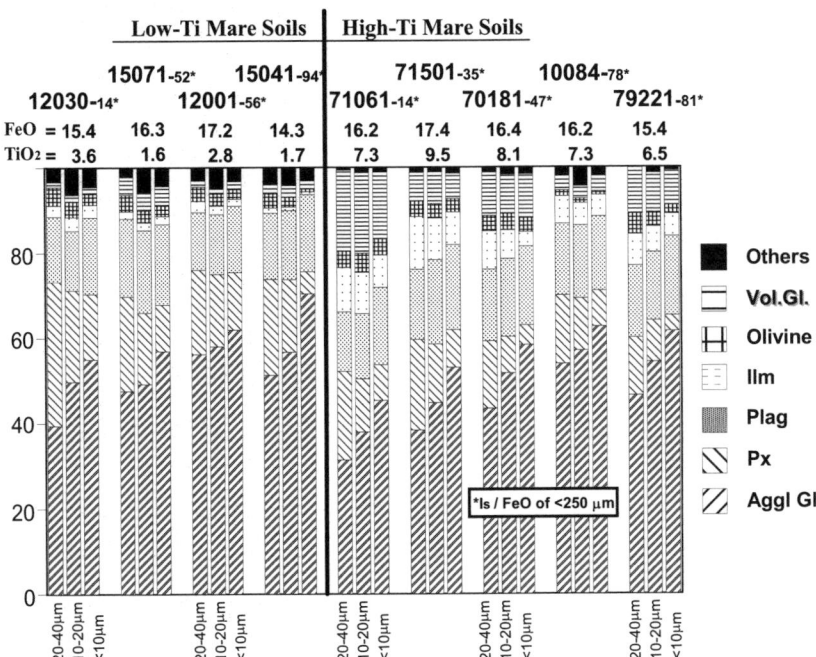

Figure 2.35. Modal analyses of the finest size fractions of Mare Soils, from Taylor et al. (2001b). The soils are divided into high- and low-Ti soils for this presentation and further arranged left to right within these two groups from least to most mature. The small number after the sample number (e.g., 12030-14) is the I_s/FeO value of the <250 μm fraction (Morris 1978), commonly referenced as the maturity value for the soil in general. The FeO and TiO$_2$ contents of the soils are shown for comparison. These soils were selected to have similar compositions, as well as contrasting maturities. Note the general increase in agglutinitic-glass and decrease in pyroxene contents as maturity increases *between* soils, as well as the same trends *within* a given soil with decrease in grain size.

Data for lunar soils are dependent upon the nature of the location. For example, the modal data of Taylor et al. (2003) for mare soils show that although the plagioclase abundance remains nearly constant with decreasing grain size, the pyroxene abundance decreases significantly (Fig. 2.35), with a net result that the plagioclase concentration in the finest fractions increases. From the detailed analyses for a suite of soils from Apollo 14 and 16, Taylor et al. (2003a,b) reported that the feldspathic component also increases significantly with decreasing particle size, consistent with the expected differential comminution of plagioclase during soil formation (e.g., Hörz et al. 1984). However, the mean composition of the agglutinitic glass from these highland soils is clearly not associated with the composition of the finest fraction. It would appear that the F^3 Model (Papike et al. 1981) does not apply to the highland samples. Instead, the data are interpreted to indicate that lateral mare-highland mixing and selective melting of soil phases are both significant parts of the lunar soil formation and evolution (Pieters and Taylor 2003).

Qualitatively the two processes, mixing and preferential comminution, lead to the same effect in mare soils—increasing plagioclase and incompatible element-rich phases and decreasing pyroxene with decreasing grain size. On the basis of a compositional mass-balance modeling, Korotev (1976) concluded that both processes were important in Apollo 17 soils. The effect is also strong, however, in the Luna 24 regolith (Laul et al. 1988), which contains only minimal nonmare material and which is not likely to contain any appreciable component of KREEP-rich impact-melt breccia or other form of KREEP on the basis of data from Lunar Prospector. Similarly, Al is enriched in the finest grain-size fractions of Apollo 12 soils (Papike et al. 1982; Taylor 2001b) and feldspathic lithologies are rare at the Apollo 12 site. Thus in general, enrichment of the finest grain-size fractions with basalt-derived mesostasis and plagioclase is probably the main cause of variation in composition with grain size in mare soils. A similar effect (Al, Ca, and incompatible-element enrichment, Fe and Mg depletion in the finest grain-size fraction; Laul and Papike 1980) is even seen in the Apollo 14 soil, which is dominated by KREEP-rich impact-melt breccias. At the Apollo 16 site, mixing effects may be more important than mineral effects.

(4) Agglutinate fractionation processes. In the 1970's, several papers addressed the issue of whether agglutinates or the glass in agglutinates differ in composition from the soils in which they were formed. If, for example, some fractionation occurs during the melting associated with a micrometeorite impact such that an agglutinate particle is systematically enriched in Fe, for example, compared to the bulk soil, then formation of an agglutinate would cause the concentration of Fe to decrease imperceptibly in finer grain-size fractions and increase in a coarser fraction. If such a process were important and the other three processes discussed here were not, then we would expect differences in Fe concentrations between fine and coarse grain-size fractions to increase as a soil matures.

In the first of the 1970's papers, Adams et al. (1975) and Rhodes et al. (1975) made magnetic separates of soils and identified these as "agglutinate fractions." The implicit assumption was that most of the metal in the soil was or had been nanophase Fe associated with agglutinate formation. They observed that the "agglutinate fractions" were consistently richer in incompatible elements and elements associated with mafic minerals and depleted in elements associated with plagioclase compared to the bulk soil. The effect was particularly strong among the Apollo 16 soils. Adams et al. (1975) and Rhodes et al. (1975) interpreted the result in terms of chemical fractionation associated with the formation of agglutinates by micrometeorite impacts. Various later works showed that the composition of glass in agglutinates, although variable in composition, does not differ significantly, on average, from the composition of the soil in which the agglutinates are formed (Marvin et al. 1971; Gibbons et al. 1976; Via and Taylor 1976; Hu and Taylor 1977). The explanation is that much of the Fe metal in lunar soils, particularly soils from Apollo 16, is not associated with maturation but instead derives from the ancient KREEP-bearing impact-melt breccias which constitute part of the soil. The magnetic separates may have been richer in agglutinates than the nonmagnetic separates, but they were also richer in the melt-breccia component (high Fe, high Th) and poorer in metal-poor anorthosite than the bulk soils.

Although the initial work on "agglutinate fractions" was a "red herring," it led to subsequent work on understanding agglutinate formation. For example, there is evidence that the glass in agglutinates more closely reflects the composition of the <10 μm grain-size fraction than the <1-mm fraction (Walker and Papike 1981; Taylor et al. 2001a,b). One explanation is that the finest material preferentially melts during micrometeorite impacts (Papike et al. 1981). However, an agglutinate particle can only be composed of material that was smaller than itself prior to the impact that formed it, so a 100-μm agglutinate will necessarily have a composition more similar to the <100-μm material in the regolith than to the >100-μm material. Large agglutinates in the 1–2 mm and 2–4 mm grain-size fractions of lunar soils have compositions, on average, indistinguishable from the soils (<1-mm) in which they form (Blanchard et al. 1975; Jolliff et al. 1991a, 1996). Little work has been done on the bulk composition of submillimeter agglutinates, where we might expect the largest differences between the compositions of agglutinate and the <1-mm soil. One such work, that of Taylor et al. (1978) on ~100-μm agglutinates, does, in fact, demonstrate that the average composition of agglutinates can differ from that of the soil (<1-mm) in which they are *found*. This work concludes that agglutinates move and that they do record the composition of the soil in which they are *formed*. However, it is likely that the six sets of average agglutinate compositions reported by G. J. Taylor et al. (1978) do, in fact, match the <100-μm grain-size fractions of the soils in which they were found much better than they do the <1-mm fraction. The high Fe and Ti concentrations of agglutinates from the South Massif compared to <1-mm soils almost certainly reflects that the agglutinates were formed from fine-grained material in which the mass proportion of pyroclastic ash ("orange glass") was much greater than in the <1-mm soil used for comparison.

The pattern and magnitude of compositional differences among grain-size fractions is not a strong function of soil maturity. The lack of dependence on maturity is particularly evident in results for Apollo 17 soils for which I_s/FeO varies considerably (Korotev 1976; Taylor et al. 2001a; Fig. 2.34). This observation suggests that the formation of agglutinates is probably not a significant cause of differences in composition with grain-size in lunar soils and, in fact, probably acts to diminish the magnitude of differences caused by other processes.

The bulk chemistry of the nine mare soils studied by Taylor et al. (2001b), representing both high-Ti and low-Ti compositions, have systematic differences relative to each other, as well as among the different grain-size fractions of a given soil (Fig. 2.34). These differences permit comparisons of their chemistries, both as individuals and collectively, as a function of different maturities. With decreasing grain size, FeO, MgO, and TiO_2 concentrations decrease, and CaO, Na_2O, and Al_2O_3 (plagioclase components) increase for all soils. By comparison with the modal values in Figure 2.35, these chemical variations parallel increases in agglutinitic glass and decreases in the oxide minerals (mainly ilmenite), pyroxene, olivine, and volcanic glass. There is also a large increase (i.e., >100%) in the agglutinitic glass contents between the 90-150 μm and the 10-20 μm size fractions (Taylor et al. 1996, 1999). The I_s/FeO values for each soil increase significantly as the grain size decreases, which is another effect of space weathering (see below).

The glass that forms the binder in an agglutinate particle is a product of several complex processes. Primarily the glass is a melt from the very small volume of soil that was directly impacted by a micrometeoroid. As mentioned above, small compositional changes must have occurred during melting, such as the escape (at least in part) of volatile elements, or the extraction of metallic Fe^0 from silicate melt, leaving the melt slightly depleted in Fe. Additional effects could have been produced by the character of the mineral grains involved. Close examination of the composition of individual agglutinates suggests that the composition of the agglutinate glass is determined by the initial shock-induced melting and is not modified by incorporation (assimilation) of unmelted grains into the glass. Seemingly pure glass is actually riddled with many submicrometer-sized vesicles and clasts. Analyses by Basu and McKay (1985) of the "pure" glass, both adjacent to and away from such clasts, both large and small, show no differences in chemical composition between different areas. This observation suggests that assimilation of the clasts by the primary impact melt has not been a major process modifying the final composition of the glass in different areas.

As determined by Taylor et al. (2001a,b), chemical compositions of individual mineral and glass phases within a soil are relatively constant for <45 μm sizes and independent of grain size. The agglutinitic glass compositions remain relatively constant with changes in grain size and soil maturity. In fact, changes in the chemical compositions of the soil size fractions are caused mainly by the abundances of the phases (e.g., pyroxene, agglutinitic glass), not by distinct changes in the chemistry of the individual minerals and/or glasses. It was originally suggested by Hu and Taylor (1977) that the composition of agglutinitic glass in a soil mimics, to large extent, the chemistry of the bulk soil. However, the compositions of the agglutinitic glasses presented here, at least for the grain sizes <45 μm, are more feldspathic than the overall bulk chemistry of the soil. Thus the composition of the agglutinitic glass is a fair representation of the bulk chemistry plus a feldspathic component.

Figure 2.36 compares the bulk chemistry of each size fraction and the composition of the corresponding agglutinitic glass for the high-Ti mare soils. The systematic changes in

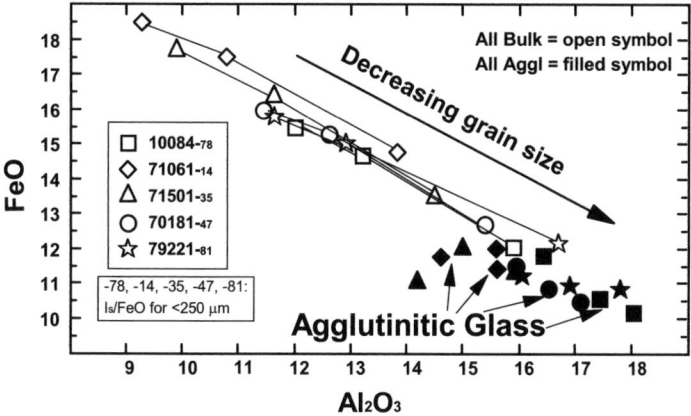

Figure 2.36. Comparisons of changes in the bulk chemistry of the three finest size fractions for the high-Ti soils versus the compositions of their agglutinitic glasses. The small number after the sample number (e.g., 10084-78) is the I_s/FeO value of the <250 μm fraction (Morris 1978). Note that the bulk compositions of decreasing-size fractions within a given soil converge on the composition of the agglutinitic glass by decreasing FeO and increasing Al_2O_3 contents.

soil composition with decreasing grain size show that the bulk chemistry of each size fraction becomes more feldspathic with increasing maturity, with the effect being most pronounced in the finest fractions (Fig. 2.34. The composition of the agglutinitic glass is relatively invariant and even more feldspathic (i.e., rich in Al_2O_3) than even the <10 μm fraction, largely a function of the "Fusion of the Finest Fraction" (F^3) model (Papike et al. 1981). However, the selective impact-produced vaporization of FeO may also play a role. Thus, it would appear that the differences in bulk compositions of the mare soil fractions, with decreasing grain size, approach the composition of the agglutinitic glasses (Taylor et al. 2001a,b).

10. REMOTE SENSING OF LUNAR SURFACE COMPOSITION

Remote sensing is the determination of the physical or compositional state of a surface without physical contact. While the parameters derived remotely are extremely limited compared to analyses that can be conducted in the laboratory, remote sensing offers a synoptic view inaccessible in any other way. Compositional remote sensing of the Moon is broadly divided in two on the basis of physical technique and spatial resolution: mineralogical remote sensing utilizes spectral analysis of reflected (and potentially, emitted) light from the lunar surface to determine mineralogical parameters with existing data at the 10's to 100's of meters scale, and elemental remote sensing exploits energy dependent variations in neutron, X-ray or gamma-ray flux to determine elemental compositions 10's to 100's of kilometers scale. The signal available to mineralogical remote sensing is very strong and enables high spatial resolution to be obtained, even from Earth. Ground-based astronomy established the requirements for spectral observations of the Moon, and the Clementine mission obtained multispectral imaging data for nearly the entire lunar surface at wavelengths sensitive to lunar mineralogy. The elemental signals are far weaker and require that spatially resolved measurements be obtained from lunar orbit. Apollo carried the first elemental sensors to the Moon that provided data for important elements over a significant portion of the lunar surface and set a baseline for future missions. Lunar Prospector collected gamma-ray and neutron flux data for the entire Moon and returned data for many important elements.

Many missions are planned for remote sensing of the Moon in the first decade of the 21st century, with the ESA SMART-1 mission underway at this writing, and the Japanese SELENE mission in an advanced state of development, the Indian mission Chandrayan and the US Lunar Reconnaissance Orbiter in development. Undoubtedly these missions will yield a new bounty of information on the Moon, but these missions rest on the lessons of Clementine and Lunar Prospector, and the brief pause the lunar science community has enjoyed to contemplate the new results.

In this section we review the major remote-sensing measurements made by the two missions Clementine and Lunar Prospector, but do not neglect the important contributions of ground-based spectroscopy, Apollo geochemical measurements, and the potential of radar and thermal infrared spectroscopy for understanding the composition of the Moon in the future.

10.1. Gamma-ray spectroscopy

The concentrations of many elements can be determined for spatially resolved portions of the lunar surface by measuring the characteristic gamma-ray line emissions of specific elements produced by either nuclear spallation reactions or radioactive decay reactions. Figure 2.37 shows a spectrum of gamma-rays coming from the lunar surface as measured by the Lunar Prospector gamma-ray spectrometer. The nuclear spallation reactions are initiated by high-energy galactic cosmic rays that continually hit the lunar surface. There are two dominant nuclear spallation reactions: neutron inelastic scatter and thermal neutron capture. Inelastic scatter reactions occur when high energy neutrons initially produced by spallation reactions lose energy to a target nucleus, leaving the nucleus in an excited state. The nucleus can then de-excite by producing a gamma-ray having an energy characteristic of that particular element. Thermal neutron capture occurs when low-energy neutrons (again the product of galactic-cosmic-ray spallation reactions) are absorbed by a target nucleus, which again leaves the nucleus in an excited state. As with the inelastic scatter reaction, the nucleus can de-excite by producing one or more gamma rays,

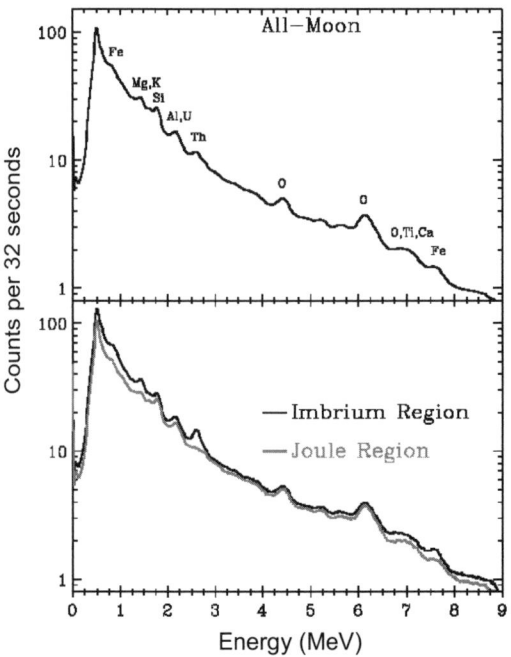

Figure 2-37. Gamma-ray spectra measured from the lunar surface using the LP-GRS. The top panel shows gamma-ray spectra from the entire Moon. The bottom panel shows spectra from two regions that have substantially different gamma-ray responses and surface compositions.

which have energies characteristic of the particular element. Finally, there are a few naturally radioactive elements (Th, U, K) that produce gamma rays having fluxes large enough such that their lunar abundances can be measured from orbit. A comprehensive description of gamma-ray spectroscopy is given by Evans et al. (1993). The most complete tabulation of gamma-ray fluxes coming from the lunar surface is given by Reedy (1978). Table 2.8 gives a summary of various gamma-ray spectroscopy observational parameters. Table 2.9 lists the most prominent gamma-ray lines that can be used for orbital elemental-composition measurements.

There have been two experiments that have carried out orbital gamma-ray measurements of the lunar surface: the Apollo Gamma-ray Spectrometer (AGRS) and the Lunar Prospector Gamma-ray Spectrometer (LP-GRS). Table 2.10 lists some of the important measurement parameters for each experiment. Further details about these experiments can be found in Metzger et al. (1993); Feldman et al. (1999), and Lawrence et al. (2004). Both experiments used scintillator-based gamma-ray detectors adequate for identifying the major gamma-ray lines seen from the lunar surface. Three of the major differences between the AGRS and LP-GRS experiments are the mission duration, coverage of the lunar surface, and direct measurement of the neutron flux. The AGRS measurements were taken from an equatorial orbit over the Apollo 15 and 16 missions, and resulted in a measurement coverage of ~18% of the lunar surface. In contrast, the LP-GRS measurements were taken from a polar orbit and resulted in global coverage from two different average altitudes of 30 and 100 km (Lawrence et al. 2000). Because the measured spatial resolution is proportional to the instrument height above the lunar surface, the two average altitudes had two different average spatial resolutions. The high altitude data has an average spatial resolution of 150–200 km^2; the low altitude data has an average spatial resolution of 60–80 km^2 (Lawrence et al. 2003) (though efforts can be made to carry out a spatial deconvolution if the gamma-ray angular emission and detection can be sufficiently well understood (Etchegaray-Ramirez et al. 1983)). The ultimate size of an analysis region, however, depends on the particular element being analyzed. For example, Th and Fe lines (Sections 10.1.1 and 10.1.2) are sufficiently resolved in the gamma-ray spectra and have a sufficiently high flux so that the intrinsic spatial resolution dominates the measured spatial resolution However, for other elements, such as Mg, the spatial resolution is more likely to be determined by the need to have a large enough region to reduce statistical uncertainties to an acceptable level.

Gamma-ray spectroscopy has unique measurement properties that distinguish it from other methods, such as multispectral imaging. Gamma-ray spectroscopy, like X-ray spectroscopy, measures elemental abundances independent of the molecular or mineralogical state of a given element. There are, however, some composition effects that need to be accounted for when deriving elemental abundances from gamma rays produced by neutron capture and inelastic scatter reactions, and which distinguish the AGRS from the LP-GRS experiment. For example, when abundances are measured using neutron-capture gamma rays, variations in thermal-neutron number density, which has a dynamic range of over a factor of three across the lunar surface, need to be taken into account (Lawrence et al. 2002). For inelastic-scatter gamma rays, variations in the fast-neutron flux need to be accounted for before deriving elemental abundances (Prettyman et al. 2002a,b). One of the distinguishing features of the LP-GRS experiment compared to the AGRS experiment is that Lunar Prospector had onboard neutron measurements that were used to make these neutron corrections.

Another characteristic of gamma-ray spectroscopy, compared to techniques such as optical, infrared and X-ray methods, is that the measured gamma-ray signal comes from an average depth of 30 cm beneath the lunar surface that avoids very superficial deposits such as small crater ray systems. A strength of gamma-ray spectroscopy is that absolute elemental abundances can be measured that are to a first approximation independent of ground-truth data (Metzger et al. 1977; Lawrence et al. 1999, 2000). This capability is important because the footprint of gamma-ray spectroscopy data is large (~60–200 km) compared to the sample-return sites and

Table 2.8. Summary of observational measurement parameters for different remote sensing techniques.

Technique	spatial resolution	depth of signal	surface effects	mineral effects	measurement specificity	absolute abund.	collection times
multispectral imaging	~100 m	~1 μm	Yes	Yes	Fe, Ti	Yes[1]	minutes
thermal imaging							
radar							
gamma-ray spectroscopy	50 – 200 km	30 cm	Minimal	No	Th, K, U, Fe, Ti, Al, Mg, Si, Ca, O	Yes[2]	minutes to hours
thermal neutrons	50 – 200 km	100 cm	Minimal	Not likely	Combinations of Fe, Ti, Gd, Sm	Yes[3]	minutes to hours
epithermal neutrons	50 – 200 km	50 cm	Minimal	No	H, Gd+Sm	Yes[3]	minutes to hours
fast neutrons	50 – 200 km	50 cm	Minimal	No	Fe, Ti, <A>	Yes[3]	minutes to hours
X-ray spectroscopy	20 – 100 km	10 μm	Cannot see through dust layer	No	Si, Mg, Al, Fe[4], Ti[4]	No[5]	minutes to hours

[1]Absolute abundances require ground truth calibration. [2]Absolute abundances have been obtained using both ground truth calibration and independent of ground truth calibration. [3]Information about ground truth abundances and/or other remotely sensed data needed to obtain absolute abundances. [4]No lunar surface measurements for Fe and Ti using XRF have been reported. Fe and Ti abundances can only be obtained using XRF during times of high solar activity. [5]Absolute abundances have not been reported using XRF.

Table 2.10. Comparison of the Apollo and Lunar Prospector gamma-ray spectrometer measurements.

Parameter	Apollo GRS	Lunar Prospector GRS
type of instrument	NaI scintillator	BGO scintillator
energy resolution	10% @ 662 keV	10.5% at 662 keV
mission duration	dates of Apollo 15 and 16	16 Jan 1998 – 31 July 1999
collection time over a 150 × 150 km² area	0.21 hours	4.7 hours at equator, 228 hours at poles
coverage of lunar surface	~18% coverage from an equatorial orbit	100% coverage from a polar orbit
spatial resolution	~200 km at 100 km altitude	~200 km at 100 km altitude, ~60 km at 30 km altitude
reported elemental measurements	Th, K, Fe, Ti	Th, K, U, Fe, Ti, Mg, Al, Si, Ca, O,

Table 2.9. List of strongest gamma-ray lines for the ten elements that have been measured using orbital gamma-ray spectroscopy (after Reedy 1978).

Element	Assumed Abundance	Energy (MeV)	Estimated Flux[1]	Reaction Type
O	43.5 wt%	2.741	5.98×10^{-3}	inelastic scatter
		3.086	4.72×10^{-3}	inelastic scatter
		3.684	1.15×10^{-2}	inelastic scatter
		3.854	6.20×10^{-3}	inelastic scatter
		4.438	2.02×10^{-2}	inelastic scatter
		6.129	4.78×10^{-2}	inelastic scatter
		6.917	1.23×10^{-3}	inelastic scatter
		7.117	1.35×10^{-2}	inelastic scatter
Mg	4 wt%	1.369	1.33×10^{-2}	inelastic scatter
Al	11 wt%	2.210	1.13×10^{-2}	inelastic scatter
		7.724	2.88×10^{-2}	neutron capture
Si	20 wt%	1.779	6.54×10^{-2}	inelastic scatter
K	1200 μg/g	1.461	3.92×10^{-2}	radioactive decay
Ca	10 wt%	1.943	3.42×10^{-3}	neutron capture
		3.737	5.77×10^{-3}	inelastic scatter
		3.904	3.87×10^{-3}	inelastic scatter
		6.420	4.02×10^{-3}	neutron capture
Ti	1.4 wt%	1.382	4.48×10^{-3}	neutron capture
		6.419	4.67×10^{-3}	neutron capture
		6.762	6.87×10^{-3}	neutron capture
Fe	9 wt%	0.847	1.92×10^{-2}	inelastic scatter
		7.631	1.00×10^{-2}	neutron capture
		7.646	9.20×10^{-3}	neutron capture
Th	1.9 μg/g	0.911	1.76×10^{-2}	radioactive decay
		0.965	3.37×10^{-2}	radioactive decay
		0.969	1.09×10^{-2}	radioactive decay
		2.615	3.66×10^{-2}	radioactive decay
U	0.5 μg/g	1.120	8.02×10^{-2}	radioactive decay
		1.765	1.06×10^{-2}	radioactive decay
		2.204	3.73×10^{-3}	radioactive decay

[1]The estimated flux is given in gamma-rays per second and assumes the abundances given in the second column. Other assumptions and details about the modeled gamma-ray fluxes are given in Reedy (1978).

the abundances sampled over the small-area landing sites may not be representative of the large areas sampled by the gamma-ray spectrometer. The ability to independently obtain absolute abundances, however, requires a detailed understanding of the gamma-ray and neutron detectors as well as the gamma-ray and neutron production processes on the lunar surface (Prettyman et al. 2002a). Finally, gamma-ray spectroscopy provides the ability to measure some key elements that cannot be measured by any other remote technique, such as Th, K, U, O, and Ca.

10.1.1. Gamma-ray spectroscopy measurements of Th and K. There are three naturally radioactive elements – Th, K, and U (Table 2.9) – that produce gamma-ray fluxes high enough to be measured from a gamma-ray detector in lunar orbit. Each of these elements are "incompatible" and are therefore expected to show similar abundance distributions (Section 3.4.2).

Of the three elements, the easiest to measure is Th. Th produces a very strong gamma-ray signal. According to the gamma-ray tabulation of Reedy (1978), the 2.61 MeV Th gamma ray is the second strongest elemental gamma ray coming from the lunar surface (the 1.46 MeV K gamma-ray is the strongest). Second, the 2.61 MeV gamma ray is one of the best separated gamma rays in the measured lunar gamma-ray spectra, which is particularly important for gamma-ray spectra measured using scintillator detectors that have low energy resolution. Finally, because the 2.61 MeV Th line has the highest energy of the radioactive and incompatible elements, it is located in a region of relatively low gamma-ray background. In contrast, the lower energy lines are contaminated by the Compton continuum background from the 2.6 MeV Th line, which makes the determination of their fluxes more difficult.

The gamma ray line with the largest flux in lunar spectra is the 1.46 MeV gamma-ray produced by ^{40}K. In spite of this large flux, however, several difficulties arise in using this line to measure K abundances. First, compared to the 2.61 MeV Th line, the 1.46 MeV K line is not as well separated in the measured gamma-ray spectra (see Fig. 2.37). It is close to a number of other large-flux gamma rays such as the 1.38 MeV Mg line. Second, since it has a relatively low energy, it sits atop a large gamma-ray background that consists in part of Compton continuum gamma rays originating from the 2.61 MeV Th line. On most of the lunar surface, the background component that originates from the 2.61 MeV Th line varies in the same way as the 1.46 MeV full-energy K line. One of the goals for measuring K abundances, however, is to find where they deviate from Th abundances, which for nonmare regions would indicate the occurrence of materials with non-KREEP-like K/Th (Section 6.2.4).

Finally, U produces a number of gamma-ray lines having energies less than 3 MeV. Because the expected fluxes of these lines are lower than either the large Th or K line and because there is a high degree of correlation between U and the elements Th and K, it is difficult to cleanly separate out the effect of the U lines from the rest of the spectrum. Because of these effects, there are no reported measurements of U independent of Th and K.

Since the gamma rays from Th and K are produced by radioactive decay reactions, it is relatively straightforward to determine absolute abundances to first order, independent of ground-truth measurements. This is because no neutron corrections need to be made to the measured gamma-ray fluxes, and the parameters that describe the radioactive decay reactions are very well known.

The first remotely measured Th abundances on the lunar surface used data from the AGRS (e.g., Bielefeld et al. 1976; Metzger et al. 1977). Maps of these data on the lunar surface are given by Spudis and Pieters (1991) and Metzger (1993). Metzger et al. (1977) used an energy band analysis ($0.55 \leq E \leq 2.75$) combined with information about the gamma-ray background and detector efficiency to determine absolute Th abundances independent of ground truth. While these data only covered 18% of the lunar surface, many features of the global Th distribution were identified. These include the near-side/far-side composition dichotomy, elevated Th abundances around Aristarchus crater, the Fra Mauro region, the Apennine mountains SE of Imbrium basin, and near Van de Graff crater in SPA basin.

The first global maps of Th abundances on the lunar surface used high-altitude LP-GRS data to derive relative abundances on 150 km × 150 km pixels (Lawrence et al. 1998). Further studies derived absolute abundances using low-altitude (Lawrence et al. 1999) and both high- and low-altitude LP-GRS data (Lawrence et al. 2000) on 60 km × 60 km pixels. Finally, the spatial resolution and information content of the low-altitude data were optimized by Lawrence et al. (2003) to produce a map (Plate 2.3) having an intrinsic spatial resolution of ~(80 km)^2 mapped onto 0.5° × 0.5° pixels. The absolute abundances in Lawrence et al. (2003) were derived from Prettyman et al. (2002a,b), who determined absolute Th abundances using a full modeling and spectral analysis of the gamma-ray spectra. The Lawrence et al. studies used an energy band analysis ($2.5 \leq E \leq 2.7$ MeV) to determine the relative number of counts in the 2.61 MeV Th peak. Other studies have used alternate forms of determining absolute abundances using either ground truth information (Gillis et al. 2004) or a more detailed analysis of the intrinsic uncertainties and background determination (Warren 2001a, 2005). Although neither of these analyses changed the abundance determinations by more than 20% for most of the abundance range, they argued that for abundances less than 2 ppm, the Lawrence et al. and Prettyman et al. concentrations may be high by up to a factor of two. The Warren studies used a combination of statistical analysis of LP-GRS and sample data to adjust the absolute Th concentrations, and the Gillis et al. studies used a combination of lunar samples, including lunar meteorite data, and AGRS regional data to make an empirical correction.

Upon inspection, the global distribution of Plate 2.3 is consistent with the partial coverage map made with AGRS data. The general features that are seen include high Th concentrations (up to 12 µg/g) on the lunar near side in and around Imbrium basin and western Oceanus Procellarum. There are also Th enhancements at a number of large craters that appear to have penetrated a mare basalt layer and exposed underlying material with high Th concentrations. These craters include Aristarchus, Mairan, Aristillus, and Kepler. Detailed descriptions of the Th abundances around some of these craters are given in a number of studies (e.g., Blewett et al. 2001; Gillis et al. 2004). Plate 2.3 also shows moderately elevated Th concentrations across much of the SPA basin.

Despite the fact that most of the lunar highlands have very low Th abundances, a few locations exist in the lunar highlands that have elevated Th abundances. One of these locations is a Th enrichment near the craters Compton and Belkovich (60°N, 100°E) that appears to be co-located with a bright albedo feature having a size of 15 × 30 km (Gillis et al. 2002). This Th enrichment likely has surface Th concentration in the range of 40–55 ppm (Lawrence et al. 2003), which suggests this region is dominated by an evolved lithology such as alkali anorthosite (Elphic et al. 2000; Lawrence et al. 2000, 2003; Gillis et al. 2002).

Measurements of K abundances were first obtained using AGRS data (Bielefeld et al. 1976; Metzger et al. 1977). As described by Metzger et al. (1977), K/Th in the western and eastern mare regions were shown to be roughly the same as for KREEP-like materials (~400, Korotev 2000). However, in the Th-poor farside regions, K/Th ranged up to 1800, which is much higher than the near-side values. While these large farside values were attributed, in part, to the low abundances causing large scatter in the data, Metzger et al. (1977) stated that some of the variation was attributable to real compositional variations.

With LP-GRS data, a preliminary global map of K abundances was reported by Lawrence et al. (1998). Using the analysis that is described in Section 10.1.4, Prettyman et al. (2002a,b) reported global maps of the absolute abundances of K. These data show a strong correlation between Th and K, with a correlation trend that is similar to what is seen in the sample data.

10.1.2. Gamma-ray spectroscopy measurements of Fe. Two dominant gamma-ray lines can be used to measure Fe abundances: the 7.6 MeV line produced by neutron-capture reactions and the 0.846 MeV line produced by inelastic-scatter reactions. With current data sets, the 7.6 MeV line is easier to analyze because (1) it is largely separated in measured gamma-ray spectra (see Figs. 2.37 and 2.38); (2) of the identified lines, it has a relatively large dynamic range (again see Fig. 2.37); and (3) because it occurs at a high energy, it sits on top of a relatively low background of gamma rays scattered from higher energies. In contrast, despite its comparable flux (Reedy 1978), the 0.846 MeV line is difficult to analyze with existing data because it is not well separated in the measured gamma-ray spectra and it sits on top of a large and variable background. In particular, variations in the Compton continuum from the higher energy Th, K, and U lines dominate the background.

Several different studies have used data from both the 7.6 and 0.846 MeV lines to derive Fe abundances on the lunar surface. The first comprehensive map of Fe abundances using the 7.6 MeV gamma-ray line was produced by Davis (1980) using AGRS data. In that study, approximately 18% of the lunar surface was mapped with a spatial resolution of ~200 km. Maps of these data can also be found in Spudis and Pieters (1991). To derive these abundances, Davis measured the total counting rate within an energy window (6.99 $E \leq 8.99$ MeV) around

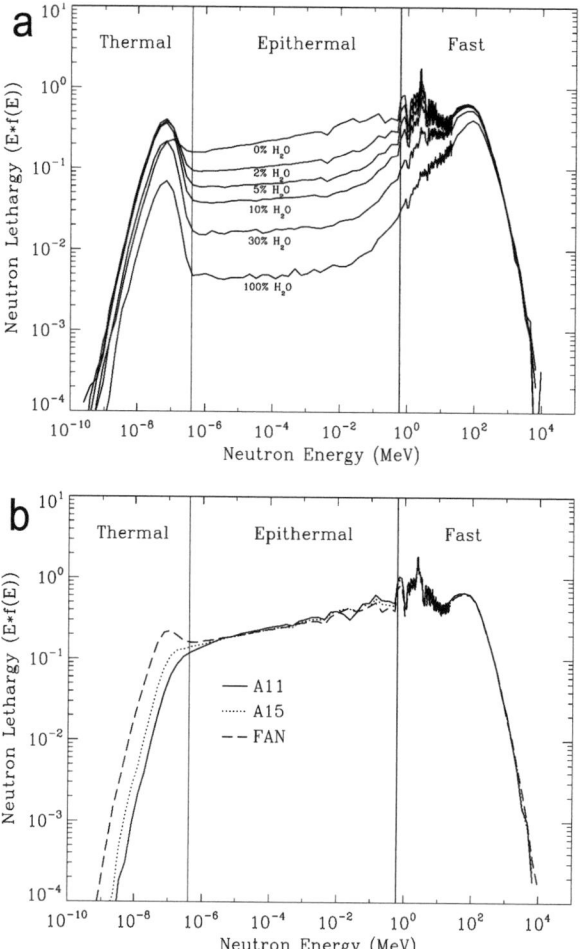

Figure 2.38. Simulated neutron flux spectra for different lunar soil compositions. Top panel shows the simulated neutron flux (or lethargy, which is flux times energy) for three different lunar soils: Apollo 11, Apollo 15, and ferroan anorthosite. Bottom panel shows simulated spectra for a ferroan anorthosite soil combined with varying amounts of water from 0 wt% to 100 wt%.

the 7.6 MeV line. Absolute abundances were obtained by normalizing the measured counting rate with ground truth abundances at the sample sites.

In a subsequent study, Davis and Bielefeld (1981) reported Fe abundances derived from variations in the 0.846 MeV gamma-ray line using AGRS data. The background from Th, K, and U was accounted for using a semi-empirical model derived from measured counting rates and assumed Th/U and Th/K abundance ratios. When the AGRS data were averaged over large regions, a good correlation was found between the inelastic scatter and the AGRS neutron-capture data of Davis (1980).

The first global map (Plate 2.4) of gamma-ray-derived Fe abundances using low-altitude LP-GRS data measured the counting-rate variations of the 7.6 MeV line (Lawrence et al. 2002). In a manner similar to the study of Davis (1980), the total counting rate within an energy window ($7.35 \le E \le 7.92$) was measured to derive the Fe abundances. These counting rates were binned onto $0.5° \times 0.5°$ pixels on the lunar surface and smoothed using a two-dimensional, equal-area Gaussian function in order to maximize the compositional contrast and minimize the scatter in the data. Finally, absolute abundances were obtained by normalizing to landing site abundances via the Clementine spectral-reflectance-derived FeO concentrations of Lucey et al. (2000). For this normalization, the Clementine spectral reflectance data were smoothed to match the LP-GRS footprint. The validity of this normalization is founded upon the assumption that the Clementine-derived data give a good representation of the true FeO concentrations within ~150 km of the landing sites (see Section 10.4.7).

Prettyman et al. (2002a,b) did a comprehensive analysis of high- and low-altitude LP-GRS data to derive abundances for ten different elements (Th, K, U, Fe, Ti, Mg, Ca, Al, Si, and O). In this study, absolute abundances were determined largely independent of ground truth; modeling and experimental verification of the gamma-ray production and detection process enabled absolute abundances to be derived from the measured LP-GRS and LP-NS data alone. Figure 2.39 shows the resulting fits from the analysis of Prettyman et al. (2002b) to measured gamma-ray spectra for a few selected regions. As part of this analysis, FeO abundances were determined on (150 km × 150 km) and (60 km × 60 km) equal area pixels using variations of both the 7.6 and 0.846 MeV lines.

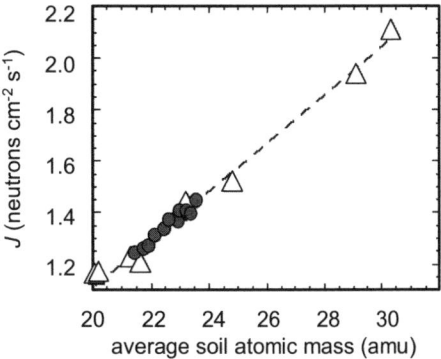

Figure 2.39. Simulated integrated neutron leakage flux from the lunar surface for typical lunar minerals (triangles) and typical lunar compositions (circles). The integration is between 600 keV and 8 MeV (taken from Gasnault et al. 2001).

When inter-comparisons are made among the various gamma-ray-derived Fe datasets, reasonably good agreement is found for all measurements. For example, the correlation coefficient between the Fe dataset of Davis (1980) and Lawrence et al. (2002) is 0.85. The correlation coefficient between the two LP-GRS datasets is 0.98.

Upon close inspection, however, significant differences occur between the datasets. For example, when the analysis of AGRS data (Davis 1980; Clark and McFadden 2000) is compared to the analysis of LP-GRS data (Lawrence et al. 2002), there are systematic differences in both mare and highlands regions. Specifically, the AGRS data tend to show higher Fe abundances in the highlands compared to the LP-GRS data, and the AGRS data tend to show lower Fe abundances in the mare regions compared to the LP-GRS data. These differences are almost certainly due to the fact that a neutron-number-density correction was done on the LP-GRS data and not on the AGRS data (Lawrence et al. 2002). Such a correction is needed because the total number of thermal neutrons in a given location dictates the number of neutron-capture gamma rays that can be produced (Lawrence et al. 2002). Because the number of thermal neutrons varies by over a factor of three across the lunar surface (see Plate 2.5), this correction is non-negligible. When a detailed comparison is made between the two LP-GRS datasets (Lawrence et al. 2002; Prettyman et al. 2002a,b), discrepancies appear to be related in part to differences in how latitude-dependent detection asymmetries were treated.

10.1.3. Gamma-ray spectroscopy measurements of Ti. The strongest grouping of Ti gamma-ray lines are neutron-capture lines that occur at 6.42, 6.56, and 6.76 MeV. In contrast to the Fe neutron-capture line at 7.6 MeV, these Ti lines are not cleanly separated in either the AGRS or LP-GRS spectra (e.g., Fig. 2.37). In particular, interferences from gamma-ray lines are produced by O (6.92 MeV), Si (6.88 MeV), and Ca (6.42 MeV). Furthermore, these lines are located on the Compton continuum background produced by the higher-energy Fe lines.

Despite these complications, Ti abundances have been reported using data from both the AGRS and LP-GRS experiments. Davis (1980) reported the first gamma-ray derived map of Ti abundances using AGRS data. As with the Fe data, the reported Ti data covered approximately 18% of the lunar surface. Davis derived absolute Ti abundances using a combination of an energy-window counting-rate technique ($4.78 < E < 6.99$) and normalization to ground-truth abundances. Furthermore, an Fe- and Al-dependent background was subtracted from the window counting rate to account for the Fe-dependent Compton-continuum-background gamma rays. A global Ti map using high-altitude LP-GRS data was reported by Prettyman et al. (2002a,b). The Ti map was derived using the same analysis used to determine Fe and other elemental abundances (Section 10.1.4).

10.1.4. Gamma-ray spectroscopy measurements of Mg, Al, Ca, Si and O. In addition to the abundances for Th, K, Fe, and Ti, preliminary results for the global, absolute abundances of Mg, Al, Ca, Si, and O were reported by Prettyman et al. (2002a,b). Prettyman et al. (2002a,b) did a comprehensive analysis of the entire gamma-ray production and detection process. Included in this analysis was detailed modeling of the gamma-ray production process, the surface-to-spacecraft gamma-ray transport process, and the instrument gamma-ray detection process. A linear least-squares fitting procedure of the gamma-ray energy spectra was used to determine the absolute elemental abundances. Finally, absolute concentrations were obtained by normalizing the gamma-ray flux to the composition of feldspathic lunar meteorites for selected highlands regions. For all elements, data from the high-altitude portion of the LP mission were used and data were mapped using pixels having a size of ~ 150 km × 150 km (or 5°×5° at the equator).

The general behavior of the elemental abundance maps shows trends that are to first order consistent with sample studies. The near-side mare basalt regions have relatively high Mg abundances and relatively low Al, Ca, and Si abundances. In contrast, the far-side highlands regions have relatively low Mg abundances and relatively high Al, Ca, and Si abundances. The O abundances fall within a narrow range near 44 wt% as is expected from sample studies (Section 6.2.4; Fig. 2.18).

It should be noted that these data are preliminary and work continues as of this writing to refine and improve the analysis of Prettyman et al. (2002a,b). Specific issues include the gamma-ray production cross sections for a number of key elements (e.g., O, Ti, Mg), the high-energy physics models that describe the proton-to-neutron (and hence gamma-ray) production, and the source, magnitude, and energy dependence of the non-lunar background gamma-rays.

10.2. Neutron spectroscopy

The technique of orbital neutron spectroscopy can measure the composition of the lunar surface by measuring broad energy ranges of neutrons produced by nuclear spallation reactions. Specifically, these neutrons are generated by interactions between galactic cosmic rays and their secondary particles, and the nuclear constituents of surface material. After production, an equilibrium velocity distribution forms as a result of multiple elastic and non-elastic collisions between the neutrons and nuclei in the surface. The intensity and velocity-space structure of resulting spectra depend on the elemental composition of surface material. Figure 2.38 shows a plot of simulated neutron energy spectra for various assumed lunar compositions (Feldman et al. 1998a).

In practice, there are three broad ranges of neutron energies that contain compositional information about the lunar surface (see Fig. 2.38). The neutrons within these energy ranges are referred to as fast neutrons (0.6–3 MeV), epithermal neutrons (0.4 eV $< E <$ 0.6 MeV), and thermal neutrons ($E <$ 0.4 eV). In general, the flux of fast neutron is related to average atomic mass of the surface material (Section 10.2.1; Gasnault et al. 2001). Since variations in the atomic mass of lunar material is dominated by variations in Fe and Ti abundances, a global map of fast neutrons looks similar to a combination of the observed Fe and Ti abundances (Maurice et al. 2000). The flux of epithermal neutrons reflects primarily the abundance of H (Lingenfelter et al. 1961; Feldman et al. 1998b, 2000a, 2001). However, epithermal neutrons are also observed to be sensitive to Gd and Sm because of a few very large (n,γ) resonances between 0.4 eV and 10 eV. Finally, because the neutron absorption cross sections for Fe, Ti, Gd, and Sm are quite large, the flux of thermal neutrons is highly sensitive to abundance variations of these elements across the lunar surface (Elphic et al. 1998; 2000; Feldman et al. 1998a; 2000b).

The only mission to conduct orbital neutron spectroscopy measurements of the lunar surface has been the Lunar Prospector mission (Binder 1998). The Lunar Prospector spacecraft carried the neutron spectrometer (LP-NS) that measured the fluxes of thermal and epithermal neutrons using two ^{3}He gas proportional counters (Feldman et al. 1999, 2004; Maurice et al. 2004). Fast neutrons and a broad energy range of epithermal neutrons (Genetay et al. 2003) were measured using signals from the scintillator anti-coincidence shield of the LP-GRS (Feldman et al. 1999, 2004).

As demonstrated by the LP-NS, there are a number of strengths to the orbital neutron spectroscopy technique. First, the signal from both epithermal and fast neutrons provides one of the only ways to measure the H content of the lunar surface (see Section 10.9). Second, the dynamic range of the fast and thermal neutron flux measurements is relatively large and provides a very sensitive measure of composition parameters across the lunar surface. Neutron measurements also provide critical corrections for a the analysis of the gamma-ray spectroscopy data (Lawrence et al. 2002; Prettyman et al. 2002a,b).

Some of the limitations of orbital neutron spectroscopy are the same as gamma-ray spectroscopy. For example, the spatial resolution of the epithermal neutron measurements is between 1 and 1.5 × the spacecraft altitude above the lunar surface (Feldman et al. 2001). In addition, with the exception of the strong sensitivity to H, orbital neutron spectroscopy measurements do not by themselves provide specific elemental abundance measurements. Other data must supplement the orbital neutron spectroscopy measurements to provide information about specific elemental abundances. Nevertheless, even with this limitation, orbital neutron spectroscopy measurements can still provide valuable constraints for the composition of the lunar surface.

10.2.1. Average atomic mass of the lunar surface. In a study of fast-neutron data from Lunar Prospector, Gasnault et al. (2001) showed that a direct relation between the fast neutron flux and the average atomic mass of the soil is generally valid for planetary surfaces having very low H abundances. The correspondence between average atomic mass and fast neutrons results from neutron production processes and the subsequent interactions of neutrons with the surrounding materials. This relationship has been characterized using numerical calculations supported by laboratory experiments of the irradiation of thick targets (materials like Be, Pb, W, and Sn) with high-energy protons.

As described by Gasnault et al. (2001), the average atomic mass is related to the fast neutron flux according to the following relation:

$$< A > = \alpha J + \beta \qquad (2.1)$$

where $<A>$ is the average soil atomic mass (amu), J is the neutron flux at the surface (neutrons cm^{-2} s^{-1}), $\alpha = 10.6 \pm 0.3$, and $\beta = 8.3 \pm 0.4$. The presence of light elements in the soil may induce modifications of this relation. However, except for H, these modifications will be small. Fast neutron spectroscopy can therefore be used to provide a measure of the average soil atomic mass of planetary surfaces through use of Equation (2.1) when the soil has typical low lunar abundances of hydrogen.

Because the Moon is mostly devoid of H, Equation (2.1) is valid for the Moon. Gasnault et al. (2001) correlated Lunar Prospector measurements with <A> of the various lunar soil samples. Although there are mixing effects due to the relatively low spatial resolution of the data, Figure 2.39 shows that the correlation is very good (0.917) and compatible with Equation (2.1). That result leads to the construction of a map of the average atomic mass for the lunar surface from measurements of Lunar Prospector (Plate 2.6). The lowest <A> values (about 21 amu) are scattered over the highlands, whereas the highest <A> values (about 24 amu) are concentrated at the west and south of the Aristarchus plateau. The uncertainties lead to a standard deviation in <A> for this map of between 15% and 20%. Finally, we note that since the abundances of Fe and Ti dominate the variation of <A> across the Moon, fast neutrons are alternatively a good indicator of the abundance and location of Fe and Ti (Maurice et al. 2000).

10.2.2. Thermal and epithermal neutron measurements of Fe , Ti, Gd, and Sm. Monte Carlo simulations of cosmic ray-induced neutron production in lunar soils and rocks indicate that the leakage flux of thermal neutrons out of the lunar surface depends strongly on the soil composition. Materials that are abundant in thermal-neutron absorbing elements, such as Fe, will have a markedly lower leakage flux than those lacking these absorbers. This is clearly seen in orbital neutron spectrometer measurements from the Moon. For example, the thermal-neutron leakage flux is much lower over the Fe-rich maria than over the Fe-poor highlands (Maurice et al. 2004, Plate 2.5). The epithermal neutron flux is much less affected by the presence of such absorbers, and simulations show that there is a linear relationship between the ratio of epithermal to thermal fluxes (or count rates) and the macroscopic absorption cross section of materials with typical lunar compositions. The macroscopic absorption cross section Σ_a is a measure of the net thermal-neutron-absorbing effect due to all elements making up the material:

$$\Sigma_a = \sum_i \sigma_{ai} f_i \frac{N_A}{A_i} \tag{2.2}$$

where σ_{ai} is the thermal neutron absorption cross section at 0.025 eV (usually expressed in barns, where 1 barn = 10^{-28} cm^2), f_i is the weight fraction, A_i is the atomic mass of element i, and N_A is Avogadro's number. For the known range of lunar soil and rock compositions, Σ_a is strongly dominated by Fe, Ti and the REEs Gd and Sm. These trace elements can have a profound effect on thermal-neutron absorption because certain of their isotopes have very large thermal-neutron-absorption cross sections.

Because the relationship between the epithermal-to-thermal flux ratio and Σ_a depends almost entirely on three constituents, Fe, Ti, and the REEs, it is in principle possible to obtain the abundance of one of these constituents if the other two can be somehow constrained. For example, if Fe abundance can be determined by either gamma-ray spectroscopy or optical spectral-reflectance techniques, and if the REEs can be constrained by a close relationship with Th as measured by gamma-ray spectroscopy, then the Ti abundance can be inferred using the neutron data. This Ti-detection technique complements that of optical spectral reflectance and gamma-ray spectroscopy. The effect in thermal neutrons is very strong for the high-Ti mare basalts, and much more readily distinguished than the (n,γ) capture line of Ti in the BGO gamma-ray spectra obtained by the LP-GRS. In contrast, given estimates of Fe and Ti abundances, the combined REE concentrations of Gd and Sm can be extracted using this technique. Finally, in feldspathic highlands regions where both REE and Ti concentrations are very low, thermal neutrons can be used to estimate Fe concentrations.

The technique suffers in two areas: first, it necessarily has the broad spatial resolution of the gamma-ray and neutron-spectrometer footprints, and second, it depends on having well-determined concentration values for Fe and the REEs, Gd and Sm, or Ti. The former issue limits the technique's effectiveness in prospecting for ilmenite-rich regions, whereas optical techniques can provide this information down to arbitrarily small scales. However there is evidence that the optical technique may respond to more than just ilmenite (see Section 10.4.6). The latter issue is a serious drawback, since uncertainties in Fe or REE determinations add to the uncertainty in inferred Ti abundance. Likewise, using Th as a proxy for REEs may be problematic, since more highly-evolved lithologies such as alkali anorthosite, monzogabbro, or granite fractionate Th and REEs differently. Finally, uncertainties in the determination of Ti cause uncertainty in the determination of REE abundances using thermal neutrons.

Regarding quantitative abundance determinations, various analyses have presented results for Fe, Ti, and the REE. An analysis of the thermal-to-epithermal neutron counting rates in feldspathic highlands materials where Ti and REE abundances are low (and therefore should make almost no contributions to the observed thermal-neutron flux) show that the weight fraction of FeO must be between 2% and 5% (Feldman et al. 2000b). Elphic et al. (2002) presented the first neutron-derived abundances of Ti on the Moon. The results disagree with the abundance estimates provided by UVVIS spectral reflectance techniques (Lucey et al. 2000; see Section 10.4.6); the neutron abundance is on average about half that of the optical estimates (see also Section 10.5.3). The disagreement between the two data sets is likely to shed light on lunar mare basalt opaque mineral abundances and compositions as can be determined from orbit. Preliminary Ti abundance estimates from the Lunar Prospector GRS support the neutron numbers, but there are spatial discrepancies (Prettyman et al. 2002b) (see above).

Finally, the first qualitative maps of REE abundance (Gd and Sm) were shown by Elphic et al. (1998). Quantitative estimates were produced using both preliminary Lunar Prospector global FeO abundance estimates based on GRS data, and TiO_2 abundance estimates derived from Clementine UVVIS spectral reflectance data (Plate 2.7, Elphic et al. 2000). In general the correlation with LP-GRS Th distributions is very good. However, the TiO_2 abundance estimates appeared to be a factor of two too high, which had the effect of reducing the REE contribution in some high-Ti mare basalts (Elphic et al. 2000; 2002).

One strength of these REE measurements is that they are an independent measure of incompatible trace elements, specifically the rare earth elements Gd and Sm. The REE abundances can then be compared with that of Th to determine the Th/REE ratio, one indicator of the degree to which igneous evolution has caused incompatible-element fractionation, insofar as it may occur on a large scale. So in principle it is possible to separate KREEP, alkali anorthosite, and granitic compositions. It may be possible to investigate this ratio in mare basalts as well. The effect of the REEs in the neutron data is profound, and can be comparable to that of Fe and Ti in mare basalts.

It should be noted that the technique described here depends on knowing Fe and Ti abundance; uncertainties in these values produce uncertainties in the inferred REE abundance. Finally, there is some debate about the values of "effective" thermal-neutron capture cross sections for Gd and Sm, which are critical to determining the elemental abundance of the REEs.

10.2.3. Epithermal neutron measurements of Gd and Sm. The REEs Gd and Sm can also be detected using epithermal neutrons (Maurice et al. 2004). In contrast to the Gd and Sm measurement with thermal neutrons, the measurement of Gd and Sm concentrations with epithermal neutrons requires fewer assumptions about the surface composition. In general, epithermal-neutron data from the LP-NS yield a map (Plate 2.8) that comprises three levels of information relevant to the surface composition. The first signal is related to the production rate of fast neutrons that generate the epithermal neutrons. The second signal is due to solar-wind-implanted H at mid latitudes and water deposits at the poles (Section 10.8; Feldman et al. 2000a; Johnson et al. 2002). The third signal is caused by a combination of Gd and Sm (Maurice et al. 2004). With appropriate corrections, maps of Sm and Gd concentrations can be derived with values that correlate well in non-mare regions with the measured Th concentrations (see Fig. 2.40 and Plate

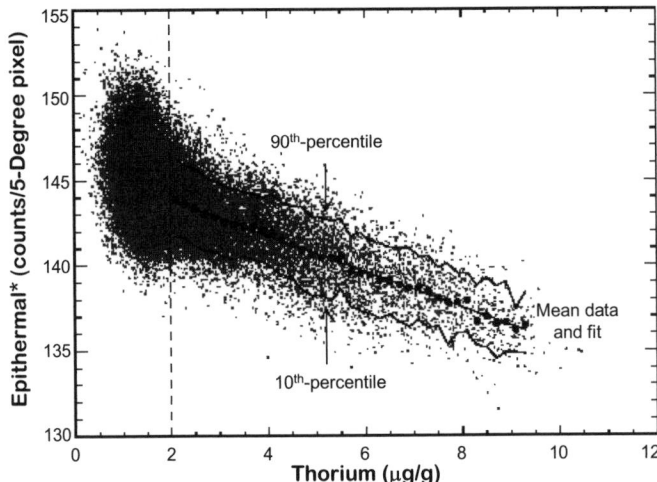

Figure 2.40. Comparison of Gd+Sm abundances measured using epithermal neutrons
with thorium concentrations as measured with the LP-GRS.

2.7). In terms of Sm concentration, epithermal-neutron measurements have a 10 ppm threshold
sensitivity, a 6.4 ppm precision, and a maximum value of ~35 ppm.

10.3. X-ray spectroscopy

The technique of planetary X-ray fluorescence (XRF) spectroscopy can measure the
chemical composition of an airless planetary body by measuring the characteristic X-ray line
emission that is excited on the planetary surface by solar X-rays. During solar quiet times, the
spectral shape and intensity of the solar X-ray flux limit the available elemental measurements
to the geophysically important elements Mg (1.254 keV), Al (1.487 keV), and Si (1.740 keV).
During times of enhanced solar activity, it is possible to measure the higher energy X-rays
from Ca (3.691 keV) and Fe (6.403 keV). Yin et al. (1993) give a detailed description of the
technique of planetary XRF spectroscopy.

Prior to the present time, the only XRF spectroscopy measurements to successfully measure
the composition of the lunar surface were the Apollo 15 and 16 XRF instruments. As described
in a variety of studies (Adler et al. 1973; Clark and Adler 1978; Adler and Trombka 1980;
Clark and Hawke 1981; Spudis and Pieters 1991; Yin et al. 1993), the Apollo 15 and 16 XRF
instruments measured the elemental ratios of Mg/Si and Al/Si over 9% of the near-side surface.
To simplify the data analysis procedures, the results were reported as elemental ratios instead
of absolute elemental abundances. Abundance maps and summaries of these measurements are
given by Andre et al. (1977), Spudis and Pieters (1991), and Yin et al. (1993).

The strengths of the XRF technique include the ability to measure elemental ratios and
absolute elemental abundances such as Mg , Al, and Si that may be difficult to measure with
other techniques. In addition, with the XRF technique, the measured elemental abundances
are measured independently of the molecular or mineralogical state of the given elements.
Furthermore, with suitable data processing, the spatial resolution of XRF measurements has
been shown to be better (~30 km) than unprocessed gamma-ray and neutron data taken from
a similar altitude above the lunar surface (~150–200 km). The spatial resolution of XRF
measurements, however, is still quite broad compared to imaging data. The depth sensitivity
in the lunar regolith of the XRF technique is on the order of tens of microns (Yin et al.
1993). Consequently, any thin deposit of excavated material can be identified with the XRF

technique. Because of the shallow depth sensitivity, the XRF technique is complementary to the gamma-ray and neutron techniques that measure compositions down to about half a meter in the regolith. One of the drawbacks of XRF is that because solar X-rays provide the incident X-ray illumination, composition measurements can only be made on the sunlit side of the Moon. This limits the duty cycle and potential surface coverage for any XRF mission.

Finally, there are two missions, one in progress and one in planning, to fly XRF instruments to measure the composition of the lunar surface (Grande et al. 2001; Okada et al. 1999). These include the European Space Agency's SMART-1 mission (presently in orbit) and the Japanese-sponsored SELENE mission. It is anticipated that both these instruments will provide larger surface coverage and have better X-ray energy resolution than the Apollo 15 and 16 instruments.

10.4. Spectral reflectance remote sensing of lunar surface composition

Optical, ultraviolet and infrared remote sensing of the Moon has a venerable history, beginning with Wood (1912) who discovered significant variations in the visible and ultraviolet characteristics of the lunar surface. A host of techniques has been applied using ground-based telescopes; the most scientifically fruitful of these for compositional mapping have been spectroscopy and multispectral imaging in the region of solar reflectance from 0.4 to 2.5 microns. Spectral reflectance measurements of the lunar surface are sensitive to the mineralogy, mineral chemistry, and physical state of the regolith, including the important optical effects of space weathering.

The foundation for remote compositional analysis lies in optical absorption physics (Burns 1993) and the linking of spectral properties of materials measured in the laboratory to well-understood mineral species and their mixtures. The pioneering work by Burns (1970), Hapke et al. (1970), McCord and Adams (1973), McCord et al. (1981), and others amply demonstrated the potential of spectral reflectance measurements for understanding lunar materials. Following this initial era, a diverse set of remote-sensing investigations made significant progress in understanding the distribution of mare and highland materials. Mare studies using spectroscopy and multispectral imaging produced maps of a diverse suite of mare basalts, including unsampled types, pyroclastic deposits and the abundance of Ti. Spectroscopy of the lunar highlands uncovered a spectral diversity that mirrors the diversity found in the pristine rock collection, but occurs in vast exposures that the tiny fragments found in the Apollo collection did not anticipate (Pieters 1986). The state of lunar spectroscopic remote sensing just prior to the new spacecraft results is summarized by Pieters (1993).

Techniques and methodologies developed during the 1970's and 1980's for lunar spectral analysis provided the foundation for understanding of the new mission data, such as using observations of craters of all sizes to probe beneath the regolith, and inferring the abundance of Ti from lunar color. These and similar approaches benefit from three key characteristics of the Clementine data set: global reach, high spatial resolution, optimal band passes, and data uniformity. The value of global data is obvious; studies previously confined to the lunar near side can now be conducted globally. High spatial resolution enables entirely new studies, such as examination of compositional properties of individual lava flows. The Clementine band passes were chosen to characterize the major features in the spectra of lunar rock-forming minerals (Fig. 2.41). Data uniformity in terms of calibration and photometric correction enable confident comparison of radiometric and derived compositional information Moon-wide and provide access to new quantities such as albedo to be applied more confidently. Continuing studies of lunar samples, especially the regolith, have provided new understanding of lunar soil formation processes and effects that strongly influence the ability to glean information from the Moon remotely.

10.4.1. Spectral properties of major lunar materials.

Pyroxene and olivine. Of the major lunar minerals, the mafic silicates pyroxene and olivine are the most readily detected using near-infrared spectroscopy. They exhibit strong and unam-

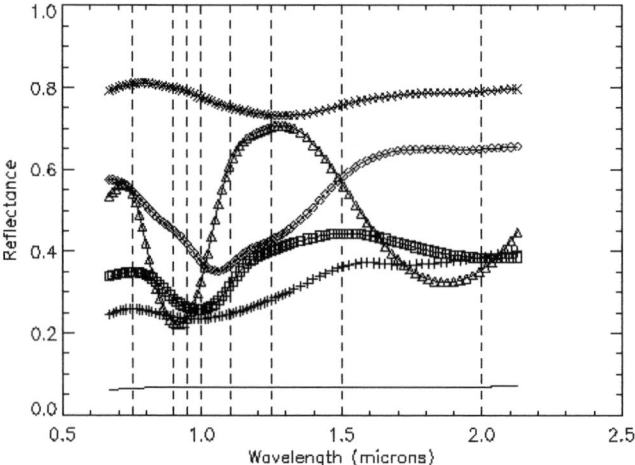

Figure 2.41. Spectra of individual lunar mineral phases: Ilmenite (solid line), glass (+), Orthopyroxene (Δ), clinopyroxene (), olivine (◊), plagioclase (×). Wavelengths of Clementine band passes for the UVVIS and NIR cameras (vertical dashed lines) are superimposed for comparison.

biguously diagnostic spectral features near 1-2 μm due to electronic transitions in ferrous iron (Burns 1970); the vast majority of lunar spectra obtained—remotely or in the laboratory—show evidence of pyroxene or olivine. Only spectra of regions essentially lacking these minerals can show clear evidence of the presence of other phases by casual inspection, such as areas identified as anorthosites (Spudis et al. 1984) and glass-rich pyroclastic deposits (Adams et al. 1974). The dependence of key spectral properties of these minerals on mineral chemistry has been well documented; the number, centers and widths of the absorption features due to ferrous iron in the mafic silicates are strong indicators of their compositions, enabling not only distinctions between minerals, but constraints on the composition within mineral classes (Fig. 2.42) (Adams 1974; Adams and Goulaud 1978; Hazen et al. 1977; Hazen et al. 1978; King and Ridley 1987; Cloutis and Gaffey 1991; Sunshine and Pieters 1998).

The strong spectral properties of pyroxene led to the first remote detection of a lunar mineral as reported in McCord and Johnson (1970). New measurement capability led to the detection of regions differing in pyroxene chemistry (McCord et al. 1981). Although many lunar highland regions show evidence for abundant low-Ca pyroxene, consistent with the sample collection, highland areas with high abundances of high-Ca pyroxene were also detected (McCord et al. 1981; Lucey et al. 1986; Pieters 1986; Lucey and Hawke 1987). In a survey of central-peak compositions, Pieters (1986) found that the frequency of occurrence of central peaks showing strong evidence of abundant high-Ca pyroxene was relatively high (18 of 77 locations, or 22%), which is in strong contrast to the scarcity of such rocks in the sample collection.

Olivine was the first unambiguously detected mineral other than pyroxene. Pieters (1982) reported that the central-peak complex of the large crater Copernicus exhibited a spectrum indicating the presence of a mineral assemblage with abundant olivine, but little or no pyroxene. Using geologic arguments, (Pieters and Wilhelms 1985) concluded that this spectrum represented troctolite, a mixture of olivine and plagioclase. Spacecraft observations would show that olivine-rich/pyroxene-poor locations occur elsewhere on the Moon (see below, and Pieters and Tompkins 1999; Tompkins and Pieters 1999; Pieters et al. 2001a,b).

Feldspar. Iron-bearing plagioclase feldspar also exhibits a diagnostic absorption feature, albeit weak, and the wavelength position of the feature is sensitive to the sodium content of the feldspar (Adams and Goulaud 1978). Subtle features in spectra of lunar materials appearing on the wings of pyroxene features are sometimes attributed to feldspar, but pyroxenes can also inherently exhibit such a feature (Sunshine and Pieters 1993), and in the absence of detailed analysis or high spectral resolution, olivine can also plausibly mimic feldspar in this respect. The plagioclase iron feature is also known to be susceptible to shock at levels frequently experienced by lunar rocks (Adams et al. 1979; Bruckenthal and Pieters 1984). Despite weakness and fragility of this feature the inherent high albedo of feldspar owing to its low ferrous iron content enhances its detectability.

Feldspar was the last of the major silicates to be unambiguously detected remotely. Spudis et al. (1984) reported the presence of anorthosite in the Inner Rook rings of the Orientale multiringed impact basin. In the strictest sense, that and subsequent similar detections of anorthite are not in themselves unambiguous. Spectra attributed to anorthite (or anorthosite) are characterized by the apparent lack of a measurable spectral feature due to ferrous iron; the spectra are featureless except for a spectrally red slope and a break in this slope near 1.5 microns (Fig. 2.43). Excepting the break in slope, the characteristics are consistent with a number of Fe-free minerals (e.g., enstatite), but of minerals and rocks present in the lunar sample collection, anorthite and anorthosite are the only plausible candidates that can account for the detected spectral properties. The remote spectra themselves are unusual in the sense that Fe-bearing lunar anorthite and anorthosite do exhibit weak ferrous-iron absorptions near 1.25 μm that the remotely-obtained data lack. The lack of this feature in the remote spectra has been attributed to the effect of shock (Spudis et al. 1984), thus these regions have been interpreted to be composed of maskelynite. However, recent modeling (Lucey 2002) has shown that space-weathering effects even on immature anorthosite surfaces can obscure the weak band and cause the observed break in slope without any shock effects (Fig. 2.43).

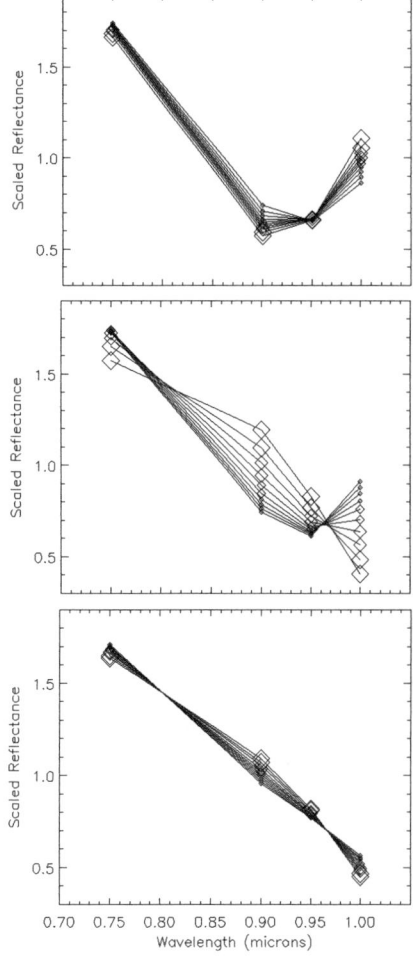

Figure 2.42. Spectra of orthopyroxene (a), clinopyroxene (b), and olivine (c) as a function of Mg-number at four Clementine wavelengths. Spectra are scaled to their mean reflectance and computed from the optical constants of Lucey 1998 at a grain size of 15 microns. The range of Mg-number in each case is 50-95, and intervals of 5 are shown, with increasing symbol size indicating increasing Mg-number. Both pyroxenes show strong variations in spectral properties with change in Mg-number (or iron content) suggesting that the chemistry of the minerals may be detectable remotely, even with the sparse spectral sampling of Clementine data. In each case the reflectance also a strong function of Mg-number and iron content.

Ilmenite. Ilmenite, $FeTiO_3$, is the most abundant oxide mineral found in lunar rocks, varying between 0 and 24 vol% (Papike et al. 1982). Spectrally, ilmenite is dark and neutral,

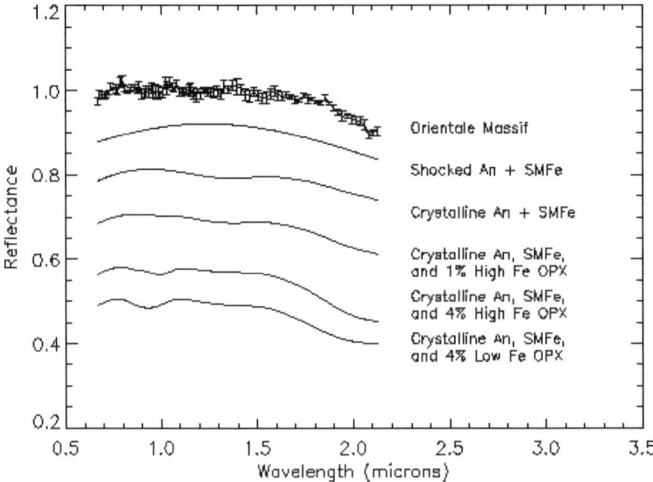

Figure 2.43. Spectra of Orientale Massif from Spudis et al. 1984 and model spectra from Lucey 2000 with straight line continua removed. Space weathered shocked anorthosite is generally consistent with the spectrum of the Orientale anorthosite, but does not account for the break in slope found in anorthosite spectra. Models containing crystalline material reproduce the shape of the lunar spectrum better, so the presence of shocked material is not required.

but not entirely featureless, exhibiting broad absorptions centered near 500 nm and 1200 nm (Adams 1975; Burns 1985). As with feldspar, the effect of the distinctive reflectance of ilmenite is arguably more diagnostic of its presence than its spectral properties. The presence of ilmenite is inferred in mare regions on the basis of its strong optical effects on both immature and mature mare soils. Mare soils with lower TiO_2 tend to exhibit steeper and "redder" UVVIS slopes than mare soils with relatively more TiO_2, which appear spectrally bluer (Section 1.5.3).

Glass. Technically not a mineral, glass is important geologically on the Moon, both as a product of space weathering and in pyroclastic deposits. Glass shows spectral features at 1 and 2 microns due to ferrous iron and these features are broad, which indicate a lack of long-range order. Laboratory experiments show the reflectance and intensity of glass absorptions is a strong function of Fe content beyond 700 nm; shortward of 700 nm the reflectance is controlled by the sum of Fe and Ti which interact by means of a charge transfer absorption (Bell and Mao 1972; Lucey et al. 1998). Iron-bearing glass has been unambiguously detected in regional pyroclastic deposits (Gaddis et al. 1985, 2003) and in the form of impact melt glass in the dark halo around Tycho (Smrekar and Pieters 1985).

Coarse-grained metallic iron. The presence of significant amounts of iron metal in some rocks has been neglected by the remote-sensing community. However, its abundance is generally below stated performance goals for uncertainties in determining modal abundance (~10%). Experiments aimed at understanding the spectral properties of metal on asteroids show that Fe metal is moderate in reflectance relative to other minerals, and spectrally red, characteristics that may confound its detection. In some rocks it is the major carrier of Fe, so its spectrally stealthy nature may be an important source of uncertainty in Fe estimation using spectral techniques in low-Fe terrains.

10.4.2. Characterization of soil modal mineralogy: validation of remote sensing. Since the return of the first lunar samples, standard operating procedure for soil petrography has been to characterize a lunar soil by "particle counting" (e.g., Simon et al. 1981; Heiken and McKay 1974). Such analyses provide detailed information about the abundances of mineral and rock fragments, volcanic glasses, impact-produced glasses, and glass-bonded aggregates, called agglutinates. Particle counting simply involves classifying a soil fragment with a title (e.g., pyroxene, basalt, breccia, agglutinate). These particle count data, however, do not provide information on the real percentages of minerals (modes) locked in rock fragments and fused-soil particles (e.g., agglutinates).

With studies of lunar-soil formational processes (e.g., Simon et al. 1981; Fischer and Pieters 1995), particle abundances are commonly reported, and each agglutinate contains 30-80% glass that binds these soil aggregates together. However, it is the absolute abundance of individual mineral phases and the agglutinitic glass that is important for chemical considerations (Hu and Taylor 1978), as well as spectral reflectance modeling, and these data do not exist in the literature. Modal analyses of the phases in the soils also permit us to address the abundances of nanophase Fe^0-bearing agglutinitic glass, as a function of grain size.

The actual amounts of the various minerals and glasses in the soil that interact with solar radiation are the important input data for the remote compositional analysis and space-weathering studies. Modal analysis, *sensu stricto*, is defined as the volume percentage (or calculated mass %) of the mineral constituents, not the particle type. It is essential that accurate quantitative modal analyses of the components of lunar soils be obtained, particularly of the <45 μm grain sizes, where the characteristics of the individual agglutinates are lost. This is accomplished using the techniques described and illustrated by Taylor et al. (1993, 1996), Chambers et al. (1995), and Higgins et al. (1996). Accurate modal analyses were performed with X-ray digital-imaging analyses on grain mounts of lunar soils as detailed by Taylor et al. (1996). In addition to the modes, the average chemical composition of each phase is determined (e.g., different types of pyroxene, plagioclase, high-Ti volcanic glass, high-Al agglutinitic glass), as related in Taylor et al. (2000, 2001a,b).

This methodology, applied to a statistical sample of lunar soils of varying composition, is essential to validate many of the results that have begun to emerge from spectral remote sensing of the Moon. Such a sample does not yet exist, but these crucial data could be obtained from the existing lunar-sample collection.

10.4.3. Optical effects of space weathering. Even before the Apollo missions, a darkening process was thought to act on the lunar surface (Gold 1955). With the return of lunar samples, it was immediately apparent that the optical properties of lunar soils are very different from those of lunar rocks powdered in the laboratory (McCord and Johnson 1970; McCord and Adams 1973). As noted by McCord and Adams (1973) and described by Fischer and Pieters (1994), the three principal manifestations of space weathering on optical properties of lunar materials are: 1) overall reduction of reflectance; 2) general attenuation of diagnostic absorption bands; and 3) development of a red-sloped continuum. These effects increase with soil maturity (i.e., surface exposure). Cassidy and Hapke (1975) first suggested that the red slope might be due to tiny inclusions of nanophase iron ($npFe^0$). Experimental (Allen et al. 1995, 1996), observational (Noble et al. 2001), and modeling (Lucey et al. 1995; Pieters et al. 2000; Hapke 2001) results have demonstrated the systematic effects of increasing amounts of small particles of $npFe^0$ (Fig. 2.44). Experimentally-produced coatings of nanophase iron cause strong reddening and darkening of samples (Allen et al. 1995, 1996). The observations of generally similar coatings rich in nanophase iron on most soil particles (Wentworth et al. 1999) confirms the Cassidy and Hapke hypothesis that lunar reddening and darkening are caused by submicroscopic metal.

The effects of nanophase iron can be broken into four stages (Noble et al. 2001): (1) Miniscule amounts of nanophase Fe^0 result in a large curvature in visible wavelengths, while leaving the longer wavelengths largely unaffected. (2) As nanophase Fe^0 accumulates, the continuum becomes less curved and significantly redder, reaching peak redness somewhere between 0.15 and 0.35 wt% Fe^0. (3) Additional nanophase Fe^0 results in an increasingly linear continuum that starts to lose redness in the visible. (4) From the experimental and modeling results, it is clear

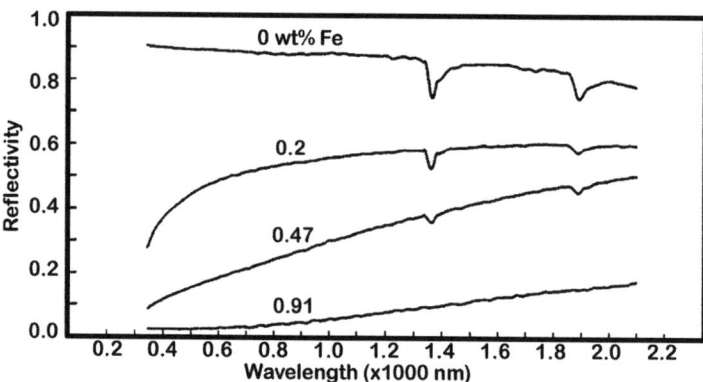

Figure 2.44. Visible-NIR reflectance spectra of silica gel particles (35-74 μm in diameter) with various amounts of nanophase Fe^0 filling pores 6 nm in diameter (Allen et al. 1996).

that if significantly more nanophase Fe^0 could be added to the soil, eventually the continuum would become dark and featureless. However, even the most mature Fe-rich <10 μm mare sample studied to date (15041) still has a significantly red slope. The natural soils achieve an equilibrium state where there is a balance between the creation and destruction of weathered rims and the influx of new material. Mature mare soils reach this equilibrium state at stage 3. Mature highland soils, having less iron available to them, reach steady state around stage 2. Soils of intermediate iron content, either inherently or due to mixing, show intermediate properties.

In addition to the demonstrated importance of nanophase iron, radiative-transfer modeling suggests that lunar soil spectral properties also require a dark neutral component to match the red slope and the low reflectance: extremely fine-grained nanophase metal does not darken sufficiently as it reddens to match both these characteristics of soils (Pieters et al. 2000; Hapke 2001). This dark component is undoubtedly the dark agglutinate glass that is so abundant in lunar soils.

Separated agglutinates are dark, but not extremely red despite being riddled with fine-grained iron (Pieters et al. 1993). However, this is consistent with observations that the nanophase metal in agglutinate glass is generally larger than coating metal (Keller and Clement 2001). Larger nanophase Fe^0 particles (those greater than ~10 nm in diameter) result in darkening of the soil (Britt and Pieters 1994; Keller et al. 1998), whereas the smaller particles (<5 nm in diameter) are largely responsible for the reddening effect. It is of note that the size range of nanophase iron particles sensed by FMR and represented in I_s/FeO measurements is much narrower than those that cause optical effects (Hapke 2001). In the optical literature the nanophase iron is commonly referred to as "submicroscopic iron" abbreviated as SMFE. The correlation between the size range sensed by FMR ($npFe^0$) and larger, but optically active sizes (SMFE), is only modest (Morris 1976, 1977; Lucey et al. 2000) so that I_s/FeO cannot even in principle be a precise predictor of optical effects of space weathering, and vice versa.

Historically, both micro-meteoroid impacts and solar-wind ion bombardment have been considered as contributors to the production of nanophase iron metal on grain surfaces that lead to the observed maturation (Keller et al. 1999; Hapke 2001). In the case of the ion bombardment, the resulting implantation of solar-wind H may act as an effective reducing agent that enhances the rate of production of nanophase metal from pre-existing silicates during micrometeorite impacts (Taylor 1988). In addition, solar-wind ion sputtering may contribute to vapor deposition of lunar patinas, which are coatings or discolorations caused by space weathering (Wentworth et al. 1999).

It has been suggested that lunar orbital magnetic field data may afford a macroscopic means of investigating the importance of the solar-wind ion bombardment in producing the optical maturation of the lunar surface (Hood and Schubert 1980; Hood and Williams 1989). Specifically, some lunar crustal field anomalies may be sufficiently strong to shield portions of the surface from the ion bombardment (Lin et al. 1998). Ideally, by comparing the optical properties of shielded surfaces with those of unshielded surfaces, one might expect to be able to estimate the relative contributions of the ion bombardment and micrometeoroid impacts in the maturation of the surface with time.

In this regard, it is of interest that the strongest local lunar magnetic field anomalies have been found to correlate in location with unusual, curvilinear albedo markings (Hood et al. 1979, 2001). For example, Figure 2.45 shows a correlation of a magnetic anomaly over western Oceanus Procellarum with unusual albedo markings northwest of the crater Reiner (6.7 N, 54.7 W). These markings are classified by lunar geologists as the "Reiner Gamma Formation" (e.g., Schultz 1976) and are also found elsewhere on the Moon in association with strong crustal fields. The higher albedos and curvilinear shapes of these markings have been suggested to be, at least partly, a consequence of deflection of the ion bombardment by strong local magnetic fields (Hood and Schubert 1980). The sources of the magnetic anomalies are

Figure 2.45. Contour map of the magnetic field magnitude over the Reiner Gamma region on western Oceanus Procellarum. The image was constructed from Lunar Prospector magnetometer data using a series of closely spaced orbit passes at altitudes ranging from 18 to 20 km. The contour interval is 3 nT and the 30-km crater Reiner is at the lower right. (after Hood et al. 2001)

proposed to be basin ejecta materials beneath the visible mare surface (Hood et al. 2001). However, there are several unresolved problems with this interpretation. First, it has not been demonstrated conclusively that lunar crustal field anomalies are sufficiently strong to completely shield portions of the surface from the solar wind. Second, the geologic origin of the Reiner Gamma Formation remains poorly understood. The alternate suggestion has been made, for example, that these albedo markings represent regions of surface scouring by a recent (<100 m.y.) cometary impact (Schultz and Srnka 1980).

In spite of remaining difficulties, the correlation of strong, local lunar magnetic-field anomalies with unusual albedo markings must be regarded as circumstantial evidence for a role of the solar-wind ion bombardment in the optical maturation of the lunar surface. Future combined orbital and surface measurements, together with returned samples from possible shielded surfaces, may therefore be very useful in establishing the precise contributions of ion and micrometeoroid bombardments to the lunar space-weathering process.

10.4.4. Grain-size effects. The presence of nanophase iron-rich coatings explains the surprising relative insensitivity of lunar soil spectra to variations in particle size. It is well known that spectra of transparent material normally brighten as particle size decreases (e.g., Adams and Filice 1967; Pieters 1983). This characteristic is not observed in particle-size separates of naturally formed lunar soils. While some increase in the albedo with decreasing particle size is observed in the near infrared along with reddening of the continuum, reflectance in the visible range (near 0.4 μm) typically has very little, if any, change as a function of particle size. Synthetic particle size separates of lunar fractions produced by crushing larger fractions of lunar soils follow the normal pattern of brightening with decreasing particle size throughout and do not duplicate the optical effects of natural particle separates of lunar soils (Pieters et al. 1993). This difference suggests that the structure of the natural surface is destroyed in the grinding process as fresh surfaces are exposed. Thus, the optical properties of the finest fractions must be due largely to surface correlated weathering products rather than simply to particle size effects.

A related property of lunar soils is that the spectral properties of bulk soil appear to be dominated by a very narrow grain size distribution. Fischer and Pieters (1996) demonstrated that the 10–20 μm and 20–45 μm sizes are optically the most similar to the bulk soil (Fig. 2.46). Larger size fractions resemble less altered material and are not representative of bulk soil optical properties (Pieters et al. 1993) whereas the <10 μm fractions appear to be highly unusual (Fischer and Pieters 1995). The dominance of the fine (but not finest) size fractions has been attributed to two factors: 1) fine particles coat larger particles, whereas, photons that enter large particles are unlikely to escape; and 2) although the <45 μm portions of lunar soils make up almost 50 wt% of the soils, this fraction constitutes over 75% of the surface area of the soil (Taylor et al. 1998). This observation has an important implication. The effect of grain size has marked effects on the spectral properties of minerals and glasses in a manner that can mimic compositional variations. The dominance of lunar soils by a narrow size fraction minimizes grain-size effects, enhancing the ability to map composition with confidence. When this spectral dominance of the fine particles begins during the evolution from rock to regolith is unstudied, so caution should be applied in assessing analyses of very immature surfaces; nevertheless, this observation supports a significant simplification in spectral analysis. A companion issue is that if spectral reflectance perceives a narrow grain size range, it is important to understand the relationship between the composition of this range and the bulk soil. In Section 9.5, it was shown that compositions of soils do change with grain size, so this effect must be borne in mind when interpreting spectral-reflectance data.

10.4.5. Mapping major mineralogy. The Clementine mission carried two multispectral imaging cameras that covered the spectral range from 0.4 to 1 μm in 5 bands (the UVVIS camera) and 1.1 to 2.7 μm in 6 bands (the NIR camera) (Fig. 2.47). The wavelengths were selected to characterize the major spectral properties of the lunar surface as understood from

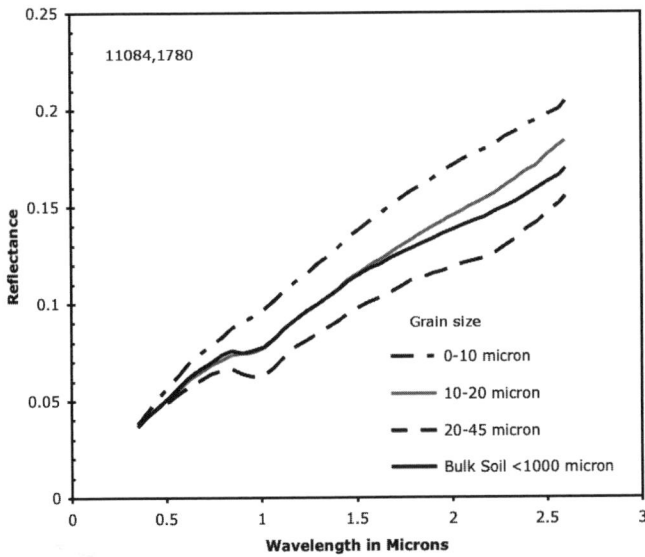

Figure 2.46. Reflectance spectra of 10084 size separates and bulk soil.

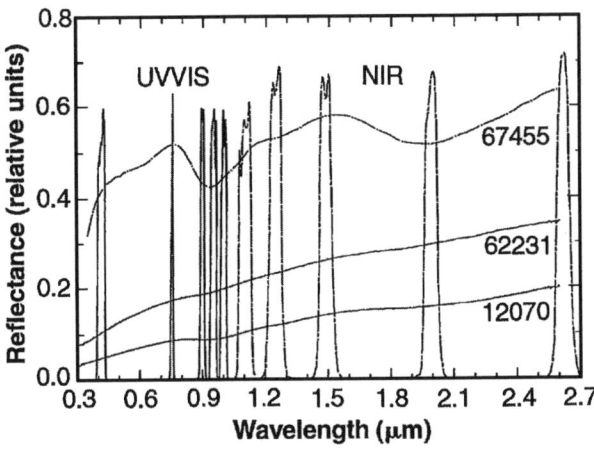

Figure 2.47. Laboratory Reflectance spectra of lunar samples superimposed on bandpasses of Clementine filters for the UVVIS and NIR cameras. Sample 67455 is a feldspathic breccia with low-calcium pyroxene, sample 62231 is a mature feldspathic soil that served as a ground-truth spectral standard for the Clementine data, and 12070 is a mature basaltic soil typical of the Apollo 12 landing site.

telescopic spectroscopy of the Moon and spectral characteristics of lunar samples. The UVVIS and NIR cameras were similar in spatial resolution and field of view and would ultimately achieve essentially 100% coverage of the lunar surface at resolutions higher than 500-m per pixel (Nozette et al. 1994; McEwen and Robinson 1997). Data from the UVVIS camera have been available for some ten years at this writing (see Plate 2.9); pilot projects have been carried out using NIR data and the global mosaic was available for analysis as of January 2004.

The spectral distinctions among lunar minerals are subtle over the wavelengths covered by the UVVIS camera, but Tompkins and Pieters (1999) developed a method to distinguish and identify spectrally dominant mineralogies from UVVIS camera data and applied this method to Clementine data of 109 globally distributed central peaks (Pieters et al. 1997, 2001;

Pieters and Tompkins 1999). They detected all the major silicates in portions of their sample of crater central peaks and in that study were able to reveal mineralogies of the lunar farside. They found evidence for both shocked and unshocked plagioclase, high- and low-Ca pyroxene, and olivine in craters on the near side where these minerals had already been detected, and they extended these discoveries to the far side. Because lunar space-weathering processes rapidly weaken spectral features, Tompkins and Pieters (1999) took care to select study areas with steep topographic slopes (clearly common in central peaks). In these locations space-weathering products are shed by mass wasting, leaving behind highly immature material that exhibits high spectral contrast, simplifying analysis. Using the high resolution of Clementine data, they were also able to illustrate spatial relationships previously unreported. Pieters and Tompkins (1999) made a detailed case study of the central peak of Tsiolkovsky.

The first publication exploiting NIR data was McEwen et al. (1994), who assembled a mosaic of an orbital swath through a portion of Aristarchus crater and surroundings, and used ground-based spectral data to calibrate the orbital swath. They found evidence for exposures of crystalline anorthite and olivine within Aristarchus crater. More extensive work with the NIR data was performed by LeMouelic et al. (1999, 2000) who further investigated Aristarchus crater and also the craters Aristillus and Kepler. They found clear evidence of variations in orthopyroxene-clinopyroxene ratio in the central peak of Aristillus and confirmed the presence of olivine on the rim and wall of Aristarchus. Their analysis showed evidence of about 25% pyroxene in the Aristarchus olivine locations and indicated that these olivine-rich locations would not have been detectable using UVVIS data alone.

In their demonstration that that mineralogy could be extracted from Clementine data, Tompkins and Pieters (1999) used a non-linear mixing model based on the work of Bruce Hapke to guide their interpretations of the spectral data. Lucey (2004) automated this analysis to produce global maps of the abundance of plagioclase, olivine, and clinopyroxene, and orthopyroxene (e.g., Fig. 2.48; see also Plates 2.9 and 2.10). Following the approach of Tompkins and Pieters, this analysis was confined to the most immature surfaces on the basis of a combination of reflectance and NIR-visible ratio, so only about 5% of the lunar surface was analyzed. Global maps of olivine, orthopyroxene, clinopyroxene, and plagioclase were produced from these sparse analyses by interpolating between the data points in a manner similar to production gridded data from sparse laser-altimeter profiles.

10.4.6. Mapping titanium. Prior to the global remote-sensing missions, many groups studied the relationship between lunar color in the ultra-violet and visible (UVVIS) and Ti content. Full-disk color photography was processed to reveal strong color differences on the lunar disk (McCord 1969; McCord and Johnson 1969; Pieters and McCord 1976; Whitaker 1972). These images show strong variations in color between the UV and visible, especially in the mare. Whitaker (1972) cited R. Strom as being the first to suggest that the strong color variations were due to the principal compositional variable in mare basalt samples, namely Ti. Shortly thereafter, Charette et al. (1974) showed that Ti content is correlated with telescopic measurements of UVVIS ratio in mature (agglutinate-rich) basaltic regolith, whereby samples with high TiO_2 concentrations (e.g., derived from Mare Tranquillitatis) exhibit relatively flatter and "bluer" UVVIS slopes than basalts with lower concentrations of TiO_2 (e.g., central Mare Serenitatis), which are spectrally redder. The Charette Relation provided quantitative comparisons of TiO_2 for sites in the near-side mare for the first time.

Failure to predict the low-Ti content of Luna 24 basalts returned from Mare Crisium indicated that a more complex relation exists between the two parameters. Subsequent revisions to the Charette Relation were made to increase the predictability of TiO_2 concentrations by improving spectral contrast (Johnson et al. 1991, 1977), and spatial resolution (Melendrez et al. 1994).

New algorithms for TiO_2 mapping were developed using Clementine and Galileo multispectral data. These algorithms included the ability to estimate TiO_2 for highland and

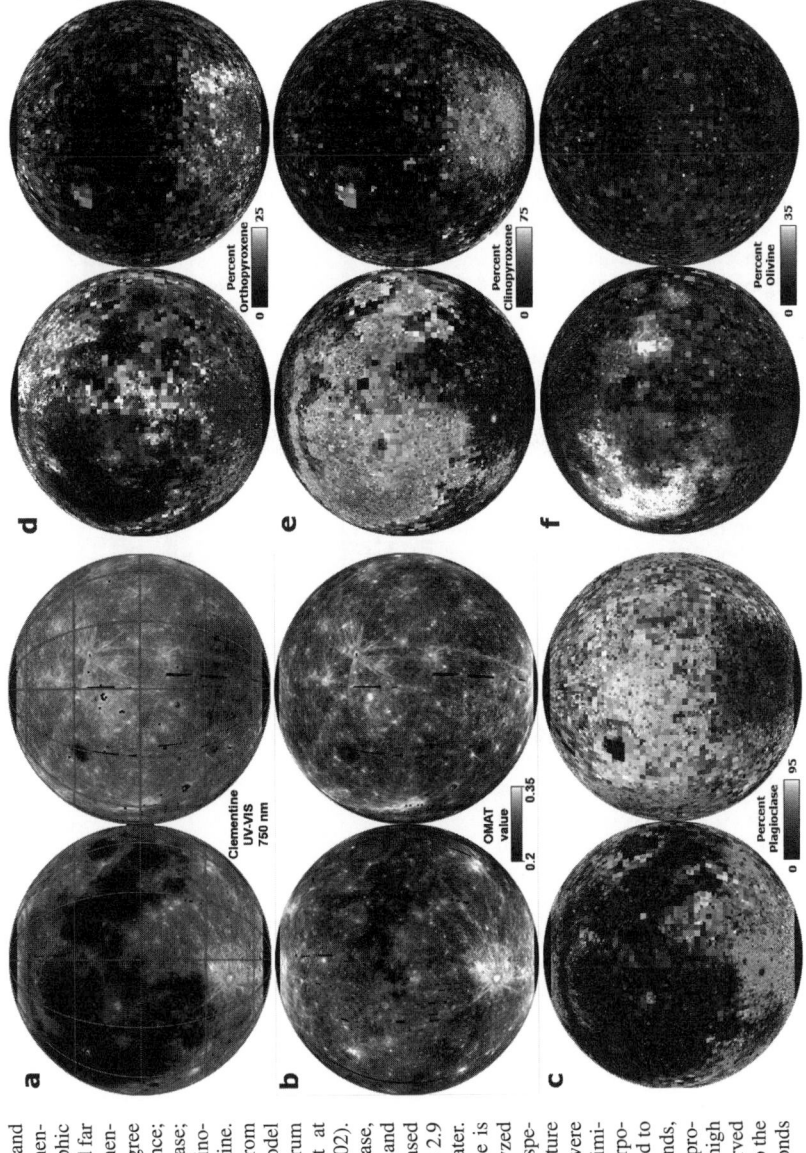

Figure 2.48 Global mineral maps and optical maturity derived from Clementine multispectral data in orthographic projection, with near side on left and far side on right of each pair. (a) Clementine 750 nm filter image with 30 degree latitude-longitude grid for reference; (b) optical maturity; (c) plagioclase; (d) orthopyroxene (low-Ca); (e) clinopyroxene (high-Ca); and (f) olivine. The mineral maps were made from 4-component nonlinear mixing model analysis of each immature spectrum in the global Clementine data set at 1 km spatial resolution (Lucey 2002). The four components are plagioclase, orthopyroxene, clinopyroxene and olivine. The maturity cutoff was based on the OMAT parameter (see Plates 2.9 and 2.10) with a value of 0.3 or greater. Only about 5% of the lunar surface is this optically immature, so the analyzed data feature large spatial gaps, especially in the highlands where immature exposures are infrequent. The data were then interpolated using algorithm similar to kriging. An artifact of the interpolation is that mare compositions tend to be over-reported, even in the highlands, as these more recent lava flows can provide abundant fresh material. The high abundance of clinopyroxene observed in South Pole-Aitken is likely due to the presence of frequent small mare ponds and cryptomare.

immature mare surfaces (Blewett et al. 1997; Lucey et al. 1998, 2000). Results using these algorithms support that the apparent gap between high- and low-Ti basalt samples is observed at the remote-sensing level for all of the Apollo and Luna landing sites (Gillis et al. 2003) (Fig 2.49). On a global scale, however, a continuum of TiO$_2$ concentrations exist for mare basalt regions (Giguere et al. 2000; Gillis et al. 2003).

Ti maps derived from multispectral imaging, however, have been shown to exhibit low precision and accuracy, on the order of 5 wt% in Ti, using sample analyses (Gillis et al. 2003) and Lunar Prospector neutron-spectrometer estimates of TiO$_2$ (Elphic et al. 1998, 2002). A complex relation between the UV/VIS ratio and TiO$_2$ was further shown to exist by comparing Lunar Prospector neutron TiO$_2$ concentrations and the UV/VIS ratio for mare regions of uniform color (Gillis et al. 2005). Data from these regions of interest reveal an apparent sigmoidal trend in the UV/VIS-TiO$_2$ relationship (Fig. 2.50) in which mare areas with <1.5 wt% TiO$_2$ trend along a shallower slope than basalts with higher TiO$_2$ contents. Also, the high-Ti trend is redder than predicted on the basis of the low-Ti trend alone by inferring that the UV/VIS values for basalts of the lower trend would extend linearly to higher values if they were to contain >3 wt% TiO$_2$.

The sigmoidal relation is notably different than the single curvilinear trend of Charette (1974), but reminiscent of large variation in color at low-Ti values (<2 wt%) that indicate higher uncertainty at low-Ti, as previous workers have noted Pieters (1993). In contrast to previous observations with sparser data, it appears that color values for high-Ti soils (>5 wt%) are also quite variable and that the lowest scatter in color is at intermediate Ti values (2-4 wt%).

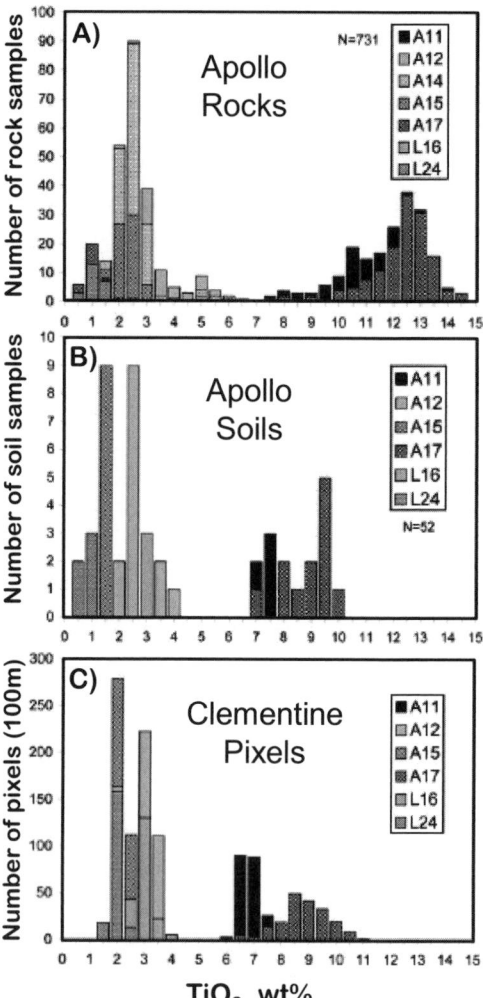

Figure 2.49. A) Histogram of TiO$_2$ concentrations for Apollo and Luna mare basalt samples. Note the "gap" in TiO$_2$ concentration between 5.5 and 9 wt% TiO$_2$: data from (BVSP 1981; Papike et al. 1998). B) Histogram of TiO$_2$ concentrations determined for representative basaltic soils from individual Apollo (A) and Luna (L) sampling sites. C) Frequency of Clementine UVVIS-based TiO$_2$ estimates determine using the modified regression for a 20×20 pixel image (100 m/pixel) that contains the respective landing site.

Focusing on predictability yields new insights regarding the utility of the spectral data. The utility of the multispectral data is in the prediction of Ti from color, rather than the spectroscopically interesting color effects at a given Ti content. At the lowest color values, (UV/VIS <0.58)

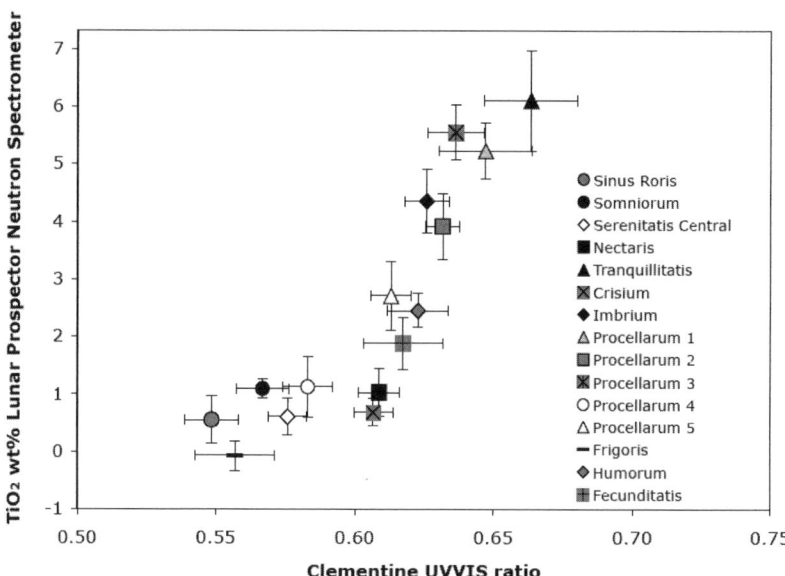

Figure 2.50. Average TiO$_2$ concentrations (Lunar Prospector neutron spectrometer) and UVVIS ratio data (Clementine) for 15 regions of interest. These data reveal an apparent non-linear, sigmoidal shape to the UVVIS-TiO$_2$ correlation. Regions of interest were selected on the basis of uniform color and composition in an effort to reduce error as a result of physical and spectral mixing. Error bars represent 1σ deviation.

all Ti values are <2 wt% (Fig. 2.50), so very low color values do indicate low Ti, however, all low Ti surfaces do not have low color values. For instance, UV/VIS ratios as high as 0.64 exist for some low-Ti basalts, <2 wt% TiO$_2$. Similarly, at high color values (UV/VIS >0.66) all Ti values are uniformly high, >5 wt%, but again high-Ti surfaces exist with lower color values. At intermediate UV/VIS ratios, Ti varies across nearly its entire range of values (Fig. 2.50). Figure 2.51 maps the distribution of surfaces with varying confidence. High confidence is assigned to UV/VIS values <0.58 or >0.66. On the basis of these arguments, one might conclude that no useful TiO$_2$ concentrations can be obtained where UVVIS values occur between 0.58 and 0.66. In this UV/VIS range, TiO$_2$ can vary unpredictably from 0 to 7.4 wt%, with a mean of 2.2 and a standard deviation of ±1.6. Perhaps UVVIS color can place a lower limit on the distribution of high- and low-Ti basalts, but it may not capture all basalts of these compositions, and little can be said with confidence about the basalts that have intermediate color values, which cover a vast majority of the maria (75%). This large area of uncertainty is in agreement with the finding of Gillis et al. (2003) who showed that TiO$_2$ concentrations for over two-thirds of maria were over- or underestimated by algorithms of Lucey et al. (2000). This result would affect the conclusion of Giguere et al. (2000) that intermediate Ti basalts (4.5–7.5 wt% TiO$_2$) are common. Basalts with intermediate color, previously interpreted as intermediate Ti, are now recognized to have poorly constrained Ti contents. The absolute mode of TiO$_2$ concentrations and/or the width of the mode might change, but it is unlikely that there is a bimodal distribution for basalt Ti contents globally (Fig. 2.52) because the Lunar Prospector neutron and gamma-ray spectrometer data (Elphic et al. 2002; Prettyman et al. 2002) do not exhibit a bimodal distribution.

Two qualitative models may explain the relationship between UVVIS color and TiO$_2$. One model, summarized by Pieters (1993), suggests that predictability in high Ti terrains should be good because the low albedo of soil components allows surface scattering to dominate; thus masking the optical properties of components. At low Ti contents, however, the complexity

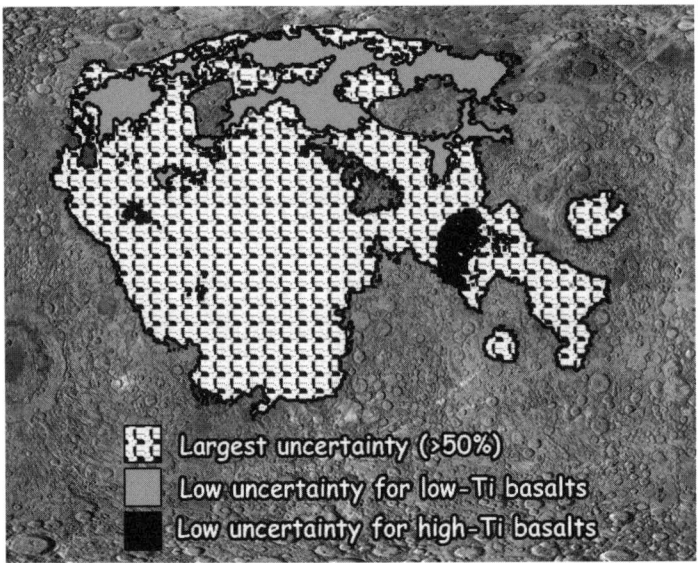

Figure 2.51. The distribution of TiO$_2$ uncertainty values based on the prediction of TiO$_2$ from color.

associated with mineralogy is increasingly perceived, breaking down the UVVIS-TiO$_2$ correlation.

Rava and Hapke (1987) emphasized the role of spectrally neutral opaques in controlling the Ti-color relationship. Ilmenite is dark and spectrally neutral relative to the spectrally red mature lunar mare soils, and must contribute to lunar UVVIS color as its volume proportion varies from essentially zero to as much as 25% (Section 4.2.1). Other oxides such as chromite, ulvöspinel, and armalcolite, however, are spectrally similar to ilmenite and thus their presence contributes to deviations from the UVVIS-TiO$_2$ correlation. In addition, Ti partitioned into non-opaque silicates (e.g., pyroxene) would also contribute to deviations from the correlation.

Recent results on the composition of agglutinate glasses (Taylor et al. 2001) show that Ti is only weakly enriched in the glasses of high-Ti basalts, reducing the effectiveness of Ti-Fe charge transfer absorptions in lowering reflectance of high-Ti soils. This finding suggests that opaques are more important than was previously assumed. The Fe content of silicate minerals and glasses in basaltic soils has a strong control on the visible reflectance that can also contribute to variations in the UVVIS ratio (Gillis et al. 2005).

Mare units show strong and consistent variations in UVVIS color that can be used to map geologic units; however, to what extent the variations in any individual mare deposit depend on Ti or other compositional parameter is not known. Clementine 2.7-µm data may help improve the accuracy of mapping TiO$_2$ because it correlates well with the UVVIS ratio, but the 2.7-µm data show a greater range in values with fewer effects from maturity than the UVVIS ratio and are not affected by Fe-Ti charge transfer as in the case of the UVVIS (Gillis et al. 2005). A full understanding the relationship between Ti and lunar color can be addressed by a focused study of lunar soils and a quantitative understanding of the optical properties of their components, but such a study has not yet been done.

10.4.7. Mapping iron. Efforts to map Fe using multispectral methods prior to Clementine were limited, but suggested that an algorithm might be derived analogous to Ti mapping algorithms (Charette et al. 1977; Fischer and Pieters 1994). Iron dominates the reflectance properties of the Moon in three ways: (1) the reflectance and absorption properties of mafic silicates and glass are proportional to Fe content; (2) the major and locally abundant opaque mineral phase, ilmenite, is Fe rich; and (3) nanophase Fe, the principal actor in space-weathering optical effects, strongly effects the lunar reflectance and is proportional to Fe content. In rocks and minerals, increasing Fe makes them darker and if the carrier of Fe is a silicate, the absorption features near 1 and 2 μm are strong, leading to a decreased ratio of near-IR to visible reflectance (Burns 1993). Nanophase Fe imparts a different behavior. Reflectance decreases with increasing nanophase Fe (and hence maturity), but the ratio of near-IR to visible reflectance increases (Fischer and Pieters 1994). A series of algorithms exploiting these observations

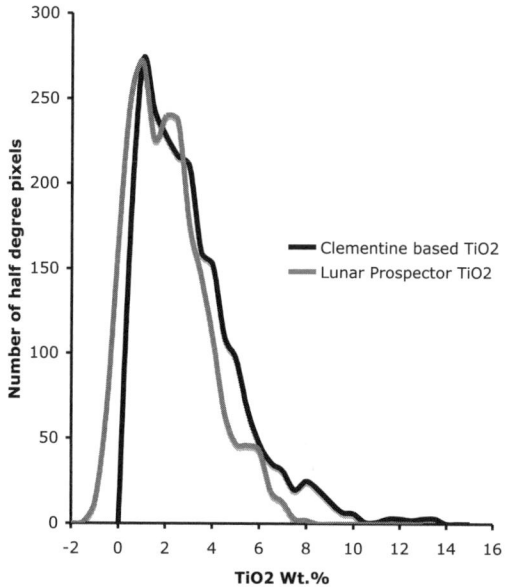

Figure 2.52. Comparison of Clementine derived TiO_2 and Lunar Prospector neutron spectrometer TiO_2 results. Differences between the two histograms highlight the uncertainty in TiO_2 distribution predicted from UVVIS color data.

were developed for predicting Fe contents from visible reflectance and near-IR to visible ratio of lunar soils. Correlations between derived Fe-sensitive spectral parameters and known FeO contents were shown to be high in several studies, with few outliers, in contrast to similar studies of Ti mapping (Lucey et al. 1995, 1998, 2000; Blewett et al. 1997; Lawrence et al. 2002; Gillis et al. 2004).

As with mapping of Ti, there are mineralogical effects that contaminate the Fe estimates (Staid and Pieters 2000). The spectral properties of silicates alter at different rates as a function of Fe content (Lawrence et al. 2002). Comparison of Lunar Prospector Fe abundances to multispectral Fe shows three major regional anomalies: Western Oceanus Procellarum, South Pole-Aitken Basin, and Mare Tranquillitatis. Western Procellarum is now thought to be olivine-rich (Staid and Pieters 2001). The absorbance of olivine is weaker than pyroxene at Clementine wavelengths for equivalent Fe (Lawrence et al. 2002). This leads to systematic underestimates of the Fe content at high olivine contents using calibration curves dominated by soils rich in pyroxene. In contrast to the FeO overestimate of the olivine-rich mare, the South Pole-Aitken (and Apollo 14) FeO abundances are overestimated. Current global mineral maps (Section 10.4.5) suggest that South Pole-Aitken basin is anomalously low in olivine relative to the rest of the Moon, which seems to carry a background olivine component of a few weight percent. If this is correct, the absence of regional olivine explains the systematic overestimate of Fe by spectral reflectance methods at South Pole-Aitken basin. Multispectral estimates of Fe content at Mare Tranquillitatis are anomalously high relative to Lunar Prospector by a few weight percent, indicating the effect of opaque ilmenite on the Fe estimates. In practice, few mare surfaces show high Ti or high olivine contents, so for mare deposits low in these components, Fe estimates are probably valid to the degree the calibration curves suggest, namely 1–2 wt% (Lucey et al. 1995). On the basis of the measured error of Tranquillitatis relative to LP-GRS Fe, even moderate Ti basalts would cause only a 1–2 wt% systematic underestimate if the error is proportional to opaque mineral content. In the highlands, no opaque rich deposits are known, and olivine contents can be estimated, suggesting that either a correction could be developed, or problematic areas could be avoided.

Much of the value of the multispectral Fe (and Ti) maps is in their ability to be applied at high spatial resolution not achievable using geochemical (e.g., GRS, XRF) methods. However, for much of the Clementine data, Fe estimates based on the method introduced by Lucey et al. (1995) are compromised by variations of shading due to variable topography observed under oblique lighting conditions, conditions that hold for most of the Clementine data set (e.g., Robinson and Jolliff 2002). Le Mouélic et al. (2000) developed a method for calculating the Fe in silicates based on the correlation of Fe with the 1 μm band depth and the visible-NIR continuum slope. Because the method uses only spectral ratios, it eliminates the influence of topography. It is likely that the Le Mouelic method shares the mineralogic overprints possessed by the Lucey et al. (1995) method, but these have yet to be evaluated.

10.4.8. Mapping maturity. Because the optical effects of space weathering and Fe variations are inversely correlated, these components can be separated. Fischer and Pieters (1996) showed that I_s/FeO correlates with the 1-μm band depth and could be estimated with multispectral images. Lucey et al. (1996) presented lunar soil and remote measurements of optical maturity and showed that compositional effects (especially mare vs. highland) can be largely suppressed, leaving optical maturity as the principal parameter (Fig. 2.48b; Plate 2.9). The range of values in this optical-maturity parameter is almost ten times the residual compositional effects that can be observed in the background, such as variations in the mare and among pyroclastic deposits. Grier et al. (2001) used this parameter to study age relationship among large Copernican and Eratosthenian craters. Like Fe estimates, this method is subject to contamination by topographic shading at high phase angles, but Le Mouélic et al. (2002) also derived a method using NIR data that does not suffer from these effects.

10.4.9. Hydrated minerals. The returned (Apollo) samples are extremely dry and in fact no indigenous lunar water has been unambiguously identified in the sample collection. However, the Moon has been impacted by comets regularly over its history so it is possible that evidence of interaction of the Moon with cometary water might be preserved. Considerable attention has been paid to water which might have accumulated in lunar cold traps (see Chapter 6) but evidence of water might be present elsewhere on the Moon.

Telescopic and space-based searches for evidence of water or hydrated minerals have been conducted. Roush and Lucey (1988) searched for evidence of a water-of-hydration band at 3 μm at the Reiner Gamma Formation, a region proposed to be the result of interaction of a comet with the lunar surface. The Near Infrared Mapping Spectrometer on Galileo obtained spectra of a portion of the Moon through the 3-μm region as well and detected no evidence of a 3-micron absorption band. Vilas et al. (1999) suggested that anomalous features in Galileo multispectral imaging of portions of the Moon in the lunar south showed evidence of phyllosilicates (Galileo used somewhat different bands than Clementine). Finally, McConnochie et al. (2002) used Clementine NIR data obtained in polar regions with no direct illumination but indirectly illuminated by reflections off crater walls to search for evidence of a 2.0-μm band, with no reported detection.

The Clementine NIR data are actually well suited to a search for water as the 2.7-micron band is centered on the strongest portion of the hydroxyl absorption. Any searches using these data must cope with the very significant thermal-emission component inherent at 2.6 and 2.7 microns, but while this component will tend to reduce spectral contrast, reflected solar radiation still dominates, so a search of the data for spectral anomalies consistent with water or other hydroxyl-bearing phase could be useful. A preliminary survey of the data at 500-m resolution between latitudes of 70°S and 70°N revealed no obvious anomalies in 2.7-μm reflectance, or ratios of 2.7 μm to 2.6 or 2.0 μm that might have indicated the presence of hydrated minerals. For example, there is no population of craters exhibiting hydration anomalies that might indicate the impacts of a cometary population, nor did inspection of the Reiner Gamma formation of other lunar swirls exhibit 2.7-μm anomalies. This survey probably was not sensitive to anomalies less than 5-10% in depth owing to the ubiquitous effect of maturity near small craters. More sensitive methods or investigations in the polar regions might bear more fruit (Pieters et al. 1988; Gillis and Lucey 2004).

10.5. Spectral remote sensing in the thermal IR

Spectroscopy in the region of thermal emission (7–14 μm) has the potential to make important contributions to lunar science (Nash et al. 1993). This technique is sensitive to silicate polymerization so in principal it should be more sensitive to the presence of feldspar than near-IR spectroscopy, and also to variations in feldspar chemistry, as well as enable searches for quartz that might be present in outcrop in rare locations on the Moon. There have been a handful of telescopic observations of the Moon in the thermal region (e.g., Sprague et al. 1992). Although none of these observations have contributed substantially to understanding the composition of the lunar surface, they are valuable in determining performance requirements for future experiments. Laboratory measurements of lunar soils by Salisbury et al. (1997) showed that spectral variations among lunar soils in the thermal IR are extremely subtle, and a successful telescopic experiment will likely require extremely high signal to noise ratios, on the order of 1000 or greater. Because of the extreme sensitivity of this technique to grain size, a possible future application might be to detect the presence of bare rock for other types of remote sensing to exploit.

10.6. Remote sensing of lunar thermal properties

10.6.1. Clementine LWIR. The long-wave infrared (LWIR) camera on Clementine measured lunar surface emission between 8 and 9.5 μm. The LWIR spatial resolution ranged from 200 m/pixel near the poles to 55 m/pixel at the equator. Contiguous pole-to-pole imaging strips were obtained with ~10% overlap between adjacent frames; however, significant longitudinal gaps exist between successive orbital passes. The LWIR local lunar noontime thermal-emission observations are unique in many ways. The high resolution of the LWIR images surpasses all other lunar temperature observations and provides substantial information on the spatial variations in temperature. The LWIR nadir-looking observations offer a unique thermal emission perspective on the lunar surface, and the measurements allow observations of variations with local incidence and emission angles that are not available to Earth-based observers. The global nature of the measurements, including the lunar farside, is unprecedented.

The LWIR global dataset demonstrates that the Lambertian temperature model of $\cos^{1/4}(i)$, where i is the solar incidence angle, is a fair approximation for nadir-looking temperatures, rather than the $\cos^{1/6}(i)$ behavior observed for ground-based measurements of the full Moon where both incidence and emission angle vary (Lawson et al. 2000). Deviations from the Lambertian model are likely due to surface roughness effects.

LWIR temperature information can also be combined with other data sets to investigate lunar surface thermophysical properties. Albedo, large-scale surface roughness, and small-scale (subresolution) surface roughness each may affect the lunar surface temperature. An ideal way to explore the effects of these properties is to compare the reflectance and the emission of an area, since they are closely connected. Each thermophysical property will result in a different relationship between reflectance and temperature, and investigating the trends as they vary with terrain and phase angle can help to determine which processes dominate the lunar surface response under varying conditions.

The lunar surface response in different highland and mare locations has been explored as a function of varying phase angle (Lawson and Jakosky 2001). An example can be seen in the Aristarchus Plateau images of Figure 2.53. The continuous LWIR mosaic is bounded

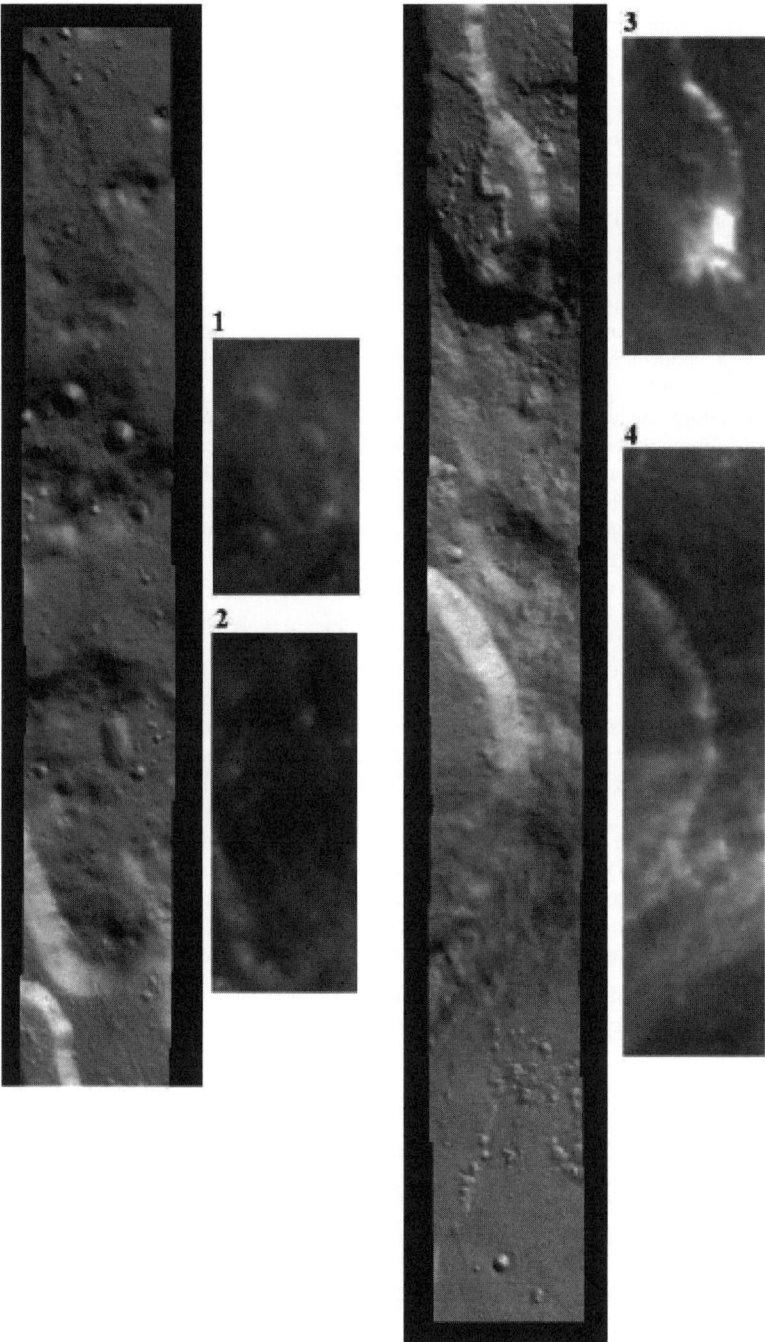

Figure 2.53. Aristarchus Plateau LWIR mosaic (strips) and corresponding UVVIS images (panels) (taken from Lawson and Jakosky 2001). Latitude range: 22°–28°N, longitude: 310.8°E.LWIR resolution: 100 m; LWIR temp range: 320-380 K. Numbered parts correspond to sections discussed in the text.

by black pixels and divided into two portions; the northernmost point of the mosaic is at the top of the left strip, and the southernmost point is at the bottom of the right strip. To the right of each LWIR mosaic strip are UVVIS 750-nm images of the same area of the lunar surface; the brightness stretch across the four UVVIS panels is constant. The Aristarchus Plateau is a 2-km elevated crustal block approximately 170×200 km which slopes downward to the north-northwest and is surrounded by the younger mare basalts of Oceanus Procellarum. The 42-km-diameter impact crater Aristarchus lies on the plateau's southwestern edge, and the older, partially embayed 35-km-diameter crater Herodotus lies just west of Aristarchus. The plateau contains the densest concentration of lunar sinuous rilles, most of which originate from cobra-head craters. A dark-mantle deposit, with the lowest reflectivity of any large lunar area blankets the entire plateau (Wilhelms 1987; McEwen et al. 1994; Weitz et al. 1998).

At very low phase angles the temperature and reflectance response is primarily governed by the variation in single-scattering albedo regardless of the presence of topography. As the phase angle increases, the influence of surface roughness grows. Finally, at moderate to high phase angles the effect of surface roughness dominates. In the absence of large-scale topography, the lunar surface temperature and reflectance response at all phase angles is governed by the variation in single-scattering albedo. LWIR-measured temperature variations yield local topographic information at high incidence angles that is unavailable via the reflectance, whereas UVVIS-measured reflectance variations yield local topographic information at low incidence angles that is unavailable via the temperature. The largest factors affecting the lunar surface daytime temperatures are the albedo and the incidence angle of solar insolation.

10.6.2. Neutron spectrometer measurements of subsurface temperature. The flux of thermal neutrons that leak out from planets depend mostly on surface composition, but also on surface temperature. The temperature dependence reflects the fact that in equilibrium, the rate of moderating the neutron flux from the epithermal range to the thermal range must exactly balance the rate of neutron absorption by elements in the soil. Because most absorption cross sections in the thermal energy range depend inversely on the neutron speed, the loss of neutron flux (density times speed) to absorption depends only on the neutron number density. This density should not, therefore, depend on temperature. Instead, the flux of neutrons in the thermal energy range should increase with increasing temperature.

With numerical simulations of this effect using the Monte Carlo Neutral Particle transport code MCNPX, Little et al. (2003) showed that the effect is most prominent in materials having a low macroscopic absorption cross section (such as the feldspathic lunar highlands) because the thermal-neutron population is relatively largest for that composition. On the other hand, the effect is insignificant for material having a large macroscopic cross section (such as the mare basalts). They also showed that the depth being sampled by leakage thermal-neutron fluxes was centered at about 30 g cm^{-2} below the surface. For a density of 2 g cm^{-2}, this depth is 15 cm. Because most models of the thermal structure of the lunar surface (e.g., Vasavada et al. 1999) indicate that this depth is below the diurnal thermal wave, the temperature of such a layer should depend only on latitude (λ). Indeed, Figure 2.54 shows that a preliminary comparison of the latitudinal variation of the flux of thermal neutrons above the feldspathic highlands were shown to be consistent with a cos$^{1/4}$ (λ) temperature law (Little et al. 2003) as was determined using the Clementine LWIR data (Lawson and Jakosky 2001). The net effect on the thermal neutron flux from equator to pole amounted to about 13.5%.

10.6.3. Infrared night-time and eclipse observations of the Moon. Ground-based telescopic infrared-scanner measurements of the Moon during eclipse, normalized to initial temperature and time of observation, show numerous thermal anomalies and that these anomalies often correlate with stratigraphic ages of craters and crater count ages of individual maria (Shorthill and Saari 1965; Fudali 1966; Saari et al. 1966). The Apollo 17 Infrared Scanning Radiometer collected high-quality temperature data for portions of both the night and

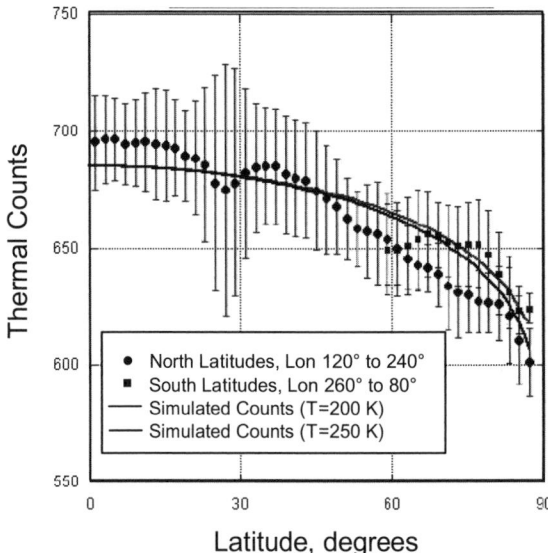

Figure 2.54. Latitude dependence of lunar surface temperature measured with thermal neutrons. The points show measured thermal neutron count rate in feldspathic highlands regions either for northern latitudes () or southern latitudes (). The solid line is a $cos^{1/4}(\lambda)$ temperature dependence for $T = 250$ K at the equator.

day side of the Moon (Mendell and Low 1975). These data showed that nighttime temperature anomalies were observed and that they often correlated with geologic features. This experiment showed that areas with high thermal inertia (indicated by high nighttime temperatures) also had higher frequencies of exposed rock. Mendell (1976) suggested, that with an appropriate dose of geologic common sense, these or similar data might be used for age dating.

Recent observations of the southern near side of the Moon in partial eclipse in the thermal infrared at high resolution show results similar in spatial resolution and character to the Apollo 17 radiometer results, but for a different portion of the Moon (Wilcox et al. 2005). As reported by Mendell and Low (1975), mare surfaces are dotted with small anomalies, small impact craters having excavated cobble-rich material from thin mare regoliths. Highland regolith almost entirely lacks anomalies owing to small craters suggesting the highland regolith is free of blocks to depths on the order of 100 m. These data, coupled with radar measurements at a range of wavelengths, should allow detailed understanding of the regolith structure at the meter scale.

10.7. Radar remote sensing of the regolith

Earth-based radar measurements, acquired using the delay-Doppler technique (Pettengill et al. 1974), provide synoptic views of the lunar surface over a range of wavelengths from 3.8 cm to 7.5 m (e.g., Zisk et al. 1974; Thompson 1987; Thompson 1978). These maps have been used to study regional pyroclastic deposits (Zisk et al. 1977), mare titanium abundance (Schaber et al. 1975; Campbell et al. 1997), and variations in large-scale regolith properties (Thompson et al. 1979). Recent work has improved considerably the effective spatial resolution of these observations through the use of focused processing (Stacy et al. 1997; Campbell et al. 2003), and permitted derivation of detailed topography using interferometric techniques (Margot et al. 2000).

The radar echo from the Moon is modulated by a number of factors: surface topography on scales of several meters to many kilometers, the abundance of wavelength-scale rocky debris on the surface or suspended within the fine-grained dust, and the bulk microwave loss properties of the regolith. Longer radar wavelengths are sensitive to the abundance of larger-diameter rocks, and can penetrate to greater depth for any particular value of the regolith loss factor. Attenuation of the radar signals with depth is controlled primarily by ilmenite in basaltic or basalt-contaminated deposits. Losses in highland-dominated materials are typically low, and may be controlled by a number of possible minerals.

Radar data can reveal properties of the regolith deeper than apparent to visible imaging. Ghent et al. (2005) found that virtually all large, young craters on the Moon have extensive "haloes" of low 70-cm radar return concentric to rugged, radar-bright materials of the continuous ejecta blanket (Fig. 2.55). It is most likely that these haloes are due to a lower abundance of 10-cm to meter-scale rocks on and within the regolith than is typical of "well-gardened" highlands and maria across the Moon. In effect, the impact process reduces the block abundance of a large surrounding region. This deep view also allows direct imaging buried mare deposits (crypto-mare), for example in the region west of Oceanus Procellarum (Campbell and Hawke 2005).

Ongoing work is opening new opportunities for Earth-based radar studies of the Moon. By using focused processing (similar to synthetic aperture radar), high spatial resolution can be achieved over a large illuminated area (Stacy et al. 1997; Campbell and Hawke 2005). At 12.6-cm radar wavelength, spatial resolution as fine as 20 meters per pixel has been achieved using these techniques. Applications include studies of the polar shadowed terrain, detailed analysis of pyroclastic deposits that may provide useful resources for eventual exploitation, and improved understanding of regional differences in regolith properties linked with basin ejecta patterns.

10.8. Hydrogen and the lunar poles

10.8.1. Importance of hydrogen. The elemental composition of planetary bodies provides a very important diagnostic of the processes that shaped their formation and evolution. The giant molecular cloud from which the Sun and its planetary system formed was very rich in H. It is generally thought that 90% by number was H, about 8% was He, and all of the heavier elements comprised the remaining 2%. Although the Sun contains all elements in these proportions, the various planetary bodies of the solar system sustained considerable chemical fractionation that depended on the location within the solar nebula where they coalesced and the mix of mechanisms that guided their evolution. The H content of the terrestrial planets is generally low compared to that of the Sun, but increases gradually with increasing heliocentric distance. Beyond Mars, the abundance of H takes a large jump in all of the gas-rich giant planets and comets that populate the outer solar system (see also Chapter 7).

An early result of the Apollo program was that the composition of the Moon was deficient in all volatile elements relative to the Earth, especially so in H. Whereas most dry basalts on

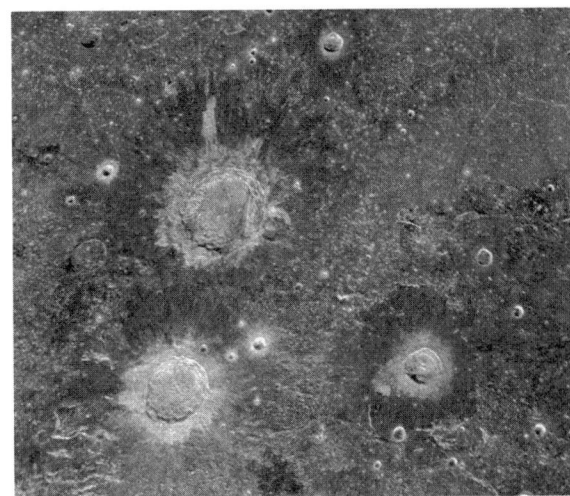

Figure 2.55. 70-cm wavelength radar image, collected using the Arecibo and Greenbank radio telescopes, of three large lunar craters (Aristoteles, at upper left, is 87 km diameter). The brightness variations reflect differences in decimeter-scale rocks on the surface or suspended within the upper 3-10 m of the regolith. Many large, young lunar craters have associated "haloes" of low 70-cm radar return that indicate rock-poor ejecta deposits outside rock-rich continuous ejecta.

Earth contain the equivalent of 1% water by weight, similar rocks on the Moon contain the equivalent of about 0.045% water. In addition, there are no reservoirs of water on the Moon such as is contained in the terrestrial oceans, glaciers, and polar ice packs. This striking difference has been used to infer that the Moon formed, or evolved through a single or series of catastrophic heating events in which most of the Moon's inventory of volatiles was either stripped or evaporated away. Hydrogen, being the lightest volatile, is thought to have been completely lost during this period. Hydrogen that has been detected within lunar samples from the present lunar surface is thought to have been implanted in the outer amorphous layers of regolith grains by the solar wind (Section 8.3).

Speculation regarding reservoirs of H that might exist on the present-day surface of the Moon is centered on the very cold regions near both lunar poles that form the floors of craters permanently shaded from the Sun (Watson et al. 1961; Arnold 1979). This H is thought to have been delivered in the form of water molecules through impacts of comets, meteoroids, and dust particles with the Moon. After delivery, the water molecules slowly migrated to the poles while undergoing loss to interplanetary space due to ionization by solar UV and pickup by the solar wind. Once at polar latitudes, the residual component of water molecules becomes trapped by the cold surfaces of the permanently-shaded polar craters (Butler et al. 1993; Butler 1997).

There are several reasons for special interest in detecting and locating enhanced reservoirs of H on the Moon. Foremost is that layered horizons within these deposits provide a history of past impacts that can be interpreted using the isotopic ratios of all volatile elements within individual layers and their thickness, to determine the size and origin of the delivery agent. Next, is that water is an essential ingredient for the support of any extended human presence on the Moon. An assay of its lunar inventory (or by proxy, that of H) is needed to conduct advanced planning for future scientific and commercial human missions to the Moon (see Chapter 6). And finally, the H content of non-polar surface soils provides a potential marker of soil maturity.

10.8.2. Epithermal neutron measurements of H. Neutrons generated by galactic cosmic rays that leak from planets provide a sensitive measure of H within near-surface regolith layers. This sensitivity stems from the fact that neutrons and protons have the same mass. The fractional energy transferred per elastic collision is thus greater for protons than for all other nuclei, which are heavier. On average, neutrons lose half their energy per collision with H and so moderate to low energies very quickly when H is present. They consequently travel very quickly from the energy range in which they are born (the fast neutron range), to the energy range in which they are absorbed (the thermal range). At equilibrium, the intensity of neutrons in the intermediate energy range (predominantly the epithermal range) is reduced in proportion to the enhancement in energy-transfer rate. The measurement signature of enhanced H in planetary leakage neutrons is therefore a reduced flux in the intermediate energy range.

Another factor that regulates the amount of neutron flux-reduction is the cross section for elastic scattering. Here again, H is outstanding because the (n,p) cross section is 20 barns, independent of energy, E, below $E = \sim 20$ keV, and decreases slowly as the neutron energy increases. It is equal to about 5 barns at $E = 1$ MeV, which is about equal to elastic cross sections for most elements heavier than H. The detectability of H using neutrons is therefore greatest in the energy range between the upper thermal energy limit, about 0.4 eV, and 20 keV. Although it diminishes steadily at higher energies, the neutron-detection technique is still useful and significant at the low energy end of the fast-neutron range (0.5 MeV $< E < 10$ MeV). The technique is not applicable in the thermal energy range because detection in this range is masked by the fact that thermal fluxes depend critically on the macroscopic absorption cross section, which varies over a range of 300% for the different compositions of lunar surface material (Feldman et al. 2000b).

Although the characteristic signature of H as a reduction in the flux of epithermal neutrons is very specific, it is not unique. Several REE such as Gd and Sm, support extremely large and sharp resonances in their neutron absorption cross sections in the energy range between 0.4 eV and 10 eV (see Section 10.2.2). The effect of these resonances on the equilibrium neutron flux spectrum can sometimes be significant, thereby masking the reduction effect due to H (Maurice et al. 2000). This effect is not present at fast neutron energies. Detection of a reduction of leakage neutrons in both ranges of energy in the proportions dictated by both the composition of the major rock-forming elements and the known elastic-scattering cross sections then provides a unique signature of H. This effect was used to verify the presence of enhanced H near the south pole of the Moon by Feldman et al. (2000b) (Plate 2.11).

Although enhanced abundances of H near both lunar poles is evident in neutron measurements made using the Lunar Prospector neutron spectrometers, identification of their host molecule as water requires more information. Arguments in this regard were provided by Feldman et al. (2001). Two models of enhanced H emplacement have been published in the literature. The first, presented earlier, is the delivery of water to the Moon from impacts of comets, meteoroids, and/or dust particles, followed by a random walk to the poles where they are permanently trapped on surfaces sufficiently cold that their characteristic evaporation time is longer than a few billion years (Watson et al. 1961; Arnold 1979; Butler et al. 1993; Butler 1997; Crider and Vondrak 2000). The second is the delivery of H to the Moon from the solar wind and related plasmas near the Earth (Crider and Vondrak 2000; Starukhina and Shkuratov 2000). Typical energies of these plasma protons are between 1 keV and 4 keV, which results in burial depth within the amorphous coatings of regolith grains to the order of 1000 Å. At sufficiently low temperatures, diffusive loss of implanted protons from these amorphous layers is sufficiently low that the H is trapped for times longer than several billion years. However, estimates of characteristic loss times for all major minerals known to be present on the Moon yield times longer than the age of the solar system for temperatures lower than about 200 K (Feldman et al. 2001). But temperatures lower than this limit are predicted (Vasavada et al. 1999) to exist within several large, flat-floored, only intermittently-shaded craters near the north pole. However these craters are observed to support H concentrations that are much lower than those within neighboring permanently-shaded craters whose temperatures are predicted (Vasavada et al. 1999) to be less than 100 K. While the difference in temperature between 200 K and 100 K cannot account for the needed difference in H diffusion times, it can account for the difference in water-ice sublimation times (Feldman et al. 2001). The presence of water molecules therefore provides a natural explanation of the observed difference in H concentration. Nevertheless, an association between variable concentrations of H in non-polar surface soils and the Clementine spectral-reflectance maturity index (Lucey et al. 2000) has been reported (Johnson et al. 2002) and shown to be in quantitative agreement with the H content of returned Apollo and Luna soil samples (Feldman et al. 2001).

10.8.3. Radar measurements of lunar poles. The discovery of deposits at the poles of Mercury consistent with water ice prompted the lunar radar community to make similar observations of the lunar poles. While experiments have detected local anomalies (Nozette et al. 1994), the most recent ground-based measurements of the poles have shown these local anomalies are not similar to the Mercury polar deposits, and that floors of permanently shaded craters that can be observed do not show evidence of thick ice deposits (Campbell et al. 2003). This places a strong constraint on the distribution and abundance of H-bearing materials at the poles; if in the form of water ice, they must be well mixed with regolith, or at most exist in the form of thin, centimeter scale layers.

11. THE LUNAR ATMOSPHERE

11.1. Introduction

The possible existence of a lunar atmosphere has both fascinated and challenged astronomers for hundreds of years. Galileo searched for evidence of clouds, and Kepler imagined an Earth-like climate. Landings during the Apollo era brought instruments that measured a weak atmospheric pressure. Decades later, new spectrographic and imaging instruments detected sodium and potassium gas that extended to surprisingly large distances, making the Moon's gaseous environment appear as comet-like. The sources of the lunar atmosphere involve transport of the materials to the surface and their release by the impact of sunlight, solar-wind plasma and meteorites upon the surface to release atoms and perhaps molecules. Solar radiation and the gravitational influence of the Earth play dramatic roles in the subsequent evolution of the lunar atmosphere.

The first studies of a possible lunar atmosphere were carried out by Galileo, who devoted considerable effort to a systematic search (Mendillo 2001). He carried out the very first stellar occultation experiment, finding no sign of an atmosphere. Sensors brought to the surface of the Moon by Apollo astronauts detected a very low atmospheric pressure and detected atoms of He and Ar (Hodges 1975; Stern 1999). The concentration of gases was found to be approximately 10^7 particles/cm^3 during the day and about 10^5 particles/cm^3 at night. Such densities are still comparable to vacuum conditions in comparison to the gas content at the surface of the Earth (10^{19} cm^{-3}). Initial concerns that the instruments were merely detecting gases that evaporated off the lunar modules were ultimately dismissed in favor of accepting a weak atmosphere produced by capture of solar-wind particles (such as He) or the radioactive decay of elements in the lunar soil (such as Ar from decay of K). As interests turned to other areas in the solar system, the Moon's minor atmosphere received little attention for many years.

11.2. Remote sensing of the Moon's atmosphere

An astonishing discovery by Potter and Morgan (1985) was the presence of very bright emission of the yellow D lines of Na at Mercury; shortly afterwards they found the near-infrared emission of K (Potter and Morgan 1986). A comprehensive summary of this work appears in Hunten et al. (1988). At that time, the much fainter emissions from the Moon had not been discovered; they were found by looking past the sunlit limb by Potter and Morgan (1988a) and Tyler et al. (1988). In the same way as comets, the rocky surfaces of Mercury, the Moon and some asteroids emit gases that escape into space, providing a transient atmosphere to study. The elements Na and K are not particularly abundant in the solar system but are relatively easy to detect because they scatter sunlight very efficiently. They are not the major constituents of the atmospheres of the Moon (or Mercury), but they do serve as excellent "tracers" of other gases presumably there but more difficult to detect. For example, the total number of Na plus K atoms detected just above the Moon's surface is barely 100 atoms/cm^3, far below the concentrations suggested for total abundances in the Apollo data.

Plate 2.12 provides a dramatic example of Na abundance profiles versus height above the limb. They confirm the discovery by Potter and Morgan (1988b) that the Na atoms extend to much higher altitudes than they would if they were at the temperature of the local surface. A small thermal component can, however, be detected near the subsolar point. Higher velocities have been suggested to be the result of sputtering by impact of ions or solar photons; micrometeoroid impact could also be involved. Evidently these other processes are more important than thermal evaporation everywhere except the warmest region at and near the subsolar point. We will return to these points below after a further discussion of the observations.

As demonstrated in Plate 2.12, it is necessary to use high-dispersion spectroscopy for studies near the limb to reduce the signals from the scattered light of the lunar limb and disk. For altitudes more than one or two lunar radii, the much more sensitive technique of imaging through narrow-band filters can be used. These new low-light-level imaging techniques, capable of taking a picture of the full extent of the atmosphere, provided the next step in understanding the Na atmosphere of the Moon. Images of the Na brightness in two dimensions show that the atmosphere extends to several times the radius of the Moon (Fig. 2.56). If the brightness levels were very much higher, the Moon's Na atmosphere would be visible to the naked eye as a large cloud (nebula) spanning several degrees of the night sky. This large extent implies that many Na atoms have relatively high speeds, approaching the escape speed (2.3 km/sec); yet, the pattern of brightness decreases with distance (d) as d^{-4} showing that there are slower Na atoms as well. To test the mechanisms responsible for sputtering processes with the required release speed distributions, several research groups are making lunar observations during meteor showers and at times when the solar wind is shielded from the surface by the Earth's magnetic field.

11.3. Lunar atmosphere generated by meteors

There have been several reports of possible enhancements in the Moon's atmosphere during a meteor shower (Hunten et al. 1992; Cremonese and Verani 1997; Verani et al. 1998, 2001). A particularly strong case occurred during the Leonids "storm" in November 1998. Using a wide-angle (180° all-sky) field of view, Smith et al. (1999) described a persistent Na "spot" seen on the nights after the peak meteor events of 17 November 1998. The unusual aspect of their observations was that they were made near new Moon phase, i.e., when the Moon cannot be observed directly owing to its location between the Earth and the Sun. Their analysis and companion computer simulations (Wilson et al. 1999) showed that Na atoms released by the meteor-regolith impacts at a rate of 7×10^{21} s^{-1} were accelerated away from the Moon by the radiation pressure of sunlight. In approximately two days (i.e., on 19 November), a cloud of Na gas swept past the Earth where it was focused into a beam by the Earth's gravitational field. Thus, in viewing the sky in the direction opposite from both the Moon and the Sun, the distant lunar tail was observed. Figure 2.57 summarizes the observational and modeling description of this effect. Subsequent observations reported by Smith et al. (2001) show that the lunar Na spot (and therefore the distant atmospheric tail) is a permanent feature of the Earth-Moon system, occasionally modulated in prominence by transient sources of sputtering.

11.4. Sputtering sources

The source of the Na gas on the Moon is a research topic very much in active debate (Stern 1999). Liberating gases from the surface material (regolith) requires the impact of micrometeorites or solar-wind ions and electrons, or sunlight. These are called sputtering agents, and laboratory experiments show that they indeed can free atoms and molecules from surfaces with sufficient energy to move away from the surface. Both hot and cold gaseous populations are possible from these processes, and the degree to which the thermal and

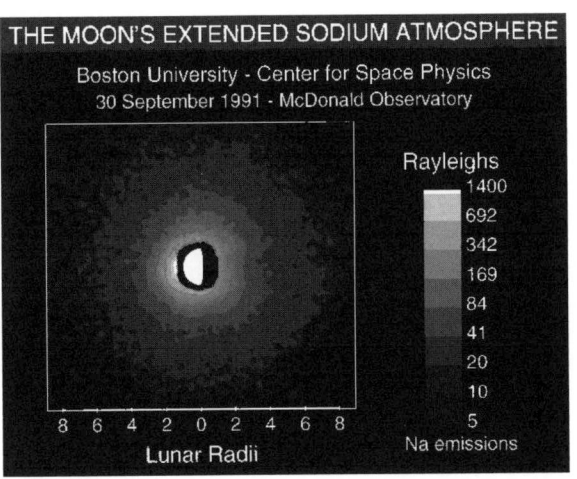

Figure 2.56. Image of the sodium atmosphere surrounding the Moon near quarter phase. The sodium brightness units shown are far below those capable of being seen by the unaided human eye. The Rayleigh unit is defined as $10^6/4\pi$ photons per square centimeter per second per steradian.

Lunar Sodium Tail

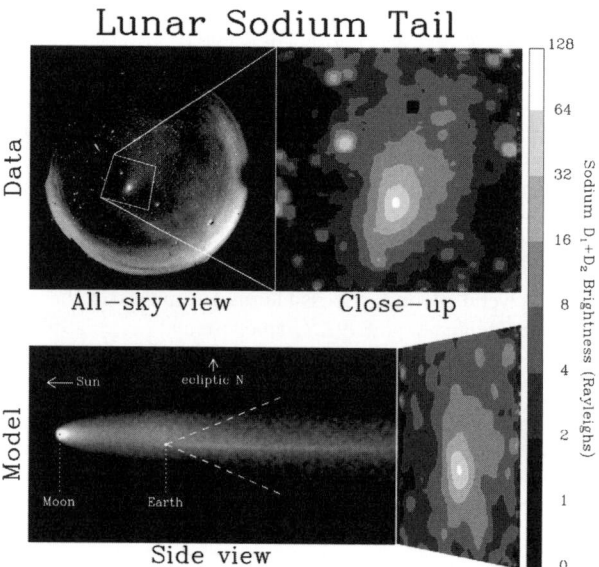

Figure 2.57. *Top panel: (Left)* all-sky image showing the sodium "spot" discovered on 19 November 1998. *(Right)* magnified portion of the feature. *Bottom panel: (Left)* computer model showing (to scale) neutral sodium originating from the Moon, accelerated by radiation pressure to the Earth where a portion is gravitationally focused into a column of enhanced sodium brightness. *(Right)* This simulated "spot" is shown as it would appear to a wide-angle camera system (dashed lines) on the night side of the Earth.

suprathermal components dominate close and distant regions is still under study (Sprague et al. 1992, 1998; Hunten et al. 1998). Sputtered gases are either pulled back to the regolith by gravity, pushed away by solar radiation pressure, or lost by photoionization and removal by the magnetic field in the solar wind (see Plate 2.12). The term surface-boundary-exosphere (SBE) is applied to an atmosphere produced by vaporization of surface material under conditions where collisions aloft are so rare that the liberated gases can have long parabolic trajectories back to the surface, or can escape directly from it.

11.5. Sources, recycling and sinks

The discussion of sources and sinks can be assisted by the illustration in Fig. 2.58. It is useful to distinguish between source processes (often very slow) that bring material to the surface and loss processes that release them from the surface into the atmosphere or space.

The need for some kind of source process is demonstrated as follows: The loss rate of Na atoms, in global mean, is 5×10^3 atoms per cm^2 per second. If material near the surface contains 1% Na by weight, all of it could be depleted in 10^7 years. Possible mechanisms for replacing it are regolith turnover, addition of meteoritic material, or diffusion from deeper layers. Hunten and Sprague (1997) and Morgan et al. (1988, 1989) suggest that regolith turnover is adequate, with an augmentation by meteoritic material.

Any atom that is released into the atmosphere must return to the surface unless it is ionized or otherwise lost to space. Half the ions find themselves in a downward electric field and are carried back to the surface. Sprague et al. (1992) discuss the fate of these recycled atoms, suggesting that the average one executes about 50 such hops, each lasting about 1000 sec.

The relation of the various source and release processes was studied by Sprague et al. (1992) with the use of a set of height profiles (0–700 km) of intensity measured at five solar

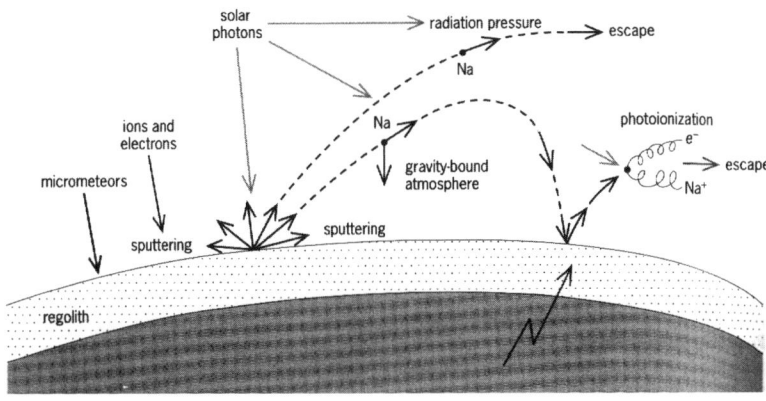

Figure 2.58. A schematic illustration of the processes that release and govern the subsequent motion of sodium atoms (Na) ejected from the regolith of the Moon.

zenith angles from 3 to 80°. The rates, which had to be assumed, have now been measured in the laboratory (Madey et al. 1998, Yakshinskiy and Madey 2000, Yakshinskiy et al. 2000). They are slower than the assumed ones by a factor of 1700 for photo-stimulated desorption (PSD) and 4000 for thermal evaporation. Thus, nearly the same results are obtained for the atmosphere if the number of atoms adsorbed per unit area of surface is increased by a factor of 3000; the value is then 3×10^{10} cm^{-2} at 9° from the subsolar point, increasing by a factor of 4 near the terminator. Thermal evaporation dominates by a factor ~ 3 near the subsolar point and rapidly becomes negligible elsewhere. Sputtering by solar-wind protons has been shown to be less important as a source (see the next section).

11.6. Lunar atmosphere seen during eclipses

For approximately four days each month, the Moon passes through the Earth's magnetic envelope (magnetosphere) thereby shielding its surface from solar-wind plasma impact. To see if the lunar atmosphere is affected by this removal of a sputtering agent, observations of the tenuous Na gas have to be made during the nights spanning the bright full Moon. This presents a serious, if not impossible, impediment to wide-angle imaging systems. Such observations can however be made during the totality phase of a lunar eclipse. Under such conditions, the bright lunar disk (and therefore scattered light) is very much reduced, and yet sunlight beyond the penumbra is still at full strength to illuminate any distant Na that may be present. Mendillo et al. (1999) reported on four such experiments and found that the robust, extended nature of the Moon's atmosphere is not affected in any drastic way by the absence of a solar-wind sputtering source. Thus, of the three proposed mechanisms for generating the extended lunar atmosphere, photon-sputtering is always present on one hemisphere and is thus considered to be the dominant source; ever-present micrometeorites are a secondary source, one certainly enhanced during meteor showers; solar wind sputtering appears to be the least significant source. Its weakness is traceable mainly to the small proton flux, 3×10^8 cm^{-2} sec^{-1}, while that of photons effective in producing PSD is 2×10^{14} cm^{-2} sec^{-1}.

11.7. Summary

Studies of the Moon's atmosphere, once a topic of only speculation, is now an active research field in comparative atmospheric science. State-of-the-art observational tools have transformed the available database and numerical simulations offer insights and tests of physical processes. Future observational work will center on attempts to identify species other than Na and K (i.e., the major constituents), probably requiring dedicated space-based ultraviolet observations. Modeling efforts will concentrate on the variability patterns associated with sources (i.e., as illustrated in Fig. 2.58); laboratory experiments will explore surface sputtering efficiencies and yields. Thus, our closest cosmic neighbor continues to fascinate us, and its role as a laboratory-in-space for the study of surface-boundary-exospheres will continue to enrich the study of primitive bodies in the solar system.

12. REFERENCES

Adams JB (1974) Visible and near-infrared diffuse reflectance spectra of pyroxenes as applied to remote sensing of solid objects in the solar system. J Geophys Res 79:4829-4836

Adams JB (1975) Interpretation of visible and near-infrared diffuse reflectance spectra of pyroxenes and other rock-forming minerals. *In*: Infrared and Raman Spectroscopy of Lunar and Terrestrial Minerals. Karr Jr C (ed) Academic Press, p 91-116

Adams JB, Charette MP, Rhodes JM (1975) Chemical fractionation of the lunar regolith by impact melting. Science 190:380-381

Adams JB, Filice AL (1967) Spectral reflectance 0.4 to 2.0 microns of silicate rock powders. J Geophys Res 72:5705-5715

Adams JB, Goulaud LH (1978) Plagioclase feldspars: visible and near infrared diffuse reflectance spectra as applied to remote sensing. Proc Lunar Planet Sci Conf 9:2901-2909

Adams JB, Hörz F, Gibbons RV (1979) Effects of shock-loading on the reflectance spectra of plagioclase, pyroxene and glass. Proc Lunar Planet Sci Conf 10:1-3

Adams JB, Pieters CM, McCord TB (1974) Orange glass: Evidence for regional deposits of pyroclastic origin on the Moon. Proc Lunar Sci Conf 5:171-186

Adler I, Trombka J, Schmadeback R, Lowman P, Blodget H, Yin L, Eller E, Podwysocki M, Weidner JR, Bickel AL, Lum RKL, Gerard J, Gorenstein P, Bjorkholm P, Harris B (1973) Results of the Apollo 15 and 16 x-ray fluorescence experiment. Proc Lunar Sci Conf 4:2783-2791

Adler I, Trombka JI (1980) Orbital chemistry-lunar surface analysis from the X-ray and gamma ray remote sensing experiments. *In:* Chemistry of the Moon: Physics and Chemistry of the Earth, Volume 10. Ahrens LH (ed) Pergamon, p 17-43

Agrell SO, Peckett A, Boyd FR, Haggerty SE, Bunch TE, Cameron EN, Dence MR, Douglas JAV, Plant AG, Traill RJ, James OB, Keil K, Prinz M (1970) Titanian chromite, aluminian chromite, and chromian ulvöspinel from Apollo 11 rocks. Proc Apollo 11 Lunar Sci Conf 1:81-86

Ali MZ, Ehmann WD (1977) Chemical characterization of lunar core 60010. Proc Lunar Sci Conf 8:2967-2981

Allen CC, Morris RV, McKay DS (1995) Experimental space weathering of lunar soils. Meteoritics 30(5):479-480

Allen CC, Morris RV, McKay DS (1996) An experimental analog to maturing lunar soil. Lunar Planet Sci 27:13-14

Allen CC, Morris RV, McKay DS (1996) Oxygen extraction from lunar soils and pyroclastic glass. J Geophys Res 101: 26085-26095

Allton JH (1989) Catalog of Apollo Lunar Surface Geological Sampling Tools and Containers. JSC-23454, NASA Johnson Space Center, Houston, p 97

Anand M, Taylor LA, Patchen A (2003) Petrologic comparisons of lunar mare basalt meteorites Dh-287A and NWA 032. Lunar Planet Sci 34:1787 (CD-ROM)

Anand M, Taylor LA, Neal C, Patchen A, Kramer G (2004) Petrology and geochemistry of LAP 02205: A new low-Ti mare-basalt meteorite. Lunar Planet Sci XXXV:1626 (CD-ROM)

Anders E, Grevesse N (1989) Abundances of the elements: meteoritic and solar. Geochim Cosmochim Acta 53:197-214

Andre CG, Wolfe RW, Adler I, Clark PE, Weidner JR, Philpotts JA (1977) Chemical character of the partially flooded Smythii Basin based on Al/Si orbital x-ray data. Proc Lunar Sci Conf 8(1):925-931

Antonenko I, Head JW, Mustard JF, Hawke BR (1995) Criteria for the detection of lunar cryptomaria. Earth Moon Planets 69:141-172

Arai T, Warren PH (1999) Lunar meteorite Queen Alexandra Range 94281: Glass compositions and other evidence for launch pairing with Yamato 793274. Meteorit Planet Sci 34:209-234

Arnold JR (1979) Ice in the Lunar Polar Regions. J Geophys Res 84:5659-5668

Basu A, McKay DS (1985) Chemical variability and origin of agglutinitic glass. Proc Lunar Planet Sci Conf 16:D87-D94

Basu A, Wentworth SJ, McKay DS (2001) Occurrence and distribution of Fe⁰-globules in lunar agglutinates. Lunar Planet Sci XXXII:1942 (CD-ROM)

Bell PM, Mao HK (1972) Crystal-field effects of iron and titanium in selected grains of Apollo 12, 14, and 15 rocks, glasses, and fine fractions. Proc Lunar Sci Conf 3(1):545-553

Bernatowicz TJ, Hohenberg CM, Podosek FA (1979) Xenon component organization in 14301. Proc Lunar Planet Sci Conf 10:1587-1616

Bernatowicz TJ, Nichols RH, Jr., Hohenberg CM (1994) Origin of amorphous rims on lunar soil grains. Lunar Planet Sci XXV:105-106

Bickel CE (1977) Petrology of 78155 - an early, thermally metamorphosed polymict breccia. Proc Lunar Planet Sci Conf 8:2007-2027

Bielefeld MJ, Reedy RC, Metzger AE, Trombka JI, Arnold JR (1976) Surface chemistry of selected lunar regions. Proc Lunar Sci Conf 7:2661-2676

Binder AB (1998) Lunar Prospector: overview. Science 281:1475-1476

Blanchard DP, Jacobs JW, Brannon JC, Brown RW (1976) Drive tube 60009: A chemical study of magnetic separates of size fractions from five strata. Proc Lunar Sci Conf 7:281-294

Blanchard DP, Korotev RL, Brannon JC, Jacobs JW, Haskin LA, Reid AM, Donaldson CH, Brown RW (1975) A geochemical and petrographic study of 1-2 mm fines from Apollo 17. Proc Lunar Sci Conf 6:2321-2341

Blewett DT, Hawke BR (2001) Remote sensing and geological studies of the Hadley-Apennine region of the Moon. Meteorit Planet Sci 36(5):701-730

Blewett DT, Lucey PG, Hawke BR, Jolliff BL (1997) Clementine images of the lunar sample-return stations: refinement of FeO and TiO_2 mapping techniques. J Geophys Res 102:16,319-16,325

Bogard DD, Morris RV, Hirsch WC, Jr. HVL (1980) Depositional and irradiational history of the Hadley Rille core 15010/11. Proc Lunar Planet Sci Conf 11:1511-1529

Bogard DD, Morris RV, Johnson P, Lauer HVJ (1982) The Apennine Front core 15007/8: irradiational and depositional history. Proc Lunar Planet Sci Conf 13:A221-A231

Borchardt R, Stöffler D, Spettel B, Palme C, Wänke H, Wacker K, Jessberger EK (1986) Composition, structure, and age of the Apollo 16 subregolith basement as deduced from the chemistry of post-Imbrium melt bombs. Proc Lunar Planet Sci Conf 17. J Geophys Res 91:E43-E54

Boynton WV, Chou C-L, Bild RW, Baedecker PA, Wasson JT (1976) Element distribution in size fractions of Apollo-16 soils; evidence for element mobility during regolith processes. Earth Planet Sci Lett 29:21-33

Britt DT, Pieters CM (1994) Darkening in black and gas-rich ordinary chondrites - the spectral effects of opaque morphology and distribution. Geochim Cosmochim Acta 58(18):3905-3919

Brownlee DE, Hörz F, Bradley J (1992) Interplanetary meteoroid debris in LDEF metal craters. Second LDEF Post-Retrieval Symposium Abstracts, NASA CP-10097, p. 47. NASA Langley Research Center

Brownlee DE, Hörz F, Laurance M, Bernhard RP, Warren J, Bradley J (1991) The composition of meteoroids impacting LDEF. 54th Annual Meeting of the Meteoritical Society, LPI Contrib No 766:38

Bruckenthal EA, Pieters CM (1984) Spectral effects of natural shock on plagioclase feldspar. Proc Lunar Planet Sci Conf 15:96-97

Burns RG (1970) Crystal field spectra and evidence for cation ordering in olivine minerals. Am Mineral 55:1608-1632

Burns RG (1985) Electronic spectra of minerals. *In*: Chemical Bonding and Spectroscopy in Mineral Chemistry. Berry FJ, Vaughan DJ (eds) Chapman and Hall, p 63-101

Burns RG (1993) Mineralogical Applications of Crystal Field Theory. Cambridge University Press

Busche FD, Prinz M, Keil K, Bunch TE (1972) Spinels and the petrogenesis of some Apollo 12 igneous rocks. Am Mineral 57:1729-1747

Butler B (1997) The migration of volatiles on the surfaces of Mercury and the Moon. J Geophys Res 102:19283-19291

Butler BJ, Muhleman DO, Slade MA (1993) Mercury; full-disk radar images and the detection and stability of ice at the North Pole. J Geophys Res 98:15003-15023

BVSP (1981) Basaltic Volcanism on the Terrestrial Planets. Pergamon Press

Cameron EN (1971) Opaque minerals in certain lunar rocks from Apollo 12. Proc Lunar Sci Conf 2:193-206

Campbell BA, Campbell DB, Chandler JF, Hine AA, Nolan MC, Perillat PJ (2003) Radar mapping of the lunar poles. Nature 426:137-138

Campbell BA, Hawke BR (2005) Radar mapping of lunar cryptomaria east of Orientale basin. J Geophys Res doi: 10.1029/2005JE002425

Campbell BA, Hawke BR, Thompson TW (1997) Long-wavelength radar studies of the lunar maria. J Geophys Res 102:19,307-19,320

Cassidy W, Hapke B (1975) Effects of darkening processes on surfaces of airless bodies. Icarus 25:371-383

Chambers JG, Taylor LA, Patchen A, McKay DS (1995) Quantitative mineralogical characterization of lunar high-ti mare basalts and soils for oxygen production. J Geophys Res 100:14391-14401

Champness PE, Dunham AC, Gibb FGF, Giles HN, MacKenzie WS, Stumpfl EF, Zussman J (1971) Mineralogy and petrology of some Apollo 12 samples. Proc Lunar Sci Conf 2:359-376

Charette MP, McCord TB, Pieters CM, Adams JB (1974) Application of remote spectral reflectance measurements to lunar geology classification and determination of titanium content of lunar soils. J Geophys Res 79(11):1605-1613

Charette MP, Taylor SR, Adams JB, McCord TB (1977) The detection of soils of Fra Mauro basalt and anorthositic gabbro composition in the lunar highlands by remote spectral reflectance techniques. Proc Lunar Sci Conf 8: 1049-1061

Cintala MJ, Hörz F (1992) An experimental evaluation of mineral-specific comminution. Meteorit Planet Sci 27:395-403

Cirlin EH, Goldberg IB, Housley RM, Weeks RA, Perhac R (1975) Ferromagnetic resonance as a method of studying the micrometeorite bombardment history of lunar fines. Lunar Sci Conf 6:146-148

Cirlin EH, Housley RM, Goldberg IB, Paton NE (1974) Ferromagnetic resonance as a method for studying regolith dynamics and breccia formation. Lunar Sci Conf 5:121-122

Clark PE, Adler I (1978) Utilization of independent solar flux measurements to eliminate nongeochemical variation in X-ray fluorescence data. Proc Lunar Planet Sci Conf 9:3029-3036

Clark PE, Hawke BR (1981) Compositional variation in the Hadley Apennine region. Proc Lunar Planet Sci Conf 12: 727-749

Clark PE, McFadden LA (2000) New results and implications for lunar crustal iron distribution using sensor data fusion techniques. J Geophys Res 105:4291-4316

Cloutis EA, Gaffey MJ (1991) Pyroxene spectroscopy revisited - spectral-compositional correlations and relationship to geothermometry. J Geophys Res 96:22809-22826

Cremonese G, Verani S (1997) High resolution observations of the sodium emission from the Moon. Adv Space Res 19:1561-1569

Crider DH, Vondrak RR (2000) The solar wind as a possible source of lunar polar hydrogen deposits. J Geophys Res 105:26773-26782

Cushing JA, Taylor GJ, Norman MD, Keil K (1999) The granulitic impactite suite: Impact melts and metamorphic breccias of the early lunar crust. Meteorit Planet Sci 34:185-195

Dalton J, Hollister LS, Kulick CG, Hargraves RB (1974) The nature of the chromite to ulvöspinel transition in mare basalt 15555. Lunar Planet Sci V:160-162

Davis Jr PA (1980) Iron and titanium on the Moon from orbital gamma-ray spectrometry with implications for crustal evolutionary models. J Geophys Res 85(B6):3209-3224

Davis PA, Bielefeld MJ (1981) Inelastic neutron scatter iron concentrations of the Moon from orbital gamma ray data. J Geophys Res 86:11919

Davis PA, Spudis PD (1985) Petrologic province maps of the lunar highlands derived from orbital geochemical data. Proc Lunar Planet Sci Conf 16. J Geophys Res 90:D61-D74

Davis PA, Spudis PD (1987) Global petrologic variations on the Moon: a ternary-diagram approach. Proc Lunar Planet Sci Conf 17:E387-E395

Delano JW (1986) Pristine lunar glasses: criteria, data and implications. Proc Lunar Planet Sci Conf 16. J Geophys Res 90:D201-D213

Delano JW, Ringwood AE (1978) Siderophile elements in the lunar highlands: Nature of the indigenous component and implications for the origin of the Moon. Proc Lunar Planet Sci Conf 9:111-159

Dence MR, Douglas JAV, Plant AG, Traill RJ (1970) Petrology, mineralogy and deformation of Apollo 11 samples. Proc Apollo 11 Lunar Sci Conf 1:315-340

Deutsch A, Stöffler D (1987) Rb-Sr analyses of Apollo 16 melt rocks and a new age estimate for the Imbrium basin: Lunar basin chronology and the early heavy bombardment of the Moon. Geochim Cosmochim Acta 51:1951-1964

Dickinson T, Taylor GJ, Keil K, Schmitt RA, Hughes SS, Smith MR (1985) Apollo 14 aluminous mare basalts and their possible relationship to KREEP. Proc Lunar Planet Sci Conf 15. J Geophys Res 90(Supplement):C365-C374

Donaldson CH, Usselman TM, Wiliams RJ, Lofgren GE (1975) Experimental modeling of the cooling history of Apollo 12 olivine basalts. Proc Lunar Sci Conf 6:843-869

Dowty E (1977) Phosphate in Angra Dos Reis: structure and composition of the $Ca_3(PO_4)_2$ minerals. Earth Planet Sci Lett 35:347-351

Dowty E, Keil K, Prinz M (1974a) Igneous rocks from Apollo 16 rake samples. Lunar Science V:174-176

Dowty E, Prinz M, Keil K (1974b) Ferroan anorthosite: a widespread and distinctive lunar rock type. Earth Planet Sci Lett 24:15-25

Dymek RF (1986) Characterization of the Apollo 15 feldspathic basalt suite. *In*: Geology and Petrology of the Apollo 15 Landing Site, LPI Tech. Rpt. 86-03. Spudis PD, Ryder G (eds) Lunar and Planetary Institute, p 52-57

Eberhardt P, Geiss J, Graf H, Groegler N, Krahenbuehl U, Schwaller H, Schwarzmuller J, Stettler A (1970) Trapped solar wind noble gases, exposure age and K/Ar-age in Apollo 11 lunar fine material. Lunar Sci Conf 1:1037-1070

El Goresy A, Prinz M, Ramdohr P (1976) Zoning in spinels as an indicator of the crystallization histories of mare basalts. Proc Lunar Sci Conf 7:1261-1279

El Goresy A, Ramdohr P, Taylor LA (1971a) The opaque minerals in the lunar rocks from Oceanus Procellarum. Proc Lunar Sci Conf 2:219-235

El Goresy A, Ramdohr P, Taylor LA (1971b) The geochemistry of the opaque minerals in Apollo 14 crystalline rocks. Earth Planet Sci Lett 13:121-129

Elphic RC, Lawrence DJ, Feldman WC, Barraclough BL, Maurice S, Binder AB, Lucey PG (1998) Lunar Fe and Ti abundances: comparison of Lunar Prospector and Clementine data. Science 281:1493-1496

Elphic RC, Lawrence DJ, Feldman WC, Barraclough BL, Maurice S, Binder AB, Lucey PG (2000) Lunar rare earth element distribution and ramifications for FeO and TiO$_2$: Lunar Prospector neutron spectrometer observations. J Geophys Res 105:20,333-20,345

Elphic RC, Lawrence DJ, Feldman WC, Barraclough BL, Maurice S, Lucey PG, Blewett DT, Binder AB (2002) The Lunar Prospector neutron spectrometer constraints on TiO$_2$. J Geophys Res 107(E4):10.1029/2000JE001460

Etchegaray-Ramirez MI, Metzger AE, Haines EL, Hawke RB (1983) Thorium concentrations in the lunar surface; IV, Deconvolution of the Mare Imbrium, Aristarchus, and adjacent regions. Proc Lunar Planet Sci Conf 13:A529-A543

Eugster O (1999) Chronology of dimict breccias and the age of South Ray crater at the Apollo 16 site. Meteorit Planet Sci 34:385-391

Eugster O, Terribilini D, Polnau E, Kramers J (2001) The antiquity indicator argon-40/argon-36 for lunar surface samples calibrated by uranium-235-xenon-136 dating. Meteorit Planet Sci 36:1097-1115

Evans LG, Reedy RC, Trombka JI (1993) Introduction to planetary remote sensing gamma ray spectroscopy. *In*: Remote Geochemical Analysis: Elemental and Mineralogical Composition. Pieters CM, Englert PAJ (eds) Cambridge Univ. Press, p 167-198

Evensen NM, Murthy VR, Coscio Jr. MR (1974) Provenance of KREEP and the exotic component: Elemental and isotopic studies of grain size fractions in lunar soils. Proc Lunar Sci Conf 5:1401-1417

Feldman W, Maurice S, Lawrence DJ, Little RC, Lawson SL, Gasnault O, Wiens RC, Barraclough BL, Elphic RC, Prettyman TH, Steinberg JT, Binder AB (2001) Evidence for water ice near the lunar poles. J Geophys Res 106(E10):23,231-23,251

Feldman WC, Ahola K, Barraclough BL, Belian RD, Black RK, Elphic RC, Everett DT, Fuller KR, Kroesche J, Lawrence DJ, Lawson SL, Longmire JL, Maurice S, Miller MC, Prettyman TH, Storms SA, Thornton GW (2004) Gamma-ray, neutron, and alpha-particle spectrometers for the Lunar Prospector mission. J Geophys Res 109:doi 10.1029/2003JE002207

Feldman WC, Barraclough BL, Fuller KR, Lawrence DJ, Maurice S, Miller MC, Prettyman TH, Binder AB (1999) The Lunar Prospector gamma-ray and neutron spectrometers. Nucl Instrum Methods Phys Res(Sect, A 422):562-566

Feldman WC, Barraclough BL, Maurice S, Elphic RC, Lawrence DJ, Thomsen DR, Binder AB (1998a) Major compositional units of the Moon: Lunar Prospector thermal and fast neutrons. Science 281:1489-1493

Feldman WC, Lawrence DJ, Elphic RC, Barraclough BL, Maurice S, Genetay I, Binder AB (2000a) Polar hydrogen deposits on the Moon. J Geophys Res 105:4175-4195

Feldman WC, Lawrence DJ, Elphic RC, Vaniman DT, Thomsen DR, Barraclough BL, Maurice S, Binder AB (2000b) The chemical information content of lunar thermal and epithermal neutrons. J Geophys Res 105:20347-20363

Feldman WC, Maurice S, Binder AB, Barraclough BL, Elphic RC, Lawrence DJ (1998b) Fluxes of fast and epithermal neutrons from lunar prospector: evidence for water ice at the lunar poles. Science 281:1496-1500

Fink D, Klein J, Middleton R, Vogt S, Herzog GF, Reedy RC (1998) [41]Ca, [26]Al, and [10]Be in lunar basalt 74275 and [10]Be in the double drive tube 74002/74001. Geochim Cosmochim Acta 62:2389-2402

Fischer EM, Pieters CM (1994) Remote determination of exposure degree and iron concentration of lunar soils using VIS-NIR spectroscopic methods. Icarus 111(2):475-488

Fischer EM, Pieters CM (1995) Lunar-surface aluminum and iron concentration from Galileo solid-state imaging data, and the mixing of mare and highland materials. J Geophys Res 100(E11):23279-23290

Fischer EM, Pieters CM (1996) Composition and exposure age of the Apollo 16 Cayley and Descartes regions from Clementine data: normalizing the optical effects of space weathering. J Geophys Res 101(E1):2225-2234

Folk RL (1980) Petrology of Sedimentary Rocks. Hemphill Publishing Co.

Fruland RM, Morris RV, McKay DS, Clanton US (1977) Apollo 17 ropy glasses. Proc Lunar Sci Conf 8:3095-3111

Fudali RF (1966) Implications of the nonuniform cooling behavior of the eclipsed Moon. Icarus 5(1-6):536-544

Gaddis LR, Pieters CM, Hawke BR (1985) Remote sensing of lunar pyroclastic mantling deposits. Icarus 61:461-489

Gaddis LR, Staid MI, Tyburczy JA, Hawke BR (2003) Compositional analyses of lunar pyroclastic deposits. Icarus 161:262-280

Garrison JR, Taylor LA (1980) Genesis of highland basalt breccias: A view from 66095. *In*: Proc. Conf. Lunar Highlands Crust. Papike JJ, Merrill RB (eds) Pergamon Press, p 395-417

Gasnault O, Feldman WC, Maurice S, Genetay I, d'Uston C, Prettyman TH, Moore KR (2001) Composition from fast neutrons: application to the Moon. Geophys Res Lett 28:3797-3800

Genetay I, Maurice S, Feldman WC, Gasnault O, Lawrence DJ, Elphic RC, d'Uston C, Binder AB (2003) Elemental content from 0 to 500 keV neutrons: Lunar Prospector results. Planet Space Sci 51:271-280

Ghent RR, Leverington DW, Campbell BA, Hawke BR, Campbell DB (2005) Earth-based observations of radar-dark crater haloes on the Moon: Implications for regolith properties. J Geophys Res 110: doi:10.1029/2004JE002366

Gibbons RV, Hörz F, Schaal RB (1976) The chemistry of some individual lunar soil agglutinates. Proc Lunar Sci Conf 7:405-422

Giguere TA, Taylor GJ, Hawke BR, Lucey PG (2000) The titanium contents of lunar mare basalts. Meteorit Planet Sci 35(1):193-200

Gillis JJ, Jolliff BL, Elphic RC (2003) A revised algorithm for calculation TiO$_2$ concentrations from Clementine **UVVIS** data: a synthesis of rock, soil, and remotely sensed TiO$_2$ concentrations. J Geophys Res 108(E2):5009, doi:10.1029/2001JE001515

Gillis JJ, Jolliff BL, Lawrence DJ, Lawson SL, Prettyman TH (2002) The Compton-Belkovich region of the Moon: Remotely sensed observations and lunar sample association. Proc Lunar Planet Sci Conf XXXIII:1967 (CD-ROM)

Gillis JJ, Korotev RL, Jolliff BL (2004) Lunar surface geochemistry: Global concentrations of Th, K, and FeO as derived from Lunar Prospector and Clementine data. Geochim Cosmochim Acta 68(18):3791-3805

Gillis JJ, Lucey PG (2004) Clementine 2.7 micron data: mapping the mare and searching for water. Proc Lunar Planet Sci Conf 35:2158

Gillis JJ, Lucey PG, Campbell BA, Hawke BR (2005) Clementine 2.7 micron data and 70-cm Earth-based radar data provide additional constraints for **UVVIS**-based estimates of TiO$_2$ content for lunar mare basalts. Proc Lunar Planet Sci Conf XXXVI:2254

Gnos E, Hofmann BA, A. A-K, Lorenzetti S, Eugster O, Whitehouse MJ, Villa IM, Jull AJT, Eikenberg J, Spettel B, Krähenbühl U, Franchi IA, Greenwood RC (2004) Pinpointing the source of a lunar meteorite: implications for the evolution of the Moon. Science 305:657-659

Gold T (1955) The lunar surface. Monthly Notices Royal Astron Soc 115:585-604

Goldstein JI, Axon HJ, Agrell SO (1972) The grape cluster, metal particle 63344,1. Earth Planet Sci Lett 28:217-224

Goldstein JI, Hewins RH, Axon HJ (1974) Metal silicate relationships in Apollo 17 soils. Proc Lunar Conf 5:653-671

Gose WA, Morris RV (1977) Depositional history of the Apollo 16 deep drill core. Proc Lunar Sci Conf 8:2909-2928

Gose WA, Strangway DW, Pearce GW (1978) Origin of magnetization in lunar breccias; an example of thermal overprinting. Earth Planet Sci Lett 38:373-384

Graf JC (1993) Lunar Soils Grain Size Catalog. NASA Reference Publication 1265, NASA

Grande M (2001) The D-CIXS X-ray spectrometer on ESA's SMART-1 mission to the Moon. Earth Moon Planets 85: 143-152

Grier JA, McEwen AS, Lucey PG, Milazzo M, Strom RG (2001) Optical maturity of ejecta from large rayed lunar craters. J Geophys Res 106(E12):32847-32862

Grove TL (1982) Use of lamellae in lunar clinopyroxenes as cooling rate speedometers: An experimental calibration. Am Mineral 67:251-268

Haggerty SE (1971) Compositional variations in lunar spinels. Nature Phys Sci 233:156-160

Haggerty SE (1972a) Luna 16: An opaque mineral study and a systematic examination of compositional variations of spinels from Mare Fecunditatis. Earth Planet Sci Lett 13:328-352

Haggerty SE (1972b) Chemical characteristics of spinels in some Apollo 15 basalts. *In*: The Apollo 15 Lunar Samples. Chamberlain JW (ed) Lunar Science Institute, p 92-97

Haggerty SE (1973a) Luna 20: mineral Chemistry of spinel, pleonaste, chromite, ulvöspinel, ilmenite, and rutile. Geochim Cosmochim Acta 37:857-867

Haggerty SE (1973b) Armalcolite and genetically associated opaque minerals in the lunar samples. Proc Lunar Sci Conf 4:777-797

Haggerty SE (1978) Luna 24: systematics in spinel mineral chemistry in the context of an intrusive petrogenetic grid. *In*: Mare Crisium: The View From Luna 24. Merrill RB, Papike JJ (eds) Pergamon, p 523-536

Haggerty SE, Boyd FR, Bell PM, Finger LW, Bryan WB (1970) Opaque minerals and olivine in lavas and breccias from Mare Tranquillitatis. Proc Apollo 11 Lunar Sci Conf 3:513–538

Hapke B (2001) Space weathering from Mercury to the asteroid belt. J Geophys Res 106(E5):10039-10073

Hapke B, Cassidy W, Wells EN (1975) Effects of vapor-phase deposition processes on the optical, chemical and magnetic properties of the lunar regolith. The Moon 13:339-353

Hapke B, Cohen A, Cassidy W, Wells E (1970) Solar radiation effects on the optical properties of Apollo 11 lunar samples. Proc of the Apollo 11 Lunar Sci Conf 3:2199-2212

Haselton JD, Nash WP (1975) A model for the evolution of opaques in mare lavas. Proc Lunar Sci Conf 6:747-755

Hashizume K, Chaussidon M, Marty B, Robert F (2000) Solar wind record on the Moon: deciphering presolar from planetary nitrogen. Science 290:1142-1145

Haskin LA (1985) Chapter 4. Petrogenetic modeling - Use of rare earth elements. *In*: Rare Earth Element Geochemistry, Developments in Geochemistry 2. Henderson P (ed) Elsevier, p 115-152

Haskin LA, Gillis JJ, Korotev RL, Jolliff BL (2000) The materials of the lunar Procellarum KREEP Terrane: a synthesis of data from geomorphological mapping, remote sensing, and sample analyses. J Geophys Res 105:20,403-20,415

Haskin LA, Korotev RL (1977) Test of a model for trace element partition during closed-system solidification of a silicate liquid. Geochim Cosmochim Acta 41:921-939

Haskin LA, Korotev RL, Rockow KM, Jolliff BL (1998) The case for an Imbrium origin of the Apollo thorium-rich impact-melt breccias. Meteorit Planet Sci 33:959-975

Haskin LA, Warren PH (1991) Lunar chemistry. *In*: Lunar Sourcebook: A User's Guide to the Moon. Heiken G, Vaniman DT, French BM (eds) Cambridge University Press, p 357-474

Hawke BR, Head JW (1978) Lunar KREEP volcanism: geologic evidence for history and mode of emplacement. Proc Lunar Planet Sci Conf 9:3285-3309

Hawke BR, Lucey PG, Bell JF, Spudis PD (1990) Ancient mare volcanism. LPI-LAPST Workshop on Mare Volcanism and Basalt Petrogenesis: Astounding Fundamental Concepts (AFC) Developed Over the Last Fifteen Years. p 5-6

Hawke BR, Peterson CA, Blewett DT, Bussey DBJ, Lucey PG, Taylor GJ, Spudis PD (2003) Distribution and modes of occurrence of lunar anorthosite. J Geophys Res 108(E6):10.1029/2002JE001890

Hazen RM, Bell PM, Mao HK (1978) Effects of compositional variation on absorption spectra of lunar pyroxenes. Proc Lunar Sci Conf 9:2919-2934

Hazen RM, Mao HK, Bell PM (1977) Effects of compositional variation on absorption spectra of lunar olivines. Proc Lunar Sci Conf 8:1081-1090

Head JW (1974) Lunar dark mantle deposits: Possible clues to the distribution of early mare deposits. Proc Lunar Sci Conf 5:207-222

Head JW (1976) Lunar volcanism in space and time. Rev Geophys Space Phys 14:265-300

Head JW, Wilson L (1992) Lunar mare volcanism: stratigraphy, eruption conditions, and the evolution of secondary crusts. Geochim Cosmochim Acta 56:2155-2175

Head JW, Wilson L, Wilhelms D (1997) Lunar mare basalt volcanism; early stages of secondary crustal formation and implications for petrogenetic evolution and magma emplacement processes. Lunar Planet Sci XXVIII:545-546

Heiken G, McKay DS (1974) Petrography of Apollo 17 soils. Proc Lunar Sci Conf 5:843-860

Heiken G, McKay DS, Fruland RM (1973) Apollo 16 soils: Grain size analysis and petrography. Proc Lunar Planet Sci Conf 4:251-265

Heiken G, Vaniman DT, French BM (1991) Lunar Sourcebook: A User's Guide to the Moon. Cambridge University Press

Heiken GH, Morris RV, McKay DS, Fruland RM (1976) Petrographic and ferromagnetic resonance studies of the Apollo 15 deep drill core. Proc Lunar Sci Conf 7:93-111

Herzog GF (1994) Applications of accelerator mass spectrometry in extraterrestrial materials. Nucl Instr Methods B 92:492-499

Hewins RH, Goldstein JI (1975) The provenance of metal in anorthositic rocks. Proc Lunar Sci Conf 6:343-362

Hidaka H, Ebihara M, Yoneda S (2000) Neutron capture effects on samarium, europium, and gadolinium in Apollo 15 deep drill-core samples. Meteorit Planet Sci 35:581-589

Higgins SJ, Taylor LA, Chambers JG, Patchen A, McKay DS (1996) X-ray digital-imaging petrography: Technique development for lunar mare soils. Meteorit Planet Sci 31(3):356-361

Hodges RJ (1975) Formation of the lunar atmosphere. The Moon 14:139-157

Hood LL, Coleman PJ, Wilhelms DE (1979) Lunar nearside magnetic anomalies. Proc Lunar Planet Sci Conf 19:99-113

Hood LL, Schubert G (1980) Lunar magnetic-anomalies and surface optical-properties. Science 208(4439):49-51

Hood LL, Williams CR (1989) The lunar swirls: distribution and possible origins. Proc Lunar Planet Sci Conf 19:99-113

Hood LL, Zakharian A, Halekas J, Mitchell DL, Lin RP, Acuña MH, Binder AB (2001) Initial mapping and interpretation of lunar crustal magnetic anomalies using Lunar Prospector magnetometer data. J Geophys Res 106:27,825-27,840

Hörz F, Cintala MJ, See TH, Cardenas F, Thompson TD (1984) Grain size evolution and fractionation trends in an experimental regolith. Proc Lunar Planet Sci Conf 15:C183-C196

Housley RM, Cirlin EH, Goldberg IB, Crowe H (1976) Ferromagnetic resonance studies of lunar core stratigraphy. Proc Lunar Sci Conf 7:13-26.

Housley RM, Cirlin EH, Goldberg IB, Crowe H, Weeks RA, Perhac R (1975) Ferromagnetic resonance as a method of studying the micrometeorite bombardment history of the lunar surface. Proc Lunar Sci Conf 6:3173-3186

Housley RM, Cirlin EH, Paton N, Goldberg IB (1974) Solar wind and micrometeorite alteration of the lunar regolith. Proc Lunar Sci Conf 5:2623-2642

Hu HN, Taylor LA (1977) Lack of chemical fractionation in major and minor elements during agglutinate formation. Proc Lunar Sci Conf 8:3645-3656

Hu HN, Taylor LA (1978) Soils from Mare Crisium: Agglutinitic glass chemistry and soil development. *In:* Mare Crisium: The View from Luna 24. Merrill RB, Papike JJ (eds) Pergamon Press, p 291-302

Hubbard NJ, Meyer CJ, Gast PW, Wiesmann H (1971) The composition and derivation of Apollo 12 soils. Earth Planet Sci Lett 10:341-350

Hunten DM, Cremonese G, Sprague AL, Hill RE, Verani S, Kozlowski RW (1998) The Leonid meteor shower and the lunar sodium atmosphere. Icarus 136:298-303

Hunten DM, Kozlowski RWH, Sprague AL (1992) A possible meteor shower on the Moon. Geophys Res Lett 18:2101-2104

Hunten DM, Morgan TH, Shemansky DE (1988) The Mercury atmosphere. *In*: Mercury. Vilas F, Chapman CR, Matthews MS (eds) University of Arizona Press, p 562-612

Hunten DM, Sprague AL (1997) Origin and character of the lunar and mercurian atmospheres. Adv Space Res 10: 1551-1560

James OB (1995) Siderophile elements in lunar impact melts: Nature of the impactors. Lunar Planet Sci XXVI:671-672

James OB (1996) Siderophile elements in lunar impact melts define nature of the of the impactors. Lunar Planet Sci XXVII:603-604

James OB, Flohr MK (1983) Subdivision of the Mg-suite noritic rocks into Mg-gabbronorites and Mg-norites. J Geophys Res 88:A603-A614

James OB, Flohr MK, Lindstrom MM (1984) Petrology and geochemistry of lunar dimict breccia 61015. Proc Lunar Planet Sci Conf 15. J Geophys Res 89:C63-C86

James OB, Hammarstrom JG (1977) Petrology of four clasts from consortium breccia 73215. Proc Lunar Sci Conf 8: 2459-2494

Jerde EA, Morris RV, Warren PH (1990) In quest of lunar regolith breccias of exotic provenance: a uniquely anorthositic sample from the Fra Mauro (Apollo 14) highlands. Earth Planet Sci Lett 98:90-108

Jerde EA, Warren PH, Morris RV, Heiken GH, Vaniman DT (1987) A potpourri of regolith breccias; "new" samples from the Apollo 14, 16, and 17 landing sites. Proc Lunar Planet Science Conf 17:E526-E536

Johnson JR, Feldman WC, Lawrence DJ, Maurice S, Swindle TD, Lucey PG (2002) Lunar Prospector epithermal neutrons from impact craters and landing sites: Implications for surface maturity and hydrogen distribution. J Geophys Res 107(E2):10.1029/2000JE001430

Johnson JR, Larson SM, Singer RB (1991a) A reevaluation of spectral ratios for lunar mare TiO_2 Mapping. Geophys Res Lett 18(11):2153-2156

Johnson JR, Larson SM, Singer RB (1991b) Remote sensing of potential lunar resources, 1. Near-side compositional properties. J Geophys Res 96(E3):18,861-18,882

Johnson TV, Saunders RS, Matson DL, Mosher JA (1977) A TiO_2 abundance map for the northern maria. Proc Lunar Sci Conf 8:1029-1036

Jolliff BL (1991) Fragments of quartz monzodiorite and felsite in Apollo 14 soil particles. Lunar Planet Sci 21:101-118

Jolliff BL (1993) A monazite-bearing clast in Apollo 17 melt breccia. Lunar Planet Sci XXIV:725-726

Jolliff BL (1998) Large-scale separation of K-frac and REEP-frac in the source regions of Apollo impact-melt breccias, and a revised estimate of the KREEP composition. Int Geology Rev 40:916-935

Jolliff BL, Floss C, McCallum IS, Schwartz JM (1999) Geochemistry, petrology, and cooling history of 14161,7373: a plutonic lunar sample with textural evidence of granitic-fraction separation by silicate-liquid immiscibility. Am Mineral 84:821-837

Jolliff BL, Gillis JJ, Haskin L, Korotev RL, Wieczorek MA (2000) Major lunar crustal terranes: surface expressions and crust-mantle origins. J Geophys Res 105:4197-4216

Jolliff BL, Haskin LA (1995) Cogenetic rock fragments from a lunar soil: evidence of a ferroan noritic-anorthosite pluton on the Moon. Geochim Cosmochim Acta 59:2345-2374

Jolliff BL, Haskin LA, Colson RO, Wadhwa M (1993) Partitioning in REE-saturating minerals: theory, experiment, and modelling of whitlockite, apatite, and evolution of lunar residual magmas. Geochim Cosmochim Acta 57: 4069-4094

Jolliff BL, Haskin LA, Korotev RL, Papike JJ, Shearer CK, Pieters C, Cohen BA (2003) Scientific expectations from a sample of regolith and rock fragments from the interior of the lunar South Pole-Aitken basin. Lunar Planet Sci XXXIV:1989 (CD-ROM)

Jolliff BL, Korotev RL, Haskin LA (1991) Geochemistry of 2-4 mm particles from Apollo 14 soil (14161) and implications regarding igneous components and soil-forming processes. Proc Lunar Planet Sci Conf 21:193-219

Jolliff BL, Korotev RL, Rockow KM (1998) Geochemistry and petrology of lunar meteorite Queen Alexandra Range 94281, a mixed mare and highland regolith breccia, with special emphasis on very-low-Ti mafic components. Meteorit Planet Sci 33:581-601

Jolliff BL, Rockow KM, Korotev RL, Haskin LA (1996) Lithologic distribution and geologic history of the Apollo 17 site: the record in soils and small rock particles from the highland massifs. Meteorit Planet Sci 31:116-145

Jull AJT, Cloudt S (1996) Evidence for recent gardening or disturbance of lunar core 76001 from solar-cosmic-ray records of ^{14}C. Lunar Planet Sci XXVII:627-628

Jull AJT, Cloudt S, Donahue DJ, J.M. S, Reedy RC, Masarik J (1998) ^{14}C depth profiles in Apollo 15 and 17 cores and lunar rock 68815. Geochim Cosmochim Acta 62:3025-3036

Jull AJT, Lal D, McHargue LR, Burr GS, Donahue DJ (2000) Cosmogenic and implanted radionuclides studied by selective etching of lunar soils. Nucl Instr Methods B 172:867-872

Kaula WM, Schubert G, Lingenfelter RE, Sjogren WL, Wollenhaupt WR (1972) Analysis and interpretation of lunar laser altimetry. Proc Lunar Sci Conf 3:2189-2204

Kaula WM, Schubert G, Lingenfelter RE, Sjogren WL, Wollenhaupt WR (1974) Apollo laser altimetry and inferences as to lunar structure. Proc Lunar Sci Conf 5:3049-3058

Keil K, Kurat G, Green JA (1972) Lithic fragments, glasses and chondrules from Luna 16 fines. Earth Planet Sci Lett 13:243-256

Keller L, McKay D (1997) The nature and origin of rims on lunar soil grains. Geochim Cosmochim Acta 61:2331-2340

Keller LP, Clemett SJ (2001) Formation of nanophase iron in the lunar regolith. Lunar Planet Sci XXXII:2097 (CD-ROM)

Keller LP, McKay DS (1993) Discovery of vapor deposits in the lunar regolith. Science 261:1305-1307

Keller LP, Wentworth S, McKay D (1998) Space Weathering: Reflectance spectroscopy and TEM analysis of individual lunar soil grains. Proc Lunar Planet Sci Conf 29:1762

Keller LP, Wentworth SJ, McKay DS, Taylor LA, Pieters C, Morris RV (1999) Space weathering in the fine size fractions of lunar soils: Soil maturity effects. Workshop on New Views of the Moon II, p. 32-34. Lunar and Planetary Institute, Houston, Texas

Keller LP, Wentworth SJ, McKay DS, Taylor LA, Pieters C, Morris RV (2000) Space weathering in the fine size fractions of lunar soils: mare/highland differences. Lunar Planet Sci XXXI:1655

Kerridge JF (2001) Isotopic variability of nitrogen in lunar regolith. Science 293:U1-U2

King TVV, Ridley WI (1987) Relation of the Spectroscopic Reflectance of Olivine to Mineral Chemistry and Some Remote-Sensing Implications. J Geophys Res-Solid Earth Planets 92(B11):11457-11469

Klein CJ, Drake JC, Frondel C (1971) Mineralogical, petrological, and chemical features of four Apollo 12 lunar microgabbros. Proc Lunar Sci Conf 2:265-284

Korotev RL (1976) Geochemistry of grain size fractions of soils from the Taurus-Littrow valley floor. Proc Lunar Sci Conf 7:695-726

Korotev RL (1983a) Geochemical study of individual 1-2 mm particles from Apollo 16 soil 67712. Lunar Planet Sci Conf XIV:399-400

Korotev RL (1983b) Geochemical study of individual 1-2 mm particles from Apollo 16 soil 65502. Lunar Planet Sci Conf XIV:397-398

Korotev RL (1987a) The nature of the meteoritic components of Apollo 16 soil. As inferred from correlations of iron, cobalt, iridium, and gold with nickel. Proc Lunar Planet Sci Conf 17. J Geophys Res 92:E447-E461

Korotev RL (1987b) The meteoritic component of Apollo 16 noritic impact melt breccias. Proc Lunar Planet Sci Conf 17. J Geophys Res 92:E491-E512

Korotev RL (1987c) Mixing levels, the Apennine Front soil component, and compositional trends in the Apollo 15 soils. Proc Lunar Planet Sci Conf 17. J Geophys Res 92:E411-E431

Korotev RL (1991) Geochemical stratigraphy of two regolith cores from the central highlands of the Moon. Proc Lunar Planet Sci Conf 21:229-289

Korotev RL (1994) Compositional variation in Apollo 16 impact-melt breccias and inferences for the geology and bombardment history of the Central Highlands of the Moon. Geochim Cosmochim Acta 58:3931-3969

Korotev RL (1996) On the relationship between the Apollo 16 ancient regolith breccias and feldspathic fragmental breccias, and the composition of the prebasin crust in the central highlands of the moon. Meteorit Planet Sci 31:403-412

Korotev RL (1997) Some things we can infer about the moon from the composition of the Apollo 16 regolith. Meteorit Planet Sci 32:447-478

Korotev RL (2000) The great lunar hot spot and the composition and origin of the Apollo mafic ("LKFM") impact-melt breccias. J Geophys Res 105:4317-4345

Korotev RL, Gillis JJ (2001) A new look at the Apollo 11 regolith and KREEP. J Geophys Res 106(E6):12,339-12,354

Korotev RL, Haskin LA (1988) Europium mass balance in polymict samples and implications for plutonic rocks of the lunar crust. Geochim Cosmochim Acta 52:1795-1813

Korotev RL, Jolliff BL (2000) Siderophile element concentrations in two metal fragments from the Apollo 16 regolith. Lunar Planet Sci XXXI:1385 (CD-ROM)

Korotev RL, Jolliff BL (2001) The curious case of the lunar magnesian granulitic breccias. Lunar Planetary Science 32:1013 (CD-ROM)

Korotev RL, Jolliff BL, Campbell AJ, Humayun M (2003a) Laser-ablation ICP-MS analyses of meteoritic metal grains in lunar impact-melt breccias. Lunar Planet Sci XXXIV:1487 (CD-ROM)

Korotev RL, Jolliff BL, Rockow KM (1996) Lunar meteorite Queen Alexandra Range 93096 and the iron concentration of the lunar highland surface. Meteorit Planet Sci 31:909-924

Korotev RL, Jolliff BL, Zeigler RA (2000) The KREEP components of the Apollo 12 regolith. Lunar Planet Sci XXXI:1363 (CD-ROM)

Korotev RL, Jolliff BL, Zeigler RA, Gillis JJ, Haskin LA (2003b) Feldspathic Lunar Meteorites and Their Implications for Compositional Remote Sensing of the Lunar Surface and the Composition of the Lunar Crust. Geochim Cosmochim Acta 67:4895-4923

Korotev RL, Kremser DT (1992) Compositional variations in Apollo 17 soils and their relationship to the geology of the Taurus-Littrow site. Proc Lunar Planet Sci Conf 22:275-301

Korotev RL, Morris RV (1993) Composition and maturity of Apollo 16 regolith core 60013/14. Geochim Cosmochim Acta 57:4813-4826

Korotev RL, Morris RV, Jolliff BL, Schwarz C (1997a) Lithological variation with depth and decoupling of maturity parameters in Apollo 16 regolith core 68001/2. Geochim Cosmochim Acta 61:2989-3002

Korotev RL, Morris RV, Lauer HVJ (1984) Stratigraphy and geochemistry of the Stone Mountain core (64001/2). Proc Lunar Planet Sci Conf 15:C143-C160

Korotev RL, Rockow KM, Jolliff BL, Haskin LA (1997b) Lithic fragments of the Cayley plains. Lunar Planet Sci XXVIII:753-754

Laul JC, Lepel EA, Vaniman DT, Papike JJ (1979) The Apollo 17 drill core: Chemical systematics of grain-size fractions. Proc Lunar Planet Sci Conf 10:1269-1298

Laul JC, Papike JJ (1980) The Apollo 17 drill core: Chemistry of size fractions and the nature of the fused soil component. Proc Lunar Planet Sci Conf 11:1395-1413

Laul JC, Papike JJ, Simon SB (1981) The lunar regolith: Comparative studies of the Apollo and Luna sites. Chemistry of soils from Apollo 17, Luna 16, 20, and 24. Proc Lunar Planet Sci Conf 12B:389-407

Laul JC, Rode OD, Simon SB, Papike JJ (1987) The lunar regolith; chemistry and petrology of Luna 24 grain size fractions. Geochim Cosmochim Acta 51:661-673

Laul JC, Vaniman DT, Papike JJ, Simon S (1978) Chemistry and petrology of size fractions of Apollo 17 deep drill core 70009-70006. Proc Lunar Planet Sci Conf 9:2065-2097

Lawrence DJ, Elphic RC, Feldman WC, Prettyman T, Gasnault O, Maurice S (2003) Small-area thorium features on the lunar surface. J Geophys Res 108(E9):doi 10.1029/2003/JE002050

Lawrence DJ, Feldman WC, Barraclough BL, Binder AB, Elphic RC, Maurice S, Thomsen DR (1998) Global elemental maps of the Moon: the Lunar Prospector gamma-ray spectrometer. Science 281:1484-1489

Lawrence DJ, Feldman WC, Barraclough BL, Binder AB, Elphic RC, Maurice S, Miller MC, Prettyman TH (2000) Thorium abundances on the lunar surface. J Geophys Res 105:20,307-20,331

Lawrence DJ, Feldman WC, Barraclough BL, Elphic RC, Maurice S, Binder AB, Miller MC, Thomsen DR (1999) High resolution measurements of absolute thorium abundance on the lunar surface. Geophys Res Lett 26(17): 2681-2683

Lawrence DJ, Feldman WC, Elphic RC, Little RC, Prettyman TH, Maurice S, Lucey PG, Binder AB (2002) Iron abundances on the lunar surface as measured by the Lunar Prospector gamma-ray and neutron spectrometers. J Geophys Res 107(E12):10.1029/2001JE001530

Lawson SL, Jakosky BM (2001) Lunar surface thermophysical properties derived from Clementine LWIR and UVVIS images. J Geophys Res 106(E11):27,911-27,932

Lawson SL, Jakosky BM, Park H-S, Mellon MT (2000) Brightness temperature of the lunar surface: Calibration and global analysis of the Clementine long-wave infrared camera data. J Geophys Res 105(E2):4273-4290

Le Mouélic S, Langevin Y, Erard S (1999) The distribution of olivine in the crater Aristarchus inferred from Clementine NIR data. Geophys Res Lett 26(9):1195-1198

Le Mouélic S, Langevin Y, Erard S, Pinet P, Chevrel S, Daydou Y (2000) Discrimination between maturity and composition of lunar soils from integrated Clementine UV-VISible/near-infrared data: Application to the Aristarchus Plateau. J Geophys Res 105(E4):9445-9455

Le Mouélic S, Lucey PG, Langevin Y, Hawke BR (2002) Calculating iron contents of lunar highland materials surrounding Tycho crater from integrated Clementine UV-VISible and near-infrared data. J Geophys Res 107(E10):5074, doi:10.1029/2000JE001484

Leya I, Lange HJ, Lüpke M, Neupert U, Daunke R, Fanenbruck O, Michel R, Rösel R, Meltzow B, Schiekel T, Sudbrock F, Herpers U, Filges D, Bonani G, Dittrich-Hannen B, Suter M, Kubik PW, Synal HA (2000a) Simulation of the interaction of galactic cosmic-ray protons with meteoroids: On the production of radionuclides in thick gabbro and iron targets irradiated isotropically with 1.6 GeV protons. Meteorit Planet Sci 35:287-318

Leya I, Neumann S, Wieler R, Michel R (2001) The production of cosmogenic nuclides by galactic cosmic-ray particles for 2p exposure geometries. Meteorit Planet Sci 36:1547-1561

Leya I, Wieler R, Halliday AN (2000b) Cosmic-ray production of tungsten isotopes in lunar samples and meteorites and its implications for Hf-W cosmochemistry. Earth Planet Sci Lett 175:1-12

Lin RP, Mitchell DL, Curtis DW, Anderson KA, Carlson CW, McFadden J, Acuña MH, Hood LL, Binder AB (1998) Lunar surface magnetic fields and their interaction with the solar wind: results from lunar prospector. Science 281:1480-1484

Lindstrom MM, Lindstrom DJ (1986) Lunar granulites and their precursor anorthositic norites of the early lunar crust. Proc Lunar Planet Sci Conf 16th in J Geophys Res 91:D263-D276

Lindstrom MM, Mittlefehldt DW, Morris RV, Martinez RR, Wentworth S (1995) QUE93069, a more mature regolith breccia for the Apollo 25th anniversary. Lunar Planet Sci XXVI:849-850

Lindstrom MM, Schwarz C, Score R, Mason B (1991) MacAlpine Hills 88104 and 88105 lunar highland meteorites: General description and consortium overview. Geochim Cosmochim Acta 55:2999-3007

Lingenfelter RE, Canfield EH, Hess WN (1961) The lunar neutron flux. J Geophys Res 66:2665-2671

Little RC, Feldman WC, Maurice S, Genetay I, Lawrence DJ, Lawson SL, Gasnault O, Barraclough BL, Elphic RC, Prettyman TH, Binder AB (2003) Latitude variation of the subsurface lunar temperature: Lunar Prospector thermal neutrons. J Geophys Res 108(E5):10.1029/2001JE001497

Love SG, Brownlee DE (1993) A direct measurement of the terrestrial mass accretion rate of cosmic dust. Science 262: 550-553

Lovering JF, Wark DA, Gleadow AJW, Britten R (1974) Lunar monazite: A late-stage (mesostasis) phase in mare basalts. Earth Planet Sci Lett 21:164-168

Lucey PG (1998) Model near-infrared optical constants of olivine and pyroxene as a function of iron content. J Geophys Res 103:1703-1714

Lucey PG (2002) Radiative transfer model constraints on the shock state of remotely sensed lunar anorthosites. Geophys Res Lett29(10):124-1 to 124-3

Lucey PG (2004) Mineral maps of the Moon. Geophys Res Lett 31(8): doi:10.1029/2003GL019406

Lucey PG, Blewett DT, Hawke BR (1998) Mapping the FeO and TiO2 content of the lunar surface multispectral imagery. J Geophys Res 103(E2):3679-3699

Lucey PG, Blewett DT, Jolliff BL (2000a) Lunar iron and titanium abundance algorithms based on final processing Clementine **UVVIS** images. J Geophys Res 105(E8):20,297-20,305

Lucey PG, Blewett DT, Taylor GJ, Hawke BR (2000b) Imaging of lunar surface maturity. J Geophys Res 105(E8): 20377-20386

Lucey PG, Hawke BR (1987) Probable outcrops of Mg-gabbronorite in the lunar highlands detected by near-infrared remote sensing. Proc Lunar Planet Sci Conf 18:578

Lucey PG, Hawke BR, Pieters CM, Head JW, McCord TB (1986) A compositional study of the Aristarchus region of the Moon using near-infrared reflectance spectroscopy. J Geophys Res-Solid Earth Planets 91(B4):D344-D354

Lucey PG, Taylor GJ, Malaret E (1995) Abundance and distribution of iron on the Moon. Science 268(5214):1150-1153

Madey TE, Yakshinskiy BV, Ageev VN, Johnson RE (1998) Desorption of alkali atoms from oxide surfaces: relevance to origins of Na and K in atmospheres of Mercury and the Moon. J Geophys Res 103:5873-5887

Manatt SL, Elleman DD, Vaughn RW, Chan SI, Tsay F-D, Huntress WT Jr. (1970) Magnetic resonance of lunar samples. Science 167:709-711

Margot J, Campbell DB, Jurgens RF, Slade MA (2000) Digital elevation models of the Moon from Earth-based radar interferometry. IEEE Geosci Rem Sens 38:1122-1133

Marvin UB (1983) The discovery and initial characterization of Allan Hills 81005; the first lunar meteorite. Geophys Res Lett 10:775-778

Marvin UB, Lindstrom MM, Holmberg BB, Martinez RR (1991) New observations on the quartz monzodiorite-granite suite. Proc Lunar Planet Sci Conf 21:119-135

Marvin UB, Wood JA, Taylor GJ, Reid JB Jr., Powell BN, Dickey JS Jr., Bower JF (1971) Relative proportions and probable sources of rock fragments in the Apollo 12 soil samples. Proc Lunar Sci Conf 2:679-699

Masarik J, Nishiizumi K, Reedy RC (2001) Production rates of cosmogenic helium-3, neon-21, and neon-22 in ordinary chondrites and the lunar surface. Meteorit Planet Sci 36:643-650

Mathew KJ, Marti K (2001) Lunar nitrogen: indigenous signature and cosmic-ray production rate. Earth Planet Sci Lett 184:659-669

Maurice S, Feldman WC, Lawrence DJ, Elphic RC, Gasnault O, d'Uston C, Genetay I, Lucey PG (2000) High-energy neutrons from the Moon. J Geophys Res 105:20,365-20,375

Maurice S, Lawrence DJ, Feldman WC, Elphic RC, Gasnault O (2004) Reduction of neutron data from Lunar Prospector. J Geophys Res 109:doi 10.1029/2003JE002208

McCallum IS, O'Brien HE (1996) Stratigraphy of the lunar highlands crust: depth of burial of lunar samples from cooling rate studies. Am Mineral 81:1166-1175

McConnochie TH, Buratti BJ, Hillier JK, Tryka KA (2002) A search for water ice at the lunar poles with Clementine images. Icarus 156(2):335-351

McCord TB (1969) Color differences on the lunar surface. J Geophys Res 74(12):3131-3142

McCord TB, Adams JB (1973) Progress in remote optical analysis of lunar surface composition. The Moon 7:453-474

McCord TB, Clark RN, Hawke BR, McFadden LA, Owensby PD, Pieters CM, Adams JB (1981) Moon: near-infrared spectral reflectance, a good first look. J Geophys Res 86(B11):10,883-10,892

McCord TB, Johnson TV (1969) Relative spectral reflectivity 0.4-1m of selected areas of the lunar surface. J Geophys Res 74(17):4395-4401

McCord TB, Johnson TV (1970) Lunar spectral reflectivity (0.30 to 2.50 microns) and implications for remote mineralogical analysis. Science 169:855-858

McEwen AS, Robinson MS (1997) Mapping of the moon by Clementine. Comparative Studies Moon Mercury 19(10): 1523-1533

McEwen AS, Robinson MS, Eliason EM, Lucey PG, Duxbury TC, Spudis PD (1994) Clementine observations of the Aristarchus Region of the Moon. Science 266(5192):1858-1862

McKay DS, Basu A, Nace G (1980) Lunar core 15010/11; grain size, petrology, and implications for regolith dynamics. Proc Lunar Planet Sci Conf 11:1531-1550

McKay DS, Basu A, Waits G (1978a) Grain size and the evolution of Luna 24 soils. *In:* Mare Crisium: The View from Luna 24. Merrill RB, Papike JJ (eds) Pergamon Press, p 125-136

McKay DS, Bogard DD, Morris RV, Korotev RL, Johnson P, Wentworth SJ (1986) Apollo 16 regolith breccias: characterization and evidence for early formation in the mega-regolith. Proc Lunar Planet Sci Conf 16. J Geophys Res :D277-D303

McKay DS, Dungan MA, Morris RV, Fruland RM (1977) Grain size, petrographic, and FMR studies of double core 60009/10: a study of soil evolution. Proc Lunar Sci Conf 8:2929-2952

McKay DS, Fruland RM, Heiken GH (1974) Grain size and the evolution of lunar soil. Proc Lunar Sci Conf 5:887-906

McKay DS, Heiken G, Basu A, Blanford G, Simon S, Reedy R, French B, Papike JJ (1991) The lunar regolith. *In:* The Lunar Source Book: A User's Guide to the Moon. Heiken G, Vaniman D, French B, (eds) Cambridge Univ. Press, p 285-356

McKay DS, Heiken GH, Waits G (1978b) Core 74001/2; grain size and petrology as a key to the rate of *in situ* reworking and lateral transport on the lunar surface. Proc Lunar Planet Sci Conf 9:1913-1932

McKay DS, Williams RJ (1979) A geologic assessment of potential lunar ores. *In:* Space Resources and Space Settlement, NASA SP-428. Billingham J, Gilbreath W, O'Leary B (eds) NASA, p 243-256

McKay GA, unnamed coauthors (1986) Topic 2: Apollo 15 KREEP basalt. *In:* Geology and Petrology of the Apollo 15 Landing Site, LPI Tech Rpt 86-03. Spudis PD, Ryder G (eds) Lunar and Planetary Institute, p 14-16

McKay GA, Weill DF (1977) KREEP petrogenesis revisited. Proc Lunar Sci Conf 8:2339-2355

McKinley JP, Taylor GJ, Keil K, Ma M-S, Schmitt RA (1984) Apollo 16; impact melt sheets, contrasting nature of the Cayley Plains and Descartes Mountains, and geologic history. Proc Lunar Planet Sci Conf 14:514-524

Melendrez DE, Johnson JR, Larson SM, Singer RB (1994) Remote sensing of potential lunar resources, 2. High spatial resolution mapping of spectral reflectance ratios and implications for near side mare TiO_2 content. J Geophys Res 99(E3):5601-5619

Mendell WW (1976) Degradation of large, period II lunar craters. Proc Lunar Planet Sci Conf 7:2705-2716

Mendell WW, Low FJ (1975) Infrared orbital mapping of lunar features. Proc Lunar Planet Sci Conf 6:2711-2719

Mendillo M (2001) The atmosphere of the moon. Earth Moon Planets 85-86:271-277

Mendillo M, Baumgardner J, Wilson J (1999) Observational test for the solar wind origin of the Moon's extended sodium atmosphere. Icarus 137:13-23

Metzger AE (1993) Composition of the Moon as determined from orbit by gamma ray spectroscopy. *In:* Remote Geochemical Analysis: Alemental and Mineralogical Composition. Pieters CM, Englert PAJ (eds) Cambridge Univ. Press, p 341-365

Metzger AE, Haines EL, Parker RE, Radocinski RG (1977) Thorium concentrations in the lunar surface. 1. Regional values and crustal content. Proc Lunar Sci Conf 8:949-999

Metzger AE, Parker RE (1979) The distribution of titanium on the lunar surface. Earth Planet Sci Lett 45:155-171

Metzger AE, Trombka JI, Peterson LE, Reedy RC, Arnold JR (1973) Lunar surface radioactivity: Preliminary results of the Apollo 15 and Apollo 16 gamma-ray spectrometer experiments. Science 179:800-803

Meyer CJ, McKay DS, Anderson DH, Butler PJ (1975) The source of sublimates on the Apollo 15 green and Apollo 17 orange glass samples. Proc Lunar Sci Conf 6:1673-1699

Michel R (1999) Long-lived radionuclides as tracers in terrestrial and extraterrestrial matter. Radiochimica Acta 87: 47-73

Michel R, Bodemann R, Busemann H, Daunke R, Gloris M, Lange HJ, Klug B, Krins A, Leya I, Lüpke M, Neumann S, Reinhardt H, Schnatz-Büttgen M, Herpers U, Schiekel T, Sudbrock F, Holmqvist B, Condé H, Malmborg P, Suter M, Dittrich-Hannen B, Kubik PW, Synal HA, Filges D (1997) Cross sections for the production of residual nuclides by low- and medium-energy protons from the target elements C, N, O, Mg, Al, Si, Ca, Ti, V, Mn, Fe, Co, Ni, Cu, Sr, Y, Zr, Nb, Ba and Au. Nucl Instr Methods B 129:153-193

Morgan TH, Zook HA, Potter AE (1988) Impact-driven supply of sodium and potassium to the atmosphere of Mercury. Icarus 74:156-170

Morgan TH, Zook HA, Potter AE (1989) Production of sodium vapor from exposed regolith in the inner solar system. Proc Lunar Planet Sci Conf 19:297-304

Morris RV (1976) Surface exposure indices of lunar soils: a comparative FMR study. Proc Lunar Sci Conf 7:315-335

Morris RV (1977) Origin and evolution of the grains-size dependence of the concentration of fine-grained metal in lunar soils: the maturation of lunar soils to a steady-state stage. Proc Lunar Planet Sci Conf 8:3719-3747

Morris RV (1978a) The surface exposure (maturity) of lunar soils; some concepts and I_s/FeO compilation. Proc Lunar Planet Sci Conf 9:2287-2297

Morris RV (1978b) *In situ* reworking (gardening) of the lunar surface: evidence from the Apollo cores. Proc Lunar Planet Sci Conf 9:1801-1811

Morris RV (1980) Origins and size distribution of metallic iron particles in the lunar regolith. Proc Lunar Planet Sci Conf 11:1697-1712

Morris RV (1985) Determination of optical penetration depths from reflectance and transmittance measurements on albite powders. Lunar Planet Sci XVI:581-582

Morris RV, Gibbons RV, Hörz F (1975) FMR thermomagnetic studies up to 900°C of lunar soils and potential magnetic analogues. Geophys Res Lett 2:461-464

Morris RV, Gose WA (1976) Ferromagnetic resonance and magnetic studies of cores 60009/60010 and 60003: Compositional and surface-exposure stratigraphy. Proc Lunar Sci Conf 7:1-11

Morris RV, Gose WA (1977) Depositional history of core section 74001; depth profiles of maturity, FeO, and metal. Proc Lunar Sci Conf 8:3113-3122

Morris RV, Korotev RL, Lauer HVJ (1989) Maturity and geochemistry of the Van Serg crater core (79001/2) with implications for micrometeorite composition. Proc Lunar Planet Sci Conf 19:269-284

Morris RV, Lauer HVJ (1980) The case against UV photostimulated oxidation of magnetite. Geophys Res Lett 7:605-608

Morris RV, Lauer HVJ, Gose WA (1979) Characterization and depositional and evolutionary history of the Apollo 17 deep drill core. Proc Lunar Planet Sci Conf 10:1141-1157

Morris RV, See TH, Hörz F (1983) Some evidence concerning the source material of large glass objects from the Moon. Proc Lunar Planet Sci Conf 14:528-529

Morris RV, See TH, Hörz F (1986) Composition of the Cayley Formation at Apollo 16 as inferred from impact melt splashes. Proc Lunar Planet Sci Conf 17:E21-E42

Naney MT, Crowl DM, Papike JJ (1976) The Apollo 16 drill core: Statistical analysis of glass chemistry and the characterization of a high alumina-silica poor (HASP) glass. Proc Lunar Sci Conf 7:155-184

Nash DB, Salisbury JW, Conel JE, Lucey PG, Christensen PR (1993) Evaluation of infrared-emission spectroscopy for mapping the Moons surface-composition from lunar orbit. J Geophys Res 98(E12):23535-23552

Neal CR, Taylor LA (1989) The nature of barium partitioning between immiscible melts: A comparison of experimental and natural systems with reference to lunar felsite petrogenesis. Proc Lunar Planet Sci Conf 19:209-218

Neal CR, Taylor LA (1992) Petrogenesis of mare basalts: a record of lunar volcanism. Geochim Cosmochim Acta 56:2177-2211

Neal CR, Taylor LA, Lindstrom MM (1988) Apollo 14 mare basalt petrogenesis: assimilation of KREEP-like components by a fractionating magma. Proc Lunar Planet Sci Conf 18:139-153

Neal CR, Taylor LA, Patchen AD (1989) High alumina (HA) and very high potassium (VHK) basalt clasts from Apollo 14 breccias. Part 1. Mineralogy and petrology: Evidence of crystallization from evolving magmas. Proc Lunar Planet Sci Conf 19:137-145.

Nehru CE, Prinz M, Dowty E, Keil K (1974) Spinel-group minerals and ilmenite in Apollo 15 rock samples. Am Mineral 59:1220-1234

Nehru CE, Warner RD, Keil K (1976) Electron Microprobe Analyses of Opaque Mineral Phases from Apollo 11 Basalts. University of New Mexico Special Publication 17

Nishiizumi K, Caffee MW (2001) Beryllium-10 from the Sun. Science 294:352-354

Nishiizumi K, Caffee MW, Jull AJT, Reedy RC (1996) Exposure history of lunar meteorites Queen Alexandra range 93069 and 94269. Meteorit Planet Sci 31:893-896

Nishiizumi K, Fink D, Klein J, Middleton R, Masarik J, Reedy RC, Arnold JR (1997) Depth profile of ^{41}Ca in an Apollo 15 drill core and the low-energy neutron flux in the Moon. Earth Planet Sci Lett 148:545-552

Nishiizumi K, Kohl CP, Arnold JR, Finkel RC, Caffee MW, Masarik J, Reedy RC (1995) Final results of cosmogenic nuclides in lunar rock 64455. Lunar Planet Sci XXVI:1055-1056

Nishiizumi K, Masarik J, Caffee MW, Jull AJT (1999) Exposure histories of pair lunar meteorites EET 96008 and EET 87521. Lunar Planet Sci XXX:1980 (CD-ROM)

Noble SK, Pieters CM, Taylor LA, Morris RV, Allen CC, McKay DS, Keller LP (2001) The optical properties of the finest fraction of lunar soil: implications for space weathering. Meteorit Planet Sci 36:31-42

Norman MD, Ryder G (1980) Geochemical constraints on the igneous evolution of the lunar crust. Proc Lunar Planet Sci Conf 11:317-331

Nozette S, Rustan P, Pleasance LP, Horan DM, Regeon P, Shoemaker EM, Spudis PD, Acton CH, Baker DN, Blamont JE, Buratti BJ, Corson MP, Davies ME, Duxbury TC, Eliason EM, Jakosky BM, Kordas JF, Lewis IT, Lichtenberg CL, Lucey PG, Malaret E, Massie MA, Resnick JH, Rollins CJ, Park HS, McEwen AS, Priest RE, Pieters CM, Reisse RA, Robinson MS, Simpson RA, Smith DE, Sorenson TC, Breugge RWV, Zuber MT (1994) The Clementine Mission to the Moon - Scientific Overview. Science 266:1835-1839

Okada T, Kato M, Fujimura A, Tsunemi H, Kitamoto S (1999) X-ray fluorescence spectrometry with the SELENE orbiter. Adv Space Res 23:1833-1836

Palme H (1977) On the age of KREEP. Geochim Cosmochim Acta 41:1791-1801

Papike JJ (1987) Chemistry of the rock-forming silicates: ortho, ring, and single-chain structures. Rev Geophys 25:1483-1526

Papike JJ (1988) Chemistry of the rock-forming silicates: multiple-chain, sheet, and framework silicates. Rev Geophys 26:407-444

Papike JJ (1998a) Planetary Materials. Rev Mineralogy, Volume 36. Mineralogical Society of America, Washington DC

Papike JJ (1998b) Comparative planetary mineralogy: chemistry of melt-derived pyroxene, feldspar, and olivine. Rev Mineral 36:7-1–7-11

Papike JJ, Bence AE, Brown GE, Prewitt CT, Wu CH (1971) Apollo 12 clinopyroxenes: exsolution and epitaxy. Earth Planet Sci Lett 10:307-315

Papike JJ, Cameron M (1976) Crystal chemistry of silicate minerals of geophysical interest. Rev Geophys Space Phys 14:37-80

Papike JJ, Hodges FN, Bence AE, Cameron M, Rhodes JM (1976) Mare basalts: crystal chemistry, mineralogy, and petrology. Rev Geophys Space Phys 14(4):475-540

Papike JJ, Ryder G, Shearer CK (1998) Lunar samples. Rev Mineral 36:5-1–5-234

Papike JJ, Simon SB, Laul JC (1982) The lunar regolith: chemistry, mineralogy, petrology. Rev Geophys Space Phys 20(4):761-826

Papike JJ, Simon SB, White C, Laul JC (1981) The relationship of the lunar regolith <10 μm fraction and agglutinates; Part I, A model for agglutinate formation and some indirect supportive evidence. Proc Lunar Planet Sci Conf 12: 409-420

Papike JJ, Spilde MN, Adcock CT, Fowler GW, Shearer CK (1997) Trace element fractionation by impact-induced volatilization: SIMS study of lunar HASP Samples. Am Mineral 82:630-634

Papike JJ, Taylor LA, Simon SE (1991) Lunar minerals. *In:* Lunar Sourcebook: A User's Guide to the Moon, Heiken G, Vaniman DT, French BM (eds) Cambridge University Press, p 121-181

Pearce GW, Strangway DW, Gose WA (1974) Magnetic properties of Apollo samples and implications for regolith formation. Proc Lunar Sci Conf 5:2815-2826

Pettengill GH, Zisk SH, Thompson TW (1974) The mapping of lunar radar scattering characteristics. The Moon 10: 3-16

Pieters CM (1982) Copernicus crater central peak - lunar mountain of unique composition. Science 215:59-61

Pieters CM (1983) Strength of mineral absorption features in the transmitted component of near-infrared reflected light - 1st results from Relab. J Geophys Res 88(NB11):9534-9544

Pieters CM (1986) Composition of the lunar highland crust from near-infrared spectroscopy. Rev Geophys 24:557-578

Pieters CM (1993) Compositional diversity and stratigraphy of the lunar crust derived from reflectance spectroscopy. *In:* Remote Geochemical Analysis: Elemental and Mineralogical Composition (Topics in remote sensing 4). Englert PAJ, Pieters CM (eds) Cambridge University Press, p 309-339

Pieters CM, Fischer EM, Rode O, Basu A (1993) Optical effects of space weathering - the role of the finest fraction. J Geophys Res 98:20817-20824

Pieters CM, Head JW, Gaddis L, Jolliff B, Duke M (2001a) Rock types of South Pole-Aitken basin and extent of basaltic volcanism. J Geophys Res 106:28001-28022

Pieters CM, Head JW, Gaddis LR, Jolliff BL, Duke M (2001b) The character and possible origin of olivine hill in South Pole-Aitken Basin. Proc Lunar Planet Sci Conf XXXII:#1810

Pieters CM, Hill PM, Magee KP, Sunshine JM (1988) Water on the Moon? Potential detection of recent cometary impacts in the Earth/Moon environment using LGO/VIMS. Lunar Planet Sci 19:935-936

Pieters CM, McCord TB (1976) Characterization of lunar mare basalt types: A remote sensing study using reflection spectroscopy of surface soils. Proc Lunar Sci Conf 7:2677-2690

Pieters CM, Taylor LA (2003) Systematic global mixing and melting in lunar soil evolution. Geophys Res Lett 30(20): 2048, doi:10.1029/2003GL018212

Pieters CM, Taylor LA, Noble SK, Keller LP, Hapke B, Morris RV, Allen CC, McKay DS, Wentworth S (2000) Space weathering on airless bodies: Resolving a mystery with lunar samples. Meteorit Planet Sci 35(5):1101-1107

Pieters CM, Tompkins S (1999) Tsiolkovsky crater: a window into crustal processes on the lunar farside. J Geophys Res 104:21935-21949

Pieters CM, Tompkins S, Head JW, Hess PC (1997) Mineralogy of the mafic anomaly in the South Pole-Aitken Basin: implication for excavation of the lunar mantle. Geophys Res Lett 24(15):1903-1906

Pieters CM, Wilhelms DE (1985) Origin of olivine at Copernicus. J Geophys Res 90:C415-C420

Pillinger CT, Davis PR, Eglinton G, Gowar AP, Jull AJT, Maxwell JR, Housley RM, Cirlin EH (1974) The association between carbide and finely divided metallic iron in lunar fines. Proc Lunar Sci Conf 5:1949-1961

Polnau E, Eugster O (1998) Cosmic-ray produced, radiogenic, and solar noble gases in lunar meteorites Queen Alexandra Range 94269 and 94281. Meteorit Planet Sci 33:313-319

Potter AE, Morgan TH (1985) Discovery of sodium in the atmosphere of Mercury. Science 229:651-653

Potter AE, Morgan TH (1986) Potassium in the atmosphere of Mercury. Icarus 67:336-340

Potter AE, Morgan TH (1988a) Extended sodium atmosphere of the Moon. Geophys Res Lett 15:1515-1518

Potter AE, Morgan TH (1988b) Discovery of sodium and potassium vapor in the atmosphere of the Moon. Science 241:675-680

Prettyman TH, Feldman WC, Lawrence DJ, McKinney GW, Binder AB, Elphic RC, Gasnault O, Maurice S, Moore KR (2002a) Library least squares analysis of Lunar Prospector gamma ray spectra. Lunar Planet Sci XXXIII: 2012 (CD-ROM)

Prettyman TH, Lawrence DJ, Vaniman DT, Elphic RC, Feldman WC (2002b) Classification of regolith materials from lunar prospector data reveals a magnesium-rich province. *In:* The Moon Beyond 2002: Next steps in Lunar Science and Exploration, LPI Contr. Abstract #1128. Lunar and Planetary Institute

Quick JE, Albee AL, Ma M-S, Murali AV, Schmitt RA (1977) Chemical composition and possible immiscibility of two silicate melts in 12013. Proc Lunar Sci Conf 8:2153-2189

Quick JE, James OB, Albee AL (1981) Petrology and petrogenesis of lunar breccia 12013. Proc Lunar Planet Sci Conf 12B:117-172

Rao MN, Garrison DH, Bogard DD, Reedy RC (1994) Determination of the flux and energy distribution of energetic solar protons in the past 2 Myr using lunar rock 68815. Geochim Cosmochim Acta 58:4231-4245

Rava B, Hapke B (1987) An Analysis of the Mariner 10 color ratio map of Mercury. Icarus 71(3):397-429

Reed SJB, Taylor SR (1974) Meteoritic metal in Apollo 16 samples. Meteorit Planet Sci 9:23-34

Reedy RC (1978) Planetary gamma-ray spectroscopy. Proc Lunar Planet Sci Conf 9:2961-2984

Reedy RC (1998) Studies of modern and ancient solar energetic particles. Proc Indian Acad Sci (Earth Planet Sci) 107: 433-440

Reedy RC (2000) Predicting the production rates of cosmogenic nuclides. Nucl Instr Methods B 172:782-785

Reedy RC, Arnold JR, Lal D (1983) Cosmic-ray record in solar system matter. Science 219:127-135

Reedy RC, Nishiizumi K (1998) Factors affecting the interpretation of solar-proton-produced nuclides and some chlorine-36 results. Lunar Planet Sci XXIX:1698 (CD-ROM)

Reid AM, Meyer C, Harmon RS, Brett R (1970) Metal grains in Apollo 12 igneous rocks. Earth Planet Sci Lett 9:1-5

Rhodes JM, Adams JB, Blanchard DP, Charette MP, Rodgers KV, Jacobs JW, Brannon JC, Haskin LA (1975) Chemistry of agglutinate fractions in lunar soils. Proc Lunar Sci Conf 6:2291-2307

Rhodes JM, Rodgers KV, Shih C, Bansal BM, Nyquist LE, Wiesmann H, Hubbard NJ (1974) The relationships between geology and soil chemistry at the Apollo 17 landing site. Proc Lunar Sci Conf 5:1097- 1117

Robinson MS, Jolliff BL (2002) Apollo 17 landing site: Topography, photometric corrections, and heterogeneity of the surrounding highland massifs. J Geophys Res 107(E11):doi 10.1029/2001JE001614

Ross M, Bence AE, Dwornik EJ, Clark JR, Papike JJ (1970) Mineralogy of lunar clinopyroxenes, augite and pigeonite. Proc Lunar Sci Conf 1:839-848

Ross M, Huebner JS, Dowty E (1973) Delineation of the one atmosphere augite-pigeonite miscibility gap for pyroxenes from lunar basalt 12021. Am Mineral 58:619-635

Roush TL, Lucey PG (1988) A search for water on the moon at the Reiner Gamma formation, a possible site of cometary coma impact. Proc Lunar Planet Sci Conf 18:397-402

Rubin AE (1997) Mineralogy of meteorite groups. Meteorit Planet Sci 32:231-247

Rutherford MJ, Dixon S, Hess P (1980) Ilmenite saturation at high pressure in KREEP basalts: origin of KREEP and high TiO_2 in mare basalts. Lunar Planet Sci XI:966-967

Ryder G (1985) Catalog of Apollo 15 Rocks. Curatorial Publication 20787. NASA Johnson Space Center

Ryder G (1994) Coincidence in time of the Imbrium basin impact and Apollo 15 KREEP volcanic flows: the case for impact-induced melting. Spec Pap Geol Soc Am 293:11-18

Ryder G, Norman MD, Score RA (1980) The distinction of pristine from meteorite-contaminated highland rocks using metal compositions. Proc Lunar Planet Sci Conf 11:471-479

Ryder G, Stoeser DB, Wood JA (1977) Apollo 17 KREEPy basalt: a rock type intermediate between KREEP and mare basalts. Earth Planet Sci Lett 35:1-13

Ryder G, Wood JA (1977) Serenitatis and Imbrium impact melts: implications for large-scale layering in the lunar crust. Proc Lunar Sci Conf 8:655-668

Saari JM, Shorthill RW, Deaton TK (1966) Infrared and visible images of the eclipsed moon of December 19, 1964. Icarus 5:635-659

Salisbury JW, Basu A, Fischer EM (1997) Thermal infrared spectra of lunar soils. Icarus 130:125-139

Salpas PA, Taylor LA, Lindstrom MM (1987) Apollo 17 KREEPy basalts: evidence for the non-uniformity of KREEP. J Geophys Res 92:E340-E348

Sands DG, De Laeter JR, Rosman KJR (2001) Measurements of neutron capture effects on Cd, Sm and Gd in lunar samples with implications for the neutron energy spectrum. Earth Planet Sci Lett 186:335-346

Sato M, Hickling NL, McLane JE (1973) Oxygen fugacity values of Apollo 12, 14, and 15 lunar samples and reduced state of lunar magmas. Proc Lunar Sci Conf 4:1061–1079

Schaal RB, Hörz F (1977) Shock metamorphism of lunar and terrestrial basalts. Proc Lunar Sci Conf 8:1697-1729

Schaber GG, Thompson TW, Zisk SH (1975) Lava flows in Mare Imbrium: An evaluation of anomalously low Earth-based radar reflectivity. The Moon 13:395-423

Schnabel C, Xue S, Ma P, Herzog GF, Fifield K, Cresswell RG, di Tada ML, Hausladen P, Reedy RC (2000) Nickel-59 in surface layers of lunar basalt 74275: implications for the solar alpha particle flux. Lunar Planet Sci XXXI: 1778 (CD-ROM)

Schonfeld E (1974) The contamination of lunar highlands rocks by KREEP: interpretation by mixing models. Proc Lunar Sci Conf 5:1269-1286

Schultz PH (1976a) Moon morphology: interpretations based on Lunar Orbiter photography. University of Texas Press

Schultz PH (1976b) Floor-fracture lunar craters. Moon 15:241-273

Schultz PH, Srnka LJ (1980) Cometary collisions on the Moon and Mercury. Nature 284:22-26

Shervais JW, McGee JJ (1998) Ion and electron microprobe study of troctolites, norite, and anorthosites from Apollo 14: evidence for urKREEP assimilation during petrogenesis of Apollo 14 Mg-suite rocks. Geochim Cosmochim Acta 62:3009-3023

Shervais JW, Taylor LA, Lindstrom MM (1988) Olivine vitrophyres: a nonpristine high-Mg component in lunar breccia 14321. Proc Lunar Planet Sci Conf 18:45-57

Shih C-Y, Nyquist LE, Bogard DD, Wooden JL, Bansal BM, Wiesmann H (1985) Chronology and petrogenesis of a 1.8 g lunar granitic clast: 14321, 1062. Geochim Cosmochim Acta 49:411-426

Shih C-Y, Nyquist LE, Wiesmann H (1993) K-Ca chronology of lunar granites. Geochim Cosmochim Acta 57:4827-4841

Shorthill RW, Saari JM (1965) Nonuniform cooling of the eclipsed Moon: a listing of thirty prominent anomalies. Science 150:210-212

Simon SB, Papike JJ (1985) Petrology of the Apollo 12 highland component. J Geophys Res 90:D47-D60

Simon SB, Papike JJ, Gosselin DC, Laul JC (1985) Petrology and chemistry of Apollo 12 regolith breccias. J Geophys Res 90:D75-D86

Simon SB, Papike JJ, Gosselin DC, Laul JC (1986) Petrology, chemistry, and origin of Apollo 15 regolith breccias. Geochim Cosmochim Acta 50:2675-2691

Simon SB, Papike JJ, Laul JC (1981) The lunar regolith: Comparative studies of the Apollo and Luna sites. Petrology of soils from Apollo 17, Luna 16, 20, and 24. Proc Lunar Planet Sci Conf 12B:371-388

Simon SB, Papike JJ, Shearer CK, Laul JC (1983) Petrology of the Apollo 11 highland component. Proc Lunar Planet Sci Conf 14:B103-B138

Sippel RF (1971) Luminescence petrography of the Apollo 12 rocks and comparative features in terrestrial rocks and meteorites. Proc Lunar Sci Conf 2:247-263

Sisterson JM, Kim K, Beverding A, Englert PAJ, Caffee MW, Vincent J, Castaneda C, Reedy RC (1997) Measuring excitation functions needed to interpret cosmogenic nuclide production in lunar rocks. Eds. *In:* Applications of Accelerators in Research and Industry, 392. Duggan JL, Morgan IL (eds) AIP Conf. Proc., p 811-814

Smith JV (1974) Lunar mineralogy: a heavenly detective story, Presidential Address, Part I. Am Mineral 59:231-243

Smith JV, Steele IM (1976) Lunar Mineralogy, a heavenly detective story. Am Mineral 61:1059-1116

Smith SM, Wilson JK, Baumgardner J, Mendillo M (1999) Discovery of the distant sodium tail and its enhancement following the Leonid meteor shower. Geophys Res Lett 26:1649-1652

Smith SM, Wilson JK, Baumgardner J, Mendillo M (2001) Monitoring the Moon's transient atmosphere with an all-sky imager. Adv Space Res 27:1181-1187

Smrekar S, Pieters CM (1985) Near-Infrared spectroscopy of probable impact melt from 3 large lunar highland craters. Icarus 63:442-452

Snyder GA, Taylor LA, Jerde EA (1994) Evolved QMD-melt parentage for lunar highlands alkali suite cumulates: Evidence from ion-probe rare-earth element analyses of individual minerals. Lunar Planet Sci XXV:1311-1312

Snyder GA, Taylor LA, Liu Y-G, Schmitt RA (1992) Petrogenesis of the western highlands of the Moon: Evidence from a diverse group of whitlockite-rich rocks from the Fra Mauro Formation. Proc Lunar Planet Sci Conf 22: 399-416

Sonett CP, Giampapa MS, Mathews MS (1991) The Sun in Time, 990 p, Univ. Arizona Press, Tucson

Spettel B, Dreibus G, Burghele A, Jochum KP, Schultz L, Weber HW, Wlotzka F, Wänke H (1995) Chemistry, petrology, and noble gases of lunar highland meteorite Queen Alexandra Range 93069. Meteorit Planet Sci 30:581-582

Sprague AL, Kozlowski WH, Hunten DM, Wells WK, Grosse FA (1992b) The sodium and potassium atmosphere of the Moon and its interaction with the surface. Icarus 96:27-42

Sprague AL, Kozlowski WH, Hunten DM, Wells WK, Grosse FA (1998) Observations of sodium in the lunar atmosphere during International Lunar Atmosphere Week, 1995. Icarus 131:372-381

Sprague AL, Witteborn FC, Kozlowski RW, Cruikshank DP, Bartholomew MJ, Graps AL (1992a) The Moon – mid-infrared (7.5-Mu-M to 11.4-Mu-M) spectroscopy of selected regions. Icarus 100(1):73-84

Spudis P, Pieters C (1991) Global and regional data about the Moon. *In:* Lunar Sourcebook: A Users Guide to the Moon. Heiken GH, Vaniman DT, French BM (eds) Cambridge University Press, p 595-632

Spudis PD, Davis PA (1986) A chemical and petrological model of the lunar crust and implications for lunar crustal origin. Proc Lunar Planet Sci Conf 17. J Geophys Res 91:E84-E90

Spudis PD, Hawke BR (1986) The Apennine Bench formation revisited. *In:* Workshop on the Geology and Petrology of the Apollo 15 Landing Site. LPI Tech. Report 86-03. Spudis PD, Ryder G (eds) Lunar and Planetary Institute, p 105-107

Spudis PD, Hawke BR, Lucey P (1984) Composition of Orientale basin deposits and implications for the lunar basin-forming process. J Geophys Res 89:C197-C210

Spudis PD, Ryder G (1981) Apollo 17 impact melts and their relation to the Serenitatis basin. Multi-Ring Basins. Proc Lunar Planet Sci Conf 12A:133-148

Spudis PD, Ryder G (1985) Geology and petrology of the Apollo 15 landing site - Past, present, and future understanding. EOS, Trans Am Geophys Union 66:721,724-726

Stacy NJS, Campbell DB, Ford PG (1997) Radar mapping of the lunar poles: A search for ice deposits. Science 276: 1527

Staid MI, Pieters CM (2000) Integrated spectral analysis of mare soils and craters: application to eastern nearside basalts. Icarus 145:122-139

Staid MI, Pieters CM (2001) Mineralogy of the last lunar basalts: results from Clementine. J Geophys Res 106:27,887-27,900

Starukhina LV, Shkuratov YG (2000) The lunar poles: water ice or chemically trapped hydrogen? Icarus 147:585-587

Stern SA (1999) The lunar atmosphere: history status current problems and context. Rev Geophys 37:453-491

Stewart DB (1975) Apollonian metamorphic rocks - the products of prolonged subsolidus equilibration. Lunar Sci Conf 6:774-776

Stöffler D, Bischoff A, Borchardt R, Burghele A, Deutsch A, Jessberger EK, Ostertag R, Palme H, Spettel B, Reimold WU, Wacker K, Wänke H (1985) Composition and evolution of the lunar crust in the Descartes Highlands, Apollo 16. Proc Lunar Planet Sci Conf 15. J Geophys Res 90:C449-C506

Stöffler D, Knoll H-D, Marvin UB, Simonds CH, Warren PH (1980) Recommended classification and nomenclature of lunar highland rock—A committee report. *In:* Proceedings of the Conference on the Lunar Highland Crust. Papike JJ, Merrill RB (eds) Pergamon Press, p 51-70

Stöffler D, Ostertag R, Reimold WU, Borchardt R, Malley J, Rehfeldt A (1981) Distribution and provenance of lunar highland rock types at North Ray Crater Apollo 16. Proc Lunar Planet Sci Conf 12:185-207

Sunshine JM, Pieters CM (1993) Estimating modal abundances from the spectra of natural and laboratory pyroxene mixtures using the modified gaussian model. J Geophys Res 98:9075-9087

Sunshine JM, Pieters CM (1998) Determining the composition of olivine from reflectance spectroscopy. J Geophys Res 103:13675-13688

Takeda H, Miyamoto M, Ishii T, Lofgren GE (1975) Relative cooling rates of mare basalts at the Apollo 12 and 15 sites as estimated from pyroxene exsolution data. Proc Lunar Sci Conf 6:987-996

Taylor GJ, Warner RD, Keil K, Ma M-S, Schmitt RA (1980) Silicate liquid immiscibility, evolved lunar rocks and the formation of KREEP. *In:* Proceedings of the Conference on the Lunar Highland Crust. Papike JJ, Merrill RB (eds) Pergamon Press, p 339-352

Taylor GJ, Warren P, Ryder G, Delano J, Pieters C, G. L (1991) Lunar Rocks. *In:* Lunar Sourcebook: A Users Guide to the Moon. Heiken GH, Vaniman DT, French BM (eds) Cambridge University Press, p 183-284

Taylor GJ, Wentworth S, Warner RD, Keil K (1978) Agglutinates as recorders of fossil soil compositions. Proc Lunar Planet Sci Conf 9:1959-1967

Taylor LA (1988) Generation of native Fe in lunar soil. Proceedings of the Space '88 Conference. American Society of Civil Engineers, Albuquerque, NM, p 67-77

Taylor LA, Chambers JG, Patchen A, Jerde EA, McKay DS, Graf J, Order RR (1993) Evaluation of lunar rocks and soils for resource utilization: detailed image analysis of raw material and beneficiated products. Lunar Planet Sci XXIV:1409-1410

Taylor LA, Kullerud G, Bryan WB (1971) Opaque mineralogy and textural features of Apollo 12 samples and a comparison with Apollo 11 rocks. Proc Lunar Sci Conf 2:855-871

Taylor LA, McCallister RH (1972) Experimental investigation of significance of zirconium partitioning in lunar ilmenite and ulvöspinel. Earth Planet Sci Lett 17:105-111

Taylor LA, McCallister RH, Sardi O (1973) Cooling histories of lunar rocks based on opaque mineral geothermometers. Proc Lunar Sci Conf 4:819-828

Taylor LA, Meek TT (2005) Microwave sintering of lunar soil: properties, theory, and practice. J Aerospace Engr 18: 188-196

Taylor LA, Morris RV, Keller LP, Pieters CM, Patchen A, Taylor DH, Wentworth SJ, McKay DS (2000) Major contributions to spectral reflectance opacity by non-agglutinitic, surface-correlated nanophase iron. Lunar Planet Sci XXXI:1842 (CD-ROM)

Taylor LA, Onorato PIK, Uhlmann DR, Coish RA (1978) Subophitic basalts from Mare Crisium: cooling rates. *In:* Mare Crisium: The View from Luna 24. Merrill RB, Papike JJ (eds) Pergamon Press, p 473-482

Taylor LA, Patchen A, Morris RV, Taylor D, Pieters CM, Keller LP, McKay DS, Wentworth S (1999) Chemical and mineralogical characterization of the 44-20, 20-10, and <10 micron fractions of lunar mare soils. Lunar Planet Sci XXX:1885 (CD-ROM)

Taylor LA, Patchen A, Taylor DS, Chambers JG, McKay DS (1996) X-ray digital imaging petrography of lunar soils: Modal analyses of minerals and glasses. Icarus 124:500-512

Taylor LA, Pieters C, Keller LP, Morris RV, McKay DS, Patchen A, Wentworth S (2001a) The effects of space weathering on Apollo 17 mare soils: petrographic and chemical characterization. Meteorit Planet Sci 36:285-299

Taylor LA, Pieters CM, Morris RV, Keller LP, McKay DS (2001b) Lunar mare soils: Space weathering and the major effects of surface-correlated nanophase Fe. J Geophys Res 106:27,985-28,000

Taylor LA, Pieters CM, Patchen A, Taylor D, Wentworth S, McKay D (1998) Optical properties and abundances of minerals and classes in the 10 to 44 size fraction of mare soils: Lunar Planet Sci XXIX:1160 (CD-ROM)

Taylor LA, Pieters CM, Patchen A, Taylor DH, Morris RV, Keller LP, McKay DS (2003) Mineralogical characterization of lunar highland soils. Lunar Planet Sci 34:1774 (CD-ROM)

Taylor LA, Shervais JW, Hunter RH, Laul JC (1983) Ancient (4.2 AE) highlands volcanism: the gabbronorite connection? Lunar Planet Sci 14:777-778

Taylor LA, Taylor DH (2000) Considerations for return to the Moon and lunar base site selection Workshops. J Aerospace Engineering 10:68-79

Taylor LA, Uhlmann DR, Hopper RW, Misra KC (1975) Absolute cooling rates of lunar rocks: theory and application. Proc Lunar Sci Conf 6:181-191

Taylor SR (1975) Lunar Science: A Post-Apollo View; Scientific Results and Insights from the Lunar Samples. Pergamon Press

Taylor SR (1982) Planetary Science: A Lunar Perspective. Lunar and Planetary Institute

Thalmann C, Eugster O, Herzog GF, Klein J, Krähenbühl U, Vogt S, Xue S (1996) History of lunar meteorites Queen Alexandra Range 93069, Asuka 881757, and Yamato 793169 based on noble gas isotopic abundances, radionuclide concentrations, and chemical composition. Meteorit Planet Sci 31:857-868

Thompson TW (1978) High resolution lunar radar map at 7.5 m wavelength. Icarus 36:174-188

Thompson TW (1987) High resolution lunar radar map at 70-cm wavelength. Earth Moon Planets 37:59-70

Thompson TW, Roberts WJ, Hartmann WK, Shorthill RW, Zisk SH (1979) Blocky craters: Implications about the lunar megaregolith. Moon and Planets 21:319-342

Toksöz MN, Dainty AM, Solomon SC, Anderson KR (1974) Structure of the Moon. Rev Geophys 12:539-567

Tompkins S, Pieters CM (1999) Mineralogy of the lunar crust: Results from Clementine. Meteorit Planet Sci 34:25-41

Tsay F-D, Chan SI, Manatt SL (1971a) Magnetic resonance studies of Apollo 11 and Apollo 12 samples. Proc Lunar Sci Conf 2:2515-2528

Tsay F-D, Chan SI, Manatt SL (1971b) Ferromagnetic resonance of lunar samples. Geochim Cosmochim Acta 35: 865-875

Tuniz C, Bird JR, Fink D, Herzog GF (1998) Accelerator Mass Spectrometry. CRC Press, Boca Raton, Florida

Tyler AL, Kozlowski RWH, Hunten DM (1988) Observations of sodium in the tenuous lunar atmosphere. Geophys Res Lett 15:1141-1144

Uhlmann DR, Onorato PIK, Yinnon H, Taylor LA (1979) Partitioning as a cooling rate indicator. Lunar Planet Sci X: 1253-1255

Usselman TM, Lofgren GE (1976) The phase relations, textures, and mineral chemistries of high titanium mare basalts as a function of oxygen fugacity and cooling rate. Proc Lunar Sci Conf 7:1345-1363

Usselman TM, Lofgren GE, Donaldson CH, Williams RJ (1975) Experimentally reproduced textures and mineral chemistries of high-titanium mare basalts. Proc Lunar Sci Conf 6:997-1020

Vaniman D, Heiken G, Mendell W, Olhoeft G, Reedy R (1991) The lunar environment. *In:* Lunar Sourcebook: A Users Guide to the Moon. Heiken GH, Vaniman DT, French BM (eds) Cambridge University Press, p 27-60

Vaniman DT, Bish DL (1990) Yoshiokaite, a new Ca,Al-silicate mineral from the Moon. Am Mineral 75:676-686

Vaniman DT, Papike JJ (1977) Very low Ti (VLT) basalts; a new mare rock type from the Apollo 17 drill core. Proc Lunar Sci Conf 8:1443-1471

Vaniman DT, Papike JJ (1980) Lunar highland melt rocks: Chemistry, petrology, and silicate mineralogy. *In:* Proceedings of the Conference on the Lunar Highland Crust. Papike JJ, Merrill RB (eds) Pergamon Press, p 271-337

Vasavada AR, Paige DA, Wood SE (1999) Near-surface temperatures on Mercury and the Moon and the stability of polar ice deposits. Icarus 141:179-193

Verani SC, Barbieri C, Benn CR, Cremonese G, Mendillo M (2001) The 1999 Quadrantids and the lunar Na atmosphere. Mon Not R Astron Soc 327:244-248

Verani SC, Benn C, Cremonese G (1998) Meteor stream effects on the lunar sodium atmosphere. Planet Space Sci 46: 1003-1006

Via WN, Taylor LA (1976) Chemical aspects of agglutinate formation; relationships between agglutinate composition and the composition of the bulk soil. Proc Lunar Sci Conf 7:393-403

Vilas F, Domingue DL, Jensen EA, McFadden LA, Coombs R, Mendell W (1999) Aqueous alteration on the Moon. Lunar Planet Sci XXX:1343

Vogt S, Herzog GF, Reedy RC (1990) Cosmogenic nuclides in extraterrestrial materials. Rev Geophys 28:253-275

Walker RJ, Papike JJ (1981) The relationship of the lunar regolith, <10 μm fraction and agglutinates; Part II, Chemical composition of agglutinate glass as a test of the "fusion of the finest fraction" (F (super 3)) model. Proc Lunar Planet Sci Conf 12:421-432

Warner JL, Phinney WC, Bickel CE, Simonds CH (1977) Feldspathic granulitic impactites and pre-final bombardment lunar evolution. Proc Lunar Sci Conf 8:2051-2066

Warren PH (1985) The magma ocean concept and lunar evolution. Ann Rev Earth Planet Sci 13:201-240

Warren PH (1988) The origin of pristine KREEP: effects of mixing between urKREEP and the magmas parental to the Mg-rich cumulates. Proc Lunar Planet Sci Conf 18:233-241

Warren PH (1990) Lunar anorthosites and the magma-ocean plagioclase-flotation hypothesis: importance of FeO enrichment in the parent magma. Am Mineral 75:46-58

Warren PH (1993) A concise compilation of petrologic information on possibly pristine nonmare Moon rocks. Am Mineral 78:360-376

Warren PH (1994) Lunar and martian meteorite delivery services. Icarus 111:338-363

Warren PH (2001a) Porosities of lunar meteorites: strength, porosity, and petrologic screening during the meteorite delivery process. J Geophys Res 106:10101-10111

Warren PH (2001b) Early lunar crustal genesis: the ferroan anorthosite epsilon-neodymium paradox. Meteorit Planet Sci 36:A 219

Warren PH (2005) "New" lunar meteorites: implications for composition of the global lunar surface, lunar crust, and the bulk Moon. Meteorit Planet Sci 40:477-506

Warren PH, Jerde EA, Kallemeyn GW (1987) Pristine Moon rocks: A "large" felsite and a metal-rich ferroan anorthosite. Proc Lunar Planet Sci Conf 17. J Geophys Res 92:E303-E313

Warren PH, Taylor GJ, Keil K, Marshall C, Wasson JT (1981) Foraging westward for pristine nonmare rocks; complications for petrogenetic models. Proc Lunar Planet Sci Conf 12:21-40

Warren PH, Taylor GJ, Keil K, Shirley DN, Wasson JT (1983) Petrology and chemistry of two "large" granite clasts from the Moon. Earth Planet Sci Lett 64:175-185

Warren PH, Wasson JT (1977) Pristine nonmare rocks and the nature of the lunar crust. Proc Lunar Sci Conf 8:2215-2235

Warren PH, Wasson JT (1978) Compositional-petrographic investigation of pristine nonmare rocks. Proc Lunar Planet Sci Conf 9:185-217

Warren PH, Wasson JT (1979a) The origin of KREEP. Rev Geophys 17:73-88

Warren PH, Wasson JT (1979b) The compositional-petrographic search for pristine nonmare rocks: Third foray. Proc Lunar Sci Conf 10:583-610

Warren PH, Wasson JT (1980) Further foraging for pristine nonmare rocks: Correlations between geochemistry and longitude. Proc Lunar Planet Sci Conf 11:431-470

Wasson JT, Boynton WV, Chou C-L, Baedecker PA (1975) Compositional evidence regarding the influx of interplanetary materials onto the lunar surface. Moon 13:121-141

Watson K, Murray BC, Brown H (1961) The behavior of volatiles on the lunar surface. J Geophys Res 66:3033

Weitz CM, Head JW, Pieters CM (1998) Lunar regional dark mantle deposits: geologic, multispectral, and modeling studies. J Geophys Res 103:22725-22759

Wentworth SJ, Keller LP, McKay DS, Morris RV (1999) Space weathering on the Moon: patina on Apollo 17 samples 75075 and 76015. Meteorit Planet Sci 34:593-603

Wentworth SJ, McKay DS, Lindstrom DJ, Basu A, Martinez RR, Bogard DD, Garrison DH (1994) Apollo 12 ropy glasses revisited. Meteorit Planet Sci 29:323-333

Whitaker EA (1972) Lunar color boundaries and their relationship to topographic features. The Moon 4:348-355

Wieczorek MA, Phillips RJ (2000) The Procellarum KREEP terrane: implications for mare volcanism and lunar evolution. J Geophys Res 105:20,417-20,430

Wieler R (1998) The solar noble gas record in lunar samples and meteorites. Space Sci Rev 85:303-314

Wieler R, Baur H (1995) Fractionation of Xe, Kr, and Ar in the solar corpuscular radiation deduced by closed system etching of lunar soils. Astrophys J 453:987-997

Wilcox BB, Lucey PG, T. CJ (2005) Space weathering and thermal properties of fresh craters on the Moon. Lunar Planet Sci XXXVI:2293 (CD-ROM)

Wilhelms DE (1987) The geologic history of the Moon. US Geol. Surv. Prof. Paper 1348

Wilhelms DE, McCauley JF (1971) Geologic map of the near side of the moon. Miscellaneous geologic investigations; USGS Map I-703, p. 1 map. U.S. Geological Survey, Washington, D.C.

Williams KL, Taylor LA (1974) Optical properties and chemical compositions of Apollo 17 armalcolites. Geology 2:5-8

Wilson JK, Smith SM, Baumgardner J, Mendillo M (1999) Modeling an enhancement of the extended lunar atmosphere during the Leonid meteor shower of 1998. Geophys Res Lett 26:1645-1648

Wlotzka F, Spettel B, Wänke H (1973) On the composition of metal from Apollo 16 fines and the meteoritic component. Proc Lunar Sci Conf 4:1483-1491

Wood RW (1912) Selective absorption of light on the Moon's surface and lunar petrography. Astrophys J 36:75

Yakshinskiy BV, Madey TE (2000a) Desorption induced by electronic transitions of Na from SiO_2: relevance to planetary exospheres. Surf Sci 451:160–165

Yakshinskiy BV, Madey TE, Ageev VN (2000b) Thermal desorption of sodium atoms from thin SiO_2 films. Surf Rev Lett 7:75-87

Yin L, Trombka JI, Adler I, Bielefeld M (1993) X-ray remote sensing techniques for geochemical analysis of planetary surfaces. *In:* Remote Geochemical Analysis: Elemental and Mineralogical Composition. Pieters CM, Englert PAJ (eds) Cambridge Univ. Press, p 199-212

Zisk SH, Hodges CA, Moore HJ, Shorthill RW, Thompson TW, Whitaker EA, Wilhelms DE (1977) The Aristarchus-Harbinger region of the Moon: Surface geology and history from recent remote sensing observations. The Moon 17:59-99

Zisk SH, Pettengill GH, Catuna GW (1974) High-resolution radar maps of the lunar surface at 3.8-cm wavelength. The Moon 10:17-50Campbell HW, Hess PC, Rutherford MJ (1978) Ilmenite crystallization in non-mare basalts. Lunar Planet Sci IX:149-151

Reviews in Mineralogy & Geochemistry
Vol. 60, pp. 221-364, 2006
Copyright © Mineralogical Society of America

3

The Constitution and Structure of the Lunar Interior

Mark A. Wieczorek[1], Bradley L. Jolliff[2], Amir Khan[3], Matthew E. Pritchard[4], Benjamin P. Weiss[5], James G. Williams[6], Lon L. Hood[7], Kevin Righter[8], Clive R. Neal[9], Charles K. Shearer[10], I. Stewart McCallum[11], Stephanie Tompkins[12], B. Ray Hawke[13], Chris Peterson[13], Jeffrey J. Gillis[13], Ben Bussey[14]

[1]*Institut de Physique du Globe de Paris, Saint Maur, France*
[2]*Washington University, St. Louis, Missouri, U.S.A.*
[3]*Niels Bohr Institute, University of Copenhagen, Denmark*
[4]*Cornell University, Ithaca, New York, U.S.A.*
[5]*Massachusetts Institute of Technology, Cambridge, Massachusetts, U.S.A.*
[6]*Jet Propulsion Laboratory, California Institute of Technology, Pasadena, CA, U.S.A.*
[7]*University of Arizona, Tucson, Arizona, U.S.A.*
[8]*Astromaterials Branch, Johnson Space Center, Houston, Texas, U.S.A.*
[9]*University of Notre Dame, Notre Dame, Indiana, U.S.A.*
[10]*University of New Mexico, Albuquerque, New Mexico, U.S.A.*
[11]*University of Washington, Seattle, Washington, U.S.A.*
[12]*Science Applications International Corporation, Chantilly, Virginia, U.S.A.*
[13]*University of Hawaii, Honolulu, Hawaii, U.S.A.*
[14]*Applied Physics Laboratory, Laurel, Maryland, U.S.A.*

e-mail: wieczor@ipgp.jussieu.fr

"This picture, though internally consistent, is subject to revision or rejection on the basis of better data." Don Wilhelms (1987) *The Geologic History of the Moon*, USGS Prof. Paper 1348.

1. INTRODUCTION

The current state of understanding of the lunar interior is the sum of nearly four decades of work and a range of exploration programs spanning that same time period. Missions of the 1960s including the Rangers, Surveyors, and Lunar Orbiters, as well as Earth-based telescopic studies, laid the groundwork for the Apollo program and provided a basic understanding of the surface, its stratigraphy, and chronology. Through a combination of remote sensing, surface exploration, and sample return, the Apollo missions provided a general picture of the lunar interior and spawned the concept of the lunar magma ocean. In particular, the discovery of anorthite clasts in the returned samples led to the view that a large portion of the Moon was initially molten, and that crystallization of this magma ocean gave rise to mafic cumulates that make up the mantle, and plagioclase flotation cumulates that make up the crust (Smith et al. 1970; Wood et al. 1970). This model is now generally accepted and is the framework that unifies our knowledge of the structure and composition of the Moon. The intention of this chapter is to review the major advances that have been made over the past decade regarding the constitution of the Moon's interior. Much of this new knowledge is a direct result of data acquired from the successful Clementine and Lunar Prospector missions, as well as the analysis of new lunar meteorites. As will be seen, results from these studies have led to many fundamental amendments to the magma ocean model.

1529-6466/06/0060-0003$15.00　　　　　　　　　　　　　　　　DOI: 10.2138/rmg.2006.60.3

Much of what we know from sample analyses has been previously summarized elsewhere, and only their most important aspects will be discussed in this chapter. The reader is referred to the relevant chapters in the books *Basaltic Volcanism on the Terrestrial Planets* (Basaltic Volcanism Study Project 1981), *The Lunar Sourcebook* (Heiken et al. 1991), and *Planetary Materials* (Papike et al. 1998) for more in-depth assessments. Fortunately, the lunar samples were curated carefully and with forethought as to the possibility that there might be no sample-return missions for a long time after Apollo. Thus, as new analytical methods are developed and old ones are improved, analyses of the samples continue to yield important results.

Much of our geophysical knowledge of the Moon comes from instruments that were deployed during the manned Apollo landings. As part of the Apollo Lunar Surface Experiments Package (ALSEP), Surface magnetometers, heat-flow probes, and seismometers operated until 1977 when the transmission of data was terminated. While re-analyses of these data occasionally yield surprises (such as the Apollo seismic data), many aspects of the pre-1990s geophysical reviews are just as relevant today as when they were written. The electromagnetic sounding data of the Apollo era is thoroughly reviewed by Sonnett (1982), much of the geophysics as reviewed by Hood (1986) is still highly relevant, and the paleomagnetism of the lunar samples has been comprehensively reviewed by Fuller and Cisowski (1987) and Collinson (1993). One suite of passive surface instruments deployed during the Apollo and Russian Luna missions that is still operational are the corner-cube retroreflectors. Laser ranging to these stations continues to the present day, and with the gradual accumulation of data, our knowledge of the deep lunar interior is slowly being revealed. The results of this experiment up to 1994 have been reviewed by Dickey et al. (1994).

The remote sensing missions of the 1990s—Galileo, Clementine, and Lunar Prospector—for the first time obtained near-global compositional data sets that enable us to assess the nature of materials at the surface of the Moon. One of the major results of global remote sensing has been to locate regions of the lunar surface that differ from the Apollo landing sites and that are difficult to explain given the known compositions of the present lunar sample collection. In particular, the floor of the South Pole-Aitken basin, which is the largest recognized impact structure in the solar system, may contain rocks that formed deep in the crust and as yet, not sampled elsewhere. The mineralogy and petrology of known rock types, nevertheless, provide constraints on lunar petrogenetic processes, which then allow informed extrapolation to those areas that were not directly sampled. Near-global topography, magnetic data, and improved models of the lunar gravity field from these missions have further offered vastly improved datasets in comparison to the spatially limited Apollo observations. By combining general characteristics of the lunar samples and surface-derived geophysical constraints with these near-global orbital data, our understanding of the Moon's interior structure has been much improved over the general sketch provided during the Apollo era.

The Apollo and Luna samples were found to contain many different rock components brought together by impact processes. Any given sample of soil or breccia thus might contain a much wider diversity of rock types than whatever the local bedrock material might be. In fact, because of the potentially widespread effects of mixing together of widely separated materials by large impact events, a general (although incorrect) argument could be made that the Moon was adequately sampled by the Apollo and Luna missions. The discovery of lunar meteorites (Warren 1994), coupled with telescopic and global remote sensing, prove this argument wrong. Furthermore, recent studies of basin-ejecta deposits and deposition processes suggest that most of the Apollo sites were strongly influenced by material derived from one or a few late basin-forming events, especially Imbrium (e.g., Haskin 1998; Haskin et al. 1998). Thus, the feldspathic lunar meteorites provide our best samples of the lunar highland crust far removed from the effects of contamination by Imbrium ejecta. Several of the lunar meteorites also provide samples of basalt that differ from any of the basalt types known from the Apollo and Luna suites.

The information presented and discussed in this chapter is set within the context of the new datasets. Key constraints based upon the samples, Apollo-era geophysical experiments, and pre-1990s remote sensing are reviewed and then coupled with recent results to reassess the makeup and structure of the lunar interior. One of the recurring themes of this chapter will be to demonstrate that the traditional "pie-wedge" cross-sectional view of the lunar interior can no longer be considered as being globally representative. Instead, the lunar crust and underlying mantle appear to be best characterized as being composed of discreet geologic terranes, with each possessing a unique composition, origin, and geologic evolution. This concept is best illustrated by the gamma-ray spectrometer data obtained from the Lunar Prospector mission, which shows that incompatible and heat-producing elements are highly concentrated in a single region of the Moon that was once volcanically very active. This recognition has led many workers to reassess the significance of the data derived from the Apollo era, particularly in light of the fact that all six Apollo landing sites straddle the border between two of the most distinctive terranes—the Procellarum KREEP Terrane and the Feldspathic Highlands Terrane (e.g., Jolliff et al. 2000a). The fact that a body as small as our Moon possesses vastly different geologic crustal provinces should be borne in mind when one attempts to assess the geologic history of a less understood and larger planetary body, such as Mercury and Mars.

Because the concept of a magma ocean is so thoroughly ingrained in modern notions of how lunar rocks types are distributed as a function of depth in the crust and upper mantle (and for good reasons), we use it as a general framework for discussion. We do not maintain that we understand how it originated or solidified in detail, nor how deep it might have been and how it might have evolved. Indeed, fundamental questions remain that relate to these topics, and some of these are addressed in more detail later in this chapter and in Chapter 4 of this book. Nevertheless, we begin this chapter with a very brief summary of some of the key implications that a magma ocean has for the makeup and structure of the Moon, and what types of question we might address within the context of this model. From this point, we sequentially review our current knowledge of the crust, mantle, and core of the Moon (Sections 2, 3, and 4). In Section 5, we emphasize the concept that the lunar crust and mantle are best understood in terms of unique geologic terranes. Following this, we address some of the fundamental unanswered questions in lunar science dealing with the lunar interior, and end with a discussion of how these may be addressed in future studies and exploration.

1.1. The magma ocean model and its many questions

Since the time of the first Apollo sample studies, the oldest lunar crust was correctly inferred to be rich in plagioclase (Smith et al. 1970; Wood et al. 1970). Assuming an average crustal thickness of ~60 km and a bulk Moon aluminum concentration similar to the Earth's mantle, a significant differentiation event must have occurred in the Moon's early evolution to have formed this crust (e.g., Warren 1985). On the basis of petrologic experiments, the mare basalts were shown to have been derived from an ultramafic pyroxene and olivine source. Trace-element geochemistry, especially rare earth element (REE) patterns, has further shown that the ultramafic sources of the mare basalts are complementary to the feldspathic crust.

To explain these observations, the global magma ocean (or magmasphere) hypothesis was developed (e.g., Smith et al. 1970; Wood et al. 1970; Warren and Wasson 1977, 1979b; Warren 1985). In its simplest form, this hypothesis postulated that the magma ocean differentiated during crystallization to form a dense, ultramafic mantle rich in olivine and pyroxene that was overlain by a buoyant, globe-encircling, plagioclase-rich crust. This hypothesis has served well because it accommodates many remotely sensed observations as well as much of the data collected on the samples returned by the Apollo and Luna missions. In addition, the magma-ocean hypothesis is consistent with a hot accretion of the Moon as would be expected if it formed as a result of a giant impact between the early Earth and a Mars-sized object (e.g., Pritchard and Stevenson 2000) (see Chapter 4).

Estimated depths of the primordial magmasphere range from whole Moon melting to thin melt layers above partially molten zones. Deep magma oceans are currently favored in the literature, and a popular hypothesis is that a seismic discontinuity located about 500 km below the surface might represent this maximum depth of melting. Whether the middle and lower mantle escaped melting and differentiation remains one of the key unknowns regarding the Moon's evolution. Regardless of its initial depth, isotopic evidence and thermal considerations suggest that the materials that make up the lunar mantle crystallized within a period of about 30 m.y. (Lee et al. 2002; Shearer and Righter 2003) to 200 m.y. (Solomon and Longhi 1977; Nyquist and Shih 1992).

Several thermal and petrologic models have been developed that track the compositional evolution of a crystallizing magma ocean (see Shearer and Papike 1999 and references therein). In general, the first mineral to crystallize and sink would have been a magnesium-rich olivine, and as crystallization proceeded the cumulus phases would have become more iron rich. After about 75% of the magma ocean had crystallized, plagioclase would have become a liquidus phase, and because of its low density, it would have been buoyant with a tendency to rise. After the magma ocean was about 90% crystallized, dense Ti-bearing phases (ilmenite, armalcolite) would have crystallized. In all models, efficient fractional crystallization would produce a chemically evolved, late-stage residual melt referred to as urKREEP (Warren and Wasson 1979b) that was initially sandwiched between the crust and mantle. This residual magma would have been extremely enriched in iron and the incompatible and heat-producing elements that are represented by the acronym KREEP (Potassium, Rare Earth Elements, and Phosphorous). Because of the high concentration of heat-producing elements, concentrations of this material at depth could remain in a molten phase for many hundreds of millions of years (Solomon and Longhi 1977). Somewhat paradoxically, though, samples directly related to urKREEP have not yet been identified, though KREEP-rich samples that are rich in magnesium have. Although the high concentrations of incompatible elements in these rocks are consistent with a late-stage magma ocean origin, their high magnesium concentrations remain enigmatic. The origin of these magnesian KREEP-rich samples remains one of the major unresolved problems in understanding the Moon's magmatic evolution.

In simple models of magma-ocean crystallization, depending upon the density of the cumulate phases, they are either sequentially laid down at the base of the magma ocean, or at the base of the crust. Because the mantle cumulates become increasingly rich in iron with the progress of crystallization (culminating in dense ferropyroxene and ilmenite-rich cumulates), this mantle cumulate pile would have been gravitationally unstable. Several studies have shown that the late-stage ilmenite cumulates should have sunk through the mantle. Others have predicted that the deep, olivine-rich mantle cumulates should have participated in the overturn as well, bringing early-crystallized magnesium-rich olivine to the upper mantle (see Chapter 4). A key question regarding these overturn models is their timing, extent, and efficiency. In particular, did this event act to homogenize the composition of the mantle? Or was mixing inefficient such that any original stratification within the magma-ocean cumulates might still be present, though in an inverted sequence? Did such overturn occur globally, or was it localized in one or several regions of the mantle? Chapter 4 considers the internal evolution of the Moon in great detail; in this chapter, we mainly focus on sample, remote sensing, and geophysical evidence for the present-day lithology and structure of the interior.

The compositional stratigraphy that results from a crystallizing magma ocean has commonly been illustrated by the use of simple circular cross sections, or 1-dimensional pie-wedge diagrams (for a recent example, see Fig. 3.1). Implicit in such drawings is the assumption that early lunar differentiation was spherically symmetrical and that any asymmetry was imparted by later basin-forming impact processes. This view of an initially laterally uniform internal composition (and hence laterally uniform internal processes) was widely accepted

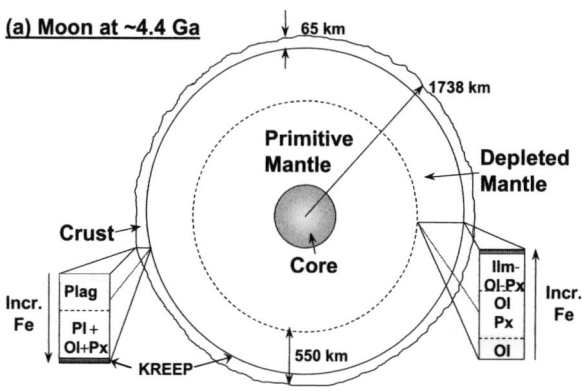

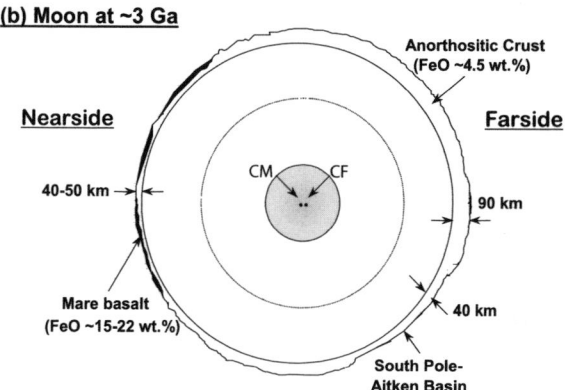

Figure 3.1. One possible interpretation of the Moon's internal structure (a) just after magma-ocean crystallization and (b) near the end of mare basaltic volcanism. With the exception of a hemispheric dichotomy in crustal thickness and mare basaltic volcanism, this model, like many pre-Lunar Prospector models, represents the crust and mantle as being laterally uniform in composition. (a) The lunar magma ocean is assumed to have a depth of 550 km, the lower mantle is composed of "primitive" unmelted materials, and a small core is assumed. The sequences of major mineralogy as a function of radius in the mantle (olivine → olivine + pyroxene → ilmenite + olivine + pyroxene) and crust (plagioclase-rich in the upper crust and plagioclase + pyroxene + olivine in the lower crust), as well as the existence of a global KREEP-rich layer at the crust-mantle boundary, are a result of a fractionally crystallizing magma ocean. Complex processes, such as mantle overturn and asymmetric solidification of the magma ocean, are not considered in this model. (b) The lunar interior at ~3 Ga, emphasizing the nearside-farside dichotomy in both the distribution of mare basalts and the thickness of the lunar crust. (CM and CF represent the center of mass and center of figure, respectively, which are offset from each other by about 2 km). Compare with the more recent interpretations presented in Figure 3.27. [Used by permission of Springer, from McCallum (2001), *Earth Moon Planets*, Vol. 85-86, Figs. 2 and 4, pp. 256 and 260.]

until the acquisition of thorium surface concentrations by the Lunar Prospector spacecraft. It now appears that the Moon possesses a fundamental asymmetry, stemming from the time of initial differentiation, with KREEP-rich rocks and mare volcanism having been concentrated on the Moon's Earth-facing hemisphere. Curiously, the elevations of the nearside hemisphere are about 2 km less than that of the far side. No matter the origin of the asymmetric distribution of KREEP-rich rocks, it appears that their associated high heat production fundamentally affected the post-magma-ocean magmatic history of the Moon's near side.

The above description of the magma-ocean model is admittedly simplified and is discussed in far more detail in Chapter 4. Nonetheless, the model in its generic form has held up remarkably well over the past thirty or so years. The concept of the Moon possessing distinct geologic terranes is the most recent contribution to the lunar magma-ocean concept, and this will be more fully discussed in Section 5.

2. THE CRUST

The lunar crust provides us with the most direct evidence for a major differentiation event early in the Moon's history. Because the crust is easily accessible to both remote-sensing techniques and sample analyses, much is now known of its composition and the processes by which it formed. Although these two lines of study are only applicable to rocks that now reside at the surface, processes associated with impact cratering have brought once deep-seated rocks to the surface, allowing the deep crust to be investigated at certain locales as well. Geophysical investigations, including analysis of the Apollo seismic data and the Moon's gravity and topography, are also sensitive to properties of the crust.

The most important questions that will be addressed in this section are related to the composition of the crust, any lateral or vertical compositional variations that might be present, and its thickness. When used in combination, these pieces of information can be used to constrain the bulk composition of the crust, and hence place constraints on its mode of origin and the extent of lunar differentiation. At the end of this section a brief review will be given of our current knowledge of the lunar crust.

2.1. Samples of the crust

Rock samples collected during the Apollo and Luna missions and the identification and characterization of lunar meteorites provide first-hand information about the mineralogy and lithology of the Moon's principal crustal rock types. These can be related to remotely sensed information, and their global distribution (lateral and vertical) can be inferred through consideration of mixing models (chemical and physical) and geologic processes. The samples include several varieties of intrusive igneous rocks, basaltic lavas and volcanic glasses, and a wide range of breccias formed by impact melting and ballistic sedimentation. Even though the samples represent a spatially restricted subset of lunar rocks, there are good reasons to think that the existing rock suites represent fairly well the global crustal rock types with but a few exceptions. First, because of the lateral and vertical redistribution of materials by large impacts, and the assembly of diverse rock components into impact breccias, many lithologies of potentially widely separated sources may be combined as clasts into a single breccia. The same argument applies to the diversity of materials and potential sources for small rock fragments found in the unconsolidated regolith. Second, because the Moon is essentially anhydrous, and there are no aqueous sedimentary or alteration processes to mechanically sort minerals, the mineralogy of lunar rocks is greatly restricted (Chapter 2). Third, even though large impacts may have exhumed deeply buried rocks, high-pressure mineralogy is restricted because of the low lunar pressure gradient (~ 5.3 MPa per km or 1 kbar per 19 km). Fourth, because the Moon cooled early in its history compared to larger bodies such as Mars, Venus, and Earth, most lunar rocks have not been extensively reprocessed (metamorphosed or melted repeatedly) except as associated with impacts. Thus, big surprises in terms of new lunar rock types are not likely to be encountered in future exploration. As discussed below, however, rocks from deep within the lunar crust (and mantle, as in mantle xenoliths) are not well represented among the samples (or not at all) and the full range of basalt types has not yet been sampled.

The Lunar Sourcebook subdivided lunar igneous rocks into "mare basaltic lavas and related volcanic rocks" and "pristine highland rocks" (Taylor et al. 1991). This scheme was generally followed in *Planetary Materials* (Papike et al. 1998). It is not our intent here to revise the classification scheme. Instead, the focus is on the makeup of the crust and mantle of the Moon and on relating rock types to geochemical or mineralogical signatures at the surface of the Moon (Chapter 2). Thus, while the grouping of major rock types in this chapter differs somewhat from groupings associated with some previous classifications, it is essentially the same as that adopted by Warren (1993). This section focuses on the *crustal* rock types, including the ferroan anorthositic, magnesian and alkalic intrusive suites, KREEP basalts, and polymict breccias produced by impact mixing of the primary rock types. Mare basalts and volcanic glasses are covered in more detail as samples of the mantle in Section 3.1 and in Chapter 4. Herein, we refer to the glasses as "volcanic glasses," but this is essentially the same as referred to elsewhere as "pyroclastic" glasses, and in those cases where MgO contents are sufficiently high (~15 wt% or greater), we also use the term "picritic" glasses.

The crustal igneous rocks, also referred to as "pristine igneous" rocks (Warren 1993; Papike et al. 1998), are almost exclusively plagioclase rich, and the plagioclase is almost exclusively calcic, with anorthite (An) contents typically much greater than 50%, and most greater than 90% (Fig. 3.2). The major mafic silicate minerals are pyroxene and olivine, with pyroxene dominated by the orthorhombic, low-Ca variety. Thus the major rock types are anorthosite (sensu stricto >90 vol% plagioclase), norite, troctolite, and rarely (at least among the samples) gabbro (see Table 3.1). The nomenclature of Stöffler et al. (1980) incorporates intermediate rock types and is followed here. Rarer rock types include spinel troctolite, feldspathic lherzolite, dunite, granite or felsite (fine-grained texture, granitic assemblage), monzogabbro, and quartz monzodiorite (QMD). Although the plagioclase anorthite (An) content typically exceeds 50%, which dictates use of the term "gabbro" (Streckeisen 1976; Le Maitre 1989), "quartz monzodiorite" has heritage in previous lunar studies, and thus the term is retained for lunar samples that have been described previously as QMD. The use of the word "pristine" refers to the set of characteristics that distinguish a rock as having originated by magmatic processes related to the intrinsic thermal properties of the Moon and not as a result of impact-induced melting. The key distinguishing characteristic is very low siderophile-element concentrations because these elements are added in high concentration by most impacting materials and provide an effective tracer of the impact process. Other characteristics include compositions that are consistent with an igneous process as opposed to mixtures of disparate rock types of known composition. The pristine designation represents a means to restrict the lunar igneous rock sets to those rocks that truly formed by internally generated magma processes. The distinction, however, becomes blurred in light of the possibility that large impact-melt bodies associated with the largest impact basins may have been many kilometers thick and may have differentiated (e.g., Sudbury, see Chapter 5). In zones that cooled slowly and where gravitational differentiation occurred, sinking metal might have effectively scavenged siderophile trace elements. Recognizing a sample as being pristine is important in order to assess processes related to the internal igneous differentiation of the Moon.

The crustal igneous rocks have been subdivided into suites according to their major mineral compositions, and in some cases other characteristics such as trace-element signatures and isotopic systematics correspond to these subdivisions. Objections to calling the subdivisions "suites" may be raised on the grounds that the term "suite" implies a petrogenetic relationship (Papike et al. 1998) and such may not be the case. Here, the term "suite" is retained for major groupings and objections are addressed in the following sections. The ferroan-anorthositic suite includes those rocks whose plagioclase is highly calcic (An content generally >94) and whose mafic silicates, either pyroxene or olivine, have relatively high Fe/Mg values (Fig. 3.2). The magnesian suite includes rock types whose mafic silicate minerals have high values of Mg/Fe and whose plagioclase compositions cover a range of An contents that are correlated with their Mg/Fe values. Rocks of the magnesian suite have high values of Mg/Fe relative to their An contents, but they are not all necessarily "magnesium rich" (see rock compositions in the online RIMG supporting materials at *www.minsocam.org*, Tables A3.1–A3.12). The

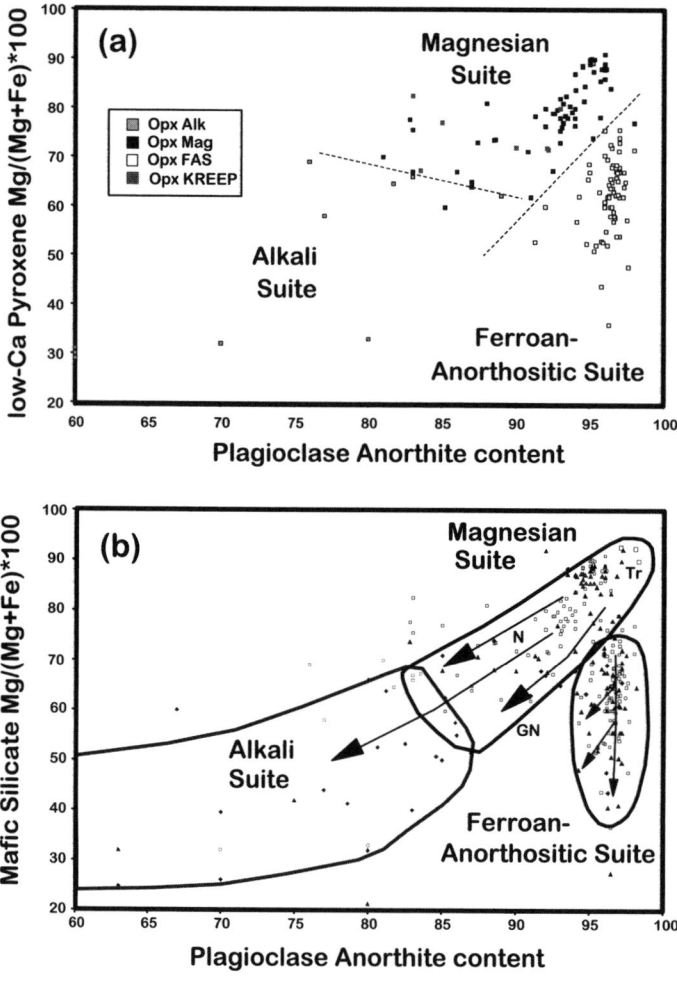

Figure 3.2. Mg/(Mg+Fe) in mafic silicates vs. anorthite content, Ca/(Ca+Na+K), of coexisting plagioclase in lunar crustal igneous rocks. (a) Orthopyroxene Mg/(Mg+Fe) vs. coexisting plagioclase An content, showing rock types divided into a ferroan-anorthositic suite (high An content and relatively ferroan orthopyroxene), magnesian suite (high Mg at high An, ranging to lower Mg at lower An), and alkali suite (plagioclase relatively enriched in Na and K, and relatively ferroan mafic silicates). Dashed lines represent values typically used to distinguish groups. Data are from the compilation by Warren (1993), supplemented with additional data from Papike et al. (1998). (b) Mg/(Mg+Fe) for all mafic silicates (olivine, low- and high-Ca pyroxene) vs. coexisting plagioclase An content for the intrusive igneous rock suites and KREEP basalts. Typical trends resulting from fractional crystallization (arrows) suggest a possible relationship between the magnesian- and alkali-suite rocks. The overall trend from upper right to lower left is similar to trends observed for rocks of terrestrial layered mafic intrusive bodies and is the general trend of fractionation of the minerals of KREEP basalt as determined from early crystallizing cores to late rims. The ferroan-anorthositic and Mg suites are difficult to relate through a common magmatic process. The near-vertical trend of the ferroan-anorthositic suite requires a more complex petrogenetic scenario (Herbert et al. 1978; Longhi and Boudreau 1979; Raedeke and McCallum 1980). Specific groups within the magnesian suite such as the norites (N) and gabbronorites (GN) appear to differ petrogenetically from each other and thus are shown associated with different fractionation trends (arrows) (e.g., James and Flohr 1983). Mineral compositions of troctolite (Tr) tend to be at the magnesian and anorthitic end of the trend.

Table 3.1. Nonmare pristine rocks according to mass and landing site, Apollo samples.

Igneous Rock Suite		grams	A11	A12	A14	A15	A16	A17
Ferroan	*sum, >1 g, pristinity >5*	9474	0	0	0	280	9186	8
Anorthosite	anorthosite	9174	0	0	0	279	8887	8
Suite	noritic anorthosite	174	0	0	0	1.1	173	0
	troctolitic anorthosite	6	0	0	0	0	6	0
	anorthositic norite	120	0	0	0	0	120	0
Magnesian	*sum, all listed samples*	2593	0	1.4	17	222	10	2342
Suite	norites (> 0.5 g)	1563	0	0	1.6	212	0	1350
	gabbronorite	310	0	0	0.4	0	8	301
	troctolite	656	0	0	14	3	0.7	636
	spinel-troctolite	9.2	0	1.3	0.01	7.5	0.4	*10*
	ultramafic	56	0	0.1	0.4	0	0	55.2
Alkali Suite	*sum, all listed samples*	14.3	0	1.7	7.8	3.7	0.5	0.6
	Alkali anorthosite	3.9	0	0.45	3.33	0.10	0	0
	alkali norite	2.8	0	0.08	1.60	0.04	0.45	0.60
	alkali gabbronorite	0.2	0	0	0.23	0	0	0
	alkali troctolitic anorth.	0.2	0	0	0.23	0	0	0
	felsite (granite)	3.6	0	1.20	2.32	0.01	0	0.02
	monzogabbro (QMD)	3.6	0	0	0.06	3.54	0	0
KREEP Basalt	sum, all listed samples	17.9	0	0	0.02	15.1	0.04	2.7

Footnotes: Values are derived from data in Planetary Materials, RiM-G Vol. 36 (Papike et al. 1998). Pristinity refers to grading scale as developed and applied by Warren (1993). Spinel troctolite in Apollo 17 samples estimated from descriptions of large clasts in 72395 and 72435 (Dymek et al. 1976).

"alkali" suite includes rock types for which Na and K contents are high compared to those of the magnesian suite. A growing body of evidence points to a relationship between KREEP basalt and members of the magnesian and alkali suites (Jolliff et al. 1993; Snyder et al. 1995b), thus these are discussed in succession. The following sections refer to Table 3.1 to illustrate the distribution of igneous rock types as a function of landing site and, broadly, their spatial location and abundance relative to major geochemical provinces or "terranes." This table contains a sample selection bias in that it derives from well-known (published) analyzed samples, thus igneous rocks that occur as clasts in breccias are under-represented. However, the compilation points to trends that appear to be robust, such as the high concentration of ferroan anorthosite at Apollo 16. When combined with global remotely sensed data and data from the lunar meteorites, information derived from this simple tabulation provides important constraints for the distribution of these important rock types.

In the following sections, descriptions of rock distribution and chemical compositions are intended to focus on those aspects relating to the integration of sample and remote-sensing information. For example, chemical compositions are discussed according to their FeO, TiO_2, and Th content so as to highlight possible relationships to global compositional maps and variations in concentrations of these relatively well-determined elements.

2.1.1. Ferroan-anorthositic suite. The ferroan-anorthositic rocks are chemically and mineralogically distinct from other lunar igneous rock groups (see also Chapter 2). Although rare in the Apollo 11, 12, 14, and 17 samples, rocks of this suite were found among the Apollo 15 samples and they dominate the Apollo 16 and Luna 20 samples. The two key aspects of the ferroan-anorthositic suite (FAS) embodied in the name provides the motivation for grouping these rocks together and for inferring a specific petrologic relationship. First, the majority of samples of FAS rocks are highly anorthositic, that is they contain very little other than plagioclase. Of the large rocks listed in the compilation of Warren (1993), only a handful of the FAS rocks contain more than 10% of the mafic silicates pyroxene and olivine. The FAS rocks contain few to none of the common minor or accessory minerals such as ilmenite, spinel, phosphates, or zircon. The modal mineralogy of the FAS rocks coupled with relatively coarse relict grain size and cumulus texture, where such texture is preserved, clearly indicate formation by a cumulus igneous process. The very high plagioclase content of the FAS rocks (>90%) is a defining characteristic of the suite. Nevertheless, at the end of this section we review the possibility that not all FAS rocks throughout the crust, laterally and with depth, are so highly anorthositic.

The second major aspect embodied in the title of "ferroan anorthositic suite" is that the mafic silicates are ferroan in composition, that is, the mafic silicates have a relatively low (atomic) $Mg/(Mg+Fe)$ ratio. This characteristic is not simply a matter of definition because among the demonstrably igneous lunar rocks, there exists a gap in the $Mg/(Mg+Fe)$ values at high plagioclase An content (see Fig. 3.2a). Furthermore, several other key geochemical signatures discussed below correspond to the low $Mg/(Mg+Fe)$ values. There is petrologic significance to the combination of highly calcic plagioclase (anorthite) and ferroan mafic silicates. Experience from terrestrial layered mafic rock bodies suggests that simple igneous differentiation that produces rocks such as troctolite, norite, and gabbro would give a trend with a positive slope on a plot of the $Mg/(Mg+Fe)$ ratio of coexisting mafic silicates vs. plagioclase An content, such as shown in Figure 3.2b. As crystallization proceeds, plagioclase compositions progress from calcic to sodic and coexisting mafic silicate compositions, from magnesian to ferroan. What sets the FAS rocks apart is the consistently high Ca content of plagioclase with little regard for variations in the $Mg/(Mg+Fe)$ values of the coexisting mafic silicates. Such a trend (or lack of a trend) requires special petrogenetic explanations. Raedeke and McCallum (1980) showed that rocks of the terrestrial Stillwater Complex, which includes thick anorthosite layers, also show a vertical trend at high An, similar to the lunar FAS rocks. They attributed these trends to a process of equilibrium crystallization of intercumulus melt in a plagioclase-rich mush. Longhi and Ashwal (1985) proposed a two-stage formation and emplacement process that would further explain the separation of plagioclase-rich rocks from co-crystallizing mafic silicates and trapped melt. In general, this characteristic of FAS rocks and the history and processes of their emplacement have been the focus of much study (e.g., Longhi 1977, 1980; Longhi and Boudreau 1979; Raedeke and McCallum 1980; Longhi and Ashwal 1985; James et al. 1989; Jolliff and Haskin 1995; Floss et al. 1998) and remain among the outstanding questions of lunar petrology.

Mineralogy and mineral modes. Most of the FAS samples are indeed rich in plagioclase, with a mass weighted average of about 96%. However, some rocks consist of over 20% pyroxene (the range for samples greater than one gram is 77–99% plagioclase). Pyroxene is typically more abundant than olivine in the FAS rocks and, in some samples, high-Ca pyroxene is present as well as pigeonite and hypersthene (see Table 3.2). Because of ferroan bulk compositions, pigeonite (or inverted pigeonite, see below) is also common. At more magnesian compositions [$Mg/(Mg+Fe) > 0.7$], high- and low-Ca pyroxenes are found, but not pigeonite. Of the large rock samples, only one is olivine rich (62237; 62 g). The major mineral assemblage is important because it constrains the composition of the melt that crystallized these minerals to lie generally along the plagioclase-pyroxene cotectic. This constraint in turn is important for petrologic models of the crystallization of magma systems to produce FAS rocks. According to models of magma-ocean crystallization, much of the magma ocean solidified as olivine plus pyroxene cumulates, but by the time plagioclase began to crystallize, the dominant mafic silicate would have been pyroxene, not olivine.

Table 3.2. List of ferroan anorthosites with mass exceeding 1 gram.

Sample	Rock type	estimated mass (g)	~modal % feldspar	weighted % feldspar		Main reference
				all	**w/o 60015**	
60015	anorthosite	4600	99	48.069		Ryder & Norman (1979, 1980)
60025	anorthosite	1836	90	17.442	33.904	James et al. (1991)
62255	anorthosite	800	97	8.191	15.922	Ryder and Norman (1979)
62275	anorthosite	443	93	4.349	8.453	Warren et al. (1983a)
60215c30	anorthosite	300	97	3.072	5.971	Rose et al. (1975)
61015	anorthosite	300	96	3.040	5.909	James et al. (1984)
65315	anorthosite	285	99	2.963	5.760	Ebihara et al. (1992)
15415	anorthosite	269	99	2.811	5.464	Ryder (1985)
60135	norite	120	77	0.975	1.896	Ryder and Norman (1980)
64435c210A	anorthosite	100	98	1.034	2.011	James et al. (1989)
65325	anorthosite	65	99	0.676	1.314	Warren and Wasson (1978)
62237	noritic anorthosite	62	85	0.560	1.088	Ebihara et al. (1992)
62236	noritic anorthosite	57	86	0.520	1.011	Nord and Wandless (1983)
67075c17	anorthosite	50	96	0.507	0.985	Haskin et al. (1973)
67915c12-1	noritic anorthosite	50	85	0.449	0.872	Taylor and Mosie (1979)
60055	anorthosite	36	98	0.367	0.714	Ryder and Norman (1980)
60515	anorthosite	17	95	0.170	0.331	Warren et al. (1983a)
60056	anorthosite	16	95	0.160	0.312	Warren et al. (1983a)
60639c19	anorthosite	10	99	0.104	0.203	Warren and Wasson (1978)
67635	anorthosite	9.1	92	0.088	0.172	Stöffler et al. (1985)
65327	anorthosite	7.0	99	0.073	0.141	Warren and Wasson (1978)
77539c15	anorthosite	6.2	99	0.065	0.125	Warren et al. (1991)
64435c239	troctolitic anorth.	6.0	83	0.053	0.102	James et al. (1989)
15295c41	anorthosite	5.3	99	0.056	0.108	Warren et al. (1990)
15362	anorthosite	4.2	98	0.043	0.084	Ryder (1985)
67636	anorthosite	3.2	97	0.033	0.064	Stöffler et al. (1985)
67035c26	gabbronoritic anorth.	2.3	80	0.019	0.038	Warren (1993)
67637	anorthosite	2.3	96	0.023	0.045	Stöffler et al. (1985)
65767c3	anorthosite	2.0	98	0.021	0.040	Dowty et al. (1974)
67016c346	anorthosite	2.0	95	0.020	0.039	Norman and Taylor (1992)
67455c30	anorthosite	1.7	95	0.017	0.033	Ryder and Norman (1978)
73217c35	anorthosite	1.7	95	0.017	0.033	Warren et al. (1983a)
67539c7	anorthosite	1.5	96	0.015	0.029	Stöffler et al. (1985)
15437	troctolitic anorth.	1.1	80	0.009	0.018	Ryder (1985)
67915c26	noritic anorthosite	1.0	85	0.009	0.017	Marti et al. (1983)
67535	anorthosite	0.99	93	0.010	0.019	Stöffler et al. (1985)
Sum		9473.8		**96.03**	**93.23**	
Sum w/o 60015		4873.8				

Source of data: Papike et al., 1998; Planetary Materials, Warren, 1993 (pristinity index >5).

Taking the modal proportion of plagioclase weighted according to mass, then of the listed large FAS rocks, the average proportion of plagioclase is about 96%. Owing to the fact that 60015 contains nearly half the mass and is estimated to contain 99% plagioclase, the mass-weighted average may be biased. Excluding 60015, the mass-weighted average proportion of plagioclase drops to 93%.

FAS summary by Appollo landing site (g):

A15	280
A16	9180
A17	8

Minor or trace minerals found in the ferroan anorthositic rocks include ilmenite, aluminous Cr spinel, troilite, Fe-Ni metal, and an Si phase (Papike et al. 1998). Minerals that would signal the presence of trapped melt such as the phosphates, zircon, and K-feldspar are not found in these rocks. Their absence indicates a cumulus formation process that very efficiently excluded late-stage trapped melt (e.g., Morse 1982). Recrystallization textures or subsolidus annealing is evident in many FAS samples, leaving open the possibility that low-melting accessory minerals might have been lost during a metamorphic episode (Phinney 1994).

Compositions. The FAS rocks are compositionally distinctive. Because most are rich in calcic plagioclase, most compositions are highly aluminous. Some of the distinguishing geochemical traits include very low incompatible-element concentrations (Taylor et al. 1991), high Sc/Sm (typically >10; Norman and Ryder 1980), 0.6-1.0 ppm Eu concentrations (Korotev and Haskin 1988; Jolliff and Haskin 1995), and low Cr/Sc (see Table A3.1 and Norman and Ryder 1980). High Sc concentrations are consistent with the abundance of ferroan pyroxene because in lunar rocks, Sc correlates closely with Fe^{2+}. Low Cr/Sc is consistent with formation of these rocks following extensive crystallization of mafic cumulates, including chromite and Cr-bearing pyroxenes (i.e., upper-mantle cumulates). Owing to the low oxygen fugacity, evidenced by the stability of Fe-metal, Eu is predominantly divalent and is enriched in plagioclase relative to the trivalent REEs. Since incompatible trace elements are so low in concentration in FAS rocks, whole-rock compositions exhibit large positive Eu anomalies.

The very low incompatible-trace-element concentrations are consistent with the virtual absence of late-stage accessory minerals. Calculated equilibrium melt compositions during the interval of plagioclase crystallization from the magma ocean have trace-element concentrations within about a factor of ten to fifty times chondritic values (Snyder et al. 1992b; Jolliff and Haskin 1995; Papike et al. 1998), indicating their formation from relatively unevolved melt (i.e., other than fractionation associated with the crystallization of olivine and pyroxene from a magma ocean beginning with trace-element concentrations at several times chondritic). Among the elements well determined by remote sensing, surface FeO concentrations span a range from the extremely low values of <1 wt% of nearly pure anorthosite to 15 wt% in ferroan norite, but the average

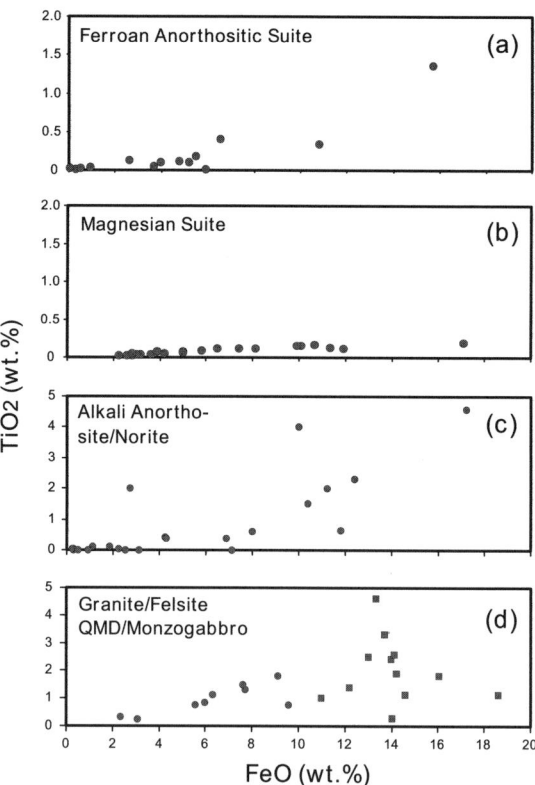

Figure 3.3. TiO_2 vs. FeO for crustal igneous rock suites. In (d), compositions with >10 wt% FeO have been arbitrarily considered to be QMD or monzogabbro (squares), even though some of those with <10 wt% FeO may have too much FeO and too little SiO_2 to be considered granite. Note different TiO_2 scales between the upper two and lower two plots.

concentration (excluding mare basalts) is around 4 wt%. Concentrations of TiO_2 and Th are extremely low in all of the FAS rocks (see Figs. 3.3 and 3.4).

Age relationships. The FAS rocks are isotopically very primitive and extremely old. Values of $^{87}Sr/^{86}Sr$ are nearly as low as primitive meteorites and indicate formation from an unfractionated reservoir. Although age determinations using different isotopic systems vary, Sm-Nd internal isochrons for two rocks yield ages of 4.44 and 4.54 Ga (see Fig. 3.5), providing a firm link between formation of the FAS rocks and the early differentiation of the Moon. Recent determinations of Sm-Nd isochron ages of 4.29 and 4.4 Ga (Norman et al. 1998; Borg et al. 1999; Norman et al. 2003) might imply that FAS rocks formed (or remained very hot) over an extended period of perhaps two hundred million years or more, or that the younger of the FAS ages were somehow disturbed or reset. Some of the FAS samples also appear to have extremely positive ε_{Nd} values ($\varepsilon_{Nd} = +3.1$ for 62236) indicating that these could have formed from reservoirs previously depleted of light rare-earth elements (standard magma-ocean crystallization models predict $\varepsilon_{Nd} \sim 0$). Alternatively, Norman et al. (2003) have suggested that the plagioclase fractions may have been affected by subsequent reequilibration, whereas the mafic fractions of the ferroan anorthosites define a crystallization age of 4.46 ± 0.04 Gyr. If the positive ε_{Nd} of 62236 is real, then its relatively young age combined with the recrystallized nature of some of the ferroan rocks may record an event such as recrystallization associated with remobilization of FAS cumulates, or perhaps underplating of anorthositic materials by later-formed magmas. The magmatic evolution of FAS rocks is addressed in detail in Section 3 of Chapter 4.

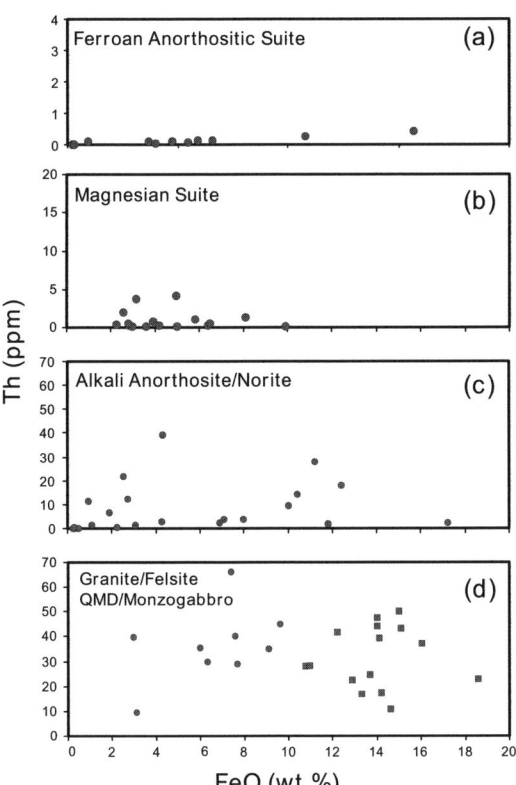

Depths of origin. The textures and mineralogy of FAS rocks indicate that they crystallized at least in part slowly and over a range of depths. Major mineral compositions are well equilibrated, and elements that can be enriched during rapid crystallization such as Al in pyroxene and Ca in olivine are low in concentration in these minerals. In terrestrial rocks, Al can also be incorporated in pyroxene as a result of high pressure, but other characteristics of the rocks would permit distinguishing such effects. Pyroxene grains in FAS rocks are typically exsolved (low-Ca and high-Ca separation), and in some samples, plagioclase contains exsolved grains of pyroxene and other minerals (e.g., Jolliff and Haskin 1995; Norman et al. 2003). Unfortunately, there are no direct and sensitive mineralogical indicators of pressure in the FAS rocks, so the next best method to assess

Figure 3.4. Th vs. FeO for crustal igneous rock suites. In (d), compositions have been arbitrarily assigned to granite for those samples with FeO < 10 wt% and QMD/monzogabbro for FeO > 10 wt% (squares). Note different Th scales.

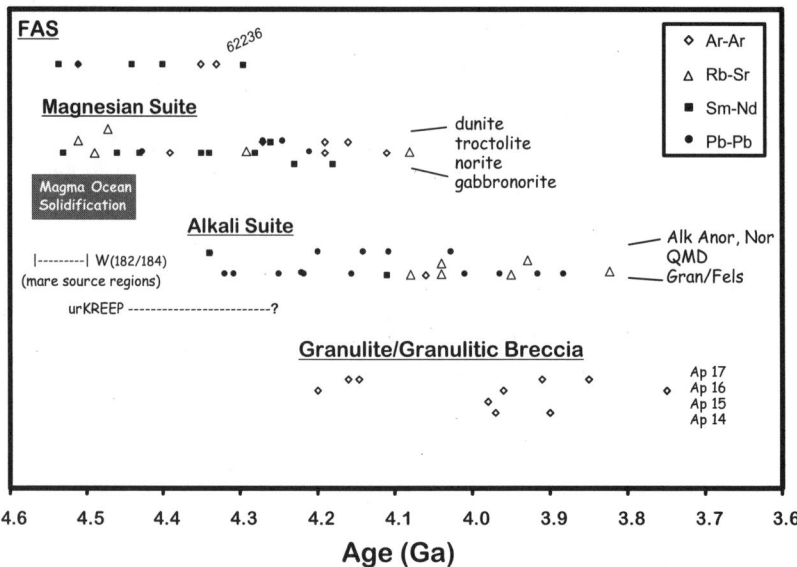

Figure 3.5. Crystallization ages of selected crustal rock suites. Time spans indicating magma-ocean solidification and the extended period for urKREEP crystallization are interpretive. The time span for development of the mare source regions based upon W(182/184) systematics is meant to indicate a rapid magma-ocean crystallization of the mare source (see Chapter 4). The Sm-Nd age of sample 62236 (Borg et al. 1999) is surprisingly young (see discussion in text). For other ages and data sources, see Chapter 5, Table 5.6.

depth of formation is to estimate cooling rates from features such as compositions and widths of exsolution lamellae or the ordering of cations in pyroxene (McCallum and O'Brien 1996). Of the samples studied so far, none indicate depths of origin exceeding 25 km (McCallum and O'Brien 1996). On the basis of exsolution features in FAS rocks from North Ray Crater, Apollo 16, Jolliff and Haskin (1995) inferred a two-stage cooling and depth history that was initially slow and deep, and later rapid and shallow, as postulated by Longhi and Ashwal (1985). McCallum and Schwartz (2001) suggested a two-stage process of solidification at 4.5 Ga and partial melting at 4.3 Ga, coupled with the removal of melt bearing a high Sm/Nd component, to explain the isotopic characteristics of rock 67215.

Distribution. Because the Apollo 16 site, where FAS rocks are so abundant, was thought early in post-Apollo studies to be representative of the feldspathic lunar highlands, a logical inference was that rocks of the ferroan-anorthositic suite should be the main component of the feldspathic highlands. Global remote sensing, mainly of FeO and Th concentrations, showed that the Apollo 16 site was, however, not as representative of the feldspathic highlands as once thought. Concentrations of FeO and Th in regolith at the Apollo 16 site are higher (5.5 wt% FeO, 1.8 ppm Th) than much of the feldspathic highlands of the lunar far side (4.5 wt% FeO, <1 ppm Th). Examination of the Apollo 16 regolith shows that this difference results largely from the presence of mafic impact melt, delivered to the site probably as part of ejecta deposits related to the Imbrium, Serenitatis, and/or Nectaris events (Haskin et al. 2002). The question remains as to just how anorthositic the pre-basin substrate at this site might have been. After subtraction of the impact-melt component and exotic mare basalt components (Korotev 1997), the remaining Cayley Plains regolith composition apparently has <4 wt% FeO and ~30 wt% Al_2O_3, but an Mg/(Mg+Fe) of ~0.7, somewhat higher than expected if the soil is composed predominantly of FAS components. Stöffler et al. (1985) did a similar mixing model and cast the results into the following rock components: 86–87% anorthosite, 6–8% magnesian gabbronorite, 4% ferrogabbro, ~2% dunite, ~1% feldspathic lherzolite, and a trace of spinel troctolite.

Samples of regolith from the ejecta deposits of North Ray Crater are poor in mafic impact-melt components and are more directly representative of typical feldspathic highlands materials (Stöffler et al. 1985; Norman et al. 1995, 2003). The North Ray Crater regolith and the large feldspathic fragmental boulders strewn about its rim appear to be more dominated by FAS components (see also Jolliff and Haskin 1995; Korotev 1996, 1997; Norman et al. 2003) than the Cayley regolith, and the most ferroan regolith compositions are very similar to the most ferroan feldspathic lunar meteorites (see also Chapter 2).

2.1.2. Magnesian suite. The magnesian-suite rocks include pristine lunar rocks for which Mg/(Mg+Fe) values (*Mg'*) of mafic silicates range from about 0.95 to 0.6, coupled with systematic variations in the An content of coexisting plagioclase from about 98% to 84% (see Fig. 3.2). As implied by the term "magnesian," these rocks have relatively magnesian compositions for their plagioclase anorthite content, compared to the other major igneous suite, the FAS rocks. All members of this suite, however, are not necessarily *magnesium rich*, especially the more anorthositic members (see Table A3.2). Although the general trend of decreasing *Mg'* with decreasing An content resembles trends observed for different rock types of terrestrial layered intrusives (Raedeke and McCallum 1980), inclusion of the diverse members within the magnesian suite does not necessarily imply a petrogenetic relationship among all members. On the contrary, subgroups exist that appear to be unrelated to one another (e.g., James 1980; Papike et al. 1998). For this reason, the magnesian-suite rocks have been considered to represent intrusions into the early lunar crust, which is presumed to have been ferroan anorthositic.

Mineralogy, mineral modes, rock types. The major minerals of the magnesian-suite rocks are plagioclase, low-Ca pyroxene, and olivine, plus minor amounts of high-Ca pyroxene, and, in some cases, pleonaste (Mg-Al) spinel. The most common accessory minerals include ilmenite, the phosphates apatite and RE-merrillite (whitlockite), troilite, zircon, baddeleyite, chromite, and rarely cordierite, silica, K-feldspar, niobian rutile, zirconian armalcolite, zirconolite, and Fe metal (see Table A3.2). The main rock types are norite, troctolite, spinel troctolite, and gabbronorite, plus more feldspathic variations, such as noritic anorthosite (or, in terrestrial rock nomenclature, leuconorite). Several "ultramafic" assemblages are found among the Apollo samples that are rich in pyroxene and/or olivine (see Table 3.3), but of these, most are tiny rock fragments with the notable exception of a dunite (72415-8, with a total mass of 55 g) collected at the Apollo 17 site. Others of the very small rock fragments listed in the table may not represent the modal mineralogy of their parent rocks. Of the samples listed by Warren (1993), norites are the most abundant according to mass, followed by troctolites and gabbronorites, however, in all cases, the sample masses are dominated by a few large rock samples collected at the Apollo 17 site (Papike et al. 1998). The proportions of rock types present at a given landing site can also be determined by coupled analyses of the soils and rock fragments contained therein (e.g., Bence et al. 1974; Jolliff et al. 1991b, 1996). Lithologic proportions determined in this way reflect the same general systematics of rock distribution as indicated by the larger samples, including the predominance of the magnesian suite among the nonmare igneous rock types found at the Apollo 17 site.

Compositions. Major-element compositions of the magnesian-suite rocks vary widely, as shown in Table A3.2. This variation results partly from the fact that most samples are small, and many are unlikely to be representative of the larger rocks from which they were derived. Among the rocks listed in Table A3.2, MgO, for example, ranges from 7 wt% (an anorthositic norite) to 45 wt% (dunite) and Al_2O_3 ranges from <2 wt% (dunite) to nearly 29 wt% (troctolitic anorthosite). Because of the small and potentially unrepresentative sample sizes, mineral assemblage and mineral chemistry have been used as the main criteria for classification, as described above. Still, numerous compositional features serve also to distinguish members of

Table 3.3. List of magnesian-suite intrusive rocks.

Sample	mass (g)	~Plag An mol%	~Low-Ca pxroxene Mg#	~Olivine Mg#	~Modal % feldspar	Main reference
Troctolite						
76335	465	96	88	87	77	Warren and Wasson (1978)
76535	155	96	86	88	50	Ryder and Norman (1979)
76536	10.3	—	86	83	70	Warren and Wasson (1979)
14321c1020	9.2	95	89	86	70	Lindstrom et al. (1984)
15455c106	3.0	95	85	83	71	Warren and Wasson (1979)
73146	3.0	95	88	86	85	Warren and Wasson (1979)
14303c194	2.0	95	—	88	70	Warren and Wasson (1980)
76255c57("U5B")	2.0	96	91	89	77	Ryder and Norman (1979)
12071c10	1.3	70				Warren et al. (1990)
14304c95("a")	0.9	94	—	87	55	Goodrich et al. (1986)
60035c21	0.7	96	89	88	57	RD Warner et al. (1980)
14172c11	0.7	94	—	87	65	Warren and Wasson (1980)
14179c6	0.7	94	—	87	70	Warren et al. (1981)
14321c1024	0.7	95	—	80	85	Warren et al. (1981)
73235c127	0.7	96	86	83	60	Warren and Wasson (1979)
14305c264	0.2	70				Warren and Wasson (1980)
14305c279	0.2	85			85	Warren et al. (1983)
sum mass	**655.5**					
Spinel-troctolite						
15445G (F?)	4.0	—	—	90	50	Ryder and Norman (1979)
15445c71-A	1.5	92?	—	92	35	Ridley et al. (1973), Ryder (1985)
12071c10	1.3	97	—	>78	70	Warren et al. (1990)
65785c4	0.3	96	84	83	65	Dowty et al. (1974b)
67435c77	0.1	97	—	92	40	Ma et al. (1981b)
14304c109("q")	0.0003	94	—	87	?	Goodrich et al. (1986)
15455 H	2?	96	91	92	25	Baker and Herzberg (1980)
15295 c	—	93	—	91	75	Marvin et al. (1989)
72435	—	96	70	73	~80	Dymek et al. (1976)
73263 particles	—	96	90	90	~70	Bence et al. (1974)
76503 particles	—	96	90	90	~70	Bence et al. (1974)
77517c disagg	—	97	90	90	—	RD Warner et al. (1978b)
sum mass	**9.2**					
Ultramafic						
72415/8 Dunite	55.2	94	87	87	4	Ryder and Norman (1979)
12033;503 Harzburgite	0.1	—	91	89	—	Warren et al. (1990)
14161;212,1 Peridotite	<0.1	—	87	85	1	Morris et al. (1990)
14161;212,4 Dunite	<0.1	—	—	85	—	Morris et al. (1990)
14304c121("d")	0.1	—	—	89	—	Warren et al. (1987)
14305c389 Pyroxenite	<0.1	—	91	90	1	Shervais et al. (1984a)
14321c1141 Dunite	0.1	—	—	89	2	Lindstrom et al. (1984)
sum mass	**55.5**					
Norite						
77075/77215	840	91	71	—	55	Ryder and Norman (1979)
78235/78255	395	93	81	—	47	Warren et al. (1979a)
15455c228	200	93	83	—	70	Ryder (1985)
77035c130	100	93	79	—	60	Warren and Wasson (1979)
15445c17("B")	10	95	82	—	63	Shih et al. (1990)
72255c42	10	93	75	—	40	Ryder and Norman (1979)
78527	5.2	93	80	77	50	Warren et al. (1983b)
14318c146	1.5	87	73	71	55	Warren et al. (1983b)
15361	0.9	94	84	—	40	Warren et al. (1990)
15360;11	0.7	93	78	—	65	Warren et al. (1990)
14318c150	0.5	83	78	74	65	Warren et al. (1986)
78236,3						Blanchard and McKay (1981)
sum mass	**1563.8**					
Gabbronorite						
76255c82("U4")	300	87	65	—	41	Warren et al. (1986)
67667	7.9	91	78	71	24	Warren and Wasson (1979a)
73255c27,45	0.9	89	74	—	53	James and McGee (1979)
61224;6	0.3	83	67	—	34	Marvin and Warren (1980)
14311c220	0.2	85	60	—	75	Warren et al. (1983c)
76255c72("U5A")	0.1	86	67	—	39	Ryder and Norman (1979)
14161,7044	<0.1	88	64	—	60	Jolliff et al. (1993)
14304c114("h")	<0.1	89	—	68	40	Goodrich et al. (1986)
sum mass	**309.5**					

Troctolites and norites listed for those samples larger than 0.5 g; for others, all samples documented by Warren (1993) and Papike et al. (1998) are listed. All of the "ultramafic" rocks except the Apollo 17 dunite are tiny rock fragments or clasts. Whether they actually represent larger ultramfic rocks is not certain. Sample 67667 also referred to as feldspathic lherzolite (Stöffler et al. 1985).

this rock suite. TiO_2 values are typically low, rarely exceeding 1 wt% and averaging less than 0.5 wt% (Table A3.2). Values of Mg/(Mg+Fe) vary systematically and correlate positively with Ca/(Ca+Na+K). Among the trace elements, the magnesian-suite rocks are distinguished from FAS rocks by low Sc/Sm and high Cr/Sc ratios. Most of the magnesian-suite rocks are cumulates that do not contain representative proportions of trapped melt, so bulk incompatible-element concentrations tend to be fairly low, albeit significantly higher than FAS rocks by 1–2 orders of magnitude. Some, however, have high incompatible element concentrations, and ion-microprobe analyses of REE concentrations in magnesian-suite silicates, especially pyroxene and plagioclase, have shown relatively high concentrations for these minerals (Papike et al. 1994, 1996; Shervais and McGee 1998).

In fact, much has been made in the past decade about the apparent trace-element enrichment of the magnesian-suite rocks (or their parent melts) despite the fact that most have very small trapped-melt components. Early studies of rocks such as troctolite 76535 (e.g., Haskin et al. 1974) recognized that the varied accessory mineral accompaniment and trace-element contents (trapped-melt components) signaled a more evolved equilibrium melt (and inferred parent melt) than the ferroan anorthosites. However, the trace-element-rich accessory minerals could also be explained as a result of metasomatic alteration or addition of the accessory components at some later time. The application of the ion microprobe to determine trace-element contents of the major silicate minerals, coupled with equilibrium mineral/melt distribution coefficients, showed that the cumulus minerals themselves had relatively high concentrations of incompatible elements. This argued against metasomatism and led to the recognition that indeed many of the magnesian-suite cumulates, even those lacking a trapped melt component, must have formed from trace-element-rich equilibrium melts (Papike et al. 1994, 1996; Shervais and McGee 1998), in particular, ~0.7 to 2 times high-K KREEP, which corresponds to ~15–45 ppm Th. Even the group 2 "trace-element poor" Apollo 17 norites cited by James (1980) were shown by Papike (1996) to have high REE concentrations in plagioclase and orthopyroxene that, when combined with distribution coefficients, indicate equilibrium melts with REE concentrations 200–1000 times CI chondrites. That many of these rocks lack abundant accessory minerals means that trapped melt was effectively excluded from them when they crystallized.

Despite finding significant trace-element enrichment in many of the magnesian-suite norites and presumably, in their parent melts, it remains to be proven that all magnesian-suite rocks had such parentage.

Age relationships. Isotopic data for the magnesian-suite rocks are considered in detail in Chapter 4, especially as they relate to the magmatic evolution of crustal rocks. Here, we summarize the age data and some of the inferences that can be made with respect to the distribution of these rocks within the Moon's crust. Several important points can be made with respect to the age data, which are shown in Figure 3.5. First is the partial overlap of the magnesian-suite rocks ages with those of the FAS rocks. In particular, some of the oldest magnesian-suite rocks are nearly as old as the oldest FAS rocks, which means that they formed at essentially the same time as the formation of crust from the magma ocean, and not necessarily as later remelting and intrusion events into an already solid ferroan-anorthositic crust. Second is the extended range of ages from >4.5 Ga to about 4.1 Ga, suggesting that the processes that led to magnesian-suite intrusive activity occurred over a prolonged interval relative to the ferroan-anorthositic rocks. We will return to this point in the discussion of ages of the alkali-suite rocks.

Depths of origin. The application of thermobarometry to lunar samples is difficult because of the low-pressure gradients in the Moon and the paucity of minerals whose compositions are sensitive to low-pressure variations. Only a few rock types have appropriate assemblages of coexisting minerals for such determinations. Among these are symplectite-bearing troctolites such as 76535 (Gooley et al. 1974) and spinel cataclasites such as breccia clasts from 72435

(Dymek et al. 1976; Herzberg 1978) and 15295 (Marvin et al. 1989). McCallum and Schwartz (2001) have recently reevaluated the depths of origin of several of the magnesian-suite rocks using updated thermodynamic data and advances in quantifying the energetics of crystalline solid solutions to calculate pressures and temperatures of equilibration. For troctolite 76535, assuming local equilibria in the vicinity of the symplectites, which consist of clinopyroxene + orthopyroxene + Cr-spinel surrounded by olivine and plagioclase, McCallum and Schwartz computed temperatures for the recrystallization of about 800–900°C and pressures of 220–250 MPa, corresponding to 42–50 km depth. For the spinel cataclasites, the relevant reaction involves olivine + cordierite and orthopyroxene + spinel (Herzberg 1978; Herzberg and Baker 1980). Temperatures are estimated from olivine-spinel equilibria (Sack and Ghiorso 1991) and Al_2O_3 in orthopyroxene (Herzberg 1978). For three samples studied by McCallum and Schwartz (15445, 77517, 73263), equilibration temperatures fall within the range of 600–900°C with minimum pressures of 100–200 MPa, corresponding to 20–40 km depth. Crustal thickness modeling that utilizes gravity, topography and seismic constraints suggests that the lunar crust is on average between 40 and 50 km thick (see Section 2.7). Thus, if the thermobarometry results are taken at face value, they imply that the magnesian-suite rocks originate from the lower half of the crust.

Distribution. Magnesian-suite rocks are found among the samples from all of the Apollo landing sites, even Apollo 11 (Korotev and Gillis 2001). They are most abundant, however, at the Apollo 14 and 17 sites (see Tables 3.1 and 3.3), especially the latter, and specific groups of rock types tend to be found mainly at one or two specific sites. For example, troctolites are common as clasts in Apollo 14 breccias and as rocks at the Apollo 17 site. Magnesian-suite norites also occur as clasts in Apollo 14 breccias and as rocks at the Apollo 15 and 17 sites. Gabbronorites occur as clasts and small rocks at the Apollo 14 site and as clasts and rocks at the Apollo 16 and 17 sites. Samples of like rock types at a given site, such as the Apollo 16 gabbronorites, are more likely related to one another by mineral and trace-element compositions than they are to the samples of like rock types from other sites (James and Flohr 1983). This similarity suggests petrogenetic links between samples at one site, but not between sites, and led to the idea that the magnesian-suite rocks represent distinct intrusive bodies into the early crust, which was presumed to be a more-or-less global ferroan-anorthositic shell. That most of the reported ages of magnesian-suite rocks are younger than the ferroan anorthosites supports this interpretation, but the overlap between the youngest ferroan anorthosites and the oldest magnesian-suite samples means that the process that formed the sampled magnesian-suite rocks must have been operative very early, even before the initial differentiation was complete globally (see also Chapter 4).

The results of studies of depths of equilibration and burial of magnesian-suite rocks indicate that at least some of them last equilibrated in the middle to lower crust. In the past, it was assumed that the process of magnesian-suite emplacement into early FAS crust was a global phenomenon, and that they were only excavated by the largest basin impact events. Early on, however, in lunar sample studies, attention was called to the so-called east-west dichotomy relative to Apollo and Luna landing sites whereby rocks of the magnesian and ferroan anorthosite suites seemed to be more abundant among nonmare samples of the western and eastern sites, respectively (e.g., Hunter and Taylor 1983; Stöffler et al. 1985; Shervais and Taylor 1986).

The assumption of a global distribution of the magnesian suite within the crust is also challenged on the basis of the components of the feldspathic lunar meteorites, which are mostly dominated by ferroan-anorthositic materials (Jolliff et al. 1991b; Korotev 1996; Korotev et al. 2003a). This observation, coupled with recent global remote sensing (FeO as indicated by Clementine and Lunar Prospector data), indicates that these meteorites may represent a vast portion of the feldspathic highlands (see also Chapter 2). One could also argue, however, that where the crust is thicker beneath the feldspathic highlands, basin impacts may have been less likely to penetrate deeply enough to exhume magnesian-suite intrusives. In this regard, analysis of the materials exhumed by the South Pole-Aitken basin, whether by return of samples or remote sensing of Mg, would be a key test of the global distribution of magnesian-suite rocks. At present, the global distribution of magnesian-suite intrusives within the crust is not known, although recent studies coupling the magnesian suite to KREEP-rich parent magmas (Snyder et al. 1995a,b; Korotev 2000) makes it likely that their primary occurrence (aside from impact redistribution) is controlled by the early distribution of KREEP, which appears to be concentrated in the regions of mare Imbrium and Oceanus Procellarum (see Section 5.1).

Although the magnesian-suite rocks are discussed here as a group, it is with recognition that various subgroups can not be related directly to one another through a common parent magma (see James and Flohr 1983). It also appears to be the case that igneous rocks containing two or more of the different rock types (e.g., a norite and a troctolite) as might be found in a sample from an igneous contact of a layered intrusion, are not found among the samples. Moreover, impact breccias in the sample collection are not found that represent simple fragmented mixtures of two or more of the magnesian-suite rock types. Thus, direct evidence that layered intrusions or other composite igneous bodies exist that contain two or more of these rock types is lacking among the samples (Papike et al. 1998). Thus, one of the key unresolved issues is how the different magnesian-suite lithologies are related to one another.

Remote sensing studies of central peaks of large craters, which could have depths of origin up to ~30 km below the surface, may provide some clues to the distribution of rock types in the crust (e.g., Tompkins and Pieters 1999; Wieczorek and Zuber 2001a). With the 100–200 m resolution of the Clementine multispectral data, distinct rock types can be discerned on the slopes of central peak mountains, which are steep enough to prevent buildup of thick regolith deposits. Rock types such as anorthosite, norite, and troctolite are observed together and a common inference is that these assemblages represent uplifted magnesian-suite intrusive rock bodies. However, the observation that mafic crustal igneous rocks (norite, gabbro, troctolite) in the sample collections are commonly of the magnesian suite and that rocks rich in olivine and pyroxene are rare among the FAS samples leads to a potentially erroneous inference (e.g., Heiken et al. 1991) that mafic crustal rocks observed remotely are necessarily of the magnesian-suite variety. Commonly, when remote sensing indicates a relatively mafic crustal rock, the assumption has been made that the exposure is of a magnesian-suite lithology and composition. On the other hand, several lines of argument support the possibility that such exposures may be mafic *ferroan* rocks instead. First is the observation that mafic varieties of FAS rocks do occur in the sample collection and share many of the distinguishing trace-element geochemical signatures of the more plagioclase-rich FAS samples. Second is the observation that most of the feldspathic lunar meteorites, which may be argued on the basis of statistical reasoning to represent the vast feldspathic highlands that constitute the northern lunar far side, are dominated by ferroan-anorthositic lithologic components. If the rocks of central peaks are uplifted from depths as great as 30 km, they could represent ferroan mafic rocks that are the deep-seated complement to the ferroan anorthosites.

2.1.3. Alkali suite. Compared to the ferroan-anorthositic and magnesian-suite rocks, the alkali suite is minor by mass (see Table 3.1) and occurs mainly as small rock fragments or clasts in impact breccia. These rocks are distinguished by their alkali-rich bulk compositions and by the combination of relatively ferroan mafic silicates and plagioclase that is less calcic than plagioclase of the magnesian suite. The alkali suite comprises a variety of assemblages including anorthosite, troctolitic anorthosite, norite, gabbronorite, gabbro, felsite (fine-grained granite), and monzogabbro. The majority of these rocks by mass occur in the Apollo 12, 14, and 15 samples, especially Apollo 14 (see Table 3.1); however, traces of alkalic igneous rocks have been reported in samples from all of the Apollo sites except Apollo 11. In most cases, rock samples are small (<1 g), with but a few exceptions (see Tables 3.4 and 3.5). The largest specimens amount to only a few grams each, such as a 1.7 g alkali anorthosite clast in 14047 (Warren et al. 1983a), a 1.8 g granite clast in 14321 (Warren et al. 1983c), and a 2.5 g quartz monzodiorite

Table 3.4. List of alkali-suite anorthosite and norite.

Sample	mass (g) (if avail.)	~Plag An mol%	~Modal % feldspar	Main reference
Alkali Anorthosite				
12003,179/210	0.10	82	100	Warren et al. (1990)
12033,425/501	0.13	83	99	Warren et al. (1990)
12033,550/532	0.02	83	96	Laul (1986), Simon and Papike (1985)
12033,97.7	0.10	88	100	Hubbard et al. (1971b)
12037,178/177	0.02	~92	~97	Laul (1986), Simon and Papike (1985)
12073c120/122	0.08	79	99	Warren and Wasson (1980)
14160,106/105	0.19	82	100	Warren and Wasson (1980)
14161,7245	0.04	83	90	Jolliff et al. (1991)
14304c122"b"	0.49	82	98	Warren et al. (1987)
14305c283WhtA	0.14	85	95	Warren et al. (1983c)
14305c400	0.67	76	99	Shervais et al. (1984a)
14321c1060WhtA	0.14	86	96	Warren et al. (1983c)
15405c181	0.10	84	99	Lindstrom et al. (1988)
Alkali Noritic or Troctolitic Anorthosite				
14047c112/113	1.65	81	84	Warren et al. (1983b)
14066c49/51	0.01	81	85	Shervais et al. (1983)
14160,197/217		~70	80	Snyder et al. (1992)
14305c91	0.14	86	90	Hunter and Taylor (1983)
67975,131N		85	85	James et al. (1987)
Alkali Norite and Gabbronorite				
12033,555/534	0.07	81	49	Laul (1986), Simon and Papike (1985)
12042,280/281		~85		Laul (1986), Simon and Papike (1985)
14303,44		86		Hunter and Taylor (1983)
14304c86"g"	0.23	82	14	Goodrich et al. (1986)
14311,96		85		Hunter and Taylor (1983)
14311,220	0.23	85	75	Warren et al. (1983c)
14313c70WhtA	0.03	83	50	Warren et al. (1983c)
14316,6/12	0.002	84	60	Warren et al. (1981)
14318,146/149	1.20	87	55	Warren et al. (1983b)
15405c170	0.04	89	70	Lindstrom et al. (1988)
67915,163	0.23	63	43	Marti et al. (1983)
67975,14		88	50	James et al. (1987)
67975,44Nm		86	21	James et al. (1987)
67975,44Nf		~70	63	James et al. (1987)
67975,62		85	38	James et al. (1987)
67975,86		82	15	James et al. (1987)
67975,117N			50	James et al. (1987)
67975,136N		~70	42	James et al. (1987)
67975,42N		85	48	James et al. (1987)
77115c19	0.60	95	70	Winzer et al. (1974)

clast in 15405 (Ryder 1985). The occurrence of an 82 g breccia composed mostly of granitic and monzogabbroic materials (12013), however, suggests that larger rocks or rock bodies composed mainly of alkali-rich materials may exist. The KREEP-rich impact-melt breccia 15405 (513 g), which contains clasts of quartz monzodiorite, indicates a spatial, if not petrogenetic, link between some of the alkali lithologies and KREEP basalt. Such relationships are also observed in smaller impact breccias, especially in samples from the Apollo 12 and 14 sites.

Mineralogy and mineral modes. Minerals of the alkali suite include plagioclase, low-Ca and high-Ca pyroxene, K-feldspar, a silica phase, apatite, merrillite, ilmenite, Cr-spinel, fayalite, zircon, baddeleyite, troilite, and Fe-Ni metal. Plagioclase typically is more sodic than

Table 3.5. List of alkaline-suite QMD/monzogabbro and granite/felsite.

Sample	mass (g) (if avail)	high-Ca pyroxene Mg#	Olivine Mg#	Plag An mol%	Modal % feldspar	Main reference
Quartz monzodiorites (monzogabbro)						
15405c56	2.50	—		—	46	Ryder (1985)
15434,12	0.62	33		80	40	Ryder and Martinez (1991)
15434,14 bx	0.15	—		80	40	Ryder and Martinez (1991)
15434,10	0.12	—		60	30	Ryder and Martinez (1991)
15459c315	0.10	29		60	59	Lindstrom et al. (1988)
15403,24,7001	0.04	32		70	50	Marvin et al. (1991)
14161,7069	0.023	35		70	41	Jolliff (1991)
14161,7264 bx	0.020	62		71	50	Jolliff et al. (1991)
14161,7373	0.018	49		70	28	Jolliff (1991)
15403,71a	0.005	30		60	50	Marvin et al. (1991)
15403,71c	0.001	—		20	60	Marvin et al. (1991)
15403,71b	<0.001	30		—	30	Marvin et al. (1991)
15403,7002	<0.001	31		60	45	Marvin et al. (1991)
15403,23c	—	—		—	—	Marvin et al. (1991)
Granite/Felsite						
12013 bx	—	—	—	—	—	Quick et al. (1977)
14321c1028	1.80	5	2	—	60	Warren et al. (1983a)
12033,507	1.20	—	8	50	55	Warren et al. (1987)
14303c204	0.17	40	42	75	60	Warren et al. (1983a)
14004,94 glass/bx	0.15	—	—	—	40	Snyder et al. (1992)
14004,96 glass/bx	0.10	—	—	—	35	Snyder et al. (1992)
14161,7269 glass/bx	0.04	50	—	67	50	Jolliff (1991)
14001,28,2 bx	0.03	—	—	80	45	Morris et al. (1990)
73215c43,3	0.02	35	19	60	50	James and Hammarstrom (1977)
14001,28,3 bx	0.02	—	—	80	45	Morris et al. (1990)
14001,28,4 bx	0.02	—	—	80	45	Morris et al. (1990)
12070,102-5	<0.01	30	13	50	—	Marvin et al. (1991)
73255,27,3	<0.01	—	—	—	—	Blanchard and Budahn (1979)
15403,71c	—	45	—	—	—	Marvin et al. (1991)

bx = brecciated, not strictly monomict

An_{86} (Warren 1993) and low-Ca pyroxene is more ferroan than En_{70} (Snyder et al. 1995b). Samples rich in plagioclase, i.e., > 80 vol%, occur primarily among Apollo 12 and 14 materials. Some of these are clasts in breccias and some are small rocks or rock fragments. A variety of more pyroxene-rich assemblages also have been found at these sites, and these have been referred to mainly as alkali norite, or in a few cases, gabbronorite or monzogabbro. Similarly alkalic and ferroan assemblages also occur in a few Apollo 15, 16, and 17 samples, but these are less common than in Apollo 12 and 14 samples. Some of the rocks that have relatively sodic plagioclase and ferroan pyroxenes also have modally significant silica and K-feldspar, and have been referred to as quartz monzodiorite (QMD) (Ryder 1976) or as monzogabbro (Jolliff et al. 1999). Some of these have high proportions of phosphates, as high as 10 wt% (Jolliff 1991), although the small size of most of these rock fragments casts uncertainty on whether their modes might be representative of larger rocks or outcrops. Rock assemblages that are rich in K-feldspar and silica, commonly occurring in a granophyric intergrowth, are found among non-mare samples from most sites. Like other alkali-suite samples, these are most common among Apollo 12 and 14 samples, and are more numerous than reflected in Tables 3.4 and 3.5. Masses of several samples exceed a gram, and among the samples in these tables, the QMD-monzogabbro and granite groups each have summed masses comparable to the alkali-anorthosite and alkali-norite groups.

Compositions. Bulk compositions of the alkali-suite rocks are distinctive (see Table A3.3). Sodic plagioclase and the presence of K-feldspar are reflected in Na- and K-rich compositions. Trace-elements follow Na and K, with enrichments in Eu, Ba, Rb, and Cs. Bulk compositions are relatively ferroan, with Mg/(Mg+Fe) of alkali anorthosites and alkali norites averaging about 0.6, and granite and QMD-monzogabbro, ~0.4.

The alkali anorthosites differ from the other rock types in this group in their alkali-element proportions and trace-element signatures. The hallmark of the alkali anorthosites is their high Na_2O concentrations, averaging about 1.6 wt%, which are not necessarily accompanied by commensurately high K_2O (their average K_2O wt% is ~0.3). Enrichment in Eu is also a signature of this rock type, with Eu ranging from ~3 to 8 ppm and averaging ~6 ppm, as compared to average concentrations in FAS rocks of ~0.8 ppm and in magnesian-suite rocks, of ~1.2 ppm. The more mafic alkali norites, on the other hand, have lower Na_2O (with an average of 1.25 wt%) and higher K_2O (~0.5 wt%). In these rocks, Eu follows Na, and both correlate to plagioclase content. Barium follows K and is typically enriched in the more mafic norites by about a factor of 3 relative to the alkali anorthosites. Other incompatible trace elements such as the REEs, Zr, and Th do likewise. These variations are consistent with the norites containing a higher proportion of trapped melt.

In terms of the three elements sensed most accurately from orbit (FeO, TiO_2, and Th), compositions of these rocks cover a broad range (see Figs. 3.3 and 3.4). The alkali anorthosites typically have concentrations of FeO <4 wt% and very low TiO_2, averaging <0.5 wt%. The alkali norites have FeO concentrations ranging to 17 wt% with an average of ~10 wt%, and TiO_2 concentrations range to nearly 5 wt% with an average of ~2 wt%. Concentrations of Th are variable, averaging 5 ppm in the alkali anorthosites and 12 ppm in the alkali norites and alkali gabbronorites, but with maximum values near 40 ppm.

The QMD-monzogabbro and granite samples have been distinguished generally on the basis of mineralogy, with most workers referring to samples with abundant silica and K-feldspar as granite, and those with abundant silica, K-feldspar, and pyroxene, as QMD or monzogabbro. The granitic samples typically have SiO_2 contents of 65–75 wt%, FeO <10 wt% and K_2O >3 wt% ranging up to ~8 wt% (Table A3.4). The monzogabbros typically have >10 wt% FeO, ranging to over 16 wt%, and K_2O <3 wt%. Concentrations of TiO_2 are higher in the monzogabbros, averaging about 2 wt% compared to ~1 wt% in the granites. The trace-element signature of the monzogabbros is an extreme enrichment in REE concentrations and high field-strength elements such as Zr and Hf. The granite compositions show extreme enrichments in the alkali elements (K, Rb, Cs) and Ba, as well as in Nb, Ta, Th, and U. They also typically have a distinctive "V"-shaped REE pattern at about 100–300 times chondritic, with enrichment of both the lightest and heaviest REEs. The incompatible elements in these rock types do not follow a typical KREEP-like pattern, and this has been attributed to fractionation involving merrillite and other late-stage trace-element-rich minerals, and immiscible silicate-liquid segregation that occurred during late-stage crystallization (Jolliff 1991, 1998; Snyder et al. 1995b).

Recent studies have supported a relationship between the alkali-suite and magnesian-suite rocks (e.g., Snyder et al. 1995a; Shervais and McGee 1999). The trend among mineral compositions shown in Figure 3.2b suggests that fractional crystallization of a KREEP basalt-like magma could produce both the magnesian- and alkali-suite rocks (see also Chapter 4, Fig. 4.20). Most of the magnesian-suite rocks appear to be cumulates that were effectively separated from the bulk of their residual-melt component. The alkali suite could represent other, more chemically differentiated products of the same magmas. Alkali anorthosites might have originated as perched or flotation cumulates within intrusive magma bodies and alkali norite and gabbronorite could represent mafic cumulates that retained their trapped-melt components. Granite and monzogabbro would represent segregations of late-stage residual melt. One of the problems with this scenario is that members of these two rock suites (magnesian- and alkali-suite) are not typically found together in crystalline rocks or breccias.

Age relationships. Isotopic characteristics are also not clear on the relationship between the alkali and magnesian suites. While crystallization ages overlap, the alkali suite extends to younger ages (see Fig. 3.5). Ages extend from a little over 4.3 Ga to nearly 3.8 Ga (Meyer et al. 1989), with the apparent cut off at ~3.8 Ga likely resulting merely from the absence of younger impact basins that were large enough to excavate significant quantities of subsurface rocks. Other isotopic systematics indicate relationships between the alkali-suite rocks and KREEP basalt. Additional details of the isotopic systematics of alkali-suite samples are given by Snyder et al. (1995b; 2000).

Depths of origin. Mineral compositions and textures of members of the alkali suite so far are consistent with fairly rapid cooling associated with shallow emplacement and crystallization (see McCallum and O'Brien 1996; McCallum 1998; Jolliff et al. 1999). Even samples containing fairly coarsely exsolved pyroxene (coarse, at least, for a lunar rock) appear to have been emplaced in the shallow crust, within one or two kilometers of the surface. This is in contrast to some members of the magnesian suite that appear to have depths of origin deeper than 20 km.

Distribution. Perhaps the key questions regarding the alkali suite are related to their extent and distribution. Clearly these rock types are more abundant at the western landing sites, Apollo 12, 14, and 15, and Lunar Prospector results suggest a general confinement of concentrations of these materials to the Procellarum KREEP Terrane. What is not known is whether the alkali-suite rocks form extensive outcrops or separate intrusive bodies. Small sizes suggest that they could simply be the broken, trapped-melt-rich parts of overall more magnesian rocks. Remote sensing, however, points to several locations as places where such rocks may be more common, if not making up large outcrops or volcanic exposures, such as the domes and other volcanic constructs observed in parts of Oceanus Procellarum (e.g., Hawke et al. 2003a), and exposed by impacts such as Aristarchus, Kepler, and Mairan. A very curious Th-rich compositional anomaly occurs at the Compton Belkovich region (Gillis et al. 2002) that is associated with a low FeO abundance (as sensed remotely) suggesting the possibility of an alkali-anorthosite exposure at this locale. Some of the more mafic rocks that would be expected to occur with such an anorthosite have not yet been observed, however.

2.1.4. KREEP basalts. Although KREEP-bearing materials occur among samples from all of the Apollo sites, volcanic KREEP basalts are found primarily among the samples from Apollo 15. Apollo 17 samples have yielded one significant occurrence of KREEP basalt and a few small clasts or rock fragments occur in the Apollo 12, 14, and 15 samples (Table 3.6). Among these, only the Apollo 15 KREEP basalts appear to have a demonstrably local source (e.g., Spudis 1978; Ryder 1994). The term "KREEP basalt" refers strictly to the rocks that can be shown to have formed as a result of internal melting, not impact processes. Nevertheless, some of the KREEP basalts are almost identical in composition to known impact-melt rocks, especially those from the Apollo 14 site. Most of the texturally and compositionally similar rocks found at the Apollo 14 site have been shown to contain clasts or to have too much plagioclase to represent basaltic liquids.

A breccia containing clasts of an apparently volcanic KREEP basalt was found at the Apollo 17 site, and these have widely been referred to in the literature as "Apollo 17 KREEP basalt." It should be emphasized, however, that these clasts all occur in a single boulder from Station 2 (72275) (Ryder et al. 1977) and that they differ compositionally from those found at the Apollo 14 and 15 sites (e.g., different FeO, Al_2O_3, and trace element concentrations; see Salpas et al. (1987) and Table A3.5). No rock fragments of this KREEP basalt type have been found in the rock fragments of the Apollo 17 regolith (Bence et al. 1974; Jolliff et al. 1996),

Table 3.6. KREEP basalts.

Sample	mass (g)	~modal % feldspar	Main reference
15386	7.50	43	Ryder (1985)
15382	3.20	41	Ryder (1985)
72275c91	2.73	40	Salpas et al. (1987)
15434,18	0.74	45	Ryder and Sherman (1989)
15434,16	0.44	45	Ryder and Sherman (1989)
15434,189	0.41		Ryder and Sherman (1989)
15434,8	0.32	42	Ryder and Sherman (1989)
15304,6	0.31	49	Ryder and Sherman (1989)
15434,29	0.30		Ryder and Sherman (1989)
15264,4	0.28		Ryder and Sherman (1989)
15434,21	0.22	32	Ryder and Sherman (1989)
15024,11	0.21		Ryder and Sherman (1989)
15434,17	0.20	48	Ryder and Sherman (1989)
15404,5	0.17		Ryder and Sherman (1989)
15564,16	0.16		Ryder and Sherman (1989)
15434,25	0.14		Ryder and Sherman (1989)
15314,34	0.11	47	Ryder and Sherman (1989)
15434192	0.10		Ryder and Sherman (1989)
15405c68	0.08	40	Ryder (1985)
15007,290/291	0.06		Warren et al. (1983c)
15434194	0.05		Ryder and Sherman (1989)
67015c310	0.043	72	Marvin et al. (1987)
15007,292/293	0.030		Warren et al. (1983c)
15007,302	0.021		Warren et al. (1983c)
14161,7048	0.020	55	Jolliff et al. (1991)
15007,304	0.015		Warren et al. (1983c)
15007,294	0.011		Warren et al. (1983c)
Sum mass	17.9		
Average modal % feldspar		49	

which is consistent with this rock type having come from elsewhere and not being a common derivative of a local source. The Station 2 boulder (72275) may well be a fragment of distal ejecta, perhaps from Imbrium and almost certainly with a provenance different from the main Apollo 17 impact-melt breccia formations thought to be from Serenitatis.

Mineralogy and mineral modes. Minerals of the KREEP basalts include pyroxene (mostly pigeonite and augite, with minor orthopyroxene) and plagioclase as the main minerals, with lesser amounts of K-feldspar, phosphates, ilmenite, zircon (and other Zr-bearing phases including baddeleyite, zirconolite, zirconian armalcolite), K-feldspar, fayalite, silica (cristobalite), troilite, and Fe-metal. Mineralogically, the absence of magnesian olivine, higher plagioclase and low-Ca pyroxene proportions, and the presence of abundant late-stage mesostasis, distinguish these basalts from the mare basalts. Mineral compositional variations are consistent with fractional crystallization such that plagioclase and pyroxene cores are calcic and magnesian, with their rims being sodic and ferroan, respectively. These mineral-compositional variations constitute one of the lines of evidence linking KREEP basalts to magnesian- and alkali-suite rocks.

Compositions. KREEP basalts are typically aluminous (13–16 wt% Al_2O_3 and 9–15 wt% FeO, Table A3.5), and their defining characteristic is an incompatible trace-element enrichment at about 100–150 times chondritic concentrations. The compositions of Apollo 15

KREEP basalts are very similar to one another, but the Apollo 17 KREEP basalt composition differs significantly from these. The 72275 compositions are more ferroan (Mg/(Mg+Fe) ~0.5 compared to 0.6), they have lower TiO_2 concentrations (see Fig. 3.6), they have incompatible-trace-element concentrations 1/2 to 1/3 of those of Apollo 15 KREEP basalt, and they have higher Sc, V, Co, and Ni concentrations. Figure 3.7 illustrates the differences in FeO and Th concentrations. Salpas et al. (1987) concluded that the Apollo 15 and Apollo 17 KREEP basalts are not related to each other in any direct petrologic way. The 72275 KREEP basalt is intermediate between Apollo 15 KREEP basalt and mare basalts, and is most similar to the Apollo 14 "group 5" aluminous basalts (Dickinson et al. 1985).

A key aspect of the geochemistry of volcanic KREEP basalts is their rather magnesian bulk composition given their strong trace-element enrichment. This characteristic seems to require mixing of mantle-derived magnesian melts with KREEP-rich magma-ocean residua (urKREEP) or sinking of urKREEP into the mantle with subsequent remelting (see Warren 1988) (summarized recently by Papike et al. 1998).

The volcanic KREEP basalts have somewhat lower incompatible-element concentrations than some of the Apollo 14 and 15 impact-melt rocks (see below). Nevertheless, the incompatible trace elements (e.g., K, Ba, Rb, Cs, trivalent REEs, Zr, Hf, Nb, Ta, Th, U) occur in remarkably similar inter-element ratios in all the KREEP basalts as well as the KREEP-bearing mafic impact-melt breccias. On the basis of remotely sensed compositions, volcanic KREEP basalt cannot be distinguished from mafic impact melt—geologic and morphologic relationships are required (Spudis 1978).

Age relationships. Crystallization ages of the Apollo 15 KREEP basalts are well determined by analyses of multiple isotopic systems to be 3.82–3.86 Ga (summarized by Papike et al. 1998), presumably shortly after the Imbrium impact event (Ryder 1994). KREEP basalt fragments from 72275 yield slightly older crystallization ages (3.93–4.08 Ga). Isotopic systematics (ε_{Nd}) suggest a relationship between KREEP basalts and magnesian- and alkali-suite rocks (Shih et al. 1992).

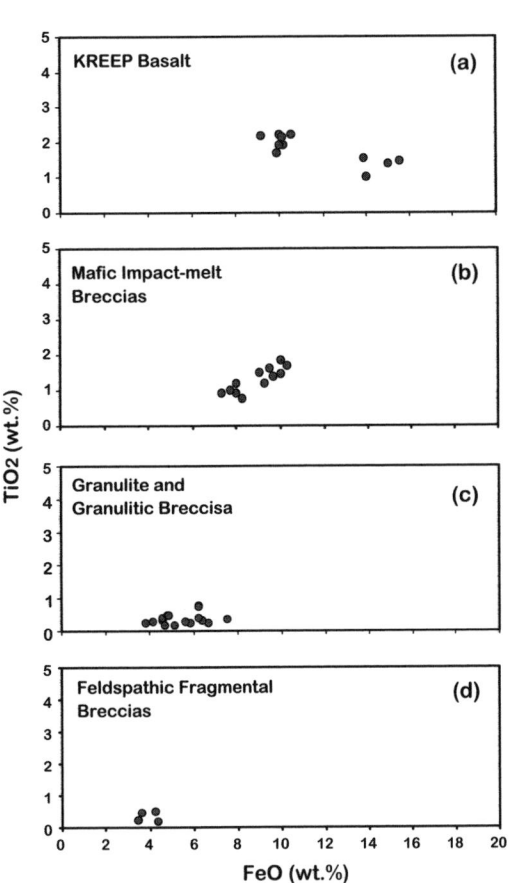

Figure 3.6. TiO_2 vs. FeO for KREEP basalts, mafic impact-melt breccia groups, granulites and granulitic breccias, and feldspathic fragmental breccias. For KREEP basalts, the cluster at ~10% FeO is from Apollo 15 samples and the cluster at 14–16% wt% FeO is from a single Apollo 17 sample, 72275. Decreasing FeO and TiO_2 concentrations from (a) to (d) is probably a complex function of compositional stratigraphy of the crust with depth as well as wide-scale lateral variations associated with major geochemical terranes (see also Fig. 3.8).

Distribution. Recent remote sensing and photogeologic analysis reveal several locations where KREEP basalts may occur on the surface the Moon. Two such places occur near the Apollo 15 site, where KREEP basalt is a common component of the regolith. The Apennine Bench has been cited as a likely location for KREEP basalt flows (Hawke and Head 1978; Spudis 1978), and Clementine and Lunar Prospector data suggest the presence of KREEP basalt in the target region of the Aristillus crater, located just to the west of the Apollo 15 site (Gillis and Jolliff 1999; Blewett and Hawke 2001). KREEP basalts do not appear to make up any of the extensive mare flows in the Procellarum-Imbrium region, but they could be a significant buried component or a component of nonmare materials in the Procellarum KREEP Terrane. The absence of KREEP basalt among rock fragments in regolith of the Apollo 17 site (Jolliff et al. 1996) indicates that KREEP basalt is most likely not a local component there, and its occurrence as a clast component in the station 2 boulder is consistent with a remote origin (e.g., Spudis and Pieters 1991).

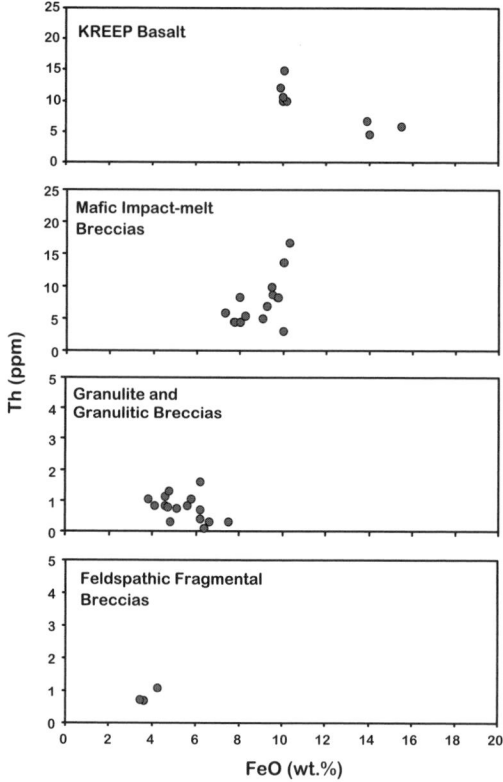

Figure 3.7. Th vs. FeO for KREEP basalts, mafic impact-melt breccias, granulitic breccias, and feldspathic fragmental breccias. Note the difference in thorium scales between the upper and lower plots.

Most of the materials at the Apollo 14 site that are mineralogically and geochemically similar to KREEP basalt are probably clast-free/poor, crystalline impact melt. The high Th concentration of KREEP basalts coupled with the observed global distribution of Th suggests that their occurrence is restricted to the Procellarum KREEP Terrane (see Section 5).

2.1.5. Mafic impact-melt rocks and breccias. Detailed descriptions of the different lunar breccia (polymict) groups can be found elsewhere and constitute the subject of a voluminous literature and extensive investigations. In particular, the major groups of polymict breccias are described and summarized in *The Lunar Sourcebook* (Heiken et al. 1991) and in Chapter 5 of *Planetary Materials* (1998). In this and in subsequent sections, we focus on key characteristics of several of the main lunar breccia types as indicators of specific crustal components or geologic provinces of the Moon. As their compositions are more mafic than typical lunar highlands surface materials, they are thought to have formed mainly as ejected impact melt from basin-sized impacts that penetrated deep enough into the lunar crust to encounter rocks more mafic than observed at the surface.

Mafic impact-melt breccias from the Apollo landing sites, particularly Apollo 14, 15, 16, and 17, are abundant and form compositionally distinct groups. These groups exhibit a range of major-element compositions and incompatible-element enrichments. While concentrations of incompatible elements span a significant range, inter-element ratios vary little and have been used to infer a common lunar KREEP component. This discussion of mafic impact melt breccias draws upon groupings defined by numerous past studies (e.g., Ryder and Wood 1977; Spudis and Ryder 1981; Korotev 1994; Jolliff et al. 1996). These groupings are the same as those presented by Korotev (1998) in which he presented average concentrations of several of the key elements sensed remotely by the Lunar Prospector gamma-ray spectrometer. These groups were also used by Jolliff (1998) and the compositions listed in Table A3.6 are from that paper.

Mineralogy and mineral modes. The mafic impact-melt breccias have also been referred to as basaltic impact melts, and depending on their specific textures, they have been called crystalline-matrix breccias, "dimict" breccias, poikilitic breccias, or aphanitic-matrix breccias (see Stöffler et al. 1980). The mineralogy of the mafic impact-melt breccias is very similar to that of the KREEP basalts, except that clasts are as variable as the known lunar rock types. Importantly, clasts tend to be of relatively calcic plagioclase and magnesian mafic silicates when compared to the matrix, and the matrix usually contains fine-grained accessory minerals that provide the incompatible-element enrichment.

Compositions. The mafic impact-melt breccia groups have concentrations of FeO >7 wt% and Al_2O_3 <22 wt% (Table A3.6). Though TiO_2 concentrations range only from ~1 to 2 wt% (Fig. 3.6), incompatible elements range from ~50 to >200 times chondritic values (Fig. 3.7). The key compositional characteristic of all the mafic impact-melt breccias is that the incompatible elements, regardless of concentration, occur in KREEP-like proportions with very little deviation.

The Apollo 14 melt breccias have a broad range of incompatible-element concentrations, but most have very similar inter-element ratios (Warren 1989; Jolliff et al. 1991b). The concentration of Th, for example, ranges from about 10–30 ppm, averaging ~17 ppm. Compositions of individual samples vary according to clast contents and the proportions of clasts and matrix. Jolliff et al. (1991b) selected a set of rock fragments of similar texture whose compositions form a very tight group to define an average impact-melt-breccia composition represented by the majority of Apollo 14 melt breccias; this composition is similar to the "average high-K KREEP" of Warren (1989).

Apollo 15 melt breccias were divided into 5 groups by Ryder and Spudis (1987) on the basis of compositional clustering of samples, and these groups have different levels of trace-element enrichment. Nevertheless, analyses of additional samples have revealed a continuum of compositions between the groups (Lindstrom et al. 1988). A common interpretation has been that distinctive group compositions indicate that members of a particular group originated in the same impact event, and that different groups require different origins. However, [40]Ar/[39]Ar dating by Dalrymple and Ryder (1993; 1996) has shown that even within groups, compositionally similar samples apparently have different ages, thus interpretations relating a group of impact-melt breccias to a specific impact are not straightforward.

Of the five Apollo 15 groups, group A has the highest level of incompatible-element concentrations at about 70% of high-K KREEP, overlapping in part the high concentrations of the Apollo 14 group. Group B compositions have incompatible-trace-element concentrations at about 50% of high-K KREEP. Group C compositions are notably more aluminous and less mafic, and may simply be the more feldspathic extreme of compositional group B (Korotev 1994). Group D consists of well-studied impact-melt rocks 15445 and 15455, taken by Ryder and Spudis (1987) to be potential candidates for samples of the Imbrium melt sheet (as opposed to ejected melt or reworked Imbrium ejecta). Concentrations of incompatible elements in group D are relatively low, being about 20% of those of high-K KREEP.

Among the Apollo 17 samples, Spudis and Ryder (1981) distinguished two texturally and compositionally distinct groups: poikilitic breccias and aphanitic breccias. The poikilitic breccias are by far the more abundant of the two and form the large breccia boulders at the base and on the slopes of the North and South Massifs. A third compositionally distinct group of breccias described by Jolliff et al. (1996) occurs among small rock fragments found in the massif soils. These small samples were distinguished as a group on the basis of their similar and high concentrations of incompatible trace elements. They are similar in trace-element composition to Apollo 15 group B and Apollo 16 group 1M, but have siderophile-element signatures more like the other Apollo 17 groups.

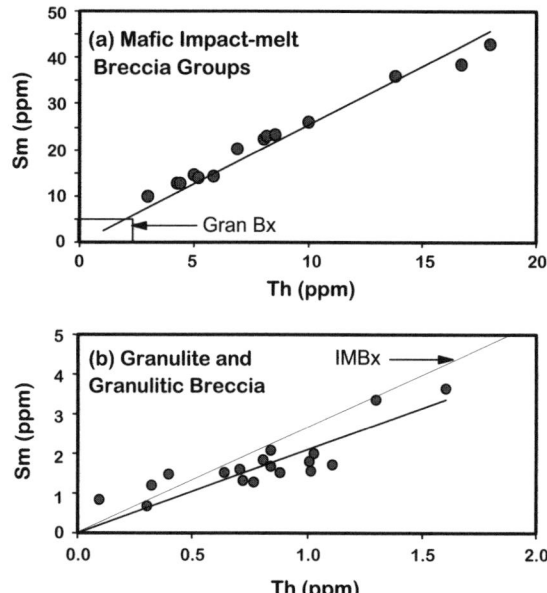

Figure 3.8. Sm vs. Th in mafic impact-melt breccias and in feldspathic granulites and granulitic breccias. Note different Sm scales between (a) and (b). Gran Bx and IMBx denote granulitic breccia and impact melt breccia, respectively.

Among all of the mafic impact-melt breccia groups, incompatible trace elements exhibit extremely tight correlations, such as between Th and Sm (Fig. 3.8). These correlations form a key line of evidence that the impacts that produced these melt breccias tapped a common, deep reservoir. An inverse correlation exists between Mg/(Mg+Fe) and Th contents (see Fig. 3.9), which suggests a relationship between Fe and incompatible-element enrichment. The Apollo 15 and 17 KREEP basalts do not adhere to this correlation.

Age relationships. The formation ages of numerous impact-melt breccias have been determined, principally using $^{40}Ar/^{39}Ar$ methods (e.g., Dalrymple and Ryder 1993, 1996).

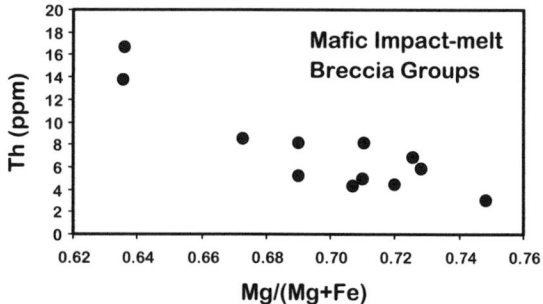

Figure 3.9. Th vs. molar Mg/(Mg+Fe) in the mafic impact-melt breccias. Data are group averages from Table A3.6.

However, the dating method is difficult because of incomplete degassing of clasts and subsequent disturbances. Some of the Ar release patterns are simply too difficult to interpret unambiguously. Still, most of the dated mafic impact-melt breccias have ages in the range of ~3–3.9 Ga, so either most of the impact basins formed at about this time, or the samples are dominated by a small number of events. This topic is considered in more detail in Chapter 5. The important point here is that a source of KREEP enrichment was apparently not widespread until this time, judging by the compositions of other key breccia groups discussed below.

Depths of origin. Depths of origin are difficult to constrain in these breccias. Minerals that might record depths of melting generally don't survive the impact-melting process, and the clasts present may have been entrained in a different part of the target as the melt left the excavation cavity. Judging by mineral compositions and textures, most of the clasts in the sampled melt breccias appear to have relatively shallow origins. An argument based on compositions is that the breccia bulk compositions, and especially the matrix material, are difficult to match by mixing known rock compositions. The implication is that the melted component includes materials from depth that are not common or not found near the surface. That component is typically richer in FeO and incompatible elements than the clasts (Ryder and Wood 1977; Ryder 1979). The broad-scale inference is that the crust must become more mafic with depth. Although the ejecta of most large basins are more mafic than the most feldspathic parts of the upper crust, we now know from global remote sensing that broad lateral variations in composition exist in the crust. Thus, any layering that might be present locally may not be representative globally.

Distribution. Studies of melt rocks and breccias from terrestrial impact craters indicate that most small impacts very effectively homogenize target materials. A notable exception is the 100 km diameter Popigai crater where distinct compositional variability in impact-melt lithologies is found that is related to differences in target rock compositions (Kettrup et al. 2003). Regardless, the common interpretation of lunar impact-melt rocks is that specific compositional groups represent individual impacts. This interpretation, coupled with the observation that impact-melt groups from a given landing site bear more resemblance to other groups from the same site than they do to groups from other sites, has led to the paradigm that the impact-melt breccias originate from the nearest large basin. This approach has been challenged in recent years on several grounds (e.g., Haskin et al. 1998). One argument is that the very large impacts that formed basins likely struck targets that were heterogeneous on a scale large enough to prevent total homogenization, especially in the clast content entrained by melt exiting different regions of a basin as has been demonstrated for the Popigai crater (Kettrup et al. 2003). Another is that isotopic age data are difficult to interpret and are not entirely consistent with the nearest-basin paradigm. Finally, impact ejecta modeling (Haskin 1998; Haskin et al. 2003) suggests that ejecta from one or two of the latest and largest of the nearside basins (Imbrium and Serenitatis) may dominate the mafic impact-melt breccias sampled at all of the Apollo sites.

A common interpretation of the KREEP trace-element similarity of the diverse mafic melt-breccia groups is that the impacts that formed these tapped a common source of very incom-patible-element-rich material, presumably a deep global KREEP layer sandwiched between the crust and mantle. Global remote sensing by Lunar Prospector shows, however, that such KREEP rich material was strongly confined to the Procellarum KREEP Terrane and was not ex-cavated by all large basins (especially notable are the Crisium and South Pole-Aitken basins).

2.1.6. Granulitic breccias. The feldspathic granulites and granulitic breccias are found at most of the landing sites and within the nonmare lunar meteorites (e.g., Dhofar 026, Cohen et al. 2004) as small rocks, rock particles in regolith, and lithic clasts in breccias. Their textures suggest heating and recrystallization, and compositional as well as shock features in some indicate a relationship to impact processes. However, in many cases, the geologic setting in which this metamorphism occurred remains unclear. These rocks have compositions that are similar to estimates of the bulk lunar crust, and many are clearly mixtures of different precursor rock types, yet they are not easily explained as chemical mixtures of the known igneous rocks (Korotev and Jolliff 2001).

Mineralogy and mineral modes. The mineralogy and textures of granulitic breccias are described in Bickel et al. (1978), James and Hammarstrom (1977), Warner and Phinney (1977), Bickel and Warner (1978), Lindstrom et al. (1986), Cushing et al. (1999), and Cohen et al. (2004). These rocks form a continuum between fine-grained igneous textures, impact-melt textures (e.g., poikilitic-poikiloblastic), and metamorphic textures (granoblastic). In terms of normative mineralogy, these rocks typically have about 80 vol% plagioclase and most contain olivine, orthopyroxene, and clinopyroxene. The magnesian granulites typically have olivine > pyroxene, and the ferroan ones have pyroxene > olivine. From their normative mineralogy, these all lie near the boundary between troctolitic anorthosite, noritic anorthosite, and anorthositic norite.

Compositions. Meteoritic siderophile-element contamination indicates impact involvement in the origin of these rocks. Average compositions cover a relatively restricted range; for example, the range of Al_2O_3 is 25–29 wt% averaging 26.8 wt%, MgO is 4–9 wt% averaging 7 wt%, and FeO is 3.8–7.5 wt% averaging 5.4 wt% (see Table A3.7). Incompatible element concentrations are extremely low; for example, the range of Th is 0.1–1.6 ppm averaging 0.8 ppm, and Sm is 0.7–3.7 ppm averaging 1.7 (see Fig. 3.8). The most common interpretation of these compositions is that they represent upper-crustal materials uncontaminated by the excavation of KREEP-rich materials from the Procellarum KREEP Terrane by impacts such as Imbrium and Serenitatis.

The granulites and granulitic breccias can be divided into magnesian and ferroan varieties (Lindstrom and Lindstrom 1986) and for some elements, such as FeO and Sc, the compositions appear to form two distinct groups. Both groups, however, are found at a given landing site and the significance of the variation and grouping in Mg/(Mg+Fe) is not known. The magnesian group has Mg/(Mg+Fe) ranging from 0.72 to 0.78 with an average of 0.75, whereas the ferroan group ranges from ~0.5 to 0.7 with an average of 0.62. It is important to note that the designation of one of these groups as "magnesian" does not necessarily imply any relationship to the magnesian-suite igneous rocks, especially those magnesian-suite rocks that have been shown to have a trace-element-rich lineage (Korotev and Jolliff 2001). At the Apollo 17 site, for example, the magnesian granulitic breccia compositions do not correspond to the magnesian-suite norites, troctolites and troctolitic anorthosites that are also found at the site.

Age relationships. Age data are difficult to determine on these rocks (see Cohen et al. 2004) but for those that have been reported (all of which are based on the $^{40}Ar/^{39}Ar$ chronometer), ages range from about 3.8 to 4.2 Ga. The younger of these ages may have been reset by the impacts that exhumed the breccias (Papike et al. 1998). The formation of most of these rocks appears to predate the formation of the large, late basins such as Imbrium and Serenitatis.

Depths of origin. While textures of the granulitic breccias indicate burial and thermal metamorphism, some disagreement exists as to the depth of burial and duration of annealing needed to produce these (Cushing et al. 1999). Although textures and mineral compositions reflect annealing processes, these do not necessarily require deep burial and slow heating/cooling. In some cases, but not all, mineral compositional zoning is minimal, suggesting equilibration. Judging by FeO concentrations, typical compositions are similar to highlands compositions sensed remotely, which suggests that the granulitic breccias might be representative of the near-surface and upper crust.

Distribution. The granulites and granulitic breccias are found at all Apollo and Luna landing sites, but they are most common in samples from the eastern nonmare sites. The low incompatible trace-element concentrations further suggest an origin away from the Procellarum KREEP Terrane. Ages and textures are consistent with formation in large impacts, burial coupled with heating beneath either a thick ejecta blanket or an impact-melt sheet, and later excavation to the surface. The clustered compositions (such as high Al_2O_3 and low FeO) of the granulitic breccias suggest that they might be fairly representative of the average compositions of significant areas of the ancient upper crust of the Moon.

2.1.7. Feldspathic fragmental breccias. Fragmental breccias are a common product of impact processes and they occur at all Apollo landing sites (e.g., Lindstrom et al. 1977; James 1981; Lindstrom and Salpas 1981; Norman 1981; Stöffler et al. 1985). Because the fragmental breccias consist of a diversity of rock types mixed together, they potentially record information about the target areas of large impacts. When a group of such breccias is found to have a common composition, that composition takes on added significance as potentially representing an especially broad area. Such is likely to be the case for the plagioclase-rich fragmental breccias, which occur most abundantly at the Apollo 16 site. In recent years, the similarity between their composition and the compositions of many of the feldspathic lunar meteorites has led to the conclusion that these may best represent vast expanses of the feldspathic highlands surface far from the influence of mixing with more mafic ejecta from the large and late nearside basins such as Imbrium and Serenitatis. Key clast types found in these breccias include granulitic breccia, cataclastic anorthosite, and fragment-laden feldspathic impact melt (Papike et al. 1998). The compositions of these breccias provide a key constraint on the composition of the Feldspathic Highlands Terrane and the feldspathic regions of the Moon's upper crust.

Compositions and mineral modes. Of the major breccia groups, the feldspathic fragmental breccias are the most aluminous, with typical Al_2O_3 concentrations of 29–31 wt%, which corresponds to noritic anorthosite (see Table A3.8). The group from Apollo 16, North Ray Crater, has ferroan bulk compositions and components are dominated by ferroan-anorthositic-suite lithologies. They have extremely low incompatible-element concentrations (e.g., TiO_2 <0.5 wt% and Th <1 ppm, see Figs. 3.6 and 3.7), and their compositions are very similar to the most feldspathic of the lunar meteorites (see Table A3.9).

Age relationships. The Apollo 16 breccias contain no clasts younger than 3.9 Ga, which is taken to be approximately the age of their formation. They have so little KREEP that it is reasonable to assume that they were assembled prior to the widespread distribution of KREEP by the Imbrium and other impact basins in the Procellarum KREEP Terrane.

Distribution. The feldspathic fragmental breccias are most common at the Apollo 16 site, specifically, in the ejecta of North Ray Crater. Stöffler et al. (1985) argued that the materials excavated from North Ray Crater represent the Descartes Highlands, which are, in turn, part of the ejecta deposit of Nectaris (see also Bussey and Spudis 2000). At the Apollo 16 landing site, the Descartes Formation is buried beneath the Cayley Formation, which is considered to be largely an Imbrium-produced deposit. Remote sensing indicates that the distinctive composition of materials excavated by North Ray crater as reflected by breccias such as 67455 (highly feldspathic and ferroan) is widespread to the east, towards and beyond Nectaris. In fact, the feldspathic fragmental breccias from North Ray Crater, along with the feldspathic lunar meteorites (see next section) provide the most plausible match for the vast anorthositic region of the Feldspathic Highlands Terrane on the lunar farside north of the equator.

2.1.8. Lunar meteorites. The lunar meteorites now constitute a large enough group of samples that they form a statistically significant data set. At the time of writing, the number of stones is in excess of 80 and the number of distinctly different sources, counting likely pairs (pieces of the same rock or different rocks likely to have come from the same crater) as a single source, is about 30 (see Tables 3.7 and 2.1). Because these are likely to represent more randomly sampled locations on the Moon than the six Apollo and 3 Luna sample locations, the distribution of rock types and compositions represented by the meteorites provide an important comparison to global compositional and mineralogical data sets. The range and distribution of compositions of the lunar meteorites contrast with those of the Apollo and Luna samples and reflect both the abundance of highly feldspathic highlands and of low-Ti basalts.

The meteorites fall into three broad groups, which are reflected on compositional plots of TiO_2 vs. FeO and Th vs. FeO (see Fig. 3.10, and Tables A3.9 and 2.1): (1) feldspathic regolith and impact-melt breccias, (2) mixed mare-highland breccias, and (3) mare basalts and gabbros. The feldspathic breccias form the most numerous group, accounting for ~62% of the samples both by number and by mass. The clustering of compositions of the feldspathic

Table 3.7. List of lunar Meteorites according to likely source pairs.

Designation	Lithology	mass (g)	mass all	TiO$_2$	Al$_2$O$_3$	FeO	MgO	Mg'	Th (ppm)	Reference
Dhofar 081 (plus 280, 490, 910, 1084)	Fsp Frg Bx	174	459	0.15	30.5	2.93	2.82	63.1		Warren et al. (2001); Russell et al. (2004)
Dhofar 026	Fsp Gr Bx	148		0.22	29.6	4.06	3.92	63.3	0.4	Cohen et al. (2004)
Northwest Africa 482	Fsp IMBx	1015		0.16	29.4	3.80	4.28	66.8	0.2	Korotev et al. (2003b)
Dhofar 302, 305-307, 309-311, 489, 730-731, 908, 909, 911, 925, 1085	Fsp Rbx		1068		29					Russell et al. (2002); Nazarov et al. (2002, 2004), and Demidova et al. (2003)
QUE 93069 (plus 94269)	Fsp Rbx	21	25	0.25	28.9	4.44	4.53	64.5	0.5	Korotev et al. (1996)
Dar al Gani 400	Fsp Rbx/IMBx	1425		0.18	28.9	3.70	4.88	70.1	0.3	Korotev et al. (2003b); simple average
Dar al Gani 262	Fsp RBx	513		0.20	28.4	4.47	5.50	68.7	0.4	Korotev et al. (2003b)
MAC 88105 (plus MAC88104)	Fsp RBx	663	724	0.23	28.1	4.28	4.05	62.8	0.4	Jolliff et al. (1991)
Yamato 86032 (plus 82192/3)	Fsp Frg Bx	648	712	0.23	27.5	4.82	5.32	66.4	0.2	Lindstrom et al. (1991a); simple average
"Achondrite 1153"	Fsp Rbx	??		0.18	27.4	5.20	3.89	57.1		Yanai and Ueda (2000)
Yamato 791197	Fsp Rbx	52		0.40	27.3	5.71	5.76	64.3	0.3	Korotev et al. (2003b)
Northeast Africa 001	Frg Bx	262			~27				0.2	Haloda et al. (2005); Korotev and Irving (2005)
Dar al Gani 996	Fsp Rbx	12								Russell et al. (2003)
Dhofar 025 (plus 301, 304, 308)	Fsp RBx	751	772	0.29	26.9	4.90	7.10	72.1	0.5	Korotev et al. (2003b)
Pecora Escarpment (PCA) 02007	Fsp RBx	22		0.28	26.5	5.90	6.69	66.9	0.4	Korotev et al. (2004)
Dhofar 733	Fsp Gr Bx	98								Russell et al. (2003)
Allan Hills 81005	Fsp Rbx	31		0.27	25.7	5.47	8.17	72.7	0.3	Korotev et al. (1983)
Meteorite Hills (MET) 01210	Anorth-bearing Bas Bx	23								Antarctic Met Newslett (Feb 2004)
Yamato 983885	Fsp RBx w/ bas	290			22.0					Kaiden and Kojima (2002); Warren and Bridges (2004)
Calcalong Creek	Fsp RBx w/ basalt	19		0.80	20.9	10.9	5.14	45.7	4.6	Hill and Boynton (2003)
Yamato 981031 (plus 793274)	Mixed Bas/Fsp RBx	186	195	0.58	17.7	12.8	8.54	54.5	1.0	Korotev et al. (2003a); simple average
QUE 94281	Mixed Bas/Fsp RBx	23		0.60	16.0	13.2	9.10	55.1	1.0	Jolliff et al. (1998)
Sayh al Uhaymir (SaU) 169	KREEP IMBx w/ Reg	206		2.21	15.9	10.7	11.1	64.9	32.7	Gnos et al. (2004)
EET 87521 & 96008	Bas/Gab Frg Bx		84	0.53	14.8	16.2	8.91	49.3	1.0	Korotev et al. (2002, 2003a); simple average
Northwest Africa 3136	Anorth-bearing Bas Bx	95				15.3			1.3	Kuehner et al.(2005); Korotev and Irving (2005)
Yamato 793169	Low-Ti Basalt	6		2.11	11.8	20.7	5.58	32.5	0.7	Korotev et al. (2003a); simple average
Northwest Africa 773 basaltic matrix	Bas RBx w/ Ol-Gab	633		0.78	10.6	17.3	13.2	57.6	2.2	Fagan et al. (2003); and Jolliff et al. (2003)
Northwest Africa 773 Ol-cumulate	Ol-Gab cumulate			0.24	3.100	18.1	27.1	72.8	1.6	
Asuka 881757	Low-Ti Basalt/gabbro	442		2.42	10.0	22.6	6.04	33.2	0.5	Korotev et al. (2003a); simple average
LAP 02205 (plus LAP xxxxx)	Low-Ti Basalt	1226	1875	3.23	9.9	22.3	6.34	33.6	2.1	Anand et al. (2004); Jolliff et al. (2004)
Northwest Africa 032 (plus 479)	Low-Ti Basalt	300	456	3.21	9.0	22.6	8.21	39.3	1.9	Fagan et al. (2002)
Dhofar 287	Basalt w/ RBx	154		2.86	8.1	22.2	12.3	49.7		Taylor et al. (2001)

Mass (g) column is the mass of the first meteorite listed. "Mass all" is the combined mass of the group if more than one meteorite is listed. Pairing as suggested by Korotev (see note below).
Fsp RBx = feldspathic regolith breccia; IMBx = impact-melt breccia; Gr Bx = granulitic breccia; Frg Bx = fragmental breccia; Bas = basalt; Gab = gabbro
EET: Elephant Moraine; LAP: LaPaz Icefield; MAC: MacAlpine Hills;QUE: Queen Alexandra Range (Antarctic locales).
An up-to-date, though subject-to-frequent-change resource at time of writing: http://epsc.wustl.edu/admin/resources/meteorites/moon_meteorites_list.html (maintained by R. Korotev)

breccias reflects a common and highly feldspathic composition that is under-represented by the samples of the Apollo and Luna missions. The mixed mare-highland breccias could correspond to areas marginal to mare basalt flows, a regolith formed by vertical mixing between a surface mare basalt and underlying feldspathic crust, or they could correspond to cryptomaria where older basalt flows were covered by younger basin ejecta. The basaltic meteorites account for some 20% of the lunar meteorites which is somewhat greater than the approximately 12% surface area of the Moon covered by basalts. They account for ~30% of the lunar meteorite mass, but this number is skewed by the abundant (1.875 kg) LaPaz icefield meteorite suite (Anand et al. 2004; Jolliff et al. 2004). The basaltic meteorites are remarkable in that they are all either of low-Ti or very-low-Ti composition, in contrast to the

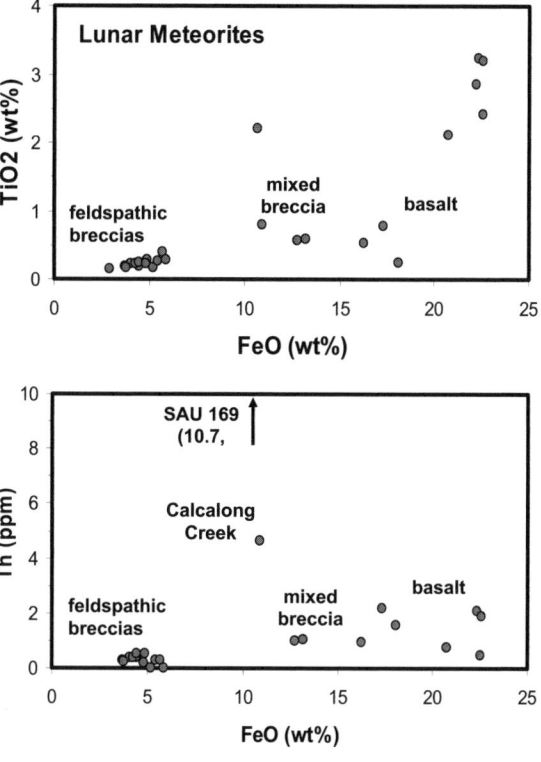

Figure 3.10. Compositions of the lunar meteorites. (a) TiO_2 vs. FeO and (b) Th vs. FeO. Data from Table A3.9.

Apollo basalt suites that were distinctly bimodal. The basaltic meteorites also include coarser-grained varieties than typical Apollo and Luna basalts, commonly exhibiting fine lamellar exsolution in pyroxene. One of the basaltic meteorites, Northwest Africa 773, contains a prominent olivine-gabbro clast component whose mineral proportions clearly indicate a cumulus origin (Fagan et al. 2003; Jolliff et al. 2003). A lack of mineral indicators of slow cooling and thus of deep origins, however, indicates that even this cumulate assemblage probably formed in a shallow intrusive setting. Interestingly (and importantly), NWA773 was shown by Ar-Ar dating (Fernandes et al. 2003) and by Sm-Nd dating (Borg et al. 2004) to be the youngest igneous rock thus far analyzed (2.91 Ga by Ar-Ar, and 2.865 Ga by Sm-Nd).

The lunar meteorite groups correspond broadly to the major geologic terranes as expressed at the Moon's surface, with one apparent exception. The feldspathic breccias correspond well to the Feldspathic Highlands Terrane, with FeO concentrations of 3–6 wt% matching well with the peak of global values, especially over the northern far-side highlands. The percentage of feldspathic breccias among the meteorites is remarkably consistent with the proportion of the Moon's surface that falls in the range of "feldspathic" according to its FeO content (i.e., less than ~6 wt%). The low and very-low-Ti basaltic meteorites correspond to the abundant low-Ti basalt regions of the Moon as indicated by the global distribution of Ti (Giguere et al. 2000; Lucey et al. 2000; Gillis et al. 2003). Even the Procellarum KREEP Terrane is now represented by at least two of the meteorites, Calcalong Creek (Hill and Boynton 2003) and Sayh al Uhaymir 169 (Gnos et al. 2004), which have elevated concentrations of incompatible elements. The only major terrane apparently not represented among the meteorites is the South Pole-Aitken basin, which is expected to yield samples that are relatively mafic but not necessarily basaltic. Although several of the meteorites (Yamato 981031, Yamato 793274 and QUE 94281) have bulk compositions that are intermediate in composition (e.g., ~13 wt% FeO), these are composed of bimodal mixtures of feldspathic and basaltic components (Lindstrom et al. 1991; Jolliff et al. 1998; Arai and Warren 1999).

2.2. Seismology

Of all the geophysical methods used to study the Earth's structure, seismology is uniquely suited to determine many parameters that are critically important to understanding its dynamic behavior. For this reason, NASA deployed seismometers on the lunar surface during each of the Apollo missions. Four of the passive seismic stations that were installed came to operate concurrently from Dec. 1972 to Sep. 1977, at which point the transmission of data was suspended because of a lack of funding. These stations were placed in an approximate equilateral triangle with distances between corners being about 1100 km and two of them, Apollo 12 and 14, placed only 180 km apart in one corner. In the 8-year period beginning with Apollo 11 when the experiment was underway, more than 12000 events were recorded (see Table 3.8). These events were found to differ in signal characteristics and were grouped into four categories: deep moonquakes, shallow moonquakes, meteoroid impacts and thermal moonquakes (see reviews by Toksöz et al. 1974, Nakamura et al. 1982, Lognonné and Mosser 1993, and Lognonné 2005).

Table 3.8. Catalogued seismic events (Nakamura et al. 1982; Nakamura 2003).

Type	Number of events
Artificial impacts	9
Meteoroid impacts	1743
Shallow moonquakes	28
Deep moonquakes	7245
Unclassified	3533
Total	**12,558**

In comparison to the Earth, the moonquakes are very small magnitude events with the largest, the shallow moonquakes, having Earth-equivalent body-wave magnitudes of about 5 (Goins et al. 1981b). The deep moonquakes are by far the most numerous events and generally have magnitudes less than 3. Stress drops associated with the shallow events are about 40 MPa (400 bars), whereas stress drops associated with the deep moonquakes are much smaller, being only about 10 kPa (0.1 bar). This manifests itself in a significant difference in the level of seismicity between the Earth and Moon. The annual seismic energy release from moonquakes was estimated to be about 10^{10} J, whereas it is ~10^{18} J for the Earth (Goins et al. 1981b). The reason that such small events were observable at all is because the Moon's surface is very quiet in comparison to the Earth, with no oceans or an atmosphere to produce micro-seismic background noise.

The types of seismicity observed on the Moon are distinguishable from those that we observe on the Earth. Subsets of the deep moonquakes have nearly identical waveforms, implying a common origin in a small hypocentral region (Nakamura 1978), and occur monthly (Lammlein et al. 1974) deep within the Moon (~700-1200 km). These events were originally classified as belonging to 109 distinct hypocentral regions (Nakamura et al. 1982), and was recently updated to at least 166 (Nakamura 2003). The monthly intervals at which deep moonquakes occur suggest a strong relationship between them and the tides raised on the Moon by the Earth and the Sun. A large percentage of the seismic events observed on the short period components of the Apollo seismic experiment are very small moonquakes that occur with great regularity. These are believed to be natural seismic occurrences generated by small near-surface moonquakes less than about 1.5 to 4 km away that are triggered by diurnal thermal variations (Duennebier and Sutton 1974). Shallow moonquakes (~50 to 220 km depth, Khan and Mosegaard 2002) are the most energetic seismic sources observed on the Moon, although they are less abundant than the other types of seismic events, with an average of only

5 events occurring per year (Nakamura 1977). Shallow moonquakes are the only events that are believed to be related to tectonic activity (Nakamura et al. 1982).

The seismic events due to meteoroid impacts provide important information on the interplanetary medium, and the impacts detected correspond to meteoroids having estimated masses in the range from 100 g to 1 t or greater (Duennebier et al. 1975; Dorman et al. 1978; Oberst 1989). In addition, Earth-crossing cometary and asteroidal objects have been identified from the seismic data with the former occurring in clusters related to known meteor showers (Oberst and Nakamura 1991).

The lunar seismic signals were found to differ from terrestrial ones in their anomalously long continuance and high frequency content. For instance, the signal from the S-IVB impact continued for up to 4 hours. First arrivals typically are small, with slowly building amplitudes obscuring much of the information usually present in terrestrial seismograms. These characteristics are believed to be due to intense scattering in the highly heterogeneous and porous lunar regolith coupled with a low level of attenuation. The lunar crustal Q-value is an order of magnitude greater than the Earth's and is ascribed as being due to the absence of water and volatiles in the Moon (Latham et al. 1970a) (Q^{-1} is proportional to the fraction of energy that is dissipated in the lunar interior per loading cycle). This apparent complexity together with the small number of usable seismic events, the paucity of stations and their locations set certain limits on the amount of information that could be extracted from this dataset.

It has, nevertheless, been possible to retrieve a set of first arriving P- and S-waves that constitute the primary source of data from which several models of the lunar velocity structure have been obtained. Using linear inversion techniques (Toksöz et al. 1974; Goins et al. 1981a; Nakamura et al. 1982), the gross structure of the lunar interior has been delineated. Khan et al. (2000) reanalyzed the same dataset using a non-linear inversion technique, obtaining a more realistic view of the inherent resolution associated with the lunar seismic velocity profile. Lognonné et al. (2003) have further performed a complete reanalysis of the original data records which, when compared with the previous analyses, offers an assessment of the ambiguities in picking lunar arrival times.

Finally, we note that the information content of the Apollo seismic data has not yet been completely exhausted. As an example, the technique of using seismic-receiver functions has recently been shown to be viable in detecting converted phases at the crust-mantle interface (Vinnik et al. 2001). Additionally, waveform cross-correlation techniques have recently been used by Nakamura (2003) as a tool to reclassify and group the lunar seismic events. This study has discovered an additional 88 deep moonquake nests, and has further shown that 31 of the original 109 deep nests were incorrectly classified. While arrival times for these new deep moonquake sources have not yet been picked, they would aid in refining the seismic velocity structure of the deep lunar interior. It is further possible that some of these new sources have a far-side origin, allowing the question of whether of not the Moon possesses a core to be addressed (Nakamura 2005).

2.2.1. The crust-mantle interface. The Earth's internal structure can be thought of as a series of concentric shells, each of which transmits seismic waves at a different speed. The boundaries between shells are marked by seismic discontinuities at which the speed of transmission makes a sudden jump. The shallowest seismic discontinuity that is global in its extent is known as the Mohorovicic discontinuity, or Moho for short, and the terrestrial crust is almost always defined as the region above this discontinuity. The Moho was defined by Steinhart (1967) as the level where the compressional wave velocity first increases rapidly or discontinuously to a value between 7.6 km/s and 8.6 km/s. In the absence of a recognizable steep gradient, it is taken as the level where the P-wave velocity first exceeds 7.6 km/s. This definition was also employed in delineating the crust-mantle interface of the Moon (i.e., the "lunar Moho").

Geometrical ray theory has proven to be the most successful technique in tackling the lunar seismograms. Using this method, the crustal structure beneath the Apollo seismic stations, in particular stations 12 and 14, has been investigated using the artificial impact data set (this includes impacts of the lunar module ascent stage and the upper stage of the Saturn rocket). Epicentral distances range from 9 to 1750 km, although not all of these events have been recorded due to their limited energy and the sequential emplacement of the stations.

Several studies have reported crustal velocity profiles beneath the Mare Cognitum region near the Apollo 12 and 14 landing sites by inverting travel times of first P-wave arrivals (S-wave arrivals from the artificial-impact events are highly indistinct and accordingly were not used in these studies). Using linear inversion techniques, these models generally found that the seismic velocity gradually increased from the surface to a depth of ~20 km, at which point a seismic discontinuity was found. Below this depth, the velocity was found to be approximately constant down to the crust-mantle interface. Estimates for the thickness of the crust at this locale varied from 65 km (Toksöz et al. 1972), over 55 km (Toksöz et al. 1974), and finally 58 ± 8 km (Nakamura et al. 1982). It should additionally be mention that Goins et al. (1981c) obtained a crustal thickness of 75 km beneath the Apollo 16 site based on their identification of secondary phases that were interpreted as reflections off the lunar Moho. An additional reflected phase suggested to them that a 20-km discontinuity was present at this site as well. The data presented as evidence of these purported reflected and refracted arrivals, however, are not entirely convincing and await a reanalysis.

Figure 3.11 shows the seismic velocity model of the lunar crust and upper mantle as obtained by Khan et al. (2000). This model was constructed using a Bayesian inversion technique, and their seismic model is presented as a series of 1-D marginal probability distributions at 1-km depth intervals. As the contours in this image correspond to the likelihood of obtaining a given velocity at a given depth, it should be noted that the most probable velocity as a function of depth does not necessarily correspond to the most probable velocity profile. The construction of this profile used two arrivals at the Apollo 15 station and one at

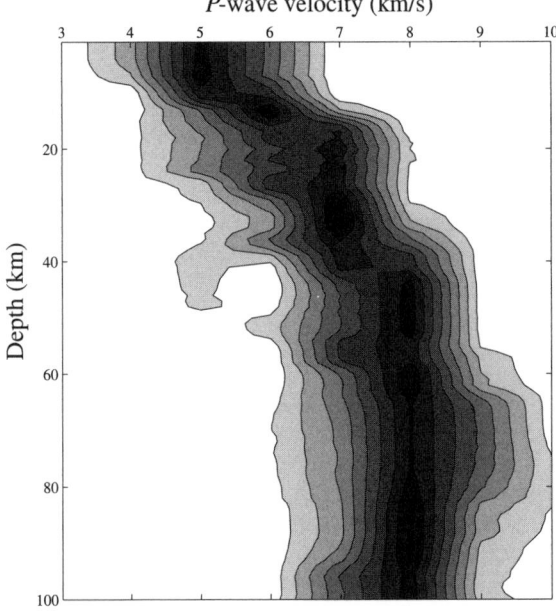

Figure 3.11. P-wave velocity model of the upper 100 km of the Moon derived from the non-linear Bayesian inversion of Khan et al. (2000). The marginal probability of the seismic velocity is plotted at one-kilometer depth intervals, and the contours define nine equally sized probability intervals.

the Apollo 16 station, with the remainder coming from stations 12 and 14. Hence, this velocity profile should be considered as representative of the Mare Cognitum region near the Apollo 12 and 14 landing sites. Of particular importance in this figure are the sharp increases in velocity that occur near 20 and 45 ± 5 km below the surface. The latter of these velocity increases is interpreted as representing the crust-mantle interface, and is designated by a P-wave velocity increase between the crust and upper mantle of ~1.0 km/s. Using Bayesian hypothesis testing to further distinguish between a thin or thick crust led to the result that a 35–45 km thick crust is highly favored to a crust 50–70 km thick. If the thin crust hypothesis is correct, then its thickness is 38 ± 3 km (Khan and Mosegaard 2002). An inversion by Lognonné et al. (2003), which is based upon an independent set of arrival times, also indicates that the crust at the Apollo 12 and 14 sites is thinner than previously assumed, possessing a thickness of only 30 ± 2.5 km. As will be shown in Section 2.6, estimates of the average crustal thickness based on inversions of gravity and topography data are more consistent with these recent thin-crust estimates than the Apollo era value of ~60 km.

2.2.2. The 20-km seismic discontinuity. The Apollo-era crustal velocity models beneath the Apollo 12 and 14 sites showed that the seismic velocity gradually increased from the surface to a depth of about 20 km, at which point a seismic discontinuity was observed. Beneath this depth to the base of the crust the seismic velocity was found to be relatively constant. The crustal velocity model of Khan et al. (2000) (as well as the preferred thin-crust model of Khan and Mosegaard (2002)) is also characterized by a sharp increase in velocity ~20 km beneath the surface. In this section, we assume that this feature is real and assess the possible origins that have been proposed for this seismic discontinuity.

The gradual increase in velocity with depth in the upper 20 km of the crust has been attributed to the closure of microfractures with increasing lithostatic pressure (e.g., Toksöz et al. 1972; Simmons et al. 1973). This interpretation is supported by the in situ measurement of seismic velocities for fractured lunar rocks as a function of increasing confining pressure, which shows a similar behavior (e.g., Todd et al. 1973; Wang et al. 1973; Simmons et al. 1975). The constant seismic velocity beneath a depth of ~20 km has been suggested to be representative of competent unfractured bedrock. While these interpretations are widely accepted, there is no consensus as to the origin of the sharp velocity increase that is observed ~20 km below the surface. Three possible explanations are that this feature is the result of (1) a fracture discontinuity, (2) a change in composition, or (3) a combination of the two.

Fracture discontinuity. It was recognized from investigations of the elastic properties of the returned lunar samples that the increase in lithostatic pressure with depth could not by itself produce a seismic discontinuity at ~20 km depth. Indeed, the confining pressure at a depth of ~60 km was shown to be insufficient to close all fractures in the lunar samples (Todd et al. 1973; Wang et al. 1973; Simmons et al. 1975). While Simmons et al. (1973) suggested that an increase in temperature with depth could have annealed fractures that were once present beneath a depth of ~20 km, this hypothesis has never been quantifiably tested.

Another hypothesis suggested by Simmons et al. (1973) for the intracrustal discontinuity is that the lower crust formed (i.e., crystallized from the magma ocean) after the major impact basins fractured the upper portion of the already solidified crust. However, since magma-ocean crystallization models at the time suggested that the entire crust probably formed in a relatively short time period of ~100 Ma (e.g., Solomon and Longhi 1977), this explanation never gained much popularity. Gamma-ray data obtained from the Apollo and Lunar Prospector missions (Lawrence et al. 1998, 2000), though, show that KREEP-rich rocks are primarily located in the regions of Oceanus Procellarum and Mare Imbrium, and this observation has suggested to some that the crystallization of the lunar magma ocean may have been prolonged beneath this region (Wasson and Warren 1980; Jolliff et al. 2000a; Korotev 2000; Wieczorek and Phillips 2000). Regardless of whenever the lower portion of the crust solidified in this region, the high abundance of incompatible elements that are present there could have given rise to atypically high crustal temperatures, possibly contributing to the progressive annealing of fractures with depth.

Compositional discontinuity. It has been suggested by some that the sharp increase in seismic velocity at 20-km depth could (partly) be the result of a change in chemical composition (e.g., Toksöz et al. 1972, 1974). The initial interpretation was that the upper 20 km of crust was basaltic in composition, whereas the underlying crust was anorthositic. Even though the basalt flows in the vicinity of the seismic network are now thought to be no more than a kilometer thick, it is important to note that seismic velocity alone can not uniquely distinguish between rocks of gabbro, anorthositic gabbro, and anorthositic compositions (Liebermann and Ringwood 1976). Nevertheless, there are some compelling reasons to suspect that the deep crust might be compositionally distinct from the upper crust. As is discussed more thoroughly throughout this section, these include the following observations: (1) the ejecta of some large impact basins is more mafic than the surrounding highlands (e.g., Reid et al. 1977; Ryder and Wood 1977; Spudis and Davis 1986), (2) the central peaks of some complex craters contain mafic lithologies (e.g., Tompkins and Pieters 1999), and (3) the floor of the South Pole-Aitken basin, interpreted to represent lower crustal materials, has a highly noritic composition (Lucey et al. 1995; Pieters et al. 1997, 2001). We emphasize that these studies cannot easily distinguish between a compositionally zoned and stratified crust.

Compositional and fracture discontinuity. Aspects of the above two interpretations can be combined in two possible ways. First, Wieczorek and Phillips (1997) suggested that if the crust was initially stratified in composition, this might conspire to generate a fracture discontinuity as well. The rationale was that during the formation of an impact crater, a density contrast at depth would cause only some portion of the shock energy to be transmitted into the lower crustal layer, possibly inhibiting the formation of shock-related microfractures there. A second possible interpretation is that the crust within the region of Oceanus Procellarum is indeed stratified, with the upper crust being anorthositic and the lower layer being composed of Mg-suite plutonic rocks. Since the Mg-suite rocks generally have high abundances of incompatible elements, it is possible that the lower crust in this region was hotter than typical, causing fractures in the lower crust to completely anneal.

2.3. Gravity and topography datasets

2.3.1. Gravity. Because of a the relative velocity between the Earth and a spacecraft, a radio signal transmitted to or from the Earth will have its frequency shifted as a result of the Doppler effect. This phenomenon was first used by Muller and Sjogren (1968) to map the gravitational field of the Moon in the line-of-sight direction between the Lunar Orbiter spacecraft and the Earth. A surprising outcome of this pioneering work was the discovery of positive gravity anomalies over the depressions associated with the nearside impact basins. Since a surface depression that is isostatically compensated should possess a near-zero gravitational anomaly, these regions were inferred to contain some quantity of material that is not compensated. These inferred "mass concentrations" have ever since been referred to as "mascons." This peculiar gravity signature has generally been interpreted as being the combined result of the uplift of the lunar crust-mantle interface beneath these impact basins, and/or the later flooding of mare basalts onto a thick and undeformable lithosphere (e.g., Wise and Yates 1970; Sjogren and Smith 1976; Phillips and Lambeck 1980; Phillips and Dvorak 1981). Mascon basins are also found on Mars (e.g., Yuan et al. 2001).

The collection of radio tracking data from the Apollo 15 and 16 sub-satellites and the Clementine and Lunar Prospector missions has subsequently greatly improved our knowledge of the lunar gravity field. The modeled gravitational-potential, U, has traditionally been expressed as a sum of spherical harmonic functions

$$U(r,\theta,\phi) = \frac{GM}{r} \sum_{i=1}^{2} \sum_{l=2}^{L_{max}} \sum_{m=0}^{l} \left(\frac{R}{r}\right)^{l} C_{ilm} Y_{ilm}(\theta,\phi) \tag{3.1}$$

where G is the gravitational constant, M is the mass of the planet, and (r, θ, ϕ) is the location of a point in space referenced to the center of mass of the body with $r > R$. Here, l and m are the degree and order of the spherical harmonic function Y_{ilm}, the corresponding spherical harmonic coefficient referenced to a radius R is C_{ilm}, and L_{max} is the maximum degree of the expansion. The spherical harmonic functions are defined by

$$Y_{ilm}(\theta,\phi) = P_{lm}(\cos\theta) \begin{cases} \cos m\phi & i = 1 \\ \sin m\phi & i = 2 \end{cases} \tag{3.2}$$

where P_{lm} is an associated Legendre function, and are normalized such that the integral of the square of each spherical harmonic over the entire sphere is 4π. A useful visualization property of the spherical harmonic functions is that they have $2m$ zeros in the ϕ direction (longitude), and $l-m$ zeros in the θ direction (co-latitude). The effective wavelength of a given spherical harmonic can be approximated by $\lambda = 2\pi R/l$. The gravitational acceleration exterior to R is given by

$$\vec{g} = \nabla U \tag{3.3}$$

and the height of an equipotential, or level, surface (the geoid) with respect to a radius R is given to first order by

$$N(r,\theta,\phi) = R \sum_{i=1}^{2} \sum_{l=2}^{L_{max}} \sum_{m=0}^{l} C_{ilm} Y_{ilm}(\theta,\phi) \tag{3.4}$$

In center-of-mass coordinates the $l = 1$ terms in the above sums are identically zero. The C_{1lm} and C_{2lm} terms are often referred to as C_{lm} and S_{lm}, respectively, and the C_{l0} coefficients are typically referred to as $-J_l$ (see Lambeck (1988) and Kaula (1967; 2000) for further details). The J_2 term is affected by the rotational flattening of the Moon, and both the J_2 and C_{22} terms are affected by deformations cause by the gravitational attraction of the Earth.

The latest model of the lunar gravity field (Konopliv et al. 2001) includes the processing of tracking data obtained from the Lunar Prospector extended mission where the spacecraft sometimes approached the surface to within 10 km. While the nearside gravity field is now known to a good accuracy of ~30 mgals, because a spacecraft cannot be tracked much beyond the limbs of the Moon, the farside gravity field is only poorly determined, being uncertain by as much as ~200 mgals. As a spacecraft in orbit about the Moon can be tracked about 20° over the limb, the poorest gravity resolution occurs in an ~60° shadow zone centered on 0°N and 180°E. Gravity anomalies that are present within this shadow zone, however, do affect the long-term evolution of an orbiting spacecraft, and because of this it is possible to infer the gravity field within this region. Globally, though, the spherical harmonic gravity model is only accurate to approximately degree 15. Interpretation of farside gravity anomalies that are present within the shadow zone must be treated with caution as it is possible that their amplitudes have been muted and/or laterally offset.

The most recent JPL gravity model (LP150Q, Konopliv et al. 2001) has been determined up to degree and order 150, and possesses an effective half-wavelength spatial resolution of ~36 km. As the gravity field becomes increasingly noisy beyond approximately degree 100, and since the topography of the Moon is only known globally to less than degree 90 (with a half-wavelength resolution of ~60 km), we plot this gravity field in Color Plate 3.1 truncated at degree 90. This figure shows the familiar gravity highs over the major impact basins and the more poorly determined and noisy farside field. Even though direct tracking data does not exist over most of the farside, many anomalies there are found to correspond to actual geologic features, such as the Hertzsprung, Korolev, and Mendeleev impact basins. Color Plate 3.2 shows the lunar geoid as determined from this gravity model. When the elevation of the Moon is referenced to the geoid (as opposed to a spheroid), lava flow directions should accurately track elevation changes (i.e., lava will flow "downhill"). We emphasize that the lunar geoid varies by almost a kilometer within the nearside maria, and that the gridded Clementine topography data is not referenced to the full geoid (see Section 2.3.2).

One of the more surprising discoveries from the Lunar Prospector based gravity fields is that some mascon basins do not appear to be associated (or only marginally) with mare basalts. A few of the most convincing non-mare mascon basins include Schiller-Zucchius (56°S, 44.5°W), Bailly (67°S, 68°W), Schrödinger (75°S, 134°E), Lorentz (34°N, 97°E), and Hertzsprung (1.5°N, 128.5°W) (e.g., Konopliv et al. 1998, 2001). While the interiors of some of these impact basins might contain mare deposits that were subsequently buried by impact ejecta from nearby craters (i.e., Schiller-Zucchius), it is unlikely that "cryptomare" is the entire origin of the gravity anomalies associated with these basins. Modeling of the thickness of the lunar crust has shed some light on the origin of the largest mascon basins. Neumann et al. (1996) found that the crust-mantle interface beneath some impact basins was uplifted above its isostatic position. Wieczorek and Phillips (1999) further found that some nearside basins were in a superisostatic state, even after the load associated with the mare basalt fill was removed. As was originally suggested by Neumann et al. (1996), this could be the result of the basin floor rebounding above its isostatic level during the impact event, and subsequently being frozen in place. This topic will be discussed in Section 5.1.2.

Finally, we note that the orientation of the Moon is controlled by differences in its principal moments of inertia, and that these moments completely determine the degree-2 gravity field. The minimum energy configuration of the Moon is a state in which the maximum moment of inertia lies along its rotational axis and where the axis of its minimum moment of inertia coincides with the Earth-Moon direction. We note that both the present orientation of the Moon, and one that is rotated by 180° about its rotation axis, are equally stable. The direction of the Moon's center-of-mass/center-of-figure offset (which is a degree-1 feature) is unrelated to its current orientation.

2.3.2. Topography. The topography of the lunar surface has been measured in a number of ways, including the use of laser altimetry, stereo-photogrammetry, radar interferometry, radar sounding, and limb profile data. Prior to the Clementine mission, though, the highest resolution datasets were primarily restricted to the Earth-facing hemisphere and equatorial ground tracks of the Apollo command and service modules (see Bills and Ferrari 1977a and references therein). Data from the Clementine Lidar have since greatly improved our knowledge of the global shape of the Moon (Zuber et al. 1994; Smith et al. 1997). While this instrument was originally designed for military purposes, through careful in-flight programming of the ranging process, and judicious filtering of the returned data, near-global surface elevation data were obtained from this mission. However, because of the short time in lunar orbit (~2 months), and the high spacecraft altitudes over the poles, significant gaps in coverage exist.

During the first month of the Clementine mission, topographic profiles were obtained from 79°S to 22°N, and during the second month from 20°S to 81°N. The Lidar had an intrinsic range resolution of ~40 m, though absolute surface elevations may be uncertain by ~130 m as a result of uncertainties in the spacecraft orbit and orientation. The laser spot size on the lunar surface when the spacecraft was at periapse is ~200 m, and the minimum along-track spacing of shots is about 20 km. Over the rougher highlands many spurious returns were ultimately discarded, yielding a more typical along-track spacing of about 100 km. The cross-track spacing of the orbit tracks is approximately 60 km at the equator (~2° of longitude; 1° of lunar longitude corresponds to 30.3 km).

Because of the non-optimal design of the Clementine Lidar, it is necessary to be aware of some possible artifacts in the data. Electronic noise and background solar radiation caused the Lidar to detect multiple returns for each laser shot. In order to determine which return (if any) was from the lunar surface, the data were filtered in two ways. First, up to four returns that occurred within an expected predefined range window were recorded. Second, a stochastic and fractal model of surface topography was used to construct a sequential filter to track the surface along a given profile (see Smith et al. 1997 for details). Filtering the data in this manner ultimately led to the acceptance of 72,548 range measurements.

All of the accepted elevations from the Clementine Lidar are plotted in Color Plate 3.3. Of particular notice are the uneven coverage and data gaps, especially over the poles. Elevation data is always reported with respect to some reference surface, and the data in this figure (as well as the $0.25° \times 0.25°$ gridded data on the PDS site) are referenced to a spheroid with an equatorial radius of 1738 km and a flattening of 1/3234.93 that corresponds to the J_2 portion of the lunar geoid. We emphasize that these data are not referenced to the full geoid; hence basalt flow directions inferred from this representation of the data may be erroneous, especially near the mascon basins.

The Clementine elevation measurements have been expanded in spherical harmonics up to degree and order 90 (GLTM2c) using the expression

$$h(\theta,\phi) = \sum_{ilm} H_{ilm} Y_{ilm}(\theta,\phi) \tag{3.5}$$

In contrast to the gridded dataset, the elevations are referenced to a sphere with a radius of 1738 km. The elevation of the lunar surface from this model, referenced to the full geoid, is plotted in Color Plate 3.4. Clearly visible in this image are the major impact basins and farside South Pole-Aitken basin. Also of note is the fact that the Moon possesses a 1.9 km displacement of the center of figure with respect to the center of mass toward the backside direction of 8°N and 157°W. The origin of this will be discussed in Section 2.7, and here we only note that this causes the average elevation of the nearside to be ~1.9 km less than that of the far side.

Another significant global feature of the Moon is its degree-2 shape. The Moon currently possesses an ~2.2 km polar flattening (related to the spherical harmonic term J_2), and a longitudinal ellipticity of 800 m (related to the spherical harmonic terms C_{22} and S_{22}) with the maximum amplitude occurring at ~40°E. As a result of the combination of tides raised by the Earth and the rotation of the Moon, if the Moon were ever in a state of hydrostatic equilibrium, its expected shape and gravity field would be composed almost exclusively of J_2 and C_{22} terms, corresponding to a tri-axial ellipsoid. The measured values, however, are many times greater than what is expected for the current Earth-Moon separation. A popular explanation for the origin of the degree-2 shape and gravity field is that these terms were "frozen" into the lithosphere when the Moon was much closer to the Earth and was rotating more rapidly (e.g., Jeffreys 1970). Lambeck and Pullan (1980) estimated that the present shape of the Moon was similar to the equilibrium shape corresponding to an Earth-Moon separation of about 25 Earth radii (the Earth-Moon separation is currently about 60 Earth-radii). This interpretation is potentially problematic as the lunar orbit should rapidly evolve to a distance of ~30 Earth radii in less than 100 My (e.g., Webb 1982). It is alternatively possible that the degree-2 shape of the Moon is a result of primordial crustal thickness variations.

The fidelity of the Clementine altimetry data can be assessed by comparing it to independently obtained measurements. Using the technique of radar interferometry, Margot et al. (1999b) determined absolute elevations for the region of Tycho crater with a spatial and vertical resolution of about 200 and 30 m, respectively. Comparing this data set with the 87 Clementine measurements that occur in this region, two important differences were noticed. First, the Clementine data for this region appear to be shifted in latitude by ~3 km.

Second, while the RMS difference between the two data sets is only ~90 m, up to three of the Clementine measurements could be in error for this region. This latter result suggests that about 3–4% of the accepted Clementine range data could be erroneous.

Finally, two techniques have been used in an attempt to fill the polar data gaps of the Clementine altimetry data. Similar to the Tycho study above, Margot et al. (1999a) used radar interferometry to determine elevations poleward of 87.5°. Because of the Earth-Moon viewing geometry, these data were primarily obtained over the lunar nearside. These measurements have been used to constrain the elevations of those regions that are permanently shadowed and that might possess near-surface deposits of ice (see Chapter 2). An alternative method for determining polar topography is to use Clementine stereo imagery. Using these data, Cook et al. (2000) constructed digital elevation models poleward of 60° that possess a 1-km spatial resolution. These relative elevation determinations were tied to the absolute Clementine altimetry data, but while relative vertical elevations could be as good as 100 m, absolute elevations far from the Clementine tie-points might be in error by as much as a kilometer. The combination of elevation measurements obtained from stereo-imagery and radar-interferometry studies should ultimately be able to improve the current 90-degree spherical harmonic representation of the shape of the Moon. This would be of particular importance in modeling the high-resolution line-of-sight nearside gravity data obtained from the extended portion of the Lunar Prospector mission.

2.4. Thickness of the mare basalts

Knowledge of the thickness of extruded mare basalts is key to understanding the Moon's thermal history, lithospheric thickness, gravity field, and the effect of vertical impact mixing on regolith composition. Basic approaches for determining mare thicknesses make use of a flooded crater's morphology, the composition of crater ejecta, and subsurface reflections observed in the Apollo 17 radar sounding experiment. The interpretation of gravity anomalies can also be used to constrain the thickness of a mare deposit, though solutions based on this method are not unique. Using orbital images, the first two methods have been applied regionally, whereas the Apollo 17 radar sounding data are only applicable to swaths across the Serenitatis and Crisium basins, and Oceanus Procellarum. In this section, we review the results of these three approaches.

By assuming that a crater forms with a characteristic shape, the thickness of mare basalts within the rim of a partially flooded crater can be estimated (e.g., Baldwin 1970). The first depth estimates were made by Eggleton et al. (1974), using wrinkle ridges that were presumed to represent the crater rims of shallowly buried craters. Later, De Hon (1974; 1977; 1979) and De Hon and Waskom (1976) systematically mapped the thickness of the nearside maria using flooded and embayed crater relationships. In doing so, they found that the eastern maria averaged 200 to 400 m in thickness and that the western maria were closer to 400 m. Isolated regions thicker than 1 km were found to occur where basins and large craters are superposed on the floor of the Procellarum region and in the central portion of some of the larger impact basins. Individual basalt flows (where recognizable) vary considerably in thickness, but are generally in the range of 30 to 60 m (e.g., Schaber 1973; Schaber et al. 1976; Hiesinger et al. 2002). Thus, the above inferred basalt thicknesses imply the superposition of multiple discreet flows.

The accuracy of mare-thickness maps using the partially flooded crater method, however, is difficult to assess because of the few partially embayed craters that are present within the maria. Additional uncertainties are also inherent in the method for individual craters. For example, Hörz (1978) concluded that the craters used to measure basalt thickness were more degraded prior to mare flooding than De Hon had originally assumed. Between the time of crater formation and mare inundation, crater profiles likely became shallower as a result of impact erosion, infilling by crater ejecta, and/or rebound of the crater floor. The net result of a shallower initial crater profile is an overestimation of mare thickness. On the basis of this argument, Hörz (1978) reduced the depth of mare deposits obtained by De Hon by a factor of two. Another limitation to the method is that it can only measure the thickness of basalts deposited since the time the impact crater formed. If mare basalts were emplaced before the crater formed, then the actual basalt thickness could be considerably greater.

The above crater morphology studies could only determine mare thicknesses up to ~2 km, as a greater basalt thickness would completely bury any medium sized crater. A similar morphological technique was applied to large impact basins by Williams and Zuber (1998) that extended the maximum measurable basalt thickness up to ~7 km. Using Clementine altimetry, these authors determined the depth/diameter relationship for unflooded impact basins. Basins that possessed maria were found to be shallower than the depth-to-diameter trend predicted, and this shallowing was attributed to the thickness of mare basalts. The inferred mare thicknesses from this study are the following: Grimaldi (3.5 km), Humorum (3.6 km), Orientale (0.6 km), Nectaris (0.8 km), Smythii (1.3 km), Crisium (2.9 km), Serenitatis (4.3 km), and Imbrium (5.2 km). These values are in general thinner than previous studies that predicted basalt thicknesses up to 10 km (e.g., Solomon and Head 1980). We note that some of these thickness estimates may still be overestimated because some impact basins may have been partially shallowed by viscous relaxation (such as Imbrium and Serenitatis; see also Section 5.1.2). Thus, it seems likely that the basalts within the large impact basins are probably no thicker than about 4 km.

Clementine multispectral data, along with the relationship between the depth and diameter of a crater's excavation cavity (e.g., Croft 1980), have additionally been used to measure the thickness of mare basalts. In this technique, if impact craters excavate highland materials beneath the maria, the ejecta composition can be utilized to estimate the depth of the mare/highland interface. Specifically, a crater that penetrates the mare/highland contact at depth should excavate feldspathic materials that are spectrally distinct from the surrounding mare basalts. In general, these measurements are more robust than those of the crater infilling model because they offer an increased number of control points with which to contour basalt isopach maps. Moreover, the crater drill-core method is not sensitive to the amount of crater degradation prior to mare flooding, or whether these craters are perched on top of older mare deposits. The method does, however, rely on the accuracy with which highland materials can be detected in the ejecta of the observed craters, and this shortcoming will cause a slight overestimate of the mare thickness. In comparing the crater excavation method of measuring mare thicknesses with those previously determined by De Hon and Hörz for similar areas—Oceanus Procellarum (Heather and Dunkin 2002), Mare Humorum (Budney and Lucey 1998), and Mare Smythii (Gillis and Spudis 2000) it appears that in most instances thickness values are closer to those of Hörz (1978), thus supporting his lower thickness estimates for the maria.

The poor image quality of the lunar farside prior to the Galileo and Clementine missions prevented the measurement of farside mare thicknesses. Recent thickness measurements have been made using the methods discussed above, along with shadow measurements of flow fronts, and partially buried topography (Yingst and Head 1997, 1998, 1999; Gillis 1998; Gillis and Spudis 2000). These measurements yield basalt thicknesses of ~200 to 300 m on average, approximately half as thick as their nearside counterparts. There are a few deposits on the farside that approach or exceed 1 km in thickness, and these occur as isolated deposits within the South Pole-Aitken basin (e.g., Ingenii, Jules Verne, Leibnitz, and Von Kármán), Mare Orientale, and Mare Moscoviense.

Another method of mapping the thickness of mare basalts comes from an analysis of data collected from the Apollo 17 radar sounding experiment (Phillips et al. 1973). After transmitting an energy pulse towards the Moon, this instrument measured the time delay between any reflected energy. In addition to detecting the initial signal reflected from the surface, reflections were often observed from subsurface horizons as well. Within the mare, these horizons were found to be approximately parallel to the surface, and were interpreted to represent a partially developed regolith that was buried by subsequent lava flows (Maxwell and Phillips 1978; Peeples et al. 1978).

Within Mare Serenitatis, two nearly continuous buried horizons were observed, one at a mean depth of 0.9 km and the other at a mean depth of 1.6 km. The regolith horizon is required to be at least 2 m thick to be detectable, and several hundred million years are probably necessary to form such a layer (Sharpton and Head 1982). This suggests that at least two distinct major phases of mare volcanism occurred within this basin. While the deepest reflector in this basin most likely represents pre-mare "bedrock" (Sharpton and Head 1982), it is also possible that it might represent the top of another thick sequence of lava flows. Unfortunately, radar-sounding data over the Serenitatis basin was only obtained for the southern portion of this basin, leaving open the possibility that the basalts could be considerable thicker in the basin center. Nonetheless, these results imply that the basalts in this basin are on average at least 1.6 km thick, which is less than half of the 4.3 km value as determined by Williams and Zuber (1998). In Mare Crisium, only one nearly continuous reflector was observed at a mean depth of 1.4 km below the surface. In contrast to the Serenitatis basin, this profile passed through the center of this basin. If this subsurface reflector represents pre-mare bedrock, then the inferred average basalt thickness of this basin is again about half of the 2.9 km value as obtained by Williams and Zuber (1998). Alternatively, it is possible that this reflector might represent the surface of an older basaltic flow (Head et al. 1978). A subsurface horizon was also found in the central portion of Oceanus Procellarum (Cooper et al. 1994). These data show that the maria are about 1 km thick just east of Grimaldi, and decrease in thickness to ~0.6 km near the crater Kepler.

Integrating the above methods for measuring mare thickness indicates that the mare basalts are a volumetrically minor component of the crust. The average mare thickness on the nearside is generally less than 400 m with the basins containing up to 4 km of basalt fill. Mare thicknesses on the far side are comparatively thinner, with the average thickness being close to 200 to 300 meters. Even in the areas where the crust is thinnest on the far side (e.g., in South Pole-Aitken basin) mare thicknesses rarely exceed 500 m. One estimate of the total volume of extruded mare basalts is 10^7 km^3 (Head and Wilson 1992). As these authors used mare thicknesses that were about twice that of De Hon's measurements, this number should probably be viewed as an upper limit. Assuming an average thickness of 0.3 km yields a total mare basalt volume of 2×10^6 km^3. Assuming that the lunar crust is on average 50 km thick, the mare basalts only represent about 0.1–0.5% by volume of the crust.

2.5. Lithospheric thickness and mascon tectonics

The elastic lithosphere defines the strong outer rind of a planetary body. Because the lithosphere must be relatively cold to maintain its strength, its thickness provides information on the temperature of the shallow portions of the Moon and thus can constrain thermal history models (see Chapter 4). In this section, we focus on the principal methods that have been used to infer lunar lithospheric thicknesses—faulting around mascons—and critically examine what work remains to be done. We briefly mention how gravity and topography studies have also been used to estimate the lithospheric thickness of the Moon.

Mascons are excess concentrations of mass that act as loads on the lunar lithosphere, causing it to flex under the applied force (e.g., Melosh 1978). If the stress within the lithosphere exceeds a critical value (the yield stress) the material will fracture, possibly resulting in visible faults on the surface of a planet. Around many of the mascon basins, faults are indeed visible as concentric or linear rilles (interpreted as grabens), and within the mare as ridges (i.e., "wrinkle" ridges interpreted to be thrust faults (e.g., Golombek 1985)). Many have assumed that these faults formed when the flexural stresses within the lunar lithosphere exceeded the yield stress, and that the location and width of the faulted zones can be related to the thickness of the lithosphere at this time (e.g., Comer et al. 1979; Solomon and Head 1979, 1980; Freed et al. 2001). It is important to note that since the lithosphere is in reality not perfectly elastic, the obtained values should be considered as an "effective" elastic thickness (Burov and Diament 1995; Kohlstedt et al. 1995). The effective elastic thickness is, furthermore, likely to differ from the lithospheric thicknesses as defined by thermal considerations (e.g., McNutt 1984).

The method of matching flexural topographic profiles from a load to the location and width of zones of faulting is well established and has been used on all of the terrestrial planets (e.g., Thurber and Toksöz 1978; Solomon and Head 1980; McGovern and Solomon 1993, 1998; Albert et al. 2000). In interpreting these results, though, it is important to consider the limitations of the present models and our ignorance of the rheology of the outer layers of planetary bodies. Plate-flexure models that have been applied to the Moon generally assume that the lithosphere behaves purely elastically (see Zhong and Zuber 2000 for a visco-elastic treatment). Once the stress exceeds the yield stress, however, the material is no longer elastic and is assumed to behave plastically. This means that fractures or faults relieve some of the elastic stress and change the stress distribution within the lithosphere (Schultz and Zuber 1994). No models of mascon tectonics to date have yet modeled this inelastic rheology correctly (Albert and Phillips 2000; Albert et al. 2000), so models are really only able to predict how the material initially fails. Pre-existing faults (perhaps caused by the basin-forming impact) might also affect the distribution of stress within the lithosphere and how the material fails. The state of stress before the mascon load was applied is additionally poorly constrained. Some indication of the state of isostasy before mare emplacement is provided by the lunar gravity and topography fields (Neumann et al. 1996; Konopliv et al. 1998, 2001 Wieczorek and Phillips 1999), and these studies suggest that while some basins might have been in a pre-mare isostatic state, others were probably not.

The plate flexure models also require specific knowledge about the load, such as the temporal and spatial distribution of the mare load, and how much of the load is from the top and bottom (e.g., mare basalts vs. volcanic intrusions, respectively). The plate flexure model also requires a knowledge of the elastic properties of the lithosphere (rigidity and Poisson's ratio), but the vertical and lateral variations of these parameters within the lunar lithosphere are not well constrained. In order to understand how the stresses induced by flexure causes faulting, a rock-failure criteria must be used, and different criteria can yield different results (Freed et al. 2001).

Many of the uncertainties mentioned above have important or unknown effects upon calculated estimates of the lithospheric thickness. Important first steps in exploring this parameter space have been made by Solomon and Head (1980), Pullan and Lambeck (1981) and Freed et al. (2001). Even with the above-mentioned uncertainties, much can be learned by modeling the locations of compressional and extensional features. Solomon and Head (1980) looked at 8 different basins and found that the extensional rilles formed when the lithosphere was thin (less than about 70 km) and that the ridges formed later when the lithosphere was thicker (more than 100 km). The ridges are mostly younger than the rilles, so the thicker lithosphere associated with the younger features is consistent with the lithosphere cooling and thickening with time. Solomon and Head (1980) also found that some basins with the same age have different lithospheric thicknesses at the time of rille formation, but that the thickness at the time of ridge formation was more uniform. They interpret this to mean that there was shallow thermal heterogeneity in the Moon during rille formation, but that with time, thermal conduction or convection evened out the shallow temperature differences.

Although simple plate flexure models can explain the origin of concentric rilles and radial ridges, a major problem is that the early models predicted that there should be strike-slip faulting between these two regimes. No significant strike-slip faulting is visible anywhere on the Moon, and this discrepancy has been called the "strike-slip paradox." Several potential resolutions of this paradox have been proposed. Solomon and Head (1980) postulated that global expansion and contraction related to the lunar thermal evolution (e.g., Solomon and Chaiken 1976) modified the local stress field around the mascons and inhibited strike-slip faulting. If true, the timing and location of faulting at the mascons provides a stringent constraint on lunar thermal histories. Golombek (1985) noted that strike-slip faulting is not likely to occur where faults actually nucleate, several kilometers beneath the surface, especially if the stress distribution in the megaregolith is not isotropic. Schultz and Zuber (1994) further noted that (1) because the load was not placed instantaneously, the history of loading and faulting must be considered, and (2) the mode of failure is non-unique in that it could be strike-slip or tensional jointing. Freed et al. (2001) have recently shown that lunar curvature, certain initial stress distributions, and certain failure criteria reduce, and perhaps eliminate the zone of strike-slip faulting.

While the strike-slip paradox has been given much attention in the literature, there is another aspect of lunar tectonics that until recently has remained just as paradoxical. Namely, plate flexure models predict that radial compressive faults should preferentially form in the central portion of the mare basalt load (e.g., Melosh 1978), whereas geologic mapping of the mascon basins shows that the majority of ridges within the mare filled basins are in fact concentrically oriented (e.g., Solomon and Head 1980). In the study of Freed et al. (2001), it was shown that by taking into account a putative global compressive stress field, as well as the temporal history of basaltic eruptions in a basin, that it is possible to form both concentric and radial ridges within the basin. While this is one possible explanation, an analysis of their figures reveals two other explanations that we consider to be equally plausible. First, the geometry of the load was shown to influence where radial and concentric ridges should form. In particular, if the mare basalts were much thicker in the central portion of the basin than near its periphery, then radial ridges were favored to form. In contrast, if the basalts in the basin had a more uniform thickness (as is the case for at least Mare Crisium; see Section 2.4), then the formation of concentric ridges would be favored. In addition, the pre-mare isostatic state of the basin was shown to influence the distribution of radial and concentric ridges. If the basin was initially sub-isostatic, and the basin floor subsequently rebounded after mare flooding, then the formation of radially oriented ridges would be favored. In contrast, if the basin was initially super-isostatic, then basin subsidence after mare volcanism would be enhanced and the formation of concentrically oriented ridges would be favored. As many basins are predicted to have been in a pre-mare super-isostatic state (Neumann et al. 1996; Konopliv et al. 1998, 2001; Wieczorek and Phillips 1999), this latter explanation is particularly compelling.

Considering the simplifications used in relating mascon tectonics to lithospheric thickness, it is useful to consider methods based on gravity and topography data for constraining this quantity. It is important to realize, though, that the lithospheric thickness recorded by the formation of rilles and ridges might be from a time that is different than those recorded by a basin's current gravity and topography fields. Using the Apollo data, Kuckes (1977) found the elastic lithospheric thickness at several basins to be around 50 to 100 km. Using Clementine gravity and topography data and a thin elastic spherical shell formulation Arkani-Hamed (1998) found the elastic thickness to range from about 20 to 60 km beneath the nearside mascon basins. Using Lunar Prospector gravity data, Aoshima and Namiki (2001) reported that for at least the Orientale and Serenitatis basins, the lithospheric thickness was not well constrained, casting doubt on whether gravity and topography will be as useful as mascon tectonics in determining lithospheric thickness at the time of mare basalt loading. In an investigation by Sugano and Heki (2004), the gravity signatures of the major basins were modeled by assuming an initial basin topographic profile, and then comparing the flexural deformation of the Moho with mass deficits inferred from their present Bouguer anomalies. Lithospheric thicknesses were generally found to lie between 20 and 60 km, with Serenitatis and Imbrium possessing smaller values. However, the employed crater diameters were taken from Wilhelms (1984), and if the excavation cavity diameters of Wieczorek and Phillips (1999) were used instead, elastic thicknesses close to zero would be obtained for most basins (T. Sugano, personal communication 2005). This would imply that these basins were in a pre-mare isostatic state before the emplacement of the mare basalts. Elastic thickness estimates were also reported in a study by Crosby and McKenzie (2005), but their model admittance spectra were not windowed in the same manner as the observations (see Pérez-Gussinyé et al. 2004). In comparing the above studies, it should be clear that there is no consensus concerning the thickness of the elastic lithosphere based on gravity and topography methods.

In the final analysis, there is still much work to be done in determining the lunar lithospheric thickness and how it varies both spatially and temporally. A more complete exploration of the parameter space of mascon tectonic models is needed and more sophisticated models that include inelastic effects need to be applied. As for inversions based on gravity and topography data, the major impediment is the lack of a suitable loading model that correctly takes into account the basin forming process (which may include acoustic fluidization of the crust (Melosh 1979) and near-instantaneous Moho uplift), the initial surface and Moho relief, the loads associated with ejecta exterior to the basin, and the spatial and temporal loading history of mare volcanism. Without such a model, gravity and topography data in this regard will probably be limited to determining the degree of compensation of a basin at the present time, and before the commencement of mare volcanism.

2.6. Gravity and topography admittance studies

If the density structure of a planet were completely known, then it would be straightforward to compute the gravitational attraction at any point in space by employing Newton's law of gravitation. Unfortunately, as a result of the non-uniqueness of potential modeling, it is conversely not possible to uniquely invert for the density structure of a body by using gravity and topography information alone. This non-uniqueness of potential modeling, however, can be overcome by invoking reasonable assumptions about the interior structure of a planet.

In this section, we describe two general techniques that have been used to place constraints on the thickness of the lunar crust. One method relies upon the relationship between the lunar geoid and topography in the space domain, whereas the other relies upon the relationship between these fields in the spectral (wavelength) domain. Both techniques are roughly consistent in suggesting that the lunar crust is on average about 50 km thick. A summary of the various methods that have been used to constrain the average thickness of the crust is given at the end of this section.

2.6.1. Geoid to topography ratios. Both spatial and spectral techniques have been used to invert for the average thickness of the crust for planetary bodies, with each technique possessing its own merits and caveats. Both methods first assume that the crust is compensated by a specific model (i.e., Airy, Pratt, or elastic support), and then adjust the parameters of this model in order to best match the observed relationship between these two fields. One of the strengths of investigating the relationship between the lunar geoid and topography in the space domain is that certain regions can be excluded from consideration if they appear to be inconsistent with the assumed compensation model. For instance, it is reasonable to assume that the lunar highland crust might be compensated by an Airy mechanism, in which surface topography is supported by a crustal root. As the lunar mascon basins are clearly inconsistent with this hypothesis, these regions can be simply ignored in the analysis.

One spatial inverse technique that relates isostatically compensated topography to its associated geoid anomaly was developed by Ockendon and Turcotte (1977) and Haxby and Turcotte (1978). If the crust is compensated by an Airy mechanism, the local geoid to topography ratio (GTR) was shown to be proportional to the average crustal thickness. A more accurate approach was developed in spherical coordinates by Wieczorek and Phillips (1997) who showed that the previous technique incurred errors on the order of 10 km for small planets such as the Moon and Mars (see also Wieczorek and Zuber 2004).

In an investigation by Wieczorek and Phillips (1997), geoid to topography ratios of the nearside highland crust were determined after excluding the maria and removing the degree-2 terms from the lunar gravity and topography fields. We have redone these calculations here using the updated gravity model of Konopliv et al. (2001) and obtain an average GTR of 26.7 ± 6.9 m/km. If the lunar crust is compensated by an Airy mechanism, then this value corresponds to a globally averaged crustal thickness of 49 ± 16 km. In comparing this value to the seismically constrained crustal thickness near the Apollo 12 and 14 sites, the difference in elevation between these sites and the mean planetary radius must be considered. If we assume that the degree-2 portion of the lunar topography is entirely the result of rotational and tidal flattening (see Section 2.3.2), and hence ignore these terms, then the average elevation of these sites is about 1.6 km below the mean planetary radius. Assuming reasonable values for the density of the crust (2710–2885 kg m^{-3}) and mantle (3200–3335 kg m^{-3}) the crustal thickness at the Apollo 12 and 14 sites based upon the above GTR analysis is predicted to lie between 16 and 56 km. (If the degree-2 topography were retained in the above analysis, then the average elevation of the Apollo 12 and 14 sites would be about 1.0 km below mean planetary radius, and the inferred crustal thickness there would be slightly greater.)

The Apollo-era seismic velocity profile beneath the Apollo 12 and 14 sites (e.g., Toksöz et al. 1974) suggested that the crust there was about 60 km thick. As this is outside the range inferred from the above study, Wieczorek and Phillips (1997) suggested that the highland crust was more consistent with being partially compensated at an intra-crustal interface, as opposed to solely at the crust-mantle interface. The simplest interpretation was that the upper crust was "floating" on a more-mafic lower crust, and in support of this hypothesis they noted that a seismic discontinuity appears to exist about 20 km beneath the surface that could be compositional in origin (see Section 2.2.2). Reanalyses of the Apollo seismic data by Khan and Mosegaard (2002) and Lognonné et al. (2003), however, forces us to re-evaluate this hypothesis. In contrast to the Apollo-era models, these recent studies imply that the crust beneath the Apollo 12 and 14 sites is between 27 and 50 km thick. As this is within the range of 16 to 56 km as predicted by the above Airy compensation model, the seismic and gravity data can now be considered consistent with the hypothesis of an Airy compensated uniform density crust.

In addition to the above single layer compensation model, Wieczorek and Phillips (1997), also considered models in which the surface topography was compensated at both intracrustal and crust-mantle interfaces. In one model, the upper crust was constrained to have a constant thickness while the lower crustal thickness varied, whereas in a second model the inverse was assumed. Using the crustal thickness constraints of Khan et al. (2000) we find that the highland GTRs and inferred Apollo 12 and 14 crustal structure are only consistent with the model in which the upper crustal thickness varies while possessing a constant thickness lower crust. This model is consistent with the expectation that impact cratering would primarily redistribute upper crustal materials, leaving the lower crust of the highlands relatively untouched.

If it is assumed that the crust is entirely compensated by lateral variations in density (i.e., topographic highs correspond to lower crustal densities as in the model of Pratt) then the average crustal thickness derived from the above calculated GTRs is constrained to lie between 53 and 96 km. As this model predicts the crustal thickness of the Apollo 12 and 14 sites to be greater than 52 km, this model is inconsistent with the recent seismic velocity models. Wieczorek and Phillips (1997) have further shown that the composition and density of the highland crust (as inferred from the Clementine-derived iron abundances of Lucey et al. (1995)) are not significantly correlated with elevation, suggesting that Pratt compensation is not of regional importance. As is discussed in Section 2.7.1, though, it is still possible that a portion of the Moon's center-of-mass/center-of-figure offset could be caused by hemispheric variations in crustal density.

We note briefly that two other studies have used geoid-to-topography ratios to constrain the thickness of the lunar crust. Using the Cartesian method of Ockendon and Turcotte (1977) and Haxby and Turcotte (1978), Arkani-Hamed (1998) obtained an average crustal thickness of 52 km (no error bar quoted) for the region encompassing the South-Pole Aitken basin. While this value is in agreement with those presented above, the assumption of Cartesian geometry may have biased this number by ~10 km. Furthermore, as will be discussed in Section 2.8.4, the floor of the SPA basin is more mafic and dense than the surrounding highlands, complicating the interpretation of this result. The gravity and topography coverage of this basin is also less well known than that of the nearside. Konopliv et al. (1998) have also attempted to place constraints on the thickness of the lunar crust by using the same Cartesian method. The average crustal thickness from their study, however, varies from less than zero to more than 200 km, and we suspect that this unphysical behavior is in part the result of not removing the mascon basins prior to computing regional geoid to topography ratios.

2.6.2. Spectral admittance studies. A different approach to modeling the relationship between the gravity and topography fields of a planet is to work in the spectral domain. In contrast to the spatial approach described above, in which a single admittance (GTR) is calculated for a given region, spectral approaches result in admittance and coherence functions that are wavelength dependent. One of the benefits of using a spectral approach is that the form of these functions can give clues to the compensation mechanism that is operating in the study region. If multiple compensation mechanisms are locally present, then it may be possible to isolate the effects of each by analyzing different wavelength bands. In addition, since an individual admittance and coherence is calculated for each wavelength under consideration, it is in principle possible to invert for multiple model parameters.

While spectral approaches are often superior to the interpretation of a single spatially derived geoid to topography ratio, they have their associated caveats as well. A primary drawback is that the methods so far employed could only investigate regions that are either square in Cartesian studies (e.g., Forsyth 1985; Simons et al. 2000) or circular in spherical studies (Simons et al. 1997; Wieczorek and Simons 2005). Hence, if one wishes to investigate a region that is compensated by a single mechanism, this restriction will often place strict constraints on the size of the study region, and hence the maximum wavelength that can be analyzed. As an example, the largest nearside highlands region that can be studied while avoiding the lunar maria and mascon basins is about 1000 km wide. In this case, the longest wavelength that could be analyzed would correspond to about degree 10. In contrast, GTR studies for the Moon are most sensitive to degrees less than this; spatial and spectral studies are thus seen to provide complementary information. We note that a recent method developed by Simons et al. (2006) allows for the spectral analysis of irregularly shaped regions on the sphere, and this somewhat mitigates the drawbacks associated with applying spectral admittance studies to the Moon.

Simons et al. (1997) have developed a technique in which spectral admittance and coherence functions are estimated for localized regions on a sphere. In this approach, the gravity and topography fields are multiplied by a windowing function (which is expanded in spherical harmonics up to a maximum degree L) that isolates a particular region. The optimal windows for this procedure are described in Wieczorek and Simons (2005). As the multiplication of two fields in the spatial domain is analogous to the convolution of their spherical harmonic coefficients in the spectral domain, a complementary tradeoff was shown to exist between spatial and spectral resolution. In particular, since the gravity and topography fields are spectrally truncated at a maximum degree L_g, the maximum degree that can be analyzed is $L_g - L$. Hence, the more that one localizes the data (i.e., the sharper the window and the greater L), the fewer the number of degree-dependent admittances that can be analyzed.

This localized spectral analysis technique was applied by Aoshima and Namiki (2001) to the nearside mascon basins in order to place limits on the average crustal thickness and elastic thickness. As the thickness of the mare basalts in these basins is not well constrained, this was estimated by assuming that these basins were in a state of Airy isostasy prior to the commencement of mare volcanism, and that these loads are presently supported by the elastic lithosphere (see also Bratt et al. 1985). Their results are presented in Table 3.9. Wieczorek and Phillips (1999) have shown that this pre-mare isostatic assumption is probably valid for those basins that formed within the Procellarum KREEP Terrane (see Section 5.1.2; Imbrium, Serenitatis, Humorum, and Grimaldi), and for Serenitatis and Humorum, an average crustal thickness of 50 ± 10 and 50 km was determined, respectively. These values are consistent with those from the highland GTR analysis presented in the previous section. The interpretation of the other basins in their study is less certain as the validity of the pre-mare isostatic constraint is questionable.

In another study, Arkani-Hamed (1998) used a Cartesian admittance modeling approach to investigate the region encompassing the South Pole-Aitken basin. Assuming that this region is compensated by an Airy mechanism, he discarded those wavenumbers that did not show a positive correlation between the gravitational potential and topography. This spectral editing procedure improved the spectral correlation between the two fields and a reference crustal thickness of 54 km (no error bar quoted) was obtained. However, as the spectral admittance function was not explicitly shown, it is difficult to assess the uncertainty associated with this number. It is also not clear as to whether the lack of topographic data poleward of 80°S, the low fidelity of the farside gravity field, or the Cartesian approximation in this analysis biased this crustal thickness determination. This value is, nevertheless, consistent with the above-mentioned studies.

2.6.3. The average thickness of the lunar crust. The above admittance studies place constraints on the average thickness of the lunar crust that are independent of the Apollo seismic data. As is summarized in Table 3.9, these studies are in general agreement. Assuming that the crust is compensated by an Airy mechanism, the highland geoid to topography ratios imply

Table 3.9. Reference thickness of the lunar Crust.

Region	Reference Crustal Thickness (km)	Comments
Extrapolating Apollo 12/14 seismic constraints to mean planetary radius		
[†]Nearside highlands	65 – 76	Airy compensation; Toksöz et al. (1974)
[†]Nearside highlands	45 – 66	Airy compensation; Khan et al. (2000)
[†]Nearside highlands	33 – 48	Airy compensation; Lognonné et al. (2003)
Geoid to topography ratios		
Nearside highlands	49 ± 16	this study
[§]South Pole-Aitken	52	Arkani-Hamed (1998)
Localized admittance modeling		
Serenitatis	50 ± 10	Aoshima and Namiki (2001)
[*]Crisium	35 – 40	Aoshima and Namiki (2001)
[*]Nectaris	50 ± 5	Aoshima and Namiki (2001)
[*]Orientale	70 ± 5	Aoshima and Namiki (2001)
Humorum	50	Aoshima and Namiki (2001)
[*]Humboldtianum	55 ± 5	Aoshima and Namiki (2001)
Cartesian spectral admittance modeling		
[§]South Pole-Aitken	54	Arkani-Hamed (1998)

[†]Includes the cases where the degree-2 topography is assumed to be compensated and uncompensated.
[*]These basins may not satisfy the pre-mare isostatic assumption.
[§]Based on farside gravity data.

that the average crust is 49 ± 16 km thick. Using localized spectral admittance functions, the two basins that are likely to satisfy the pre-mare isostatic assumption give a reference crustal thickness of 50 ± 10 km and 50 km for the Serenitatis and Humorum basins, respectively (Aoshima and Namiki 2001). A Cartesian admittance analysis of the South Pole-Aitken basin also suggests that the average crustal thickness is about 54 km (Arkani-Hamed 1998).

As was discussed in Section 2.2, the crust beneath the Apollo 12 and 14 sites was determined to be about 60-km thick during the Apollo era, whereas the reanalyses of Khan et al. (2000), Khan and Mosegaard (2002) and Lognonné et al. (2003) suggest a thinner crustal thickness of 45 ± 5 km, 38 ± 8 km, and 30 ± 2.5 km, respectively. We can now compare these thick and thin seismic estimates with the results of the above gravity-topography admittance studies. In doing so, we will assume that the Apollo 12 and 14 sites are compensated by an Airy mechanism and extrapolate the seismically derived crustal thickness to the mean planetary radius. For the following calculations, we will include the cases where the degree-2 topography of the Moon is compensated and where it represents the Moon's equilibrium shape that was frozen into the lithosphere when the lunar orbit was closer to the Earth (e.g., Lambeck and Pullan 1980). If the crust at this locale was 60-km thick, then using a plausible range of crustal and mantle densities (see Section 2.6.1) the average thickness at mean planetary radius would lie between 65 and 76 km. We note that this value is inconsistent with the estimates inferred from the gravity and topography admittance studies. In contrast, if we assume that the thickness of the crust beneath the Apollo 12 and 14 sites is 45 ± 5 km, then the average crustal thickness at mean planetary radius would lie between 45 and 66 km. Assuming yet a thinner value of 30 ± 2.5 km yields the range of 33–48 km. The limits of the above two thin-crust extrapolations are nearly identical to those of the GTR analysis of Section 2.6.1, and we therefore advocate an average lunar crustal thickness of 49 ± 16 km.

2.7 Crustal thickness models

It was recognized very early that the lunar gravity and topography fields are highly correlated with the large impact basins. The large positive gravity anomalies that were found over these topographic depressions led to the hypothesis that these were the result of both uncompensated basalt flows and relief along the crust-mantle interface (e.g., Wise and Yates 1970; Sjogren and Smith 1976; Phillips and Lambeck 1980; Phillips and Dvorak 1981). Based on higher resolution gravity and topography data, the mascons associated with some basins have been suggested to be the result of superisostatic uplift of the lunar mantle (Neumann et al. 1996; Konopliv et al. 1998, 2001; Wieczorek and Phillips 1999).

A number of attempts have been made to constrain globally the crustal thickness variations of the lunar crust using both gravity and topography data (e.g., Bills and Ferrari 1977b; Thurber and Solomon 1978; Bratt et al. 1985; Zuber et al. 1994; Neumann et al. 1996; von Frese et al. 1997; Arkani-Hamed 1998; Wieczorek and Phillips 1998). While the approach of each study is slightly different, a commonality among them is the assumption that the observed gravity field is the result of surface topography, mare basalts, and relief along the crust-mantle interface. Some of these studies have additionally attempted to take into account the effects of impact brecciation (Phillips and Dvorak 1981; von Frese et al. 1997), or the effects of density stratification within the crust (Wieczorek and Phillips 1998). As gravity modeling is non-unique, these crustal thickness models have traditionally been anchored to the seismically constrained crustal thickness of 60 km (e.g., Toksöz et al. 1974) at the Apollo 12 and 14 sites. Recent seismic inversions, however, suggest that the crustal thickness at these two sites is probably considerably thinner than once thought; in particular 45 ± 5 km (Khan et al. 2000), 38 ± 8 km (Khan and Mosegaard 2002), and 30 ± 2.5 km (Lognonné et al. 2003). In this section, we construct new crustal thickness models that utilize these new seismic constraints and the latest gravity model of Konopliv et al. (2001).

While there are many complications in constructing a crustal thickness model, the general approach is relatively straightforward. First the gravity field that results from the surface topography is computed (the Bouguer correction). This contribution is subtracted from the observed free-air gravity field of the planet, resulting in the Bouguer anomaly. Finally, the Bouguer anomaly is interpreted as relief along a subsurface density interface. The first complication is related to how one computes the gravity field of relief along a density interface. Standard first-order potential theory predicts the spherical-harmonic gravity and topography coefficients to be linearly related. Neumann et al. (1996), however, emphasized that the large crustal thickness variations inferred for the Moon invalidate this approximation. Using the higher-order Cartesian method of Parker (1972), they showed that the first-order method under-predicted the crustal thickness of large impact basins by about 10 to 20 km. Using spherical geometry, Wieczorek and Phillips (1998) derived an analogous higher-order method for calculating gravity anomalies due to finite-amplitude relief on a sphere. (See Thurber and Solomon (1978), Bratt et al. (1985) and von Frese et al. (1997) for alternative approaches in the space domain).

The next complication is related to how one inverts for the subcrustal relief given the Bouguer anomaly. In doing so, the Bouguer anomaly must be downward continued to the average depth of the crust-mantle interface, and this is a process that amplifies noise. The amplification of noise in the lunar gravity field is particularly destructive in the spectral domain as a result of the lack of farside tracking data. To counteract this problem on the Moon, Neumann et al. (1996) used the Cartesian downward continuation filter of Phipps Morgan and Blackman (1993) to stabilize their inversion, and Wieczorek and Phillips (1998) developed an analogous filter for use in the spherical-harmonic domain. The amount of filtering to be applied is a subjective decision, being controlled by the degree at which the filter is assigned a value of 0.5. For the crustal thickness inversions presented below, these were found to be unstable if this degree was greater than 40. In this study, the filter was chosen to have a value of 0.5 at degree 30 in order to offer a subjective compromise between retaining a high nearside spatial resolution versus the damping of unreasonable farside short-wavelength undulations.

Our individual inversions, described separately below, differ only slightly from those of Wieczorek and Phillips (1998). As was originally noted by Neumann et al. (1996), the short-wavelength power in the gravity fields of Lemoine et al. (1997) and Konopliv et al. (1998) was found to be inconsistent with predictions based upon the observed topography. This artifact was attributed to the stabilization procedures used to counteract the lack of farside tracking data in constructing these gravity models. Because of this, it was necessary to filter the Bouguer correction so that high frequency noise would not be introduced into the Bouguer anomaly. The LP150Q gravity field of Konopliv et al. (2001) used here, however, is consistent with predictions from the surface topography up to about degree 65. Hence, in this study, we truncate both the gravity and topography fields at this degree.

We describe below three plausible crustal thickness models of the Moon. The first two models assume that the crust is uniform in composition (excluding a thin veneer of mare basalts), whereas the third model assumes that the crust is stratified into upper anorthositic and lower noritic layers. The salient features unique to each model are discussed below, whereas the general aspects of the crustal models are discussed in the following section. We emphasize here that none of these models assume that the lunar crust is isostatically compensated.

Model 1 (LP150Q). This is the canonical model in which the lunar gravity field is assumed to be solely the result of surface topography, a thin veneer of dense mare basalts, and relief along the crust-mantle interface. For this model, as well as the two that follow, the gravitational attraction of the mare basalts within the nearside mascon basins was taken into account using the mare thickness model of Solomon and Head (1980) modified by the maximum thickness constraints of Williams and Zuber (1998).

A density of 2900 and 3300 kg m^{-3} was assumed for the crust and mare basalts, respectively (see Table 3.10). Given the known mass of the Moon, and a crust that is on average 55-km thick, the density of the mantle can be no greater than 3400 kg m^{-3}. For this model, a density of 3320 kg m^{-3} was assumed for the upper mantle, which results in an approximately zero-km thick crust (excluding the mare fill) beneath the Crisium basin. Because the density of the mare basalts and mantle are approximately the same, the assumed mare thickness model does not significantly affect the inverted total thickness of the crust. The crustal thickness at the Apollo 12 and 14 sites was constrained to be 45 km, and the globally averaged crustal thickness for this model is 53.4 km, consistent with the limits derived from gravity and topography admittance studies. The total crustal thickness for this model is shown in Color Plate 3.5.

Model 2 (LP150Q). This model is similar to the previous one, but with one major exception. Here we allow for the possibility that the degree-1 shape of the Moon (i.e., the 1.9 km center-of-mass/center-of-figure offset) might not be the result of crustal thickness variations. Particularly, after computing the Bouguer anomaly, the degree-1 terms were set to zero. In this model, the degree-1 Bouguer anomaly could either be the result of lateral density variations within the crust and/or mantle. We chose the density of the mantle for this model to be 3400 kg m^{-3} in order to give a zero crustal thickness beneath the Apollo basin that lies within the larger farside South Pole-Aitken basin. The globally averaged crustal thickness for this model is 43.4 km, consistent with the limits derived from the gravity and topography admittance studies. The total crustal thickness is shown in Color Plate 3.6.

Model 3 (LP150Q). As is discussed throughout Section 2, there are some compelling reasons to suspect that the lunar crust might either be vertically zoned or stratified in composition. In particular, a model in which the upper crustal thickness varies, while the lower crustal thickness remains constant, was found to satisfy the highland geoid to topography ratios (see Section 2.6.1). Here, we construct a dual-layered crustal thickness model of the Moon using this general structure. By strictly employing this model, however, it is not possible to satisfy the lunar gravity field over many of the large impact basins. Wherever this occurred, we allowed the lower crust to vary in thickness as well. The densities of the upper crust, lower crust, and mantle in this model were assumed to be 2820, 3040, and 3350 kg m^{-3}, respectively (see Section 2.8.4 for a discussion of the density of the lower crust). The upper crustal and total crustal thicknesses for this model are displayed in Color Plate 3.7. The globally averaged thickness of the crust for this model is 52 km, consistent with the limits derived from the gravity and topography admittance studies, and the average thickness of the upper crust is 26.9 km. We note that this dual-layered model is different in detail than the favored dual-layered model presented in Wieczorek and Phillips (1998).

Table 3.10. Crustal thickness model parameters.

Parameter	Model 1 (single-layered)	Model 2 (single-layered)	Model 3 (dual-layered)
Reference crustal thickness, km	53.4	43.4	52.0
Reference upper crustal thickness, km	—	—	26.9
Hemispheric crustal thickness difference, km	16.7	2.0	13.1
Density of upper crust, kg m^{-3}	2900	2900	2820
Density of lower crust, kg m^{-3}	—	—	3040
Density of mantle, kg m^{-3}	3320	3400	3350
Density of mare basalts, kg m^{-3}	3300	3300	3300

2.7.1. Implications

The center-of-mass/center-of-figure offset. One noticeable feature of the shape of the Moon is that its center of figure is displaced 1.9 km away from its center of mass in the direction of 8°N and 157°W. A popular explanation for this observation is that it could be the result of the lunar farside crust being significantly thicker than that of the nearside (e.g., Kaula et al. 1972; Kaula et al. 1974). In our crustal thickness models that assume constant density crustal layers, we indeed find that the farside crust is on average 17 km thicker than the nearside for our single-layered model (Model 1), and 13 km thicker for our dual-layered model (Model 3).

A thicker farside crust, however, is not the only possible explanation for the center-of-mass/center-of-figure offset of the Moon. In particular, it is possible that this feature could be the result of the degree-1 variations in either crustal and/or mantle density, as is implicitly assumed in our Model 2. As an example, in the strict case of Pratt compensation, the crust-mantle interface lies at a constant depth below the surface; in order for there to be a constant pressure along this interface, the crust must be denser than average at low elevations, and less dense than average at high elevations. Solomon (1978) and Thurber and Solomon (1978) emphasized that some portion of the lunar crust could be regionally or globally compensated by crustal density variations. This was based in part on observed correlations between the remotely sensed surface composition (Al/Si, Th, and Fe) and elevation in the limited Apollo data. Using the near-global iron abundance data of Lucey et al. (1995) derived from the Clementine mission, though, Wieczorek and Phillips (1997) have since shown that no significant regional correlation exists between density and elevation for the highlands. While this implies that Pratt compensation is regionally unimportant, they could not rule out a Pratt origin for the global degree-1 shape of the Moon. For example, the 1.9 km center-of-mass/center-of-figure offset could be accounted for if the nearside crust was on average 115 kg m^{-3} denser than that of the farside. This would correspond to about an ~4 wt% difference in iron content between the two hemispheres (Wieczorek and Phillips 1997). Additional studies of how crustal composition varies with elevation are needed before a Pratt origin can be ruled out for the long-wavelength shape of the Moon. Even if it turns out that density differences do not exist between the near- and far-side crust, it is possible that they might yet exist within the lunar mantle (e.g., Wasson and Warren 1980). In support of this, we note that the high-Ti basalts erupted exclusively on the lunar nearside, implying a higher TiO_2 content of the nearside mantle.

Impact basins. A feature common to all three crustal thickness models is that the crust is significantly thinned beneath many of the large impact basins. This dramatic feature was noted in previous models (e.g., Bratt et al. 1985; Neumann et al. 1996; Wieczorek and Phillips 1998) and has been interpreted as being the result of the large amount of material that is excavated in the cratering process. Most of these basins are also surrounded by an annulus of thickened crust, and this feature is most likely attributed to the thick ejecta deposit of the basin and/or to the lateral displacement of crustal materials that occurs during the excavation stage of the cratering process (Neumann et al. 1996; Wieczorek and Phillips 1999).

The large impact basins also exhibit some notable asymmetries based on their inferred crustal structure (see online RIMG supporting material at *www.minsocam.org* for crustal thickness images of the major basins). For instance, the excavation cavity of the Imbrium basin is not located at the geometric center of its main topographic rim, but rather is displaced towards the northwest (see also Spudis 1993). Furthermore, many nearside impact basins appear to possess a symmetry axis that might be indicative of having formed by an oblique impact (e.g., Serenitatis, Humorum, Crisium, and Nectaris). The crustal structure of the Nectaris basin is particularly odd with its northwest-southeast elongation. This structure could be the result of two closely spaced and unrelated impact basins, a double impact basin formed by two related bolides, or the result of a single highly-oblique impact.

Exposures of the mantle. Even though the crust is inferred to have been thinned dramatically beneath many of the large impact basins, these crustal thickness models predict that exposures of the lunar mantle should be rare to non-existent. Depending upon the assumed crustal thickness model, there are only two possible regions were we might expect to find evidence for mantle materials exposed near the surface. If we assume that the degree-1 shape of the Moon is a result of lateral density variations in the crust and/or mantle (Model 2) then this model predicts the mantle to be exposed beneath the Apollo basin that lies within the larger South Pole-Aitken basin. In contrast, if we assume that the degree-1 shape of the Moon is a result of crustal thickness variations (Models 1 and 3) then the only place where the crust is predicted to be entirely absent (excluding the mare fill) is beneath the Crisium basin. The possibility that the Crisium basin might have excavated into the lunar mantle was originally noted by Wieczorek and Phillips (1998), and the thinner seismic crustal thickness constraints (Khan and Mosegaard 2002; Lognonné et al. 2003) only strengthen the case that this indeed might have happened.

Given that the Apollo and Crisium basins have been partially flooded by mare basalts, it is unclear whether these putative mantle deposits would be presently visible at the surface. It is possible, though, that a mantle geochemical signature might be detectable in the ejecta blanket surrounding these basins. Alternatively, a mantle signature might be associated with younger craters that formed within the mare deposits of these basins. Two particularly interesting candidates are the craters Pierce and Picard that are located within Mare Crisium. X-ray fluorescence data obtained from lunar orbit over these craters have shown that they possess anomalously high concentrations of magnesium, and Andre et al. (1978) originally interpreted this as being the result of these craters having excavating into a buried Mg-rich basalt flow. In support of this interpretation, they noted that some of the Luna 24 samples from eastern Mare Crisium (which were presumed to be volcanic) similarly had high abundances of magnesium. Wieczorek and Phillips (1998) alternatively suggested that the high magnesium concentrations associated with these two craters might be the result of having excavated into either the underlying mantle, or an impact melt sheet that was partially derived from the mantle. A reassessment of the Mg-rich Luna 24 samples might be worthwhile in light of this suggestion.

Exposures of the lower crust. If the lunar crust is vertically stratified into upper anorthositic and lower noritic layers, then our dual-layered crustal thickness model (Model 3) makes specific predications as to where we might find exposures of the lower crust. Similar to the dual-layered crustal thickness model of Wieczorek and Phillips (1998), our model predicts that the entire upper crust should be absent beneath many of the large impact basins, including Imbrium, Serenitatis, Crisium, Smythii, Humboldtianum, Orientale, Humorum, Nectaris, and South Pole-Aitken. Additionally, the upper crust is predicted to be less than 5 km thick beneath Mendel-Rydberg, Moscoviense, Freundlich-Sharonov, and a portion of eastern Mare Frigoris. Unfortunately, mare basalts have extensively flooded most of these regions, obscuring these putative lower crustal exposures. A major exception to this is the South Pole-Aitken basin where mare deposits are relatively scarce (e.g., Wilhelms 1987; Yingst and Head 1997). The remotely-sensed composition of the floor of this basin appears to be composed of relatively mafic noritic materials (Pieters et al. 1997; 2001), in agreement with the predictions of this dual-layered crustal thickness model. As will be discussed more in Section 2.8.4, this model can also be used to predict whether the central peak of a complex crater is derived from either the upper or lower crust.

Other features. The crustal thickness models possess other notable features, a few of which will be mentioned here. First, the crust beneath the Aristarchus Plateau and Marius Hills is thickened by ~10 km with respect to the surrounding maria. Since these regions have been inferred to be unique long-lived volcanic centers (e.g., Wilhelms 1987), this may be related to a thick shield-like volcanic pile of extrusive and intrusive volcanic rocks. Second, the crust beneath the Apennine Bench is also seen to be thickened by ~15 km. This could similarly be attributed to extensive volcanism, or perhaps to some process associated with the deposition of the Imbrium and Serenitatis ejecta. Though some portion of this unit may have formed by post-Imbrium KREEP-basaltic volcanism (e.g., Hawke and Head 1978; Spudis 1978), it is not clear whether the bulk of this thickened crust formed before, after, or during the Imbrium impact. Finally, a localized region of highly thickened crust is seen to lie west of the Orientale basin and northeast of the South Pole-Aitken basin. As this region is located on the farside of the Moon, this feature might (partially) be an artifact of the poor resolution of the gravity field in this region.

2.8. Crustal stratigraphy

2.8.1. Spectral reflectance studies of central peaks. Probing the composition of the crust at depth can be accomplished by using a unique property of complex impact craters. Typically, lunar craters larger than 35 km in diameter have peaks at their center that are believed to represent rocks that have been uplifted from beneath the surface during crater formation. While the exact depths from which these central peaks are uplifted are not exactly known, most estimates range between 0.1 and 0.2 times the crater diameter (e.g., Dence 1968; Roddy 1977; Melosh 1989). As an example, the crater Copernicus with a diameter of ~98 km may contain rocks from pre-impact depths of ~10–20 km exposed in its central peak. Most recently, Cintala and Grieve (1998) have made use of models that take into account both the excavation of crustal materials and the extent of impact melting to predict the amount of uplift in lunar central peaks. Assuming that the materials that constitute these peaks are derived from beneath the maximum depth of impact melting, their depths of origin were found to be approximately given by $0.109\,D^{1.08}$, where D is the final rim diameter of the crater in kilometers. This depth of peak origin corresponds to almost twice the maximum depth of excavation of the crater.

Although no rocks were collected during the Apollo missions that directly sampled the central peak of a crater, spectral-reflectance remote sensing techniques can provide estimates of their composition. In particular, at visible through near-infrared wavelengths, crystal field transitions of Fe and Ti cause diagnostic absorption bands in reflectance spectra (see Chapter 2) allowing mafic minerals such as olivine and pyroxene to be identified. Until the 1990s, though, this type of remote sensing was confined to the acquisition of individual spectral measurements from Earth-based telescopes. While telescopic based spectra offer excellent spectral resolution, each measurement is an average over an area approximately 2 to 10 km in diameter.

Over the past 30 years, a database of high-resolution, telescopic near-infrared reflectance spectra has been acquired for numerous locations across the lunar nearside (e.g., McCord et al. 1981; Pieters 1993). Spectra of impact craters have been particularly fruitful as they expose fresh material from beneath the surface. Pieters' (1986) summary of the compositions of nearside craters based on these telescopic spectra revealed a crust with a surprisingly varied composition. Central peak rocks spanned the range of compositions found among pristine lunar highland rock samples, including troctolites, gabbros, norites, and anorthosites. These initial central peak studies led to several key conclusions for lunar geology. First, the peak compositions indicated potential gaps in the lunar sample collection (suggesting, for example, a greater abundance of gabbroic rocks than found among the samples (Lucey and Hawke 1989). Second, several of the central peaks might have tapped mafic plutons, potential sources for the Mg-suite samples (Lucey et al. 1986; Pieters 1991). And third, central peaks in many locations exposed anorthosite, allowing this rock type to be mapped across the lunar nearside and compared to the predictions of the magma-ocean theory.

Although the high spectral resolution of telescopic measurements allow detailed assessments of mineralogy, only spatial variations on the order of kilometers could be detected. Data collected from lunar orbiting spacecraft, in contrast, offer increased spatial resolution, but significantly less spectral information. In 1994, the Clementine mission returned global multispectral images of the Moon, including large portions of the lunar farside whose composition was previously unknown. Clementine's UVVIS camera provided high-spatial resolution (100–400 m/pixel) images at five wavelengths, allowing first-order mineralogical determinations (see Chapter 2) at scales sufficient for detailed geologic mapping. Rather than measuring a single average composition for a central peak as with telescopic studies, multiple compositions within the peaks were distinguishable.

Using the Clementine data, Tompkins and Pieters (1999) conducted a global survey of central peaks in order to estimate the compositional variability of the lunar crust beneath the megaregolith. The craters ranged in diameter from 40 to 180 km, and the depths of origin of their central peaks were believed to range from approximately 5 to 30 km. Representative five-color spectra from spectrally and spatially distinct areas within the peaks were selected and classified on a relative scale, from which mineralogical abundances were estimated. These were then translated to rock types based on the classification scheme of Stöffler et al. (1980) (see Fig. 3.12). The lithologies found in these central peaks are shown in Color Plate 3.8.

From this survey, a more complete picture of the global trends emerged. Given the observed central peak compositions, the lunar crust is observed to be extremely anorthositic, gradually increasing in mafic content with depth. The crustal composition ranges from predominantly "pure" anorthosite to anorthositic norite. In general, these results are considerably less mafic than earlier estimates and predictions. Isolated mafic outcrops occur throughout the Moon, though true mafic rocks are rare among the surveyed craters; only ~10% of the central peaks contain mafic rocks such as gabbro or norite, and about half of the peaks are composed of

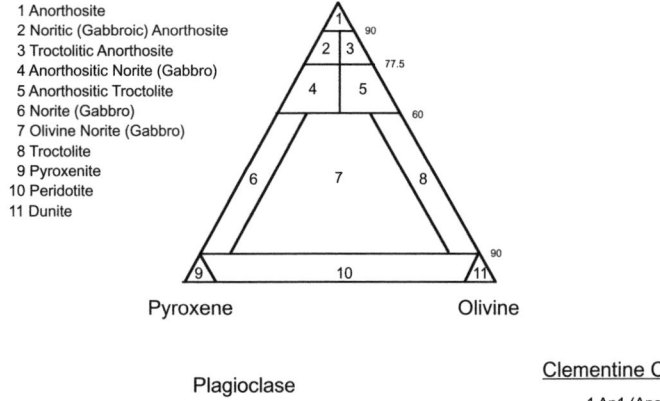

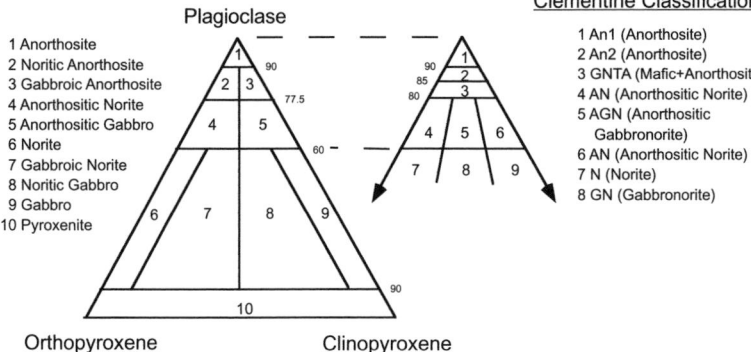

Figure 3.12. Classification scheme of lunar rocks based on Stöffler et al. (1980), and its relationship to the classification scheme used in Tompkins and Pieters (1999). Lower figure after Tompkins and Pieters (1999).

rocks that contain more than 85% plagioclase. The mafic lithologies identified among the central peaks are typically more anorthositic than the average value for large Mg-suite samples (Warren 1990). Nonetheless, central peaks that were identified as gabbro, norite, troctolite, gabbronorite, anorthositic troctolite, anorthositic gabbro, and anorthositic gabbronorite (following the Stöffler et al. classification scheme) were considered to be potential candidates for magnesian-suite plutons. This method, however, has no way of distinguishing Mg', so identifications of more mafic rock types could also reflect exhumation of relatively mafic ferroan anorthositic rocks.

2.8.2. The distribution of anorthosite.

During the 1980s and early 1990s, analyses of telescopic near-infrared reflectance spectra revealed outcrops or patches of anorthosite on the order of tens of square kilometers in plan dimension on the lunar nearside, principally associated with uplifted blocks on the inner rings of impact basins such as the inner Rook Mountains of the Orientale basin and in parts of the inner ring of the Grimaldi basin (Spudis et al. 1984; Hawke et al. 1991, 1993, 2003b). Other exposures of anorthosite were found near the outer rings of basins such as Grimaldi, Humorum, and Nectaris, but in these cases, the anorthosites were found in central peaks and walls of superposed impact craters. (e.g., Hawke et al. 1993). Anorthosite has further been identified in the central peaks of Alphonsus and Petavius (Pieters 1986; Coombs et al. 1990) which are located near major rings of ancient impact basins, and in plains units of the northern highlands that were affected by large Copernican craters (Thales and Anaxagoras). In general, these anorthosite deposits appear to have been exposed from beneath a slightly more mafic surface layer. Locations where anorthosite has been identified using Earth-based reflectance spectra are shown in Figure 3.13 (see also Color Plate 3.9).

Clementine UVVIS spectra have subsequently been used to search for anorthosite globally using remotely sensed estimates of FeO abundances (Hawke et al. 2003b). In these studies, a surface iron abundance less than 4 wt% FeO, which likely corresponds to more than 90% plagioclase, was used as a criterion for detecting this rock type. In addition to exposures

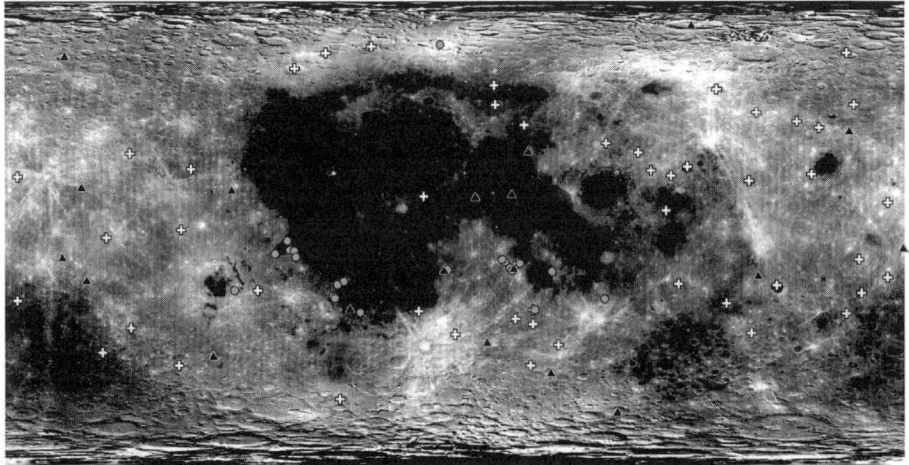

Figure 3.13. Anorthosite distribution superposed on the Clementine 750 nm global albedo basemap. Circles represent locations of outcrops of anorthosite observed with telescopic near-infrared spectra (Hawke et al. 2003b), triangles represent locations within central peaks of craters where Tompkins and Pieters (1999) identified anorthosite (An1), and crosses represent locations within central peaks of craters where anorthosite plus one or more additional rock types were identified. The base map is projected in a simple cylindrical projection. See also Color Plate 3.9.

identified previously, anorthosites were found in the inner rings of basins such as Hertzsprung, Korolev, Grimaldi, Humorum, Nectaris, and Orientale. Tompkins and Pieters (1999) further found numerous central peaks with exposed anorthosite which were presumably derived from depths ranging from 5 to 20 km below the surface (these occurrences are listed in Table A3.10). One should be aware, however, that as the mafic absorption features of these spectra rely on ferrous iron-bearing silicates, the possibility exists that feldspathic magnesian rocks could be misclassified as anorthosite, even if they contained more than 10% mafic silicates.

The exposures of anorthosite reported by Hawke and coworkers are especially noteworthy because many appear to contain less than 2 wt% FeO, which likely corresponds to a rock being composed of nearly 95% or more plagioclase. Some of these exposures, which can extend for more than 150 km, are even more anorthositic than the farside feldspathic highlands terrane, whose composition is typically 3.5–5% wt% FeO (Jolliff et al. 2000a; Lawrence et al. 2002). These nearly pure anorthosite exposures are particularly significant as they may represent remnants of the ancient anorthositic upper crust. As such, their distribution may provide an indication of the extent and stratigraphic position of positively buoyant anorthositic cumulates that formed during the primary crustal differentiation event of the Moon.

Anorthosite exposures are most common in the feldspathic highlands terrane where basin impacts have excavated through a slightly more noritic megaregolith that is on the order of a few kilometers to ~10 km thick (Spudis et al. 1984; Hawke et al. 1991, 2003b; Bussey and Spudis 2000). For example, the Hertzsprung and Korolev basins in the farside highlands probably excavated as deep as ~20-30 km, and abundant anorthosite exposures are found in the ring mountains of these basins. Such deep excavations would probably have been sufficient to breach a putative thick ejecta deposit from the South Pole-Aitken basin, which appears to be more mafic than the underlying anorthosites (Hawke et al. 2003b). Anorthosite is also exposed within the South Pole-Aitken basin in blocks uplifted by the Apollo and Ingenii basins, as well as in the Leibnitz and von Kármán craters (e.g., Pieters et al. 2001). It is possible that these deposits represent ancient feldspathic crustal materials that underlie the South Pole-Aitken impact melt sheet.

Anorthosite exposures are less common in the highlands of the southern nearside and in the region encompassing the Imbrium basin and Oceanus Procellarum (the Procellarum KREEP Terrane). Their paucity in these regions may partially result from obscuration by basalt flows and thick mafic ejecta deposits from the large nearside impact basins, or alternatively, a continuous anorthositic layer may never have existed within this region of the crust. While anorthosites have been identified in the central peaks of Eratosthenes and Aristarchus (McEwen et al. 1994; Tompkins and Pieters 1999; Hawke et al. 2003b), elevated thorium concentrations associated with these craters suggest that these exposures might be of an alkali variety, possibly associated with KREEP-rich intrusives, and may not be related to a primary ferroan anorthositic crust. Moreover, impact breccias found at the Apollo 14 site, thought to be predominantly Imbrium ejecta or remnants of older upper crust of the Procellarum KREEP terrane (Haskin et al. 2000), mostly lack ferroan anorthosite as clasts or chemical components (Jolliff et al. 1991b; Spudis et al. 1991). Nevertheless, exceptions to this general rule do exist (Shervais et al. 1983; Warren et al. 1983a). Finally, we note that anorthosites are also present in the inner ring of the Humorum basin which lies on the boundary between the Procellarum KREEP and Feldspathic Highlands Terranes (Hawke et al. 1993; Bussey and Spudis 2000).

A key question that has yet to be resolved is whether or not these anorthosite deposits are the remnants of a once globe-encircling layer as is predicted by simple models of magma-ocean crystallization. Even if ferroan anorthosites never existed within the Procellarum KREEP terrane, their lateral and depth distribution within the Feldspathic Highlands terrane could be used to test more complex anorthosite formation scenarios (e.g., Longhi and Ashwal 1985; Jolliff and Haskin 1995; Longhi 2000).

2.8.3. Composition of basin ejecta. Just as exposures of anorthosite in the rings of impact basins and central peaks of complex craters represent materials uplifted from depth, the composition of ejecta from impact basins similarly can provide a window into the subsurface crust. While the current depth of most lunar impact basins is generally fairly shallow (a few kilometers at most), this structure is the result of crustal rebound following the crater excavation stage as well as a variety of post-impact degradational processes (e.g., Melosh 1989). In contrast to the shallow observed depths of the large impact basins, ejecta blankets are derived from that portion of the crater that was ballistically excavated during the cratering process (i.e., the excavation cavity), and this is known to be appreciably deeper.

A wide variety of sources—including computational studies, laboratory experiments, and the inferred structure of lunar and terrestrial impact craters (e.g., Croft 1980; O'Keefe and Ahrens 1993; Wieczorek and Phillips 1999)—suggest that the excavation cavity is roughly parabolic in shape, possessing a depth/diameter ratio of approximately 1/10. Unfortunately, diameters of the excavation cavity of craters are in general not known, as post-impact slumping of the rim destroys its initial morphology. Nevertheless, for small craters the observed diameter and excavation diameter are comparable, and if desired, various methods can be employed that relate one to the other (e.g., Croft 1980). For the largest basins, the issue is more complex as it is sometimes difficult to determine which basin ring most closely corresponds to its excavation cavity (e.g., Wieczorek and Phillips 1999).

As two examples, consider the Imbrium and Serenitatis basins, whose excavation cavity diameters have been estimated at 744 km and 657 km, respectively, based on crustal thickness modeling (Wieczorek and Phillips 1999). These two basins should have excavated to a maximum depth of ~70 km, likely including the entire crustal column and portions of the upper mantle. However, because the excavation cavity has a parabolic shape, most of the material ejected would come from the upper parts of the crust. In particular, the upper 30% by depth contributes about half of the ejecta, whereas the deepest 10% by depth contributes only ~1% (e.g., Spudis 1993; Haskin et al. 2003). Finally, it should be noted that when the primary ejecta is ultimately deposited on the surface, it will mix with the local substrate (e.g., Haskin et al. 2003).

Caveats notwithstanding, basins of different sizes can be used as probes into the lunar crust, allowing an investigation of how the composition varies both laterally and vertically (e.g., Spudis et al. 1984; Spudis and Davis 1986). With the acquisition of global compositional datasets, Clementine and Lunar Prospector data have been used to estimate FeO concentrations of ejecta deposits (Spudis et al. 1996; Bussey and Spudis 1997, 2000; Blewett and Hawke 2001; Jolliff et al. 2002). Using these data, Bussey and Spudis (2000) advocated a general three-layer compositional model of the crust based on a study of the ejecta deposits of Orientale, Humorum, Nectaris, and Crisium (see also Hawke et al. 2003b). Ejecta-deposit compositions reveal a top layer of mixed feldspathic megaregolith that may extend to depths of ~10–20 km, and uplift structures in the form of basin inner-ring massifs exposed highly anorthositic rocks. While this anorthositic "layer" was presumed to overlie a more mafic lower crust, such as is likely exposed in the South Pole-Aitken basin (Pieters et al. 1997, 2001), for the most part, these basins did not excavate materials that were significantly more mafic than the upper feldspathic crust. In particular, a lack of mafic ejecta at the Orientale basin suggests the presence of a thick anorthositic upper crust at this locale. This case appears to be representative for most of the basins that lie within the Feldspathic Highlands Terrane (Jolliff et al. 2000a) where gravity and topography data imply a relatively thick upper crust (Wieczorek and Phillips 1998).

2.8.4. The composition of the upper and lower crust. The central peaks that were investigated in the Tompkins and Pieters (1999) study are approximately randomly distributed across the lunar surface (see Color Plate 3.8) making it is possible to investigate systematically lateral and vertical variations in crustal composition. These authors originally noted that the central peaks of complex craters located within larger impact basins were generally more mafic than those that formed within the highlands. This observation was taken as being consistent with the hypothesis that large impacts excavate through the upper crust, bringing more mafic lower-crustal materials to surface. This was further quantified by Wieczorek and Zuber (2001a) who utilized a geophysically based crustal thickness model, and here we update their results using the updated model of this paper.

The dual-layered crustal thickness model presented in Section 2.7 assumes that the crust is stratified into upper anorthositic and lower noritic layers, with the thickness of the lower crust being approximately constant. By utilizing the composition of a complex crater's central peak and its associated pre-impact depth of origin, the validity of this geophysical model can be assessed. Specifically, if the depth of origin of a peak is predicted to be greater than the thickness of the upper crust at a given locale, then the peak should be composed of lower crustal (and hence noritic) materials. Using this approach, 91 of the 108 peaks in the Tompkins and Pieters (1999) study are predicted to be derived from the upper crust, whereas only 17 are predicted to have a lower crustal origin. Consistent with the geophysical crustal thickness model, those peaks that are exclusively composed of rocks containing more than 85% plagioclase (there are 51) are derived from the upper crust, whereas those peaks that contain some norite or gabbro-norite (there are only 6) have a lower crustal origin. Five of these mafic peaks are located within the South Pole-Aitken basin, whereas the remaining peak is from the crater Bullialdus, which lies near the Nubium basin (Pieters 1991). Peaks that have intermediate compositions do not show any clear correlation with being exclusively derived from either the upper or lower crustal layers.

If it is assumed that these central peaks are randomly distributed across the lunar surface and with depth, and that the dual-layered crustal thickness model is accurate, the bulk composition of the upper and lower crustal layers can be estimated. First, the average proportion of rock types that are present in the upper and lower crustal layers is summarized in Table 3.11 where it is seen that the upper crust is composed of 38% pure anorthosite. Furthermore, 89% of the upper crust is predicted to be composed of rocks that contain more than 80% plagioclase. The majority of the remaining material is composed of intermediate lithologies such as anorthositic norite, anorthositic gabbro-norite, anorthositic gabbro, and anorthositic troctolite.

Since only 17 central peaks are predicted to be derived from the lower crust of the Moon, its derived composition will be statistically less robust. Furthermore, because of uncertainties associated with the adopted crustal thickness model, as well as the modeled depth of origin

Table 3.11. Mineralogy of the upper and lower crust.

Rock type and assumed volume % plagioclase	Modal Abundance (%)		
	Upper Crust	**Lower Crust**	**Most Mafic Lower Crust**
anorthosite (95 ± 5)	38.3	22.1	0
An2 (87.5 ± 2.5)	34.3	17.6	4.2
GNTA (82.5 ± 2.5)	16.2	24.2	17.2
anorthositic gabbro (70 ± 10)	2.7	1.5	4.2
anorthositic gabbro-norite (70 ± 10)	2.9	4.6	7.5
anorthositic norite (70 ± 10)	3.0	15.4	29.7
anorthositic troctolite (70 ± 10)	1.6	1.5	0
gabbro (50 ± 10)	0.5	1.5	4.2
gabbronorite (50 ± 10)	0	6.1	17.2
norite (50 ± 10)	0	5.6	15.8
troctolite (50 ± 10)	0.4	0	0

of central peak materials, it is possible that a few upper crustal central peaks could have been misclassified as having a lower crustal origin. Recognizing this sampling problem, Wieczorek and Zuber (2001a) chose to analyze a subset of the lower-crustal central peaks that contain some gabbro and norite, and that might be representative of the "most-mafic lower crust." This end-member composition (which is based only on six craters) lacks pure anorthosite and is composed of 33% norite and gabbro-norite and 30% anorthositic norite. When compared to the upper crust, the most-mafic lower crust is seen to contain a greater relative abundance of norite with respect to gabbro. For example, in the upper crust, anorthositic norite, anorthositic gabbro-norite, and anorthositic gabbro all occur in roughly equal proportions. In the most-mafic lower crust, though, anorthositic norite is approximately seven times more abundant than anorthositic gabbro and norite is approximately four times more abundant than gabbro.

Using the data in Table 3.11, the upper crust is found to be composed of 87 ± 4% plagioclase by volume and to have an Al_2O_3 content that lies between 28.5 and 32.2 wt%. The most-mafic lower crust is found to be composed of 65 ± 8% plagioclase and to have an Al_2O_3 content that lies between 18.2 and 24.7 wt%. The computed density of these crustal layers is listed in Table 3.12.

This bulk composition of the upper crust is comparable with, though slightly more feldspathic than, previous estimates. Pre-Clementine and Lunar Prospector estimates for the Al_2O_3 content of the upper crust were in the range of 26 to 28 wt% (e.g., Spudis and Davis 1986). Using Clementine derived iron abundances, Lucey et al. (1998a) found the lunar highland soils to be slightly more feldspathic, ranging from 27 to 29 wt% Al_2O_3. Using the five most feldspathic lunar meteorites, Korotev (2000) estimated an Al_2O_3 abundance of ~28 wt% for the uppermost portion of the lunar crust. Finally, Jolliff et al. (2000a) obtained an Al_2O_3 abundance of ~29 wt% for the most feldspathic portion of the highlands crust based on an empirical correlation between aluminum and iron concentrations in the Apollo sample collection. Central peaks, which are likely derived from beneath the megaregolith (Bussey and Spudis 2000; Hawke et al. 2003b), indicate that the feldspathic component may in some places be even greater, up to 32 wt% Al_2O_3.

Table 3.12. Bulk properties of the upper and lower crust.

Parameter	Upper Crust	Lower Crust	Most Mafic Lower Crust
Plagioclase, vol%	87.4 ± 4.3	79.0 ± 5.8	65.4 ± 8.4
Al_2O_3, wt%	28.5–32.2	24.4–29.2	18.2–24.7
density, kg m^{-3}	2858 ± 35	2927 ± 47	3038 ± 69

2.9. Summary

Geophysical, remote sensing, and sample analyses are all converging on a picture of the lunar crust that varies both laterally and vertically in composition. To a good approximation, the lunar highland crust can be thought of as being either compositionally zoned or stratified. The average composition of the upper portion of the crust is extremely anorthositic (here estimated at ~87 ± 4% plagioclase or 28.5–32.2 wt% Al_2O_3), and pure exposures of anorthosite are commonly found beneath a surficial layer of mixed, slightly more mafic regolith materials. The anorthositic lunar samples have ferroan compositions and ancient ages (~4.45 Ga) consistent with being derived by plagioclase flotation in a near-global magma ocean. The lower portion of the crust, in contrast, appears to be more noritic in comparison to the upper crust. This is evident from both the noritic composition of some central peaks, as well as the noritic composition for the floor of the giant South Pole-Aitken basin. Unfortunately, we do not possess samples of these important rock types, leaving open the question of whether the lower crust is "ferroan" or "magnesian" in composition.

Geophysical and seismic analysis both suggest that the crust is thinner than previously assumed, being about 30–45 km thick at the Apollo 12 and 14 sites in comparison to the Apollo-era value of ~60 km. When this value is extrapolated globally, the crust is found to be on average about 49 ± 16 km thick. While the gravity and topography fields of the Moon are consistent with a crust that is homogenous in composition, remote-sensing considerations suggest that this view is too simplistic. A model that satisfies both the geophysical and remote-sensing constraints is one in which the crust is stratified into upper anorthositic and lower noritic layers. The thickness of the upper anorthositic crust in this model varies, whereas the thickness of the lower crust is constant except beneath the large impact basins. Geophysical and remote-sensing techniques constrain each of these crustal layers to be on average ~25 km thick. While crustal thickness models predict that exposures of the lunar mantle should be extremely rare to absent, the lower crust should have been excavated beneath many of the large impact basins, and is probably currently exposed within the South Pole-Aitken basin.

One group of rocks in the Apollo sample collection that remains enigmatic are those that comprise what is commonly referred to as the magnesian-suite. These rocks have high abundances of incompatible elements (i.e., KREEP), and much higher magnesium numbers than those of the ferroan anorthositic rocks. Global thorium abundances collected from orbit suggest that the KREEP-rich rocks of this suite are confined to the region encompassing Oceanus Procellarum and Mare Imbrium (see Section 5 for further details). Most of these rocks appear to have formed as crustal intrusions at various depths within the crust. The large span of crystallization ages (from ~4.45 to 3.85 Ga) attest to an extended period of Mg-suite magmatism in this region. As the youngest of these plutonic rocks were likely excavated and brought to the surface by the Imbrium impact, it is likely that Mg-suite plutonism may have continued for a much longer time than the ages of these samples suggest.

3. THE MANTLE

As of yet, no samples of the lunar mantle have been unambiguously identified in either the sample collection or by remote sensing techniques. Thus, unlike the lunar crust, the mantle can only be studied through less direct means. One method described below includes the chemical and isotopic analysis of mare basalts and volcanic glasses. Although they are now exposed at the surface of the Moon, the origin of these materials is ultimately related to partial melting deep within the lunar mantle, and their compositions reflect that of the mantle at the time they formed. Basalt and volcanic glass compositions combined with experimental petrology can infer the depths of origin and compositions of the mantle source regions. The only other fruitful method that can be used to infer the composition and structure of the mantle comes from an analysis of the Apollo seismic data.

Key questions that are addressed below include the major and trace element geochemistry of the mare source regions, the depths of origin of the volcanic samples, and whether there is evidence for chemical heterogeneities within the mare basalt source regions. Seismic analysis suggest that a major discontinuity may exist about 500 km below the surface, and the origin of this feature is crucial to understanding the Moon's thermal and magmatic evolution. Of key importance is whether this discontinuity represents the maximum depth of the magma ocean, or instead the maximum depth of melting within the mare source region.

3.1. Mare basalts and volcanic glasses

Almost all of the prominent mare basalts and related pyroclastic deposits are associated with the large nearside basins and Oceanus Procellarum. Basalts are comparatively rare on the lunar farside but do occur in Mare Moscoviense, Mare Ingenii, Mare Orientale, and in patches within other basins such as Apollo and the South Pole-Aitken basin. However, the distribution of mare basalts is not simply a case of lavas flooding topographic lows or thinned crust as the deepest basin on the Moon, the South Pole-Aitken basin, was demonstratively less volcanically active than regions of thinned nearside crust (Lucey et al. 1994; Smith et al. 1997; Wieczorek et al. 2001). Although mare basalts are prominent in lunar images, they form a rather thin veneer on the older crust. Thicknesses are typically less than 1 km exterior to impact basins and probably not more than 4 km within the largest basins (see Section 2.4). While the ages of large mare basalt samples dated radiometrically range from ~3.9 Ga to 3.2 Ga, the duration of basaltic volcanism appears to be longer. An age of 4.2 Ga has been reported for aluminous basalt clasts in a nonmare impact breccia (Taylor et al. 1983), and crater-counting methods indicate that basalts in some unsampled areas have ages approaching 1 Ga (Schultz and Spudis 1983; Hiesinger et al. 2000; Hiesinger and Head 2003). Although the concentration of radiogenic heat sources appears to be enhanced within the nearside crust and mantle, the existence of basalts on the farside indicates that heat sources there were also capable of raising the mantle temperature locally above the solidus. Experimental evidence for the depth of mare source regions is discussed in Section 3.3.1 below.

3.1.1. Basalt and volcanic glass compositional groups—major elements. Many volumes and hundreds of research papers have been written about the lunar mare basalts. Our intent here is not to describe the groups per se; instead it is to relate key aspects of the mare basalts to what can be inferred from orbital remote sensing and to discuss new implications for lateral and vertical heterogeneity of mantle source regions. For many years, the remote-sensing community has been aware that despite the wide variety of basalt compositions known from the direct analysis of Apollo and Luna samples, many basalt types exist for which we have no samples (e.g., Pieters 1978). For example, crater statistics imply relatively young ages of ~1 Ga for basalt flows in some locations, but so far, the youngest radiometrically dated ages are ~3 Ga. Remote sensing has revealed the occurrence of extensive buried mare basalt regions, referred to as cryptomare, indicating that ancient mare volcanism (>3.8 Ga) may be more common than would be inferred from the samples. Studies of basalts in the Western Procellarum region suggest an olivine rich mineralogy (Staid and Pieters 2001) coupled with higher Fe and Th contents (Jolliff et al. 2001). Clearly, much is yet to be learned about the extents of mare basaltic volcanism in space and time, especially regarding their petrogenesis and geochemical variability.

The first-order classification of lunar mare basalts is traditionally based upon TiO_2 concentrations since the mare basalts and volcanic glasses span a huge range. Basalts with TiO_2 concentrations <1 wt% are called very-low Ti or VLT, those with TiO_2 of 1–4 wt% are low-Ti, those with 4–8 wt% are intermediate, and >8 wt%, high-Ti. A second order classification has been made according to other compositional parameters or major mineralogy. The aluminous basalts, found mainly as clasts in breccias at Apollo 14, have >11 wt% Al_2O_3. High-K basalts from Apollo 11 have ~0.3 wt% K_2O, compared to <0.1 % in most others. The Apollo 14 very-high K basalts, however, have 0.6–0.8 wt% K_2O and are also aluminous. At Apollo 12, the three main basalt groups were distinguished on the basis of mineralogy: olivine basalts, pigeonite basalts, and ilmenite basalts.

The major mafic mineralogy (olivine, pyroxene, Fe-Ti oxides) of basalts can be determined from orbit and the concentrations of FeO and TiO_2 can be estimated as has been done with the Clementine UVVIS data (Chapter 2). FeO and TiO_2 as well as other elements such as Th have been determined by the Lunar Prospector gamma-ray spectrometer, but at a much lower spatial resolution than the Clementine data. The two key compositional parameters useful for distinguishing mare basalts in the lab or from orbit are FeO and TiO_2. FeO concentrations of the main mare basalt groups from Apollo and Luna samples range from ~15.5 wt% in the aluminous basalts to ~23 wt% in the Luna 24 ferrobasalts and Apollo 15 olivine basalts, with most being in the range of 18–21 wt%. Concentrations of FeO in the volcanic glasses are slightly higher, with a range of 16–24 wt%, and averaging ~21 wt%. Concentrations of TiO_2 range from <1 wt% to 13 wt% in the high-Ti basalts and to 17 wt% in the high-Ti volcanic glasses (Tables A3.11 and A3.12). Because the high-Fe and high-Ti basalts are endmember compositions, they are seen readily from orbit. Where mare surfaces contain admixed nonmare materials, the ejecta of small craters can be used to discern the degree of mixing and the composition of the basalt flows (Staid and Pieters 2000).

Using global compositional data derived from Clementine spectra, Giguere et al. (2000) showed that the TiO_2 contents of mare basalts form a unimodal distribution, with a peak at about 4 wt% and a long tail to high concentrations. This is in contrast to the samples, which show a distinctly bimodal distribution, with intermediate-TiO_2 compositions being rare. Gillis et al. (2003) showed that this is not an artifact of the remotely sensed data, but is related instead to the non-representativeness of the landing sites. In particular, Clementine data for just the basaltic landing sites reproduce the bimodal distribution as is seen in the samples. In comparing landing site and sample TiO_2 and FeO, Gillis et al. also showed that remotely sensed data were consistent with significant mixing of basalt and non-volcanic materials relative to the returned rock samples.

3.1.2. Trace-element signatures. Trace-element contents and signatures of mare basalts have been studied at length and are described in much detail elsewhere (see Basaltic Volcanism Study Project 1981; Papike et al. 1998; Shearer and Floss 2000 for summaries). Among the trace elements, the REEs have proven to be the most diagnostic and useful for discriminating relationships and petrogenetic processes. Recognition of the complementary patterns of chondrite-normalized REE contents of mare basalts and nonmare crustal materials was key to understanding the basic relationship between the crust and mantle of the Moon (e.g., Taylor 1975). The common characteristics of depletion of the light REEs (LREE), the relative enrichment of the heavy REEs (HREE), and the negative Eu anomaly were found to be consistent with the partitioning behavior between the minerals olivine and pyroxene, and melt. Basaltic melts represented by almost all the lunar basalt types and the picritic glasses share these characteristics, which were inherited from their deep mantle sources. The negative Eu anomaly has been attributed to the fractionation of divalent Eu from the trivalent REEs during plagioclase crystallization and extraction to form the crust, but studies of experimentally determined distribution coefficients show that the formation of a negative Eu anomaly can also occur from the crystallization of mafic silicates (e.g., McKay et al. 1990). Thus the shallow negative Eu anomaly of the picritic glasses does not necessarily imply plagioclase separation, but likely reflects the olivine and pyroxene cumulate nature of their source region. The general variations in REE patterns between the mare basalts, feldspathic crustal lithologies, and KREEP-rich materials are consistent with the simple magma-ocean scenario. Significant variations in REE concentrations and patterns in the mare basalts are taken to reflect variations in their sources, but the degree of melting associated with formation of a specific basalt, assimilation of crustal material during emplacement, and fractionation during solidification can impart additional variations. These factors must be taken into account before appealing to source-region variations to explain differences between basalt groups. Because minerals fractionate trace elements when they crystallize, the volcanic glasses have proven to be the most useful for study of mantle source-region variations. Such variations are discussed further in Section 3.3.1 and in Chapter 4, and were reviewed recently by Papike et al. (1998).

Elements other than the REEs that behave incompatibly in lunar basaltic systems include the large alkalis (K, Rb, Cs), Sr, Ba, Zr, Hf, Nb, Ta, Th, and U. In general, these elements occur in low concentrations in lunar basalts and volcanic glasses either because their source regions were intrinsically low (undifferentiated or depleted) or because the source regions are cumulates composed largely of minerals that do not readily incorporate them. Some of these elements are geochemically similar pairs that are difficult under common petrologic processes to fractionate from one another, such as Rb-Cs, Zr-Hf, Nb-Ta, and Th-U. In such cases, significant differences in trace-element abundance patterns signal specific processes. For example, the presence of garnet, if a component of a mantle source residue following partial melting, would have the effect of retaining preferentially the HREEs and so would impart a negative slope to the REE pattern of the basaltic liquid formed during partial melting. The lack of such patterns among the crystalline mare basalts was taken early on as evidence that garnet was not a significant component of basalt source regions, although it may be in some volcanic glasses (Neal 2001). Another example is the formation of ilmenite and other Ti-rich oxides, which can fractionate Hf from Zr. Scandium, which is incompatible in olivine, is enriched in pyroxene (especially high-Ca pyroxene), thus variations between Sc and the compatible trace element Cr correspond well to variations associated with the proportions of the major silicate minerals. Taken together, the patterns of trace elements, which are usefully compared between groups when normalized to an average composition or to chondritic compositions, are very useful discriminators of process and source variations.

Not all mare basalts are depleted of incompatible trace elements. Several groups such as the Apollo 14 very-high-K basalts, some of the Apollo 14 and Luna 16 high-alumina basalts, and the Apollo 11 high-K basalts have incompatible-trace-element enrichments that indicate either assimilation of some trace-element rich crustal component or enrichment of their source regions in incompatible elements. These processes can be difficult to distinguish and can be aided by isotopic studies. Recent investigations comparing trace-element patterns of the volcanic glass groups, enabled by secondary-ion mass spectrometry (SIMS), provide the best look at trace-element variations that relate to source regions (Shearer et al. 1996; Shearer and Floss 2000).

The divalent and trivalent trace transition metals (ferromagnesian trace elements) are generally compatible in mare basalts. Chromium tends to be enriched in basalts that have the highest Mg/(Mg+Fe) and that contain magnesian olivine and spinel, and its concentration decreases in basalts with progressively higher Fe/Mg ratios. Concentrations of Cr are higher in the volcanic glasses (2000-8000 ppm) than in the crystalline basalts (1000-6500 ppm) (Tables A3.11 and A3.12). Nickel and Co tend to correlate inversely with the incompatible elements and are in highest concentration in the most "primitive" olivine-rich basalts, reaching concentrations of ~90 and 70 ppm, respectively (Karner et al. 2003). The volcanic glasses have higher concentrations on average, reaching ~100 and 200 ppm for Ni and Co, respectively. Vanadium follows Ti and is enriched in the high-Ti basalts (up to 270 ppm). The range of concentrations of V in the crystalline basalts and the volcanic glasses is similar. Scandium also correlates with Ti and is enriched most in pyroxene-rich basalts. It typically occurs in higher concentration in crystalline basalts (30-90 ppm) than in volcanic glasses (30-60 ppm). Although oxygen fugacities are low and some divalent Cr may be present in the upper mantle source regions, Fe and Ni are mostly divalent. Reduction of iron and crystallization of Fe metal in basalts occurs late and such metal is low in Ni and Co content.

The Lunar Prospector mission results opened up a new frontier in lunar trace-element research with important implications for basalt generation. Two instruments provided data relevant to trace-element concentrations of surface materials, the gamma-ray spectrometer and the neutron spectrometer. The gamma-ray spectrometer provided data on the distribution of Th, U, and K (see Chapter 2), whereas the neutron spectrometer provided a means to infer the concentrations of Sm and Gd, two REEs with high neutron-absorption cross sections. In areas where vast expanses of basalt occur, the concentration of these elements in the regolith can be determined. Then, using simple mixing models with components derived from local geologic relationships, the concentration of these trace elements in the underlying basalts can be inferred. Using this approach, some basalts of the western Procellarum region, which were not sampled directly by any of the sample return missions, have been inferred to be relatively enriched in the naturally radioactive, heat producing trace elements Th, U, and K. For example, whereas most of the Fe-rich mare basalts have Th concentrations less than 2 ppm (Korotev 1998), the gamma-ray data suggest that some of the basalts in western Procellarum may have Th concentrations as high as 6 ppm (Haskin et al. 2000; Jolliff et al. 2001). As these basalts cover vast areas, this trace-element signature is not likely to be a local phenomenon affected by assimilation or by crystallization processes, and their high iron content argues against vertical mixing with an underlying KREEP-rich substrate. If the mixing models are correct, this signature would represent a characteristic of the mantle source regions for these basalts, the significance of which is discussed below.

3.1.3. Mare basalt source regions. Basalts serve as probes of the composition, structure, dynamics, and thermal history of planetary interiors, and for the Moon and Mars, they represent the only samples derived from their respective mantles. The composition and depth of the source regions for mare basalts have been tied to models of lunar differentiation (i.e., the lunar magma ocean) and post-LMO mantle dynamics (cumulate overturn and assimilation). The mineralogy of the source region and the composition of the source, whether differentiated or undifferentiated, have tremendous consequences for the bulk composition and thermal history of the planet. In this section, we focus on observations that potentially constrain the depth and composition of the mare basalt source regions. Thermal evolution is the subject of Chapter 4.

Depth and mineralogy of the source. Two contrasting models for the depths of origin of the mare basalts are as follows. One is that mare magmatism represents melting over a wide range of depths from just below the lunar crust to deep in the upper mantle. The second model is that most of the mare basalts represent partial melts initiated deep within the lunar mantle and below a fairly rigid and cool lithosphere (>350 km). Both models have profound implications concerning the structure and thermal evolution of the Moon between 3.9 and 3.0 Ga when most of the mare basalts now visible at the surface erupted. Several lines of observation can be used to address this issue: high-pressure experiments on mare basalts, trace element characteristics of mare basalts, and geophysical observations.

Studies of high-pressure phase relations of mare basalts provide information about conditions under which melting occurred, including temperature, pressure (and thus depth), and the minerals involved in melting. A common method is to determine the pressure at which a given basaltic magma is multiply saturated; that is, possessing two minerals such as olivine and orthopyroxene on the liquidus. High-pressure experiments (e.g., Kesson and Lindsley 1976; Longhi 1992) indicate that (1) multiple saturation (predominantly olivine + orthopyroxene but also plagioclase ± spinel ± pigeonite ± augite ± ilmenite ± armalcolite) for the mare basalts and volcanic glasses occurs over an extremely wide range of temperature and pressure conditions (<1200°C to 1460°C and <5 to 25 kbar), (2) the temperature and pressure of multiple saturation of the volcanic glasses (1410–1500°C and 17–25 kbar, corresponding to depths of 350–520 km) generally exceed those of the crystalline mare basalts (1200–1380°C and <5 to 12 kbar, corresponding to depths less than 250 km) (Longhi 1992, 1995), and (3) the pressure of multiple saturation for both the high-Ti and very low-Ti basalts overlap, suggesting that the mantle is not uniformly stratified in terms of composition, but maintains compositionally diverse regions at a given depth.

It is important to note that the depths of multiple saturation correspond to the depth of melting only if (1) two or more phases were indeed present in the source residue after melting, and (2) melting occurred at a single pressure. In the case of extensive melting, leaving just olivine in the residue, the depth inferred from the pressure of multiple saturation would be in error. Arguments based on observed volumes of basalt and the preservation of diverse compositional signatures, however, suggest that small degrees of partial melting (<20%) were probably the norm in the lunar mantle (Shearer and Papike 1993). Trace element and isotopic modeling also support modest degrees of melting (Hughes et al. 1989; Nyquist and

Shih 1992; Beard et al. 1998; Shearer and Papike 1999). Furthermore, Hess (1991; 1993) and Hess and Finnila (1997) showed that most of the basaltic magma compositions represented by the pyroclastic glasses surround and follow the olivine-orthopyroxene cotectic boundary, indicating multiple saturation with these two minerals. For crystalline mare basalts that have undergone changes in composition by assimilation or crystallization, multiple saturation experiments represent a minimum depth of melting, and the potentially wide range of depths that are inferred could be misleading. An important consideration for depth-of-melting experiments is that melting may have occurred over a range of depths as a partially molten diapir rose through the mantle.

That the mare basalts and the picritic glasses have different pressures of multiple saturation may be interpreted in two ways: (1) the parental magmas for the picritic glasses and mare basalts may be unrelated and in fact produced by melting at two different pressure regimes in the lunar mantle (parental magmas for the crystalline basalts at depths less than 350 km and the volcanic glasses at depths greater than 350 km), or (2) the multiple saturation depths for the mare basalts are an underestimation of their true depths of melting, and that the parental magmas for both the mare basalts and picritic glasses were generated at depths greater than 350-400 km. At issue are the depths at which melting occurred, as well as the compositional variations within the mantle both laterally and with depth. The volcanic glass experimental data may be interpreted as indicating that sources for high-Ti basalts and very low-Ti basalts both occurred at depths greater than 250 km. Such an arrangement would require disruption or overturn of the simple layered cumulate sequence that might result from static crystallization of a magma ocean (e.g., Hughes et al. 1988; Ryder 1991; Spera 1992; Shearer and Papike; Hess and Parmentier 1995). Given the limited spatial sampling of mare basalts, compositional variations within the lunar mantle remain poorly known.

Composition of mantle source regions. Mare basalts exhibit numerous distinct mineralogical, chemical and isotopic characteristics that reveal many characteristics of their mantle source. Within the context of terrestrial basalt classification, the normative mineral assemblage of mare basalts ranges from quartz to olivine. Most importantly, all of these mare basaltic magmas are non-tholeiitic (Longhi 1981; Hess 1998). The mineralogical (Fe metal) and chemical (reduced valence states for Fe, Ti, Cr) signatures of the mare basalts reflect extremely low oxygen fugacities (~IW buffer). They are depleted in alkali, volatile, and siderophile elements, and there is a complete absence of water and hydrous phases. The crystalline mare basalts and pyroclastic glasses have a ubiquitous negative Eu anomaly. In the crystalline mare basalts, the depth of the negative Eu anomaly increases with increasing TiO_2. Also, from very low TiO_2 to the high TiO_2 mare basalts there is an increase in the abundances of incompatible-elements (REEs, Zr, Ba, Nb), and a decrease in the abundances of compatible elements (e.g., Ni). Sm-Nd and Lu-Hf isotopic compositions indicate that the mare basalt sources were fractionated with respect to chondritic compositions early in the evolution of the lunar mantle and have since remained isolated (Nyquist and Shih 1992; Beard et al. 1998). High-pressure experiments on near-primary basalts indicate that phases such as ilmenite, clinopyroxene, plagioclase, and spinel are not found on the high-pressure liquidus and therefore must have been consumed during melting. Olivine and orthopyroxene occur as high-pressure liquidus phases.

These observations constrain the mineralogical and chemical characteristics of the lunar interior. First, the isotopic data, incompatible-element enrichments and fractionation, the negative Eu anomaly, and the absence of plagioclase as a high-pressure liquidus phase for the near-primary basalts indicate that the source region had experienced early differentiation that resulted in a non-chondritic mantle, and that the individual source regions remained isotopically and mineralogically distinct. Isotopic closure of the systems could correspond to either crystallization of the magma ocean or subsolidus overturn of the LMO cumulate pile. Whether the non-chondritic nature of the lunar mantle is solely a product of lunar differentiation or is instead a product of early differentiation of a non-chondritic bulk Moon is debatable. Second, the source region of the mare basalts is highly reduced, anhydrous and depleted in volatile and siderophile elements. Third, mineral assemblages making up the lunar mantle are dominated by olivine and orthopyroxene. This indicates that the source regions for the mare basalts are predominately harzburgitic rather than lherzolitic. Other phases such as ilmenite, clinopyroxene, plagioclase, spinel and perhaps garnet are heterogeneously distributed in the lunar mantle and were in most cases totally consumed during melting.

Trace element constraints on source regions. Understanding the trace-element contents of mare source regions is key for several reasons. One is that the naturally radioactive elements K, U, and Th provide long-lived sources that contribute to the buildup of heat needed to remelt mantle materials long after solidification of the magma ocean. Another relates to the trace elements whose abundance patterns can help to distinguish primitive unfractionated sources from sources that were fractionated by magma-ocean solidification processes. Related to these are the volatile elements that appear to have been associated with the eruption of volcanic glasses. Differences in the style of eruption of the volcanic glasses and mare basalts may be related to their volatile content, depth of origin, and whether or not the sources were undifferentiated and thus volatile-rich relative to the cumulate upper mantle. Volcanic volatiles are discussed in Section 3.1.4.

The low incompatible-element concentrations of most sampled basalts are consistent with being derived from generally depleted sources. Some portion of the trace elements in their source was likely incorporated as a trapped melt component in cumulates of the crystallizing magma ocean. However, small amounts of a component with a KREEP-like trace-element signature are indicated for some of the basalts and picritic magmas, and the main question is whether this signature is inherent to the mantle or is rather due to contamination during magma ascent.

The potential contamination of basaltic melts by trace-element-rich components is significant in relation to models of dynamic overturn of mantle cumulates. Early in lunar history, near the last stages of magma-ocean solidification, density instabilities may have mixed dense Fe-rich, late-stage cumulates formed near the crust-mantle boundary with deeper mantle cumulates. This mixing could have occurred either by sinking of the dense high-level cumulates or by diapiric rise of low-density magnesian mantle cumulates (e.g., Hess and Parmentier 1995). In either case, the end result could have been the incorporation of incompatible-trace-element enriched residua in various unrelated mantle source regions (see Chapter 4). In places where such processes occurred, the timing of mixing of different materials and the crystallization of late-stage minerals that can fractionate key trace elements may have produced some of the fundamental variations in composition observed between basalt groups. The most important example is the timing of crystallization of ilmenite and perhaps armalcolite. Titanium is an incompatible element in olivine, orthopyroxene, and plagioclase, yet it was effectively decoupled in lunar magma systems from the other incompatible elements and is not strongly enriched, for example in KREEP basalt and other materials with KREEP-like compositions. Likewise, the separation of a low-density Si-K-enriched melt following the crystallization of Ca-REE-phosphate in very late-stage LMO residua may have decoupled the REEs from K. Such a separation might lead to different mixing reservoirs with which rising basaltic liquids could interact.

Low-Ti mare basalt compositions exhibit generally chondritic ratios of high field strength elements (HFSE; Zr, Nb, Hf, Ta, Th) and Y, even though melting and crystallization of the magma ocean have processed their source regions. The ratios of these elements should not change significantly under low-pressure fractional crystallization processes (i.e., <18 kb), because unlike the terrestrial mantle, the lunar mantle has not experienced repeated processing after magma-ocean crystallization, which would tend to enhance the effects of slight differences in HFSE mineral partition coefficients. However, Ti-rich mare basalts and KREEP basalt generally exhibit non-chondritic ratios of these elements, which is probably a result of Fe-Ti oxide addition (Ti-rich basalts) or removal (KREEP); the complement to the KREEP pattern is seen in the compositions of the high-Ti basalts (Fig. 3.14). Mare basalt compositions are consistent with an overturn model of the cumulate pile that would have incorporated excess ilmenite into the low-Ti basalt source (or possibly a KREEP component, as in the case of the Apollo 11 high-K basalts), thus disturbing a source region that had originally developed with chondritic HFSE (and Y) ratios (Neal 2001).

High-pressure phases in mantle source regions. Recent work suggests that some volcanic glasses may have source regions at depths below the differentiated cumulate mantle, and trace-element measurements suggest the possibility that garnet may occur in their residue (Neal 2001). In particular, if garnet were retained in the residue during partial melting, its influence on

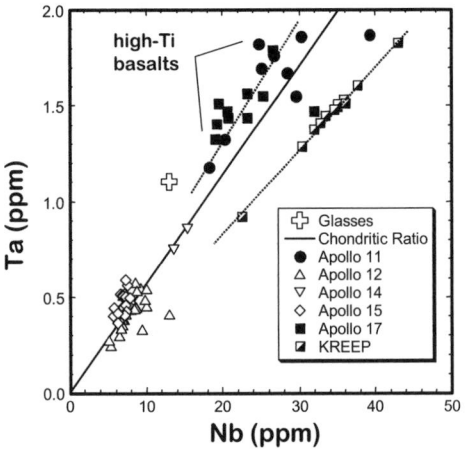

Figure 3.14. Plot of Ta vs. Nb for the mare basalts. Only one datum is available for the volcanic glasses (Shearer et al. 1990). Low-Ti basalts cluster around the chondritic ratio for Nb/Ta, whereas the high-Ti basalts have Ta/Nb ratios generally less than chondritic. Conversely, Ta/Nb ratios for KREEP are superchondritic. Modified after Neal (2001).

magmatic compositions would be most noticeable in the heavy REE and Y concentrations because garnet has high mineral/melt partition coefficients for these elements. Retention of garnet in the residue would thus result in elevated LREE/HREE or incompatible-element/Y ratios in the corresponding partial melts. However, assimilation of late-stage cumulates and KREEP-rich materials to produce the high-Ti contents of some glasses, as well as the ubiquitous negative Eu anomaly (cf., Longhi 1992; Wagner and Grove 1997; Elkins et al. 2000; Neal 2001), can also lead to elevated values for these ratios. Therefore careful examination of the data is required to distinguish between the incorporation of fractionated material and the presence of residual garnet in the mantle source region.

The mare basalts generally exhibit chondritic Zr/Y ratios, whereas the volcanic glasses and KREEP-rich samples are suprachondritic (see Fig. 3.15). The Apollo 11 high-K and Apollo 14 basalts tend to exhibit suprachondritic Zr/Y, but these are probably due to the incorporation of KREEP (cf., Neal et al. 1988, 1989). While the Zr/Y database is small, the mare basalt and picritic glass groups are clearly distinguishable. In particular, the mare basalts and some glasses have Zr/Y ratios <4, KREEP forms a tight group with Zr/Y ~4.5, and some volcanic glasses have ratios in excess of the KREEP values (Fig. 3.15a). This is further emphasized when the slope of the REE profiles is considered as measured by the Sm/Yb ratio (Fig. 3.15b). Some volcanic glasses exhibit the steepest slope (highest Sm/Yb ratio of ~2), whereas the mare basalts and remaining glasses display Sm/Yb ratios between 1 and 1.25, with KREEP being intermediate. It is the glasses with Sm/Yb ratios greater than KREEP that are considered evidence that garnet was a residual component after the cessation of partial melting (Neal et al. 1988, 1989; Neal 2001). This result further requires these glasses to be derived from depths where Al-bearing garnet would be stable.

The geochemical evidence for garnet in the lunar mantle is consistent with both the predicted range of thermodynamically stable phase assemblages with depth and the lunar seismic data (e.g., Anderson 1975; Hood 1986; Hood and Jones 1987; Mueller et al. 1988). As an ex-

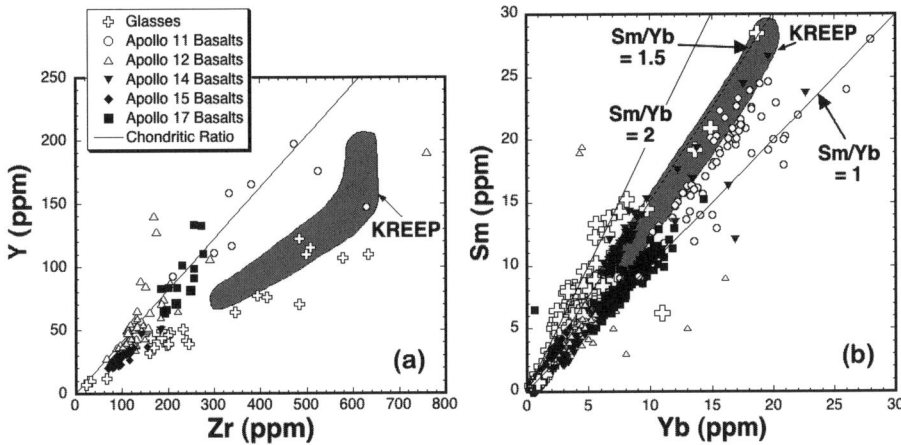

Figure 3.15. Two diagrams showing evidence for garnet being retained in the source regions of some volcanic glasses through increased Y/Zr (a) and Sm/Yb (b) ratios relative to KREEP. Modified after Neal (2001).

ample, Kuskov and Fabrichnaya (1994), Kuskov (1995; 1997), and Kuskov and Kronrod (1998) calculated the stable phases of a variety of compositional models by minimizing the Gibbs free energy. In comparing their predicted seismic velocities with the Apollo-era seismic velocity profiles, it was found that garnet should be a significant phase (8-13 mol%) in the lower mantle between depths of ~500 and 1260 km. For comparison, Neal (2001) modeled the high Zr/Y and Sm/Yb glasses using a source containing 5 vol% garnet. Although the coupled thermodynamic-seismic models imply that garnet is present in the lower mantle, garnet by itself is probably not responsible for the apparent 500 km seismic discontinuity, as is discussed in Section 3.2.3. We further note that garnet has been predicted to be thermodynamically stable in the mantle above the 500 km seismic discontinuity for certain compositional models and temperature profiles (Kuskov 1995, 1997; Khan et al. 2006). Based on the modeling of Kuskov (1995), up to 4 mol% garnet could be stable at a depth of 400 km, whereas Khan et al. (2006) find that ~15 wt% garnet may be present between depths of ~200 and 600 km. Thus, a garnet signature by itself should not be used as evidence for an origin beneath this discontinuity.

Despite the evidence articulated above for the presence of garnet in the lower mantle, whether this region escaped processing by the lunar magma ocean remains unknown. If the deep garnet-bearing source was "primitive" and "unprocessed" then it might have elevated siderophile- and chalcophile-element concentrations (e.g., Ir, Au, Cu, and Zn) for two reasons. First, these elements would not have been as efficiently scavenged from the primitive Moon during a core formation event as they would have been from a magma ocean. Second, the siderophile and chalcophile elements scavenged from the magma ocean by an immiscible sulfide melt (Neal and Ely 2002) would have to permeate through the solid primitive interior of the Moon, possibly enriching this region with these elements in the process. Simple elemental relationships show that at least some of the volcanic glasses possess higher siderophile and chalcophile abundances than the mare basalts, supporting the hypothesis that they were derived from a deep undifferentiated source (see Fig. 3.16). Although the database is not extensive, the glasses for which data are available (Anders et al. 1971; Morgan et al. 1972a; Morgan et al. 1972b; Ganapathy et al. 1973; Morgan et al. 1974) also have generally higher Ir and Au abundances relative to the crystalline basalts (see Neal 2001). Perhaps most significantly, the Cu and Zn data demonstrate that there are two groups of glasses; one with abundances greater than the basalts, and the other with abundances comparable to the mare basalts (Fig. 3.16). The former group contains glasses with chondrite normalized Sm/Yb >1.5, consistent with being derived from a garnet-bearing, primitive source. The latter group contains glasses with lower chondrite normalized Sm/Yb (<1.5) and higher Sc/Sm (>30) ratios, suggestive of being derived from a garnet-free source region in the lunar magma ocean cumulate pile, similar to the mare basalts (Neal 2001).

3.1.4. Volatile element content of the lunar mantle.
Volatiles play a fundamental role in the evolution of a planet and exert an important influence on mantle processes (e.g., melting, dynamics, and heat transfer) and characteristics (e.g., mineralogy and melt composition). Unlike the Earth where estimates of juvenile volatiles can be made using a variety of approaches, the volatile budget of the lunar mantle can, at the present time, only be reconstructed from the record preserved in mare basalts and volcanic glasses. Several lines of observation can be used to reconstruct the volatile budget of the lunar mantle from mare basalt samples: mineral assemblage and composition, bulk volatile content, volatile coatings on

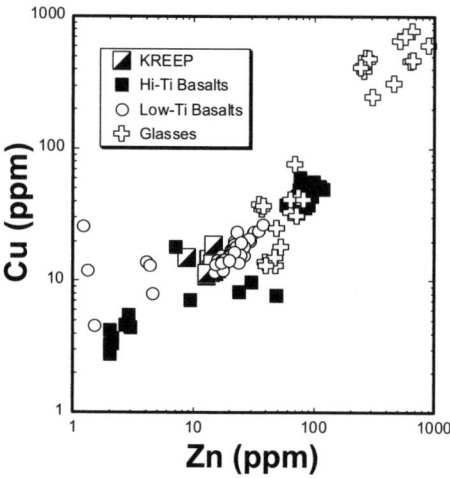

Figure 3.16. Plot of Cu vs. Zn for the volcanic glasses and mare basalts. Those glasses with elevated Cu and Zn abundances are inferred to originate in a deep primitive lunar mantle that escaped magma-ocean differentiation. Samples with elevated Cu and Zn abundances generally correlate with elevated Zr/Y and Sm/Yb ratios. [Used by permission of the American Geophysical Union, from Neal (2001), *Journal of Geophysical Research*, Vol. 106, Fig. 10, p. 27,881.]

mineral or glass grains, volatile element profiles in volcanic glasses, volatile element contents of volcanic glass beads within a well defined pyroclastic deposit stratigraphy, and distribution of volatile-driven pyroclastic eruptions. As in the terrestrial case, reconstructing the volatile content of the lunar mantle from basaltic melts is compromised by volatile degassing at the time the lavas erupted and subsequent contamination from external sources. Unlike terrestrial basalts where atmospheric and non-magmatic water add to the complexity, lunar-volatile contamination may be attributed to low-pressure condensation following impact, solar wind implantation, and assimilation or sublimation of cometary or meteoritic material.

To a first approximation, the presence of metallic iron or Ni-Fe alloys in mare basalts indicates that they crystallized at oxygen fugacities below the iron-wüstite buffer. Identification of an assemblage of Fe-rich olivine, cristobalite, and metal in the mesostasis of Apollo 14 basalts (El Goresy et al. 1972) constrain the f_{O_2} at the end of crystallization to be lower than the IW buffer and near the quartz-fayalite-iron buffer (QFI). These mineralogical observations are further corroborated by the total absence of ferric iron in both silicates and oxides, and the presence of reduced valance states of Cr and Ti (Haggerty et al. 1970). Intrinsic f_{O_2} determinations as well as those based on thermodynamic gas equilibria calculations show that the range of temperature and f_{O_2} for mare basalt crystallization is relatively small, varying from 10^{-13} at 1200°C to 10^{-16} at 1000°C (Wellman 1970; Sato and Helz 1971; Sato et al. 1973). Relative to standard buffer curves, these values are approximately 0.2 to 1.0 log units below iron-wüstite (IW) and above the univariant curve that defines the stability of ilmenite (see Fig. 3.17). An exception to this is the pyroclastic high-Ti basaltic glass, e.g., 74220. Sato (1979) showed that at temperatures between 1000 and 1100°C, the measured f_{O_2} for the high-Ti glass was approximately 0.15 to 0.3 log units above the IW buffer. However, extrapolating this relationship to the liquidus temperature of the high-Ti glass (>1325°C) shifts the estimate of f_{O_2} to approximately 0.3 log units below the IW buffer. Some lunar basalts show textural and mineralogical evidence for further subsolidus reduction (summarized by El Goresy et al. 1976) taking place at a range of f_{O_2} below both IW and FIQ buffers.

Although the f_{O_2} conditions under which the mare basalts crystallized is restricted to below IW, it is unclear whether these reducing conditions reflect reducing crustal and mantle environments or reduction processes during magma transport and eruption. The oxidation state of the deep lunar mantle depends on the specific reactions that control the redox state. If the dominant redox controlling reaction in the lunar mantle were the uv-I-il buffer or a similar reaction, then the oxygen fugacity at constant temperature would rise with depth in the Moon at a rate of about 1 log-unit per 10–15 kbar (200–300 km). On the other hand, if the oxidation state in the Moon was controlled by the graphite-$(CO+CO_2)$ system, the oxygen fugacity of the lunar mantle would increase at a greater rate with increasing depth, such that it would be near the fayalite–magnetite–quartz (FMQ) buffer at 400–500 km (Delano 1990).

Ringwood et al (1981) suggested that the reduced nature of the mare basalts was a surface phenomenon

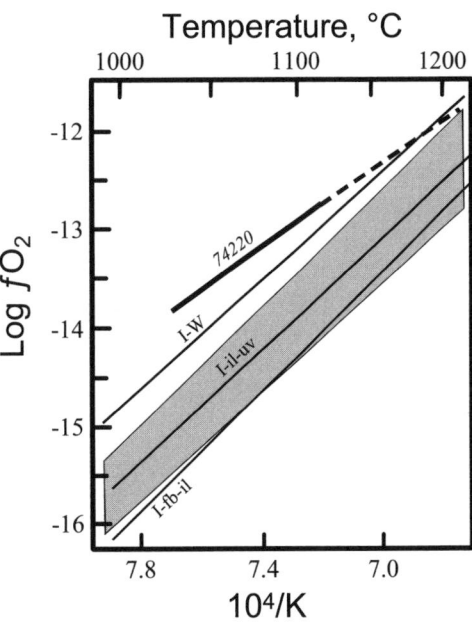

Figure 3.17. Oxygen fugacity as a function of temperature for several buffers relevant to the lunar interior: I-W, iron-wüstite; I-il-uv, iron-ilmenite-ulvöspinel; I-fb-il, iron-ferrobustamite-ilmenite. Although the intrinsic f_{O_2} curve for the Apollo 17 orange glass (74220) is at more oxidizing conditions than IW below 1175°C, the curve intersects its liquidus temperature at an f_{O_2} more reducing than IW. Patterned region represents the range of f_{O_2} corresponding to the crystallization of mare basalts.

and that the lunar mantle was more oxidized (between the IW and Co-CoO buffers). They sited the late crystallization of metal in mare basalts and the depleted behavior of siderophile elements (W, Re) as lines of evidence. However, several direct and indirect lines of evidence do support a reduced mantle with f_{O_2} at or below the IW buffer. First, small metal inclusions of Fe-Ni metal in olivine phenocrysts in the Apollo 12 basalts indicate that the basalts were at or near metal saturation prior to extensive crystallization at the lunar surface (Brett et al. 1971). Second, Delano (1990) showed that the Cr content of liquids coexisting with spinel varies considerably with f_{O_2}, such that Cr is low (~1000 ppm) at the FMQ buffer and high (>2700 ppm) at the IW buffer. This is due to the lower solubility of Cr^{2+} in spinel at reducing conditions and therefore remaining in the melt. Delano pointed out that the Cr content of primitive lunar basalts (2700–6300 ppm) is significantly higher than similar terrestrial basalts (<1500 ppm). Therefore, if spinel was a residual phase in the lunar mantle and the Cr content was approximately chondritic, the high Cr content of the mare basalts suggests a reducing lunar mantle. Third, carbon solubility modeling by Fogel and Rutherford (1995) showed that the solubility of C is low at conditions 0.5 log units below the IW buffer and increases substantially above the IW buffer (>300 ppm). Therefore, the low concentration of dissolved C in mare basalts (~30 ppm) and volcanic glasses (<100 ppm) implies reducing conditions in the lunar mantle. Fourth, Weitz et al. (1997), in their examination of inclusions (olivine, metal, spinel) in Apollo 17 orange glass beads, concluded that metal-melt equilibria and the Cr content of the olivine and spinel indicated a pre-eruptive f_{O_2} of 1.3 log units below IW. They also concluded that an episode of magma oxidation occurred during later stages of eruption. Finally, Fe metal is also present in intrusive igneous lunar rocks that bear no indication of reduction due to subsolidus reequilibration or volatile loss. By analogy to terrestrial basalts, the oxidation state of extrusive basalts and their intrusive equivalents ought to be similar (Basaltic Volcanism Study Project 1981).

Unlike terrestrial volcanic gases, those associated with lunar basaltic magmatism are devoid of H_2O. Judging from the samples, the lunar magmas also contained lower concentrations of gases. This finding has been interpreted as indicating that the lunar mantle is extremely dry and volatile poor. Hydroxyl contents for the Apollo 15 green and yellow volcanic glasses are below the detection limits of Fourier transform infrared spectroscopy (FTIR; ~10 to 50 ppm), and the absence of hydrous silicates such as amphibole and phosphates reflect their intrinsic low H_2O content. The bulk carbon content of analyzed mare basalt samples lies between 15 and 67 ppm (Gibson 1977), and this agrees with the more recent study by Fogel and Rutherford (1995) which demonstrated that the dissolved C species in lunar volcanic glasses is less than FTIR detectability (<50–100 ppm). The ratio of $CO:CO_2$ in Apollo 15 basalts ranges from 1:2 to 1:4 and appears to be dependent upon concentration (Gibson et al. 1975). Wellman (1970) calculated that the $CO:CO_2$ ratio should be approximately 1:2 at crystallization conditions experienced by lunar basalts in contrast to terrestrial volcanic glasses that have a CO_2 content that ranges from 0.05 to 0.43 wt% (Basaltic Volcanism Study Project 1981). The low C content of mare basalts has been attributed to the low C content of their source, the low solubility of C at reducing conditions (below the IW buffer), and volatile loss during eruption owing to diffusion and evaporation (Fogel and Rutherford 1995). Sato (1978; 1979), Wilson and Head (1981), Spera (1992) and Fogel and Rutherford (1995) suggested that the oxidation of graphite at relatively shallow depths (<4 km) was the mechanism for producing pyroclastic deposits on the Moon. In order for this process to be effective, though, graphite would need to be transported from the lunar interior by the basaltic magmas, and thus far, graphite has not been identified in mare basalts associated with pyroclastic deposits.

If CO was a major volcanic gas in lunar magmas, then carbonyl species such as COS, COCl, $COCl_2$, $FeCO_5$ and others would have played a role in elemental transport (Colson 1992). The presence of carbonyls would favor the deposition of sulfides and possibly Fe metal with decompression, while liberating CO. Reactions involving carbonyls also may have played a role in the formation of vapor-deposited sulfides on volcanic glasses.

The sulfur concentrations of crystalline mare basalts range from an average of 400 ppm in some of the Apollo 15 low-Ti basalts to 2600 ppm in the high-Ti basalts from Apollo 17 (Delano 1986b, and references therein). The relationship between Ti and S content in the mare basalts, as shown in Figure 3.18, is partially attributed to increases in S saturation with increasing Fe and Ti content, with calculated S saturation in mare basalts ranging from 1500 to 3400 ppm. For comparison, pre-eruptive concentrations of S in terrestrial basalts range from 800 to 2500 ppm and generally exhibit a strong correlation with Fe (Basaltic Volcanism Study Project 1981). The S content of the bulk volcanic glasses ranges from 550 to 750 ppm for the Apollo 17 high-Ti glasses and averages 400 ppm for the very low Ti glasses. The individual glasses have been analyzed by both microprobe and ion probe methods (Delano et al. 1994; Shearer et al. 1997; Klein and Rutherford 1998) and the S concentrations of the individual glass beads are lower than both the crystalline basalts and the bulk glass beads (Fig. 3.18). The higher S content of the bulk beads relative to the individual beads reflects volatile-rich coatings on the glasses. Unlike the crystalline mare basalts, the glasses do not show a clear relationship between Ti and S content. The lack of the relationship between Ti and S and the overall lower S concentrations of the individual glass beads indicate substantial loss of S during pyroclastic eruptions. Fogel and Rutherford (1995) calculated that S loss by diffusion during fire fountaining would be significantly less than either CO_2 or Cl loss.

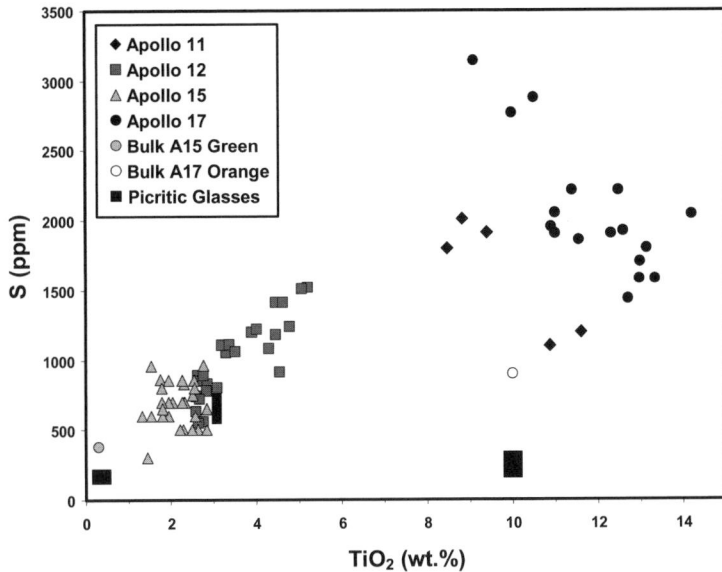

Figure 3.18. Sulfur concentrations in lunar picritic glasses and mare basalts. Data from Shearer et al. (1998), Agrell et al. (1970), Beaty and Albee (1978), Gibson et al. (1975, 1977), Gibson and Moore (1974), Rees and Thode (1972, 1974), and Wänke et al. (1975).

Clearly, abundant evidence suggests that the lunar magmas were reduced prior to eruption at the lunar surface (below IW-1) and that they were derived from a fairly reduced lunar mantle (below IW). All indications are that the lunar mantle was also anhydrous. It is more difficult to extract information concerning the nature of C and S in the lunar mantle. These volatiles were lost to various degrees during eruption of basaltic magmas at the lunar surface (highest loss during pyroclastic eruptions) and their solubility in lunar basalts was controlled by variables such as temperature, pressure, composition, and f_{O_2}. Although the solubility of C is low at the reducing conditions of lunar magmas, most models for lunar fire fountaining evoke the shallow oxidation of mantle-derived graphite that was entrained in the mare basalts.

3.2. Seismology

Besides the analysis of mare basalts, which are ultimately derived from the mantle, the only other detailed source of information about the lunar mantle comes through an analysis of the Apollo seismic data. However, because of the limited depth distribution of the deep moonquakes and the absence of identifiable quakes on the hemisphere opposite that of the seismic network, seismic velocity information was precluded from being inferred below a depth of ~1150 km.

Figure 3.19 depicts the seismic velocity model of Khan et al. (2000), which is based upon the first P- and S-wave arrival times of Nakamura and coworkers. This model was constructed using a Monte Carlo sampling of parameters within 56 variable thickness homogeneous layers, and the results were interpreted in terms of a Bayesian probabilistic framework. Plotted contours correspond to the probability of obtaining a given velocity at any given depth (see the discussion in Section 2.2). This inverse approach most faithfully addresses the inherent resolution of the data, as well as errors on estimated parameters (i.e., velocities, seismic discontinuities, etc.). However, as a relatively large number of layers were used, the range of acceptable velocities at any depth is fairly large, and thus not entirely useful for placing

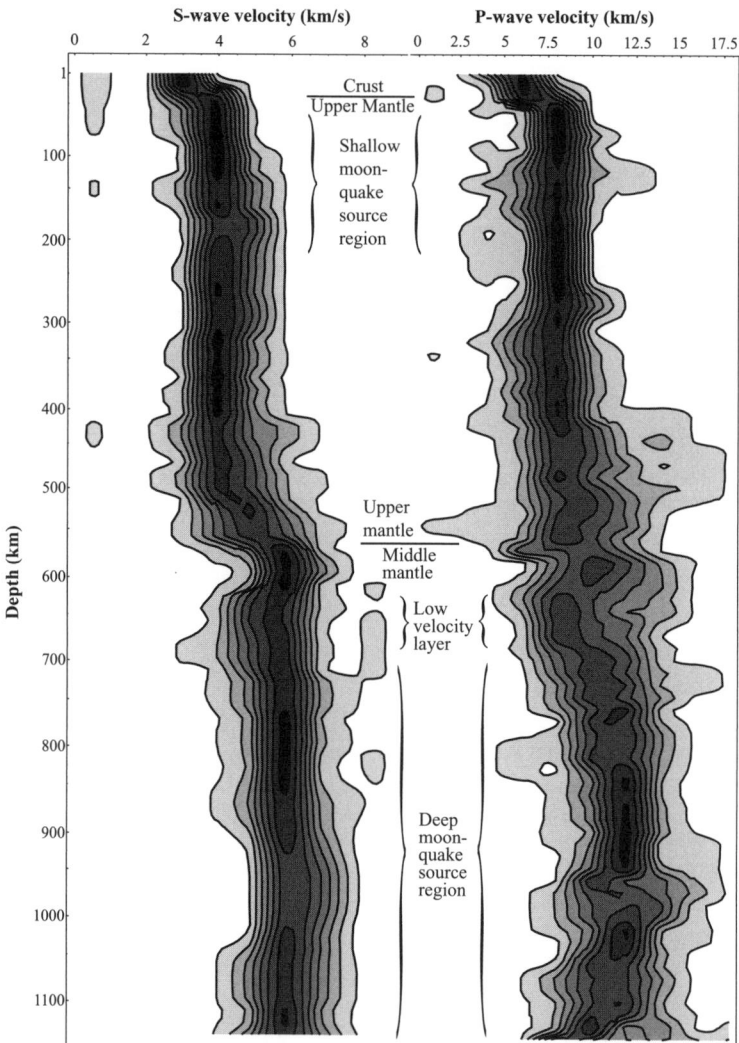

Figure 3.19. P and S-wave velocity profiles of the lunar mantle based on the Bayesian inversion of Khan et al. (2000). The marginal probability of the seismic velocity is plotted at one-kilometer depth intervals, and the contours define nine equally sized probability intervals. [Used by permission of the American Geophysical Union, from Khan et al. (2000), *Geophysical Research Letters*, Vol. 27, Fig. 1a, p. 1592.]

quantitative constraints on the geochemistry of the lunar interior. For this purpose, the models of Nakamura et al. (1982) Goins et al. (1981a) and Lognonné et al. (2003) would be more useful (see also, Khan et al. 2006).

   Whereas earlier studies (e.g., Goins et al. 1981a; Nakamura et al. 1982) could not resolve the depth dependence of any velocity variations in the mantle (those present were introduced for computational convenience), quantifiable velocity variations are found in the Khan et al. (2000) model providing evidence for compositional stratification within the mantle. Specifically, their results suggest an almost constant velocity upper mantle from the base of the crust down to a depth of roughly 500 km. A slight decrease of S-wave velocity with depth is indicated over this depth range (Khan and Mosegaard 2002, Fig. 14), consistent with the findings of previous studies (Nakamura et al. 1976; Goins et al. 1981a). Below a depth of 500 km both the P- and S-wave velocity structure indicate a gradual velocity increase toward the transition between the upper and middle mantle. At the base of this transition zone, at a depth of 560 ± 15 km, a substantial increase in velocity of the order of 1 km/s is found to occur (Khan and Mosegaard 2002, Figs. 9 and 13). A discontinuity at this depth was previously suggested by the Nakamura et al. (1982) velocity model. As a result of the large uncertainties of arrival times associated with the deep moonquakes, velocity variations within the middle mantle could not be unambiguously identified.

Relocalization of the shallow moonquake hypocenters showed these to be distributed in the depth range from 50 to 220 km (Khan et al. 2000), confirming earlier suggestions that the upper mantle is a source of tectonic activity (Nakamura et al. 1979). The high frequency content of seismic signals from the shallow moonquakes indicates large stress concentrations at their source. Stress drops associated with these quakes might exceed 100 MPa (Oberst 1987), consistent with a model in which thermoelastic strain within a thick lithosphere is released by an occasional shallow quake as the Moon cools. However, as only 28 shallow events were observed, definitive conclusions as to their origin cannot be drawn. Relocalization of the deep moonquake hypocenters indicates a well-defined source region with all quakes situated in the depth range of ~700 to 1150 km (see Fig. 3.20).

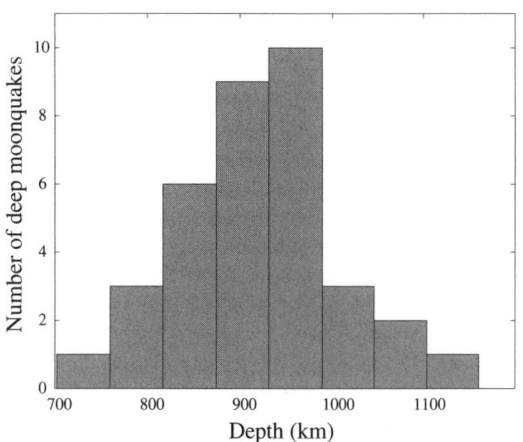

Figure 3.20. Depth distribution of the deep moonquakes. Data from Khan and Mosegaard (2002)

We note that some unresolved differences exist among the velocity models of Goins et al. (1981a), Nakamura et al. (1982), Khan et al. (2000), and Lognonné et al. (2003). Perhaps the most important is that the study of Goins et al. (1981a) found a decrease in velocity below a depth of ~480 km, in contrast to a velocity increase as found by Nakamura et al. (1982) and Khan et al. (2000). As these latter two studies relied upon the same data set, it should not be too surprising that they are in agreement. However, a complete independent analysis and inversion of the Apollo data by Lognonné et al. (2003) has found a significant P-wave velocity decrease to occur below a depth of 488 km, in agreement with the Goins et al. (1981a) model (see Fig. 3.21). Possible causes of this discrepancy include differences among these three datasets, as well as the different inversion techniques employed. We note that the inversions based upon the Nakamura et al. dataset have relied upon 41 deep moonquake sources whereas both Goins et al. (1981a) and Lognonné et al. (2003) have used 24. Lognonné et al. (2003) have also demonstrated that their lower mantle velocities are not robust to small changes in the number of utilized moonquakes. Indeed, when five of the deepest moonquakes were excluded in their inversion, a velocity increase below 500 km was obtained. We remark that in the coupled thermodynamic-seismic inversion of Khan et al. (2006), which is based on the Lognonné et al. (2003) first arrival data set, a seismic velocity increase below a depth of ~500 km has been found, consistent with the studies of Nakamura et al. (1982) and Khan et al. (2000).

3.2.1. Evidence for a partially molten lower mantle. Information on the deepest structure of the Moon is relatively indirect since the seismic studies mentioned so far were limited to a depth of ~1150 km. The lack of knowledge deeper than this arises because of the distribution of the deep moonquake sources, where until recently, the only farside moonquakes that had been located came from a single focus that lies just over the eastern limb of the Moon (A33, Nakamura et al. 1973). Analysis of the signal characteristics from this sole farside focus showed clear onset of shear-wave arrivals at the two stations situated closest to the source, whereas at the furthest two stations none could be discerned. This is particularly telling, as the characteristic feature used in identifying the deep moonquakes is their prominent shear-wave and weak P-wave arrivals. At the two stations where S-wave arrivals were visible, their ray paths were found to have bottomed at depths of roughly 1000 km, while rays traveling to the two farther stations bottomed at about 1300 km depth. This observation led to the suggestion that the absence of shear wave arrivals at the two stations

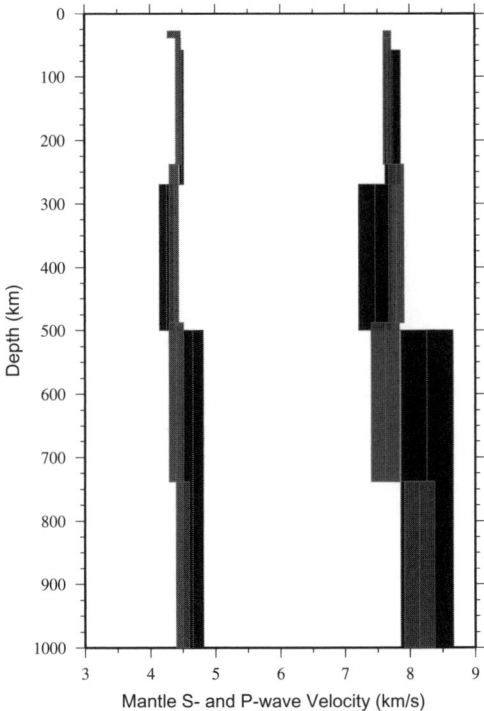

Figure 3.21. P- and S-wave velocity profiles of the lunar mantle based on the studies of Lognonné et al. (2003) (gray) and Nakamura et al. (1982) (black) with 1-σ uncertainties. Note the different velocity scale with respect to Figure 3.19.

farthest from the source was due to the rays having probed an attenuating region somewhere below a depth of 1000 km (or a radius less than ~740 km). As P-wave arrivals were observed, the attenuation must be relatively greater for shear waves. These observations are most easily explained by the presence of a high attenuation region, such as the presence of a partial melt, below a depth of ~1000 km (Nakamura et al. 1973).

A recent analysis by Nakamura (2005) has identified and located seven new moonquake nests that are either clearly on the farside, or, to within error, on the lunar limb. The distribution of these nests shows that there is a lack a moonquakes occurring within a radius of ~40° antipodal to the sub-Earth point. Given the observed asymmetric distribution of nearside moonquakes, as well as the asymmetric thermal evolution of the Moon (e.g., Wieczorek and Phillips 2000; Zhong et al. 2000; Parmentier et al. 2002), it is possible that this region might truly be aseismic. Nevertheless, of these seven nests, two were found to possess signal characteristics similar to the previously known farside nest, with clear S-wave arrivals being absent at the farthest stations in the seismic array. This strengthens the original suggestion of Nakamura (1973) that the deep lunar interior is highly attenuating to S-waves.

Lunar rotation studies indicate that the lunar interior is very effective in dissipating energy (e.g., Yoder 1981), with current estimates of the solid body tidal Q being very low (<90, Williams et al. 2001b). Since the seismic Q of the mantle has been found to be relatively high (>4000 in the upper mantle, e.g., Nakamura and Koyama 1982) and about 1400–1500 in the lower mantle

(Dainty et al. 1976; Nakamura et al. 1976), the deeper interior must possess a significantly lower Q, possibly as a result of the presence of a partial melt. As noted by Nakamura (2005), a region possessing a seismic Q of 100 would attenuate P-wave amplitudes by a factor of 3 over a distance of 500 km, and those of S-waves by a factor of 6. Such a highly attenuating region could easily explain the near absence of detectable farside deep moonquakes, for even if they did exist, direct shear-wave arrivals would not be expected for these events.

3.2.2. Mineralogy and composition of the lunar mantle.

A few studies have used the lunar seismic data to constrain the composition and mineralogy of the lunar interior. Initial investigations were limited to comparing the seismic velocity models of Goins et al. (1981a) and Nakamura et al. (1982) to those expected from a restricted suite of mantle mineral assemblages (Buck and Toksöz 1980; Hood and Jones 1987; Mueller et al. 1988). Later studies by Kuskov and coworkers (Fabrichnaya and Kuskov 1994; Kuskov and Fabrichnaya 1994; Kuskov 1995, 1997; Kuskov and Kronrod 1998, 2001; Kuskov et al. 2002) employed a thermodynamic approach to calculate stable mineral phases for a range of mantle compositions and temperature profiles, and then compared the obtained seismic velocities to the model of Nakamura et al. (1982). Finally, a recent study by Khan et al. (2006) used a similar thermodynamic approach, but instead of just comparing their results to a seismic velocity model, they inverted directly the first arrival time data set of Lognonné et al. (2003) for mantle composition and temperature using a Monte Carlo Bayesian inverse method.

Kuskov and Kronrod (1998) assumed that the mantle composition was constant in the three layers defined by the Nakamura et al. (1982) seismic velocity model (58–270 km, 270–500 km, and greater than 500 km), and parameterized the mantle temperature profile by an exponential function. For a range of chemical compositions and temperature profiles, thermodynamically stable mineral phases were determined, and the seismic velocity was calculated as a function of depth. In general, they found that the Nakamura et al. (1982) velocity model was best explained by an ~500 km orthopyroxenite upper mantle, and a garnet-bearing olivine and clinopyroxene mantle below this depth. Two best-fit models were presented, with each being constrained by a prescribed mantle density just beneath the crust. Concerning their upper mantle, one model contained about 95 mol% orthopyroxene and 4 mol% clinopyroxene between the crust and 270 km depth, and about 92 mol% orthopyroxene, 4 mol% clinopyroxene, 4 mol% olivine and 1 mol% garnet between 270 and 500 km. For this model, the magnesium number decreased from 84 at the base of the crust to 73.5 at a depth of 270 km. Their second model was somewhat less silica-rich above a depth of 270 km, possessing about 75 mol% orthopyroxene, 5 mol% clinopyroxene, and 19 mol% olivine. Between 270 and 500 km depth the composition was similar to the first model, being composed of approximately 96 mol% orthopyroxene and 4 mol% clinopyroxene. For this second model the magnesium number decreased from 87 at the base of the crust to 75 at a depth of 270 km. Both models gave nearly identical results for the mantle below 500 km. Here, a magnesium number of 86 was obtained and the mineral phases olivine, clinopyroxene, and garnet were found to be present in the molar proportions 56:35:9.

A recent study by Khan et al. (2006) similarly employed a thermodynamic approach and obtained results that are somewhat consistent with those of Kuskov and coworkers. As for the inverse approach, the most important difference between these studies is that Khan et al. (2006) directly modeled the first seismic arrivals, whereas Kuskov and coworkers compared the seismic velocities obtained from a thermodynamic model to the seismic velocity profiles of Nakamura et al. (1982), which is in itself a model. The first arrival data set of Lognonné et al. (2003) was used, and the Moon was assumed to be composed of a variable thickness and constant composition crust, upper mantle, and lower mantle. The bulk crustal composition was found to be consistent with the model of Taylor (1982), and the transition between the upper and lower mantle occurred near a depth of 600 km. The upper mantle was found to be an orthopyroxenite containing ~75 wt% orthopyroxene, a relatively unconstrained amount of clinopyroxene, trace amounts of olivine, and depending upon depth, either plagioclase or garnet. Above a depth of about 200 km, plagioclase was stable with abundances less than about 10 wt%, and below this depth, garnet was present with abundances near 15 wt%. Below ~600 km depth, the mantle was found to be composed of approximately 60 wt% olivine and 40 wt% garnet. As for the magnesium number, this was found to increase from ~66 in the crust, to ~75 in the upper mantle, and finally to ~89 beneath a depth of 600 km.

While the overall mantle mineralogy of the Kuskov and coworkers and Khan et al. studies are similar, they differ in detail. Perhaps most importantly is that Khan et al. find relatively higher abundances of garnet in the upper mantle (~15 wt%) between depths of ~200 and 600 km, whereas Kuskov and coworkers only found approximately 1 mol% over a similar depth range. The lower mantle of Khan et al. is also more garnet rich, being composed of about 40% garnet by weight in contrast to about 9 mol% as obtained by Kuskov and coworkers. As a result of these differences, the Khan et al. mantle contains about twice as much aluminum as that of Kuskov and coworkers. Furthermore, clinopyroxene is predicted to be a rather minor phase by Khan et al. in the lower mantle, whereas this mineral is present with an abundance of ~35 mol% in the model of Kuskov and coworkers. While the magnesium number of the lower mantle is similar for these two models, this number is not easily comparable for the upper mantle as Kuskov and coworkers employed two compositional layers where Khan et al. used one.

One potential concern with the above studies is that neither has fully explored the consequences of possible titanium-rich phases in the mantle. While Kuskov and Kronrod (2001) and Kuskov et al. (2002) included titanium in their most recent thermodynamic modeling, its concentration was fixed to chondritic values. All other studies have neglected this element. As high-titanium basalts are somewhat common on the lunar nearside (with some containing up to 13 wt% TiO_2), high abundances of titanium (most likely present as ilmenite or ulvöspinel) are at least locally required in the nearside mantle beneath the Apollo seismic network. The presence of titanium-rich phases could possibly affect either the thermodynamic mineral stability fields, and/or have an effect on the seismic velocity of the mineral assemblage. It remains to be demonstrated as to whether this effect is significant or negligible.

3.2.3. Origin of the 500 km seismic discontinuity. Several seismic velocity models of the lunar mantle are characterized by a substantial velocity increase near a depth of ~500 km (Nakamura et al. 1982; Nakamura 1983; Khan et al. 2000; Khan and Mosegaard 2002). According to the Nakamura (1982) model, P-wave velocities here increase from 7.46 ± 0.25 km/s to 8.26 ± 0.40 km/s, whereas S-wave velocities increases from 4.25 ± 0.10 km/s to 4.65 ± 0.16 km/s. The recent thermodynamic-seismic inversion of Khan et al. (2006), which is based on the first arrival dataset of Lognonné et al. (2003), finds similar velocity increases across an interface located approximately 600 km below the surface.

As reviewed by Hood and Zuber (2000), previous work indicates that the large velocity increases near 500 km depth most probably require a change in composition in addition to a possible mineralogical phase transition (Hood and Jones 1987; Mueller et al. 1988; Kuskov 1997). When the allowed ranges in mantle temperature and the stability fields of appropriate mineral assemblages are considered, the spinel to garnet phase transition may occur at depths between approximately 300 km (20 kbar) to 550 km (25 kbar) (e.g., Green and Ringwood 1967; Kuskov 1995, 1997), or at depths as shallow as ~200 km (Khan et al. 2006). Although a spinel to garnet phase transition is consistent with the depth of the observed seismic discontinuity, this phase transition alone would result in only a small velocity increase (≤0.1 km/s for S-waves) for plausible mantle compositions (e.g., Hood and Jones 1987). In contrast, the Nakamura (1982) velocity model estimates probable S-wave velocity increases of 0.4 km/s, and such a large velocity increase can not be explained solely in terms of this, or any other known, phase transition. A change in composition is required and possibilities include a more aluminous composition below 500 km depth, a higher Mg number of the lower mantle, or both (e.g., Hood 1986; Hood and Jones 1987; Mueller et al. 1988; Kuskov and Kronrod 1998; Khan et al. 2006).

Four possible explanations for the implied change in composition at 500 km depth may be considered. First, it is possible that the lunar mantle was initially compositionally homogeneous, but that melting and differentiation of a magma ocean occurred only to a depth of ~500 km. This would effectively transfer most of the aluminum from the upper mantle to the crust while the lower portion of the mantle would remain compositionally pristine. However, for plausible maximum bulk aluminum abundances (which determine the allowed amount of garnet in the lower mantle), the implied change in composition at this depth is still insufficient to explain the observed velocity increases (e.g., Hood and Jones 1987; Mueller et al. 1988). Furthermore, as was noted by Wieczorek and Phillips (2000), the gravitationally unstable nature of the post-magma-ocean mantle may have given rise to a global redistribution of mantle materials (Herbert 1980; Spera 1992; Hess and Parmentier 1995), obscuring any such putative compositional interface.

A second interpretation of the 500-km discontinuity is that the lunar mantle may have initially been compositionally zoned with more aluminous and/or magnesium-rich phases being present in the lower portion of the mantle. If the formation and differentiation of a lunar magma ocean to a depth of 500 km were to occur for this initial structure, then such a model could in principle produce velocity increases that are compatible with the Nakamura (1982) seismic velocity model (Hood and Jones 1987; Mueller et al. 1988). A lower mantle unaffected by magma-ocean differentiation containing higher abundances of aluminous and Mg-rich mineral phases could possibly be a consequence of the giant-impact origin of the Moon (see review of Hood and Zuber 2000). In particular, this model predicts that the Moon formed from a hot circumterrestrial silicate vapor cloud (e.g., Canup and Esposito 1996; Cameron 1997; Canup 2004), and fractional condensation of such a vapor would favor the earliest condensates to have a more aluminous and Mg-rich composition (Mueller et al. 1988). If these early refractory condensates accreted to form a small proto-Moon, then their subsequent remelting would have been inhibited because of their higher mean melting temperatures. The later accretion of less refractory condensates might thus have melted only the outer portion of the Moon.

While the 500-km seismic discontinuity is often cited as possibly marking the base of the lunar magma ocean, it is becoming increasing clear from giant impact simulations that the Moon could have formed in a completely molten state (e.g., Pritchard and Stevenson 2000; Canup 2004). A third interpretation of this feature is that it is directly related to the crystallization products of a deeper, near-global magma ocean. Though the exact crystallization sequence of a magma ocean is dependent upon several factors (such as its assumed bulk composition and the relative importance of fractional and equilibrium crystallization), it is generally agreed upon that the first ~40% of crystallization is dominated by olivine, and that the subsequent ~40% is dominated by orthopyroxene (e.g., Snyder et al. 1992b; Hess 2000; Shearer and Floss 2000). The magnesium number of these cumulates should further progressively decrease with increasing crystallization. As the mantle below a depth of 500 km corresponds to about 35% by volume of the Moon, it is conceivable that the 500-km discontinuity could represent the boundary between early olivine-rich and later orthopyroxene-rich magma-ocean cumulates. This mineralogical stratification is generally consistent with the thermodynamic-seismic inversions of Kuskov and coworkers (e.g., Kuskov 1995, 1997; Kuskov and Kronrod 1998), and more recently Khan et al. (2006) (see Section 3.2.2). The results of Khan et al. (2006) further imply that the magnesium number increases from ~75 in the upper mantle to ~89 in the lower mantle, consistent with the sequential emplacement of cumulates at the base of the magma ocean. Nevertheless, a potential problem with this scenario is that the post-magma-ocean cumulate pile is expected to be gravitationally unstable as a result of its progressively increasing iron content with radius (Hess and Parmentier 1995). If a large scale overturn were to occur, then the magma-ocean cumulates might possibly end up in an inverted sequence, with high magnesium-number, olivine-rich cumulates residing in the upper mantle, and with low magnesium-number, orthopyroxene-rich cumulates in the lower mantle.

A fourth interpretation of the 500-km discontinuity is that it represents the maximum depth of melting of the mare source region. Motivated by the recognition that the crust beneath Oceanus Procellarum and Mare Imbrium is a unique geochemical province with high abundances of KREEP (e.g., Haskin 1998; Jolliff et al. 2000a; Korotev 2000; Wieczorek and Phillips 2000), Wieczorek and Phillips (2000) constructed a thermal-evolution model of the Moon that contained a large quantity of heat-producing elements within the Procellarum KREEP Terrane. They found that melting within the mantle primarily occurred beneath this province, and that the depth of melting increased with time, ultimately achieving a maximum depth somewhere between ~200 and 600 km. Based on this observation, as well as the fact that none of the picritic glasses have depths of multiple saturation (commonly interpreted to be depths of melting) in excess of 540 km (e.g., Longhi 1992; Elkins et al. 2000; Elkins-Tanton et al. 2003), they interpreted the 500-km discontinuity as possibly representing the maximum depth of melting beneath this nearside province. Because of the extraction of aluminous basaltic melts from the mare source region above ~500 km depth, the mantle below this depth would be expected to be relatively enriched in aluminous phase by perhaps up to 1 wt% Al_2O_3. If this interpretation is correct, then the 500-km discontinuity is not a global feature of the Moon, but is rather only locally present beneath the Procellarum KREEP Terrane. Since three of the four seismic stations are located within this geochemical province, it is presently not a simple task to delineate the lateral extent of this feature.

3.3. Summary

As samples of the lunar mantle have not been identified within the lunar sample collection, information concerning its composition and structure can only come from indirect sources, such as the analyses of the mare basalts and volcanic glasses, and the Apollo seismic data. The sample data show that the lunar mantle is heterogeneous in composition, with low- to high-Ti sources at similar depths being required to account for the wide range of basaltic compositions. The mantle is depleted in alkali, volatile, and siderophile elements, and further possesses an extremely low oxygen fugacity near or below the iron-wüstite buffer. Some picritic glasses show evidence for garnet being present in their source, and high pressure experiments demonstrate that the mare basaltic magmas and picritic glasses are multiply saturated, usually with olivine and orthopyroxene, at pressures corresponding to depths less than 560 km. While these depths are often quoted as corresponding to depths of melting, other interpretations are possible. Crater counting studies suggest that mare volcanism extended over a large portion of lunar history (from >4 Ga to ~1 Ga), but was most active between about 3.9 and 3 Ga.

Most investigations of the lunar seismic data agree that there is a major seismic discontinuity at a depth between 500 and 600 km. It also appears that the deepest mantle (below a depth of ~1000 km) might be partially molten. Investigations of thermodynamically stable mineral phases that are consistent with the Apollo seismic data indicate that the upper ~500 km of the mantle is predominantly composed of orthopyroxene, with smaller abundances of olivine, clinopyroxene, plagioclase and garnet. In contrast, the lower mantle is predominately composed of olivine, with lesser quantities of garnet and possibly clinopyroxene. Whereas garnet is predicted to exist with high abundances in the lower mantle, it is also likely to be stable in the upper mantle as well between depths of ~200 and 600 km. Thus, the garnet signature seen in some picritic glasses does not necessarily imply an origin beneath the ~500 km seismic discontinuity.

The magnitude of the ~500-km seismic discontinuity requires that a change in composition occurs at this depth, and several hypothesis have been put forth to explain this feature. One possibility is that this depth represents the base of melting of the lunar magma ocean. However, in order to explain the magnitude of the seismic discontinuity, the bulk composition of the Moon below this depth must be more Mg- and/or Al-rich than above it. Alternatively, if the lunar magma ocean was global in extent, and its cumulates were laid down sequentially and not subsequently disturbed, then this depth could mark the transition between early olivine-rich cumulates and later orthopyroxene-rich cumulates that are predicted to crystallize from a lunar magma ocean. If this scenario is true, then a large-scale post-magma-ocean overturn of the lunar mantle probably did not occur. Finally, it is possible that this discontinuity represents the maximum depth of melting in the mare source region. This is supported by thermal modeling which takes into account the high abundances of heat sources in the Procellarum KREEP terrane, as well as the fact that all mare basalts and picritic glasses possess multiple saturation depths less than 560 km. If this interpretation is correct, then the ~500-km seismic discontinuity might only be a local feature beneath the nearside mantle.

4. THE CORE

All of the terrestrial planets and many of the icy satellites have undergone a major differentiation event that resulted in the formation of a metallic core. Seismic data demonstrate that the central portion of the Earth is composed of a large iron-rich core whose size is about 55% of its radius. Venus is inferred to have approximately the same internal structure based on its similarity in size and mass, whereas the high bulk density of Mercury is suggestive of an iron-rich core that is about ~75% of its radius. Moment of inertia and mass constraints for Mars (e.g., Sohl and Spohn 1997; Bertka and Fei 1998) and Jupiter's moon Io (Anderson et al. 1996b) suggest that these bodies have iron-rich cores that are about half of their radius. By similar means, Ganymede and Europa are each inferred to have an iron-rich core, though their absolute size is not well constrained (Anderson et al. 1996a; 1998). In addition to the terrestrial planets and icy satellites, some of the asteroids apparently differentiated early forming iron cores that are sampled by the iron meteorites.

Because of the above observations, it is natural to suspect that the Moon should also possess a sizeable metallic core. However, as we describe below, several lines of evidence imply that if the Moon does have a core, it must be small (<460 km radius). The existence, size and composition of such a core is of fundamental importance in deciphering many aspects of the origin and evolution of this body. For example, it is widely believed that the origin of the Moon is related to a collision between the Earth and a Mars-sized object that ejected debris and vapor into circumterrestrial orbit (see Cameron 2000 and references therein). While the debris that accreted to form the Moon is predicted to be derived primarily from the silicate mantle of the impactor, current models do not uniquely constrain the amount of iron that would be entrained in this material. Knowledge of the size of the lunar core could thus be used to constrain the many unknown parameters associated with these models (Canup and Asphaug 2001). Secondly, if the Moon does possess an iron-rich core, then it is possible that it could have at one time generated a magnetic field. A lunar dynamo might help explain the curious magnetizations that have been measured in some of the Apollo samples, and the crustal magnetic fields that have been mapped from orbit. Thirdly, knowledge of the physical state of the core (liquid vs. molten) would help constrain its composition and temperature, and hence the thermal evolution of the Moon. Finally, the formation of an iron core may have had a noticeable effect on the composition of the lunar mantle, and subsequently on the composition of the mare basalts that erupted at the surface.

While most studies to date that have attempted to constrain the size of the lunar core presume that it is composed of metallic iron, in this section, we address the possibility, as suggested by Wieczorek and Zuber (2002), that the Moon might instead possess a molten, dense, iron- and titanium-rich *silicate* "core." Two reasons argue for the plausibility of this hypothesis. First, ilmenite is predicted to crystallize near the terminal stages of magma-ocean crystallization, and because of its relatively high density, sink through the lunar mantle. This could plausibly result in the formation of a small dense silicate core possessing high concentrations of titanium and iron (Hess and Parmentier 1995). A second reason is tied to buoyancy considerations of lunar basaltic magmas. Because of the high iron and titanium concentrations that are found in some of these basalts, Delano (1990) realized that some magmas might have been negatively buoyant with respect to the deep lunar mantle. Specifically, he showed that if a basaltic melt with a titanium concentration greater than about 16 wt% was produced deep within the Moon, this melt would probably sink (see also, Circone and Agee 1996; Agee 1998). The absence of basalts with titanium abundances greater than this cut-off value was used as evidence in favor of this hypothesis.

Below we review the relevant information that bears on the size, physical state, and composition of the lunar core. Constraints on the core size come from the moments of inertia of the Moon, lunar laser ranging data, magnetic induction studies, and the abundance of siderophile elements in the mare basalts. We also discuss whether the lunar paleomagnetic data require the existence of a core dynamo, and whether the physical state of the core can be used to constrain its composition and current temperature.

4.1. Mass and moments of inertia

The mass and moments of inertia of the Moon are bulk properties that any lunar density model must satisfy. Differences in the principal moments of inertia are uniquely related to the second-degree gravitational harmonics of the Moon (e.g., Lambeck 1988) which have been measured to high precision by orbiting spacecraft. The rotation of the Moon is affected by its moments of inertia, and analyses of ranges to the lunar laser retroreflectors determine the independent librational parameters, which depend upon ratios of the principal moments (Dickey et al. 1994; Konopliv et al. 1998). By combining these pieces of information, the entire moment of inertia tensor of the Moon is now known to good accuracy. For studies of the lunar interior, one is generally only interested in the mean moment of inertia, which can be calculated by the expression

$$I = \frac{2}{3} \int_V \rho(r,\theta,\phi) \, r^2 \, dV \qquad (3.6)$$

where r is radius, ρ is the position dependent density, and dV is the differential volume element.

While the moments of inertia do not uniquely determine the density structure of a body, they do give qualitative information as to whether or not the density increases with depth. For a homogeneous sphere, the moment of inertia normalized by MR^2 (where M and R are its mass and radius, respectively) is equal to 0.4. In contrast, the Earth, which has an iron core that is about half of its radius, has a normalized moment of inertia of about 0.33. Most recently, Konopliv et al. (1998) obtained a value of $I/MR^2 = 0.3931 \pm 0.0002$ for the Moon. This value was computed using a reference radius of 1738 km, and if the mean planetary radius were used instead, the normalized moment of inertia would be increased by 0.0004 (see Table 3.13).

As the mean moment of inertia of the Moon is seen to be very close to the value of a uniform density sphere, if the Moon possesses an iron core, it must be small. For instance, if one assumes that the density of the crust, mantle and core are uniform, then the lunar mass and moment of inertia can both be satisfied by either an ~330-km radius solid iron core (ρ~8.1 g cm^{-3}) or an ~460-km radius liquid eutectic Fe-FeS core (ρ~5.2 g cm^{-3}). In contrast, if the "core" were instead composed of a dense silicate material having a density less than 5.2 g cm^{-3}, then its radius would be larger. If the density of the lunar mantle increases with depth, as is suggested by the existence of the 500-km seismic discontinuity, then the above estimated core sizes should be viewed as upper limits. Indeed, plausible compositional models of the

Table 3.13. Properties of the Moon and its orbit.

Parameter	Value	Note
Semi-major axis (km)	384399	(1)
Orbit eccentricity	0.0549	(2)
Obliquity to orbit plane	6.688°	(1)
Orbit inclination	5.145°	(1)
Inclination of equator to ecliptic	1.543°	(1)
Orbital period (days)	27.321582	(2)
Mean planetary radius (km)	1737.103 ± 0.015	(3)
Polar flattening (km)	2.17 ± 0.11	(3)
Center-of-mass/center-of-figure offset (km)	1.90 ± 0.01 (8.1°N, 156.6°W)	(3)
GM_{Moon} (10^9 m^3 s^{-2})	4902.801076 ± 0.000081	(4) (5)
Gravitational constant (G) (10^{-11} kg m^3 s^{-2})	6.67259 ± 0.00030	(6)
Mass (10^{21} kg)	73.4767 ± 0.0033	
Mean density (kg m^{-3})	3346.45 ± 0.17	(7)
Normalized polar moment of inertia (C M^{-1} R^{-2})	0.3932 ± 0.0002	(8) (9)
	0.3936 ± 0.0002	(8) (10)
Normalized mean moment of inertia (I M^{-1} R^{-2})	0.3931 ± 0.0002	(8) (9)
	0.3935 ± 0.0002	(8) (10)
k_2	0.026 ± 0.003	(4)
	0.025 ± 0.003	(11)

Notes: (1) Williams et al. 2001b; (2) Yoder 1995a; (3) Smith et al. 1997; (4) Konopliv et al. 2001; (5) model LP150Q; (6) Bursa 1992; (7) assuming a sphere of mean planetary radius; (8) Konopliv et al. 1998; (9) R = 1738 km; (10) R = mean planetary radius; (11) Williams et al. 2001a

lunar interior can be constructed that do not require the existence of an iron-rich core (e.g., Hood 1986). As one simple example, the measured moment of inertia and mass could be satisfied by a mantle that was linearly zoned in composition, with the top of the mantle and center of the Moon being composed of Fo$_{95}$ and Fo$_{61}$, respectively.

The measured moment of inertia of the Moon by itself does not place any firm constraints on the existence or non-existence of a lunar core. However, as will be seen below, this parameter does offer important constraints on the interior composition when used in conjunction with the Apollo seismic data (Hood and Jones 1987; Mueller et al. 1988; Kuskov and Kronrod 1998, 2001; Khan and Mosegaard 2001) and k_2 Love number (Khan et al. 2006).

4.2. Seismology

The materials that might be expected to comprise a lunar core are characterized by relatively low P-wave seismic velocities (~5.5 km/s for solid iron, ~4 km/s for liquid iron, ~3 km/s for an Fe-FeS eutectic liquid, and ~2.5 km/s for a silicate magma). An ideal approach toward determining the size and composition of a putative lunar core would thus be to determine the seismic velocity structure of the deep interior. Some evidence for a 170 to 360 km radius low-velocity (3.7 to 5.1 km/s) core was tentatively suggested by Nakamura et al. (1974) on the basis of a single P-wave arrival from a farside meteoroid impact, but this observation is not definitive (e.g., Sellers 1992). Sellers (1992) further tentatively identified PKP arrivals from two additional farside impact events, and if these arrivals are ultimately deemed to be reliable, they are consistent with a 400 to 450 km radius core possessing a P-wave velocity of ~5 km/s (the sensitivity of the radius to the core velocity was not addressed). While these farside meteorite impacts may ultimately shed light on the existence and size of a putative lunar core, we consider these studies to be more provocative than definitive at the present time.

One indirect approach for investigating the size of a dense metallic core is to construct compositional models of the lunar crust and mantle that are consistent with the Apollo seismic velocity models. The mass of material below the limits of the seismic velocity profile can then be inferred from a knowledge of the known lunar mass, and the core size can be constrained by use of the known moment of inertia (Hood and Jones 1987; Mueller et al. 1988; Kuskov and Kronrod 1998, 2001; Kuskov et al. 2002). Because of the uncertainties in the velocity models, the mantle density profiles are only weakly constrained by the seismic data. Furthermore, these density profiles depend upon the assumed thermal structure of the mantle, which is also uncertain. Nevertheless, because of the relatively small density changes that are expected to occur in the mantle for plausible bulk compositions and temperature profiles, application of this approach has consistently indicated the likely existence of a small dense core. Hood and Jones (1987) found that a core radius of 200–450 km was implied for an iron composition, representing about 1–4% of the lunar mass. Mueller et al. (1988) similarly concluded that a metallic core at least 150 km in radius was necessary to reconcile the mass and moment of inertia constraints with the Nakamura (1982) seismic model. Kuskov and Kronrod (2001) have estimated the radius of the core to be 310–320 km for a pure iron composition and 430–440 km for a eutectic Fe-FeS composition, and the more recent analysis of Kuskov et al. (2002) gives a range between 330 and 530 km. We note that all of the above core sizes would be larger if the core was instead assumed to be composed of a less dense titanium- and iron-rich silicate composition.

4.3. Induced dipole moment

An alternate approach toward investigating the deep lunar interior consists of estimating limits on the electrical conductivity as a function of depth using surface and/or orbital magnetometer data (see the reviews of Sonnett 1982, Hood 1986, and Hood and Zuber 2000). This approach can, in principle, yield limits on the size of a high electrical-conductivity core. However, the composition of the core (e.g., molten silicate, metallic iron, metallic Fe-FeS) is not easily constrained by electromagnetic sounding data alone.

Limits on the mantle electrical conductivity profile were obtained using time-dependent measurements from a high-altitude orbiting magnetometer to monitor the input field and a surface magnetometer to monitor the sum of the input and induced fields. Application of this technique yielded bounds on the mantle conductivity profile and an upper bound of ~435 km on the radius of a high electrical-conductivity core (Sonnett et al. 1972; Dyal et al. 1976; Hood et al. 1982; Hobbs et al. 1983). The mantle conductivity was found to continuously increase from 10^{-4}–10^{-3} S/m at a depth of 300 km to 10^{-3}–10^{-2} S/m at a depth of 700 km to 10^{-2}–10^{-1} S/m at 1000-km depth (e.g., Hood 1986). For this experiment, the electrical conductivity of a molten silicate core (~10 S/m) cannot be distinguished from that of a metallic iron core (~10^5 S/m) using Apollo data records with typical lengths under 100 hours (Hood et al. 1982). One difficulty with this time-dependent sounding technique is that two separate magnetometer data records are used and a very accurate intercalibration is required if the weak core signal is to be detected (Daily and Dyal 1979). Because the Explorer 35 orbital magnetometer and the Apollo 12 surface magnetometer that were employed were not perfectly intercalibrated, this error source essentially precluded an accurate determination of the core size via this technique.

A method for sounding the deep lunar interior that requires only data from a single magnetometer involves the measurement of the induced magnetic dipole moment of the Moon as it passes through the geomagnetic tail of the Earth (Goldstein et al. 1976). This method, while not capable of determining electrical conductivity bounds as a function of depth, has the advantage of being free of intercalibration errors and therefore allows more sensitive measurements of the weak induced fields expected from a small core. The method exploits the quasi-vacuum (i.e., nearly plasma-free) environment experienced by the Moon during its monthly traversals of the geomagnetic tail lobes. During these periods, the Moon is occasionally exposed to a nearly spatially uniform magnetic field for durations ranging from hours to several days. This steady magnetic field slowly diffuses into the Moon, with the rate of diffusion being controlled by the material's electric conductivity. That portion of the lunar interior that is unaffected by the external field will give rise to an induced magnetic dipole moment oriented opposite to the applied field. Initially, the induced moment originates from electrical currents set up in the lunar mantle. However, after a decay period of ~5 hours or less, the external field diffuses through the mantle and induces currents on the surface of a high electrical-conductivity core, if one is present. At this point, the amplitude and time-dependence of the dipole moment is directly relatable to the core radius and its conductivity. If the core is a good conductor ($>10^2$ S/m), then the induced dipole moment should not significantly change during the few days time that the Moon spends in the geomagnetic tail lobe. However, if the core conductivity were representative of a silicate magma (~10 S/m) then the time-variability of the induced magnetic moment could be significant.

The first measurements of the lunar induced magnetic dipole moment in the geomagnetic tail were obtained using data from the Apollo 15 and 16 subsatellite magnetometers (Russell et al. 1981). After eliminating intervals when significant tail lobe plasma densities were present, the final estimated induced moment amplitude was $(-4.23 \pm 0.64) \times 10^{23}$ A m^2 T^{-1}. Assuming that mantle contributions to the induced moment were negligible, the corresponding radius of a high electrical-conductivity core is 439 ± 22 km. Additional measurements of the induced moment were reported by Hood et al. (1999) using data from the Lunar Prospector magnetometer. The selected data were obtained when the Lunar Prospector orbit plane was in an optimal orientation for induced moment measurements. Editing and averaging of individual orbit segments over a duration of about a day and a half yielded an estimate for the induced moment of $(-2.4 \pm 1.6) \times 10^{23}$ A m^2 T^{-1}. Assuming that mantle currents were negligible, the core radius was found to lie between 250 and 430 km for an electrical-conductivity typical of metallic iron. Alternatively, if the electrical-conductivity of the core were typical of a basaltic melt, then we find here that a core radius between 361 and 538 km would satisfy the average induced dipole moment.

Although the error estimates of the induced dipole moment of Hood et al. (1999) and Russell et al. (1981) overlap, the difference in the mean core radius estimates of nearly 100 km is an indicator of the difficulty of performing this measurement using a single magnetometer. Nevertheless, the fact that both sets of measurements yielded evidence for a small conducting core increases the likelihood that such a core actually exists in the Moon. In fact, a slightly larger core radius estimate would result if the positive paramagnetic and ferromagnetic induced moments of the crust and mantle were taken into account in the above calculations (Rochette 2000). However, this effect is small given the large measurement uncertainties associated with the induced dipole moment (Hood 2000).

4.4. Lunar laser ranging

The analysis of lunar laser ranges from stations on the Earth to corner-cube retroreflectors on the Moon can be used to determine the time-varying rotation of the Moon. The robustness of the measured lunar rotation depends upon the geometrical spread of the retroreflectors that were emplaced at the Apollo 11, 14, and 15 sites, and the Russian Lunakhod 2 site. At present, the accuracy of ranges to these retroreflectors is less than 2 cm. A review of the lunar laser ranging program up to 1994 is given by Dickey et al. (1994).

The rotation of the Moon can be characterized by two angles that describe the orientation of its polar principal axis, and a third angle that describes the rotation about this axis. Currently, the polar axis is inclined 6.69° to the orbit normal (the obliquity), the orbit plane is inclined 5.14° to the ecliptic plane, and the polar axis is tilted 1.54° from the normal to the ecliptic plane. However, in the past when the Moon was much closer to the Earth (~34 Earth radii away at ~4 Ga) the lunar obliquity could have been as high as 77° (Ward 1975).

Both the polar axis and the orbit normal precess with the same 18.6-year period about the normal to the ecliptic plane, but they are out of phase by nearly 180°. This rotational configuration, known as a Cassini state, describes the first-order rotational state of the Moon. Superposed on this configuration are small rotational oscillations about the polar axis, as well as small oscillations in the pole direction, that can be detected from an analysis of the lunar laser ranging (LLR) data. These deviations, referred to as physical librations, are mainly due to the differences between the principal moments of inertia and the low-degree static gravity field of the Moon. As described below, these and other rotational signatures can be used to infer the elastic properties of the Moon and the presence of a fluid core. In particular, a slight misalignment of the spin axis and orbit normal from the Cassini state has been observed, and this can be modeled as a result of energy dissipation within both the solid body of the Moon and along a liquid-core/solid-mantle interface.

4.4.1. Tides and the solid mantle. Solid-body tides are raised on the Moon by the gravitational attraction of the Earth, and the response of the Moon to these tides provides an opportunity to sample the elastic properties of its interior. The familiar response to the second-degree Earth tide is football shaped, with the elongated axis lying nearly along the Earth-Moon line. Higher-degree tides are also raised by the Earth, but the strength of these decrease by about two orders-of-magnitude for each increase in degree. For a spherically symmetrical body, the tidal response at each degree is proportional to three degree-dependent Love numbers. Here we concern ourselves only with the degree-2 terms. One Love number describes the change in the gravitational field of the Moon as a response to the Earth's tidal potential (k_2), another describes the vertical change in elevation of the surface (h_2), and the third describes the horizontal displacement of the surface (l_2). The Love numbers and their associated Qs (where Q^{-1} is a measure of the amount of dissipation that is incurred during a loading cycle) are bulk properties of the Moon that depend upon the interior structure and are therefore useful as constraints on models of the lunar interior. In particular, the degree-2 Love numbers depend most strongly upon the radial profile of the shear modulus, μ, or equivalently the S-wave velocity via the relationship $v_s = (\mu/\rho)^{1/2}$. In comparison to the shear modulus, the bulk modulus and P-wave velocity have relatively minor effects on the degree-2 love numbers. The presence of a dense liquid or solid core would change the k_2 Love number by a few percent.

The surface displacements of time-varying tides on the Moon are only about 0.1 m in height and about half of that horizontally. The time variations of these tides, which may be represented by a sum of periodic components, arise from variations in the distance and direction of the Earth as seen from the Moon. The two largest components have periods of 27.555 days (the anomalistic period, which is the time it takes for the Moon to go from perigee to perigee) and 27.212 days (the period with respect to the ascending node). Some additional tidal components have periods of 1/3 month, 1/2 month, 7 months, and 1 year. The precession periods for the lunar orbit are 6 years for the argument of perigee, 9 years for the longitude of perigee, and 18.6 years for the longitude of ascending node. In addition, the rotation of the Moon has motions such as the 75-year wobble of the pole direction. Because of these long dynamical time scales, the accurate interpretation of the rotation and orbit of the Moon via laser ranges requires the analysis of many years of data.

Time variations in the tidal field give rise to torques on the Moon, affect its moments of inertia, and influence its rotation. Tidal distortion of the second-degree gravity potential and moment of inertia tensor is proportional to the Love number k_2. As the phase of the tidal response depends on Q^{-1}, tidal dissipation is proportional to k_2/Q (e.g., Segatz et al. 1988). However, when analyzing the rotation data, the mantle and core contributions to these numbers are not immediately isolated from one another. The oblateness of the core-mantle boundary influences the determination of the Love number k_2 and dissipation at this interface must be taken into account when the solid-body tidal Q is estimated. Fortunately, these core and mantle effects can be separated as their influence on different periodic rotation terms is different. (The effects of core dissipation will be discussed separately in the following section.) A recent analysis of the LLR data yields a value of $k_2 = 0.025 \pm 0.003$ (Williams et al. 2001a), where most of the uncertainty in this value is a result of the uncertainty associated with the core flattening. (If the core flattening were set to zero, then LLR determination of k_2 would be larger by 14%.) LLR analyses also detect the tidal displacement Love numbers h_2 and l_2, though k_2 is the most accurately determined among these quantities. If the displacement Love numbers could be improved sufficiently, then they would provide two additional constraints on the interior properties of the Moon. Time variations of the gravity field are measurable with accurate tracking of lunar orbiting spacecraft, and analysis of these data yield a concordant estimate of the Love number $k_2 = 0.026 \pm 0.003$ (Konopliv et al. 2001). Unfortunately, the uncertainty in these numbers is close to that which is required to distinguish between different models of the deep lunar interior. In particular, the uncertainty of k_2 is ~10%, whereas the presence of an iron core would only affect this number by ~5%.

The Apollo seismic experiment placed good constraints on the elastic properties of the upper layers of the Moon, but gave less definitive constraints for the lower mantle and possible core. When Love numbers are computed from these S-wave seismic profiles (Dickey et al. 1994), they are found to be similar (within about 20%), but not equal to the above values. Thus, the k_2 Love number offers the possibility of better constraining the S-wave velocity (and corresponding shear modulus) of the deep lunar interior where the seismically obtained values are less well determined (see Khan et al. 2006).

Imperfect elastic properties result in dissipation of energy whenever the Moon is flexed. Just as the k_2 Love number depends upon the radial profiles of elastic properties, the parameter k_2/Q for the whole Moon depends on the radial distribution of dissipation. In principle, the k_2/Q value may also depend on tidal frequency. Dissipative effects in the Moon cause a slight phase shift in the precessing polar axis with respect to the orbit normal axis from which the tidal Q can be determined (Yoder 1981; Williams et al. 2001b). In particular, a small advance in the polar axis of 0.26" from the Cassini configuration has been measured (e.g., Dickey et al. 1994; Williams et al. 2001b). The tidal solid-body Qs obtained from this number are extremely low and imply that a large amount of tidal dissipation is presently occurring within the Moon. For instance, the tidal Q for one month is found to be 37 whereas the annual Q is 60. While the bulk tidal Q of the Earth is smaller at ~12 (e.g., Burns 1986), this is primarily a result of dissipation within its oceans. Mars has been inferred to have a somewhat larger (though still relatively small) tidal Q between 50 and 150 from the acceleration of its satellite Phobos (e.g., Burns 1986). In contrast to the low whole-moon tidal Q, at seismic frequencies the Moon has a local Q of about 3000 to 3600 for the upper crust (Latham et al. 1970a,b), 4000 to more than 7000 for the upper mantle (Dainty et al. 1976; Nakamura et al. 1976; Nakamura and Koyama 1982), and 1400 to 1500 for the deep moonquake source region (Dainty et al. 1976; Nakamura et al. 1976). The Q has been inferred to decrease dramatically below the deep moonquake source region and this has been interpreted as possibly indicating the presence of a partial melt below this depth (Nakamura et al. 1973). Thus, given the high seismic Qs above a depth of ~1150 km, it appears likely that most of the solid body dissipation in the Moon occurs below this depth. In fact, if the tidal and seismic Qs are considered to be equivalent, then the average Q below ~1150 km depth is required to be less than 4 (Yoder 1981).

4.4.2. The core. While the presence of any significantly sized dense core will increase the Love numbers by a few percent, there are influences on the rotation of the Moon that are unique to the presence of a molten core. In particular, two torques arise from interactions at a solid-mantle/liquid-core boundary, and detection of the corresponding perturbations on the rotation would provide evidence for a fluid core. Because a fluid core will rotate independently from the solid mantle, a peak velocity difference of about 2 cm/s at the interface between the two should exist. This velocity difference results in a local force, causes energy dissipation, and leads to a net torque over the whole surface of the core-mantle boundary. If this boundary is oblate, there will be a second force due to flow along this aspherical boundary. These two torques have different directions and their influences on the rotation are in principle distinguishable.

Analysis of the lunar laser ranging data has detected dissipation-caused phase shifts at four different rotation frequencies (Williams et al. 2001b). Solid-body tidal dissipation acting alone provides a poor match to these results, whereas dissipation due to both tides and an independently rotating core are able to satisfy these results. This analysis of the LLR data implies that about 34% of the dissipation-caused phase shift from the Cassini state is due to a liquid core. To obtain the core size from the strength of the core dissipation requires a model of the dissipation at the core-mantle boundary, and for this Yoder's (1981; 1995b) turbulent boundary layer theory has been used. In this model, the core size is primarily dependent upon its density, and only secondarily on its viscosity. The effect of viscosity on the core size would be more important if dissipation was a result of laminar flow, but laminar flow is only to be expected if the viscosity is in excess of ~10 Pa s (Yoder 1981). For the pressures and temperatures expected at the center of the Moon, the viscosity of a molten Fe-FeS metallic core is on the order of 10^{-2} Pa s (Dobson et al. 2000; Vocadlo et al. 2000), whereas the viscosity of a basaltic silicate magma would be slightly larger at ~10^{-1} Pa s.

Assuming a kinematic viscosity of 10^{-6} m^2 s^{-1} (7×10^{-3} Pa s for a density of 7 g cm^{-3}), the LLR analysis constrains the radius of a molten iron core to lie between 314 and 352 km. For a molten Fe-FeS eutectic composition ($\rho \sim 5.2$ g cm^{-3}) the core radius is constrained to lie between 334 and 375 km. Alternatively, if the core is composed of a dense iron- and titanium-rich silicate magma ($\rho \sim 3.5$ g cm^{-3}), then we find here that the core would be slightly larger, having a radius that lies between 363 and 407 km. These computed core radii are only slightly dependent upon the assumed core viscosity, where an order of magnitude increase in viscosity would only result in a decrease in core size by about 20 km. If the Moon possesses a solid inner and outer liquid core, then these core radii estimates would be overestimates as there would be two surfaces at which dissipation could occur.

Oblateness of the liquid-core/solid-mantle interface also influences the rotation of the Moon, and at present, the detection of the core flattening effect is about twice its uncertainty. This measurement provides an independent line of evidence for the fluid state of the core, and as the k_2 Love number is anticorrelated with the core flattening, its detection gives rise to a slightly smaller value of k_2 (see Williams et al. 2005). When used in combination with the core moment of inertia, this parameter can allow a computation of the resonant period of the free core nutation. If there is a solid inner core in addition to a fluid outer core, then this might also influence the rotation of the Moon. An aspherical inner core would experience gravitational torques from both the Earth and the lunar mantle, but the associated rotational signatures have not yet been tested against the available data.

Much of the rotation-derived lunar science information comes through forced terms, but there are also free rotational modes. The three solid-body free libration modes have been detected and their amplitudes and phases have been determined (Newhall and Williams 1997). Dissipation from flexing and fluid core interactions should cause lunar free librations to damp with geologically short time scales, so the finite amplitudes imply recent or active stimulation. Of particular interest to understanding the interior is the 75-year polar wobble, analogous to the Earth's Chandler wobble. It has been suggested by Yoder (1981) that this mode may be stimulated by eddies at the solid-mantle/liquid-core boundary. Both damping and active excitation may be observable in the future, and active excitation should cause temporal irregularities in this mode. There are also free rotational modes for a fluid core, and any core-mantle boundary flattening would give rise to a resonant free core nutation frequency.

Finally, we note that dissipation due to both tides and the core-mantle interaction deposits heat in the Moon. At present these are minor heat sources, but both could have significantly heated the Moon when it was closer to the Earth (Peale and Cassen 1978; Williams et al. 2001b). The amount of heating depends on the rate of expansion of the early lunar orbit due to the tidal dissipation that occurs in the Earth and its oceans. This early expansion rate is uncertain, and for a slow evolution, the early dynamical heating could have been comparable to the internal radiogenic heating. Furthermore, this energy might have been able to promote convection, and if the core was metallic, a dynamo. However, a faster orbital evolution would have less dramatic consequences. In addition to tidal heating, a large amount of heat would have been generated within a fluid core (if present) during a spin axis transition between two distinct Cassini states when the semi-major axis of the Moon passed through ~34 Earth radii (Ward 1975). At that time, the obliquity of the Moon is predicted to have switched from ~26° to a maximum of 77°, and then back down to ~49° on the order of ~10^5 years.

4.5. Lunar free oscillations

When a planet is abruptly perturbed, it oscillates with a distinct set of frequencies. As the amplitudes of these free oscillations depend upon various rheologic properties of the planet, they can be used to constrain its deep interior structure. An advantage of applying this method to a planetary body is that it is applicable even if seismic recordings from only one station are available. Since the state of the core affects the fundamental modes of free oscillations, identifying these are relevant to addressing the existence and physical state of the lunar core. Since the spheroidal free oscillations involve movement of the whole body, they are particularly sensitive to the density structure of the planetary body.

In a recent study, Khan and Mosegaard (2001) claimed to have identified excitations of lunar free oscillations associated with five large meteoroid impacts occurring in 1976 while the long-period instrumental response was extended to low frequencies. Only spheroidal modes were considered because of non-seismic interference in the horizontal component of the seismometers. Although a signal to noise ratio of ~1.9 was stated for the stacked normal mode amplitude spectra following these impacts, this has remained somewhat contentious. Gudkova and Zharkov (2002) have shown that the expected amplitudes from such events should be about two orders of magnitude below the detection limit of current broad-band seismometers. Furthermore, based on a synthetic seismogram of an impact with approximately the same energy as those used in the Khan and Mosegaard study, Gagnepain-Beyneix et al. (2006) have shown that the signal to noise ratio should be less than 0.01 over the frequency band that was investigated. They further noted that the stacking of individual amplitude spectra does not noticeably increase the signal to noise ratio as a result of the loss of phase information that is inherent in such a procedure.

While the detection of fundamental spheroidal lunar normal modes is unlikely to be realized with the current Apollo seismic data set, these should in principle be detectable following large shallow moonquake events using superconducting gravimeters (Lognonné 2005). Low-order torsional modes are also in principle detectable using modern broad-band seismometers (Gudkova and Zharkov 2002).

4.6. Thermal constraints on the physical state of the core

If one knew the composition of the lunar core, estimates of its temperature over time would place constraints on its current and past physical state. Thermal evolution models of the Moon, however, are highly variable, and this limits the uniqueness of this approach. Based on a variety of published thermal evolution models, the temperature at the center of the Moon at ~4 Ga may have been between about 1280 and 1750°C, and at present is predicted to be between about 1000 to 1480°C (see Chapter 4). Despite these large uncertainties, a few generalizations can be made.

For the pressures that would be encountered in a lunar core (~4 GPa), the melting temperature of pure iron is ~1690°C (e.g., Presnall 1995). If the core were composed of pure iron, then it could possibly have been in a molten state early in the geologic history of the Moon. However, current estimates of the internal temperature of the Moon all predict that such a core would have subsequently completely frozen. It is generally believed that the crystallization of the Earth's core is a contributing factor towards generating a dynamo (e.g., Roberts and Glatzmaier 2000). Therefore the gradual crystallization of a lunar iron core may similarly have powered a dynamo, possibly explaining the curious paleomagnetic signatures of the lunar rocks and magnetic anomalies as observed from orbiting magnetometers (see Section 4.7). However, as a presently solid core is incompatible with the LLR data, the hypothesis of a pure iron lunar core can probably be dismissed.

The addition of either carbon or sulfur to iron will act to lower its melting temperature. The Fe-C system exhibits eutectic melting behavior, possessing a eutectic composition of about 3.5 wt% carbon, and a eutectic melting temperature of ~1175°C (Hirayama and Fujii 1993). If only a small quantity of carbon was sequestered in the lunar core, then it is likely that some portion of it would have remained molten for most to all of lunar history. If the current temperature of the core were at or above the Fe-C eutectic temperature, then the Moon would currently possess both a solid iron inner core and a liquid Fe-C outer core. The Fe-FeS system similarly exhibits eutectic melting. In this system the eutectic composition at lunar core pressures is about 25 wt% sulfur, and the eutectic melting temperature is about 950°C (Fei et al. 1997). Thus, if a small amount of sulfur were present in the core, some portion of it would almost certainly be molten at the present time. If the core of the Moon is predominantly composed of metallic iron, then these temperature considerations highly suggest that it currently possesses both a solid inner and liquid outer core.

As an alternative to metallic compositions, it is also possible that the "core" of the Moon might be composed of dense iron- and titanium-rich silicate materials. The composition of such a material is not well constrained, and hence its liquidus and solidus temperature are also uncertain. A plausible composition might be similar to that of the Apollo 14 high-titanium black glass, as a magma of this composition is predicted to be more dense than the deep lunar mantle (Delano 1990; Circone and Agee 1996; Agee 1998). At core pressures, this composition has a liquidus temperature of ~1570°C (Wagner and Grove 1997). Though the solidus temperature is unknown, it seems likely that some portion of this material could remain in a molten state at the present time.

4.7. Paleomagnetism of the lunar samples

It has been known for some time from Apollo surface magnetometers and subsatellite data that the lunar crust possesses an intrinsic remnant magnetization. The central question of lunar magnetism is to determine the intensity, duration and, ultimately, the origin of the fields that magnetized the crustal rocks of the Moon. The answer to these questions would dramatically affect our understanding of the lunar interior and the physical environment of the early solar system. Unfortunately, while there have been a number of reviews of the subject (e.g., Collinson 1985, 1993; Fuller and Cisowski 1987; Hood 1995), little new paleomagnetic data on lunar samples has been published since the early 1980s (e.g., Collinson 1984; Chowdhary et al. 1987). Major advances have instead come in our understanding of the reliability and applicability of the various kinds of paleointensity experiments as well as in the growth and understanding of the terrestrial paleointensity database. The overall quality of the terrestrial data far exceed that of the Moon, and the analysis of these data have accompanied advances in our understanding of the dynamics of dynamo generation. In light of these advances, it is useful to reexamine the nature and quality of the lunar paleointensity experiments of two decades ago.

4.7.1. Paleointensity experiments on lunar samples. Paleomagnetic conglomerate tests on single Apollo breccias (Banerjee and Swits 1974) demonstrate that the remnant magnetization of the lunar samples is of lunar origin and is billions of years old. However, as no outcrops of lunar bedrock were unambiguously sampled during the Apollo missions, the geometry of the paleomagnetic field of the Moon cannot be determined from the Apollo samples alone. As a result, nearly all work has focused on paleointensity measurements of the ancient field that magnetized the lunar rocks. Unfortunately, since the bulk of the these experiments were completed in the late 1970s, it has become clear that most of these methods give results that could be inaccurate. (For critiques of the various paleointensity methods see Levi and Banerjee (1976), Bailey and Dunlop (1977), Collinson and Stephenson (1977), Sugiura (1979), Sugiura and Strangway (1980), Kono (1987), Tauxe (1993), Dunlop and Ozdemir (1997), Goguitchaichvili et al. (1999), and Herrero-Bervera and Valet (2000).)

Today, the Thellier-Thellier double-heating method (Coe et al. 1978) sets the standard for paleointensity studies because it progressively thermally demagnetizes the sample. In doing so, it provides an internal check that the magnetization is a primary thermoremanence, produces multiple independent field measurements for each sample, and can often yield accurate results even when the sample is altered during heating. A major improvement in the Thellier-Thellier experiment during the last twenty years has been the widespread incorporation of partial thermoremanent magnetism (pTRM) checks that can diagnose alteration of the ferromagnetic minerals during the experiment (e.g., oxidation of iron, change in grain sizes of magnetic minerals, and quenched phase transitions in kamacite). Since such alteration can lead to dramatically inaccurate paleointensities, the incorporation of pTRM checks leads to a significant improvement in the success rate of the Thellier-Thellier method. Although nearly all Thellier-Thellier experiments on terrestrial rocks today employ pTRM checks, these have never been used in any published lunar experiment.

Selkin and Tauxe (2000) have developed six criteria for judging the quality of Thellier-Thellier paleointensity experiments. Unfortunately, none of the paleointensity experiments ever performed on lunar rocks pass all six of these. In fact, even the most successful lunar experiments pass no more than three or four. Note, however, that these criteria are highly restrictive and were meant to select for data capable of demonstrating field intensity changes of a few tens of percent, whereas even an order-of-magnitude accuracy would be valuable for lunar studies. The reasons that the lunar paleointensity experiments do not pass these criteria are severalfold. To begin with, many samples were found to hold weak and/or unstable natural remanent magnetization. Secondly, fewer than twenty lunar experiments were investigated with the Thellier-Thellier method, and none of these used pTRM checks. Thirdly, magnetostatic interactions between ferromagnetic grains in the samples often resulted in wild swings in pTRM intensities at several different temperatures. Fourthly, heating-induced changes in the mineralogy destroyed the primary minerals carrying the remanent magnetizations. Finally, the scarcity of lunar materials has drastically limited access to samples and made it difficult to make repeated measurements on the same individual samples.

4.7.2. Evidence for time-varying paleointensities. The most significant development in lunar paleointensity experiments in the last twenty years was the saturation isothermal remanent magnetism (IRMs) normalization studies of Cisowski and coworkers (Cisowski et al. 1983; Cisowski and Fuller 1986; Fuller and Cisowski 1987). The basis of this method is to determine the ratio of the NRM to laboratory induced IRMs for each sample, which is intended to quickly measure the relative intensity of the paleofield by normalizing the sample's moment by the mass and mineralogy of its ferromagnetic material. Although these studies employed an empirically-derived calibration factor for the ratio between the IRMs and the absolute paleointensity, it has since been suggested that this factor may not be constant, but instead may be sample dependent because of its intrinsic dependence on crystal size (Fuller and Cisowski 1987; Tauxe 1993). On the other hand, recent work by several groups suggests that perhaps the factor relating paleointensity to the NMR/IRMs ratio may be constant after all (Kletetschka et al. 2003, 2004; Gattacceca and Rochette 2004).

The most reliable absolute and IRMs normalization paleointensity measurements are shown in Figure 3.22 as a function of age. Even with this select dataset there are a number of problems. (1) At any given age, the inferred paleofield often has a range of intensities. (2) The ages in this figure are either upper limits or are highly uncertain. This is because detailed ^{40}Ar/^{39}Ar thermochronology was not performed on any of the samples, and worse, some of the ages are based on the U/Pb isotopic system that is usually reset at much higher temperatures than the magnetization. Although metamorphic Ar-Ar ages were used for some samples (e.g., for breccias that were heated during lithification), there was little attempt to put thermal constraints on the samples using the Ar release pattern during stepped heating experiments. (3) In the absence of an external magnetic field, impact shock effects will tend to demagnetize a sample over time (Cisowski and Fuller 1978), which implies that paleointensity values from shocked samples are underestimates. And finally, (4) the temporal sampling of lunar paleointensities is very nonuniform with a substantial data gap occurring between 3.0 and 1.5 Ga, and few existing measurements from before 3.9 Ga.

While considerable care must be exercised when interpreting the lunar paleointensity data, the data in Figure 3.22 have been interpreted by some as evidence for a weak magnetic field prior to ~4 Ga, a rapid increase in intensity between 4.1–3.9 Ga to a value of ~100 μT (or 1 Gauss in cgs units), followed by a return after 3.6 Ga to a value of ~10 μT (e.g., Cisowski et al. 1983; Runcorn 1994, 1996). Much speculation has been published linking the temporally variable nature of these data to the evolution of a putative lunar dynamo. In particular, Runcorn (1994; 1996), and most recently Stegman et al. (2003), have argued that the apparent 4.1–3.9 Ga rise and subsequent fall in intensities could be the result of a late turn-on of the lunar dynamo and its subsequent demise. If true, such a late dynamo turn-on would constrain models of the initial thermal state of the Moon (e.g., Pritchard and Stevenson 2000; Stegman et al. 2003).

An alternative to the lunar dynamo hypothesis is that the lunar samples acquired a remanent magnetization by some process associated with impact cratering. Hood and Huang (1991) have shown that during a large impact, an expanding impact-generated plasma cloud could amplify ambient fields near the antipode of the basin. The rocks in these regions could then be magnetized by the process of shock remanent magnetization when seismic waves from the impact converged at this region (e.g., Hughes et al. 1977; Hood and Huang 1991; Watts

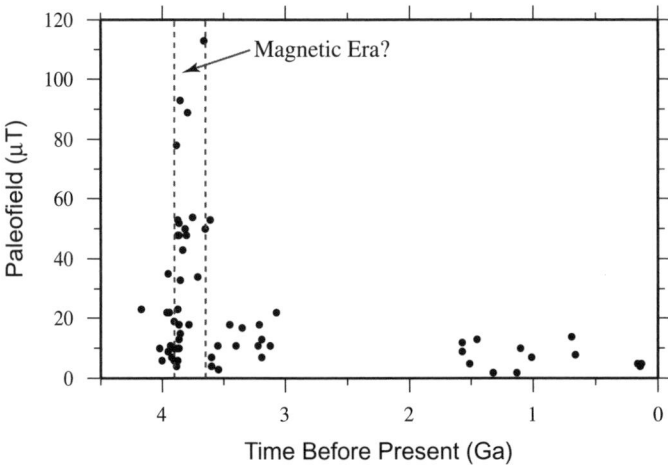

Figure 3.22. Absolute and IRMs normalization paleointensities of the lunar samples as a function of time. Data taken from Fuller and Cisowski (1987, Fig. 53).

et al. 1991; Wieczorek and Zuber 2001b). Consistent with this theory, the largest magnetic anomalies on the lunar surface as measured from orbit are in fact antipodal to many of the large basins (Lin et al. 1988, 1998; Hood et al. 2001). The field strengths in these regions may be as large as ~300 nT, whereas more typical fields of the Moon are only ~10 nT or less.

It is also possible that transient electrical currents set up in a plasma cloud close to an impact crater could have themselves generated a significant magnetic field (e.g., Martelli and Newton 1977; Srnka 1977; Srnka et al. 1979; Cerroni and Martelli 1982; Hood and Vickery 1984; Crawford and Schultz 1988, 1991; Hood and Huang 1991; Crawford and Schultz 1999). Measurements of surface fields by the Lunar Prospector electron reflectometer show that while basins possess low field strengths, basin ejecta is often strongly magnetized (Halekas et al. 2001, 2003; Richmond et al. 2003), consistent with such a process. As a lunar dynamo is unlikely to have operated during the past few Ga, the high paleofields measured from young lunar samples are most easily explained by impact processes. Finally, it is perhaps noteworthy that a large number of the high paleofield measurements occur near ~3.9 Ga, a time at which some models of the lunar cratering flux predict a large spike in the cratering rate (e.g., Cohen et al. 2000; Hartmann et al. 2000; Ryder et al. 2000).

While some temporal variations in the lunar paleointensity data may reflect real changes in field strength, it is important to assess the reliability and quality of the IRMs and absolute paleointensity data. Even the most reliable terrestrial Thellier-Thellier pTRM check paleointensity datasets show quite a bit of scatter. The Selkin and Tauxe (2000) dataset for just the last 300 My of Earth history has a ratio of variance to mean field equal to 0.7, which is actually not much less than that of the lunar IRMs and absolute paleointensity datasets (which have ratios of 1.0 and 1.6, respectively). Given that the terrestrial paleointensities show only a factor of two variation in mean intensity over the last 300 My (Selkin and Tauxe 2000), this begs the question of whether the apparent order-of-magnitude rise and fall of lunar paleointensity values with time is at all statistically meaningful.

An answer to this question can be obtained by use of the Kolmogorov-Smirnov (KS) test (Press et al. 1992, pp. 617-622), which assesses the probability that two datasets are drawn from the same underlying distribution. It does not make any assumptions about the nature of the underlying distribution, and is appropriate for even the smallest sample sizes (Lindgren 1993). It has also been used previously to analyze the significance of changes in the paleointensities of the Earth (Selkin and Tauxe 2000). We reject the null hypothesis that two sets of data are drawn from the same underlying distribution when the probability of this occurring is less than 5%.

We have applied the KS test to various subsets of the paleointensity database. These data were divided into three time periods (pre-3.9 Ga, 3.9–3.65 Ga, and post-3.65 Ga), and three different paleointensity techniques were considered (IRMs, the Thellier-Thellier subset of the absolute measurements, and all absolute measurements). These data were taken from the figures in Fuller and Cisowski (1987) and our results are summarized in Table 3.14. Using the IRMs database, it is seen that there is some statistical basis for saying that the IRMs data provide evidence for an increase in field strength around 3.9 Ga (there is a 3.8% chance that the pre-3.9 Ga and 3.9–3.65 data sets are drawn from the same distribution). There is an even better statistical basis for saying that the field strength decreased after ~3.65 Ga. It is also clear that there is no statistical difference between the pre-3.9 Ga and post-3.65 Ga IRMs normalization datasets.

Unfortunately, the absolute paleointensity measurements are not as definitive. Using all the absolute paleointensity measurements, we can similarly say that there is a good statistical basis for claiming that the field strength decreased after 3.65 Ga. However, the same is not true when only the Thellier-Thellier absolute paleointensity measurements are used (these are probably the most robust subset of the absolute measurements). Although we cannot reject the hypothesis that the 3.9–3.65 Ga and post-3.65 Ga periods of the Thellier-Thellier paleointensities are

Table 3.14. Kolmogorov-Smirnov (KS) tests on lunar paleointensity data.

Paleointensity Data Subsets	P	Reject Null Hypothesis?
Early (IRM) vs. Mid (IRM)	0.038	YES
Early (IRM) vs. Late (IRM)	0.25	NO
Mid (IRM) vs. Late (IRM)	1.0×10^{-5}	YES
Mid (TT) vs. Late (TT)	0.18	NO
Mid (A) vs. Late (A)	0.030	YES

P = Probability of the null hypothesis that the two distributions are drawn from the same underlying distribution.
IRM = Saturation isothermal remanent magnetism normalization dataset.
TT = Thellier-Thellier dataset.
A = Entire Absolute paleointensity dataset.
Early = pre-3.90 Ga, Mid = 3.90-3.65 Ga, Late = 3.65 Ga to present.

drawn from the same distribution, this discrepancy with the other datasets might be explained by the very small number of Thellier-Thellier measurements (the effective number of degrees of freedom for this test is significantly lower than that of any of the others).

4.7.3. What we know about lunar paleomagnetism. So what can we confidently take away from the lunar paleointensity work of the last thirty years? First, it is near certain that many of the lunar samples retain ancient (billions of years old) magnetizations. Second, it is likely that there were at least transient magnetic fields on the Moon that reached 10 µT and possibly as much as 100 µT. In particular, it is difficult to dismiss the 100 µT value from 62235 that has been reproduced in four separate experiments (two of which used the Thellier-Thellier technique) by two independent laboratories (Collinson et al. 1973; Stephenson et al. 1974; Cisowski et al. 1983; Sugiura and Strangway 1983). Given the uneven sampling of the lunar surface during the Apollo missions, it is highly probable that some lunar rocks might show evidence for an even higher field strength. Third, the lunar paleointensity data give statistically significant evidence for a strong rise in paleointensities around 3.9 Ga, followed by a weakening sometime after ~3.6 Ga. This evolution in intensity could be reflective of (1) a rising and weakening dynamo field, or (2) a time period in which impact-generated magnetic fields were more common, possibly as a result of a spike in the lunar cratering rate. And finally, the fact that (a) the paleointensities from more than a dozen samples younger than 1.5 Ga give fairly high values between ~1–10 µT (including a Thellier-Thellier value from the <200 Ma impact glass of 70019) and (b) large magnetic anomalies are associated with ejecta and the antipodes of large impact basins, we strongly suspect that impact events have had at least some role in magnetizing the lunar crust. Given the possibility that such impacts may have merely amplified a preexisting field, this does not exclude the possibility of an ancient lunar dynamo.

4.7.4. A lunar dynamo? If the observations of lunar paleomagnetism can be believed, then the intensity of the magnetic field was at one point as large as 100 µT at the lunar surface. If this were a result of an internal dynamo, then assuming a 375 km radius metallic core (which is the largest radius allowed by the LLR data), the field strength at the core-mantle boundary would be about 100 times greater (~10 mT). This is more than 20 times greater than the field present at the core-mantle boundary of the Earth, and larger field strengths would be required if the lunar core was smaller. Is it possible for a convecting lunar dynamo to produce a field intensity of this magnitude? Such a question is important, because if a convecting lunar dynamo was present early in lunar history, it has implications for the early thermal state of the Moon and could potentially constrain lunar formation scenarios.

There is much uncertainty in using dynamo theory to estimate the strength of magnetic fields, but our current theoretical understanding (and using current planetary and satellite dynamos as analogs) makes a convecting lunar dynamo seem unlikely. A dynamo is formed when an electrically conducting fluid flows in the presence of a magnetic field, regenerating a self-sustaining magnetic field by induction (e.g., Busse 2000; Roberts and Glatzmaier 2000). Here we consider a fluid metallic iron core whose flow is generated by convective cooling. However, other conductive fluids (e.g., liquid silicates) or sources of flow (e.g. tidal effects) could possibly be relevant for the Moon.

When a dynamo is operating, its internal field strength can be estimated by use of the Elsasser number

$$\Lambda = \frac{B^2}{2\,\rho\,\mu_0\,\lambda\,\Omega} \tag{3.7}$$

where λ is the magnetic diffusivity (~2 m^2 s^{-1} for liquid iron), Ω is the angular rotation rate of the body, ρ is the density of liquid iron, μ_0 is the magnetic permeability of free space, and B is the internal intensity of the magnetic field at the core-mantle boundary (e.g., Stevenson 2003). Convection in the presence of rotation and the generation of a magnetic field is most efficient when the Elsasser number is close to unity, or perhaps between 1–10 (Zhang and Jones 1994; Stevenson 2003). Dynamos might exist for $\Lambda < 1$, but cannot exist when Λ is much greater than unity because large fields eliminate those aspects of the fluid motion that are favorable for dynamo generation. When the Elsasser number is near unity, the Coriolis and Lorentz forces are roughly balanced so that the buoyancy that promotes convection is less inhibited by the other two forces. Notice that the Elsasser number does not require either the core size or the relative importance of compositional versus thermal convection to be known. Dynamos that do not require convection (e.g., tidally or nutationally driven flows) are also expected to be limited by the condition $\Lambda \sim 1$, although this is less certain.

Letting Λ vary between 1 and 10, the present lunar rotation rate implies a maximum field strength between 300–1000 μT at the lunar core. When the Moon was closer to the Earth, say 30 Earth radii away, the field at the core would have been slightly larger with a strength of 500–1600 μT. Upward continuing the maximum of these values to the surface yields a maximum field strength of ~16 μT, which is smaller by a factor of six than the maximum paleointensity implied by the lunar samples. Moreover, this maximum estimate is based upon the strength of the internal toroidal field of the core, and it is generally understood that the poloidal field external to the core is generally smaller, possibly by a large factor (e.g., Hide and Roberts 1979).

The above analysis suggests that a lunar core dynamo would not have been able to generate a magnetic field of ~100 μT at the lunar surface. However, considering the limitations in the above analysis, the existence of an early lunar core dynamo cannot be definitively ruled out. A better understanding of lunar magnetism will have to wait until the paleointensities of lunar samples are reassessed using modern measurement techniques. Advances in dynamo theory will further help elucidate whether or not 100 μT magnetic fields can be generated at the lunar surface.

4.8. Geochemical constraints on the presence of a lunar core from siderophile elements and short-lived isotopes

The siderophile (metal-seeking) elements comprise a suite of nearly 30 elements that are sensitive indicators to metal-silicate partitioning in planetary bodies. Because these elements should be extracted into metallic phases according to their metal/silicate partition coefficients during accretion, these elements may represent a chemical "fingerprint" of a planetary core formation. Over the past 30 years, estimates of siderophile elements in the lunar mantle have been used to argue for the presence of a small metallic core (0.1–5.5 lunar wt%; see Table 3.15),

Table 3.15. Summary of lunar core sizes based upon siderophile element concentrations.

Study	Core Mass Fraction (%)	Core Radius (km)[†]	Silicate Mantle Degree of Melting (%)	Core Ni Abund. (wt%)	Bulk Moon Comp.[*]
Newsom (1984)	2.0 – 5.5	369 – 517	2 – 9	12 – 25	CI
O'Neill (1991)	~1	~ 293	0	35 – 55	PUM, CI, H
Ringwood & Seifert (1986)	0.4	216	0	40	PUM
Righter & Drake (1996)	1	293	100	43	PUM/CI/H
Righter & Drake (1996)	5	500	100	8.3	PUM/CI/H
Righter (2002)	0.7 – 1.0	260 – 293	100	20.0 –25.7	Proto-Earth/ Impactor

*CI (CI chondrite); PUM (Primitive upper mantle); H (H chondrite).
†Assuming a core density of 7 g cm^{-3}

based mainly upon experimental partition coefficients determined at low temperatures and pressures (1200–1300 °C and 1 bar; e.g., Ringwood 1979; Drake et al. 1984; Newsom 1984; O'Neill 1991). Recent experimental work has been carried out at higher pressures and temperatures, and this allows a re-examination of siderophile partitioning behavior for conditions that are likely to have endured shortly after the formation of the Moon. This section will describe how siderophile elements are estimated for the lunar mantle, outline the conditions under which the lunar core may have formed, and finally discuss the evidence and time interval of core formation as implied by short-lived radioisotopes.

4.8.1. Siderophile elements in lunar samples and mantle. Samples of the lunar mantle have not yet been recognized in the Apollo sample collection (even as xenoliths), so estimates of its siderophile-element concentrations must be derived entirely from crustal samples. Because impact processes have affected the lunar surface, surface materials often contain impact melts, and their original (or pristine) composition and mineralogy have thus been compromised. Impacting materials are commonly chondritic, and chondrites generally contain more than 10 times the siderophile-element concentrations of typical crustal and mantle rocks. As a result, distinguishing pristine from compromised siderophile-element concentrations can be challenging and has been contentious in the past (e.g., Anders 1978; Delano and Ringwood 1978).

The approach of looking for correlations between a siderophile element and a refractory lithophile element of equal compatibility (or incompatibility) has proven the most reliable and straightforward way to estimate siderophile-element concentrations in the lunar mantle. Siderophile elements, however, can be difficult to determine at the low concentrations typical of lunar basaltic rocks. For example, Mo and Sn are present at low ppb levels in lunar samples, and some of the highly siderophile elements (HSEs) such as Re, Os and Ir are in ppt levels. Data for highly siderophile elements in lunar rocks obtained to date have come almost exclusively from the highly sensitive technique of radiochemical neutron activation analysis (RNAA) (e.g., Wolf et al. 1979; Warren et al. 1986). Many early data, even from otherwise reputable labs, have been discredited (Wolf et al. 1979) as suspiciously high in comparison to later analyses of the same or closely similar samples. The most frequent cause of spurious data is probably laboratory contamination, a problem that is enhanced for the lunar basalts by their low concentrations of the HSEs, and by the small sample masses available for analysis.

When the lunar mantle is partially melted, elements can exhibit compatible, $D > 1$, or incompatible, $D < 1$, behavior, where D is the partitioning coefficient between the solid and liquid phases for that element. The relative depletions of an element in comparison to CI chondrites can be used to estimate the original siderophile element abundances in the lunar and terrestrial mantle, and the abundances discussed below have all been estimated in this manner (see Newsom 1995 for a summary of this approach). Because this is an enormous field of literature, we do not review it here, but rather direct the reader to key articles by Jagoutz et al. (1979) and Palme and Nickel (1986) and recent reviews by Walter et al. (2000) and Righter et al. (2000).

A summary of siderophile-element concentrations and depletions are presented here in order to highlight differences between the Earth and Moon. Because the origin of the Earth and Moon are likely to be directly linked, the siderophile-element abundances for each body are compared and plotted in Figure 3.23 for the compatible elements and Figure 3.24 for the incompatible elements. The importance of these differences is discussed in the following paragraphs.

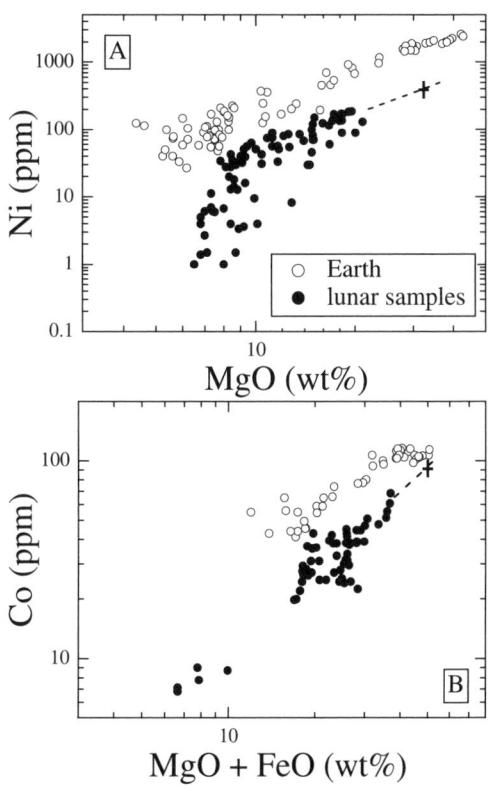

Ni and Co. Since both of these elements exhibit compatible behavior due to olivine and chromite fractionation, their concentrations in planetary mantles can be estimated by correlations with MgO and FeO (e.g, Wänke and Dreibus 1986) (see Fig. 3.23). Both Ni and Co for the lunar mantle are found to be approximately 5 times more depleted with respect to the terrestrial mantle.

Mo, W, Re. In many basalts from diverse planetary bodies, moderately siderophile elements (such as Mo, W, and Re) are incompatible and thus are positively correlated with other incompatible, refractory lithophile elements (such as Pr, Ba, Yb, La or Nd; see Fig. 3.24). This correlation line is

Figure 3.23. Compatible-element depletion diagrams for Ni-MgO and Co-(MgO+FeO) data. Data from Delano (1986a) and references therein. [Used by permission of Elsevier Science, from Righter (2002), *Icarus*, Vol. 158, Fig. 1, p. 2.]

found to be well below chondritic values, and this "depletion" is most likely due to metal-silicate equilibration (core formation) in that particular body (see the schematic illustration in Fig. 3.24a). The lunar mantle is substantially more depleted in Mo and Re than the terrestrial mantle (e.g., Newsom and Palme 1984; Newsom et al. 1996; Righter et al. 1998). However, the W depletions found in the lunar mantle are very similar to those found both in the terrestrial mantle and the eucrite parent body (Vesta) (e.g., Righter and Drake 1996), indicating that W may not be a very sensitive indicator of the pressure and temperature conditions of core formation.

P and Ga. Besides being siderophile, many elements such as P and Ga are also volatile. As a result of this, a measured depletion can be due to either core formation or volatility-controlled processes. In order to estimate the depletion of P and Ga due to core formation, these two elements must first be normalized to lithophile elements of nearly equal volatility. Good candidates for such a comparison are the alkali elements Na, K, Rb and Cs. Though it

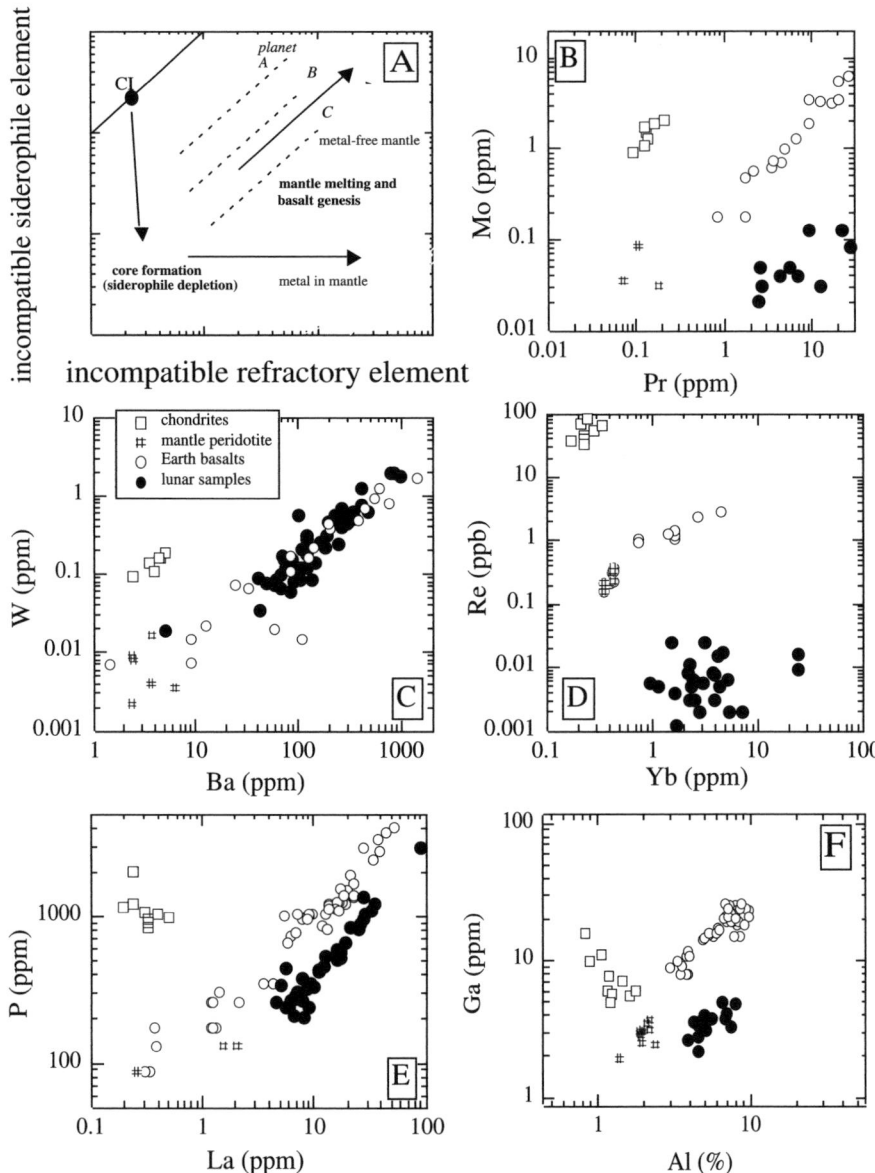

Figure 3.24. Siderophile-element depletion diagrams for Mo-Pr, W-Ba, P-La, Re-Yb, and Ga-Ti. Data sources: Lunar Mo and Pr: Taylor et al. (1971), Newsom and Palme (1984); terrestrial Mo and Pr: Newsom et al. (1986); Lunar P, La, Ga, Ti, W and Ba: BVSP (1981), Warren et al. (1986), Wänke et al. (1972), Wänke et al. (1974), Wänke et al. (1977), Palme et al. (1978); Terrestrial W and Ba: Newsom et al. (1996); terrestrial P and La: BVSP (1981), Newsom and Drake (1983) and references therein; Terrestrial Ga and Ti: Dickey et al. (1977), BVSP (1981); Norman and Garcia (1999), Frey et al. (1985). [Used by permission of Elsevier Science, from Righter (2002), *Icarus*, Vol. 158, Fig. 2, p. 3.]

has been recognized for some time that these elements are depleted in the lunar mantle relative to the Earth (e.g., Kreutzberger et al. 1986), Ga has been found to be no more depleted than Na in lunar highland feldspars (Norman et al. 1995). In fact, neither P nor Ga are depleted relative to the lithophile, volatile elements (Li, Na, K, Rb, Cs; see Fig. 3.25), suggesting that neither was depleted substantially by lunar core formation (e.g., Righter 2002).

Overall the siderophile elements show increasing depletion with increasing siderophility. Ni, Co, Mo and Re exhibit a further depletion over that in the Earth, whereas W depletions are similar to those in the Earth and eucrites. These observations argue qualitatively that the material that makes up the Moon has at some point undergone a metal-silicate differentiation event. We emphasize here, however, that this evidence does not necessarily imply that the Moon currently possesses a metallic core. In particular, if the giant-impact origin of the Earth-Moon system is correct, then the Moon is predicted to accrete from the material that was put in circumterrestrial orbit following this event. If the Earth and impacting body previously underwent a core-forming event, and if the core of the impacting body accreted to the Earth during this event, then a core-free Moon could form that nonetheless possessed the siderophile fingerprint of metal-silicate equilibration. In contrast, if a substantial amount of metallic iron was incorporated in the circum terrestrial disk during the giant impact, then it becomes possible that this material could have been reprocessed by a subsequent lunar core-forming event at lower pressures. Separating the consequences of these two putative core-forming events is not a simple matter, and depends upon knowing the composition and thermal history of the Moon, proto-Earth and impactor (Righter 2002).

4.8.2. Core formation models. Temperature, pressure, the fugacities of oxygen and sulfur, and silicate melt structure and composition all affect the partitioning behavior of an element between metal and silicate phases. Because of the relatively low pressures associated with the lunar core (~4 GPa), partition coefficients for typical pressures encountered at the core-mantle boundary have been obtained in the laboratory (Walker et al. 1993; Hillgren et al. 1994, 1996; Ito et al. 1998). From these experiments, metal/silicate partition coefficients have been parameterized for Fe, Ni, Co, Mo, W, P, Ga, Sn and Cu (Righter and Drake 1997, 1999, 2000). Using these expressions it is possible, for instance, to calculate metal/silicate partition coefficients along an adiabatic gradient in a deep magma ocean system (see Fig. 3.26). While the pressure at the base of a lunar magma ocean would only be ~4 GPa (or 40 kbar), if the Moon was derived from a much larger planet that collided with the Earth, the basal pressures of a magma ocean in this body could have been considerably greater.

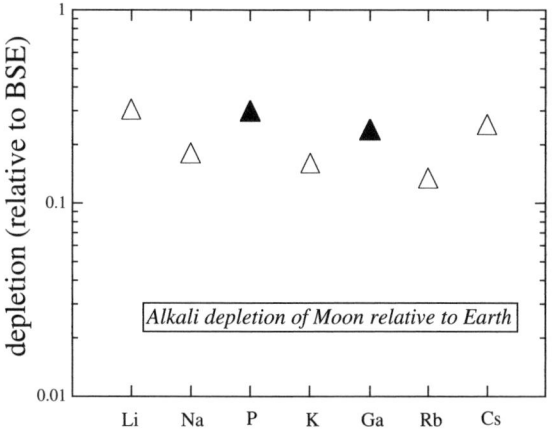

Figure 3.25. Depletion of P and Ga in the lunar mantle relative to the alkali elements, Li, Na, K, Rb and Cs, normalized to the composition of the bulk silicate Earth (data from Jones and Palme 2000). [Used by permission of Elsevier Science, from Righter (2002), *Icarus*, Vol. 158, Fig. 7, p. 9.]

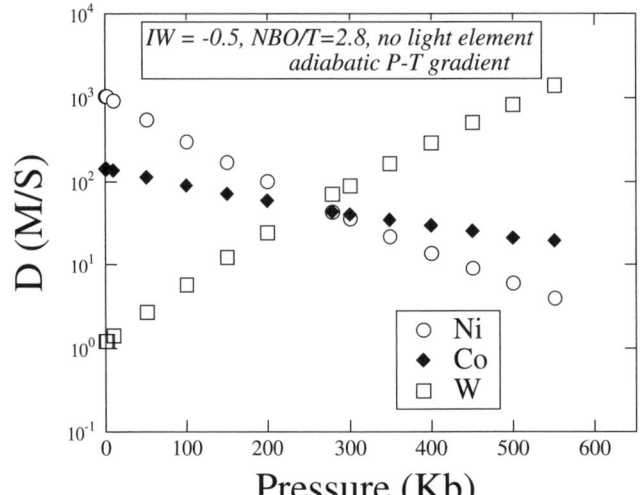

Figure 3.26. Effect of pressure and temperature on metal-silicate partition coefficients for Ni, Co and W using predictive expressions presented by Righter and Drake (1999). Conditions are an adiabatic temperature gradient with the oxygen fugacity fixed at 0.5 log f_{O_2} units below the IW buffer, a peridotite magma (nbo/t = 2.8), and no light elements in the metallic liquid.

Most previous work on lunar siderophile elements has concluded that the Moon has undergone metal-silicate equilibration, and therefore likely possesses a small metallic core. These models differ in the size of the core and its composition (e.g., Brett 1973; Wänke et al. 1977; Ringwood 1979; Drake et al. 1984; Newsom 1984, 1986; Ringwood and Seifert 1986; Wänke and Dreibus 1986; Hillgren 1991; O'Neill 1991; Righter and Drake 1996) and a summary of some of these models is presented in Table 3.15. Early models were based on the siderophile-element data reported by Wänke et al. (1977). Subsequent Mo and Re analyses of lunar materials were considered in the modeling of Newsom (1984), who accounted for the lunar Re and Mo depletions with a 2–5 lunar wt% metallic core if the Moon accreted directly from the solar nebula, or a 0.5–1 lunar wt% core if it formed from material coming from the terrestrial mantle. Later work by Ringwood and Seifert (1986), O'Neill (1991) and Hillgren (1991) all argued that the 3 times larger depletion of Ni in the lunar mantle in comparison to the terrestrial mantle could be explained by a Ni-rich lunar core. These conclusions were based primarily on the assumption that the Moon was made of material from the primitive upper mantle of the Earth. This latter assumption is now considered to be at odds with giant-impact models in which the Moon is predicted to be primarily derived from the object that collided with the proto-Earth.

While the thermal state of the Moon during and after its formation is not well known, it is likely that a large portion of it was at one time completely molten. Recognizing that metal-silicate partition coefficients change as both a function of pressure and temperature, Righter and Drake (1996) considered high pressure and temperature scenarios for the depletions of the lunar siderophile elements. They found that the siderophile-element depletions of the mare source region are consistent with a scenario in which a small metallic core (1–5 lunar wt%) equilibrated with an overlying molten mantle. Using a peridotite magma ocean, a best fit to the lunar depletions was obtained by using metal-silicate partition coefficients calculated for $T = 2200$ K, $P = 35$ kbar, $\Delta IW = -1$, and a core that contained 15 mol% sulfur. Again, these conclusions were based on the assumption that the Moon was derived from material like the primitive terrestrial mantle.

The issue of whether the Moon has a small metallic core has been re-examined by Righter (2002) in light of improved dynamical modeling of the giant-impact Moon-forming event and high temperature and pressure metal-silicate partition coefficients. In particular, he assumed that both the proto-Earth and impactor underwent an initial high-pressure core-forming event.

End-member initial conditions for these bodies assumed either that a molten core equilibrated with a totally molten mantle ("hot" initial conditions), or that metal equilibration occurred at the base of a magma ocean that coincided with the top of the perovskite stability field ("warm" initial conditions). Following the giant-impact event, the Moon was assumed to have formed exclusively from the mantle of either of these bodies. After the Moon accreted from this material, a second stage of metal-silicate equilibration was assumed to have occurred, but this time at the low pressures associated with the lunar molten-core/molten-silicate-mantle interface. Given reasonable compositions for the impactor and proto-Earth, a small metallic core (0.7–2 wt%) is predicted to have segregated and equilibrated with the lunar mantle following the giant impact event. The scenario in which the Moon is made from the mantle of a "hot" proto-Earth is the least likely of these models because the lunar mantle is predicted to be more depleted in W, P and Re than is observed. Discarding this scenario, the Moon is predicted to have an Fe-rich core that is 0.7–1 % of its mass. The results from this latest study eliminate previous geochemical objections to the Moon having a composition primarily derived from the impactor that collided with the proto-Earth.

While the recent modeling of Righter (2002) is consistent with the Moon possessing a small metallic core, he also investigated the possibility of whether core formation in the impactor alone could account for the observed lunar siderophile element abundances. In this scenario, if the core of the impactor completely accreted to the Earth, then the siderophile element data would not require the existence of a lunar core. It was found that if core formation in the impactor occurred at low temperature, pressure and f_{O_2} conditions, the abundances of Ni, Mo, P and Re could be accounted for. The abundances of Ga, Co, and W, however, were found to be slightly lower than observed. Given the uncertain composition, size, and thermal history of the impactor, it is possible that this might not be a fundamental objection to the hypothesis that the Moon does not possess an iron core.

Finally, we note that several factors have yet to be considered in the above models, including the kinetics of metal-silicate equilibration, the dynamics of metal-silicate differentiation, and the time scale of magma-ocean crystallization. In a study by Rubie et al. (2003) for the terrestrial magma ocean, it was shown that the timescale of metal-silicate equilibration between a layer of molten iron and an overlying silicate magma was about two orders of magnitude greater than the timescale of magma-ocean crystallization. This is particularly important because as soon as a solid silicate layer forms between the two, metal-silicate equilibration will effectively cease. If this result were applicable to the Moon, then metal-silicate equilibration would be favored to occur by the sinking of small metal droplets through a silicate magma ocean. This process, however, is sensitive to the dynamics of metal segregation from the silicate magma, and the "effective" pressure of equilibration could either over- or under-estimate the pressure at the core-mantle interface (Rubie et al. 2003).

4.8.3. Isotopic studies bearing on core formation. The short half life of ^{182}Hf (~9 My) as it decays to ^{182}W enables Hf-W isotope data to be used in examining planetary differentiation processes early in the history of the solar system. With Hf being lithophile and W siderophile, core formation should have dramatically increased the Hf/W ratio of the silicate mantle and reduced it in the core. If core formation occurred while ^{182}Hf was alive, differentiation would further produce positive and negative ε^{182}W anomalies relative to chondrites in the silicate mantle and core, respectively (Lee and Halliday 1995; Halliday et al. 1996; Jacobsen and Harper 1996). (ε^{182}W = $[(^{182}W/^{183}W)_{sample}/(^{182}W/^{183}W)_{standard} - 1] \times 10^4$). Early differentiation in the solar system is witnessed by negative ε^{182}W anomalies in metal from iron meteorites and extremely positive anomalies in the eucrites (Lee and Halliday 1997).

Early tungsten isotopic measurements of mantle-derived terrestrial basalts showed that while they did not possess distinct ε^{182}W anomalies (ε^{182}W~0) (Halliday et al. 1996; Halliday 2000), they did possess a suprachondritic Hf/W ratio. As chondritic materials were thought to have near-zero $\varepsilon^{182}W$ anomalies, these data were broadly interpreted to indicate that core formation for the Earth was "late" (>60 My). Even if a late veneer of chondritic material was added to the Earth, as may be required to account for the highly siderophile element abundances in the mantle (e.g., Morgan 1986; Morgan et al. 2001 though see Frost et al. 2004 for an alternative scenario), this would not be able to completely dilute a positive early-core forming $\varepsilon^{182}W$ signature back to zero (Halliday and Lee 1999). Recent studies, however, have found that the $^{182}Hf/^{180}Hf$ ratio at the start of the solar system was about three times lower than previously suggested, and this implies that the bulk silicate Earth has a resolvable ^{182}W excess of ~1.9ε units relative to chondrites (Kleine et al. 2002; Schoenberg et al. 2002; Yin et al. 2002). These data imply that core formation occurred early for the Earth, within 33 ± 2 My of solar system formation.

Apollo samples analyzed for W isotopes show $\varepsilon^{182}W$ values ranging from zero to more than +6, although the Moon appears to have an approximately chondritic Hf/W ratio (Lee et al. 1997). Because of the highest measured $\varepsilon^{182}W$ anomalies, it was recognized that these analyses needed to be corrected for cosmic ray-induced neutron capture on ^{181}Ta (e.g., Jones and Palme 2000). It was demonstrated that the measured ^{182}W excess in mineral separates from some lunar basalts (specifically the high-Ti basalts) correlated with their Ta/W ratios, and this confirmed the theoretical predictions that the $^{181}Ta(n,\gamma)^{182}Ta(\beta^-)^{182}W$ reaction due to cosmic irradiation was indeed the cause of some of the excess ^{182}W. Lee et al. (2002) demonstrated that although such a correction reduced the magnitude of the $\varepsilon^{182}W$ anomalies, it did not remove them. In addition to this process, some of the excess ^{182}W can be attributed to silicate fractionation effects, since ilmenite, clinopyroxene, and garnet all fractionate Hf from W (Righter and Shearer 2003; Shearer and Righter 2003). Model ages based on these data suggest that lunar core formation occurred ~53 ± 4 My after the formation of the solar system (Lee et al. 1997, 2002; Halliday and Lee 1999) and less than 70 My after solar system formation for the source of the high-Ti mare basalts (Lee et al. 2002). If the Earth and Moon possess identical Hf/W ratios, then the results of Kleine et al. (2002) suggest that core formation occurred earlier, between 24 and 35 My after solar system formation.

4.9. Joint inversions of several datasets

As has been highlighted in the preceding sections, the size, composition and physical state of the lunar core cannot be uniquely determined by the analysis of individual datasets in isolation. Each measurement can be interpreted in multiple ways, and both a metallic and dense molten-silicate core are defensibly permissible. One method that might further constrain the deep interior structure of the Moon would be to construct models that are not only consistent with a single measurement, but several. An investigation in this spirit has recently been performed by Khan et al. (2006) who jointly modeled the measured mass, moment of inertia, k_2 Love number, and tidal dissipation quality factor Q. As these four measurements are affected by the depth dependence of the lunar density, shear modulus, and local quality factor q, a joint inversion of these should be able to better resolve the attributes of the lunar core.

The approach taken in the aforementioned study was to first address the question as to whether the lunar core is either completely solid or molten, and then to use the most likely answer to place limits on its density and shear modulus. Assuming that the structure of the Moon could be adequately approximated by a series of five concentric homogeneous shells, a Monte Carlo method was used to sample values of the density, shear modulus, q, and thickness for each layer. The crust was restricted to have an average thickness between 30 and 60 km, and the shear modulus of the lowermost layer was constrained to be representative of either completely molten or solid materials. Using these sampled parameters, the mass, moment of inertia, Love number and tidal dissipation factor were computed, from which the misfit between the data and model was quantified. All results were then interpreted in terms of a Bayesian probabilistic framework.

Using the method of Bayesian hypothesis testing (by use of the Bayes factor), it was shown that the hypothesis of the Moon possessing a completely molten core is highly favored to that of a solid core by a very large factor. For the molten core scenario, its radius was found to have a nearly uniform probability distribution between about 70 and 450 km, and to be nearly zero exterior to this range. The probability of the core density was found to linearly increase from a value near zero at ~3.5 g cm^{-3} to a maximal value at a density near ~7.5 g cm^{-3} (densities greater than this were not sampled). While central densities typical of molten iron, an Fe-S eutectic melt, and a dense molten silicate core can be found that are consistent with the data, their results favor the denser side among these possibilities.

A few considerations should be borne in mind when interpreting these results. First, the value adopted for the k_2 Love number was taken from a recent LLR study (see Williams et al. 2005) that is smaller by one standard deviation than the independent determination based on radio tracking data of lunar spacecraft (Konopliv et al. 2001). Second, the above model results are based upon the premise that the Moon can be adequately described by five homogeneous layers. While this is a reasonable assumption, models with more layers are also defensible. In particular, if the Moon possesses a solid inner core (see Section 4.6), then a minimum of six homogenous layers would be required to describe its interior density and rheological structure (i.e., crust, upper mantle, middle mantle, lower attenuating mantle, liquid outer core and solid inner core; see Fig. 3.27). The inclusion of extra layers would, at a minimum, increase the variance of the solution parameters, and could affect the magnitude of the Bayes factor.

Finally, we note a possible difference in interpretation that could arise between Bayesian and non-Bayesian inverse approaches when the physical model is highly non-unique. Let us presume that the mass, moment of inertia, k_2 Love number, and quality factor of the Moon were perfectly known. (It is unlikely that higher precision measurements of at least the mass and moment of inertia would significantly improve models of the lunar interior.) As only four scalar numbers are being used to constrain how the lunar density, shear modulus, and quality factor varies with depth, a potentially large range of models might exist that could exactly fit these observations. In the Bayesian framework, the posterior probability distribution would be

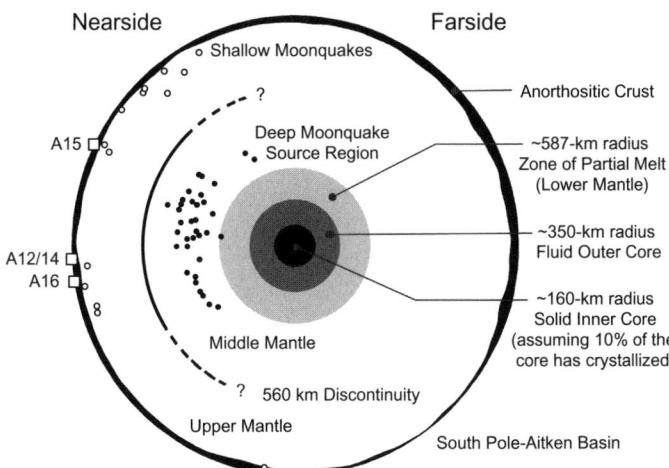

Figure 3.27. Schematic diagram of the internal structure of the Moon as revealed by various geophysical methods. The thickness of the crust is shown for a pole-to-pole profile at 0 and 180° longitude, and the shallow and deep moonquakes have been projected onto the nearside hemisphere as a function of depth and latitude. While the radius of a solid inner core is completely undetermined, thermal considerations suggest that at least some portion of the core has crystallized.

obtained simply by multiplying this model space by the prior probability distribution, as each model has the same zero misfit (i.e., the same likelihood, see Ulrych et al. 2001). In contrast to the situation where the likelihood function is somewhat peaked, the posterior probability distribution would be critically dependent upon the assumed prior probability distribution. A more conservative inverse approach might be to instead treat all physically reasonable models that satisfy perfectly known measurements as being possible solutions. If this criterion were to be used, it is unlikely that the data would be able to distinguish between the case of a pure liquid iron core and that of a dense molten silicate core.

4.10. Summary

The geophysical evidence presented above implies that the central portion of the Moon is molten (with or without a solid inner core), has a high electrical conductivity, and is denser than the overlying mantle. The precise composition of this "core," however, is not well constrained by these studies. Indeed, the moment of inertia, the k_2 Love number, the quality factor Q, and lunar rotation and magnetic induction studies are all consistent with the core being composed of either a molten Fe-FeS-C metallic alloy with a radius less than 375 km, or a molten silicate composition that is slightly larger. If the lunar core is composed of molten silicate materials, then this composition must be enriched in both iron and titanium in order to be denser than the overlying silicate mantle. Thermal considerations suggest that some portion of the core, regardless of its composition, should have crystallized over the past 4.5 Ga, thus predicting that the Moon currently possesses a solid inner core as well.

Although the above geophysical methods for inferring the properties of the lunar core are not very sensitive to its composition, two additional pieces of information could be used to favor one hypothesis over the other. Firstly, modeling of the lunar siderophile-element abundances is consistent with the Moon possessing an iron-nickel rich core with a radius between 260 and 293 km (Righter 2002). While this range of core sizes is consistent with the range of core radii as implied by the geophysical techniques, it remains possible that core-free models could be constructed that are consistent with the measured siderophile-element abundances. Secondly, the paleomagnetic signature of the lunar samples, as well as the crustal fields as measured from orbit, imply that the magnetic field intensity of the lunar environment was once much greater than is currently observed. While this might be taken as being consistent with the Moon once possessing a dynamo (which would require the core to be metallic), these field strengths might also be consistent with transient magnetic fields generated during impact events. Indeed, the largest magnetic anomalies observed from orbit appear to be associated with the antipodes of the largest basins. Current dynamo scaling relations further suggest that a convecting iron core would not be able to generate the magnetic field intensities that are required by the paleomagnetic data.

A schematic illustration of what is known of the internal structure of the Moon is presented in Figure 3.27. This figure illustrates (1) the lateral variations in crustal thickness that are the result of impact cratering events (i.e., the SPA basin on the farside), (2) the existence of a seismic discontinuity at a depth of ~560 km, (3) the spatial distribution of the deep and shallow moonquakes, (4) the likely existence of a partially molten region beneath a depth of 1150 km, (5) and the existence of a molten outer core with a radius of ~350 km. As some portion of the liquid core is likely to have crystallized, a solid inner core is most likely present as well. An arbitrary core crystallization of 10% by volume would result in an inner core radius of about 260 km. While the 560-km seismic discontinuity may be a global feature of the Moon, this presumption cannot be tested using the limited spatial distribution of the Apollo seismic network.

5. CRUSTAL PROVINCES

The lunar maria and highlands have long been known to be fundamentally different based upon contrasts in their albedo, morphology, and spectral properties. For this reason it was, and still is, common to broadly divide the lunar surface into these two distinctive terrains—where the term *terrain* refers to "an area of ground with a particular physical character" (Allaby and Allaby 1991). With the collection of geophysical and remote-sensing data from orbit, as well as the analysis of surface samples, it has become apparent that there are limitations with this mare/highland classification scheme. Equatorial gamma-ray data obtained from orbit during the Apollo missions showed that the highest abundances of thorium, and by inference KREEP, are in the Procellarum and Imbrium regions on the lunar nearside (e.g., Metzger et al. 1973). Laser altimetry data has shown that the Moon possesses a 2-km center-of-mass/center-of-figure offset (e.g., Kaula et al. 1972; Zuber et al. 1994), suggesting that the farside crust might be significantly thicker than that of the nearside. Furthermore, analyses of the Apollo samples suggest that the eastern and western landing sites are significantly different in composition (e.g., Shervais and Taylor 1986). Although the Apollo data were limited in coverage, some workers have suggested, or hinted at, the possibility that differences in geologic evolution between the near and far sides of the Moon may be more fundamental than the traditional dichotomy between the maria and highlands (e.g., Wasson and Warren 1980; Warren and Rasmussen 1987; Ryder 1994; Haskin 1998).

Global gamma-ray and neutron-flux data from the Lunar Prospector mission have since convincingly shown that the surface concentrations of Th and other incompatible elements are highly localized only within the Imbrium and Procellarum regions (Lawrence et al. 1998, 2000; Elphic et al. 2000). This distribution has led to the idea that the Imbrium-Procellarum region is a unique geochemical crustal province that possesses a distinctive geochemistry and thermal evolution (e.g., Haskin et al. 2000; Jolliff et al. 2000a; Korotev 2000; Wieczorek and Phillips 2000). Jolliff et al. (2000a) have further advocated dividing the Moon into three major crust-mantle *terranes*: the Procellarum KREEP Terrane (PKT), the Feldspathic Highlands Terrane (FHT), and the South Pole-Aitken Terrane. In the terrestrial vernacular, a terrane is defined as a "fault bounded area or region which is characterized by a stratigraphy, structural style, and geologic history distinct from those of adjacent areas, and which is not related to those areas by unconformable contacts or facies changes" (Allaby and Allaby 1991). Although the fault-bounded aspect of this definition does not apply to the Moon because of its lack of plate tectonics, as will be seen in the following sections, the remainder of this definition is probably an accurate description for at least the Procellarum KREEP and Feldspathic Highlands Terranes. Though the surface properties of the region encompassed by the South Pole-Aitken basin are unique for the Moon, we consider it likely that this region simply represents an exposure of lower crustal rocks that may well be typical of the Feldspathic Highlands Terrane.

In this section, we first give a general description of the major terranes that have been proposed by Jolliff et al. (2000a). Our discussion here closely follows this paper as well as that of Wieczorek and Phillips (2000), Haskin et al. (2000) and Korotev (2000). Finally, using the inferred compositions of these lunar terranes, we place constraints on the bulk composition of the Moon.

5.1. The Procellarum KREEP Terrane

The Procellarum KREEP Terrane differs from other regions of the Moon primarily by its KREEP-rich geochemistry and extended volcanic history. Data from the Lunar Prospector mission shows that the surface abundance of incompatible elements (Th, K, Gd, Sm, and by inference the other elements that make up KREEP) are highly concentrated in a single region that encompasses Oceanus Procellarum, Mare Imbrium, and the adjoining mare and highlands (Lawrence et al. 1998, 2000; Elphic et al. 2000). It is by this surface enhancement of KREEP that the confines of this terrane are tentatively delineated in Color Plate 3.10. Whereas the mare exterior to the PKT (as derived from Lunar Prospector data) have Th abundances of only ~2 ppm or less, the mare regolith within the PKT has concentrations that lie between ~3 and 7 ppm. Highland regions that reside in the PKT are inferred to be dominated by Imbrium ejecta and possess Th abundances that range up to ~12 ppm, consistent with samples that are believed to be derived from this impact basin (e.g., Korotev 1998).

Another defining characteristic of the Procellarum KREEP Terrane is its unique volcanic history. Outside of this terrane, the majority of mare basalts erupted within the confines of large impact basins where the low-density upper crust has been substantially thinned. In contrast to this behavior, Oceanus Procellarum represents the largest contiguous expanse of mare basalts, even though this region possesses a relatively thick crust and is not in an obvious basin setting (e.g., Wieczorek et al. 2001). While the PKT encompasses only about 16% of the surface area of the Moon, more than 60% of the mare basalts by area are located there. The only recognized lunar volcanic complexes are found within Oceanus Procellarum (e.g., Marius Hills, the Aristarchus Plateau, Rümker Hills, and Hansteen Alpha; see Whitford-Stark and Head 1977, 1980; Wilhelms 1987; Hawke et al. 2003a), and the vast majority of the young Eratosthenian and Copernican basalts appear to have erupted close to these complexes. Furthermore, even though only ~20% of the maria have TiO_2 abundances greater than 5 wt%, more than half of the basalts within the Procellarum KREEP Terrane have TiO_2 abundances that are at least this great (Haskin et al. 2000). The youngest lunar basaltic lava flows erupted within Oceanus Procellarum, possibly as recently as 900 ± 400 My ago (Schultz and Spudis 1983), and the oldest lunar basalt that has been dated (~4.2 Ga) came from a clast within an Apollo 14 breccia that was likely derived from this terrane (Taylor et al. 1983). Most basaltic eruptions exterior to this terrane, in contrast, occurred primarily during the Imbrian period.

Although Lunar Prospector has demonstrated that the Procellarum KREEP Terrane has a high surface abundance of incompatible elements, we caution that the gamma-ray and neutron spectrometers only sense the upper meter of the regolith. A major question is whether the KREEP component within this terrane is solely surficial (as in a veneer of KREEP-rich mare basalts and Imbrium ejecta), or whether it extends deep into the underlying crust and/or mantle. In either case, the KREEP-rich geochemistry and extensive volcanic history of this terrane suggest that these two phenomena are genetically related (Wieczorek and Phillips 2000). In the following two sections we first present evidence that a large portion of the crust within the PKT is enriched in incompatible elements. Following this, we consider whether the mare basalts within this terrane have high abundances of incompatible elements, or whether this is merely the result of vertical mixing with an underlying KREEP-rich substrate. Finally, we speculate on the origin and evolution of this unique region of the Moon.

5.1.1. Crustal composition. On the basis of Apollo gamma-ray data, Haskin (1998) made the suggestion that the Imbrium basin may have formed in a unique KREEP-rich geochemical province and that the primary ejecta from this basin should hence be enriched in incompatible elements. By modeling the deposition of KREEP-rich Imbrium ejecta onto a KREEP-poor substrate, he argued that the distribution and concentration of thorium exterior to this province was consistent with having an Imbrium origin. In particular, the concentration of thorium in the highlands was found to gradually decrease with distance from the Imbrium basin and then to slightly increase near its antipode. Global thorium data obtained from Lunar Prospector have since confirmed this prediction (Lawrence et al. 1998). It is important to note that this distribution of Th would not be obtained if the Imbrium bolide impacted a target that only possessed a near-surface KREEP enhancement. Instead, quantitative modeling seems to require an average crustal composition that lies somewhere between 4 and 9 ppm Th (Haskin et al. 1999).

The uniqueness of the Imbrium target composition is illustrated by a comparison with the composition of ejecta from other large lunar impact basins. With the possible exception of Serenitatis (Wieczorek and Zuber 2001b), which is adjacent to Imbrium, no other impact basin possesses a Th-enriched ejecta blanket. Because other large impact basins are known to have excavated deep into the crust (e.g., Wieczorek and Phillips 1999), this implies that only the crust near the Imbrium basin is significantly enriched in incompatible elements. As the confines of the Procellarum KREEP Terrane are not symmetric about the Imbrium basin, the ejecta from this basin cannot be the sole source of the Th enhancement within this terrane.

Modeling of the thickness of Imbrium's ejecta blanket has shown that a significant quantity of Imbrium-derived materials should be present at each of the Apollo landing sites (Haskin 1998; Haskin et al. 2003). On the supposition that only ejecta from this basin would be rich in incompatible elements, Haskin et al. (1998) have reevaluated the old hypothesis that all KREEP-rich impact melts in the Apollo sample collection have an Imbrium origin (e.g., Evensen et al. 1974; Tera et al. 1974; Reid et al. 1977; Schaeffer and Schaeffer 1977). If this were indeed true, then these samples would place a strict constraint on the composition of the Imbrium target. As an example, Korotev (2000) has shown that the composition of these mafic impact melts can be modeled to a good approximation as being a mixture of three components: (1) typical feldspathic upper crustal materials (82.5 vol% plagioclase or 28.4 wt% Al_2O_3), (2) a material very similar in composition to Apollo 15 KREEP basalt, and (3) highly forsteritic dunite (Fo_{90}). Though there exists considerable variability in the composition of the mafic impact-melt rocks, the average proportion of these components is 58% KREEP basalt, 29% feldspathic upper crust, and 13% dunite. As a basin the size of Imbrium should have excavated through the entire crust (e.g., Wieczorek and Phillips 1999), the dunite component may represent the upper mantle in this locale (Korotev 2000).

If the above model results can be extrapolated to the rest of the Procellarum KREEP Terrane, and if the dunite component has a mantle origin, then about 67% of the crust within this terrane has a composition similar to that of KREEP basalt, or its differentiated equivalent. Since KREEP basalt has an average Th concentration of about 12–13 ppm, this implies that a large portion of the incompatible elements of the Moon (~30%, Jolliff et al. 2000a) are sequestered within this relatively small region of the crust. The origin of the feldspathic upper-crustal component in these impact-melt breccias is less certain. One possibility is that this component was similarly derived from the Imbrium target. Under this scenario, the crust of the PKT might best be described as possessing compositionally distinct feldspathic and KREEP basaltic materials, perhaps stratified into the upper and lower crust, respectively. Consistent with this interpretation is the observation that many central peaks in this terrane have broadly anorthositic compositions (~60–90 vol% plagioclase, see Tompkins and Pieters 1999 and Section 2.8). We caution, however, that spectral studies cannot distinguish between ferroan and alkali lithologies.

Petrological studies of the mafic impact melts, however, suggest that ferroan-anorthositic materials might have been absent from both the Imbrium (Spudis et al. 1991) and Serenitatis targets (Ryder et al. 1997). If this is true, then the feldspathic upper-crustal component in these melt rocks as inferred by Korotev (2000) might instead represent locally derived substrate that was melted and mixed with the primary ejecta of these basins. In either case, this feldspathic chemical component would have been present within the confines of the Procellarum KREEP Terrane in some form or another. Alternatively, it is possible that the three-component mixing model of Korotev (2000) overly simplifies the composition of the Imbrium target, and that anorthositic materials may truly be absent from this region of the crust.

5.1.2. Viscous relaxation of impact basins. In support of the hypothesis that the Procellarum KREEP Terrane possesses a high abundance of heat-producing elements, of the two sites where the lunar heat flow was determined (Apollo 15 and 17) the heat flow of the site within the PKT was found to be the larger (Langseth et al. 1976). Unfortunately, these two landing sites are situated close to the boundary of this terrane and it is not certain if these measurements are representative of either the PKT or FHT. Nevertheless, other pieces of evidence seem to suggest that the crust was hotter than typical within this region, causing impact-induced topography to viscously relax faster than in the surrounding highlands.

The size and shape of the excavation cavity of the young large nearside impact basins was investigated by Wieczorek and Phillips (1999) using a gravity-derived crustal thickness model. In concordance with craters that are orders of magnitude smaller in size (e.g., Croft 1980; O'Keefe and Ahrens 1993), it was found that the maximum depth of excavation was approximately 1/10th of their transient crater diameter. The Imbrium and Serenitatis basins, however, were found to be anomalous in that their reconstructed excavation cavities were tens of kilometers shallower than expected. One likely explanation for this observation is that these basins originally formed in concordance with the general scaling relationship, but that they were modified during, or after, the impact event. One possible form of post-impact modification is viscous relaxation, whose rate is highly dependent on temperature. Since the Imbrium basin lies within the confines of the Procellarum KREEP Terrane, and Serenitatis is just on its border, a higher than typical crustal temperature at these locales is a reasonable expectation as a result of the high abundance of radioactive elements found there. It is also possible that the apparent shallow structure of these basins is a result of their having being filled by lavas that possess a density similar to that of the lower crust. The only volcanic rock that meets this criterion is KREEP basalt (mare basalts have densities of ~3.3 g cm^{-3} which is similar to the underlying mantle). While rare in the sample collection, KREEP basalts have been identified at the Apollo 15 and 17 sites. The composition of the mafic impact-melt breccias further implies that KREEP basalt may be a major chemical component within the crust of this region.

The second piece of evidence for higher crustal temperatures in the PKT comes from the isostatic state of large impact basins. As originally noted by Neumann et al. (1996), the inferred crustal structure of the large impact basins suggests that some of these are currently superisostatic (i.e., the crust-mantle interface has been uplifted above its isostatic level). This appears to be true even when the load of the mare basalts is accounted for (Wieczorek and Phillips 1999), and the existence of mascon basins that do not possess significant mare fill is similarly suggestive of their being in a superistostatic state (Konopliv et al. 1998, 2001). Only those basins that formed near the PKT (Imbrium, Serenitatis, Grimaldi, and Humorum) appear to have been in an isostatic state prior to the emplacement of their mare fill (Wieczorek and Phillips 1999), consistent with the suggestion that viscous stress relaxation was important only within the Procellarum KREEP Terrane.

The final piece of evidence for significant viscous relaxation within the PKT is a qualitative one. Haskin et al. (2000) recognized that there are very few highland structures that breach lava flows within this region, which is somewhat surprising as the thickness of the mare is inferred to be relatively thin (see Section 2.4). While these lavas would completely cover some ancient impact structures, if the pre-mare topography within the PKT were similar to that of the surrounding highlands, then highland exposures would still be expected. Since this is not the case, they reasoned that this initially high standing surface topography might have viscously relaxed as a result of high crustal temperatures.

5.1.3. Are the Procellarum basalts thorium rich? The evidence presented in the previous sections seems to suggest that a large portion of the crust within the Procellarum KREEP Terrane is enriched in incompatible elements. Lunar Prospector gamma-ray data further show that the upper meter of the mare regolith in this region has a moderate Th enhancement as well (~3 to 7 ppm in the most FeO-rich regions). Somewhat surprisingly, though, almost all of the crystalline mare basalts in the Apollo sample collection have Th abundances less than ~2 ppm, including those from the Apollo 12 mare site that lies within this terrane (e.g., Jolliff et al. 2001). All of the basaltic lunar meteorites for which data exist similarly have a low abundance of Th (recall that more than 60% of the mare basalts by area erupted within the 16% surface area of the PKT). A major question is thus whether some of the basalts in Oceanus Procellarum are enriched in a KREEP component, or if the surface enrichment as observed from orbit is merely the result of vertical mixing with an underlying KREEP-rich substrate. The resolution of this question has very fundamental implications for lunar magmatic processes. If it turns out that these basalts are indeed enriched in KREEP, then either the mare source region was

KREEP-rich as well, or these basalts assimilated a substantial quantity of incompatible elements as they traversed the crust. Alternatively, if the basalts in this region are similar to those collected by Apollo, then the heat source for their generation may have been from the overlying crust (Wieczorek and Phillips 2000).

Either side of this debate can be reasonably defended (see Haskin et al. 2000; Jolliff et al. 2000a; Wieczorek and Phillips 2000), and it may turn out that reality lies between these two extremes. Evidence in favor of the eruption of Th-poor basalts in this region includes the apparent paucity of Th-rich mare basalts in the Apollo, Luna, and lunar meteorite sample collections. Perhaps more telling, though, is the nature of samples returned from the Apollo 12 landing site. Orbital gamma-ray data imply a surface Th concentration of ~6 ppm for this site (Jolliff et al. 2000b), which is consistent with the soils that were collected there (see also Chapter 2). However, these soils were found to represent a mixture between locally derived Th-poor mare basalts and a KREEP-rich nonmare component (e.g., Hubbard and Gast 1971; Hubbard et al. 1971; McKay et al. 1971; Meyer et al. 1971; Evensen et al. 1974; Jolliff et al. 2000a). As the mare basalts at this locale are relatively thin, it is possible that local impact craters could have excavated this KREEP-rich component from the underlying crust (e.g., Wasson and Baedecker 1972), or incorporated material through vertical mixing of buried ejecta horizons produced by large nearby craters such as Lansberg and Reinhold (Jolliff et al. 2000b). Part of the KREEP component at the Apollo 12 site might also be derived from a ray of Copernicus (Wentworth et al. 1994).

Alternatively, two pieces of evidence give support to the view that the mare basalts in the Procellarum KREEP Terrane possess high abundances of incompatible elements. Firstly, remotely sensed data indicate that some of the basalts in this region have both high concentrations of FeO (~20–22 wt%) and Th (~5 ppm). Such a composition could in principle be derived by mixing atypically high FeO basalts (>23 wt%) with KREEP-rich materials such as monzogabbro (QMD) and/or granite (35–40 ppm Th). However, QMD and granite are rare in the Apollo sample collection and it is unlikely that they are abundant in any significant quantity underlying the expansive mare basalts in Oceanus Procellarum. Secondly, although crystalline KREEP-rich mare basalts are not common in the sample collection (with the exception of the Apollo 11 high-K group (3.3–4.8 ppm Th, see Section 3.1), this fact can not be used to conclude that basalts with higher concentrations are not common elsewhere. Some Th- and FeO-rich volcanic and impact glasses with basaltic compositions do occur in the Apollo samples (Delano et al. 1982; Jolliff et al. 1991b) suggesting that this putative crystalline rock might be more common than the Apollo and lunar meteorite collections imply.

5.1.4. Origin and evolution of the Procellarum KREEP Terrane. The enrichment of incompatible elements and the unique volcanic history of the Procellarum KREEP Terrane bear strongly on the origin and evolution of this province. In particular, the observation that KREEP-rich materials are only abundant within this region of the Moon seems to suggest that the final dregs of the lunar magma ocean ultimately accumulated within this region. While it is difficult to dispute that this happened, the process by which this occurred is not forthcoming (see Chapter 4). One possible explanation is that the crust in this region might have been thinner than typical, perhaps because of a now unrecognizable Procellarum impact event. Hydrostatic forces would then have caused a KREEP-rich magma sandwiched between the crust and mantle to accumulate beneath this region (Warren 2001). Alternatively, the last remaining dregs of the lunar magma ocean may have accumulated above a degree-1 downwelling of dense ilmenite cumulates (Parmentier et al. 2002).

If the mafic impact-melt breccias are representative of the crust within the PKT, and if we take the mixing-model results of Korotev (2000) literally, then one third of the crust in this region is composed of anorthositic upper crustal materials, with the remainder being of a composition similar to Apollo 15 KREEP basalt. Assuming an average crustal thickness of

40 km for this region, this translates to the equivalent of ~13 km of feldspathic upper crust and ~27 km of KREEP basalt. As a result of the high abundance of incompatible elements in KREEP basalt (~12.4 ppm Th), this material could have remained partially molten up until at least the time of the Imbrium impact (e.g., Solomon and Longhi 1977; Wieczorek and Phillips 2000; Hess and Parmentier 2001). This is consistent with the observation of Ryder (1994) that the Imbrium impact event induced KREEP-basaltic volcanism in this region, most likely by denuding a subcrustal KREEP-basalt magma chamber. The slow crystallization of this subcrustal KREEP-basalt layer could have given rise to the Mg- and alkali-suite rocks that were most likely exhumed by the Imbrium impact event (e.g., Snyder et al. 1995a,b). Additionally, the large range of crystallization ages for these rocks (~3.9 to 4.4 Ga e.g., Nyquist and Shih 1992; Snyder et al. 2000) are consistent with such a protracted period of crystallization. As the Imbrium basin was the last basin forming event to have excavated deeply into the crust of this region, we would not expect to find Mg- and alkali-suite rocks any younger than the ~3.84 Ga age of this basin. Indeed, it is very likely that Mg- and alkali-suite plutonism extended beyond ~3.9 Ga.

The high abundance of radioactive elements in the Procellarum KREEP Terrane also appears to be related to the voluminous and protracted volcanic history of this province. Wieczorek and Phillips (2000) demonstrated that the inferred heat production of this province was more than sufficient to melt the underlying mantle over the duration of observed mare volcanism. Similarly, Parmentier et al. (2002) have shown that if the magma-ocean ilmenite cumulates sank beneath the PKT, carrying with them a significant quantity of incompatible elements, melting of the mantle beneath this region would naturally occur as well. These models are discussed further in Chapter 4.

Finally, we note that extensive melting of the mantle beneath the Procellarum KREEP Terrane would have drastically altered its composition. In the most extreme case, massive melting would leave a residuum composed solely of olivine, with its Mg-number depending upon the degree of melt extraction. This appears to be consistent with the inference of Korotev (2000) that the mantle beneath this region is composed of a highly forsteritic dunite. Alternatively, it is possible that this high-magnesium dunite could be related to a possible overturn of magma-ocean cumulates (Hess and Parmentier 1995). Figure 7.5 (in Chapter 7) shows an interpretative and highly idealized cross section of the crust and mantle beneath the Procellarum KREEP Terrane.

5.2. The Feldspathic Highlands Terrane

The Feldspathic Highlands Terrane differs from the Procellarum KREEP Terrane primarily by its ancient age, feldspathic composition, limited post magma-ocean magmatism, and low abundances of incompatible elements (Jolliff et al. 2000a). Other representative characteristics include both its high albedo and extensively cratered terrain. The confines of this terrane are based primarily upon the Clementine-derived iron abundance data and, as is shown in Color Plate 3.10, span more than 60% of the surface area of the Moon. In contrast to the PKT, less than 10% of this area is covered by mare basalts, and those mare deposits that are present erupted primarily during the Imbrian period (Wilhelms 1987). With the exception of these minor mare deposits, the feldspathic Apollo samples and lunar meteorites that make up this terrane have crystallization ages of about 4.4–4.5 Ga (e.g., Nyquist and Shih 1992).

Based on surface FeO abundances, Jolliff et al. (2000a) have distinguished between a central anorthositic portion, and an outer slightly more mafic region of the Feldspathic Highlands Terrane. The central anorthositic portion has some of the lowest FeO abundances on the Moon, averaging ~4.2 wt% FeO, and is centered on the farside near 40°N and 180°E where the crust is inferred to be its thickest (see Section 2.7). Flanking this central region to both the west and east, the outer portion of this terrane is seen to be slightly more mafic, having an average FeO abundance of ~6 wt%. Since the outer portion of the Feldspathic Highlands

Terrane is inferred to have a thinner crust than the central portion, the ejecta blankets of large impact basins in the outer region are likely to contain a greater quantity of material derived from the lower crust, which is expected to be somewhat more mafic.

The composition of the upper crust of the Feldspathic Highlands Terrane has been estimated from both sample and remote sensing studies. Firstly, an average of the five most feldspathic meteorites, which are presumably derived from this terrane, imply that the upper crust is composed of ~28 wt% Al_2O_3, or 82.5% plagioclase (Korotev 2000). Secondly, based on the Clementine-derived ~4.2 wt% FeO abundance for the central portion of this terrane, as well as an empirical correlation between FeO and Al_2O_3 in the Apollo sample collection (Korotev 1999), Jolliff et al. (2000a) estimated an Al_2O_3 abundance of ~29 wt%. Finally, based on the composition of central peaks, which are inferred to have an upper-crustal origin beneath a slightly more mafic surficial regolith (see Section 2.8.4; Wieczorek and Zuber 2001a), the abundance of Al_2O_3 is constrained to lie between 28.5 and 32.2 wt%, corresponding to a modal abundance of 87 ± 4% plagioclase.

The lower portion of the Feldspathic Highlands Terrane is likely to be more mafic than its upper portion. This inference is supported primarily by the somewhat more mafic composition of ejecta from large impact basins (e.g., Bussey and Spudis 2000), the mafic compositions of some central peaks (e.g., Tompkins and Pieters 1999) and the noritic floor of the South Pole-Aitken basin (e.g., Pieters et al. 2001). Based on central peak compositions and crustal thickness modeling, the lower portion of the crust appears to have a high abundance of noritic materials and a relatively low abundance of gabbroic materials (see Section 2.8.4, Wieczorek and Zuber 2001a). This implies that the lower portion of crust in this terrane does not contain a significant quantity of mare basaltic intrusions, which is consistent with the scarcity of basaltic lavas found there. The most-mafic portion of the lower crust from this study is inferred to have a composition of 18.2–24.7 wt% Al_2O_3, corresponding to a modal abundance of 65 ± 8% plagioclase. Figure 7.4 (in Chapter 7) shows an interpretative and highly idealized cross section of the crust and mantle beneath the Feldspathic Highlands Terrane.

5.2.1. South Pole-Aitken basin. The South Pole-Aitken basin on the lunar farside is unique in terms of its surface composition and geophysical character, and it looms large in debates about the structure of the lunar crust. Although its precise age is not known, superposition relationships mark it as the oldest of the topographically well-preserved lunar impact basins. With a 2100–2500 km diameter and 12–14 km total vertical relief (Spudis et al. 1994; Wieczorek and Phillips 1999), it is the largest recognizable impact structure in the solar system.

The first studies using Clementine derived FeO and TiO_2 concentrations (Lucey et al. 1998b) indicated that upper-mantle materials were likely to be a part of the mixture making up the basin floor. However, absolute FeO abundances based on Lunar Prospector gamma-ray data (which lie between ~6 and 12 wt% FeO) have since shown that the Clementine derived abundances for this basin were overestimated. This is likely a result of this region possessing a mineralogy that differs from the Apollo and Luna sites, which were employed in calibrating the Clementine-based FeO determinations. In particular, it appears that higher proportions of orthopyroxene are present within this basin when compared to most of the Apollo sites where clinopyroxene and olivine are more significant (see Chapter 2). Investigations of South Pole-Aitken basin rock types using combined remotely sensed mineralogy and compositions derived from Lunar Prospector data indicate that the basin, despite its great size, probably penetrated mainly into a noritic lower crust (Pieters et al. 1997, 2001) and that if mantle rocks are present, or incorporated into the melt sheet, they are not abundant. However, the precise mineralogy of the basin floor is still debatable, as mineralogical mapping using Clementine multispectral data imply that clinopyroxene might be significantly more abundant than previously thought (Lucey 2004). If true, this may be suggestive of a widespread basaltic component mixed with noritic lower crustal materials.

Questions remain as to the nature of the lower crust of the Moon and its relationship to the materials exposed in the South Pole-Aitken basin. While this basin clearly differs from the Th-rich Procellarum KREEP Terrane, its abundance of Th is slightly enhanced compared to the surrounding highlands (~2–3 ppm vs. <1 ppm). Whether the exposed crust there is typical of the deep crust of the Feldspathic Highlands Terrane is not known, but we think that this is likely for several reasons. Firstly, geophysical crustal thickness modeling suggests that this basin did not excavate all the way into the mantle as approximately 25 km of crustal materials are predicted to underlay the floor of this basin (see Section 2.7). Secondly, the high-standing topography and inferred thickened crust northeast of this basin has been interpreted as representing the ejecta from an oblique SPA impact (Zuber et al. 1994; Schultz 1997). As this putative ejecta deposit has some of the lowest iron abundances on the Moon, it is unlikely that there is a significant mantle component within it (Wieczorek and Phillips 1999). Thirdly, the dual-layered crustal thickness model presented in Section 2.7 implies that the floor of this basin should be almost exclusively composed of lower crustal materials. And finally, the central peak of Bullialdus, which is inferred to have a lower crustal origin (see Section 2.8.4), has a composition similar to the floor of this basin.

While we can not dismiss the possibility that this basin might have excavated into the mantle (e.g., Lucey et al. 1998b), we think that the evidence is currently against this hypothesis (see Lucey 2004). Nevertheless, even if the material that is exposed in this basin is derived largely from the lower crust, we recognize that this impact event would likely have reprocessed this material in some manner. In particular, the floor of this basin could represent a giant impact melt sheet that might have partially differentiated (Morrison 1998). Even though the mantle was probably not excavated in this impact event, it is still possible that impact melting of the underlying mantle may have occurred.

5.2.2. Origin and evolution of the Feldspathic Highlands Terrane. The feldspathic composition and ancient age of materials that comprise the Feldspathic Highlands Terrane is consistent with it having a magma-ocean flotation-cumulate origin. Unlike the Procellarum KREEP Terrane, which has been heavily modified by post-magma-ocean magmatism, extrusive and intrusive volcanic products are comparatively rare within this terrane. Furthermore, the duration of mare volcanism within the Feldspathic Highlands Terrane seems to have been limited primarily to a relatively short period of time within the Imbrian period. Because this terrane dates to the time of primary differentiation, its surface has been affected heavily by numerous large impact events, including the event that created the giant South Pole-Aitken basin. Regardless, this terrane probably represents the best example of a primary differentiation crust within our solar system.

The available evidence seems to suggest that the crust of the Feldspathic Highlands Terrane becomes more mafic and noritic with depth. Furthermore, if the floor of the South Pole-Aitken basin is representative of lower crustal materials, then thorium abundances grade from <1 ppm in the upper crust to ~2–3 ppm in the lower crust. A major question regarding this terrane is the origin of this compositional zonation, and several hypotheses have been advanced to explain this observation. One hypothesis advocated by Head and Wilson (1992) is that the mafic character of the lower crust could be a consequence of numerous basaltic intrusions. However, as gabbroic lithologies are relatively rare in the lower portion of the crust (<5% by volume, see Section 2.8.4), this interpretation is probably not correct. Many investigators have also suggested that the deep crustal rocks of this terrane could be composed of Mg-suite plutonic rocks (e.g., Reid et al. 1977; Ryder and Wood 1977). However, most Mg-suite rocks appear to be genetically related to the petrogenesis of KREEP, and these rocks are most likely to reside exclusively within the Procellarum KREEP Terrane (e.g., Jolliff et al. 2000a; Korotev 2000; Wieczorek and Phillips 2000).

An alternative explanation for the mafic character of the lower crust is that this might simply be a result of cumulate flotation in a chemically evolving lunar magma ocean (Wieczorek and Zuber 2001a). In particular, as the magma ocean crystallized, it should have become both increasingly FeO rich and dense (Warren 1990). In order for an assemblage of minerals to float in this magma and accrete to the underlying crust, its bulk density must be less than that of the surrounding magma. Thus, as the residual magma ocean became more dense, and the crust grew downward, the proportion of mafic cumulates that could be incorporated into floating mineral assemblages would have increased. Furthermore, since the magma ocean would have become increasingly enriched in incompatible elements as crystallization proceeded, the deepest flotation cumulates would be expected to contain a greater quantity of thorium and other incompatible elements.

Finally, it is possible that the observed chemical zonation of the crust could have been achieved by the mechanical separation of feldspathic and mafic phases after the crust was largely formed. As a result of high crustal temperatures early in lunar history, if anorthosites were present in extensive discontinuous bodies, they could have buoyantly rose through the crust while their mafic complements would have sunk (e.g., Longhi 1980; Longhi and Ashwal 1985).

5.3. Bulk composition: Th and Al mass balance

Based upon a number of assumptions that were not too controversial at the time, Taylor (1982) argued that the Moon is enriched in refractory elements when compared to the terrestrial primitive mantle by about a factor of two. In particular, Al_2O_3 and Th abundances of 6 wt% and 125 ppb were obtained, respectively, for the Moon, in contrast to 3.6 wt% and 64 ppb for the primitive terrestrial mantle. Though a similar conclusion was obtained by Drake (1986), Warren and Rasmussen (1987) later argued for a more terrestrial-like composition based upon revised globally-averaged heat-flow estimates. One commonality among these studies is that they are strongly dependent upon estimates of the average thickness of the crust, the composition of the highlands crust, and the present day mean heat flow. In light of results from the Clementine and Lunar Prospector missions, as well as reanalyses of the seismic data which point to a thinner crust, it is clear that some of the assumptions employed in these models may need revision. Indeed, due to the added complexity of the inferred structure of the Moon as portrayed in Color Plate 3.10, the bulk composition of the Moon is perhaps less agreed upon today than it was a decade or more ago.

The most significant recent development concerning the bulk composition of the Moon was the realization that the crust and mantle are not as uniform in composition as once thought. Some recent mass-balance calculations based on surface remote sensing data, and using the concept of distinct crustal terranes (Jolliff et al. 2000a; Jolliff and Gillis 2002; Taylor et al. 2002), have come to similar conclusions as Taylor (1982). On the other hand, others (Warren 2000; Warren and Humphrys 2003) have argued for a more nearly chondritic or terrestrial mantle-like composition. Given the lack of direct samples of the lunar mantle, our limited spatial sampling of lunar basalts, our current uncertainty in the average thickness of the crust, and the unfortunate locations of the Apollo heat flow measurements, it is perhaps not at all surprising that such divergences exits. Many of the parameters that are input into these bulk compositional models are educated guesses, and variations of nearly a factor of two can often be justified.

Given these caveats, it is probably more important to understand the greatest sources of uncertainty inherent in these calculations than it is to advocate one bulk compositional model over another. As the bulk Th and Al abundances are key to deciphering the bulk composition and origin of the Moon, a few simple examples will be given here that demonstrate how fundamental uncertainties associated with key model parameters affect these calculations.

5.3.1. Aluminum. Since a significant portion of the lunar aluminum budget resides in the crust, its bulk concentration will necessarily be dependent upon the assumed average crustal thickness. Recent reanalyses of the Apollo seismic data have revised the crustal thickness at the Apollo 12 and 14 sites down from 60 km to 45 ± 5 km (Khan et al. 2000), 38 ± 3 km (Khan and Mosegaard 2002), and 30 ± 2.5 km (Lognonné et al. 2003). A global extrapolation of these numbers in geophysical crustal thickness modeling then requires the assumption that the density of the crust and mantle are uniform. Thus, any mean crustal thickness estimate should probably be considered to be uncertain by at least ±10 km. For illustrative purposes, if we assume the crust to be composed of 29 wt% Al_2O_3, then a reduction in the average crustal thickness of 10 km will reduce the bulk lunar Al_2O_3 content by about 0.4 wt%.

While the concentration of Al_2O_3 in the upper crust is somewhat agreed upon, it appears that the lower crust of the Moon is less feldspathic, and its composition is not as well known (see Section 2.8.4). Taking this into account, a reasonable uncertainty of 3 wt% for the bulk Al_2O_3 content of the crust (see for example Table 3.12) would correspond to an additional uncertainty of about 0.2 wt% in terms of the bulk lunar Al_2O_3 composition. If we assume a uniform composition crust with an average thickness of 30 and 60 km, then the contribution of the crust to the lunar bulk Al_2O_3 abundance could lie anywhere between about 1.2 and 2.2 wt%, respectively.

An additional complexity regarding the bulk Al concentration of the Moon concerns the deep lunar interior. Mantle Al_2O_3 abundances have been estimated by the thermodynamic-seismic investigations of Kuskov and coworkers (e.g., Kuskov and Kronrod 1998) and Khan et al. (2006), and we here take the differences between these studies as an indication of the uncertainty in this value. Kuskov and Kronrod (1998) presented two best-fit models (see Section 3.2.2) whose chemical compositions were defined to be constant within the three mantle seismic velocity layers of Nakamura et al. (1982) (58–270 km, 270–500 km, and greater than 500 km). Their abundances correspond to 2.5 and 2 wt% Al_2O_3 for the upper mantle, 3.3 and 3.1 wt% for the middle mantle, and 5.1 and 5.3 wt% for the lower mantle. In contrast, Khan et al. (2006) found abundances that are higher by a factor of ~2, with a most probable value of ~6 wt% for the upper 600 km of the mantle and ~11 wt% for the lower mantle.

For the calculations to follow, we assume that the Moon possesses a 50 km thick crust, a metallic core of radius 350 km, and a constant density mantle. We find that the difference between the two mantle compositions of Kuskov and Kronrod (1998) leads to an uncertainty in the bulk lunar Al_2O_3 concentration of only ~0.3 wt%. In contrast, the difference between the models of Kuskov and Kronrod (1998) and Khan et al. (2006) is considerably larger at about 3.5 wt%. For the models of Kuskov and Kronrod (1998), the contribution of the mantle to the bulk Al_2O_3 abundance is about 3.3–3.6 wt%, whereas for the model of Khan et al. (2006) it is about 6.8 wt%.

These examples demonstrate that the largest uncertainty in calculating the bulk aluminum concentration of the Moon resides not in the thickness and composition of the crust, but in the composition of the mantle. Nevertheless, tentative conclusions can be made. By employing the minimum mantle Al_2O_3 abundance of Kuskov and Kronrod, as well as the minimum crustal abundance noted above, the minimum bulk lunar Al_2O_3 concentration should be ~4.5 wt% This is greater than the value of ~3.6 wt% for the bulk silicate Earth (Taylor 1992), but is less than the value of 6 wt% that was advocated by Taylor (1982) for the Moon. If the results of Khan et al. (2006) are correct, then the Moon is likely to have a bulk concentration of Al_2O_3 that is even greater than the model of Taylor (1982).

5.3.2. Thorium. As a large portion of the lunar incompatible elements is sequestered in the crust, bulk compositional models for these elements are dependent upon the average crustal thickness and composition. As an example, consider the average thorium abundance for the Feldspathic Highlands Terrane. While the surface abundance of this element is well known as a result of the Lunar Prospector measurements, the thorium concentration with depth in this terrane is less certain. If the floor of the South Pole-Aitken basin is representative of the lower crust, then the thorium concentration grades from less than ~1 ppm in the upper crust to about

2 or 3 ppm in the lower crust. If we assume that the lower crust is uniform in composition with a thickness of 25 km, a 1 ppm uncertainty in its composition would result in a bulk Moon uncertainty of 30 ppb. Furthermore, if we assume an average 1 ppm thorium abundance for the entire crust, then a change in crustal thickness of 10 km would correspond to a change in the bulk Th abundance of ~10 ppb.

The situation is considerably worse when one considers the uncertainties associated with the Procellarum KREEP Terrane as both its areal extent and crustal composition are highly uncertain. While many of the highland exposures within this region are likely to be composed of Imbrium ejecta, it is not clear how representative this material is in terms of the average crustal composition there. A plausible range of bulk Th abundances for this terrane is between 4 ppm (similar to that used in Jolliff et al. 2000a) and ~7 ppm (corresponding to a 50-50 mixture of KREEP basalt and feldspathic materials). Assuming that the crust of the PKT is on average 40 km thick, and that it occupies 10% of the surface area of the Moon, an uncertainty of 3 ppm Th for the PKT would correspond to an uncertainty in the bulk Th abundance of the Moon of ~20 ppb. An uncertainty in the crustal thickness of 10 km would correspond to an additional uncertainty of ~5–10 ppb. In addition, if we were to assume that the PKT had a slightly greater surface area of 15%, then the bulk lunar Th abundance would be increased by about 10–20 ppb.

Although direct samples of the lunar mantle have not been found among the sample collections, its geochemistry can be estimated from the composition of the lunar basalts. If primary melts of the mantle could be identified, then melt-solid distribution coefficients could be employed to back out the composition of its mantle source. In practice, though, the abundance of Th in a typical primary melt (as obtained from the picritic glasses), as well as the appropriate distribution coefficients, are both uncertain by a factor of at least two. While mantle Th abundances are expected to be low, the large volume of the mantle compensates this. In particular, the difference between a mantle possessing 40 ppb (e.g., Jolliff et al. 2000a) vs. 25 ppb Th (e.g., Warren and Wasson 1979b) simply corresponds to a bulk Moon difference of ~15 ppb.

It is thus clear that the bulk Th abundance of the Moon, which is principally based upon remote-sensing measurements, could be uncertain by a large amount. Although it is probably unreasonable to presume that all of the above sources of error would add constructively, giving rise to a maximum error of ~100 ppb Th, it is probably equally unlikely to assume that these uncertainties are distributed in a Gaussian manner, in which case the expected error would be about 50 ppb. In any case, this exercise demonstrates the difficulties involved in estimating the bulk thorium concentration of the Moon, and suggests that such studies cannot distinguish unambiguously between Taylor's enriched model (125 ppb) and that of a terrestrial primitive mantle composition (64 ppb). Nevertheless, as both Al and Th are refractory lithophile elements, if the Moon is enriched in Al, as appears to be the case, then it would be reasonable to also expect a corresponding enrichment in Th.

6. MAJOR RECENT ADVANCES IN LUNAR SCIENCE, UNANSWERED QUESTIONS, AND FUTURE DEVELOPMENTS

Prior to the Apollo missions, little was known of the composition and internal structure of the Moon. The wealth of data collected during Apollo led to a rapid increase of understanding in this regard, showing that, like the Earth, the Moon also possesses a crust, a mantle, and possibly even a small metallic core. The feldspathic composition of the crust quickly gave rise to the idea that a large portion of the Moon was initially molten, and this magma-ocean concept has since been the guiding paradigm under which most data are now interpreted.

The Clementine and Lunar Prospector missions have similarly led to great advances in understanding of the structure, composition, origin, and evolution of the Moon. Though one might presume that this was the result of collecting new types of data, these missions merely completed the orbital reconnaissance that was initiated during Apollo. Topography, gravity, gamma-ray, magnetometer, electron-reflectometer, and spectral reflectance data were all acquired during the Apollo era, but these were either limited to swaths near the equator or to the Moon's Earth-facing hemisphere. During the 1990s, these measurements were extended globally for the first time and have shown that the traditional dichotomous mare-highland classification is inadequate in describing the structure and geologic evolution of this body. Large regions were found to possess distinctive geological characteristics, and these are inferred to be the result of either asymmetries associated with a crystallizing magma ocean, or later large impact events.

Much of what has been reported on in this chapter is the result of data collected from Clementine and Lunar Prospector, the continued analysis of the lunar samples, and the reanalysis of key geophysical datasets. Below we list some of the key advances in lunar science that have been made since the Apollo era, and some of the key scientific questions that have yet to be resolved.

6.1. Major advances in lunar science

1. The origin and evolution of the lunar crust and interior can be best understood by dividing the crust and underlying mantle into three major geological provinces: the Procellarum KREEP Terrane, the Feldspathic Highlands Terrane and the South Pole-Aitken basin.

2. The Procellarum KREEP Terrane encompasses Mare Imbrium and Oceanus Procellarum and is characterized by its high abundance of incompatible and heat-producing elements, its temporally extended volcanic history (from ~4.2 Ga to <900 Ma), the nearly complete basaltic resurfacing of this region, and the prevalence of Ti-rich lava flows.

3. The Feldspathic Highlands Terrane is characterized by its ancient age (>4.3 Ga) and highly feldspathic composition. It probably represents the primary magma-ocean flotation crust of the Moon, and has been altered only by subsequent impact processes and minor quantities of mare volcanism. The available evidence suggests that the crust of this terrane becomes increasing mafic with depth.

4. The South Pole-Aitken basin is not only the largest impact structure recognized on the Moon, but also in the solar system, with a diameter slightly larger than that of the Hellas basin on Mars. However, even though this basin has a diameter of over 2100 km, it does not appear to have excavated all the way through the lunar crust. The basin floor is noritic in composition, and this likely reflects the composition of the deep crust within the Feldspathic Highlands Terrane.

5. Seismic data demonstrate that the Moon possesses substantial internal structure. At the Apollo 12 and 14 sites within the Procellarum KREEP Terrane, the crust is about 30 to 40 km thick, and there is some evidence for an intracrustal seismic discontinuity ~20 km below the surface. The upper 500 km of the mantle beneath the Procellarum KREEP Terrane appears to have a nearly constant seismic velocity, suggesting that this region has a relatively uniform composition. A major seismic discontinuity is located about 500 km below the surface that most likely represents a change in mantle composition. Beneath a depth of ~1150 km, the mantle may be partially molten.

6. Many pieces of evidence suggest that the Moon possesses some form of a dense, molten, and high-electrical-conductivity core whose radius is less than 400 km.

6.2. Unanswered questions

1. *Is the surface distribution of KREEP representative of the underlying crust?* Although gamma-ray data collected from orbit shows that the surface abundance of incompatible and heat-producing elements are concentrated within the Procellarum KREEP Terrane, it is possible that the subcrustal distribution of KREEP could be more widespread than these data suggest.

2. *What is the provenance of the magnesian-suite rocks?* Since most of the Mg- and alkali-suite rocks appear to have some form of genetic relationship to KREEP, these samples probably originated within the Procellarum KREEP Terrane. Nevertheless, using the available remote-sensing data, it is uncertain as to whether or not this suite of rocks (possibly with lower abundances of incompatible elements) will be found at other locales on the Moon.

3. *What is the composition and origin of the lower crust?* Though many lines of evidence suggest that the lower portion of the feldspathic crust is more mafic and noritic than its upper portion, we do not have any samples of this important rock type. Though a more mafic lower crust is compatible with cumulate flotation in a chemically evolving magma ocean, its origin can only be definitively answered by chemical analyses of these crustal rocks.

4. *What is the composition of the South Pole-Aitken basin?* The floor of the South Pole-Aitken basin has been inferred to be largely composed of noritic materials, though recent mineralogical mapping suggests that clinopyroxene abundances may be significantly higher than previously though. Its composition might be representative of the lower crust of the Feldspathic Highlands Terrane, although at present it cannot be ruled out that its composition is unique, and that it might contain components from the upper mantle. As suggested by the recent NAS Decadal Survey (National Research Council 2003), a mission to sample and return materials from the SPA Terrane would probably answer this question. Although lunar meteorites provide several samples of the Feldspathic Highlands Terrane, none yet appear to sample materials of the SPA basin.

5. *What is the composition and depth of origin of the farside and young nearside basalts?* Many different types of mare basalts are represented within the sample collections, yet key mare deposits remain unsampled. The volcanic history of the Procellarum KREEP Terrane differs dramatically from that of the Feldspathic Highlands Terrane, and samples of basalts that erupted on the lunar farside would help elucidate at what depths melting there occurred, as well as the composition of the farside mantle. Samples of the youngest basalts within the Procellarum KREEP Terrane would help constrain how mare volcanic processes have evolved with time within this geologic province.

6. *What is the origin and lateral extent of the 500-km seismic discontinuity?* Seismic velocity profiles beneath the Procellarum KREEP Terrane imply that a major seismic discontinuity occurs ~500 km below the surface. Models of the lunar interior imply that this discontinuity is compositional in origin, with the deeper mantle containing a higher abundance of Al and/or Mg. One possibility is that this boundary could represent the maximum depth of the lunar magma ocean. Alternatively, if the magma ocean was global in extent, then this depth might mark the transition between early olivine- and later orthopyroxene-rich cumulates. Finally, it is possible that this discontinuity might correspond to the maximum depth of melting within the post-magma-ocean mare source. If the latter explanation is correct, then the discontinuity may only be locally present beneath the Procellarum KREEP Terrane.

7. *What is the size and composition of the core, and was there ever a core dynamo?* While the available data imply that the Moon has a dense, molten, and high electrical-conductivity core, we do not know the composition of this core. Though it may be composed of metallic iron (with some amount of alloying Ni, S, and C), the geophysical data can also be reconciled with a core composed of a dense molten Ti-rich silicate magma. In support of this latter hypothesis, some magmatic evolution models of the Moon suggest that late Ti-rich magma-ocean cumulates may have sunk through the mantle, possibly to the center of the Moon. It is presently unclear if the crustal magnetizations are related to an internal core dynamo, or to processes associated with impact cratering.

The above lists demonstrate that even though we appear to have a good understanding of the composition and internal structure of the Moon, fundamental questions of first-order importance remain unanswered. While the answers to some of these questions might be inferred by the continuing analyses of existing datasets, the next big leap forward in lunar science will certainly occur with the collection of additional data from both the surface and orbit.

Below we list what we think are the most important science objectives that should guide the future scientific exploration of the Moon. We divide these into two main categories. The primary science objectives will almost certainly lead to a dramatic, and possibly revolutionary, understanding of the Moon. These objectives will require the collection of samples from the lunar surface, and the emplacement of a network of geophysical stations. While these goals might be achievable by robotic means, human exploration would be desirable. In contrast, many of the other scientific objectives can be achieved either from lunar orbit or from advances in technology on Earth.

6.3. Primary science objectives

1. *Determine the composition of the South Pole–Aitken Basin.* Knowledge of the composition of the floor of this basin will address many questions: Did this basin excavate into the mantle? Is the basin floor composed of ferroan or magnesian rocks? Is the floor of this basin typical of lower crustal materials? If so, how does this composition relate to the lunar magma ocean? Although some progress in deciphering the character of the South Pole-Aitken basin could be achieved by acquiring higher precision spectral reflectance, gamma-ray, and X-ray fluorescence data, the questions above will only be definitively resolved by the collection and analysis of samples. Samples from locations interior to the basin may consist of impact-melt rocks formed by the SPA event, perhaps reworked and excavated from depth by later cratering, whereas samples from outside the basin may represent materials ejected from deep within the crust and/or mantle.

2. *Determine the global seismic velocity structure of the lunar interior.* The Apollo seismic network only spanned a small region of the Moon that straddled the Procellarum KREEP and Feldspathic Highlands Terranes. Thus, it is not clear as to whether the inferred seismic velocity structure beneath this network is representative of either of these regions. A network of about 10 short-period seismometers operating for close to a decade would provide data on which to base much improved estimates of the crustal thickness at diverse sites, constrain the size, composition and physical state of the core, and determine whether the seismic velocity structure of the mantle differs beneath the major geologic terranes. A single long-period seismometer or superconducting gravimeter operating for about a year might also be able to measure lunar free oscillations excited by shallow moonquakes and meteoroid impacts. Such data would increase our knowledge of both the density and rigidity structure of the Moon.

3. *Determine the heat flow at various locations within the major terranes.* The heat flow of the Moon was measured at the Apollo 15 and 17 sites. Unfortunately, these

two sites straddle the Procellarum KREEP and Feldspathic Highlands Terranes, and it is thus not certain if these values are representative of either region. Some thermal models suggest that the heat flow within the heart of the Procellarum KREEP Terrane may be many times greater than was measured at the Apollo 15 site. Heat flow determinations within representative regions of each of these terranes would dramatically improve our understanding of how radioactive elements are distributed within both the crust and mantle, and lead to a better understanding of how the magma ocean crystallized. Multiple heat flow probes in and around the Procellarum KREEP Terrane would additionally help delineate the boundaries of this terrane, and hence place better constraints on the bulk composition of the Moon.

A mission concept currently exists to return samples from the South Pole-Aitken Basin (Duke et al. 2000; Pieters et al. 2003) that follows from the recommendations made in the NAS Decadal Survey (National Research Council 2003). This mission would collect additional remote sensing data over the basin and would return samples from two locations, one interior to, and the other exterior to its excavation cavity. The main objectives would be to date the age of the basin, and determine the composition and mineralogy of the basin impact melt sheet, the lower crustal and/or upper mantle, and the basalts that subsequently erupted there. Samples from relatively homogeneous regions of the surface in this basin would also provide a ground truth for remote sensing studies within this major lunar terrane.

6.4. Other science objectives

1. *Obtain global high precision and high resolution major-element concentrations of the lunar surface.* The primary major elements that are needed to understand the makeup of the surface are Fe, Mg, Ca, and Al. These and other elements can either be measured by an orbiting gamma-ray spectrometer (GRS) or an X-ray fluorescence (XRF) spectrometer. The D-CIXS X-ray fluorescence spectrometer (Grande et al. 2003) onboard the European Space Agency's (ESA) SMART-1 mission is currently collecting data. Furthermore, both XRF and GRS instruments are slated for the Japanese SELENE mission that should be launched in either 2006 or 2007. The XRF instruments will obtain the first such global datasets with a spatial resolution of ~30 km. While the SELENE gamma-ray spectrometer has an approximately six times better energy resolution than the respective Lunar Prospector instrument (~2 keV in comparison to 17.6 keV), as a result of the ~100 km mapping orbit of SELENE, these data will only have a spatial resolution of ~100 km in contrast to ~30 km for the Lunar Prospector data.

2. *Obtain global high spectral-precision spectral reflectance data in order to more accurately infer surface mineralogy.* In combination with global absolute atomic abundances, high spectral-resolution reflectance data will help to more uniquely determine the mineralogy of the surface. An infrared spectrometer on the SMART-1 mission (Keller et al. 2001) currently in orbit about the Moon, is obtaining spectra from 900 to 2400 nm, with a 60 nm spectral resolution and 300 m spatial resolution. The Spectral Profiler onboard SELENE would collect similar data from 500 to 2600 nm, with a 6–8 nm spectral resolution and a 500 m spatial resolution. The Indian Chandrayan-1 mission, planned for launch in 2007, and the US Lunar Reconnaissance Orbiter (LRO), scheduled for launch in 2008, would collect additional multispectral data.

3. *Obtain global high-resolution topography.* Whereas the Clementine altimetry data greatly improved knowledge of the shape of the Moon, significant gaps remain in this dataset, especially over the lunar poles. A laser altimeter is expected to accompany the SELENE mission and is designed to acquire data with an along- and cross-track spacing of approximately 2 and 5 km, respectively. Such data would represent a significant improvement over the Clementine dataset that only has an along- and

cross-track shot spacing of ~60 and 20–100 km, respectively. High-resolution topographic data are necessary to interpret the high-resolution gravity data that are currently available over the nearside of the Moon, as well as to delineate where deposits of ice may be found at the poles. The LRO mission would also employ a laser altimeter.

4. *Obtain gravity tracking over the lunar farside.* Currently, there is no direct spacecraft tracking data within ~60° of the center of the lunar farside. This lack of tracking data greatly limits the resolution the global gravity field, lunar crustal thickness investigations, and gravity-topography admittance studies. The SELENE mission is expected to carry a relay subsatellite transponder that would for the first time obtain direct tracking data over the central farside region. Using these tracking data the inferred gravity field is expected to be globally determined to about spherical harmonic degree 70 (in comparison to approximately degree 15 for the current gravity field).

5. *Determine the thickness of the mare basalts.* Knowledge of the thickness of basaltic lava flows on the Moon is necessary to understand its thermal evolution. In addition, by measuring the thickness of basalts within an impact basin, the underlying basin structure is revealed as well. The SELENE Lunar Radar Sounder instrument should accomplish this goal.

6. *Improve knowledge of the lunar core and internal properties by continued LLR ranging and the emplacement of additional retroreflectors on the lunar surface.* With the exception of the collection of additional seismic data, the analysis of lunar laser ranging (LLR) data offers the next best opportunity to further our understanding of the deep lunar interior. As modeling of the lunar ranges requires long spans of data (on the order of six years), and the most accurate of these data were collected over the past few years, continued ranging to the lunar retroreflectors will continue to improve models of the deep mantle and core. The emplacement of additional retroreflectors (~4) would significantly improve the analysis of these data. A ranging station that is currently under assembly should be able to range to lightweight retroreflectors that are smaller than the Apollo arrays (either individual corner cubes or small arrays of corner cubes) with a much higher precision. Another approach using optical transponders could produce a bright signal that would be detectable by the many existing Earth satellite ranging stations that cannot presently range to the Moon. In the future, lightweight retroreflectors could be carried to the surface by small spacecraft while larger payload capabilities could accommodate the transponders needed for stronger signals.

7. *Sample the youngest volcanic rocks within the Procellarum KREEP Terrane and the farside volcanic rocks.* Sampling farside volcanic rocks would enable us to determine if magmatic processes varied beneath the two major geologic terranes of the Moon. At present, it is not certain if farside magmas were derived from a similar source composition and depth as the nearside basalts. Sampling the youngest basalts within the Procellarum KREEP Terrane would additionally elucidate how volcanic processes varied as a function of time within this terrane. Other sampling targets would be regions that possess "extreme" compositions, such as those present within the Aristarchus plateau, Apennine bench, and Compton-Belkovich thorium anomaly.

8. *Obtain global imagery at multiple spatial resolutions and multiple illumination angles.* The primary image data that is used for geologic mapping of the Moon is still the Lunar Orbiter photographs from the 1960s. These images are, however, nonuniform in both spatial coverage and resolution. The Terrain Camera of the SELENE mission is designed to globally image the Moon with spatial resolutions up to 10 meters. Images will be taken under low sun illumination (<30°) and stereo imagery will be useful for constructing high spatial resolution digital elevation

models. Narrow- and wide-angle imaging planned for the LRO mission will provide 0.5-m resolution panchromatic images over selected 5-km swaths and 100-m resolution multi-band imaging over a 100-km swaths, respectively.

Successful SMART-1, SELENE, Chandrayan, and LRO missions[†] may accomplish many of the above scientific objectives within the next few years. Within five years time, it is also likely that technological advances on Earth will improve the collection of laser-ranging data, although the emplacement of additional ranging sites is still highly desirable. The remaining major tasks in lunar exploration will thus be the construction of a global geophysical network that includes both seismometers and heat flow probes, as well as the robotic sampling of a few key rock types (SPA, farside basalts, young basalts) (see also Crawford 2004). With the accomplishment of these tasks, the next logical steps in lunar exploration will likely involve human return to the nearest celestial object in our solar system.

7. ACKNOWLEDGMENTS

We thank the various NASA programs, especially Planetary Geology and Geophysics, and Cosmochemistry, that have supported the research of individual co-authors. Klaus Mosegaard and Philippe Lognonné are thanked for comments and reviews that improved this manuscript.

8. ONLINE SUPPORTING MATERIALS AT *WWW.MINSOCAM.ORG*

Table A3.1. FAS compositions.
Table A3.2. Magnesian-suite compositions.
Table A3.3. Alkali-suite compositions: alkali anorthosite/norite.
Table A3.4. Alkali-suite compositions: granite/felsite & QMD/monzogabbro.
Table A3.5. KREEP basalt compositions.
Table A3.6. Impact-melt rocks & breccia compositions.
Table A3.7. Granulitic breccia compositions.
Table A3.8. Feldspathic fragmental breccia compositions.
Table A3.9. Lunar meteorite compositions.
Table A3.10. Remote sensing identifications of arthorthosite.
Table A3.11. Mare basalt group compositions.
Table A3.12. Volcanic glass group compositions.
Crustal thickness archive.

9. REFERENCES

Agee CB (1998) Crystal-liquid density inversions in terrestrial and lunar magmas. Phys Earth Planet Inter 107:63-74
Agrell SO, Scoon JH, Muir ID, Long JVP, McConnell JDC, Peckett A (1970) Observations on the chemistry, mineralogy and petrology of some Apollo 11 lunar samples. Proc Apollo 11 Lunar Sci Conf 1:93-128
Albert RA, Phillips RJ (2000) Paleoflexure. Geophys Res Lett 27:2385-2388
Albert RA, Phillips RJ, Dombard AJ, Brown CD (2000) A test of the validity of yield strength envelopes with an elastoviscoplastic finite element model. Geophys J Int 140:399-409
Allaby A, Allaby M (eds) (1991) The Concise Oxford Dictionary of Earth Sciences. Oxford Univ. Press
Anand M, Taylor LA, Nazarov MA, Patchen A (2003) Petrologic comparisons of lunar mare basalt meteorites Dh-287A and NWA 032. Lunar Planet Sci XXXIV:1787
Anand M, Taylor LA, Neal C, Patchen A, Kramer G (2004) Petrology and geochemistry of LAP 02 205: A new low-Ti mare-basalt meteorite. Lunar Planet Sci XXXV:1626

[†] Definition of the LRO and Chandrayan missions has occurred only recently and is on-going, thus we have not attempted to incorporate fully the contributions expected from these missions and their instrument payloads.

Anders E (1978) Procrustean science: indigenous siderophiles in the lunar highlands according to Delano and Ringwood. Proc Lunar Planet Sci Conf 9:161-184

Anders E, Ganapathy R, Keays RR, Laul JC, Morgan JW (1971) Volatile and siderophile elements in lunar rocks: Comparison with terrestrial and meteoritic basalts. Proc Lunar Sci Conf 2:1021-1036

Anderson DL (1975) On the composition of the lunar interior. J Geophys Res 80:1555-1557

Anderson JD, Lau EL, Sjogren WL, Schubert G, Moore WB (1996a) Gravitational constraints on the internal structure of Ganymede. Nature 384:541-543

Anderson JD, Schubert G, Jacobson RA, Lau EL, Moore WB, Sjogren WL (1998) Europa's differentiated internal structure: Inferences from four Galileo encounters. Science 281:2019-2022

Anderson JD, Sjogren WL, Schubert G (1996b) Galileo gravity results and the internal structure of Io. Science 272: 709-712

Andre CG, Wolfe RW, Andler I (1978) Evidence for a high-magnesium subsurface basalt in Mare Crisium from orbital X-ray flourescence data. In: Mare Crisium: The View from Luna 24. Merrill RB, Papike JJ (eds) Pergamon Press, p 1-12

Aoshima C, Namiki N (2001) Structures beneath lunar basins: Estimates of Moho and elastic thickness from local analysis of gravity and topography. Lunar Planet Sci XXXII:1561

Arai T, Warren PH (1999) Lunar meteorite Queen Alexandra Range 94281: Glass compositions and other evidence for launch pairing with Yamato 793274. Meteorit Planet Sci 34:209-234

Arkani-Hamed J (1998) The lunar mascons revisited. J Geophys Res 103:3709-3739

Bailey ME, Dunlop DJ (1977) On the use of anhysteretic remanent magnetization in paleointensity determination. Phys Earth Planet Inter 13:360-362

Baker M, Herzberg CT (1980) Spinel cataclasites in 15445 and 72435: Petrology and criteria for equilibrium. Proc Lunar Planet Sci Conf 11:535-553

Baldwin RB (1970) A new method of determining the depth of lava in the lunar mare. Publ Astron Soc Pacific 82: 857-864

Banerjee SK, Swits G (1974) Natural remanent magnetization studies of a layered breccia boulder from the lunar highland region. Moon 14:473-481

Basaltic Volcanism Study Project (1981) Basaltic Volcanism on the Terrestrial Planets. Pergamon Press

Beard BL, Taylor LA, Scherer EE, Johnson CM, Snyder GA (1998) The source region and melting mineralogy of high-titanium and low-titanium lunar basalts deduced from Lu-Hf isotope data. Geochim Cosmochim Acta 62: 525-544

Beaty DW, Albee AL (1978) Comparative petrology and possible petrogenetic relations among the Apollo 11 basalts. Lunar Planet Sci Conf 9:359-463

Bence AE, Delano JW, Papike JJ, Cameron KL (1974) Petrology of the highlands massifs at Taurus Littrow: An analysis of the 2-4 mm soil fraction. Proc Lunar Sci Conf 5:785-827

Bertka CM, Fei Y (1998) Implications of Mars Pathfinder data for the accretion history of the terrestrial planets. Science 281:1838-1840

Bickel CE, Warner JL (1978) Survey of lunar plutonic and granulitic lithic fragments. Proc Lunar Planet Sci Conf 9: 629-652

Bills BG, Ferrari AJ (1977a) A harmonic analysis of lunar topography. Icarus 31:244-259

Bills BG, Ferrari AJ (1977b) A lunar density model consistent with topographic, gravitational, librational, and seismic data. J Geophys Res 82:1306-1314

Blanchard DP, Budahn JR (1979) Remnants from the ancient crust: Clasts from Consortium breccia 73255. Proc Lunar Planet Sci Conf 10:803-816

Blanchard DP, McKay GA (1981) Remnants from the ancient lunar crust III: Norite 78236. Lunar Planet Sci 12:83-85

Blewett DT, Hawke BR (2001) Remote sensing and geological studies of the Hadley-Apennine region of the Moon. Meteor Planet Sci 36:701-730

Borg L, Norman M, Nyquist L, Bogard D, Snyder G, Taylor L, Lindstrom M (1999) Isotopic studies of ferroan anorthosite 62236: A young lunar crustal rock from a light-rare-earth-element-depleted source. Geochim Cosmochim Acta 63:2679-2691

Borg LE, Shearer CK, Asmerom Y, Papike JJ (2004) Prolonged KREEP magmatism on the Moon indicated by the youngest dated lunar igneous rock. Nature 432:209-211

Bratt SR, Solomon SC, Head JW, Thurber CH (1985) The deep structure of lunar basins: implications for basin formation and modification. J Geophys Res 90:3049-3064

Brett R (1973) A lunar core of Fe-Ni-S. Geochim Cosmochim Acta 37:165-170

Brett R, Butler PJ, Meyer C, Jr., Reid AM, Takeda H, Williams R (1971) Apollo 12 igneous rocks 12004, 12008, 12009, and 10022: a mineralogical and petrological study. Proc Lunar Sci Conf 2:301-318

Buck WR, Toksöz MN (1980) The bulk composition of the Moon based on geophysical constraints. Proc Lunar Planet Sci Conf 11:2043-2058

Budney CJ, Lucey PG (1998) Basalt thickness in Mare Humorum: the crater excavation method. J Geophys Res 103: 16,855-16,870

Burns JA (1986) The evolution of satellite orbits. *In:* Satellites. Burns JA, Mathews MS (eds) Univ. Arizona Press, p 117-158

Burov EB, Diament M (1995) The effective elastic thickness (T_e) of continental lithosphere: What does it really mean? J Geophys Res 100:3905-3927

Bursa M (1992) Parameters of common relevance of astronomy, geodesy and geodynamics. Bull Geod 66:193-197

Busse FH (2000) Homogeneous dynamos in planetary cores and in the laboratory. Annu Rev Fluid Mech 32:383-408

Bussey DBJ, Spudis PD (1997) Compositional analysis of the Orientale basin using full resolution Clementine data: Some preliminary results. Geophys Res Lett 24:445-448

Bussey DBJ, Spudis PD (2000) Compositional studies of the Orientale, Humorum, Nectaris, and Crisium lunar basins. J Geophys Res 105:4235-4243

Cameron AGW (1997) The origin of the Moon and the single impact hypothesis, V. Icarus 126:126-137

Cameron AGW (2000) Higher-resolution simulations of the giant impact. *In:* Origin of the Earth and Moon. Canup RM, Righter K (eds) Univ. Arizona Press, p 133-144

Canup R (2004) Simulations of a late lunar-forming impact. Icarus 168:433-456

Canup RM, Asphaug E (2001) Origin of the Moon in a giant impact near the end of the Earth's formation. Nature 412: 708-712

Canup RM, Esposito LW (1996) Accretion of the Moon from an impact-generated disk. Icarus 119:427-446

Cerroni P, Martelli G (1982) Magnification of pre-existing magnetic fields in impact-produced plasmas, with reference to impact craters. Planet Space Sci 30:395-398

Chowdhary SK, Collinson DW, Stephenson A, Runcorn SK (1987) Further investigations into lunar paleointensity determinations. Phys Earth Planet Inter 49:133-141

Cintala MJ, Grieve RAF (1998) Scaling impact melting and crater dimensions: Implications for the lunar cratering record. Meteorit Planet Sci 33:889-912

Circone S, Agee CB (1996) Compressibility of molten high-Ti mare glass: Evidence for crystal-liquid density inversions in the lunar mantle. Geochim Cosmochim Acta 60:2709-2720

Cisowski SM, Collinson DW, Runcorn SK, Stephenson A, Fulle M (1983) A review of lunar paleointensity data and implications for origin of lunar magnetism. Proc 13th Lunar Planet Sci Conf, Part 2, J Geophys Res, suppl. 88: A691-A704

Cisowski SM, Fuller M (1978) The effect of shock on the magnetism of terrestrial rocks. J Geophys Res 83:3441-3458

Cisowski SM, Fuller M (1986) Lunar paleointensities via the IRMs normalization method and the early magnetic history of the Moon. *In:* Origin of the Moon. Hartmann WK, Phillips RJ, Taylor GJ (eds) Lunar and Planetary Institute, p 411-424

Coe RS, Grommé S, Mankinen EA (1978) Geomagnetic paleointensities from radiocarbon-dated lava flows on Hawaii and the question of the Pacific nondipole low. J Geophys Res 83:1740-1756

Cohen BA, James OB, Taylor LA, Nazarov MA, Baruskova LD (2004) Lunar highland meteorite Dhofar 026 and Apollo sample 15418: Two strongly shocked, partially melted, granulitic breccias. Meteor Planet Sci 39:1419-1447

Cohen BA, Swindle TD, Kring DA (2000) Support for the lunar cataclysm hypothesis from lunar impact melt ages. Science 290:1754-1756

Collinson DW (1984) On the existence of magnetic fields on the Moon between 3.6 Ga ago and the present. Phys Earth Planet Inter 34:102-116

Collinson DW (1985) Primary and secondary magnetizations in lunar rocks: Implications for the ancient magnetic-field of the Moon. Earth Moon Planets 33:31-58

Collinson DW (1993) Magnetism of the Moon: A lunar core dynamo or impact magnetization? Surv Geophys 14: 89-118

Collinson DW, Stephenson A (1977) Paleointensity experiments using alternating field demagnetization. Phys Earth Planet Inter 13:380-385

Collinson DW, Stephenson A, Runcorn SK (1973) Magnetic studies of Apollo 15 and 16 rocks. Proc Lunar Sci Conf 4:2963-2976

Colson RO (1992) Mineralization on the Moon? Theoretical consideration of Apollo 16 "rusty rocks," sulfide replacement in 67016, and surface-correlated volatiles on lunar volcanic glasses. Proc Lunar Planet Sci Conf 22: 427-436

Comer RP, Solomon SC, Head JW (1979) Elastic lithospheric thickness on the Moon from mare tectonic features: a formal inversion. Proc Lunar Planet Sci Conf 10:2441-2463

Cook AC, Watters TR, Robinson MS, Spudis PD, Bussey DBJ (2000) Lunar polar topography derived from Clementine stereoimages. J Geophys Res 105:12,023-12,033

Coombs CR, Hawke BR, Lucey PG, Owensby PD, Zisk SH (1990) The Alphonsus region: A geologic and remote sensing persepective. Proc Lunar Planet Sci Conf 20:161-174

Cooper BL, Carter JL, Sapp CA (1994) New evidence for graben origin of Oceanus Procellarum from lunar sounder optical imagery. J Geophys Res 99:3799-3812

Crawford DA, Schultz PH (1988) Laboratory observations of impact-generated magnetic fields. Nature 336:50-52

Crawford DA, Schultz PH (1991) Laboratory investigations of impact-generated plasma. J Geophys Res 96:18,807-18,817

Crawford DA, Schultz PH (1999) Electromagnetic properties of impact-generated plasma, vapor and debris. Int J Impact Engng 23:169-180

Crawford I (2004) The scientific case for renewed human activities on the Moon. Space Policy 20:91-97

Croft SK (1980) Cratering flow fields: Implications for the excavation and transient expansion stages of crater formation. Proc Lunar Planet Sci Conf 11:2347-2378

Crosby A, McKenzie D (2005) Measurements of the elastic thickness under ancient lunar terrain. Icarus 173:100-107

Cushing JA, Taylor GJ, Norman MD, Keil K (1999) The granulitic impactite suite: Impact melts and metamorphic breccias of the early lunar crust. Meteor Planet Sci 34:185-195

Daily WD, Dyal P (1979) Magnetometer data errors and lunar induction studies. J Geophys Res 84:3313-3326

Dainty AM, Toksöz MN, Stein S (1976) Seismic investigation of the lunar interior. Proc Lunar Sci Conf 7:3057-3075

Dalrymple GB, Ryder G (1993) $^{40}Ar/^{39}Ar$ age spectra of Apollo 15 impact melt rocks by laser step-heating and their bearing on the history of the lunar basin formation. J Geophys Res 98:13,085-13,095

Dalrymple GB, Ryder G (1996) Argon-40/argon-39 age spectra of Apollo 17 highlands breccia samples by laser step heating and the age of the Serenitatis basin. J Geophys Res 101:26,069-26,084

De Hon RA (1974) Thickness of mare material in the Tranquilitatis and Nectaris basins. Proc Lunar Sci Conf 5:53-59

De Hon RA (1977) Mare Humorum and Mare Nubium: Basalt thickness and basin-forming history. Proc Lunar Sci Conf 8:633-641

De Hon RA (1979) Thickness of the western mare basalts. Proc Lunar Planet Sci Conf 10:2935-2955

De Hon RA, Waskom JD (1976) Geologic structure of the eastern mare basins. Proc Lunar Sci Conf 7:2729-2746

Delano JW (1986a) Abundances of cobalt, nickel, and volatiles in the silicate portion of the Moon. *In:* Origin of the Moon. Hartmann WK, Phillips RJ, Taylor GJ (eds) Lunar and Planetary Institute, p 231-248

Delano JW (1986b) Pristine lunar glasses: criteria, data, and implications. Proc 16th Lunar Planet Sci Conf, Part 2, J Geophys Res, suppl. 91:D201-D213

Delano JW (1990) Buoyancy-driven melt segregation in the Earth's Moon, I, Numerical results. Proc Lunar Planet Sci Conf 20:3-12

Delano JW, Hanson BZ, Watson EB (1994) Abundance and diffusivity of sulfur in lunar picritic magmas. Lunar Planet Sci 25:325-326

Delano JW, Lindsley DH, M.-S. M, Schmitt RA (1982) The Apollo 15 yello impact glasses: Chemistry, petrology, and exotic origin. Proc 13th Lunar Planet Sci Conf, Part 1, J Geophys Res, suppl. 87:A159-A170

Delano JW, Ringwood AE (1978) Siderophile elements in the lunar highlands: Nature of the indigenous component and implications for the origin of the Moon. Proc Lunar Planet Sci Conf 9:111-159

Dence MR (1968) Shock zoning at Canadian craters: Petrography and structural implications. *In:* Shock Metamorphism of Natural Materials. French BM, Short NM (eds) Mono Book Corp., p 169-184

Dickey JO, Bender PL, Faller JE, Newhall XX, Ricklefs RL, Ries JG, Shelus PJ, Veillet C, Whipple AL, Wiant JR, Williams JG, Yoder CF (1994) Lunar laser ranging: a continuing legacy of the Apollo program. Science 265:482-490

Dickey JS, Frey FA, Hart SR, Watson EB, Thompson G (1977) Geochemistry and petrology of dredged basalts from the Bouvet triple junction, South Atlantic. Geochim Cosmochim Acta 41:1105-1118

Dickinson T, Taylor GJ, Keil K, Schmitt RA, Hughes SS, Smith MR (1985) Apollo 14 aluminous mare basalts and their possible relationship to KREEP. Proc 15th Lunar Planet Sci Conf, in J Geophys Res 90:C365-C374

Dobson DP, Crichton WA, Vocadlo L, Jones AP, Wang Y, Uchida T, Rivers M, Sutton S, Brodholt JP (2000) In situ measurement of viscosity of liquids in the Fe-FeS sytem at high pressures and temperatures. Am Mineral 85:1838-1842

Dorman J, Evans S, Nakamura Y, Latham G (1978) On the time varying properties of the lunar seismic meteoroid population. Proc Lunar Planet Sci Conf 9:3615-3626

Dowty E, Keil K, Prinz M (1974) Igneous rocks from Apollo 16 rake samples. Lunar Science V:174-176

Drake DM (1986) Is lunar bulk material similar to Earth's mantle? *In:* Origin of the Moon. Hartmann WK, Phillips RJ, Taylor GJ (eds) Lunar and Planetary Institute, p 105-124

Drake MJ, Newsom HE, Reed SJ, Enright MC (1984) Experimental determination of the partitioning of gallium between solid iron metal and synthetic basaltic melt: electron and ion microprobe study. Geochim Cosmochim Acta 48:1609-1615

Duennebier F, Dorman J, Lammlein D, G. L, Nakamura Y (1975) Meteoroid flux from passive seismic experiment data. Proc Lunar Planet Sci Conf 6:2417-2426

Duennebier F, Sutton GH (1974) Thermal moonquakes. J Geophys Res 79:4351-4363

Duke M, Gage C, Bogard D, Carrier W, Coombs C, Gaddis L, Head JI, Jolliff B, Lofgren G, Papanastassiou D, Papike J, Pieters C, Ryder G (2000) South Pole-Aitken basin sample return mission. *In:* Proceedings of the Fourth International Conference on Exploration and Utilization of the Moon SP-462: European Space Agency, ESTEC, Noordwijk

Dunlop DJ, Ozdemir O (1997) Rock Magnetism: Fundamentals and Frontiers. Cambridge Univ. Press

Dyal P, Parkin CW, Daily WD (1976) Structure of the lunar interior from magnetic field measurements. Proc Lunar Sci Conf 7:3077-3095

Dymek RF, Albee AL, Chodos AA (1976) Petrology and origin of Boulders #2 and #3, Apollo 17 Station 2. Proc Lunar Sci Conf 7:2335-2378

Ebihara M, Wolf R, Warren PH, Anders E (1992) Trace elements in 59 mostly highland Moon rocks. Proc Lunar Planet Sci 22:417-426

Eggleton RE, Schaber GG, Pike RJ (1974) Photogeologic detection of surfaces buried by mare basalts. Lunar Science V:200-2002

El Goresy A, Prinz M, Ramdohr P (1976) Zoning in spinels as an indicator of the crystallization histories of mare basalts. Proc Lunar Sci Conf 7:1261-1279

El Goresy A, Taylor LA, Ramdohr P (1972) Fra Mauro crystalline rocks: mineralogy, geochemistry, and subsolidus reduction of opaque minerals. Proc Lunar Sci Conf 3:333-349

Elkins LT, Fernandes VA, Delano JW, Grove TL (2000) Origin of lunar ultramafic green glasses: Constraints from phase equilibrium studies. Geochim Cosmochim Acta 64:2339-2350

Elkins-Tanton LT, Chatterjee N, Grove TL (2003) Experimental and petrologic constraints on lunar differentiation from the Apollo 15 green picritic glasses. Meteorit Planet Sci 38:515-527

Elphic RC, Lawrence DJ, Feldman WC, Barraclough BL, Maurice S, Binder AB, Lucey PG (2000) Lunar rare earth element distribution and ramifications for FeO and TiO_2: Lunar Prospector neutron spectrometer observations. J Geophys Res 105:20,333-20,345

Evensen NM, Murthy VR, Coscio Jr. MR (1974) Provenance of KREEP and the exotic component: Elemental and isotopic studies of grain size fractions in lunar soils. Proc Lunar Sci Conf 5:1401-1417

Fabrichnaya OB, Kuskov OL (1994) Constitution of the Moon: 1. Assessment of thermodynamic properties and reliability of phase relation calculations in the $FeO-MgO-Al_2O_3-SiO_2$ system. Phys Earth Planet Inter 83:175-196

Fagan TJ, Taylor GJ, Keil K, Bunch TE, Wittke JH, Korotev RL, Jolliff BL, Gillis JJ, Haskin LA, Jarosewich E, Clayton RN, Mayeda TK, Fernandes VA, Burgess R, Turner G, Eugster O, Lorenzetti S (2002) Northwest Africa 032: Product of lunar volcanism. Meteorit Planet Sci 37:371-394

Fagan TJ, Taylor GJ, Keil K, Hicks TL, Killgore M, Bunch TE, Wittke JH, Mittlefehldt DW, Clayton RN, Mayeda TK, Eugster O, Lorenzetti S, Norman MD (2003) Northwest Africa 773: Lunar origin and iron-enrichment trend. Meteor Planet Sci 38:529-554

Fei Y, Bertka CM, Finger LW (1997) High-pressure iron-sulfer compound, Fe_3S_2, and melting relations in the Fe-FeS system. Science 275:1621-1623

Fernandes VA, Burgess R, Turner G (2003) [40]Ar-[39]Ar chronology of lunar meteorites Northwest Africa 032 and 773. Meteor Planet Sci 38:555-564

Floss C, James OB, McGee JJ, Crozaz G (1998) Lunar ferroan anorthosite petrogenesis: Clues from trace element distributions in FAN subgroups. Geochim Cosmochim Acta 62:1255-1283

Fogel RA, Rutherford MJ (1995) Magmatic volatiles in primitive lunar glasses: FTIR and EMPA analyses of Apollo 15 green and yellow glasses and revision of the volatile assisted fire-fountaining theory. Geochim Cosmochim Acta 59:201-215

Forsyth DW (1985) Subsurface loading and estimates of the flexural rigidity of continental lithospheres. J Geophys Res 90:12,623-12,632

Freed AM, Melosh HJ, Solomon SC (2001) Tectonics of mascon loading: Resolution of the strike-slip faulting paradox. J Geophys Res 106:20,603-20,620

Frey FA, Suen CJ, Stockman HW (1985) The Ronda high temperature peridotite: Geochemistry and petrogenesis. Geochim Cosmochim Acta 49:2469-2491

Frost DJ, Liebske C, Langenhorst CA, McCammon RG, Trønnes RG, Rubie DC (2004) Experimental evidence for the existence of iron-rich metal in the Earth's lower mantle. Nature 428:409-412

Fuller M, Cisowski SM (1987) Lunar paleomagnetism. In: Geomagnetism. Jacobs JA (ed) Academic Press, p 307-455

Gagnepain-Beyneix J, Lognonné P, Chenet H, Lombardi D, Spohn T (2006) Seismic model of the Moon mantle and their constraints on the mantle temperature and mineralogy. Phys Earth Planet Inter (in press)

Ganapathy R, Morgan JW, Krähenbühl U, Anders E (1973) Ancient meteoritic components in lunar highland rocks: clues from trace elements in Apollo 15 and 16 samples. Proc Lunar Sci Conf 4:1239-1261

Gattacceca J, Rochette P (2004) Toward a robust normalized magnetic paleointensity method applied to meteorites. Earth Planet Sci Lett 227:377-393

Gibson EKJ (1977) Volatile elements, carbon, nitrogen, sulfur, sodium, potassium, and rubidium in the lunar regolith. Phys Chem Earth 10:57-62

Gibson EKJ, Brett R, Andrawes F (1977) Sulfur in lunar mare basalts as a function of bulk composition. Proc Lunar Sci Conf 8th 1417-1428

Gibson EKJ, Chang S, Lennon K, Moore GW, Pearce GW (1975) Sulfur abundances and distributions in mare basalts and their source magmas. Proc Lunar Sci Conf 6:1287-1301

Gibson EKJ, Moore GW (1974) Sulfur abundances and distributions in the valley of Taurus-Littrow. Proc Lunar Sci Conf 5:1823-1837

Giguere TA, Taylor GJ, Hawke BR, Lucey PG (2000) The titanium contents of lunar mare basalts. Meteor Planet Sci 35:193-200

Gillis JJ (1998) The composition and geologic setting of mare deposits on the far side of the Moon. Ph.D. thesis, Rice University, Houston

Gillis JJ, Jolliff BL (1999) Lateral and vertical heterogeneity of thorium in the procellarum KREEP terrane; as reflected in the ejecta deposits of post-imbrium craters. *In:* Workshop on New views of the Moon II. Flagstaff, AZ, p 18-19

Gillis JJ, Jolliff BL, Elphic RC (2003) A revised algorithm for calculating TiO$_2$ from Clementine UVVIS data: A synthesis of rock, soil, and remotely sensed TiO$_2$ concentrations. J Geophys Res 108:10.1029/2001JE001515

Gillis JJ, Jolliff BL, Lawrence DJ, Lawson SL, Prettyman TH (2002) The Compton-Belkovich region of the Moon: Remotely sensed observations and lunar sample association. Lunar Planet Sci XXXIII:1967

Gillis JJ, Spudis PD (2000) Geology of the Smythii and Marginis region of the Moon: using integrated remotely sensed data. J Geophys Res 105:4217-4233

Gnos E, Hofmann BA, A. A-K, Lorenzetti S, Eugster O, Whitehouse MJ, Villa IM, Jull AJT, Eikenberg J, Spettel B, Krähenbühl U, Franchi IA, Greenwood RC (2004) Pinpointing the source of a lunar meteorite: Implications for the evolution of the Moon. Science 305:657-659

Goguitchaichvili AT, Prevot M, Camps P (1999) No evidence for strong fields during the R3-N3 Icelandic geomagnetic reversal. Earth Planet Sci Lett 167:15-34

Goins NR, Dainty AM, Toksöz MN (1981a) Lunar seismology: The internal structure of the Moon. J Geophys Res 86: 5061-5074

Goins NR, Dainty AM, Toksöz MN (1981b) Seismic energy release of the Moon. J Geophys Res 86:378-388

Goins NR, Dainty AM, Toksöz MN (1981c) Structure of the lunar crust at highland site Apollo station 16. Geophys Res Lett 8:29-32

Goldstein BE, Phillips RJ, Russell CT (1976) Magnetic evidence concerning a lunar core. Proc Lunar Sci Conf 7: 3321-3341

Golombek MP (1985) Fault type predictions from stress distributions on planetary surfaces: Importance of fault initiation depth. J Geophys Res 90:3065-3074

Goodrich CA, Taylor GJ, Keil K, Kallemeyn GW, Warren PH (1986) Alkali norite, troctolites, and VHK mare basalts from Apollo 14 breccia 14304. Proc 16th Lunar Planet Sci Conf, J Geophys Res 91:D305-D318

Gooley R, FBrett R, Warner J, Smyth JR (1974) A lunar rock of deep crustal origin: 76535. Geochim Cosmochim Acta 38:1329-1339

Grande M, Browning R, Waltham N, Parker D, Dunkin SK, Kent B, Kellett B, Perry CH, Swinyard B, Perry A, Feraday J, Howe C, McBride G, Phillips K, Huovelin J, Muhli P, Hakala PJ, Vilhu O, Laukkanen J, Thomas N, Hughes D, Alleyne H, Grady M, Lundin R, Barabash S, Baker D, Clark PE, Murray CD, Guest JE, Casanova I, d'Uston LC, Maurice S, Foing B, Heather DJ, Fernandes VA, Muinonen K, Russell SS, Christou A, Owen C, Charles P, Koskinen H, Mato M, Sipila K, Nenonen S, Holmstrom M, Bhandari N, Elphic R, Lawrence D (2003) The D-CIXS X-ray mapping spectrometer on SMART-1. Planet Space Sci 51:427-433

Green DH, Ringwood AE (1967) The stability field of aluminous pyroxene peridotite and garnet peridotite and their relevance in upper mantle structure. Earth Planet Sci Lett 3:151-160

Gudkova TV, Zharkov VN (2002) The exploration of the lunar interior using torsional oscillations. Planet Space Sci 50: 1037-1048

Haggerty SE, Boyd FR, Bell PM, Finger LW, Bryan WB (1970) Opaque minerals and olivine in lavas and breccias from Mare Tranquillitatis. Proc Apollo 11 Lunar Sci Conf 1:513-538

Halekas JS, Lin RP, Mitchell DL (2003) Magnetic fields of lunar multi-ring impact basins. Meteorit Planet Sci 38: 565-578

Halekas JS, Mitchell DL, Lin RP, Frey S, Hood LL, Acuña MH, Binder AB (2001) Mapping of crustal magnetic anomalies on the lunar near side by the Lunar Prospector electron reflectometer. J Geophys Res 106:27,841-27,852

Halliday AN (2000) Terrestrial accretion rates and the origin of the Moon. Earth Planet Sci Lett 176:17-30

Halliday AN, Lee D-C (1999) Tungsten isotopes and the early development of the Earth and Moon. Geochim Cosmochim Acta 63:4157-4179

Halliday AN, Rehkämper M, Lee D-C, Yi W (1996) Early evolution of the Earth and Moon: New constraints from Hf-W isotope geochemistry. Earth Planet Sci Lett 142:75-90

Haloda J, Irving AJ, Tycova P (2005) Lunar meteorite Northeast Africa 001: An anorthositic regolith breccia with mixed highland/mare components. Lunar Planet Sci XXXVI:1487

Hartmann WK, Ryder G, Dones L, Grinspoon D (2000) The time-dependent intense bombardment of the primordial Earth/Moon system. *In:* The Origin of the Earth and Moon. Canup RM, Righter K (eds) Univ. Arizona Press, p 493-512

Haskin LA (1998) The Imbrium impact event and the thorium distribution at the lunar highlands surface. J Geophys Res 103:1679-1689

Haskin LA, Gillis JJ, Jolliff BL, Korotev RL (1999) On the distribution of Th in lunar surface materials. Lunar Planet Sci XXX:1858

Haskin LA, Gillis JJ, Korotev RL, Jolliff BL (2000) The materials of the lunar Procellarum KREEP Terrane: A synthesis of data from geomorphological mapping, remote sensing, and sample analyses. J Geophys Res 105: 20,403-20,415

Haskin LA, Helmke PA, Blanchard DP, Jacobs JW, Telander K (1973) Major and trace element abundances in samples from the lunar highlands. Proc Lunar Sci Conf 3:1275-1296

Haskin LA, Korotev RL, Gillis JJ, Jolliff BL (2002) Stratigraphies of Apollo and Luna highland landing sites and provenances of materials from the perspective of basin impact ejecta modeling. Lunar Planet Sci XXXIII:1364

Haskin LA, Korotev RL, Rockow KM, Jolliff BL (1998) The case for an Imbrium origin of the Apollo thorium-rich impact-melt breccias. Meteorit Planet Sci 33:959-975

Haskin LA, Moss BE, McKinnon WB (2003) On estimating contributions of basin ejecta to regolith deposits at lunar sites. Meteorit Planet Sci 38:13-33

Haskin LA, Shih C-Y, Bansal BM, Rhodes JM, Weismann H, Nyquist LE (1974) Chemical evidence for the origin of 76535 as a cumulate. Proc Lunar Sci Conf 5:1213-1225

Hawke BR, Head JW (1978) Lunar KREEP volcanism: Geologic evidence for history and mode of emplacement. Proc Lunar Planet Sci Conf 9:3285-3309

Hawke BR, Lawrence DJ, Blewett DT, Lucey PG, Smith GA, Spudis PD, Taylor GJ (2003a) Hansteen Alpha: A volcanic construct in the lunar highlands. J Geophys Res 108:doi:10.1029/2002JE002013

Hawke BR, Lucey PG, Taylor GJ, Bell JF, Peterson CA, Blewett DT, Horton K, Smith GA, Spudis PD (1991) Remonte sensing studies of the Orientale region of the Moon: A pre-Galileo view. Geophys Res Lett 18:2141-2144

Hawke BR, Peterson CA, Blewett DT, Bussey DBJ, Lucey PG, Taylor GJ, Spudis PD (2003b) Distribution and modes of occurrence of lunar anorthosite. J Geophys Res 108:5050, doi:10.1029/2002JE001890

Hawke BR, Peterson CA, Lucey PG, Taylor GJ, Blewett DT, Cambell BA, Coombs CR, Spudis PD (1993) Remote sensing studies of the terrain northwest of the Humorum basin. Geophys Res Lett 20:419-422

Haxby WF, Turcotte DL (1978) On isostatic geoid anomalies. J Geophys Res 83:5473-5478

Head JW, III, Adams JB, McCord TB, Pieters CM, Zisk SH (1978) Regional stratigraphy and geologic history of Mare Crisium. *In:* Mare Crisium: The view from Luna 24. Merrill RB, Papike JJ (eds) Pergamon Press, p 43-74

Head JW, Wilson L (1992) Lunar mare volcanism: Stratigraphy, eruption conditions, and the evolution of secondary crusts. Geochim Cosmochim Acta 56:2155-2175

Heather DJ, Dunkin SK (2002) A stratigraphic study of southern Oceanus Procellarum using Clementine multispectral data. Planet Space Sci 50:1299-1309

Heiken GH, Vaniman DT, French BM (eds) (1991) The Lunar Sourcebook: a user's guide to the Moon. Cambridge Univ. Press

Herbert F (1980) Time-dependent lunar density models. Proc Lunar Planet Sci Conf 11:2015-2030

Herbert F, Drake MJ, Sonnett CP (1978) Geophysical and geochemical evolutioon of the lunar magma ocean. Proc Lunar Planet Sci Conf 9:249-262

Herrero-Bervera E, Valet J-P (2000) Paleointensity experiments using alternating field demagnetization. Earth Planet Sci Lett 177:43-58

Herzberg CT (1978) The bearing of spinel-cataclasites on the crust-mantle structure of the Moon. Lunar Planet Sci Conf 9:319-336

Herzberg CT, Baker MB (1980) The cordierite- to spinel-cataclasite transition: Structure of the lunar crust. *In:* Proc. Conf. Lunar Highlands Crust. Merrill RB (ed) Pergammon, p 113-132

Hess PC (1991) Diapirism and the origin of high TiO_2 mare glasses. Geophys Res Lett 18:2069-2072

Hess PC (1993) Ilmenite liquidus and depths of segregation of high-Ti picritic glasses. Lunar Planet Sci XXIV:649-650

Hess PC (1998) Source regions to lunar troctolite parent magmas. Lunar Planet Sci XXVIIII:1225

Hess PC (2000) On the source regions for mare picrite glasses. J Geophys Res 105:4347-4360

Hess PC, Finnila A (1997) Depths of segregation of high-TiO_2 picrite mare glasses. Lunar Planet Sci 28:559-560

Hess PC, Parmentier EM (1995) A model for the thermal and chemical evolution of the Moon's interior: Implications for the onset of mare volcanism. Earth Planet Sci Lett 134:501-514

Hess PC, Parmentier EM (2001) Thermal evolution of a thicker KREEP liquid layer. J Geophys Res 106:28,023-28,032

Hide R, Roberts PH (1979) How strong is the magnetic field in the Earth's liquid core? Phys Earth Planet Inter 20: 124-126

Hiesinger H, Head JW, III (2003) Ages and stratigraphy of mare basalts in Oceanus Procellarum, Mare Nubium, Mare Cognitum, and Mare Insularum. J Geophys Res 108:5065, doi:10.1029/2002JE001985

Hiesinger H, Head JW, III, Wolf U, Jaumann R, Neukum G (2002) Lunar mare basalt flow units: Thicknesses determined from crater size-frequency distributions. Geophys Res Lett 29:10.1029/2002GL014847

Hiesinger H, Jaumann R, Neukum G, Head JW, III (2000) Ages of mare basalts on the lunar nearside. J Geophys Res 105:29,239-29,275

Hill DH, Boynton WV (2003) Chemistry of the Calcalong Creek lunar meteorite and its relationship to lunar terranes. Meteorit Planet Sci 38:595-626

Hillgren VJ (1991) Partioning behavior of Ni, Co, Mo, and W between basaltic liquid and Ni-rich metal: Implications for the origin of the Moon and lunar core formation. Geophys Res Lett 18:2077-2080

Hillgren VJ, Drake MJ, Rubie DC (1994) High-pressure and high-temperature experiments on core-mantle segregation in the accreting Earth. Science 264:1442-1445

Hillgren VJ, Drake MJ, Rubie DC (1996) High-pressure and high-temperature metal-silicate partitioning of siderophile elements: The importance of silicate liquid composition. Geochim Cosmochim Acta 60:2257-2263

Hirayama Y, Fujii T (1993) The melting relation of the system iron and carbon at high pressure and its bearing on the early stage of the Earth. Geophys Res Lett 20:2095-2098

Hobbs BA, Hood LL, Herbert F, Sonnett CP (1983) An upper bound on the radius of a highly electrically conducting lunar core. Proc 14th Lunar Planet Sci Conf, Part 1, J Geophys Res, suppl. 88:B97-B102

Hood LL (1986) Geophysical constraints on the lunar interior. *In:* Origin of the Moon. Hartmann WK, Phillips RJ, Taylor GJ (eds) Lunar and Planet. Inst., p 361-410

Hood LL (1995) Frozen fields. Earth Moon Planets 67:131-142

Hood LL (2000) Reply to comment on "Initial measurements of the lunar induced magnetic dipole moment using Lunar Prospector magnetometer data" by Hood et al. Geophys Res Lett 27:1079

Hood LL, Herbert F, Sonnett CP (1982) The deep lunar electrical conductivity profile: Structural and thermal inferences. J Geophys Res 87:5311-5326

Hood LL, Huang Z (1991) Formation of magnetic anomalies antipodal to lunar impact basins: Two-dimensional model calculations. J Geophys Res 96:9837-9846

Hood LL, Jones JH (1987) Geophysical constraints on the lunar bulk composition and structure: A reassessment. Proc 17th Lunar Planet Sci Conf, Part 2, J Geophys Res, suppl. 92:E396-E410

Hood LL, Mitchell DL, Lin RP, Acuna MH, Binder AB (1999) Initial measurements of the lunar induced magnetic dipole moment using Lunar Prospector magnetometer data. Geophys Res Lett 26:2327-2330

Hood LL, Vickery A (1984) Magnetic field amplification and generation in hypervelocity meteoroid impacts with application to lunar paleomagnetism. Proc 15th Lunar Planet Sci Conf, Part 1, J Geophys Res, suppl. 89:C211-C223

Hood LL, Zakharian A, Halekas J, Mitchell DL, Lin RP, Acuña MH, Binder AB (2001) Initial mapping and interpretation of lunar crustal magnetic anomalies using Lunar Prospector magnetometer data. J Geophys Res 106:27,825-27,840

Hood LL, Zuber MT (2000) Recent refinements in geophysical constraints on lunar origin and evolution. *In:* Origin of the Earth and Moon. Canup RM, Righter K (eds) Univ. of Arizona Press, p 397-412

Hubbard NJ, Gast PW (1971) Chemical composition and origin of nonmare lunar basalts. Proc Lunar Sci Conf 2: 999-1020

Hubbard NJ, Meyer Jr. C, Gast PW, Wiesmann H (1971) The composition and derivation of Apollo 12 soils. Earth Planet Sci Lett 10:341-350

Hughes HG, App FN, McGetchin TR (1977) Global seismic effects of basin-forming impacts. Phys Earth Planet Inter 15:251-263

Hughes SS, Delano JW, Schmitt RA (1988) Apollo 15 yellow-brown volcanic glass: Chemistry and petrogenetic relations to green volcanic glass and olivine-normative mare basalts. Geochim Cosmochim Acta 52:2379-2391

Hughes SS, Delano JW, Schmitt RA (1989) Petrogenetic modelling of 74220 high-Ti orange volcanic glasses and the Apollo 11 and 17 high-Ti mare basalts. Proc Lunar Planet Sci Conf 19:175-188

Hunter RH, Taylor LA (1983) The magma ocean from the Fra Mauro shoreline: An overview of the Apollo 14 crust. Proc. 13th Lunar Planet Sci Conf, J Geophys Res 88:A591-A602

Hörz F (1978) How thick are the lunar mare basalts? Proc Lunar Planet Sci Conf 9:3311-3331

Ito E, Katsura T, Suzuki T (1998) Metal/silicate partitioning of Mn, Co, and Ni at high pressures and high temperatures and implications for core formation in a deep magma ocean. *In:* Properties of Earth and Planetary Materials at High Pressure and Temperature. Manghnani MH (ed) American Geophysical Union, p 215-225

Jacobsen SB, Harper CL (1996) Accretion and early differentiation history of the Earth based on extinct radionuclides. *In:* Earth Processes: Reading the Isotope Code. Basu A, Hart S (eds), American Geophysical Union, p 47-74

Jagoutz E, Baddenhausen H, Blum K, Cendales M, Dreibus G, Spettel B, Lorenz V, Wänke H (1979) The abundance of major, minor, and trace elements in the Earth's mantle as derived from primitive ultramafic nodules. Proc Lunar Planet Sci Conf 10:2031-2050

James OB (1980) Rocks of the early lunar crust. Proc Lunar Planet Sci Conf 11:365-393

James OB (1981) Petrologic and age relationsof Apollo 16 rocks: Implications for subsurface geology and the the age of the Nectaris basin. Proc Lunar Planet Sci Conf 12B:209-233

James OB, Flohr MK (1983) Subdivision of the Mg-suite noritic rocks into Mg-gabbronorites and Mg-norites. Proc 13th Lunar Planet Sci Conf, Part 2, J Geophys Res, suppl. 88:A603-A614

James OB, Flohr MK, Lindstrom MM (1984) Petrology and geochemistry of lunar dimict breccia 61015. Proc 15th Lunar Planet Sci Conf, J Geophys Res 89:C63-C86

James OB, Hammarstrom JG (1977) Petrology of four clasts from consortium breccia 73215. Proc Lunar Sci Conf 8: 2459-2494

James OB, Lindstrom MM, Flohr MK (1987) Petrology and geochemistry of alkali gabbronorites from lunar breccia 67975. Proc 17th Lunar Planet Sci Conf, J Geophys Res 89:E314-E330

James OB, Lindstrom MM, Flohr MK (1989) Ferroan anorthosite from lunar breccia 64435: Implications for the origin and and history of lunar ferroan anorthosites. Proc Lunar Planet Sci Conf 19:219-243

James OB, McGee JJ (1979) Consortium breccia 73255: Genesis and history of two coarse-grained "norite" clasts. Proc Lunar Planet Sci Conf 10:713-743

Jeffreys H (1970) The Earth. Cambridge Univ. Press

Jolliff BJ, Gillis JJ, Lawrence DJ, Maurice S (2001) Thorium content of mare basalts of the western Procellarum region. Lunar and Planetary Science XXXII:2144

Jolliff BL (1991) Fragments of quartz monzodiorite and felsite in Apollo 14 soil particles. Lunar and Planetary Science 21:101-118

Jolliff BL (1998) Large-scale separation of K-frac and REEP-frac in the source regions of Apollo impact-melt breccias, and a revised estimate of the KREEP composition. Int Geol Rev 40:916-935

Jolliff BL, Floss C, McCallum IS, Schwartz JM (1999) Geochemistry, petrology, and cooling history of 14161,7373: A plutonic lunar sample with textural evidence of granitic-fraction separation by silicate-liquid immiscibility. Am Mineral 84:821-837

Jolliff BL, Gillis JJ (2002) Lunar Crustal and Bulk Composition. *In:* Workshop on The Moon Beyond 2002: Next Steps in Lunar Science and Exploration. Abstract 3056

Jolliff BL, Gillis JJ, Haskin L, Korotev RL, Wieczorek MA (2000a) Major lunar crustal terranes: Surface expressions and crust-mantle origins. J Geophys Res 105:4197-4216

Jolliff BL, Gillis JJ, Haskin LA (2002) Eastern Basin Terrane and South Pole-Aitken Basin ejecta: Mid-level Crust? Lunar Planet Sci XXXIII:1157

Jolliff BL, Gillis JJ, Korotev RL, Haskin LA (2000b) On the origin of nonmare materials at the Apollo 12 landing site. Lunar and Planetary Science XXXI:1671

Jolliff BL, Haskin LA (1995) Cogenetic rock fragments from a lunar soil: Evidence of a ferroan noritic-anorthosite pluton on the Moon. Geochim Cosmochim Acta 59:2345-2374

Jolliff BL, Haskin LA, Colson RO, Wadhwa M (1993) Partitioning in REE-saturating minerals: Theory, experiment, and modelling of whitlockite, apatite, and evolution of lunar residual magmas. Geochim Cosmochim Acta 57: 4069-4094

Jolliff BL, Korotev RL, Haskin LA (1991a) A ferroan region of the lunar highlands crust as recorded in meteorites MAC88104 and MAC88105. Geochim Cosmochim Acta 55:3051-3071

Jolliff BL, Korotev RL, Haskin LA (1991b) Geochemistry of 2-4 mm particles from Apollo 14 soil (14161) and implications regarding igneous components and soil-forming processes. Proc Lunar Planet Sci Conf 21:193-219

Jolliff BL, Korotev RL, Rockow KM (1998) Geochemistry and petrology of lunar meteorite Queen Alexandra Range 94281, a mixed mare and highland regolith breccia, with special emphasis on very-low-Ti mafic components. Meteorit Planet Sci 33:581-601

Jolliff BL, Korotev RL, Zeigler RA, Floss C (2003) Northwest Africa 773: Lunar mare breccia with a shallow-formed olivine-cumulate component, inferred very-low-Ti (VLT) heritage, and a KREEP connection. Geochim Cosmochim Acta 24:4857-4879

Jolliff BL, Rockow KM, Korotev RL, Haskin LA (1996) Lithologic distribution and geologic history of the Apollo 17 site: The record in soils and small rock particles from the highland massifs. Meteor Planet Sci 31:116-145

Jolliff BL, Zeigler RA, Korotev RL (2004) Petrography of lunar meteorite LAP 02205, a new low-Ti basalt possibly launch paired with NWA 032. Lunar Planet Sci XXXV:1438

Jones JH, Palme H (2000) Geochemical constraints on the origin of the Earth and Moon. *In:* Origin of the Earth and Moon. Canup RM, Righter K (eds) Univ. Arizona Press, p 197-216

Kaiden H, Kojima H (2002) Yamato 983885: A second lunar meteorite from the Yamato 98 collection. Antarctic Meteorites XXVII, Tokyo, Nat Inst Polar Res:49-51

Karner J, Papike JJ, Shearer CK (2003) Olivine from planetary basalts: Chemical signatures that indicate planetary parentage and those that record igneous setting and process. Am Mineral 88:806-816

Kaula WM (1967) Theory of statistical analysis of data distributed over a sphere. Rev Geophys 5:83-107

Kaula WM (2000) Theory of Satellite Geodesy: Applications of Satellites to Geodesy. Dover

Kaula WM, Schubert G, Lingenfelter RE, Sjogren WL, Wollenhaupt WR (1972) Analysis and interpretation of lunar laser altimetry. Proc Lunar Sci Conf 3:2189-2204

Kaula WM, Schubert G, Lingenfelter RE, Sjogren WL, Wollenhaupt WR (1974) Apollo laser altimetry and inferences as to lunar structure. Proc Lunar Sci Conf 5:3049-3058

Keller HU, Mall U, Nathues A (2001) Mapping the Moon with SIR and infrared spectrometer for SMART-1. Earth Moon Planets 85:545

Kesson SE, Lindsley DH (1976) Mare basalt petrogenesis-A review of experimental studies. Rev Geophys Space Phys 14:361-373

Kettrup B, Deutsch A, Masaitis VL (2003) Homogeneous impact melts produced by a heterogeneous target? Sr-Nd isotopic evidence from the Popigai crater, Russia. Geochim Cosmochim Acta 67:733-750

Khan A, Connolly JAD, Maclennan J, Mosegarrd K (2006) Joint inversion of swismic and gravity data for luanr composition and thermal state. Geoophys J Int (in press)

Khan A, Mosegaard K (2001) New information on the deep lunar interior from an inversion of lunar free oscillation periods. Geophys Res Lett 28:1791-1794

Khan A, Mosegaard K (2002) An enquiry into the lunar interior--A non-linear inversion of the Apollo lunar seismic data. J Geophys Res 107:10.1029/2001JE001658

Khan A, Mosegaard K, Rasmussen KL (2000) A new seismic velocity model for the Moon from a monte carlo inversion of the Apollo lunar seismic data. Geophys Res Lett 27:1591-1594

Khan A, Mosegaard K, Williams JG, Lognonné P (2004) Does the Moon possess a molten core? Probing the deep lunar interior using results from LLR and Lunar Prospector. J Geophys Res 109:doi:10.1029/2004JE002294

Klein N, Rutherford MJ (1998) Volcanic gas formed during eruption of Apollo 17 orange glass magma: evidence from glassy melt inclusions and emperiments. Lunar Planet Sci XXVIIII:1448

Kleine T, Münker C, Mezger K, Palme H (2002) Rapid accretion and early core formation on asteroids and the terrestrial planets from Hf-W chronometry. Nature 418:952-955

Kletetschka G, Acuña MH, Kohout T, Wasilewski PJ, Connerney JEP (2004) An empirical scaling law for acquisition of thermoremanent magnetization. Earth Planet Sci Lett 226:521-528

Kletetschka G, Kohout T, Wasilewski PJ (2003) Magnetic remanence in the Murchison meteorite. Meteor Planet Sci 38:399-405

Kohlstedt DL, Evans B, Mackwell SJ (1995) Strength of the lithosphere: Constraints imposed by laboratory experiments. J Geophys Res 100:17,587-17,602

Kono M (1987) Changes in TRM and ARM in a basalt due to laboratory heating. Phys Earth Planet Inter 46:1-8

Konopliv AS, Asmar SW, Yuan DN (2001) Recent gravity models as a result of the Lunar Prospector mission. Icarus 150:1-18

Konopliv AS, Binder AB, Hood LL, Kucinskas AB, Sjogren WL, Williams JG (1998) Improved gravity field of the Moon from Lunar Prospector. Science 281:1476-1480

Korotev R (1998) Concentrations of radioactive elements in lunar materials. J Geophys Res 103:1691-1701

Korotev RL (1994) Compositional variation in Apollo 16 impact-melt breccias and inferences for the geology and bombardment history of the Central Highlands of the Moon. Geochim Cosmochim Acta 58:3931-3969

Korotev RL (1996) On the relationship between the Apollo 16 ancient regolith breccias and feldspathic fragmental breccias, and the composition of the prebasin crust in the central highlands of the moon. Meteor Planet Sci 31:403-412

Korotev RL (1997) Some things we can infer about the moon from the composition of the Apollo 16 regolith. Meteor Planet Sci 32:447-478

Korotev RL (1999) A new estimate of the composition of the feldspathic upper crust of the Moon. Lunar Planet Sci XXX:1303

Korotev RL (2000) The great lunar hot spot and the composition and origin of the Apollo mafic ("LKFM") impact-melt breccias. J Geophys Res 105:4317-4345

Korotev RL, Gillis JJ (2001) A new look at the Apollo 11 regolith and KREEP. J Geophys Res 106:12,339-12,354

Korotev RL, Haskin LA (1988) Europium mass balance in polymict samples and implications for plutonic rocks of the lunar crust. Geochim Cosmochim Acta 52:1795-1813

Korotev RL, Irving AJ (2005) Compositions of three lunar meteorites: Meteorite Hills 01210, Northeast Africa 001, and Northwest Africa 3136. Lunar Planet Sci XXXVI:1220

Korotev RL, Jolliff BL (2001) The curious case of the lunar magnesian granulitic breccias. Lunar Planet Sci XXXII:1013

Korotev RL, Jolliff BL, Zeigler RA, Gillis JJ, Haskin LA (2003a) Feldspathic lunar meteorites and their implications for compositional remote sensing of the lunar surface and the composition of the lunar crust. Geochim Cosmochim Acta 67:4895-4923

Korotev RL, Jolliff BL, Zeigler RA, Haskin LA (2003b) Compositional constraints on the launch pairing of three brecciated lunar meteorites of basaltic composition. Antarctic Meteor Res 16:152-175

Korotev RL, Jolliff BL, Zeigler RA, Haskin LA (2003c) Compositional evidence for launch pairing of the YQ and Elephant Moraine lunar meteorites. Lunar Planet Sci XXXIV:1357

Korotev RL, Lindstrom MM, Lindstrom DJ, Haskin LA (1983) Antarctic meteorite ALHA81005—Not just another lunar anorthositic norite. Geophys Res Lett 10:829-832

Korotev RL, Zeigler RA, Jolliff BL (2004) Compositional constraints on the launch pairing of LAP 02205 and PCA 02007 with other lunar meteorites. Lunar Planet Sci XXXV:1416

Kreutzberger ME, Drake MJ, Jones JH (1986) Origin of Earth's Moon: Constraints from alkali volatile trace elements. Geochim Cosmochim Acta 50:91-98

Kuckes AF (1977) Strength and rigidity of the elastic lunar lithosphere and implications for present-day mantle convection in the Moon. Phys Earth Planet Inter 14:1-12

Kuehner SM, Irving AJ, Rumble D III, Hupé AC, Hupé GM (2005) Mineralogy and petrology of lunar meteorite NWA 3136: A glass-welded mare regolith breccia of mixed heritage. Lunar Planet Sci XXXVI:1228

Kuskov OL (1995) Constitution of the Moon: 3. Composition of middle mantle from seismic data. Phys Earth Planet Inter 90:55-74

Kuskov OL (1997) Constitution of the Moon: 4. Composition of the mantle from seismic data. Phys Earth Planet Inter 102:239-257

Kuskov OL, Fabrichnaya OB (1994) Constitution of the Moon: 2. Composition and seismic properties of the lower mantle. Phys Earth Planet Inter 83:197-216

Kuskov OL, Kronrod VA (1998) Constitution of the Moon: 5. Constraints on composition, density, temperature, and radius of a core. Phys Earth Planet Inter 107:285-306

Kuskov OL, Kronrod VA (2001) Core sizes and internal structure of Earth's and Jupiter's satellites. Icarus 152:204-227

Kuskov OL, Kronrod VA, Hood LL (2002) Geochemical constraints on the seismic properties of the lunar mantle. Phys Earth Planet Inter 134:175-189

Lambeck K (1988) Geophysical geodesy: The slow deformations of the Earth. Clarendon Press

Lambeck K, Pullan S (1980) The lunar fossil bulge hypothesis revisited. Phys Earth Planet Inter 22:29-35

Lammlein DR, Latham GV, Dorman J, Nakamura Y, Ewing M (1974) Lunar seismicity, structure and tectonics. Rev Geophys Space Phys 12:1-21

Langseth MG, Keihm SJ, Peters K (1976) Revised lunar heat-flow values. Proc Lunar Sci Conf 7:3143-3171

Latham G, Ewing M, Dorman J, Press F, Toksöz MN, Sutton G, Meissner F, Duennebier F, Nakamura Y, Kovach R, Yates M (1970a) Seismic data from man-made impacts on the Moon. Science 170:620-626

Latham G, Ewing M, Press F, Sutton G, Dorman J, Nakamura Y, Toksöz MN, Wiggins R, Derr J, Duennebier F (1970b) Passive seismic experiment. Science 167:455-457

Laul JC (1986) Chemistry of the Apollo 12 highland component. Proc 16th Lunar Planet Sci Conf, J Geophys Res 91: D251-D261

Lawrence DJ, Feldman WC, Barraclough BL, Binder AB, Elphic RC, Maurice S, Miller MC, Prettyman TH (2000) Thorium abundances on the lunar surface. J Geophys Res 105:20,307-20,331

Lawrence DJ, Feldman WC, Barraclough BL, Binder AB, Elphic RC, Maurice S, Thomsen DR (1998) Global elemental maps of the Moon: The Lunar Prospector gamma-ray spectrometer. Science 281:1484-1489

Lawrence DJ, Feldman WC, Elphic RC, Little RC, Prettyman TH, Maurice S, Lucey PG, Binder AB (2002) Iron aabundances on the lunar surface as measured by the Lunar Prospector gamma-ray and neutron spectrometer. J Geophys Res 107:5130, doi:10.1029/2001JE001530

Le Maitre RW, editor (1989) A Classification of Igneous Rocks and Glossary of Terms. Blackwell Scientific Publications

Lee D-C, Halliday AN (1995) Hafnium-tungsten chronometry and the timing of terrestrial core formation. Nature 378: 771-774

Lee D-C, Halliday AN (1997) Core formation on Mars and differentiated asteroids. Nature 388:854-857

Lee D-C, Halliday AN, Leya I, Wieler R, Wiechert U (2002) Cosmogenic tungsten and the origin and earliest differentiation of the Moon. Earth Planet Sci Lett 198:267-274

Lee D-C, Halliday AN, Snyder GA, Taylor LA (1997) Age and origin of the Moon. Science 278:1098-1103

Lemoine FG, Smith DE, Zuber MT, Neumann GA, Rowlands DD (1997) A 70th degree lunar gravity model (GLGM-2) from Clementine and other tracking data. J Geophys Res 102:16,339-16,359

Levi S, Banerjee SK (1976) On the possibility of obtaining relative paleointensities from lake sediments. Earth Planet Sci Lett 29:219-226

Liebermann RC, Ringwood AE (1976) Elastic properties of anorthite and the nature of the lunar crust. Earth Planet Sci Lett 31:69-74

Lin RP, Anderson KA, Hood LL (1988) Lunar surface magnetic field concentrations antipodal to young large impact basins. Icarus 74:529-541

Lin RP, Mitchell DL, Curtis DW, Anderson KA, Carlson CW, McFadden J, Acuña MH, Hood LL, Binder AB (1998) Lunar surface magnetic fields and their interaction with the solar wind: results from lunar prospector. Science 281:1480-1484

Lindgren BW (1993) Statistical theory. Chapman and Hall

Lindstrom MM, Knapp SA, Shervais JW, Taylor LA (1984) Magnesian anorthosites and associated troctolites and dunite in Apollo 14 breccias. Proc 15th Lunar Planet Sci Conf, J Geophys Res 89:C41-C49

Lindstrom MM, Lindstrom DJ (1986) Lunar granulites and their precursor anorthositic norites of the early lunar crust. Proc 16th Lunar Planet Sci Conf, J Geophys Res 91:D263-D276

Lindstrom MM, Marvin UB, Vetter SK, Shervais JW (1988) Apennine front revisited: Diversity of Apollo 15 highland rock types. Proc Lunar Planet Sci Conf 18:169-185

Lindstrom MM, Mittlefehldt DW, Martinez RR, Lipschutz MJ, Wang M-S (1991) Geochemistry of Yamato-82192, -86032 and -793274 lunar meteorites. Proc NIPR Symp, Antarct Meteorit Res 4:12-32

Lindstrom MM, Nava DF, Lindstrom DJ, Winzer SR, Lum RKL, Schuhmann PJ, Schuhmann S, Philpotts JA (1977) Geochemical studies of the white breccia boulders at North Ray Crater, Descartes region of the lunar highlands. Proc Lunar Sci Conf 8:2137-2151

Lindstrom MM, Salpas PA (1981) Geochemical studies of rocks from North Ray Crater, Apollo 16. Proc Lunar Planet Sci Conf 12B:305-322

Lognonné P (2005) Planetary seismology. Annu Rev Earth Planet Sci 33:571-604

Lognonné P, Gagnepain-Beyneix J, Chenet H (2003) A new seismic model for the Moon: Implications for structure, thermal evolution and formation of the Moon. Earth Planet Sci Lett 211:27-44

Lognonné P, Mosser B (1993) Planetary seismology. Surv Geophys 14:239-302

Longhi J (1977) Magma oceanography 2: Chemical evolution and crustal formation. Proc Lunar Sci Conf 8:601-621

Longhi J (1980) A model of early lunar differentiation. Proc Lunar Planet Sci Conf 11:289-315

Longhi J (1981) Preliminary modeling of high pressure partial melting. Implications for early lunar differentiation. Proc Lunar Sci Conf 12B:1001-1018

Longhi J (1992) Experimental petrology and petrogenesis of mare volcanics. Geochim Cosmochim Acta 56:2235-2251

Longhi J (1995) Liquidus equilibria of some primary lunar and terrestrial melts in the garnet stability field. Geochim Cosmochim Acta 59:2375-2386

Longhi J (2000) Anorthosite petrogenesis revisited. Lunar and Planetary Science XXXI:1592

Longhi J, Ashwal LD (1985) Two-stage models for lunar and terrestrial anorthosites: Petrogenesis without a magma ocean. Proc Lunar Planet Sci Conf 15:C571-C584

Longhi J, Boudreau AE (1979) Complex igneous processes and the formation of the primitive lunar crustal rocks. Proc Lunar Planet Sci Conf 10:2085-2105

Lucey PG (2004) Mineral maps of the Moon. Geophys Res Lett 31:L08701, doi:10.1029/2003GL019406

Lucey PG, Blewett DT, Hawke BR (1998a) Mapping the FeO and TiO_2 content of the lunar surface with multispectral imagery. J Geophys Res 103:3679-3699

Lucey PG, Blewett DT, Jolliff BL (2000) Lunar iron and titanium abundance algorithms based on final processing Clementine UVVIS images. J Geophys Res 105:20,297-20,305

Lucey PG, Hawke BR (1989) A remote mineralogical perspective on gabbroic units in the lunar highlands. Proc Lunar Planet Sci Conf 19:355-363

Lucey PG, Hawke BR, Pieters CM, Head JW, III, McCord TB (1986) A compositional study of the Aristarchus region of the Moon using near-infrared reflectance spectroscopy. J Geophys Res 91:D344-D354

Lucey PG, Spudis PD, Zuber M, Smith D, Malaret E (1994) Topographic-compositional units on the Moon and the early evolution of the lunar crust. Science 266:1855-1858

Lucey PG, Taylor GJ, Hawke BR, Spudis PD (1998b) FeO and TiO_2 concentrations in the South Pole-Aitken basin: Implications for mantle composition and basin formation. J Geophys Res 103:3701-3708

Lucey PG, Taylor GJ, Malaret E (1995) Abundance and distribution of iron on the Moon. Science 268:1150-1153

Ma M-S, Schmitt RA, Taylor GJ, Warner RD, Keil K (1981) Chemical and petrographic study of spinel troctolite in 67435: Implication for the origin of Mg-rich plutonic rocks. Lunar Planet Sci 12:640-642

Margot J-L, Cambell DB, Jurgens RF, Slade MA (1999a) Topography of the lunar poles from radar interferometry: A survey of cold trap locations. Science 284:1658-1660

Margot J-L, Cambell DB, Jurgens RF, Slade MA (1999b) Topography of Tycho crater. J Geophys Res 104:11,875-11,882

Martelli G, Newton G (1977) Hypervelocity cratering and impact magnetisation of basalt. Nature 269:478-480

Marti K, Aeschlimann U, Eberhardt P, Geiss J, Grogler N, Jost DT, Laul JC, Ma M-S, Schmitt RA, Taylor GJ (1983) Pieces of the ancient lunar crust: Ages and composition of clasts in consortium breccia 67915. Proc 14th Lunar Planet Sci Conf, J Geophys Res 88:B165-B175

Marvin UB, Carey JW, Lindstrom MM (1989) Cordierite-spinel troctolite, a new magnesium-rich lithology from the lunar highlands. Science 243:925-928

Marvin UB, Lindstrom MM, Bernatowicz TJ, Podosek FA, Sugiura N (1987) The composition and history of breccia 67015 from North Ray Crater. Proc 17th Lunar Planet Sci Conf, J Geophys Res 92:E472-E490

Marvin UB, Lindstrom MM, Holmberg BB, Martinez RR (1991) New observations on the quartz monzodiorite-granite suite. Proc Lunar Planet Sci Conf 21:119-135

Marvin UB, Warren PH (1980) A pristine eucrite-like gabbro from Descartes and its exotic kindred. Proc Lunar Planet Sci Conf 11:507-521

Maxwell TA, Phillips RJ (1978) Stratigraphic correlation of the radar-detected subsurface interface in Mare Crisium. Geophys Res Lett 5:811-814

McCallum IS (1998) The stratigraphy and evolution of the lunar crust. *In:* Workshop on New Views of the Moon: Integrated Remotely Sensed, Geophysical, and Sample Datasets. Lunar and Planetary Institute, p 54-55

McCallum IS (2001) A new view of the Moon in light of data from Clementine and Prospector missions. Earth Moon Planets 85-85:253-269

McCallum IS, O'Brien HE (1996) Stratigraphy of the lunar highlands crust: Depth of burial of lunar samples from cooling rate studies. Am Mineral 81:1166-1175

McCallum IS, Schwartz JM (2001) Lunar Mg suite: Thermobarometry and petrogenesis of parental magmas. J Geophys Res 106:27,969-27,983

McCord TB, Clark RN, Hawke BR, McFadden LA, Owensby PD, Pieters CM, Adams JB (1981) Moon: Near-infrared spectral reflectance, a good first look. J Geophys Res 86:10,883-10,892

McEwen AS, Robinson MS, Eliason EM, Lucey PG, Duxbury TC, Spudis PD (1994) Clementine observations of the Aristarchus region of the Moon. Science 266:1858-1861

McGovern PJ, Solomon SC (1993) State of stress, faulting, and eruption characteristics of large volcanoes on Mars. J Geophys Res 98:23,533-23,579

McGovern PJ, Solomon SC (1998) Growth of large volcanoes on Venus: Mechanical models and implications for structural evolution. J Geophys Res 103:11,071-11,101

McKay DS, Morrison DA, Clanton US, Ladle GH, Lindsay JF (1971) Apollo 12 soil and breccia. Proc Lunar Sci Conf 2:755-773

McKay GA, Wagstaff J, Le L (1990) REE distribution coefficients for pigeonite: Constraints on the origin of the mare basalt europium anomaly. LPI Tech Rep 90-02:48-49

McNutt MK (1984) Lithospheric flexure and thermal anomalies. J Geophys Res 89:11,180-11,194

Melosh HJ (1978) The tectonics of mascon loading. Proc Lunar Planet Sci Conf 9:3513-3525

Melosh HJ (1979) Acoustic fluidization: A new geologic process? J Geophys Res 84:7513-7520

Melosh HJ (1989) Impact Cratering: A Geologic Process. Oxford Univ. Press

Metzger AE, Trombka JI, Peterson LE, Reedy RC, Arnold JR (1973) Lunar surface radioactivity: Preliminary results of the Apollo 15 and Apollo 16 gamma-ray spectrometer experiments. Science 179:800-803

Meyer C, Jr., Brett R, Hubbard NJ, Morrison DA, McKay DS, Aitken FK, Takeda H, Schonfeld E (1971) Mineralogy, chemistry, and origin of the KREEP component in soil samples from the Ocean of Storms. Proc Lunar Sci Conf 2:393-411

Meyer C, Jr., Williams IS, Compston W (1989) Uranium-lead ages for lunar zircons:Evidence for prolonged period of granophyre formation from 4.32 to 3.88 Ga. Meteor Planet Sci 31:379-387

Morgan JW (1986) Ultramafic xenoliths: Clues to Earth's late accretionary history. J Geophys Res 91:12,375-12,387

Morgan JW, Ganapathy R, Higuchi H, Krähenbühl U, Anders E (1974) Lunar basins: tentative characterization of projectiles from meteoritic elements in Apollo 17 boulders. Proc Lunar Sci Conf 5:1703-1736

Morgan JW, Krähenbühl U, Ganapathy R, Anders E (1972a) Trace elements in Apollo 15 samples: implications for meteorite influx and volatile depletion on the Moon. Proc Lunar Sci Conf 3:1361-1376

Morgan JW, Laul JC, Krähenbühl U, Ganapathy R, Anders E (1972b) Major impacts on the Moon: characterization from trace elements in Apollo 12 and 14 samples. Proc Lunar Sci Conf 3:1377-1395

Morgan JW, Walker RJ, Brandon AD, Horan MF (2001) Siderophile elements in Earths's upper mantle and lunar breccias: Data sythesis suggests manifestations of the same late influx. Meteorit Planet Sci 36:1257-1275

Morris RW, Taylor GJ, Newsom HE, Keil K, Garcia SR (1990) Highly evolved and ultramafic lithologies from Apollo 14 soils. Proc Lunar Planet Sci Conf 20:61-75

Morrison DA (1998) Did a thick South Pole-Aitken basin melt sheet differentiate to form cumulates? Lunar Planet Sci XXIX:1657

Morse SA (1982) Adcumulus growth of anorthosite at the base of the lunar crust. Proc 13th Lunar Planet Sci Conf, J Geophys Res 87:A10-A18

Mueller S, Taylor GJ, Phillips RJ (1988) Lunar composition: A geophysical and petrological synthesis. J Geophys Res 93:6338-6352

Muller PM, Sjogren WL (1968) Masons: lunar mass concentrations. Science 161:680-684

Nakamura Y (1977) HFT events: Shallow moonquakes? Phys Earth Planet Inter 14:217-223

Nakamura Y (1978) A₁ moonquakes: Source distribution and mechanism. Proc Lunar Planet Sci Conf 9:3589-3607

Nakamura Y (1983) Seismic velocity structure of the lunar mantle. J Geophys Res 88:677-686

Nakamura Y (2003) New identification of deep moonquakes in the Apollo lunar seismic data. Phys Earth Planet Inter 139:197-205

Nakamura Y (2005) Farside deep moonquakes and deep interior of the Moon. J Geophys Res 110:E01001, doi:10.1029/2004JE002332

Nakamura Y, Duennebier F, Latham GV, Dorman HJ (1976) Structure of the lunar mantle. J Geophys Res 81:4818-4824

Nakamura Y, Koyama J (1982) Seismic Q of the lunar upper mantle. J Geophys Res 87:4855-4861

Nakamura Y, Lammlein D, Latham G, Ewing M, Dorman J, Press F, Toksöz MN (1973) New seismic data on the state of the deep lunar interior. Science 181:49-51

Nakamura Y, Latham G, Lammlein D, Ewing M, Duennebier F, Dorman J (1974) Deep lunar interior inferred from recent seismic data. Geophys Res Lett 1:137-140

Nakamura Y, Latham GV, Dorman HJ (1982) Apollo lunar seismic experiment Final summary. Proc 13th Lunar Planet Sci Conf, Part 1, J Geophys Res 87:A117-A123

Nakamura Y, Latham GV, Dorman HJ, Ibrahim AK, Koyama J, Horvath P (1979) Shallow moonquakes: Depth, distribution and implications as to the present state of the lunar interior. Proc Lunar Planet Sci Conf 10:2299-2309

National Research Council SSES, Space Studies Board (2003) New Frontiers in the Solar System: An Integrated Exploration Strategy. National Academies Press

Nazarov MA, Demidova SI, Patchen A, Taylor LA (2002) Dhofar 301, 302 and 303: Three new lunar highland meteorites from Oman. Lunar Planet Sci XXXII:1293

Nazarov MA, Demidova SI, Patchen A, Taylor LA (2004) Dhofar 311, 730 and 731: New lunar meteorites from Oman. Lunar Planet Sci XXXV:1233

Neal CR (2001) Interior of the Moon: The presence of garnet in the primitive deep lunar mantle. J Geophys Res 106: 27,865-27,885

Neal CR, Ely JC (2002) Sulfide immiscibility in the lunar magma ocean: evidence for a primitive lunar lower mantle and the origin of high-μ mare basalts. Lunar Planet Sci XXXIII:1821

Neal CR, Taylor LA, Lindstrom MM (1988) Apollo 14 mare basalt petrogenesis: Assimilation of KREEP-like components by a fractionating magma. Proc Lunar Planet Sci Conf 18:139-153

Neal CR, Taylor LA, Schmitt RA, Hughes SS, Lindstrom MM (1989) High-alumina (HA) and very high potassium (VHK) basalt clasts from Apollo 14 breccia, part 2, whole rock geochemistry: Further evidence for combined assimilation and fractional crystallization within the lunar crust. Proc Lunar Planet Sci Conf 19:147-161

Neumann GA, Zuber MT, Smith DE, Lemoine FG (1996) The lunar crust: Global structure and signature of major basins. J Geophys Res 101:16,841-16,843

Newhall XX, Williams JC (1997) Estimates of the lunar physical librations. Celest Mech Dyn Astron 66:21-30

Newsom H (1984) The lunar core and the origin of the Moon. EOS 65:369-370

Newsom H (1986) Constraints on the origin of the Moon from the abundance of molybdenum and other siderophile elements. *In:* Origin of the Moon. Hartmann WK, Phillips RJ, Taylor GJ (eds) Lunar and Planetary Institute, p 203-229

Newsom H, Drake MJ (1983) Experimental investigations of the partioning of phosphorus between metal and silicate phases: Implications for the Earth, Moon, and eucrite parent body. Geochim Cosmochim Acta 47:93-100

Newsom HE (1995) Composition of the solar system, planets, meteorites, and major terrestrial reservoirs. *In:* Global Earth Physics: A Handbook of Physical Constants. Ahrens TJ (ed) American Geophysical Union p 159-189

Newsom HE, Palme H (1984) The depletion of siderophile elements in the Earth's mantle: New evidence from molybdenum and tungsten. Earth Planet Sci Lett 69:354-364

Newsom HE, Sims KWW, Noll PD, Jaeger WL, Maehr SA, Beserra TB (1996) The depletion of tungsten in the bulk silicate Ear: Constraints on core formation. Geochim Cosmochim Acta 60:115-1169

Newsom HE, White WM, Jochum KP, Hofmann AW (1986) Siderophile and chalcophile element abundances in oceanic basalts, Pb isotope evolution and growth of the Earth's core. Earth Planet Sci Lett 80:299-313

Nord GL, Wandless M-V (1983) Petrology and comparative thermal and mechanical histories of clasts in breccia 62236. Proc 13th Lunar Planet Sci Conf, J Geophys Res 88:A645-A657

Norman M, Borg L, Nyquist LE, Bogard D, Snyder G, Taylor L, Lindstrom M (1998) Composition and age of the lunar highlands: Petrogenesis of ferroan noritic anorthosite 62236. Lunar Planet Sci XXVIIII:1551

Norman MD (1981) Petrology of suevitic lunar breccia 67016. Proc Lunar Planet Sci Conf 12B:235-252

Norman MD, Borg LE, Nyquist LE, Bogard DD (2003) Chronology, geochemistry, and petrology of a ferroan noritic anorthosite clast from Descartes breccia 67215: Clues to the age, origin, structure, and impact history of the lunar crust. Meteor Planet Sci 38:645-661

Norman MD, Garcia MO (1999) Primitive magmas and source characteristics of the Hawaiian plume: Petrology and geochemistry of shield picrites. Earth Planet Sci Lett 168:27-44

Norman MD, Keil K, Griffin WL, Ryan CG (1995) Fragments of ancient lunar crust: Petrology and geochemistry of ferroan noritic anorthosites from the Descartes region of the Moon. Geochim Cosmochim Acta 59:831-847

Norman MD, Ryder G (1980) Geochemical constraints on the igneous evolution of the lunar crust. Proc Lunar Planet Sci Conf 11:317-331

Norman MD, Taylor SR (1992) Geochemistry of lunar crustal rocks from breccia 67016 and the composition of the Moon. Geochim Cosmochim Acta 56:1013-1024

Nyquist LE, Shih C-Y (1992) The isotopic record of lunar volcanism. Geochim Cosmochim Acta 56:2213-2234

O'Keefe JD, Ahrens TJ (1993) Planetary cratering mechanics. J Geophys Res 98:17,011-17,028

O'Neill HSC (1991) The origin of the Moon and the early history of the Earth—chemical model. Part 1: The Moon. Geochim Cosmochim Acta 55:1143-1158

Oberst J (1987) Unusually high stress drops associated with shallow moonquakes. J Geophys Res 92:1397-1405

Oberst J (1989) Meteoroids near the Earth-Moon system as infered from temporal and spatial distribution of impacts detected by the lunar seismic network. Ph.D. thesis, University of Texas, Austin

Oberst J, Nakamura Y (1991) A search for clustering among the meteoroid impacts detected by the Apollo lunar seismic network. Icarus 91:315-325

Ockendon JR, Turcotte DL (1977) On the gravitational potential and field anomalies due to thin mass layers. Geophys J R Astron Soc 48:479-492

Palme H, Baddenhausen H, Blum K, Cendales M, Dreibus G, Hofmeister H, Kruse H, Palme C, Spettel B, Vilczek E, Wänke H, Kurat G (1978) New data on lunar samples and achondrites and a comparison of the least fractionated samples from the Earth, Moon, and the eucrite parent body. Proc Lunar Planet Sci Conf 9:25-57

Palme H, Nickel KG (1986) Ca/Al ratio and composition of the Earth's primitive upper mantle. Geochim Cosmochim Acta 49:2123-2132

Papike JJ, Fowler GW, Shearer CK (1994) Orthopyroxene as a recorder of lunar crust evolution: Am ion microprobe investigation of Mg-suite norites. Am Mineral 79:796-800

Papike JJ, Fowler GW, Shearer CK, Layne GD (1996) Ion microprobe investigation of plagioclase from lunar Mg-suite norites: Implications for calculating parental melt REE concentrations and for assessing post-crystallization REE distribution. Geochim Cosmochim Acta 60:3967-3978

Papike JJ, Ryder G, Shearer CK (1998) Lunar samples. Rev Mineral 36:5.1-5.234

Parker RL (1972) The rapid calculation of potential anomalies. Geophys J R Astron Soc 31:447-455

Parmentier EM, Zhong S, Zuber MT (2002) Gravitational differentiation due to initial chemical stratification: origin of lunar asymmetry by the creep of dense KREEP? Earth Planet Sci Lett 201:473-480

Peale SJ, Cassen P (1978) Contribution of tidal dissipation to lunar thermal history. Icarus 36:245-269

Peeples WJ, Sill WR, May TW, Ward SH, Phillips RJ, Jordan RL, Abbott EA, Killpack TJ (1978) Orbital radar evidence for lunar subsurface layering in Maria Serenitatis and Crisium. J Geophys Res 83:3459-3468

Phillips RJ, Adams GF, Brown WE, Jr., Eggleton RE, Jackson P, Jordan R, Peeples WJ, Porcello LJ, Ryu J, Schaber GG, Sill WR, Thompson TW, Ward SH, Zelenka JS (1973) The Apollo 17 lunar sounder. Proc Lunar Sci Conf 4: 2821-2831

Phillips RJ, Dvorak J (1981) The origin of lunar mascons: analysis of the Bouguer gravity associated with Grimaldi. *In:* Multi-ring Basins. Schultz PH, Merrill RB (eds) Pergamon Press, p 91-104

Phillips RJ, Lambeck K (1980) Gravity fields of the terrestrial planets: long-wavelength anomalies and tectonics. Rev Geophys Space Phys 18:27-76

Phinney WC (1994) FeO and MgO in plagioclase of lunar anorthosites: Igneous or metamorphic? Lunar Planet Sci XXV:1081-1082

Phipps Morgan J, Blackman DK (1993) Inversion of combined gravity and bathymetry data for crustal structure: A prescription for downward continuation. Earth Planet Sci Lett 119:167-179

Pieters CM (1978) Mare basalt types on the front side of the Moon: A summary of spectral reflectance data. Proc Lunar Planet Sci Conf 9:2825-2849

Pieters CM (1986) Composition of the lunar highland crust from near-infrared spectroscopy. Rev Geophys 24:557-578

Pieters CM (1991) Bullialdus: Strengthening the case for lunar plutons. Geophys Res Lett 18:2129-2132

Pieters CM, Duke M, Head JWI, Jolliff B (2003) Science options for sampling South Pole-Aitken Basin. Lunar Planet Sci XXXIV:1366

Pieters CM, et al. (1993) Crustal diversity of the Moon: Compositional analyses of Galileo solid state imaging data. J Geophys Res 98:17,127-17,148

Pieters CM, Head JW, III, Gaddis L, Jolliff BL, Duke M (2001) Rock types of the South Pole-Aitken basin and extent of basaltic volcanism. J Geophys Res 106:28,001-28,022

Pieters CM, Tompkins S, Head JW, Hess PC (1997) Mineralogy of the mafic anomaly in the South Pole-Aitken Basin: Implications for excavation of the lunar mantle. Geophys Res Lett 24:1903-1906

Presnall DC (1995) Phase diagrams of Earth-forming minerals. *In:* Mineral Physics and Crystallography: A Handbook of Physical Constants (AGU reference shelf 2). Ahrens TJ (ed) American Geophysical Union, p 248-268

Press WH, Teukolsky SA, Vetterling WT, LFlannery BP (1992) Numerical Recipes in Fortran 77: The Art of Scientific Computing. Cambridge Univ. Press

Pritchard ME, Stevenson DJ (2000) Thermal aspects of a lunar origin by giant impact. *In:* Origin of the Earth and Moon. Canup R, Righter K (eds) Univ. Arizona Press, p 179-196

Pullan S, Lambeck K (1981) Mascons and loading of the lunar lithosphere. Proc Lunar Planet Sci Conf 12:853-865

Pérez-Gussinyé M, Lowry AR, Watts AB, Velicogna I (2004) On the recovery of effective elastic thickness using spectral methods: Examples from synthetic data and from the Fennoscandian Shield. J Geophys Res 109:B10409, doi:10.1029/2003JB002788

Quick JE, Albee AL, Ma M-S, Murali AV, Schmitt RA (1977) Chemical composition and possible immiscibility of two silicate melts in 12013. Proc Lunar Sci Conf 8:2153-2189

Raedeke LD, McCallum IS (1980) A comparison of fractionation trends in the lunar crust and the Stillwater Complex. *In:* Proc. Conf. Lunar Highlands Crust. Papike JJ, Merrill RB (eds) Pergamon Press, p 133-153

Rees CE, Thode HG (1972) Sulfur concentrations and isotope ratios in lunar samples. Proc Lunar Sci Conf 3:1479-1485

Rees CE, Thode HG (1974) Sulfur concentrations and isotope ratios in Apollo 16 and 17 samples. Proc Lunar Sci Conf 5:1963-1973

Reid AM, Duncan AR, Richardson SH (1977) In search of LKFM. Proc Lunar Sci Conf 8:2321-2338

Richmond NC, Hood LL, Halekas JS, Mitchell DL, Lin RP, Acuña MH, Binder AB (2003) Correlation of strong lunar magnetic anomaly with a high-albedo region of the Descartes mountains. Geophys Res Lett 30:1395, doi: 10.1029/2003GL016938

Ridley WI, Hubbard NJ, Rhodes JM, Wiesmann H, Bansal BM (1973) The petrology of lunar breccia 15445 and petrogenetic implications. J Geol 81:621-631

Righter K (2002) Does the Moon have a metallic core? Constraints from giant-impact modeling and siderophile elements. Icarus 158:1-13

Righter K, Drake M (1996) Core formation in Earth's Moon, Mars, and Vesta. Icarus 124:513-529

Righter K, Drake MJ (1997) Metal-silicate equilibrium in a homogeneously accreting Ear: New results for Re. Earth Planet Sci Lett 146:541-553

Righter K, Drake MJ (1999) Effect of water on metal-silicate partioning of siderophile elements: A high pressure and temperature terrestrial magma ocean and core formation. Earth Planet Sci Lett 171:383-399

Righter K, Drake MJ (2000) Metal-silicate equilibrium in the early Ear: New constraints from volatile moderately siderophile elements Ga, Sn, Cu, P. Geochim Cosmochim Acta 64:3581-3597

Righter K, Hervig RL, Kring D (1998) Accretion and core formation in Mars: Molybdenum contents of melt inclusion glasses from three SNC meteorites. Geochim Cosmochim Acta 62:2167-2177

Righter K, Shearer CK (2003) Magmatic fractionation of Hf and W: constraints on the timing of core formation and differentiation in the Moon and Mars. Geochim Cosmochim Acta 67:2497-2507

Righter K, Walker RJ, Warren PH (2000) The origin and significance of highly siderophile elements in the lunar and terrestrial mantles. *In:* Origin of the Earth and Moon. Canup R, Righter K (eds) Univ. Arizona Press, p 291-322

Ringwood AE (1979) Origin of the Earth and Moon. Springer-Verlag

Ringwood AE, Kesson SE, Hibberson W (1981) Rhenium depletion in mare basalts and redox state of the lunar interior. Lunar Planet Sci XII:891-893

Ringwood AE, Seifert S (1986) Nickel-cobalt abundance systematics and their bearing on lunar origin. *In:* Origin of the Moon. Hartmann WK, Phillips RJ, Taylor GJ (eds) Lunar and Planetary Institute, p 249-277

Roberts PH, Glatzmaier GA (2000) Geodynamo theory and simulations. Rev Mod Phys 72:1081-1123

Rochette P (2000) Comment on "Initial measurements of the lunar induced magnetic dipole moment using Lunar Prospector magnetometer data" by Hood et al. Geophys Res Lett 27:1077-1078

Roddy DJ (1977) Large-scale impact and explosion craters: Comparisons of morphological and structural analogs. *In:* Impacts and Explosion Cratering. Roddy DJ, Pepin RO, Merrill RB (eds) Pergamon Press, p 815-841

Rose HJ Jr., Baedecker PA, Berman S, Christian RP, Dwornik EJ, Finkelman RB, Schnepfe MM (1975) Chemical composition of rocks and soils returned by the Apollo 15, 16, and 17 missions. Proc Lunar Sci Conf 6:1363-1373

Rubie DC, Melosh HJ, Reid JE, Liebske C, Righter K (2003) Mechanisms of metal-silicate equilibration in the terrestrial magma ocean. Earth Planet Sci Lett 205:

Runcorn SK (1994) The early magnetic-field and primeval satellite of the Moon: Clues to planetary formation. Philos Trans R Soc London Ser A 349:181-196

Runcorn SK (1996) The formation of the lunar core. Geochim Cosmochim Acta 60:1205-1208

Russell CT, Coleman PJ, Goldstein BE (1981) Measurements of the lunar induced magnetic moment in the geomagnetic tail: Evidence for a lunar core. Proc Lunar Planet Sci Conf 12:831-836

Russell SS, Folco L, Grady MM, Zolensky ME, Jones R, Righter K, Zipfel J, Grossman JN (2004) The Meteoritical Bulletin No. 88, 2004, July. Meteorit Planet Sci 39:A215-A272

Russell SS, Zipfel J, Folco L, Jones R, Grady MM, McCoy T, Grossman JN (2003) The Meteoritical Bulletin, No. 87, 2003, July. Meteorit Planet Sci 38:A194

Russell SS, Zipfel J, Grossman JN, Grady MM (2002) The Meteoritical Bulletin, No. 86, 2002, July. Meteor Planet Sci 37:A157-A184

Ryder G (1976) Lunar sample 15405: Remnant of a KREEP basalt-granite differentiated pluton. Earth Planet Sci Lett 29:255-268

Ryder G (1979) The chemical components of highlands breccias. Proc Lunar Planet Sci Conf 10:561-581

Ryder G (1985) Catalog of Apollo 15 Rocks. Curatorial Publication 20787. NASA Johnson Space Center, Houston

Ryder G (1991) Lunar ferroan anorthosites and mare basalt sources: The mixed connection. Geophys Res Lett 18:2065-2068

Ryder G (1994) Coincidence in time of the Imbrium basin impact and Apollo 15 KREEP volcanic flows: The case for impact-induced melting. Spec Paper Geol Soc Am 293:11-18

Ryder G, Koeberl C, Mojzsis SJ (2000) Heavy bombardment of the Earth at ~3.85 Ga: The search for petrologic and geochemical evidence. *In:* Origin of the Earth and Moon. Canup R, Righter K (eds) Univ. Arizona Press, p 475-492

Ryder G, Martinez RR (1991) Evolved hypabyssal rocks from Station 7, Apennine Front, Apollo 15. Proc Lunar Planet Sci 21:137-150

Ryder G, Norman MD (1978) Catalog of Pristine Nonmare Materials Part 2. Anorthosites. NASA Johnson Space Center

Ryder G, Norman MD (1979) Catalog of Pristine Non-mare Materials Part 1, Non-Anorthosites (Revised). NASA Johnson Space Center Curatorial Facility

Ryder G, Norman MD (1980) Catalog of Apollo 16 Rocks. NASA Johnson Space Center Curatorial Facility

Ryder G, Norman MD, Taylor GJ (1997) The complex stratigraphy of the highland crust in the Serenitatis region of the Moon inferred from mineral fragment chemistry. Geochim Cosmochim Acta 61:1083-1105

Ryder G, Sherman SB (1989) The Apollo 15 coarse fines (4-10 mm). National Aeronautics and Space Administration Lyndon B. Johnson Space Center, TM-101934

Ryder G, Spudis PD (1987) Chemical composition and origin of Apollo 15 impact melts. Proc 17th Lunar Planet Sci Conf, J Geophys Res 92:E432-E446

Ryder G, Stoeser DB, Wood JA (1977) Apollo 17 KREEPy basalt: A rock type intermediate between KREEP and mare basalts. Earth Planet Sci Lett 35:1-13

Ryder G, Wood JA (1977) Serenitatis and Imbrium impact melts: Implications for large-scale layering in the lunar crust. Proc Lunar Sci Conf 8:655-668

Sack RO, Ghiorso MS (1991) Chromian spinel as petrogenetic indicators: Thermodynamic and petrologic applications. Am Mineral 76:827-847

Salpas PA, Taylor LA, Lindstrom MM (1987) Apollo 17 KREEPy basalts: Evidence for the non-uniformity of KREEP. Proc 17th Lunar Planet Sci Conf, J Geophys Res 92:E340-E348

Sato M (1978) Oxygen fugacity of basaltic magmas and the role of gas forming elements. Geophys Res Lett 5:447-449

Sato M (1979) The driving mechanism of lunar pyroclastic eruptions inferred from the oxygen fugacity behavior of the Apollo 17 orange glass. Proc Lunar Planet Sci Conf 10:311-325

Sato M, Helz RT (1971) Oxygen fugacity values of Apollo 12 basaltic rocks. Lunar Sci II:144-145

Sato M, Hickling NL, McLane JE (1973) Oxygen fugacity values of Apollo 12, 14 and 15 lunar samples and reduced state of lunar magmas. Proc Lunar Sci Conf 4:1061-1079

Schaber GG (1973) Lava flows in Mare Imbrium: geologic evaluation from Apollo orbital photography. Proc Lunar Sci Conf 4:73-92

Schaber GG, Boyce JM, Moore HJ (1976) The scarcity of mappable flow lobes on the lunar maria: Unique morphology of the Imbrium flows. Proc Lunar Sci Conf 7:2783-2800

Schaeffer GA, Schaeffer OA (1977) ^{39}Ar-^{40}Ar ages of lunar rocks. Proc Lunar Sci Conf 8:2253-2300

Schoenberg R, Kamber BS, Collerson KD, Eugster O (2002) New W-isotope evidence for rapid terrestrial accretion and very early core formation. Geochim Cosmochim Acta 66:3151-3160

Schultz PH (1997) Forming the South-Pole Aitken basin: the extreme games. Lunar Planet Sci XXVIII:1787

Schultz PH, Spudis PD (1983) Beginning and end of lunar mare volcanism. Nature 302:233-236

Schultz RA, Zuber MT (1994) Observations, models, and mechanisms of failure of surface rocks surrounding planetary surface loads. J Geophys Res 99:14,691-14,702

Segatz M, Spohn T, Ross N, Schuber G (1988) Tidal dissipation, surface heat flow, and figures of viscoelastic models of Io. Icarus 75:187-206

Selkin PA, Tauxe L (2000) Long-term variations in palaeointensity. Philos Trans R Soc London Ser A 358:1065-1088

Sellers PC (1992) Seismic evidence for a low-velocity lunar core. J Geophys Res 97:11,663-11,672

Sharpton VL, Head JW III (1982) Stratigraphy and structural evolution of southern Mare Serenitatis: A reinterpretation based on Apollo lunar sounder experiment data. J Geophys Res 87:10,983-10,998

Shearer CK, Floss C (2000) Evolution of the Moon's mantle and crust as reflected in trace-element microbeam studies of lunar magmatism. *In:* Origin of the Earth and Moon. Canup R, Righter K (eds) Univ. Arizona Press, p 339-359

Shearer CK, Papike JJ (1993) Exploring volcanism on the Moon: A perspective from volcanic picritic glass beads. Geochim Cosmochim Acta 57:4785-4812

Shearer CK, Papike JJ (1999) Magmatic evolution of the Moon. Am Mineral 84:1469-1494

Shearer CK, Papike JJ, Layne GD (1996) The role of ilmenite in the source region for mare basalts: Evidence from niobium, zirconium, and cerium in picritic glasses. Geochim Cosmochim Acta 60:3521-3530

Shearer CK, Papike JJ, Simon SB, Galbreath KC, Shimizu N (1990) Ion microprobe studies of REE and other trace elements in Apollo 14 'volcanic' glass beads and comparison to Apollo 14 mare basalts. Geochim Cosmochim Acta 54:851-867

Shearer CK, Righter K (2003) Behavior of tungsten and hafnium in silicates: a crystal chemical basis for understanding the early evolution of the terrestrial planets. Geophys Res Lett 30:doi:10.1029/2002GL015523

Shearer CK, Weidenbeck MG, Fowler GW, Papike JJ (1997) Volatiles in planetary mantles. The behavior of sulfur in lunar picritic magmas and the Moon's mantle. Geol Soc Amer Abstr with Prog 29:A-192

Shearer CK, Weidenbeck MG, Fowler GW, Papike JJ (1998) S and other volatiles in lunar picritic magmas and the lunar mantle. An approach using secondary ion mass spectrometry. Lunar Planet Sci XXVIIII:1284

Shervais JW, McGee JJ (1998) Ion and electron microprobe study of troctolites, norite, and anorthosites from Apollo 14: Evidence for urKREEP assimilation during petrogenesis of Apollo 14 Mg-suite rocks. Geochim Cosmochim Acta 62:3009-3023

Shervais JW, McGee JJ (1999) KREEP in the western lunar highlands: Ion and electron microprobe study of alkali anorthosites and norites from Apollo 14. Am Mineral 84:806-820

Shervais JW, Taylor LA (1986) Petrologic constraints on the origin of the Moon. *In:* Origin of the Moon. Hartmann WK, Phillips RJ, Taylor GJ (eds) Lunar and Planetary Institute, p 173-201

Shervais JW, Taylor LA, Laul JC (1983) Ancient crustal components in Fra Mauro breccias. Proc 14th Lunar Sci Conf, J Geophys Res 88:B177-B192

Shervais JW, Taylor LA, Laul JC, Smith MR (1984) Pristine highlands clasts in consortium breccia 14305: Petrology and geochemistry. Proc 15th Lunar Planet Sci Conf, J Geophys Res 89:C25-C40

Shih C-Y, Nyquist LE, Bansal BM, Wiesmann H (1992) Rb-Sr and Sm-Nd chronology of an Apollo 17 KREEP basalt. Earth Planet Sci Lett 108:203-215

Shih C-Y, Nyquist LE, Bogard DD, Bansal BM, Wiesmann H (1993) Ages of pristine noritic clasts from lunar breccia 15445 and 15455. Geochim Cosmochim Acta 57:915-931

Simmons G, Siegfried R, Richter D (1975) Characteristics of microcracks in lunar samples. Proc Lunar Sci Conf 6: 3227-3254

Simmons G, Todd T, Wang H (1973) The 25-km discontinuity: Implications for lunar history. Science 182:158-161

Simon SB, Papike JJ (1985) Petrology of the Apollo 12 highland component. Proc 16th Lunar Planet Sci Conf, J Geophys Res 90:D47-D60

Simons FJ, Dahlen FA, Wieczorek MA (2006) Spatiospectral concentration on a sphere. SIAM Rev (in press)

Simons FJ, Zuber MT, Korenaga J (2000) Isostatic response of the Australian lithosphere: Estimations of effective elastic thickness and anisotropy using multitaper spectral analysis. J Geophys Res 105:19,163-19,184

Simons M, Solomon SC, Hager BH (1997) Localization of gravity and topography: constraints on the tectonics and mantle dynamics of Venus. Geophys J Int 131:24-44

Sjogren WL, Smith JC (1976) Quantitative mass distribution models for Mare Orientale. Proc Lunar Planet Sci Conf 7:2639-2648

Smith DE, Zuber MT, Neumann GA, Lemoine FG (1997) Topography of the Moon from Clementine lidar. J Geophys Res 102:1591-1611

Smith JV, Anderson AT, Newton RC, Olsen EJ, Wyllie PJ, Crewe AV, Isaacson MS, Johnson D (1970) Petrologic history of the Moon inferred from petrography, mineralogy, and petrogenesis of Apollo 11 rocks. *In:* Proceedings of the Apollo 11 Lunar Science Conference. Pergamon Press, p 897-925

Snyder GA, Borg LE, Nyquist LE, Taylor LA (2000) Chronology and isotopic constraints on lunar evolution. *In:* Origin of the Earth and Moon. Canup R, Righter K (eds) Univ. Arizona Press, p 361-395

Snyder GA, Neal CR, Taylor LA (1995a) Processes involved in the formation of magnesian-suite plutonic rocks from the highlands of the Earth's Moon. J Geophys Res 100:9365-9388

Snyder GA, Taylor LA, Halliday A (1995b) Chronology and petrogenesis of the lunar highlands alkali suite: Cumulates from KREEP basalt crystallization. Geochim Cosmochim Acta 59:1185-1203

Snyder GA, Taylor LA, Liu Y-G, Schmitt RA (1992a) Petrogenesis of the western highlands of the Moon: Evidence from a diverse group of whitlockite-rich rocks from the Fra Mauro Formation. Proc Lunar Planet Sci 22:399-416

Snyder GA, Taylor LA, Neal CR (1992b) A chemical model for generating the sources of mare basalts: Combined equilibrium and fractional crystallization of the lunar magmasphere. Geochim Cosmochim Acta 56:3809-3823

Sohl F, Spohn T (1997) The interior structure of Mars: Implications from SNC meteorites. J Geophys Res 102:1613-1635

Solomon SC (1978) The nature of isostasy on the Moon: How big of a Pratt-fall for Airy methods. Proc Lunar Planet Sci Conf 9:3499-3511

Solomon SC, Chaiken J (1976) Thermal expansion and thermal stress in the Moon and terrestrial planets: Clues to early thermal history. Proc Lunar Sci Conf 7:3229-3243

Solomon SC, Head JW (1979) Vertical movement in mare basins: Relation to mare emplacement, basin tectonics, and lunar thermal history. J Geophys Res 84:1667-1682

Solomon SC, Head JW (1980) Lunar mascon basins: lava filling, tectonics, and evolution of the lithosphere. Rev Geophys Space Phys 18:107-141

Solomon SC, Longhi J (1977) Magma oceanography, 1, Thermal evolution. Proc Lunar Sci Conf 8:583-599

Sonnett CP (1982) Electromagnetic induction in the Moon. Rev Geophys Space Phys 20:411-455

Sonnett CP, Smith BF, Colburn DS, Schubert G, Schwartz K (1972) The induced magnetic field of the moon: Conductivity profiles and inferred temperature. Proc Lunar Sci Conf 3:2309-2336

Spera FJ (1992) Lunar magma transport phenomena. Geochim Cosmochim Acta 56:2253-2265

Spudis P, Pieters C (1991) Global and regional data about the Moon (Chapter 10). *In:* Lunar Sourcebook: A Users Guide to the Moon. Heiken GH, Vaniman DT, French BM (eds) Cambridge University Press, p 595-632

Spudis PD (1978) Composition and origin of the Apennine Bench Formation. Proc Lunar Planet Sci Conf 9:3379-3394

Spudis PD (1993) The Geology of Multi-Ring Impact Basins. Cambridge Univ. Press

Spudis PD, Davis PA (1986) A chemical and petrological model of the lunar crust and implications for lunar crustal origin. Proc 17th Lunar Planet Sci Conf, Part 1, J Geophys Res suppl. 91:E84-E90

Spudis PD, Hawke BR, Lucey P (1984) Composition of Orientale basin deposits and implications for the lunar basin-forming process. Proc 15th, Lunar Planet Sci Conf, Part 1, J Geophys Res, suppl. 89:C197-C210

Spudis PD, Hawke BR, Lucey PG, Taylor GJ, Stockstill K (1996) Composition of the ejecta deposits of selected lunar basins from Clementine elemental maps. Lunar Planet Sci XXVII:1255-1256

Spudis PD, Reisse RA, Gillis JJ (1994) Ancient multiring basins on the Moon revealed by Clementine laser altimetry. Science 266:1848-1851

Spudis PD, Ryder G (1981) Apollo 17 impact melts and their relation to the Serenitatis basin. Multi-Ring Basins, Proc Lunar Planet Sci Conf 12A:133-148

Spudis PD, Ryder G, Taylor GJ, McCormick KA, Keil K, Grieve RAF (1991) Source of mineral fragments in impact melts 15445 and 15455: Toward the origin of Low-K Fra Mauro basalt. Proc Lunar Planet Sci Conf 21:151-165

Srnka LJ (1977) Spontaneous magnetic field generation in hypervelocity impacts. Proc Lunar Sci Conf 8:785-792

Srnka LJ, Martelli G, Newton G, Cisowski SM, Fuller MD, Schaal RB (1979) Magnetic field and shock effects and remanent magnetization in a hypervelocity impact experiment. Earth Planet Sci Lett 42:127-137

Staid MI, Pieters CM (2000) Integrated spectral analysis of mare soils and craters: Application to eastern nearside basalts. Icarus 145:122-139

Staid MI, Pieters CM (2001) Mineralogy of the last lunar basalts: Results from Clementine. J Geophys Res 106:27887-27900

Stegman DR, Jellinek AM, Zatman SA, Baumgardner JR, Richards MA (2003) An early lunar core dynamo driven by thermochemical mantle convection. Nature 421:143-146

Steinhart J (1967) Mohorovicic discontinuity. *In:* International Dictionary of Geophysics. Runcorn S (ed) Pergamon Press, p 991-994

Stephenson A, Collinson DW, Runcorn SK (1974) Lunar magnetic field paleointensity determinations on Apollo 11, 16, and 17 rocks. Proc Lunar Planet Sci Conf 10:2859-2871

Stevenson DJ (2003) Planetary magnetic fields. Earth Planet Sci Lett 208:1-11

Streckeisen A (1976) To each plutonic rock, its proper name. Earth Sci Rev 12:1-33

Stöffler D, Bischoff A, Borchardt R, Burghele A, Deutsch A, Jessberger EK, Ostertag R, Palme H, Spettel B, Reimold WU, Wacker K, Wänke H (1985) Composition and evolution of the lunar crust in the Descartes Highlands, Apollo 16. Proc 15th Lunar Planet Sci Conf, Part 2, J Geophys Res 90:C449-C506

Stöffler D, Knoll H-D, Marvin UB, Simonds CH, Warren PH (1980) Recommended classification and nomenclature of lunar highland rock—committee report. *In:* Proc. Conf Lunar Highland Crust. Papike JJ, Merrill RB (eds) Pergamon Press, p 51-70

Sugano T, Heki K (2004) Isostasy of the Moon from high-resolution gravity and topography data: Implication for its thermal history. Geophys Res Lett 31:L24703, doi:10.1029/2004GL022059

Sugiura N (1979) ARM, TRM, and magnetic interactions: concentration dependence. Earth Planet Sci Lett 42:451-455

Sugiura N, Strangway DW (1980) Comparisons of magnetic paleointensity methods using a lunar sample. Proc Lunar Planet Sci Conf 9:1801-1813

Sugiura N, Strangway DW (1983) Magnetic paleointensity determination on lunar sample 62235. Proc 13th Lunar Planet Sci Conf, Part 2, J Geophys Res, suppl. 88:A684-690

Tauxe L (1993) Sedimentary records of relative paleointensity of the geomagnetic field: Theory and practice. Rev Geophys 31:319-354

Taylor GJ, Hawke BR, Spudis PD (2002) Bulk composition of the Moon: Importance, uncertainties, and what we need to know. *In:* Workshop on Moon Beyond 2002: Next Steps in Lunar Exploration abstract 3049. Lunar and Planetary Institute, Houston, Taos, New Mexico

Taylor GJ, Warren P, Ryder G, Delano J, Pieters C, G. L (1991) Lunar Rocks. *In:* Lunar Sourcebook, A User's Guide to the Moon. Heiken GH, Vaniman DT, French BM (eds) Cambridge University Press, p 183-284

Taylor LA, Mosie AB (1979) Breccia Guidebook #3, 67915. NASA Johnson Space Center

Taylor LA, Nazarov MA, Cohen BA, Warren PH, Barsukova LD, Clayton RN, Mayeda TK (2001) Bulk chemistry and oxygen isotopic compositions of lunar meteorites Dhofar 025 and Dhofar 026: A second-generation impact melt. Lunar Planet Sci XXXII:1985

Taylor LA, Shervais JW, Hunter RH, Shih CY, Bansal BM, Wooden J, Nyquist LE, Laul LC (1983) Pre-4.2 AE mare-basalt volcanism in the lunar highlands. Earth Planet Sci Lett 66:33-47

Taylor SR (1975) Lunar Science: a Post-Apollo View; Scientific Results and Insights from the Lunar Samples. Pergamon Press

Taylor SR (1982) Planetary Science: A Lunar Perspective. Lunar and Planetary Institute

Taylor SR (1992) Solar System Evolution: A New Perspective. Cambridge Univ. Press

Taylor SR, Rudowski R, Muir P, Graham A, Kaye M (1971) Trace element chemistry of lunar samples from the Ocean of Storms. Proc Lunar Sci Conf 2:1083-1099

Tera F, Papanastassiou DA, Wasserburg GJ (1974) Isotopic evidence for a terminal lunar cataclysm. Earth Planet Sci Lett 22:1-21

Thurber CH, Solomon SC (1978) An assessment of crustal thickness variations on the lunar near side: models, uncertainties, and implications for crustal differentiation. Proc Lunar Planet Sci Conf 9:3481-3497

Thurber CH, Toksöz MN (1978) Martian lithospheric thickness from elastic flexure theory. Geophys Res Lett 5:977-980

Todd T, Richter DA, Simmons G, Wang H (1973) Unique characterization of lunar samples by physical properties. Proc Lunar Sci Conf 4:2639-2662

Toksöz MN, Dainty AM, Solomon SC, Anderson KR (1974) Structure of the Moon. Rev Geophys 12:539-567

Toksöz MN, Press F, Dainty A, Anderson K, Latham G, Ewing M, Dorman J, Lammlein D, Sutton G, Duennebier F (1972) Structure, composition, and properties of lunar crust. Proc Lunar Sci Conf 3:2527-2544

Tompkins S, Pieters CM (1999) Mineralogy of the lunar crust: Results from Clementine. Meteorit Planet Sci 34:25-41

Ulrych TJ, Sacchi MD, Woodbury A (2001) A Bayes tour of inversion: A tutorial. Geophysics 66:55-69

Vinnik L, Chenet H, Gagnepain-Beyneix J, Lognonné P (2001) First seismic receiver functions on the Moon. Geophys Res Lett 28:3031-3034

Vocadlo L, Alfè D, Price GD, Gillan MJ (2000) First principles calculations on the diffusivity and viscosity of liquid Fe-S at experimentally accessible conditions. Phys Earth Planet Int 120:145-152

von Frese RRB, Tan L, Potts LV, Kim JW, Merry CJ, Bossler JD (1997) Lunar crustal analysis of Mare Orientale from topographic and gravity correlations. J Geophys Res 102:25,657-25,676

Wagner TP, Grove TL (1997) Experimental constraints on the origin of lunar high-Ti ultramafic glasses. Geochim Cosmochim Acta 61:1315-1327

Walker D, Norby L, Jones JH (1993) Superheating effects on metal/silicate partitioning of siderophile elements. Science 262:1858-1861

Walter MJ, Newsom H, Ertel W, Holzheid A (2000) Siderophile elements in the Earth and Moon: Metal/silicate partitioning and implications for core formation. *In:* Origin of the Earth and Moon. Canup R, Righter K (eds) Univ. Arizona Press, p 265-290

Wang H, Todd T, Richter D, Simmons G (1973) Elastic properties of plagioclase aggregates and seismic velocities in the Moon. Proc Lunar Sci Conf 4:2663-2671

Wänke H, Baddenhausen H, Balacescu A, Teschke F, Spettel B, Dreibus G, Palme H, Quijano-Rico M, Kruse H, Wlotzka F, Begemann F (1972) Multi-element analyses of lunar samples and some implications of the results. Proc Lunar Sci Conf 3:1251-1268

Wänke H, Baddenhausen H, Blum K, Cendales M, Dreibus G, Hofmeister H, Kruse H, Jagoutz E, Palme C, Spettel B, Thacker R, Vilczek E (1977) On the chemistry of lunar samples and achondrites. Proc Lunar Planet Sci Conf 8: 2191-2213

Wänke H, Dreibus G (1986) Geochemical evidnece for the formation of the Moon by impact-induced fission of the proto-Earth. *In:* Origin of the Moon. Hartmann WK, Phillips RJ, Taylor GJ (eds) Lunar and Planetary Institute, p 649-672

Wänke H, Palme H, Baddenhausen H, Dreibus G, Jagoutz E, Kruse H, Palme C, Spettel B, Teshke F, Thacker R (1975) New data on the chemistry of lunar samples: Primary matter in the lunar highlands and the bulk composition of the Moon. Proc Lunar Sci Conf 6:1313-1340

Wänke H, Palme H, Baddenhausen H, Dreibus G, Jagoutz E, Kruse H, Spettel B, Teschke F, Thacker R (1974) Chemistry of Apollo 16 and 17 samples: Bulk composition, late stage accumulation and early differentiation of the Moon. Proc Lunar Sci Conf 5:1307-1335

Ward WR (1975) Past Orientation of the lunar spin axis. Science 189:377-379

Warner JL, Phinney WC, Bickel CE, Simonds CH (1977) Feldspathic granulitic impactites and pre-final bombardment lunar evolution. Proc Lunar Sci Conf 8:2051-2066

Warner RD, Taylor GJ, Keil K (1980) Petrology of 60035: Evolution of a polymict ANT breccia. *In:* Proc. Conf. Lunar Highlands Crust. Papike JJ, Merrill RB (eds) Pergamon, p 377-394

Warner RD, Taylor GJ, Mansker WL, K. K (1978) Clast assemblage of possible deep-seated (77517) and immiscible-melt (77538) origins in Apollo 17 breccias. Proc Lunar Planet Sci Conf 9:941-958

Warren P (1994) Lunar and martian meteorite delivery services. Icarus 111:338-363

Warren PH (1985) The magma ocean concept and lunar evolution. Annu Rev Earth Planet Sci 13:201-240

Warren PH (1988) The origin of pristine KREEP: Effects of mixing between urKREEP and the magmas parental to the Mg-rich cumulates. Proc Lunar Planet Sci Conf 18:233-241

Warren PH (1989) KREEP: Major-element diversity, trace-element uniformity (almost). *In:* Workshop on Moon in Transition: Apollo 14, KREEP, and Evolved Lunar Rocks LPI Tech. Report. 89-03:149-153. Lunar and Planetary Institute

Warren PH (1990) Lunar anorthosites and the magma-ocean plagioclase-flotation hypothesis: importance of FeO enrichment in the parent magma. Am Mineral 75:46-58

Warren PH (1993) A concise compilation of petrologic information on possibly pristine nonmare Moon rocks. Am Mineral 78:360-376

Warren PH (2000) Bulk composition of the Moon as constrained by Lunar Prospector Th data. I. Application of ground truth for calibration. Lunar Planet Sci XXXI:1756

Warren PH (2001) Early lunar crustal genesis: The ferroan anorthosite epsilon-neodymium paradox as a possible result of crustal overturn. Meteorit Planet Sci 36(supplement):A219

Warren PH, Bridges JC (2004) Lunar meteorite Yamato-983885: A relatively KREEPy regolith breccia not paired with Y-791197. *In:* 67th Annual Meteoritical Society Meeting #5095. Meteorit Planet Sci, August 2-6, 2004, Rio de Janeiro, Brazil

Warren PH, Humphrys TL (2003) Bulk composition of the moon as constrained by thorium data: Comparison of Lunar Prospector versus Apollo GRS results. Lunar Planet Sci XXXIV:2034

Warren PH, Jerde EA, Kallemeyn GW (1987) Pristine Moon rocks: A "large" felsite and a metal-rich ferroan anorthosite. Proc 17th Lunar Planet Sci Conf, J Geophys Res 92:E303-E313

Warren PH, Jerde EA, Kallemeyn GW (1990) Pristine moon rocks: An alkali anorthosite with coarse augite exsolution from plagioclase, a magnesian harzburgite, and other oddities. Proc Lunar Planet Sci Conf 20:31-59

Warren PH, Jerde EA, Kallemeyn GW (1991) Pristine Moon rocks: Apollo 17 anorthosites. Proc Lunar Planet Sci 21: 51-61

Warren PH, Rasmussen KL (1987) Megaregolith insulation, internal temperatures, and bulk uranium content of the Moon. J Geophys Res 92:3453-3465

Warren PH, Shirley DN, Kallemeyn GW (1986) A potpourri of pristine lunar rocks, including a VHK mare basalt and a unique, augite-rich Apollo 17 anorthosite. Proc 16th Lunar Planet Sci Conf, Part 2, J Geophys Res, suppl. 91: D319-D330

Warren PH, Taylor GJ, Kallemeyn GW, Cohen BA, Nazarov MA (2001) Bulk-compositional study of three lunar meteorites: Enigmatic siderophile element results for Dhofar 026. Lunar Planet Sci XXXII:2197

Warren PH, Taylor GJ, Keil K, Kallemeyn GW, Rosener PS, Wasson JT (1983a) Sixth foray for pristine nonmare rocks and an assessment of the diversity of lunar anorthosites. Proc 13th Lunar Planet Sci Conf, in J Geophys Res 88: A615-A630

Warren PH, Taylor GJ, Keil K, Kallemeyn GW, Shirley DN, Wasson JT (1983b) Seventh foray: Whitlockite-rich lithologies, a diopside-bearing troctolitic anorthosite, ferroan anorthosites, and KREEP. Proc 14th Lunar Planet Sci Conf, J Geophys Res 88:B151-B164

Warren PH, Taylor GJ, Keil K, Marshall C, Wasson JT (1981) Foraging westward for pristine non-mare rocks: Complications for petrogenetic models. Proc Lunar Planet Sci 12B:21-40

Warren PH, Taylor GJ, Keil K, Shirley DN, Wasson JT (1983c) Petrology and chemistry of two "large" granite clasts from the Moon. Earth Planet Sci Lett 64:175-185

Warren PH, Wasson JT (1977) Pristine nonmare rocks and the nature of the lunar crust. Proc Lunar Sci Conf 8:2215-2235

Warren PH, Wasson JT (1978) Compositional-petrographic investigation of pristine nonmare rocks. Proc Lunar Planet Sci Conf 9:185-217

Warren PH, Wasson JT (1979a) The compositional-petrographic search for pristine nonmare rocks: Third foray. Proc Lunar Sci Conf 10:583-610

Warren PH, Wasson JT (1979b) The origin of KREEP. Rev Geophys 17:73-88

Warren PH, Wasson JT (1980) Further foraging for pristine nonmare rocks: Correlations between geochemistry and longitutde. Proc Lunar Planet Sci Conf 11:431-470

Wasson JT, Baedecker PA (1972) Provenance of Apollo 12 KREEP. Proc Lunar Sci Conf 3rd:1315-1326

Wasson JT, Warren PH (1980) Contribution of the mantle to the lunar asymmetry. Icarus 44:752-771

Watts AW, Greeley R, Melosh HJ (1991) The formation of terrains antipodal to major impact basins. Icarus 93:159-168

Webb DJ (1982) Tides and the evolution of the Earth-Moon system. Geophys J R Astron Soc 70:261-271

Weitz CM, Rutherford MJ, Head JW (1997) Oxidation states during ascent and eruption of the volcanic glasses as inferred from metal-melt equilibria in the 74001/2 core. Geochim Cosmochim Acta 61:2765-2775

Wellman TR (1970) Gaseous species in equilibrium with Apollo 11 holocrystalline rocks during their crystallization. Nature 225:716-717

Wentworth SJ, McKay DS, Lindstrom DJ, Basu A, Martinez RR, Bogard DD, Garrison DH (1994) Apollo 12 ropy glassses revisited. Meteorit Planet Sci 29:323-333

Whitford-Stark JL, Head JW (1977) The Procellarum volcanic complexes: Contrasting styles of volcanism. Proc Lunar Sci Conf 8:2705-2724

Whitford-Stark JL, Head JW III (1980) Stratigraphy of Oceanus Procellarum basalts: Sources and styles of emplacement. J Geophys Res 85:6579-6609

Wieczorek MA, Phillips RJ (1997) The structure and compensation of the lunar highland crust. J Geophys Res 102: 10,933-10,943

Wieczorek MA, Phillips RJ (1998) Potential anomalies on a sphere: Applications to the thickness of the lunar crust. J Geophys Res 103:1715-1724

Wieczorek MA, Phillips RJ (1999) Lunar multiring basins and the cratering process. Icarus 139:246-259

Wieczorek MA, Phillips RJ (2000) The Procellarum KREEP Terrane: Implications for mare volcanism and lunar evolution. J Geophys Res 105:20,417-20,430

Wieczorek MA, Simons FJ (2005) Localized spectral analysis on the sphere. Geophys J Int in press:

Wieczorek MA, Zuber MT (2001a) The composition and origin of the lunar crust: Constraints from central peaks and crustal thickness modeling. Geophys Res Lett 28:4023-4026

Wieczorek MA, Zuber MT (2001b) A Serenitatis origin for the Imbrium grooves and South Pole-Aitken thorium anomaly. J Geophys Res 106:27,825-27,840

Wieczorek MA, Zuber MT (2002) The "core" of the Moon: Iron or titanium rich? Lunar Planet Sci XXXIII:1384

Wieczorek MA, Zuber MT (2004) Thickness of the Martian Crust: Improved Constraints from geoid-to-topography Ratios. J Geophys Res 109:E01009, doi:10.1029/2003JE002153

Wieczorek MA, Zuber MT, Phillips RJ (2001) The role of magma buoyancy on the eruption of lunar basalts. Earth Planet Sci Lett 185:71-83

Wilhelms DE (1984) Moon. *In:* The Geology of the Terrestrial Planets. Carr MH (ed) NASA SP-469, 106-205

Wilhelms DE (1987) The Geologic History of the Moon. US Geol Surv Spec Pap 1348

Williams JG, Boggs DH, Ratcliff JT (2005) Lunar fluid core and solid-body tides. Lunar Planet Sci XXXVI:1503
Williams JG, Boggs DH, Ratcliff JT, Yoder CF, Dickey JO (2001a) Influence of a fluid lunar core on the Moon's orientation. Lunar Planet Sci XXXII:2028
Williams JG, Boggs DH, Yoder CF, Ratcliff JT, Dickey JO (2001b) Lunar rotational dissipation in solid body and molten core. J Geophys Res 106:27,933-27,968
Williams KK, Zuber MT (1998) Measurement and analysis of lunar basin deposits from Clementine altimetry. Icarus 131:107-122
Wilson L, Head JW, III (1981) Ascent and eruption of basaltic magma on the Earth and Moon. J Geophys Res 86: 2971-3001
Winzer SR, Nava DF, Schuhmann S, Kouns CW, Lum RKL, Philpotts JA (1974) Major, minor, and trace element abundances in samples from the Apollo 17 Station 7 boulder: Implications for the origin of early crustal rocks. Earth Planet Sci Lett 23:439-444
Wise DU, Yates MT (1970) Mascons as structural relief on a lunar 'Moho'. J Geophys Res 75:261-268
Wolf R, Woodrow A, Anders E (1979) Lunar basalts and pristine highland rocks: Comparison of siderophile and volatile elements. Proc Lunar Planet Sci Conf 10:2107-2130
Wood JA, Dickey JS, Marvin UB, Powell BN (1970) Lunar anorthosites and a geophysical model of the Moon. *In:* Proceedings of the Apollo 11 Lunar Science Conference. Pergamon Press, p 965-988
Yanai K, Ueda M (2000) Achondrite polymict breccia 1153: A new lunar meteorite classified to anorthositic regolith breccia. Lunar Planet Sci XXXI :1101
Yin Q, Jacobsen SB, Yamashita K, Blichert-Toft J, Télouk P, Albarède F (2002) A short timescale for terrestrial planet formation from Hf-W chronometry of meteorites. Nature 418:949-952
Yingst RA, Head JW (1999) Geology of mare deposits in South Pole-Aitken basin as seen by Clementine UV/VIS data. J Geophys Res 104:18,957-18,979
Yingst RA, Head JW III (1997) Volumes of lunar lava ponds in South Pole-Aitken and Orientale basins: Implications for eruption conditions, transport mechanisms, and magma source regions. J Geophys Res 102:10,909-10,931
Yingst RA, Head JW III (1998) Characteristics of lunar mare deposits in Smythii and Marginis basins: implications for magma transport mechanisms. J Geophys Res 103:11,135-11,158
Yoder CF (1981) The free librations of a dissipative Moon. Philos Trans R Soc London Ser A 303:327-338
Yoder CF (1995a) Astrometric and geodetic properties of Earth and the solar system. *In:* Global Earth Physics: A Handbook of Physical Constants (AGU reference shelf 1). Ahrens TJ (ed) American Geophysical Union, p 1-31
Yoder CF (1995b) Venus' free obliquity. Icarus 117:250-286
Yuan D-N, Sjogren WL, Konopliv AS, Kucinskas AB (2001) Gravity field of Mars: A 75th degree and order model. J Geophys Res 106:23,377-23,401
Zhang K, Jones CA (1994) Convective motions in the Earth's fluid core. Geophys Res Lett 21:1939-1942
Zhong S, Parmentier EM, Zuber MT (2000) A dynamic origin for the global asymmetry of lunar mare basalts. Earth Planet Sci Lett 177:131-140
Zhong S, Zuber MT (2000) Long-wavelength topographic relaxation for self-gravitating planets and implications for the time-dependent compensation of surface topography. J Geophys Res 105:4153-4164
Zuber MT, Smith DE, Lemoine FG, Neumann GA (1994) The shape and internal structure of the Moon from the Clementine mission. Science 266:1839-1843

Plate 1.1

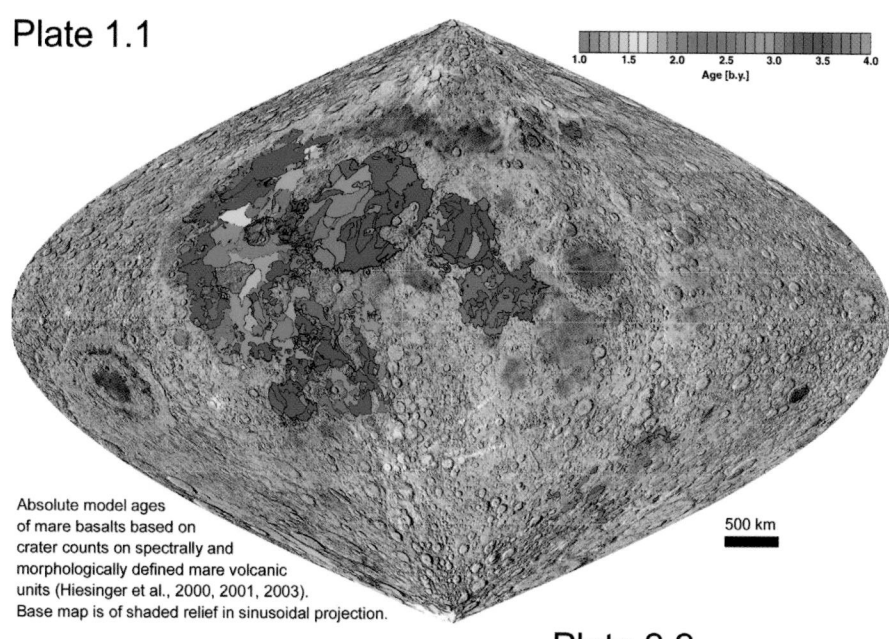

Age [b.y.]
1.0 1.5 2.0 2.5 3.0 3.5 4.0

500 km

Absolute model ages
of mare basalts based on
crater counts on spectrally and
morphologically defined mare volcanic
units (Hiesinger et al., 2000, 2001, 2003).
Base map is of shaded relief in sinusoidal projection.

Plate 2.1

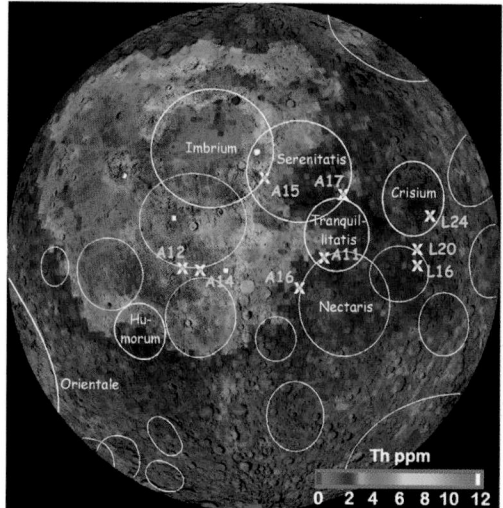

Th ppm
0 2 4 6 8 10 12

Map of the concentration of Th on the lunar near-side
surface (Lawrence et al., 2000; Gillis et al., 2000) with the
locations of the Apollo and Luna landing sites (A11, L24,
etc.) and nearside basins (from Fig. 2.3 of Vaniman et al.
(1991). The seven youngest basins, from oldest to
youngest, are Tranquillitatis, Nectaris, Humorum, Crisium,
Serenitatis, Imbrium, and Orientale (Wilhelms, 1987).

Plate 2.2

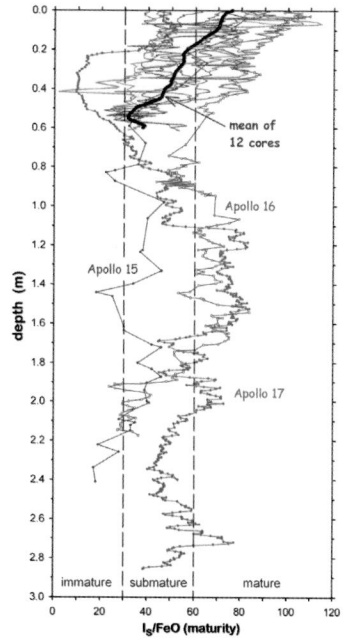

Variation in the regolith maturity parameter
Is/FeO with depth in 12 Apollo regolith cores.
See Fig. 2.29 caption for additional details.

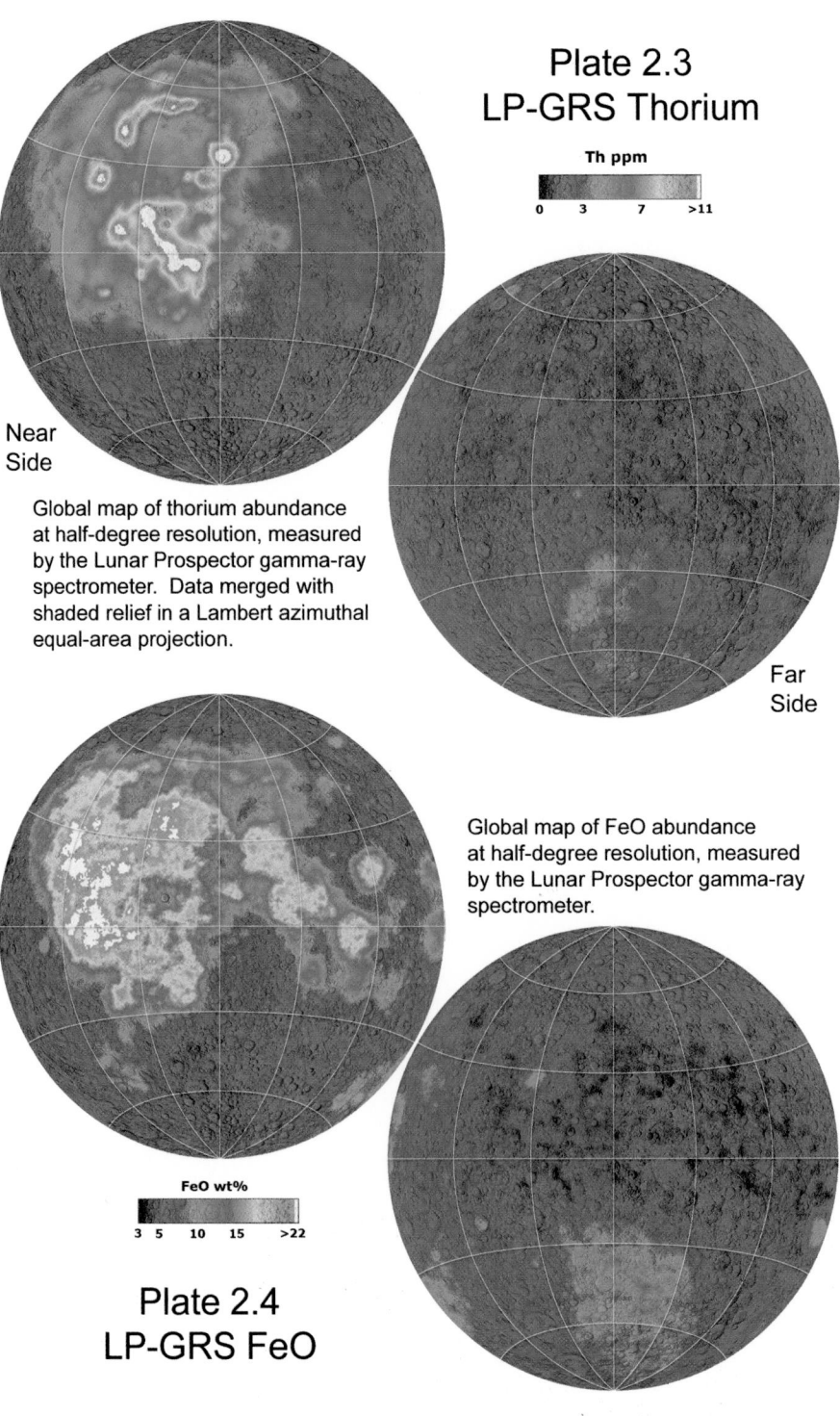

Plate 2.3
LP-GRS Thorium

Th ppm

0   3   7   >11

Near
Side

Global map of thorium abundance
at half-degree resolution, measured
by the Lunar Prospector gamma-ray
spectrometer. Data merged with
shaded relief in a Lambert azimuthal
equal-area projection.

Far
Side

Global map of FeO abundance
at half-degree resolution, measured
by the Lunar Prospector gamma-ray
spectrometer.

FeO wt%

3   5   10   15   >22

Plate 2.4
LP-GRS FeO

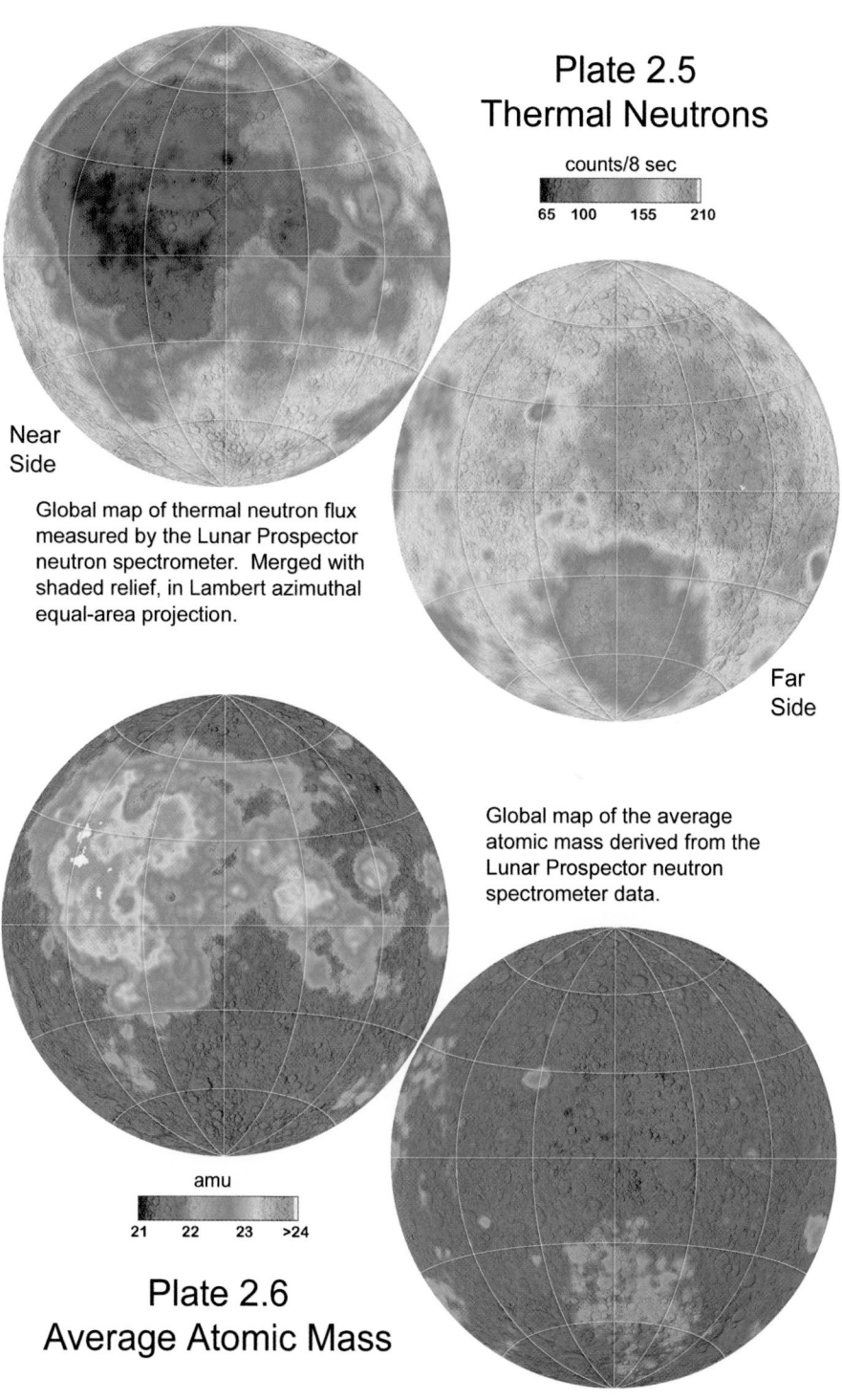

Plate 2.5
Thermal Neutrons

counts/8 sec

65 100 155 210

Near Side

Global map of thermal neutron flux measured by the Lunar Prospector neutron spectrometer. Merged with shaded relief, in Lambert azimuthal equal-area projection.

Far Side

Global map of the average atomic mass derived from the Lunar Prospector neutron spectrometer data.

amu

21 22 23 >24

Plate 2.6
Average Atomic Mass

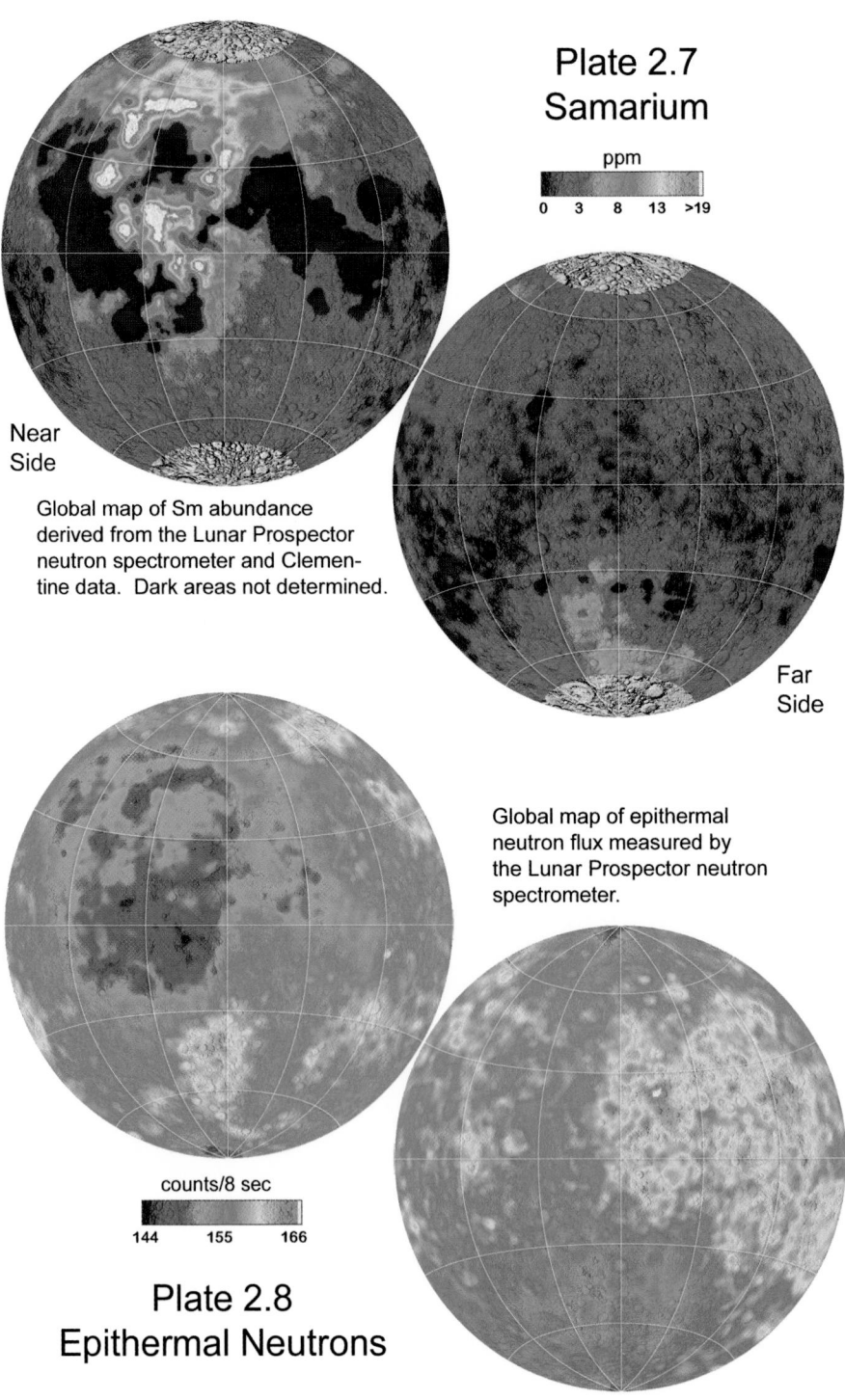

Plate 2.7
Samarium

ppm

0   3   8   13   >19

Near
Side

Global map of Sm abundance
derived from the Lunar Prospector
neutron spectrometer and Clementine data. Dark areas not determined.

Far
Side

Global map of epithermal
neutron flux measured by
the Lunar Prospector neutron
spectrometer.

counts/8 sec

144   155   166

Plate 2.8
Epithermal Neutrons

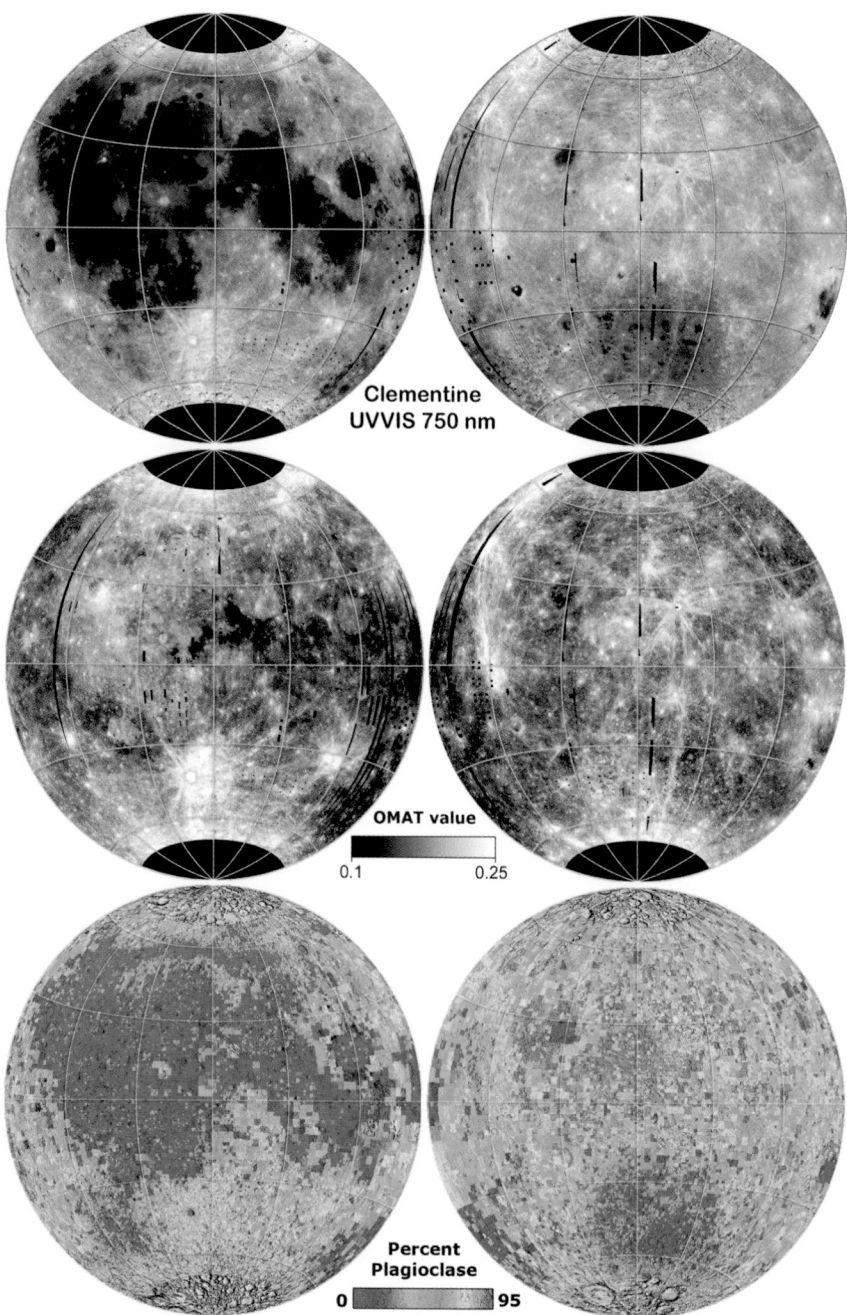

Clementine UVVIS 750 nm

OMAT value
0.1   0.25

Percent Plagioclase
0   95

Plate 2.9. Global images in Lambert projection showing (top) the Clementine UVVIS 750 nm mosaic with lat/lon grids at 30° spacing; (middle) the OMAT (optical maturity) parameter (see caption to Fig. 2.48 for details); (bottom) plagioclase abundance derived from Clementine multispectral datasets as described in the text.

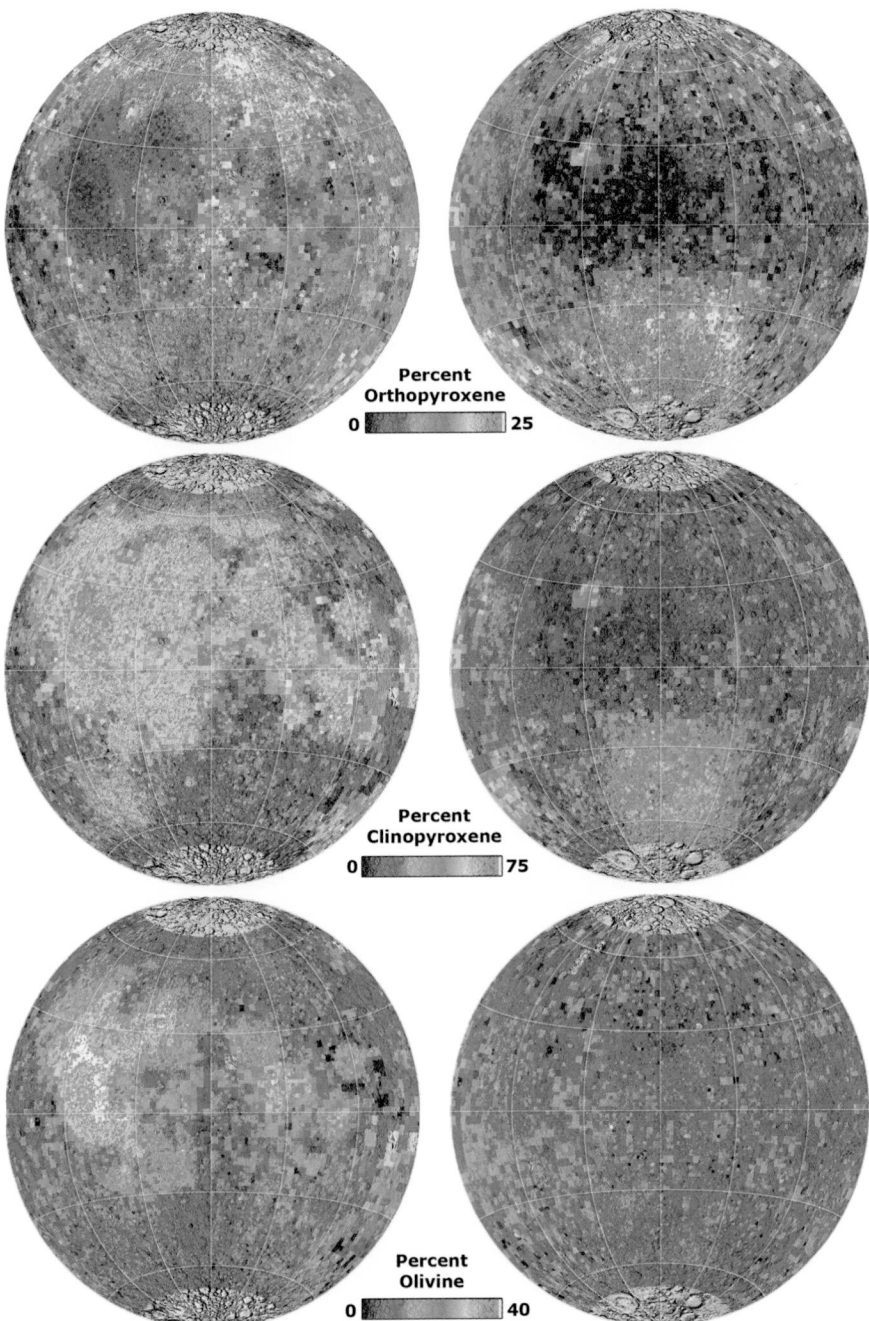

Percent
Orthopyroxene
0   25

Percent
Clinopyroxene
0   75

Percent
Olivine
0   40

Plate 2.10. Mineral maps of orthopyroxene (top), clinopyroxene (middle), and olivine (bottom) derived from Clementine multispectral data (see caption to Fig. 2.48 and explanation in Chapter 2 text). Maps in Lambert azimuthal equal-area projection merged with shaded relief.

Plate 2.11 Maps of epithermal neutrons for the north and south poles measured by the Lunar Prospector neutron spectrometer.

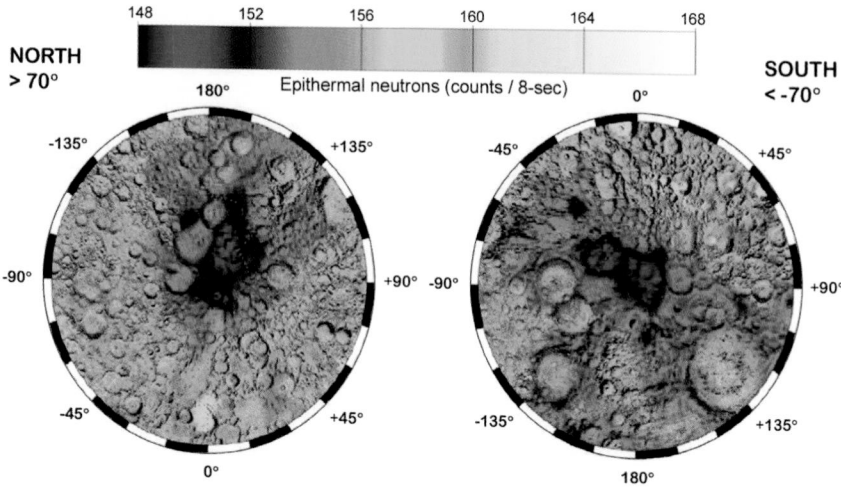

Epithermal neutrons (counts / 8-sec)

NORTH > 70°

SOUTH < -70°

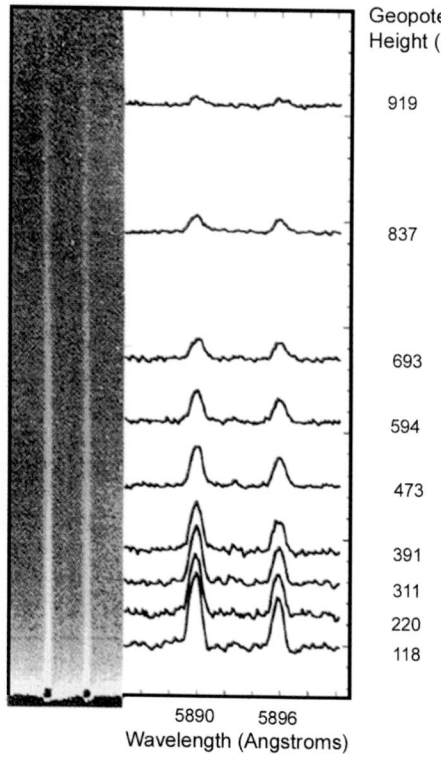

Geopotential Height (km)

919

837

693

594

473

391

311

220

118

5890   5896
Wavelength (Angstroms)

Plate 2.12

A spectrograph image showing Na emission lines extending from the lunar surface to 1800 km in altitude (~900 km geopotential height) above the limb. Brightness close to the surface is ~ 600 R for D2 line and ~ 400 R for D1 line on 20 September 1995 (after Sprague et al. 1998).

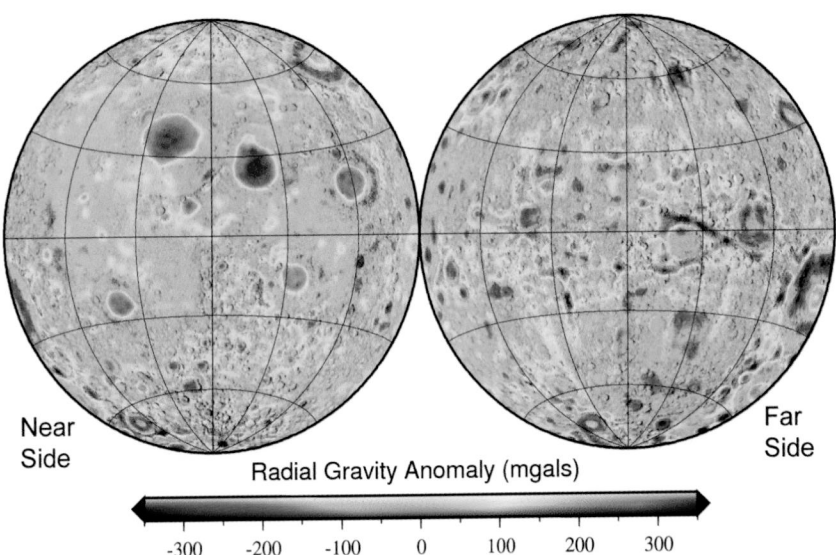

Near
Side

Far
Side

Radial Gravity Anomaly (mgals)

-300   -200   -100   0   100   200   300

Plate 3.1. Radial gravity anomalies of the Moon as determined from the model LP150Q, expanded to degree 90 with the degree-zero and J_2 terms removed. The gravity anomalies (determined at 1738 km radius) range from a minimum of -545 mgal to a maximum of 604 mgal. Images are in Lambert azimuthal equal area projection.

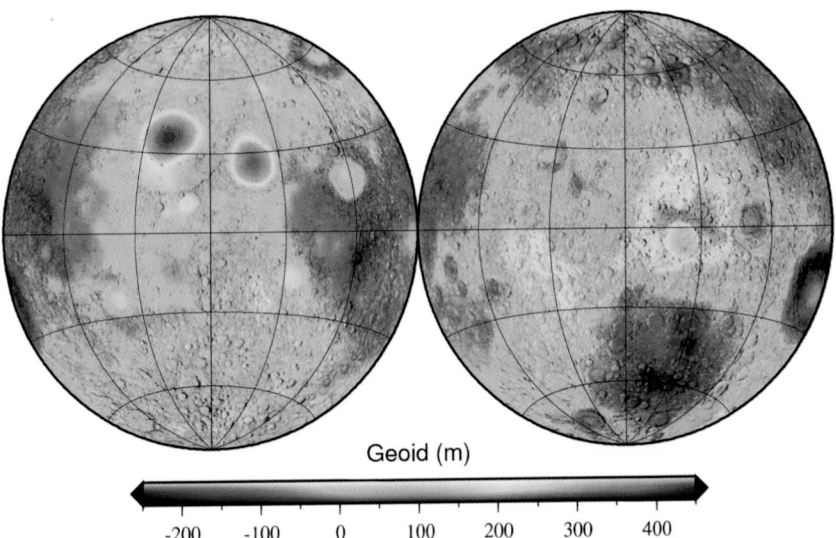

Geoid (m)

-200   -100   0   100   200   300   400

Plate 3.2. The geoid of the Moon as determined from the model LP150Q, expanded to degree 150 with the J_2 term removed. The maximum and minimum geoid elevations in this figure are 461 and -296 meters, respectively. Images are in Lambert azimuthal equal area projection.

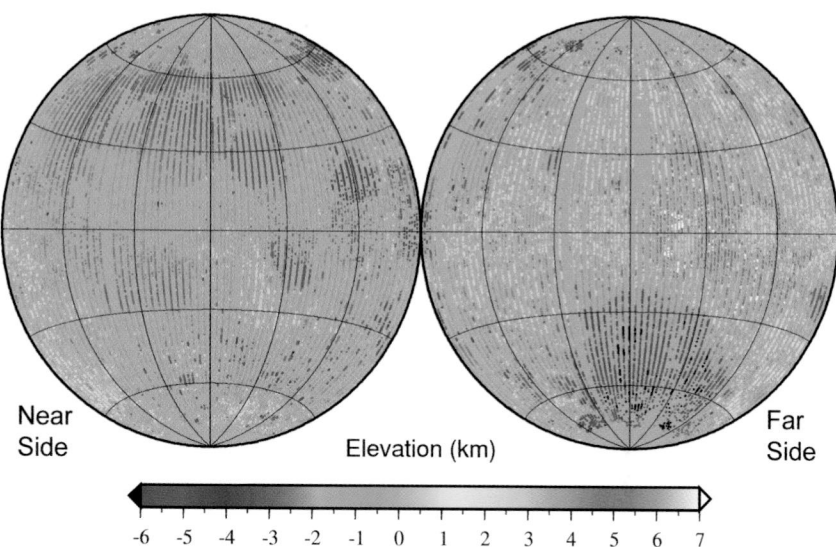

Plate 3.3. Topography of the Moon as determined by the individual range measurements of the Clementine Lidar. The elevation is measured with respect to a spheroid of radius 1738 km at the equator and with a flattening of 1/3234.93 (which corresponds to the J_2 portion of the lunar geoid). Images are in Lambert azimuthal equal area projection.

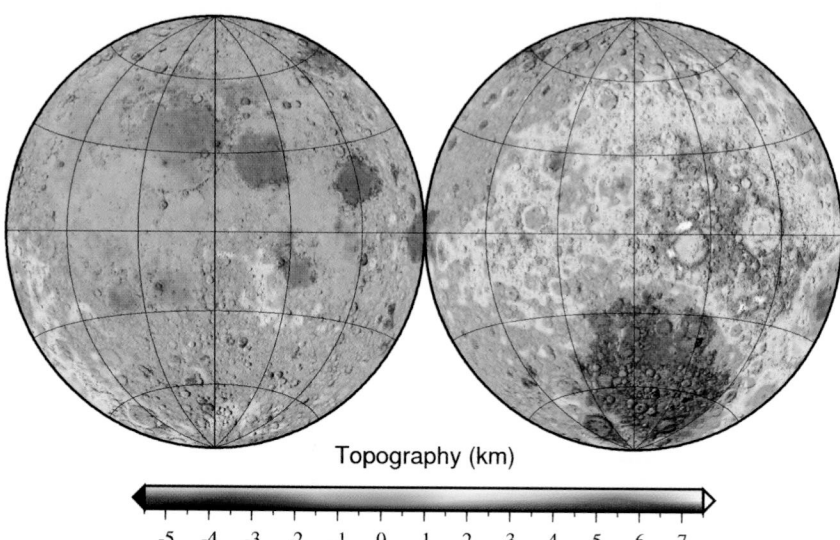

Plate 3.4. Topography of the Moon (GLTM2c) referenced to the full lunar geoid (LP150Q) expanded to degree 90. Maximum and minimum elevations are -6.3 and 8.4 km respectively. Images are in Lambert azimuthal equal area projection.

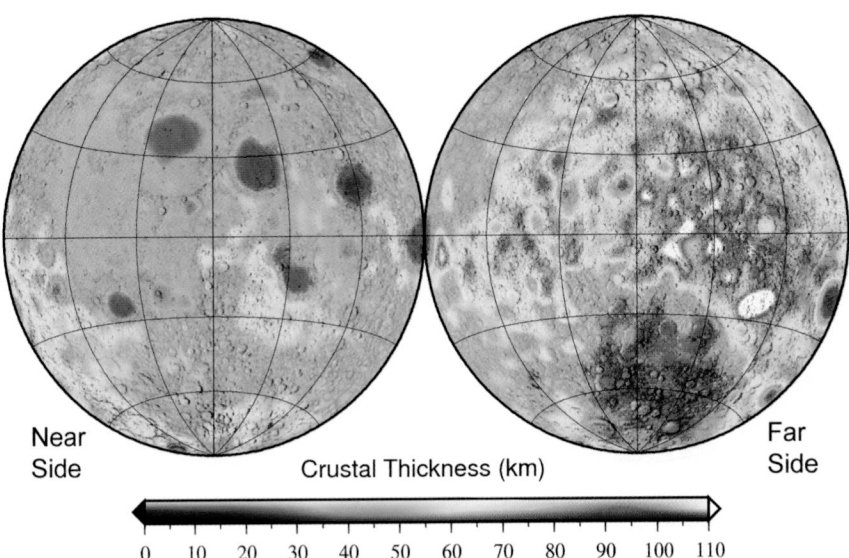

Plate 3.5. Total crustal thickness of the Moon (Model 1). Crustal thicknesses range from ~3 km beneath the Crisium basin to 154 km northeast of the South Pole-Aitken basin. Images are in Lambert azimuthal equal area projection.

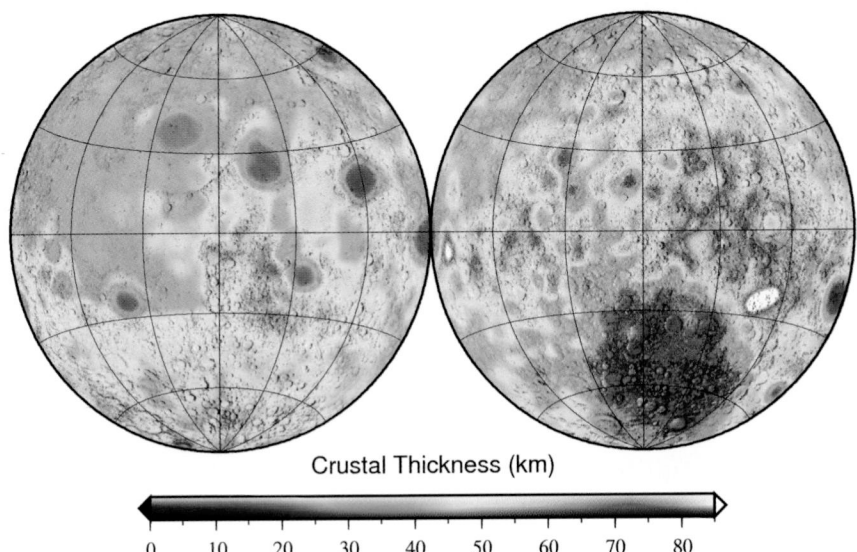

Plate 3.6. Total crustal thickness of the Moon (Model 2). Crustal thicknesses range from ~0 km beneath the Apollo basin to 104 km northeast of the South Pole-Aitken basin. This model assumes that the degree-1 Bouguer anomaly is a result of lateral density variations. Images are in Lambert azimuthal equal area projection.

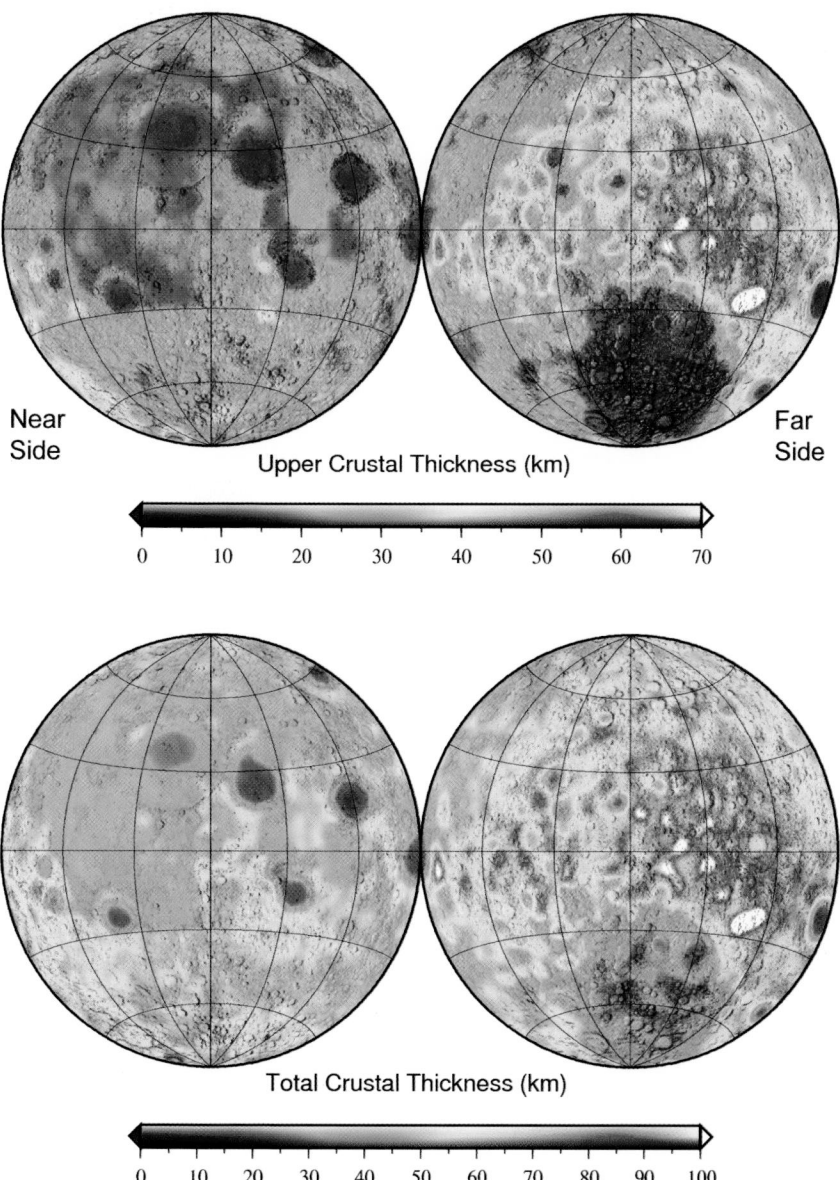

Near
Side

Far
Side

Upper Crustal Thickness (km)

0   10   20   30   40   50   60   70

Total Crustal Thickness (km)

0   10   20   30   40   50   60   70   80   90   100

Plate 3.7. Crustal thickness of the Moon (Model 3). Upper image is the thickness of the upper crust (excluding mare fill), whereas the lower image is the total thickness of the crust. The total thickness of the crust for this model ranges from ~3 km beneath the Crisium basin to 134 km northeast of the South Pole-Aitken basin. Images in Lambert azimuthal equal area projections.

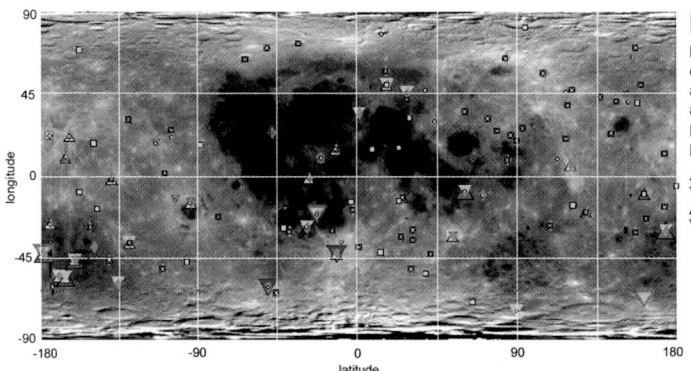

Plate 3.8

Map showing locations of craters in the Tompkins and Pieters (1999) study, and the inferred lithologies present in their central peaks.

▲ T (Troctolite)
▼ N (Norite)
▲ GN (Gabbronorite)
▼ G (Gabbro)
▲ AT (Anorth. Troc)
▽ AN (Anorth. Norite)
△ AGN (Anorth. Gabnor)
▽ AG (Anorth. Gabbro)
◆ GNTA (Gab Nor Troc Anor)
▫ An1 (Anorthosite 1)
◇ An2 (Anorthosite 2

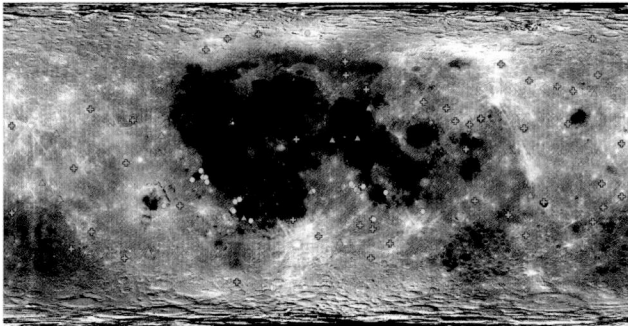

Plate 3.9

Anorthosite distribution shown on the Clementine 750 nm global basemap. Circles show locations where anorthosite is observed with telescopic near-infrared spectra (Hawke et al. 2003b); cyan triangles represent locations on central peaks of craters where Tompkins and Pieters (1999) identified anorthosite (An1); and crosses are locations on central peaks of craters where anorthosite plus one or more additional rock types were identified. Base map is in simple cylindrical projection.

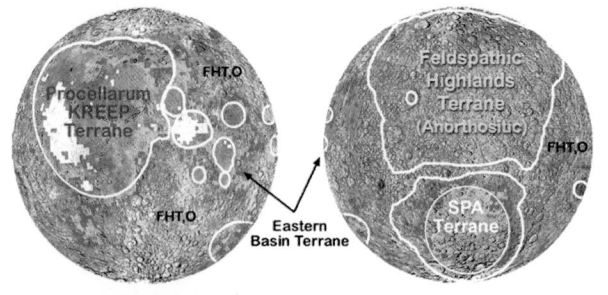

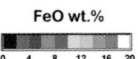

Plate 3.10

Surface expression of major lunar terranes as inferred from the distribution of FeO and Th on the lunar surface. After Jolliff et al. 2000.

FeO wt.%

0   4   8   12   16   20

FeO derived from Clementine data, convolved to 2 degree resolution

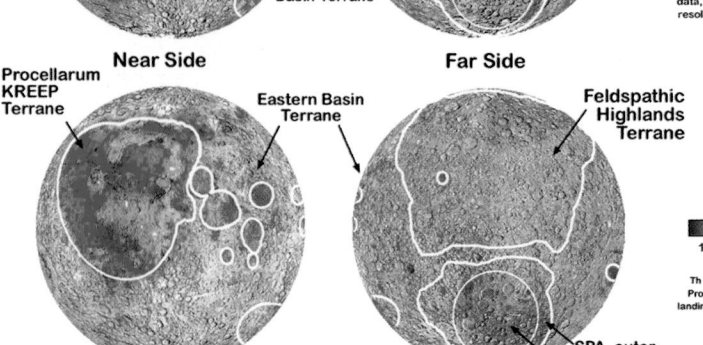

Th, ppm

1   2   4   6   8   10   12

Th concentrations from Lunar Prospector data, calibrated to landing site soils (Gillis et al., 2000)

Reviews in Mineralogy & Geochemistry
Vol. 60, pp. 365-518, 2006
Copyright © Mineralogical Society of America

4

Thermal and Magmatic Evolution of the Moon

Charles K. Shearer[1], Paul C. Hess[2], Mark A. Wieczorek[3],
Matt E. Pritchard[4], E. Mark Parmentier[2], Lars E. Borg[1], John Longhi[5],
Linda T. Elkins-Tanton[2], Clive R. Neal[6], Irene Antonenko[7],
Robin M. Canup[8], Alex N. Halliday[9], Tim L. Grove[10],
Bradford H. Hager[10], D-C. Lee[11], Uwe Wiechert[12]

[1]*Inst. of Meteorites, University of New Mexico, Albuquerque, New Mexico, U.S.A.*
[2]*Dept. of Geol. Sci., Brown University, Providence, Rhode Island, U.S.A.*
[3]*Institut de Physique du Globe de Paris, Paris, France*
[4]*Dept. of Earth & Atmospheric Sci., Cornell University, Ithaca, New York, U.S.A.*
[5]*Lamont-Doherty Earth Observatory, Palisades, New York, U.S.A.*
[6]*Dept. of Civil Eng. & Geol. Sci., Univ. of Notre Dame, Notre Dame, Indiana, U.S.A.*
[7]*University of Toronto, Toronto, ON, Canada*
[8]*Dept. of Space Studies, Southwest Research Institute, Boulder, Colorado, U.S.A.*
[9]*Dept. of Earth Sciences, University of Oxford, Oxford, United Kingdom*
[10]*Dept. of Earth, Atmospheric, & Planetary Sci., MIT, Cambridge, Massachusetts, U.S.A.*
[11]*Academica Sinica, Institute of Earth Sciences, Taipei, Taiwan*
[12]*Eidgenossische Technische Hochschule, Zurich, Switzerland*

Corresponding author e-mail: Charles K. Shearer <cshearer@unm.edu>

1. INTRODUCTION

As with all science, our continually developing concepts of lunar evolution are firmly tied to both new types of observations and the integration of these observations to the known pool of data. This process invigorates the intellectual foundation on which old models are tested and new concepts are built. Just as the application of new observational tools to lunar science in 1610 (Galileo's telescope) and 1840 (photography) yielded breakthroughs concerning the true nature of the lunar surface, the computational and technological advances highlighted by the Apollo and post-Apollo missions and associated scientific investigations provided a new view of the thermal and magmatic evolution of the Moon.

1.1. Pre-Apollo view of the thermal and magmatic evolution of the Moon

Many of the early views of the Moon manifested in mythology and art throughout the world were primarily tied to lunar and terrestrial cycles and the relationships between the Sun and the Moon. Prophetically, myths involving the lunar deities Mwuetsi from Zimbabwe and Coyolxauhqui from Mexico told of rather violent or catastrophic events in which the Moon was expunged from the Earth. Numerous ancient scientific observations were made about the nature of the Moon ranging from those uncovered in early Neolithic sites that correctly identified mare Crisium and mare Humorum to the insights made by Greek philosophers such as Anaxagoras (ca. 500-428 B.C.) and Democritus (ca. 460-370 B.C.), who attached terrestrial analogues to its character (stone, mountains).

With the advent of the telescope (1610) and photography (1840) as scientific tools for lunar exploration, semiquantitative data could be collected that would provide an intellectual foundation for scientific interpretation. Initially, modern terrestrial geological analogs were extended to the Moon (lunar highlands, volcanic craters, seas). Combined with the rigors of

1529-6466/06/0060-0004$15.00

DOI: 10.2138/rmg.2006.60.4

computational modeling, these observational data were extended to predict the original thermal state of the Moon and its thermal and magmatic history. Its proximity to the Earth made the Moon a prime candidate for the source of a wide range of meteorites (chondrites, eucrites) (Urey 1962, 1965; Duke and Silver 1967) and potential extraterrestrial materials (tektites) (Verbeek 1897). Numerous models were made for the ejection of material from the lunar surface (Arnold 1965). The possibility that the Moon was the source for these materials erroneously added "lunar sample" mineralogical and geochemical observations to the pre-Apollo computational models for its thermal and magmatic history.

1.1.1. Pre-Apollo view of the initial thermal state of the Moon. The conditions and processes under which the Moon formed had profound implications for its initial thermal state and its subsequent thermal and magmatic evolution. Pre-Apollo models for its origin fall within three groups (1) accretion or condensation along with the Earth as a double planetary system, (2) fission from a rapidly rotating Earth, and (3) capture by the earth of a fully formed body that was assembled elsewhere in the solar system.

The co-accretion of the Earth and Moon had been a fairly popular model for the origin of this double planet system. In its simplest form it suffered from two major problems. It failed to explain the orbital relationship between the two bodies and their contrasting densities. More elaborate models appealed to fractionation during accretion or slightly different accretional environments to account for the differences in density. In the 19th century it was anticipated that planetary bodies were formed by the accretion of incandescent matter. Therefore, the Moon and the Earth were once molten and differentiated during the crystallization of these molten oceans (Thomson 1864). In the co-accretion model, hot or cold accretion of the Moon dictated its initial thermal state.

In the fission model initially proposed by Darwin (1879), the Moon separated from the Earth by solar tidal forces and the Pacific Ocean basin was the resulting scar. The rationale for this model was that it accounted for the density differences between the Earth and the Moon because it implied that this event occurred following core formation and involved material derived from the upper mantle and crust of the Earth. The similarity of the composition of tektites to the terrestrial crust and the erroneous conclusion that were derived from the Moon (Verbeek 1897; Nininger 1943, 1947; O'Keefe 1963) added inaccurate geochemical evidence for this model. The fission model was initially criticized because it did not account for the orbital dynamics of the Earth-Moon system and because tides of the magnitude required to extract the Moon were thought to be mechanically impossible. The thermal consequences of the fission model for the Moon were not widely explored. The incorporation of a terrestrial crustal component enriched in U, Th, and K into the Moon would have resulted in melting of the lunar interior over a substantial period of time due to the release of radioactive heat. Although abandoned by many prior to Apollo missions, the primordial extraction of the Moon from the Earth by other mechanisms proved to be attractive following the Apollo missions.

The perceived similarities of moons in the solar system, the uniqueness of the Earth-Moon system, and the dynamics of the Moon's orbit around the Earth were the philosophical basis for the capture model. Capture of the Moon by the Earth was long advocated by Urey (1952, 1957, 1959) and the theoretical basis was provided by Gerstenkorn (1955), MacDonald (1964), and Goldreich (1966). The capture model in its pre-Apollo incarnations implies that a relatively undifferentiated and cool Moon was produced elsewhere in the solar system with a primitive composition similar to chondritic meteorites. The Moon as a source for chondritic meteorites was advocated by many (Urey 1959, 1962, 1965; Arnold 1965; Öpik 1966) until the return of the first Apollo mission to the Moon. Most of the capture models suggested that the Moon was captured during the very early history of the Earth. However, in a pamphlet that was privately printed in 1908, F.B. Taylor suggested that the Moon was captured as late as the Cretaceous. Fitting within this model, Urey (1952) calculated that the initial interior of the

Moon was less than 600 °C and more likely was 300 °C. Consequently, magmatic evolution of the Moon primarily involved surface melting caused by impacts and basin formation.

1.1.2. Pre-Apollo view of the thermal and magmatic evolution of the Moon. The first observations related to the thermal and magmatic evolution of the Moon were made by Renaissance scientists in the 17th century. Drawing analogies to crater-producing volcanic processes on Earth, these early observers understandably attached a volcanic origin to the lunar craters. Dana (1846) bolstered these earlier views of a volcanic origin for lunar craters. Gilbert (1893) and Baldwin (1949) challenged this view by concluding that the large lunar craters were of impact origin. The debate continued until the Apollo missions.

Besides the debate on the origin of lunar craters, the interpretation of the early magmatic history of the Moon hinged upon the nature of the lunar maria and highlands. Originally considered seas by the first lunar explorers during the Renaissance, 20th century observers prior to Apollo speculated that they were asphalt lakes (Wilson 1962), dust (Gold 1955), sedimentary rocks (Gilvarry 1968), impact derived melts (Urey 1952), and flood basalts (Baldwin 1949; Kuiper 1954; Fielder 1963). Each of these origins implies a distinctive thermal and magmatic history for the Moon. Proposed ages for the mare ranged from 4 billion years to tens of millions of years (Baldwin 1949; Hartmann 1965; Gault 1970). Drawing on the initial terrestrial analogy made by Galileo in 1610 for the lunar terra, the brighter reflectivity and the presumed lunar origin for tektites, most observers prior to Apollo equated the lunar highlands to terrestrial continental masses. Although their composition and origin were far less debated than the maria, the highlands were suspected to be largely volcanic or to represent more sialic rocks such as granites and rhyolites.

Gilbert (1893) predicted that the Moon formed cold and remained cold throughout its history. The mathematical problem of the cooling of a sphere radioactively heated was solved by Lowan (1933) and first applied to the thermal history of the Moon in the 1950s (Urey 1952, 1955) and MacDonald (1959). There was increasingly a perception (Runcorn 1962; Tozer 1967; Schubert et al. 1969) that solid-state convection has been important in controlling the thermal history of the Moon and other planetary bodies. Urey's early work on the thermal history of the Moon suggested the possibility that internal temperature achieved by radioactive heating could be sufficient to melt at least a part of the lunar interior. Urey (1960) remained skeptical of the extent of lunar melting predicted by this model and anticipated that the Moon accreted cold and perhaps the interior experienced only moderate heating. The melt produced in the lunar interior through radioactive decay was predicted to behave as a highly viscous liquid. Its convection in the lunar interior was modeled by Kopal (1961) and Runcorn (1962, 1963). Prior to the Apollo missions, the basic assumptions underlying Runcorn's approach, namely that the Moon is internally hot enough for a large part of its mass to behave as a molten globe, was challenged by several observations. A rigid crust and a mostly solid interior were implied by relief of the lunar surface and a significant free libration of the Moon in longitude (Koziel 1964).

Looking back from the vantage point of one of mankind's genuine scientific adventures, many of these observations and models for the Moon appear naive and ambiguous. The status of understanding the thermal and magmatic history of the Moon is perhaps best depicted in a lament made by Cook (1967) only two years prior to Apollo and quoted by Taylor (1975): "No conclusions can be drawn other than that the interpretation of the existing experimental data leads to many ambiguities."

1.2. Apollo view of the thermal and magmatic evolution of the Moon

The Apollo missions and associated science had a profound impact on our understanding of the thermal and magmatic history of the Moon. A detailed analysis of the view of the Moon immediately following the Apollo missions is presented by Taylor (1975). Critical to this immense intellectual step forward was the collection of lunar samples and their integration into both Earth-, orbital-, and surface-based observations made prior to and during the Apollo missions. Prior vanguard missions (Luna, Ranger, and Surveyor) provided invaluable guidance for planning the Apollo missions, but the remotely sensed data were mostly ambiguous at answering important questions concerning the thermal history of the Moon.

The observational data derived from the Apollo missions resolved the nature and age of many of the basic lithologies and landforms making up the Moon that added critical constraints in interpreting the magmatic and thermal evolution of the Moon. Some of the fundamental observations made by Apollo-related science were:

1. Large impacts produced all of the large lunar craters and a variety of impact lithologies.

2. Lunar magmatism was essentially basaltic in nature.

3. The maria represent impact basins that were flooded with fluid basalts, produced by melting of the lunar mantle and a majority were erupted to the lunar surface in the interval between 3.1-3.9 billion years ago. The mare basalts exhibit a wide range in composition and are depleted in europium compared to other rare earth elements.

4. The terra are complex, ancient (>3.9 billion years old) and dominated by plagioclase-rich lithologies such as ferroan anorthosite that is enriched in refractory elements. Several other plutonic lithologies were also identified. The trace element characteristics of the primary lunar crust are complementary to the mare basalts and their mantle sources.

5. The Moon is depleted in volatile and siderophile elements and enriched in refractory elements compared to the Earth, yet oxygen isotopes indicate that the Earth and Moon have an ancestral relationship.

6. The Moon is highly differentiated and therefore is not the source for primitive meteorites such as chondrites. This initial differentiation event occurred soon after lunar accretion and was completed by ≈ 4.4 Ga.

7. The coefficient of moment of inertia and seismic data are consistent with nearly uniform density with depth and a small lunar core (diameter <700 km). Conductivity, heat-flow, and tectonic observational data added additional constraints for the internal temperature and the bulk composition of the Moon.

The synthesis of these observational data led to numerous fundamental models for the thermal and magmatic evolution of the Moon immediately following Apollo. The Moon also provided a window to the early thermal evolution of terrestrial planetary bodies that is inaccessible due to eradication of early planetary surfaces by tectonics or by the remoteness of other planetary surfaces. Early Apollo-era results led to the realization that immediately following accretion the Moon underwent extensive, Moon-wide melting. The resulting expression of this event was eloquently referred to as the lunar magma ocean (LMO). Models involving a LMO held that the early lunar crust of ferroan anorthosite was produced by plagioclase flotation and that the mantle source for the mare basalts was a product of the sinking and accumulation of more dense mafic silicates. Early crystallization sequences were calculated for the LMO from olivine-dominated early cumulates to ilmenite-rich late stage cumulates. The final dregs of this Moon-wide melting event produced a widespread, uniformly distributed, incompatible-element-rich and isotopically uniform material referred to as KREEP (Potassium, Rare Earth Elements, Phosphorous). Crystallization of the LMO was prior to 4.4 Ga. and models for the depth of the LMO ranged from 100 km to whole-Moon melting. Models that advocate melting of only the outer portion of the Moon imply that primitive lunar mantle lies below the cumulate pile. The extent of early lunar melting was tied to the potential sources for the primordial heat. The composition of a given mare basalt reflected the composition of the cumulate horizon that was melted. Therefore, it was expected that basalt compositions reflected depth of melting. Estimates for the depth of origin for the mare basalts ranged from 100 to 400 km. The mantle sources for the mare basalts were volatile-poor and most were thought to be very reduced. Even so, pyroclastic deposits found at the Apollo 15 and 17 landing sites were interpreted to be products of volatile driven fire-fountaining. The existence of basaltic volcanism older than the basin-filling basalts was speculative. Most of the highland basaltic rocks such as the Fra Mauro basalts were concluded to have been produced by impacts. The magmatic rocks in the lunar highlands were characterized into several distinct plutonic suites. It was concluded that these plutonic rocks were not partial melts from the same sources as the mare basalts. Whether all the highland plutonic rocks were products of the same primordial lunar differentiation or a broader range of lunar processes (LMO, post-LMO magmatism, impacts) was a point of debate.

On the basis of observations from the Apollo missions, numerous thermal history models for the Moon appeared in the literature prior to 1976. The LMO concept was prominent in many of these early models. However, most avoided treating this episode of lunar differentiation in detail. The early history of the Moon immediately following initial differentiation was dominated by an internal thermal environment that resulted in significant magmatic and tectonic activity. Thermal models both with and without solid convection were suggested and were equally able to satisfy many of the observations made during the Apollo missions. Many of the models concluded that convection played an important role in lunar differentiation, dynamics, and the introduction of lateral heterogeneity. There was reasonable agreement that the range of present-day temperatures in the lunar interior is rather low to a depth of 500 km and rises to about 1000 °C at 700 km depth. Below this depth, there was less agreement in the models, with estimates for temperatures at the center of the Moon ranging between 1000-1600 °C. Some models predicted a partially molten lower mantle and partially molten core. Most of these thermal models were dependent upon the mechanism (primary accretional zoning or early differentiation) for the distribution of heat-producing elements in the outer part of the Moon and the existence of an undifferentiated lunar mantle.

The observations and the resulting science related to the Apollo missions established a robust observational and intellectual framework to provide context for future lunar exploration. Since the high-water mark of the 1970s, new inroads have been made owing to new analytical-computational-technological advances, new missions during the last decade of the twentieth century, and a retrospective view of the pioneering science attributed directly to Apollo. In this chapter, we present new insights into the thermal and magmatic evolution of the Moon afforded us by these new approaches and views.

2. LUNAR ORIGIN AND FORMATION

The origin of the Moon is of fundamental importance in providing a context for lunar and terrestrial studies in a variety of fields, and has received much attention, particularly over the past 50 years. Numerous theories of lunar origin have been advanced, and have been well summarized in a book (*Origin of the Moon 1985*) and several review papers (Boss 1985; Wood 1986; Stevenson 1987). Almost 20 years ago, at a Conference in Kona, Hawaii, (that led to the publication of the *Origin of the Moon* book), a single hypothesis of lunar origin came to the forefront: the giant impact (e.g., Brush 1996). In the intervening time, scenarios that might lead from a giant impact to the Moon as we known it have been explored and summarized in another volume, *Origin of the Earth and Moon* (2000). A variety of physical properties of the Moon must be explained, and in this section, we summarize how consistent the giant impact model is with the real Moon. We first review why the giant impact model is currently the favored hypothesis, and then explore how well it explains observed lunar characteristics, such as early lunar dynamics, composition, initial thermal state, and timing of formation.

2.1. Mechanisms of lunar formation

There are four general types of lunar origin models: fission, capture, co-accretion, and giant impact. Characteristics of these models, as well as their strengths and weaknesses, are covered by Boss (1985), Wood (1986), and Stevenson (1987). The fission model postulates that the early Earth was spinning so rapidly that it became rotationally unstable, causing material in the equatorial regions to be flung into orbit to yield the Moon. In the capture model, the Moon formed elsewhere in the solar system, but it eventually suffered a close encounter with the Earth and was able to enter a bound and stable circum-terrestrial orbit. During Earth accretion, a circum-terrestrial disk might have formed through a variety of mechanisms, which could have led to the formation of the Moon. The giant impact model supposes that such a disk formed from the ejecta of a giant collision of a projectile and the Earth, while the co-accretion model supposes that the circum-terrestrial disk was present throughout Earth's accretion, and was built up of smaller heliocentric projectiles.

The key properties of the Earth/Moon system that serve to separate the models from each other are the ability of each to explain the large angular momentum of the Earth-Moon system and the observation that the Moon is depleted in iron. There is no single observation that rules out the capture model, but there is only a very narrow range of orbital parameters that allow capture to occur (Wood 1986). The dynamical problem with the capture model can be summarized as follows: a dissipative process near the Earth (e.g., extended atmosphere or accretion disk) must operate to allow the object to be captured into Earth orbit, but such a process must disappear soon after the object is in orbit so that it will not continue to lose orbital energy and crash into the Earth. Compositionally, it is also difficult to explain why a captured Moon would not have a bulk iron content similar to the terrestrial planets. The fundamental objection to the fission and co-accretion models is the same; they cannot readily explain the angular momentum of the Earth-Moon system. The fission model requires much more angular momentum than the system currently has, and there is no viable mechanism to remove a sufficient amount of angular momentum. For example, the amount of angular momentum lost to solar tides is likely small (e.g., Canup et al. 2001).

It is difficult, but not impossible, for co-accretion to produce a sufficiently high angular momentum Earth-Moon system. Thus, on the basis simply of the angular momentum, co-accretion must be considered a viable alternative to the giant impact (Weidenschilling et al. 1986; Morishima and Watanabe 2001). Another challenge for the co-accretion model is to explain how the Moon could form side-by-side with the Earth but be deficient in iron. Weidenshilling et al. (1986) proposed that the iron deficiency can be created during co-accretion if the circum-terrestrial disk acts as a compositional filter, where mechanically weak silicates break up and are retained in the disk, while strong iron objects pass through the disk and do not become accreted to the proto-Moon. Recent work has strengthened the co-accretion model by showing that the proto-Earth could have gravitationally altered the orbits of heliocentric impactors such that their trajectories would become spatially non-uniform and the right amount of angular momentum could be deposited into the Earth-Moon system if accretion occurred in the presence of nebular gas (Ohtsuki and Ida 1998). The recent co-accretion model of Morishima and Watanabe (2001) makes use of the spatially non-uniform distribution of heliocentric impactors to get the right angular momentum, but requires the co-accretion to happen before dissipation of the solar nebula or early in Earth accretion. As the solar nebula probably dissipated within the first 10 m.y., it is unlikely that such an early formation is consistent with the proposed formation age of the Moon within the first 50 m.y. (see below). Furthermore, co-accretion during the early formative stages of Earth requires a delicate balance between tidal evolution and mass addition to prevent the Moon from crashing into the Earth. The required balance between tidal evolution and mass addition is physically unlikely (Sleep and Fujita 1998).

The Moon's lack of a large iron core together with planet accretion models that predict that large impacts would be common in the final stages of terrestrial accretion led Hartmann and Davis (1975) to propose that the Moon formed from iron-depleted mantle material ejected from the Earth during a large-scale impact. They also hypothesized that material subject to such conditions might be depleted in volatile elements relative to the Earth. An independent and contemporaneous investigation by Cameron and Ward (1976) recognized that the oblique impact of a roughly Mars-sized planet could account for the rapid initial terrestrial rotation rate implied by the current angular momentum of the Earth-Moon system, and suggested that vaporization might provide a physical mechanism to emplace material into bound orbit. The concepts described in these two works form the basis for what is now the leading theory for lunar origin, the giant impact hypothesis.

A viable model for lunar formation must create a Moon that is consistent with myriad observations, from studies of lunar composition provided by the Apollo and Luna missions, to detailed properties of the lunar orbit. Before exploring how well models of the giant impact can explain these observations, a definition is in order. What makes a giant impact different than a common large impact? Non-giant impacts would have ejected material on ballistic trajectories that would re-impact the proto-Earth, or on hyperbolic ones that would leave the system altogether, but not on trajectories that would place the ejecta in stable orbits that would allow them to accrete into the Moon. A giant impact is not only different in size than other impacts, it involves fundamentally different physics that allows material to enter orbit through impact jets, vapor clouds, gravitational torques, or gravitational interactions among the ejected debris (Stevenson 1987). One might ask whether a giant impact is likely? Were there vagrant protoplanets roaming the early solar system big enough to cause such a collision? The answer appears to be yes. Many models of accretion of the terrestrial planets indicate that there could be objects of the necessary size formed near the Earth (e.g., Wetherill 1986; Agnor et al. 1999; also the review by Canup and Agnor 2000). The results of calculations used to determine the size distribution of the protoplanets in the terrestrial neighborhood must, however, be considered cautiously because they neglect the effects of fragmentation (Stewart 2000) and still struggle to reproduce the low inclination and eccentricity orbits of Earth and Venus (e.g., Chambers and Wetherill 1998; Agnor et al. 1999).

Accepting that a giant impact is possible, other questions arise: can a giant impact eject enough mass into Earth orbit such that it can accrete into a single Moon with the observed properties? With recent computational advances, it has become possible to more fully explore the parameter space of plausible impact scenarios and determine which are most consistent with the Moon. Dynamical models of lunar formation, discussed in the next section, are important because they not only address whether a giant impact could lead to formation of the Moon, but also provide key information for geochemical models, such as the likely mass of the impactor and proto-Earth at the time of the giant impact, the relative amount of material from each that goes into forming the Moon, and the thermal state of the proto-Earth and ejecta following the impact.

2.2. Dynamical models

2.2.1. Overview. Beginning in the mid-1980's (e.g., Benz et al. 1986, 1987; Melosh and Kipp 1989), numerical hydrocode simulations of planet-scale collisions sought to determine 1) whether an impact could eject sufficient iron-depleted material into a bound orbit to form the Moon, and 2) what specific type of impact would be required. Although early simulations were hindered by coarse resolutions and very long computational run times (see also Chapter 5), results generally supported the impact hypothesis (Benz et al. 1986, 1987, 1989; Cameron and Benz 1991). Calculations relating to the subsequent evolution of a protolunar disk composed of solids suggested that the disk could spread extremely rapidly via disk particle collisions, with a characteristic time scale of only about a month (Ward and Cameron 1978). More detailed models that considered the thermodynamic behavior of the disk suggested that disk material would be a mixture of fluid and vapor, whose overall evolution time would be regulated by the radiative cooling time of the disk, ~ 100 years (Stevenson 1987; Thompson and Stevenson 1988).

In the mid-1990's, models describing the accumulation of impact-ejected material were developed and provided new constraints on the distribution of orbiting material required to yield a lunar-size moon (Canup and Esposito 1996; Ida et al. 1997; Kokubo et al. 2001). In particular, N-body accretion models predicted the rapid accretion of a single moon at a typical radial distance of 3-4 Earth radii ($R_{\oplus}$) and revealed a relationship between the angular momentum and mass of the protolunar disk and the size of the resulting moon (Ida et al. 1997). In cases where multiple moons accumulated initially, analytic and numerical models suggested that further evolution of the system due to tidal interaction with the Earth and satellite-satellite interactions would still yield a single moon system, as the moons collided with each other, or as inner moons collided with the Earth (Canup et al. 1999).

Recent works have attempted to identify specific impact scenarios that can account for the current Earth-Moon system, in essence attempting to link the predictions of impact simulations with those of the lunar accumulation models (Cameron 1997, 2000, 2001; Cameron and Canup 1998; Canup and Asphaug 2001). A common difficulty has been producing a massive enough protolunar debris disk with an impact that leaves the Earth-Moon system with its final mass and angular momentum. An inability to produce simultaneously the main dynamic features of the Earth and Moon with a single impact led to proposals that multiple impacts might have been required, or that the Moon-forming impact occurred when Earth was only about half-accreted (Cameron and Canup 1998; Cameron 2000, 2001). While both of these possibilities are not inconsistent with the character of impacts predicted during the final stages of terrestrial accretion (Agnor et al. 1999), they are more restrictive and problematic than the original single-impact hypothesis. However, recent work (Canup and Asphaug 2001) suggested that such restrictions may be unnecessary, as high-resolution simulations predict that an impactor containing about 10% of the Earth's mass appears capable of both producing the Moon and an Earth-Moon system with approximately the current mass and angular momentum.

Below we discuss recent impact simulation results; for more detail, the reader is referred to the chapters by Cameron and Kokubo, and Canup and Ida in the "Origin of the Earth and Moon" (2000) volume.

2.2.2. Constraints on the Moon-forming impact. The angular momentum contained in both the lunar orbit and the Earth's spin has been nearly conserved throughout the history of the Earth-Moon system, with some loss due to tidal interaction with the sun. Tidal interaction between the Earth and Moon causes the lunar orbit to expand and the Earth's rotation to slow, implying that the Moon initially formed very close to a more rapidly rotating Earth. A minimum orbital distance for lunar formation would be the Earth's Roche limit ($a_{Roche} = 2.9R_{\oplus}$ for lunar density material); significant accretion interior to this distance would be frustrated by planetary tides (e.g., Canup and Esposito 1995).

The current angular momentum of the Earth-Moon system, $L_{E-M} = 3.5 \times 10^{41}$ g-cm^2/s, is quite large, and implies that the initial length of the terrestrial day was only about 4-5 hours when the Moon was located near the Roche limit. A lunar mass satellite ($M_L = 7.349 \times 10^{25}$ g or $0.012M_{\oplus}$) orbiting on a circular orbit with $a = a_{Roche}$ would contain an orbital angular momentum of $M_L(GM_{\oplus}a_{Roche})^{1/2} \approx 0.18L_{E-M}$, providing a lower limit on both the mass and angular momentum of orbiting material necessary to yield the Moon assuming completely efficient accumulation.

In reality, much of the orbiting material initially within a_{Roche} will re-impact the Earth in the course of angular momentum exchange within the protolunar disk and through interaction with the forming Moon; some may also be ejected from the system entirely. Lunar accumulation simulations (Ida et al. 1997; Kokubo et al. 2001; also review by Kokubo et al. 2000) predicted a characteristic initial satellite orbital radius ~ $1.2a_{Roche}$, or a ~ $3.5R_\oplus$. A basic conservation argument can be made to estimate the mass of a satellite, M_M, forming with $a = 1.2a_{Roche}$ from an initial protolunar disk containing mass M_D and angular momentum L_D (Ida et al. 1997):

$$M_M \approx \frac{1.9L_D}{\sqrt{GM_\oplus a_{Rocha}}} - 1.1M_D - 1.9M_{esc} \qquad (4.1)$$

where M_{esc} is the amount of escaping material during accretion.

In the simplest impact scenario, the Moon-forming impact would produce an Earth-Moon system consistent with the current one, requiring no significant subsequent dynamic modification. In this case, a candidate impact should place material containing sufficient mass and angular momentum to yield the Moon, typically at least 1.5-2 lunar masses, and produce an Earth with approximately its final mass and an Earth-Moon system with somewhat more than its current angular momentum in order to account for later loss due to interactions with the sun. In addition, the ejected material must be appropriately iron-depleted to account for the Moon's low bulk density. The exact fraction of iron contained in the Moon is uncertain. The upper limit on the mass fraction of the lunar core is estimated to be ~ 3% (e.g., Hood and Zuber 2000); additional iron is incorporated in silicates in the lunar mantle and crust (e.g., Lucey et al. 1995). From the bulk lunar density, an upper limit on the total percent by mass of iron in the Moon is estimated to be about 8% in the limit that all of the iron is contained in low-density silicates (Wood 1986).

2.2.3. Numerical impact simulation method. The general approach in performing numerical impact experiments has been to consider four primary impact variables: impact parameter, b, [defined here as $b \equiv \sin(\xi)$, where ξ is the impact angle], impact velocity, v_{imp}, [whose minimum value is the mutual escape velocity of the colliding objects, $v_{esc} \equiv \{2G(M_I + M_{Tar})/(R_I + R_{Tar})\}^{1/2}$], the mass ratio of the impactor to the total mass, γ, and the total colliding mass, M_T. A series of simulations are examined to determine what set of impact conditions yields the most favorable results; however, the possible parameter space is large and multi-dimensional, and individual impact simulations are computationally intensive.

Most works modeling potential Moon-forming impact have utilized a method known as smooth particle hydrodynamics, or SPH (e.g., Lucy 1977; Benz 1990). The SPH technique requires no underlying grid and is well suited to intensely deforming systems evolving within mostly empty space. Moreover, its Lagrangian nature allows for tracking of material history during a simulation. In SPH, an object is represented as a large number of spherical overlapping 'particles' whose evolution is tracked as a function of time. Each particle represents a quantity of mass of a given composition, whose 3-dimensional spatial extent is specified by a probability density function known as the kernel, and the characteristic spatial width of the particle, known as the smoothing length. The functional form of the kernel does not change during a simulation, but the smoothing length of each particle is adjusted to maintain an overlap with a minimum number of other particles, thus insuring that even low-density regions are smoothly resolved. For modeling planet-scale impacts (i.e., those involving objects 1000's of kilometers in radius or larger), the evolution of each particle's kinematic (position and velocity) and state (internal energy, density) variables are evolved due to 1) gravity, 2) compressional heating and expansional cooling, and 3) shock dissipation. The chosen form for the equation of state relates a particle's specific internal energy and local density to pressure as a function of specified material constants. Commonly used equations of state range from simple analytic forms (e.g., Tillotson 1962) to complex semi-analytic methods (e.g., ANEOS; Thompson and Lauson 1972).

A critical element in the accuracy of SPH is numerical resolution. In early works, simulations with $N = 3000$ particles were done in which a lunar mass (and thus the typical amount of ejected material) was represented by only a few tens of SPH particles. Computational advances now allow for much greater resolution (Cameron 2000, 2001; Canup and Asphaug 2001), and recent simulations have $N = 10^4\text{-}10^5$. With $N > 10,000$ particles, comparisons (Cameron 2000; Canup and Asphaug 2001) indicate that predictions for the amount of orbiting material are consistent to within about 10-15% across simulations with varying resolution. However, resolving the amount of orbiting iron remains a challenge, as modern resolutions describe a lunar mass using hundreds to thousands of SPH particles and the upper limit on the amount of lunar iron with typically several tens of particles.

2.2.4. Simulating lunar-forming impacts. SPH simulations performed in the 1990s by Cameron were reviewed in detail in Cameron (2000, 2001). Since the work of Cameron and Benz (1991), increasingly larger impactors relative to the targets were considered in an effort to increase the yield of material placed into bound orbit, with Cameron (2000, 2001) considering collisions that all involved impactors containing 30% of the total colliding mass, or $\gamma = 0.3$.

A prevailing trait of these simulations has been an apparent difficulty in placing sufficient mass into orbit to yield the Moon for a total system mass and impact angular momentum close to that of the current Earth-Moon system (e.g., Cameron 1997, 2000, 2001; Cameron and Canup 1998). Two classes of impacts, both with $\gamma = 0.3$, were identified as being capable of producing the Moon. The first involved an impact by an object with about 3× the mass of Mars, with an impact angular momentum, L, much greater than $L_{E\text{-}M}$, typically by a factor of 2. A significant dynamic event—such as another giant impact—would then be required to decrease the Earth-Moon system angular momentum by this large amount subsequent to the Moon-forming event. The second class of impacts involved an impact of an object with roughly twice the mass of Mars with $L \sim L_{E\text{-}M}$, but with a total mass (impactor plus target) of only $M_T \sim 0.65 M_\oplus$. In this "early-Earth" scenario, the Earth is only partially accreted when the Moon forms and must subsequently gain $\sim 0.35 M_\oplus$, with the later growth involving sufficiently small and numerous impacts so that the system angular momentum is not significantly altered.

Both of these impact scenarios are more restrictive than the original single impact-hypothesis. A specific problem with the early-Earth scenario is the potential for the Moon to become contaminated by iron-rich material during the period when the Earth was accumulating the final $\sim 35\%$ of its mass (e.g., Stewart 2000). If, during this period, the Moon also accumulated even an approximately proportionate amount of material, it would have gained excessive amounts of iron. The ratio of impact rate onto the Moon vs. that with the Earth is $N_I = f_M R_M^2 / f_\oplus R_\oplus^2$, where $(f_M/f_\oplus)$ is the ratio of the gravitational focusing factor of the Moon to that of the Earth, and R_M and $R_\oplus$ are lunar and terrestrial radii. Reasonable values of impactor velocity yield $0.03 < N_I < 0.074$ (Stewart 2000); assuming an impacting population with a terrestrial abundance of iron implies that the Earth could accrete $0.10 M_\oplus$ before impacts with the Moon delivered an amount of iron equal to the upper limit on the mass of lunar iron. There are several factors that might serve to mitigate the accretion of iron by the Moon, including a less than perfect accretion efficiency (e.g., Morishima and Watanabe 2001). However, in general, as the amount of material which must be added to the Earth after the Moon-forming impact in a given impact scenario increases, difficulties with the Moon becoming compositionally more similar to the other terrestrial planets during the period of subsequent terrestrial growth increase. The simplest explanation for the Moon's unusual compositional characteristics is that it is the result of an impact that occurred near the very end of terrestrial accretion.

Work has been ongoing to identify other impact scenarios that can more closely produce the Earth-Moon system. An important step in this regard has been the realization that the results of SPH simulations display common trends when viewed in terms of scaled quantities. Canup et al. (2001) re-examined the results of Cameron's (2000) simulations and found that the ratio of the mass of the orbiting material to the total colliding mass, (M_D/M_T) is a function of the impact angular momentum (L) scaled by a quantity L^*. Here L^* is the critical angular momentum for rotational stability of a spherical body with mass M_T. The quantity (L/L^*) for a $v_{imp} = v_{esc}$ impact is given by

$$\frac{L}{L^*} = \frac{\sqrt{2}}{K} \gamma(1-\gamma)\sqrt{\gamma^{1/3} + (1-\gamma)^{1/3}} \sin\xi \qquad (4.2)$$

where K is the gyration constant (2/5 for a uniform sphere), γ is the impactor-to-total mass ratio, and ξ is impact angle. The protolunar disk masses and angular momenta produced by SPH impact simulations performed by Cameron (2000) with $\gamma = 0.3$ were found to be well approximated by power-laws of the form

$$\frac{M_D}{M_T} = C_M \left[\frac{L}{L^*}\right]^{S_M} ; \quad \frac{L_D}{L^*} = C_L \left[\frac{L}{L^*}\right]^{S_L} \qquad (4.3)$$

with $C_M = 0.054$, $s_M = 3.9$, and $C_L = 0.38$, $s_L = 4.4$. For $\gamma = 0.3$, the maximum yield of orbiting material resulted for an impact angular momentum about 70-80% of that of a grazing impact, independent of the total colliding mass M_T.

2.2.5. Recent simulations involving smaller impactors. Using the scaling analysis in Canup et al. (2001), Canup and Asphaug (2001) estimated what other impactor sizes might be better able to produce the Earth-Moon system. They predicted that the maximum yield for an $L \approx L_{E\text{-}M}$ and $M_T \approx M_\oplus$ impact should be achieved when $b \approx L_{E\text{-}M}/L_{graz} \approx 0.8$, with the angular momentum of a grazing impact, L_{graz}, given by (again assuming $v_{imp} = v_{esc}$):

$$L_{graz} = [3/(4\pi\rho)]^{1/6} \sqrt{2G} f(\gamma) M_T^{5/3} \qquad (4.4)$$

where ρ is the average target/impactor density, and $f(\gamma) \equiv \gamma(1-\gamma)[\gamma^{1/3} + (1-\gamma)^{1/3}]^{1/2}$. Impactors with $\gamma < 0.12$ had been ruled out as lunar-forming candidates in early low-resolution studies (Benz et al. 1987), because they appeared to produce overly iron-rich disks. However, for those $N = 3000$-particle simulations, a single iron particle contained a mass comparable to the upper limit on the lunar core.

In a survey of 36 simulations, Canup and Asphaug (2001) found a variety of successful candidate impacts involving impactor-to-target mass ratios $\gamma \sim 0.1$, or an impactor mass of $\sim 6 \times 10^{26}$ g. This is essentially equivalent to the mass of Mars. Such impacts are attractive in that they appear to require little or no dynamical modification of the system after the Moon-forming impact, thus avoiding problems associated with a period of extended terrestrial growth after lunar formation. Nominally, a smaller impactor than those proposed by Cameron (2000, 2001) would also be more likely, because in collisional populations small objects are more common than large ones (the number of objects, dN, in a mass range dm is typically proportional to d$N \propto m^{-q}$dm, with $q \sim 1.5\text{-}1.8$). The Mars-mass impactor proposed as optimal by Canup and Asphaug (2001) is essentially the same as that originally proposed a decade prior to the first Moon-forming impact simulations by Cameron and Ward (1976).

2.2.6. Discussion and open issues. Recent work simulating potential Moon-forming impacts indicate that a wide variety of oblique, low-velocity impacts are capable of placing material into bound orbit where it would then accumulate into a satellite(s). Two specific impact scenarios have recently been promoted: one involves the impact of an object with about 2× the mass of Mars with an Earth than is only about 65% formed after the impact (Cameron 2000, 2001), while the other invokes the impact of a Mars-mass object with a fully formed Earth (Canup and Asphaug 2001). The former involves a very massive impactor relative to the target protoearth, with the impactor containing 30% of the total mass; the latter involves a somewhat smaller impactor containing about 12% of the total mass. Although not yet investigated with simulation, we predict that there will be a class of impacts intermediate to these two cases involving progressively decreasing impactor-to-total mass ratios combined with increasing total colliding masses, which would also yield a lunar-sized Moon and an Earth-Moon system with an angular momentum close to L_{E-M}. The range of these impacts that could be viable Moon-forming candidates would then depend upon on how much the Earth could accrete in the presence of the Moon without the Moon becoming contaminated with iron or volatile-rich material. Other types of impacts with $v_{imp} > v_{esc}$ may also be promising candidates.

To date, impact simulations predict that the majority of the material placed in orbit originates from the impactor, with contributions from the target proto-Earth of up to tens of percent by mass (e.g., Canup and Asphaug 2001). Recent work suggests that this may be compatible with the lunar and terrestrial siderophile-element profiles (Righter 2002), although the identical O-isotope signatures in the Moon and Earth (see Section 4.2.5) then require the impactor and the target to have very similar compositions.

The choice of equation of state (EOS) on impact simulation outcome has an important effect and has yet to be fully assessed for potential Moon-forming impacts. Both early simulations (Benz et al. 1986, 1987) and Canup and Asphaug (2001) utilized the Tillotson EOS (Tillotson 1962), while those of Benz et al. (1989), Cameron and Benz (1991), and Cameron (1997, 2000, 2001) used ANEOS (Thompson and Lauson 1972). Tillotson is a simple analytical EOS that lacks an actual treatment of phase changes and mixed phase states. Unlike Tillotson, ANEOS handles mixed phases in a thermodynamically consistent manner; however, in its standard rendition, ANEOS treats all vapor as monatomic species (e.g., Melosh and Pierazzo 1997). The entropy required for vaporization of molecular species such as mantle material is therefore greatly overestimated by this feature of ANEOS, which may be responsible for the apparent lack of vapor production in the simulations of Cameron (e.g., 2000, 2001). An extension to ANEOS to allow for molecular vapor was undertaken by Melosh (2000), and simulations have now been done with this EOS (Canup et al. 2002). Open questions include the relative role of gas pressure gradients vs. gravitational torquing in placing material into bound orbit, and the predicted initial physical state (solids-melts-vapor) of the ejected material.

2.3. Lunar accretion time scales and the early lunar orbit

Assuming that a giant impact occurred and ejected sufficient mass into Earth orbit, the question then becomes whether the material would accrete into a single Moon that is consistent with observed physical and orbital properties. Recent numerical simulations that track the evolution of a swarm of 10^3-10^4 moonlets in a circum-terrestrial disk using direct N-body simulation indicate that a single Moon would form under a variety of initial conditions on a timescale of order 1 month to 1 year (Ida et al. 1997; Kobuko et al. 2000, 2001). Lunar accretion seems to be somewhat inefficient (only 10-55% of the ejected material goes into forming the Moon) so an important conclusion from these simulations is that more than 1 lunar mass must be ejected into orbit from the giant impact (Canup and Esposito 1996; Kobuko et al. 2000, 2001). Many different initial conditions (size and radial distribution of particles, total mass in the disk, etc.) are put into the N-body simulations because the exact initial disk conditions are unknown. There is uncertainty both because the distribution of ejecta in the circum-terrestrial disk from the giant impact remains an active area of research (see Section 2.2) and because the amount of time the disk evolves between the giant impact and the onset of accretion is not well constrained. It is unlikely that accretion could begin immediately following the giant impact because the ejected material is mostly melt and vapor, and significant cooling would be required before accretion could occur. In fact, N-body simulations assume that all particles in the circum-terrestrial disk are solid and behave like infinitely strong particles during collisions (e.g., no fragmentation). Thompson and Stevenson (1988) estimated that 10-100 years might elapse between the giant impact and when the material is cool enough to accrete, but coupled thermal and dynamic models will be necessary to constrain the exact time (e.g., Stewart 2000; Kokubo et al. 2000). Using a hybrid N-body/fluid disk model, Canup and Ward (2000) suggested that accretion exterior to the Roche limit could occur rapidly, while the Roche interior disk remains hot and evolves on the longer, 10-100 year time scale. Thus one important unanswered question about the giant impact is that of timing. How soon after the giant impact did accretion begin and over what time interval did the Moon form? These issues are important for understanding whether the Moon was ever completely molten, with major implications for the lunar thermal evolution and composition (see Sections 2.4 and 2.6).

Simulations suggest that the Moon would accrete near the Earth's Roche limit, at around 3-4 Earth radii, and would subsequently tidally evolve outward to its present location, where it is still moving slowly away from the Earth at about 3 cm/yr (Dickey et al. 1994). However, lunar accretion simulations also predict that the initial inclination of the lunar orbit relative to the Earth's equator would be of order 1 degree, while integrations of the lunar orbit back in time indicate that this initial inclination was at least 10 degrees (Goldreich 1966; Touma and Wisdom 1994). There are currently two different proposed explanations for how the Moon's orbit could have become significantly inclined to the Earth's equator. (1) If a remnant of the protolunar disk continued to exist interior to the Roche limit for some time after the accretion of a Moon outside a_{Roche}, resonant interaction between the Moon and this disk through the 3:1 inner vertical resonance could increase lunar orbital inclination to the required value (Ward and Canup 2000). A bending mode of the disk is excited, which coherently perturbs the lunar orbit at each midplane crossing, progressively increasing the tilt of the lunar orbit. This mechanism would occur during the earliest evolution of the Moon's orbit, within a few hundred years of the Moon's accretion. Its effectiveness is dependent upon the Roche interior disk containing at least ~0.5 lunar masses and having a characteristic viscous lifetime $\geq$ 50-100 years; for such values, a small initial inclination would be increased to ~10-15 degrees. The recoil of the Moon destroys the disk by causing it to decay and be accreted by the Earth. (2) If a Roche interior disk did not interact with the early Moon, later resonant interactions of the Moon with the Sun might have been able to increase the lunar inclination to appropriately high values instead (Touma and Wisdom 1998; Touma 2000). Assuming the protolunar disk has been removed, the Moon's passage through a secular resonance with the Sun, known as the evection resonance (where the apsidal precession rate of the Moon equals the orbital period of the Earth, 1 year), would occur at about 4.6 Earth radii (Kaula and Yoder 1976; Touma and Wisdom 1998). Capture into this resonance would lead to an increase in the lunar eccentricity, which if followed by capture in a second inclination-eccentricity resonance (termed the eviction resonance by Touma and Wisdom) can produce the needed inclination. This mechanism's effectiveness depends upon having the tidal dissipation "Q" factor of the Moon change between passage through evection and eviction, such that an early low rate of dissipation allows for eccentricity growth at the evection resonance, whereas a later enhanced rate of dissipation allows for capture into the eviction resonance. Such a change could occur if an initially cold Moon was tidally heated during the eccentricity excitation phase (Touma and Wisdom 1998).

In summary, the difficulty of reconciling the large initial lunar inclination with many lunar origin models, including the giant impact model, appears to have been mitigated by plausible dynamical scenarios that could increase the inclination of an initially equatorial Moon to ~10 degrees. The next section explores whether other observations about the Moon (particularly from geochemistry and geology) can constrain whether the Moon formed hot or cold, and might be used to infer the timescale between the giant impact and accretion.

2.4. The initial lunar thermal state

A giant impact is a major thermal event that could have consequences for numerous properties of the Moon that we observe today. Recent N-body simulations of lunar formation (Section 2.3) predict that once accretion began, it proceeded very rapidly, leaving little time to lose heat of accretion. If accretion occurred less than 100 years after the giant impact, it is likely that the Moon formed almost completely molten (Pritchard and Stevenson 2000). Uncertainties in the physical state of the ejecta from a giant impact (Section 2.2) and the interval between the giant impact and accretion (Section 2.3) make it difficult to state precisely the range of initial lunar temperatures. Therefore in this section we consider the extreme case: do any observations rule out a completely molten Moon? For more complete analysis see Pritchard and Stevenson (2000).

The most widely cited observational constraint on the initial temperature profile within the Moon is the radius constraint, which relates surface observations of faulting on the Moon to internal thermal evolution (MacDonald 1960; Solomon and Chaiken 1976). If the Moon formed almost completely molten and cooled monotonically throughout geologic history, then the lunar radius would contract, building up stress in the rigid outer layers of the Moon through the relation (Solomon 1986),

$$\frac{\Delta R}{R} = \frac{(1-\nu)\sigma}{E} \tag{4.5}$$

where $\Delta R/R$ is the fractional change in radius, ν is Poisson's ratio, E is Young's modulus, and σ is the horizontal stress in the lithosphere. If the stress exceeds a certain level, faulting would occur. The fact that there are no global scale contractional or extensional features on the Moon constrains the amount of radius change (and therefore temperature change) within the Moon since the creation of the stable lunar surface at 3.5 Ga, after the end of heavy bombardment. By comparison, global thrust faults on Mercury have been used as evidence of global contraction (e.g., Strom et al. 1975). That no such features are seen on the Moon indicates that there has been a delicate balance between heating of the lunar interior and cooling of the exterior throughout lunar history, such that the radius has remained nearly constant (Solomon and Chaiken 1976; Solomon 1977, 1986; Kirk and Stevenson 1989). The implication for lunar origin scenarios is that the interior of the Moon must have started cold, and certainly not molten. However, there are several scenarios that complicate the relation between internal thermal evolution and surficial tectonics such that hotter conditions deep within the Moon would be possible (Pritchard and Stevenson 2000). First, as Equation (4.5) indicates, the Young's modulus must be used to relate stress to strain. The outer layers of the Moon are highly cracked due to impact processes, as indicated by Apollo seismic experiments (e.g., Toksöz et al. 1974), and this type of material can withstand more strain before failing than solid rock (He and Ahrens 1994; Pritchard and Stevenson 2000). Mathematically, this means that by using a lower E, a higher strain is allowed before the stress exceeds failure criteria. Another possible factor that could obscure the relation between initial temperature and faulting is if the lunar thermal state has not decreased monotonically. For example, a change in convective regime within the interior could lead to episodes when the interior heated up. Other factors that complicate the relation between initial temperatures and lunar faulting are mentioned in Pritchard and Stevenson (2000) and below. The conclusion of these works is that a hot initial Moon cannot be ruled out.

An initially molten Moon has geochemical consequences. One of the first and most unexpected results of the Apollo missions was that the outer several 100 km of the Moon were probably initially completely molten (see Section 4.3). The giant impact and subsequent rapid accretion of the Moon naturally leads to a LMO, whereas the slower accretion envisioned in co-accretion models may or may not (e.g., Wood 1986). Early proponents of the giant impact also claimed that vaporization during the giant impact could lead to the observed volatile depletion of the Moon (Wood 1986), and more recently some have noted that accretion of the Moon within 1 month could itself lead to elemental fractionation (Abe et al. 1999). Alternatively, the existence of volatiles in samples thought to be from deep within the lunar interior has been cited as evidence that the Moon was never totally molten (Dickenson et al. 1988). Another line of evidence against vaporization of proto-lunar material is the relative abundance of alkali elements. Although the Moon is depleted in absolute abundances of volatile alkalis, their relative abundances (e.g., Li/Na, Na/K, K/Rb, Rb/Cs) do not show a volatile-depletion pattern (i.e., Li < Na < K < Rb < Cs) relative to the Earth (McDonough and Sun 1995; Jones and Drake 1993; Taylor 1982). Turcotte and Kellogg (1986) also claimed that the Moon could not have begun molten because of the inferred lack of mixing and vigorous convection in the lunar interior. Although the issues of lunar composition are discussed in more detail in Section 2.6, it is emphasized here that the geochemical data have multiple interpretations, and there is not necessarily a simple relation between a hot/molten state and a complete depletion in volatiles or homogenization. The existence of volatiles deep in the lunar interior could be consistent with a molten Moon if the process of volatile extraction was inefficient. Similarly, the existence of undepleted xenoliths from the terrestrial mantle (Jones and Palme 2000) cannot rule out a terrestrial magma ocean and the existence of a giant impact (Stewart 2000). Furthermore, the volatile depletion of the Moon might be inherited from the giant impactor and have little to do with the giant impact event itself (Humayan and Cassen 2000). It is also possible that the Moon could have begun nearly molten and be poorly mixed if the mantle heterogeneity was formed during differentiation, and if mantle convection was weak or non-existent.

Two sets of geophysical data could provide information on the initial thermal state of the Moon, but they require further investigation. First, is the possibility that a record still exists from the time when the Moon was close to the Earth and it rotated faster. The fundamental shape of the lunar surface and gravitational field (as recorded in the low order spherical harmonics of those fields) might be interpreted as a "fossil" of the earlier rotation state, which might be inconsistent with a totally molten Moon (Hood and Zuber 2000). The other set of geophysical data is paleomagnetism of the lunar samples (discussed in Chapter 3), which might indicate the presence of an early convecting lunar dynamo, although there are theoretical difficulties with this interpretation. If there really was a lunar dynamo, then core formation occurred early in the Moon, and this has implications for the early thermal state, but a totally molten Moon is not required (Section 6.3.9).

2.5. Age of the Moon

The age of the Moon represents more than just a point in time in early solar system history. The various theories for the origin of the Moon as well as those for the accretion of the terrestrial planets more generally are directly testable by the precise determination of the Moon's age. Furthermore, without a properly tested theory for how the Moon formed we cannot evaluate the reasons for its chemical composition and what that composition reveals about the early stages of planetary development. The age of the Moon is now known extremely well; consequently, we can confirm certain theories and refute others. In this section the background to these issues is briefly explained and the state of understanding at the end of the Apollo missions is summarized. Then the powerful new constraints from hafnium-tungsten (Hf-W) cosmochemistry are discussed.

The earliest dynamic and chemical evolution of the terrestrial planets is now relatively well understood in terms that are consistent with both isotopic and chemical data as well as theoretical modeling. The process of runaway gravitational growth should build Moon-sized objects over periods of roughly 10^5 years (Lissauer 1987). Therefore if the Moon formed as such a planetary embryo there are obvious implications for its expected age. The importance of protracted accretionary processes involving collisions between such embryos were not widely appreciated until the 1970's (Safronov 1974; Wetherill 1976, 1986), and the concept of the origin of the Moon as a result of a late collision between two planets came about in the 1980's (Cameron and Benz 1991). The most serious alternate theory was capture. However, this theory makes no predictions that the Moon itself should form late. The Giant Impact theory predicts that the age of the Moon should post-date the origin of the solar system by some considerable time period; probably tens of millions of years if Wetherill's calculations are correct.

Attempts to date the Moon were initially focused on dating the oldest rocks and therefore providing a lower limit. These studies mainly involved precise Sr, Nd and Pb isotopic constraints

(Tera et al. 1973; Wasserburg et al. 1977; Hanan and Tilton 1987; Carlson and Lugmair 1988; Shih et al. 1993; Alibert et al. 1994). At the end of the Apollo era, Wasserburg and colleagues wrote "The actual time of aggregation of the Moon is not precisely known, but the Moon existed as a planetary body at 4.45 Ga, based on mutually consistent Rb-Sr and U-Pb data. This is remarkably close to the ^{207}Pb-^{206}Pb age of the Earth and suggests that the Moon and the Earth were formed or differentiated at the same time." (Wasserburg et al. 1977). Such constraints on the age of the Moon still leave considerable scope (>100 m.y.) for an exact age.

Some of the most precise and reliable early ages for lunar rocks are given in Chapter 5 (Table 5.7). They provide considerable support for an age of >4.42 Ga. Perhaps the most compelling evidence comes from ferroan anorthosite 60025, which defines a relatively low first-stage μ and an age of about 4.5 Ga. Carlson and Lugmair (1988) reviewed all of the most precise and concordant data and concluded that the Moon had to have formed in the time interval 4.44-4.51 Ga, which is consistent with the estimate of 4.47 ± 0.02 Ga of Tera et al. (1973). The anomalous ages and initial Nd isotopic compositions of selected ferroan anorthosites is discussed in Section 3.4.

Model ages can provide upper and lower limits on the age of an object. Halliday and Porcelli (2001) reviewed the Sr isotope data for early solar system objects and showed that the initial Sr isotopic compositions of early lunar highlands samples are all slightly high relative to the best estimates of the solar system initial ratio (Fig. 4.1). Even the most conservative assessment of the Sr isotope data would conclude that the difference between the ^{87}Sr/^{86}Sr of the bulk solar system initial at 4.566 Ga = 0.69891 ± 2 and the Moon at (roughly) 4.515 Ga = 0.69906 ± 2 is resolvable. A Rb-Sr model age for the Moon can be calculated by assuming that objects formed from material that separated from a solar nebula reservoir with the Moon's current Rb/Sr ratio. As the Rb/Sr ratios of the lunar samples are extremely low, the uncertainty in formation age does not affect the calculated initial Sr isotopic composition, hence model age, significantly. The CI chondritic Rb/Sr ratio (^{87}Rb/^{86}Sr = 0.92) is assumed to represent the solar nebula. This model provides an upper limit on the formation age of the object, because the solar nebula is thought to represent the most extreme Rb/Sr reservoir in which the increase in Sr isotopic composition could have occurred. The Sr isotopic composition, however, probably evolved in a more complex manner over a longer time period. The calculated time required to generate the difference in Sr isotopic composition in a primitive solar nebula

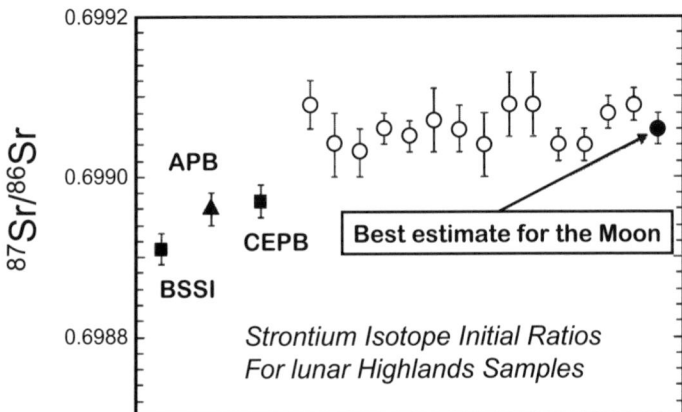

Figure 4.1. The initial Sr isotopic compositions of early lunar highlands samples are clearly resolvable from the initial Sr isotopic composition of the solar system. CAI: calcium-aluminum refractory inclusions; APB: angrite parent body; EPB: eucrite parent body.

environment is 11 ± 3 m.y. (Halliday and Porcelli 2001). Unless the Rb/Sr of the bulk solar system or solar nebula is not that of CI chondrites, this is the earliest point in time at which the Moon could have formed. This alone establishes that the Moon acquired its Rb/Sr ratio (or chemical composition) at a stage that is later than predicted from modeling the accretion dynamics of planetary embryos by runaway gravitational growth (Lissauer 1987).

A similar model-age approach can be used with Hf-W cosmochemistry. The W isotopic compositions of bulk rock lunar samples range from chondritic to $\varepsilon_W > 10$ (Lee et al. 1997, 2001). This was originally interpreted as the result of radioactive decay of formerly live ^{182}Hf within the Moon, which has a variable but generally high Hf/W mantle (Lee et al. 1997; Halliday 2000). We now know that a major portion of the ^{182}W excess in lunar samples is cosmogenic and the result of the reaction ^{181}Ta(n,γ) $\rightarrow$ ^{182}Ta(β^-) $\rightarrow$ ^{182}W while these rocks were exposed on the surface of the Moon (Leya et al. 2000; Lee et al. 2002). This can be corrected using (1) estimates of the cosmic ray flux from Sm and Gd compositions, (2) the exposure age and Ta/W ratio or (3) internal isochrons of W isotopic composition against Ta/W (Lee et al. 2002). The best current estimates for the corrected compositions are shown in Figure 4.2. The spread in the data is reduced and the stated uncertainties are greater relative to the raw W isotopic compositions (Lee et al. 1997). Most data are within error of chondritic but a small excess ^{182}W is still resolvable for some samples.

From these data, the most obvious and clear implication of the chondritic W isotopic compositions is that the Moon, a high Hf/W object, must have formed late. Otherwise the very high Hf/W would have resulted in a large excess in ^{182}W produced from the relatively large amount of (still) live ^{182}Hf. On this basis Halliday (2000) pointed out that the chondritic W isotopic composition was hard to explain if the Moon formed before about 50 Myrs after the start of the solar system. Subsequent to these studies it was shown that the present-day W isotopic composition of the average solar system, as measured in chondrites is offset to ε_W −1.9 ± 0.1 (Kleine et al. 2002; Schönberg et al. 2002; Yin et al. 2002). This change, as well as the corresponding change in the initial ^{182}Hf/^{180}Hf of the solar system shortens the model timescales. Kleine et al. (2002) and Yin et al. (2002) argued that the Moon had to have formed within the first 30 Ma of the start of the solar system. More recent modeling of the magnitude of the radiogenic effect on the Moon indicates a somewhat younger age of 40 to 50 Ma (Halliday 2003, 2004).

Of course the radiogenic excesses found in some lunar samples could be either the result of inherited isotopic heterogeneity or radioactive decay of with formerly live ^{182}Hf. High precision oxygen isotope data (Wiechert et al. 2001) provide no evidence of inherited heterogeneities. Although significant meteoritic material might have been incompletely admixed into the Moon after it formed, (perhaps equivalent to the late veneer on Earth) and early large impacts clearly struck the Moon, there is no indication of incompletely admixed meteoritic components in the lunar interior. This does not exclude addition of meteoritic material completely but limits the amount to a few percent. Admixing 3% H-chondritic material, for example, would be detectable when 0.016‰ deviation (in Δ^{17}O) is considered significant (Fig. 4.3). Even less material from L, LL or carbonaceous chondrites would be detectable because these groups are further displaced from the terrestrial fractionation line (TFL). Although CI-chondrites overlap with the TFL, they characterized by high δ^{18}O values above 16‰. Any proportion larger than 5% would increase the δ^{18}O value by at least 0.5‰.

Furthermore, the oxygen data indicate that the Earth and Moon have compositions that are identical ($\Delta^{17}O_{MOON}$ = 0.000003 ± 0.000005, 99.7% confidence). Therefore, if the giant impact theory is correct, the proto-Earth and the impacting planet were constructed from an identical mix of inner-solar-system material and provenance. Only if the Δ^{17}O values of the proto-Earth and impactor were identical to within 0.03‰ would it be possible that the average Δ^{17}O value of the Moon plots within 0.005‰ on the TFL. This calculation is conservative, assuming the Moon is made of equal proportions from the proto-Earth and the impactor planet. Improved

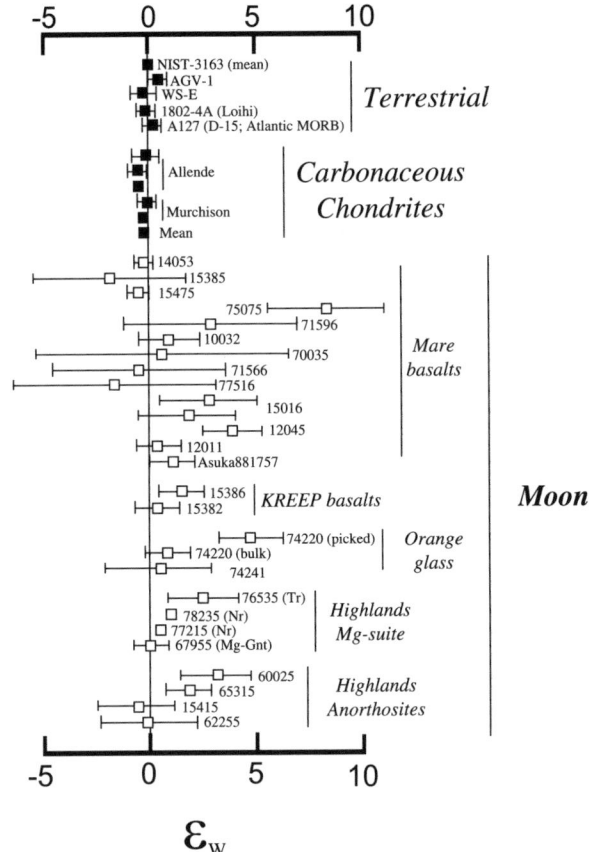

Figure 4.2. Tungsten isotopic compositions for terrestrial samples, whole rock carbonaceous chondrites and lunar samples expressed in ε units as deviations from the terrestrial value. Data are from Lee and Halliday (1996, 1997), Lee et al. (1997, 2002), Kleine et al. 2002, Schönberg et al. 2002, and Yin et al. 2002. Note that the chondrite data for Allende and Murchison reported in Lee and Halliday (1996) are now known to be in error by 2 ε units.

models of the Moon-forming giant impact indicate that a large portion (i.e., 70-90%) derived from the impactor (Asphaug and Canup 2001). If the material from impactor and the proto-Earth were incompletely mixed, it would not be detectable in the oxygen isotopes. However, all giant impact models involve an extraordinarily energetic start to the Moon and preservation of pre-accretion heterogeneities is unlikely.

2.6. Compositional constraints on the origin of the Moon

Estimates for the composition of the Moon have been made in numerous studies (Anders 1977; Wänke et al. 1977; Taylor 1982; Ringwood 1986; Jones and Delano 1989; O'Neill 1991; summarized by Newsom 1995) and in Chapter 3 (Section 5.3). A comparison between the estimated bulk composition of the Earth's mantle and the Moon indicates that there are both similarities and differences, but the differences are such that a lunar origin by fission can be ruled out because it predicts that the two should have identical compositions (Drake 1986). Estimating planetary bulk composition is difficult because samples of the terrestrial and lunar lower mantles are lacking. Furthermore, the compositions of the objects(s) that would form a Moon through co-accretion, capture, or the giant impact are not known. A particular problem for the giant impact is that the exact proportion of material from the Earth and impactor that form the Moon is unknown, although recent numerical simulations of the giant impact indicate that 70-90% of the Moon would consist of material from the impact (see Section 2.2). Still, identical oxygen isotopic compositions of the Earth and Moon, and differences from other solar system materials provides a strong constraint (Wiechert et al. 2000). The similarity in the oxygen isotopes suggests that the material that formed the Moon came from nearly the same location within the solar nebula as Earth, and although consistent with current simulations of the giant impact (Wiechert et al. 2000), it is also consistent with other lunar formation scenarios such as co-accretion and capture.

2.7. Summary

We cannot yet answer the question of whether the Moon as we know it is completely consistent with formation by a giant impact into the early Earth, because realistic models of the giant impact, evolution of the protolunar disk, and lunar accretion are only beginning to be explored. However, during the past 20 years, sophisticated (albeit incomplete) numerical models have revealed that a giant impact can eject sufficient mass into Earth orbit to yield a Moon consistent with the observed orbital and compositional properties, providing strong support for the giant-impact hypothesis. These models

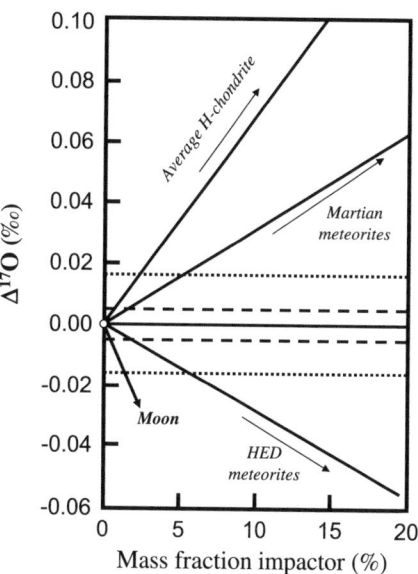

Figure 4.3. The $\Delta^{17}O$ values for lunar samples plot within a 2σ error of $\pm$ 0.016 ‰ (dashed lines) on the terrestrial fractionation line. If the impactor would have formed from the same raw material as Mars or the HED parent body then all lunar samples must have formed within 2% the same portion from the impactor and proto-Earth than the Earth using the triple standard error of the mean ($3\sigma_{MEAN}$). The H-chondrites plot on average 0.7‰ above the terrestrial fractionation line allowing a maximum of 3% chondritic material mixed into any of the studied lunar samples (2σ). Other chondrite groups like L, LL or carbonaceous chondrites show an even larger deviation from the terrestrial fractionation line and, therefore, even less of these primitive materials can be mixed into the lunar samples. For the mixing lines in this figure identical oxygen abundances have been assumed for all objects.

predict that the Moon can accrete very quickly (within a year or so), and further work that couples thermal and dynamical evolution is needed to determine plausible initial thermal conditions for the Moon. The long-standing constraint on initial lunar temperatures from the lack of contractional faults on the Moon should be re-examined using a wider range of plausible lunar thermal histories and rheological conditions for the lunar lithosphere. In terms of geochemistry, some now argue that because a large contribution to lunar composition might come from the (mostly unknown) composition of the impactor, lunar formation from the giant impact might not be "testable" in extreme cases (Spudis 1996; Jones and Palme 2000). Along with more sophisticated and complete models of the giant impact, co-accretion models that could be consistent with the angular momentum constraint, bulk iron deficiency, and the timing of terrestrial accretion, should be explored. In the end, we may only be able to say that one scenario is more plausible than another, but that many possibilities are allowed.

3. INITIAL MAGMATIC DIFFERENTIATION

3.1. The magma ocean: rationale, origin, size and final outcome

3.1.1. The rationale for a lunar magma ocean. Ever since petrographic studies revealed anorthositic lithologies in Apollo 11 soils (Wood et al. 1971), researchers have appealed to buoyant segregation of relatively low-density plagioclase in an extensive, primordial melting event to explain the plagioclase-rich crust of the Moon (e.g., Solomon and Töksoz 1973; Taylor and Jakes 1974). That event, which came to be known as the lunar magma ocean, has received additional affirmation through the chemistry and petrology of mare basalts and geophysical modeling.

The abundance, distribution, characteristics, and apparent antiquity of the plagioclase-rich lunar crust remain the most compelling arguments for a globe-encircling magma ocean. Remote-sensing data from the Apollo missions for Fe, Mg, Al/Si, and Mg/Si (Adler and Trombka 1977) were interpreted as indicating that the highlands are at least as anorthositic as the soils and polymict breccias (75% plagioclase) from the Apollo 16 site (Warren 1985). Estimates for the bulk composition of the veneer of megaregolith on the surface of the Moon range from anorthositic gabbro (Spudis and Davis 1986) to ferroan-anorthositic norite (Spudis et al. 1999). The bulk composition of the deep lunar crust remains obscure, but material in large basin floors and materials ejected from large impact basins indicate that the lower crust is basaltic in composition. Estimates for its average composition range from norite to KREEP basalt. Clearly, the bulk composition of the lunar crust is not that of anorthosite ($Al_2O_3 \approx 35$ wt%) as predicted by most LMO models, but approaches that of anorthositic norite ($Al_2O_3 \approx$ 25 wt%). This is not, however, an impediment to the LMO concept. The lunar crust does not represent simply a primordial lithology produced during early differentiation, but a composite planetary feature consisting of LMO-produced anorthosites, post-LMO igneous assemblages derived from melting of the lunar mantle and emplaced into the crust, and megaregolith mixtures of these petrogenetically distinct magmatic assemblages. Longhi (1982) estimated that the lunar crust (assuming a mean thickness of 60 km) contains the equivalent of at least 5-6 km of anorthosite. Wood (1986) calculated that a minimum thickness of 20 km of pure anorthite must reside in the lunar crust. Crustal models by Spudis et al. (1999) and Hawke et al. (2003) suggested that this primordial anorthosite crust resides primarily in the middle crust (15-35 km) and is exposed in some areas of the northern far side and within inner rings of multiring basins (see also Chapter 3).

The antiquity of the ancient lunar crustal rocks has been authenticated in numerous studies (reviewed by Carlson and Lugmair 1981; Nyquist 1982; Nyquist and Shih 1992; Snyder et al. 2000; Nyquist et al. 2002). Very old ages were postulated initially for the ferroan anorthosites because of their primitive $^{87}Sr/^{86}Sr$ ratios. Unfortunately, ferroan anorthosites are extremely poor in Rb and REE, so no Rb-Sr or Sm-Nd internal isochron ages were measured prior to the late 1980s (Carlson and Lugmair 1988). The new results for the ferroan anorthosites (Carlson and Lugmair 1988; Alibert et al. 1994; Borg et al. 1997, 1999) confirm ancient ages, although their implied crystallization ages point to a much more complex picture of the initial stages of lunar differentiation or an early disturbance of the isotopic systematics. This will be discussed later in this section.

Other chemical attributes of the ferroan anorthositic crust provide further evidence for their origin as LMO flotation cumulates. Differences in mineral compositions, such as molar ratios of Ca/(Ca+Na) (or An content) in plagioclase and Mg/(Mg+Fe) in mafic minerals (Warner et al. 1976) and trace element signatures, such as Sr, Eu/Al, Eu/Sm, Sc/Sm and Ti/Sm, between the ferroan anorthosites and spatial associated mafic lithologies (Norman and Ryder 1979; Raedeke and McCallum 1980; Warren and Wasson 1980; James and Flohr 1983; Warren 1986) are not consistent with a complementary petrogenetic relationship. Thus, there appears to be an absence of mafic cumulates complementary to the ferroan anorthosites in the lunar crust. This implies that the differentiation event that produced the early lunar crust was of a large scale and that the bulk of mafic cumulates associated with the ferroan anorthosites remain in the lunar mantle.

The notion of a complementary, plagioclase-depleted lunar interior did not gain wide acceptance until after studies of the Apollo 12 mare basalts. Initially, the nearly ubiquitous negative Eu-anomalies in rare- earth-element (REE) plots of mare basalt analyses were interpreted as indicating residual plagioclase in the basalt source region. Experimental work by Green et al. (1971a,b) demonstrated that plagioclase was absent from the liquidi of primitive mare basalts at both low and high pressure. The inability of plagioclase to crystallize in the early stages of low-pressure differentiation or to remain a residual phase in the mare basalt source region meant that the widespread signal of plagioclase fractionation in mare basalts was indicated that an earlier event depleted the mare-basalt source region of plagioclase (Helmke et al. 1972; Philpotts et al. 1972). In that the ferroan anorthosites have a substantial positive Eu anomaly, the simple inference is that the mare basalt source regions were the complementary mafic cumulates of the same magma that produced the ferroan-anorthositic crust.

As reviewed by Warren and Wasson (1979) and Warren (1985), the ratios of incompatible elements among lunar rocks rich in such elements virtually all conform to a uniform pattern throughout the sampled portion of the Moon. The key characteristic of this pattern is that it is high in concentration of incompatible elements such as **K**, **R**are **E**arth **E**lements, **P** (KREEP), Th, U, Zr, Hf, and Nb, and exhibits a uniform fractionation among the incompatible elements. For example, it is uniformly enriched in the light REE relative to the heavy REE as indicated by precisely determined La/Lu and Sm/Nd elemental ratios and ^{143}Nd/^{144}Nd isotopic ratios. The widespread uniformity and the extreme incompatible-element enrichment of the KREEP signature relative to bulk Moon can not be produced by small-scale fractional crystallization processes (Taylor 1975 1982; Warren and Wasson 1979; Warren 1985). Model ages for KREEP also suggest its origin during the early stages of lunar differentiation (Nyquist and Shih 1992). This uniformity and enrichment was viewed as being produced from the residuum (urKREEP) of a single, global magma during primary lunar differentiation. Although recent remotely sensed data indicate that this geochemical signature is localized in the Procellarum-Imbrium region on the Earth-facing side of the Moon, it is still a major planetary signature that requires a large-scale planetary process for its origin. Warren and Wasson (1979) and Walker et al. (1979) cautioned that most rocks with a KREEP signature are too mafic in major-element composition and mineralogy to be simple, unaltered samples of LMO residuum. More likely, the KREEP signature was incorporated into the lunar crust through remobilization, assimilation, and mixing.

Numerous attempts have been made to demonstrate the former existence of a LMO, gauge its size, and predict its effects. Warren (1985) provided a thorough review and analysis of the problem. Here, we focus on the formation of the LMO and its effect on lunar thermal history, the extent of early differentiation as a gauge of its size and duration of crystallization, and the relation between ferroan anorthosites and the LMO.

3.1.2. Thermal constraints for a magma ocean.
Although several heating mechanisms have been proposed (see summaries in Warren 1985, Shearer and Papike 1999), the only plausible way to melt a large portion of the Moon in a short time is by rapid ($< 10^3$ years) accretion. So when Wetherill's (1976, 1980) dynamic modeling of the early solar system showed that the planets accreted on a time scale of 10^7 to 10^8 years, serious doubts arose about the plausibility of the LMO. Several alternate hypotheses that did not involve an LMO were advanced to explain the Eu-anomaly in mare basalts (e.g., Walker 1983) and the abundance of ferroan anorthosite without cogenetic mafic rocks in the lunar crust (e.g., Longhi and Ashwal 1984). However, the development of the giant impact hypothesis alleviated many of the concerns about rapid heating. As discussed earlier, models of the dynamics of debris orbiting Earth after the giant impact predicts very short accretion times ranging from several months to

a year. Furthermore, collisional heating of particles within the disc might maintain the inner portion of the disc in a liquid to partially vaporized condition (Cameron and Benz 1991). Melting a substantial portion of the Moon no longer seems to be a serious obstacle.

Perhaps, the most insightful constraints on the size of the LMO came from the work of Solomon and Chaiken (1976), who modeled the change in planetary volume through time as a function of initial temperature distribution (Section 4.2). Their thermal modeling was consistent with the outer 200 km of the Moon being initially molten. Solomon (1977) revised the depth of the molten zone to 300 km. Kirk and Stevenson (1989) repeated the calculations and found that if one allowed for heating and melting of the part of the Moon below the LMO that the volume increase produced by partial melting through time could offset the contraction produced by solidification of a 400 km deep LMO. To provide sufficient expansion, the amount of melt required was ~100× the volume of the mare basalts. This magma need not have erupted or even intruded the crust; however, the production of this magma does coincide with the eruption of the mare basalts. So unless the Moon supported two coeval, but distinct melting systems throughout its history, it is reasonable to equate the melting predicted by Kirk and Stevenson (1989) with mare basalt magmatism. Head and Wilson (1992) calculated that mare basalt eruption geometries and rates implied a system of dikes and sills 50× the volume of the erupted mare basalts. Thus the 100:1 ratio of melt to erupted mare basalt envisioned by Kirk and Stevenson (1989) is certainly plausible. As shown below, however, the mare basalts could not have formed from melting the undifferentiated lunar interior. Moreover, there is no evidence of melts of primitive lunar material ever erupting as lavas. Thus it appears from thermal modeling that 400 km is an upper limit to the thickness of the LMO.

It may be that some additional thermal process, not incorporated in the thermal models described above, counteracted the contraction expected if the initial depth of melting exceeded 400 km. One such process is asymmetric heating of the crust. Haskin (1998) called attention to the enrichment of heat-producing, incompatible elements in a region of thinned crust on the lunar near side, the Procellarum-KREEP terrane (PKT; see Chapter 3). Perhaps higher temperatures in the PKT expanded and/or weakened the crust sufficiently to mask the effects of global contraction of a LMO deeper than 400 km. Another such process is deep burial of heat-producing elements. Revising a suggestion of Kesson and Ringwood (1975), Hess and Parmentier (1995) proposed that overturn in the late-stage cumulates of an LMO would have sent plumes of ilmenite and ferropyroxene sinking into the lunar interior, carrying with them trapped melt enriched in heat-producing elements. Even though the buildup of heat released by radioactive decay would eventually lead to melting, the presence of ilmenite would constrain the melts to be denser than their surroundings at depths >500 km (Delano 1990). Such a deep-seated heat source was not considered in the thermal calculations described in Section 2.4. It is unclear whether sufficient abundances of incompatible elements could have been sequestered in the deep interior to accommodate a deep (>400 km) LMO; this remains a fruitful area of investigation.

3.1.3. Duration of magma-ocean crystallization. The duration over which the LMO crystallized is uncertain. Part of the problem is the uncertainty in the age of the ferroan-anorthositic crust. The anorthosites are among the oldest lunar rocks, yet their crystallization ages are difficult to interpret (see Section 3.4). Low $^{87}Sr/^{86}Sr$ ratios for ferroan anorthosites approach those of primitive meteorites, implying that they have been separated from any significant concentrations of Rb for a length of time comparable to the age of the Moon (Nyquist and Shih 1992). Model ages for nonpristine rocks using U-Pb isotopes indicate a major differentiation event occurred at about 4.47-4.56 Ga (Oberli et al. 1979; Carlson and Lugmair 1988; Hanan and Tilton 1987; Premo and Tatsumoto 1991). Relatively young ages for some ferroan anorthosites (Borg et al. 1998) may be explained by prolonged subsolidus cooling or open system behavior, but their derivation from light REE-depleted sources (Borg et al. 1998) is still open to interpretation (Section 3.4).

The idea that the KREEP component in the lunar crust was the residuum from the crystallization of the LMO has played a key role in estimating the duration of lunar differentiation (Papanastassiou et al. 1970; Tera and Wasserburg 1974; Lugmair and Carlson 1978). The average model age for KREEP is 4.42 ± 0.07 Ga for Sm/Rb = 2.3 (Nyquist and Shih 1992). The interpretation of this model age is open to speculation in that it may represent a true crystallization age, a cooling or disturbance age, or a mixing event. It does, however, represent the maximum duration over which the LMO crystallized. Sm-Nd model ages for the mare basalts are interpreted as indicating that their source region (i.e., mafic cumulates from LMO crystallization) were isotopically closed by 4.4 Ga.

Thermal models for the duration of LMO crystallization are highly dependent on the size of the LMO and the nature and thermal conductivity of the early crust. Thermal models predict that before plagioclase became a stable liquidus phase, parts of the LMO may have crystallized over a period of time ranging from 10^2 years to less than 10^7 years (Solomon and Longhi 1977; Binder 1976; Minear 1980). The discrepancy between these durations result primarily from assumptions about the chilled margin or proto-lunar crust (thickness of proto-crust and boundary layers, viscosity, and composition). Assuming development of a thick (50-100 km), stable and continuous plagioclase crust following substantial LMO crystallization, Solomon and Longhi (1977) and Minear (1980) predicted total LMO crystallization times of 6×10^7 to 2×10^8 years. Shorter time scales for crystallization are anticipated for a thinner lunar crust or an early crust destabilized by large impacts. Minear (1980) estimated that the LMO crystallized over a period of 3×10^7 to 8×10^7 years in a multi-stage crustal growth model that anticipated an early lunar proto-crust and a later ferroan-anorthosite dominated crust disrupted by impacts and regrown by conductive cooling. In models in which the ferroan-anorthositic crust formed after LMO crystallization (Longhi 2003), the LMO could have crystallized over a timeframe of less than 200 years to 10 m.y. (Longhi 2003).

Although the excess ^{182}W produced by the early decay of ^{182}Hf to ^{182}W is significantly smaller than initially reported (Lee et al. 1996), there still appear to be mantle reservoirs for the mare basalts that have excess ^{182}W ($\varepsilon_W^{182} > 0$). As the high ε_W^{182} appears to be associated with the mare basalts that were generated by melting of a clinopyroxene- and ilmenite-bearing mantle, Shearer and Newsom (1998, 2000) suggested that the radiogenic W sources were produced during LMO crystallization and not core formation. From the extent of Hf/W fractionation by ilmenite and high-Ca clinopyroxene determined by Righter and Shearer (2003) and Shearer and Righter (2003), over 95% of the LMO crystallized less than 60 million years after solar-system formation. Depending on the accretion age of the Moon, this inference suggests that most of the LMO crystallized over a period of 10-40 million years. Such a short period of time to crystallize the LMO contrasts with models suggesting durations from 100 m.y. to greater than 200 m.y. (Solomon and Longhi 1977; Longhi 1980). Rapid LMO crystallization also implies that the transition from LMO (characterized by anorthosite formation) to serial magmatism (with the emplacement of Mg-suite plutonic rocks) occurred earlier, in agreement with age constraints from other studies (Shih et al. 1993).

3.1.4. Chemical and petrological constraints on early lunar differentiation. The concept of the LMO began with the recognition of an anorthositic crust, so it is useful to examine some mass-balance constraints involving the crust. Figure 4.4 illustrates the results of some simple calculations that determine the amount of Al_2O_3 residual to the formation of crusts of different thickness and composition for 3 potential whole-Moon compositions. Following the gravity modeling of Haines and Metzger (1980), Taylor (1982) adopted 2.93 and 73.4 km for the average density and thickness, respectively, of the lunar highlands crust; he also derived 26 wt% Al_2O_3 as the average crustal composition. The 50:1 ratio for intrusive to extrusive mare basalt implies approximately 20 wt% of the crust consists of mare basalt (Head and Wilson 1992). Taking the average Al_2O_3 concentration of mare basalts to be 10 wt%, then removing the

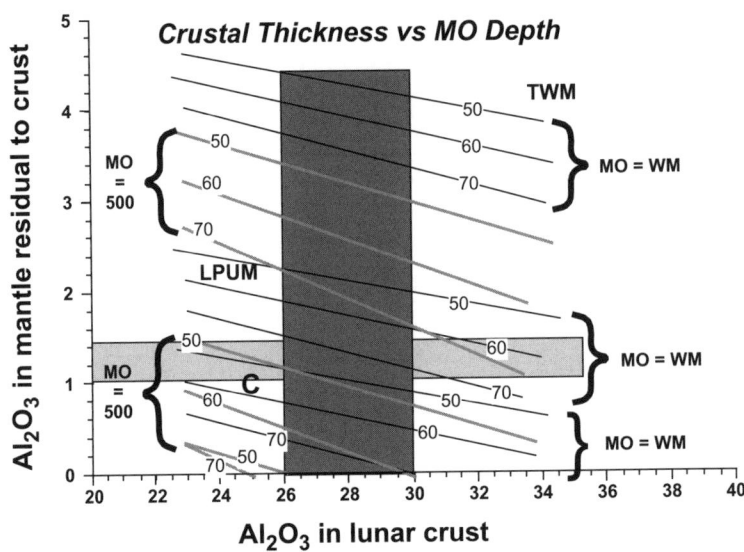

Figure 4.4. Calculations that determine the amount of Al_2O_3 residual to the formation of crusts of different thickness and composition for 3 potential whole-Moon compositions: a refractory element-enriched composition with 6 wt% Al_2O_3 (TWM or *Taylor Whole Moon* – Taylor 1982), an alkali-depleted version (LPUM or *Lunar Primitive Upper Mantle*) of an estimated terrestrial upper mantle composition (Hart and Zindler 1986) with 4 wt% Al_2O_3, and an approximate chondritic composition (C) with 3 wt% Al_2O_3.

mare basalt component from the crust increases the Al_2O_3 to 30 wt% and decreases the average thickness to ~58 km. These values are likely to be extremes because most of the intrusive mare magmas are likely to be situated beneath maria not modeled by Haines and Metzger (1980). Global crustal mass balance based on global FeO and using basin ejecta as probes of the middle to lower crust indicate average crustal Al_2O_3 of 24-25 wt% (see also Chapter 3, Section 5.3). According to this type of calculation, the Al_2O_3 concentration of the pre-mare-basalt lunar crust probably lies within the yellow-shaded area bounded by 26 and 30 wt% Al_2O_3 (Fig. 4.4). This total Al_2O_3 includes LMO flotation cumulates as well as post-LMO plutonic rocks.

Figure 4.4 also shows that the chondritic (C) composition cannot satisfy the Al_2O_3 requirement of the Moon unless nearly the entire Moon was initially molten and the extraction of Al_2O_3 was nearly perfect. In Figure 4.4, the TWM and LPUM compositions both contain sufficient Al_2O_3 for total melts of the Moon, but LPUM does not for an LMO <500 km deep. We may constrain the composition-depth relations further by considering constraints imposed by mare-basalt petrogenesis. For example, the pressures of multiple saturation on the liquidus of the picritic green glasses provide a crude estimate of the minimum depth of the mare basalt source region and, therefore, the minimum depth of early differentiation. Elkins-Tanton and Grove (2000) showed multiple saturation pressures from 1.3 to 2.5 GPa, which correspond to depths of 260 to 700 km. If decompression provided most of the heat for mantle melting on the Moon, as it does on the Earth, then pressures of multiple saturation represent average depths of melting (Klein and Langmuir 1987). Longhi's (1992) modeling of green-glass petrogenesis by polybaric melting, which is updated below, constrained the green glass source region to lie at depths ≥1000 km and to contain ~1.4 wt% Al_2O_3 (horizontal blue shading). Figure 4.4 indicates that the TWM composition contains too much Al_2O_3 to produce the depth and composition of the green glass source region, whereas LPUM agrees well for an LMO depth between 500 km and the entire Moon.

3.1.5. Petrologic constraints on the extent of early lunar melting and differentiation. The compositions of basalts reflect the compositions and depths of their source regions, although not as directly as we would like. Not only do magma compositions change prior to eruption in response to low-pressure crystallization and assimilation, but also the melting process intrinsically obscures itself. Melting in the Earth's mantle occurs primarily by the release of pressure during convective upwelling. Adiabatic decompression permits approximately 10% melting per GPa (Hess 1992). Low porosity and density differences lead to separation of melt from matrix during the melting process (McKenzie 1984). Consequently, most primitive magmas are blends of low-degree melts accumulated over a range of depths from a progressively depleted source. An important corollary is that a magma composition cannot be related to a specific depth by major elements. Nonetheless, there have several reasonably successful calculations of primitive magma composition based upon parameterization of the P-T-X data obtained from melting experiments (Kinzler and Grove 1992a; Klein and Langmuir 1987; Longhi 1992).

In order to understand the implications of mare basalt compositions for lunar structure, it is useful to compare them to terrestrial basalts in the same format that the melting calculations are presented. Accordingly, Figure 4.5 compares the compositions of low-Ti mare basalts with those of the most abundant terrestrial basalts—mid-ocean ridge (MORB), ocean island (OIB), and continental flood basalts (CFB). Each set of compositions is projected against appropriate olivine-saturated, low-pressure liquidus boundaries calculated by the algorithms presented by Longhi (1991). The terrestrial basalts have higher proportions of plagioclase (Pl) to pyroxene (Opx, Di) components reflecting higher concentrations of both Al$_2$O$_3$ and alkalis. The high relative proportions of the Pl component lead to early crystallization of plagioclase and augite and the relatively late appearance of low-Ca pyroxene as pigeonite. This is the typical tholeiitic trend. The lower proportion of the Pl component in the mare basalts leads to the early appearance of pigeonite. The differences in the moderately volatile alkalis between lunar

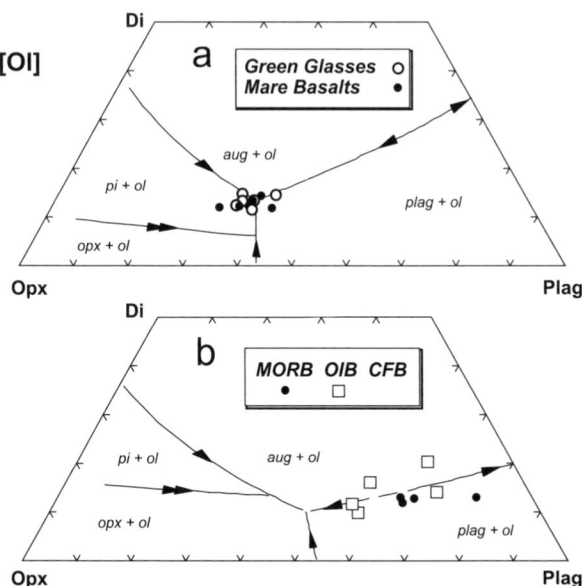

Figure 4.5. Comparison of the compositions of low-Ti mare basalts with those of the most abundant terrestrial basalts—mid-ocean ridge (MORB), ocean island (OIB), and continental flood (CFB). Each set of compositions is projected against appropriate olivine-saturated low-pressure liquidus boundaries calculated by the algorithms presented by Longhi (1991).

and terrestrial basalts reflect different accretion histories, whereas the differences in Al_2O_3 are more likely to reflect different magmatic histories.

Figure 4.6 shows a series of batch and polybaric fractional melts of a source with the composition of depleted terrestrial upper mantle from Kinzler and Grove (1992b) calculated with the programs BATCH and FFTHERM. BATCH (Longhi 2002), which calculates equilibrium crystallization and partial melting, is a new version of MAGPOX (Longhi 1991, 1992). FFTHERM is a modification that balances the amount of melting in each pressure step (0.1 GPa) against the temperature drop along an adiabat of 15°/GPa, the calculated temperature drop along the solidus, and the temperature equivalent of the heat of fusion. Any melt that exceeds the assumed background porosity fraction of 0.008 is withdrawn and accumulated. In Figure 4.6 the calculated batch and accumulated fractional melts overlap a significant portion of the natural compositions in both projections. The more extreme natural compositions with negative Opx components probably reflect the influence of CO_2 in the melting process. The batch melts all represent liquids that are saturated with or nearly saturated with an aluminous phase (plagioclase, spinel, or garnet). Saturation with an aluminous phase at high pressure ensures that a liquid will have a high ratio of plagioclase to pyroxene components.

The whole-Moon composition of Taylor (1982) contains 6 wt% Al_2O_3—1.5× the estimated concentration of the Earth's upper mantle (Hart and Zindler 1986). It should not be surprising, therefore, that 10% batch melts of TWM have relatively high ratios of plagioclase to pyroxene components, a generally tholeiitic aspect, and do not overlap the compositions of the mare volcanics (Fig. 4.7). At 30% melting at 3.0-3.5 GPa, TWM melts begin to approach the component proportions of the mare volcanics, because the garnet is melted out, but Mg/(Mg+Fe) (Mg') values are much higher than those of the green glasses (0.72 vs. 0.61). Also, there is no plausible way to generate such high degrees of melting at depths (>700 km) corresponding to these pressures. Similar arguments apply to batch melts of LPUM.

Figure 4.8 illustrates the track of accumulated polybaric partial melts of TWM, LPUM, and GGS (a modified version of the green glass source region of Longhi (1992) that was constructed

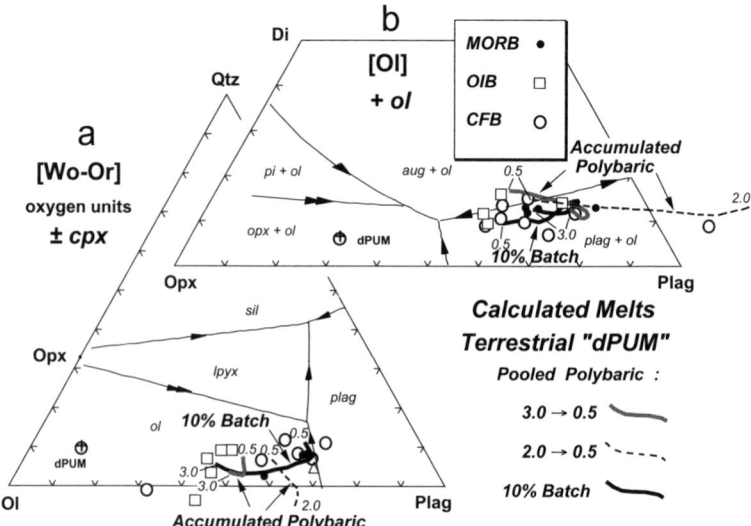

Figure 4.6. A series of calculated batch and polybaric fractional melts of a source with the composition of depleted terrestrial upper mantle (Terrestrial "dPUM") from Kinzler and Grove (1992b).

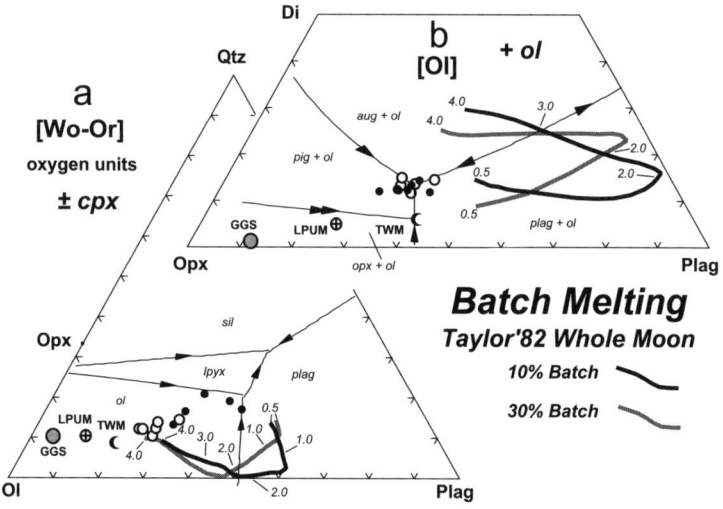

Figure 4.7. Calculated batch melts (10% and 30%) at different pressures for Taylor's (1982) whole-Moon composition (TWM).

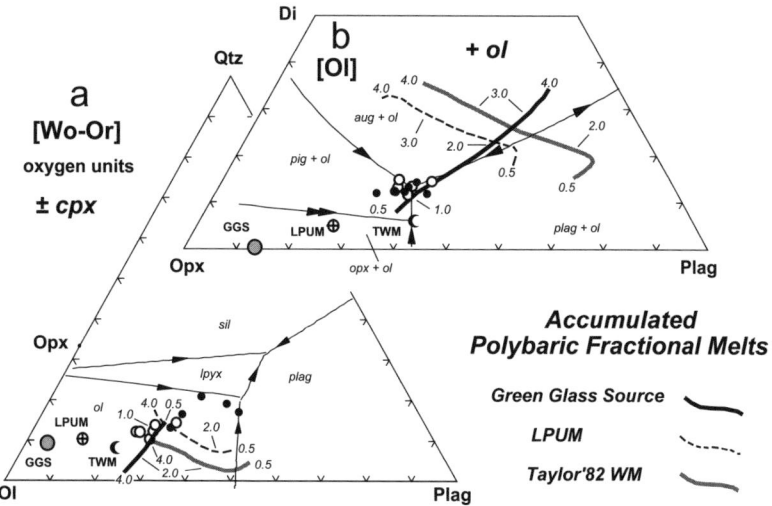

Figure 4.8. Plot illustrates the track of calculated accumulated polybaric partial melts of TWM, LPUM, and GGS (modified version of the *Green Glass Source* region of Longhi 1992).

by adding forsteritic olivine, ferroan clinopyroxene, and ilmenite in various proportions). Each composition has garnet at the solidus at 4.0 GPa, but GGS has only a trace, and following the removal of 0.2% liquid from the source, garnet does not appear at lower pressures. The high-P melts of TWM have markedly lower quartz coordinates than melts of LPUM or GGS because of different solidus assemblages. Orthopyroxene is not present at the TWM solidus above 3.2 GPa because of a peritectic reaction with the liquid. At these pressures, melting of TWM begins in the presence of olivine, augite, garnet, and, possibly spinel. At ~7% melting, orthopyroxene begins to crystallize, but the polybaric melting process never gets beyond 1-2% melting at any time. The orthopyroxene-free liquids lie on a boundary curve that is not tightly constrained by experimental data. However, the data that do exist constrain the olivine + clinopyroxene + garnet curve to trend toward low silica (Longhi 1995). The LPUM accumulated melts cut across the green glass array in the Wo (wollastonite, $CaSiO_3$) projection in Figure 4.8a, but are far removed in Figure 4.8b. The GGS melts do cut across the field of the green glasses in both projections. Some sample compositions are listed in Table 4.1. These compositions are by no means unique—it is possible to spread the track of the calculated liquids out by varying the initial composition or the porosity (lower porosity means more melt is extracted at each step, so the source is depleted faster and the melts become less aluminous).

The fact that neither batch nor polybaric melting of bulk lunar compositions can produce the picritic green glasses or mare basalts has several implications for lunar differentiation. The simplest are that: a) the primitive isotopic and trace-element patterns identified in the green glasses (e.g., Delano 1986; Neal 2001) are not intrinsic, but result from some sort of hybridization or assimilation; b) the post-LMO deep melting of the Moon envisioned by Kirk and Stevenson (1989) could not have produced the mare volcanics, at least not directly.

The most significant implications relate to the scale and extent of the primordial differentiation. To the extent that the model is accurate, the results imply that the differentiation event must have been sufficiently fluid to deliver differentiated material to depths of ~1200 km. Conceivably, this transport could have been accomplished by crystallization of an LMO of modest depth (400-500 km) followed either by plumes of dense Fe-rich pyroxene ± ilmenite that penetrated the deep interior or by large-scale overturn that placed the initially unmelted portion of the Moon above part or all of the crystallized LMO. Indeed, the presence of primitive components in some of the volcanic glasses (Delano 1986) suggests that some undifferentiated material survived at least until the time of mare volcanism whereas the *P-T* path of the depleted green glass source implies that only modest amounts of undifferentiated Moon could have been present during mare volcanism.

Figure 4.9 shows the *P-T* path of the GGS that produced the compositional trend in Figure 4.8 in relation to its own solidus and the solidii of TWM and LPUM. Clearly, the picritic glass magmas would have melted and partially assimilated any primitive material with which they came into contact. More important, however, is the observation that the temperatures along the GGS solidus exceed the solidus temperatures of undifferentiated material. Thus it is not likely that any significant fraction of undifferentiated lunar material was present above ~1200 km at the time of the green glass magmatism, otherwise it would have melted preferentially.

Table 4.1. Whole Moon, Green Glass Source, and model magma compositions.

	TWM	LPUM	GGS	FF11*	Ap15G	Ap14	VLT
SiO_2	44.4	46.1	43.9	45.8	44.1	46.0	48.56
TiO_2	0.31	0.17	0.07	0.37	0.37	0.55	0.79
Al_2O_3	6.14	3.93	1.44	7.24	7.81	9.30	16.42
Cr_2O_3	0.61	0.50	0.42	0.43	0.33	0.58	0.32
FeO	10.9	7.62	16.15	20.7	21.0	18.2	7.05
MnO	0.15	0.13	0.18	0.28	0.10	0.21	0.13
MgO	32.7	38.3	36.26	17.9	16.7	15.9	12.48
CaO	4.60	3.18	1.55	7.12	8.41	9.24	14.99
Na_2O	0.09	0.05	0.03	0.16	0.13	0.11	0.33
K_2O	0.009	0.003	0.03	0.02	0.03	0.07	0.02
Mg'	0.84	0.90	0.80	0.61	0.59	0.61	0.76

*accumulated melt composition: GGS source, 4.0 GPa initial melting, 0.1 GPa steps, 0.008 porosity

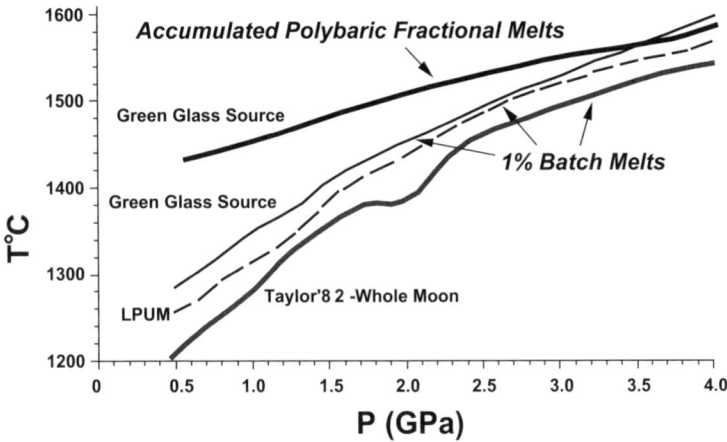

Figure 4.9. Calculated pressure-temperature path of the GGS that produced the compositional trend in Figure 4.8 in relation to its own solidus and the solidii of TWM and LPUM.

Such melts would have been lunar tholeiites in which high-Ca clinopyroxene was the dominant pyroxene. Such rocks have not been found among the samples. Therefore, ~1200 km is a reasonable approximation of the depth of the primordial differentiation.

The recent interpretation of seismic data contrasts with the petrologic interpretation of the extent of the LMO to ≈ 1200 km depth. Several of the Apollo-era seismic models have been analyzed to infer constraints on the structure and composition of the lunar mantle (Buck and Toksöz 1980; Hood and Jones 1987; Mueller et al. 1988; Hood and Zuber 2000; Khan et al. 2000). In general, the preferred interpretation of the increase in seismic velocity at 500 km depth is a change in composition to more aluminous and Mg-rich silicates. This inferred change in composition and increase in the degree of heterogeneity has been interpreted as indicating that the Moon was initially melted and differentiated only to a depth of 500 km. If this interpretation is correct, it implies that an Al- and Mg-rich primitive mantle exists below 500 km, that all the Al in the primordial lunar crust must have been extracted from the upper mantle, and that much of the mare basaltic magmatism must have involved melting above 500 km.

A final consideration is the composition of the green glass source itself. Figure 4.8 illustrates the bulk compositions TWM, LPUM, and GGS. The three symbols are widely spaced, reflecting their different Al_2O_3 concentrations, 6, 4, and 1.4 wt%, respectively. Estimates of the difference in Al_2O_3 between primitive and deleted upper mantle on the Earth are on the order of 0.1 wt% (Kinzler and Grove 1992b). This difference is smaller than the size of the symbols. If the Moon has a bulk composition similar to that of the Earth's upper mantle, then differentiation by extraction of basalt would imply that roughly twice the mass of the GGS had been extracted. Such material is not in evidence on the surface, and if buried would have to have a remarkably cryptic history, inasmuch as any basalt would be much less refractory than either GGS or undifferentiated Moon. Although the GGS composition was developed to fit the green glasses, it nonetheless is consistent with mixing of early- and late-stage components of an LMO. These various lines of evidence are thus consistent not only with an LMO, but with one that involved at least 2/3 of the Moon's radius. From the mass balance relations in Figure 4.8 and the need for a low-Al_2O_3 source for the green glasses, the bulk composition of this ocean was very likely similar to an alkali-depleted version (LPUM) of the Earth's upper mantle.

3.1.6. Final outcome. The LMO model predicts several outcomes regarding the current nature of the Moon. An appealing aspect of these LMO models is that they provide (1) processed mantle cumulate assemblages that are appropriate sources for subsequent periods of lunar magmatism and (2) a primary lunar crust consisting of plagioclase-rich flotation cumulates.

Different cumulate rock types have been identified as potential sources for a wide range of parental magmas for both pre-mare and mare basalts. For example, the highly fractionated KREEP component in the cumulate pile is considered to have been remobilized and emplaced into the lunar crust. Many of the pre-mare basaltic magmas are thought to have acquired a KREEP signature through assimilation or mixing. The late-stage, ilmenite-bearing cumulates have been considered as possible sources for the high-Ti mare basalts (Taylor and Jakes 1974; Taylor 1982; Snyder et al. 1992), whereas early olivine-orthopyroxene dominated cumulates have been proposed as the mantle source for the very-low-Ti picritic magmas (Taylor and Jakes 1974; Taylor 1982).

Chemical signatures within the LMO cumulate stratigraphy are illustrated in Figure 4.10. For example, early cumulates should be enriched in elements such as Ni and perhaps Co, and have REE patterns with virtually no Eu anomaly (Shearer and Papike 1989). Later cumulates should develop an REE pattern with negative Eu anomalies, be systematically enriched in incompatible elements and depleted in compatible elements such as Ni.

The crystallization history of the LMO has been summarized by Schnetzler and Philpotts (1971), Taylor and Jakes (1974), Taylor (1982), Longhi (1977, 1981), and Snyder et al. (1992). Schematic diagrams for the cumulate pile produced by LMO crystallization are presented in Figures 4.10 and 4.11. The sequence of crystallization is highly dependent on LMO bulk composition and the pressure and flow regimes under which crystallization occurred, and is therefore difficult to predict. In a dynamically simple LMO, the crystallization sequence advocated by most of these models is olivine $\rightarrow$ orthopyroxene $\pm$ olivine $\rightarrow$ olivine + clinopyroxene $\pm$ plagioclase $\rightarrow$ clinopyroxene + plagioclase $\rightarrow$ clinopyroxene + plagioclase + ilmenite. On the basis of estimated bulk compositions of the LMO (Taylor and Bence 1975; Ringwood and Kesson 1976; Buck and Toksöz 1980; Warren 1986; Hughes et al. 1988, 1989), olivine crystallizes first until the olivine-orthopyroxene boundary line is reached. The extent of the olivine cumulate assemblage in the LMO cumulate pile is estimated to be between 30 and 40%. The effect of the olivine-orthopyroxene field boundary on cumulate assemblages depends on both the pressure of crystallization and the efficiency of crystal accumulation (equilibrium versus fractional crystallization). As pressure increases, this boundary moves toward the olivine apex of the olivine-anorthite-SiO$_2$ pseudoternary, which has two effects. First, this shift effectively decreases the volume of the monomineralic olivine assemblage in the cumulate pile. Second, the olivine-orthopyroxene boundary becomes a cotectic surface, resulting in coprecipitation of olivine + orthopyroxene regardless of the efficiency of crystal separation. Within the context of the flow regimes proposed by Spera (1992), olivine and orthopyroxene precipitated from magmas at fairly high-pressure near the base of the cumulate zone. At the upper LMO boundary, significant radiative heat loss resulted in rapid olivine and orthopyroxene crystallization. These early crystallization products of the upper boundary layer should sink and be incorporated into the inertial inner region of the LMO (a nearly isothermal region dominated by high convective velocities detailed in Section 3.2). It appears unlikely that these early phases from the upper boundary would have been incorporated into the basal cumulate layer. The appearance of low-Ca clinopyroxene and high-Ca clinopyroxene followed the precipitation of orthopyroxene and olivine. The exact sequence of the appearance of these pyroxenes is dependent upon bulk composition. The relationship among clinopyroxenes and plagioclase in the LMO crystallization sequence is discussed in detail by Longhi (1980). In the bulk composition used by Snyder et al. (1992), high-Ca clinopyroxene crystallized after plagioclase.

The appearance of plagioclase in the crystallization sequence is fundamentally important to understanding the development and evolution of the early lunar crust. In simple dynamic models, the lunar crust is thought to have formed by plagioclase crystallization and flotation

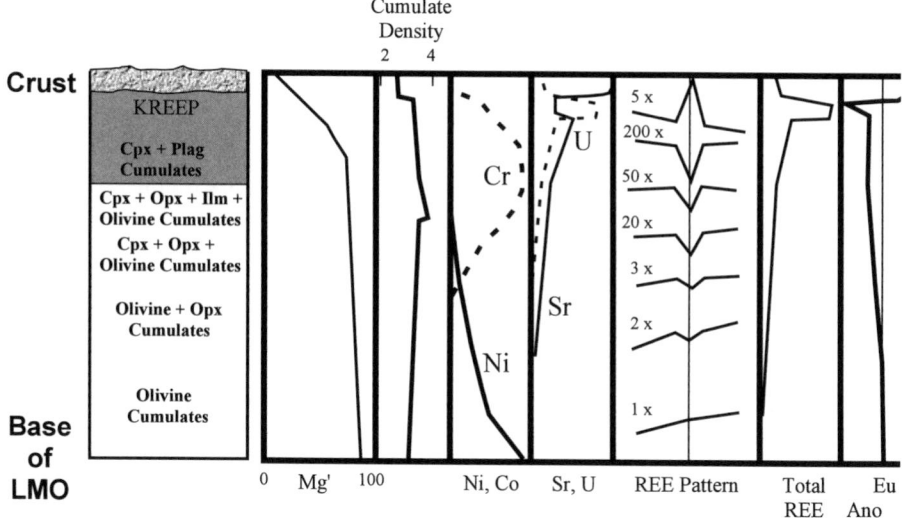

Figure 4.10. Mineral and chemical characteristics of the LMO cumulate pile (modified after Shearer and Papike 1999).

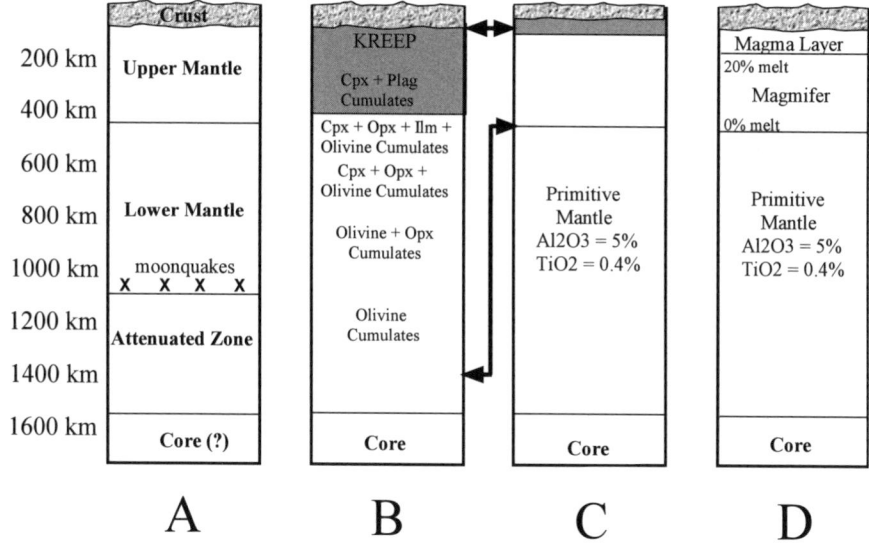

Figure 4.11. Different types of lunar magma oceans (LMO). (A) Geophysical structure of the Moon. (B) Deep LMO that consumed all of the primitive lunar mantle. (C) Shallow LMO that preserves the primitive lunar mantle at depth. D. Magmifer processing of the primitive lunar mantle (modified after Taylor 1982).

after substantial amounts of the LMO had crystallized. Using the initial bulk composition suggested by Warren (1986) in which the Al_2O_3 was 7 wt%, Snyder et al. (1992) calculated that plagioclase would be a liquidus phase after 57% LMO crystallization. Using estimates of bulk compositions containing <5 wt% Al_2O_3, Snyder et al. (1992) estimated that plagioclase would appear in the crystallization sequence after 70-80% crystallization. The ability of plagioclase to float and accumulate in "rockbergs" at the surface also depends on the bulk Mg′ of the LMO.

Flotation of anorthositic rocks (plagioclase + minor mafic minerals) may have been impossible until the "magmasphere" became more Fe-rich after substantial crystallization of mafic silicates (Warren and Wasson 1980). High degrees of crystallization prior to plagioclase flotation are also implied by the trace-element characteristics of the ferroan anorthosites (Palme et al. 1984, 1997). On the basis of these models, a large portion of the cumulate pile has either no Eu anomaly or only a minor one (Taylor 1982; Shearer and Papike 1989; Snyder et al. 1992).

In more complex models for LMO dynamics that will be discussed in the subsequent section, it could be expected that plagioclase was a liquidus phase during a substantial period of crystallization (Longhi 1980). Due to rapid heat loss of the upper LMO boundary, plagioclase might have become a liquidus phase earlier in the upper boundary layer than in the lower one. This may have resulted in formation of a protocrust prior to 60% LMO crystallization. In addition, this early plagioclase crystallization may have imposed a negative Eu anomaly on early basal cumulates (olivine + orthopyroxene) by sinking of residual melt into the central zone of the LMO.

Ilmenite-bearing cumulates precipitated after 90% crystallization of olivine + pyroxene + plagioclase. Ilmenite made up between 3-12% of the cumulus minerals. Incompatible trace elements excluded from the crystal structures of olivine, pyroxene, and plagioclase, were concentrated in the residual melt (KREEP). The signature of this KREEP component was later incorporated into the feldspathic highland crust through remobilization, assimilation, and mixing.

Longhi (2003) suggested a stratified, post-LMO lunar mantle that is different from that illustrated in Figures 4.10 and 4.11. This difference is not due to a dramatic difference in crystallization sequence of the LMO, but the efficiency by which plagioclase could be removed and accumulated from the LMO. As discussed below, the efficiency of plagioclase removal is dependent upon the evolution of different flow regimes in the crystallizing LMO. Under an end-member model in which the ferroan-anorthosite crust is not a product of LMO crystallization, Longhi (2003) proposed that the post-LMO Moon consists of four density-stratified layers: (1) an undifferentiated zone beneath the LMO cumulate pile that is initially denser than the overlying olivine-rich cumulates because of its lower Mg' and the presence of garnet, (2) an intermediate ultramafic zone of dunites grading upward into harzburgites, (3) an upper mafic zone composed of norites and gabbronorites, and (4) a thin protocrust consisting of dense LMO residua, quenched LMO liquids, and perhaps a small amount of anorthosite. These differences in densities and thermal characteristics would drive subsequent cumulate overturn resulting in the post-LMO formation of the ferroan-anorthosite crust, formation of mare-basalt mantle sources by fertilizing the lower LMO cumulate pile with upper LMO cumulates, and the derivation of mantle plumes from the undifferentiated zone to form Mg-suite magmas.

Complementary to the LMO cumulates that constitute at least a portion of the lunar mantle are the flotation cumulates that make up the primordial lunar crust. They represent the only remnant in the lunar sample collection that is presumably a product of the first stage of lunar differentiation. Unlike most of the other terrestrial planets, the primary lunar crust is preserved and provides significant information concerning the initial differentiation of the Moon. A summary of the petrography, geochemistry, and distribution of the ferroan anorthosites that comprise this primary crust is given by Papike et al. (1998) and in Chapter 3.

Apart from high Al_2O_3 and CaO, and the low Mg' documented in Chapter 3, the ferroan anorthosites are characterized by low incompatible element abundances with the trivalent REE generally less than chondritic. Their REE patterns are typified by a large positive Eu anomaly with $(Eu/Sm)_N$ as high as 100. This large anomaly is attributed to the highly reducing conditions during the early stages of lunar differentiation ($\approx$ IW−1) which resulted in a high proportion of divalent Eu relative to terrestrial magmatic environments. Estimates of parental magma compositions from bulk rock and plagioclase core compositions suggest that the parent magmas had refractory incompatible-element ratios close to chondritic (e.g., Ti/Sm, Sc/Sm) (e.g., Hubbard et al. 1971; James et al. 1991). These parental melts are estimated to contain REE concentrations 10-50× chondritic abundances with fairly flat REE patterns (Floss 1991; Floss et al. 1991, 1998; Jolliff and Hsu 1996; Papike et al. 1997b). The calculated parent melts with lower abundances of REE lack distinct Eu anomalies and are presumably earlier and more primitive, whereas those with higher REE have small negative Eu anomalies. This is consistent with a model magma ocean originally with 5 wt% Al_2O_3 crystallizing plagioclase from ~78-90% solid (Papike et al. 1997b). However, the general lack of correlation between An contents of the plagioclase with the Mg' of coexisting mafic minerals does not fit the simple crystallization model implied by the REE. In general, this lack of correlation between mineral compositions is attributable to variable amounts of adcumulus growth (Morse 1982; Ryder 1982) or more complex processes (e.g., Haskin et al. 1981; Jolliff and Haskin 1995).

Although in the general model of anorthosite petrogenesis the anorthosites are complementary to the mare-basalt source regions, the relationship is complex. Some of the mare sources are too magnesian, contain too much high-Ca pyroxene, and too much Ni to be directly complementary to ferroan anorthosites (e.g., Ryder 1982, 1991). These lines of evidence suggest modification of the LMO cumulates by mixing during solid-state convection. More details are presented in Section 3.3.

3.2. Physics of the lunar magma ocean

Most models for the thermal evolution of the LMO are based on the assumption that the outer portion of the Moon was totally molten (Wood et al. 1970; Wood 1972, 1975; Taylor and Jakes 1974; Hubbard and Minear 1975; Walker et al. 1975; Longhi 1977, 1981; Solomon and Longhi 1977; Minear and Fletcher 1978; Tonks and Melosh 1990; Spera 1992). Early models focused on the geochemical and petrologic implications of this early differentiation event.

Among the more sophisticated of these early models were those developed by Solomon and Longhi (1977). Their thermal models were based on the numerical solutions of the finite-difference analog of the equation for energy conservation with spherical symmetry, as originally described by Toksöz et al. (1972). Although these thermal models are one dimensional and did not describe convective flow in the LMO, they accounted for heat loss from the base and top of the LMO and approximated the average thermal state as a function of time. The models proposed by Solomon and Longhi (1977) shared several parameters. The thermal conductivity was assumed to be constant at 4×10^5 erg/cm-sec-K and the specific heat was set to 1.2×10^7 erg/g-K. The surface temperature was fixed at −20 °C. Initial density was set at the mean lunar density of 3.34 g/cm³. In more complex models the cumulate layers were assigned appropriate densities on the basis of mineral assemblage. The heat of fusion (4×10^9 erg/g) was taken to be distributed uniformly with temperature over the solidus-liquidus interval. Convective heat transport with the liquid part of the LMO was simulated by the assumption that heat in excess of that necessary to maintain a given layer along a specific temperature profile is transferred to the overlying layer at each time step. The average abundances of heat-generating elements were taken from Langseth et al. (1976) and these elements were modeled to be strongly partitioned into the liquid phase. Models proposed by Solomon and Longhi (1977) included both crystal-liquid fractionation and the expected variation of the physical properties of the liquid and solids with the extent of LMO crystallization. The starting solidus and liquidus were taken from the data of Ringwood (1976) for the whole-Moon composition of Taylor and Jakes (1974) and the effects of pressure were ignored.

In the more sophisticated models of Solomon and Longhi (1977), cooling and crystallization would occur primarily at the top of the LMO in response to radiative cooling. The base of the LMO consists of primitive lunar mantle. This model assumes a protocrust that is a Moon-wide chilled margin at least a few kilometers thick (Solomon and Longhi 1977). The protocrust of this scale forms a cold thermal boundary layer that affects the overall heat loss from the LMO. A thinner protocrust on the order of meters predicted by other models (e.g., Hofmeister 1983) would provide less impedance to heat loss. Individual blocks of this chilled margin are recycled into the LMO by a combination of mini-plate tectonics and impact disruption. The early crystallizing minerals, dominated by olivine, sink to the base of the LMO. Initial cooling is efficient, crustal growth is slow, and the lower mafic cumulate layer grows rapidly. Once plagioclase becomes a liquidus phase and is buoyant, crustal growth accelerates and the LMO cooling rate decreases. Herbert et al. (1977) envisioned the formation of the initial plagioclase crust as surface expressions of convection-cell subsidence. On the basis of these models, the time for the crystallization of approximately 90% of the LMO would have been 100 to 200 m.y.. The duration of crystallization of the LMO is virtually independent of initial depth as long as it is covered by a 70 km thick insulating layer of flotation cumulates.

More complex models of the dynamics of a totally molten LMO have been explored by Tonks and Melosh (1990) and Spera (1992). A workshop on the physics and chemistry of magma oceans from 1 bar to 4 mbar contributed significantly to understanding the complexities of the LMO (1992, LPI Tech. Report 92-03). Studies by Solomatov and Stevenson (1993a,b,c), Abe (1997), and Solomatov (2000) have application for the LMO as well as for other terrestrial planets. At the high Rayleigh numbers relevant to the LMO, convection changes to a regime referred to as "hard" turbulent convection. At even higher Rayleigh numbers, convection is expected to enter a new regime of turbulent convection ("ultrahard"), but recent experimental studies by Glazier et al. (1999) suggested that hard turbulence is probably the pertinent regime for the LMO. Tonks and Melosh (1990) used the turbulent mixing-length theory of Kraichnan (1962) to describe a hard-turbulent LMO regime that initially convected vigorously with an initial Rayleigh number (Ra) on the order of $10^{25\text{-}27}$. Circulation in a hard-turbulent regime is characterized by small correlation scales for velocity, temperature, and composition, both spatially and temporally (Spera 1992). The evaluation by Spera (1992) used experimental and simulation studies combined with a scale analysis. Spera adopted the experiments and arguments of Castaing et. al. (1989) to support a $Nu \sim Ra^{2/7}$ rather than the traditional $Nu \sim Ra^{1/3}$ scaling. The Nusselt number (Nu) is the ratio of the actual heat flux at the surface relative to the conductive flux giving the same drop in temperature across the horizontal boundaries of the system. The difference in scaling adopted by these models is critical. As shown by Spera (1992), calculation of the LMO surface temperature and cooling rate from the two different scaling laws leads to significantly different results. Adopting the 2/7 law yields an LMO surface temperature (T_s) of 580 °C, whereas the traditional 1/3 law yields an LMO T_s of 1100 °C. Since the radiative cooling rate is proportional to T_s^4, cooling is nearly a factor of 10 faster for the 1/3 law scaling. A better understanding of the relationship between $Nu\text{-}Ra$ relations for high Ra convection of fluids with high Prandtl number (= kinematic viscosity/thermal diffusivity) is required to better constrain models for the duration of LMO cooling and crystallization (Spera 1992).

In their comparison of the nature of magma oceans on the Earth and Moon, Tonks and Melosh (1990) concluded that the effect of gravitational acceleration on the slopes of the solidus, liquidus, and adiabatic temperature profiles was one factor in the divergent geochemical evolutionary paths of the two bodies. They proposed that the adiabat of the LMO lies entirely between the liquidus and solidus. Therefore, nearly the entire LMO would have been at subliquidus conditions. An implication of this is that crystals suspended in the LMO would remain below the liquidus temperature. They concluded that individual crystals would continue to grow no matter where they were transported in the LMO and would eventually become large enough to overcome convective stirring and settle into the cumulate pile. This conclusion is a simplification of the relationship between the adiabat and the liquidus. It would appear that crystals that are liquidus phases in cooler portions of the LMO could be transported to and absorbed in hotter regions that are still below the liquidus.

In high Ra convection regimes where $Ra^{1/3} > 35\sigma^{1/2}$ (Eqn. 4.6) (σ = Prandtl number), Tonks and Melosh (1990) and Spera (1992) defined a complex LMO system that could be divided conceptually into several thermal-mechanical regimes on the basis of rheological properties. Immediately adjacent to either the floor or top of the LMO is a thin thermo-mechanical boundary layer where heat is transported by conduction. The thickness of this layer is defined by Spera (1992) as follows:

$$\delta_T \approx 4 L Ra^{-\frac{2}{7}}$$ (4.7)

where L is the time dependent depth of the LMO. The surface heat flux supported by the high Rayleigh number heat flux is defined by Spera (1992) as follows:

$$q_s \approx \left(\frac{k\Delta T}{4L}\right) Ra^{\frac{2}{7}}$$ (4.8)

where ΔT is the temperature difference across the LMO from its base, at or near the solidus, to its surface and k is the thermal conductivity ($k \approx 3$ Wm^{-1}K^{-1}). The thickness of the surface conductive crust and the basal cumulate pile increase through time.

Adjacent to both boundary layers are sublayers where heat is transferred primarily by convection but in which viscosity dominates. The thickness of the layer is as follows (Spera 1992):

$$\delta_v \approx 4 L Ra^{-\frac{2}{7}}$$ (4.9)

and the rms convective velocity within the layer is:

$$u_v \approx \left(\frac{\kappa}{L}\right) Ra^{\frac{3}{7}}$$ (4.10)

The separation of crystals from melt is restricted to this flow regime near the floor region or along the base of the crust. Melt trapped within the basal cumulates communicates very slowly with the adjacent layers because of the large Darcy friction within the crystal mush (Oldenburg and Spera 1990, 1991; Spera 1992).

The interior layer of the LMO is dominated by inertial forces. The temperatures of this part of the LMO are nearly isothermal and typical convective velocities are on the order of:

$$u_v \approx \left(\frac{\nu}{L}\right) \sigma^{-\frac{2}{3}} Ra^{\frac{3}{7}}$$ (4.11)

This layer consists of any material above its liquidus, along with partly crystalline regions where crystal fraction is below a value $\Phi_{critical}$, too low for crystals in the liquid to form networks. One of the important features of hard turbulent convection in this zone is the existence of large-scale turbulent eddies (Tonks and Melosh 1990; Solomatov 2000). Tonks and Melosh (1990) evaluated crystal suspension and settling in this zone using empirical studies of sediment transport. Settling and flotation only take place for Rouse numbers (terminal velocity of the crystal/turbulent friction velocity) greater than 1. Results show that crystals up to 1-2 cm in diameter can be suspended from the onset of cooling. As viscosity increases in the later stages of cooling, crystals of 10 cm to >1 m in diameter may be suspended in the inertial layer. The critical crystal size for equilibrium crystallization to be an important process during the crystallization of early magma oceans was estimated by Solomatov (2000) as:

$$D_{crit} \approx 10^{-3} \left(\frac{\eta_l}{0.1 \text{ Pa·s}}\right)^{\frac{1}{2}} \left(\frac{F}{10^6 \text{ W·m}^{-2}}\right)^{\frac{1}{2}} \text{ m}$$ (4.12)

The dimensions and depths of the various rheological zones during the evolution of the LMO depend on the liquidus and solidus slopes, and the intersections of the adiabat with the liquidus and solidus and the line of $\Phi_{critical}$ (see Solomatov 2000; Fig. 4.9). The structure of a magma ocean may thus depend significantly on planet size, a relationship also noted and discussed by Tonks and Melosh (1990). Determination of $\Phi_{critical}$ is significant for the modeling process. An example of the thermal structure and evolution implied by many of the above idealizations is shown in Figure 4.12, which also shows two potential relationships between the lunar adiabat and the solidus and liquidus (Tonks and Melosh 1990; Spera 1992).

Spera (1992) reconstructed the thermal and physical history of the LMO using the above parameters. During the early stages of the LMO with $Ra \approx 1025$, $\sigma \approx 102$, and $L \approx 500$ km, the conductive and discontinuous crust had a thickness of $\delta_T \approx 0.1$ m through which enormous heat flux, $q_s \approx 2 \times 10^4$ W/m^2, is carried. Solomatov (2000) estimated a heat flux of ~10^6 W/m^2 during the initial stages of MO crystallization. Hofmeister (1983) argued that the thickness of the chilled crust would reach a steady-state thickness of no more that a few meters. As long as frequent impacts and shear stresses generated by convection kept the surface discontinuous, heat transfer would remain high and be governed by the balance between heat brought to the surface by convection and radiative heat transferred from the hot LMO surface. During this time, the viscous sublayer has a limited thickness of $\delta_y \sim 0.5$ km. Based on the above equations, the magma convection velocities within this layer are on the order of 10^{-2} m/s.

Within the inertial layer, convective velocities are extremely high (~10-40 m/s; Spera 1992; Solomatov 2000). The temperature gradient across the layer is small due to the very rapid advective transport of heat. Cooling rates are extraordinarily high during early stages of LMO crystallization because of the significant radiative heat loss through the discontinuous and thin

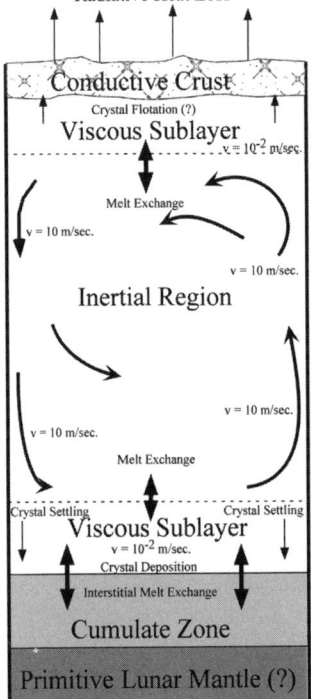

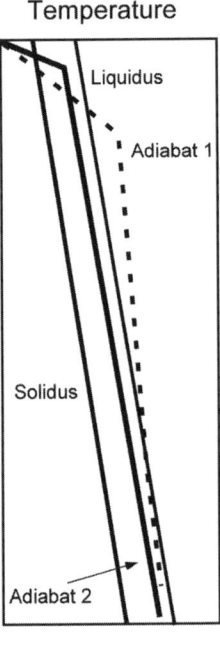

Figure 4.12. Schematic diagrams illustrating LMO structure for high Rayleigh number convection (after Spera 1992) and relationships between two possible lunar adiabats and the solidus and liquidus. Adiabat 1 is constructed based on the regime diagram of Spera (1992). A portion of this adiabat lies above the liquidus. In this relationship crystals nucleate in relatively narrow zones near the top and bottom of the LMO. Crystals that are swept into the main body of the LMO remelt, thus keeping the mean crystal size small. During cooling the adiabat will shift to lower temperatures. Adiabat 2 is from Tonks and Melosh (1990) and lies between the liquidus and the solidus for the Moon. In this scenario, crystals continue to grow in size until they are large enough to settle out or float. Tonks and Melosh (1990) suggested that the LMO crystallized predominantly under the condition of the adiabat lying entirely between the solidus and liquidus.

protocrust. Estimates of crystal size during the early stages of LMO crystallization are very close to the critical size separating fractional and equilibrium crystallization (Solomatov 2000), which means that both equilibrium and fractional crystallization are equally acceptable within the uncertainties of the physical parameters.

As the LMO continued to cool, the thicknesses of the boundary layers and the viscous sublayers continued to grow at the expense of the interior region. The development of the former results in the substantial decrease in radiative heat loss. Tonks and Melosh (1990) examined the longevity of the inertial flow layer using Equation (4.6). They calculated that the inertial flow zone disappeared at $\Phi_{critical}$ = 46% in a 400 km LMO and 49% in a 1000 km terrestrial magma ocean. This is similar to the disappearance of the inertial flow zone estimated by Murase and McBirney (1973). Most theoretical and experimental investigations based on spherical particles estimate a $\Phi_{critical}$ value between 50 and 60% (e.g., Campbell and Forgacz 1990; Lejeune and Richet 1995; Marsh 1995). Philpotts et al. (1999) found that in natural basaltic systems at low pressures, plagioclase laths form three-dimensional networks at crystal fractions as low as 25%. Therefore, the liquidus phases and their typical crystal forms may be a highly significant parameter. The propagation of this rheological front from the base of the terrestrial magma ocean toward the surface may take as little as 10^3 years (Solomatov 2000). Melosh (1992) suggested that at this crystal "lock up" point the effective viscosity increases by many orders of magnitude and melt moves by percolation through the crystals. In this environment, chemical differentiation of the magma ocean takes place as the liquid percolates through the solid and may be driven by density differences (Melosh 1992). At this stage, fractional crystallization may be the more dominant process. Magma-ocean cooling and further crystallization within this rheological regime may be as fast as 10^7-10^8 years (Davies 1990). However, if surface recycling was inefficient, thereby slowing convection beneath the lunar crust, the crystallization time would be longer (Solomatov and Moresi 1996; Solomatov 2000).

Using initial bulk compositions of 5-7 wt% Al_2O_3, Snyder et al. (1992) calculated that plagioclase would be a liquidus phase after 57% to 80% of the LMO crystallized. On the basis of these estimates, the onset of plagioclase crystallization would occur when the crystal fraction is well above any reasonable value of $\Phi_{critical}$ and the inertial zone of the LMO has disappeared. Therefore, plagioclase flotation depends not only on the buoyancy of plagioclase in highly evolved liquids, but also on the ability of plagioclase to separate buoyantly from a crystal mush. Upon the appearance of plagioclase on the liquidus and the development of a well-defined plagioclase crust, the rate of LMO cooling would drop precipitously owing to thermal blanketing effects of stable crust. The crust and cumulate zone continue to grow until they eventually merge at the base of the crust.

Tracking incompatible and heat-producing elements trapped in interstitial liquids is one critical step to understanding the magmatic source regions that would result from LMO crystallization. Interstitial liquids are trapped in solidifying crystal mushes when solidification proceeds more quickly than compaction and percolation. The bottom of the partially solidified zone moves to shallower depth due to cooling by heat loss at the surface. Therefore, the upward velocity V of this solidification front will influence the amount of trapped, interstitial liquid in the LMO cumulate pile. Buoyant percolation of liquid that is less dense than solid would occur at a pore velocity given by:

$$u = \frac{K}{\phi\mu}\Delta\rho g \qquad (4.13)$$

where K is permeability, $\Delta\rho$ is the density difference between the crystals and the liquid, μ is the liquid viscosity and Φ is the melt fraction. If liquid percolation occurs through tube-like channels along mineral grain edges, then

$$K \approx \frac{b^2 \phi^3}{200} \tag{4.14}$$

where b is the grain size (cf. Wark and Watson 2002). The amount of melt trapped is given by the value of Φ for which $V \cong u$. Escape of interstitial melt requires compaction of the solid. This estimate neglects the resistance of deformation of the solid that would further increase the trapped melt fraction (cf. Shirley 1986). A surface heat flux by radiation through an H_2-H_2O iron-wüstite buffered atmosphere (Abe et al. 2000), at a surface temperature equal to the potential temperature of the liquid, would give trapped liquid fractions on the order of 1-10% for a grain size of 1 mm and a liquid viscosity of 1-10 Pa·s. If a conductive lid forms at the surface, the resulting much lower rates of cooling would give a smaller trapped melt fraction. Even a trapped melt fraction less than 1% may be significant in controlling the incompatible-element content of LMO cumulates.

3.3. Physics of cumulate overturn

3.3.1. Introduction and relevance. As discussed in previous parts of Section 3, crystallization of the LMO produced a wide compositional range of cumulates that eventually served as mantle sources for subsequent periods of lunar basaltic magmatism. Models for the stagnant crystallization of an LMO predict a cumulate stratigraphy (Fig. 4.10) that is inconsistent with the observation that there is a limited relationship between basalt composition and its depth of origin (Section 4.5). Also, it appears that basaltic magmas represented by the volcanic glasses, and probably most of the mare basalts, were a product of melting initiated deep in the lunar mantle (Delano 1986; Shearer and Papike 1993, 1998; Elkins-Tanton et al. 2003a). Since Ti-bearing cumulates and plagioclase do not crystallize until near the top of a solidifying magma ocean (Fig. 4.10) and assimilation of Ti by low-Ti magmas appears not to be a viable mechanism (Ringwood and Kesson 1976; Van Orman and Grove 2000; Elkins-Tanton et al. 2001), the source materials needed for mare magmatism requires some mechanism of overturn in which late magma-ocean cumulates are mixed downward into the underlying mantle of earlier cumulates to a depth of at least 400 km.

Solidification of a magma ocean would define initial planetary compositional differentiation, but could lead to an unstable cumulate density stratification, which overturns to a stable configuration. The nature of the final, stable compositional stratification can have important implications for a planet's subsequent evolution by delaying or suppressing thermal convection and by controlling the initial temperature distribution and the distribution of heat-producing elements in the mantle. Processes occurring during the first few hundred million years of magma-ocean solidification and overturn surely shaped the evolution of the planet on much longer time scales.

For the Moon, recent progress on LMO crystallization (e.g., Snyder et al. 1992; Hess and Parmentier 1995) point to unexplored aspects of cumulate formation and overturn that may be critical to understanding lunar evolution. The origin of the source materials and the processes responsible for mantle melting that generated lunar basalts (mare basalts, Mg-suite rocks) and perhaps the temporal evolution of lunar magnetic field may all have been controlled by the initial formation and disruption of the LMO cumulate pile.

As discussed above, LMO crystallization models suggest that highly fractionated, late-stage liquids were trapped between deep mafic cumulates and a flotation anorthositic crust. Depending on the LMO composition used, Ti-oxides (i.e., ilmenite) would begin to crystallize from the late-stage liquid with clinopyroxene and plagioclase when 89-95% of the LMO had solidified. Depending on the thickness of anorthosite crust assumed, crystallization of ilmenite would begin at a depth between 150 and 100 km at a temperature between 1180 and 1125 °C (Hess and Parmentier 1995; Van Orman and Grove 2000). Assuming efficient separation from plagioclase, the high-Ti cumulates would have a density of 3700-3800 kg/m^3, compared to the underlying olivine + pyroxene mantle density of about 3300 kg/m^3. Ringwood and Kesson (1976) proposed that because the solid ilmenite + clinopyroxene cumulate layers were denser than the underlying, less evolved olivine- and pyroxene-bearing cumulates, the ilmenite + clinopyroxene cumulates sank into the underlying cumulates. Ringwood and Kesson (1976) and other investigators (e.g., Shearer et al. 1992, 1993; Hess and Parmentier 1995; Zhong et al. 2000) hypothesized that the sunken high-Ti cumulates subsequently remelted and contributed to the formation of high-Ti mare basalts. Other investigators have challenged the hypothesis of deeply foundered, remelted high-Ti cumulates on the basis that high-Ti melts are not buoyant below 200-300 km depth in the Moon (Circone and Agee 1996; Wagner and Grove 1997), or on the basis that the high viscosities in the cold planetary lid would prevent or delay their sinking (Elkins-Tanton et al. 2002).

3.3.2 Evolution of the mantle immediately following magma ocean crystallization. As discussed above, crystallization of the LMO would generate an unstably stratified mantle that should overturn to a gravitationally stable state. Dense cumulates that were added to the top of the already solidified mantle would have been gravitationally unstable and would tend to sink. The relative timescales of LMO solidification and cumulate overturn also influence the final cumulate stratigraphy. If overturn occurred while the LMO was crystallizing, deeper already solidified mantle would be brought to the solidification front thus affecting the composition of the remaining liquid. Overturn therefore involves a competition between the rate of thickening of the solidified layer and its Rayleigh-Taylor time scale (Elkins-Tanton et al. 2003b).

The time scale of Rayleigh-Taylor instability and overturn of an unstably stratified viscous layer of thickness d and solid-state viscosity η with stress-free top and bottom boundaries is

$$t_{overturn} = \frac{4\pi^2\eta}{\gamma g d^2} \tag{4.15}$$

where γ is the compositional density gradient, assumed constant with depth, and g is gravity (Hess and Parmentier 1995). As the thickness of the solidified layer d increases, its overturn time decreases. Gravity at the solidification front also increases as solidification proceeds, further decreasing overturn time as the solidified layer thickens. At the beginning of solidification near the bottom of a 1000 km deep LMO on the Moon, g is about 0.7 m/s^2, overturn of a 100 km thick layer with $\gamma = 10^{-4}$ kg/m^3/m in olivine and olivine-orthopyroxene cumulates (below a depth of about 100 km in Fig. 4.13) and $\eta = 10^{19}$ Pa·s, is about 2 million years. This value of viscosity is taken to be comparable to values near the bottom of the Earth's oceanic lithosphere, in the range 10^{19}-10^{20} Pa·s. The creep rate of mantle silicates, however, depends strongly on the presence of small amounts of water. In the absence of water, laboratory deformation studies (Hirth and Kohlstedt 1996) indicate that the viscosity is larger by at least a factor of 100. In this case the overturn time of a 100 km thick layer is more than 200 m.y. longer than estimates for the solidification time of the LMO. Near the end of solidification when d is 10× thicker, the overturn time is only a few million years.

Overturn rates therefore do not become less than thickening rates until near the end of LMO solidification, thus the LMO is expected to crystallize almost completely before density instability initiates cumulate overturn. Based on the simple model of very rapid, turbulent convective heat transfer to the surface, assuming that a solidified layer at the surface partici-pates in this convection rather than forming a conductive lid, solidification may be more rapid than this estimated overturn time for the olivine-orthopyroxene cumulates. Once a plagioclase flotation crust forms a conductive lid, the last stages of solidification will be much longer. Therefore overturn of the mafic cumulates may occur while the evolved portion of the LMO is still liquid.

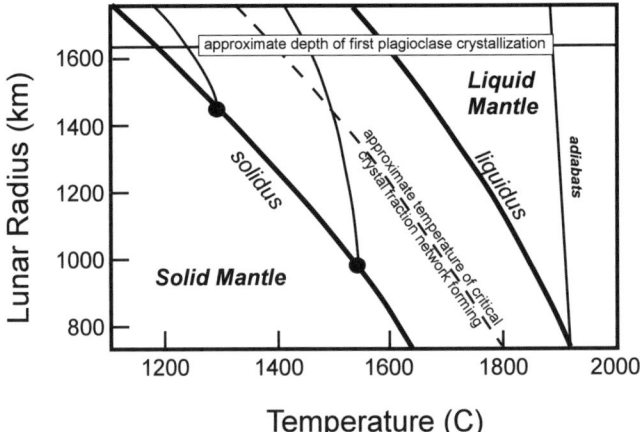

Figure 4.13. Solidus and liquidus for a lunar bulk mantle composition from Ringwood and Kesson (1976), with approximate adiabats (thin lines) for the liquid and partly crystalline magma oceans to demonstrate the possibility of crystalline, crystalline network, and liquid and crystal mush zones in the solidifying mantle. The lunar adiabats have slopes such that during almost the entirety of magma ocean crystallization the adiabat lies between the solidus and liquidus, as discussed in Tonks and Melosh (1990). Plagioclase begins to crystallize when the magma ocean is about 80% solid, at which point the magma ocean is likely to have a higher than critical crystal fraction, making complete plagioclase flotation impossible (modified after Tonks and Melosh (1990).

The simplified overturn scenario is where the entire cumulate pile is simply reordered with a monotonically decreasing density upwards and assuming that plagioclase separates perfectly by flotation. The resulting compositional profiles before and after overturn are shown in Figure 4.14. Prior to overturn, primitive olivine-orthopyroxene cumulates with progressively increasing Fe/Mg are overlain by an ilmenite-rich layer. After overturn, this dense layer forms the base of the cumulate pile with a high-Ti and low-Si composition (Fig. 4.14).

During overturn, hot material rising from depth could cross its solidus through decompression and melt (Fig. 4.14). The hottest cumulates that rise to just below the plagioclase flotation crust are olivine with an Mg′ of 95-96. The cumulates just below are a mixture of high Mg′ olivine and orthopyroxene. The approximate solidus for these materials is shown as a dashed line, indicating that the overturned cumulates in this model should not remelt through decompression. In other models, overturn can create up to 50 km of mafic material that could form a lower lunar crust. The very old ages of the Mg-suite (Papike et al. 1998, and references therein) suggest that a mechanism such as remelting high-Mg cumulates during overturn may be required to explain their formation.

Despite gravitational instability, however, the rate and occurrence of overturn depend on complex rheologies of solids deforming by thermally activated creep. The calculations presented in Elkins-Tanton et al. (2002) suggest that, with a non-linear stress-dependent rheology appropriate for a pyroxene-dominated cumulate, the late-forming high-Ti cumulates may creep at rates too low to allow descent via Rayleigh-Taylor instabilities on a time scale appropriate to form the high-Ti basalts. Wieczorek and Phillips (2000) presented a model with a ten-km thick layer of material of a KREEP-basalt composition at the base of the anorthositic crust that predicts that melting should eventually extend to depths of about 600 km and persist through the entire evolution of the Moon to the present day. Hess and Parmentier (2001) pointed out that such a high concentration of heat sources might prevent the crystallization of the LMO over much of the history of the Moon, but in particular they question whether the

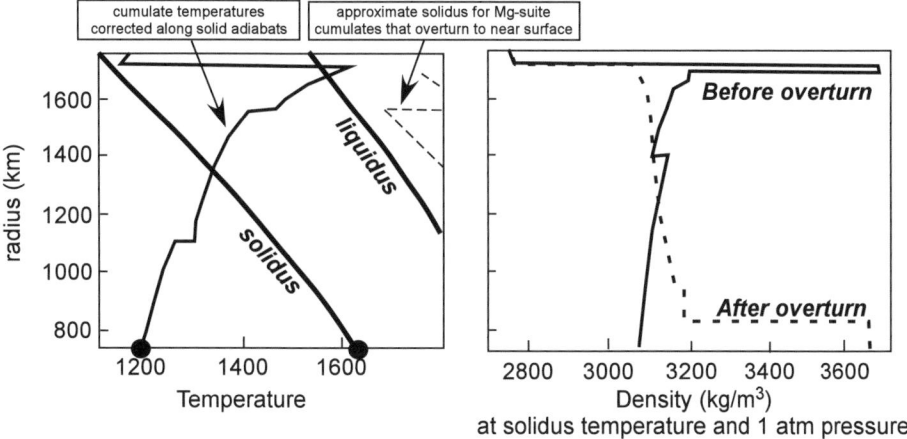

Figure 4.14. On the left, temperature of the lunar cumulate pile after overturn. When hot material rises from depth, it may cross its solidus through depressurization and melt. In the simple plagioclase flotation model, the hottest cumulates that rise to just below the plagioclase flotation crust are olivine with an Mg# of 95 to 96 (where Mg# = 100 × molar Mg/(molar Mg + molar Fe)). The cumulates just below are a mixture of high Mg# olivine and orthopyroxene. The approximate solidus for these materials is shown as a dashed line, indicating that the overturned cumulates cannot remelt through depressurization in this model. The solidus and liquidus for the original bulk mantle are shown in gray for reference. Bold circles mark the temperature at the bottom of the cumulate pile before and after overturn. Solid lines are temperature corrected along solid adiabats. On the right, the density profile of the cumulate pile before and after overturn is shown calculated at the solidus temperatures of the minerals and at one atmosphere pressure, for comparison of layers during overturn. The dense, spinel-rich layer just under the plagioclase crust during initial crystallization falls to the bottom of the cumulate pile after overturn, and the middle of the initial pile roughly inverts to make the final profile, bringing the most magnesian minerals to the shallowest depths under the anorthosite crust.

high temperatures at the base of the crust generated in such a model would allow the presence of an elastic lithosphere thick enough to support the mascons at the time of mare basalt eruptions (Solomon and Head 1980).

Longhi (2003) suggested a somewhat different outcome resulting from density and temperature stratification of an LMO cumulate pile that contained an upper, plagioclase-bearing mafic zone, formed in the absence of a thick, floating anorthositic crust (Section 3.1). He proposed that because of unstable density profiles in the mafic and ultramafic zones (lower, olivine- and pyroxene-bearing cumulates) produced by the upward decrease in Mg′, two-layer convection occurred. Overturn in each zone not only brings lower-density mineral assemblages upward but also advects heat, resulting in transportation of olivine cumulates at 1800 °C to the base of the mafic cumulate zone that contains mineral assemblages with solidii of 1000-1250 °C. Melting in the mafic cumulate zone ensues and produces thick suspensions of plagioclase in ferroan-anorthosite-like liquid that ascend buoyantly. The denser, complementary pyroxenitic component (both Ti-poor and Ti-rich) sinks into the lunar interior to form the source region for the mare basalts through mixing of early LMO cumulates. Overturn steepens and eventually reverses the density profile in the LMO cumulate pile. This change in the density profile along with radioactive and conductive heating of undifferentiated mantle beneath the LMO cumulate pile may activate plumes. These plumes may eventually melt to produce additional post-LMO magmas such as the Mg-suite.

3.3.3. Spatial scales of mantle overturn. The spatial scale of gravitational instability is especially important for testing the overturn hypothesis and for understanding the subsequent evolution of the Moon. The simplest outcome of the above idealized overturn would be for all the ilmenite cumulates to move to the bottom of the LMO cumulates, or perhaps the base of the mantle. This dense material, rich in heat-producing elements, would heat up with time allowing thermal expansion to offset compositional density. If the deep layer is pure ilmenite cumulate, a temperature increase of several thousand °C is required for thermal expansion to offset the high intrinsic density, which would result in a completely liquid layer. This melt would be expected to be denser than the solid (Circone and Agee 1996), and so might remain at depth in a molten state. Thus deep, compositionally dense layers may also be difficult to remobilize through convective instabilities, and may form a reservoir sequestering material from the rest of the mantle (Alley and Parmentier 1998; Elkins-Tanton et al. 2003b). Alternatively, thermally driven convective upwelling in the overlying mantle might entrain small amounts of such dense melt creating a possible mare-basalt source. If the ilmenite cumulates mixed with olivine-pyroxene cumulates during overturn, then a smaller increase in temperature might allow the solid mixture to become buoyant at a more modest temperature increase, rise, decompress, and melt (Zhong et al. 2000; Stegman et al. 2003).

The final stages of this heterogeneous overturn would thus involve the simultaneous solidification and sinking of the ilmenite-rich cumulates. The ilmenite-rich layer before overturn is relatively thin but contains the densest mantle material (Fig. 4.15). The wavelength or spacing of dense diapirs would depend on the thickness of the layer from which they form and thus on the balance between the rate of solidification and the rate of diapiric sinking. Estimated diapir sizes for reasonable ranges of solidification rate are invariably less than 10 km (Hess and Parmentier 1995). The low sinking velocity of small diapirs means that they would sink to depths only on the order of 100 km over the time for final solidification of the LMO, thus generating a layer consisting of a mixture of ilmenite-poor cumulates with ilmenite-rich inclusions (Beard et al. 1998 suggested deeper sinking of the pyroxene-bearing cumulate). This mixed layer is denser than underlying cumulates and its thickness may be large enough that it would be most unstable at long wavelengths (cf. Parmentier et al. 2002). The observed hemispheric asymmetry in the distribution of mare basalts on the surface and Th in surface materials (Lawrence et al. 1998) might thus be explained. Alternatively, ilmenite cumulates that initially sink to the bottom of the mantle might rise, decompress, and melt in a spherical-harmonic mode-one pattern (Zhong et al. 2000) generating mare basalts in only one region of the surface. Stegman et al. (2003) suggest that such a process may also explain the generation and subsequent decay of a lunar core magnetic dynamo based on available paleomagnetic data.

3.4. Challenges to the LMO hypothesis

3.4.1. Introduction. The LMO model has been the cornerstone of lunar petrology since the examination of the first samples returned by the Apollo missions (Smith et al. 1970; Wood et al. 1970). One of the first challenges to the LMO model was the identification of energy sources capable of producing extensive planetary-scale melting. Initial models for lunar accretion and other energy-producing processes were found to generate inadequate amounts of heat. As illustrated earlier in this chapter (Section 2), with the acceptance of new lunar accretion models that predict the storage of enough energy in the Moon to induce extensive lunar melting, the necessity for alternative, low melting differentiation models is diminished. The potential problems with forming a primordial lunar crust through the separation and flotation of plagioclase in an LMO following moderate degrees of crystallization (40-60%) were discussed in detail in Secton 3.2. Another potential problem that needs to be factored into LMO models are new isotopic observations for the ferroan anorthosites.

3.4.2. Ages of early crustal rocks. Isochron ages have been determined for a growing number of lunar crustal rocks belonging to the alkali, magnesian, and ferroan-anorthositic suites. Notable inconsistencies exist between the measured ages and initial isotopic compositions of many lunar crustal rocks and the ages and isotopic compositions expected

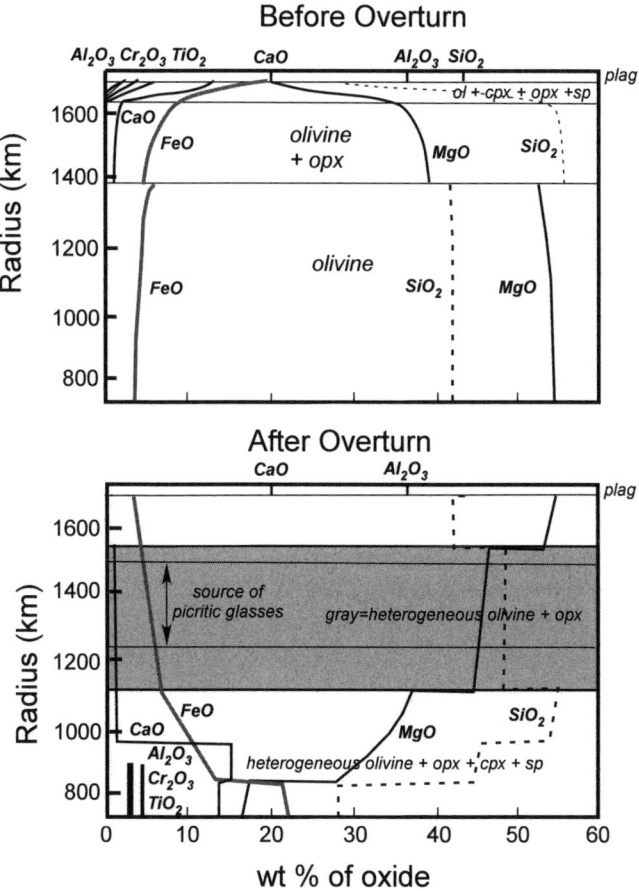

Figure 4.15. Compositional stratification resulting from (a) initial fractional crystallization of a 1000 km-deep lunar magma ocean, and (b) overturn of cumulates to a gravitationally stable configuration. "Spinel" here refers to spinel-group minerals, including ilmenite and ulvöspinel. The phase assemblages and volumes are from calculations and personal communication with John Longhi. This simplified model assumes that all the plagioclase separates from the magma ocean as it crystallized, forming a crust 37 km thick. The remaining liquid crystallized below this crust. After overturn, the spinel-rich late cumulates have fallen to the bottom of the magma ocean (though had they fallen, they may have fallen still further, through the primordial mantle beneath the magma ocean in this model). Note that high viscosities due to cold near-surface temperatures, which inhibit overturn of shallow cumulates, are not considered in this model (Elkins Tanton et al. 2002). The region from which the picritic glasses and mare basalts later formed is a mixture of olivine and orthopyroxene, appropriate for the most magnesian of those volcanic products, but titanium, chromium, and KREEP components must be explained with a further process, possibly re-eruption of deeply fallen cumulates as proposed by Hess and Parmentier (1995) and Zhong et al. (2000).

according to LMO models. Such inconsistencies suggest that the LMO model, in its simplest form, cannot explain all of the isotopic systematics observed in the lunar rocks.

Ages for alkali-suite rocks range from ~3.80 to 4.37 Ga, for Mg-suite rocks, from ~4.18 to 4.46 Ga, and for ferroan anorthosites, from 4.29 to 4.56 Ga (Fig. 4.16; see also Chapter 3, Fig. 3.5 and Chapter 5, Table 5.6). Numerous Rb-Sr ages determined on Mg-suite rocks range from 3.90 to 4.61 Ga and also overlap with the Sm-Nd ages determined on the ferroan anorthosites (see summaries in Nyquist and Shih 1992; Nyquist et al. 2001). Sm-Nd isotopic

Ages of Lunar Crustal Rocks

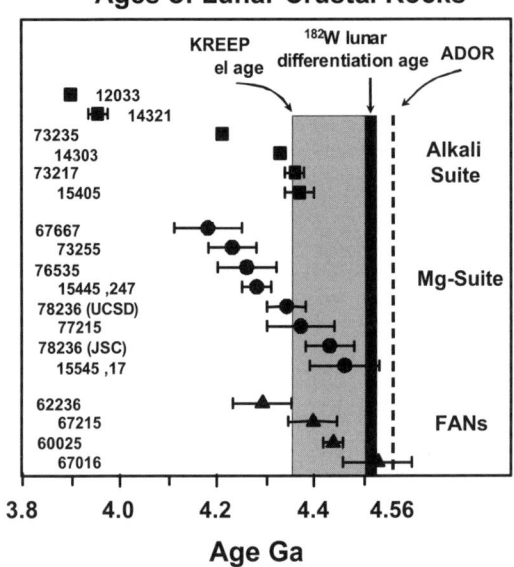

Figure 4.16. Compilation of Sm-Nd isochron and Pb-Pb zircon ages of the lunar crust modified from Borg et al. (1999). These techniques are probably the least susceptible to resetting by impact metamorphism, and so best represent the age of lunar crustal rocks. Data from Lugmair et al. (1976), Nakamura et al. (1976), Carlson and Lugmair (1981), Nyquist et al. (1981c), Compston et al. (1984a; 1984b), Carlson & Lugmair (1988), Shih et al. (1993), Alibert et al. (1994), Meyer et al. (1996), and Norman et al. (2000). This figure demonstrates that the ages of the ferroan anorthosites and Mg-suite rocks overlap, suggesting that by about 4.4 Ga both types of magmas were being produced on the Moon. Even the oldest ages of the ferroan anorthosites and Mg-suite rocks are within error of one another requiring the magma ocean to have existed for only a short period of time.

analyses of the ferroan anorthosites are very difficult owing to their monomineralic, highly shocked, and commonly polymict character, and their very low concentrations of REE. Thus the ferroan anorthosites that have been dated tend to have higher abundances of mafic minerals than typical ferroan anorthosites (in order to obtain a wider range of parent/daughter ratios to give a more precise isochron) and may therefore not be representative of the ferroan-anorthositic suite as a whole. Furthermore, the Sm-Nd isotopic systematics of the ferroan anorthosites could somehow be disturbed (see discussion in Borg et al. 1999), and the Sm-Nd ages of the ferroan anorthosites may not represent crystallization ages. For the purpose of this initial discussion, the ferroan anorthosite ages are taken at face value. We then explore the implications of plausible mechanisms that have been proposed to disturb the Sm-Nd isotopic systematics of the ferroan anorthosites.

There is significant overlap between the ages of the alkali, magnesian, and ferroan-anorthositic suites of rocks, and particularly the latter two. The overlap is not dependent solely on the relatively young ages of ferroan anorthosites such as 62236. In fact, Mg-suite clast 15445,247 (Shih et al. 1993) is within analytical uncertainty of the age of the oldest ferroan anorthosite (Fig. 4.16). The ages of the ferroan anorthosites are younger than the age of LMO differentiation estimated from the short-lived chronometer $^{182}Hf \rightarrow ^{182}W$ ($t_{1/2} = 9$ Ma). Although this chronometer only provides model ages for parent/daughter fractionation it suggests that lunar differentiation occurred between 4.50 and 4.52 Ga (Lee et al. 1997; Shearer and Newsom 1999; Righter and Shearer 2003; Shearer and Righter 2003). Furthermore, the age of urKREEP formation (i.e., the late-stage residual melt of LMO crystallization) has also been estimated by a variety of isotopic methods. Nyquist and Shih (1992) suggested that an average model age of 4.42 ± 0.07 Ga is probably the most representative of the age of urKREEP formation. This model age is as old or older than the Sm-Nd age of ferroan anorthosite 62236 (4.29 ± 0.06 Ga; Fig. 4.16), suggesting that the age of 62236 could post-date urKREEP formation.

Taken at face value, these age relationships suggest that the earliest Mg-suite magmatism was contemporaneous with the magmatism that produced at least some ferroan anorthosite. Contemporaneous crystallization of the ferroan anorthosites and Mg-suite rocks is not consistent with the standard LMO model in which ferroan anorthosites are the oldest crust and are intruded by younger Mg-suite plutonic rocks. The crystallization ages of at least some ferroan anorthosites post-date the age estimates for crystallization of the LMO, therefore the age relationship between the ferroan anorthosites, Mg-suite rocks, and LMO crystallization are more consistent with lunar crustal genesis by intrusion of multiple magma bodies, i.e., serial magmatism (e.g., Walker 1983; Longhi and Ashwal 1985; Longhi 2003).

3.4.3. Initial Nd isotopic composition. Six ferroan anorthosites have been dated by the Sm-Nd isotopic system: 60025, 67016, 62236, 67215, and clasts from lunar meteorites MAC88105 and Yamato 86032 (Carlson and Lugmair 1988; Alibert et al. 1994; Borg et al. 1999; Norman et al. 2000, 2003; Nyquist et al. 2002). It is apparent that all of the ferroan anorthosites from the Apollo 16 site analyzed thus far have positive initial ε_{Nd}^{143} values (Fig. 4.17), which is consistent with derivation from LREE-depleted sources. However, the ferroan-anorthosite clasts from the lunar meteorites have negative initial ε_{Nd}^{143} values. Data on the MAC88105 clast is inconclusive for negative initial ε_{Nd}^{143} because it may be contaminated by Mg-suite materials (Nyquist et al. 2002). Thus, the presence of an Mg-suite component could lower the initial Nd value of MAC, making it negative. Conversely, the clast from Y86032 appears to be a ferroan-anorthosite breccia without contamination from other lithologies.

The $^{147}Sm/^{144}Nd$ ratio of potential ferroan-anorthosite sources are estimated from their ages and initial ε_{Nd}^{143} values, assuming that their sources formed at ~4.56 Ga. The model illustrates potential relationships between the four analyzed ferroan anorthosites from Apollo 16. All of the ferroan anorthosites lie within error of a single growth line with a $^{147}Sm/^{144}Nd$ ratio of 0.279 ± 0.17. Thus, although differences in crystallization and exposure ages indicate that these four ferroan anorthosites are unlikely to be comagmatic, they maybe derived from the same source (Borg et al. 1999). Conversely, they may simply be derived from different sources with similar $^{147}Sm/^{144}Nd$ ratios. In any case the fact that the Apollo 16 ferroan anorthosites have positive initial ε_{Nd} values, and consequently appear to be derived from LREE-depleted sources, is not consistent with the simple LMO model. This inconsistency stems from the fact that if the initial conditions are chondritic, the LMO is not expected to dramatically fractionate the Sm/Nd ratio, and hence ε_{Nd} values, throughout crystallization.

The preceding discussion presumes that the Sm-Nd isotopic systematics of the samples have not been disturbed by shock processes. Although most Sm-Nd isotopic systematics appear fairly well behaved, this is not a forgone conclusion (Norman et al. 2003).

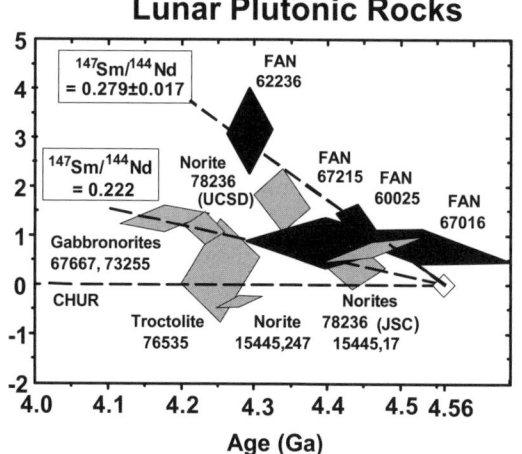

Figure 4.17. Time - ε_{Nd}^{143} diagram of ferroan-anorthosites and magnesian-suite rocks (sources of data same as Fig. 4.16). Note that all ferroan anorthosites (FAN) have positive initial ε_{Nd}^{143} values, consistent with derivation from LREE-depleted sources. The $^{147}Sm/^{144}Nd$ ratio estimated for the sources of FAN, assuming the source formed 4.56 Ga, is 0.279 ± 0.017. This value is typical of highly depleted sources from terrestrial planets.

3.4.4. Comparison between modeled anorthositic cumulates and ferroan anorthosites.
Snyder et al. (1992) estimated the mineralogy and major and trace element compositions of the LMO cumulates using a combination of phase equilibria and equilibrium/fractional crystallization models. Using these models as a basis, the Sm-Nd abundances of anorthositic cumulates may be calculated in a similar fashion, but using the modal mineralogy observed in actual samples. This approach permits the modeled compositions to be directly compared to analyzed ferroan anorthosites. The ferroan anorthosites are assumed to have formed with their present modal mineralogy during the interval of crystallization from 78-86% (Snyder et al. 1992). The results of the model are not affected if the 86-95 interval is used instead. From these data and model results, the ^{147}Sm/^{144}Nd ratios of the anorthositic cumulates can be calculated.

The isotopic composition of the cumulates depends on when they formed. One estimate of the time of LMO crystallization comes from the 4.42 ± 0.07 Ga model age for KREEP (Nyquist and Shih 1992). This age is comparable, but slightly younger than the age estimated from the Hf-W isotopic system of 4.50 and 4.52 Ga (Lee et al. 1997; Shearer and Newsom 1999). For the models presented below the LMO is assumed to have occurred at 4.50 Ga. The ε_{Nd}^{143} values of the anorthositic cumulates that are modeled are not, however, strongly dependent on this age because of the long half life of the ^{147}Sm → ^{143}Nd ($t_{1/2}$ = 106 Ga) chronometer.

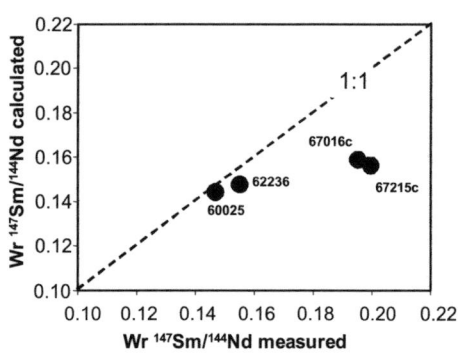

Figure 4.18 is a plot of the ε_{Nd}^{143} of whole-rock values for ferroan anorthosites 60025, 62236, 67215, and 67016 versus the ε_{Nd}^{143} calculated for anorthositic cumulates with the modal mineralogy of the ferroan anorthosites. It is apparent from this figure that the measured ε_{Nd}^{143} whole-rock values are substantially below the 1:1 line predicted by the model. The ε_{Nd}^{143} calculated for the anorthositic cumulates are dependent on the time of LMO crystallization, the Sm/Nd ratio of the bulk Moon, and the Nd isotopic composition of the bulk Moon. However, the crystallization of the LMO must be significantly younger than 4.35 Ga (3.85 Ga) and the bulk Moon is required to have a ^{147}Sm/^{144}Nd ratio significantly higher than the chondritic value of 0.1967 (0.207) and ε_{Nd}^{143} value significantly higher than 0 (+4.1) to account for the differences between the modeled and measured ε_{Nd}^{143} whole rock values of 60025. The situation is worse

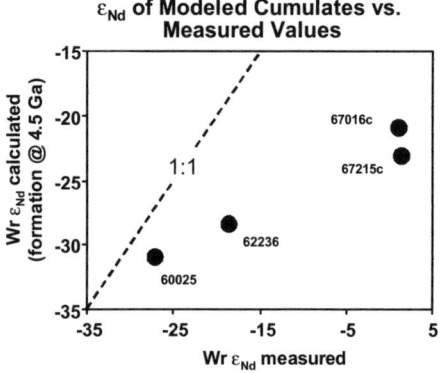

Figure 4.18. Present-day ε_{Nd}^{143} of modeled anorthositic cumulates versus present-day whole-rock values. Source of whole-rock values same as Figure 4.17. Whole rock value for 60025 calculated from data from mineral fractions assuming a mode of plag:ol:pig = 80:10:5. The composition of the modeled anorthositic cumulates are calculated using the LMO model of Snyder et al. (1992) and the modal mineralogy of the individual ferroan anorthosites assuming formation at 4.50 Ga. The present-day ε_{Nd}^{143} of values of FAN are more radiogenic than the calculated values suggesting that they are not simple floatation cumulates of a magma ocean.

for samples that lie farther from the 1:1 line on Figure 4.18. Another possibility is that the ferroan anorthosites that have been analyzed isotopically contain an LREE-depleted component that was not considered in the models. Although the presence of this component would shift a sample to the right on Figure 4.18, it is unlikely because it would have to be a major source of REEs in the ferroan anorthosites. A final possibility is that the analyzed ferroan anorthosites are mixtures of materials derived from a LREE-depleted source. This also seems unlikely because most mineral phases from the ferroan anorthosites lie on isochrons. The simplest explanation for the differences between the modeled and measured ε_{Nd}^{143} whole-rock values is that these ferroan anorthosites are derived from a source that is more LREE-depleted than is hypothesized for the LMO.

The LREE-depleted source region postulated for the ferroan anorthosites is confirmed, at least for sample 62236, by the presence of a small positive ε_{Nd}^{142} anomaly of $+0.25 \pm 0.11$ (Borg et al. 1999). Positive ε_{Nd}^{142} anomalies, such as that observed in 62236, are the result of the decay of $^{146}Sm \rightarrow {}^{142}Nd$ ($t_{1/2} = 103$ Ma), in an LREE-depleted source region (i.e., Sm/Nd ratio that is greater than the chondritic value). Although the size of the ε_{Nd}^{142} anomaly observed in 62236 is relatively small, it is significantly larger than that predicted by the models. For example, a modeled anorthositic cumulate forming at 4.50 Ga is expected to have a negative ε_{Nd}^{142} value of -0.45. As with whole-rock ε_{Nd}^{143} values, the ε_{Nd}^{142} is not dependant on isochron determinations and is expected to be unaffected by post-crystallization shock metamorphism of the sample.

3.4.5. Ramifications of isotopic data for the LMO. Both new and resurrected models have been employed to account for the production of ferroan anorthosites with younger ages and derived from LREE-depleted mantle sources (positive ε_{Nd}^{142}). Wetherill (1975) envisioned a differentiation model in which accretional melting of the Moon was restricted, creating isolated magma chambers that crystallized to form large layered intrusions consisting of mafic cumulates (source for mare basalts) capped with anorthosites (lunar highlands). An extensive primitive lunar mantle remained unprocessed during these events. Walker (1983) and Longhi and Ashwal (1985) suggested that subsequent heating of the Moon due to the decay of long-lived radionuclides initiated global convection that heated the stack of intrusions envisioned by Wetherill (1975). This resulted in melting and allowed the separation, growth and ascent of anorthositic diapirs. The coeval mafic component was buried in the mantle at depths greater than the sampling depth of multiring impact basins (30-60 km). At such depths, these mafic cumulates could have mixed with primitive lunar mantle assemblages that did not participate in the localized melting. This model accounts for the overlapping ages of ferroan anorthosites and Mg-suite rocks, the young age of some of the ferroan anorthosites, and perhaps even the LREE-depleted mantle sources for the ferroan anorthosites.

This type of model, however, has several weaknesses (Hess 1989; Shearer and Papike 1993). First, it is clear that substantial early melting occurred rather than insignificant melting. Second, this model does not adequately explain the diversity of mantle sources for the mare basalts (e.g., TiO_2 variability). Third, in comparison with terrestrial analogs, most of the larger layered intrusions do not have such large anorthosite units. Fourth, there is little evidence for the existence of a primitive lunar mantle that was involved in lunar magmatism. Fifth, the apparent consistency of the KREEP-like interelement ratios in almost all nonmare materials is not compatible with basalt crystallization in relatively small, isolated magma chambers (Warren 1985).

One way to reconcile the isotopic results discussed above is to have a chondritic LMO crystallizing early and very quickly and for the ferroan anorthosites from the Apollo 16 landing site not be floatation cumulates of the LMO. Rather they have to be derived from mafic sources characterized by LREE-depletion. Multi-stage growth modeling of ε_{Nd}^{142} and ε_{Nd}^{143} values in 62236 by Borg et al. (1999) suggest that the source region of this sample formed at 4.45 ± 0.01 Ga. This age is younger than some estimates of LMO crystallization of (e.g., 4.50-4.52 Ga; Lee et al. 1997; Shearer and Newsom 1999) and therefore does not require the source of 62236 to be a cumulate of the LMO. A post-LMO melting model was advanced by Longhi (2003) to explain the isotopic observations made by Borg et al. (1999) for the Apollo 16 ferroan-anorthosite. His calculations implied that the LMO was capable of producing plagioclase-bearing mafic cumulate sources as a result of the inefficient flotation-separation of plagioclase during crystallization. Remelting of these sources at relatively low-pressures (5-10 kb) is capable of generating ferroan-anorthosite-like magmas. These post-LMO magmas would be derived from LREE-depleted sources and would have ages that post-date ferroan anorthosites produced during LMO crystallization and overlap with the Mg-suite.

3.4.6. Crystallization and impact resetting ages of ferroan anorthosites. An alternative to distinct periods of ferroan-anorthosite production is that post-crystallization, open-system processes affected the Nd isotopic composition and/or Sm/Nd ratios of the ferroan anorthosites. These subsolidus processes have been explored by Shearer et al. (2002), Nyquist et al. (2002), and Norman et al. (2003). Numerous studies have demonstrated that the ferroan anorthosites have undergone significant post-crystallization cooling, reheating to the point of remelting, brecciation, and mixing (i.e., Nord and Wandless 1983). Shearer et al. (2002) showed that 62236 consists of several presumably ferroan-anorthositic lithologies that have distinct trace-element signatures and that have been metamorphosed to fairly high temperatures (800-1000 °C). They suggested that during these high temperature events, isotopic reequilibration may have occurred among the different lithologies in the breccia. Nyquist et al. (2002) showed that the plagioclase matrix surrounding the Yamato 86032 ferroan-anorthosite clast had a distinctly different Nd isotopic composition and Sm/Nd. The higher REE diffusion rates in plagioclase compared to pyroxene and the significantly different REE patterns of the two phases suggest that open system processes may have affected plagioclase more severely than primary pyroxene. Using isotopic data derived from the mafic minerals from the Apollo 16 ferroan anorthosites, Norman et al. (2003) derived an age of 4.456 ± 0.04 Ga and an ε_{Nd}^{142} of $+0.8 \pm 1.4$. Using isotopic data derived from mafic minerals from the Apollo 16 ferroan anorthosites and bulk data from an anorthositic clast from Yamato 86032, Nyquist et al. (2002) derived an age of 4.49 ± 0.09 Ga and an ε_{Nd}^{142} of 0.0 ± 0.8. The scale and mechanism of REE distribution among phases and their effect on the Nd isotopic composition and Sm/Nd of plagioclase are still in need of validation. Further isotopic measurements of ferroan anorthosites are required to better understand the observations and diametrically opposing conclusions made by Borg et al. (1999) and Norman et al. (2003).

4. LUNAR (PRE-MARE) MAGMATISM 4.5 TO ~3.85 Ga

4.1. Introduction

4.1.1. Definition of pre-mare magmatism. Based on Earth observations and a cursory view of the Apollo and Luna sample suites, to a first approximation the lunar crust consists predominately of an ancient highland crust of ferroan anorthosite, younger basalts that fill many impact basins, and impact-generated lithologies. However, a closer examination of the sample suite indicates the existence of a wide compositional breadth of plutonic and volcanic rocks that represent lunar magmatic events that presumably followed ferroan-anorthosite formation and preceded the eruption of basin-filling basalts (Fig. 4.19). Although referred to as pre-mare magmatism, it is highly likely that mare magmatism is a continuation of early episodes of basaltic magmatism. For example, on the basis of lunar meteorites, KREEP basaltic magmatism is known to extend to at least 2.9 Ga (Borg et al. 2004). In addition, the sampling of the plutonic rocks discussed below is biased because their availability is a product of impact events that were curtailed significantly after 3.9 Ga. These plutonic and volcanic lithologies have been arranged into several petrologic groups: Mg-suite plutonic rocks, alkali-rich plutonic rocks, and highly evolved highland plutonic rocks, KREEP basalts, high-Al basalts, and high-K basalts (Hubbard et al. 1971; Brown et al. 1972; Warner et al. 1976; Meyer

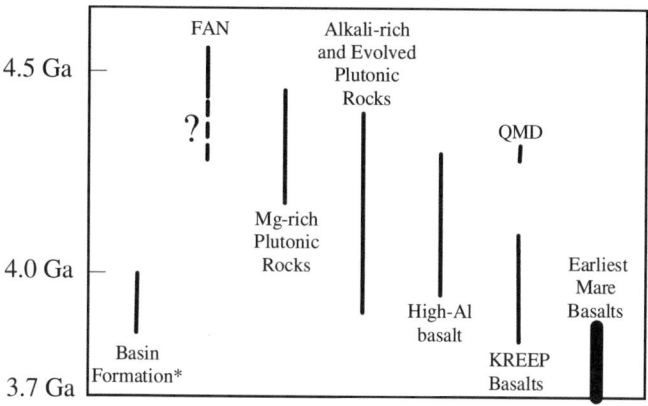

Figure 4.19. Bar diagram illustrating the ages of various events during the early evolution of the Moon (modified after Shearer and Papike 1999).

1977; Warren and Wasson 1977, 1979, 1980; Norman and Ryder 1979; Shervais et al. 1984, 1985; Warren 1989). Many of these lithologies occur as clasts in highland soils and breccias, and have been modified by post-crystallization processes to varying degrees. Of the 384 kg of samples returned from the Moon, less than 0.75% (2.88 kg) are coherent fragments of these pre-mare lithologies (see Chapter 3, Section 2.1). Of this mass, approximately 85% is represented by only 7 rock fragments.

The existence of large volumes of pre-mare basaltic lithologies has also been postulated based on remotely sensed data (Head 1974; Metzger and Parker 1979; Schultz and Spudis 1979; Bell and Hawke 1984; Davis and Spudis 1985, 1987; Hawke et al. 1990; Head and Wilson 1992). A series of low-albedo surface units of varying distribution and areal extent have been interpreted as pre-mare volcanic deposits (Head 1974). These deposits are concentrated in and near upland areas adjacent to maria, appear to have a smooth surface that gives the appearance of mantling underlying topography, and have an age that is post-Imbrium and pre-mare (Head 1974). Head and Hawke (1992) have suggested that perhaps up to a third of the erupted basalts at the lunar surface were this type of "cryptomaria volcanism." The distribution of plutonic rocks has been examined through windows to the lower lunar crust provided by central peaks of impact craters (Pieters and Wilhelms 1985; Hawke et al. 1986; Lucey et al. 1986; Pieters 1991, 1993; Tompkins and Pieters 1999; see also Chapter 3). The relationship of older dark-mantle deposits and central peak lithologies to the volcanic and plutonic lithologies represented by the sample suite is equivocal.

4.1.2. New approaches to exploring episodes of lunar magmatism. The relatively old magmatic lithologies are critical for reconstructing the magmatic and thermal history of the Moon between 3.85 and 4.5 Ga. Are these lithologies products of melting resulting from internal (mantle) or external heat (impact)? If endogenous, are these lithologies related to one another? Is younger mare magmatism really a continuation of these early stages of magmatism? Understanding their composition, distribution, and origin is important for reconstructing the structure of the lunar crust and calculating the bulk composition of the crust and the whole Moon.

New data on these lithologies have focused on three sets of interpretive problems: lack of new samples, small sample size, and lack of geological context in which to place the sample suite. "New" samples of these early stages of lunar magmatism have been "mined" as clasts and mineral fragments from breccias. Because of their complex history (magmatic, extended cooling, impact, potential mixing) and in many cases small, potentially unrepresentative sample size, recent work on these samples has focused upon isotopic and microbeam studies. Many of these samples have little or no geological control. In order to better use the sample data to reconstruct the magmatic, crustal, and mantle evolution of the Moon, their relationship to remotely sensed lithologies must be established and they must be put within the context of the global data sets.

4.2. Plutonic rocks, magnesian suite

4.2.1. Introduction. Rocks that belong to the Mg-suite (also referred to as the Mg-suite, magnesium-rich plutonic rocks, highland Mg-suite or HMS) can be differentiated from other plutonic-hypabyssal lithologies making up pristine lunar highland rocks with a variety of mineralogical and chemical criteria (Chapter 3). They are distinguished from ferroan anorthosites and alkali-suite rocks by the An content of plagioclase vs. the Mg/(Mg+Fe) of coexisting mafic minerals (Fig. 4.20). Compared to the ferroan anorthosites, they generally have higher abundances of mafic minerals and elevated incompatible-element concentrations (Chapter 3 and references therein). The REE contents of rocks from the Mg-suite vary widely (≈ 0.4-400× chondrite) and in most cases are greater than those of ferroan anorthosites (Fig. 4.21). The mineralogical and chemical differences between ferroan anorthosites and the Mg-suite have been interpreted as indicating formation of these two crustal rock types under different petrogenetic circumstances (i.e., they cannot be simply related along a common liquid line of descent).

Commonly, it is assumed that the Mg-suite plutonic rocks intruded preexisting LMO-derived ferroan anorthositic crust and that they formed intrusive complexes analogous to terrestrial layered mafic intrusions. The evidence for the former is that early age dates suggested that the ferroan anorthosites are older than the Mg-suite rocks. The evidence for the latter is tied to the sampled lithologies that are analogous to terrestrial layered intrusions and mimic a liquid line of descent (dunite → troctolite → norite) predicted in the olivine-anorthite-SiO$_2$ pseudoternary system. However, orbital and field evidence, although far from complete, calls for a much more complex scenario. In fact, as noted above, the interpretation of some recent isotopic data cast some doubt on the temporal relationship between the ferroan anorthosites and the Mg-suite. Also, there is a spatial dissociation between ferroan anorthosite (Apollo 16) and Mg-suite (Apollo 14, 15, and 17) samples suggesting that lateral and vertical separations dominate the distributions rather than the intrusion of one into another (Ryder et al. 1995; Papike et al. 1998).

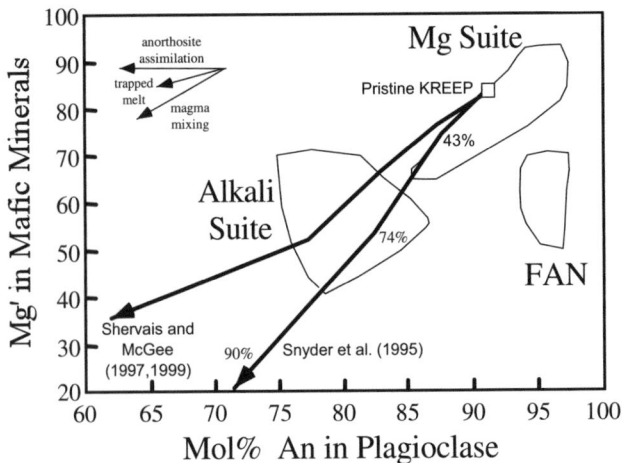

Figure 4.20. Anorthite content (An) in plagioclase vs. Mg′ of coexisting mafic silicates for lunar highland plutonic rocks (modified after Warren 1986 and Snyder et al. 1995).

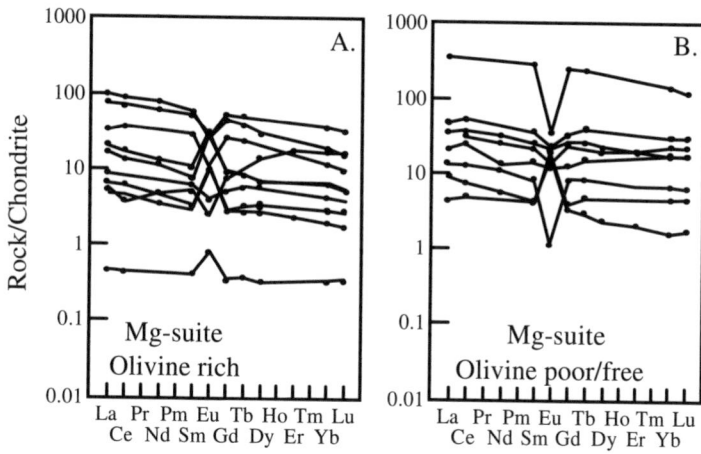

Figure 4.21. Chondrite-normalized rare earth element plots for magnesian-suite plutonic rocks (modified after Papike et al. 1998).

As with the ferroan anorthosites, the petrologic history of the Mg-suite is obscured to varying degrees because of the effect of intense meteorite bombardment and prolonged subsolidus cooling. The first isotopic data were interpreted as indicating that Mg-suite magmatism followed the generation of the ferroan-anorthositic lunar crust and extended over a period of 400 m.y. (Carlson and Lugmair 1981a;b; Carlson 1982; Dasch et al. 1989) (Figs. 4.16 and 4.19). This is also implied by the initial Sr isotopic compositions of the Mg-suite plutonic rocks that are higher than those of ferroan anorthosites (Carlson and Lugmair 1981a,b). The duration of Mg-suite magmatism is also problematical. Some of the younger ages for the Mg-suite may not be crystallization ages at all, but may be more closely related to subsolidus closure (Carlson and Lugmair 1981a,b; Carlson 1982). Conversely, the decrease in the intensity of meteorite bombardment after ~3.9 Ga limited the natural sampling of these deep crustal lithologies.

4.2.2. Recent petrologic observations. Many of the basic petrologic observations concerning the Mg-suite were made during the early- to mid-1970s. Since that time numerous petrologic and geochemical observations have been made that reflect on the origin of the Mg-suite and their place in the evolution of the lunar crust.

In an attempt to reconstruct the geology of the lunar highlands crust, Spudis et al. (1991) and Ryder et al. (1997) collected electron-microprobe data on mineral fragments in poikilitic melt breccias produced by the Serenitatis and Imbrium impact events. They held that the nature and distribution of the mineral fragments could be used to reconstruct the stratigraphy of the highland crust and thereby provide additional insight into crust-building episodes related to the emplacement of the Mg-suite. Ryder et al. (1997) concluded that the highland lithologies at this site were extremely complex, consisting of numerous separate igneous intrusions that crystallized at various crustal depths. Mg-suite plutons were emplaced at depths of 40-50 km in the lunar crust (McCallum and Schwartz 2001). Some of the mineral fragments were from previously unknown Mg-suite lithologies (i.e., troctolites with olivine with even higher Mg'). The deepest crust sampled by this impact event was basaltic in composition, consisting of KREEPy gabbroic rocks with limited olivine. Mg-suite lithologies consisting of norites and troctolites overlaid the gabbroic rocks. The intrusive relationship between the Mg-suite and ferroan anorthosites generally predicted by more simple crust-building models was lacking in the Serenitatis region. Ferroan anorthosites are essentially absent in the mineral fragments.

The studies of Jolliff et al. (1993) and Snyder et al. (1995) are examples of a similar approach for investigating "new" Mg-suite plutonic rocks. They compiled major-trace element and isotopic data for relatively pristine rocks that occurred as small (<100 g) clasts from Apollo 14 breccias. Using these data, they calculated parental melts and placed these samples within the context of large Mg-suite samples. The study by Snyder et al. (1995) concluded that these remnants of the Mg-suite were cumulates derived from the crystallization of 0 to 43% of pristine KREEP basalt (Fig. 4.20). They also suggested models for the incorporation of the KREEP component and elevated large-ion lithophile elements into the parent magmas.

One of the more insightful petrologic-experimental observations concerning the origin of the Mg-suite was made by Longhi (1981) and was reemphasized by Warren (1985) and Hess (1998). They concluded that the crystallization sequence of the Mg-suite, like that of the mare basalts (orthopyroxene followed by clinopyroxene), indicated that lunar basaltic magmatism was non-tholeiitic. Longhi (1981) and Hess (1998) interpreted this observation as indicating that the mantle source for the parental magmas for the Mg-suite had a sub-chondritic Ca/Al ratio and consisted of mineral assemblages dominated by harzburgites rather than lherzolites. Alternatively, this could be a product of very high degrees of partial melting (Longhi 1981) or assimilation of a ferroan-anorthosite crust (Warren 1985).

4.3.2. Isotopic studies. Nyquist and Shih (1992) and Snyder et al. (2000) compiled ages for the Mg-suite rocks. These ages ranged from 4.61 ± 0.07 to 4.17 ± 0.02. This range is perhaps a little deceiving in that it includes ^{40}Ar-^{39}Ar and Rb-Sr ages that are clearly disturbed by impact-related processes. Still, from these data many of the early studies concluded that the Mg-suite rocks post-dated the ferroan anorthosites. U-Pb and Sm-Nd data have more precisely defined the duration of Mg-suite magmatism. A compilation of U-Pb and zircon ages for the Apollo 14 Mg-suite rocks yields a range of 4.136 to 4.320 Ga. Shih et al. (1993) determined Sm-Nd ages for two norites from the Apollo 15 site of 4.46 ± 0.07 and 4.28 ± 0.03 Ga. It is interesting to note that these two ages come from the same clast in sample 15445 from locations separated by one cm (Nyquist and Shih 1992). These data seem to indicate that Mg-suite magmatism was initiated soon after the formation of the primitive ferroan-anorthosite crust and extended for 100s of millions of years. Snyder et al. (1995) suggested that Mg-suite magmatism on the lunar near side occurred first in the northeast and then swept to the southwest over a period of 300 to 400 million years. However, it is possible that in reality all of these intrusions were solidified before 4.3 Ga and represent a significant episode of crust building (Papike et al. 1998).

Nyquist and Shih (1992), Shih et al. (1993), and Snyder et al. (1995) have shown that many of the Mg-suite rocks exhibit enriched Nd isotopic signatures (i.e., negative ε^{143}_{Nd}) and a $^{147}Sm/^{144}Nd$ ratio of 0.169 that approximates that of KREEP (Fig. 4.22). This signature contrasts with what would be expected for early Mg-rich LMO cumulates, which should have a depleted Nd isotopic signature (i.e., LREE depleted with a positive ε^{143}_{Nd}). The similarity of $^{147}Sm/^{144}Nd$ ratio implies that the KREEP basalts are potentially the volcanic equivalent to the Mg-suite and Mg-rich KREEP basalts could be analogous to the parental magmas for the Mg-suite (Snyder et al. 1995).

A dunite from the Mg-suite has a very subchondritic $^{187}Os/^{188}Os$ value of 0.1045 (Walker et al. 2004). This means that the mantle source of its parent melt could not have been highly fractionated with regard to Re/Os prior to its formation. This and other siderophile-element data indicate that the lunar mantle (at least the sources for Mg-suite magmas) had much lower abundances of highly siderophile elements relative to the terrestrial and martian mantles.

4.2.4. Trace element mineralogy. One of the difficulties of interpreting bulk rock geochemical data from the Mg-suite is that they represent cumulates, not melts, and many of the samples are small clasts that are non-representative of the cumulate lithologies from which they were derived. Because of this, numerous studies have approached the interpretation of the Mg-

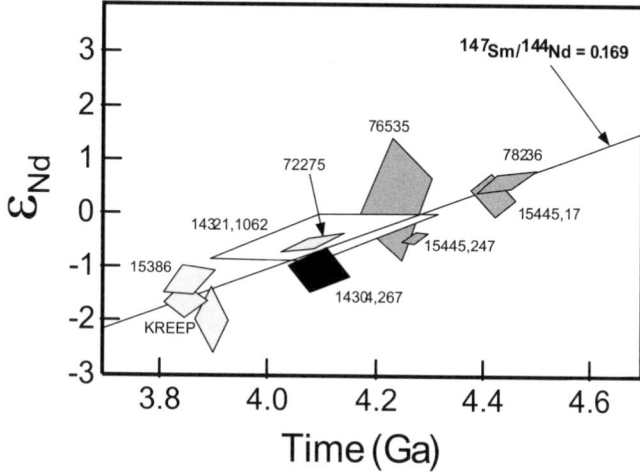

Figure 4.22. Initial ε_{Nd} values vs. crystallization ages for Mg-suite rocks, evolved highland plutonic rocks and KREEP basalts. These rocks form a linear array corresponding to an LREE-enriched source with $^{147}Sm/^{144}Nd = 0.169$ (modified after Nyquist and Shih 1992).

suite by trace-element analysis of individual phases using the ion microprobe. These studies have used this approach to interpret magmatic history by comparing and inverting elemental data of individual phases and for interpreting age dates by investigating on a mineralogical scale the effects that mixing, cooling, and reheating have on isotopic systematics. To date, trace-element analyses by ion microprobe have been reported for pyroxene, olivine, plagioclase and phosphates for most of the lithologies making up the Mg-suite (Jolliff et al. 1993; Papike et al. 1994, 1996; Shervais and McGee 1997, 1999; Shearer and Floss 2000; Shearer and Papike 2000; Shearer et al. 2001). Based on these analyses and the assumption that the trace-element compositions of mineral cores could be inverted to determine melt compositions, these investigators determined the trace element characteristics of Mg-suite magmas.

As implied by the bulk rock data, the plutonic rocks of the Mg-suite appear to have crystallized from magmas with REE and incompatible trace-element concentrations equal to or greater than the high-K KREEP estimated by Warren (1985) (Fig. 4.23). In addition, relative to high-K KREEP, the calculated parental magmas for the norites are depleted in Eu (Papike et al. 1994, 1996). The trace-element data provide additional insight from the distribution of the incompatible elements in the lithologies that cannot be extracted from the whole-rock data. That is, phases that crystallized early from the melt (Figs. 4.23 and 4.24) (cores of olivine, orthopyroxene) are enriched in incompatible elements indicating that this enrichment was a characteristic of the parental magmas, not a product of late-stage magmatic processes or subsolidus mixing or mingling of a KREEP component.

Several other insights are offered by the ion-probe data regarding the characteristics of the Mg-suite parental magmas and the petrogenetic relationships among highland lithologies. A puzzling characteristic of the Mg-suite is that most lithologies have primitive major-element characteristics (high Mg′) yet are enriched in incompatible elements. With such a high Mg′, compatible elements such as Ni should also be high. For example, Ni in olivine (Mg′ ≈ 0.92) from terrestrial komatiites is approximately 3700 ppm (e.g., Karner et al. 2001) and the Ni in olivine (Mg′ ≈ 0.75) from the Apollo 12 olivine basalts ranges from 400 to 500 ppm (Papike et al. 1999; Karner et al. 2001). Surprisingly, Ni in olivine from the Mg-suite ranges from 300 to 100 ppm for the ultramafic lithologies (Mg′ for olivine ≈ 0.90 to 0.85) and 160 to 40 ppm

Shearer et al.

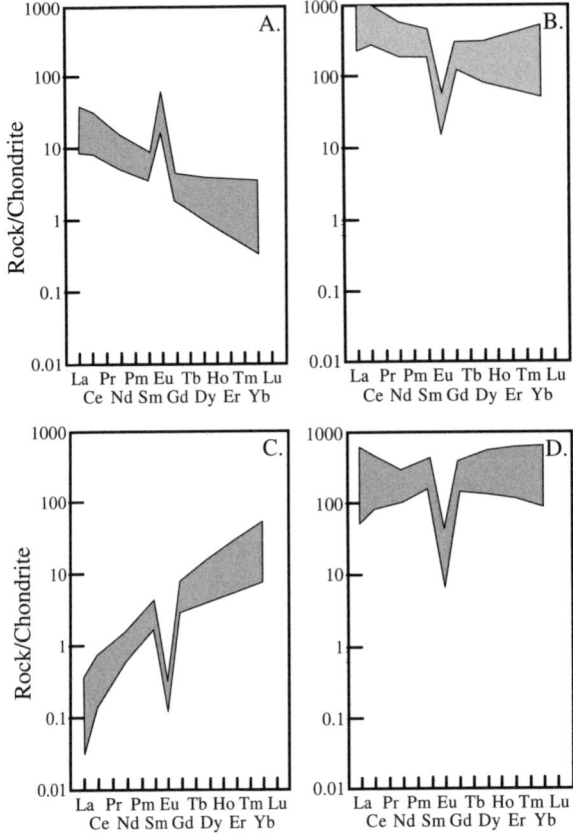

Figure 4.23. Chondrite-normalized REE plots for plagioclase and pyroxene in magnesian-suite norites and pyroxenes (analyzed by ion microprobe) and parental melts calculated from them by inversion using appropriate distribution coefficients (modified from Papike et al. 1998).

Figure 4.24. Ni vs. Co for olivine from the various lithologies making up the magnesian-suite, ferroan-anorthosites and Select mare basalts (modified from Shearer and Papike 2005).

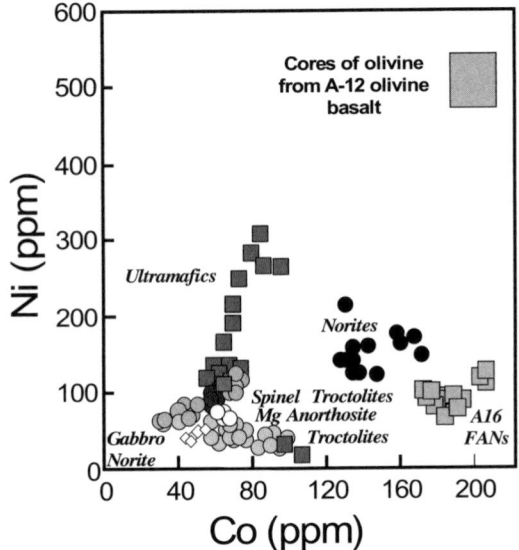

for the troctolites and norites (Mg' for olivine ≈ 0.94 to 0.71) (Shearer et al. 2001) (Fig. 4.24). Therefore, the only "primitive" signature of these magmas is their high Mg'.

Among the Mg-suite lithologies thus far analyzed, there is clearly a relationship between Ni and Co in the olivine (Fig. 4.24). Nickel and to some degree Co decrease from the ultramafic rocks → troctolites, spinel troctolites, Mg anorthosites → gabbronorites. Yttrium in olivine and incompatible elements in plagioclase and the bulk rock increase with decreasing Ni. However, the olivine-bearing norites plot at higher Co. Compositions of the olivine in these lithologies are offset toward ferroan-anorthosite olivine that plot at much higher Co.

4.3. Plutonic rocks, evolved highland rocks

4.3.1. Introduction. Here, included in this discussion of evolved lunar highland rocks are the alkali suite, consisting of anorthosites, norites, gabbros, and monzodiorites, a suite of quartz monzodiorites (monzogabbro), and a range of siliceous plutonic rocks referred to as granites, rhyolites, and felsites (Hubbard et al. 1971; Brown et al. 1972; Warren and Wasson 1980; Jolliff 1991; Snyder et al. 1995). The sample population and the petrography of these evolved highland rocks are described in Chapter 3 and by Papike et al. (1998). The small size and potential non-representative nature of these small samples is an obvious obstacle to their interpretation.

Early studies of the lunar highlands identified plutonic rocks with anomalously high alkali-element contents (Hubbard et al. 1971) and highly evolved mineral and chemical signatures (Brown et al. 1972). These alkali-rich lunar rocks are generally defined as containing greater than 0.1 wt% K_2O and 0.3 wt% Na_2O (Hubbard et al. 1971; Brown et al. 1972; Warren and Wasson 1980). Although this is low compared with terrestrial magmatic rocks, the bulk Moon is highly depleted in alkali and volatile elements (Taylor 1982). Although the alkali rocks have an elevated incompatible-element contents (La = 20 to >1000× chondrite), they rarely have the typical KREEP incompatible-element abundance pattern (Fig. 4.25) nor the high K_2O contents observed in the other evolved plutonic rocks such as the quartz monzodiorites and granites/felsites (Jolliff 1991; Papike et al. 1998). Characteristic features of these rock types are discussed in Chapter 3.

The evolved highland lithologies appear to make up a much smaller portion of the lunar crust relative to the ferroan anorthosites and Mg-suite rocks. However, this conjecture is based on their limited abundance in the lunar sample suite and speculation concerning their origin. Most of the early ages of tiny fragments of these evolved rocks indicate that this style of magmatism followed the production of ferroan-anorthosite and overlapped with episodes of Mg-suite magmatism.

4.3.2. Isotopic studies. Nyquist and Shih (1992) and Snyder et al. (2000) compiled ages for the alkali-suite rocks. Ages tabulated at that time range from 4.370 ± 0.03 to 3.796 ± 0.01. Like the Mg-suite rocks, this range is deceptive in that it includes ^{40}Ar-^{39}Ar and Rb-Sr ages that are clearly disturbed. Zircon (U-Pb) ages and Sm-Nd data have more precisely defined the duration of alkali-suite magmatism. A compilation of zircon ages yields a range of 4.028 to 4.370 Ga for alkali anorthosite (Apollo 14), gabbronorite (Apollo 16), and quartz monzodiorite (Apollo 15). Zircon ages for the lunar granites (Apollo 12, 14, 17) range from 3.883 to 4.360 Ga. Sm-Nd ages for an alkali anorthosite (Snyder et al. 1995) and a granite (Shih et al. 1985) are 4.108 ± 0.053 Ga and 4.110 ± 0.200 Ga, respectively.

Shih et al. (1985) and Snyder et al. (1995) have shown that granite and alkali anorthosite exhibit enriched initial-Nd isotopic signatures (ε_{Nd} = −0.6 and −1.0, respectively) and a $^{147}Sm/^{144}Nd$ ratio of 0.169 that approximates that of KREEP (Fig. 4.22). The similarity of $^{147}Sm/^{144}Nd$ ratio implies that KREEP, KREEP basalts, Mg-suite rocks, and the evolved plutonic rocks have a common petrologic link, which is consistent with the inference from mineral compositions (e.g., Fig. 4.20).

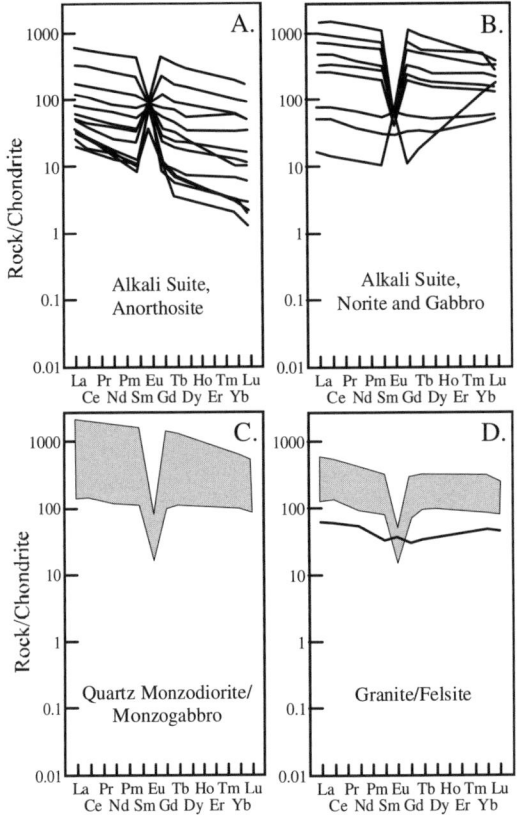

Figure 4.25. Chondrite-normalized rare earth element plots for evolved, highland plutonic rocks. (A) alkali-suite, anorthosites, (B) alkali suite, norites and gabbros, (C) quartz monzodiorites, (D) granites/felsites (modified after Papike et al. 1998).

4.3.3. Trace element mineralogy. Snyder et al. (1994) and Shervais and McGee (1997, 1999) presented analyses of plagioclase and pyroxene from norites and anorthosites. Based on these analyses, it appears that the alkali norites crystallized from magmas with REE and incompatible trace-element concentrations equal to or slightly greater than "high K-KREEP" estimated by Warren (1985). Relative to high K-KREEP, the calculated magmas for the anorthosites and norites of the alkali suite are slightly depleted in HREE and enriched in Eu (Shervais and McGee 1997, 1999). From ion microprobe analyses in phases from an alkali gabbro (in 14318), Shervais and McGee (1997) concluded that the alkali suite samples they analyzed were derived from parental magma with trace-element characteristics similar to those for many of the Mg-suite lithologies. Snyder et al. (1994) deduced that the parent magma was similar to the composition of the quartz monzodiorite lithologies and more evolved than those for the Mg-suite.

Jolliff et al. (1993) analyzed the REE in apatite and merrillite in a suite of samples including alkali-suite lithologies. Although the phosphates are likely to represent products of very late-stage trapped melt, the coexisting silicate mineral assemblages were used to model the geochemical evolution of residual melt, constrained by bulk trace-element analyses by INAA, and thus to back out likely parent-melt compositions. Parent-melt REE concentrations inferred in this way were found to be consistent with fractionation of KREEP-like parent melts. Jolliff et al. (1999) reported ion-microprobe analyses of pyroxene and plagioclase in sample 14161,7373, a well-preserved phosphate- and alkali-rich monzogabbro that likely crystallized in a shallow intrusive body. Although this sample has the highest bulk

REE concentrations of any lunar rock, the trace-element concentrations of its minerals are consistent with fractionation of a KREEP-like parent magma. Segregation of phosphates from late-stage melts, especially RE-merrillite, coupled with the effects of late-stage silicate-liquid immiscibility to produce felsic segregations, imposes a trace-element signature that leads to the unusual enrichment pattern seen in granite/felsite.

4.4. Remotely sensed data on lunar plutonic rocks

If the Mg-suite and alkali-suite lithologies crystallized at depths of 10-30 km (Herzberg and Baker 1980) or deeper, they should rarely be exposed at the surface, unless excavated by large basin- forming events or uplifted in the central peaks of large craters. Their identification through remote sensing methods is further complicated because characteristics such as Mg′, which is used to distinguish Mg-suite rocks, cannot be determined from current spectral data (e.g., Clementine) that are available at high spatial resolution. However, the relative abundance of mafic minerals can be obtained from such data and may be used to distinguish candidate Mg-suite rocks from anorthosites and basalts (see Chapter 2).

The study of rock types exposed in the central peaks of large craters by Tompkins and Pieters (1999) and described in Chapter 3 classified central peaks according to their major mineralogy. Exposures containing less than 80% plagioclase were identified as gabbro, norite, troctolite, gabbronorite, anorthositic troctolite, anorthositic gabbro, and anorthositic gabbronorite following the classification scheme of Stöffler et al. (1980). These were inferred to be Mg-suite lithologies because these rock types are known from the samples. Central peaks containing anorthositic norite, on the other hand, could indicate either Mg-suite rocks or ferroan-anorthositic rocks (Jolliff and Haskin 1995; Norman et al. 1995; Tompkins and Pieters 1999; Hawke et al. 2003; Norman 2003) (Fig. 4.26). Because Mg′ is not determined, however, the assignment to the ferroan, magnesian, or alkali suites remains an inference.

In one location, Tycho, the suggestion of Mg-suite plutonic material in the central peaks is perhaps more conclusive. The identification of gabbroic material in the peaks of Tycho (Tompkins and Pieters 1999) is supported by previous studies, including the analysis of

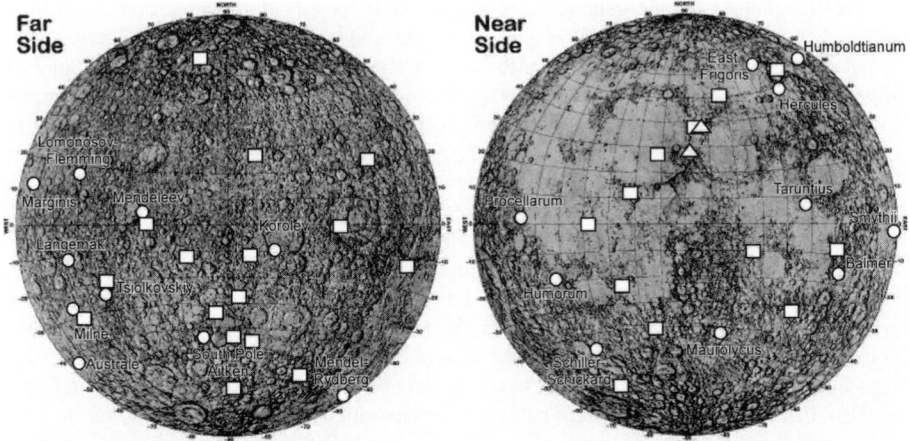

Figure 4.26. Locations of proposed ancient volcanic units. Hidden mafic deposits are shown with circles, and the names corresponding to Table 4.3 are given. Potential Mg-suite materials are indicated with squares. Possible locations of KREEP volcanism are identified with triangles. Basemap is a downsampled version of the Shaded Relief Map of the Lunar Near Side and Far Side Hemispheres, L 10M, I-2769, Lambert azimuthal equal area projection, US Geological Survey 2002.

telescopic hyperspectral data (Hawke et al. 1986; Lucey and Hawke 1989; Pieters 1993). The central peaks of Tycho exhumed material from a depth of approximately 15 km (Tompkins and Pieters 1999), which is too deep to represent buried extrusive materials. Thus, it is likely that Tycho crater tapped into an Mg-suite pluton or a gabbroic intrusion petrogenetically related to the mare basalts. From the composition of surrounding crater peaks, and from other studies (Hawke et al. 1986), the deposit is thought to be vertically and laterally extensive, on the order of tens of kilometers (Tompkins and Pieters 1999).

The mafic central-peak lithologies are not constrained to the western lunar near side, i.e., the Procellarum KREEP terrane (PKT). If these are indeed exhumed Mg-suite plutonic rocks, this is an important observation that would constrain petrogenetic models for both the origin of the Mg-suite and the nature of asymmetric lunar magmatism prior to mare volcanism. More studies are needed to determine the volumetric significance of these types of deposits and the role they played in early lunar magmatism. In particular, it is important to combine UVVIS spectral data (mineralogy) with Lunar Prospector compositional data (KREEP signature: Th, Sm) to fingerprint the nature of Mg-suite lithologies both inside and outside the PKT, although in practice this is difficult because of the coarse spatial resolution of the LP data (Chapter 2).

4.5. Models for early lunar plutonism

Age dates and reconstructions of the lunar crust using remotely sensed data and mineral fragments paint a complex picture of crust-building by the emplacement of post-LMO magmas. The Mg-suite appears to mark the transition between magmatism associated with the LMO and serial magmatism. This transition period may have occurred as early as 30 m.y. (Shearer and Newsom 2000) to as late as 200 m.y. after LMO formation (Solomon and Longhi 1977; Longhi 1980). The duration of Mg-suite emplacement into the crust is also unclear. The absence of Mg-suite rocks with ages that correspond to the younger episodes of KREEP-basaltic magmatism is possibly a result of a lack of deep sampling because of a decrease in impact flux after ~3.9 Ga. Although it seems logical to assume that the Mg-suite was emplaced into the older ferroan-anorthosite crust, neither crustal reconstructions such as those of Ryder et al. (1997) nor analysis of central peaks confirm this relationship. The study by Ryder et al. (1997) seems to suggest a near-side crust consisting of multiple episodes of Mg-suite emplacement.

Several models for the generation of the Mg-suite have been dismissed based on more recent data: impact origin, LMO origin, remelting and remobilization of late-stage LMO cumulates, and metasomatism of Mg-suite rocks by a incompatible-element-enriched fluid. An impact origin for the Mg-suite had been proposed to explain the chemical paradox of primitive and evolved chemical signatures in the same rocks (Wänke and Dreibus 1986; Taylor et al. 1993). The model proposed by Taylor et al. (1993) combined late accretion impactors (~bulk Moon) as a source for the primitive component with LMO crystallization products (anorthosite, KREEP) as a source for the evolved component. The impact of this material into the Moon during the end of LMO crystallization would have mixed this primitive material with remelted ferroan-anorthosite and residual KREEP liquid. The resulting magmas pooled beneath the ferroan anorthositic crust and subsequently intruded the crust. There are several difficulties with this model. The trace-element fingerprints for impactors, such as elevated siderophile-element abundances, are not found in the Mg-suite. For example, the Cr/Ni ratios for the Mg-suite show a typical lunar value (Cr/Ni > 5) in contrast with primitive cosmic abundances (Cr/Ni = 0.25). There is also a mass-balance problem with this process. The ratio of the mass of the impactor to the mass of the impact melt is too small (O'Keefe and Ahrens 1977) to make a substantial contribution to the high Mg' observed in the Mg-suite. Formation of the Mg-suite by impact also circumvented the problem of a heat source capable of producing large volumes of primitive, high-Mg magmas following LMO crystallization (Taylor et al. 1993). However, several studies have identified viable processes that could have triggered melting in the deep lunar mantle (Ryder 1991; Spera 1992; Warren and Kallemeyn 1993; Shearer and Papike 1993). Hess (1994) explored the possibility that the Mg-suite was generated by impact melting of the plagioclase-rich lunar crust and olivine-rich cumulates of the LMO. This eliminates the mass-balance problem in the model of Taylor et al. (1993). Hess (1994) postulated that large impact-melt sheets that were superheated and sufficiently insulated could cool slowly and differentiate to produce the troctolite-norite-gabbro sequence observed in the Mg-suite. Variable incorporation of KREEP and crustal components during shock melting could explain the variation in the evolved component in the Mg-suite plutonic rocks. The siderophile-element signature of the impactor would have been significantly diluted during these processes. If an impact origin for the Mg-suite is correct, one would not expect that the ferroan-anorthosite and Mg-suite chemical trends to be distinctly different (Fig. 4.20). The observed trends cannot be rationalized in terms of mixing of crustal and LMO cumulate components. Hess (1994) also pointed out that impact melts involving a substantial upper-mantle component should have olivine on the liquidus and olivine-rich mineral assemblages are expected. However, dunites are not well represented in the Mg-suite, at least in the returned sample suite and on the lunar surface. Finally, if a substantial upper-mantle component was incorporated into surface impact-melt sheets, it should be expected that upper mantle lithologies would have been excavated and incorporated into the lunar regolith. No such samples have been found.

Wood (1975), Longhi and Boudreau (1979) and Raedeke and McCallum (1980) proposed models in which the Mg-suite was produced during LMO crystallization and evolution. Wood (1975) suggested that the Mg-suite and ferroan anorthosites were simply contemporaneous products of crystal accumulation and trapped melts. In this scenario, the Mg-suite intrusions consist of cumulus olivine/pyroxene plus plagioclase, whereas the ferroan anorthosites consisted of cumulus plagioclase with mafic crystals produced from intercumulus melts. In addition to the differential incorporation of cumulates and trapped melts, Longhi and Boudreau (1979) proposed that the cumulus minerals in both rock types were produced by different styles of crystallization. The plagioclase in the ferroan anorthosites were products of equilibrium crystallization. The cumulate mafic silicates of the Mg-suite precipitated at approximately the same time, but under conditions of fractional crystallization. Raedeke and McCallum (1980) demonstrated that minerals from the banded zone in the Stillwater Complex showed two fractionation trends remarkably similar to the fractionation of Mg/(Mg+Fe) of mafic minerals versus An of plagioclase exhibited by the lunar highland plutonic rocks (see Figure 4.20). They attributed the bimodality of the Stillwater rocks to differences in the style of crystallization (fractional crystallization accompanied by crystal accumulation versus equilibrium crystallization of trapped intercumulus liquid in a plagioclase-rich crystal mush). A contemporaneous relationship between members of these suites is also suggested by the overlapping ages for some of the Mg-suite rocks with the ferroan anorthosites (Shih et al. 1993).

The above models are not consistent with all observations made for the Mg-suite. First, the ages for many Mg-suite rocks postdate the ferroan anorthosites that may be derived from LMO crystallization (Papanastassiou and Wasserburg 1975; Carlson and Lugmair 1981a,b; Nyquist and Shih 1992; Shih et al. 1994). Some of the apparent overlap of ages between the ferroan anorthosites and the Mg-suite is the combined result of an early transition between the LMO and serial magmatism (Shearer and Newsom 2000) and an inaccurate interpretation of the crystallization ages for these early crustal rocks (Papike et al. 1998). Second, detailed geochemical characteristics are not consistent with a near contemporaneous origin of the two magmatic suites from the LMO. For example, the Mg-suite parental magmas have high initial Sr, Sm, and Eu/Al, but low Sc/Sm and Ti/Sm relative to the melts parental to the ferroan anorthosites (Norman and Ryder 1979; Raedeke and McCallum 1980; Warren and Wasson 1980; James and Flohr 1983; Warren 1986). These KREEPy signatures indicate that the Mg-suite cumulates crystallized after the development of KREEP.

The above difficulties prompted the development of numerous models that advocate a post-LMO igneous origin for the Mg-suite that involves the melting of LMO cumulates or even an undifferentiated lunar mantle. These models call upon either the melting of shallow, evolved LMO cumulates, or melting of the deep lunar mantle (LMO cumulates, or primitive lunar mantle) with incorporation of KREEP or crustal component through assimilation, mixing or metasomatism. Hess et al. (1978), Hess (1989), Snyder et al. (1995), Papike et al. (1994, 1996, 1997) and Korotev (2000) explored the possibility that the parental magmas for the Mg-suite originated by partial melting of LMO cumulates at a shallow depth. Although they did not draw a genetic connection between the Mg-suite and KREEP basalts, Hess et al. (1978) suggested that the KREEP-like highland basalts were generated by partial melting of LMO cumulates that crystallized soon after extensive ilmenite crystallization, but prior to the formation of KREEP (between 95 and 99% crystallization of the LMO). These magmas may have assimilated KREEP. Alternatively, Hess (1989) explored the possibility that the KREEP-rich magmas were a product of partial melting of a lower lunar crust that had been altered metasomatically by KREEP. Snyder et al. (1995) and Papike et al. (1994, 1996, 1997) suggested that pressure release melting and remobilization of these rock types may be related to catastrophic impacts on the lunar surface. However, pressure-release melting is unlikely to be important for the deep lunar crust and very shallow mantle as the pressures in those environments are rather low (i.e., 5-10 kb).

Neal and Taylor (1991) suggested that a metasomatic event occurred after emplacement and crystallization of the Mg-suite magmas in the lunar crust. Korotev (2000) proposed that the Mg-suite magmas are the product of dissolution of anorthosite and the most primitive olivine cumulates into the late-stage, KREEP-enriched liquid of the LMO. The KREEP-rich liquid would become multisaturated with both plagioclase and olivine. Although Hess and Parmentier (2001) discounted this model based on geochemical rationale, they did provide a thermal model for this dissolution process. They suggested that because the KREEP layer contains more than 200× chondritic heat-producing elements, it would undergo reheating and would grow by dissolving adjacent lithologies.

In most cases, these models do not account for the substantial observational data for the Mg-suite. Models that advocate remobilization of the KREEP horizon generally cannot account for the extraordinarily high incompatible-element content and the high Mg' of the Mg-suite (Shearer and Papike 1999; Hess and Parmentier 2001). Nor can they account for higher Ni abundance and lower Cr/Ni of the Mg-suite. The mechanism in which a large melt layer is produced just below the crust would retard the transport of later mare magmas to the lunar surface and impede the preservation of mascons associated with large impact basins (Hess and Parmentier 2001). The subsolidus addition of a "KREEP-like" component to the Mg-suite lithologies is not likely because ion probe studies indicate that the cores of large olivine and orthopyroxene grains in the Mg-suite have a KREEP signature. There is no evidence that these components diffused across these grains from a metasomatic fluid.

Models that require the Mg-suite to represent crystallization products of high Mg' magmas generated in the deep lunar mantle were developed to resolve some of the problems with a shallow cumulate source (James 1980; Warren and Wasson 1980; Longhi 1981, 1982; Morse 1982; Smith 1982; Hunter and Taylor 1983; James and Flohr 1983; Shirley 1983; Shervais et al. 1984; Warren 1986; Ryder 1991; Hess 1994). Hess (1994) demonstrated that magmas with Mg' appropriate for Mg-suite parent magmas could be generated by melting of early LMO cumulates. An LMO with an Mg' value equivalent to the bulk Moon (80 to 84; Jones and Delano 1989; O'Neill 1991) upon crystallization at high pressures would produce early cumulates of olivine having Mg' greater than 91. Subsequent melting of these cumulates would produce magmas with Mg' equivalent to that of the parental Mg-suite magmas. Because of the pressure dependence of the FeO-MgO exchange equilibrium between olivine and basaltic melt, crystallization of these high-pressure melts near the lunar surface would result in liquidus olivine that is slightly more magnesian than residual olivine in the mantle source. Higher Mg' values for the melt may result from the reduction of small amounts of FeO to Fe in the source (Hess 1994).

Melting of the early LMO cumulates initially could have been triggered by either radioactive decay (Hess 1994) or cumulate overturn (Hess and Parmentier 1994), or both. The deep lunar mantle materials would have been less dense than the overlying cumulates and tend to move upward and be subjected to pressure-release melting. This would have resulted in relatively high degrees of melting (>30%; Ringwood 1976; Kesson and Ringwood 1977; Herbert 1980). Partial melting of early LMO cumulates could have produced primitive melts with high Mg', but these magmas would not have the same geochemical characteristics as the Mg-suite. For example, these primitive magmas would not possess the high incompatible-element enrichments, fractionated Eu/Al and Na/(Na+Ca), low Ni and plagioclase as a liquidus phase until the Mg' of the melt was <42. Two types of processes have been proposed to resolve this problem: assimilation of KREEP and melting of a hybrid KREEP-early cumulate sources.

For KREEP assimilation, Warren (1986) calculated that if these high-Mg magmas assimilated ferroan-anorthosite and KREEP, they would have reached plagioclase saturation at values of Mg' appropriate for Mg-suite magmas. Such magmas also would have inherited a fractionated incompatible-element signature (high REE, fractionated Eu/Al). Hess (1994) explored the thermal and chemical implications of anorthosite melting and plagioclase dissolution by high-Mg basaltic magmas. In his analysis of anorthosite melting and mixing as a mechanism to drive high Mg' magmas to plagioclase saturation, Hess concluded that the resulting crystallization of olivine and the mixing of relatively Mg-poor cotectic melts produced from the anorthosites would lower the Mg' of the hybrid melt below that expected for the Mg-suite parental magmas. In addition, diffusion rates for Al_2O_3 in basaltic melts are extremely slow (Finnila et al. 1994) and indicate that the time scales to dissolve even a small amount of plagioclase are of the same order as the characteristic times of solidification of a large magma body. Similar thermal constraints are less severe for the shallow melting-KREEP assimilation by primitive Mg-suite magma. Although KREEP assimilation does not dramatically drive the Mg-rich magmas to plagioclase saturation, this mechanism may account for the evolved trace-element signatures in the Mg-suite. However, mixing of relatively viscous melts of KREEP composition with more-fluid Mg-rich magmas could be prohibitive (Finnila et al. 1994). In addition, melt compositions for the Mg-suite norites that were calculated from pyroxene trace element data have a KREEP component equivalent to or slightly higher than KREEP (Papike et al. 1997). Simple mass-balance calculations indicate that it may be impossible for primitive Mg-suite magma to assimilate such an abundant amount of KREEP (Shearer and Floss 1999). However, it is possible that the lower Mg' of the norites along with their elevated REE may reflect assimilation of KREEP and fractional crystallization of a more primitive basalt that was parental to the troctolites. Thus-far unevaluated aspects of assimilation models are the assimilation of a KREEP horizon with a temperature elevated by heat-producing elements, a lunar mantle capable of producing parental magma with a higher Mg', and the effect of crustal processes on the subsolidus redistribution of elements in the Mg-suite lithologies (i.e., pyroxene reequilibration with intercumulus melt, subsolidus reequilibration of primary olivine or pyroxene compositions; McCallum and Schwartz 2001).

As an alternative to assimilation at shallow mantle levels, is it possible that the KREEP component was added to the deep mantle source for the Mg-suite? Hess (1994) proposed that the source for the Mg-suite may be a hybrid mantle consisting of early LMO cumulates (dunite) and a bulk Moon component that may be either primitive lunar mantle or an early quenched LMO crust. Polybaric fractional melting of a 50-50 mixture would produce melts with appropriate Mg' and reasonable Al_2O_3. However, these mixing components would not produce some of the incompatible-element signatures exhibited by the Mg-suite. Shearer et al. (1991) suggested a cumulate overturn mechanism to transport a KREEP component to the deep lunar mantle, thus explaining the evolved KREEP signature imprinted on selected picritic glasses associated with mare basaltic magmatism. A similar process may have produced the KREEP signature in the Mg-suite. A potential pitfall of this model is the Ni content of the olivine in Mg-suite rocks. Although the Mg′ of olivine (Fo_{92-88}) in the Mg-suite is high and suggests a primitive magma derived from deep within the LMO cumulate pile, the Ni content of the olivine determined by electron microprobe (Ryder 1982) and ion microprobe (Shearer and Papike 2000; Shearer et al. 2001) ranges from 10 to 350 ppm. This is fairly low for primitive lunar basaltic magma. In comparison, ion microprobe analyses of olivine cores (Fo_{75-72}) from Apollo 12 olivine basalts yield Ni concentrations of 450-515 ppm (Papike et al. 1999). The calculated olivine composition in equilibrium with mare basalts more primitive than the Apollo 12 olivine basalts (i.e., Apollo 15 green glass) is approximately Fo_{85} and Ni ≈ 1500 ppm. This poses a dilemma for interpretation of the high Mg′ of the Mg-suite and hints at the involvement of a metal phase either during melting of the source or during evolution of the basaltic magma (Papike et al. 1997). Alternatively, it could suggest that the parental magmas for the Mg-suite are not derived from deep within the LMO cumulate pile, but from a lunar mantle that was not part of the LMO cumulate lithologies.

The most reasonable models for the generation of the parental magmas for the Mg-suite and other highland plutonic and volcanic rocks involve the incorporation of a KREEP component. How the KREEP component is incorporated into these magmas is the point of debate. How would the asymmetrical distribution of KREEP on the Moon, as indicated by remotely sensed data, influence the formation and distribution of the Mg-suite magmas under these two petrogenetic scenarios (shallow assimilation versus deep melting of a hybrid mantle)? If magmas parental to the Mg-suite where produced by a two-stage process involving mantle melting followed by KREEP assimilation, there should be two types of high-Mg plutonic rocks: those with a KREEP signature spatially associated with the PKT on the near side of the Moon and those without a KREEP signature emplaced into ferroan anorthosites in non-KREEPy lunar terranes. In the scenario in which KREEP in the mantle source drives melting, it should be anticipated that only KREEP-enriched Mg-suite rocks will exist and they will be closely associated with the PKT on the near side of the Moon. Combining the remote sensing approach taken by Tompkins and Pieters (1999) with gamma-ray chemical signatures of KREEP may provide addition insight into distinguishing between these two petrogenetic models.

The presence of a trace-element KREEP signature and Nd isotopic systematics implies a link not only among the rocks making up the Mg-suite, but also among the Mg-suite, alkali suite, evolved plutonic rocks, and the KREEP basalts. As a first approximation, Warren (1988) and Snyder et al. (1995, 2000) suggested that fractional crystallization of a KREEP basalt could potentially produce the cumulate lithologies making up the Mg-suite. The KREEP basalts in the lunar collection have bulk rock Mg′s of approximately 61 to 66. This range in Mg′ is significantly lower than would be expected for parental magmas of the Mg-suite cumulates (mafic minerals with Mg′ ≥ 90). However, it is possible that more Mg-rich KREEP basalts exist but have not yet been sampled.

The Mg-suite and evolved plutonic rocks may be either part of a continuum of crystallization products of parental basaltic magmas similar to the KREEP basalts (Ryder 1976; James 1980; Warren and Wasson 1980; James et al. 1987; Snyder et al. 1995; Shervais and McGee 1999) or separate episodes of basaltic magmatism (Warren and Wasson 1980). Using a KREEP basalt (from the Apollo 15 site) as a starting parental magma, Snyder et al. (1995) demonstrated that fractional crystallization and the accumulation of mineral phases and trapped KREEP-like residual liquid (2-15%) could produce the range of mineral and rock compositions observed in the highland Mg- and alkali suites. The sequence of crystallization and accumulation that they proposed is Mg-suite (0-43% crystallization) → alkali anorthosites, alkali norites (43-74% crystallization) → alkali gabbros, alkali norites (74-90% crystallization) → quartz monzodiorites (90-99.8% crystallization).

Shervais and McGee (1997, 1999) envisioned a more complex process to account for the REE characteristics of the alkali-suite magmas and the difference in the albite content of the plagioclase between the calculated liquid lines of descent for KREEP magmas and alkali-suite magmas. They concluded that in addition to fractional crystallization, assimilation of ferroan anorthosites, magma mixing, and local equilibrium crystallization were important processes relating these magmas.

An alternative model is one in which the Mg-suite and alkali-suite magmas represent contemporaneous, but separate, episodes of basaltic magmatism (James 1980; Warren and Wasson 1980; James et al. 1987). There is some compositional evidence to suggest genetically distinct highland rock types. For example, James et al. (1987) subdivided many of these highland rock types into various groups on the basis of their mineral chemistry and mineral associations. Whether these subdivisions are artificial or petrologically significant is unclear. Within this scenario, however, differences between the two suites may be attributed to the depth of initial melting prior to assimilation. For example, Mg-suite magmatism would be a product of deep mantle melting followed by KREEP assimilation just below the lunar crust, whereas the more evolved magmas would involve initial melting at shallower mantle levels followed by assimilation. Hunter and Taylor (1983) suggested that the parental magmas of the alkali suite could represent the "dregs" of the LMO.

Materials of granitic composition occur as immiscible glasses within the mesostasis of mare basalts, in melt inclusions, as glasses in selective lunar soils, as crystalline clasts in breccias, and as segregations in monzogabbro (Papike et al. 1998; Jolliff et al. 1999). In all petrologic models, the granite, rhyolite, and felsite clast are considered to be products of extensive fractional crystallization of alkali-suite parental magma. Marvin et al. (1991) proposed that these evolved granitic rock types were produced through the crystallization and removal of phosphates and zircon from parental magmas equivalent to the quartz monzodiorite of the alkali suite. This fractional crystallization mechanism accounted for the enrichment of REE, Zr, and Hf in the quartz monzodiorites relative to the more evolved lunar granites. It has been also proposed that the lunar granites are a product of liquid immiscibility following extensive fractional crystallization of a KREEP basalt. Experimental studies by Hess et al. (1975) illustrated that after 90-95% of crystallization of a variety of lunar basaltic magmas, spherules of granitic magma exsolved from the residual ferrobasaltic liquid. Chemical signatures of these lunar granites have been attributed to primarily immiscibility (Quick et al. 1977; Taylor et al. 1980; Warren et al. 1983; Neal and Taylor 1989; Snyder et al. 1993). On the other hand, Jolliff (1991), Jolliff et al. (1993, 1999) and Shearer et al. (1998) similarly argued that many of these chemical signatures were a result silicate-liquid immiscibility, but after phosphate crystallization.

4.6. Volcanic rocks, KREEP basalts

4.6.1. Introduction. KREEP basalts were returned primarily by the Apollo 15 and Apollo 17 missions. Small and rare fragments of KREEPy basalts from the Apollo 14 and 16 and Luna 20 sites have been documented (Salpas et al. 1987; Jolliff et al. 1991; Nyquist and Shih 1992), but only the Apollo 15 and 17 basalts lack elevated siderophile-element abundances attributed to meteorite contamination. The KREEP basalts occur primarily as numerous small fragments and particles among the rake material, fines and as clasts in breccias. Only a few samples of the KREEP basalts are over 1 g, 72275 (clast 91, 2.73 g), 15382 (3.2 g), and 15386 (7.5 g). Most are subophitic to intersertal in texture. The mineralogy is dominated by pyroxene and plagioclase. In the Apollo 15 basalts many pyroxenes have cores of orthopyroxene, overgrown with pigeonite and then augite. The pyroxene in the Apollo 17 basalts is primarily pigeonite with minor augite. Olivine is extremely rare. The KREEP basalts are distinguished from the mare basalts by their higher plagioclase abundances and their highly enriched concentrations of incompatible minor and trace elements that have elemental ratios similar to the "type" KREEP samples (soils and breccias from Apollo 14) (Fig. 4.27). Compared to the Apollo 15 KREEP basalts, those from Apollo 17 are more iron-rich, lower in incompatible elements and more depleted in heavy REE. A list of KREEP basalts in the Apollo collection and more detailed description of the petrography are presented by Papike et al. (1998) and in Chapter 3.

The KREEP basalts sampled from the Apollo missions have crystallization ages of 3.84 to 4.08 Ga (Nyquist and Shih 1992). The Apollo 15 KREEP basalts (3.84 ± 0.02 Ga) are younger than the KREEP basalts from the Apollo 17 site (≈ 4.01 to 4.08 Ga). Based strictly on sample mineralogy, geochemistry and isotopic data, the relationship among KREEP basaltic volcanism, mare basaltic volcanism, and basin formation is equivocal. The Apollo 15 KREEP basalts may represent volcanic eruptions nearly contemporaneous with the formation of the Imbrium basin and predated the earliest stages of Imbrium-filling mare magmatism. The KREEP volcanism represented by the Apollo 17 samples predates the formation of Serenitatis basin and associated basin-filling basalts. The oldest mare basalts overlap with the ages of some of these KREEP basalts. High-Ti basalts as old as 3.88 ± 0.06 Ga are found at the Apollo 11 site (Papanastassiou et al. 1977).

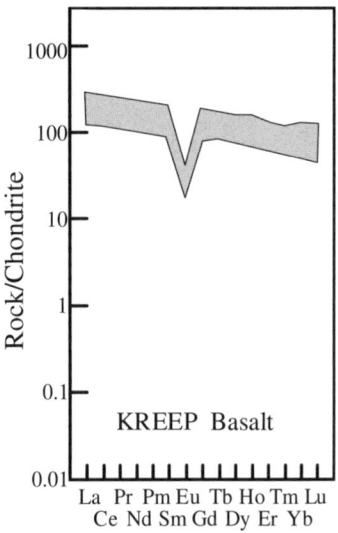

Figure 4.27. Chondrite-normalized rare-earth-element pattern of KREEP basalts (modified after Papike et al. 1998).

More recently, an olivine-gabbro cumulate clast with a significant KREEP signature was identified in lunar meteorite Northwest Africa 773 (Fagan et al. 2001; Jolliff et al. 2003, Borg et al. 2004). The Ar-Ar age of the cumulate lithology is ~2.7 Ga (Fernandes et al. 2002). The Sm-Nd age for the clast is ~2.9 Ga (Borg et al. 2004). This sample indicates that the duration of KREEP basaltic magmatism continued for at least another 1 billion years after the eruption of the Apollo 15 KREEP basalts.

4.6.2. Isotopic studies. Isotopic data may be able to answer questions concerning the age relationship between KREEP basalts and plutonic highland rocks and mare basalts and the petrogenesis of the KREEP basalts. As was stated earlier, KREEP basalts erupted to the lunar surface between 3.84 and 4.08 Ga (Nyquist and Shih 1992). This overlaps with the eruption of mare basalts. Papanastassiou et al. (1977), Snyder et al. (1996) and Shih et al. (1999) confirmed that some of the Apollo 11 high-Ti basalts were as old as 3.88 Ga. Misawa et al. (1993) determined a Sm-Nd age of 3.87 ± 0.06 Ga on basaltic lunar meteorite Asuka 881757 that had a composition that are is similar to the low-Ti to very low-Ti basalts. Compared to the ages for KREEP basalt volcanism, emplacement of Mg-suite magmas into the lunar crust is considerably older (Sm-Nd ages range from 4.18 ± 0.07 Ga to 4.46 ± 0.07 Ga). Figure 4.22 shows the initial ε_{Nd} for KREEP basalts compared to highland Mg-suite rocks. Data for most of the KREEP basalts plot along the same evolution line as do the data for the Mg-suite (Nyquist and Shih 1992). NWA 773 deviates from this trend. It has an initial ε_{Nd} of −7.84 and a $^{147}Sm/^{144}Nd$ ratio for its source region of 0.161 (Borg et al. 2004). This indicates that its source is more LREE-enriched than the sources of the other KREEP basalts and Mg-suite rocks.

4.6.3. Trace-element mineralogy. Trace-element data of mineral phases in KREEP basalts are essentially absent in the literature. Most of the available ion microprobe data is from the olivine-gabbro clast in NWA 773 (Jolliff et al. 2004; Shearer and Borg 2004). The mineral phases such as plagioclase and pyroxene exhibit substantial incompatible-element enrichments relative to similar phases in mare basalts. This is consistent with the isotopic measurements that suggest that a substantial KREEP component has been added to this basalt. Olivine in the clast has Ni and Co concentrations exceeding that found in the Mg-suite and high-Al basalts (Fig. 4.24). The Ni and Co concentrations in olivine overlap or exceed that found in olivine in low-Ti mare basalts.

4.7. Volcanic rocks, high-Al and K-rich basalts

4.7.1. Introduction. The high-Al basalts (>11 wt% Al_2O_3) are enriched in Al relative to other lunar basalts, but are not "high-Al" relative to terrestrial basalts. They were returned by the Apollo 14 and Luna 16 missions and bear a close resemblance to low-Ti mare basalts based on similar contents of TiO_2 (2-3 wt%), FeO (14-18 wt%) and CaO (9-11 wt%) (Table 4.2). Feldspathic basalts from other collection sites (e.g., the one Apollo 12 feldspathic basalt 12038 and the evolved Luna 24 VLT basalts) are not considered in this discussion because they may be either unrepresentative samples, fractional crystallization products of melts that originally contained lower abundances of Al_2O_3, or may not have been derived from the location where they were collected. As with other vestiges of older lunar magmatism, the study of these rocks has been hampered by sample size and the lack of geological context. The importance of both the Apollo 14 and Luna 16 rocks is that they perhaps represent the oldest record of lunar volcanism thus far recognized. The high-Al basalts clasts from the Apollo 14 site have ages that range from 4.0 to 4.3 Ga (Taylor et al. 1983; Dasch et al. 1987; Nyquist and Shih 1992; Shih et al. 1992), up to 200 m.y. older than the oldest known KREEP basalt. It is still a point of debate whether the high-Al basalts represent non-mare basaltic magmatism (Hubbard and Gast 1972), high-Al mare basaltic magmatism (Ridley et al. 1975), or impact melts (Snyder et al. 2000).

In addition to the high-Al basalts, Shervais et al. (1985b), Goodrich et al. (1986) and Neal et al. (1988b, 1989a,b) described basaltic clasts with relatively high potassium contents (i.e., > 0.5 wt% K_2O) in polymict lunar breccias from the Apollo 14 site. These were described as high-K (Goodrich et al. 1986) or very high-K basalts (Shervais et al. 1985b) which we will collectively call "K-rich" basalts. A description of these basalts is presented in Chapter 3.

4.7.2. Trace element studies. Although high-Al basalts were first reported during the early days of lunar sample studies (i.e., Helmke et al. 1972; Hubbard et al. 1972), it was not

Table 4.2. Comparison of bulk compositions of high alumina basalts and low-Ti mare basalts.

	High-Alumina Basalt			Low-TiO$_2$ Mare Basalt		
	14053	**14072**	**14305**	**12052**	**12020**	**15475**
SiO_2	46.4	45.2	45.3	46.40	44.57	48.15
TiO_2	2.64	2.57	2.2	3.28	2.76	1.77
Al_2O_3	13.6	11.1	13.0	10.16	7.77	9.44
Cr_2O_3	0.40	0.51	0.59	0.52	0.61	0.63
FeO	16.8	17.8	16.0	20.15	20.98	19.98
MnO	0.26	0.27	0.20	0.27	0.27	0.30
MgO	8.48	12.2	9.9	8.22	14.40	8.85
CaO	11.2	9.8	10.6	10.80	8.60	10.58
Na_2O	0.44	0.32	0.80	0.27	0.22	0.27
K_2O	0.10	0.08	0.80	0.07	0.06	0.06
Total	100.3	99.9	99.0	100.1	100.2	100.0

Note that the low Ti mare basalts have higher FeO but lower Al_2O_3 contents. From Papike et al. 1997.

until detailed studies of clasts in lunar breccias (i.e., 14303, 14304, 14305, and 14321) during the 1980s that the importance and complexity of these possible remnants of early basaltic volcanism were first recognized (Taylor et al. 1983). The elemental data from these Apollo 14 clasts reported during the mid-1980s to the early 1990s indicated an extremely wide variation in incompatible elements both in overall concentration and REE slope with limited variation in major element whole rock chemistry (i.e., Mg' = 57-51) and mineral composition (Dickinson et al. 1985; Shervais et al. 1985 a,b). The database for these basalts was expanded considerably by Neal et al. (1988, 1989, 2003). The compilation of all the analyses of Apollo 14 high-Al basalts confirm the earlier studies showing a strong positive correlation between La and Hf over an extremely wide range in composition (Fig. 4.28a).

The K-rich basalts have the same major-element composition as the high-Al basalts, but with elevated contents of K, Ba, and Rb (Shervais et al. 1985b). They exhibit the same general positive correlation between La and Hf as the high-Al basalts, although in some samples La and Hf is significantly greater. The concentration of Sm and Th overlap or exceed the range observed in the high-Al basalts and generally have higher Th/Sm ratios (Fig. 4.28b). The K/La ratios for most mare basalts and even KREEP basalts typically fall between 30 and 100 whereas the VHK basalts have ratios that exceed 1000. Chondrite-normalized trace-element patterns show a negative Eu anomaly, a flat to negatively sloped REE pattern with REE abundances up to ~40× chondritic, and with low Sc (≤60 ppm).

4.7.3. Isotopic studies. Isotopic data for the high-Al and K-rich basalts have been summarized by Nyquist and Shih (1992), Shearer and Papike (1998), and Snyder et al. (2000). Since the recognition that the high-Al basaltic clasts may represent old episodes of basaltic volcanism (Taylor et al. 1983), subsequent studies have confirmed the initial observations (Dasch et al. 1987; Neal and Taylor 1990). The work by Dasch et al. (1987) extended the age of the high-Al basalts back to 4.29 ± 0.13 Ga. The combination of all age dates up to that time led Neal and Taylor (1990) to conclude that distinct periods of high-Al basaltic volcanism occurred at 4.3, 4.1, and 3.95 Ga.

Figures 4.29 and 4.30 show initial $^{87}Sr/^{86}Sr$ and ε_{Nd} for the Apollo 14 high-Al basalts as a function of their ages. Dasch et al. (1987) noted that the data for the basalts defined two distinct

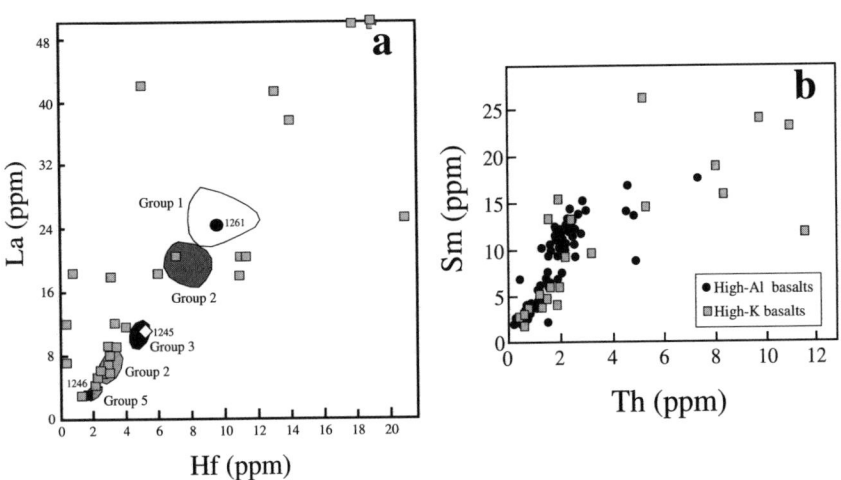

Figure 4.28. (a) Hf vs. La for the high-Al and high-K basalts. Groups outlined are from Dickinson et al. (1984). Also plotted are the high-K basalts (compiled by Clive Neal) (b) Sm versus Th for the high-Al and high-K basalts.

reservoirs with different Rb/Sr. Groups 1, 2, 3, and 5 were consistent with radiogenic growth of ^{87}Sr in a reservoir of uniform Rb/Sr = 0.021, whereas groups 4 and 5′ are consistent with radiogenic growth in a reservoir with Rb/Sr a factor of two lower. In terms of Sm-Nd isotopes, the high-Al basalts plot at ε_{Nd} values greater than 0 and do not plot along the ^{147}Sm/^{144}Nd evolution line of 0.169 defined by the Mg-suite, the evolved plutonic rocks, and the KREEP basalts.

The K-rich basalts are characterized by ε_{Nd}~0 implying derivation from a long-lived chondritic reservoir (Shih et al. 1986). Because of large uncertainties, the (^{87}Sr/^{86}Sr)$_I$ and the implied Rb/Sr ratios are not well defined, but appear to be roughly comparable to that of Apollo 14 high-Al basalts. The initial ^{87}Sr/^{86}Sr ratio is low (0.6995 ± 0.0005) and significantly lower than KREEP basalts (>0.700).

4.7.4. Trace-element mineralogy. One of the obvious problems with understanding the petrogenesis of these rocks is whether the compositional variations observed and modeled are a product of magmatic or impact processes, or a function of unrepresentative sampling. To this end, Hagerty et al. (2002, 2003, 2004) analyzed selected olivine-bearing high-Al basalt clasts representing incompatible-element enriched and depleted basaltic lithologies as defined by Shervais et al. (1985a), Dickinson et al. (1985, 1989), Neal et al. (1988a, 1989a,b), and Hughes et al. (1990). Plagioclase in these clasts is enriched in REE (Fig. 4.31), Ba, and Sr in the same whole-rock enrichment sequence observed by Dickinson et al. (1985) (Fig. 4.28a). In addition,

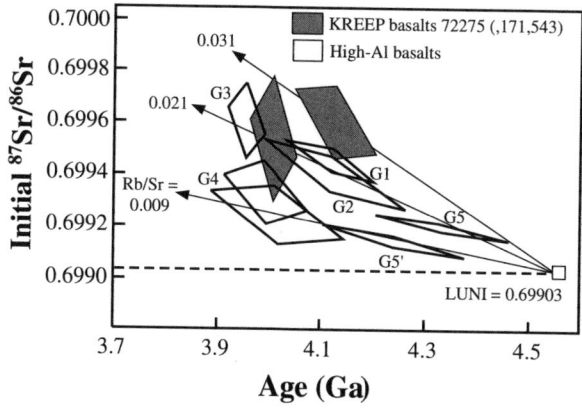

Figure 4.29. Initial Sr values vs. ages for the high-Al basalts and KREEP (modified after Nyquist and Shih 1992).

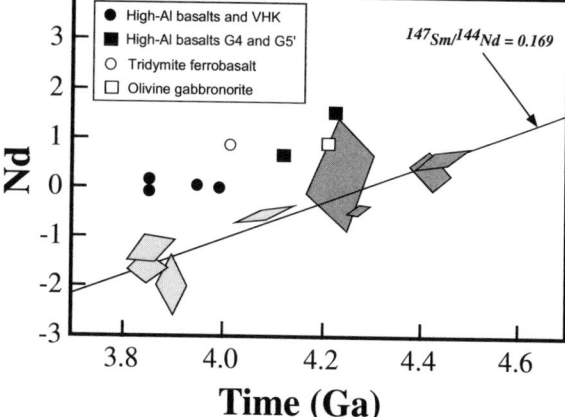

Figure 4.30. Initial ε_{Nd} values vs. ages for the high-Al basalts (modified after Nyquist and Shih 1992).

the Ce/Yb ratios for plagioclase increase (~5 → 7 → 19) with increasing REE content (Fig. 4.31a). These indicate that the basaltic melts from which these plagioclase grains crystallized exhibited different REE contents and Ce/Yb (i.e., LREE enrichments). Assuming D's for the plagioclase do not significantly change, the REE concentration in the melt increased by ~10× between the depleted and enriched basaltic clasts. This incompatible-element enrichment is on the same scale as predicted by whole-rock analyses of the individual clasts.

Although whole-rock analyses show limited variation in compatible elements (i.e., Co, Ni), the olivine from clasts show that each fragment exhibits different Ni-Co concentrations and behavior (Fig. 4.32). The basaltic clasts with the lowest incompatible element concentrations in the bulk rock and in plagioclase contain olivine with cores exhibiting the highest Ni abundances (360 ppm). Conversely, the clasts with the highest incompatible-element concentrations contain olivine with cores exhibiting the lowest Ni concentrations (80 ppm). Nickel decreases from the olivine cores to their rims.

The ion-microprobe analyses of olivine and plagioclase in Apollo 14 high-Al basalt clasts indicate that both compatible and incompatible elements exhibit variability consistent with compositional differences observed in the bulk samples. This indicates that the 10- to 20-fold increase in incompatible elements observed in the bulk clasts is real and not simply a product of unrepresentative sampling (i.e., higher proportions of incompatible-element-enriched mesostasis). Simple fractional crystallization of a single basaltic magma or melting of a single mantle source cannot account for the range of incompatible-element enrichments observed in the olivine-bearing high-Al basalts. The range of ages exhibited by these basalts also argues against simple fractional crystallization. Instead, the range observed in the high-Al clasts represents distinct pulses of basaltic magma (Neal and Taylor 1990; Hagerty et al. 2002, 2003; Kramer and Neal 2003) that were produced by the melting of distinct mantle source or different degrees of assimilation.

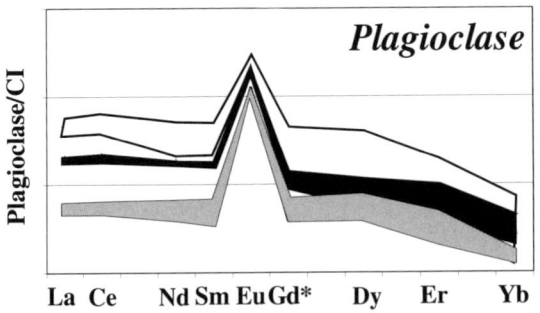

Figure 4.31. Chondrite-normalized rare earth element patterns for plagioclase from the high-Al basalts (Black = sample 14321, 1245; Gray = sample 14321,1246; White = sample 14321,1261). (Hagerty et al. 2002).

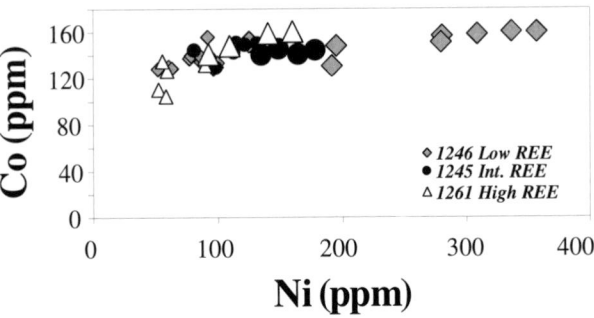

Figure 4.32. Ni vs. Co for olivine from the high-Al basalts (Hagerty et al. 2002).

4.8. Remotely sensed data on pre-mare volcanic rocks

4.8.1. Introduction. The global distribution of pre-mare lithologies on the Moon is difficult to describe, because to date, no contiguous deposits of non-mare volcanism have been unequivocally identified. This may result from several factors. Current remote sensing techniques are not able to distinguish the pre-mare rock types from highland or mare materials, or mixtures thereof. Additionally, these ancient lithologies may correspond to deposits that have been buried or obscured by subsequent events or that are intrusive in nature, making them difficult to observe remotely. Despite these obstacles, some advances have been made.

4.8.2. Hidden mafic deposits. The presence of ancient basalt clasts in the Apollo sample record, some as old as 4.23 Ga (Taylor et al. 1983), contrasts sharply with the age range of observed mare deposits, which are only 3.9-2.5 Ga (Head 1976). When Schultz and Spudis (1979) first recognized the existence of hidden mafic deposits, they proposed that these represented the contiguous bodies from which such ancient basalt clasts originated. Since then, many hidden mafic deposits, termed "cryptomaria" by Head and Wilson (1992) have been proposed and identified.

By definition, cryptomaria are strictly mare basalt deposits whose low albedo signature has been hidden or obscured by superposed high albedo material (Antonenko et al. 1995; Fig. 4.33). The thickness of the overlying ejecta can range from a few kilometers near large impact basins to only several meters near small craters. Thus, the smooth surface topography of hidden deposits is not always guaranteed, since thick basin ejecta may be very textured (Head 1974). Criteria for the detection of cryptomare deposits include the presence of dark-haloed craters (craters that penetrate through the overlying ejecta to excavate underlying mafic material) and the identification of spectral or geochemical anomalies (Antonenko et al. 1995). However, mare basalts can be difficult to distinguish from other mafic materials when only remote sensing tools are available, and especially if the deposits in question are obscured by other materials. As a result, the term "cryptomare" has been used to include all hidden mafic deposits, regardless of specific chemical or mineralogical composition (Antonenko et al. 1995).

A global census of all previously identified and suspected hidden mafic deposits, based on a survey of existing literature, was conducted by Antonenko (1999). A rough estimate of the potential surface area and volume was calculated for each deposit, and a possible age range determined (Fig. 4.26 and Table 4.3). This list provides a good start for understanding the importance of these deposits and their importance for early lunar volcanism. Not all such deposits are necessarily included in the list of Table 4.3. Additional hidden mafic deposits may be found upon further investigation. Also, Nectarian or pre-Nectarian ejecta deposits

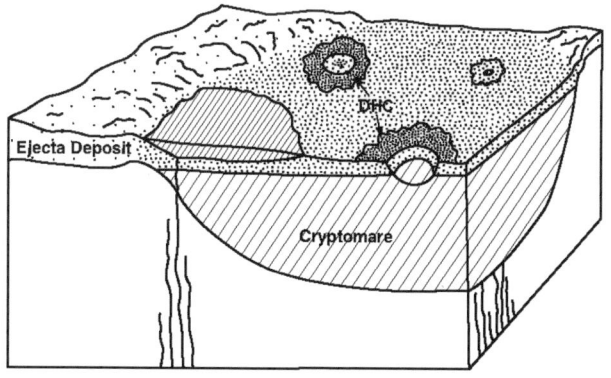

Figure 4.33. Illustration showing the formation of a cryptomare. High albedo basin ejecta is emplaced on top of a pre-existing low albedo mafic deposit. These hidden mafic may be sampled by dark halo impact craters, which penetrate the regional ejecta to excavate mafic material, emplacing it in a halo around the crater.

Table 4.3. List of suspected cryptomare deposits and their characteristics.

Cryptomare Deposit	Area (×10⁵ km²)	Thickness (m)	Volume (×10⁵ km³)	Age (Ga)
Schiller-Schickard	3.6	400	1.5	3.8 - 3.92
Humorum	2.3	450	1.0	3.8 - 3.92
Procellarum	0.35	400	0.15	3.8 - 3.92
Australe	6.4	500	3.2	3.8 - 3.92
				3.92 - 4.0
Balmer	1.7	400	0.68	3.87 - 4.1
East Frigoris	3.7	400	1.5	3.8 - 3.92
Langemak	4.0	400	1.6	3.92 - 4.1
Smythii	2.0	400	0.80	3.8 - 3.84
Taruntius	0.87	600	0.52	~3.8
South Pole-Aitken	2.5	400	1.0	3.63 - 3.9
				3.9 - 4.1
Mendel-Rydberg	0.31	400	0.12	3.8 - 3.92
Marginis	0.29	400	0.12	3.8 - 4.1
	0.67	400	0.27	3.87 - 4.1
Lomonosov-Flemming	0.52	400	0.21	3.8 - 4.0
Hercules	0.62	400	0.25	>3.2
Tsiolkovsky	0.50	400	0.20	3.2 - 4.0
Korolev	0.95	400	0.38	3.4 - 4.04
Humboldtianum	0.21	400	0.084	3.8 - 3.87
	1.12	400	0.45	3.87- 3.92
Maurolycus	1.6	400	0.64	3.87- 3.92
Milne	0.92	500	0.46	3.92 - 4.0
Mendeleev	0.53	400	0.21	3.2 - 3.9
Totals	35.7		15.3	

that obscure even older mafic units may lie beneath the circular maria (Antonenko 1999). The totals given in Table 4.3 are therefore considered minimum estimates for the area and volume of all hidden mafic deposits on the Moon.

A comparison between these hidden mare deposits and known maria can be made on the basis of such totals. Maria have a total surface area of approximately 6.3×10^6 km² (Head 1975b) and a total volume of approximately 5×10^6 km³ (Hörz 1978; Budney and Lucey 1998; Antonenko 1999). Hidden mafic deposits may, therefore, cover as much of the lunar surface as half of the known maria and have a volume that is equal to 30% of the known mare volume (Antonenko 1999). It is clear that hidden mafic deposits represent a significant contribution to global lunar volcanism.

The age ranges given in Table 4.3 can be used to assess the volcanic flux of these mafic materials. Considering only units that are unambiguously older than 3.8 Ga, the data indicate that volcanic flux prior to 3.8 Ga may have been as high as 60% of the flux after 3.8 Ga (Antonenko 1999). It is possible that early volcanic fluxes were comparable to those in the late Imbrian period when volcanism is believed to have peaked (Head and Wilson 1992).

The specific nature of these hidden mafic deposits is not well understood. Owing to their proximity to mare deposits, they are generally assumed to represent an earlier phase of mare volcanism. Because of their presence close to the surface, they are generally believed to be extrusive. The likelihood of shallow intrusive sills is often ruled out because the current models of mare emplacement do not allow for them (Head and Wilson 1992). However, although no clasts of intrusive mare basalts have been unambiguously identified in the returned lunar samples (e.g., Heiken et al. 1991), there appear to be examples of intrusive equivalents of mare basalts among the lunar meteorites (e.g., NWA 773, Asuka 881757, and clasts in QUE 94281 and EET 87521 and 96008; see Chapter 2). Such deposits could represent any of the more mafic materials that have been identified in the sample record, not just basalts. Some hidden mafic deposits could, for example, represent shallow plutons of Mg-suite rocks or KREEP basalt flows.

4.8.3. KREEP basaltic volcanism. Two related instances of potential KREEP volcanism have been proposed. The Apennine Bench Formation near the Apollo 15 landing site has long been suspected of being composed of extrusive KREEP basalts (Hawke and Head 1978; Spudis 1978; Metzger et al. 1979; Spudis and Hawke 1986). This formation may be underlain by an intrusive equivalent of the KREEP basalts (Hawke and Head 1978; Blewett and Hawke 2001). The Apennine Bench Formation was initially mapped as a light-plains deposit (Wilhelms and McCauley 1971). However, photogeologic studies showed evidence of volcanic features, suggesting that extrusive volcanism produced the deposit (Hawke and Head 1978; Spudis 1978). Furthermore, Apollo gamma-ray spectrometer data (Davis 1980; Metzger et al. 1977) indicated that the FeO, TiO_2, and Th values for the Apennine Bench were consistent with the composition of Apollo 15 KREEP basalt samples (Hawke and Head 1978; Spudis 1978; Metzger et al. 1979; Spudis and Hawke 1986). FeO and TiO_2 abundances derived from Clementine remotely sensed data (Lucey et al. 2000) and Lunar Prospector Th abundances (Lawrence et al. 2000) are in good agreement with those derived from the Apollo data, and are consistent with Apollo 15 KREEP basalts (Blewett and Hawke 2001). Telescopic hyperspectral reflectance spectra indicate the presence of low-Ca pyroxenes such as those found in KREEP basalts (Spudis et al. 1988; Blewett et al. 2001).

Adjacent to the Apennine Bench, the craters Aristillus and Autolycus have excavated KREEP-like material. Clementine data shows the ejecta of Aristillus to have FeO values that are lower than the nearby mare basalts, while TiO_2 values are greater than the surrounding mare, highlands and even the Apennine Bench (Blewett and Hawke 2001). Lunar Prospector and Apollo Th data from Aristillus also show elevated Th concentrations (Metzger et al. 1979; Gillis and Jolliff 1999). Similar FeO, TiO_2, and Th concentrations are noted in the ejecta of Autolycus crater (Metzger et al. 1979; Blewett and Hawke 2001). Such evidence suggests that these two craters excavated KREEP-rich material (Metzger et al. 1979; Gillis and Jolliff 1999). Considering the depth of excavation and the diameter of the crater, Aristillus ejecta originate from a depth of <5 km, which may be too deep to represent extrusive volcanism that was buried by later events, but which could be Imbrium impact melt. Blewett and Hawke (2001) suggested that the Aristillus and Autolycus craters excavated a KREEP plutonic complex. This interpretation of the provenance of Aristillus and Autolycus ejecta supports an earlier stratigraphic model suggested for the region (Hawke and Head 1978). In that model, the lowest unit consists of the "highland" basement, which was impacted to form the Imbrium basin. Thorium-rich KREEP plutonic materials and/or KREEP basalts were emplaced following basin formation, and then covered by Imbrium mare basalts. Lastly, the Aristillus and Autolycus impacts penetrated into the KREEP-rich substrate. Detailed studies of stratigraphic relations exposed by these and other craters are needed to confirm the nature of the proposed KREEP magmatic units and to determine their extents and volumes.

4.9. Models for early lunar volcanism

4.9.1. KREEP basalts. The early partial melting models for KREEP basalts were inadequate because they did not explain the constancy of relative trace-element abundances and ratios in KREEP. Therefore, most of the current models for the petrogenesis of KREEP basalts are tied to the role played by the residual dregs ("urKREEP") of the LMO during lunar magmatism. Two general types of models are currently considered: assimilation of urKREEP by magmas undergoing fractional crystallization and generation of a melt from an urKREEP-enriched cumulate source.

Using trace-element data generated in the early 1980s and assimilation models that had been previously proposed for the Mg-suite plutonic rocks, Warren (1988) modeled the chemical compositions of Apollo 15 and 17 KREEP basalts as a product of AFC (assimilation and fractional crystallization) of urKREEP by Mg-rich parental magmas. Snyder et al. (1995) quantitatively modeled the AFC generation of parental KREEP basalts using a model and end-member components similar to that of Warren (1988).

Several recent models have suggested that perhaps the KREEP basalts and the high-Al basalts were products of melting of a hybrid lunar mantle consisting of mixtures of evolved (KREEP) and early LMO cumulates. This type of model is partially based on the LMO cumulate-overturn models of Ringwood and Kesson (1976), Spera (1992), and Hess and Parmentier (1995), and perhaps demonstrated by chemical and experimental data for the lunar pyroclastic glasses (Hughes et al. 1988, 1989, 1990; Shearer et al. 1989, 1990, 1991; Shearer and Papike 1993). Hess and Parmentier (1995), Hess (1994) and Shearer and Papike (1999) suggested the possibility that the KREEP component was transported into the deep (and Mg-rich) lunar mantle where it was incorporated into the source for magmas parental to the KREEP basalts and Mg-suite magmas. An alternative mechanism to produce a hybrid source and to initiate melting was proposed by Nyquist and Shih (1992) who suggested that mixing and melting was the product of large impacts. Clearly, the interrelationships among KREEP volcanism, highland plutonism, mare volcanism, and basin formation are complex. Are the KREEP basalts the volcanic equivalents of both the highland Mg-suite and the highland alkali suite? Is there a transition between mare basalts and the KREEP basalts?

As discussed earlier, Warren (1988) and Snyder et al. (1995) suggested that primitive KREEP basaltic magmas were parental to the plutonic cumulates making up the Mg- and alkali rock suites. This is consistent with the fractional crystallization modeling of Snyder et al. (1995) and the Nd isotopic data compiled by Nyquist and Shih (1992). There is a significant difference in ages between Mg-suite plutonic rocks (4.14-4.46 Ga) and KREEP basalts (2.9-4.08 Ga) that suggests they represent different episodes of lunar magmatism. However, this difference could be the result of sampling bias attributed to changes in the meteorite impact flux that would change with time both the efficiency of sampling of the lower crust and the preservation of old KREEP basalt flows. On the basis of lunar meteorite NWA 773 there appears to be an age overlap between the production of KREEP basalts and that of the most of the mare-filling basalts. The lower initial ε_{Nd} for this one example of rather young KREEP basaltic magmatism suggests that a substantial KREEP component was in the mantle source and that KREEP in the lunar mantle played an important part in extending the duration of lunar magmatism in the Procellarum basin on the near side of the Moon (Jolliff et al. 2003; Borg et al. 2004).

4.9.2. High-Al basalts. Models for the generation of the high-Al basalts and the VHK basalts range from melting of hybridized and nonhybridized mantle sources (Reid and Jakes 1974; Taylor and Jakes 1974; Binder 1976; Kurat et al. 1976; Ma et al. 1979; Dickinson et al. 1989; Hughes et al. 1990; Shervais and Vetter 1990) to assimilation of an evolved LMO component (KREEP, lunar granite rock types) by mantle-derived basaltic magmas (Dickinson et al. 1985; Shervais et al. 1985a; Neal et al. 1988a,b, 1989a,b) to impact melting of crustal lithologies (Snyder et al. 2000). Many of these various models are similar to those proposed for the KREEP basalts, but involve different sources and assimilants.

Are the basalt represented by these fragments a product of mantle melting or impact melting? Snyder et al. (2000) suggested that the mixing of components to produce the high-Al basalts was a surface process resulting from impact. According to this hypothesis, the high-Al basalts are not true basalts but are instead impact-melt rocks. The low siderophile contents (Dickinson et al. 1985; Shervais et al. 1985a), however, are not consistent with addition of a meteorite component to the target or melt by the potential impactor. Moreover, the narrow range of major-element bulk compositions reinforces constraints provided by phase equilibria (e.g., Neal et al. 1989b). Impact melts might be expected to show a wider range of compositions that do not respect phase boundaries, unless the impact melt has subsequently differentiated. Irving (1975) used the cotectic composition of KREEP basalts to argue for a magmatic origin. The major-element compositions of the high-Al basalts overlap those of low-Ti mare basalts, except that the CaO/Al_2O_3 weight ratios (0.8-0.9) of the former are slightly less than those of the latter ($CaO/Al_2O_3{\sim}1$). However, given the small samples available, slight discrepancies such as these might be expected. The convex REE patterns and the initial ε_{Nd} vs. age plots are both consistent with magmas having been derived from cumulate mantle sources. We tentatively agree that these basalts are related to mare volcanism and have significant petrogenetic implications.

Early studies, based primarily on high-Al basalts returned by the Luna 16 mission, placed their mantle source just below the crust at depths of 40 to 100 km (Reid and Jakes 1974; Taylor and Jakes 1974; Binder 1976; Kurat et al. 1976). In these studies, the source region was thought to be a late-stage LMO cumulate that was rich in clinopyroxene and plagioclase (Reid and Jakes 1974; Taylor and Jakes 1974; Binder 1976; Ma et al. 1979) and that was LREE-enriched (Ma et al. 1979). Ridley et al. (1975) used mineralogical and trace-element justifications to argue that plagioclase was not a required phase in the source and that a slight increase in silica activity in an olivine + pyroxene source would result in a stronger partitioning of Al into the melt. Alternatively, he suggested that the high-Al basalts could simply be residual liquids. Clearly, some of the younger and first-discovered plagioclase-rich basalts (e.g., the Apollo 12 feldspathic basalt, evolved Luna 24 VLT basalts) may have been produced by this latter process.

The wide range in incompatible trace-element abundances of high-Al basalt clasts from the Apollo 14 site was shown to be inconsistent with these models (Shervais et al. 1985a; Dickinson et al. 1985, 1989; Neal et al. 1988a,b, 1989a,b; Hughes et al. 1990). Although this variation could be a function of different amounts of mesostasis in the individual small fragments, ion microprobe data imply that the variation in the fragments is related to significant differences in melt composition. This variability primarily in incompatible trace elements was ascribed to different degrees of partial melting and KREEP assimilation (Dickinson et al. 1985; Shervais et al. 1985; Neal et al. 1988a,b, 1989a,b; Kramer and Neal 2003). The best-fit model by Neal et al. (1988, 1989a,b) and Neal and Taylor (1990) employed cyclical assimilation of KREEP by a LREE-depleted olivine basalt. They concluded that assimilation of up to 15% KREEP and 70% fractional crystallization would produce the array of compositions observed in a suite of high-Al basalts from the Apollo 14 site. Based upon Sr and Nd isotopic data first reported by Dasch et al. (1987), Nyquist and Shih (1989) modified this model by suggesting that there were multiple batches of olivine basaltic magmas produced by different degrees of partial melting of at least two mantle sources. They also concluded that the parental olivine basalt assimilated KREEP prior to extensive fractional crystallization. The latter is consistent with the ion-microprobe data of Hagerty et al. (2001-2003) that show that the cores of olivine in several samples are relatively enriched in incompatible elements such as Kramer and Neal (2003) concluded, on the basis of trace-element data, that three groups of Apollo 14 high-Al basalts existed representing different degrees of partial melting of a common source at three separate times (3.95, 4.1, 4.3 Ga). Each magma experienced variable degrees of KREEP assimilation and fractional crystallization.

Although assimilation-fractional crystallization (AFC) models involving basaltic magmas and KREEP are consistent with major- and trace-element data, they are not entirely compatible with the isotopic data. Nyquist and Shih (1992) interpreted the Sr and Nd isotopic data as follows: (1) Source regions for the high-Al basalts with low incompatible-element concentrations were established early in lunar evolution (LMO cumulates) and had depleted LREE abundance patterns. (2) Coincident with the onset of earliest episodes of high-Al magmatism, these sources were contaminated with REE-rich, LREE-enriched magmas having

^{147}Sm/^{144}Nd similar to that of KREEP and the Mg-suite crustal rocks. (3) Radiogenic growth continued in these source regions during episodes of high-Al magmatism.

As an alternative to AFC, several models have been suggested to account for the incorporation of a highly evolved component into the source for the high-Al basalts. Hughes et al. (1990) suggested that the incompatible trace-element variability could be attributed to partial melting of a hybridized source consisting of a mixture of early- and late-stage LMO cumulates. This model appears to agree with the Ge abundances of the high-Al basalts (Dickinson et al. 1989) and variations in initial isotopic compositions (Nyquist and Shih 1992). Whereas Hughes et al. (1990) suggested that the mechanism for mantle mixing and source hybridization was the gravitational destabilization of the cumulate pile, Nyquist and Shih (1992) suggested that the mechanism for mixing was the impact of a large bolide. Nyquist and Shih (1992) based their interpretation on the seemingly cyclical and localized nature of high-Al basaltic magmatism (Dasch et al. 1987; Neal and Taylor 1990; Nyquist and Shih 1992). These two cumulate mixing models are not only different in the mechanism of mixing, they also differ with respect to the depth of the mantle cumulate source and the area covered by high-Al basalt flows. The gravitational destabilization of the cumulate pile implies a deep (>400 km) mantle source for the high-Al magmas, whereas impact mixing implies a shallow source (<100 km). The two models should also differ with respect to timing of mixing. Gravitational destabilization of the cumulate should be an older event (>4.3 Ga) than impact mixing (<4.3 Ga). Approaches using remotely sensed compositional constraints may help to distinguish high-Al basalts from mixtures of mare basalt and feldspathic highlands, and early results suggest a more widespread distribution of high-Al basalts (Kramer et al. 2004 a,b).

The K-rich basalts with their elevated K, Ba, and Rb concentrations require a complex petrogenesis because simple partial-melting models are difficult to reconcile with the phase equilibria; primitive partial melts must be in equilibrium with plagioclase but high pressure experiments on compositions similar to K-rich basalts have only olivine and orthopyroxene on the liquidus (Walker et al. 1972). AFC models have been proposed for the origin for the K-rich basalts; however, rather than assimilating a KREEP component, their petrogenesis may have involved the assimilation of a very small amount (1-3%) of a "granitic component" (Shervais et al. 1985b; Shih et al. 1986, 1987; Neal et al. 1988b, 1989b). Initial Sr and Nd isotopic compositions of these basalts lie on the growth curves defined by the high-Al basalts indicating that Sr and Nd were not assimilated significantly from a granitic component. These results are not unique since our granite sample data base is sparse, although such a process does account for the elevated K, Rb, and Ba concentrations and elevated K\La.

One important conclusion inferred from these older basalts is that the crystallization of the LMO, convective-overturn of the LMO cumulate pile, and reheating/melting of cumulate sources were achieved by 4.3 Ga. These results are consistent with constraints provided by the short-lived isotopes (e.g., Shearer and Newsom 2000; Righter and Shearer 2003) and the physics of the cumulate overturn model (e.g., Hess and Parmentier 1999).

5. LUNAR (MARE) MAGMATISM 3.85 TO ~1.0 Ga

5.1. Introduction

Successful models for the internal evolution of the Moon must consider the volume, distribution, timing, composition and, ultimately the petrogenesis of mare basaltic volcanism. Indeed, given the paucity of geophysical data, the evolution of the lunar interior can be gleaned only by unraveling the petrogenesis of the mare basalts. Mare basalts as represented by the dark smooth areas on the Moon, cover roughly 17% of the surface but are primarily relegated to fills of multiringed impact basins and irregular depressions (Head 1976). The basalts form low albedo, generally smooth surfaces with characteristics that give evidence for highly effusive, flood-type eruptions (Head and Wilson 1992). Since these flood basalts completely fill most of the near-side basins, the early stages of basin-filling volcanism are largely erased. Partly filled basins, however, such as Orientale and Smythii afford a glimpse at the beginning stages of basin volcanism (Gillis and Spudis 2000). Even earlier volcanic episodes appear as dark-mantling deposits or "dark haloes" around impact craters that excavated pre-basin deposits (Head and Wilson 1992), but their volumes and links to mare volcanism are not yet fully documented (Antonenko 1999). Thickness of the mare flood basalts are estimated to vary widely, ranging from a few hundred meters in Smythii (Gillis and Spudis 2000) to 4.5 km in the central portions of multiringed basins (Head and Wilson 1992; see also Chapter 3). In addition to the eruption of large volumes of lava, variously colored volcanic glass beads (or picritic glasses) with mare basalt affinities were produced by fire-fountaining (Head and Wilson 1992 and references therein). These glass beads provide some of the most important constraints to models of mare-basalt petrogenesis.

Mare volcanism on the lunar far side is limited to a few volcanic deposits in comparatively small and young craters and lacks the extensive basaltic plains so characteristic of the near-side crust. This asymmetry in mare volcanism has generally been attributed to the thicker anorthosite crust on the lunar far side, where it is roughly 70 km thick (Zuber et al. 1994, 1999), on average and may act as a low density barrier preventing the eruption of the denser mare magmas. However, the largest and possibly the most ancient lunar basin, the South Pole-Aitken Basin, occurs on the lunar far side. Despite its great size, it contains only sporadic patches of mare volcanic deposits (Pieters et al. 2001; Petro and Pieters 2004). The basin is floored by lunar crust that appears to be no thicker than crust flooring the multiringed, basalt-filled craters on the lunar near side (Wieczorek and Phillips 1999). It is clear that the creation of a major multiringed basin over thinned anorthosite crust is not directly correlated to the appearance of large volumes of mare volcanism.

Radiometric crystallization ages of common mare-basalt compositions range from 3.6-3.9 Ga for the high-Ti basalts at the Apollo 11 and Apollo 17 sites to 3.16-3.4 Ga for low and very-low-Ti basalts at the Apollo 12, Apollo 15 and Lunar 24 sites (Nyquist and Shih 1992; Snyder et al. 2000). Hiesinger et al. (2000), using crater size frequency ages on well defined volcanic units, found that mare volcanism on the lunar near side was active from about 3.9-4.0 Ga to about 2.0 Ga, a span of almost 2 Ga. Younger patches of mare volcanism apparently are as young as 1.2 Ga (Hiesinger et al. 2003). Most magma production, however, occurred in the late Imbrian at about 3.6-3.8 Ga (Fig. 4.34). A second weaker pulse of volcanism between 3.3-3.5 Ga is observed in the younger multiring basins. No single basin records 2.0 Ga years of volcanism (Hiesinger et al. 2000, 2003), and the older basins such as Tranquillitatis contain only a "single" pulse of volcanism between 3.3-3.9 Ga. If all the volcanic effusive activity is considered together, the 3.5-3.8 Ga period marks the peak in volcanic output. Mare volcanism on the lunar far side, while comparatively sparse, was also a long-lived process (Pieters et al. 2001).

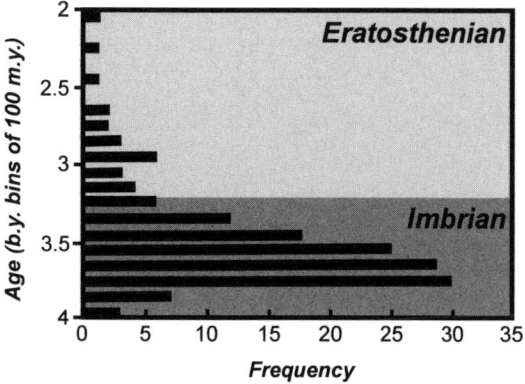

Figure 4.34. Histogram showing model ages for mare basalts on the lunar nearside (modified after Hiesinger et al. 2000).

5.2. Composition of the mare basalts and source mineralogy

Mare basalts and picritic glasses are unique to the Moon; neither the Earth, Mars, or the basaltic meteorites have basalts that even approach the major-element composition of some of the lunar basalts. The major-element composition of mare volcanic rocks have been summarized in numerous publications (e.g., Papike et al. 1976, 1997, 1998; BVSP 1981; Delano 1986; Neal and Taylor 1992; Shearer and Papike 1993, 1999). Our aim here is to show how the composition of the mare volcanics require that the mantle source is not part of the primitive Moon but instead is a highly differentiated product of the LMO.

Relative to terrestrial basalts, the major-element compositions of the mare volcanic suite have widely varying and extreme compositions (Table 4.4). The mare volcanics and particularly the picritic glasses have significantly lower CaO, Al_2O_3, and Na_2O contents but higher FeO and MgO contents than terrestrial basalts from the ocean floor and ocean islands. The low Na_2O contents probably reflect, in part, the paucity of Na_2O and other volatile elements on the Moon (Taylor 1982), but the low CaO and Al_2O_3 contents record an unusual mantle source and/or petrogenesis. The low CaO and Al_2O_3 contents are not characteristics of magmas in equilibrium with lherzolite mantle. The bulk compositions of picritic glasses are more similar to the compositions of komatiites than basalts; the mare "basalts" are not basalts but products of ultramafic magmas. We will, nevertheless, continue the tradition and use mare "basalts" to characterize both crystalline basalts and the volcanic picritic glasses. The reader should be cautioned not to jump to conclusions; unlike the terrestrial experience, ultramafic magmas on the Moon do not necessarily imply that such magmas are products of large degrees of melting.

Another major difference in bulk composition between terrestrial and lunar mafic/ultramafic magmas are revealed in the very high FeO contents of the mare basalts and picritic glasses. The FeO contents of the picritic glasses are roughly twice those of primitive ocean-floor basalts (Delano 1986; Hess 1992). Only highly differentiated ocean-floor basalts and rare Archean basalts or komatiites attain FeO values of the high-Ti picritic glasses. The MgO contents of the lunar picritic glasses are more typical of terrestrial picritic basalts or komatiites but given their high FeO values, the Mg' values of the lunar mare suite are significantly lower. Komatiite/picrite terrestrial magmas, for example, coexist with olivine with Mg' values in the low 90's (Hess 1989, 1992) but the most magnesian picritic glasses crystallize olivine with Mg' values in the low 80's (Delano 1990). These differences are very significant and speak to material differences in the compositions of the lunar and terrestrial mantles.

Table 4.4. Compositions of (1) very low-Ti picrite glass (2) high-Ti picrite glass (Shearer and Papike 1993) (3) ocean floor basalt (4) ocean island basalt and (5) komatiite (Hess 1989).

	Picrite Glasses		Terrestrial Magmas		
	1	2	3	4	5
SiO_2	47.6	34.1	49.7	48.8	47.3
TiO_2	0.3	17.0	0.7	2.5	0.4
Al_2O_3	7.8	3.8	16.5	11.9	7.8
Cr_2O_3	0.6	0.7	0.1	0.1	0.4
FeO	16.3	23.5	8.3	11.4	10.8
MnO	0.3	0.4	0.2	0.2	0.2
MgO	18.6	13.6	9.0	9.7	25.8
CaO	8.0	6.4	13.8	10.5	7.6
Na_2O	0.2	0.5	1.9	2.0	—
K_2O	0.0	0.2	0.1	0.5	—

The range of TiO_2 contents in the picritic glasses is most noteworthy, varying from very low values of 0.22 wt% in Apollo 15 green glass to 17.0 wt% TiO_2 in Apollo 14 black glass (Shearer and Papike 1993). Terrestrial basalts, in comparison, typically have between about 0.5% and 6.0% TiO_2. Primitive MORB, for example, have TiO_2 contents between 0.5-1.2% TiO_2 (Presnall and Hoover 1987). Only highly magnesian alkaline lavas, as for example the meimechites, which, based on their elevated incompatible trace element content, are products of small degrees melting of a primitive or an enriched mantle source, have more extreme TiO_2 values that reach only 4-6% (Arndt et al. 1995). Indeed, it is easy to show as is done below that even very small degrees of melting of a typical primitive terrestrial mantle cannot give rise to the very high-Ti contents of the most enriched lunar picrite mare glasses.

The bulk partition coefficient for TiO_2 between lherzolite mineral assemblages and basalt is strongly affected by the high-Ca clinopyroxene and garnet content. The high-Ca clinopyroxene-basalt TiO_2 partition coefficient varies from about 0.3-0.5 (Skulski et al. 1994; Jones 1995). The partition coefficients for garnet and orthopyroxene are estimated from the TiO_2 partition coefficients between these phases and coexisting high-Ca clinopyroxene (e.g., Seitz et al. 1999). While there is considerable scatter, the garnet-basalt partition coefficient is roughly twice that of high-Ca clinopyroxene, whereas that of orthopyroxene is roughly half that of clinopyroxene. In support of these estimates, experiments done by Xirouchakis et al. (2001) indicate a $D(TiO_2)$ ~0.1-0.2 for orthopyroxene. Using $D = 0$ for olivine, a typical fertile spinel lherzolite would have a bulk $D > 0.1$. A depleted harzburgite, in contrast, would have $D \leq 0.1$.

Terrestrial fertile mantle contains roughly 0.2 wt% TiO_2, depleted mantle significantly less (Taylor 1980). Basalts derived by small degrees of melting of fertile lherzolite cannot have more than 2 wt% TiO_2 assuming $D \geq 0.1$ (the maximum enrichment of TiO_2 in the melt is given by $1/D \approx 10$). Depleted mantle approaching harzburgite in composition has a smaller bulk partition coefficient for TiO_2 but also a much smaller TiO_2 content. It follows that terrestrial basalts are constrained to have less than about 2 wt% TiO_2; those highly alkaline primitive basalts with higher TiO_2 concentrations probably reflect the influence of CO_2 on the melting process. It comes as no surprise that the TiO_2 enriched mare basalts and glasses cannot be derived from primitive mantle on a volatile-depleted Moon, but must originate from mantle that is both differentiated and TiO_2 enriched.

The sources for the very low-Ti mare basalts (VLT) and glasses are also unlikely to be primitive mantle. Consider, as an illustrative example, the petrogenesis of a VLT glass with 0.4 wt% TiO_2 (Table 4.5). Assuming a mantle with 0.2 wt% TiO_2, such glasses can be generated by roughly 50% melting of fertile terrestrial lherzolite. At this extent of melting all the accessory phases and high-Ca clinopyroxene are consumed leaving harzburgite in the mantle. Indeed, even 20-25% melting is sufficient to create a harzburgite residue (Hess 1992). Under these conditions, both CaO and Al_2O_3 are incompatible and their contents should be enriched, roughly doubled, in the basalt melt relative to the contents in the mantle. At first glance, these expectations are satisfied as the VLT glass has roughly 8 wt% each of Al_2O_3 and CaO or double that in fertile mantle (Table 4.5). However, for this explanation to be generally accepted, the TiO_2 contents of other VLT glasses should correlate with their Al_2O_3 and CaO contents. Such correlations do not exist. Whereas VLT glasses have TiO_2 contents that range from 0.22 to 1.0 wt%, the corresponding CaO and Al_2O_3 each are within about 10% from their mean values (Delano 1986; Shearer and Papike 1993). The highest TiO_2 contents, moreover, are not necessarily associated with the highest CaO and Al_2O_3 contents. These arguments show that primitive lherzolite mantle is not an appropriate source for VLT basalts and glasses (See also Section 4.3.1). Furthermore, large degrees of melting of a peridotite source are not a requisite to account for the low-TiO_2 contents. A more reasonable hypothesis is that the source for the VLT mare magmas is differentiated and depleted in the incompatible oxides TiO_2, Al_2O_3 and CaO.

Table 4.5. Selected analyses of picritic glasses from the Apollo missions.

	1	2	3	4	5	6	7	8	9	10	11	12
SiO_2	48.68	44.10	44.41	46.53	44.5	46.3	45.15	41.92	39.8	39.28	36.69	33.55
TiO_2	0.22	0.40	0.63	0.78	0.47	0.55	1.00	4.82	6.84	9.07	13.69	16.75
Al_2O_3	7.06	7.79	7.23	8.30	7.65	9.82	6.87	6.52	8.50	7.55	6.87	4.03
Cr_2O_3	0.60	0.40	0.50	0.37	0.45	0.46	0.64	0.40	0.54	0.84	0.80	0.97
FeO	16.30	21.82	19.27	18.54	21.3	19.5	23.00	24.10	21.7	23.35	20.59	22.43
MnO	0.34	0.25	0.28	0.32	0.32	0.30	0.33	0.32	0.36	0.27	0.23	0.26
MgO	18.66	17.00	18.75	16.42	17.6	14.1	15.68	13.74	11.8	11.90	11.84	15.18
CaO	8.43	8.55	8.05	8.48	8.37	9.41	7.93	7.90	8.60	8.02	7.68	6.30
Na_2O	0.17	0.10	0.25	0.14	0.28	0.16	0.21	0.68	0.31	0.35	0.69	0.20
KO_2	0.00	0.00	n.a.	0.08	0.00	0.01	0.00	0.14	0.00	0.03	n.a.	0.08
Total	100.5	100.4	99.4	100.0	100.9	100.6	100.8	100.5	98.5	100.5	99.1	99.8

1=A15 Green C, 2=A16 Green B, 3=A14 Green B Type, 4=A14VLT, 5=Apollo11 Green, 6=A17 VLT, 7=A14 Green A, 8=A14 Yellow, 9=A17 Yellow, 10=A17 Orange 2, 11=A15 Red, 12=A 14 Red-Black (Shearer and Papike 1993) Glasses selected to show the range in glass population and most common composition.

With the exception of one or two VLT picritic glass compositions, it is noteworthy that the contents of both Al_2O_3 and CaO in VLT glasses appear to be buffered to near constant values (Table 4.5). What these results suggest is that the Al_2O_3 and CaO are not strongly incompatible and/or the extent of melting is roughly constant. The mantle phases that are capable of buffering the CaO and Al_2O_3 contents are plagioclase, high-Ca clinopyroxene, garnet and orthopyroxene. The first two phases listed are poor candidates since the CaO and Al_2O_3 contents of the VLT picritic glasses are too low—the VLT melts are undersaturated with respect to plagioclase and high-Ca clinopyroxene at all lunar pressures. The same is true for garnet except at depths greater than 500 km. In addition, the HREE contents of most of the picritic glasses are not consistent with significant quantities of garnet occurring as a residual phase in the source region (Papike et al. 1997). Of the possible phases listed above, only orthopyroxene has the characteristics of the desired residual phase.

The VLT picritic glasses have orthopyroxene as a liquidus phase only at high pressures (Chen et al. 1982; Chen and Lindsley 1983; Elkins-Tanton et al. 2000) so that the source lies deep within the Moon. The orthopyroxene-basalt partition coefficient for Al is pressure dependent; this dependency is approximated by $D_{Al} = 0.11 + 0.10 \times GPa$. Assuming that the pressures are 2-3 GPa it follows that $D_{Al} = 0.3$-0.4. The bulk partition coefficient, of course, is a function of the olivine content in the source; assuming that the sources for mare basalts are cumulates to the LMO, the source could range from an olivine-bearing orthopyroxenite to an orthopyroxene-bearing dunite (e.g., Hughes et al. 1989; 1990; Delano 1980; Longhi 1982; Hess 2000 among others). Therefore, for the source to buffer the Al_2O_3 and CaO contents to approximately constant values for varying amounts of melting, an orthopyroxene-rich source is clearly favored. This conclusion follows from the fact that the contents of more compatible elements in partial melts are less sensitive to varying degrees of partial melting than are incompatible elements. Both CaO and Al_2O_3 in an olivine-rich mantle would behave much more incompatibly than in a pyroxene-rich mantle.

The FeO contents of VLT picritic glasses range from 16-22 wt% (Table 4.5). This range in FeO contents requires either that the FeO contents of the source region vary a comparable amount and/or the mineralogy of the source is dominated by orthopyroxene rather than olivine. The FeO olivine/basalt partition coefficient for VLT to high-Ti mare basalts is approximately unity (Delano 1980; Elkins-Tanton et al. 2000). It follows that partial melting of an olivine-rich source to produce mare basalts or fractionation of olivine from such basaltic magmas will leave the FeO contents of the melt largely unchanged. The FeO orthopyroxene/basalt partition coefficient is approximately 0.6-0.7 for VLT (Elkins-Tanton et al. 2000) and hi-Ti mare basalt (Delano 1980). Partial melting of an orthopyroxene-rich source from 0 to 40% is required to vary the FeO contents by the observed amounts. Based on our discussions detailing the near constancy of Al_2O_3 and CaO in VLT glasses, such a large range in the degree of melting is unlikely. It follows that much of the variation in FeO contents must be ascribed to variations of the source composition.

We conclude from this analysis that VLT basalts were derived from heterogeneous mantle dominated by orthopyroxene but with more than trace amounts of olivine. The need to include olivine in the source is argued more fully in forthcoming sections. The mantle source was depleted in CaO and Al_2O_3 relative to primitive terrestrial lherzolite; the mantle residue to the VLT cannot have contained high-Ca clinopyroxene, plagioclase, ilmenite or garnet because the melts are undersaturated with respect to these phases at lunar pressures below 3.0 GPa (Elkins-Tanton et al. 2000; Chen et al. 1982; Shearer et al. 2003; Draper et al. 2004). This conclusion does not exclude the possibility that such phases were originally in the source region albeit in small amounts and were consumed by melting or that the mantle source containing such phases was in contact with the VLT during some stage of their evolution (Longhi 1992; Elkins-Tanton et al. 2000). In addition, high pressure studies by Shearer et al. (2003) and Draper et al. (2004) hint at the possibility that garnet may be a liquidus phase at pressures above 3.0 GPa.

The high-Ti picritic glasses must also come from a similar source given that the CaO and to a lesser degree the Al_2O_3 contents are similar to those of VLT glasses. What is different, of course, are the extraordinary levels of TiO_2 and the concomitant increase in FeO contents (Table 4.5). It is apparent from the preceding discussions that the higher TiO_2 and FeO contents must in large part reflect the higher TiO_2 and FeO contents of the mantle source. The most extreme compositions as represented by the Apollo 14 black glasses contain the highest TiO_2 (17 wt%) but the lowest Al_2O_3 contents of only 3.8 wt%. These low Al_2O_3 contents are roughly half of those of the VLT picritic glasses yet the CaO contents (6.5 wt%) of the high-Ti glass are not reduced a comparable amount. Whereas the Al_2O_3 contents in high-Ti glasses vary widely from 3.8-8.5 wt%, the spread in CaO contents is much more subdued (6.3-8.2 wt%). Note that the CaO/Al_2O_3 ratio in VLT is much less variable than that of the high-Ti picritic glasses. Perhaps the explanation centers on the pressure dependence of the Al_2O_3 partition coefficient of orthopyroxene. In comparison, the partition coefficient for CaO does not have strong pressure dependency.

Picritic mare glasses are the most primitive mare liquid compositions yet returned from the Moon and are the best candidates for primary magmas (Delano 1986). Crystalline mare basalts, on the other hand, are more evolved by comparison and their compositional variations within each suite of high-Ti, low-Ti and even VLT basalts appear to have been controlled by low-P crystal fractionation of olivine and possibly other phases (BSVP 1981). It is not clear, however, whether the more primitive parental liquids are themselves products of fractional crystallization or approximate primary compositions. We tentatively make the assumption that the most primitive mare basalts are comparable to picritic glasses in petrogenetic importance. An instructive comparison between the two groups of magmas is given in Figure 4.35 where the TiO_2 and Al_2O_3 contents are plotted versus the Mg' value. According to Papike et al. (1976) the Mg' values of the most primitive high- and low-Ti basalts are roughly 0.50. Those basalts with larger values appear to contain cumulates of olivine. Certainly, the low-Ti olivine basalts with such elevated Mg' have suffered significant additions of olivine (Walker et al. 1975). If this analysis is correct, then the Mg' values of the most primitive high- and low-Ti mare basalt liquids and the picritic glasses are very similar. The same can be said of the VLT basalts and their glasses. If we now compare the major-element compositions of the picritic glasses with mare basalts with the same Mg' values, it is evident that the mare basalts have higher CaO and Al_2O_3 contents than the picritic glasses (Fig. 4.35 and Table 4.6). High-Ti picritic glasses, for example, have Al_2O_3

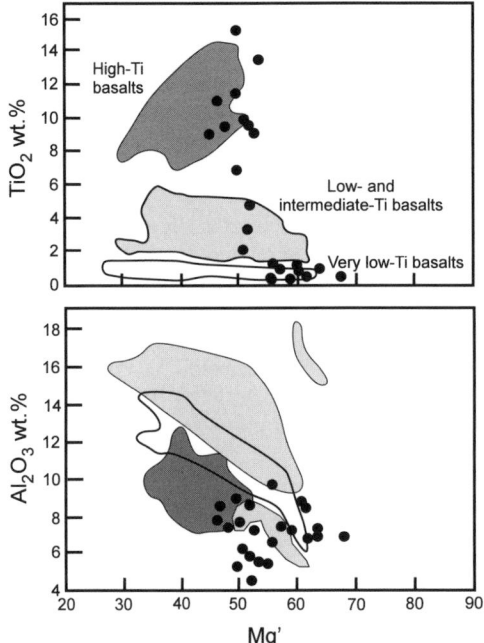

Figure 4.35. Mg' values vs. TiO$_2$ and Al$_2$O$_3$ for mare basalts (enclosed fields) and picrite glasses filled circles (modified after Shearer and Papike 1993).

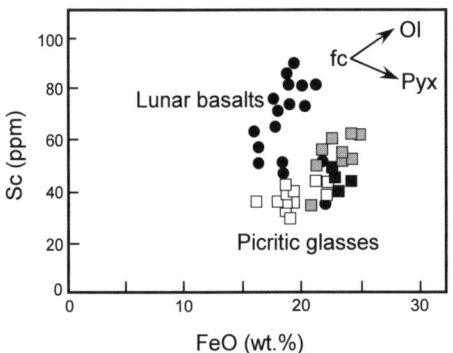

Figure 4.36. Sc vs. FeO of lunar picrite glasses (squares) and more basalts (filled circles) (from data of Shearer and Floss 2000; Delano 1990).

contents in the range from 4-8 wt% whereas the primitive high-Ti mare basalts with Mg' ≅ 0.5 have Al$_2$O$_3$ contents from 7-12 wt%. Of course, more evolved mare basalts have even higher Al$_2$O$_3$ contents because Al$_2$O$_3$ behaves incompatibly with olivine and such basalts typically are undersaturated with respect to plagioclase.

Low-pressure experimental studies show that the most primitive mare basalts and picritic glasses have only olivine (± minor chromite) as their liquidus phase (Longhi 1992). Calculated low-pressure fractionation paths for a representative set of picritic glasses, show that the differentiates of these more primitive magmas are similar to typical crystalline mare basalts (Longhi 1987). This fact suggests that picritic glasses are parental magmas to the more evolved mare basalts. What is a puzzling result for some workers is that none of the 24 picritic glass compositions listed by Delano (1986) has exactly the right composition to be a parent to one of the known varieties of mare basalts with the one exception of the Apollo 17 green glass (Longhi 1987). The simplest explanation is that the collection of mare basalts is incomplete—we simply have a sampling problem. Remote-sensing data suggest that there is indeed a continuum of mare basalts across the TiO$_2$ spectrum, and the lunar meteorites are providing new and somewhat random samples with which to test the global distribution (see Chapters 2 and 3).

Delano (1990) suggested, however, that a "sampling problem" may not be a sufficient explanation for the lack of direct-lineage relationships between the picritic glasses and mare basalts. Picritic glasses, including those analyzed subsequently by Shearer and Papike (1993), have FeO/Sc weight ratios that range from about 4000 to 7000, whereas crystalline mare basalts have ratios from about 2000 to 6000 (Fig. 4.36) (Delano 1990; Shearer and Floss 2000). Fractionation of olivine leaves the FeO contents largely unchanged and thereby decrease the FeO/Sc ratios of differentiated melts. Fractionation of pyroxene, however, increases the FeO content and decreases Sc. Delano (1990) argued, however, that crystal fractionation processes could not reconcile the two data sets.

Table 4.6. Comparison of the bulk composition of high TiO_2, low TiO_2 and VLT mare basalts (B) with corresponding glass compositions (G) (from Papike et al. 1997).

	B 10003	B 70215	G Ap 14 Orange	G Ap 17 Orange	B 12052	G Ap 15 Yellow	B 70008	B 24077	G Ap 16 Green	G Ap 14 Green
SiO_2	39.72	37.79	37.88	39.28	46.40	43.05	48.1	46.0	44.10	45.71
TiO_2	10.50	12.97	12.50	9.07	3.28	3.58	0.36	0.75	0.40	0.89
Al_2O_3	10.43	8.85	5.73	7.55	10.16	8.58	11.2	10.1	7.79	8.50
Cr_2O_3	0.25	0.41	0.70	0.84	0.52	0.52	0.60	0.42	0.40	0.50
FeO	19.80	19.66	20.83	23.35	20.15	21.22	18.2	22.4	21.82	18.90
MnO	0.30	0.27	0.31	0.27	0.27	0.52	0.26	0.26	0.25	0.20
MgO	6.69	8.44	13.90	11.90	8.22	13.06	11.0	10.5	17.00	16.16
CaO	11.13	10.74	6.94	8.02	10.80	8.35	10.2	10.8	8.55	8.74
Na_2O	0.40	0.36	0.48	0.35	0.27	0.34	0.15	0.21	0.10	0.14
K_2O	0.06	0.05	0.15	0.03	0.07	n.a.	0.01	0.016	0.00	0.02
Total	99.3	99.5	99.4	100.5	100.1	99.2	100.1	101.5	100.4	100.0

The conclusion that picritic glasses are distinct from mare basalts is supported also by the contrasting concentrations of large-ion lithophile elements in these lithologies (Shearer et al. 1994). Specifically, the Li/Be ratios in mare basalts (3-5) are consistently lower and are offset from the value of these ratios in picritic glasses (14-34) (Fig. 4.37). Since Li is moderately incompatible in olivine and pyroxene (partition coefficients are 0.15 in clinopyroxene and 0.33 in olivine) and Be is strongly incompatible ($D\sim0$ in olivine and $D\sim0.06$ in clinopyroxene), the Li/Be ratio should decrease in the cumulate pile with fractionation of the LMO. Indeed, KREEP-like components have the lowest Li/Be ratios. The sharp differences in Li/Be between mare basalts and picritic glasses probably record distinct source regions for the two groups of mare magmas. Neither partial melting nor fractional melting models can reconcile these data (Shearer et al. 1994).

5.3. Petrography

The petrographic characteristics of mare basalts have been thoroughly reviewed in BVSP (1981) , Longhi (1992), Papike et al. (1997), Chapter 3 of this book and references therein; only the briefest summary is presented here. We focus mainly on those features that are most relevant to our understanding of mare-basalt petrogenesis. Many of the textural features and especially the modes are controlled by the vagaries of cooling and crystal differentiation. In addition, many studies (e.g., Papike et al. 1976; Rhodes et al. 1976) have concluded that the wide range of modal and chemical compositions is a result of both magmatic differentiation and the small non-representative sample size of many of the basalt lithologies.

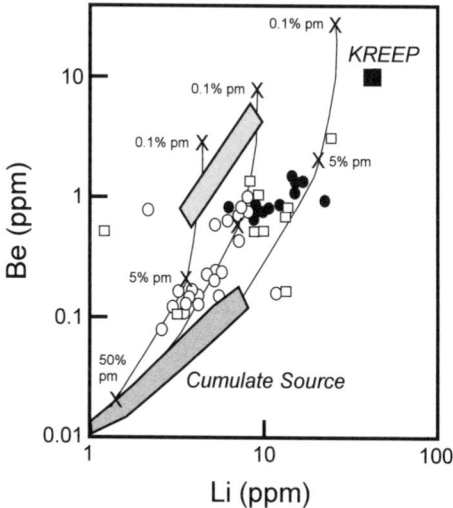

Figure 4.37. Be vs. Li for picrite glasses (Symbols), crystalline mare basalts (rectangle) and KREEP (square). Trajectories produced by melting cumulate source by various degrees (pm) (modified after Shearer et al 1994).

Table 4.7. Average modes for high-TiO$_2$ mare basalts (Apollo 11, 17) and low-TiO$_2$ mare basalts (Apollo 12, 15). (From Papike et al. 1997)

Apollo	11	17	12	15
Olivine	2.2	4.5	20	7
Pyroxene	49.0	47.0	53	63
Plagioclase	31.0	23.0	19	24
Ilmenite	14.0	24.0	7	5

The average modes of high-Ti Apollo 11 and Apollo 17 mare basalts are given in Table 4.7. The crystallization sequence inferred from petrochemical analysis and experimental investigations (Longhi 1992; Papike et al. 1997) is

$$olivine \rightarrow armalcolite \rightarrow pyroxene \rightarrow plagioclase + ilmenite$$

where pyroxene is derived from a reaction relation between olivine and melt and ilmenite from a reaction relation between armalcolite ([Fe,Mg]Ti$_2$O$_5$) and melt. The appearances of plagioclase and to a lesser extent pyroxene are strongly affected by cooling rate (see Longhi 1992) so that their position in the crystallization sequence is variable but, in general, plagioclase, augite, and pigeonite appear after armalcolite. As such, the most magnesian samples have olivine ± armalcolite as a liquidus phases whereas more evolved magmas typically have only ilmenite, pyroxene, and plagioclase as early crystallizing phases. Experiments have shown that high-Ti mare basalts are multisaturated with 2 or more phases within 25 °C of their one bar liquidus (Longhi 1992). These are not characteristics of primary or even primitive magmas unless the mantle sources are shallow and differentiated.

Low-Ti mare basalts exhibit a wide range of textures and crystal sizes ranging from vitrophyric to porphyritic gabbros (BVSP 1981). The crystallization sequence is similar to that of high-Ti mare basalts, except that ilmenite, not armalcolite, is the primary TiO$_2$ phase and ilmenite is generally the last major phase to appear (the natural consequence of lower TiO$_2$ contents). Pigeonite typically forms cores to augite mantles so that the general crystallization sequence is

$$spinel + olivine \rightarrow pigeonite \rightarrow augite \rightarrow plagioclase \rightarrow ilmenite$$

Iron metal, phosphates, cristobalite and quenched immiscible melts are all late appearing phases (Roedder and Weiblen 1970). A feature that distinguishes mare basalts from terrestrial basalts is the prominence of low-Ca clinopyroxene, mainly pigeonite, in the crystallization sequence. Pigeonite precedes or is largely coeval with augite in the crystallization sequence whereas in terrestrial MORBs augite is by far the most dominant pyroxene (Longhi 1981). Indeed, the tholeiitic silicate crystallization sequence on Earth is

$$olivine \rightarrow plagioclase \rightarrow augite \rightarrow pigeonite$$

whereas in mare basalts, and in other lunar magmas, the sequence is

$$olivine \rightarrow plagioclase \rightarrow pigeonite \rightarrow augite$$

where pigeonite and plagioclase may switch positions in mare basalts (Longhi 1992). These petrographic data alone clearly distinguish the mantle source on Earth, typically fertile lherzolite, from that of the sources for magmas on the Moon.

5.4. Trace elements

The REE patterns of mare basalts and picritic glasses provide useful but, in some cases, somewhat confusing constraints on their petrogenesis. We focus first on some features most basalts have in common. A persistent feature is that nearly all mare basalts and picritic glasses have negative europium (Eu) anomalies (Fig. 4.38) (Papike et al. 1997). The negative Eu anomaly is typically ascribed to the nature of the source region because primitive mare basalts and picritic glasses are far from being saturated with respect to plagioclase at any pressures (BVSP 1981; Longhi 1992). At the liquidus temperatures appropriate for basalts at 1 bar, plagioclase saturated melts would contain more than 15% Al_2O_3 (Hess et al. 1977); in contrast, most picritic glasses contain less than 10 wt% Al_2O_3 (Delano 1986; Table A3.12) and primitive mare basalts typically less than 12 wt% (Neal and Taylor 1992; Table A3.11). Mare basalts with greater Al_2O_3 contents are differentiated, have suffered accumulation of plagioclase, and/or reflect a characteristic of a mantle source (high-Al mare basalts). The Eu anomaly is therefore intrinsic to the source and is commonly attributed to the fractionation of plagioclase to form the floated anorthosite crust. (BVSP 1981). Pyroxene-bearing cumulates complementary to the anorthosite would bear the negative Eu signature. Small Eu anomalies recorded in some VLT glasses and basalts may not require plagioclase fractionation but reflect the very low distribution coefficient for Eu^{+2} in low-

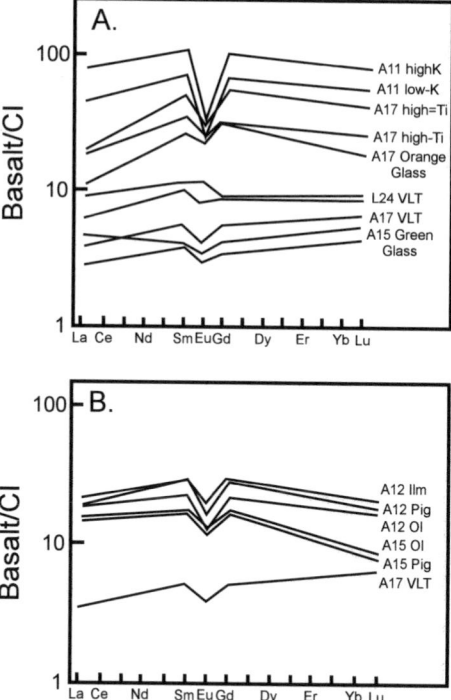

Figure 4.38. (A) Chondrite normalized REE for crystalline mare basalt and one glass (modified after Papike et al. 1999). (B) Chondrite normalized REE for Apollo 12 and Apollo 15 crystalline mare basalts and one VLT mare basalt (modified after Papike et al. 1997).

Ca pyroxene (McKay et al. 1990). It is also possible that the Eu anomalies are products of source hybridization from Eu depleted liquids. Europium anomalies in high-Ti magmas are, on average, much larger than those in VLT and low-Ti magmas (BVSP 1981; Papike et al. 1997) and can be generated only from liquids that have lost significant quantities of plagioclase. That the depth of the Eu anomaly increases with total HREE contents is consistent with a source hybridized by various amounts of trace element-rich components that formed from liquids which had lost substantial plagioclase.

Most high-Ti and low-Ti mare basalts have a distinctive chondrite-normalized REE pattern that distinguishes lunar from terrestrial basalts. The mare basalts typically possess positive LREE slopes and negative HREE slopes, together forming a convex-up REE pattern (Fig. 4.38) (Papike et al. 1997, 1998). Terrestrial ocean-floor basalts (N-MORB), which cover 70% of the Earth, contain positive LREE slopes but generally flat HREE (e.g., Hess 1989). The N-MORB are products of partial melting of somewhat depleted lherzolite, peridotites that still contain roughly 10% (±) of high-Ca clinopyroxene. The primary melts parental to N-MORB were equilibrated with olivine, orthopyroxene, high-Ca clinopyroxene, spinel and/or garnet. The REE patterns of N-MORBs appear to be largely controlled by high-Ca clinopyroxene; in fact, the shape of D_{REE} patterns of high-Ca clinopyroxene is very similar to the REE patterns of N-MORBs (Fig. 4.39). The distinctive convex-upward REE patterns of mare basalts, therefore, give us a clue to their petrogenesis.

The convex-upward REE patterns observed in mare basalts cannot be derived by partial melting of primitive lherzolite nor of highly depleted harzburgite mantle. Partial melts from fertile lherzolite have negatively sloped REE patterns if garnet is in the residue and negatively sloped LREE and flat HREE patterns if a garnet-free lherzolite assemblage is in the residue. Melting of high-Ca clinopyroxene-free harzburgite produces melts with positively sloped REE patterns, whereas if the harzburgites are high-Ca clinopyroxene-bearing, the HREE pattern will tend to be flat. A source formed of harzburgitic cumulates from a LMO with a flat chondrite-

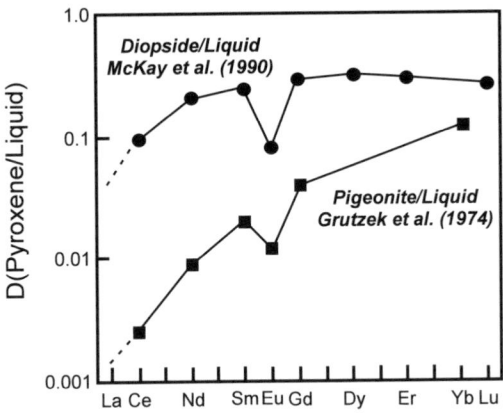

Figure 4.39. Mineral/Melt distribution coefficients (*D*) for REE in pyroxenes and basaltic melt (modified after Jones 1995).

normalized REE pattern largely mimic the REE patterns of harzburgite created by large degrees of melting of primitive mantle. High-Ca clinopyroxene-free harzburgites have positively sloped REE patterns (e.g., BVSP 1981; Hess 1989). Partial melts generated from these cumulates will also have positively sloped REE patterns for relatively small amounts of melting.

The distinctive mare-basalt convex REE pattern must, in part, reflect a fractionated non-chondritic source, most likely the mafic cumulates of the LMO (BVSP 1981). The REE content of the cumulates is controlled by the REE pattern of the magma, the nature of the crystallizing phases and the amount of trapped intercumulus liquid. Multiplication of the bulk distribution coefficient by the abundances of the REE in the LMO existing at the time the source accumulated gives the REE content of the adcumulate. The source REE is a weighted average of the REE content of the adcumulate and intercumulus liquid. Nyquist et al. (1979) completed this exercise and concluded that the negatively sloped HREE pattern was also a feature of the LMO parental to the cumulates (i.e., the LMO did not have a flat chondrite-normalized HREE pattern). The REE patterns of low-Ti mare basalts could be reproduced by small degrees of melting of an adcumulate that retained significant quantities of high-Ca clinopyroxene in the mantle residue.

Hughes et al. (1988) argued instead, that the REE patterns in low-Ti mare basalts were produced by small degrees of melting of sources in which olivine and orthopyroxene alone were residual phases. The adcumulates were crystallized from a chondritic LMO (i.e., flat REE pattern). The negatively sloped HREE pattern was created through the hybridization of the adcumulate source through addition of late-stage LMO cumulates and their intercumulus liquid. Hybridization resulted from the convective overturn of the gravitationally unstable cumulates. Similar models were devised to produce an adequate source for the high-Ti mare basalts (Hughes et al. 1989). Unfortunately, such models are not always internally consistent when major-element constraints are included in the analysis.

Hughes et al. (1989) in their attempts to model the petrogenesis of 74220 orange glass, for example, concluded that the trace elements could be duplicated by melting 4-7% of a hybridized source formed from 95.8% of cumulate olivine and 4.2% late-stage cumulates composed of 24% augite, 40% plagioclase, 35% ilmenite, plus small but critical amounts of apatite and highly evolved KREEPy intercumulus liquid. Whereas the trace elements in these melts were similar to those of the orange glass, the major elements were not. A very powerful constraint on the petrogenesis is that ilmenite, clinopyroxene, and plagioclase (or spinel) are not found on the high-pressure liquidus of the orange glass (Green et al. 1971) and must therefore be consumed and eliminated from the source. The liquid produced by 5% melting and leaving only olivine in the residue has about 16 wt% TiO_2, 11 wt% Al_2O_3 and 9 wt% CaO. In comparison, the orange glass has about 9 wt% TiO_2, 6 wt% Al_2O_3 and 7 wt% CaO. The melt produced by 7% melting is a better fit for the major elements (11 wt% TiO_2, 7 wt% Al_2O_3, and 6 wt% CaO) but is a poorer fit for the trace elements.

The chondrite-normalized REE patterns of the picritic glasses are more varied and more clearly express the effects of source hybridization and/or assimilation (Fig. 4.40) (Shearer and Papike 1993). Whereas a few of the high-Ti picritic glasses have the characteristic convex-upward mare basalt REE pattern, the Apollo 15 and 17 orange glasses for example, some REE patterns are negatively sloped. Apollo 14 and 15 high-Ti picritic glasses have such negatively sloped REE patterns. Some VLT picritic-glass REE patterns are also negatively sloped in direct contrast to the REE patterns in crystalline VLT basalts, which generally have a positive slope. These VLT glasses have LREE abundances that are more than 30× chondritic values whereas the VLT mare basalts have LREE abundances at less than 10× chondritic. The negatively sloped patterns appear to record some incorporation of KREEP-like components in the source (Shearer and Papike 1993). Apollo 16 green glass has low total REE, almost no negative Eu anomaly and a flat REE pattern. Only the VLT Apollo 14 green glass has the convex upward REE pattern.

The Ni and Co contents of VLT glasses vary within a factor of 2 from their mean values (Shearer and Papike 1993). More significantly, the Ni/Co ratio does not vary widely and, with a few exceptions, is constrained to values between 2 and 3 (Shearer et al. 1996; Table A3.12). Both Ni and Co are compatible in olivine and pyroxene; the D_{Ni} is roughly 10 and 3 and D_{Co} is 1.7 and 1.1 for olivine and orthopyroxene, respectively. Using the batch melting equation, the ratio of Ni/Co in melts derived by relatively modest degrees of melting ($\leq 20\%$) is approximately the ratio of the bulk distribution coefficients, D_{Ni}/D_{Co} multiplied by the initial Ni/Co ratio, $C°_{Ni}/C°_{Co}$ of the source.

It follows that the constrained variations in the Ni/Co content of the VLT glasses required that the product of bulk distribution coefficients and the initial Ni/Co contents of the source vary only slightly. It also requires that the most primitive VLT glasses experienced only very little olivine fractionation subsequent to their separation from the mantle. As little as 10% fractional crystallization of olivine will change the Ni/Co ratio in the melt by a factor of more than 2.

Similarly, the Ni/Co contents of the cumulates of the LMO will vary not only with the extent of fractionation but also as a function of whether olivine or orthopyroxene is the dominant liquidus phase. Forty percent fractionation of olivine, for example, lowers the Ni/Co ratio by more than an order of magnitude (Fig. 4.41) (Shearer et al. 1996). The effects of pyroxene fractionation are directly related to the degree of fractionation (F) and the Ni/Co ratio decreases with F.

The limited range of Ni/Co ratios in VLT picritic glasses sets important limits on the Ni/Co contents of the source region. Since Co and Ni are both strongly compatible, the Ni/Co ratio is a robust indicator of the cumulate mantle and is little affected by subsequent hybridization from highly evolved cumulates and KREEP-enriched liquids, provided that the latter occur in relatively small amounts. First, the sources must be distributed over a limited "stratigraphic section" of the cumulate pile (Shearer et al. 1996). The stratigraphic section represents roughly 10% of the volume of the cumulate pile. Second, this same ratio characterizes VLT glasses from Apollo 11, 14, 15 and 17 sites, implying that the source mantle for VLT picritic glasses is widely distributed across the lunar near-side mantle.

The Co and especially the Ni contents of high-Ti picritic glasses are lower than those in the VLT glasses. Nickel contents typically are, with a few exceptions, less than 50 ppm and some

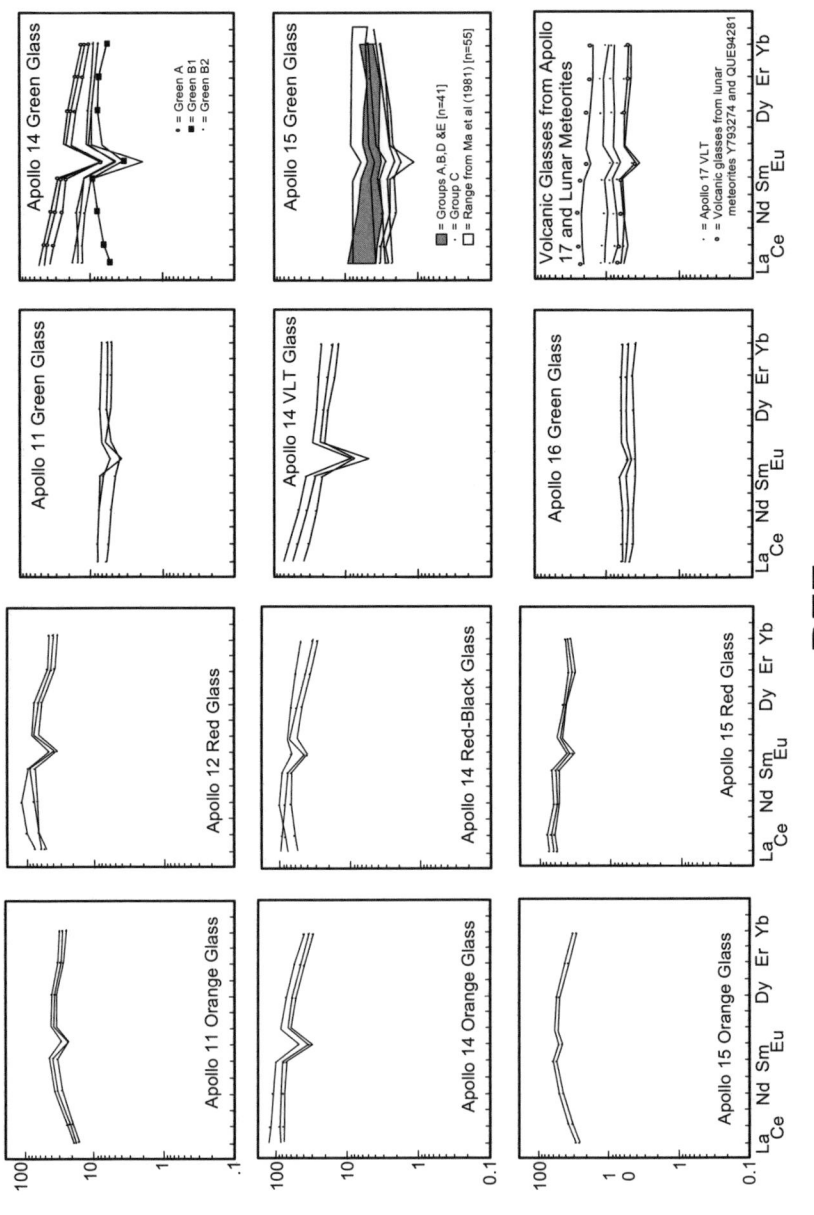

Figure 4.40. Chondrite normalized REE in lunar picrite glasses (modified after Papike et al. 1998).

glasses have Ni below detection limits. Cobalt contents are within the same range or somewhat lower than those in VLT glasses. The Ni/Co ratios are, therefore, comparatively low and range from about 0.5-1.1 (Shearer and Papike 1993; Table A3.12). Taken at face value, these data suggest that the cumulate source of high-Ti picritic glasses is located in a somewhat more evolved part of the cumulate stratigraphy. The same conclusion may apply to the crystalline mare equivalents. The Ni/Co ratios in crystalline high-Ti mare basalts range to values as low as 0.1 with Ni contents of a few ppm (Table A3.11). Both Co and Ni contents are lower in crystalline high-Ti mare basalts than in the corresponding glasses and possibly reflect the fractionation of olivine from the parent magmas.

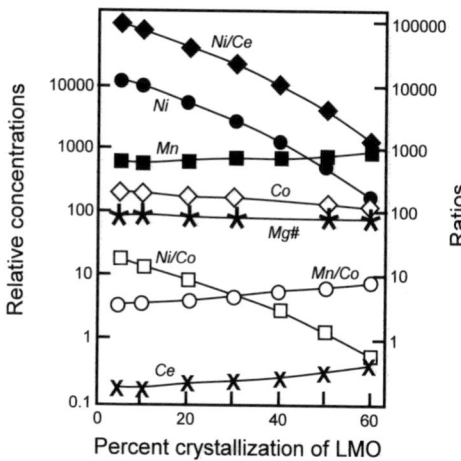

Figure 4.41. Ni, Co and Ni/Co contents of cumulates to a lunar magma ocean (LMO) (modified after Shearer et al. 1996a).

Accumulation of ilmenite in the high-Ti mare and picritic glass source(s) can be examined from the concentration of certain trace elements that partition into ilmenite over olivine and pyroxene. The transition elements Hf, Zr, Nb, and Ta tend to favor Ti-rich phases like ilmenite, armalcolite, sphene and rutile (McCallum and Charette 1978; Jones 1995; Horng and Hess 2000). Ilmenite/basalt partition coefficients for Zr ($D_{Zr} = 0.33$) and Nb ($D_{Nb} = 0.8$) are at least an order of magnitude greater than those for olivine and pyroxene (Table 4.8). During most of the crystallization of the LMO where olivine and orthopyroxene are the liquidus phases, interelement ratios between the above elements should remain constant, presumably at chondritic values. However, when ilmenite and, to a lesser extent, high-Ca pyroxene join the crystallization sequence, interelement ratios between the high-valence transition elements and trace elements

Table 4.8. Approximate crystal/melt partition coefficients for basalts at 1200 °C.

	OLV	OPX	CPX	ILM	GNT
La	0.0001	0.0001	0.05	0.002	0.003
Nd	0.0004	0.0005	0.15	0.007	0.03
Sm	0.001	0.002	0.2	0.01	0.20
Lu	0.02	0.04	0.4	0.07	7
Co	1.7	1.1	1.5	4.3	9
Ni	5	2	4	3.5	—
Ti	0.01	0.1	0.3	—	0.3
Hf	0.01	0.05	0.2	0.4	0.07
Zr	0.01	0.05	0.1	0.3	0.1
Nb	0.001	0.001	0.002	0.8	0.002
Ta	0.005	0.005	0.01	4.0	0.02
W	0.01	0.02	0.2	0.1	0.003
Sc	0.2	0.3	1.0	1.5	2

The *D* values are not constants but are functions of *T, P,* melt composition and crystal composition, particularly the wollastonite content of pyroxene (Forsyth et al. 1994; Skulski et al. 1994; Shearer and Papike 1986; Shearer et al. 1995; Green et al. 2000; McCallum and Charette 1978; Jones 1995).

strongly incompatible with these phases, such as Ce, will become fractionated. Shearer et al. (1996) concluded that the source regions for both VLT and high-Ti picritic magmas were fractionated with respect to lunar abundances for Nb/Ce and Zr/Ce. The high-Ti source region, in particular, was interpreted to contain small additions of ilmenite, but no more than a few percent. Neal (2001) found that the Hf/Th and Ta/Th are decidedly suprachondritic for high-Ti mare basalts, supporting the hypothesis that ilmenite was added to the source region. Perhaps even more compelling is the suprachondritic ratio of Ta/Nb in high-Ti mare basalts because the partition coefficient for Ta is about twice that for Nb in TiO_2-rich minerals (Horng and Hess 2000).

5.5. Isotopic observations

Isotopic data for mare basalts are presented as initial ε_{nd} vs. crystallization age in Figure 4.42, and initial ε_{Hf} vs. initial ε_{Nd} diagrams in Figure 4.43 (Nyquist and Shih 1992; Snyder et al. 1994; 1997; 2000; Beard et al. 1998). The discussion will focus first on the Nd data. All mare basalts have positive ε_{Nd} bounded by the chondritic evolution line, by definition $\varepsilon_{Nd} = 0$, and an ε_{Nd} growth curve corresponding to a source with $^{147}Sm/^{144}Nd$ ~0.21-0.28 (Nyquist et al. 1995). It follows that mare basalts are products of melting of sources with long-term LREE-depleted signatures as would be expected of pyroxene/olivine cumulates from an LMO with near chondritic Sm/Nd. Furthermore, the isotopic systematics record varying degrees of Sm/Nd fractionation in the source mantle. Apollo 12 ilmenite basalts and Apollo 17 high-Ti basalts

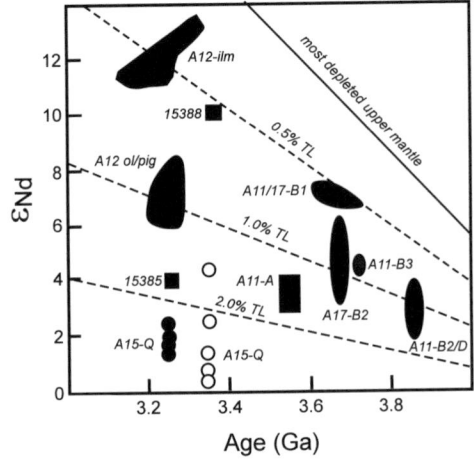

Figure 4.42. Initial ε_{Nd} vs. crystallization ages of mare basalts and array for a depleted cumulate mantle. TL represents trapped liquid component (modified after Snyder et al. 2000).

derived from the most depleted sources, whereas Apollo 15 mare basalts came from the least depleted sources, and one sample is almost chondritic. These characteristics are difficult to duplicate in an LMO because the Sm/Nd ratio is roughly constant over more than 90% of its crystallization (Snyder et al. 1992; Beard et al. 1998). The source cannot be composed solely of primitive cumulates from the LMO.

The cumulate source is assumed to consist of two components. One component is an orthopyroxene-rich adcumulate; the cumulate would have a fixed Sm/Nd, as long as it derived from the main stages of crystallization of the chondritic LMO. A second component must have smaller Sm/Nd ratios and be enriched in both Sm and Nd relatively to the adcumulate so only small masses of this component are required. Snyder et al. (1994) suggested that this is a trapped-liquid component derived from more advanced stages of crystallization of the LMO. In their model, only a few percent of this component is needed to explain the observed data. Other enriched components are possible candidates but some form of hybridization is required to recreate the isotopic systematics of mare basalts (e.g., Hughes et al. 1989).

Interestingly, some of the Apollo 12 low-Ti mare basalts and Apollo 17 high-Ti mare basalts lay along the growth curve corresponding to the most extreme LREE depletion in the mantle source (Nyquist and Shih 1992). Other Apollo 12, Apollo 17 and Apollo 11 high-Ti mare basalts define different LREE depleted growth curves (Snyder et al. 1996). This commonality of sources from widely separated landing sites suggests that the source regions

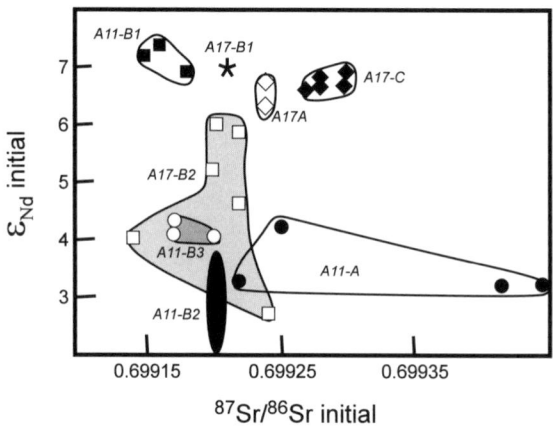

Figure 4.43. Initial ε_{Nd} vs. initial $^{87}Sr/^{86}Sr$ for high-Ti mare basalts (modified after Snyder et al. 2000).

extend Moon-wide and the timing of the depletion event was synchronous. These features imply a global process, no doubt the solidification of a global LMO (Nyquist and Shih 1992).

It is dangerous, however, to carry these conclusions much further. A commonality of sources is not as apparent in the Sr isotope data (Fig. 4.43). Initial ε_{Hf} and ε_{Nd} values for low-Ti and high-Ti mare basalts imply that the source regions for these lunar magmas are distinct (Fig. 4.44). Low-Ti mare basalts have higher initial ε_{Hf} values, at a given ε_{Nd}, than the initial ε_{Hf} values for high-Ti mare basalts (Beard et al. 1998). Indeed, the data reflect sources for low-Ti basalts that have chondrite-normalized Lu/Hf ratios approximately 4× greater than chondrite-normalized (Sm/Nd) ratios; high-Ti sources had roughly equal chondrite-normalized ratios of Lu/Hf and Sm/Nd. The two source arrays appear to intersect at KREEP basalt, perhaps indicating that these arrays are mixing lines emanating from the KREEP component (Snyder et al. 2000).

As was discussed earlier (Sections 4.2 and 4.3) it appears the some of the mare basalts on the Moon have a significant radiogenic ε_W signature, i.e., $\varepsilon_W > 0$ (Fig. 4.2). It is generally agreed that mare basalts were produced by melting of the cumulates from the LMO, thus the ε_W signature is unlikely to have been inherited from the accreting bodies that formed the Moon (Shearer and Newsom 2000; Righter and Shearer 2003; Shearer and Righter 2003). Now if the lunar reservoirs received their W signature during core formation, it would be difficult to imagine a physical process that would not imprint all crystallization products with the same W depletion event. It is more likely that if this radiogenic W signature truly exists in mare basalts, it was imparted through the fractionation of high-Ca clinopyroxene and ilmenite. Those sources that have non-radiogenic W simply did not contain sufficient concentrations of the late-stage cumulates. High-Ti source regions

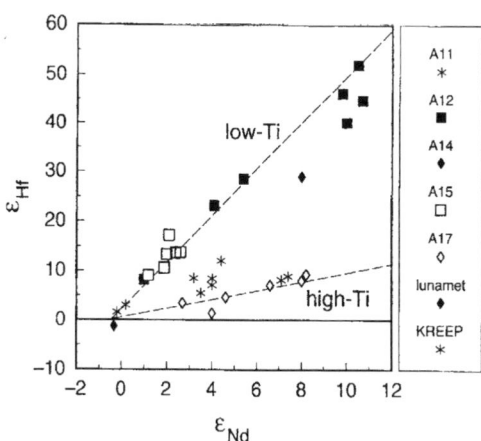

Figure 4.44. ε_{Hf} vs. ε_{Nd} for lunar mare basalts and KREEP basalt (modified after Snyder et al. 2000).

must be enriched in ilmenite and therefore should have $\varepsilon_W > 0$ as is observed. This places constraints on the age (Section 4.3) and composition of mare-basalt sources.

One difficulty with this model is that crystallization of ilmenite and high-Ca pyroxene from late-stage magmas would have created KREEP-rich liquids with depleted Hf/W. However, there are no igneous rocks on the Moon with $\varepsilon_W \leq 0$; all measured rocks have $\varepsilon_W \geq 0$. One possible resolution of this paradox is that fraction of Hf/W is due to fractionation of small amounts of metal and as well as late-stage high-Ca clinopyroxene and ilmenite. In this scenario, the early LMO would have $\varepsilon_W \geq 0$ so that late-stage depletion of Hf into high-Ca clinopyroxene and ilmenite would not produce $\varepsilon_W < 0$. Resolution of these questions awaits more data.

5.6. Experimental phase equilibria

Phase equilibria experiments on mare basalts and picritic glass compositions are ably reviewed by BVSP (1981) and Longhi (1992) so only the most relevant highlights and more recent experimental results are presented in this section. Low-pressure phase equilibria on VLT glass compositions have only olivine as a liquidus phase for a temperature interval of about 200 °C (Fig. 4.45) (Chen et al. 1982; Chen and Lindsley 1983; Elkins-Tanton et al. 2000). Liquidus temperatures are very high, reaching almost 1400 °C, comparable to those of some of the less magnesian terrestrial komatiites (Hess 1989). Pigeonite is the second liquidus phase, followed by plagioclase and then augite (Longhi 1992). The late appearance of plagioclase and augite contrary to their comparatively earlier crystallization in terrestrial ocean-floor tholeiites is a consequence of low CaO and Al$_2$O$_3$ contents. Indeed, VLT and also high-Ti picritic glasses are strongly undersaturated with respect to these phases at all pressures (Longhi 1992).

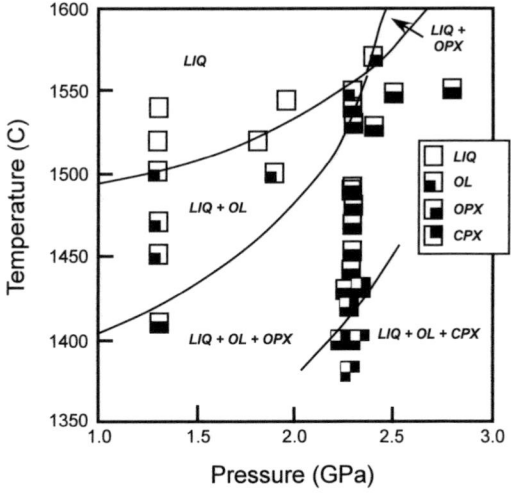

Figure 4.45. Phase equilibria of Apollo 15 green glass (modified after Elkins et al. 2000).

High-Ti picritic glasses have olivine on or near the 1-bar liquidus but the most Ti-rich compositions (TiO$_2$ > 18 wt%) also include minor amounts of Ti-rich chrome-spinel as a liquidus phase (Delano 1980; Wagner and Grove 1997). Olivine is the sole liquidus phase in the less Ti-rich Apollo 74220 orange glass (TiO$_2$ ~9 wt%), however (Green et al. 1975). Liquidus temperatures are somewhat be-low those of the VLT glasses but are still high ($T \sim 1300$ °C). Armalcolite ([Fe,Mg]Ti$_2$O$_5$) is the next major phase to appear but develops a reaction relation with liquid after ilmenite appears at only slightly lower temperatures. Pigeonite, plagioclase and high-Ca clinopyroxene appear in that order at temperatures around 1200-1150 °C.

In summary, the common fea-tures of the 1-bar phase diagrams of the picritic glasses are that they are characterized by (1) high liquidus temperatures, (2) olivine (± Cr spinel) as the sole liquidus phase, and (3) the remaining silicate and oxide phases are relegated to much lower temperatures. These features suggest that the picritic glasses are primitive magmas generated at considerable depths in the lunar mantle. These conclusions are even more robust if the source of these magmas contained both olivine and orthopyroxene. The expansion of the olivine field with decreasing pressure brings only olivine to the 1-bar liquidus: the greater the reduction in pressure, the larger the olivine liquidus field. The size of the olivine liquidus field in picritic glasses is more like those of terrestrial komatiites than of basalts, again an indication of the depths at which such magmas were generated.

The high-pressure phase equilibria are relatively simple. Olivine is the liquidus phase until orthopyroxene joins and then replaces olivine at high pressures. The pressures and temperatures where the picritic glasses are multisaturated with these two phases are given in Table 4.9. It is possible that garnet appears on the liquidus at higher pressures (Chen et al. 1982; Shearer et al. 2003; Draper et al. 2004). High-Ca pyroxene, plagioclase or ilmenite are not stable on the liquidus. The ilmenite liquidus, for example, is more than 50-100 °C below the temperatures of multiple saturation even in the most Ti-rich glasses (Delano 1980; Wagner and Grove 1997). Chrome spinel is a high-pressure liquidus phase in some picritic melts but the stability of this phase is very sensitive to Cr_2O_3 and oxygen fugacity, so the petrogenetic significance of this phase is unclear. Although low CaO pyroxene is the liquidus phase at high pressure, the pyroxene becomes progressively more calcic with decreasing temperatures or with increasing pressures at fixed temperatures (Green et al. 1971).

Table 4.9. The pressures and temperatures of olivine + orthopyroxene multi-saturation of phase equilibria experiments on picritic glass compositions.

Picrite Glass	TiO$_2$ (wt%)	Pressure (GPa)	Temp (°C)	Ref.
Apollo 14B	0.45	2.4	1560	(a)
Apollo 14 VLT	0.66	1.9	1500	(b)
Apollo 17 VLT	0.71	1.8	1500	(b)
Apollo 74220	8.9	2.0	1490	(c)
Apollo 15 Red	13.8	2.5	1460	(d)
Apollo 14 Black	16.4	1.5	1430	(e)

*Average of data tabulated in Clauser and Huenges (1995). *Ref:* (a) Elkins et al. 2000; (b) Chen et al. 1982; (c) Green et al. 1975; (d) Delano 1980; (e) Wagner and Grove 1997.

The phase-equilibria data for crystalline mare basalts are more variable because we are dealing with magmas of varying crystallinity and at different states of fractionation. The 1-bar liquidus of high-Ti mare basalt 70215, for example, is multiply saturated with olivine, armalcolite, and spinel, a feature not characteristic of primitive magmas generated at depth (Kesson 1975). If we compare these results with those of the high-Ti picritic glasses, then 70215 could be the derivative melt of a picritic glass evolved to some 100 °C below the liquidus (Longhi 1992). Other mare basalts have only olivine on the liquidus but appear to be compositions that accumulated excess olivine (Green et al. 1971; Walker et al. 1976). The petrogenetic significance of the phase equilibria of crystalline mare basalts is considered in the next section.

In the subsolidus, high-Ti mare basalts are ilmenite gabbros containing ilmenite, High-Ca clinopyroxene, and plagioclase as major phases (O'Hara et al. 1970; Ringwood and Essene 1970). With increasing pressure, the gabbro assemblage undergoes a continuous phase transition where garnet gradually replaces plagioclase as one of the major phases, rutile replaces part or all of the ilmenite. An eclogite or garnet-pyroxenite assemblage is stable at pressure greater than 1 GPa depending on the temperature and bulk composition. The eclogite assemblage is garnet + high-Ca clinopyroxene + rutile ± ilmenite ± quartz (O'Hara et al. 1971; Ringwood and Essene 1971).

5.7. Depth of the mare basalt source

High-pressure phase-relation experiments on mare basalts or their analogs record pressures of olivine plus orthopyroxene multiple saturation, which range from about 0.5 to 2.5 GPa (Longhi 1992). Multiple saturation pressures for picritic glasses, in contrast, are generally higher and define the range from 1.5-2.4 GPa (Table 4.9). If the pressures of multiple saturation have petrogenetic significance, the data imply that mare basalts come from sources that are distinct from those that gave rise to the picritic glasses. The mare- basalt source region would exist in the upper lunar mantle at depths from 100 to 500 km and the mantle would consist of cumulates more evolved than those parental to the picritic glasses. The Mg' values of fine-grained mare basalts, for example, are typically much lower than comparable picritic glasses (Delano 1986; Shearer and Papike 1993). However, before the implications of these ideas to lunar evolution are considered, we must be certain that the mare basalts pass several important tests; not only must the mare basalt compositions represent quenched liquid but these compositions must represent primary or at least, very primitive liquids.

The major-element compositional diversity of mare basalts is, in part, the result of fractionation of olivine and other near-liquidus phases (Papike et al. 1976, 1997, 1998 and references therein). It follows that only the most primitive basalt of any batch of magmas related by crystal fractionation qualifies as a potential primitive or possibly primary magma. Only a few of the mare basalts survive this filter. It is useful to illustrate how rare such primitive basaltic magmas are in the terrestrial database. The example below considers only terrestrial ocean-floor basalts because the lithospheric barriers to eruption are minimized in this tectonic environment.

The following criteria are used to identify primary ocean-floor basalts (Hess 1992). A primary basalt must have only olivine on the 1-bar liquidus, an Mg' value of 68 or greater, and an MgO content of 9 wt% or greater (Hess and Head 1990). Out of a total of more than 900 chemical analyses of ocean-floor vitrophyres, where glass assures that the basalt represents a liquid, only 21 samples had the appropriate MgO and Mg' values, and then only marginally. At best, less than 2.5% of the liquids erupted on the ocean floor can be considered a primitive liquid. These primitive basalts are probably not primary because a primary liquid derived from appropriate depths in the terrestrial mantle should have MgO values 10-30 wt% higher than recorded in these samples. Given that mare basalts must pass through a thick crust and lithosphere, it is unlikely that most crystalline mare basalts represent even primitive compositions. If such exist, they are found only in the most radically quenched compositions, particularly those compositions that are equivalent to liquids.

The identification of quenched-liquid compositions for crystalline mare samples is not a trivial task. Fine-grained to vitrophyric basalts produced by rapid cooling will most likely have preserved the original liquid composition. Aphyric samples are relatively rare, however most contain microphenocrysts and/or complex textural domains (BVSP 1981). Walker et al. (1976) found a strong positive correlation between grain size and normative olivine content in Apollo 12 olivine basalts. They argued convincingly that most samples were differentiated by simple olivine settling perhaps in the basal portion of a cooling unit. It follows that there are sets of complementary basalts that have differentiated by losing olivine.

Experimental petrology also provides a number of tests to help identify phenocryst-enriched compositions, such as the "phenocryst-matching" test. This test compares the compositions of the earliest phenocrysts in fine-grained basalt to the composition of the liquidus phase determined by experiment at 1 bar. Provided that the containers used in the experiments do not change the bulk composition of the basalt (the use of Fe, Pt or even Mo, containers can alter the FeO content of basaltic liquids), the composition of the phenocryst and liquidus phase must be identical if the basalt in question represents a liquid composition. Green et al. (1971) used this technique to examine the petrogenesis of Apollo 12 olivine basalts.

High-pressure multiple-saturation experiments on ilmenite basalt 12022 have olivine and low-Ca pyroxene on the liquidus at pressures of about 15 kb (Green et al. 1971). Olivine of composition Mg' = 77 is the sole liquidus phase at 1 bar pressure and 1290 °C. Pigeonite does not appear until about 1150 °C just above the temperatures of the incoming of ilmenite and plagioclase. The natural rock contains olivine phenocrysts with core compositions of Mg' = 69, a composition attained only at about 1170 °C in the experiments. These disparate results are most readily interpreted by assuming that the basalt sample 12022 is a partial cumulate of olivine of Mg' = 69. Because olivine is much richer in MgO than the coexisting liquid, concentrating olivine in the magma increases both the normative olivine and Mg' of the rock so that when it is totally melted, both its liquidus temperature and Mg' exceed the values of the true liquid parent. Because the liquid is enriched in normative olivine, the high pressure multiple-saturation point will exist at pressures that exceed the multiple-saturation point of the corresponding liquid without the added olivine. Clearly, only basalts that can be equated to liquids have phase equilibria with direct petrogenetic significance, but difficulties in interpretation exist even for phase equilibria performed on liquids.

The depths to the mare source regions are difficult to constrain uniquely notwithstanding the consensus among most lunar petrologists that the source regions must lie within the deep lunar mantle. One of the most important constraints is provided by the high-pressure-liquidus phase relations. The liquids corresponding in bulk composition to the VLT and high-Ti picritic glasses have both olivine and orthopyroxene on their liquidi at pressures between 1.6-2.5 GPa (Table 4.9) (Longhi 1992; Wagner and Grove 1997; Elkins-Tanton et al. 2000) but have only olivine and orthopyroxene at lower and higher pressures respectively (Fig. 4.45). Neither ilmenite, nor high-Ca clinopyroxene, nor plagioclase have stability fields that are within 100 °C of the liquidi at high pressure.

The traditional interpretation of these phase equilibria is that these melts were derived from a source in which olivine and orthopyroxene were restite phases at depths indicated by the pressures of the point of multiple saturation. A number of criteria and assumptions both stated and unstated must be satisfied before such broad conclusions are warranted. One of the most basic assumptions is that the mare-basalt sample represents a liquid composition. Even if the sample represents a melt composition, the melt must represent a primary or at least a very primitive basalt. To be "primary," the liquid composition must not have been substantially modified since the liquid was equilibrated and segregated from its source and then erupted onto the lunar surface. This requirement is a tall order on any planet. Recall that primitive ocean-floor basalts on the Earth are rare even in a tectonic environment where the lithospheric is thin to absent. Mare basalts, in contrast, are dense magmas that must pass through up to 60 km or more of anorthositic and mafic crust plus a substantial thickness of mantle lithosphere to erupt onto the lunar surface. Significant modifications to the original magma composition may result through crystallization of olivine and other near liquid phases and through assimilation of crustal materials. Even if mare liquids survive this transit intact, there remains the relevant question whether true primary magmas really exist in nature.

Longhi (1992) models the green glass VLT composition as the product of polybaric fractional melting. The final composition of melt erupted onto the lunar surface has not equilibrated with a well defined mantle residue at a specific pressure and temperature, but rather is a composite of melts generated over a range of pressures, temperatures, and mantle sources. Under the most favorable circumstances, the final melt composition reflects only an average depth of melting from some weight-averaged mantle. However, if the mantle is heterogeneous on various scales, this composite melt cannot simply be inverted to reproduce even the average conditions of melting. In any case, the concept of a primary melt loses its meaning and significance. Indeed, one of the most serious challenges facing lunar petrologists is to understand and model the physics of the generation, segregation, and transport of melt within the lunar mantle.

These caveats notwithstanding, high-pressure phase relations, particularly those that record multiple-saturation points of olivine and orthopyroxene, provide important constraints for models of petrogenesis. The significance of an olivine-orthopyroxene multiple-saturation point for terrestrial basalts has long been appreciated (BVSP 1981) because melting in the terrestrial mantle proceeds by first eliminating high-Ca clinopyroxene and the aluminous accessory phases from the mantle restite (e.g., Hess 1992). After about 25% melting, only olivine and orthopyroxene are left in the mantle residue and much higher degrees of melting are required to consume orthopyroxene. It follows that, at a minimum, primitive or near primary terrestrial basalts must have both olivine and orthopyroxene on their liquidus at their point of separation and isolation from the mantle.

No such arguments are relevant to the lunar mantle. The mare sources are believed to be cumulates composed of dunite, harzburgite, or olivine pyroxenite hybridized to various degrees by additions of high-level cumulates with small, but variable quantities of intercumulus liquid. Small but variable amounts of intercumulus liquids may also derive from the cumulates themselves and through the process of migration upwards in the pile (Snyder et al. 1992). It follows that additional arguments must be developed to argue that primitive mare basalts and picritic glasses indeed are multisaturated with respect to olivine and orthopyroxene.

One curious feature of the picritic glass compositions is their limited range of Mg'. At 1 bar these melts are in equilibrium with olivine with compositions between Mg' = 77-86 (Delano 1990). High-Ti picritic glasses have Mg' that range from 48 to 54 and are not correlated with their TiO_2 contents (Delano 1986; Shearer and Papike 1993). The VLT glasses are more primitive but are confined to Mg' numbers from 57-64 with one outlier at 68. As expected, the VLT glasses coexist with the most magnesian olivines with Mg' between 81-86. If hybridized cumulates are the source regions to mare basalts what physical processes constrain the Mg' of the mantle to only 9 Mg' units?

Depending on the Mg' number of the bulk Moon, cumulates derived from crystallization of the first half of the LMO would have Mg' greater than that observed for the source regions. In fractionating LMOs with initial Mg' = 88-79, olivine cumulates at the 50% solidification point have Mg' ranging from 89-86 respectively (Hess 1994). One mechanism to remove such mafic cumulates from participating in mare magma genesis is through overturn of the cumulate pile and then sequestering the most magnesian cumulates in a growing lithosphere (Ryder 1991; Hess and Parmentier 1995). At the time of mare basalt volcanism, the elastic lithosphere as given by the 800 °C isotherm is about 100-150 km thick and thickens at a rate of 100 km/Ga (Hess and Parmentier 2001). Isotherms of 1100-1200 °C reach deeper into the cumulate mantle, perhaps to 200-300 km depths depending on model parameters. Since the liquidus temperatures of picrite and basalt magmas exceed 1300 °C at these depths, it follows that cumulates at 200-300 km depths are not thermally conditioned to be the mare source region. A 300 km mantle section on the Moon includes roughly 50% of the volume of the Moon. If the LMO encompassed the whole Moon, then 50% of the cumulates would be contained in this cold trap and would be lost to magma genesis. With mantle overturn, these ultramafic cumulates would sequester phases with Mg' ≥ 86, explaining the high-end limits for mare sources. But how do we rationalize the absence of more Fe-rich mare sources?

The simplest explanation is that the more Fe-rich members of mare basalts are products of this iron-rich mantle. This conclusion is not warranted, however, because many, if not all, mare basalts are derivative melts produced by various amounts of fractionation and, thereby, have lowered Mg' values. In fact, most of the high-Ti crystalline mare basalts have little to no modal olivine and cannot be samples of the primary melts from an olivine-bearing cumulate mantle. These basalts cannot be used to constrain source depths as argued below. If the most evolved basalts are eliminated the composition of the mantle sources could possibly be extended by

5-10 Mg' units to olivine Mg' values of 65 (Delano 1990) but not much lower. Where then are the really Fe-rich cumulates?

In a fractionating LMO, the Mg' of the resulting cumulates vary dramatically only near the terminus of crystallization. In the model of Snyder et al. (1992), the Mg' of the mantle cumulates produced in the last 10% of fractional crystallization range from the mid 80's to 0 (Fig. 4.46). For a whole Moon LMO, these cumulates represent a 100 km thick layer in the upper mantle but cumulate overturn would sequester such cumulates in the lower 600 km of the lunar mantle. Perhaps the small volume of iron-rich cumulates is sufficient reason not to expect large volumes of mare basalts to come from such sources. Such Fe-rich cumulates are also very dense and would resist mobilization by thermal convection. Pressure-release melting would be ineffective. Lastly, melts derived from such sources may never make it to the surface in their pristine state. These magmas, instead, might act as metasomatizing agents, which hybridized more mafic and more voluminous mare mantle sources.

These arguments suggest that the cumulate sources for mare magmas are fractionation products of the LMO between 50 and about 90% solidification. If these estimates are correct, then the cumulate sources for mare basalts are likely to be olivine-bearing orthopyroxenites (see also discussion in Hess 2000), but for arguments sake let's assume that the cumulates are dunites rather then orthopyroxenites. For a chondritic bulk Moon and LMO, the lower half of the LMO cumulate pile would consist of dunites (Fig. 4.46) (e.g., Snyder et al. 1992). Such cumulates would be magnesian - an LMO with an initial Mg' of 82 would produce dunite cumulates with Mg' no lower than 88; a more magnesian Moon would produce even more magnesian dunites. Since the Mg' of olivine that would be in equilibrium with the picritic glasses is between 77-86, it follows that picritic glasses do not originate in the dunite layers of the Moon unless their parent liquids suffered significant olivine fractionation. Arguments developed by Delano (1986) and Shearer and Papike (1993) suggest that such olivine fractionation was limited only to a few percent. Dunites are, therefore, very unlikely to be the sources of the primitive mare basalts and glasses.

Evidence for orthopyroxene in the mare source region is given by the Sm/Nd isotopic systematics of mare basalts. For an assumed value for the initial ε_{Nd} (= 0 CHUR) and the age of the Moon (4.55 Ga), the initial ε_{Nd} and the time of volcanism for a particular mare basalt gives the Sm/Nd ratio of the source (e.g., Nyquist et al. 1979; Beard et al. 1998). The present day ε_{Nd} or the Sm/Nd of the mare basalt determines whether Sm was fractionated from Nd during the melting of the source mantle. Because for small amounts of melting, pyroxene (but

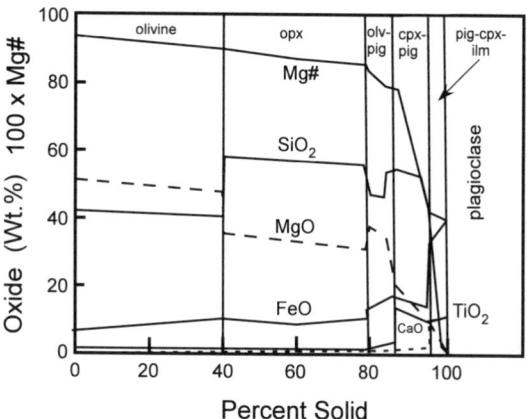

Figure 4.46. Composition of cumulates to a model magma ocean (modified after Snyder et al. 1992 and Shearer and Papike 1999).

not olivine) is capable of fractionating Sm from Nd, orthopyroxene is indicated in the mantle residue whenever (Sm/Nd) mare basalt < (Sm/Nd) source. Beard et al. (1998) and others have used these criteria to argue that orthopyroxene is required in the source mantle of low- and high-Ti mare basalts. Sources containing olivine and orthopyroxene are also indicating by the bulk compositions of the picritic glasses.

The compositions of the picritic mare glasses projected onto the MgO-SiO$_2$-TiO$_2$ phase diagram at 2 GPa (MacGregor 1969) define and follow the olivine-orthopyroxene boundary curve (Fig. 4.47). The scatter is not surprising given the approximations required to project the picritic glass compositions into the MgO-SiO$_2$-TiO$_2$ system. The fact that the pressures of multiple saturation of high-Ti and low-Ti picritic glasses vary from 1.6 to 2.4 GPa indicate that the picritic glasses were probably generated over a range of pressures (Longhi 1992). It is noteworthy, nevertheless, that the distribution of projected points certainly reproduces the expected trend of the cotectic. These results suggest that the picritic glasses were generated

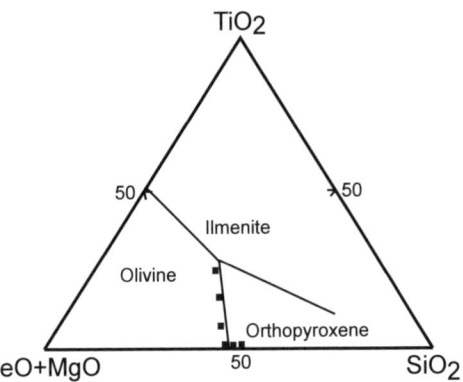

Figure 4.47. Olivine-orthopyroxene boundary curve projected from diopside and plagioclase onto the (Mg,Fe)O-SiO$_2$-TiO$_2$ system at 1.5 GPa. Shaded boxes represent the projected compositions of some picrite glasses (modified after Hess 2000).

by melting (polybaric?) processes buffered by the olivine-orthopyroxene cotectic surface. We conclude therefore that the pressures of multiple saturation are petrologically meaningful and constrain the depths of melting to regions that straddle the apparent depths of multiple saturation. Mare picritic glasses were derived by melting of olivine-orthopyroxene cumulates initiated at average depths of 400 km or more.

The petrogenetic significance of high-pressure liquidus phase relations and, in particular, the saturation pressures and temperatures of non-primary crystalline mare basalts are more questionable. The main variable that determines such pressures is the normative olivine/low CaO pyroxene ratio, although other compositional parameters such as normative feldspar and ilmenite contents are also important. The most primitive mare basalts have olivine on the 1-bar liquidus, so the principal effects of near-surface differentiation is the subtraction of olivine from differentiated liquids and the addition of olivine to partial cumulates from such liquids. The former has the effect of lowering the depths of multi-saturation, the latter of increasing such depths. Subtracting roughly 5 mol% olivine from an Apollo 15 green glass composition lowers the pressure of multiple saturation from about 2.5 GPa to 1.5 GPa; the MgO melt composition, however, is reduced by only 1 wt% (from 16 to 15, Longhi personal communication). Increasing the MgO contents of high-Ti mare basalt 70215 by 4 wt%, the equivalent of adding 10 wt% olivine, increases the depth of multiple saturation by 200 km. It is clear from these examples that the effects of even modest degrees of low-pressure olivine fractionation can have dramatic effects on the apparent depths of multiple saturation. Longhi (1992) noted that most mare basalts have apparent depths of multiple saturation between 0.5 and 2.0 GPa. These values could be achieved by as little as 10 wt% olivine fractionation from more primitive parent liquids or the accumulation of olivine into more evolved basalts.

5.8. Remelting of cumulates

Source regions for mare basalts are generally modeled as cumulates from the differentiation of the LMO. Partial melting of these cumulates, some of which are hybridized to various degrees by incompatible-element-rich components, generates the complex chemical signatures that characterize the mare-basalt suite. Of particular interest is the partitioning of TiO_2 and other incompatible or compatible trace elements during the initial cumulate-forming process and then in the subsequent partial-melting event that generated the basalts.

For simplicity, assume that the trace elements exist in chondritic proportions in the LMO. First, consider cumulates formed without intercumulus liquid. The trace-element content in the cumulates is given by the distribution coefficients for each element multiplied by the concentrations of the element in the liquid. The first partial melt from this cumulate will contain the same relative and absolute concentrations of these elements. Additional melting will dilute the incompatible elements and enrich the compatible elements in the liquid. As an example, the following describes the evolution of the TiO_2 content during partial melting of a system that is undersaturated with respect to ilmenite. TiO_2 is moderately incompatible in pyroxene and highly incompatible in olivine. Thus, the maximum amount of TiO_2 that can be generated in mare basalts is equal to the TiO_2 content of the LMO parental to the cumulate source. If we assume that the LMO has chondritic TiO_2 (about 0.2 wt%; Taylor 1980), the cumulates formed by 95% fractional crystallization coexist with a melt with about 4 wt% TiO_2. Depending on P,T and bulk composition of the magma, ilmenite crystallization will terminate the TiO_2 enrichment in the residual liquids. We therefore consider only those cumulates formed prior to ilmenite saturation. Partial melting of these cumulates cannot produce melts with more than about 4 wt% TiO_2. Indeed, if the degree of melting is comparable to the bulk distribution coefficient, say $F = 0.1$, then the TiO_2 content in the resulting basalts will be less than 2 wt%. Therefore, it is doubtful that low-Ti mare basalts containing 2-6 wt% TiO_2 can be produced from only olivine-orthopyroxene cumulates of the LMO. Certainly, low-Ti basalts at the high end of this range require a TiO_2-rich phase in the source.

Possible TiO_2-rich phases include intercumulus liquid, cumulate ilmenite, or late stage high-Ca clinopyroxene. Although cumulate with a small amount of intercumulus liquid will produce partial melts with somewhat elevated TiO_2 contents, they cannot produce TiO_2-rich mare basalts unless the intercumulus liquid makes up a significant fraction of the rock. Isotopic and trace-element considerations (Nyquist and Shih 1992; Snyder et al. 1994, 1997), however, limit the amount of acceptable intercumulus component to no more than a few percent. Becuase mare volcanism occurred more than hundreds of millions of years after the cumulates were created, the intercumulus component should have been fully crystallized and the compatible and incompatible elements would be partitioned among the existing cumulate phases, particularly if the cumulate was pyroxene-rich. The net effect will be to increase the contents of the highly incompatible elements but hardly affect the contents of the more compatible elements. Returning to the example, the TiO_2 content of a pyroxene cumulate would be increased by only a few percent of the amount present; a partial melt from this cumulate would contain little more TiO_2 than the melt from the barren cumulate. In an olivine cumulate, however, the crystallized intercumulus liquid may contain trace amounts of ilmenite depending on the composition and abundance of liquid. The first melts would then be ilmenite-saturated and contain appropriately elevated content of TiO_2. Ilmenite saturated melts contain more than 20 wt% TiO_2 at high T,P (Beck and Hess 2001).

The required ingredients to effectively hybridize the cumulate source are TiO_2-rich components, but without the large amounts of trace-element-rich components that compromise the isotopic systematics. The late-stage ilmenite-rich cumulates have some of the desired properties. These cumulates contain high-Ca clinopyroxene, ilmenite, orthopyroxene, olivine, intercumulus melt and whatever plagioclase is not lost to the growing anorthositic crust. For simplicity, we consider only the clinopyroxene and ilmenite cumulates; these cumulates should exist in roughly a 5/1 ratio (Van Orman and Grove 2000). The cumulates would contain about 7-8% TiO_2.

Ilmenite cumulates added to the olivine-rich adcumulate produce a source that contains excess high-Ca clinopyroxene and ilmenite. Small degrees of melting of this source to produce TiO_2-rich mare basalts must leave high-Ca clinopyroxene in the residual melt because the high-Ca clinopyroxene/ilmenite ratio is much greater than one. Such TiO_2-rich mare basalts must then be saturated with high-Ca clinopyroxene, contrary to what is observed in the most primitive mare basalts and picritic liquids, which are far from saturated in CaO-pyroxene. An olivine-rich cumulate with added high-level clinopyroxene-ilmenite cumulates is not an appropriate source for high-Ti mare basalts.

A different, more acceptable result follows if the cumulates are rich in orthopyroxene because orthopyroxene is a potential reservoir for both TiO_2 and the wollastonite ($CaSiO_3$) components. For simplicity, assume that the orthopyroxene is sufficiently primitive and contains very little normative $FeTiO_3$ or $CaSiO_3$. From the experiments of Delano (1980), orthopyroxene saturated with ilmenite contains about 2-3 wt% TiO_2. Since the high-level ilmenite cumulates contain 7-8 wt% TiO_2, it follows that significant amounts of this cumulate can be incorporated into orthopyroxene. Suppose that the hybridized mixture contains ilmenite cumulates and orthopyroxene in a 1/9 ratio. The remainder of the cumulate is formed from olivine. The orthopyroxene is strongly undersaturated with respect to high-Ca clinopyroxene and ilmenite, thus the hybridized orthopyroxene would contain about 1 wt% TiO_2 and 3 wt% $CaSiO_3$. Small degrees of melting of this hybridized orthopyroxene would produce TiO_2-enriched mare basalts, which are not saturated with respect to either ilmenite or Ca-pyroxene as required by the phase-equilibria constraints.

An alternative method to hybridize the orthopyroxenite cumulates follows a more indirect path. Hess and Parmentier (1995) suggested that only a small fraction of the dense ilmenite cumulates would be convectively mixed into the underlying cumulates in various proportions. Most of the ilmenite cumulates would ultimately collect in or around the lunar core. This model was guided by the constraint that none of the known mare basalts were saturated with respect to ilmenite and high-Ca clinopyroxene yet all appear to have been produced by relatively small degrees of melting. To achieve both these ends, the source for mare basalts could not have attained large additions of ilmenite cumulates. This delicate balancing act might be acceptable on Earth were continuous convective mixing of the mantle would dilute mantle inhomogeneities an appropriate amount. However, on the Moon convective mixing is not an efficient mechanism to homogenize the mantle.

It is more likely that the lunar cumulate mantle remains inhomogeneous on a number of scales. The downwelling ilmenite cumulates would form discrete pods of mantle strongly enriched in ilmenite and high-Ca clinopyroxene (Ringwood and Kesson 1976). Such pods or stringers are analogous to the pyroxenite or eclogite layers in the Earth's mantle (Hirshman et al. 1996). These ilmenite cumulates are less refractory than the pyroxene cumulate mantle and have solidus temperatures a few hundreds of degrees lower than an olivine-orthopyroxene assemblage (Wyatt 1997). The ilmenite-high-Ca clinopyroxene cumulates would melt first and create melts initially saturated with high-Ca clinopyroxene and ilmenite. Such melts would contain more than 11 wt% TiO_2 (Van Orman and Grove 2000). Buoyancy and perhaps surface energy forces would drive these melts into neighboring, more refractory cumulates where they would hybridize orthopyroxene cumulates. Partial melting of these variously hybridized zones would create the spectrum of mare basalts that exist on the Moon.

It is worth emphasizing that the partial melts of the deep-seated ilmenite cumulates cannot give rise to observed mare basalts. First, such melts would be saturated with high-Ca clinopyroxene and perhaps even ilmenite. Second, near solidus temperatures are several hundred degrees lower than the liquidus temperatures of mare basalts. Even the equilibrium liquidus of the ilmenite cumulates is at a lower temperature than the corresponding mare basalts at the same depths. Finally, some of the high-Ti mare basalts and picritic glasses have

TiO_2 contents that are greater than the composition of the minimum partial melts derived from these cumulates (Van Orman and Grove 2000). We conclude that the ilmenite cumulates alone at depth are not the appropriate source for high-Ti mare basalts.

5.9. Volatile contents of mare magmas

Picritic mare glasses are generally believed to be the products of fire-fountaining (Delano 1979 and references therein; see also Chapter 3, Section 3.1.4). The forces driving the fountaining process appear to be related to the exsolution of volatiles, a record of which now coats the glass beads. Although volatile-driven eruption is certainly a viable and likely hypothesis, questions remain as to the origin of these volatiles, the nature of the forces driving fire fountains, and the extent of degassing and volatile loss. Fogel and Rutherford (1995) and Sato (1979) before them gave a detailed analysis of the most likely eruptive mechanisms. They concluded that graphite oxidation alone is capable of generating sufficient CO/CO_2 gas to account for the pyroclastic effect. The oxidation of graphite can proceed by a number of steps perhaps via the reduction of various oxides of Cr_2O_3, TiO_2 or even FeO. The exsolution of the CO gas induces other volatiles such as S, Cl etc. to partition into the gas and be lost to the surroundings. Indeed, glassy inclusions of orange glass trapped in olivine phenocrysts contain 600 ppm S compared to 200 ppm S in the glass bead, indicating at least 400 ppm loss of S from the degassing melt sphere (Weitz et al. 1999). Delano et al. (1994) and Shearer et al. (1998) determined the sulfur contents in the interior of 14 varieties of picritic glasses varying in composition from VLT to high-Ti affinities. The mean S values were from 170 ppm to 540 ppm S. In comparison, bulk sulfur contents of the glasses which included the volatile-rich coatings were more than $4\times$ greater. The S contents of the interiors of picritic glasses are also significantly lower than those in crystalline mare basalts. If both crystalline mare basalts and picritic glasses initially had the same S content, then up to 90% of the original S was lost from these picritic melts.

There are two most likely sources for the gases. One is primitive, undegassed lunar interior. The other is the late-stage dregs of the LMO. If most of the Moon was processed by the LMO, then the volatile elements, which in many cases are also incompatible, would have been concentrated into the KREEPy remains of the LMO. Some of these materials would have been swept back into and hybridized the mantle. Partial melting of the hybridized mantle concentrates the volatiles into the melt. The volatile contents of mare magmas would then be controlled by the volatile contents of the source region and the degree of partial melting.

5.10. Buoyancy of mare basalts

Magma compressibility studies have shown that mare basalts, especially the intrinsically dense high-Ti mare basalts, eventually become more dense than the lunar mantle residue in which these melts are generated (Delano 1990; Circone and Agee 1996). The depths at which such magmas are neutrally buoyant depend on the TiO_2 content of the melt and the olivine/orthopyroxene ratio of the lunar mantle. The density of molten Apollo 14 black glass ($TiO_2 =$ 16.4 wt%), for example, is the same as the density of olivine (Fo = 80) at 2.0 GPa but equals the density of orthopyroxene (En = 80) at 0.66 GPa. It follows that the Apollo 14 black glass is neutrally buoyant at depths from 120 km to 400 km for orthopyroxenite and dunite mantles respectively. Densities of high-Ti mare basalts of more modest TiO_2 contents, as for example, the Apollo 17 orange glass ($TiO_2 = 9.3$ wt%), have depths of neutral buoyancy that range from 400 km (relative to orthopyroxene) to more than 800 km (relative to olivine) (Delano 1990). Several authors have cited these as significant if not fatal barriers to deriving mare basalts from deep with the interior of the Moon. Indeed, melts that are denser than their local surroundings will sink into the interior until neutral buoyancy is achieved rather than rise to the surface (e.g., Delano 1990; Wagner and Grove 1997).

Hess (1991) has argued that even negatively buoyant melts will rise to the lunar surface within ascending lunar diapirs, provided that the ascent velocities of the diapir exceed the velocity of the downward percolating melt. In this case the melt will either be carried past the depths of neutral buoyancy and/or be trapped above the lower thermal boundary of the diapir. If the rate of melt segregation initially exceeds the rate of diapiric ascent, then the melt will descend not only relative to the diapir but also relative to the surrounding mantle. A melt that progresses downward along the mantle adiabat to higher pressures must crystallize and therefore halt its descent. The crystallized products would then be carried upward and then remelted. The net effect is that the mean velocity of downward segregation is much less than the value calculated for the melt for which no crystallization is allowed. It thus appears that the depths of neutral buoyancy are not insurmountable barriers to deriving dense magmas from the lunar interior.

The negative buoyancy of high-Ti magmas is a key physical asset in the models of Elkins-Tanton et al. (2003). According to their models, the simple sinking of solidified high-Ti cumulates is unlikely because at the temperatures of solidification at around 1100 °C the viscosity of pyroxene-rich mantle is high, too high to allow these cumulates to sink. Instead, Ti-rich hybridized mantle is created by partial melting of these cumulates. At the "eutectic" of about 1125 °C at 100 km depth, the composition of the melt is in a ratio of two parts clinopyroxene and 1 part ilmenite (Van Orman and Grove 2000). These melts are Al_2O_3-poor (near solidus melts contain only 4-5 wt% Al_2O_3), and the densities of these melts at about 3.2 gm/cm^3 are slightly denser than the pyroxene-rich cumulates below the ilmenite cumulate layer. If the mantle is permeable to the Ti-rich melts, these negatively buoyant liquids would percolate into the underlying mantle, producing a fertile and hybridized mantle. These and other models are critically reviewed in the "Petrogenesis" section (4.5.12).

5.11. Insights from remotely sensed data

While lunar samples provide valuable insights into the character of the lunar mantle, absolute ages of mare basalt flows, and magmatic processes, remotely sensed data provide a planetary-scale context for the sample data. Remote-sensing observations have provided insights into the distribution of mare basalts, the distribution of mare-basalt compositions, and the relative ages of basalt flows. These observations and resulting conclusions are summarized in Chapters 1-3. The remote sensing observations also provide further interpretation of the mare-basalt sample suite. The sample-return data suggest that crystalline mare basalts are bimodally distributed between high-Ti mare basalts ($TiO_2 \geq 9$ wt%) and low-Ti basalts ($TiO_2 \leq 6$ wt%) with peaks at about 2.5 wt% TiO_2 and 12-13 wt% TiO_2 (see Chapter 2, Section 10.5.3). These data are of limited value, however, because they are not representative of mare volcanism given the statistically small sample. These data, moreover, are not mass weighted and simply represent the total number of basalt samples that were analyzed.

Remotely sensed TiO_2 abundances for all near-side lunar mare are area-weighted and thus more statistically representative (Giguere et al. 2000; Chapter 2, Section 10.5.3). The remote sensing technique is reasonably well calibrated in the 0.5 to 10.0 wt% TiO_2 range but still includes a generous ± 1.1 wt% standard deviation. The TiO_2 contents are not for mare basalts but rather for lunar mare soils. A comparison of TiO_2 values of rocks and regolith soils suggests that the TiO_2 values are diluted by the addition of low-TiO_2 highland materials (Gillis et al. 2003). The true TiO_2 values are between 10-30% higher than the observed values. The remotely sensed TiO_2 data, nevertheless, quite clearly demonstrate that there is a continuous distribution of TiO_2 values with no hint of the bimodality recorded in the returned sample set. Some of the continuity may be ascribed to mixing of soils under the influence of impacts. Nevertheless, high-Ti mare samples represent only a small fraction of the area-weighted lunar surface. Low-Ti basalts in the 2-3 wt% range are the dominant mare lithology. Even VLT mare basalts far outweigh the high-Ti mare basalt compositions. Far-side data are more sparse, but they too are dominated by VLT mare samples, where $TiO_2 < 1.0$ wt%. Basaltic lithologies found among the lunar meteorites support the remote-sensing result that low-TiO_2 basalts are the dominant mare volcanic product on the Moon.

This fact can be rationalized from several perspectives. High-Ti mare basalt magmas are 5-10% more dense than low-Ti counterparts (Delano 1990). Both magmas are denser than either the anorthositic crust or the lower mafic crust. Thus, buoyancy considerations alone favor the eruption low-Ti mare magmas. High-Ti mare basalts require a source enriched with variable amounts of normative ilmenite. The amount of normative ilmenite in a chondritic Moon is less than 0.5 wt% (Taylor 1980). The paucity of high-Ti mare basalts may simply reflect the low TiO_2 content of the bulk Moon.

5.12. Petrogenesis

Models for mare petrogenesis must integrate the vast amount of geochemical information, the phase equilibria, the physical settings of mare volcanism and their tectonic environment, the timing, volume and distribution of mare volcanism, the prolonged existence of mascons, and the evolution with time of the cumulate mantle and elastic lithosphere. This task is formidable; not only are some of the input parameters model dependent, but they are limited by the quality and abundance of data. This section concludes with an overview of what critical data and criteria most effectively place limits on petrogenesis. First, we give a brief critical overview of existing petrogenetic models.

Early and still influential models for mare petrogenesis called for the production of low-Ti mare basalts from deep within the lower mantle whereas high-Ti mare basalts were to have formed by melting of upper-level (shallow mantle) ilmenite-rich cumulates (Taylor and Jakes 1974). The hypothesis that high-Ti mare liquids are the direct melt products of pyroxene-ilmenite cumulates residing below the anorthositic crust still finds some support (Manga and Arkani-Hamed 1991). These authors suggested that partial melting of the ilmenite-cumulates beneath the near-side basins occurred because of the thermal insulating effect of ejecta blankets from giant impacts. Unfortunately, the model faces some considerable difficulties. Primitive high-Ti mare basalts and their ultramafic parents have liquidus temperatures more than 1300 °C at subcrustal pressures (Delano 1980; Wagner and Grove 1997). Reheating the subcrustal mantle to these temperatures in the 3.8-3.9 Ga time period conflicts with the strong and cool upper mantle needed to support lunar mascons (Hess and Parmentier 2001). A more direct argument is that primitive high-Ti mare liquids are not saturated with respect to high-Ca clinopyroxene and ilmenite at any pressures. The shallow cumulates contain high-Ca clinopyroxene and ilmenite in roughly a 5:1 ratio. Near-eutectic melts with more than 13 wt% TiO_2 are saturated with both ilmenite and high-Ca clinopyroxene (Van Orman and Grove 2000). Melts formed at higher temperatures appropriate to the liquidus temperatures of primitive mare liquids become undersaturated with respect to ilmenite, develop lower TiO_2 contents, and retain high-Ca clinopyroxene at the liquidus. These characteristics are unlike any known lunar mare basalt. These results effectively rule out a high-Ti magma source composed mainly of clinopyroxene and ilmenite.

The extended period of mare volcanism also precludes a shallow subcrustal source for high-Ti mare basalts. Radiometric ages place high-Ti mare volcanism in the early Imbrian Period (3.3-3.8 Ga) (Nyquist and Shih 1992; Snyder et al. 2000), but remote sensing observations extend the periods of volcanism to roughly 1.1 Ga ago (Shultz and Spudis 1983; Hiesinger et al. 2000, 2003). Some of the youngest basalts have 6-12 wt% TiO_2. The latter are surely high-Ti mare basalts. Given that the elastic lithosphere was thickening at about 100 km/Ga after the crystallization of the LMO (Parmentier and Hess 1998; Hess 2000; Hess and Parmentier 2001), it follows that the upper few hundred kilometers of the cumulate mantle was relatively cool, $T \leq 800$ °C, and would not be a source of volcanism in a young Moon, particularly high-Ti volcanics that have liquidus temperatures of 1200 °C or higher.

One set of models for the origin of high-Ti mare basalts involves the assimilation of shallow ilmenite-high-Ca clinopyroxene cumulates located below the lunar crust by low or very low-Ti ultramafic liquids similar in composition to the corresponding picritic glasses

(Hubbard and Minear 1975; Wagner and Grove 1997). In these models the cumulate pile is gravitationally stable and only the Ti-poor mare basalts are derived by melting of the cumulate pile at great depths. One difficulty with the assimilation model is that to make high-Ti mare basalts from their low-Ti counterparts, the proportions of ilmenite and clinopyroxene dissolved must be in a 3:1 ratio by weight. In contrast, the abundance of ilmenite relative to clinopyroxene in the late stage cumulates is roughly 1:5 by weight (Snyder et al. 1992; Hess and Parmentier 1995). Wagner and Grove (1997) suggested that faster dissolution rates of ilmenite relative to high-Ca clinopyroxene would produce the needed assimilant. Their experiments appeared to confirm that ilmenite dissolution rates were about 3× faster than diopside; however, subsequent experiments by Van Orman and Grove (2000) found that the dissolution rate of diopside in synthetic analogues in VLT mare basalts (not in alkali basalts as in Wagner and Grove 1997) were actually faster than the dissolution rates of ilmenite at comparable temperatures. They concluded that non-equilibrium and selective dissolution of the ilmenite and high-Ca clinopyroxene layer by VLT picritic magmas could not produce high-Ti mare basalts.

The strong version of the assimilation hypothesis seems to be discredited; nevertheless it is still worthwhile to discuss the consequences of assimilation of shallow cumulates since the layer may later be disrupted during cumulate overturn and inhomogeneous deposits of this layer might be scattered throughout the cumulate pile. These deposits might be selectively assimilated by migrating VLT picritic melts to produce low- and even high-Ti mare basalts and picritic glasses.

The constraints provided by Wagner and Grove (1997) are still relevant to this discussion; ilmenite and high-Ca pyroxene must enter the VLT picritic magmas in roughly a 3:1 ratio to produce high-Ti mare basalts. This assimilant contains about 42 wt% TiO_2. In order to produce a high-Ti mare basalt with 13 wt% TiO_2, about 30% of the assimilant must be added to the VLT basalt parent minus whatever olivine crystallized from the assimilating magma to provide some of the heat of dissolution. The geochemical consequences of this process should impart a distinctive geochemical signature on the high-Ti mare magma.

Some of the geochemical criteria for mare basalt petrogenesis were discussed in Hess (2000), but are repeated here for completeness. Additional observations relevant to the origin of mare basalts included:

1) Shallow-mantle ilmenite of the ilmenite cumulate layer should contain significant concentrations of incompatible elements such as Hf, Zr, Nb, and Ta because these elements are only slightly incompatible to compatible in Ti-rich phases (McCallum and Charette 1977; Jones 1995; Horng and Hess 2000). The assimilant should have Nb/Ce≈10^4 and Nb/Zr≈50 as well as high concentrations of these elements (Shearer et al. 1996; Shearer and Papike 1999). The ratios of these elements in high-Ti mare basalts should far exceed those in VLT picritic glasses. This effect is not observed. High-Ti glasses have only slightly higher Nb/Ce and Nb/Zr ratios than low-Ti magmas (Shearer et al. 1996). Moreover, Zr/Ce ratios in VLT picritic glasses and in high-Ti picritic glasses are roughly the same, contrary to expectation.

2) In the same vein, variations in the initial ε_{Hf} and ε_{Nd} show that the high- and low-Ti mare basalts formed separate, long-lived mantle source regions (Beard et al. 1998). The different ε_{Hf} and ε_{Nd} trends require that the Lu/Hf and Sm/Nd compositions of the source regions be distinctly different, inconsistent with the hypothesis that these magmas were produced by assimilation of various amounts of ilmenite-rich cumulates.

3) All mare basalts have positive initial ε_{Nd} and are derived from variably depleted mantle sources. The initial ε_{Nd} vs. age systematics are not consistent with the assimilation of incompatible-rich, KREEP-like masses (Snyder et al. 2000). Moreover, the general increase in initial ε_{Nd} with Ti (A15 green glass → A12 olivine basalts, A12 pigeonite basalts → A12 ilmenite basalts, A17 high-Ti basalts) further rules out the assimilation of KREEP to produce

these magmas and indicates an association between the ilmenite-clinopyroxene cumulates and extreme long-lived Sm/Nd source depletion. Furthermore, although the ilmenite-clinopyroxene cumulates and KREEP were produced after extensive LMO crystallization, they generally remained isolated from one another.

4) If assimilation of ilmenite cumulates by VLT picrite magmas was a common process, why was the KREEP layer spared? Conversely, why did the parent magmas to the pre-mare Mg-suite not assimilate ilmenite-rich cumulates? The lack of assimilation of ilmenite cumulates is explained by an overturn model that brings the dunite cumulates to the upper mantle and the shallow ilmenite-rich cumulates to the deeper mantle (Hess and Parmentier 1995).

5) Chromium is strongly partitioned into ilmenite under lunar conditions; the partition coefficient at 1 bar and 1180-1200 °C is >6 and the ilmenite Cr_2O_3 contents exceed 2 wt% in high-Ti mare basalts (Delano 1980). The Cr_2O_3 contents of VLT picritic glasses are in the range 0.39-0.64 wt% whereas the range for high-Ti picritic glasses is generally between 0.54-0.82 but with three values 0.93-1.77 for melts with 16.0 wt% or more TiO_2 (Shearer and Papike 1992). Only the highest Cr_2O_3 values are consistent with the assimilation of ilmenite into VLT picritic glasses.

These geochemical arguments support Van Orman and Grove's (2000) rejection of the assimilation hypothesis for the origin of high-Ti mare basalts; VLT mare basalts are not the parent magmas of high-Ti mare basalts. This conclusion, however, does not exclude the role of minor assimilation or hybridization at depths, since some of these petrogenetic processes probably played roles of varying significance in the generation of mare volcanics. Elkins-Tanton and Grove (1999) argued that generation of low-Ti green glass occurred by melting of "reshuffled" cumulates, followed by fractionation and assimilation of small amounts of high-Ti cumulates. Assimilation in this model would occur at depth and reflects the movement of melts through a heterogeneous cumulate mantle.

A second set of models for the evolution of the mare basalt source, particularly the source for high-Ti mare magmas, involve the overturn of the gravitationally unstable cumulate pile of the LMO (Ringwood and Kesson 1976; Herbert 1980; Hess 1991; Ryder 1991; Shearer et al. 1991; Spera 1992; Hess and Parmentier 1995). The overturn creates a hybrid mantle composed of mixtures of various amounts of high-Ca clinopyroxene-ilmenite cumulates and olivine + orthopyroxene cumulates. Heat-producing elements from the intercumulus liquid carried down with the shallow cumulates, moreover, provide the energy to subsequently partially melt the hybridized cumulates and produce the wide range of mare basalt magmas observed on the lunar surface. These models have several attractive features that account for key apsoects of mare-basalt petrogenesis.

The emerging view of mare volcanism is that it was long lived. Mare volcanism, peaked in the 3.6-3.8 Ga period in several Nectarian and pre-Nectarian basins and then went into a smooth decline, interrupted by short periods of vigorous activity, until volcanism devolved into more or less sporadic activity in periods younger than 2.6 Ga (Hiesinger et al. 2000, 2003). Rare outbursts of volcanic activity apparently continued to 1.1 Ga (Hiesinger et al. 2001, 2003). If we now include the earliest, pre-basin mare volcanics as recorded in the cryptomare and the high-Al basalts from Apollo 14, then mare volcanism was active for more than 3.2 b.y. or 70% of lunar history. Aluminous mare basalts show that mare volcanism initiated at least 300 m.y. earlier and the lunar mantle produced basaltic magmas that were emplaced in the lunar crust as early as 4.45 Ga. These observations suggest that the energy source for basaltic magmatism was hardly transitory (i.e., impact driven) or geographically limited.

Given the longevity of mare volcanism, it is significant that the sum of mare volcanism represents only 1% of the volume of the lunar crust and only 0.1% of the volume of the Moon (Head and Wilson 1992). It is also noteworthy that the distribution of mare volcanism is largely concentrated in the near side lunar basins. South Pole-Aitken basin, the largest and likely the most ancient lunar basin, contains only a small volume of mare eruptions. This observation means that the existence of major basins is not sufficient by itself to induce eruption of mare volcanism (Hess and Parmentier 1999). The asymmetry of mare volcanism must therefore reflect the deep-seated distribution not only of the heat-producing elements but also of TiO_2-rich source materials. The varied thickness of lunar crust is not the sole or even principal determinant for controlling mare volcanism.

The features described above are consistent with the overturn of the cumulate stratigraphy, where dense ilmenite cumulates and incompatible element-rich intercumulus materials sank into the lunar interior (Hess and Parmentier 1995). This model has several important implications. (1) It sequesters heat producing elements into the lunar interior providing for a long-lived energy source. Depending on their concentrations, an energy source with 3× (or perhaps less) lunar abundances of the heat producing elements would not only persist throughout most, if not all, of lunar history but might actually generate and maintain a liquid core. (2) The mare source region, particularly for the high-Ti mare volcanics, is provided by the mixing of these late-stage cumulates with a refractory olivine-orthopyroxene cumulate. The volume of mare volcanism is therefore limited by the fertility of this hybrid cumulate. (3) The cumulate stratigraphy produced after the "overturn" is gravitationally stable and produces a heterogeneous and relatively stable mantle (Parmentier and Hess 1998). Large and localized heating of the lower mantle must be invoked to induce the compositionally dense lower mantle to rise diapirically (thermal buoyancy) and generate mare volcanism by pressure-release melting (Hess and Parmentier 1995). This stable stratigraphy is consistent with the isotopic studies, including the most recent Hf-W results (Lee and Halliday 1997). The cumulate stratigraphy was established within the first 50 Ma or so of lunar history and has remained relatively unhomogenized throughout the magma-generating stage. This stable stratigraphy is a formidable barrier to magma production and offers one explanation for the paucity of mare volcanism.

Conductive cooling through the lunar crust and heating from below from a convecting mantle controls the rate of thickening of the elastic lithosphere (Parmentier and Hess 1996, 1997). In all models, the lithosphere is 100-150 km thick at the time of formation of the near-side lunar basins and subsequent filling of these basins with mare basalts. Since the lithosphere extends well below the lunar crust, plumes undergoing pressure-release melting are trapped at a rheological boundary well below the lunar crust (Hess 2000). This constraint on mare volcanism has several important consequences. (1) Large dikes must exist to carry mare volcanics to the lunar surface from depths of 100-200 km or more. (2) The low-density anorthositic crust probably is not a critical density filter for such magmatism. (3) The rheological boundary at the bottom of the lithosphere provides another physical barrier for mare eruptions. (4) The existence of lunar basins, even large ones, cannot strongly influence the eruption of mare basalts from such deep reservoirs.

The gravitational-overturn model rationalizes many of these geochemical and physical features of mare petrogenesis. The model, however, is not without problems. Van Orman and Grove (2000) argued that the shallow cumulates may have been physically too strong to allow the overturn to occur in a timely manner. Using a viscosity estimated from the dry flow laws of olivine and a temperature corresponding to the solidus of the LMO, diapirs 40 km in diameter are calculated to sink only 20 km in 400 million years. Only diapirs larger than 180 km in diameter would descend several hundreds of kilometers in the mantle.

Alternative approaches are needed to overcome these difficulties because there is little doubt that shallow ilmenite cumulates could not be partially melted to produce high-Ti magmas. Ti-rich materials must have sunk deep into the lunar mantle to produce suitably hybridized sources. Van Orman and Grove (2000) suggested that mixing of olivine-rich cumulates with the Ti-rich cumulates could lower the viscosity sufficiently to promote gravitational instabilities.

Parmentier and Hess (1999) showed that small ilmenite-rich diapirs would sink small distances into the subjacent cumulates but then the mixed layer would itself become gravitationally unstable. The diapirs resulting from these mixed sources would be large enough to sink deep into the Moon. Elkins-Tanton et al. (2003) suggested that evolved liquids of the LMO assimilated 10-20% of magnesian olivine from overturned lower cumulates thereby increasing the liquidus by more than 100 °C and bringing Ti-rich spinel to the liquidus. It is possible that such cumulates, being hotter and less viscous, could sink. Elkins-Tanton et al. (2003) also investigated the process of remelting the ilmenite-clinopyroxene-cumulate, which produces a high-Ti, negatively buoyant liquid that would percolate into and hybridize at least the upper mantle. However, these liquids would crystallize as they descend adiabatically. Moreover, the energy source for remelting these cumulates is not clear. If it is assumed that energy from radioactive decay in intercumulus liquids was sufficient to remelt the cumulates at 3.9 Ga, then it is hard to understand how the liquid would have crystallized to form the cumulates in the first place (Hess and Parmentier 2001). Impact-induced melting is also unlikely because it must have been a global event so as to produce widely separated (Procellarum vs. Tranquillitatis) high-Ti mare sources.

Arkani-Hamed and Pentecost (2001) argued that the SPA impact generated vigorous mantle circulation that stripped away the KREEP layer from beneath the lunar far-side crust. This circulation would have distributed radioactive elements throughout the lunar mantle, and generally cooled the mantle beneath the SPA basin, perhaps explaining the lack of substantial volcanism there. The SPA impact might have softened the upper mantle sufficiently to destabilize the ilmenite-rich layer leading to the needed gravitational instability. The gravitational instability would not be Moon-wide and would be limited to the far-side hemisphere, a result which could explain the asymmetry of basaltic volcanism and heat-producing elements on the Moon.

Certainly, a number of alternative models exist that are designed to carry products from shallow-mantle late-stage cumulates to the lunar interior. The geochemical characteristics and phase equilibria of low- and high-Ti mare magmas seem to demand such transfers. Now, there are additional geophysical data to support the convective overturn hypothesis. New information regarding the densities of the lunar interior was recently obtained from the inversion of lunar free oscillations (Khan and Mosegaard 2001). In their density model of the lunar mantle, the density increases from about 3.1 ± 0.2 gm/cm^3 near the lunar crust to values of 3.7 gm/ cm^3 below 950 km. The increase of density with depth is consistent with the profile of a lunar cumulate mantle established after a gravitationally induced convective overturn. Moreover, the density in the central region of the Moon assumes a value of 4.7 gm/cm^3, which is not consistent with a large Fe core. These densities are more appropriate to a sulfide-rich core and/or a core formed from Fe-rich high-Ca clinopyroxene-ilmenite cumulates.

5.13. Synthesis and questions

Geochemical features that best constrain petrogenetic models for mare magmatism are (1) strong, negative Eu anomalies, (2) positive initial ε_{Nd} values, (3) very large range of TiO$_2$ contents, particularly the extreme values for high-Ti picritic glasses, (4) very low CaO and Al$_2$O$_3$ contents of picritic magmas relative to values observed in terrestrial basalts, and (5) positive initial ε_w and ε_{sm}^{142}, which place the evolution of most mare source regions in the earliest 100 m.y. of lunar history. Less constraining, yet important, are phase equilibria data that require the average mantle source to be undersaturated with respect to ilmenite, high-Ca pyroxene and plagioclase. The heterogeneity of the mare sources is ecident in the range of bulk compositions, the varied contents not only of incompatible trace elements, but also in the complex isotopic signatures that characterize mare basalts. Compatible elements, however, appear to place the mantle sources to a limited stratigraphic section of an olivine-orthopyroxene cumulate pile; cumulates with very high or very low Mg′ values appear to be excluded from the mare-source mantle.

The mare source regions are best modeled as orthopyroxene-rich, olivine-orthopyroxene cumulates crystallized from an evolved plagioclase saturated LMO. Ryder (1991) pointed out that the mare basalt sources cannot be simply cumulates from plagioclase saturated LMO because these cumulates are more magnesian than the FANs and they contain too much Ni and clinopyroxene. This may be a product of cumulate mixing (Ringwood and Kesson 1976; Ryder 1991; Shearer et al. 1991; Shearer and Papike 1993). In addition, these cumulates have been fertilized to varying degrees by TiO_2-rich and incompatible element-rich phases. The incompatible element-rich components are constrained to less than about 1% of the cumulate. The characteristic trace-element composition of mare basaltic magmas is, therefore, achieved by small degrees of melting of a hybridized source, but the extent of melting must be sufficient to deplete the source of all minor phases leaving only olivine and orthopyroxene in the residue. This requirement suggests that the TiO_2-rich components must also be added to the source in relatively small amounts. The role of garnet in the source appears to be required by the Lu/Hf data (Beard et al. 1997) and trace-element ratios, such as Sm/Yb and Zr/Y exceeding those in KREEP basalts (Neal 2001). The imprint of garnet implies that melting was polybaric (Longhi 1992) and that garnet-bearing lithologies were distributed throughout the cumulate mantle, albeit in relatively minor amounts. Garnet-pyroxenite lenses may originate from plagioclase-bearing mafic cumulates that were carried deep into the mantle through convective overturn. Partial melts from these lenses would then metasomatize the mantle and create not only the garnet signatures, but possibly the requisite TiO_2 enrichments for the low-Ti and high-Ti mare magmas. If the LMO only encompassed the outer 500 km of the Moon, garnet may not be stable in the cumulate pile, only in the unmelted portion of the Moon (Neal 2001, and references therein). More data are needed to model the implications for mare-basalt magma petrogenesis in the case of the latter scenario.

This discussion briefly encapsulates the broad framework of petrogenesis advocated by most lunar researchers. A number of major questions, however, remain outstanding. First, what are the relationships between crystalline mare basalts and the picritic glasses? The geochemical data reviewed here suggest that the mantle sources of the basalts and the picritic glasses are not the same and that their petrogeneses are not directly related. Alternatively, these differences between picritic glasses and the crystalline basalts may reflect either sampling or differences in eruptive processes. Other important questions relate to the lunar asymmetry in mare volcanism, the mechanisms by which mare basalts erupt from deep within the Moon, the nature of melting and melt segregation, the role of assimilation and possible zone refining processes during magma transport, the evolution of the thermal lithosphere and the tectonic processes by which shallow-level cumulates were distributed throughout the mantle. The answers to these questions will have profound implications for models that deal with the size and evolution of the LMO and the composition and dynamics of the lunar mantle.

6. PHYSICAL PROCESSES GOVERNING THE THERMAL, CHEMICAL, AND MAGMATIC EVOLUTION SUBSEQUENT TO INITIAL PLANETARY DIFFERENTIATION

6.1. Introduction

The focus of this section is to incorporate new observations to better define the physical processes that influence lunar evolution subsequent to initial lunar differentiation (discussed in Section 4.3). In many cases, the latter evolution of the Moon must be placed within the context of accretion and primary lunar differentiation (i.e., LMO and core formation) because these primary processes have an intrinsic role in controlling later events. Fluid dynamics govern many of these processes, but in a wide variety of regimes and scales (length scales, time scales, viscosities, etc.). For example, the physics of solid-state convection is different than that of magma migration, thus these processes are discussed separately. This section is divided into five parts: (1) Basic physics used in modeling lunar thermal and magmatic history in order to facilitate cross-disciplinary communication and so that new workers can be made aware of the assumptions and uncertainties involved. (2) Observational constraints that must be met by lunar thermal evolution models. (3) A synthesis of the thermal history models that have been calculated for the Moon. (4) Mechanisms involved in the migration of mare basalts. (5) Effects of very large impacts on the thermal and chemical state of the lunar mantle.

6.2. Physical processes governing thermal and magmatic evolution

By definition, thermal history models track temperature as a function of space and time within a planetary body by considering the various sources of heat, methods of heat loss (e.g., thermal conduction, convection, magmatism) and conservation of energy. A detailed tutorial on the physics of this process is given in Chapter 9 of BVSP (1981). Here, developments over the past 20 years are emphasized.

6.2.1. What heat to lose? The Moon is a small body, but it still has significant amounts of primordial and radioactive heat. Current models of lunar formation have the Moon accreting within 1-10 years of a giant impact between the proto-Earth and a Mars-size impactor (Section 4.2). This rapid accretion following an energetic giant impact leads to a Moon that begins hot. However, as outlined in Section 4.2, there are uncertainties about the timescale between the giant impact and lunar formation, such that lunar material might have significantly cooled before accreting. No matter how cold the proto-lunar material started, the process of accretion itself caused heating within the proto-Moon. The nature of the accreting material (whether small or large impacts dominated) constrains how much heat is stored deep within the lunar interior and how much remains near the surface (Melosh 1990). Given uncertainties about the initial temperatures of proto-lunatesimals, their size distribution, and the timescale for accretion, it is difficult to constrain the accretional heat within the Moon. One generalization that can be made is that impact energy increased as the planet grew in size and mass (e.g., Kaula 1979; Ransford and Kaula 1980) such that if accretion took less than 100 years, melting of the outer several 100 km of the Moon as a magma ocean would have been an unavoidable result (Pritchard and Stevenson 2000).

The late heavy bombardment had important thermal consequences for the outer layers of the Moon and may account for some of the properties of mare basalts, as well as later eruptions to produce the volcanic glasses (discussed in Section 6.6). Large impacts produce melt through three stages: surface shock melting, instantaneous decompression melting, and pressure-release melting in response to later mantle convection under the site of impact (Section 6.6). Though movement of mantle materials in the Moon is very slow, convection and a hot selenotherm continued under the large craters for 300 to 500 Ma, a scenario that could explain the longevity of mare basalt eruptions (O'Keefe and Ahrens 1993, 1999; Cintala and Grieve 1998).

Following accretion and large impacts, the long lived radioactive isotopes (^{235}U, ^{238}U, ^{232}Th, and ^{40}K) are the most important heat source. Models normally only cite a present-day bulk U content (estimates vary between 20-46 ppb; Hood 1986) and determine concentrations of the other elements using the ratios K/U = 2000, and Th/U = 3.6 (e.g., Toksöz and Solomon 1973; Korotev 1998). The vertical and lateral distribution of these elements have important implications for thermal history. For example, the radioactive elements are concentrated strongly in the crust, and within the crust, they are further concentrated in the Procellarum KREEP Terrane (Haskin 1998; Korotev 2000; Lawrence et al. 1998, 2000; Jolliff et al. 2000). This concentration would increases the amount of melting in this region (Wieczorek and Phillips 2000; Hess and Parmentier 2001). How radioactive elements were concentrated in this region is discussed in Section 7.

6.3. Observational constraints

The chemical and isotopic composition of the *Apollo* and *Luna* igneous samples provide critical constraints for thermal models. The samples provide four different types of constraints on thermal models listed below.

6.3.1. Age of mare basalts. The majority of sampled mare basalts were extruded between about 3.9-3.1 Ga, although some volcanism clearly predates the mare, and might have continued to 1 Ga or later (Schultz and Spudis 1983). A fundamental question is whether the time span of mare extrusion was mainly governed by the duration of partial melting in the lunar interior or the ability of melts to reach the surface. For example, some workers have considered whether the apparent delay between lunar formation at 4.5 Ga and the onset of mare formation at 3.9 Ga provides important information on mantle heat sources (e.g., Hess and Parmentier 1995). There may also have been extensive volcanism between 4.5 and 3.9 Ga, but the late, heavy bombardment or younger lava flows obscured or destroyed the record. Others have suggested that the end of extensive mare volcanism at 3.1 Ga might be linked to global thermal stresses from lunar cooling and contraction (e.g., Solomon and Head 1980). Other explanations for the termination of mare volcanism are possible, such as an increase in depth of melting with time so that the path length of melt to the surface became prohibitive, or diminishment of heat sources with time so that less partial melting occurred. Owing to uncertainties in the duration and volume of mare magmatism, it is best to consider the mare ages a weak constraint on thermal models. Models should have melting and basalt production peaking between 3.9 and 3.1 Ga, and models that generate melt before 3.9 Ga or after 3.1 Ga should not be rejected. Most models that have a magma-ocean initial thermal condition in the outer layer of the Moon have no problem generating melt between 3.9 and 3.1 Ga (Cassen et al. 1979).

6.3.2. Temporal variations observed in mare basalts. Most lunar thermal models involve cooling from above and all predict that the melting zone should deepen with time, except under special circumstances (e.g., Zhong et al. 2000). If the composition and/or source depth of the mare basalts changed with time, that would constrain the location of melting in thermal models. There are, however, few observational constraints on the relationship between time and depth of mantle source. For example, there is no firm evidence that the source depth determined from multiple-saturation points increased as a function of time. Early models sought to relate age of mare basalts, their Ti content, and depth of melting; however, remote sensing studies indicate that some young mare basalts also have high Ti. The oldest examples of lunar magmatism (Mg-suite, high-Al basalts) are relatively low in Ti content. Recent studies of the ages of lunar mare meteorites, which are all low-Ti or VLT, however, have crystallization ages at both the high end [3.8-3.9 Ga (Asuka 881757, Misawa et al. 1993; Yamato 793169, Torigoye-Kita et al. 1993)] and low end [2.8-3.0 Ga (NWA 032, NWA 773, Fernandes et al. 2003; Borg et al. 2004)] of the main range of mare-basalt sample ages. These observations indicate significant mantle heterogeneity and confound any simple relationship between Ti content, age, and depth of melting.

6.3.3. Isotopic systematics observed in mare basalts. Several isotopic systems have been used to determine the lunar differentiation interval and can constrain processes related to LMO crystallization and subsequent lunar differentiation. For example, the U-Pb isotopic system for the mare basalt source appears to have closed between 4.42 Ga and mare basalt extrusion hundreds of My later (e.g., Tera and Wasserburg 1974, 1975, 1976; Papanastassiou et al. 1977; BVSP 1981) and has important consequences for thermal models, although others have questioned whether such isotopic closure occurred (Unruh and Tatsumoto 1977). Brett (1977) used U-Pb isotopic closure and arguments about atomic diffusion to constrain temperatures within the mare source region and initial mantle temperatures. Conversely, Pritchard and Stevenson (2004) argued that because diffusion is inefficient, isotopic closure instead places limits on the amount of melting in the mare source between 4.42 and mare-basalt extrusion and says little about the initial thermal state of the Moon. However, the ability to use isotopic closure to constrain the extent of melting is limited by inadequate knowledge of the partition coefficients for U and Pb under appropriate lunar conditions and by the limited data used to define isotopic closure. Further work is needed on the U-Pb system and applying results from other isotopes like Sm-Nd (e.g., Nyquist et al. 1995) to quantitative thermal-history models.

6.3.4. Mare basalts compositional heterogeneity and lunar asymmetry. The mare basalts (crystalline mare basalts, pyroclastic glasses) that erupted between 3.8 and 3.1 Ga exhibit both compositional and isotopic variation suggesting that the near-side lunar mantle is laterally and vertically heterogeneous in both temperature and composition (e.g., BVSP 1981; Nyquist and Shih 1992). Remotely sensed compositional data, such as Ti content inferred from Clementine spectral reflectance data, also indicate significant lateral variability in basalts (Gillis et al. 2003). Although it is difficult to constrain the size of the mantle heterogeneities (Spera 1992), overturn models proposed by Parmentier and co-workers have predicted the potential size range of high-Ti cumulate pods (see Section 3.3). The heterogeneous lunar mantle indicates that it was not well mixed and that mantle convection was not vigorous enough to completely homogenize the cumulate remnants of the LMO (Turcotte and Kellogg 1986).

The majority of lunar thermal-history models have been one dimensional, representing a global average of temperature and melting as a function of time. However, new global datasets confirm older results that the distribution of mare basalts, KREEP and crustal thickness are not globally uniform. Mare basalts cover about 30% of the near side but only about 1% of the far side, KREEP is concentrated in the Procellarum region, and the crust is generally thicker on the far side but spatially variable on the near side (see Chapter 3). Any thermal history should explain the distribution of mare basalts and KREEP, but the crustal dichotomy is not understood. In order to explain lunar asymmetries, two- and three- dimensional models (e.g., Wieczorek and Phillips 2000; Zhong et al. 2000; Spohn et al. 2001) incorporating lateral variations in the distribution of radioactive elements and melting are needed. For example, the concentration of KREEP in the Procellarum region could have enhanced melting in that region, and the near side in general, partially explaining the hemispheric dichotomy of mare basalts (e.g., Wieczorek and Phillips 2000). However, extremely large amounts of melt would form an impenetrable barrier to the eruption of basalts (Hess and Parmentier 2001). Hess and Parmentier (2001) also suggested that the associated high crustal temperatures might pose problems with the long-term support of the mascons. Zhong and Zuber (2000) concluded that on a planetary body the size of the Moon the existence of mascons provided no evidence for the thickness of the lunar lithosphere. More generally, two- and three-dimensional models of thermal evolution are required to understand the importance of large impact basins and their spatial distribution on thermal evolution. Recent work on the global effects of impacts on the thermal evolution on Mars (Reese et al. 2000) and regional effects upon the Moon (Arkani-Hamed and Pentecost 2001; Elkins-Tanton et al. 2003) explored the possible effects of impacts on planetary asymmetries.

6.3.5. Lunar magma ocean. Thermal histories require an initial condition. Most models begin immediately following crystallization of the LMO and consider only the thermal effects of that event—meaning that temperatures in the region of the magma ocean start at the local solidus. There are weak bounds on the thickness of the LMO with the range of 300-1000 km favored (Warren 1985; Longhi 1992). Besides thermal effects, the LMO has important effects on chemical layering. Among the possible effects are: stable stratification that would inhibit mantle convection, or unstable stratification that promotes mantle overturn with concomitant movement of radioactive elements to deeper layers and pressure-release melting (Hess and Parmentier 1995; Alley and Parmentier 1998). More detailed models of LMO crystallization are needed to better understand the thermochemical parameters used in thermal models.

6.3.6. Surface geology. The lack of a global system of thrust faults on the lunar surface has been used to constrain the amount of volume change within the Moon and the initial temperature profile (e.g., Solomon and Chaiken 1976). As mentioned in Section 4.2, it is more difficult to relate surface strain to internal processes than previously thought so the lack of thrust faults does not strongly constrain lunar volume change. Another way that surface faults have been related to volume change is through the formation of rilles (interpreted to be grabens) around the periphery of mascons. In Chapter 3, a discussion is given of rilles formed through flexural stresses related to the mass concentration in the mascon. Solomon and Head (1979, 1980) also attributed the formation of the rilles to the global stresses related to internal thermal evolution. In particular they noted that the apparent global cessation in rille formation at about 3.6 Ga (Lucchitta and Watkins 1978) required the global stress pattern to change from extension to contraction. This constraint on the timing of the change in global stress reduces the number of viable thermal models. However, given the uncertainties in predicting faulting around mascons (Chapter 3) and in dating the time of formation of rilles, it is not clear whether the timing of this change in global stress can be so precisely dated, is necessary to explain the global cessation of rilles, or if such a global cessation really occurred (since rilles are not globally distributed).

6.3.7. Current temperature profile. Hood (1986) reviewed the geophysical evidence for the current thermal state of the Moon from heat flow, seismic data and electromagnetic sounding. The evidence indicates that the outer layers of the Moon are subsolidus, but that temperatures below about 1000 km are likely warmer and possibly at the melting point (see Chapter 3). The fact that brittle failure in the form of moonquakes occurs to about 1000 km means that the lunar mantle is subsolidus to at least that point. However, because we cannot directly measure temperature within the Moon, many assumptions must be made to relate measurements to interior temperatures (BSVP 1981; Hood 1986). For example, the two heat flow measurements have been used to infer the bulk U content of the Moon (very important for thermal models), but depending on the assumptions made, inferred values range from 20-46 ppb (Langseth et al. 1976; Hood 1986; Warren and Rasmussen 1987).

6.3.8. Gravity field. Two aspects of the lunar gravity field bear upon temperatures within the Moon. (1) Mascons (discussed in Chapter 3) are mass concentrations associated with large impact craters that are clearly not in isostatic equilibrium. This means that the strength of the lithosphere has been able to support the excess mass of these structures since they were formed at ~4 Ga. However, the ability of a planet to support long-wavelength loads increases as planetary radius decreases. In other words, because of the small size of the Moon, the persistence of mascons does not strongly constrain the thickness of the lithosphere or temperatures in the upper mantle (Zhong and Zuber 2000). (2) Many workers have claimed that the low-order terms of the spherical harmonic expansion of lunar topography and the gravity field are anomalous and require an explanation separate from the other terms (Hood and Zuber 2000; Stevenson 2001 and references therein). The most common special explanation is that the low-order gravity and topography terms are a relic of faster rotation in the past, typically when the radius of the Moon's orbit was 13-16 Earth radii. The fast evolution of the Moon away from the Earth early in its history (e.g., Goldreich 1966), requires that the Moon would have to cool quickly in order for the rotational bulge from this early period to be preserved. This rapid cooling would have important implications for the timing of cooling/heating of the LMO and the initial thermal state of the Moon following differentiation (Hood and Zuber 2000). However, other explanations for the low order gravity and topography exist (Stevenson 2001), and so the thermal constraints are not unique.

6.3.9. Magnetic field. As outlined in Chapter 3, several lines of evidence support the possibility of a small, Fe-rich, conducting, and liquid lunar core. Maintaining a liquid core for the duration of lunar history provides a weak constraint on interior temperatures since the eutectic temperature of an Fe-S core under lunar conditions is only 1250 K. If a lunar dynamo existed early in lunar history, as suggested by some paleomagnetic studies (see Chapter 3), it constrains temperatures within the lunar core during that time interval. There are theoretical difficulties with a convecting lunar dynamo generating the field strengths recorded in the samples (Chapter 3), and so perhaps some other mechanism for stirring the lunar core is needed (e.g., tidal effects; Williams et al. 2001). If true, constraining the role of tidal effects would be important for understanding early lunar thermal/orbital evolution. The small size of the lunar core results in the process of core formation not having large thermal consequences for the mantle, and might even be able to coexist with a cold deep mantle (Stevenson 1980).

6.3.10. Mantle layering. The four seismometers deployed by Apollo were used to study mantle structure on the near side, and indicated a possible discontinuity at 500 km (see Chapter 3). Khan and Mosegaard (2001) used lunar free oscillations to determine that the lower mantle is denser than the upper mantle, meaning that whole-mantle convection would be inhibited by compositional stratification. If such a layer is truly global in nature it suggests that mantle convection (if it occurred) could not penetrate the 500 km discontinuity and was therefore layered, which decreases its vigor. However, because of the limited distribution of lunar seismometers and the difficulties in interpreting lunar seismograms, the sharpness, importance, and extent of the 500 km discontinuity is questionable.

6.3.11. Summary. Thermal models must agree with the available observational constraints, although in the discussion above we have highlighted the vagueness and non-uniqueness involved in each of these. Most observational constraints should not be used to rigidly divide which models are acceptable and unacceptable, although some requirements are more secure—there was an LMO and a high proportion of post-LMO mantle melting occurred between ~4.45 and 3.1 Ga. In fact, as noted above, the first requirement practically guarantees the second in most models, and so there really are few unique constraints. However, the situation is not hopeless and with future work, at least two observational constraints and two computational approaches could become more robust: (1) Additional seismometers on the lunar surface, especially on the far side, could determine if the lateral extent of the 500 km discontinuity is real and global. (2) New sample studies of lunar paleomagnetism can help to determine if there is a lunar dynamo and when it operated. (3) A mostly unexplored area is directly linking geochemical sample studies (isotopes, trace and major elements) with quantitative thermal models. A much better understanding of melt migration within the Moon and how samples were contaminated in the journey from source to surface will be required to make this link (4) Coupled thermal and dynamic models of early lunar evolution could provide better understanding of the low-order gravity and topography.

6.4. Synthesis of thermal history models

Over the past 50 years, roughly 50 thermal-history models have been published in peer-reviewed journals (Tables 4.10 and 4.11). Chapter 9 in the BSVP (1981) provides an excellent overview of the history and physics of thermal models as well as a summary of lunar thermal histories calculated before 1980. More than 80% of all thermal histories were calculated before that date. However, the lack of papers published since 1980 should not be interpreted as a lack of questions to be answered or new developments. As outlined below, there have been many developments that show how previous models were incomplete and too simplistic, and there are many fundamental observations about the Moon that are poorly understood (e.g., the asymmetric distribution of mare and KREEP). Instead, most thermal modelers have followed the data—as the number of lunar missions declined, the modelers applied their craft to explain new data from the other terrestrial planets and the galilean satellites. With the recent return to the Moon and anticipation of further new datasets, this is a good place to consider the state of this field. In this section is a summary of the types of models that have been used, simplifications and limitations of each type of model, and then synthesize the important results.

6.4.1. Types of models. Many thermal models prior to the return of Apollo samples were purely conductive and did not include the effects of melting. For example, Urey's models (see

Table 4.10. Summary of primary types of thermal models that have been used and the deficiencies with each. See text for more discussion.

Model	Deficiencies
Purely **Conductive** or purely **Convective**	Has to include melting.
Conductive – test for instability to **Convection**	Old models didn't include compositional buoyancy which can be as important as thermal buoyancy.
Conductive or **Convective** with melting but no melt migration	Melt migration, particularly segregation of radioactive elements into melt is important. Parameterization of melting is not well constrained.
Conductive with melting and melt migration	Parameterizations of melt migration are crude-- New models must include physics of melt migration.
Conductive/Convective including chemical effects	Again, must include physics of melt migration. Must further explore plausibility of initial conditions.

Table 4.11. A summary of the thermal models from the past 50 years. The criteria for putting papers in this list is that they had to be published in a peer-reviewed journal, and that the model had to quantitatively track temperatures in the lunar interior as a function of time.

Type of Model	Examples
Purely **Conductive**	Urey (1951,1952,1957,1962); MacDonald (1959,1962,1963); Phinney and Anderson (1967); Gilvarry (1970); Hanks and Anderson (1972)
Purely **Convective**	Turcotte and Oxburgh (1969); Tozer (1972,1974); Schubert et al. (1977,1979,1980), Cassen and Young (1975)
Conductive, and test for **convection**	Reynolds et al. (1972); Cassen and Reynolds (1973,1974); Toksöz et al. (1978); Cassen et al. (1979)
Conductive with melting – No melt migration	Levin (1962,1966); Levin and Majeva (1960); Majeva (1964); Solomon and Chaiken (1976); Solomon (1977); Solomon and Head (1979); Wieczorek and Phillips (2000)
Convective with melting – No melt migration	Turcotte et al. (1972,1979); Chacko and De Bremaecker (1982); Konrad and Spohn (1997); Spohn et al. (2001)
Conductive with melting and melt migration	Fricker et al. (1967); Lee (1968); McConnell and Gast (1971); Wood (1972); Solomon and Toksöz (1973); Toksöz and Solomon (1973); Toksöz and Johnston (1974,1977); Strangway and Sharpe (1975); Binder and Lange (1977, 1980); Ornatskaya et al. (1977); Kirk and Stevenson (1989)
Conductive/Convective including chemical effects	Hess and Parmentier (1995); Alley and Parmentier (1998); Zhong et al. (2000)

Table 4.11 for references) favored a cold Moon by relying upon three arguments later found to be flawed. (1) Urey and others assumed that in order to maintain structures in the lunar crust and the non-hydrostatic lunar shape, there could be little to no partial melting during lunar history. Thus, many early workers used models that were purely conductive and did not use latent heat to calculate the extent of partial melting (Table 4.10). To minimize any partial melting, these models required the Moon to start extremely cold, nearly 273 K at all lunar

radii, a result that is inconsistent with the LMO (e.g., Wood 1972). Recent work has shown that only the outermost lunar layers must remain cold to maintain long-wavelength lunar shape, and that interior temperatures are not constrained (Zhong and Zuber 2000). (2) Many early workers assumed a chondritic abundance of radioactive elements, whereas sample analysis has revealed that the lunar samples are depleted in K and enriched in U relative to chondrites (e.g., Toksöz et al. 1972; BVSP 1981). (3) Urey said that partial melting might not be required to form the mare and that they could be explained by impact melting. Sample analysis indicates that the mare were formed by partial melting in the lunar interior.

Even before analysis of lunar samples conclusively demonstrated partial melting in the lunar interior, many conductive models attempted to quantify melting in the lunar interior (e.g., Levin 1962) and the effects of moving melt to the surface (e.g., Fricker et al. 1967; Lee 1968). The physics of melt migration are complicated and there are many uncertainties about how the processes operated in the Moon (Section 6.5). Therefore, models crudely parameterize complex properties by instantaneously moving all of the melt from the source region to the near surface. Some models move melt to the top of the partial melting region (e.g., Wood 1972), while others distribute it exponentially with depth (e.g., Toksöz et al. 1972; Kirk and Stevenson 1989). The most important consequence of the melt migration is to remove the radioactive elements (which are incompatible and preferentially go into the melt) from the melting region to shallower, cooler layers. Different parameterizations of exactly how the radioactive elements segregate into the melt have been presented, but they all have the effect of limiting the total amount of melting. As an example, Figure 4.48 shows the total amount of melting as a function of depth and time within a one dimensional conductive thermal history, both with and without melt migration. The principal results of these calculations is that melt migration limits the total amount of melting that occurs at any given depth, and expands the region of melting at a given time to slightly shallower depths. Thus, melt migration is able

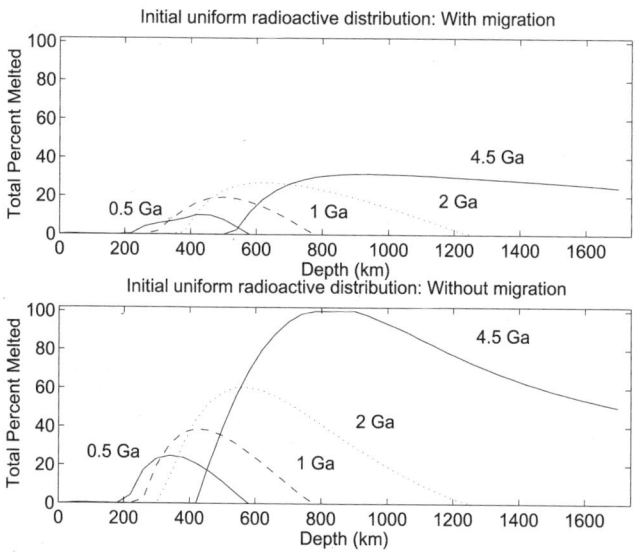

Figure 4.48. The total percent melted within the lunar interior as a function of depth and time for a one dimensional conductive model both with and without melt migration. The model uses the method and parameters of Kirk and Stevenson (1989). Although the radioactive elements are not initially distributed uniformly throughout the interior (they are concentrated in the center and in the crust), this does not affect the result that including melt migration reduces the total amount melted at a given depth.

to reduce possibly unrealistic amounts of melting that exist in some thermal models (e.g., Wieczorek and Phillips 2000).

Some thermal modelers have considered the effects of mantle convection within the Moon, first suggested to be important by Runcorn (1962). There is no observational evidence indicating convection has occurred or is occurring within the Moon (Cassen et al. 1978; Phillips and Ivins 1979; BVSP 1981). In fact, there are three observational arguments that indicate that mantle convection has been weak or nonexistent. (1) An analysis of the Moon's free oscillations by Khan and Mosegaard (2001) found that while the density of the upper ~500 km of the mantle was approximately constant, a sharp increase in density occurred below this depth. While the identification of lunar free oscillations remains controversial (see Chapter 3), the existence of a 500-km seismic discontinuity in the Apollo seismic data is consistent with a change in composition (and possibly density) below this depth (e.g., Hood and Jones 1987; Mueller et al. 1988). If these results are globally representative of the mantle (all seismometers were located in a small region on the lunar near side), then a 500-km density discontinuity would presently act to inhibit whole-mantle convection. While the time of formation and the origin of the 500-km discontinuity are presently being debated (see Section 2.3), if it is similarly ancient, then convection (if it occurred) may have only been confined to the region above or below the 500 km discontinuity. If so, this significantly reduces the length scale and vigor of convection. Such a density stratification within the lunar mantle could have been achieved as a result of the overturn of a gravitationally-unstable magma-ocean cumulate pile (e.g., Herbert 1980; Spera 1992; Hess and Parmentier 1995). (2) The distribution of heat sources in the Moon may also have acted to inhibit thermal convection within the lunar mantle. Using Lunar Prospector Th abundances at the surface (e.g., Lawrence et al. 1998, 2000), Jolliff et al. (2000) estimated the quantity of heat-producing elements in the mantle and crust. They found that about 75% of the Moon's incompatible elements were located in the crust, and furthermore, that about 40% of the crustal incompatible elements were located within the regions of Oceanus Procellarum and Mare Imbrium. Such a strong crustal enhancement of heat-producing elements could act to heat the mantle from the top down, and hence lessen the thermal buoyancy contrast between the upper and lower mantle. Estimates of the abundance of Th in the mantle are as low as 5-25 ppb based on the composition of the picritic glasses (e.g., Warren and Wasson 1979; Jolliff et al. 2000), and the low end of this range is considerable less than is usually assumed for most thermal-convection models (e.g., 20-29 ppb in the models of Spohn et al. 2001). A low concentration of heat-producing elements in the mantle would be a natural expectation if the mantle is composed of cumulates from a near-global LMO and would further act to limit the vigor of mantle convection. (3) The existence of compositional heterogeneity in the mare-basalt source region (Section 4.5) indicates that the lunar mantle was not well mixed, and that vigorous mantle convection did not occur (Turcotte and Kellogg 1986).

Consistent with the observations cited above, theoretical calculations that consider realistic viscosities also indicate that lunar-mantle convection might have been weak or non-existent. Many workers have tested various thermal models for convective instabilities that form when thermally induced positive buoyancy overcomes viscous resistance and the Rayleigh number overcomes its critical value (see Chapter 3 and Section 6.2 of this chapter). Early models that assumed a constant viscosity within the lunar interior found that the Moon and other terrestrial planets would be convectively unstable (e.g., Schubert et al. 1969). Models that started with a hot exterior (LMO) and cold interior, did not become convectively unstable until the unfavorable initial temperature profile could be reversed by the interior heating up and the exterior cooling (e.g., Cassen and Reynolds 1973). However, viscosity is a very sensitive function of temperature such that viscosity is not constant as a function of depth within the lunar mantle. Including the effects of temperature-dependent viscosity, Cassen and Reynolds (1974) calculated that the lunar mantle is currently unstable to convection. Recent progress has been made in understanding when and how convection occurs on planets with immobile lithospheres, such as the Moon (see Section 6.2). A primary result is that convection can only occur within a region where the viscosity changes by about an order of magnitude. This severely limits the amount of positive buoyancy that can form convective instabilities, and makes convection more unlikely and less vigorous on planets with stagnant lids than was previously thought. Considering the uncertainties in lunar mantle viscosity (Section 6.2), several thermal evolution scenarios are possible, including no convective instabilities or weak convection (Pritchard and Stevenson 1999).

As the Moon could be in the regime where convection is only marginally viable, compositional effects on buoyancy must be considered as well as thermal effects. Both stable and unstable compositional layering must be considered. During LMO crystallization, dense Fe-rich cumulates are placed over less dense Mg-rich cumulates, forming an unstable arrangement that causes overturn to a more stable layering that inhibits thermal convection (Hess and Parmentier 1995; Alley and Parmentier 1998). As most of the Fe- and Mg-rich cumulates crystallized in the LMO are less dense than the deeper lunar interior, overturn of the cumulates is likely to be limited to be above the bottom of the LMO (Hess and Parmentier 1995). Conversely, late-stage cumulates that include ilmenite would be more dense than all underlying material and under certain conditions, these cumulates and other late-stage materials rich in radioactive elements could sink deep into the lunar interior as Rayleigh-Taylor instabilities, perhaps all the way to the center of the Moon (Ringwood and Kesson 1976; Hess and Parmentier 1995; Zhong et al. 2000).

Alternatively, the denser material could sink into the lunar interior via percolative flow instead of as diapirs. Elkins-Tanton et al. (2001) argue that the cumulates crystallize at low temperature (about 1150 °C), and the viscosity under these conditions is very high and augmented by the viscosity law of pyroxene, which is stiffer than olivine. Following numerical modeling with a realistically cooled Moon and a solid high-Ti layer, they find the process of sinking instabilities unlikely due to viscosity constraints. If the ilmenite cumulate layer could not fall as large instabilities into the deep Moon, there is still the need to move high-Ti material deeper into the Moon in a heterogeneous manner, to create the source regions required by the picritic glasses.

Van Orman and Grove (2000) also recognized that the high viscosity of the ilmenite + pyroxene cumulate, could inhibit subsolidus flow, and suggested that mixing olivine-rich cumulate with the high-Ti cumulate could lower its viscosity enough to allow the formation of instabilities. Another possible mechanism for redistributing high-Ti material into underlying cumulates is melting by a large impact, and then sinking high-Ti liquids into underlying cumulates. As high-Ca clinopyroxene and ilmenite melt in a ratio of ~2:1 (Wyatt 1977; Van Orman and Grove 2000), remelted high-Ti cumulates would be negatively buoyant at depths as shallow as 20-80 km, and would sink as liquids, percolating downward through the underlying mantle and beginning to recrystallize ilmenite at 200 km depth, making a hybrid, heterogeneous mantle (Elkins-Tanton et al. 2001).

Subsequent to crystallization of the LMO, another process could form a stable compositional gradient that would inhibit or weaken convection. Nearly all thermal models include a deepening of the melting region as a function of time and lessening of the degree of melting with depth related to cooling from above. As the residual material left over after melting is less dense than the original material, the variable extraction of melt as a function of depth forms a stable compositional gradient that can be large enough to offset thermal buoyancy (Pritchard and Stevenson 1999). However, there is great uncertainty in the amount of melting as a function of depth and time in the mantle, and there are certainly lunar histories (especially if there is heating from below by an ilmenite/radioactive enriched core) when the variable extraction of melt with depth is not important.

6.4.2. Synthesis of results.

1.) *Conduction vs. convection.* As we emphasized above, there is no conclusive observational or theoretical requirement for lunar-mantle convection, and several reasons to believe that convection might have been weak or nonexistent. The primary difference between conductive and the incorrect constant-viscosity convective thermal histories is that the latter cool too quickly and efficiently, causing a more rapid disappearance of partial melt, thinner lithosphere, and lower mantle temperatures (BSVP 1981). Models that more appropriately account for the effects of variable-viscosity convection (e.g., Spohn et al. 2001) are more similar to conduction models in terms of mantle temperatures, lithosphere thickness, and duration of partial melt.

2.) *Initial condition.* When a thermal model begins with an accretionary thermal profile (hot exterior and cold interior, see Section 6.3), it is stable to convection. Therefore, it takes hundreds of millions to billions of years to heat up the interior before it can become thermally unstable to convection (e.g., Cassen and Young 1975; Turcotte et al. 1979; Spohn et al. 1999), although because chemical buoyancy is also important, this does not mean convection will occur. As 4.5 Ga is long enough to allow an initially cold interior to heat up, but not long enough for a hot interior to cool down, the present thermal profile does not well record the initial thermal profile (Cassen et al. 1979).

3.) *Timing of partial melting.* It is not difficult to account for partial melting in the lunar mantle during the primary era of mare volcanism (3.9-3.1 Ga) when the outer layers of the Moon begin at the solidus as expected for conditions following a LMO (Cassen et al. 1979). Most models also generate melt before and after this era. The record before 3.9 Ga is largely obscured so it is difficult to say much about that time period, but an unanswered question is why the majority of the visible mare seemed to have erupted between about 3.9 and 3 Ga, with minor additions occurring up to ~1 Ga. Proposed explanations include an inability of the melt generated to reach the surface because of global compressive stresses or an increase in the depth of melting, or a decrease in the amount of melt generated because the source regions were depleted in heat sources (due to cooling or loss of radioactive elements) or an increase in the solidus (due to prior melting or loss of volatiles; BVSP 1981).

4.) *Chemical versus thermal buoyancy.* As discussed in Section 7.5 (below), compositional gradients formed during LMO crystallization or from the variable removal of melt as a function of depth have two effects. First, stable compositional gradients can inhibit the development of convection. Second, unstable compositional gradients might overcome thermal buoyancy effects and cause sinking of dense ilmenite cumulates, if certain rheological conditions are met. Therefore, it is important to consider both chemical and thermal buoyancy when evaluating what will happen in the lunar interior.

5.) *Melt migration matters.* Melt migration in thermal history models is either ignored or parameterized in a simplistic manner, as discussed above. However, the extraction of melt with its preferential removal of volatiles and radioactive elements has important consequences for the possibility and extent of further melting (Section 4.6.5).

6.) *Lateral variations.* The distribution of heat sources, whether they are radioactive elements or large impacts, affects when and where melting occurs. While most thermal models have been one-dimensional, two- and three-dimensional models of lunar thermal evolution are needed to account for the laterally and vertically variable distributions of heat sources.

6.5. Processes governing migration of lunar basalts

6.5.1. Dynamic models of magma transportation. How the amount of melting is calculated from the conductive or convective temperature profiles was discussed in the previous sections. The processes by which melt could be transported to the surface are discussed in this section. Much progress has been made over the past two decades in modeling the dynamics of magma

transportation; however, significant gaps remain in understanding this process. A full treatment of this subject would include the effects of melt segregating from its matrix, porous flow, formation of dikes, and thermodynamics of melt interacting with the surrounding material. In this section we briefly review two aspects of magma transport that may help to decipher the complex geochemistry of the lunar basalts; porous flow and dike propagation. The reader is referred to the review by Rubin (1995), the papers in the book *Magmatic Systems* (Ryan 1994a) and the papers by Lister and Kerr (1991) and Sleep (1988) for more detailed accounts.

6.5.2. Porous flow. Porous flow of magma may be an important process deep in the lunar interior where melt would be generated. The migration of deep magma depends on the buoyancy of the melt with respect to the matrix, the permeability of the solid, viscosity of the melt, and deformation and compaction of the matrix (e.g., McKenzie 1984). As compaction should only be important for scales on the order of 100 m within the lunar mantle (e.g., Spera 1992), melt migration can be adequately modeled by Darcy's law:

$$v_l - v_s = \frac{k\,(1-\phi)\,\Delta\rho\,g}{\eta\,\phi}; \qquad k = k_0\,\phi^n \qquad (4.16)$$

where v_l and v_s are the velocity of the melt and solid, k is the permeability, ϕ is the porosity of the matrix (i.e., melt fraction), $\Delta\rho$ is the density difference between the melt and matrix, g is the gravitational acceleration, and η is the melt viscosity. Inserting parameters that might be appropriate for the lunar mantle ($\phi = 0.05$, $g = 1.6$ m s^{-2}, $\Delta\rho = 300$ kg m^{-3}, $\eta = 1$ Pa s and $k_0 = 10^{-8}$ m^2, $n = 2.5$ (e.g., Sparks and Parmentier 1994; Spera 1992)) the vertical velocity of the melt should be on the order of 1 m/year. Given uncertain knowledge of such parameters as the melt fraction and permeability, this estimate could be more than an order of magnitude too small or large. Indeed, the formation of high-porosity channels could significantly increase these estimated velocities (e.g., Dick 1999; Kelemen et al. 1999).

If the mare source was located about 500 km below the surface, it would take on the order of a million years for this melt to reach the surface via porous flow. Because of this long duration it might be expected that the melt would always be in chemical equilibrium with its surrounding matrix. Furthermore, because of the slow ascent velocities, if the temperature of the matrix ever dropped below the melt's solidus, the magma would freeze, inhibiting further vertical melt migration. Subsequent to initial freezing, melt would begin to accumulate below this low-permeability barrier, possibly leading to the formation of a magma chamber and/or to the initiation of magma transport via dike propagation.

6.5.3. Static dikes. Numerous papers have been written on the subject of determining the shape of a fluid-filled dike in an elastic medium (e.g., Sneddon 1946; Weertman 1971; Pollard and Holzhausen 1979; Pollard 1987; Rubin and Pollard 1987). In these models the shape of the dike is determined by the elastic stresses in the host rock, the hydrostatic pressure in the dike, and the resistance to fracture at the dike tips. Exact solutions exist for disk-shaped as well as 2-D blade-like fractures. These studies show that in order for a blade-like dike to be completely static, with no tendency to propagate, the vertical length of the crack must be less than

$$l_{max} = 2\left(\frac{K_c}{\Delta\rho\,g}\right)^{2/3} \qquad (4.17)$$

and have a width less than

$$w_{max} = \frac{1-\nu}{\mu}\left(\frac{K_c^4}{\Delta\rho\,g}\right)^{1/3} \qquad (4.18)$$

where K_c is the fracture toughness, μ is the shear modulus, and ν is Poisson's ratio of the rock (e.g., Lister and Kerr 1991). The fracture toughness is a material property that describes the distribution of stress near a dike tip that is necessary for a rock to fracture. Using typical values that might be appropriate for the mantle (K_c = 1-5 MPa m$^{1/2}$, μ = 60 GPa, v = 0.25), the maximum length of a static dike is ~300 m to 1 km, and its maximum width is ~0.1 mm to 1 mm.

If the length of a static dike were to exceed l_{max} by any amount, then the elastic stresses would act to open the upper tip of the dike and to close the lower tip. This quasi-static result has prompted many studies to assume that an individual dike will propagate upwards once this marginally unstable height has been achieved (e.g., Head and Wilson 1992; Wilson et al. 2001). However, since this formulation does not take into account any hydrodynamic effects (namely, the pressure gradients that are required for the magma to flow) it is inaccurate to use this model to describe a propagating dike. Furthermore, this static model can not estimate how fast such a dike would propagate. Two other fundamental problems exist using such a dike propagation model (Lister and Kerr 1991). First, owing to the extreme narrowness of the dikes, they should freeze rapidly as they traverse the lithosphere. Secondly, the bottom portion of the dike will not be able to completely close because an infinite pressure gradient is required to squeeze viscous magma out of a narrowing gap. Thus, a dike is likely to propagate only if magma is continually supplied.

6.5.4. Vertical dike propagation. In modeling the propagation of a fluid-filled dike in an elastic medium, several sources of pressure within the dike must be considered. These include the elastic stresses set up by deforming the surrounding medium, the stresses required to fracture the rock at the dike tips, hydrostatic pressures within the fluid (i.e., buoyancy), the pressure gradients required to drive fluid flow, external tectonic stresses, and internal overpressurization. Lister and Kerr (1991) considered the relative magnitudes of these stresses for a vertically propagating dike and determined that the balance between buoyancy forces and pressure gradients due to flow in a viscous fluid were the most important. Elastic stresses were found to play a moderate role, and the stresses required to fracture the medium were found to be negligible compared to the other sources of stress.

Two models of a vertically-propagating fluid-filled dike were developed by Lister and Kerr (1991) that might be applicable to the transport of magma within the Moon. In one model, magma was continuously supplied from a point source, and the equilibrium shape of the dike was determined. Neglecting the fracture strength of the rock, the breadth, b, and central width of the dike, w, were calculated as:

$$b(z) = 5.25 \left(\frac{Q\eta\mu^3 z^3}{g^4 \Delta\rho^4 (1-v)^3} \right)^{1/10} \quad ; \quad w(z) = 1.808 \left(\frac{Q^3\eta^3 (1-v)}{\mu g^2 \Delta\rho^2 z} \right)^{1/10} \quad (4.19)$$

where Q is the volumetric magma production rate, and z is the height above the point source. This model of magma transport might be valid for determining the width and length of a dike that erupts at the surface from a deep mantle source. Assuming that such a dike initiates 500 km below the surface and has a low magma production rate of 10 m^3/s, the length and width of the dike at the lunar surface would be ~50 km and ~2 cm, respectively. For a high magma production rate of 10^6 m^3/s, this model predicts a dike length of ~170 km and a width of ~0.7 m. As discussed by Lister and Kerr (1991), freezing of magma near the dike margins would tend to decrease the breadth of such a dike and increase its width.

In another model, Lister and Kerr (1991) obtained an analytical solution for a vertically propagating dike being fed by linear source at depth. It was found that the fracture strength of the rock only slightly influenced the shape of the dike near its tip. For a magma production rate q per unit length, the dike width at depth, w_∞, and the speed of dike propagation, c, were shown to be given by

$$w_\infty = \left(\frac{3\, \eta\, q}{2\, g\, \Delta\rho} \right)^{1/3} ; \qquad c = \frac{q}{2\, w_\infty} \qquad (4.20)$$

The linear magma production rate can be approximated from the first of these models. If a total magma production of 10^6 m³/s is assumed to operate over a dike length of 170 km, then q is ~6 m²/s. This gives a dike width at depth of ~0.3 m and a propagation speed of 10 m/s. If the mare source were located 500 km below the surface, then it would only take about half of an Earth day for this dike to reach the surface. Alternatively, assuming a total magma production of 10 m³/s operating over a length of 50 km, q is ~2 × 10^{-4} m²/s, w_∞ ~ 1 cm, and c is ~0.01 m/s. In this case the dike would be able to traverse 500 km in just over a year. In modeling the transport of primary basaltic magmas (represented by the VLT pyroclastic glasses) from the deep lunar mantle, Spera (1992) calculated the requirements needed to move that magma a distance of 400 km with no or very limited degrees of fractional crystallization. He concluded that such a melt must travel at 10 m/s through a crack of 40 m in diameter.

These qualitative results show that the transportation of magma via dikes is much faster than by porous flow. These short ascent times require that a dike should be able to traverse the cool lithosphere without completely crystallizing. Furthermore, the magma is unlikely to come into chemical equilibrium with the surrounding rock, thus potentially preserving isotopic and compositional information from the mare source. In that some of the above requirements may be unlikely (400 km × 40 m dike) for the Moon, alternative models have been described to transport these magmas to the lunar surface.

Wilson and Head (2003) used the dike propagation analysis of Rubin (1993) to derive a modified dike transportation model for these near-primary magmas. They proposed a model in which the dike propagates rapidly from the magma source to the surface through gas build-up in a low-pressure micro-environment near the tip the magma-filled crack. The gas-rich region consists of a free gas cavity overlying a basaltic foam extending vertically for ~20 km. It is certain that the lunar mantle is volatile-depleted. In such a volatile-poor environment, it has been proposed that the driving mechanism for lunar fire-fountaining is the oxidation of C to produce CO (Fogel and Rutherford 1995). The depth at which this reaction occurs (Fogel and Rutherford 1995) appears to be shallower than that proposed by the Wilson and Head (2003) model. In this model, the gas-producing reaction must occur in the deep lunar mantle. Also, rapid transport will aid the preservation of the primary composition of these magmas. However, to what degree volatile exsolution and heat-loss to the adjacent wall rock affect the primary nature of these melts is unknown.

Longhi (1992) and Hess (1991) acknowledged the importance of transport of these magmas through vertical dikes from the base of the lunar lithosphere (for specifics see Section 6.6). However, they suggested that these magmas were transported to the base of the lithosphere from the deep lunar mantle within ascending lunar diapers. In this model, pools of melt are generated over a range of pressures. Therefore, the multiple saturation pressures calculated for these near-primary magmas are an average and melting was initiated at greater depths (~1000 km). In these models, porous flow would have an important role in pooling magmas in a deforming and compacting crystalline matrix in the diaper and might be able to account for the transport of magmas that have a negative buoyancy if the diaper velocity was significantly large (see Section 5.9).

6.6. Processes governing the distribution of basaltic magmas on the Moon

It has long been apparent that the lunar maria are distributed in a highly non-uniform manner. On the Moon's Earth-facing hemisphere, basaltic flows comprising the maria are commonly located within the confines of large impact basins. The only major exception is the vast expanse of volcanic flows that make up Oceanus Procellarum. In contrast to the near side, lava flows on the far side are relatively scarce, even though large impact structures are located there (e.g., South Pole Aitken Basin). One of the fundamental questions of lunar geology is the cause of this near side/far side asymmetry of basaltic eruptions (Section 6.3.4) and of the preference for lavas to erupt within large impact structures.

There are two possible end-member processes that could lead to this distribution of mare basalts on the Moon. The first is that the eruption of basaltic magmas might be controlled by a magma-transport processes that inhibits eruptions on the far side and promotes eruptions within basins. This might include factors such as crustal thickness variations, volatile exsolution, and/or magma buoyancy. Alternatively, it is possible that the current distribution of maria is the result of spatial variations in magma production. In this case the thickness of the maria would be related to the quantity of melt that was produced in the underlying mantle, and the scarcity of far-side lava flows would be a consequence of a low magma production within this hemisphere. Of course, both magma production and magma-transport processes probably play some role in affecting the distribution of basaltic eruptions on the Moon. This section addresses factors that may influence whether a basaltic magma would erupt at the surface or form a crustal intrusion instead.

6.6.1. Hydrostatic head models. Until recently, the distribution of mare basalts on the Moon was thought to be solely the result of some magma transport process. The high concentrations of Fe and Ti in most lunar magmas means they are generally more dense than their terrestrial counterparts (see the following section for more details). In fact many (though not all) basalts are more dense than the Moon's anorthositic crust (e.g., Solomon 1975). These dense lunar magmas would not have been able to erupt merely on their inherent buoyancy—some other process must have "pushed" them dense magmas to the surface.

The collection of altimetry data from the *Apollo* and *Clementine* missions led to the discovery of the Moon's ~2 km center-of-mass/center-of-figure offset (e.g., Kaula et al. 1972, 1973, 1974; Zuber et al. 1994; Smith et al. 1997). The higher elevations of the far side have commonly been interpreted as a result of this hemisphere possessing a thicker crust, and the topographic depressions associated with the large impact basins have been attributed to the crust there being thinner than usual. The coincidence of regions of inferred thinned crust with the distribution of lava flows immediately suggested that these two phenomena were somehow genetically related. Based on the *Apollo* data, Kaula et al. (1973), Runcorn (1974), and Solomon (1975) all suggested that the distribution of maria could be interpreted in terms of the concept of "hydrostatic head." In this model a dike is envisioned to extend all the way from the surface to the mare source region in the mantle. Even though the magma is denser than the crust, it is less dense than the underlying mantle. If the walls of this dike were completely rigid, a simple force balance shows that a basaltic eruption could occur at any place on the Moon's surface if the mare source was deeper than ~250 km below the surface (e.g., Wieczorek et al. 2001). If the mare source was located above this level, then eruptions would only be able to occur below a critical elevation.

A modified version of the hydrostatic-head model was proposed by Head and Wilson (1992). They recognized that a basaltic magma would buoyantly rise through the mantle until it reached a level in which it was neutrally buoyant. Since they assumed all lunar basaltic magmas were denser than the crust, this level was located about 60 km below the surface at the crust-mantle interface. The hydrostatic pressure of a magma chamber at this level is insufficient for a dike to reach the surface so they postulated that an eruption would only occur if the magma chamber became over-pressurized. For a given amount of over-pressurization, they showed that basaltic eruptions would preferentially occur where the crust is thin. (Note that if the crust is isostatically compensated by an Airy mechanism, that crustal thickness is linearly related to elevation.)

The above hydrostatic-head models appeared to be in concordance with the limited *Apollo* data and hence were (and still are) widely accepted. Near-global altimetry data obtained from the *Clementine* mission, however, have since shown that there are substantial problems with hydrostatic-head being the sole factor in determining whether an eruption will or will not occur. In particular, the full depth and extent of the giant far-side South Pole-Aitken basin was delineated by the *Clementine* data. This basin's floor was found to have the lowest elevations of the entire Moon, and yet only a small number of lava ponds are found in this topographic depression (e.g., Yingst and Head 1997). While the maria are generally located at low to moderate elevations, the paucity of lava flows in the SPA basin is clearly in conflict with the predictions of the hydrostatic-head models (e.g., Lucey et al. 1994; Smith et al. 1997).

The thickness of the lunar crust also does not appear to be the sole factor controlling whether a basaltic eruption will or will not occur. Wieczorek et al. (2001) correlated the spatial distribution of maria with a geophysically derived crustal-thickness model. In seeming agreement with this model, they found that there was an abundance of lava flows where the crust was extremely thin (<30 km). However, they noted a paucity of flows within the SPA basin where the crust is about 40 km thick, and a relative abundance of flows in Oceanus Procellarum where the crust is thicker (~50 km).

Besides these observational inconsistencies with the hydrostatic-head models, some of the assumptions with these models have been called into question by more sophisticated fluid-dynamic models of dike propagation. In particular, the walls of a dike are unlikely to be completely rigid, and hence a dike should not by able to overshoot the neutral buoyancy level by any significant amount. An experimental model of dike propagation performed by Lister and Kerr (1991) showed that a dike would propagate vertically though an elastic medium whenever the magma is positively buoyant. When this dike encountered a level of neutral buoyancy, though, vertical propagation was found to nearly cease and the dike instead propagated primarily horizontally along the interface. This behavior results from the fact that vertical propagation beyond this level has to work against gravity, whereas horizontal propagation does little gravitational work. Nonetheless, if this horizontal bladed dike was continuously fed by magma, it would slowly overshoot the neutral buoyancy level, but at a rate much lower than that of the horizontal propagation.

Using the Lister and Kerr (1991) theoretical model of a growing bladed dike Wieczorek et al. (2001) tested whether a dike stalled at the crust-mantle interface on the Moon would be able to overshoot this level by ~60 km and erupt at the surface. Even by using extreme parameters, they found that an eruption would be unlikely to occur. In order for such a dike to reach the surface, the length of the dike would have to exceed ~1000 km and more reasonable parameters gave rise to dike lengths that exceeded the circumference of the Moon.

6.6.2. The role of magma buoyancy. From the study of magmatic processes on Earth, it is apparent that a magma's buoyancy plays a dominant role in its storage and transportation. Specifically, terrestrial magma chambers are located at the level in which they are neutral buoyant, and bladed dikes propagate along this level as well (e.g., Rubin and Pollard 1987; Ryan 1987, 1994; Walker 1989). As the near-surface layer of the Earth is highly fractured, this neutral buoyancy horizon is typically located only a few kilometers below the surface. Eruptions from shallow depth can then occur sporadically as a result of a bladed dike overshooting its neutral buoyancy horizon, from the exsolution of volatiles, or by temporal changes in the pressure of the magma body. Thus, the crust acts as a density filter, allowing only those magmas that are less dense to erupt. It is thus reasonable to suspect that magma buoyancy would play a dominant role in the storage and transport of magma on the Moon as well.

The density of a magma can be calculated for a given composition and temperature. The volume of a melt is well represented by the following equation:

$$V(T) = \sum_i X_i \, V_i(T) + X_{Na_2O} X_{TiO_2} V_{Na_2O-TiO_2} \qquad (4.21)$$

where X_i is the number of moles of oxide component i, $V_i(T)$ is the partial molar volume of oxide i at temperature T, and the last term takes into account a non-linear interaction between Na_2O and TiO_2. Partial molar volumes of the common oxides can be found in Lange and Carmichael (1987), and the density of a given magma is obtained by dividing the molar volumes by the total mass of the oxides. The liquidus temperature for a given composition can be approximated by using an empirical correlation between temperature and the partitioning behavior of MgO between olivine and melt (Delano 1990) or empirical relations defined by experimental studies (i.e., MELTS; e.g., Ghiorso and Sack 1995).

Using this approach, Wieczorek et al. (2001) calculated the liquidus densities of typical lunar basaltic magmas (BVSP 1981; Delano 1986). Their results are plotted in Figure 4.49 as a function of composition, and it is seen that those magmas possessing the highest concentration of Ti are generally the most dense (Delano 1990; Circone and Agee 1996), whereas those magmas possessing high concentrations of Al are the least dense. A large number of the mare-basalt magmas are more dense than typical anorthositic crustal materials ($\rho_c \approx 2.86$ g/cm³) and should not have been able to erupt owing solely to buoyancy. Nonetheless, some of the VLT, high-Al, and low-Ti basalts are seen to be less dense than the anorthositic crust and could have erupted based on their inherent buoyancy.

As is summarized in Chapter 3, a number of pieces of evidence suggest that the Moon's crust becomes increasingly mafic (and dense) with increasing depth below the surface. These include (1) the possible existence of a seismic discontinuity ~20 km below the surface that might be compositional in origin (e.g., Toksöz et al. 1974), (2) the observation that some central peaks of complex craters are highly noritic (e.g., Tompkins and Pieters 1999), (3) the observation that the ejecta of large impact basins is generally more mafic than the surrounding highlands (e.g., Reid et al. 1977; Ryder and Wood 1977; Spudis and Davis 1986; Bussey and Spudis 2000), and (4) the inference that the floor of the South Pole-Aitken basin is likely composed of noritic lower-crustal materials (Lucey et al. 1995; Pieters et al. 1997; Wieczorek and Phillips 1998). Assuming that the Moon's crust is stratified into upper anorthositic and

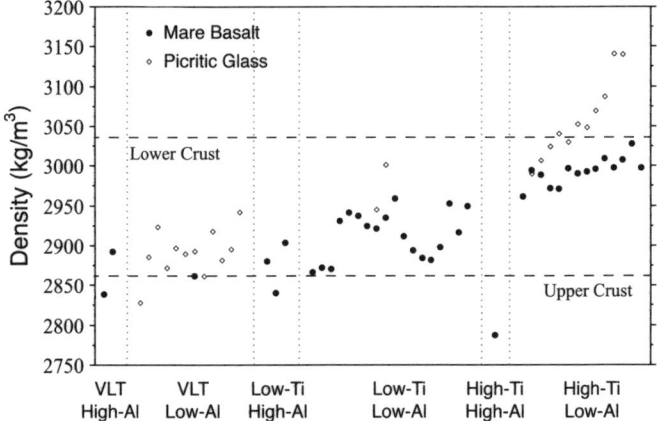

Figure 4.49. Plot of the liquidus density of mare basaltic and picritic magmas as a function of composition. Titanium concentrations increase to the right in each compositional subset and the horizontal dotted lines are the approximate density of the upper anorthositic and lower noritic crust (modified after Wieczorek et al. 2001).

lower noritic layers, Wieczorek and Phillips (1998) constructed a dual-layered geophysically-derived crustal thickness model of the Moon. Correlating this model with the composition of central peaks (Tompkins and Pieters 1999), (Wieczorek and Zuber 2001) constrained the density of the lower crust to be ~3.04 g/cm^3.

All of the mare basaltic magmas are predicted to be less dense than the Moon's lower noritic crust (Fig. 4.50). Thus, if the upper anorthositic crust was completely excavated during an impact event, any basaltic magma could have later erupted within this basin based solely on buoyancy considerations. One of the more interesting aspects of the Wieczorek and Phillips (1998) dual-layered crustal thickness model (updated and slightly modified in Chapter 3) is that the upper crust is predicted to be completely absent beneath many of the large impact basins. These basins include Imbrium, Serenitatis, Crisium, Symthii, Humboldtianum, Orientale, Humorum, Nectaris, and South Pole-Aitken. Additionally, the upper crust is predicted to be less than 5-km thick beneath Mendel-Rydberg, Moscoviense, Freundlich-Sharonov, and a portion of eastern Mare Frigoris. All of these regions are seen to possess some mare basalts, and these basaltic eruptions are hence consistent with the hypothesis that density alone determines whether a basaltic magma will erupt or form a crustal intrusion instead.

The basalts that erupted wherever the upper anorthositic crust is predicted to be present must be analyzed separately. Gillis and Spudis (1998) showed that those basalts that erupted within the far side highlands, the Australe basin, and the eastern limb have lower than typical abundances of Fe as inferred from Clementine remote-sensing data (Lucey et al. 1998). Using an empirical correlation among a magma's liquidus density, Fe and Ti content, Wieczorek et al. (2001) showed that these magmas were likely to have been less dense than the upper anorthositic crust, and hence are consistent with erupting based solely on buoyancy considerations.

The only other mare basalts that need to be reconciled with the magma buoyancy hypothesis are those that erupted within Oceanus Procellarum. The dual-layered crustal thickness model of Wieczorek and Phillips (1998) predicts that this region possesses an anorthositic crust that is ~20 km thick, yet some of the highest Ti basalts erupted in this region. There are two possible ways out of this apparent conundrum. First, it is possible that the crust in this region of the Moon might be denser than is predicted by the geophysical model.

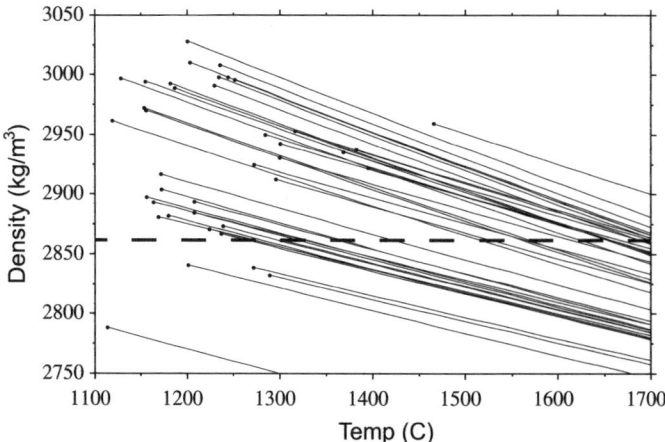

Figure 4.50. Plot of the density of mare basaltic magmas as a function of temperature. The dot at the left end of each line corresponds to the liquidus density of the magma and the horizontal line corresponds to the approximate density of the Moon's upper anorthositic crust (modified after Wieczorek et al. 2001).

Since Oceanus Procellarum lies within an anomalous KREEP-rich geochemical province, the crust there may in fact be quite different in composition from more typical highland regions (see Chapter 3). Voluminous intrusions of basaltic magma over time in this region may have substantially altered the bulk composition of the upper crust as well. Secondly, it is possible that the magmas in this region erupted with superliquidus temperatures. Figure 4.50 shows the density of the basaltic magmas from Figure 4.49 as a function of temperature. A few hundred degrees of superheating would make many of these basaltic magmas less dense than the upper anorthositic crust. Thermal models that take into account the high abundance of incompatible elements within and/or beneath the crust of Oceanus Procellarum, in fact, do predict extremely high temperatures that might be sufficient to make these magmas positively buoyant in the crust (Parmentier et al. 2000; Wieczorek and Phillips 2000; Zhong et al. 2000). Analysis of the textures in samples from this area may someday be used to test the superliquidus hypothesis.

Finally, we note that the near-surface layer of the Moon has low density as a result of impact brecciation. Thus, most basaltic magmas would probably become neutral buoyant a few kilometers below the surface and accumulate at this level. Though this low-density layer should inhibit an eruption from occurring, a number of mechanisms could lead to an eruption onto the surface. First, as mentioned above, a growing bladed dike can easily overshoot its neutral buoyancy horizon by a few kilometers. Secondly, the exsolution of gas from a magma at low pressures would help propel an eruption. Carbon monoxide gas, in particular, is predicted to exsolve from lunar magmas less than about 4 km below the surface (e.g., Wilson and Head 1981). Other possibilities are S and Cl. Because of this near-surface low-density layer, it seems likely that the quantity of intrusive mare basaltic magmas is probably much more voluminous than those that are visible on the surface.

6.6.3. Relationships between large basin impacts and mare volcanism. Before artificial satellites ever imaged the far side of the Moon, it was noted that the largest near side impact basins were filled by the relatively smooth "maria." The hypothesis was born that the two phenomena were genetically related, and furthermore that the basin-forming process was responsible for the generation of the mare basalts (e.g., Ronca 1966). Many pieces of evidence have since cast doubt upon this hypothesis. In particular, images of the far side demonstrate that there many large impact basins are not filled with mare basalts (or at least much less so than the near side basins), and radiometrically determined ages of mare samples at the Apollo 15 and 17 sites were found to be considerably younger than the basins in which they reside. Even if basins and basaltic eruptions appear to be somewhat correlated in places, it has recently been suggested that this might solely be related to large impacts removing the low density upper crust, allowing basaltic magmas to rise buoyantly through the more dense lower crust and erupt (see Wieczorek et al. 2001 and Section 6.6.2).

Based on the indistinguishable radiometric ages of the Imbrium basin and Apollo 15 KREEP basalts, as well as the inference that KREEP basalts are stratigraphically younger than the Imbrium basin (Spudis 1978), Ryder (1994) argued that the Imbrium impact event induced KREEP basaltic volcanism within this region of the Moon. A variety of pieces of evidence (e.g., Jolliff et al. 2000; Korotev 2000; Wieczorek and Phillips 2000) suggest that KREEP basalt is confined to the Procellarum KREEP Terrane (which encompasses the Imbrium basin), and thermal modeling (Wieczorek and Phillips 2000) has shown that this material could have remained molten up until at least the time of the Imbrium impact. As such, this isolated case of impact-induced volcanism might only be a result of this impact event having denuded an already molten KREEP basaltic magma chamber (Ryder 1994; Wieczorek and Phillips 2000).

The case for a genetic relationship between mare volcanism and large impact basins has recently been revived by Elkins-Tanton et al. (2004). They distinguished between two processes that could give rise to large volumes of magma. First, the excavation of large quantities of crustal material might give rise to near-instantaneous decompression melting in the underlying mantle. Second, as a result of lateral temperature gradients beneath an impact crater, small-scale convection could have led to further melting over a much longer period of time. The total amount of melt generated in their model was found to be about 10× greater than the volume of the extrusive mare basalts. Although subsequent convection beneath the crater only accounted for 0-2% of the total melt volume, this later process was found to operate for up to 350 m.y. following the impact event (see also Arkani-Hamed and Pentecost 2001).

The most debatable aspect of the Elkins-Tanton et al. (2004) model concerns the process of instantaneous decompression melting beneath the impact basin. Ivanov and Melosh (2003) showed that impact-generated volcanism by decompression melting on the Earth is improbable. In the Elkins-Tanton et al. model, the magnitude of the pressure drop beneath the crater was calculated by assuming that the entire mantle column between the crater floor and core-mantle boundary was uplifted by an amount equal to the depth of excavation of the crater. While this might approximate the pressure conditions just beneath the crater, this assumption probably overestimates the pressure drop in the deeper mantle. Following the excavation stage of the impact event, mantle rebound and crater collapse does not occur by strictly vertical displacements; inward lateral flow of mantle materials beneath the central portion of the basin partially offset the mass deficit caused by the excavation of near-surface materials (e.g., Melosh 1989; Ivanov and Melosh 2003). As illustrated in Ivanov and Melosh (2003), the amount of central uplift beneath a crater decreases substantially with increasing depth.

Although realistic considerations of the magnitude of the pressure drop beneath impact basins suggest that the quantity of decompression melt would be much less than that found by Elkins-Tanton et al. (2004), this question deserves further investigation. The study of Ivanov and Melosh (2003) did not explicitly model impact craters with sizes typical of the largest lunar basins, nor did they attempt to employ mantle temperature profiles that might be representative of the Moon at about 4 Ga. In addition, as discussed in Section 7, the Moon's thermal evolution appears to be highly asymmetric, and the possibility exists that decompression melting might have occurred beneath some craters and not others. Hydrodynamic simulations of large impact events on the Moon for a variety of mantle temperature profiles should be able to delimit which conditions could plausibly give rise to decompression melting. Subsequent impact-generated convection and melting has only been investigated by a few papers and warrants further investigation as well.

7. ORIGIN OF CRUSTAL ASYMMETRIES

7.1. Introduction

Prior to the *Clementine* and *Lunar Prospector* missions, all published thermal-evolution models possessed one commonality: they all assumed that the distribution of heat-producing elements depended solely upon radius. As such, the amount of partial melt generated beneath the surface should have been independent of location. As it is well known that the majority of the lunar basalts erupted on the near side, the scarcity of visible far-side basalts was commonly attributed to magma transport processes. The most popular explanation was based on the assumption that the mare basaltic magmas were more dense than the lunar crust, and relied upon magma-chamber over-pressurization for these magmas to reach the surface (e.g., Head and Wilson 1992). Altimetry of the South Pole-Aitken basin obtained during the *Clementine* mission, however, is inconsistent with this hypothesis and it now appears that magma buoyancy may have been the primary factor determining whether a basaltic eruption would or would not occur (for a more in-depth discussion, see Section 6 and Wieczorek et al. 2001).

In addition to differences in crustal structure affecting eruptive flux as discussed in Section 4.6.6, the scarcity of mare basaltic eruptions on the lunar far side (particularly within the SPA basin) is also a result of a difference in magma production rates between the near- and far-side hemispheres. It was not until after the *Lunar Prospector* mission in 1998, however, that such asymmetric thermal-evolution models began to be quantitatively investigated. The rational for these models is clearly evident in the global Th abundance maps generated by the LP gamma-ray spectrometer (Lawrence et al. 1998, 2000), which demonstrated that Th (and by inference KREEP) is highly concentrated within a small region of the Moon encompassing Mare Imbrium and Oceanus Procellarum. This possibility had been previously suggested on the basis of the equatorial Apollo gamma-ray data (e.g., Warren and Wasson 1979, 1980; Ryder 1994; Haskin 1998; Haskin et al. 1998), but was not widely accepted or fully appreciated.

On the basis of these and other data, Jolliff et al. (2000) divided the lunar crust and underlying mantle into distinct terranes that possess unique geochemical, geophysical, and geological histories. The two most extensive terranes are the Procellarum KREEP Terrane (PKT) and the Feldspathic Highlands Terrane (FHT), and the rational for these divisions, as well as their properties and evolutionary histories are discussed in detail in Chapter 3 (see also Haskin et al. 2000; Korotev 2000; Wieczorek and Phillips 2000). Below, we list only their most important characteristics that are necessary for deciphering the large-scale asymmetric magmatic evolution of the Moon.

The Feldspathic Highlands Terrane:

1. This terrane encompasses ~60% of the Moon's surface area and is composed primarily of ancient ferroan-anorthositic rocks that crystallized from an LMO between about 4.5 and 4.4 Ga.

2. Crustal Th concentrations range from less than 1 ppm for near surface materials, to possibly 2-3 ppm thorium for deep crustal rocks.

3. Less than 10% of this terrane by area has been resurfaced by mare basaltic lava flows, with the vast majority being emplaced almost exclusively during the Imbrian period. Remote-sensing data of central peaks in this terrane further suggest that basaltic intrusions within the underlying crust are volumetrically minor (~5% by volume).

4. Basaltic flows in this terrane generally contain low to moderate concentrations of TiO_2 (<6 wt%) as well as low concentrations of Th (~1-2 ppm).

The Procellarum KREEP Terrane:

1. This terrane encompasses the regions of Oceanus Procellarum and Mare Imbrium, and is principally characterized by high Th concentrations (3-12 ppm), as well as the abundance of mare basaltic lava flows.

2. Although this terrane only comprises ~16% of the Moon's surface area, >60% of the Moon's extrusive basalts by area occur there. Only those regions of high-standing topography (such as the rim of the Imbrium basin) have not been volcanically resurfaced.

3. Volcanism in this terrane appears to have extended from at least 4.2 Ga to about 1 Ga.

4. The majority of Moon's high-Ti mare lava flows occur within this terrane.

5. Judging from the composition of Imbrium ejecta, ~50% of the crust may have a composition similar to that of Apollo 15 KREEP basalt (~12 ppm Th).

6. Global Th mass-balance calculations suggest that ~30% of the Moon's heat-producing elements may be sequestered within the Procellarum KREEP Terrane. Furthermore, about 40% of the total amount of Th in the lunar crust resides within the PKT.

7. The heat flow at the Apollo 15 site (which lies within the PKT) is ~50% greater than that measured at the Apollo 17 site (which lies just outside of this province). In addition, only those basins that formed within, or on the edge of, this province appear to have been modified by enhanced rates of viscous relaxation.

8. On the basis that most magnesian- and alkali-suite rocks are defined by their high incompatible element contents (i.e., KREEP signature), it is unlikely that these types of rocks formed outside the PKT.

9. From orbital gamma-ray data, surface Th concentrations of the mare-basalt-dominated regoliths in this region range from ~3 to 7 ppm. It remains unclear, however, whether these Th concentrations are representative of the basalts themselves, or if they result from vertical mixing with an underlying Th-rich substrate.

The high abundances of heat-producing elements within the Procellarum KREEP Terrane are in all likelihood related to the magmatic productivity and volcanic longevity associated with this region. In order to have a complete understanding of the Moon's thermal evolution two important questions need to be addressed: (1) How did KREEP become concentrated in a single geologic province? (2) How did this enhancement in heat-production within the PKT affect its magmatic evolution? Of course, these two questions are highly coupled, but it may nonetheless be useful to initially separate these two problems. In the remainder of this section we discuss the different models that have been advocated to date in order to explain these and other characteristics. These models are somewhat preliminary; more detailed calculations are needed to fully explore their feasibility and implications.

7.2. Effects of a thicker KREEP layer within the Procellarum KREEP Terrane

Prior to the *Lunar Prospector* mission, most researchers assumed that the fractional crystallization of an LMO would have given rise to a KREEP layer between the crust and mantle that was nearly global in extent. Estimates for the thickness of this layer depended upon many factors, including the initial depth of the LMO, but was commonly assumed to be approximately two kilometers (assuming an LMO that was 360 km deep; e.g., Warren and Wasson 1979). If instead KREEP were to reside mainly within the Procellarum KREEP Terrane, as appears to be the case, then how much thicker would this layer be? Assuming this terrane encompasses ~16% of the Moon's surface area, its equivalent thickness would be ~10 km. If the LMO was deeper than originally assumed (e.g., Pritchard and Stevenson 2000), then the KREEP layer within the Procellarum KREEP Terrane would be even thicker.

Setting aside the question of how KREEP came to be concentrated within a single lunar province, Wieczorek and Phillips (2000) quantitatively modeled the thermal consequences of such a localized enhancement in heat-producing elements. On the basis of geochemical and geophysical data, they argued that up to 50% of the crust within the PKT (~30 km) might be composed of a material similar in composition to KREEP basalt. This rock possesses REE concentrations ~300× chondritic (12.4 ppm Th; Korotev 2000), which is just less than that of Warren and Wasson's (1979) urKREEP (18 ppm Th). In order to approximate the thermal state of the mantle just after the bulk of the LMO crystallized, its initial temperature profile was assumed to be adiabatic with the top of the mantle set to its solidus. The KREEP basalt layer was approximated as a spherical cap having an angular radius of 40° and was initially placed between the crust and mantle. Mantle heat-producing abundances were estimated from the mare basalts and appropriate distribution coefficients (e.g., 25 ppb Th; Warren and Wasson 1979), whereas crustal abundances were taken from orbital gamma-ray data (0.53 ppm Th). The bulk U concentration of this model Moon is just larger than that of the Earth's primitive mantle, and is about twice that of ordinary chondrites.

Whereas many lunar thermal evolution models have assumed that internal heat was partly transported by convection, a purely conductive model was employed in this study. As is discussed in Section 6, this assumption is at least plausible given that (1) the crystallization of an LMO would likely result in a mantle that was severely depleted in heat-producing elements, (2) the crystallization of an LMO could have resulted in a density-stratified mantle that was stable against convection, and (3) the mantle beneath the PKT would have been heated primarily from the top down, counteracting any tendency for large scale mantle convection. While melting was accounted for in this model using approximate latent heats and phase diagrams, melt transport was not modeled (i.e., all melt remained in the mantle).

The principal results of this model are illustrated in Figure 4.51 which shows the spatial extent of melting beneath the Procellarum KREEP Terrane as a function of time. Melting within the mantle is predicted to occur only directly below this terrane, with the depth of melting increasing with time. Owing to the lateral transport of heat near the edges of the terrane, the lateral extent of partial melting is further predicted to diminish with time. For the particular initial conditions of this model, melting should have extended ~500 km below the surface and volcanism would have spanned the entire geologic history of the Moon (though with exponentially decreasing melt-production rates with time).

In Figure 4.52 the predicted present-day heat flux is plotted as a function of distance from the center of the PKT. The heat flux decreases across the boundary between the PKT and FHT. Unfortunately, even though the measured heat-flow values at the Apollo 15 and 17 sites are consistent with this model, these landing sites are located very close to this boundary where the heat flow is predicted to vary significantly. This leaves open the possibility that these measurements are not representative of either the PKT or FHT. Knowledge of the heat flux within the heart of the Procellarum region, as well as far from it, would thus dramatically help to constrain these types of models.

The first-order result of melting within the mantle directly beneath the PKT appears to be an inevitable consequence if the equivalent of 10 km of KREEP basalt resides there. For

Figure 4.51. Maximum spatial extent of melting beneath the Procellarum KREEP Terrane at 5× before the present for the thermal model of Wieczorek and Phillips (2000) This cross section spans ±45° of latitude, whereas the PKT covers ±40° of latitude. Image from Wieczorek and Phillips (2000).

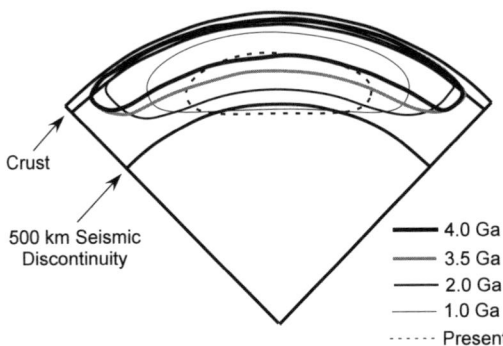

Crust

500 km Seismic
Discontinuity

—— 4.0 Ga
—— 3.5 Ga
—— 2.0 Ga
—— 1.0 Ga
······ Present

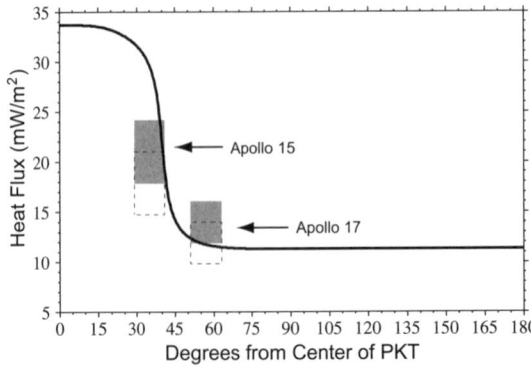

Figure 4.52. Modeled present day heat flux as a function of distance from the Procellarum KREEP Terrane for the thermal model of Wieczorek and Phillips (2000). Also shown are the Apollo heat flow measurements (gray boxes) at their approximate distance from the center of the PKT. The estimated heat flow at these sites after correction for a possible enhancement due to their location at a mare/highlands boundary (Warren and Rasmussen 1987) is shown by the dotted boxes. Image from Wieczorek and Phillips (2000).

instance, if all the heat-producing elements were removed from the mantle, then the maximum depth of partial melting beneath the PKT would still be ~340 km and span ~4 Ga of lunar history. Conversely, if the entire complement of heat-producing elements within a 10 km KREEP basalt layer were uniformly distributed over the 60 km thick crust in this province, then melting would have reached a maximum depth of ~180 km and would have lasted for ~1.5 Ga. The first-order results of this model were confirmed by Hess and Parmentier (2001) using slightly different parameters, though they ultimately considered these model predictions to be inconsistent with petrological and geophysical constraints.

7.2.1. Model implications. Although the thermal model of Wieczorek and Phillips (2000) is admittedly simple, it nevertheless offers a coherent framework in which to understand a large variety of previously unrelated aspects of the Moon's geologic and thermal evolution. Below we list those as advocated by Wieczorek and Phillips (2000). Most of these still remain to be more thoroughly explored.

1. The fact that the maria are primarily located on the lunar near side is a natural consequence of the high abundance of heat-producing elements within the crust of the Procellarum KREEP Terrane.

2. The fact that the measured heat flow is larger at the Apollo 15 site than at the Apollo 17 site is a direct consequence of this site being closer to the center of the PKT (e.g., Langseth et al. 1976; Warren and Rasmussen 1987).

3. The range of crystallization ages of the Mg- and alkali-suite rocks (Table 3.3 in Chapter 3) is the result of the slow crystallization of discreet intracrustal KREEP-basalt magma chambers (e.g., Snyder et al. 1995a, b). This extended range of KREEP-related magmatism is also indicated by the crystallization age of Apollo 15 KREEP basalts and their coincidence with the Imbrium formation event (Ryder 1994).

4. The reason that an igneous protolith to the Low-K Fra Mauro ("LKFM") mafic impact-melt breccias (commonly interpreted as representing Imbrium ejecta) has not been identified is that this impact melt is a mixture, of which one component was initially molten (Spudis et al. 1991).

5. One of the interpretations of the 500-km seismic discontinuity observed beneath the near side crust is that it is the base of the LMO cumulate pile. Based on this interpretation, mare magmatism (a product of cumulate melting) represents melts produced above this discontinuity.

7.2.2. Model shortcomings. Even though this model appears to explain the major first-order features of the Moon's post-LMO magmatic evolution, there are a few issues that still need to be critically addressed. Perhaps the most worrisome (Wieczorek and Phillips 2000; Hess and Parmentier 2001) is that the KREEP layer in these types of models generally remains partially molten for more than a billion years. Thus, even though partial melting in the mantle underlying the Procellarum KREEP Terrane might give rise to mare basaltic magmas, in order for such magmas to erupt at the surface, they must have first passed through this molten layer. Even if this were possible, one might have expected these basaltic melts to have become strongly contaminated by the KREEP component in the process, something that has not been observed in the Apollo sample collection.

There are two possible ways to resolve this conundrum. First, it is possible that the KREEP layer beneath the PKT was not a continuous giant sill, but rather that this magma resided in numerous bladed dikes and/or laterally restrictive sills. Basaltic eruptions would then be possible wherever there was a continuous solid pathway from the mare source to surface. Admittedly, this explanation is difficult to test in the absence of global heat-flow data. Alternatively, it is possible that the KREEP layer cooled more quickly than predicted in this model, solidifying around 3.9 Ga when most of the visible maria erupted. Recent reanalysis of the Apollo seismic data have shown that the crust is probably much thinner than previously assumed (~38 km (Khan and Mosegaard 2002) as opposed to ~60 km (Toksöz et al. 1974)), and this would indeed favor more rapid cooling in this province (e.g., Hess and Parmentier 2001). It remains to be tested, though, as to how this would affect the conclusions of Wieczorek and Phillips (2000).

Another potential problem noted by Wieczorek and Phillips (2000) and Hess and Parmentier (2001) is that if the crust were indeed much hotter than typical beneath the PKT, then the lithosphere might not have been able to support the excess load associated with the Imbrium "mascon." While this question deserves further scrutiny, it is probably not a fatal flaw to the hypothesis that the PKT is the host to a KREEP layer that is on the order of 10 km thick. By modeling the deformation associated with a load on a visco-elastic self-gravitating sphere, Zhong and Zuber (2000) have demonstrated that the Moon can support considerably greater stresses in comparison to that of the Earth as a result of its smaller planetary radius. In particular, for the case where the lunar lithosphere was approximated as having a temperature profile corresponding to 100 Ma terrestrial oceanic lithosphere, the degree of compensation of the surface load was found to be less than 0.3 for all wavelengths.

Another criticism mentioned by Hess and Parmentier (2001) concerns the stability of the anorthositic crust overlying a KREEP basaltic magma. In particular, molten KREEP basalt has a density of ~2.7 g/cm^3 , which would be approximately 4-7% less dense than that of the overlying anorthositic crust (2.8-2.9 g/cm^3 neglecting fractures). Thus, either this magma would have buoyantly risen through the crust with some portion of it erupting at the surface, or the crust would have catastrophically foundered. While pristine KREEP basalts are rare in the Apollo sample collection in apparent contradiction with these scenarios, the much younger visible mare basalts would have certainly buried any such ancient KREEP basaltic flows. Even presuming that one of these two scenarios did indeed occur, Wieczorek and Phillips (2000) have shown that if the heat production of a 10 km KREEP basalt layer was spread over the entire crust in this province that volcanism would still be predicted to occur there for ~1.5 Ga.

Finally, we note that the volume of melt produced in this model is predicted to be somewhere between 5× and 56× that of the visible basalts, depending upon the model assumptions. The lower end of this range might be consistent with the crust possessing voluminous basaltic intrusions, but the upper bound would imply that the entire crust is composed of mare basalts. While these large volumes of basaltic melt are problematic, it is likely that this is only an artifact of the simplified model that was employed. For instance, if melt transport were taken into account then the amount of melt generated would have been reduced. The amount of melt depends upon bulk composition and the phase relationships for the mantle source, which are rarely taken into account in thermal evolution models. While Wieczorek and Phillips (2000) used the bulk-Moon liquidus and solidus of Ringwood (1976), these are expected to only crudely approximate reality. Indeed, the effects of mantle overturn (e.g., Hess and Parmentier 1995) might have given rise to a refractory and Mg-rich upper mantle that would drastically reduce the amount of melt generated in these models.

7.3. Degree-1 upwelling of ilmenite-rich cumulates

One consequence of the Moon possessing an LMO is that ilmenite-bearing mineral assemblages should precipitate near its terminal stages of crystallization (e.g., Snyder et al. 1992). As a result of their relatively high-Ti and Fe contents, these cumulates should be relatively dense (~3.7 g/cm^3) and hence might have sunk through the lunar mantle (e.g., Ringwood and Kesson 1976; Herbert 1980; Spera 1992; Hess and Parmentier 1995) (Section 3.3). Remelting of these cumulates in the deep lunar interior is widely thought to be implicated in the origin of the high-Ti mare basalts (Section 5).

Alternative explanations for the hemispheric asymmetry in the distribution of mare basalts are related to the fate of these ilmenite-rich cumulates. Two models have been advocated that rely upon the development of a long-wavelength (degree-1) convective instability in order to concentrate these materials within one lunar hemisphere and to thus give rise to a higher than typical magma production rate. The starting premise of the model proposed by Zhong et al. (2000) is that late-stage LMO ilmenite-rich cumulates were somehow able to sink through the entire mantle, forming a global layer above the Moon's iron core. It was assumed that these cumulates were mixed in some proportion with more typical mantle materials (decreasing their density in the process), and that this material further had an enhanced abundance of heat-producing elements owing to the presence of some proportion of trapped intragranular KREEP-rich liquid. While these cumulates would have initially been gravitationally stable at the base of the mantle, radioactive heating within this layer would reduce its buoyancy with time, and at some point might have caused it to have become buoyantly unstable. The question is thus to determine under which conditions this layer would rise predominantly within a single hemisphere of the Moon, or more precisely, under which conditions a degree-1 instability would have the largest growth rate.

This question was addressed from the perspective of both a Rayleigh-Taylor linear stability analysis, as well as through finite-element modeling of thermo-chemical convection. From the Rayleigh-Taylor stability analysis, it was shown that the wavelength dependent growth rate depended primarily upon both the radius of the core and of that of the mixed ilmenite-rich cumulate layer, with the viscosity contrast between the mantle and ilmenite cumulates playing only a secondary role (Fig. 4.53). In this plot, a degree-1 instability would be the most favored when the normalized growth rate has a value of unity. The different curves within Figure 4.53 correspond to different assumed radii of the Moon's core. The most important result is that a degree-1 instability would only have the largest growth rate if the radius of the core was less than 250 km.

These results were confirmed and expanded upon by axis-symmetric thermo-chemical convection modeling. As one example, Figure 4.54 shows how both the temperature and compositional fields evolve with time. In this particular simulation, a degree-1 upwelling structure is seen to develop after ~400 Ma. While the onset time for this upwelling to occur depends upon the assumed model parameters, it should lie between ~250 and 650 Ma for reasonable choices. Partial melting could occur within this upwelling plume either as a result of the higher temperatures found there, or as a result of decompression melting.

7.3.1. Model implications and shortcomings. The main successes of the Zhong et al. (2000) model are the following:

1. Melting within a single ilmenite-rich "plume" could have given rise to a hemispheric asymmetry in the distribution of mare basaltic lava flows.

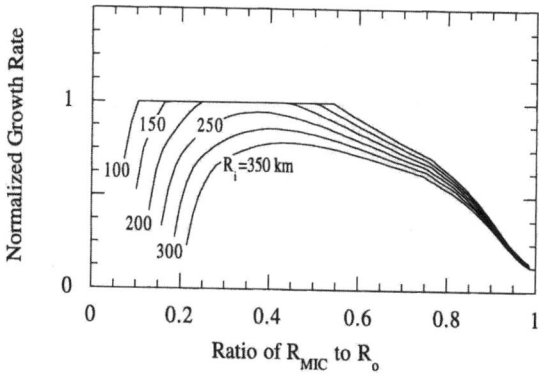

Figure 4.53. Plot showing spherical-harmonic degree-1 growth rate as a function of the radius of mixed ilmenite-rich cumulate layer (R_{MIC}). R_i is lunar core radius, and R_0 represents the lunar radius. [Reprinted with permission of Elsevier from Zhong et al. (2000) *Earth & Planetary Science Letters*, Vol. 177, p. 131-140.]

2. Some portion of these basaltic magmas should have been enriched in Ti, consistent with the abundance of high-Ti lava flows on the lunar near side.

3. The onset time for the ilmenite-rich materials to become buoyant and rise is on the order of 500 Ma, which is in general agreement with the observation that the majority of the visible mare were emplaced after ~3.9 Ga.

4. The removal of an ilmenite-rich "thermal blanket" just above the core around 4 Ga could have possibly given rise to a geodynamo, potentially explaining the paleomagnetic signatures observed in some lunar rocks (Stegman et al. 2003) (See also Chapter 3, Section 4.6).

Even though this type of model has some attractive attributes, there are a few critical (and perhaps fatal) concerns that need to be addressed. For example, this model requires the lunar core to have a radius less than 250 km, whereas most geophysical estimates suggest that the lunar core is in fact larger. For instance, an analysis of the Moon's magnetic induction response suggests that it possesses a metallic core having a radius between 250 and 430 km (Hood et al. 1999). In addition, an analysis of the Moon's rotation from three decades of lunar laser ranging data (LLR; Williams et al. 2001) strongly implies the existence of both solid-body tidal dissipation, as well as dissipation at a liquid-core/solid-mantle interface. Assuming that the core is composed of pure iron, its radius is constrained to be between 314 and 352 km. A sulfur-rich core would possess a larger radius. The possibility exists, however, that the LLR study has overestimated the core radius by neglecting the effects of dissipation at a solid-inner core interface. Nevertheless, the fact that the two most stringent core-radii estimates are currently in excess of the 250 km maximum radius required by Zhong et al. (2000) is not encouraging for this model.

An additional concern is that while this model can explain the hemispheric asymmetry of mare basaltic lava flows, it does not address why KREEP is concentrated solely within the PKT where most of these mare basalts happen to reside. Although partial melting within an ilmenite-rich plume might have given rise to Th-rich

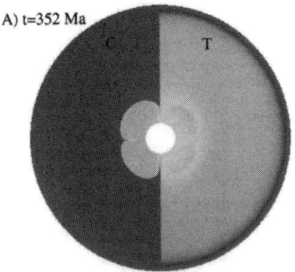

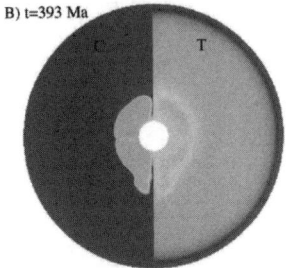

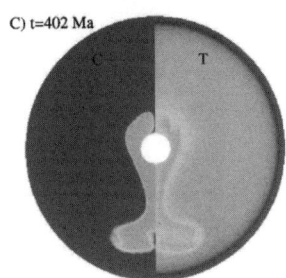

Figure 4.54. Plot showing the temperature (*T*, right) and compositional (*C*, left) fields at three different times after lunar formation. In this model, the mantle possesses a temperature dependent viscosity. Blue corresponds to mantle materials (*C*) or cold temperatures (*T*). The white central region represents a small iron core. [Reprinted with permission of Elsevier from Zhong et al. (2000) *Earth & Planetary Science Letters*, Vol. 177, p. 131-140.]

mare basaltic magmas, such samples are extremely rare, or possibly even non-existent, in the Apollo sample collection (see Section 3.5 for discussion of this issue). Even supposing that the 3-7 ppm surface Th concentrations of the basalts within the PKT were a result of such Th-rich mare basaltic lavas, it still would remain unexplained as to why the underlying crust has even higher concentrations. As such, while this model appears to be consistent with the record of mare volcanism, the spatial correlation that exists between the basaltic flows and the high-thorium underlying crust remains an unexplained coincidence.

7.4. Degree-1 downwelling of ilmenite-rich cumulates

In contrast to a degree-1 upwelling of ilmenite-rich cumulates from the base of the mantle, it has been proposed by Parmentier et al. (2002) that the ilmenite-rich cumulates could have become concentrated within a single hemisphere as they sank through the mantle in a degree-1 manner. The model's starting premise is that ilmenite-rich cumulates crystallized from the LMO forming a globe-encircling layer between the crust and mantle. As a result of the high density of this initially thin layer, small diapirs should have descended a small distance into the underlying mantle, forming a much thicker (though still global) layer of mixed ilmenite and mantle cumulates. The problem addressed in this study was to determine under which conditions this mixed layer might descend into the mantle solely within a single hemisphere.

Like the Zhong et al. (2000) study, this problem was addressed both from the perspectives of a Rayleigh-Taylor stability analysis as well as thermo-chemical convection simulations. From the Rayleigh-Taylor stability analysis it was shown that a predominantly degree-1 downwelling would occur only if (1) the thickness of the mixed ilmenite cumulate layer was greater than about 50 km, and (2) if the underlying mantle had a viscosity more than ~$10^3 \times$ that of the mixed cumulate layer (Figs. 4.55 and 4.56). The absolute growth rates were found to be proportional to the density contrast between the cumulates and mantle, and inversely proportional to the corresponding viscosity contrast. If it was assumed that the mantle viscosity increased with depth, then short-wavelength instabilities were always found to be favored. Nevertheless, as these short wavelength instabilities act to increase the thickness of the mixed layer, a degree-1 instability would eventually represent the dominant flow pattern at later times.

Thermo-chemical convection modeling has confirmed that such a degree-1 downwelling of dense cumulates could indeed occur. The timescales associated with the descending plume depend on factors such as the mantle and layer viscosity, whereas its associated temperature depends on the concentration of incompatible elements. Although these two factors are not well constrained, parameter values can be found that give rise to a plume that descends ~500 km, and starts to partially melt, in about 500 Ma. Although not explicitly modeled, if a liquid KREEP-rich layer still existed at this time, it would be expected to pool above this downwelling. In this manner mare volcanism as well as a crustal KREEP enhancement would be predicted to occur in the same region.

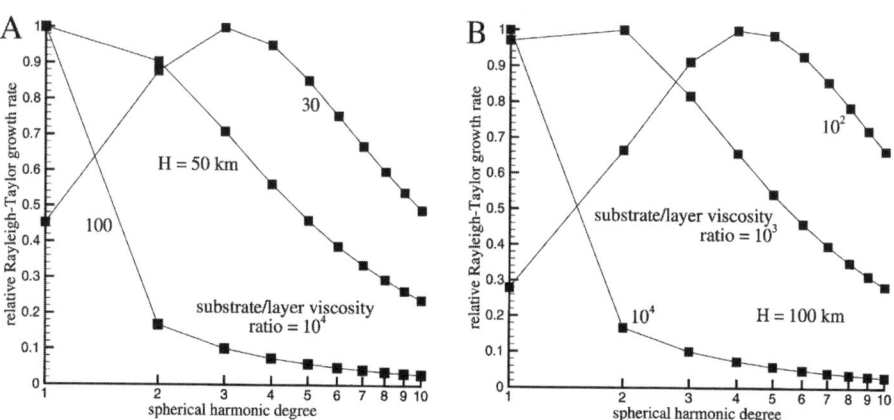

Figure 4.55. Rayleigh-Taylor instability analysis showing growth rates (normalized to the maximum growth rate) as a function of spherical harmonic degree. (A) Growth rates as a function of layer thickness for a mantle that is $10^4 \times$ more viscous than the cumulate layer. (B) Growth rates as a function of viscosity contrast for a layer thickness of 100 km. [Reprinted with permission of Elsevier from Parmentier et al. (2002) *Earth & Planetary Science Letters*, Vol. 201, p. 473-480.]

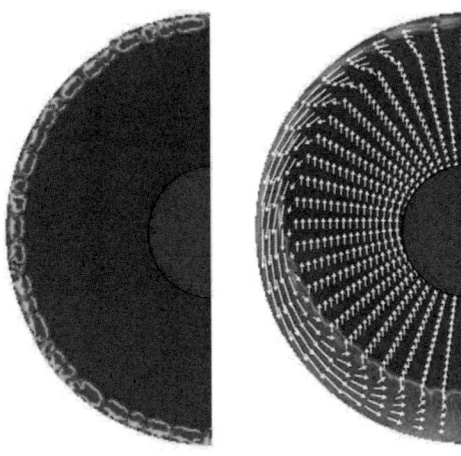

Figure 4.56. Thermo-chemical convection simulations showing the development of a degree-1 mixed ilmenite-cumulate downwelling. Initially (left), the ilmenite cumulate layer is thin, and instabilities develop with short wavelengths. These short wavelength instabilities act to thicken the mixed cumulate layer, at which point a degree-1 instability becomes the most prominent flow pattern (right). Arrows represent the final velocity pattern. [Reprinted with permission of Elsevier from Parmentier et al. (2002) *Earth & Planetary Science Letters*, Vol. 201, p. 473-480.]

7.4.1. Model implications and shortcomings. The main successes of this model include:

1. A dense and global magma-ocean cumulate layer could descend through the mantle in a degree-1 pattern, concentrating KREEP and ilmenite-rich materials within one hemisphere of the Moon.

2. Melting within this downwelling would be promoted by its high concentration of heat-producing elements and could conceivably account for the commencement of the "main phase" of mare volcanism at ~4 Ga on the lunar near side.

3. Melting within this mixed ilmenite cumulate layer could generate high Ti mare basaltic magmas.

4. If a liquid KREEP-rich layer was still present while the ilmenite cumulates were sinking through the mantle, it should have concentrated above this dense downwelling.

The main shortcoming of this model concerns the requirement that the mixed cumulate layer be some $10^3 \times$ less viscous than the underlying mantle. This is problematic because the late-stage ilmenite-rich magma-ocean cumulates would have been rich in clinopyroxene, which is about $100 \times$ more viscous than olivine. (See Elkins-Tanton et al. (2002) for a discussion of some of the problems associated with generating models in which ilmenite-rich cumulates sink into the mantle). The high-viscosity of this mineral could be partially compensated by inclusion of about 5-10% interstitial liquid within the mixed layer. While this might initially be plausible, the lower density of the liquid component would act to reduce the material's bulk density, increasing the timescales associated with the downwelling event. Additionally, compaction of the partially molten matrix might be able to expel the interstitial liquid on a timescale associated with this downwelling plume.

Another issue is whether melting within a mixed ilmenite-rich downwelling would give rise to magmas that are geochemically similar to the high- and low-Ti mare basalts. In particular, melting within this downwelling would be facilitated by high concentrations of incompatible and heat-producing elements associated with the late-stage magma-ocean cumulates, yet the sampled mare basalts for the most part possess low concentrations of these elements.

7.5. The effect of crustal thickness variations during LMO crystallization

A conceptually simple mechanism to concentrate KREEP within a single region of the Moon is based on the premise that crustal thickness variations might have existed at the time the LMO was crystallizing. It is likely that the Moon's center-of-mass/center-of-figure offset is an ancient feature dating from this time. This feature is most popularly explained by the far-side crust being thicker than that of the near side. Warren and Wasson (1979) and Wasson and Warren (1980) originally noted that if the anorthositic crust was indeed thicker than usual for the far side, then the thickness of the LMO there would have been locally thinner. As a result of the reduced heat production beneath this region of thickened crust, LMO crystallization would have proceeded at a greater rate, resulting in a feedback process leading to a yet thinner LMO. As crystallization in this region progressed towards completion, the far side crust would have eventually "grounded" on the underlying mantle, effectively removing the KREEP layer from this hemisphere. This model was reiterated and slightly modified by Warren and Kallemeyn (1998) (Fig. 4.57).

The question thus remains as to why the crust, at the time of LMO crystallization, was thinnest beneath what is now the Procellarum KREEP Terrane. Although there are plausible explanations, they are not easily testable. For instance, if an ancient and giant "Procellarum" impact occurred, forming a crater larger than the South-Pole Aitken basin, the crust would have been thinned, causing accumulation of molten KREEP. However, there is no unequivocal evidence that such a "Procellarum impact basin" (Neumann et al. 1996) exists, though admittedly such an ancient feature would have been heavily modified by the Moon's subsequent impact history. An alternative hypothesis is that a large number of statistically independent intermediate-sized impact events might have caused a net redistribution of crustal materials from one side of the Moon to the other. The expected magnitude of this process, while ultimately statistical in nature, has yet to be worked out. Finally, aspects related to convection within the LMO may have somehow caused large-scale crustal-thickness variations.

7.6. The effect of large impacts and mantle convection on the distribution of KREEP

A final model of note is that advocated by Arkani-Hamed and Pentecost (2001) for creating lateral variations in the thickness of an originally global KREEP-rich layer. This model rests upon the premise that large impact basins should have created localized near-surface thermal anomalies that subsequently affected the pattern of mantle convection in their vicinity. Convection simulations showed that the KREEP layer close to the crater should have become entrained in the mantle circulation and should have been displaced both radially outward and into the underlying mantle. For the case of a South Pole-Aitken sized impact, the convection cells were

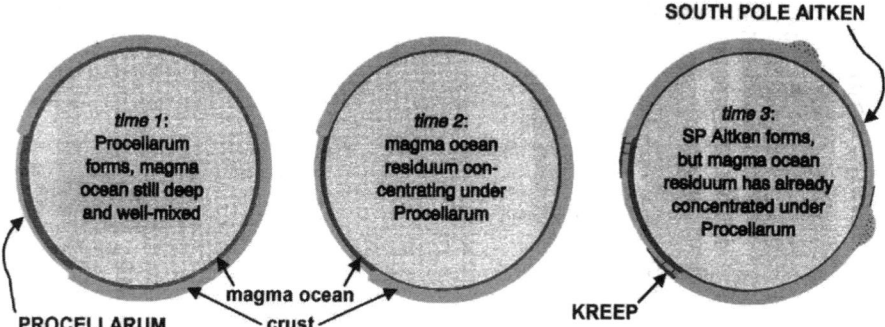

Figure 4.57. Conceptual manner in which KREEPy magma-ocean residua might have become concentrated within a single region of the Moon. Formation of a thinned crust beneath what is now the Procellarum KREEP Terrane gives rise to a relatively thicker magma ocean in this region. This could have occurred either by a "Procellarum impact" or by regional statistical fluctuations in crustal thickness. By the time the South Pole-Aitken basin formed on the far side, the majority of KREEP had already been concentrated beneath the PKT.

shown to strip away the KREEP layer within one crater radius and to transport it into the underlying mantle. In contrast, for an Imbrium-sized impact, the convective cells were weaker and
acted to concentrate all KREEP within 1 crater radius into a narrow annular ring 1 crater radius
away. From the work of Manga and Arkani-Hamed (1991), it was suggested that the combination of a low-conductivity ejecta blanket, as well as a KREEP enhancement about 1-crater radii
away, could produce partial melting in the upper mantle. Favorable stress conditions might then
have led to lateral transport of this magma from beneath surrounding highlands into the basin.

Although it is plausible to expect large impact basins to have influenced localized
convection cells, the redistribution of KREEP in this model may be an artifact of assuming that
its density was the same as that of the mantle. If the progenitor of the Mg-suite rocks was an
Mg-rich KREEPy magma (such as KREEP basalt) sandwiched between the crust and mantle,
its lower density would have resisted entrainment in the mantle circulation. Indeed, it may be
that this layer was still molten at the time of the Imbrium impact (e.g., Hubbard and Minear
1976; Solomon and Longhi 1977). It is also unlikely to expect that a low-conductivity ejecta
blanket surrounding a basin would significantly influence thermal evolution because (1) the
upper layer of the Moon had already been extensively brecciated by numerous impacts, and
(2) thermal conductivity depends primarily upon rock porosity, which is mostly a function of
lithostatic pressure (i.e., depth below the surface).

7.7. Summary and conclusions

Thermal models that attempt to explain the asymmetric magmatic evolution of the Moon
are still in their infancy. At present, two scenarios seem most plausible for the formation of the
Procellarum KREEP Terrane and the long duration of magmatic activity there. First, KREEP
may have become concentrated within the PKT as a result of the crust there being thinner than
typical. Subsequently, radiogenic heating within this layer would have heated the underlying
mantle giving rise to a wide compositional range of basalts. Secondly, a degree-1 downwelling
of ilmenite-rich cumulates may have concentrated both the residual magma-ocean KREEP
layer and subsequent magmatism in one hemisphere as has been proposed by Parmentier et
al. (2000, 2001). Although the degree-1 upwelling model of Zhong et al. (2000) can not be
dismissed, this seems to be the least plausible model as current estimates of core size would
disallow such a hemispheric upwelling. Moreover, this model does not address the origin of
the KREEP enhancement found within the Imbrium and Oceanus Procellarum region.

8. NEW VIEWS OF THE THERMAL AND MAGMATIC
EVOLUTION OF THE MOON

8.1. Introduction

The pioneering science of the Apollo missions created an observational and conceptual
framework on which to develop a comprehensive understanding of the Moon and to construct
scientifically well-founded lunar missions. In the two decades immediately following the
Apollo missions, lunar exploration was based on new analytical-computational-technological
advances and a retrospective view of the science established during Apollo. Not until the
1990s did the Galileo, Clementine and Lunar Prospector missions provide additional direct
observational data. This chapter presented new insights into the thermal and magmatic
evolution of the Moon afforded by the integration of new approaches and observation. A
summary of some of the more important new views follows.

8.2. New views

The observational data derived from the Clementine and Lunar Prospector missions
further resolved the asymmetrical nature of the Moon. Although it had been long recognized
that the distribution of mare volcanism was asymmetrical, the observation that a substantial

enrichment of the Procellarum terrane in incompatible and heat-generating elements (KREEP) was a decisive advance in understanding lunar evolution and early planetary differentiation. Although our understanding of the mechanism behind the distribution of KREEP is in its infancy, the potential ramifications are profound. The asymmetry of the thermal character of the lunar crust and mantle appears to affect not only the ability of the mantle to generate basalts but it also affects distinct evolutionary pathways for the lunar mantle.

New data on short-lived isotopic systems (Nd-Sm, Hf-W) and computational modeling of both large impacts and accretional processes has set limits on the timetable for the formation and differentiation of not only the Moon but all the terrestrial planets. Building on the derivation of the Moon from the violent encounter of the Earth with a Mars-size body, rapid accretion of the Moon is predicted by models and is consistent with isotopic systematics. An important consequence of rapid accretion is that substantial heat is stored in the early Moon. This provides a solution for the problem of not having enough energy to generate a magma ocean. Although early thermal modeling of the LMO suggests that the majority of the magma ocean crystallized in about 150 to 250 m.y., interpretation of Hf-W isotopic data and the old ages of Mg-suite plutonic rocks suggests that most of the LMO event (99%) may have been very short lived (10s of millions of years). Rapid, early differentiation of the Moon is also consistent with evidence for other planetary bodies in the solar system (i.e., Mars).

Since the realization that the Moon and potentially other terrestrial planets may have experienced rapid differentiation via a magma ocean, the concept of the LMO has become increasing both more important and complex. The increasingly complex modeling of the dynamics of a global magma ocean have yielded considerable insights and a conceptual framework on which to attach petrologic and geochemical observations. Yet to be accomplished is integrating these complex theoretical models with more simplistic petrologic perceptions of LMO crystallization.

The recognition that the pyroclastic glasses represent the best approximations of primary lunar basalts has lead to insights such as the depth of the LMO, dynamics of the lunar mantle, composition of the lunar mantle, and the depth of origin of the mare basalts. The prevalent opinion soon after the Apollo missions was that most mare basalts were derived from the shallow lunar mantle (< 400 km) and that only the outer portion of the Moon participated in the LMO. In contrast, studies of the pyroclastic glasses strongly suggest that some aspects of mare magmatism reflect melting initiated in the deep mantle. The conclusion reached from experimental studies of the pyroclastic glasses that both the high-Ti and low-Ti magmas were generated from the deep lunar mantle added credence to the Ringwood and Kesson models that were proposed in the mid-1970s that the LMO cumulate pile was gravitationally unstable. More sophisticated modeling of cumulate instability has provided a whole new perspective on both mare magmatism and earlier periods of lunar magmatism (i.e., Mg-suite).

The integration of global, remotely sensed spectral data with lunar samples has corrected early misconceptions concerning mare magmatism. Early sample analysis indicated that mare basalts were bimodal in composition with regard to Ti, that there was a temporal relationship between Ti content and eruption, and that the duration of basaltic magmatism was fairly short-lived (4.4-3.0 Ga). In contrast, current integrated data indicate that (1) low-Ti basalts are the dominate basalt type, (2) there appears to be no or limited relationship between Ti content and age (although there appears to be a relationship between age and KREEP component in lunar basalts), (3) there appears to be a relationship between mare basalt composition and its regional distribution, and (4) basaltic magmatism extended from approximately 4.5 to 1.0 Ga. These new observations have dramatic implications for the thermal history of the Moon, the thermal character of the lunar mantle, and the generation of lunar basalts.

Lunar exploration has many lessons for exploring the thermal and magmatic evolution of a planet. Exploration of the Moon has provided a paradigm for exploring a planetary body.

Achieving the scientific goals for planetary exploration requires a coordinated and integrative program involving orbital global mapping, *in situ* generation of geophysical, mineralogical, and geochemical data, and sample return and analysis. Sample return is not the last step in an extensive planetary exploration program. Rather, it is critical that data obtained from sample return missions be integrated into the analytical and scientific goals of subsequent global mapping and *in situ* missions to provide a broad planetary perspective.

8.3. Goals for future science initiatives

Lunar scientific exploration has demonstrated that for a "simple" planetary body, the early stages of planetary evolution so well preserved on the Moon are very complex. Planetary-scale remotely sensed observations and the lunar meteorites indicate that materials collected by the Apollo and Luna missions, however, sampled only a restricted subset of the Moon's complexity.

First and foremost, an understanding of the Moon's internal structure is needed to significantly advance our understanding of its thermal and magmatic evolution. New missions designed to collect new geophysical data are necessary to gain further insight into the size and composition of the lunar core, the depth of the LMO, and the existence and nature of the lower lunar mantle.

Key to understanding the early differentiation of the Moon and other planetary bodies is gaining a rigorous theoretical understanding of the workings of a magma ocean. Models have become much more sophisticated since the Apollo missions, but they are not yet well integrated with observational data. Integrating complex modeling within a strict petrologic framework is essential. The Moon provides the best preserved remnant of early planetary evolution and therefore provides a natural laboratory to test and constrain complex models. Unfortunately, much of our perspective of the Moon comes from biased sampling. Sampling of other regions of the Moon will surely provide additional constraints and undoubtedly many surprises.

9. ACKNOWLEDGMENTS

The authors of this chapter would like to thank the numerous science programs in NASA that funded both the original research summarized in this chapter and the organization and preparation of this manuscript. We would also like to thank the Lunar and Planetary Institute for their assistance in organizing many of the New Views of the Moon workshops and special LPSC sessions and contributing to the preparation of this manuscript. This manuscript was vastly improved by reviews provided by Marc Norman, Jim Papike, and Brad Jolliff. The team leader (CKS) is very appreciative for all the unselfish contributions made by team members, their valuable insights expressed in this manuscript and internal reviews that they provided. CKS would also like to thank members of the Institute of Meteoritics for the invaluable input they provided during the preparation of this manuscript.

10. REFERENCES

Abe Y, Zahnle KJ, Hashimoto A (1999) Elemental fractionation during rapid accretion of the Moon. *In:* Proc 32 ISAS Lunar Planet Symp. H Mizutani, M Kato (eds), Institute of Space and Aeronautical Science, Sagamikara, Japan, p 52-55

Agnor CB, Canup RM, Levison HF (1999) On the character and consequences of large impacts in the late stage of terrestrial planet formation. Icarus 142:219-237

Alibert C, Norman MD, McCulloch MT (1994) An ancient Sm-Nd age for a ferroan noritic anorthosite clast from lunar breccia 67016. Geochim Cosmochim Acta 58:2921-2926

Alley KM, Parmentier EM (1998) Numerical experiments on thermal convection in a chemically stratified viscous fluid heated from below; implications for a model of lunar evolution. Phys Earth Planet Interiors 108:15-32

Anders E (1977) Chemical compositions of the Moon, Earth and eucrite parent body. Phil Trans Royal Soc London Ser A 285:23-40

Antonenko I (1999) Global estimates of cryptomare deposits: Implications for lunar volcanism. Lunar Planet Sci XXX: 1703

Antonenko I (1999) Volumes of cryptomafic deposits on the Western Limb of the Moon: Implications for lunar Volcanism. PhD Dissertation, Brown University, Providence, RI

Antonenko I, Head JW, Mustard JF, Hawke BR (1995) Criteria for the detection of lunar cryptomaria. Earth Moon Planets 69:141-172

Arkani-Hamed J (1998) The lunar mascons revisited. J Geophys Res 103:3709-3739

Arkani-Hamed J, Pentecost A (2001) On the source region of the lunar mare basalts. J Geophys Res 106:14691-14700

Arndt N, Lehnert K, Vasilev Y (1995) Meimechites: highly magnesian lithosphere-contaminated alkaline magmas from deep subcontinental mantle. Lithos 34:41-59

Baker MB, Grove TL, Kinzler RJ, Donnelly-Nolan JM, Wandless GA (1991) Origin of compositional zonation (high-alumina basalt to basaltic andesite) in the Giant Crater lavafield, Medicine Lake volcano, northern California. J Geophys Res 96:21819-21842

Baldwin RB (1949) The face of the Moon. University of Chicago Press

Beard BL, Taylor LA, Scherer EE, Johnson CM, Snyder GA (1998) The source region and melting mineralogy of high-titanium and low-titanium lunar basalts deduced from Lu-Hf isotope data, Geochim Cosmochim Acta 62: 525-544

Beck AR, Hess PC (2002) The ilmenite saturation surface for high TiO_2 mare basalts. Lunar Planet Sci XXXII:1939

Bell JF, Hawke BR (1984) Lunar dark-haloed impact craters: Origin and implications for early mare volcanism. J Geophys Res 89:6899-6910

Benz W (1990) Smooth particle hydrodynamics—a review. *In:* Proc NATO Adv Res Workshop on Numerical Modelling of Nonlinear Stellar Pulsations. JR Buchler (ed), Kluwer Academic, p 1-54

Benz W, Cameron AGW, Melosh HJ (1989) The origin of the Moon and the single impact hypothesis III. Icarus 81: 113-131

Benz W, Slattery WL, Cameron AGW (1986) The origin of the Moon and the single impact hypothesis I. Icarus 66: 515-535

Benz W, Slattery WL, Cameron AGW (1987) The origin of the Moon and the single impact hypothesis II. Icarus 71: 30-45

Binder AB (1976) On the compositions and characteristics of the mare basalt magmas and their source regions. The Moon 16:115-150

Blewett DT, Hawke BR (2001) Remote sensing and geological studies of the Hadley-Apennine region of the Moon. Meteorit Planet Sci 36:701-730

Borg L, Norman M, Nyquist LE, Bogard DD, Snyder G, Taylor L, Lindstrom MM (1999) Isotopic studies of ferroan anorthosite 62236: a young lunar crustal rock from a light rare-earth-element-depleted source. Geochim Cosmochim Acta 63:2679-2691

Borg L, Nyquist LE, Taylor L, Wiesman H, Shih C-Y (1997) Constraints on martian differentiation processes from Rb-Sr and Sm-Nd isotopic analyses of the basaltic shergottite QUE 94201. Geochim Cosmochim Acta 61:4915-4931

Borg LE, Shearer CK, Asmerom Y, Papike JJ (2004) Prolonged KREEP magmatism on the Moon indicated by the youngest dated lunar igneous rock. Nature 432:209-211

Boss AP (1985) The origin of the Moon. Science 231:341-345

Brat SR, Solomon TSC, Head JW, Thurber CH (1985) The deep structure of lunar basin: Implications for basin formation and modification. J Geophys Res 90:3049-3064

Brett R (1977) The case against early melting of the bulk of the Moon. Geochim Cosmochim Acta 41:443-445

Brown GM, Emeleus CH, Holland JG, Peckett A, Phillips R (1972) Mineral-chemical variations in Apollo 14 and Apollo 15 basalts and granitic fractions. Proc Lunar Sci Conf 3:141-157

Brush SG (1996) Fruitful encounters: The origin of the solar system and of the moon from Chamberlin to Apollo. Cambridge University Press

Buck WR, Toksoz NM (1980) The bulk composition of the Moon based on geophysical constraints. Proc Lunar Planet Sci Conf 11:2043-2058

Budney CJ, Lucey PG (1998) Basalt thickness in Mare Humorum: The crater excavation method. J Geophys Res 103 E7:16855-16870

BVSP (Basaltic Volcanism Study Project) (1981) Basaltic Volcanism on the Terrestrial Planets. Pergamon

Cameron AGW (1997) The origin of the Moon and the single impact hypothesis V. Icarus 126:126-137

Cameron AGW (2000) Higher-resolution simulations of the giant impact. *In:* Origin of the Earth and Moon. Canup RM, Righter K (eds), University of Arizona Press, p 133-144

Cameron AGW (2001) From interstellar gas to the Earth-Moon system. Meteor Planet Sci 36:9-22

Cameron AGW, Benz W (1991) The origin of the Moon and the single impact hypothesis IV. Icarus 92:204-216

Cameron AGW, Ward WR (1976) The origin of the Moon. Lunar Sci VII:120-122

Canup RM, Agnor CB (2000) Accretion of the terrestrial planets and the Earth-Moon system. *In:* Origin of the Earth and Moon. Canup RM, Righter K (eds), University of Arizona Press, p 113-129

Canup RM, Asphaug E (2001) Origin of the Moon in a giant impact near the end of the Earth's formation. Nature 412: 708-712

Canup RM, Asphaug E, Pierazzo E, Melosh HJ (2002) Simulations of Moon forming impacts. Lunar Planet Sci XXXIII:1641

Canup RM, Esposito LW (1995) Accretion in the Roche Zone: co-existence of rings and ringmoons. Icarus 113:331-352

Canup RM, Esposito LW (1996) Formation of the Moon from an impact-generated disk. Icarus 119:427-446

Canup RM, Levison HF, Stewart GR (2001) Stability of a terrestrial multiple moon system. Astron J 117:603-620

Canup RM, Ward WR, Cameron AGW (2001) A scaling law for satellite-forming impacts. Icarus 150:288-296

Carlson RW (1982) Chronologic and isotopic systematics of lunar highland rocks. *In:* Workshop on Pristine Lunar Highland Rocks and the Early History of the Moon. Lunar and Planetary Institute, p 31-35

Carlson RW, Lugmair GW (1981a) Time and duration of lunar highlands crust formation. Earth Planet Sci Lett 52: 227-238

Carlson RW, Lugmair GW (1981b) Sm-Nd age of lherzolite 67667: Implications for the processes involved in lunar crustal formation. Earth Planet Sci Lett 56:1-7

Carlson RW, Lugmair GW (1988) The age of ferroan anorthosite 60025: Oldest crust on a young Moon? Earth Planet Sci Lett 90:119-130

Casse PM, Reynolds RT (1973) Role of convection in the Moon. J Geophys Res 78:3203-3215

Cassen P, Reynolds RT, Graziani F, Summers A, McNellis J, Blalock L (1979) Convection and lunar thermal theory. Phys Earth Planet Int 19:183-196

Cassen PM, Young RE (1975) On the cooling of the Moon by solid convection. The Moon 12:361-368

Chambers JE, Wetherill GW (1998) Making the terrestrill planets: N-body integrations of planetary embryos in three dimensions. Icarus 136:304-327

Chen JK, Delano JW, Lindsley DH (1982) Chemistry and phase relations of VLT volcanic glasses from Apollo 14 and Apollo 15. J Geophys Res 87:A171-A181

Chen JK, Lindsley DH (1983) Apollo 14 very low titanium glasses: melting experiments in iron-platinum alloy capsules. J Geophys Res 88:B335-342

Chevrel SD, Pinet PC, Daydou Y, Maurice S, Feldman WC, Lawrence DJ, Lucey PG (2000) Fe, Ti, and Th abundances of the lunar surface at global scale from UV-VIS spectral Clementine and gamma-ray lunar Prospector data. Lunar Planet Sci XXXI:1629

Cintala MJ, Grieve RAF (1998) Scaling impact melting and crater dimensions: Implications for the lunar cratering record. Meteorit Planet Sci 33:889-912

Circone S, Agee CB (1996) Compressibility of molten high-Ti mare glass: Evidence for crystal-liquid density inversions in the lunar mantle. Geochim Cosmochim Acta 66:2709-2720

Compston W, Williams IS, Meyer C (1984a) Age and chemistry of zircon from late-stage lunar differentiates. Lunar Planet Sci XV:182-183

Compston W, Williams IS, Meyer C (1984b) U-Pb geochronology of zircons from breccia 73217 using a sensitive high mass-resolution ion microprobe. Proc Lunar Sci Conf 14:B525-B534

Cook JJ (1967) A summary of lunar geology. USGS Interagency Report NASA-95

Dana JD (1846) On the volcanoes of the Moon. Amer J Sci 2:335-355

Darwin G (1879) Phil Trans Roy Soc 170:447

Dasche EJ, Shih C-Y, Bansal BM, Wiesmann H, Nyquist LE (1987) Isotopic analysis of basaltic fragments from lunar breccia 14321: Chronology and petrogenesis of pre-Imbrium mare volcanism. Geochim Cosmochim Acta 51: 3241-3254

Dasche, EJ, Ryder G, Shih C-Y, Wiesmann H, Bansal BM, Nyquist LE (1989) Time of crystallization of a unique A15 basalt. Lunar Planet Sci XX:218-219

Davis PA (1980) Iron and titanium distribution on the Moon from orbital gamma-ray spectrometry with implications for crustal evolutionary models. J Geophys Res 85:3209-3224

Davis PA, Spudis PD (1985) Petrologic province maps of the lunar highlands derived from orbital geochemical data. Proc Lunar Sci Conf 16, *In:* J Geophys Res 90:D61-D74

Davis PA, Spudis PD (1987) Global petrologic variations on the Moon: A ternary-diagram approach. Proc Lunar Sci Conf 17, *In:* J Geophys Res 92:E387-E395

Delano J (1990) Buoyancy-driven melt segregation in the Earth's Moon. I. Numerical results. Proc Lunar Planet Sci Conf 20:3-22

Delano JW (1979) Apollo 15 green glass: chemistry and possible origin. Proc Lunar Planet Sci Conf 10:275-300

Delano JW (1980) Chemistry and liquidus phase relations of Apollo 15 red glass: Implications for the deep lunar interior. Proc Lunar Planet Sci Conf 11:251-288

Delano JW (1986) Pristine lunar glasses: Criteria, data and implications. Proc Lunar Planet Sci Conf 16, *In:* J Geophys Res 91:D201-D213

Delano JW (1990) Pristine mare glasses and mare basalts: Evidence for a general dichotomy of source regions. *In:* Workshop on Lunar Volcanic Glasses: Scientific and Resource Potential; LPI Tech Rpt 90-02, Lunar and Planetary Institute, p 30-31

Delano JW, Hanson BZ, Watson EB (1994) Abundance and diffusivity of sulfur in lunar picritic magmas. Lunar Planet Sci XXV:325-326

Dick HJB (1999) A review of melt migration processes in the adiabatically upwelling mantle beneath oceanic spreading centers. *In:* Mid-ocean ridges: Dynamics of Processes Associated with Creation of New Oceanic Crust. Cann JR, Elderfield H, Laughton A (eds) Cambridge University Press, p 67-102

Dickey JO, Bender PI, Faller JE, Newhall XX, Ricklefs RL, Ries JG, Shelus PJ, Veillet C, Whipple AL, Wiant JR, Williams JG, Yoder CF (1994) Lunar laser ranging – A continuing legacy of the Apollo program. Science 265: 483-490

Dickinson T, Taylor GJ, Keil K, Bild RW (1988) Germanium abundances in lunar basalts: Evidence of mantle metasomatism? Proc Lunar Planet Sci Conf 19:189

Dickinson T, Taylor GJ, Keil K, Schmitt RA, Hughes SS, Smith MR, Apollo 14 aluminous basalts and their possible relationship to KREEP. Proc Lunar Planet Sci Conf 15. *In:* J Geophys Res 90:C365-C374

Drake MJ (1986) Is the lunar bulk material similar to Earth's mantle? *In:* Origin of the Moon. Hartmann WK, Phillips RJ, Taylor GJ (eds) Lunar and Planetary Institute, p 105-124

Duke MB, Silver LT (1967) Petrology of eucrites, howardites and mesosiderites. Geochim Cosmochim Acta 31:1637-1665

Elkins LT, Fernandes VA, Delano JW, Grove TL (2000) Origin of Lunar ultramafic glasses: constraints from phase equilibrium studies. Geochim Cosmochim Acta 64:2339-2350

Elkins LT, Grove PL (1999) Origin of lunar ultramafic green glass: Constraints from phase equilibrium studies. Lunar Planet Sci XXX:1035

Elkins Tanton L, Grove TL (2001) Lunar mantle composition and thermal history: Constraints from phase equilibrium studies. Lunar Planet Sci XXXII:1791

Elkins-Tanton LT, Chatterjee N, Grove TL (2003) Experimental and petrological constraints on lunar differentiation from the Apollo 15 green picritic glasses. Meteorit Planet Sci 38:515-527

Elkins-Tanton LT, Hager BH, Grove TL (2004) Magmatic effects of the lunar late heavy bombardment. Earth Planet Sci Lett 222:17-27

Elkins-Tanton LT, Van Orman JA, Hager BH, Grove TL (2002) Reexamination of the lunar magma ocean cumulate overturn hypothesis: Melting or mixing is required. Earth Planet Sci Lett 196:249-259

Elphic RC, Maurice S, Lawrence DJ, Feldman WC, Barraclough BL, Binder AB, Lucey PG (2000) Lunar Prospector measurements of incompatible elements gadolinium, samarium, and thorium. J Geophys Res 105: 20,333-20,346

Fernandes VA, Burgess R, Turner G (2003) ^{40}Ar-^{39}Ar chronology of lunar meteorites Northwest Africa 032 and 773. Meteorit Planet Sci 38(4):555-564

Finnila AB, Hess PC, Rutherford MJ (1994) Assimilation by lunar mare basalts: Melting of crustal material and dissolution of anorthite. J Geophys Res 99:14677-14690

Floss C, James OB, McGee JJ, Crozaz G (1991) Lunar ferroan anorthosites; rare earth element measurements of individual plagioclase and pyroxene grains. Proc Lunar Planet Sci Conf 22:391-392

Floss C, James OB, McGee JJ, Crozaz G (1998) Lunar ferroan anorthosite petrogenesis; clues from trace element distributions in FAN subgroups. Geochim Cosmochim Acta 62:1255-1283

Fogel RA, Rutherford MJ (1995) Magmatic volatiles in primitive lunar glasses: I FTIR and EPMA analyses of Apollo 15 green and yellow glasses and revision of the volatile-assisted fire-fountain theory. Geochim Cosmochim Acta 59:201-216

Fricker PE, Reynolds RT, Summers AL (1967) On the thermal history of the moon. J Geophys Res 172:2649-2663

Gault DE (1970) Saturation and equilibrium conditions for impact cratering on the surface of the Moon: Criteria and implications. Radio Sci 5:273-291

Gerstenkorn H (1970) The early history of the Moon. The Moon 1, 509

Ghiorso MS, Sack RO (1995) Chemical mass tansfer in magmatic process. IV. A revised and internally consistent thermodynamic model for the interpolation and extrapolation of liquid-solid equilibria in magmatic systems at elevated temperatures and pressures. Contrib Mineral Petrol 119:197-212

Giguere TA, Taylor GJ, Hawke BR, Lucey PG (2000) The titanium contacts of lunar mare basalts. Meteoritics Planet Sci 35:193-204

Gilbert GK (1893) The Moon's face: A study of the origin of its surface features. Bull Philo Soc Wash 12:241-292

Gillis JJ, Jolliff BL (1999) Lateral and vertical heterogeneity of thorium in the Procellarum KREEP terrane: As reflected in the ejecta deposits of post-Imbrium craters. *In:* Workshop on New Views of the Moon II, LPI Contribution No. 980. Lunar and Planetary Institute, p 18-19

Gillis JJ, Jolliff BL, Elphic RC (2003) A revised algorithm for calculating TiO2 from Clementine UVVIS data: A synthesis of rock, soil, and remotely sensed TiO$_2$ concentrations. J Geophys Res 108(E2) doi:10.1029/2001JE001515

Gillis JJ, Spudis PD (2000) Geology of the Smythii and Marginis region of the Moon; using integrated remotely sensed data. J Geophys Res, E, Planets 105:4217-4233

Gilvarry JJ (1968) Observational evidence for sedimentary rocks on the moon. Nature (London) 218:336-341

Glazier JA, Segawa T, Naert A, Sano M (1999) Evidence against "ultrahard" thermal turbulence at very high Rayleigh numbers. Nature (London) 398:307-310

Gold T (1955) The lunar surface. Royal Astron Soc Monthly Notices 115:585-604

Goldreich P (1966) History of the lunar orbit. Rev Geophys 4:411-439

Göpel C, Manhès G, Allègre CJ (1991) Constraints on the time of accretion and thermal evolution of chondrite parent bodies by precise U-Pb dating of phosphates. Meteoritics 26:73

Green DH, Blundy JD, Adam J, Yaxley GM (2000) SIMS determination of trace element partition coefficients between garnet, clinopyroxene and hydrous basaltic liquids at 2-7.5 GPa and 1080-1200 °C. Lithos 53:165-187

Green DH, Ringwood AE, Hibberson WO, Ware NG (1975) Experimental petrology of Apollo 17 mare basalts. Proc Lunar Sci Conf 6:871-893

Green DH, Ringwood AE, Ware NG, Hibberson WO, Major A, Kiss E (1971a) Experimental petrology and petrogenesis of Apollo 12 basalts. Proc Lunar Sci Conf 2:601-615

Green DH, Ware NG, Hibberson WO, Major A (1971b) Experimental petrology of Apollo 12 mare basalts. Part I, sample 12009. Earth Planet Sci Lett 13:85-96

Greenwood J, Hess PC (1996) Congruent Melting Kinetics: Constraints on chondrule formation. *In:* Chondrules and the Protoplanetary Disk. Hewins RH, Jones RH, Scott ERD (eds) Cambridge Univ Press, p 205-212

Greenwood J, Hess PC (1997) Congruent melting kinetics of albite: theory and experiment. J Geophy Res 103:29815-29828

Grove TL, Vaniman DT (1978) Experimental petrology of very low Ti (VLT) basalts. *In:* Mare Crisium: The View From Luna 24. Merrill RB, Papike JJ (eds) Pergamon Press, p 445-471

Gunnarsson B, Marsh BD, Taylor HP Jr. (1998) Generation of Icelandic rhyolites: silicic lavas from the Torfajokull central volcano. J Volcanol Geotherm Res 83:1-45

Hagerty J, Shearer CK, Papike JJ (2001) Trace element variability of the Apollo 14 high-Al basalts. A result of igneous processes or sample size? Lunar Planet Sci XXXII:1235

Haines EL, Metzger AE (1980) Lunar highland crustal models based on iron concentrations: isostasy and center-of-mass displacements. Proc Lunar Sci Conf 11:698-718

Halliday AN (2000) Terrestrial accretion rates and the origin of the Moon. Earth Planet Sci Lett 176:17-30

Halliday AN (2003) The origin and earliest history of the Earth. *In:* Treatise on Geochemistry. Vol. 1 Meteorites, Comets and Planets. Davis AM (vol. ed), Holland HD, Turekian KK (series eds) Elsevier-Pergamon, p 509-557

Halliday AN (2004) Mixing, volatile loss and compositional change during impact-driven accretion of the Earth. Nature 427:505-509

Halliday AN, Lee D-C (1999) Tungsten isotopes and the early development of the Earth and Moon. Geochim Cosmochim Acta 63:4157-4179

Halliday AN, Lee D-C, Jacobsen SB (2000) Tungsten isotopes, the timing of metal-silicate fractionation and the origin of the Earth and Moon. *In:* Origin of the Earth and Moon. RM Canup, K Righter (eds), Univ Arizona Press, p 45-62

Halliday AN, Lee D-C, Porcelli D, Wiechert U, Schönbächler M, Rehkämper M (2001) The rates of accretion, core formation and volatile loss in the early solar system. Phil Trans R Soc 359, in press

Halliday AN, Porcelli D (2001) In search of lost planets – the paleocosmochemistry of the inner solar system. Earth Planet Sci Lett 192(4):545-559

Halliday AN, Rehkämper M, Lee D-C, Yi W (1996) Early evolution of the Earth and Moon: new constraints from Hf-W isotope geochemistry. Earth Planet Sci Lett 142:75-89

Hanan BB, Tilton GR (1987) 60025: Relict of primitive lunar crust? Earth Planet Sci Lett 84:15-21

Hart SR, Zindler A (1986) In search of a bulk-Earth composition. Chem Geol 57:247-267

Hartmann WK (1965) Terrestrial and lunar flux of large meteorites in the last two billion years. Icarus 4:157-165

Hartmann WK (1980) Dropping stones in magma oceans: Effects of early lunar cratering. *In:* Proc Conf Lunar Highlands Crust. Papike JJ, Merrill RB (eds) Pergamon Press, p 155-171

Hartmann WK, Davis DR (1975) Satellite-sized planetesimals and lunar origin. Icarus 24:504-515

Haskin LA (1998) The Imbrium impact event and the thorium distribution at the lunar highlands surface. J Geophys Res 103:1679-1689

Haskin LA (1998) The Imbrium impact event and the thorium distribution at the lunar highlands surface. J Geophys Res 103:1679-1689

Hawke BR, Giguere TG, Blewett DT, Lucey PG, Peterson CA, Taylor GJ, Spudis PD (1999) Remote sensing studies of ancient mare basalt deposits. Lunar Planet Sci XXX:1956

Hawke BR, Head JW (1978) Lunar KREEP volcanism: Geologic evidence for history and mode of emplacement. Proc Lunar Planet Sci Conf 9:3285-3309

Hawke BR, Lucey PG, Bell JF (1986) Spectral reflectance studies of Tycho Crater: Preliminary results. Lunar Planet Sci XVII:999-1000

Hawke BR, Lucey PG, Bell JF, Spudis PD (1990) Ancient mare volcanism. LPI-LAPST Workshop on Mare Volcanism and Basalt Petrogenesis: Astounding Fundamental Concepts (AFC) Developed Over the Last Fifteen Years. Lunar and Planetary Institute, p 5-6

He H, Ahrens TJ (1994) Mechanical properties of shock-damaged rocks. Intl J Rock Mech Min Sci Geomech Abstr 31:525-533

Head JW (1974) Lunar dark mantle deposits: Possible clues to the distribution of early mare deposits. Proc Lunar Sci Conf 5:207-222

Head JW (1975) Lunar mare deposits: Areas, volumes, sequence, and implication for melting in source areas. *In:* Origins of Mare Basalts and their Implications for Lunar Evolution. Lunar Science Institute, p 66-69

Head JW (1976) Lunar volcanism in space and time. Rev Geophys Space Phys 14:265-300

Head JW (1998) Lunar mare basalt volcanism: stratigraphy, flux and implications for petrogenetic evolution. *In:* New Views of the Moon. Jolliff BL, Ryder G (eds), LPI Contrib 958:38-40

Head JW, Hawke BR (1992) The distribution and modes of occurrence of impact melt at lunar craters. International conference on large meteorite impacts and planetary evolution. LPI Contrib 790:37-38

Head JW, Wilson L (1992) Lunar mare volcanism: stratigraphy, eruption conditions, and the evolution of secondary crusts. Geochim Cosmochim Acta 56:2155-2175

Heiken GH, Vaniman DT, French BM (1991) Lunar Sourcebook: A User's Guide to the Moon. Cambridge University Press

Helmke PA, Haskin LA, Korotev RL, Ziege KE (1972) Rare earths and other trace elements in Apollo 14 samples. Proc Lunar Sci Conf 3:1275-1292

Herbert F (1980) Time dependent lunar density models. Proc Lunar Planet Sci Conf 11:2015-2030

Herbert F, Drake MJ, Sonett CP, Wiskerchen MJ (1977) Thermal history of lunar magma ocean. U.S. NASA Technical Memorandum 3511-31-33

Herzberg CT, Baker MB (1980) The cordierite- to spinel-transition: Structure of the lunar crust. *In:* Proc Conf Lunar Highlands Crust. Papike JJ, Merrill RB (eds) Pergamon, p 113-132

Hess PC (1989a) Origins of Igneous Rocks. Harvard Univ Press

Hess PC (1989b) Highly evolved liquids from the fractionation of mare and nonmare basalts. *In:* Workshop on the Moon in Transition. Taylor GJ, Warren PH (eds), LPI Tech Report 89-03, Lunar and Planetary Institute, p 46-52

Hess PC (1991) Diapirism and the origin of high TiO_2 mare glasses. Geophys Res Lett 18:2069-2072

Hess PC (1992) Phase equilibria constraints on the origin of ocean floor basalts. *In:* Mantle Flow and Melt Generation at Mid-Ocean Ridges. Geophys Monogr Series, Vol. 71. Morgan JP, Blackman DK, Sinton JM (eds) American Geophysical Union, p 67-102

Hess PC (1994) Petrogenesis of lunar troctolites. J Geophys Res 99:19,083-19,093

Hess PC, Parmentier EM (1995) A model for the thermal and chemical evolution of the Moon's interior: Implications for the onset of mare volcanism. Earth Planet Sci Lett 134:501-514

Hess PC, Parmentier EM (1999) Asymmetry and timing of mare volcanism. Lunar Planet Sci XXX:1300,

Hess PC, Parmentier EM (2001) Thermal evolution of a thicker KREEP liquid layer. *In:* New Views of the Moon II, Part 3. J Geophys Res 106:28023-28032.

Hess PC, Rutherford MJ, Campbell HW (1978) Ilmenite crystallization in non-mare basalt: Genesis of KREEP and high-Ti mare basalts. Proc Lunar Sci Conf 9:705-724

Hess PC, Rutherford MJ, Guillemette RN, Ryerson FJ, Tuchfeld HA (1975) Residual products of fractional crystallization of lunar magmas: An experimental study. Proc Lunar Sci Conf 6:895-910

Hiesinger H, Head JW, III, Wolf U, Jaumann R, Neukum G (2002) Lunar mare basalt flow units: Thicknesses determined from crater size-frequency distributions. Geophys Res Lett 29(8). doi:10.1029/2002GL014847

Hiesinger H, Head JW, Wolf U, Neukum G (2001) Lunar mare basalts: Mineralogical variations with time. Lunar Planet Sci XXXII:1826

Hiesinger HJ, Jaumann R, Neukum G, Head JW (2000) Ages of mare basalts on the lunar nearside. J Geophys Res 105: 29259-29275

Hirschmann MM, Stolper EM (1996) A possible role of garnet pyroxenite in the origin of the garnet signature in MORB. Contrib Mineral Petrol 124:185-208

Hirth G, Kohlstedt DL (1996) Water in the oceanic upper mantle; implications for rheology, melt extraction and the evolution of the lithosphere. Earth Planet Sci Lett 144:93-108

Hodges EN, Kushiro I (1974) Apollo 17 petrology and experimental determination of differentiation sequences in model Moon compositions. Proc Lunar Sci Conf 5:505-520

Hofmeister AM (1983) Effect of a Hadean terrestrial magma ocean on crust and mantle evolution. J Geophys Res B 88:4963-4983

Hood LL (1986) On lunar origins and planetary physics; impact model featured on film. Geotimes 31:15-17

Hood LL, Jones JH (1987) Geophysical constraints on lunar bulk composition and structure; a reassessment. Proc Lunar Planet Sci Conf 17. *In:* J Geophys Res B92:E396-E410

Hood LL, Zuber MT (2000) Recent refinements in geophysical constraints on lunar origin and evolution. *In:* Origin of the Earth and Moon. Canup RM, Righter K (eds), Univ Arizona Press, p 397-412

Horng WS, Hess PC (2000) Partition coefficients of Nb and Ta between rutile and anhydrous haplogranite melts. Contrib Mineral Petrol 138:176-185

Hörz F (1978) How thick are lunar mare basalts? Proc Lunar Planet Sci Conf 9:3311-3331

Housen KR, Schmidt RM, Holsapple KA (1983) Crater ejecta scaling laws: Fundamental forms based on dimensional analysis. J Geophys Res 88:2485-2499

Hubbard NJ, Gast PW (1972) Chemical composition and origin of nonmare lunar basalts. Proc Lunar Sci Conf 2: 999-1020

Hubbard NJ, Gast PW, Meyer C, Nyquist LE, Shih C, Wiesmann H (1971) Chemical composition of lunar anorthosites and their parent liquids. Earth Planet Sci Lett 13:71-75

Hubbard NJ, Gast PW, Rhodes JM, Bansal BM, Wiesmann H, Church SE (1972) Nonmare basalts: Part II. Proc Lunar Planet Sci Conf 3:1161-1179

Hubbard NJ, Minear JW (1975) A physical and chemical model of early lunar history. Proc Lunar Planet Sci Conf 6: 1057-1085

Hughes SS, Delano JW, Schmitt RA (1988) Apollo 15 yellow-brown volcanic glass: Chemistry and petrogenetic relations to green volcanic glass and olivine-normative mare basalts. Geochim Cosmochim Acta 52:2379-2391

Hughes SS, Delano JW, Schmitt RA (1989) Petrogenetic modeling of 74220 high-TiO_2 orange volcanic glasses and the Apollo 11 and 17 high-Ti mare basalt. Proc Lunar Planet Sci Conf 19:175-188

Hughes SS, Neal CR, Taylor LA (1990) Petrogenesis of Apollo 14 high-alumina parental magma. Lunar Planet Sci XXI:540-541

Humayan M, Cassen P (2000) Processes determining the volatile abundances of the meteorites and terrestrial planets. *In:* Origin of the Earth and Moon. Canup RM, Righter K (eds) Univ Arizona Press, p 3-23

Hunter RH, Taylor LA (1983) The magma ocean from the Fra Mauro shoreline: An overview of the Apollo 14 crust. Proc Lunar Planet Sci Conf 13:A591-A602

Ida S, Canup RM, Stewart G (1997) Formation of the Moon from an impact-generated disk. Nature 389:353-357

Ivanov BA, Melosh HJ (2003) Impacts do not initiate volcanic eruptions: Eruptions close to the crater. Geology 31: 869-872

James OB (1980) Rocks of the early lunar crust. Proc Lunar Planet Sci Conf 11: 365-393

James OB, Flohr MK (1983) Subdivision of the Mg suite noritic rocks into Mg-gabbronorites and Mg-norites. J Geophys Res 88:A603-A614

James OB, Lindstrom MM, Flohr MK (1987) Petrology and geochemistry of alkali gabbronorites from lunar breccia 67975. Proc Lunar Planet Sci Conf 17:E314-E330

James OB, Lindstrom MM, McGee JJ (1991) Lunar ferroan anorthosite 60025; petrology and chemistry of mafic lithologies. Proc Lunar Planet Sci Conf 21:63-87

Jolliff BL (1991) Fragments of quartz monzodiorite and felsite in Apollo 14 soil particles. Proc Lunar Planet Sci Conf 21:101-118

Jolliff BL (1998) Large-scale separation of K-frac and REEP-frac in the source regions of Apollo impact-melt breccias, and a revised estimate of the KREEP composition. Intl Geol Rev 40:916-935

Jolliff BL, Floss C, McCallum IS, Schwartz JM (1999) Geochemistry, petrology, and cooling history of 14161,7373: A plutonic lunar sample with textural evidence of granitic-fraction separation by silicate-liquid immiscibility. Am Mineral 84:821-837

Jolliff BL, Gillis JJ, Haskin LA, Korotev RL, Wieczorek MA (2000) Major lunar crustal terranes: Surface expressions and crust-mantle origins. J Geophys Res 105:4197-4216

Jolliff BL, Haskin LA, Colson RO, Wadhwa M (1993) Partitioning in REE-saturated minerals: Theory, experiment, and modelling of whitlockite, apatite, and evolution of lunar residual magmas. Geochim et Cosmochim Acta 57: 4069-4094

Jolliff BL, Hsu W (1996) Geochemical effects of recrystallization and exsolution of plagioclase of ferroan anorthosite. Proc Lunar Planet Sci Conf 27:611-612

Jones JH (1995) Experimental trace element partitioning. *In:* Rock Physics and Phase Relations. A Handbook of Physical Constants. Ahrens TJ (ed) American Geoophysical Union, p 73-104

Jones JH, Delano JW (1989) A three component model for the bulk composition of the Moon. Geochim Cosmochim Acta 53:513-527

Jones, JH, Palme H (2000) Geochemical constraints on the origin of the Earth and Moon. *In:* Origin of the Earth and Moon. Canup RM, Righter K (eds) Univ Arizona Press, p 197-216

Karner JM, Papike JJ, Shearer CK (2001) Chemistry of olivine from planetary basalts: Earth-Moon comparisons emphasizing Mn/Fe and Co/Ni systematics. Lunar Planet Sci XXXII:1017

Kaula WM (1979) Thermal evolution of Earth and Moon growing by planetesimal impacts. J Geophys Res 84:999-1008

Kaula WM, Schubert G, Lingenfelter RE, Sjogren WL, Wollenhaupt WR (1973) Lunar topography from Apollo 15 and 16 laser altimetry. Proc Lunar Planet Sci Conf 4:2811- 2819

Kaula WM, Yoder CF (1976) Lunar orbit evolution and tidal heating of the Moon. *In:* Lunar Sci VII:440-442

Kelemen PB, Hirth G, Shimizu N, Spiegelman M, Dick HJB (1999) A review of melt migration processes in the adiabatically upwelling mantle beneath oceanic spreading centers. *In:* Mid-Ocean Ridges, Dynamics of Processes Associated with Creation of New Oceanic Crust. Cann JR (ed) Cambridge Univ Press, p 67-102

Kesson SE (1975) Mare basalts: Melting experiments and petrogenetic interpretation. Proc Lunar Sci Conf 6:921-944

Khan A, Mosegaard K (2001) New information on the deep lunar interior from an inversion of lunar free oscillation periods. Geophys Res Lett 9:1791-1794

Khan A, Mosegaard K, Rasmussen KL (2000) A new seismic velocity model for the Moon from a Monte Carlo inversion of the Apollo lunar seismic data. Geophys Res Lett 27:1591-1594

King SD, Raefsky A, Hager BH (1990) ConMan: Vectorizing a finite element code for incompressible two-dimensional convection in the Earth's mantle. Phys Earth Planet Int 59:195-207

Kinzler RJ, Donnelly-Nolan JM, Grove TL (2000) Late Holocene hydrous mafic magmatism at the Paint Pot Crater and Callahan Flows, Medicine Lake Volcano, N. California and the influence of H_2O in the generation of silicic magmas. Contrib Mineral Petrol 138:1-16

Kinzler RJ, Grove TL (1992a) Primary magmas of mid-ocean ridge basalts 1. Experiments and methods. J Geophys Res 97:6907-6926

Kinzler RJ, Grove TL (1992b) Primary magmas of mid-ocean ridge basalts 2. Applications. J Geophys Res 97:6907-6926

Kirk RL, Stevenson DJ, (1989) The competition between thermal contraction and differentiation in the stress history of the Moon. J Geophys Res 94:12133-12144

Klein EM, Langmuir CH (1987) Global correlation of ocean ridge basalt chemistry with axial depth and crustal thickness. J Geophys Res 92:8089-8115

Kleine T, Münker C, Mezger K, Palme H (2002) Rapid accretion and early core formation on asteroids and the terrestrial planets from Hf-W chronometry. Nature 418:952-955

Kokubo E, Canup RM, Ida S (2000) Lunar accretion from an impact-generated disk. *In:* Origin of the Earth and Moon. Canup RM, Righter K (eds) Univ Arizona Press, p 145-163

Kokubo E, Makino J, Ida S (2001) Evolution of a circumterrestrial disk and formation of a single Moon. Icarus 148:419-436

Korotev RL (1998) On the history and origin of LKFM. Workshop on New views of the Moon; integrated remotely sensed, geophysical, and sample datasets. LPI Contribution, Report 958:47-49

Korotev RL (2000) The great lunar hot spot and the composition and origin of the Apollo mafic ("LKFM") impact-melt breccias. J Geophys Res 105:4317-4345

Kuiper GP (1954) On the origin of lunar surface features. P.N.A.S.U. 40:1096-1112

Kurat G, Kracher A, Keil K, Warner R, Prinz M (1976) Composition and origin of Luna 16 aluminous mare basalts. Proceed Lunar Sci Conf 7:1301-1321

Langseth MG, Keihm SJ, Peters K (1976) Revised lunar heat flow values. Proc Lunar Sci Conf 7:3143-3171

Lawrence DJ, Feldman WC, Barraclough BL, Binder AB, Elphic RC, Maurice S, Thomsen DR (1998) The Lunar Prospector gamma-ray spectrometer. Science 281:1484-1498

Lawrence DJ, Feldman WC, Barraclough BL, Binder AB, Ephic RC, Maurice S, Miller MC, Prettyman TH (2000) Thorium abundances on the lunar surface. J Geophys Res 105:20,307-20,331

Lee D-C, Halliday AN, Leya I, Wieler R (2001) Cosmogenic tungsten on the Moon. Meteoritics 36:A11

Lee D-C, Halliday AN, Snyder GA, Taylor LA (1997) Age and origin of the Moon. Science 278:1098-1103

Lee D-C, Halliday AN, Snyder GA, Taylor LA (2000) Lu-Hf systematics and the early evolution of the Moon. Lunar Planet Sci XXXI:1288

Lee WHK (1968) Effects of selective fusion on the thermal history of the Earth's mantle. Earth Planet Sci Lett 4:270-276

Lejeune A, Richet P (1995) Rheology of crystal-bearing silicate melts; an experimental study at high viscosities. J Geophys Res B 100:4215-4229

Leya I, Wieler R, Halliday AN (2000) Cosmic-ray production of tungsten isotopes in lunar samples and meteorites and its implications for Hf-W cosmochemistry. Earth Planet Sci Lett 175:1-12

Lissauer JJ (1987) Timescales for planetary accretion and the structure of the protoplanetry disk. Icarus 69:249-265

Lister JR, Kerr RC (1991) Fluid-mechanical models of crack propagation and their application to magma transport in dykes. J Geophys Res B 96:10,049-10,077

Longhi J (1980) A model of early lunar differentiation. Proc Lunar Planet Sci Conf 11:289-315

Longhi J (1981) Preliminary modeling of high pressure partial melting: Implications for early lunar differentiation, Proc Lunar Planet Sci 12B:1001-1018

Longhi J (1987) On the connection between mare basalts and picritic volcanic glasses. Proc Lunar Planet Sci Conf 17:E349-E360

Longhi J (1991) Comparative liquidus equilibria of hypersthene-normative basalts at low pressure. Am Mineral 76:785-800

Longhi J (1992a) Origin of green glass magmas by polybaric fractional fusion. Proc Lunar Planet Sci Conf 22:343-353

Longhi J (1995) Liquidus equilibria of some primary lunar and terrestrial melts in the garnet stability field. Geochim Cosmochim Acta 59:2375-2386

Longhi J (2002) Some phase equilibrium systematics of lherzolite melting, I. Geochem Geophys GeoSys 3(3) doi: 10.1029/2001GC000204

Longhi J, Ashwal LD (1985) Two-stage models for lunar and terrestrial anorthosites: Petrogenesis without a magma ocean. Proc Lunar Planet Sci Conf 15:C571-C584

Longhi J, Boudreau AE (1979) Complex igneous processes and the formation of the primitive lunar crustal rocks. Proc Lunar Planet Sci Conf 10:2085-2105

Longhi J, Walker D, Grove TL, Stolper EM, Hays JF (1974) The petrology of Apollo 17 mare basalts. Proc Lunar Sci Conf 5:447-469

Lucey PG, Blewett DT, Jolliff BL (2000) Lunar iron and titanium abundance algorithms based on final processing of Clementine. J Geophys Res 105:20,297-20,305

Lucey PG, Hawke BR (1989) A remote mineralogical perspective on gabbroic units in the lunar highlands. Proc Lunar Planet Sci Conf 19:355-363

Lucey PG, Taylor GJ, Malaret E (1995) Abundance and distribution of iron on the Moon. Science 268:1150-1153

Lucy LB (1977) A numerical approach to the testing of the fission hypothesis. Astron J 82:1013-1024

Lugmair GW, Carlson RW (1978) The Sm-Nd history of KREEP. Proc Lunar Planet Sci Conf 9:689-704

Lugmair GW, Galer SJG (1992) Age and isotopic relationships between the angrites Lewis Cliff 86010 and Angra dos Reis. Geochim Cosmochim Acta 56:1673-1694

Lugmair GW, Marti K, Kurtz JP, Scheinin NB (1976) History and genesis of lunar troctolite 76535. Proc Lunar Planet Sci Conf 7:2009-2033

Lugmair GW, Shukolyukov A (1998) Early solar system timescales according to ^{53}Mn-^{53}Cr systematics. Geochim Cosmochim Acta 62:2863-2886

Ma M-S, Schmitt RA, Nielsen RL, Taylor GJ, Warner RD, Keil K (1979) Petrogenesis of Luna 16 aluminous mare basalts. J Geophys Res Lett 6:909-912

MacDonald GJF (1959) Calculations on the thermal history of the earth. J Geophys Res 64:1967-2000

MacDonald GJF (1960) Stress history of the Moon. Planet Space Sci 2:249-255

MacGregor ID (1969) The system MgO-SiO_2-TiO_2 and its bearing on the distribution of TiO_2 in basalts. Am J Sci 267A:342-363

Manga M, Arkani-Hamed J (1995) Remelting mechanisms for shallow source regions of mare basalts. Phys Earth Planet Inter 68:9-31

Marvin UB, Lindstrom MM, Holmberg BB, Martinez RR (1991) New observations on the quartz monzodiorite-granite suite. Proc Lunar Planet Sci 21:119-135

McCallum IS (1983) Formation of Mg-rich pristine rocks by crustal metasomatism. Lunar Planet Sci XIV:473-474

McCallum IS, Charette M (1978) Zr and Nb partition coefficients: implications for the genesis of mare basalts, KREEP and sea floor basalts. Geochim Cosmochim Acta 42:859-870

McCallum IS, O'Brien HE (1996) Stratigraphy of the Lunar Highlands Crust: depths of burial from cooling rate studies. Am Mineral 81:1166-1175

McKay GA, Wagstaff J, Le L (1990) REE distribution coefficients for pigeonite: Constraints on the origin of the mare basalt europium anomaly. Lunar Planet Sci XXI:773-774

McKay GA, Weill DF (1976) Petrogenesis of KREEP. Proc Lunar Sci Conf 7:2339-2355

McKay GA, Weill DF (1977) KREEP petrogenesis revisited. Proc Lunar Science Conf 8:949-999

McKenzie D (1984) The generation and compaction of partially molten rock. J Petrol 25:713-765

Melosh HJ (1989) Impact Cratering: A Geologic Process. Oxford Univ Press

Melosh HJ (1990) Giant impacts and the thermal state of the early Earth. *In:* Origin of the Earth. HE Newsom, JH Jones (eds), Oxford Univ Press, p 69-83

Melosh HJ (2000) A new and improved equation of state for impact computations. Lunar Planet Sci XXXI:1903

Melosh HJ, Ivanov BA (1999) Impact crater collapse. Ann Rev Earth Planet Sci 27:385-415

Melosh HJ, Kipp ME (1989) Giant impact theory of the Moon's origin: First 3-D hydrocode results. Lunar Planet Sci XX:685-686

Melosh HJ, Pierazzo E (1997) Impact vapor plume expansion with realistic geometry and equation of state. Lunar Science XXIIX:935

Melson WG, Vallier TL, Wright TL, Byerly G, Nelen J (1976) Chemical diversity of abyssal volcanic glass erupted along Pacific, Atlantic, and Indian Ocean sea-floor spreading centers. *In:* The Geophysics of the Ocean Basin and Its Margins. Am Geophys Union Geophysical Monograph 19:351-368

Metzger AE, Haines EL, Etchegaray-Ramirez MI, Hawke BR (1979) Thorium concentrations in the lunar surface: III. Deconvolution of the Apenninus region. Proc Lunar Planet Sci Conf 10:1701-1718

Metzger AE, Parker RE (1979) The distribution of titanium on the lunar surface. Earth Planet Sci Lett 45:155-171

Meyer C (1977) Petrology, mineralogy, and chemistry of KREEP basalt. Phys Chem Earth 10:239-260

Meyer C, Williams IS, Compston W (1996) Uranium-lead ages for lunar zircons: Evidence for a prolonged period of granophyre formation from 4.32 to 3.88 Ga. Meteoritics 31:370-387

Minear JW, Fletcher CR (1978) Crystallization of a lunar magma ocean. Proc Lunar Planet Sci Conf 9:263-283

Misawa K, Tatsumoto M, Dalrymple GB, Yanai K (1993) An extremely low U/Pb source in the Moon: U-Th-Pb, Sm-Nd, Rb-Sr, and $^{40}Ar/^{39}Ar$ isotopic systematics and age of lunar meteorite Asuka 881757. Geochim Cosmochim Acta 57:4687-4702

Morishima R, Watanabe S-I (2001) Two types of co-accretion scenarios for the origin of the Moon. Earth, Planets, Space 53:213-231

Morse SA (1982) Adcumulus growth of anorthosite at the base of the lunar crust. J Geophys Res 87:A10-A18

Murase T, McBirney AR (1973) Properties of some common igneous rocks and their melts at high temperatures. Geol Soc Am Bull 84:3563-3592

Nakamura N, Tatsumoto M, Nunes P, Unruh DM, Schwab AP, Wildeman TR (1976) 4.4 by old clast in Boulder 7, Apollo 17: A comprehensive chronological study by U-Pb, Rb-Sr, and Sm-Nd methods. Proc Lunar Planet Sci Conf 7:2309-2333

Neal CR (2001) Interior of the Moon: The presence of garnet in the primitive deep lunar mantle. J Geophys Res 106: 27865-27886

Neal CR, Taylor LA (1989) The nature of barium partitioning between immiscible melts: A comparison of experimental and natural systems with reference to lunar felsite petrogenesis. Proc Lunar Planet Sci Conf 19:209-218

Neal CR, Taylor LA (1992) Petrogenesis of mare basalts: A record of lunar volcanism. Geochim et Cosmochim Acta 56:2177-2211

Neal CR, Taylor LA, Lindstrom MM (1988) Apollo 14 mare basalt petrogenesis: Assimilation of KREEP-like components by a fractionating magma. Proc Lunar Planet Sci Conf 18:139-153

Neal CR, Taylor LA, Patchen AD (1989a) High alumina (HA) and very high potassium (VHK) basalt clasts from Apollo 14 breccias, Part 1-Mineralogy and petrology: Evidence of crystallization from evolving magmas. Proc Lunar Planet Sci Conf 19:137-145

Neal CR, Taylor LA, Schmitt RA, Hughes SS, Lindstrom MM (1989b) High alumina (HA) and very high potassium (VHK) basalt clasts from Apollo 14 breccias, Part whole rock geochemistry: Further evidence of combined assimilation and fractional crystallization within the lunar crust. Proc Lunar Planet Sci Conf 19:147-161

Neukum G, Dietzel H (1975) On the development of the crater population on the Moon with time under meteoroid and solar wind bombardment. Earth Planet Sci Lett 12:59-66

Neumann GA, Zuber MT, Smith DE, Lemoine FG (1996) The lunar crust: global structure and signature of major basins. J Geophys Res 101:16841-16843

Newsom HE (1995) Composition of the solar system, planets, meteorites, and major terrestrial reservoirs. In: Global Earth Physics: A handbook of physical constants. Ahrens TJ (ed) Am. Geophys. Union, p 159-189

Nininger HH (1943) The moon as a source of tektites. Sky and Telescope 2:12-15

Nininger HH (1947) Chips from the Blasted Moon. Desert Press

Nord GL, Wandless MV (1983) Petrology and comparative thermal and mechanical histories of clasts in breccia 62236. Proc Lunar Planet Sci Conf 13, Part 2. In: J Geophys Res B 86:A645-A657

Norman M, Nyquist L, Bogard D, Borg L, Wiesmann H, Garrison D, Reese Y, Shih C-Y, Schwandt C (2000) Age and origin of the highlands on the moon: Isotopic and petrologic studies of ferroan noritic anorthosite clast from Descartes breccia 67215. Lunar Planet Sci XXXI:1552

Norman MD, Borg LE, Nyquist LE, Bogard DE (2003) Chronology, geochemistry, and petrology of a ferroan noritic anorthosite clast from Descartes breccia 67215; clues to the age, origin, structure, and impact history of the lunar crust. Meteorit Planet Sci 38:645-661

Norman MD, Ryder G (1979) A summary of the petrology and geochemistry of pristine highlands rocks. Proc Lunar Planet Sci Conf 10:531-59

Nyquist LE, Bogard DD, Shih C-Y, Wiesman H (2002) Negative ε_{Nd} in Anorthositic clasts in Yamato 86032 and MAC88105: Evidence for the LMO? Lunar Planet Sci 33:1289

Nyquist LE, Reimold WU, Bogard DD, Wooden JL, Bansal BM, Wiesmann H and Shih C-Y (1981) A comparative Rb-Sr, Sm-Nd, and K-Ar study of shocked norite 78236: Evidence of slow cooling in the lunar crust. Proc Lunar Planet Sci Conf 12B:167-197

Nyquist LE, Shih C-Y (1992) The isotopic record of lunar volcanism. Geochim Cosmochim Acta 56:2213-2234

Nyquist LE, Smith C-Y, Wooden JL, Bansal BM, Wiesmann H (1979) The Sr and Nd isotopic record of Apollo 12 basalts: Implications for lunar geochemical evolution. Proc Lunar Planet Sci Conf 10:77-114

Nyquist LE, Wiesmann H, Shih C-Y, Keith JE, Harper CL (1995) ^{146}Sm-^{142}Nd formation interval in the lunar mantle. Geochim Cosmochim Acta 59:2817-2837

O'Hara MJ, Biggar GM, Richardson SW, Ford CE, Jamieson BG (1970) The nature of seas, mascons, and the lunar interior in the light of experimental studies. Proc Apollo 11 Lunar Sci Conf 7:2449-2467

O'Keefe JA (1963) Tektites. Univ Chicago Press

O'Keefe JD, Ahrens TJ (1977) Impact-inducing energy partitioning, melting, and vaporization on terrestrial planets. Proc Lunar Planet Sci Conf 8:3357-3374

O'Keefe JD, Ahrens TJ (1993) Planetary cratering mechanics. J Geophys Res 98:17011-17028

O'Keefe JD, Ahrens TJ (1999) Complex craters: Relationship of stratigraphy and rings to impact conditions. J Geophys Res 104:27,091-27,104

Ohtsuki K, Ida S (1998) Planetary rotation by accretion of planetesimals with non-uniform spatial distribution formed by the planet's gravitational perturbation. Icarus 131:393-420

Oldenburg CM, Spera FJ (1990) Numerical experiments of magmatic solidification and convection. Eos Transactions, American Geophysical Union 71:1660

O'Neill HSC (1991) The origin of the Moon and the early history of the Earth—A chemical model. Part 1: The Moon. Geochim Cosmochim Acta 55:1143-1158

Öpik EJ (1966) The cometary origin of meteorites. Adv Astron Astrophys 4:301-336

Papanastassiou DA, Wasserburg GJ (1971) Rb-Sr ages of igneous rocks from the Apollo 14 mission and the age of the Fra Mauro formation. Earth Planet Sci Lett 12:36-48

Papanastassiou DA, Wasserburg GJ (1975) Rb-Sr study of a lunar dunite and evidence for early lunar differentiates. Proc Lunar Sci Conf 6:1467-1489

Papike JJ (1996) Pyroxene as a recorder of cumulate formational processes in asteroids, Moon, Mars, Earth: Reading the record with the ion microprobe. Am Mineral 81:525-544

Papike JJ, Fowler GW, Adcock CT Shearer CK (1999) Systematics of Ni and Co in olivine from planetary melt systems: Lunar mare basalts. Am Mineral 84:392-399

Papike JJ, Fowler GW, Shearer CK (1994) Orthopyroxene as a recorder of lunar crust evolution: An ion microprobe investigation of Mg suite norites. Am Mineral 79:790-800

Papike JJ, Fowler GW, Shearer CK (1997a) Evolution of the lunar crust: SIMS study of plagioclase from ferroan anorthosites. Geochim Cosmochim Acta 61:2343-2350

Papike JJ, Fowler GW, Shearer CK, Layne GD (1996) Ion microprobe investigation of plagioclase and orthopyroxene from lunar Mg suite norites: Implications for calculating parental melt REE concentrations and for assessing post-crystallization REE redistribution. Geochim Cosmochim Acta 60:3967-3978

Papike JJ, Hodges FN, Bence AE, Cameron M, Rhodes JM (1976) Mare basalts: Crystal chemistry, mineralogy, and petrology. Rev Geophys Space Phys 14:475-540

Papike JJ, Ryder G, Shearer CK (1998) Lunar samples. Rev Mineral 36:5-1 - 5-234

Papike JJ, Spilde MN, Adcock CT, Fowler GW, Shearer CK (1997b) Trace-element fractionation by impact-induced volatilization: SIMS study of lunar HASP samples. Am Mineral 82:630-634

Parmentier EM, Zhong S, Hess PC (1999) Asymmetric evolution of the Moon. A possible consequence of chemical differentiation. Lunar Planet Sci XXX:1289

Parmentier EM, Zhong S, Zuber MT (2001) Gravitational differentiation of an initially unstable chemical stratification: origin of lunar asymmetries. Lunar Planet Sci 32:1329

Parmentier EM, Zhong S, Zuber MT (2002) Gravitational differentiation due to initial chemical stratification; origin of lunar asymmetry by the creep of dense KREEP? Earth Planet Sci Lett 201:473-480

Phillips RJ, Ivins ER (1979) Solid convection in the terrestrial planets. Phys Earth Planet Int 19:107-148

Phillpotts JA, Schnetzler CC, Nava DF, Bottino ML, Fullagar PD, Thomas HH, Schumann S, Kouns CW (1972) Apollo 14: Some geochemical aspects. Proc Lunar Sci Conf 3:1293-1305

Philpotts S, Anthony R, Brustman CM, Shi J, Carlson WD, Denison C (1999) Plagioclase-chain networks in slowly cooled basaltic magma. Am Mineral 84:1819-1829

Phinney WC (1991) Lunar anorthosites, their equilibrium melts, and the bulk Moon. Proc Lunar Planet Sci 21:29-49

Pierazzo E, Vickery AM, Melosh HJ (1997) A reevaluation of impact melt production. Icarus 127:408-423

Pieters CM (1993) Compositional diversity and stratigraphy of the lunar crust derived from reflectance spectroscopy. *In:* Remote Geochemical Analysis: Elemental and Mineralogical Composition. Pieters CM, Englert P (eds), Cambridge University Press, p 309-336

Pieters CM, Head JW III, Gaddis L, Jolliff B, Duke M (2001) Rock types of South Pole Aitkin basin and the extent of basaltic volcanism. J Geophys Res 106:28,001-28,002

Pieters CM, Tompkins S (1999) Tsiolkovsky crater: A window into crustal processes on the lunar farside. J Geophys Res 104:21935-21949

Pieters CM, Tompkins S, Head JW, Hess PC (1997) Mineralogy of the mafic anomaly in the South Pole-Aitken Basin: Implications for excavation of the lunar mantle. Geophys Res Lett 24:1903-1906

Presnall DC, Hoover JD (1987) High pressure phase equilibrium constraints on the origin of mid-ocean ridge basalts. *In:* Magmatic Processes: Physiochemical Principles. Mysen BO (ed), Geochemical Society, p 75-89

Pritchard ME, Stevenson DJ (1999) How has the Moon released its internal heat? Lunar Planet Sci Conf XXX:1981

Pritchard ME, Stevenson DJ (2000) Thermal implications of a lunar origin by giant impact. *In:* Origin of the Earth and Moon. Canup RM, Righter K (eds) Univ Arizona Press, p 179-196

Quick JE, Albee AL, Ma M-S, Murali AV, Schmitt RA (1977) Chemical composition and possible immiscibility of two silicate melts in 12013. Proc Lunar Sci Conf 8:2153-2189

Raedeke LD, McCallum IS (1980) A comparison of fractionation trends in the lunar crust and the Stillwater Complex. *In:* Proc Conf Lunar Highlands Crust. Papike JJ, Merrill RB (eds) Pergamon, p 133-154

Ransford GA, Kaula WM (1980) Heating of the Moon by heterogeneous accretion. J Geophys Res B 85:6615-6627

Reid AM, Jakes P (1974) Luna 16 revisited: The case for aluminous mare basalts. Lunar Planet Sci V:627-629

Rhodes JM, Hubbard NJ, Wiesmann H, Rodgers KV, Brannon JC, Bansal BM (1976) Chemistry, classification, and petrogenesis of Apollo 17 mare basalts. Proc Lunar Sci Conf 7:1467-1489

Ridley WI (1975) On high-alumina mare basalts. Proc Lunar Planet Sci Conf 6:131-145

Righter K (2002) Does the Moon have a metallic core? Constraints from giant-impact modeling and siderophile elements. Icarus 158(1):1-13

Righter K, Shearer CK (2003) Magmatic fractionation of Hf and W; constraints on the timing of core formation and differentiation in the Moon and Mars. Geochim Cosmochim Acta 67:2497-2507

Ringwood AE (1976) Limits on the bulk composition of the Moon. Icarus 28:325-349

Ringwood AF, Essene E (1970) Petrogenesis of Apollo 11 basalts, internal constitution, and origin of the Moon. Proc Apollo 11 Conf, Geochim Cosmochim Acta, Supp 1, 1:769-799

Ringwood AF, Kesson SE (1976) A dynamic model for mare basalt petrogenesis, Proc Lunar Sci Conf 7:1697-1722

Roedder EP, Weiblen W (1970) Lunar petrology of silicate melt inclusions, Apollo 11 rocks. Proc Apollo 11 Lunar Sci Conf, Geochim Cosmochim Acta, Supp 1, 1:801-837

Ronca LB (1966) Meteorite impact and volcanism. Icarus 5:515-520

Runcorn SK (1962) Convection in the Moon. Nature (London) 195:1150-1151

Runcorn SK (1963) Satellite gravity measurements and convection in the mantle. Nature (London) 200:628-630

Runcorn SK (1974) On the origin of mascons and moonquakes. Proc Lunar Planet Sci Conf 5:3115-3126

Ryan MP (1994) Neutral-buoyancy controlled magma transport and storage in mid-ocean ridge magma reservoirs and their sheeted-dike complex; a summary of basic relationships. *In:* Magmatic Systems. MP Ryan (ed), International Geophysics Series 57:97-138

Ryder G (1976) Lunar sample 15405: Remnant of a KREEP-granite differentiated pluton. Earth Planet Sci Lett 29:255-268

Ryder G (1991) Lunar ferroan anorthosites and mare basalt sources: The mixed connection. Geophys Res Lett 18:2065-2068

Ryder G (1994) Coincidence in time of the Imbrium basin impact and Apollo 15 KREEP volcanic flows: The case for impact-induced melting. Spec Pap Geol Soc Am 293:11-18

Ryder G, Norman MD, Taylor GJ (1997) The complex stratigraphy of the highland crust in the Serenitatis region of the Moon inferred from mineral fragment chemistry. Geochim Cosmochim Acta 61:1083-1105

Safronov VS (1954) On the growth of planets in the protoplanetary cloud. Astron Zh 31:499-510

Salpas PA, Taylor LA, Lindstrom MM (1987) Apollo 17 KREEPy basalts: Evidence for the non-uniformity of KREEP. Proc Lunar Planet Sci Conf 17, J Geophys Res 92:E340-E348

Sato M (1979) The driving mechanism of lunar pyroclastic eruptions inferred from the oxygen fugacity behavior of Apollo 17 orange glass. Proc Lunar Planet Sci Conf 10:311-325

Schönberg R, Kamber BS, Collerson KD, Eugster O (2002) New W isotope evidence for rapid terrestrial accretion and very early core formation. Geochim Cosmochim Acta 66:3151-3160

Schubert G, Lingenfelter RE, Peale SJ (1969) The distribution and morphology of lunar sinuous rilles. The Moon 1:138-159

Schultz PH, Spudis PD (1979) Evidence for ancient mare volcanism. Proc Lunar Planet Sci Conf 10:2899-2918

Schultz PH, Spudis PD (1983) The beginning and end of mare volcanism on the Moon. Lunar Planet Sci XIV:676-677

Seitz H-M, Altherr R, Ludwig T (1999) Partitioning of transition elements between orthopyroxene and clinopyroxene in peridotite and websterite xenoliths: New empirical geothermometers. Geochim Cosmochim Acta 63:3967-3982

Shearer CK, Floss C (2000) Evolution of the Moon's mantle and crust as reflected in trace-element microbeam studies of lunar magmatism. *In:* Origin of the Earth and Moon. Canup RM, Righter K (eds) Univ Arizona Press, p 339-360

Shearer CK, Layne GD, Papike JJ (1994) The systematics of light lithophile elements in lunar picritic glasses: Implications for basaltic magmatism on the Moon and the origin of the Moon. Geochim Cosmochim Acta 58:5349-5362

Shearer CK, Newsom HE (2000) W-Hf abundances and the early origin and evolution of the Earth-Moon system. Geochim Cosmochim Acta 64:3599-3613

Shearer CK, Papike JJ (1993) Basaltic magmatism on the Moon: A perspective from volcanic picritic glass beads. Geochim Cosmochim Acta 57:4785-4812

Shearer CK, Papike JJ (1999) Magmatic evolution of the Moon. Am Mineral 84:1469-1494

Shearer CK, Papike JJ (2000) Compositional dichotomy of the Mg suite. Origin and implication for the thermal and compositional structure of the lunar mantle. Lunar Planet Sci XXXI:1405

Shearer CK, Papike JJ, Galbreath KC, Shimizu N (1991) Exploring the lunar mantle with secondary ion mass spectrometry: A comparison of lunar picritic glass beads from the Apollo 14 and Apollo 17 sites. Earth Planet Sci Lett 102:134-147

Shearer CK, Papike JJ, Hagerty J (2001) Chemical dichotomy of the Mg-suite. Insights from a comparison of trace elements in silicates from a variety of lunar basalts. Lunar Planet Sci XXXII:643

Shearer CK, Papike JJ, Layne GD (1996a) Deciphering basaltic magmatism on the Moon from the compositional variations in the Apollo 15 very low-Ti picritic magmas. Geochim Cosmochim Acta 60:509-528

Shearer CK, Papike JJ, Layne GD (1996b) The role of ilmenite in the source regions for mare basalts: Evidence from niobium, zirconium, and cerium in picritic glasses. Geochim Cosmochim Acta 60:3521-3530

Shearer CK, Papike JJ, Simon SB, Galbreath KC, Shimizu N (1989) A comparison of picritic glass beads from the Apollo 14 and Apollo 17 sites: Implications for basalt petrogenesis and compositional variability in the lunar mantle. Lunar Planet Sci XX:996-997

Shearer CK, Papike JJ, Simon SB, Galbreath KC, Shimizu N, Yurimoto Y, Sueno S (1990) Ion microprobe studies of REE and other trace elements in Apollo 14 'volcanic' glass beads and comparison to Apollo 14 mare basalts. Geochim Cosmochim Acta 54:851-867

Shearer CK, Papike JJ, Spilde MN (2001) Trace element partitioning between immiscible melts. An example from naturally occurring lunar melt inclusions. Am Mineral 86: 238-246

Shearer CK, Righter K (2001) Hafnium and tungsten partitioning in silicates. A key to understanding the early evolution of both the Moon and Mars. Lunar Planet Sci XXXII:1620

Shearer CK, Righter K (2003) Behavior of tungsten and hafnium in silicates; a crystal chemical basis for understanding the early evolution of the terrestrial planets. Geophys Res Lett 30:1007-1011

Shervais JW, McGee JJ (1997) Alkali suite anorthosites and norites: Flotation cumulates from pristine KREEP with magma mixing and the assimilation of older anorthosite. Lunar Planet Sci XXIX:1699

Shervais JW, McGee JJ (1999) Ion and electron microprobe study of troctolites, norite, and anorthosites from Apollo 14: Evidence for urKREEP assimilation during petrogenesis of Apollo 14 Mg-suite rocks. Geochim Cosmochim Acta 62:3009-3024

Shervais JW, Taylor LA, Laul JC, Shih C-Y, Nyquist LE (1985b) Very high potassium (VHK) basalts: Complications in lunar mare basalt petrogenesis. Proc Lunar Planet Sci Conf 16, J Geophys Res 90: D3-D18

Shervais JW, Taylor LA, Lindstrom MM (1985a) Apollo 14 mare basalts: Petrology and geochemistry of clast from consortium breccia 14321. Proc Lunar Planet Sci Conf 15, J Geophys Res 89:C375-C395

Shervais JW, Vetter SK (1990) Lunar mare volcanism: Mixing of distinct mantle source regions with KREEP-like component. Lunar Planet Sci XXI:1142-1143

Shih C-Y, Nyquist LE Bogard DD, Wooden JL, Bansal BM, Wiesmann H (1985) Chronology and petrogenesis of a 1.8 g lunar granitic clast: 14321, 1062. Geochim Cosmochim Acta 49:411-426

Shih C-Y, Nyquist LE, Bansal BM, Wiesmann H (1992) Rb-Sr and Sm-Nd chronology of an Apollo 17 KREEP basalt. Earth Planet Sci Lett 108:203-215

Shih C-Y, Nyquist LE, Bogard DD, Bansal BM, Wiesmann H, Johnson P, Shervais JW, Taylor LA (1986) Geochronology and petrogenesis of Apollo 14 very high potassium mare basalts. Proc Lunar Planet Sci Conf 16, J Geophys Res 91:D214-D228

Shih C-Y, Nyquist LE, Bogard DD, Dasch EJ, Bansal BM, Wiesmann H (1987) Geochronology of high-K aluminous mare basalt clasts from Apollo 14 breccia 14304. Geochim Cosmochim Acta 51:3255-3271

Shih C-Y, Nyquist LE, Bogard DD, Reese Y, Wiesmann H, Garrison D (1999) Rb-Sr, Sm-Nd and ^{40}Ar-^{39}Ar isotopic studies of an Apollo 11 group D basalt. Lunar Planet Sci XXX:1787

Shih C-Y, Nyquist LE, Dasch EJ, Bogard DD, Bansal BM, Wiesmann H (1993) Age of pristine noritic clasts from lunar breccias 15445 and 15455. Geochim Cosmochim Acta 57:915-931

Shirley DN (1983) A partially molten magma ocean model. J Geophys Res 88:A519-A527

Skulski TWM, Watson EB (1994) High-pressure experimental trace-element partitioning between clinopyroxene and basaltic melts. Chem Geol 117:127-147

Sleep NH (1988) Tapping of melt by veins and dikes. J Geophys Res B 93:10255-10,272

Sleep NH, Fujita K(1998) Principles of Geophysics. Blackwell Science

Smith JA, Anderson AT, Newton RC, Olsen EJ, Wyllie PJ, Crewe AV, Isaacson MS, Johnson D (1970) Petrologic history of the Moon inferred from petrography, mineralogy, and petrogenesis of Apollo 11 rocks. Proc Lunar Planet Sci Conf 1:1149-1162

Smith JV (1982) Heterogeneous growth of meteorites and planets, especially the Earth and Moon. J Geol 90:1-48

Smith JV, Anderson AT, Newton RC, Olsen EJ, Wyllie PJ, Crewe A, Issachson MS, Johnson D (1970) Petrologic history of the Moon inferred from petrography, mineralogy, and petrogenesis of Apollo II rocks. Proc Apollo 11 Lunar Sci Conf, Geochim Cosmochim Acta, Supp 1, 1:897-925

Snyder GA, Borg LE, Nyquist LE, Taylor LA (2000) Chronology and isotopic constraints on lunar origin and evolution. *In:* Origin of the Earth and Moon, RM Canup, K Righter (eds), University of Arizona Press, p 361-396

Snyder GA, Hall CM, Lee D-C, Taylor LA, Halliday AN (1996) Earliest high-Ti volcanism on the Moon: ^{40}Ar-^{39}Ar, Sm-Nd, and Rb-Sr isotopic studies of Group D basalts from the Apollo 11 landing site. Meteorit Planet Sci 31: 328-334

Snyder GA, Lee D-C, Taylor LA, Halliday AN, Jerde EA (1994) Evolution of the upper mantle of the Earth's Moon. Neodymium and strontium isotopic constraints on high-Ti mare basalts. Geochim Cosmochim Acta 58:4795-4808

Snyder GA, Neal CR, Taylor LA, Halliday AN (1997) Anatexis of lunar cumulate mantle in time and space: Clues from trace-element, strontium and neodymium isotopic chemistry of parental Apollo 12 basalts. Geochim Cosmochim Acta 61:2731-2748

Snyder GA, Taylor LA, Crozaz G (1993) Rare earth element selenochemistry of immiscible liquids and zircon at Apollo 14: An ion probe study of evolved rocks on the Moon. Geochim Cosmochim Acta 57:1143-1149

Snyder GA, Taylor LA, Jerde EA, Riciputi LR (1994) Evolved QMD-melt parentage for lunar highlands alkali suite cumulates: evidence from ion-probe rare-earth element analyses of individual minerals. Lunar Planet Sci XXV: 1311-1312

Snyder GA, Taylor LA, Neil CR (1992) A chemical model for generating the sources of mare basalts: Combined equilibrium and fractional crystallization of the lunar magmasphere. Geochim Cosmochim Acta 56:3809-3823

Solomatov VS (2000) Fluid dynamics of a terrestrial magma ocean. *In:* Origin of the Earth and Moon. Canup RM, Righter K (eds) Univ Arizona Press, p 323-338

Solomatov VS, Moresi LN (1996) Three regimes of mantle convection with non-Newtonian viscosity; theory and observations. Eos, Transactions, American Geophysical Union 77:750

Solomatov VS, Stevenson DJ (1993a) Suspension in convective layers and style of differentiation of a terrestrial magma ocean. J Geophys Res E 98:5375-5390

Solomatov VS, Stevenson DJ (1993b) Nonfractional crystallization of a terrestrial magma ocean. J Geophys Res E 98: 5391-5406

Solomatov VS, Stevenson DJ (1993c) Kinetics of crystal growth in a terrestrial magma ocean. J Geophys Res E 98: 5407-5418

Solomon S, Head JW (1979) Vertical movement in mare basins: Relations to mare emplacement, basin tectonics, and lunar thermal history. J Geophys Res 84:1667-1682

Solomon S, Head JW (1980) Lunar mascon basins: Lava filling, tectonics, and evolution of the lithosphere. Rev Geophys Space Phys 18:107-141

Solomon SC (1975) Mare volcanism and lunar crustal structure. Proc Lunar Planet Sci Conf 6:1021-1042

Solomon SC (1977) The relationship between crustal tectonics and internal evolution in the Moon and Mercury. Phys Earth Planet Inter 15:135-145

Solomon SC (1986) On the early thermal state of the Moon. *In:* Origin of the Moon. Hartmann WK, Phillips RJ, Taylor GJ (eds), Lunar and Planetary Institute, p 311-329

Solomon SC, Chaiken J (1976) Thermal expansion and thermal stress in the Moon and terrestrial planets: Clues to early thermal history, Proc Lunar Sci Conf 7:3229-3243

Solomon SC, Longhi J (1977) Magma oceanography 1: Thermal evolution. Proc Lunar Sci Conf 8:583-599.

Solomon SC, Töksoz MN (1973) Internal constitution and evolution of the Moon. Phys Earth Planet Inter 7:15-38

Spera FJ (1992) Lunar magma transport phenomena. Geochim Cosmochim Acta 56:2253-2266

Spohn T, Konrad W, Breuer D, Ziethe R (2001) The longevity of lunar volcanism: Implications of thermal evolution calculations with 2D and 3D mantle convection models. Icarus 149:54-65

Spudis PD (1978) Composition and origin of the Apennine Bench Formation, Proc Lunar Planet Sci Conf 9:3379-3394

Spudis PD (1996) The Once and Future Moon. Smithsonian Press

Spudis PD, Hawke BR (1986) The Apennine Bench formation revisited. *In*: Workshop on the Geology and Petrology of the Apollo 15 Landing Site, LPI Tech Report 86-03. Spudis PD, Ryder G (eds), Lunar and Planetary Institute, p 105-107

Spudis PD, Hawke BR, Lucey PG (1988) Materials and formation of the Imbrium basin. Proc Lunar Planet Sci Conf 18:155-168

Spudis PD, Reisse RA, Gillis JJ (1994) Ancient multiring basins on the Moon revealed by Clementine laser altimetry. Science 266:1848-1851

Staid MI, Pieters CM (2001) Mineralogy of the last lunar basalts. Results from Clementine. J Geophys Res 106:27,887-27,900

Stegman DR, Jellinek AM, Zatman SA, Baumgardner JR, Richards MA (2003) An early lunar core dynamo driven by thermomechanical mantle convection. Nature (London) 421:143-146

Stevenson DJ (1980) Lunar asymmetry and paleomagnetism. Nature (London), 287:520-521

Stevenson DJ (1987) Origin of the Moon: The collision hypothesis. Ann Rev Earth Planet Sci 15 p 271-315

Stewart GR (2000) Outstanding questions for the giant impact hypothesis. *In:* Origin of the Earth and Moon. Canup RM, Righter K (eds) Univ Arizona Press, p 217-226

Stöffler D, Knöll HD, Marvin UB, Simonds CH, Warren PH (1980) Recommended classification and nomenclature of lunar highland rocks. *In:* Proc Conf Lunar Highlands Crust. Papike JJ, Merrill RB (eds) Pergamon Press, p 51-70

Strom RG, Trask NJ, Guest JE (1975) Tectonism and volcanism on Mercury. J Geophys Res 80:2478-2507

Taylor GJ, Warner RD, Keil K, Ma M-S, Schmitt RA (1980) Silicate liquid immiscibility, evolved lunar rocks and the formation of KREEP. *In:* Proc Conf Lunar Highlands Crust. Papike JJ, Merrill RB (eds) Pergamon Press, p 339-352

Taylor LA, Shervais JW, Hunter RH, Laul JC (1983a) Ancient (4.2 AE) highlands volcanism: The gabbronorite connection? Lunar Planet Sci XIV:777-778

Taylor LA, Shervais JW, Hunter RH, Shih C-Y, Bansal BM, Wooden J, Nyquist LE, Laul JC (1983) Pre 4.2 AE mare-basalt volcanism in the lunar highlands. Earth Planet Sci Lett 66:33-47

Taylor SR (1975) Lunar Science: A Post-Apollo View. Pergamon Press

Taylor SR (1980) Refractory and moderately volatile element abundances in the earth, moon and meteorites. Proc Lunar Planet Sci Conf 11:333-348

Taylor SR (1982) Planetary Science: A Lunar Perspective. Lunar and Planetary Institute

Taylor SR, Jakes P (1974) The geochemical evolution of the Moon. Proc Lunar Sci Conf 5:1287-1305

Taylor SR, Norman MD, Esat TM (1993) The Mg-Suite and the highland crust: An unsolved enigma. Lunar Planet Sci XXIV:1413-1414

Taylor, S.R., P. Jakês (1974) The geochemical evolution of the Moon. Proc Lunar Sci Conf 5:1287-1305

Tera F, Papanastassiou DA, Wasserburg GJ (1973) A lunar cataclysm at ~3.95 AE and the structure of the lunar crust. Lunar Planet Sci IV:723-725

Thompson C, Stevenson DJ (1988) Gravitational instability of two-phase disks and the origin of the Moon. Astrophys J 333:452-481

Thompson SL, Lauson HS (1972) Improvements in the chart-D radiation-hydrodynamic code III: Revised analytical equation of state. Tec. Rep. SC-RR-61 0714. Sandia Nat. Laboratories, Albuquerque, NM.

Tillotson JH (1962) Metallic equations of state for hypervelocity impact. Rep. GA-3216, July 18, Gen At., San Diego, California

Toksöz MN, Dainty AM, Solomon SC Anderson KR (1974) Structure of the Moon. Rev Geophys 12:539-567

Tompkins S, Pieters CM (1999) Mineralogy of the lunar crust: Results from Clementine. Meteorit Planet Sci 34:25-41

Tonks WB, Melosh HJ (1993) Magma ocean formation due to giant impacts. J Geophys Res 98:5319-5333

Torigoye-Kita N, Misawa K, Dalrymple GB, Tatsumoto M (1993) Further evidence for a low U/Pb source in the Moon: U-Th-Pb, Sm-Nd, and Ar-Ar isotopic systematics of lunar meteorite Yamato 793169. Geochim Cosmochim Acta 59:2621-2632

Touma J (2000) The phase space adventure of the Earth and Moon. *In:* Origin of the Moon. Hartmann WK, Phillips RJ, Taylor GJ (eds), Lunar and Planetary Institute, p 165-178

Touma J, Wisdom J (1994) Evolution of the Earth-Moon system. Astron J 108:1943-1961

Touma J, Wisdom J (1998) Resonances in the early evolution of the Earth-Moon system. Astron J 115:1653-1663

Tozer DC (1967) Towards a theory of thermal convection in the mantle. *In:* The Earth's Mantle. Gaskell TF (ed), Academic Press, p 325-353

Turcotte DL, Kellogg LH (1986) Implications of isotope data for the origin of the Moon. *In:* Origin of the Moon. Hartmann WK, Phillips RJ, Taylor GJ (eds), Lunar and Planetary Institute, p 311-329

Turcotte et al. (1979) Parameterized convection within the Moon and the terrestrial planets. Proc Lunar Planet Sci Conf 10:2375-2392

Unruh DM, Tatsumoto M (1977) Evolution of mare basalts; the complexity of U-Th-Pb system. Proc Lunar Sci Conf 8:1673-1696.

Urey HC (1952) The planets: Their origin and development. Yale Univ. Press

Urey HC (1957) Origin of tektites. Nature (London) 179:556-557

Urey HC (1959) Primary and secondary objects. J Geophys Res 64:1721-1737

Urey HC (1960) Criticism of the melted Moon theory. J Geophys Res 65:358-359

Urey HC (1962) Origin of tektites. Science 137:746-748

Urey HC (1965) Meteorites and the moon. Science 147:1262-1265

Van Orman JA, Grove TL (2000) Origin of lunar high-titanium ultramafic glass: Constraints from phase relations and dissolution kinetics of clinopyroxene-ilmenite cumulates. Meteorit Planet Sci 35:783-794

Verbeek RDM (1897) Over Glaskogels van Billiton. Verslagen van der vergadering der Wissen Natuurkundig Afdeeling. K Ned Akad Wet (Koninklijke Nederlandse Akademie van Weteschappen) 5:421-425

Wagner TP, Grove TL (1997) Experimental constraints on the origin of lunar high-Ti ultramafic glasses. Geochim Cosmochim Acta 61:1315-1328

Walker D (1983) Lunar and terrestrial crust formation. Proc Lunar Planet Sci Conf 14:B17-B25

Walker D, Longhi J, Hays JF (1972) Experimental petrology and origin of Fra Mauro rocks and soil. Proc Lunar Sci Conf 3:797-817

Walker D, Longhi J, Lasaga AC, Stolper EM, Grove TL, Hays JF (1977) Slowly cooled microgabbros 15555 and 15065. Proc Lunar Planet Sci Conf 8:1521-1547

Walker D, Longhi J, Stolper EJ, Grove TL, Hays JF (1975) Origin of titaniferous lunar basalts. Geochim Cosmochim Acta 39:1219-1235

Walker D, Longhi J, Stolper EM, Grove TL, Hays JF (1976) Differentiation of an Apollo 12 picrite magma. Proc Lunar Sci Conf 7:1365-1389

Wänke H, Baddenhausen H, Blum K, Cendales M, Dreibus G, Hofmeister H, Kruse H, Jagoutz E, Palme C, Spettel B, Thacker R, Vilczek E (1977) On the chemistry of lunar samples and achondrites. Proc Lunar Planet Sci Conf 8: 2191-2213

Wänke H, Dreibus G (1986) Geochemical evidence for the formation of the moon by impact-induced fusion of the proto-Earth. *In:* Origins of the Moon. Hartmann WK, Phillips RJ, Taylor GJ (eds), Lunar and Planetary Institute, p 649-672

Wark DA, Watson BE (2002) Effect of grain size on the distribution and transport of deep-seated fluids and melts. Geophys Res Lett 27:2029-2032

Warner JL, Simonds CH, Phinney WC (1976) Genetic distinction between anorthosites and Mg-rich plutonic rocks: new data from 76255. Lunar Sci VII:915-917

Warren PH (1985) The magma ocean concept and lunar evolution. Ann Rev Earth Planet Sci 13:201-240

Warren PH (1986) Anorthosite assimilation and the origin of the Mg/Fe related bimodality of pristine Moon rocks: Support for the magmasphere hypothesis. Proc Lunar Planet Sci Conf 16, J Geophys Res 91:D331-D343

Warren PH (1988) The origin of pristine KREEP: Effects of mixing between urKREEP and the magmas parental to the Mg-rich cumulates. Proc Lunar Planet Sci Conf 18:233-241

Warren PH (1989) KREEP: Major-element diversity, trace element uniformity (almost) (abstract). *In:* Workshop on Moon in Transition: Apollo 14, KREEP, and Evolved Lunar Rocks, LPI Tech Rpt 89-03, Lunar and Planetary Institute, p 149-153

Warren PH (1993) A concise compilation of petrologic information on possible pristine nonmare Moon rocks. Am Mineral 78:360-376

Warren PH, Kallemeyn GW (1993) The ferroan-anorthositic suite, the extent of primordial lunar melting, and the bulk composition of the moon. J Geophys Res 98:5445-5455

Warren PH, Taylor GJ, Keil K, Shirley DN, Wasson JT (1983) Petrology and chemistry of two "large" granite clasts from the Moon. Earth Planet Sci Lett 61:7484

Warren PH, Wasson JT (1977) Pristine nonmare rocks and the nature of the lunar crust. Proc Lunar Sci Conf 8:2215-2235

Warren PH, Wasson JT (1979) The origin of KREEP. Rev Geophys Space Phys 17:73-88

Warren PH, Wasson JT (1980) Early lunar petrogenesis, oceanic and extraoceanic. *In:* Proc Conf Lunar Highlands Crust. Papike JJ, Merrill RB (eds), Pergamon, p 81-99

Warren PH, Wasson JT (1980) Further foraging of pristine nonmare rocks: Correlations between geochemistry and longitude. Proc Lunar Planet Sci Conf 11:431-470

Warren PW, Rasmussen KL (1987) Megaregolith insulation, internal temperatures, and bulk uranium content of the Moon. J Geophys Res 92:3453-3465

Wasserburg GJ, Papanastassiou DA, Tera F, Huneke JC (1977) Outline of a lunar chronology. Phil Trans R Soc London A 285:7-22

Weidenschilling SJ, Greenberg R, Chapman CR, Herbert F, Davis DR, Drake MJ, Jones J, Hartmann WK (1986) Origin of the Moon from a circumterrestrial disk. *In:* Origins of the Moon. Hartmann WK, Phillips RJ, Taylor GJ (eds), Lunar and Planetary Institute, p 17-55

Weitz CM, Rutherford MJ, Head JW, McKay DS (1999) Ascent and eruption of a lunar high-titanium magma as inferred from the petrology of the 74001/2 drill core. Meteorit Planet Sci 34:527-540

Wentworth S, Taylor GJ, Warner RD, Keil K, Ma M-S, Schmitt RA (1979) The unique nature of Apollo 17 VLT mare basalts. Proc Lunar Planet Sci Conf 10:207-223

Wetherill GW (1976) The role of large bodies in the formation of the Earth and Moon. Proc Lunar Sci Conf 7:3245-3257

Wetherill GW (1980) Formation of the terrestrial planets. Ann Rev Astron Astrophys 18:77-113

Wetherill GW (1986) Accumulation of the terrestrial planets and implications concerning lunar origin. *In:* Origins of the Moon. Hartmann WK, Phillips RJ, Taylor GJ (eds), Lunar and Planetary Institute, p 519-550

Wiechert UH, Halliday AN, Lee D-C, Snyder GA, Taylor LA, Rumble D (2000) Oxygen-and tungsten- isotopic constraints on the early development of the Moon. Meteorit Planet Sci 35:A169

Wieczorek MA, Phillips RJ (1999) Lunar multi-ring basins and the cratering process. Icarus 139:246-259

Wieczorek MA, Phillips RJ (2000) The "Procellarum KREEP Terrane": Implications for mare volcanism and lunar evolution. J Geophys Res 105:20417-20420

Wieczorek MA, Zuber MT (2001) The composition and origin of the lunar crust: Constraints from central peaks and crustal thickness modeling. Geophys Res Lett 28:4023-4026

Wieczorek MA, Zuber MT, Phillips RJ (2001) The role of magma buoyancy on the eruption of lunar basalts. Earth Planet Sci Lett 185:71-83

Wilhelms DE (1972) Geologic map of the Taruntius quadrangle of the Moon, 1-722 (LAC-61). US Geol Surv Washington, DC

Wilhelms DE (1987) The Geologic History of the Moon. US Geological Survey Spec. Pap. 1348

Wilhelms DE, McCauley J (1971) Geologic map of the near side of the Moon. US Geological Survey Map I-703

Wilson AT (1962) Origin of petroleum and the composition of the lunar maria. Nature (London) 196:11-13

Wood J (1986) Moon over Mauna Loa: A review of hypotheses of formation of Earth's Moon. *In:* Origins of the Moon. Hartmann WK, Phillips RJ, Taylor GJ (eds), Lunar and Planetary Institute, 17-55

Wood JA (1975) Lunar petrogenesis in a well-stirred magma ocean. Proc Lunar Sci Conf 6: 1087-1102

Wood JA, Dickey JS, Marvin UB, Powell BN (1970) Lunar anorthosites and a geophysical model of the Moon. Proc Lunar Planet Sci Conf 1:965-988

Wyatt BA (1977) The melting and crystallization behavior of a natural clinopyroxene-ilmenite intergrowth. Contrib Min Petrol 61:1-9

Xirouchakis D, Hirshemann MM, Simpson JA (2001) The effect of titanium on the silica content and on mineral-liquid partitioning of mantle-equilibrated melts. Geochim Cosmochim Acta 65:2201-2217

Yin Q, Jacobsen SB, Yamashita K, Blichert-Toft J, Télouk P, Albarède F (2002) A short timescale for terrestrial planet formation from Hf-W chronometry of meteorites. Nature 418:949-952

Zhong S, Parmentier EM, Zuber MT (2000) A dynamic origin for the global asymmetry of lunar mare basalts. Earth Planet Sci Lett 177:131-140

Zhong S, Zuber MT (2000) Long-wavelength topographic relaxation for self-gravitating planets and implications for the time-dependent compensation of surface topography. J Geophys Res E 105:4153-4164

Zuber MT, Smith DE, Lemoine FG, Neumann GA (1994) The shape and internal structure of the Moon from the Clementine mission. Science 266:1839-1843

Reviews in Mineralogy & Geochemistry
Vol. 60, pp. 519-596, 2006
Copyright © Mineralogical Society of America

5

Cratering History and Lunar Chronology

Dieter Stöffler

Institut für Mineralogie, Museum für Naturkunde
Humboldt Universität zu Berlin
Invalidenstrasse 43, 10099 Berlin, Germany
e-mail: Dieter.Stoeffler@MUSEUM.HU-Berlin.de

Graham Ryder

Lunar and Planetary Institute
3600 Bay Area Blvd., Houston, Texas 77058, U.S.A.

Boris A. Ivanov and Natalia A. Artemieva

Institute for Dynamics of Geospheres
Leninsky Prospect, 38, Bldg. 1, 119334 Moscow, Russia

Mark J. Cintala

NASA Johnson Space Center
SN2, Houston, Texas, 77058, U.S.A.

Richard A. F. Grieve

Natural Resources Canada
588 Booth Street, Ottawa, Ontario, K1A 0Y7, Canada

1. INTRODUCTION

The Moon is exceptional and important because it is the only planetary body besides the Earth for which we have both a detailed stratigraphic history and datable rock samples that can be related to specific geomorphologic units (Fig. 5.1). The Moon has preserved much of its magmatic and impact record of at least the last 4 billion years. While its endogenic history is of great interest for the fundamentals of planetary interiors and surfaces, the Moon has become a calibration plate for the cratering record of the Earth-Moon system, and by extrapolation, of the entire inner solar system if one assumes a heliocentric origin for impactor populations. These populations range from asteroids through long and short period comets to interplanetary dust, and cover a size range from hundreds of kilometers to micrometers.

This chapter reviews the presently available data sets in support of this paradigmatic assumption, as follows: (1) the phenomenology of lunar impact craters, (2) the terrestrial record of the impact cratering process and the interpretation of terrestrial impactites as far as this "ground truth" is relevant for the interpretation of lunar impact craters and datable lunar impact breccias and melt rocks, (3) the theory and numerical simulation of the cratering process and the characteristics of the Earth-Moon crossing population of impactors (asteroids and comets), (4) the principles of relative age dating of lunar surface units and the general lunar stratigraphy, (5) the stratigraphic significance and ages of lunar samples (impactites and basalts) and, based on this data set, the absolute ages of lunar surface units, (6) the cratering rate of the Moon as a function of time, and (7) the time calibration of this cratering rate based on the most recent data for the ages of multiring basins, mare basalt surfaces, and post-

1529-6466/06/0060-0005$10.00   DOI: 10.2138/rmg.2006.60.05

Figure 5.1. Telescopic view of the nearside of the Earth's Moon
with landing sites of the Apollo and Luna missions.

Eratosthenian impact craters such as Copernicus, Tycho, North Ray, Cone, and South Ray. The present state of the art confirms the concept of an early heavy bombardment of the Moon before about 3.7 Ga and a more or less constant cratering flux since then, which is compatible with the relatively restricted terrestrial cratering record and with astronomical observations. It does, however, not allow firm conclusions about the existence of a terminal lunar cataclysm. In fact, there are serious but not yet final arguments against this concept. A major obstacle to solve this question is the lack of absolute ages for heavily cratered highland regions older than about 4 Ga. Future sample return missions are required to clarify this issue, which is fundamental to understanding the collisional history of the inner solar system.

2. THE IMPACT CRATERING PROCESS: OBSERVATION AND MODELING

2.1. Morphology and morphometry of lunar impact craters

2.1.1. Morphology. Lunar impact craters exhibit a spectrum of size-dependent morphologies (e.g., Smith and Sanchez 1973; Howard 1974; Head 1976; Schultz 1976; Wilhelms et al. 1987). The basic morphologic subdivisions of lunar impact craters with increasing rim diameter are (1) simple craters, (2) complex craters, and (3) impact basins. As there can be variations in morphology even within restricted size ranges (e.g., Smith and Sanchez 1973; Howard 1974, Cintala et al. 1977), it can be difficult to choose a "typical" member of a given size class of lunar craters. A detailed discussion and classification of lunar crater morphologies can be found in Schultz (1976).

Simple craters. The classic "bowl shape" of a simple crater is typified by the 10 km diameter (*D*) crater Alfraganus C (Fig. 5.2). Fresh bowl-shaped craters are actually trapezoidal in profile, with walls possessing nearly constant slopes and small, essentially flat floors (Ravine and Grieve 1986). Wall failure is generally limited to small units commonly associated with the floor hummocks and to scree emplaced after solidification of the thin impact melt deposits on the crater floor. Hummocks and blocks are common on the floors of these craters, but central peaks do not emerge until diameters >10 km (Smith and Sanchez 1973; Howard 1974; Head 1976).

Complex craters. Complex craters are highly modified with respect to simple crater morphology. When viewed in the context of the full spectrum of crater morphologies, the transition from simple to complex craters is abrupt (e.g., Pike 1974). Inspection of the individual transitional craters, however, reveals that the changes in morphology are more gradual and less than systematic (e.g., Smith and Sanchez 1973; Howard 1974; Head 1976). For example, Lalande ($D = 25$ km) displays

Figure 5.2. Alfraganus C (10 km in diameter) in the lunar central highlands. This crater is representative of the class of lunar simple craters, which are characterized by smooth walls, relatively flat floors, and large depth/diameter ratios. This view is to the north-east (portion of Apollo 16 Panoramic Camera Frame 4615).

features that are similar to those in the smaller, simple craters, but also includes precursors of structures and units that are better developed in complex craters. Portions of Lalande's wall show only minor evidence of slumping. Overall, however, Lalande exhibits scalloped walls that begin to exhibit the complexity of the more intricately terraced complex craters. Its central peaks, on the other hand, are only emerging from the floor and are not the major topographic features that are characteristic of larger complex craters. Floor hummocks are more imposing and widespread than those in the simple craters.

Uplift of the crater floor and wall failure are well established in craters the size of Tycho ($D = 85$ km, Fig. 5.3). Terraced walls in the rim area are the rule, as are abundant floor hummocks. A crater of Tycho's size, or larger, typically exhibits a massive central peak or a cluster of peaks (Hale and Head 1979). The relative heights and volumes (Hale and Head 1979; Pike 1980a,b; Hale and Grieve 1982) of these peaks increase as a function of size until diameters of ~80 km, after which both values begin to decrease. Roughly simultaneous with this change, a ring of roughening on the floor, composed of hummocks arranged quasi-concentrically with the central structure, begins to appear (Croft 1981a,b; Hale and Grieve 1982). This represents the transition to impact *basins*.

Peak-ring basins. Central-peak basins, such as Compton ($D = 75$ km) are relatively small basins with a fragmentary ring of peaks surrounding a central peak. They are transitional to peak-ring basins. Peak-ring basins, which have a well-developed ring but lack a central peak, are found in the 175–450 km size range.

Relatively undegraded peak-ring basins on the Moon are rare, with the freshest example being the 320 km Schrödinger basin (Fig. 5.4). The interpretation of the interior morphologies of such basins is usually complicated by impact erosion, subsequent volcanic activity, or both. Nevertheless, the relevant observations can be made by inspecting a number of examples (Wilhelms et al. 1987). Additional descriptions of Schrödinger can be found in Hartmann and Wood (1971), Schultz (1976), and Spudis (1993).

Peak-ring basins are relatively shallow features for their size. Although their depths can be decreased by erosion or infilling, the fact that details of many interior features are visible (e.g., the peak ring and floor hummocks in Schrödinger) indicates that the relative shallowness is a

Figure 5.3. Tycho (85 km in diameter) in the southern lunar highlands. This is a classic complex lunar crater, with central-peak cluster, extensive wall terracing. North is toward the top of the frame (Lunar Orbiter V 125M).

Figure 5.4. The peak-ring basin Schrödinger (320 km in diameter) near the lunar south pole. Note the fractured floor and the dark-haloed volcanic vent inside the peak ring, which indicates that even this relatively fresh basin has undergone some modification from a variety of sources. This view is to the east-southeast. Antoniadi (140 km in diameter) is the peak-ring basin with the small central peak near the top left corner of the frame (Lunar Orbiter IV 9M).

primary characteristic. Wall terraces in the rim area are highly developed, and the ratio of floor diameter to rim-crest diameter is somewhat greater than in complex craters (Pike 1980a).

Multiring basins. The largest basins are multiring basins. Some researchers find as many as six concentric rings in the largest basins (e.g., Spudis 1993). Multiring basins are generally more than 400 km in diameter. The best example is the Orientale Basin, which has been only partly flooded by post-impact lavas. The definition of diameters for the various basin features varies among workers, depending on the exact criteria and data sets examined. A comprehensive treatment of lunar multiring basins can be found in Spudis (1993), which is also an excellent source of the primary literature on multiring basins. Wieczorek and Phillips (1999) pointed out that the definition of the crater diameter of multiring basins in the older literature (Wilhelms et al. 1987; Spudis 1993) is problematic and should no longer be used. The definition of the final "rim to rim" diameter of a multiring basins is difficult and continues to be a matter of dispute. Wieczorek and Phillips (1999) argue for smaller diameters than previously proposed. Their geophysical modeling demonstrates how gravity-field anomalies measured from satellites may help to improve the interpretation of multiple rings.

The most complete classification of lunar crater shapes is that of Wood and Andersson (1978), which is based on observations from Lunar Orbiter IV photographs and provides a compilation of morphologic and morphometric data for 11,462 craters. Crater shape classification is more detailed than that outlined above, and 18 different crater types are recognized. Some, however, represent rare variants. Disagreement exists as to whether all observable crater forms represent distinct morphologic types, whether some forms are transitional stages or erosional states between accepted end members (Ravine and Grieve 1986), or whether some forms are the result of effects of varying target properties on crater shape (Cintala et al. 1977).

The catalog of Wood and Andersson (1978) also includes criteria for the classification of progressively eroded and degraded impact structures. Estimates of "degradation" (extent of erosion) are based on such features as rim continuity, rim sharpness, and infilling of the crater cavity by mass wasting. Class 1 craters are the freshest and least eroded; class 5 craters are the most degraded and are only marginally recognizable as impact features. Degradational state becomes important when evaluating the relative formation ages of specific craters or crater populations (Soderblom 1970; Wilhelms 1984; Wilhelms et al. 1987; see discussion in Section 4).

2.1.2. Morphometry. Morphometry describes the fundamental diameter-dependent variations in crater topographic features, such as the statistical variation of crater depth with rim diameter. Detailed measurements of lunar imagery have made it possible to represent the different geometrical characteristics of lunar impact craters by equations of a power law form:

$$y = aD^b \qquad (5.1)$$

where y is a given crater characteristic (e.g., depth, rim height), D is the diameter of the crater (measured from rim to rim), and a and b are constants. Specific morphometric relations for what are considered fresh lunar craters are summarized in Table 5.1. It should be noted that the definition of the rim diameter for multiring basins is still an unsolved problem and there is an ongoing debate on this issue (e.g., Wieczorek and Phillips 1999).

2.2. Terrestrial impact structures

2.2.1. Structural characteristics and comparison with lunar impact craters. As exemplified above, impact craters on the Moon are recognized by their characteristic morphology. The terrestrial record of impacts, however, has been severely modified by active geologic processes and most recognized terrestrial impact craters are far from pristine in appearance. In this regard, they are better referred to as impact structures as opposed to impact craters, which implies a specific morphology. Nevertheless, terrestrial impact structures provide the major observational

Table 5.1. Morphometric relations for fresh lunar impact craters.

Crater Characteristic	D^*, km	N	Exponent (b)	Coefficient (a)	Source
Simple Craterforms					
Depth	<15	171	1.010	0.196	(1)
Rim height	<15	124	1.014	0.036	(2)
Rim diameter	<15	117	1.011	0.257	(2)
Floor diameter	<20	38	1.765	0.031	(2)
Interior volume	<13	47	3.00	0.040	(3)
Complex Craterforms					
Depth	12-275	33	0.301	1.044	(1)
Rim height	15-375	38	0.399	0.236	(2)
Rim diameter	15-375	46	0.836	0.467	(2)
Floor diameter	20-125	53	1.249	0.187	(2)
Diameter (central peak)	17-175	175	1.05	0.016	(4)
Basal area central peak	17-136	19	2.19	0.09	(5)
Height central peak	17-51	15	1.969	0.589×10^{-3}	(5)
Central peak volume	17-51	15	5.078	0.987×10^{-7}	(5)
Central peak volume	80-136	4	3.599	0.387×10^{-5}	(5)
Interior volume	19-150	21	2.31	0.238	(3)
Basins**					
Diameter ring in central peak and peak ring basins	140-435	12	1.125	0.245	(6)
Diameter inner ring in multi-ring basins	420-1160	13	0.943	0.708	(7)
Diameter intermediate ring in multi-ring basins	420-1160	13	0.970	0.845	(7)
Depth	200-630	7	0.15	2.03	(8)***

Notes: *Range of rim diameter values (D) used to establish relations for other topographic features. N is the number of craters. Volumes (central peaks, crater interiors, etc.) are in km^3 (from Heiken et al. 1991). ** Note that the "rim to rim diameter" of multi-ring basins is controversial as pointed out by Wieczorek and Phillips (1999) who propose smaller values. ***Power law in Pike's form. Original fit by Williams and Zuber (1998) is given as: $\log_{10}$ (depth)=0.41*[$\log_{10}(D)$]$^{0.57}$

Sources: (1) Pike 1974; (2) Pike 1977a; (3) Croft 1978; (4) Hale and Head 1979; (5) Hale and Grieve 1982; (6) Head 1977; (7) Pike and Spudis 1987; (8) Williams and Zuber 1998

constraints of the characteristics of natural impact craters, particularly with respect to the third dimension, i.e., subsurface characteristics, which are not evident in lunar imagery or sampling.

For example, studies at terrestrial impact structures indicate that, at simple craters, the rim consists of structurally uplifted target rocks and includes an overturned and inverted flap of near-surface target materials, which is in turn overlain by ejecta. The bowl-shaped depression observed in lunar images is only the surface manifestation of a simple crater. This bowl-shaped depression is sometimes referred to as the "apparent" crater. Terrestrial data indicate that it is actually the uppermost surface of an underlying allochthonous breccia lens, which is parabolic in cross-section and contained by fractured but allochthonous and parautochthonous target rocks (Fig. 5.5). The crater defined by the parautochthonous target rocks is referred to as the true crater.

At larger diameters, the crater structure evolves, as on the Moon, into complex structures, which consist of a structurally complex rim, a down-faulted annular trough, and a structurally uplifted central area (Fig. 5.6). As with simple structures, complex structures are partly filled by allochthonous material, such as breccias and impact-melt rocks, and an apparent and true crater can be defined (Fig. 5.6). The uplifted central area has initially the topographic form of a central peak, which rises above the floor of the structure and has a height that generally does

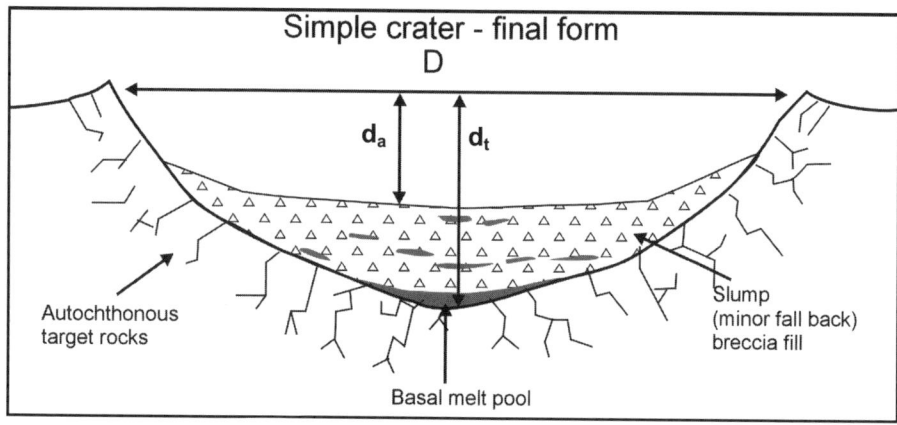

Figure 5.5. Schematic cross-section of a simple crater. *D* is the diameter and d_a and d_t are the depths of the apparent and true crater, respectively. See text for details.

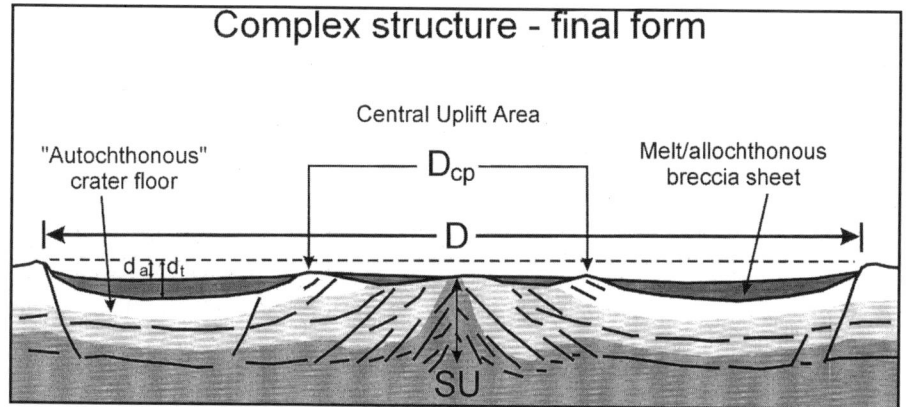

Figure 5.6. Schematic cross-section of a complex impact structure. Notation as in Figure 5.5 with SU corresponding to structural uplift and D_{cp} to the diameter of the central uplift. Note preservation of beds in outer annular trough of the structure with excavation limited to the central area. See text for details.

not exceed the depth from the rim to the floor (Fig. 5.6). With increasing diameter, the central peak is accompanied by a fragmentary ring (a central-peak basin).

Most large, terrestrial complex structures are eroded to varying degrees. There are, however, a number of complex impact structures, which were buried by post-impact sediments almost immediately after formation (e.g., Chicxulub, Mexico; Montagnais, Canada; Puchezh-Katunki, Russia; Ries, Germany), and presumably have a nearly pristine form. They can, however, only be delineated by drill-hole and geophysical data, thus the exact details of their morphologies are generally not well known except for the Ries, which was exhumed as late as Pleistocene. Only the largest terrestrial impact structures have the potential to be peak-ring basins or multiring basins. Unfortunately, the largest structures - Chicxulub, Mexico; Sudbury, Canada; and Vredefort, South Africa - are either buried, tectonically modified, or eroded. Their original detailed morphology cannot be defined with confidence, although they are assumed to represent multiple-ring or peak-ring basins (e.g., Sharpton et al. 1993; Stöffler et al. 1994; Hildebrand et al. 1995; Spray and Thompson 1995; Grieve and Therriault 2000). In the case of Chicxulub, reflection seismic data have imaged a faulted rim area and a topographic peak ring. Closer to the center, however, there is a loss of coherent seismic reflections and structural details are not known (Morgan and Warner 1999).

There is a desire to compare terrestrial impact structures with lunar impact craters (e.g., Pike 1985), and to assume a greater equivalence in detailed morphology than the observational data may support. Planetary environments result in important differences. For example, secondary target effects on Earth include the transition from simple to complex forms at diameters of ~2 km and ~4 km, depending on whether the target rocks are sedimentary or crystalline, respectively. Some complex impact structures in mixed or largely sedimentary targets do not appear to develop topographically high central peaks. For example, Ries (Germany) and Haughton (Canada) are of similar size ($D = $ ~25 km) and age, and have no emergent central peak. In contrast, Boltysh (Ukraine), which is of a similar size but in a crystalline target, has a central peak that is emergent from the surrounding ~300 m of impact lithologies filling the structure. All these structures have been affected by only minor erosion, and at this time, there is no clear explanation for this difference in their morphologies. However, there is some structural uplift of the central crater basement in all types of complex craters. Therefore, we must assume that the target properties control the morphological expression of this uplift.

Planetary gravity also has an effect on cratering mechanics and, thus, morphologies. The lower lunar gravity (1.62 m^{2}s^{-1}, or ~1/6 of the average terrestrial value for gravitational acceleration) results in deeper impact structures on the Moon compared with structures of an equivalent size on Earth because gravity acts against both the excavation of material and the formation of topography. The various forms of impact structures and their diameter ranges appear to be an inverse function of planetary gravity (Pike 1985). Moreover, although gravity is a variable in cratering mechanics; it is not a variable in determining the volume of target material melted in a specific impact event. Thus, an impact at a high velocity (e.g., 15 to 20 km s^{-1}) into crystalline target rocks generates ~2.5× more impact melt (relative to the total volume of displaced rocks) in a terrestrial than a lunar event resulting in an impact structure of equivalent size (Cintala and Grieve 1994, 1998). This additional melt, which in large part is retained within the impact structure, also has the effect of reducing observed topographic variations at terrestrial impact structures.

Owing to erosion, few terrestrial impact structures have sufficient topographic information to define morphometric relations. The most recent set of morphometric relations for terrestrial impact structures can be found in Grieve and Pilkington (1996). While erosion may be detrimental to establishing morphometries, it does result in terrestrial impact structures being exposed to different erosional levels. This, combined with on-site geologic investigations and drilling data, clearly indicate that the central peaks of complex craters are due to the uplift of deeper parautochthonous target lithologies. The amount of stratigraphic uplift at terrestrial complex impact structures is:

$$SU = 0.086D^{1.03} \quad (N = 24) \tag{5.2}$$

where $N = 24$ is the number of structures, with diameters ranging from 4 to 250 km, SU is the amount of stratigraphic uplift of the originally deepest lithology now exposed at the surface, and D is rim diameter, both in km (Grieve and Pilkington 1996).

Attempts to relate these terrestrial data to the lunar case have resulted in a minimum depth of stratigraphic uplift of:

$$SU = 0.022D^{1.45} \quad (N = 12) \tag{5.3}$$

This relation holds for a range of diameters from 17 to 136 km for the lunar case (Cintala and Grieve 1998). There are, however, a number of caveats and ambiguities, the resolution of which awaits better data. Nevertheless, the characteristic of sampling and bringing to the surface originally deeper lithologies through uplift in complex structures has been used to provide some measure of compositional variation with depth, of the lunar highland crust (Tompkins and Pieters 1999).

2.2.2. Principal impact formations and their geologic setting.

Classification of impactites. Terrestrial impact craters are the only source for a complete data base on the effects of hypervelocity impact on rocks of planetary crusts since they allow us to relate these effects to the cratering process and the final geological setting of the impact-metamorphosed rocks (reviews of these phenomena are in French and Short 1968; Roddy et al. 1977; Stöffler et al. 1979, 1988a; Melosh 1989; Grieve 1987, 1991; and French 1998). In principle, the target rocks are affected by the passage of a shock wave which propagates in a spherical geometry from the point of impact. The material engulfed by the shock wave is not only compressed and heated on an extremely short time scale but also caused to flow behind the shock with supersonic velocity. Depending of the position relative to the point of impact, rocks undergo vaporization, melting, phase transformations in a quasi-solid state, and mechanical deformation before part of rock volume affected in this way is transported (ballistically or in a ground surge mode), mixed, and deposited inside and outside the crater cavity thereby forming proximal, distal, and global deposits. The occurrence, or not, of the latter depends on the size of the cratering event.

The products of impact processes and associated nomenclature are summarized in Table 5.2 and Figures 5.7–5.10. The systematic nomenclature has been derived by the "Subcommission on the Nomenclature and Classification of Metamorphic Rocks" of the International Union of Geological Sciences (IUGS), Subgroup on Impactites (Stöffler and Grieve 1994, 1996, 2006). The proposed Systematics of Impactites apply in principle to all planetary impact formations and form a basis also for the interpretation of shock-metamorphosed lunar rocks and lunar impact formations.

Types and characteristics of impact formations. Four basic textural types of breccias are observed in terrestrial impact craters. This observation generally holds independently of the

Table 5.2. Classification of impactites (recommended by IUGS; Stöffler and Grieve 2006; see also Chapter 1 of this volume).

I. CLASSIFICATION OF IMPACTITES FROM SINGLE IMPACTS
  1. Shocked rocks
  2. Impact melt rocks*
      2.1. clast-rich
      2.2. clast-poor
      2.3. clast-free
  3. Impact breccias
      3.1. Monomict breccia
      3.2. Lithic breccia (clastic matrix breccia without melt particles)**
      3.3. Suevite (breccia with melt particles and particulate matrix)**

II. CLASSIFICATION OF IMPACTITES FROM MULTIPLE IMPACTS
  1. Impact regolith*** (unconsolidated clastic debris)
  2. Shock lithified impact regolith*** (consolidated clastic debris)
      2.1 Regolith breccias *** (breccia with *in situ* formed matrix melt and melt particles)
      2.2 Lithic breccias *** (breccia without matrix melt and melt particles)

---

*   may be subclassified into glassy, hypocrystalline, and holocrystalline varieties
** generally polymict but can be monomict in a single lithology target
*** generally polymict but can be monomict in a single lithology target

PROXIMAL IMPACTITES   DISTAL IMPACTITES

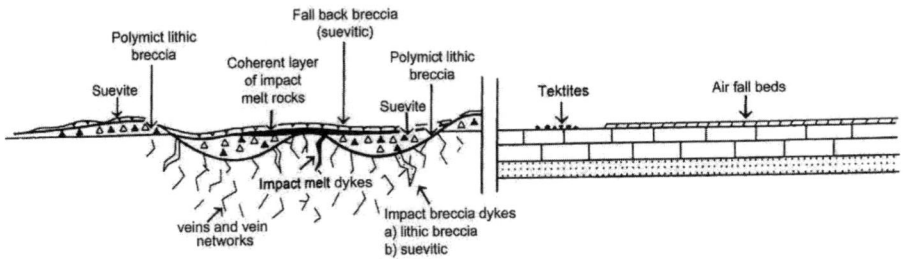

Figure 5.7. Simplified cross section of a complex terrestrial impact crater with proximal and various distal impact formations.

type of target rock and of the geological setting (Stöffler et al. 1979; Stöffler and Grieve 2006), although targets consisting exclusively of sedimentary rocks of high porosity or of evaporite composition lead to somewhat different impact formations. The four types are (Figs. 5.8–5.10; Stöffler and Grieve 2006):

1. Monomict breccias

2. Polymict breccias with particulate matrix and cogenetic melt particles ("suevite")

3. Polymict lithic (fragmental) breccias with clastic matrix (lacking melt inclusions)

4. Impact-melt rocks with variable contents of lithic and mineral clasts in a crystalline or glassy matrix (clast-laden types may be called impact-melt breccias)

These textural types occur in different geologic settings with respect to the parent crater. This is important for the correct interpretation of lunar impact breccias, which come with little or no definitive information concerning the parent crater or impact formation. Therefore, it is useful to discuss the definition of the types and characteristics of the different impact formations identified at terrestrial impact craters (Figs. 5.8 and 5.10) as context for lunar impact materials.

Impact formations may be divided into three structural subgroups:

I. Layered, allochthonous breccias

II. Autochthonous and parautochthonous breccias and shocked basement rocks

III. Breccia dikes (including melt veins and vein networks)

Following Pohl et al. (1977), we may distinguish between (1) *inner* and (2) *outer impact formations,* the latter comprising all deposits beyond the final crater rim. The outer impact formations are "layered" allochthonous breccias (Type I, which may contain Type III, mostly in large megablocks), whereas the inner impact formations include Types I, II, and III. "Layered"

Figure 5.8. *caption continued from facing page...*
Lappajärvi crater, Finland (sample La 41), white to gray lithic and mineral clasts in a dark gray aphanitic crystalline matrix, scale = cm; (f) clast-bearing lunar impact melt rock 14311 with aphanitic crystalline matrix (Apollo 14), scale = cm; (g) polymict lithic breccia with clastic matrix from the continuous ejecta blanket (Bunte breccia) of the Ries crater (Bschor quarry near Ronheim), Germany, scale = cm; (h) lunar polymict fragmental (lithic) breccia 67015 with clastic matrix from the rim of North ray crater (Apollo 16) with dark (impact melt) and light (anorthositic and granulitic rocks) clasts; scale = cm.

Figure 5.8. Basic types of impact breccia textures and macroscopic images of the main types of terrestrial and lunar breccias (in part from Stöffler et al. 1979); (a) sketches of textural types of impact breccias; (b) suevite breccia with particulate matrix and melt inclusions (black) and crystalline rock clasts (gray to white) from Mien crater, Sweden; scale = cm; (c) monomict granite breccia from Schmähingen, Ries crater, Germany; (d) lunar monomict anorthositic gabbro breccia 77017 (Apollo 17); lower part contains intruded impact melt (black), scale = 1 cm; (e) Clast-bearing impact melt rock from Kanta Ahveniemi, Kärnä island,

caption continued on facing page

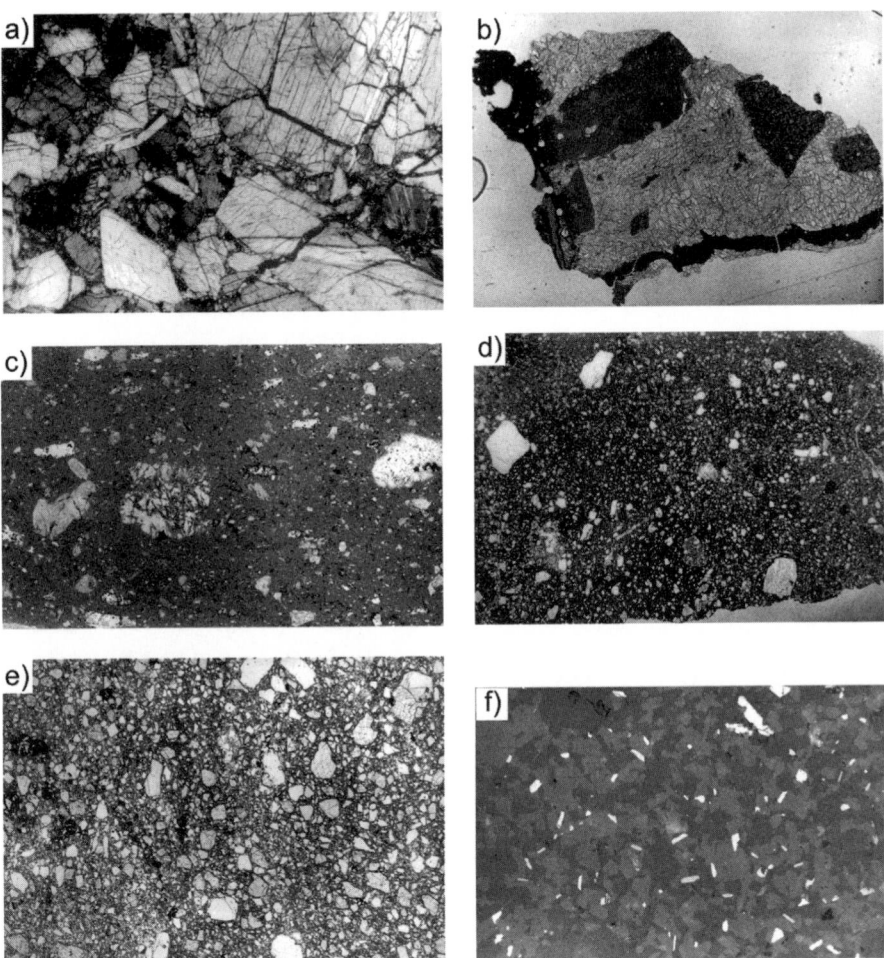

Figure 5.9. Microphotographs of typical textures of the main types of terrestrial and lunar impact breccias (in part from Stöffler et al. 1979); (a) monomict lunar anorthosite breccia 65015,16 (Apollo 16), photomicrograph using crossed polarizers, width of field is 0.25 mm, note intergranular brecciation; (b) lunar dike breccia consisting of gray intrusive impact melt with aphanitic crystalline matrix penetrating into monomictly brecciated anorthosite, younger "pseudotachylite" veins (black) occur on two sides of the sample, width of field = 1.25 mm; (c) clast-bearing impact melt rock with mineral clasts embedded in a fine-grained crystalline matrix from Lappajärvi crater, Finland (sample La 26), width of field = 33.7 mm; (d) clast-bearing lunar impact melt rock 72215,193 (Apollo 17) with lithic and mineral clasts in a fine-grained crystalline matrix, width of field = 15.1 mm; (e) lunar polymict lithic ("fragmental") breccia 76255,69 with clastic matrix consisting mainly of mineral clasts (mainly plagioclase); (f) matrix section of the lunar clast-bearing impact melt rock with crystalline matrix, thin section 14066,46, dark gray = plagioclase, light gray = pyroxene and olivine, white = ilmenite and iron, width of field = 0.23 mm, reflected light.

allochthonous impact formations occur at the top of the section, parautochthonous shocked and monomictly brecciated impact formations below the crater floor, and autochthonous impact formations (monomictly brecciated) at some depth in the crater basement, which in part may have been affected by structural uplifting. In addition, breccia dikes (Type III) occur in the parautochthonous and autochthonous crater basement. The allochthonous outer impact formations contain clasts of local substrate produced by the mechanism of "secondary mass

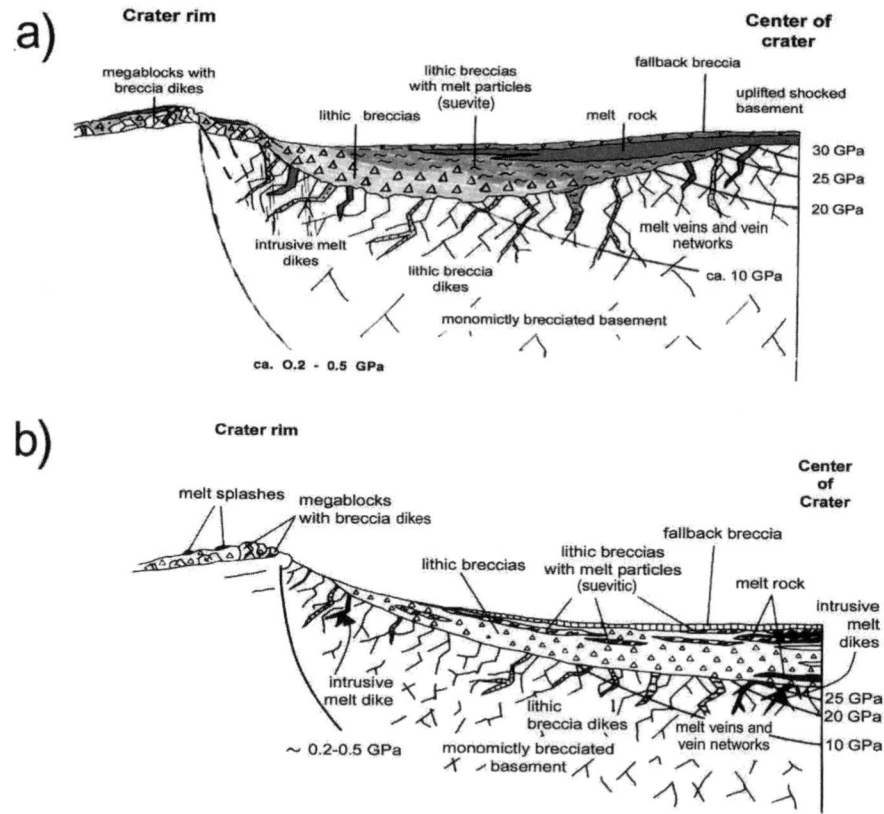

Figure 5.10. Geological setting of impact formations and types of breccias at a complex (a) and simple (b) terrestrial impact crater; note shock pressure isobars given in GPa.

wasting" (Oberbeck 1975) and whose fraction increases with radial distance from the crater rim (Hörz et al. 1983). This is important for the interpretation of lunar breccias taken from the ejecta blanket of multiring basins.

For lunar applications, it is important to distinguish between *proximal ejecta* and *distal ejecta* (Fig. 5.7; Stöffler and Grieve 2006). Proximal ejecta include the allochthonous inner impact formations and the *continuous ejecta blanket* as the innermost part of the outer impact formations. Distal ejecta comprise *global air fall beds* and *tektite glass,* which occur in strewn fields at some distance from the parent craters and represent exclusively shock fused melt from the very top section of the target (see Section 2.3) in contrast to the impact melt residing in melt sheets, suevite breccias, or as glass spherules in global airfall beds (Fig. 5.7). This melt is derived from the deep, inner melt zone of the crater, which develops in the compressed target some distance below the stagnation point of the projectile. Secondary craters—not really known from terrestrial craters—are the result of high velocity distal ejecta and are common around lunar craters, usually outside of the continuous ejecta blanket.

In conclusion, the material ejected from the crater (ejecta of the outer impact formations) forms successively: continuous deposits, discontinuous deposits, and rays (not observed so far at terrestrial craters) with increasing radial distance from a crater. Generally, the velocity of the ejecta decreases with increasing radial distance from the point of impact and with increasing depth in the target. This leads to the following characteristics of the continuous ejecta blanket: the average size of the ejecta (rock fragments) decreases with increasing distances; the final range of the ejecta is inversely proportional to their original depth in the target (the deepest rocks excavated from the crater form the rim deposits); rock fragments from deeper sections of the target are deposited later than those from higher sections, leading to an *inverted stratigraphy* in the continuous deposits; beyond about 1.5–2 crater radii, the ejecta have velocities sufficient on landing to rework local rock strata and form a radial ground surge of material (secondary mass wasting, Oberbeck 1975) by which a large fraction (e.g., 70–90%) of local rock may be incorporated into the continuous deposits.

The principles outlined above, which have been documented at terrestrial impact structures (e.g., Pohl et al. 1977; Hörz et al. 1983), are key for any geologic exploration of the lunar surface (Shoemaker and Hackmann 1962; Gault et al. 1968; and many others). Firstly, most of the rock fragments in breccias of the distal part of the continuous deposits are from the local bedrock, which is essential for the interpretation of the Imbrium-basin-related Fra Mauro Formation (Apollo 14) and the Descartes Formation (Apollo 16) related to Nectaris basin (Deutsch and Stöffler 1987; Stöffler and Ryder 2001). Secondly, a series of craters, with increasingly larger diameters may thus be used to probe progressively deeper formations in a given geologic terrain, thus enabling reconstruction of first-order stratigraphic and structural relationships at depth from simple surface observations. Such considerations affect our current perception of the lithologic make-up of the lunar crust, based on samples and remotely sensed data (Spudis et al. 1984; Wilhelms 1984; Wilhelms et al. 1987).

Isotope dating of the age of impact events. Determining the absolute age of lunar impact craters and basins and their related ejecta formations is essential for lunar stratigraphy and chronology because of the lack of other datable stratigraphic boundaries. The message for lunar studies from isotope dating to derive the impact age of terrestrial craters is fundamentally important and provides two essential implications: (1) allochthonous impact formations contain rock fragments covering the complete age range from the age of the oldest displaced target rock to the actual age of the impact crater, (2) complete resetting of the ages of the target rocks is only achieved in impact events by vaporization and whole-rock melting. That is, impact-melt rocks from the parent crater (of whatever geologic setting) are the only type of impactite that can reliably be used to date the time of impact (Deutsch and Schärer 1994; Staudacher et al. 1982; Stephan and Jessberger 1992; Bogard et al. 1988). Even then, complications may arise from unequilibrated lithic and mineral clasts of the target rocks in the melt rocks (Bottomley et al. 1990). At large impact craters where impact melt lithologies have characteristically long cooling times, partial resetting of primary ages of the target rocks is commonly observed in lithic clasts included in "hot" impact formations, such as impact melt sheets, suevite layers (Staudacher et al. 1982; Bogard et al. 1988), and thermally annealed bedrock sections of these hot impact formations if the affected bedrock breccias are completely recrystallized (e.g., Footwall breccia at the Sudbury structure, Lakomy 1990). Thus, for lunar applications, impact-melt lithologies should be the first choice in any dating effort. This may result in the direct dating of a crater or in an indirect dating on the basis of the principle that the youngest clast in a polymict impact formation is closest to the actual age of the parent crater. For further details, see Sections 5 and 6.

2.2.3. Fundamentals of progressive shock metamorphism.

Shock metamorphism. Shock-metamorphic effects in rocks and minerals as observed in many lunar samples, particularly from the lunar highlands and the regolith, are well studied and described for terrestrial impact structures in papers in French and Short (1968), Roddy et al. (1977), Stöffler (1972, 1974, 1984), Bischoff and Stöffler (1992), Stöffler and Langenhorst (1994), Grieve et al. (1996), and French (1998). The degree of shock metamorphism produced by a given shock pressure depends on a material's behavior, the so-called equation of state, which relates such parameters as compressibility, specific energy, entropy, specific volume, and phase changes. The transition from elastic to plastic behavior in dynamically loaded rocks and minerals occurs at relatively high stresses, typically on the order of 5–12 GPa. At pressures between roughly 10 and 60 GPa, mechanical deformation and transitions to high-pressure phases are typical for the common rock-forming minerals. Above about 40 to 100 GPa, thermal effects begin to dominate, and whole-rock melting begins (> ~60 GPa for felsic rocks and > ~80 GPa for mafic rocks, > ~40 GPa for porous siliceous rocks). Pressures exceeding 150 GPa cause vaporization, and ionization occurs at a few hundred GPa. The criteria for the definition of progressive stages of shock metamorphism of various rocks have been defined and classification schemes have been proposed first by papers in French and Short (1968) and later in Kieffer et al. (1976), Schaal and Hörz (1977 1980), Reimold and Stöffler 1978, Schaal et al. (1979), Snee and Ahrens (1975), Bauer (1979), Ostertag (1983), Stöffler (1984), Stöffler et al. (1986, 1988b, 1991), Bogard et al. (1987), Kitamura et al. (1977, 1992), Schmitt (2000), and Xie et al. (2001) for various terrestrial rocks (felsic rocks, basalt, dunite) and for planetary rocks and planetary analog materials such as basalt, dunite, anorthosite, lunar regolith, and chondrites (see also Bischoff and Stöffler 1992, Stöffler and Grieve 2006, and Chapter 1 of this volume).

For lunar crustal material, knowledge of residual shock effects is only essential for a few rock-forming minerals such as plagioclase, olivine, and pyroxene, and for some mafic and feldspathic igneous rocks such as basalt/gabbro/norite, dunite, and anorthosite, as well as regolith. The typical shock effects and the required formation shock pressures are summarized in Table 5.3, which is based on the specialized literature on these materials listed above and a summary in Chapter 1 of this volume.

Impact melt lithologies. Impact-melt rocks constitute a prominent rock type in the Apollo sample suite. Material identified as impact melt composes some 30–50% of all hand-specimen-sized rocks returned from highland landing sites and some 50% of all soil materials, including mare collections (Ryder 1981). Detailed studies of terrestrial impact melt sheets (e.g., Dence et al. 1977; Phinney and Simonds 1977; Whitehead et al. 2002; Dressler and Reimold 2001; and many others) show that the diverse melts derived from the various target rocks in an impact tend to be homogenized, and that the resulting glasses or crystalline rocks, depending on cooling rate, represent remarkably homogenized mixtures of the original target lithologies. In simple terms, impact melts are chemical mixtures of preexisting but now melted target rocks, although there are limitations to the degree of homogenization, particularly at the lower and upper end of the size range of impact craters (e.g., Kettrup et al. 2003). In many cases, the mixed compositions of impact melts have unique chemical characteristics that cannot be produced by conventional internal melting processes, which involve the partial melting of a compositionally restricted source rock.

The spectral composition of impact melts believed to be related to large lunar basins, therefore, has been used to make inferences regarding the composition of the lunar crust (e.g., Pieters et al. 2001). The terrestrial constraints regarding the nature of impact-melt rocks holds for craters up to the 100 km size range. It may not apply directly to the much larger lunar basins. There is an additional complication in the terrestrial environment; namely, the largest known terrestrial impact melt sheet, the Sudbury Igneous Complex at the 250 km diameter Sudbury structure, differentiated on cooling (Therriault et al. 2002). At this time, it is not known if this is a valid analog for the low gravitational environment of the Moon. The volume of impact melt normalized by the total volume of displaced rock masses does not increase linearly with crater diameter, it increases exponentially (e.g., Melosh 1989; Cintala and Grieve 1998). This is important for the interpretation of impact melt lithologies in the lunar highlands (see further discussion in Section 2.3.1)

Most impact-melt rocks contain lithic and mineral clasts from the target (e.g., Stähle 1972; references in Dressler and Reimold 2001; Figs. 5.8 and 5.9). These clasts frequently show distinct shock and thermal effects (Bischoff and Stöffler 1984). Partial digestion of clasts

Table 5.3. Shock effects in rock-forming minerals and whole rocks
with shock pressure calibration.

Shock effect	Pressure (GPa)
Shock wave barometry for non-porous felsic rocks	
Kink bands in biotite	> 0.5–1
Shatter cones	> 2
Pf's in quartz: (0001) and $\{10\bar{1}1\}$	> 5–10
Pdf's in quartz: $\{10\bar{1}3\}$	> 10
Pdf's in quartz: $\{10\bar{1}2\}$	> 20
Stishovite	> 12–15
Coesite	> 30
Diaplectic plagioclase glass	28/34* to 45
Diaplectic quartz glass	34–50
Melting of feldspar	> 45
Whole rock melting	> 60
Shock wave barometry for mafic rocks and anorthosites	
Olivine, undulatory extinction	4–5 to 10–15
mosaicism	10–15 to 60–65
planar fractures	15–20 to 60–65
planar deformation features	35–40 to 60–65
melting and recrystallization	> 60–65
Plagioclase, undulatory extinction	5–10 to 10–12
mosaicism	10–12 to 28/34*
diaplectic glass	28/34* to 45
melting	> 45
Orthopyroxene, undulatory extinction	5–10 to 20–30
mechanical twinning	> 5
mosaicism	20–30 to 75–80
planar deformation features	30–35 to 75–80
incipient melting	> 75–80
Whole rock melting, basalt/gabbro	> 75–80
Whole rock melting, dunite	> 60–70
Whole rock melting, anorthosite	> 45–50

Notes: Pf's = planar fractures, Pdf's = planar deformation features; *increasing with
increasing An-content

Data from: Müller and Hornemann 1969; Hornemann and Müller 1971; Stöffler and
Hornemann 1972; Stöffler 1972, 1974; Snee and Ahrens 1975; Kieffer et al. 1976;
Schaal and Hörz 1977; Stöffler and Reimold 1978; Schaal et al. 1979; Bauer 1979;
Ostertag 1983; Stöffler et al. 1986, 1991; Bischoff and Stöffler 1992; Stöffler and
Langenhorst 1994; Schmitt 2000; see also Chapter 1 of this volume

by the melt typically results in texturally heterogeneous glasses or fine-grained melt rocks.
Better crystallized impact melts display an increased tendency to digest clastic material. In
completely crystallized impact melts, clastic material may no longer be observed on the scales
of thin section and even hand specimens (millimeter to centimeter), yet larger lithic clasts may
be observed in the field (Phinney and Simonds 1977). Detailed studies of terrestrial impact-
melt rocks have also demonstrated that the clast population in impact-melt rocks does not
necessarily constrain the progenitor target rocks that were the source of the impact-melt rocks,
as the clasts are acquired as the melt sweeps across the expanding crater cavity (McCormick et
al. 1989). This observation has been used in attempts to constrain the source of certain Low-K
Fra Mauro rocks in the lunar collection from Apollo 15 (Spudis et al. 1991).

2.3. Impact cratering mechanics

2.3.1. Empirical observations and basic physics of cratering. The basic phenomenology of impact cratering has been well known for decades. Gault et al. (1968) divided the process into three stages, an approach that remains useful and valid today (Fig. 5.11). When a projectile first contacts another object, shock waves are generated in both the impactor and the target, causing the material to flow behind the shock front with supersonic velocity. The material flow field depends on the velocity and angle of the projectile as well as the physical and chemical properties of projectile and target materials.

The earliest part of an impact event, *the compression stage*, encompasses the time from initial contact between the impactor and target to the time that the impactor is completely engulfed by the shock (Fig. 5.11). In a vertical or near-vertical impact, energy transfer from the projectile to the target ceases at that time, as the compressed impactor and target are traveling at essentially the same velocity. This stage, however, rapidly becomes more complex with increasingly oblique impacts, as impactor shearing and highly nonlinear effects become major considerations. The reader is referred to Schultz and Gault (1990), Schultz (1996), Sugita and Schultz (1999), Pierazzo and Melosh (2000a,b), and Ivanov and Artemieva (2002) for

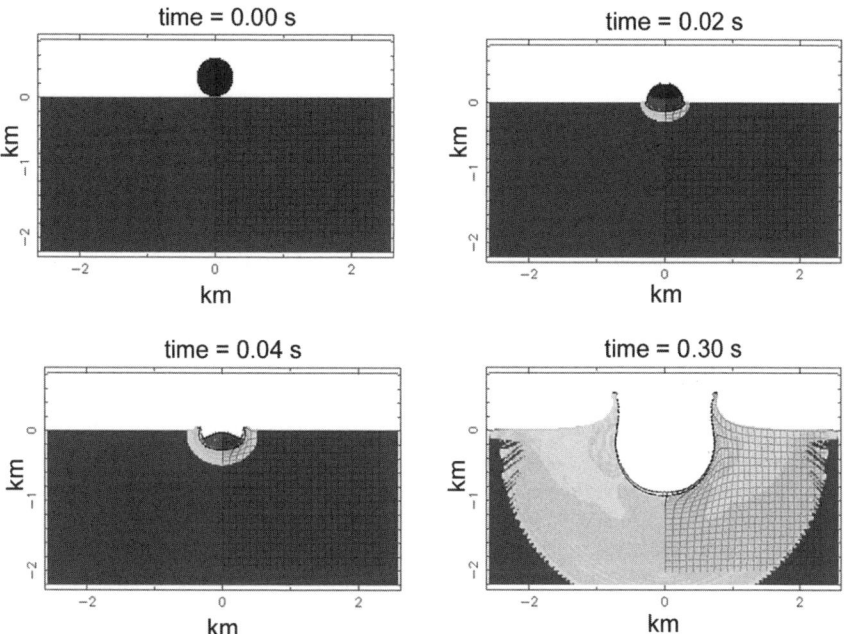

Figure 5.11. Initial stages of impact cratering, illustrated by a numerical model of a vertical impact of a granite impactor into a dunite target with an impact velocity of 15 km/s (granite and dunite are described with ANEOS—analytical equation of state—see Thomson and Lauson 1972). At $t = 0$ both the spherical projectile and the target are presented as intact brittle media. At $t = 0.02$ s, the projectile is partially penetrated into the target. Relatively less dense gray shading presents damaged zones. The progress of penetration is clearly seen at $t = 0.04$ s, the top of projectile is beneath the pre-impact surface and ejection of target material is just beginning. At these early moments the material failure occurs at the shock front. Later, at $t = 0.3$ s, the excavation stage is clearly seen. The projectile is smashed in a thin layer along the transient cavity surface. The shock wave is detached from the cratering flow area. Here the failure zone is far behind the shock front. Near the surface, the failure zone is growing as separated cracks. The mesh of dots in the target is constructed of massless Lagrangian tracers, which follow the material motion (only each 5th row and column are shown).

discussions of the effects of impact obliquity and for additional references regarding aspects of such impacts. The compression stage associated with impacts at interplanetary velocities (up to tens of km per second) is characterized by extreme stresses, rapid entropy generation, very high temperatures, and exceedingly short timescales (e.g., Ahrens and O'Keefe 1977; O'Keefe and Ahrens 1977; Melosh 1989). Pressure is so great that both the impact and target can be treated as strengthless fluids; fused or vaporized material jetted from the interface between the two can travel great distances or escape the target body completely (Kieffer 1977; Sugita and Schultz 1999). The first material ejected by the impact occurs at the time of contact and shortly thereafter, when a combination of very hot impactor and target material is squirted, or "jetted," from the contact between the two (e.g., (Kieffer 1977; Gault et al. 1968). In the case of a large target planet such as the Earth or Venus, most if not all of this jetted phase probably has velocities that exceed escape velocity. However, atmospheric drag may decelerate jets before escape (e.g., Artemieva and Ivanov 2004).

The shock in the impactor has a short but complicated history, reflecting from the irregular surface, spalling pieces of the projectile, reinforcing or attenuating interior stresses, and deceleration of the impactor. Once the shock encounters a free surface, however, it is reflected by a decompression (or "rarefaction") front. Depending on its equation of state and other physical and chemical properties, the impactor will be largely melted or vaporized, and mixed with similarly affected target materials. Conversely, if the impact velocity were sufficiently low, parts of the projectile could survive the impact relatively unscathed and spalled off as solid fragments (Melosh 1984). For oblique impacts, ricochet may occur either with the projectile remaining intact, rupturing into several large fragments, or fragmenting into a myriad of small fragments (Schultz and Gault 1990). This depends on the projectile strength, and the intensity of the effect increases with increasing impact velocity. A common misconception is that the projectile somehow explodes upon impact and that explosion creates the crater.

The *excavation stage* of the event begins after the shock completely engulfs the impactor, thus ending sensible transfer of energy from the projectile to the target (Fig. 5.11). The energy supplied to the target is partitioned into kinetic, thermal and mechanical energy (Gault and Heitowit 1963; Braslau 1970; O'Keefe and Ahrens 1977), but the total energy is constant from the end of the compression stage (energy contained within the target and the ejecta). The intensity of the shock decreases as it encompasses more mass while propagating into the target (Gault and Heitowit 1963). The rate of decrease, however, is also modulated by the irreversible nature of the shock process; the formation of a shock front in any material represents a profligate use of energy. It is highly irreversible in a thermodynamic sense, increasing the entropy of the material through which it passes (e.g., Ahrens and O'Keefe 1972), with greater changes in entropy associated with higher shock stresses. The manifestation of the entropy increase in the shocked material can range from solid-state phase changes through fusion and vaporization to the formation of ionized vapor. As a consequence of this production of waste heat, the energy available to propagate the shock is reduced, and the decay in shock stress with distance is therefore most rapid near the impact point (Gault and Heitowit 1963; Ahrens and O'Keefe 1977; Robertson and Grieve 1977; Cintala et al. 1979; Orphal et al. 1980; Croft 1982; Cintala 1992).

The shape of the shock front in the target at any given time can be thought of as a hemisphere or a truncated sphere, particularly at points far from the impact site (e.g., Gault and Heitowit 1963; Gault et al. 1968; Holsapple and Schmidt 1987; Holsapple 1993). As the shock attenuates, its effects on the target materials decrease from major phase changes through plastic deformation to fracturing. In elastic materials the shock finally propagates as a set of compressional and shear waves dissipating by non-elastic processes. In all cases, however, passage of the shock front imparts a velocity to the target material in a direction parallel to the movement of the front itself with a magnitude roughly proportional to the square root of the shock stress at that point. Because the shock is nearly hemispherical in a reasonably homogeneous target, the initial motion of the target material is radial from the impact site. This pattern changes as rarefactions propagate into the shocked region from the free surfaces encountered by the shock, including the target surface itself and the boundary of the growing crater or "transient cavity." The net result is an overall motion of the target that is first downward and outward from the impact site, eventually turning upward as individual trajectories are redirected toward the target's surface by the decompression process (Fig. 5.12). The resulting series of trajectories constitute the *flow field*; the best known descriptions of the flow field are the "*z*-model" of (Maxwell 1973, 1977) and modifications thereof (e.g., Croft 1980, 1981).

Material defining the volume of the transient cavity can be divided into two components: ejected and displaced. One of the idealized trajectories described above delineates the volume of material ejected from the transient cavity from the volume mobilized by the shock but not ejected. The former lies above the "hinge streamline" referred to in Croft's (1980) adaptation of the *z*-model, whereas the latter lies below it. Superimposing such a flow field on the estimated shape of the transient cavity shows that the ejected fraction comes from the uppermost part of the cavity's volume; the remainder, roughly half the volume of the cavity, is material that is simply pushed out of the way by the shock. This combined process of ejection and displacement has been confirmed by small-scale cratering experiments (Stöffler et al. 1975). At small natural craters, this effect is frozen into the final simple crater form, as documented at the 3.4-km terrestrial crater Brent (Robertson and Grieve 1977), but it is mostly erased in craters large enough to exhibit rebound phenomena. The z-model predicts that the maximum depth of excavation is on the order of a tenth of the diameter of the cavity formed directly by the cratering flow field, i.e., the transient cavity. While this is consistent with the few observations at terrestrial impact craters, it is not well constrained due to the general lack of preserved ejecta as the result of erosion.

As described above, material ejected from locations near the impact site in planetary-scale events is typically both hot and fast. Shocked material farther from the impact site acquires lower velocities, and therefore is ejected at lower speeds. Nevertheless, the faster fraction of

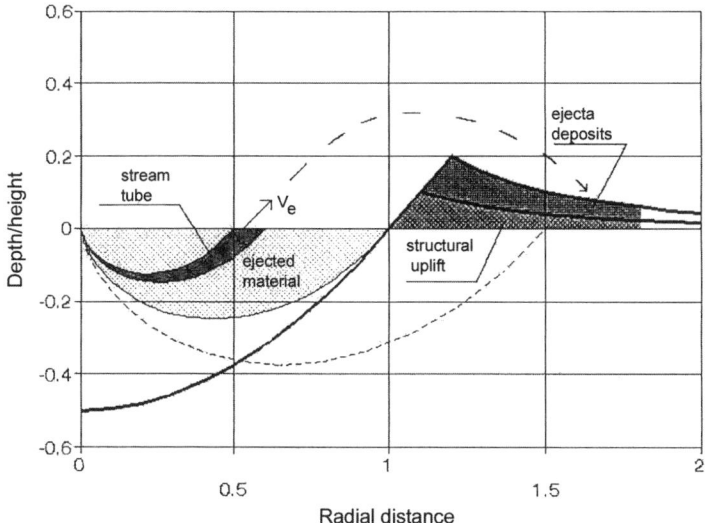

Figure 5.12. Cross section of a transient crater with flow field and excavation cavity; radial distance normalized to the diameter of the apparent crater.

this solid ejecta can be spread over much of the target planet, but because this highly shocked material originated near the impact point, it was almost certainly subjected to high stress gradients and therefore would be severely comminuted (Öpik 1971). The model by Schultz et al. (1981) records the shock state of material as a function of ballistic range and predicts that a wide range of peak pressures is present in the ejecta at a given range, although the peak pressure generally increases with increasing distance from the impact point. Ejecta originating farther from the impact site would be cooler, slower, and less fragmented. The net result of this progression is reflected in the ejecta deposits. Moving away from the crater radially, they grade from a relatively smooth, continuous formation that represents the uppermost component of the crater's rim into a textured deposit that gives way to dense fields of secondary craters. These fields fragment into clusters of secondaries at the extreme limits of the deposit (see Oberbeck et al. 1974; Oberbeck 1975). In dealing with intercrater comparisons of ejecta deposits, it must be kept in mind that the use of scaled distances can be highly misleading.

Deposition of ejecta is obviously an important factor in the mixing of lunar surface materials, but it is dependent on a variety of different factors. Because this process is so complex on the scale of individual samples, agreement among investigators on the nature and absolute extent of such mixing has been difficult to obtain (e.g., Oberbeck et al. 1975; Oberbeck and Morrison 1976; Schultz and Gault 1985). In a terrestrial context, however, a drilling program in the ejecta deposits of the Ries crater in Germany demonstrated unequivocally that local reworked material increases beyond 1 crater radius and composes between 70 and 90 % of the total clast population of the breccia deposits at 2 to 3 crater radii (Hörz et al. 1983). Perhaps the continuing improvement in remote-sensing capabilities will soon provide observational data permitting calibration of the various estimates of mixing ratios as functions of distance from the primary impact on bodies other than Earth. Rationales for this "re-calibration" have been recently presented in the form of a model of global basin ejecta "stratigraphy" by Haskin and coworkers (Haskin 1998; Haskin et al. 2003). The recent quantitative model by Haskin et al. (2002, 2003), based on impact cratering scaling equations (Housen et al. 1983; Holsapple 1993) and the concept of ballistic sedimentation (Oberbeck 1975) predicts characteristics of ejecta deposits, resulting from basin-sized cratering events. These characteristics include deposit thickness at a given distance and fraction of primary ejecta and pre-existing substrate incorporated into the ejecta deposit. The model can be useful for suggesting provenance of sampled lunar material. However, predicted secondary crater densities are at least one order of magnitude greater than observed secondary crater densities surrounding the Imbrium and Orientale basins. The proponents suggest that mutual obliteration erases essentially all secondary craters associated with the debris surge during ballistic sedimentation. If so, a process other than ballistic sedimentation is needed to produce observable secondary craters. This model does not take into account oblique impacts for which the distal ejecta deposits are strongly asymmetric (see discussion of an oblique impact in Section 2.3.2). Some concerns about the Oberbeck model have been expressed by Schultz and Gault (1985).

An attempt to take obliquity into consideration was made by Wieczorek and Zuber (2001) to test the idea of the Imbrium origin for the Imbrian grooves and South Pole-Aitken basin thorium anomaly. Their analysis suggests that the initial material ejected in an oblique impact may be qualitatively modeled by adding a constant velocity tangential to the surface in the impactor's direction to the ejecta velocities determined from the vertical impact-scaling relationships. In reality the nature of oblique impact is much more complex (see Section 2.3.2). The origin of the "grooves" antipodal to Imbrium also remain plausibly related to antipodal convergence of seismic waves (Schultz and Gault 1975).

The *modification stage* of the event begins as soon as the maximum depth of the transient cavity is attained. For a small crater, the floor is raised slightly by elastic and/or gravity-driven rebound, but in a large cavity, the rebound can be on a massive scale (e.g., Grieve et al. 1981; Ivanov 1994) to the extent that some investigators have suggested that some of the rings of multiring basins are frozen-in "tsunamis" generated by oscillating central rebounds (e.g., Baldwin 1981; O'Keefe and Ahrens 1999). Investigation of the modification stage of the impact-cratering process is much more difficult than the earlier stages for a number of reasons. To a large degree, the compression and excavation stages of large cratering events can be treated as being independent or, at most, only weakly dependent on the target's strength. Cavity modification, however, takes place under conditions that are neither static nor unaffected by material strength (e.g., Melosh and Ivanov 1999). A complication is the poorly understood role of thermal energy and its effects on the material strength of the volume of the target in the cavity's immediate proximity (Cintala and Grieve 1998; O'Keefe and Ahrens 1999). Historically, it has been difficult to include the modification stage in numerical simulations of impacts because of inadequate computational resources. Calculations of the earliest portions of the event are extremely intensive in terms of the number of computer cycles required and the time needed to complete them. The excavation stage is so long relative to the earlier phenomena that until relatively recently it had been addressed only by extrapolating the earlier particle motions ballistically (e.g., Orphal et al. 1980) which is an unappealing but nonetheless unavoidable solution to the problem. Improvements in hardware and coding made calculations to these later times possible, and in recent years, concerted efforts have been made to investigate modification phenomena through computer simulations (e.g., Ivanov 1994).

The rim crest of the transient cavity relaxes from a maximum height that is supported by dynamic stresses during the cavity's growth to a level that is dependent on the magnitude of the event. Small cavities undergo little modification from their transient-cavity form. For larger cavities, wall failure resulting from collapse of the cavity and its subsequent enlargement significantly decreases rim heights of complex craters (e.g., Pike 1977b; Cintala 1979). Mass movement of wall material into the crater has added consequences for the final crater form. Outward-moving impact melt that lined the transient cavity during its growth is transported back into the crater on the slump blocks, to be trapped in the interior as veneers, ponds, or as a coherent melt sheet (Howard and Wilshire 1975; Hawke and Head 1977; Cintala and Grieve 1998). The moving melt mixes with the shocked, brecciated, and highly comminuted material from the cavity walls and floor to form the polymict allochthonous breccias of the crater interior (compare Section 2.2).

Crater scaling. Scaling is a term typically used to denote the means by which initial conditions of an impact or explosion can be used to predict the final crater's dimensions. Because the magnitude of even a relatively small planetary impact is beyond anything that can be simulated with current technology, methods must be devised to relate the characteristics of the impactor, the target, the impact velocity, and any other relevant factors with the properties of the resulting crater. The approach used almost exclusively for this task is based on the principles of dimensional analysis (Buckingham 1914; Bridgman 1922). The first extensive use of dimensional analysis in the investigation of cratering phenomena was made by Chabai (1965) in studies of explosion craters; many subsequent investigators applied his work directly to impact events by equating explosive energy to the kinetic energy of the impactor. More than a decade later, however, the lack of a velocity (or momentum) dependence in Chabai's explosion relationships was noted by Holsapple and Schmidt, who developed an extensive suite of scaling relationships for application to impact cratering (Schmidt 1977, 1980; Schmidt and Holsapple 1978, 1982; Holsapple 1980, 1987, 1993; Holsapple and Schmidt 1980; 1982; Housen et al. 1983; Holsapple and Schmidt 1987; Schmidt and Housen 1987). Continuing work by other investigators has tested the scaling relationship predictions, suggested improvements, and incorporated more variables (e.g., Croft 1985; Schultz 1988; Cintala and Hörz 1990; Cintala et al. 1999). A brief discussion of the approach and limitations of dimensional analysis as applied to crater scaling follows, and a few examples are given to illustrate the form that these relationships typically take and to provide a reference for a few of the more common calculations.

In formulating a potential scaling relationship by dimensional analysis, all of the variables needed to describe the process are collected and, following the rules of dimensional analysis as pioneered by Bridgman (1922), groups of those variables are constructed, with the requirement being that no group can possess physical dimensions. For instance, the group P/F, where the variables P and F represent pressure and force, respectively, has the dimensions of length^{-2}, and thus is not a dimensionless group. The group at/v, on the other hand, where a represents acceleration, t time, and v velocity, is dimensionless. The number of such groups are determined by the number of dimensions (e.g., length, mass, time, temperature, etc.) and the number of variables included in the analysis. Once the dimensionless groups (typically called "π-groups") are established, any one of them can be treated as a function of the others; the real task is to determine the nature of that functionality and, in most cases, this can only be done by examining the data.

Laboratory-scale impact craters formed in wet sand probably represent the best analogs available to their large counterparts in rock in terms of scaling studies (Schmidt and Housen 1987). Dimensional analysis has been applied to a set of such craters formed by a variety of impactors over a range of velocities. Using the wet-sand craters as a basis for their analysis, the general form for the rim-crest diameter of the transient cavity D_{tc} as given by (Schmidt and Housen 1987) can be stated as follows:

$$D_{tc} = 1.16 \left(\frac{\rho_p}{\rho_t}\right)^{0.39} D_p^{0.78} v_i^{0.44} g^{-0.22} \tag{5.4}$$

in which ρ_p and ρ_t are the densities of the projectile and target, respectively, D_p is the diameter of the projectile, v_i is the impact velocity, and g is the gravitational acceleration. Most transient craters approach a parabolic shape for which transient crater depth H_{tc} is roughly 1/4 to 1/5 of the diameter D_{tc}. The diameter of the transient cavity can be used as a reasonable approximation of the diameter of a simple crater. Models show that appreciable enlargement of simple craters through slumping of their walls would reduce their depths to values lower than are observed (Cintala 1979; Grieve and Garvin 1984; Grieve et al. 1989). The diameters of complex craters, on the other hand, are significantly greater than those of their parent transient cavities because of the modification phenomena described above, and this effect must be included in the scaling relationship for large craters. Croft (1985) derived an expression relating the final rim-crest diameter D_R of a complex crater to the rim-crest diameter of the transient cavity, which can be written as

$$D_R = D_Q^{-0.18} D_{tc}^{1.18} \tag{5.5}$$

where D_Q is the diameter at which the simple-to-complex transition occurs. The value of D_Q for the Moon, for example, is generally taken as 18.7 km (Pike 1988), whereas the corresponding value for crystalline targets on Earth is 4 km (Grieve et al. 1981). Thus, Equation (5.4) can be used for craters with diameters below the simple-to-complex transition on the Moon (Pike 1988); above this size, Croft's (1985) Equation (5.5) must be used with Equation (5.4) to yield

$$D_R = 1.87 \times 10^{-2} \left(\frac{\rho_p}{\rho_t}\right)^{0.39} D_p^{0.92} v_i^{0.52} \tag{5.6}$$

where cgs units have been assumed and a lunar value for g is 162 cm s^{-1}. Equation (5.6) is applicable only to complex lunar craters. In the event that the crater diameter is known and an estimate of the projectile dimensions is desired, Equation (5.6) can be used to derive D_p for complex lunar craters, viz.,

$$D_p = 75.6 \left(\frac{\rho_p}{\rho_t}\right)^{-0.42} D_R^{1.09} v_i^{-0.56} \tag{5.7}$$

Small impact craters also depend on the target strength. If the main cratering flow energy is spent as plastic work against strength and friction, this is the so-called "strength regime of cratering," i.e., a "strength crater." For large craters most of the energy is converted into the potential energy of the excavated cavity. This case is in the "gravity regime of cratering" and the crater is termed a "gravity crater." Equation (5.4) is valid for so-called "gravity craters." Figure 5.13 illustrates the general character of the impact-crater scaling by showing the dependence of the rim-crest crater-diameter ratio to the projectile diameter, D/D_p, on the rim-crest crater diameter D for lunar gravity and the impact velocity of 18 km s^{-1} (Ivanov et al. 2001). This figure shows how D/D_p decreases for larger craters.

Melt production scaling. Standard representation of the melt volume V_m in impact craters is $V_m/V_p \sim (v_i^2/E_m)^\mu$, where V_p is the projectile volume, v_i is the projectile velocity and E_m is internal energy of melting. O'Keefe and Ahrens (1977) concluded from hydrocode calculations that above a certain limit the total volume of melt and vapor produced in an impact event scaled with the energy of the impactor, resulting in a relation including the factor μ, which they found to be 0.67. However, Bjorkman and Holsapple (1987) tested the point source limit scaling law for melt volumes and found $\mu = 0.55$–0.6. Detailed numerical modeling (Pierazzo et al. 1997) proposed the scaling of melt production for a wide range of impact velocities and target-projectile types in the form:

$$\log \frac{V_m}{V_p} = a + \frac{3}{2}\mu \log \frac{v_i^2}{E_m} \qquad (5.8)$$

with $\mu = 0.708 \pm 0.039$ as an average value for all materials. In an oblique impact the melt production is substantially lower (Pierazzo and Melosh 2000b and discussion below).

The only way to prove these multi-stage estimates is to use numerical modeling for estimates of the impact melt volume. Values for the impact melt volume are estimated from geological investigations of terrestrial impact craters (Grieve and Cintala 1992). This test gives a good result, i.e., theoretical estimates reproduce the impact melt volume within of factor of 2 for a given crater diameter (Fig. 5.14).

Excavation depth. Excavation depth is not to be confused with the depth of the transient cavity as the ejecta do not include material excavated from the full depth of the crater (see

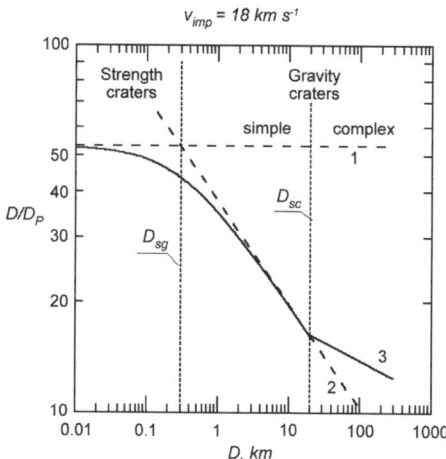

Figure 5.13. The ratio of the crater diameter, D, to the impactor diameter, D_P for various regimes of impact cratering. 1- the "strength branch" of the scaling law, where the D/D_P ratio is constant for a constant effective target material strength; 2 – the "gravity branch" of the scaling law: D/D_P ratio decreases as the crater diameter increases; 3 – the "gravity collapse" branch for complex craters, while the transient crater size is still following the "gravity branch" (2), the final crater diameter is larger than the transient crater diameter due to the crater widening during the transient cavity collapse. The figure is constructed for the impact velocity of 16 km s^{-1} assuming equal projectile and target density, lunar gravity of 1.62 m s^{-2}, strength-to-gravity regime transition at $D_{sg} = 300$ m (as assumed by Neukum and Ivanov 1994, from the onset of regular continuous ejecta blankets around lunar craters found by Moore et al. 1974), and simple-to-complex crater transition at $D_{sc} = 15$ km.

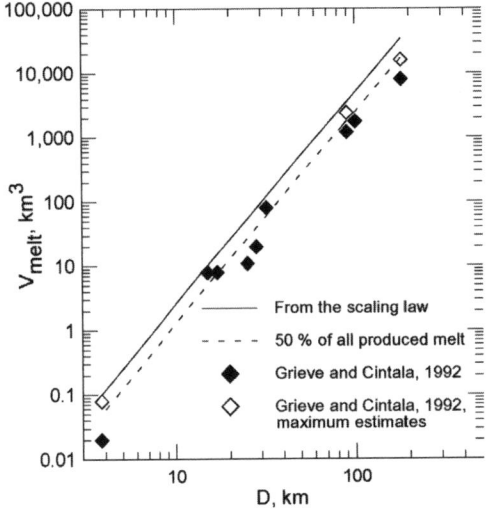

Figure 5.14. Comparison of impact melt volumes for terrestrial impact craters (Grieve and Cintala 1992) with estimates from the scaling law. Scaling laws have been used to estimate the coupling parameter. Following numerical modeling with a given coupling parameter estimates the impact melt volume for typical Earth's crust rocks (granite, gneiss). For an example of these estimates, see Pierazzo et al. (1997).

Fig. 5.12) Material corresponding to the volume of the transient cavity is *displaced* material not *ejected* material. The empirical estimate for the maximum depth of excavation H_{exc} is about 1/3 of the transient crater depth or one-tenth of the transient diameter: $H_{ex} \approx 1/3H_{tr} \approx 1/10D_{tr}$.

Ejecta deposit thickness. When the strength can be neglected, the ejection velocity v_e depends upon the ratio of ejection position r within the crater to a final crater radius R (Housen et al. 1983):

$$\frac{v_e}{\sqrt{gR}} = C_e\left(\frac{r}{R}\right)^\varepsilon \tag{5.9}$$

where C_e is an empirical coefficient. The power ε can be determined either experimentally or derived from the coupling parameter model. In the latter case it ranges from 1.9 for water to 2.4 for sand (Melosh 1989). Although the thickness of the ejecta deposit may scale similarly, morphological differences are expected as a function of crater size, because the velocity of ejecta that land at a given scaled range depends on a crater size. The ejecta blanket thickness δ falls off as a power of the distance r from the crater center: $\delta = f(R)(r/R)^{-3\pm1}$. A traditional description of the ejecta blanket thickness (McGetchin et al. 1973) $\delta = 0.14R^{0.74}(r/R)^{-3\pm1}$ is from a compilation for data mainly from explosion cratering experiments. Experimental data for impacts are only known from small-scale impact cratering experiments in sand (Stöffler et al. 1975). These data yield a somewhat different function between δ and r, namely $\delta = 0.06R(r/R)^{-3.26}$. However, any scaling is correct only for a vertical impact, and impact obliquity leads to substantial asymmetry in the ejecta blankets, especially at large distances from the crater, outside the "continuous" ejecta blanket (beyond about one crater radius). Experimental and observational results show substantial progress in the understanding ejecta distribution around craters made by oblique impacts (e.g., Anderson at al. 2003; Herrick and Forsberg-Taylor 2003; Schultz and Mustard 2004). However, the general scaling of ejecta blankets at natural impact craters is still an open issue.

2.3.2. Numerical modeling of crater formation by computer code calculations. The first numerical cratering simulation—modeling of the Meteor Crater impact—was done in 1961 by Bjork. Since then, numerical computations began to expand with steadily improving computers and codes capable of providing ever more accurate simulations of the cratering process. Many of the early results achieved by these methods are compiled in papers in Impact and Explosion Cratering (Roddy et al. 1977).

In recent years, numerical models of impact cratering have been applied mainly to Earth, Mars, and small bodies. The main application specific to the Moon has concerned the problem of the Moon's origin in a giant impact (Melosh and Sonett 1986; Cameron and Benz 1991; Cameron 2001; Canup and Asphaug 2001). Recently progress in numerical modeling in impact mechanics has been mainly in the development of more sophisticated models, creation of new three-dimensional codes, and testing on intensively studied terrestrial craters. The time to apply these improved models to problems in lunar cratering is now ripe, as new data sets and increasingly sophisticated observations of the Moon yield new constraints for models and for reevaluation of lunar stratigraphy, which is based mainly on impact events (see Sections 4 and 5).

The physical foundation. The dynamics of an impact, i.e., the displacement and deformation of materials under pressure, stress, and strain as well as the dynamics of any continuous material, are described with a set of differential equations established through the application of the principles of conservation of mass, momentum and energy (e.g., Landau and Lifshitz 1987). These equations are usually coupled with an equation of state, relating the density and internal energy of the material with pressure. A constitutive equation relates the stress in the material to the amount of strain to which the material is subjected. This set of coupled, nonlinear equations can be solved analytically only for a few problems for which certain simplifying restrictions are invoked. Only numerical techniques provide a method of obtaining solutions without appreciable restrictions and simplifications, and thus are widely used in cratering mechanics. They provide important insight into processes and phenomena observed in cratering experiments, and they are the only resource available to study impacts at cosmic velocities, which are unreachable with existing laboratory techniques.

Hydrocodes and modeling of oblique impacts. Computer programs that handle the propagation of shock waves and compute velocities, strain, and stress as a function of time are called "hydrocodes," as initially they did not include strength effects (i.e., materials were treated as liquids). Modern computer codes use sophisticated strength models, such as gradual shear and tensile-damage accumulation, thermal softening, and acoustic fluidization. Many papers and textbooks treat the main principles of hydrocode construction: Eulerian and Lagrangian descriptions, dimensionality, discretization of time and space, numerical schemes, stability, etc. (e.g., Anderson 1987; Zukas 2004 and references therein).

In 1994, hydrocodes were tested for a strengthless material in a giant, natural experiment, the collision of the Shoemaker-Levy 9 comet with Jupiter. Numerical results from various hydrocodes correctly predicted a wide range of impact-related phenomena, including light generated during entry into the Jovian atmosphere, plume formation, and fallout of ejected material (Ahrens et al. 1994; Crawford et al. 1994; Zahnle and MacLow 1994). The results also allowed interpretation of astronomical data and definition of fragment-size distribution (Crawford 1997; Nemtchinov et al. 1997).

Improvements in computer capabilities over the past few years have provided access to the next level of resolution and complexity in impact simulations. Two areas of numerical modeling of impact cratering have received considerable attention recently: (1) oblique impact simulations, which require three-dimensional capability, to model some specific features of non-vertical impacts (Fig. 5.15), and (2) modeling of terrestrial impacts with accurate target lithologies to compare the numerical results with available geological and geophysical data (Figs. 5.16–5.19).

All natural impacts are oblique to some degree, with the most probable impact angle being 45° (Gilbert, 1893; Shoemaker 1962). Although crater rims may appear circular down to low impact angles (15°), the distribution of ejecta around the crater is more sensitive to the angle of impact and currently serves as the best guide to the obliquity of the impact. Numerical modeling has reproduced asymmetric ejection in oblique impacts, providing a means of examining the production of distal ejecta (tektites and martian meteorites, see below), and

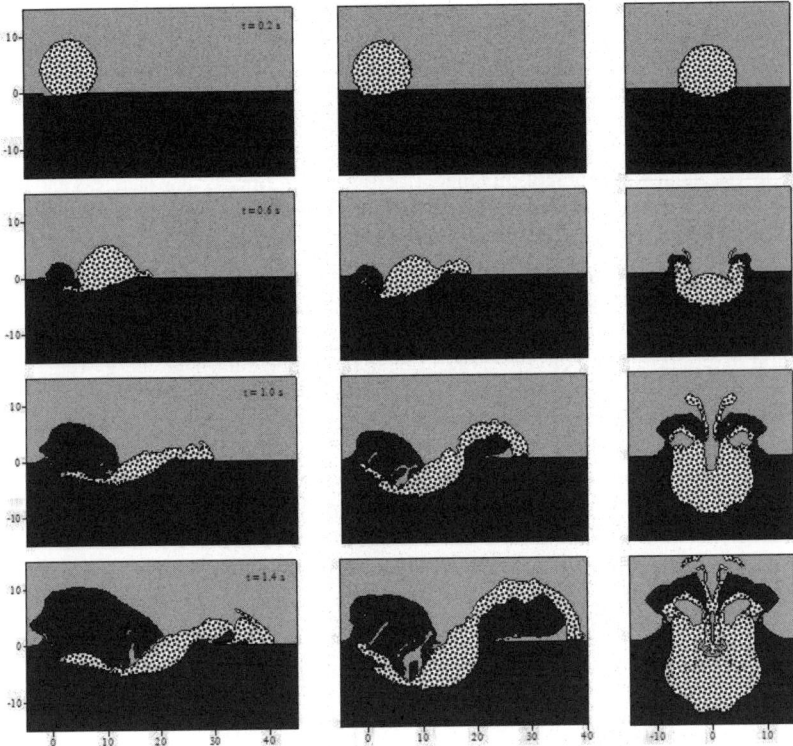

Figure 5.15. Early phases of oblique impact into a target. Material distribution in the plane of impact during the first 1.4 s after the contact (each 0.4 s). A 10 km diameter asteroid moves from the left with velocity of 20 km/s and strikes a solid surface at the point $x = 0$, $z = 0$. The impact angle is 30° in the left column, 45° in the center, and 90° (vertical impact) in the right column. The target is dark gray, atmosphere is light gray and the projectile is stippled.

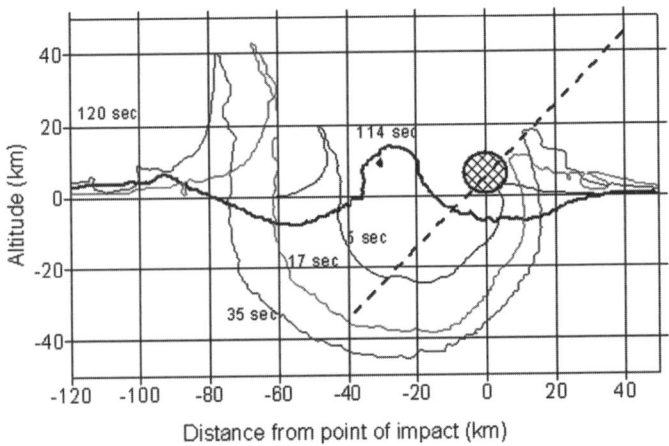

Figure 5.16. Late stage of crater modification by transient cavity collapse. Evolution of the transient cavity for a 45° impact. Patterned circle shows the projectile position at the moment of the first contact with the target. Modeling is for a Chicxulub-scale impact (Ivanov and Artemieva 2002).

revealing substantial differences between vertical and oblique impacts with regard to melt production and the fate of the projectile.

The shock wave generated by an impact weakens with decreasing impact angle (Pierazzo and Melosh 2000a; Artemieva and Ivanov 2001; Stöffler et al. 2002; Fig. 5.17). As a result, both the isobaric core of the initially shocked target material and the volume of melt are asymmetric, both concentrating downrange of the impact point and lying at shallower levels in the target relative to the vertical impact case. The ratio of impact-melt (or vapor) volume to projectile volume has a maximum for vertical (90°) impacts and decreases with impact angle (Fig. 5.17). The decrease is slow for impact angles down to 45°, corresponding to a drop in this ratio of about 20%. For impacts at 30°, however, the ratio decreases by about 50%, whereas at 15°, it is less than 10% of the vertical case (Pierazzo and Melosh 2000a). By incorporating the vertical component of impact velocity (Chapman and McKinnon 1986) standard scaling laws originally proposed by Schmidt and Housen (1987) allow recalculation of these numbers for oblique impacts relative to constant crater volume. In such a case, the ratio of impact-melt volume to crater volume is greatest for a 30° impact (Artemieva and Ivanov 2001). In general, however, scaling laws for oblique impact remain poorly understood for high-velocity planetary impacts. Initial estimates show that high-velocity (>15 km s^{-1}) oblique impacts have practically the same efficiency and produce craters with the same volumes as vertical impacts (Hayhurst et al. 1995; Burchell and MacKay 1998; Ivanov and Artemieva 2001), while laboratory impacts (5–7 km s^{-1}) lose their efficiency rather quickly (Gault and Wedekind 1978; Schultz and Gault 1990). In principle, dimensions of low-velocity impact craters formed in the laboratory depend only on the vertical component of the impact velocity.

Obliquity also influences the fate of the projectile. In particular, the amount and velocity of ricochet are strong functions of impact angle (Schultz and Gault 1990; Pierazzo and Melosh 2000c; Artemieva and Shuvalov 2001). The mass of shock melt or vapor in the projectile material decreases drastically for low impact angles as a result of the shock weakening for decreased impact angles (Fig. 5.18). For asteroidal impacts, the amount of projectile vaporized is limited to a small fraction of the projectile mass. Most of the projectile in cometary impacts, however, is vaporized even at low impact angles. A large fraction of the projectile material

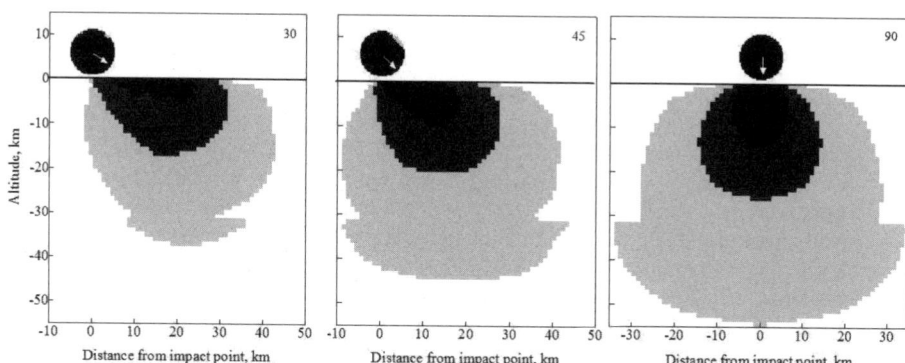

Figure 5.17. Maximum shock pressures in the projectile and the target created with oblique impacts (the central plane through the point of impact is shown). 10 km diameter projectile strikes a Chicxulub-like target (3 km of sediments and 30 km of granite crust on top of a dunitic mantle) with the velocity of 20 km/s. Impact angles are 30° (left plate), 45° (central plate), and 90° (right plate). Light gray color is for shock modified material (20 GPa<P<55 GPa), gray is for melt zone (55 GPa<P<180 GPa), black corresponds to 0 to 30% of granite vaporization (180 GPa < P < 1000 GPa). Most of the projectile is melted or partially vaporized after the shock release. The irregular shape of shock metamorphosed zone in the target is due to refraction of shock waves at the crust/mantle boundary at the depth of 33 km.

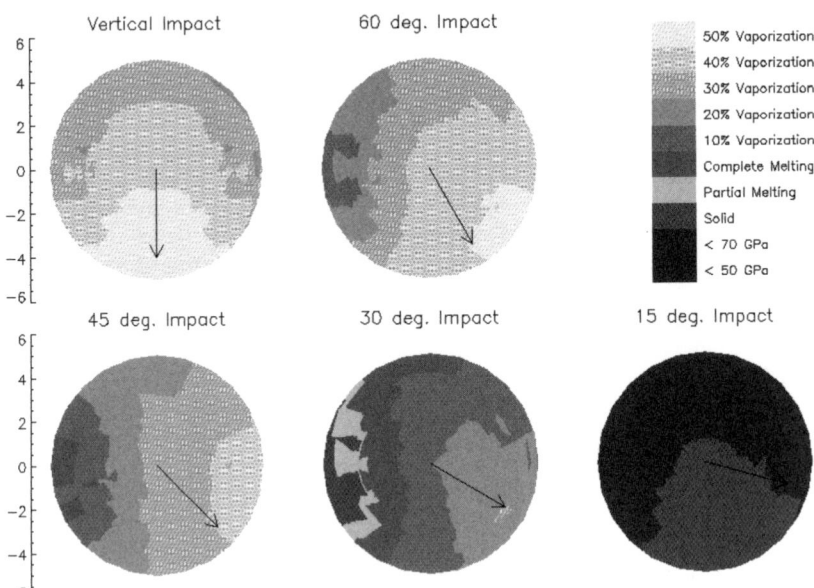

Figure 5.18. Shock pressure distribution in the projectile as a function of impact angle (from Pierazzo and Melosh 2000c).

in oblique impact simulations retains a net downrange motion and a significant amount of it may reach velocities close to or exceeding the escape velocity of Earth. Finally, most of the projectile is ejected from the growing crater in the early stages of an oblique impact. This "ricochet" could explain the absence or very low content of extraterrestrial material in some large impact craters (Palme et al. 1978; Tagle et al. 2004).

Numerical models suggest that oblique impact is the only mechanism giving rise to very special types of impact ejecta including tektites (glassy, cm-sized bodies, deposited within the strewn fields, hundreds of km from the crater, but connected geochemically with the target rocks of the crater) and meteorites from Mars and the Moon (Fig. 5.19). Full-scale, 3D modeling (Artemieva 2001; Artemieva and Ivanov 2002) suggests that no specific conditions are needed to produce these ejecta. The most conducive impacts likely have impact angles in the interval from 30° to 60° and impact velocities typical for Mars (10 km s^{-1}) or Earth (18 km s^{-1}). The obvious deficiency of tektite strewn fields (only four known strewn fields for more than 160 impact craters) may be explained by erosion and weathering and/or a paucity of young craters with diameters larger than 10 km. Martian meteorites may be launched by small asteroid impacts with final crater diameters of 1–3 km (Head et al. 2002), under the restriction that solid fragments larger than 20 cm in diameter are ejected at or above escape velocity from the upper few meters of the target (Artemieva and Ivanov 2002). Detailed numerical simulations with high spatial resolution of this upper layer do not support the idea of meteorite launch without substantial (~10 GPa) compression. This numerical result agrees well with petrological data (Fritz et al. 2003), but conflicts with simplified estimates by Mileikowsky et al. (2000) and is only marginally compatible with magnetic data for ALH84001 (Weiss et al. 2000). Similar conclusions can be drawn for lunar meteorites for which the depth of the subsurface source layer and the range of shock pressures is somewhat lower that in the martian case.

Verification of computer code by direct comparison with laboratory impact experiments, such as those by Schultz and coworkers (e.g., Schultz 1996; Dahl and Schultz 2001), remains an

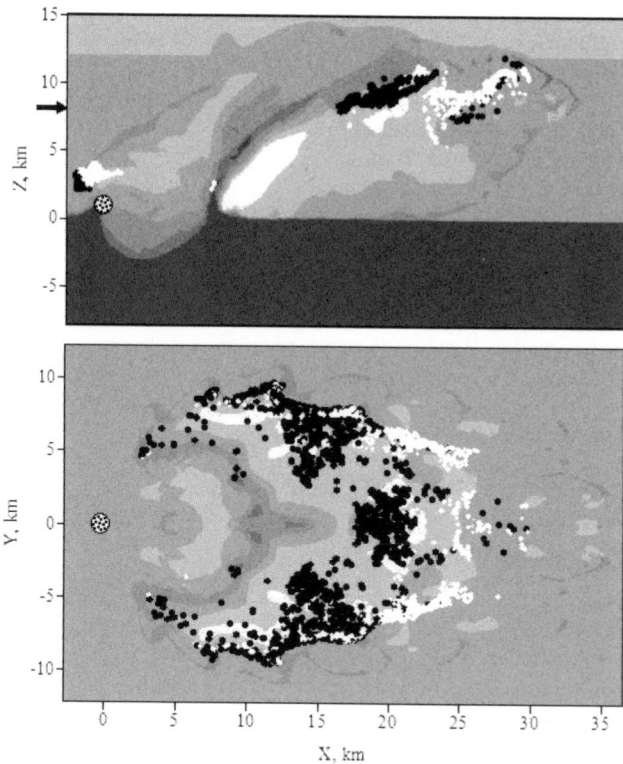

Figure 5.19. Ejection of tektites from the Ries crater. Vertical slice XZ in the plane of impact is on the upper plate, horizontal slice XY at an altitude of 8 km (see arrow) is on the bottom plate. 1.5 km in diameter asteroid (shown as dotted circle) strikes the target at the point $x = 0$, $y = 0$, $z = 0$ with velocity of 20 km/s and impact angle of 30°. Density distribution two seconds after the contact is shown in gray scale, microtektites (<0.1 cm) are shown as white, real tektites (>1 cm) are shown as black.

open issue. Problems in comparing experiments and numerical modeling include lower impact velocities and small scale in impact experiments; however, limitations also exist in the computer simulations. Notwithstanding their fundamental role in the investigation of oblique impacts, laboratory experiments cannot achieve the impact velocities and scale typical of planetary impacts. These limitations necessitate the use of numerical modeling for the study of impact events.

3. THE LUNAR IMPACT FLUX AND CRATER PRODUCTION FUNCTIONS

3.1. Crater production functions

Several decades of lunar exploration have accumulated enough data to present an approximate lunar chronology based on the ages of returned samples. The study of the *size-frequency distribution* (SFD) of lunar impact craters forms a solid basis to show that the SFD has been relatively stable since the end of the early heavy bombardment and that the process of crater formation is still going on. Impact craters represent therefore a record of small body evolution and Solar System chronology.

Despite a comprehensive study of the Solar System cratering record, the question of the "exact" form of the size-frequency distribution of impact craters created on a fresh geologic unit (i.e., totally rejuvenated at the beginning of cratering with no further obliteration of craters) is far from a final answer. Partially this is a consequence of the fact that it is difficult to identify large "fresh" surfaces because, in most cases, planetary surfaces have a complex geologic history with simultaneous crater accumulation and degradation (e.g., Hartmann 1995). However, some general conclusions can be drawn. Today there are two independent evaluations of the production SFD, named "Hartmann's SFD" and "Neukum's SFD" after the names of the main proponents.

Three main forms of SFD presentations are used. The *cumulative* form presents the number of craters with diameter equal or greater than D: $N = N(>D)$. The *differential* (incremental) distribution (i.e., the Hartmann production function) is the derivative dN/dD of the cumulative SFD $N(>D)$. It gives the number of craters in equal diameter bins: $N(D, D + \Delta D)$. The incremental distribution is a log-incremental distribution where bin boundary diameters have a constant ratio (e.g., from 1 to 2 km, from 2 to 4 km, etc.). As the crater SFD is close to a power law, the *relative* distribution R (or R-plot) presents the deviation of the SFD from the "base" power law:

$$R = D^3 \, (dN/dD) \tag{5.10}$$

In practice, the R distribution is calculated on the basis of finite increments of D.

3.2. Hartmann production function — HPF

The tabulated HPF is an assemblage of data selected by Hartmann to present the production function for one specific moment of time, the average time of lunar mare surface formation (see Neukum et al. 2001). In this case the condition for a fresh surface is satisfied by the fact that most lunar mare basalt samples have a narrow range of ages around 3.2 to 3.5 Ga (Stöffler and Ryder 2001), which restricts the lunar maria to an age range within a factor of 1.1. As the tabulated HPF is the result of some averaging of individual crater counts in different areas, it should be treated as a relatively reliable model approach to the construction of the power function. Hartmann uses a log-incremental SFD representation with a standard diameter bin size. The number of craters per km^2, N_H, is calculated for craters in the diameter bin $D_{LFT} < D < D_{RGT}$, where D_{LFT} and D_{RGT} are the left and right bin boundary and the standard bin width is $D_{RGT} = 2^{1/2} D_{LFT}$. Hartmann et al. (2000) approximated the tabulated HPF in the form of a piece-wise, three-segment power law (coefficients first published jointly by Hartman and Neukum et al. 2001):

$$\log N_H = -2.616 - 3.82 \log D_{LFT}; \; D_{LFT} < 1.41 \text{ km} \tag{5.11a}$$

$$\log N_H = -2.920 - 1.80 \log D_{LFT}; \; 1.41 \text{ km} < D_{LFT} < 64 \text{ km} \tag{5.11b}$$

$$\log N_H = -2.198 - 2.20 \log D_{LFT}; \; D_{LFT} > 64 \text{ km} \tag{5.11c}$$

This function is shown in Figure 5.20. Hartmann's choice of power law segments was made in the 1960s when he began this work and was made for historical reasons. At that time, only the shallow branch 1.41 km < D < 64 km was well established, while the preexisting literature had suggested such laws for asteroids and meteorites, which Hartmann was attempting to relate to the lunar data.

3.3. Neukum production function (NPF)

In a series of publications, Neukum proposed an analytical function to describe the cumulative SFD of lunar impact craters (Neukum 1983; Neukum and Ivanov 1994). Neukum showed that the production function had been more or less stable from Nectarian to Copernican epochs (i.e., from more than 4 Ga until the present). By this time the full size spectrum of craters was known, and in contrast to the piecewise exponential equations used for the HPF, Neukum computed a polynomial fit to the cumulative number of craters, N, per km^2 with diameters larger

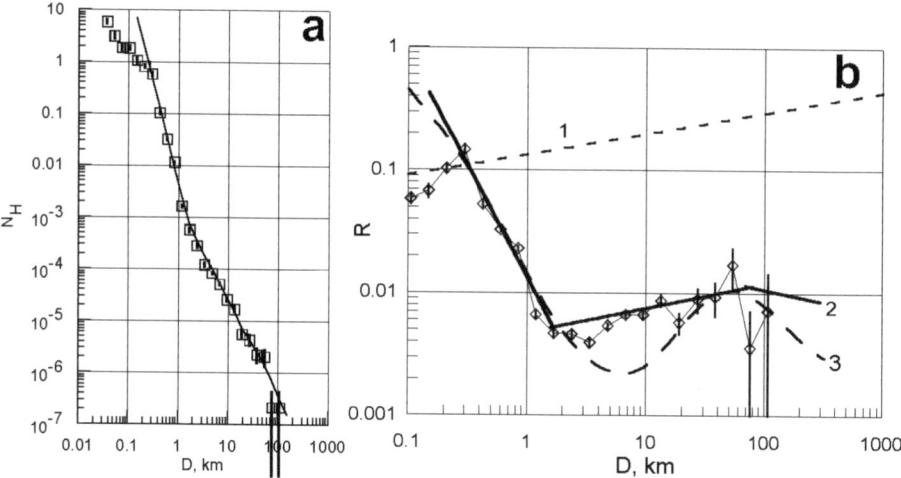

Figure 5.20. (a) Incremental representation of the Hartmann production function (HPF). The HPF, in a direct sense, is the set of points shown in the plot as squares. Straight lines represent the piece-wise power law fitting of the data (Eqn. 5.11). (b) Comparison of production functions derived by Hartmann (HPF, shown as diamonds) and Neukum (NPF) in the R plot representation. The maximum discrepancy between HPF (2) and NPF (3) (roughly a factor of 3) is observed in the diameter bins around $D \sim 6$ km. Below $D \sim 1$ km and in the diameter range of 30–100 km, the HPF and NPF give the same or similar results. Fitting the HPF to NPF Equation (5.12) yields a model age of 3.4 G.y. The NPF, which is fit to the wide range count of impact craters in the Orientale Basin, yields a model age of ~3.7 G.y. The dashed line 1 represents the approximate saturation level estimated by Hartmann (1995). The HPF power law segments (solid line labelled 2 in part b) correspond to Equation (5.11a) for $D < 1.41$ km, to Equation (5.11b) for 1.41 km $< D < 64$ km, and to Equation (5.11c) for $D > 64$ km.

than a given value D. For the time period of 1 Ga, $N(D)$ may be expressed (Neukum 1983) as:

$$\log_{10}(N) = a_0 + \sum_{n=1}^{11} a_n \left[\log_{10}(D)\right]^n \tag{5.12}$$

where D is in km, N is the number of craters with diameters $>D$ per km² per Ga; the coefficients a_n are given in Table 5.4. Equation (5.12) is valid for D from 0.01 km to 300 km. Recently, the NPF was slightly reworked toward the largest craters by re-measuring in that size range (Ivanov et al. 1999, 2001; Neukum et al. 2001). The time dependence of the a_0-coefficient is discussed in the following subsection. A similar equation is used to characterize the projectile SFD derived in Ivanov et al. (2001). Coefficients for this projectile SFD are also listed in Table 5.4. In the projectile SFD column, the first coefficient a_0 has been set to zero for simplicity. This coefficient determines the absolute number of projectiles. The absolute value of a_0 for projectiles may be found by fitting to observational data (see Section 3.7).

3.4. Towards a unified production function

In Figure 5.20b, the NPF and HPF are shown in an R-plot together with representative data for crater counts on the lunar maria and the Orientale basin. The NPF was fit to the crater counts using an assumed age of average lunar maria of 3.2 to 3.5 Ga. We find that both the HPF and NPF are a good match to the observational data below $D \sim 1$ km. However, for $D > 1$ km, the HPF lies well above the NPF, meeting again the NPF at crater diameters $D \sim 40$ km. A maximum discrepancy of a factor of 3 between HPF and NPF is observed in the diameter bins around $D \sim 6$ km. Below $D \sim 1$ km and in the $30 < D < 100$ km range, the HPF and NPF give the same or similar results.

Table 5.4. Coefficients of the analytic production function (Eqn. 5.12) for cumulative number of craters $N(D)$ with diameters larger than D, and relative number of projectiles $R(D_P)$[*] assumed from measured $N(D)$.

a_n	"Old" $N(D)$ (Neukum 1983)	"New" $N(D)$ (Neukum et al. 2001)	"New" $N(D)$ sensibility[**]	Projectile $R(D_P)$ (Ivanov et al. 2000)
a_0	−3.0768	−3.0876		—
a_1	−3.6269	−3.557528	± 3.8 %	+1.375
a_2	+0.4366	+0.781027	± 3.9 %	+0.1272
a_3	+0.7935	+1.021521	± 2.5 %	−1.2821
a_4	+0.0865	−0.156012	± 1.6 %	−0.3075
a_5	−0.2649	−0.444058	± 0.88 %	+0.4149
a_6	−0.0664	+0.019977	± 1.3 %	+0.1911
a_7	+0.0379	+0.086850	± 0.78 %	−0.04261
a_8	+0.0106	−0.005874	± 1.8 %	−0.03976
a_9	−0.0022	−0.006809	± 1.8 %	-3.1802×10^{-3}
a_{10}	-5.18×10^{-4}	$+8.25 \times 10^{-4}$	± 5.6 %	$+2.799 \times 10^{-3}$
a_{11}	$+3.97 \times 10^{-5}$	$+5.54 \times 10^{-5}$	± 24.1 %	$+6.892 \times 10^{-4}$
a_{12}			—	$+2.614 \times 10^{-6}$
a_{13}			—	-1.416×10^{-5}
a_{14}			—	-1.191×10^{-6}

[*] Relative number of objects (craters or projectiles) is defined as (e.g., Hartmann et al. 1981) $R(D) = D_x^{-3}(dN/dD_x)$ where N is the cumulative number of objects with size larger D_x (crater diameter, D, or projectile diameter, D_P, correspondingly).

[**] "Sensibility" is the coefficient variation which changes the $N(D)$ value by a factor of 2 up and down.

Although Figure 5.20b shows that the HPF and NPF share some similarities, the factor of 3 discrepancy for $2 < D < 20$ km craters requires further investigation. The discrepancy between HPF and NPF reflects the discrepancy in observational data used to construct the production function. A real "unification" of two types of production functions would require a re-counting of craters in critical areas. In general, one should be cautious in interpreting data in this range, particularly since different data sets show somewhat different SFD curvatures. Additional studies of lunar mare data are needed to further refine the accuracy of the main production-function curve.

To use a production function, one should first select a portion of the lunar surface where all the accumulated craters since the last resurfacing event can be counted. Examples of such "time slices" are:

a) The Orientale basin, which erased a large area near the base of the Imbrian stratigraphic horizon.

b) The emplacement of mare basalts (Hartmann 1970; Hartmann et al. 1981);

c) Eratosthenian-ages craters, which mostly have good stratigraphic dates (Wilhelms et al. 1987)

Figure 5.21 shows crater counts for lunar areas that differ by a factor of 100 in the density of craters per unit area. An examination of these "time slices" suggests that we cannot rule out the simple hypothesis that the lunar production function had a constant shape from ~4 Ga (lunar highland formation) to ~1 Ga (ray craters). Thus, in the limits of data accuracy, we can assume that the projectile SFD was stable over this interval. To test this hypothesis, we compare lunar data to the cratering records found on other planets and asteroids (Sections 3.5–3.7).

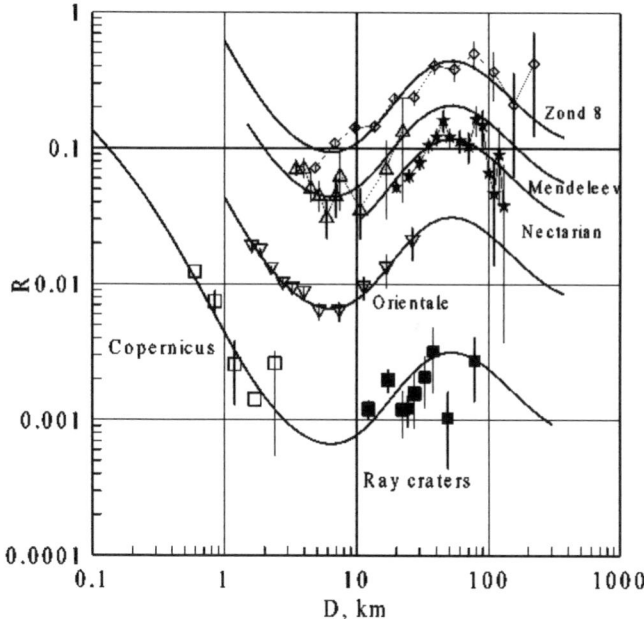

Figure 5.21. R plot for several "time slices" of the lunar impact chronology fitted with the NPF curve (Eqn. 5.11b). See data description in Neukum et al. (2001).

Although the HPF and NPF are somewhat different, both assume that the general shape of the SFD of impactors striking the Moon over the last 4 Ga was the same. A different point of view is given by R. Strom (Strom and Neukum 1988; Strom et al. 1992), who claims that the "modern" (post-mare) production function is quite different from that produced during the epoch of the early heavy bombardment. A more extensive treatment of this subject can be found in Strom and Neukum (1988).

3.5. SFD for craters in comparison with asteroids

Recently improved data in astronomical observations of asteroids permit comparison of SFD for lunar craters and asteroids. Using the impact cratering scaling laws and average impact velocity for asteroids (see review by Ivanov et al. 2001), one can convert the lunar production function into a "projectile" size-frequency distribution. This "crater-derived" SFD may be compared directly with astronomical observations of main belt asteroids (Ivanov et al. 2003).

Deviations from a simple power-law SFD of craters considered above suggest that the SFD of asteroids also deviates from a simple power law. A possible mechanism for producing such deviations is based on modeling results describing impact evolution in the main belt (i.e., a "wavy" SFD according to Campo Bagatin et al. 1994a,b; Durda et al. 1998; Davis et al. 2002). For bodies with diameters near a few hundred meters, self-gravity helps prevent catastrophic disruption events by allowing fragments to re-accumulate with the target asteroid (e.g., Love and Ahrens 1996; Melosh and Ryan 1997; Benz and Asphaug 1999). As bodies get stronger via gravity, more projectiles of that size are available to disrupt larger asteroids, ultimately leading to a wave in the shape of the SFD.

The Spacewatch data (Jedicke and Metcalfe 1998) and data of the Sloan digital sky survey (Ivezic et al. 2001) allow estimation of the asteroid SFD down to diameters of approximately 300 m. Observational data agree reasonably well with the "projectile" size frequency distribution derived from the lunar crater curve (Ivanov et al. 2003). This similarity supports the idea that lunar craters may result from bombardment by collisionally evolved families of projectiles, similar to those in the modern Main Asteroid Belt.

3.6. Near-Earth Asteroids

An SFD similar to the one of the Main Belt is found for Near-Earth Asteroids (NEAs). Figure 5.22 summarizes several recent SFDs estimated from astronomical observations and models of the NEA population. These results are compared to the projectile SFD derived from lunar cratering records with HPF and NPF. The similarity between the following SFDs, (1) crater-forming projectiles derived from 1 to 4 Ga old surfaces on the Moon, (2) the observed main belt asteroid population, and (3) NEAs (Fig. 5.22), suggests a common connection, namely that the main belt is the predominant source of both the current NEA population and those projectiles that have struck the Moon over the last several Ga.

3.7. Craters on Earth and other terrestrial planets

Having the "projectile" SFD estimates on the basis of lunar crater counts, it is possible to construct model crater SFD's for Mercury, Venus, Earth, and Mars, and to compare the model SFD with real observation. The construction of model SFD's takes into account different

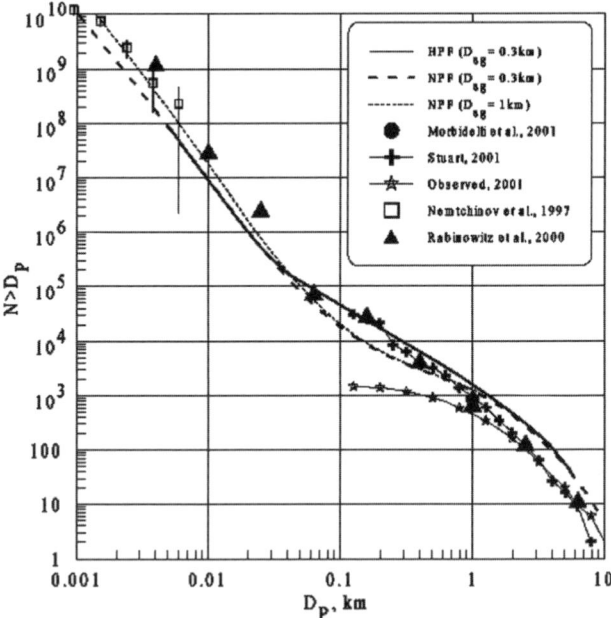

Figure 5.22. Estimates of a cumulative, $N > D_P$, size-frequency distribution for NEAs. The solid, dashed, and dotted lines are model distributions derived from the HPF and NPF for various assumed strength-to-gravity transition diameters for lunar craters. The absolute position of these curves corresponds to the lunar chronology (Eqn. 5.12) combined with estimated average impact probability for Earth-crossing asteroids (ECAs). The number of NEAs (defined as bodies with $q < 1.3$ AU) is larger. For observed bodies with $H < 15$, the ECA to NEA ratio is ~0.57. Recent astronomical estimates by Rabinowitz et al. (2000), Morbidelli et al. (2001), and Stuart (2001) are generally consistent with these estimates. Satellite observations of bolides entering Earth's atmosphere (Nemtchinov et al. 1997) are consistent with our results for small ($D_P < 10$ m) bodies, although we caution that it is problematic to convert light flashes detected in the atmosphere into projectile sizes. The average probability of ECA impacts is used to estimate the total number of projectiles and the impact (or atmospheric entrance) rate.

gravity values and different average impact velocities for these planets (see details in Ivanov 2001; Ivanov et al. 2001; Neukum et al. 2001).

A recent review of the recalculation of the lunar SFD for the conditions of other planets is given by Ivanov et al. (2003). A tentative conclusion from the review is that within the limits of data accuracy, the size-frequency distribution of craters and, hence, of crater-forming projectiles, appears similar in time (for older and younger craters) and space (for terrestrial planets). Variations in the projectile SFD cannot be excluded (e.g., the role of comets is not yet well defined). However, the assumption of a constant shape of the crater production function (SFD) as a function of time is a reasonable starting point for further discussion. The constant shape of the production function means that the systematic deviation of the lunar crater SFD from a simple power law is stable and is the same for the early heavy bombardment projectiles and for the current near-Earth asteroids. Within model constraints, crater counts can be used as a measure of a relative crater retention age for various areas on the Moon. The absolute ages have been inferred from the dating of lunar returned samples (Sections 5.5–5.7).

Details of the crater SFD for other terrestrial planetary bodies will not be discussed here. However, we briefly present the cratering record of the Earth to represent the Earth-Moon system as a whole. A comprehensive discussion of the lunar chronology from cratering statistics is given in the following sections.

Hartmann (1965, 1966) pointed out that large terrestrial craters reflect an older population, while smaller craters are continually removed by erosion, producing an observed SFD that differs from the production function. The inspection of data from the North American and European cratons (Grieve and Shoemaker 1994) suggests that it is possible to distinguish two populations of craters:

1) 8 craters with diameters from 24 to 39 km, the oldest being ~115 Ma

2) 8 craters with diameters from 55 to 100 km, with the oldest being ~370 Ma

The oldest age in each set gives an estimate of the accumulation time. For a proper balance between crater diameter bin width and the number of craters per bin, only two bins for each age sub-population are used to represent the crater production rate. We assume that craters smaller than ~20 km in the younger set and smaller than ~45 km in the older set are depleted by erosion. The poor statistics for terrestrial craters do not help to resolve the production function's shape; however, the terrestrial craters do help to constrain variations in the impact rate.

Figure 5.23 shows cumulative curves recalculated from lunar data for terrestrial craters for crater retention ages of 0.125, 0.36, and 1 Ga. Subsets of the database from Grieve and Shoemaker (1994) for smaller younger craters (the oldest one is 115 Ma) and larger older craters (the oldest one is ~370 Ma) are in a good agreement with properly scaled lunar isochrones. Recently Hughes (2000) published estimates for the terrestrial cratering rate averaged for the last 125 Ma using the "nearest neighbor" approach. This technique results in estimates that are also near (within error limits) model "lunar analogue" curves (Fig. 5.23). Thus, within a factor of 2 (which is close to error bars shown in Fig. 5.23), a reliable first approximation is to assume a constant cratering rate after the end of the heavy bombardment. Previously, the Phanerozoic cratering rate has been assumed to be larger than the average cratering rate for the last 3 Ga (in contrast to approximately constant impact rate). Partially this contradiction is due to a widely accepted concept that the cumulative number of terrestrial craters may be approximated with a simple power law (e.g., Grieve and Shoemaker 1994; Hughes 2000). The simple power law approximation used, for example, by Hughes (2000), with a cumulative exponent of −1.86, is shown as a dashed line in Fig. 5.23. Projecting the data from the crater diameter range of 10 to 50 km to the normative diameter $D = 1$ km can produce misleading results of $N(D = 1)$. Using the lunar like non-power law curve gives less deviation from a constant cratering rate on Earth (and on the Moon) for the last 3 Ga.

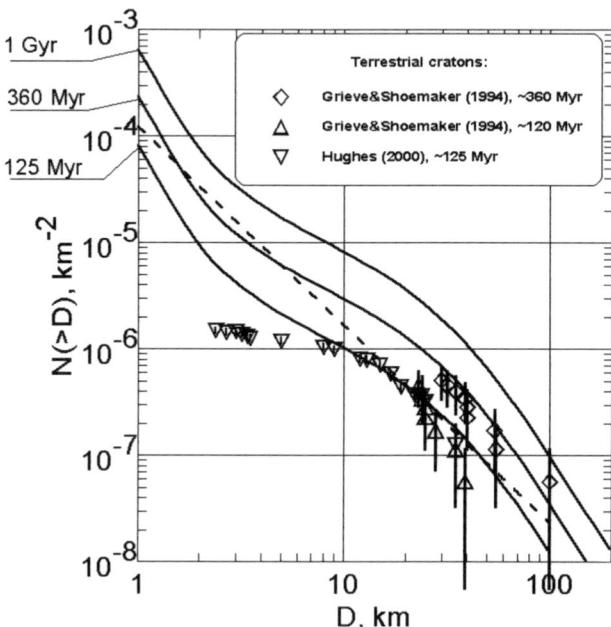

Figure 5.23. Terrestrial cratering data in cumulative form. Data for terrestrial cratons from Grieve and Shoemaker (1994)—area of 17.6 10⁶ km²—are shown as diamonds (for craters younger than ~ 360 Ma) and upward triangles (for craters younger than ~120 Ma). Downward triangles are for estimates by Hughes (2000). Solid curves are isochrones recalculated from lunar data with an account for different gravity and average impact velocity on the Earth. Dashed line illustrates a power law approximation of younger data with an exponent −1.86 used by Hughes (2000).

4. RELATIVE AGE DATING OF LUNAR SURFACE UNITS AND THE LUNAR STRATIGRAPHY

4.1. Principles of relative age dating

Baldwin (1949) provided strong arguments for the impact origin of lunar craters and for the volcanic nature of the mare plains. He also introduced general time relationships based on crater densities and superposition. Following the principles of Gilbert (1893), which were based on telescopic observations, geological mapping of the Moon was pioneered by Shoemaker and Hackman (1962) in preparation of the Apollo program. Lunar Orbiter and Apollo images greatly expanded the telescopic observations and permitted establishment of rock-stratigraphic units.

Lunar mapping and stratigraphy depends on morphologic characteristics, superposition relationships, albedo, and remotely sensed chemical composition of surface units rather than on a detailed examination of rock units and their mutual boundary characteristics. On the Moon, rock-stratigraphic units are primarily understood as morphologically distinct entities formed at a specific time by a defined geological process. Some units, however, remain of uncertain origin, and may have formed over a period of time rather than in a specific event (e.g., the "light plains" in the highlands).

The basic methods and the results of lunar mapping and stratigraphic analysis as derived from telescopic and spacecraft imagery and from the analysis of returned lunar rocks are described comprehensively in Wilhelms et al. (1987). Work on defining the relative ages of units, their geological and chemical definition, and their formative processes has continued since and progressed with global data obtained from the Clementine and Lunar Prospector missions (e.g., Nozette et al. 1994; Staid et al. 1996; Binder 1998; Jolliff et al. 2000b; Staid and Pieters 2001; Feldman et al. 2002). These missions have also helped to understand more of the third dimension of the lunar crust.

The fundamental method used for relative age dating is the application of the law of superposition. It was apparent that older units recognized this way had more craters than younger units, consistent with the craters being of impact origin. The technique of using crater density as a method of deriving relative ages when superposition relationships were lacking (e.g., non-contacting units) became standard for the Moon and for other planets. The use of size-frequency distributions as measures of relative ages has been long established (reviewed in Hartmann et al. 1981; Wilhelms 1984; Wilhelms et al. 1987; Fig. 5.24).

The degree of crater degradation is also an indicator of relative age. For a given size, a fresh crater is younger than a degraded one. The overall morphology of craters indicates their relative ages; for instance, older craters are systematically shallower and smoother than younger craters of the same size because of impact erosion. Some erosion-based method

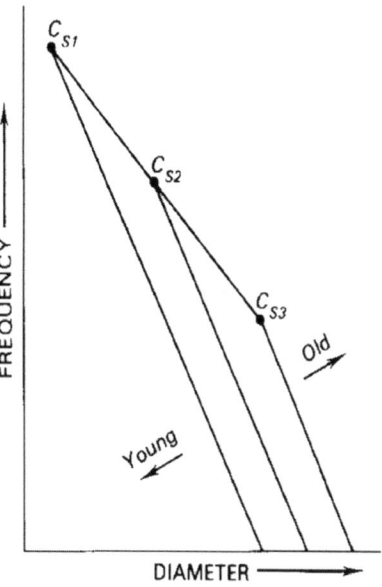

Figure 5.24. Principal graph for age dating of planetary surfaces by cumulative crater frequencies as a function of crater diameter; the kinks in the curves define the parameter C_S which is the transition diameter between the crater production curve (steep) and the crater saturation curve (flat); C_{S1}, C_{S2}, and C_{S3} represent increasing relative crater retention ages; from Wilhelms et al. (1987) (courtesy of the U.S. Geological Society).

is especially needed where the geologic unit of interest is too small for significant crater size-frequency determination. However, several different ways of addressing the morphology of a crater exist. Numerical values such as D_L relate to the size of the largest crater that is nearly destroyed (Soderblom and Lebofsky 1972; Boyce and Dial 1975; Wilhelms 1980). On small surfaces, such a crater will not necessarily be apparent, and D_L is defined as the diameter of craters with the shallow wall slope of 1° (Wilhelms et al. 1987). Despite some pitfalls, this method has been successful and used extensively (Wilhelms et al. 1987; see Table 5.5).

4.2. The lunar stratigraphy

Lunar stratigraphy establishes geologic units and arranges them into a relative time-sequenced column of global significance. A pre-requisite is the identification of rock units or morphological units formed in a single-stage process. Such units are preferably impact basins, impact craters, and lava flows. Morphological units, rather than the exposed bedrock, are necessarily used in photo-based stratigraphy. These units are assembled into higher-order systems, and Rock System boundaries so defined are intended to be the same absolute age everywhere. This chronostratigraphic division (Systems, Series) can be converted into a chronometric division (Periods, Epochs), and with application of radiogenic ages, into absolute time. Figure 5.25 shows these stratigraphic columns, following Wilhelms et al. (1987).

4.2.1. Pre-Nectarian System. The pre-Nectarian System comprises all landforms *older* than the Nectaris basin, and includes about 30 recognized impact basins. Some of these

Table 5.5. Stratigraphic criteria for lunar time-stratigraphic units (after Wilhelms et al. 1987).

System or Series	Crater frequency (number per km²)		C_S (m)	D_L (m)
	≥ 1 km	≥ 20 km		
Copernican System	$< 7.5 \times 10^{-4}$ (mare) $< 1.0 \times 10^{-3}$ (crater)	n/a	?	< 165 (mare) $<$ ca. 200 (crater)
Eratosthenian System	$< 7.5 \times 10^{-4}$ to $\sim 2.5 \times 10^{-3}$ (mare)	n/a	< 100 (mare)	145–250 (mare)
Upper Imbrian Series	$\sim 2.5 \times 10^{-3}$ to $\sim 2.2 \times 10^{-2}$ (mare)	2.8×10^{-5}	80–300 (mare)	230–550 (mare)
Lower Imbrian Series	~ 2.2–4.8×10^{-2} (basin)	1.8–3.3×10^{-5}	320–860 (basin)	n/a
Nectarian System	n/a	2.3–8.8×10^{-5}	800–4000(?) (basin)	n/a
Pre-Nectarian System	n/a	$> 7.0 \times 10^{-5}$	> 4000(?) (basin)	n/a

D_L: diameter of largest crater eroded to 1° interior slopes; C_S: limiting crater diameter for the steady state crater frequency distribution (Fig. 5.24) and from the approximate formula; $D_L = 1.7\, C_S$; n/a = not applicable.

landforms directly underlie deposits of Nectaris, and others are recognized as pre-Nectarian by the size-frequency curves of superposed craters (Fig. 5.26; Table 5.5). The oldest recognized basin is Procellarum, but this basin may not be of impact origin. The oldest basin of clear impact origin is South Pole-Aitken, which is also the deepest and the largest (Spudis 1993). The pre-Nectarian landforms are dominantly of impact origin; no volcanic landforms or tectonic features have been recognized. Pre-Nectarian terrain is predominant on the lunar farside (Fig. 5.27). Any rocks from such ancient terrains in the Apollo or Luna sample collections have been reworked as fragmental material into later impact-breccia deposits.

4.2.2. Nectarian system. The Nectarian System includes all landforms produced between the formation of the Nectaris impact basin and the formation of the Imbrium impact basin. Nectaris itself has deposits over a fairly wide area. Eleven other Nectarian basins have been recognized, including Serenitatis and Crisium (Fig. 5.26). Direct superpositional relationships allow some definition of their stratigraphic sequence, but some crater frequency distributions have been affected by later basins, e.g., Serenitatis ejecta is badly degraded by Imbrium ejecta. Nectarian "light plains" are more evident than are pre-Nectarian ones, and some of these have been suggested to be volcanic in origin (Fig. 5.28; Wilhelms et al. 1987). The Nectarian System has been masked by the Imbrian basins and later volcanic activity; thus, it is much more common on the lunar farside (Fig. 5.27). More recent studies of the relative and absolute ages of lunar mare basalts (Hiesinger et al. 2003) suggest that mare volcanism started already during the Nectarian Period (assuming a 3.92 Ga age of Nectaris, see below). Basaltic volcanism continued throughout all later Periods and ended around 1.2 Ga ago (Hiesinger et al. 2003) within the Eratosthenian or even Copernican Period depending on the assumed age for the Eratosthenian-Copernican boundary, which is either around 1.5 to 2 Ga or near 0. 8 to 1 Ga (see discussion in Stöffler and Ryder 2001 and Hiesinger et al. 2003, and in Section 6).

4.2.3. Lower Imbrian Series. The Lower Imbrian Series comprises all landforms produced between the formation of the Imbrium impact basin and the formation of the Orientale impact basin (Table 5.5). The deposits of these two basins constitute extensive, laterally continuous

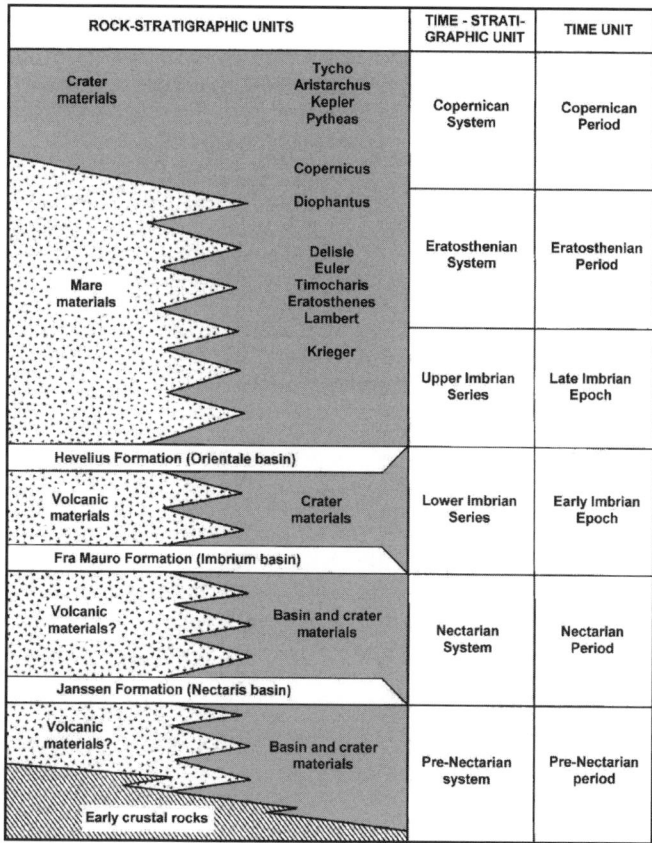

Figure 5.25. The lunar stratigraphic column with rock-stratigraphic, time-stratigraphic and time units (from Wilhelms et al. 1987, courtesy of the U.S. Geological Society).

horizons, although much of the Imbrium basin itself was later flooded with mare lavas. Crater counts as well as its topographic freshness suggest that the Schrödinger basin and its ejecta is Lower Imbrian, but no other basins are in the Lower Imbrian. The size-frequency distribution curve for Orientale deposits lies slightly below that for Imbrium, and both are distinctly below the curve for Nectaris basin deposits (Fig. 5.26). Many "light plains", including the Cayley plains on which Apollo 16 landed, have a Lower Imbrian age (Fig. 5.28), and many may be related to the Imbrium basin. A volcanic origin has also been proposed for at least some light plains, despite the absence of volcanic rocks in the Apollo 16 samples. The Apennine Bench formation, a plains unit within the Imbrium basin, may be a volcanic unit, as it correlates chemically with samples of volcanic KREEP basalt collected at the Apollo 15 landing site.

4.2.4. Upper Imbrian Series. The Upper Imbrian Series includes the landforms produced between the formation of Orientale, the youngest impact basin, and an upper boundary that is defined on the basis of D_L values (Table 5.5). The Upper Imbrian rock units are distinct from older ones: basin deposits are lacking, and two-thirds of the mare volcanic plains are in the Upper Imbrian (Fig. 5.29). The Upper Imbrian was emplaced over a much longer time period than the Lower Imbrian. The extensive mare lavas forming Maria Serenitatis, Tranquillitatis, Crisium, Nectaris, Fecunditatis, Humorum, Nubium, Cognitum, eastern Imbrium, and western Oceanus Procellarum, and several other areas including all the farside mare plains, are part

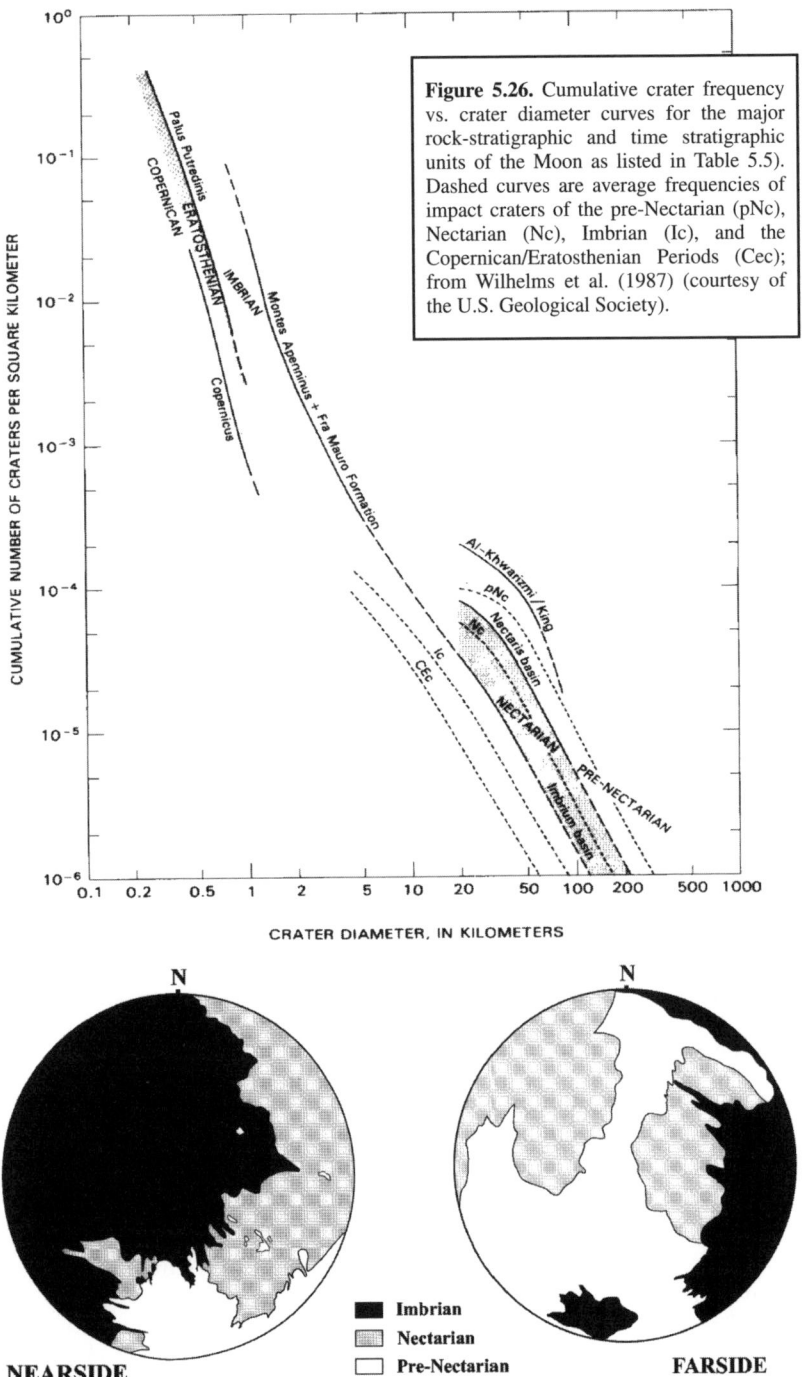

Figure 5.26. Cumulative crater frequency vs. crater diameter curves for the major rock-stratigraphic and time stratigraphic units of the Moon as listed in Table 5.5). Dashed curves are average frequencies of impact craters of the pre-Nectarian (pNc), Nectarian (Nc), Imbrian (Ic), and the Copernican/Eratosthenian Periods (Cec); from Wilhelms et al. (1987) (courtesy of the U.S. Geological Society).

Figure 5.27. Geologic map of the nearside and farside of the pre-Eratosthenian Moon showing the Imbrian, Nectarian, and pre-Nectarian Systems (compiled from Plates 3A and 3B of Wilhelms et al. 1987, courtesy of the U.S. Geological Society).

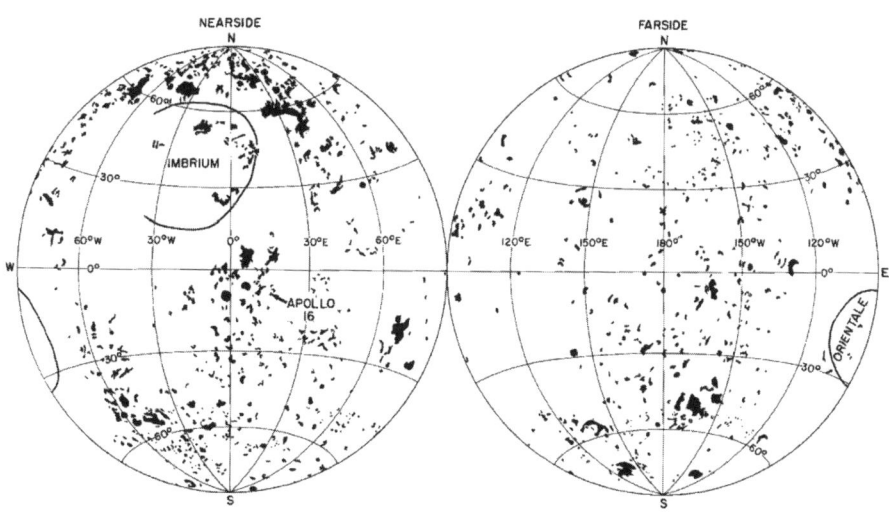

Figure 5.28. Geologic map of the distribution of "light plains" on the nearside and farside of the moon (from Wilhelms et al. 1987, courtesy of the U.S. Geological Society).

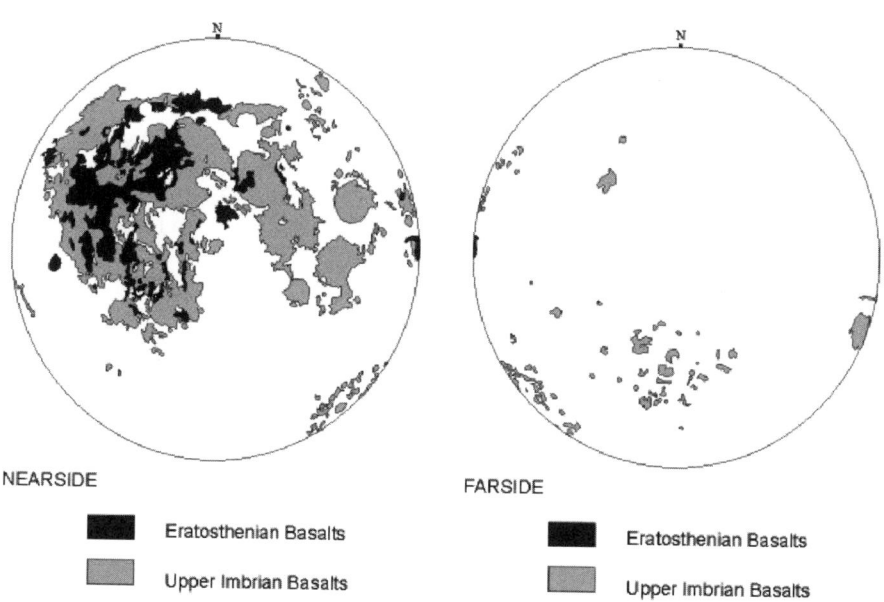

NEARSIDE

FARSIDE

■ Eratosthenian Basalts

■ Eratosthenian Basalts

▨ Upper Imbrian Basalts

▨ Upper Imbrian Basalts

Figure 5.29. Geologic map of mare basalts of the Upper Imbrian Epoch and the Eratosthenian Period on the nearside and farside of the moon; compiled from Plates 9 A,B and 10 A,B of Wilhelms et al. (1987) (courtesy of the U.S. Geological Society).

of the Upper Imbrian Series (Fig. 5.29). Their relative ages have been established on the basis of crater frequencies and by superposition, and their stratigraphic relationships have been elucidated with mineral-chemical data derived from Earth-based spectral reflectance observations and from the Clementine and Lunar Prospector orbital data (e.g., Staid et al. 1996; Jolliff et al. 2000b; Staid and Pieters 2001; Hiesinger et al. 2003). The Upper Imbrian also contains "dark mantling deposits" that have been correlated with volcanic glass of fire-fountain origin, examples of which have been sampled by the Apollo missions, notably the Apollo 15 green glass and the Apollo 17 orange glass. With the exception of the Apollo 12 mission, all mare plains sampled by Apollo and Luna were Upper Imbrian, and no samples were from the Lower Imbrian Series.

4.2.5. Eratosthenian System. The Eratosthenian System is less clearly defined than other systems (Table 5.5). Its upper boundary is even more ambiguous than the lower boundary. Formally the distinction between the Eratosthenian and the subsequent Copernican systems was made according to whether a crater was non-rayed (Eratosthenian) or bright-rayed (Copernican). Eratosthenes itself lacks rays, and Copernicus has bright rays, but according to Stöffler and Ryder (2001) neither crater defines either the base or the top of its eponymous system, instead lying within it. However, the presence of rays depends not only on age, that is degree of impact erosion, but also on crater size and on compositional differences between the impacted target and the region of ejecta deposition. For such reasons a distinction should be made between "compositional" and "maturity" rays which may have different lifetimes (Grier et al. 2001). In any case, superposition on rays of Copernicus establishes a Copernican age. Crater counts on ejecta blankets of Copernican craters are of limited use because most craters are too small for good statistics. The same limited-area constraint applies to D_L (~140 m) as well, and there is the further problem that D_L requires a flat surface, restricting measurements to the small, interior impact-melt region of a crater. Other criteria have been used including infrared (Copernican craters are hotter) and radar (blockier). A new parameter of optical maturity derived from global orbital spectral reflectance measurements from the Clementine spacecraft (Chapter 2) is a promising candidate for a more rigorous determination of relative ages of Copernican and upper Eratosthenian craters.

The Eratosthenian System includes mare plains that are much less extensive than Upper Imbrian plains. They are absent from the lunar farside (other than Mare Smythii on the limb; Fig. 5.29). The plains include those sampled at the Apollo 12 landing site in Oceanus Procellarum. None of the dark mantling volcanic-glass deposits appears to be Eratosthenian. Eratosthenian mare and crater deposits interfinger on much of the central and western nearside of the Moon, enabling a detailed relative stratigraphy to be derived. However, these show that some mare plains traditionally mapped as Eratosthenian embay bright-rayed craters, which must then be assigned to the Eratosthenian, meaning that not all rayed craters must be Copernican.

4.2.6. Copernican System. The Copernican System was first recognized by the rays of its craters, which were shown to be the youngest of lunar features because the are superposed on all other terrains (Table 5.5). Despite the difficulty of using rays in defining a lower boundary (see above), most rayed craters are indeed Copernican, and such craters are scattered all over the Moon. The upper boundary of the Copernican is the present day. Only a very small proportion of the Moon's face is Copernican, although the landscape effects can be global: The rays of Tycho stretch almost around the entire Moon. There are only half as many Copernican craters in any given size range as there are in the Eratosthenian (44 larger than 30 km, cf. 88 in the Eratosthenian). Some patches of mare basalt do overlap rayed craters, and on the basis of craters counts also seem to be Copernican in age (e.g., Schultz and Spudis 1983), mainly in northern Oceanus Procellarum. This interpretation depends, of course, on the definition and absolute age of the lower boundary of the Copernican system (e.g., Hiesinger et al. 2003, and discussion above).

5. GEOLOGIC PROVENANCE AND RADIOMETRIC AGES
OF LUNAR ROCKS

5.1. Geologic provenance and stratigraphic significance of lunar samples

Six manned Apollo missions and three robotic Luna missions returned samples from different geologic settings on the Moon (Fig. 5.1, Table 1.1). More than 30 distinct meteorites are recognized as being of lunar origin (Meteoritical Bulletin 2004; <*http//epsc.wustl.edu/admin/ resources/meteorites/moon_meteorites_list.html*>). The missions provided samples of basaltic rocks and glasses from mare terrain (Apollo 11, 12, 15, and 17; Luna 16, 24) and impactites (polymict breccias, impact-melt lithologies, and granulitic lithologies) from highland terrain (Apollo 14, 15, 16, 17; Luna 20). The lunar meteorites are derived from unknown locations in the mare as well as highland regions. The proportions of mare and highland derived meteorites indicates that they most likely represent the whole surface of the Moon.

It is obvious that the highland samples are highly processed by multiple impact cratering events. In relation to the primary igneous highland rocks (anorthosite, norite, gabbro, troctolite, dunite) the vast majority of returned samples represent second or third generation rocks. Moreover, thermally processed rocks occur as granulitic rocks and granulitic breccias whose compositions indicate polymict (anorthositic-noritic) precursors.

According to the different types of rocks—igneous rocks, crystalline impact-melt rocks and impact glasses, thermally metamorphosed rocks (granulitic lithologies) and polymict clastic matrix breccias (Table 5.6–5.8; Stöffler et al. 1980; Heiken et al. 1991)—different types of ages can be obtained by radiogenic isotope dating. These include (1) *crystallization ages* defining either magmatic, impact melting or recrystallization events, (2) *impact breccia formation ages* defining the time of the assembly and deposition of a polymict breccia, and (3) *exposure ages* defining the time since which an impact-displaced rock fragment has been exposed to cosmic rays. Concerning methods of radiogenic isotope dating, we refer to the literature (e.g., Faure 1986; Dalrymple 1991). General reviews of the results of lunar chronology have been provided by Turner (1977), Dalrymple (1991), Nyquist and Shih (1992), and Snyder et al. (2000), among others.

Radiometric ages of the different types of lunar samples are of very different geological significance which can only be derived by a careful textural analysis of the samples and their geologic setting with respect to a specific formation defined by photogeologic techniques. This hold particularly for all types of polymict rocks (breccias, impact melts). In principle, direct age dating of a parent geologic formation is only possible for mare basalts. Radiometric ages of igneous highland rocks cannot be related to any corresponding geologic surface formation on the Moon. There is the additional problem that none of the rock samples was collected directly from a bedrock unit as the entire lunar surface is covered with impact-produced regolith at least several meters thick. Consequently, the interpretation problems are different for the different types of rocks:

Volcanic rocks of mare provenance. Even for the comparatively simple case of a volcanic rock, it is not necessarily easy to relate that rock to a mapped geological unit. At any given mare collection site there is a range of basalt types—brought to the surface by multiple reworking of the regolith—that in some cases covers a distinct range of ages. While the youngest of these is most likely the age of the surface unit, if that unit is thin or discontinuous it might not be the surface that is mapped and which retains the crater density/crater degradation characteristics used to define the age of the unit in question.

Impact melt and clastic breccia lithologies of highland provenance. Radiometric age dating of impact-melt rocks is generally possible by direct dating of the glassy or crystalline matrix. However, since datable impact-melt rocks are either displaced individual clasts

Table 5.6. Radiogenic crystallization ages (Gyr) for igneous lunar highland rocks.

Sample[1]		^{40}Ar-^{39}Ar*	Rb-Sr	Sm-Nd	U-Pb, Pb-Pb
Ferroan	22013,9002	4.51 ± ?			
Anorthosite	60025			4.44 ± 0.02	4.51 ± 0.01
	67016 cl			4.56 ± 0.07	
	62236			4.36 ± 0.03	
	67435;33a cl	4.35 ± 0.05			
	67435;33b cl	4.33 ± 0.04			
Magnesian	*Troctolite* 76535	4.19 ± 0.02	4.51 ± 0.07	4.26 ± 0.06	4.27 ± ?
Suite Plutonic		4.16 ± 0.04			
Rocks		4.27 ± 0.08			
	14306,150 (?)				4.245 ± 0.075
	Dunite 72417		4.47 ± 0.10		
	Norite 14305;91 (?)				4.211 ± 0.005
	15445;17			4.46 ± 0.07	
	15445;247			4.28 ± 0.03	
	15455;228		4.49 ± 0.13	4.53 ± 0.29	
	72255		4.08 ± 0.05		
	73215;46,25	4.19 ± 0.01			
	77215		4.33 ± 0.04	4.37 ± 0.07	
	78235				4.426 ± 0.065
	78236	4.39 ± ?	4.29 ± 0.02	4.43 ± 0.05	
		4.11 ± 0.02		4.34 ± 0.04	
	Gabbro- 67667			4.18 ± 0.07	
	norite 73255c			4.23 ± 0.05	
Alkali	14066;47 (?)				4.141 ± 0.005
Rocks	14304 cl b			4.34 ± 0.08[a]	4.108 ± 0.053
	14306;60 (?)				4.20 ± 0.03
	14321;16 c				4.028 ± 0.006
	67975;131				4.339 ± 0.005
KREEP	*A15 KB* 15382	3.84 ± 0.05			
Basalt (KB)		3.85 ± 0.04	3.82 ± 0.02		
and Quartz	15386		3.86 ± 0.04	3.85 ± 0.08	
Monzogabbro	*A17 KB* 15434 particle		3.83 ± 0.05		
(QM)	72275		3.93 ± 0.04		
			4.04 ± 0.08	4.08 ± 0.07	
	QM 15405,57				4.297 ± 0.035
	15405,145				4.309 ± 0.120
Granite and	12013		> 4.08		
Felsite	12033,507				3.883 ± 0.003
	12034,106				>3.916 ± 0.17
	14082,49				4.216 ± 0.007
	14303 cl				4.308 ± 0.003
	14311, 90				4.250 ± 0.002
	14321 B1 (cl?)	K-Ca:			4.010 ± 0.002
	14321 cl	4.060 ± 0.071	4.04 ± 0.03	4.11 ± 0.20	3.965 ± 0.025
	72215 mix		3.95 ± 0.03		
	73215,43		3.82 ± 0.05		
	73235,60				4.218 ± 0.004
	73235,63				4.320 ± 0.002
	73235,73				>4.156 ± 0.003

Notes: [1] including split number if given by author; * only given if suggestive of crystallization age; [a] disturbed and suspect; cl = clast; (?) uncertain split number

Decay constants from Steiger and Jäger (1977); for references see Papike et al. (1998); from Stöffler and Ryder (2001).

within the lunar regolith or displaced clasts residing in polymict breccias, it is not obvious what geologic unit they were excavated from and what impact crater they represent. For polymict clastic impact-breccia deposits, *the age can only be constrained to be younger than that of the youngest clast.* For ancient clastic breccia deposits produced at times when the impact rate was high, the youngest clast is likely to be very close to the assembly age. Complete or partial resetting of clasts is possible for clasts residing in impact-melt breccias. In this case, the oldest clast gives a lower limit for the age of the precursor rocks of the impact melt unit (e.g., Jessberger et al. 1977). For all types of highland rocks, a meaningful interpretation of their geologic provenance and the correlation with a time-stratigraphic unit can only be made on the basis of photogeologic models of their parent geologic formations (see Section 6).

5.2. Radiometric ages of lunar rocks

Radiogenic isotope ages of lunar rocks have been determined on numerous rocks collected at the Apollo and Luna landing sites as individual fragments or as clasts within polymict breccias, and for lithic and mineral fragments extracted from lunar meteorites. The data are from several different methods, especially Rb-Sr and Sm-Nd isochron and Ar^{40}-Ar^{39} stepwise-heating methods. The data of relevance are those which directly date or indirectly constrain the age of morphological units. In particular these are crystallization ages for volcanic rocks and impact melts. It is beyond the scope of this paper to provide a complete compilation of all available data. The reader is referred to review papers and data compilations (e.g., Dalrymple 1991; Heiken et al. 1991; Nyquist and Shih 1992; Papike et al. 1998; Nyquist et al. 2001; Snyder et al. 2000). The use of these ages in dating specific surfaces and units is discussed in Section 6.

Ages of specific plutonic and volcanic ancient highlands igneous rocks are listed in Table 5.6, which is a fairly complete list of available data, including ancient mare basalts. Most of these ages are inferred to be crystallization ages. They can rarely be used to directly date a geologic unit, but can provide a lower limit in some cases.

Crystallization and recrystallization ages for clasts in specific polymict highlands rocks, mainly impact melts and granulitic breccias, are listed in Table 5.7. This is a representative list, but not complete. Most of these samples are fine grained, leading to the dominance of Ar^{40}-Ar^{39} ages. Few of these ages directly date specific geologic units as discussed in Sections 2 and 6. Some impact-melt rocks, for example at Apollo 17, may directly date impact basins on the basis of geological relationships, in this case, Serenitatis. Clasts in breccia units constrain the ages of such units by being enclosed in them. Some of the samples have been inferred to be ejecta from large craters such as Copernicus (e.g., at Apollo 12), and reheated by those events, providing a means of dating them.

Ages for groups of mare basalts are listed in Table 5.8. Unlike Tables 5.6 and 5.7, these ages are not for specific samples but are best estimates for groups, because chemistry, isotopes, and petrography allow the definition of multiple samples as from a single event and single units. More than one basalt group, with at least some differences in age, exists at each landing site; thus geological arguments are needed to establish which represents best the age of the surface on which crater counts have been established.

A separate group of ages consists of exposure ages, dating when the surface of a sample was exposed to cosmic and solar rays. Such dating is limited to rim deposits of young Copernican craters at the Apollo landing sites, such as North Ray and South Ray Craters (Apollo 16), and Cone Crater (Apollo 14), or to the landslide and secondary cratering at the Apollo 17 site inferred to be from Tycho. Others again reflect purely local events of no great stratigraphic significance in the context of this paper.

Table 5.7. Representative radiogenic crystallization ages (Gyr) for polymict lunar highland rocks.

Sample*		Description	$^{40}Ar–^{39}Ar$	Rb–Sr
Fragmental	14064,31	KREEP melt clast	3.81 ± 0.04	
Breccias	67015,320	feldspathic melt blobs	3.90 ± 0.01	
	67015,321	VHA melt blob	3.93 ± 0.01 (K-Ar)	
Glassy	61015,90	coat	1.00 ± 0.01	
Breccias	63503 particle 1m	glass fragment	2.26 ± 0.03	
and Glass	67567,4	slaggy bomb	0.84 ± 0.03	
	67627,11	slaggy bomb	0.46 ± 0.03	
	67946,17	slaggy bomb	0.37 ± 0.04	
Crystalline	14063,215	poikilitic impact melt	3.89 ± 0.01	
Melt Breccias	14063,233	aphanitic impact melt	3.87 ± 0.01	
	14167,6,3	melt	3.82 ± 0.06	
	14167,6,7	melt	3.81 ± 0.01	
	15294,6	poikilitic, Gp. Y	3.87 ± 0.01	
	15304,7	ophitic, Gp. B	3.87 ± 0.01	
	15356,9	poikilitic, Gp. C	3.84 ± 0.01	
	15356,12	poikilitic, Gp. C	3.87 ± 0.01	
	60315,6	poikilitic	3.88 ± 0.05	
	63503 particle 1c	very high-Al?	3.93 ± 0.04	
	65015	poikilitic	3.87 ± 0.04	3.84 ± 0.02
	65785	ophitic	3.91 ± 0.02	
	72215,144	aphanite, felsite melts	3.83 ± 0.03	
	72255	aphanite, felsite melts	3.85 ± 0.04	
	72215,238b	aphanite	3.87 ± 0.02	
	73215	aphanite, felsite melts		3.84 ± 0.05
	77075,18	veinlet (Serenitatis)	3.93 ± 0.03	
	72395,96	poikilitic (Serenitatis)	3.89 ± 0.02	
	72535,7	poikilitic (Serenitatis)	3.89 ± 0.02	
	76055	magnesian, poikilitic	3.92 ± 0.05	
	76055,6	magnesian, poikilitic	3.78 ± 0.04	3.78 ± 0.04
Clast-poor	14073	subophitic 14310 group		3.80 ± 0.04
Impact Melts	14074	subophitic 14310 group	3.80 ± 0.04	
	14276	subophitic 14310 group		3.80 ± 0.04
	14310	subophitic	3.88 ± 0.05	3.79 ± 0.04
	14310	subophitic; plag	3.82 ± 0.04	
	65795	subophitic; very feldspathic		3.81 ± 0.04
	60635	subophitic 68415 group		3.75 ± 0.03
	65055	subophitic 68415 group	3.89 ± 0.02	
	67559	subophitic 68415 group		3.76 ± 0.04
	68415	subophitic	3.80 ± 0.06	3.76 ± 0.04
	68416	subophitic 68415 group		3.71 ± 0.02
Granulitic	14063,207		3.90 ± 0.02	
Breccias and	14179,11	clast	3.97 ± 0.01	
Granulites	15418,50		3.98 ± 0.06	
	67215,8		3.75 ± 0.11	
	67415		3.96 ± 0.04	
	67483,13,8		4.20 ± 0.05	
	72255,235b	clast	3.85 ± 0.02	
	77017,46		3.91 ± 0.02	
	78155		4.16 ± 0.04	
	78527		4.15 ± 0.02	
	79215		3.91 ± ?	

* including split number if given by authors

Decay constants from Steiger and Jäger (1977); for references see Papike et al. (1998); from Stöffler and Ryder (2001).

Table 5.8. Best estimates of crystallization ages of mare basalt flows at the Apollo and Luna landing sites.

Landing Site	Basalt Group	Absolute Age (Gyr)
Apollo 11	High-K basalts	**3.58 ± 0.01**
	High-Ti basalts, groups B1,3	3.70 ± 0.02
	High-Ti basalts, group B2	**3.80 ± 0.02**
	High-Ti basalts, group D	3.85 ± 0.01
Apollo 12	Olivine basalt	3.22 ± 0.04
	Pigeonite basalt	3.15 ± 0.04
	Ilmenite basalt	3.17 ± 0.02
	Feldspathic basalt	3.20 ± 0.08
Apollo 15	Ol-normative basalt	**3.30 ± 0.02**
	Qz-normative basalt	3.35 ± 0.01
	Picritic basalt	3.25 ± 0.05
	Ilmenite basalt (15388)	3.35 ± 0.04
	Green glass	~3.3 – 3.4
	Yellow glass	3.62 ± 0.07
Apollo 16	Feldspathic basalt	3.74 ± 0.05
Apollo 17	High-Ti basalt, group A	**3.75 ± 0.01**
	High-Ti basalt, group B1,2	3.70 ± 0.02
	High-Ti basalt, group C	**3.75 ± 0.07**
	High-Ti basalt, group D	3.85 ± 0.04
	Orange glass	~3.5 – 3.6
Luna 16	Aluminous basalt	3.41 ± 0.04
Luna 24	Very-low-Ti basalt (VLT)	**3.22 ± 0.02**
Lunar meteorite Asuka 881757	Basalt (gabbroic)	3.87 ± 0.06

Data compiled from various sources; see especially Snyder et al. (2000), Burgess and Turner (1998), Nyquist and Shih (1992); Dalrymple (1991), Spangler et al. (1984), and references therein. Proposed ages for surface flows (crater retention ages) are given in bold (see Table 5.10); from Stöffler and Ryder (2001).

6. GEOLOGY AND ABSOLUTE AGES OF LUNAR SURFACE UNITS

6.1. Pre-Nectarian Period

The pre-Nectarian Period as a time unit is the time span between the origin of the Moon and the formation of the Nectaris basin, which is most plausibly ~3.92 Ga old (see next section). Since the oldest plausible age of solid lunar surface material is 4.52 Ga (Lee et al. 1997; Halliday 2000), a duration of the pre-Nectarian Period of ~600 Ma is suggested. The pre-Nectarian system is recorded by (1) the impact formations of some 30 multiring basins and their ejecta deposits identified photogeologically, and (2) returned samples of rocks whose absolute ages are older than Nectaris (see Table 5.6). The suite of "plutonic" pre-Nectarian rocks comprises ferroan anorthosites, alkali anorthosites, and rocks of the magnesian suite (troctolites, norites, dunites, and gabbronorites). Clasts of aluminous mare basalts, rare clasts of impact-melt rocks, and granulitic lithologies also display pre-Nectarian ages. All these rock types document the existence of magmatic, thermal metamorphic, and impact processes throughout the pre-Nectarian Period. None of the dated pre-Nectarian rock clasts can be directly related to the geologic unit (formation) in which they formed or to any specific pre-Nectarian surface unit because they were all displaced after their formation by multiple impacts.

The relative ages of most of the pre-Nectarian multiring basins are documented on the basis of crater counts on their ejecta formations (Table 5.9). Wilhelms et al. (1987) distinguished nine age groups in which the density of craters > 20 km per 10^6 km² increases from 79 (Nectaris) to 197 (Al-Khwarizimi/King). In the Wilhelms et al. (1987) scenario, no multiring basins older than 4.2 Ga are unequivocally recorded, thus implying that the oldest basins, South Pole-Aitken and Procellarum, and some 14 obliterated basins, formed between 4.2 Ga and 4.1 Ga. However, these ages are speculative as long as the actual age of the oldest recognizable multiring basin (i.e., South Pole-Aitken) is not determined by isotope dating of returned samples.

6.2. Nectarian Period

Twelve multiring basins of Nectarian age have been identified (Wilhelms et al. 1987; Spudis 1993; Table 5.9). The superposed crater densities (craters > 20 km per 10^6 km²) on the ejecta formations of these basins range from 31 for Bailly to 79 for Nectaris. Ejecta are inferred to have been sampled at Apollo and Luna landing sites for Nectaris, Crisium, and Serenitatis (Apollo 16, Luna 20, and Apollo 15 and 17, respectively). Attempts to assign absolute ages to these basins are based on samples from these landing sites (Table 5.10).

6.2.1. Age of the Nectaris impact basin. The age of the Nectaris basin is derived from radiometric ages of Apollo 16 samples (Table 5.10). The landing site was on the Cayley Formation, representing subdued smooth "light plains," Lower Imbrian Series, sculpted by the Imbrium event (Figs. 5.26, 5.28, and 5.30), and most likely part of its discontinuous ejecta. The site is 60 km west of the Kant Plateau, which is part of the Nectaris basin rim deposits (Fig. 5.30). It is near to the hilly and furrowed Descartes Formation (Muehlberger et al. 1980), which is probably related to the Nectaris ejecta blanket. Light plains similar to the Cayley Formation are common around the Imbrium basin outside of the Fra Mauro Formation (Fig. 5.28).

The local stratigraphy of the Apollo 16 landing site defines two major superimposed formations (Ulrich et al. 1981): the older *Descartes Formation* and the younger surficial *Cayley Formation* exposed as reworked regolith at the whole landing site. The sampling took advantage of two young, fresh craters, North Ray (1 km wide, 230 m deep) and South Ray (680 m, 135 m deep), as well as Stone Mountain and the subdued plains, to obtain materials from both major formations (Fig. 5.30). North Ray Crater is inferred to have excavated rocks that are part of the continuous ejecta blanket of Nectaris (Stöffler et al. 1981, 1985; Wilhelms et al. 1987). The samples collected are dominantly friable feldspathic fragmental breccias and impact melt lithologies with variable textures and compositions from very feldspathic to mafic (aluminous basaltic). Anorthosites, mostly cataclastically brecciated, are common as both individual rocks and as clasts in polymict breccias; feldspathic granulitic rocks and breccias are common, mainly as clasts within breccias. The basin ejecta model of Haskin (Haskin et al. 2002) suggests that the Nectaris, Serenitatis, and Imbrium events each would have contributed significant ejecta deposits to the Apollo 16 site. They estimated that the last of the deposits, i.e., the one produced by the Imbrium event, would have consisted on average of sub-equal proportions of Imbrium and Serenitatis ejecta, less of Nectaris ejecta, plus pre-Nectarian substrate, with only minor contributions from Humorum, Crisium, and later, Orientale. Additional complexities and observations from terrestrial craters that bear upon this model are discussed in Section 7.

In the ejecta of North Ray Crater, highly feldspathic fragmental breccias are most abundant. Lithic clasts, both individual rock fragments of the regolith and clasts within feldspathic fragmental breccias, provide the most reliable age constraints for the Descartes Formation and hence for the age of Nectaris basin (Maurer et. 1978; Jessberger 1983; Wacker et al. 1983; Stöffler et al. 1985). Their ages range from 3.84 Ga to 4.14 Ga. Since the youngest clast determines the age of the polymict impact breccia forming the basement of North Ray Crater, an age as young as 3.85 ± 0.05 Ga has been proposed for the Nectaris basin (Stöffler et al. 1985; Table 5.10). This age may also be supported by the age distribution of lithic clasts of the

Table 5.9. Tabulation of lunar time-stratigraphic units with multi-ring basins (rock-stratigraphic units) and correlated values for the frequency of superimposed craters; modified from Wilhelms et al. (1987).

Time-stratigraphic unit	Rock-stratigraphic unit			Crater frequency
	Basin	**Diameter (km)**	**Age group**	**Number of craters > 20 km per 10^6 km²**
Pre-Nectarian System	Procellarum	3,200	1	---
	South Pole-Aitken	2,500	1	---
	Tsiolkovsky-Stark	700	2	---
	Grissom-White	600	2	---
	Insularum	600	2	---
	Marginis	580	2	---
	Flamsteed-Billy	570	2	---
	Balmer-Kapteyn	550	2	---
	Werner-Airy	500	2	---
	Pingré-Hausen	300	2	---
	Al-Khwarizimi / King	590	2	197
	Fecunditatis	990	3	---
	Australe	880	3	(> 212)
	Tranquillitatis	800	3	---
	Mutus-Vlacq	700	3	225
	Nubium	690	3	---
	Lomonosov-Fleming	620	3	177
	Ingenii	650	4	162
	Poincare	340	4	(190)
	Keeler-Heaviside	780	4	186
	Coulomb-Sarton	530	5	(145)
	Smythii	840	5	166
	Lorentz	360	6	159
	Amundsen-Ganswindt	355	7	(108)
	Schiller-Zucchius	325	7	(112)
	Planck	325	7	(110)
	Birkhoff	330	7	127
	Freundlich-Sharonov	600	8	129
	Apollo	505	9	119
	Grimaldi	430	9	(97)
Nectarian System	Nectaris	860	1	79
	Mendel-Rydberg	630	(1)	(73)
	Moscoviense	445	1	87
	Korolev	440	1	79
	Mendeleev	330	(2)	63
	Humboldtianum	700	2	62
	Humorum	820	2	56
	Crisium	1,060	2	53
	Serenitatis	740	(2)	(83)
	Hertzsprung	570	2	58
	Sikorsky-Rittenhouse	310	(2)	(27)
	Bailly	300	(2)	(31)
Lower Imbrian Series	Imbrium	1160	---	28
	Schrödinger	320	---	(20)
	Orientale	930	---	22

( ) uncertain values; note that the definition of the diameter for impact basins is controversial (values proposed by Wieczorek and Phillips (1999) are smaller than those given here).

Table 5.10. Cumulative crater frequencies, crater degradation values D_L, and absolute ages of lunar surface units derived from isotope ages of lunar rocks; data are taken from literature from Stöffler and Ryder (2001).

Formation	D_L (m)(1,2)	Crater density, normalized to ave. Mare[a] (2)	$N=10^{-3}$ craters >4 km/km² (2)	$N=10^{-4}$ craters >1 km/km² (1)	$N=10^{-4}$ craters >1 km/km² (3)	$N=10^{-6}$ craters >10 km/km² (3)	Age (Gyr) (2,3)	Age (Gyr) new set a (4)	Age (Gyr) new set b (5,6)
Ancient highlands	1150	10–36	564–677		3600	920	4.3–4.55	?	?
(older crust)	±200				±1100		4.35±0.10		
Uplands		7–30	132–564				4.0-4.4	?	?
Nectaris Basin	?	16			1200	310	4.10±0.10	3.92±0.03	3.92±0.03
					±400				3.85±0.05
A16/Descartes Fm					340±70	87	3.90±0.10	3.92±0.03	3.92±0.03
									3.85±0.05
Crisium Basin	?	?			570[#]	145[#]		3.89±0.02	3.84±0.04
Serenitatis Basin	?	?			?	?	3.98±0.05	3.89±0.01	3.87±0.03
A16/Cayley Fm.	550±50	4.0	34.7					3.85±0.02	3.77±0.02
Imbrium Apennines	350±30	3.0		250–480	?	89	3.91±0.10	3.85±0.02	
A14/Fra Mauro Fm.	350±30	2.8–3.0	47.7	250–480	370±70	94[#]	3.91±0.10	3.85±0.02	3.77±0.02
Imbrium Basin							3.91±0.10	3.85±0.02	3.77±0.02
Orientale ejecta blanket	500±100	2.50	ND	220	220±?			ND	ND
Orientale Basin								3.72–3.85?	3.72–3.77?
Oldest Mare (Nubium)	315	2.5	ND	ND	ND	ND		ND	
Mare Nectaris	ND	ND	ND	ND	ND	ND		3.74	
M. Tranq., old (A11)	390	ND?	26.2?	200	90±18	23	3.72±0.10	3.80±0.02	
M. Serenitatis (A17)	330–390	1.20		90	100±30	26[#]		3.75±0.01	
M. Tranq., young (A11)	280–390	1.39	15?	34	64±20	16	3.53±0.05	3.58±0.01	
M. Fecunditatis (L16)	240–300	0.93	15.3		33±10	8.4	3.40±0.05	3.41±0.04	
M. Imbrium (A15)	255–285	0.43	8.01	26	32±11	8.2	3.28±0.10	3.30±0.02	
M. Crisium (L24)		0.43	8.17	26	30±10	7.6	3.30±0.10	3.22±0.02	
O. Procellarum (A12)	210–215	0.72	13.6	24	36±11	9.2	3.18±0.10	3.15±0.04	
Autolycus	160–200	ND	ND	ND	ND	ND		2.1±?	
Copernicus	88–112	0.30	0.06		13±3	3.3	0.85±0.20	0.8±0.015	
Tycho, A17	ND?	0.10	0.019	ND	0.9	0.23	0.109	0.109±0.004	
					±0.18		±0.004		
Tycho	10–20							0.109±0.004	
North Ray Crater	4–5	ND	ND	ND	0.44	0.11	0.05	0.053±0.008	
					±0.11		±0.0014		
Cone Crater	ND	ND	ND	ND	0.21	0.05	0.026	0.025±0.012	
					±0.05		±0.0008		
South Ray Crater								0.002±0.0002	
Terrestrial craters (Phanerozoic)					3.6±1.1	9.2	0.375 ±0.075	0.375±0.075	

(1) = Wilhelms et al. (1987); (2) Hartmann et al. (1981); (3) Neukum and Ivanov (1994); (4) Stöffler and Ryder (2001), Ryder and Spudis (1987), Wilhelms et al. (1987); (5) Stöffler and Ryder (2001), Deutsch and Stöffler (1987), Stadermann et al. (1991); (6) previous proposals, 3.85: Stöffler et al. (1985), 3.87: Jessberger et al. (1977); # from Neukum (1983); ave. = average; ND = not determined; A = Apollo; Fm. = Formation; L = Luna; M. = Mare; O. = Oceanus; Tranq. = Tranquillitatis; a: average mare = 1.88 × 10⁻⁴ craters >4 km/km². For mare and small, young craters (lower half of table), age assignments by (4,5,6) are the same (rightmost column).

Fra Mauro Formation excavated by Cone Crater at the Apollo 14 site (Stadermann et al. 1991). Other proposed ages (Table 5.10) are 3.92 ± 0.03 Ga (Deutsch and Stöffler 1987; Wilhelms et al. 1987) and 3.95 Ga (James 1981). The data used for deriving these ages do not support older ages for Nectaris such as the 4.1 Ga age proposed by Neukum (1983) as referenced in Neukum and Ivanov (1994). In part, the arguments for ages older than 3.85 Ga reflect the 3.85 Ga age of Imbrium, which is younger than Nectaris. Wilhelms et al. (1987) suggests 3.92 Ga for Nectaris only because this age is most compatible with his assumption of a constant cratering rate in the pre-Nectarian and Nectarian time since about 4.2 Ga (with 30 multiring basins formed between 4.2 and 3.92 Ga and 12 basins formed between 3.92 and 3.85 Ga, his inferred age of Imbrium). However, an assumed constant cratering rate is not a valid age constraint and ages younger than 3.85 for Imbrium have also been proposed (see Section 6.3.1).

6.2.2. Age of the Crisium impact basin. The absolute age of the Crisium basin is tentatively inferred from radiometric ages of a few small particles from the Luna 20 regolith (Wilhelms et al. 1987; Spudis 1993), collected from ejecta deposits of Crisium. Luna 20 landed on the southern rim deposits of the Nectarian Crisium basin (Fig. 5.30), about 35 km north of the mare plains of Fecunditatis (Vinogradov 1973). A core of ~50 g of fine-grained light gray regolith was collected by drilling analogous to Luna 16. Most of the rock fragments are feldspathic granulites, although the bulk soil is somewhat less aluminous than Apollo 16 soil or the highlands meteorites.

Most of the dated fragments are feldspathic, KREEP-poor impact-melt rocks not unlike some of the characteristic melt rocks at the Apollo 16 landing site, although one dated sample (22007,1; 3.87 Ga; Podosek et al. 1973) is similar to the more KREEP-rich, Apollo 17 crystalline melt rocks, which are interpreted as Serenitatis impact melt. One sample of the KREEP-poor impact melt lithology (22023,3,F) was dated at 3.895 ± 0.017 Ga (Swindle et al. 1991) which is proposed as a consistent age for the Crisium basin. Wilhelms et al. (1987) suggested an age of 3.84 ± 0.04 Ga for Crisium (Table 5.10). It remains uncertain whether any of the dated lithic clasts represent Crisium melt or even the youngest clasts of the continuous deposits of Crisium. Its actual age could be younger than 3.89 Ga (see Table 5.10) and nearly as young as the next younger dated basin (Serenitatis).

The relative ages of the Crisium and Serenitatis basins are not definitely clear. The crater density value for superimposed craters > 20 km per 10^6 km² is higher for Serenitatis (83?) than for Crisium (53) although it is based on very poor statistics (Table 5.9; Fig. 5.26), and Serenitatis has been extremely modified by Imbrium. Wilhelms et al. (1987) argues on the basis of superposition and morphology characteristics that Serenitatis is younger than Crisium. This would set an age of 3.87 ± 0.012 Ga, the proposed age for Serenitatis (see below), as the lower limit for the age of Crisium.

6.2.3. Age of the Serenitatis impact basin. The Apollo 15 and 17 landing sites are close to (though just outside of) the main rim of the Serenitatis basin, thus samples from both sites appeared to be suitable for dating the Serenitatis event (Fig. 5.30). Apollo 17 samples were collected from massifs of the Taurus-Littrow region, which are part of the eastern main rim of Serenitatis. The massifs, rising to 2 km above the floor, are dominantly of Serenitatis origin (and therefore Nectarian), and consist of autochthonous and/or allochthonous pre-Serenitatis material. This region is relatively undisturbed and only slightly modified by deposits of younger basins (Wilhelms et al. 1987; Spudis 1993). In contrast, the Imbrium basin-forming event destroyed and buried the western rim formations of Serenitatis.

Most of the sampled boulders, which are most likely derived from the massifs, and large rocks are impact-melt breccias. The most common type of melt breccia is a mafic poikilitic variety. One boulder is composed of an aphanitic, chemically more diverse melt breccia. Fragments of old igneous rocks (dunites, norites etc.) are present as clasts in these melt

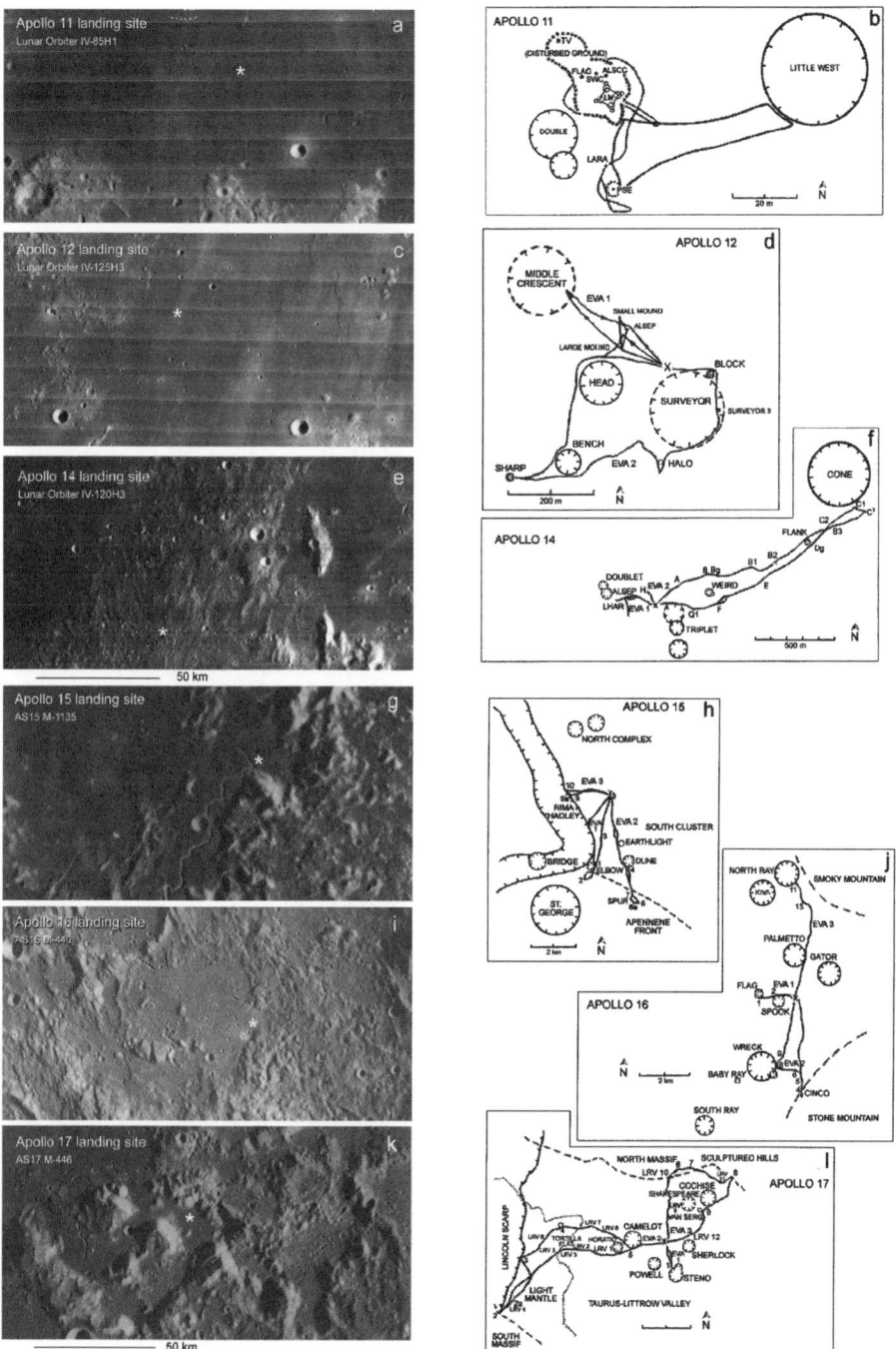

Figure 5.30. (a) Apollo 11 landing area, Mare Tranquillitatis, (b) Map and sampling traverses at the Apollo 11 landing site, (c) Apollo 12 landing area, Oceanus Procellarum, (d) Map and sampling traverses at the Apollo 12 landing site, (e) Apollo 14 landing area, Fra Mauro Formation, (f) Map and sampling traverses at the Apollo 14 landing site, (g) Apollo 15 landing area, Palus Putredinis, Mare Imbrium, and Hadley Delta,

caption and figure continued on facing page

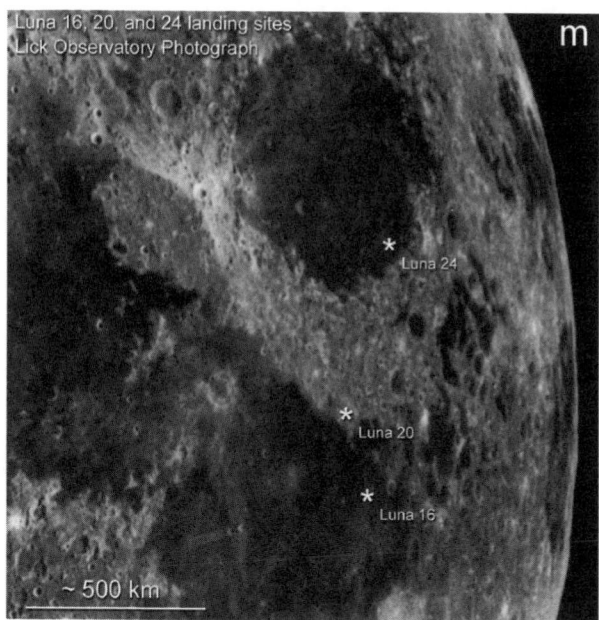

Figure 5.30 (*continued from facing page*). (h) Map and sampling traverses at the Apollo 15 landing site, (i) Apollo 16 landing area, Descartes region, (j) Map and sampling traverses at the Apollo 16 landing site, (k) Apollo 17 landing area, Taurus-Littrow region, Mare Serenitatis, (l) Map and sampling traverses at the Apollo 17 landing site, (m) Landing areas of the Luna 16, 20, and 24 missions, Mare Crisium and Mare Fecunditatis.

breccias. Abundant small fragments of feldspathic granulite suggest that much of the massif material is composed of this lithology (see also Jolliff et al. 1996). The most widespread lithology is represented by poikilitic, fragment-laden impact melt of uniform composition. This is generally inferred to be Serenitatis melt (e.g, Spudis and Ryder 1981). It provides a tightly constrained age of 3.893 ± 0.009 Ga (Dalrymple and Ryder 1996 and references therein; Table 5.10). One boulder and a few smaller fragments from the South Massif are aphanitic fragment-laden impact melts. These have a chemical composition distinct from the poikilitic melts, and they are more varied in both chemistry and fragment population. Inferred ages range between 3.86 and 3.93 Ga, but are on average, younger than those of the poikilitic rocks. These rocks might be a variant of Serenitatis melt, or even from the Imbrium event, although this would be in conflict with the 3.77 ± 0.02 Ga age of Imbrium for which arguments are presented below. Previously proposed ages of 3.86 ± 0.04 or 3.87 ± 0.03 Ga for Serenitatis are based on these younger ages (Jessberger et al. 1977; Wilhelms et al. 1987; Deutsch and Stöffler 1987; Table 5.10). Although the 3.89 Ga age of Serenitatis is well constrained, it remains open whether Serenitatis is indeed 3.89 ± 0.1 or 3.87 Ga because of the uncertainties in assigning unequivocally either one of the two types of impact melt to the Serenitatis event.

6.3. Early Imbrian Epoch

6.3.1. Age of the Imbrium impact basin. Imbrium basin deposits have been sampled at three Apollo landing sites (Apollo 14, 15, and 16) where different facies of Imbrium ejecta were deposited as indicated by photogeological interpretations (Wilhelms et al. 1987; Spudis 1993) and by cratering models (Oberbeck 1975; Schultz and Merrill 1981; Melosh 1989). *Apollo 15* sampled ejecta deposits (most likely including impact melt) at the main rim of the

Imbrium basin, *Apollo 14* sampled polymict breccias and impact-melt rocks of the continuous ejecta blanket (Fra Mauro Formation), and *Apollo 16* sampled a zone of distal discontinuous ejecta (polymict breccias and impact-melt rocks of the Cayley Formation).

Two major proposals for the age of Imbrium have been published in recent years after an age of 3.85–3.90 Ga had been generally accepted before 1980. The ages proposed more recently (Table 5.10) are 3.85 ± 0.02 Ga (Wilhelms et al. 1987; Ryder 1990a, 1994; Spudis 1993; Hartmann et al. 2000) and 3.77 ± 0.02 Ga (Deutsch and Stöffler 1987; Stadermann et al. 1991). The originally accepted age was based mainly on the measured ages of lithologies that were in some way supposedly reset by the Imbrium event and that displayed a peak of their frequency distribution within the 3.85–3.90 Ga age range (e.g., Taylor 1975). This approach is incorrect in light of the foregoing discussion, yet continues to exist. Wilhelms et al. (1987) stated: "The time of the Imbrium impact seems to be well constrained at from 3.82 to 3.87 Ga; the average and well represented age of 3.85 ± 0.03 Ga is tentatively adopted here." This incorrect view persists; for example, Hiesinger et al. (2003) referred to an age of the Imbrium basin as old as 3.91 ± 0.1 Ga. The more recent proposals discussed herein are not based on "histogram" approaches, but on age constraints that apply to relevant geological units.

Arguments for a 3.85 ± 0.02 Ga age. The continuous Imbrium ejecta (Fra Mauro Formation) was directly sampled at the Apollo 14 site. Sampling was from both Cone Crater ejecta and the smooth terrain (Fig. 5.30), both representing the Fra Mauro Formation. At the rim of Cone Crater, feldspathic fragmental breccias were sampled. Melt fragments within these breccias, have a range of ages from about 3.95 to 3.85 Ga (a few older fragments are not impact melts) (Stadermann et al. 1991 and others). Samples collected outside of the Cone Crater ejecta blanket include melt samples with younger ages, down to nearly 3.7 Ga (especially belonging to a single chemical group exemplified by 14310). However, these are not necessarily from the Fra Mauro Formation (a view which is questioned by Deutsch and Stöffler 1987). The Cone Crater samples suggest an age for the Imbrium ejecta blanket of 3.85 ± 0.02 Ga.

Melt samples at the Apennine Front (Apollo 15, Fig. 5.30) must be dominantly pre-Imbrian or contemporaneous with it, as no major impact events later affected the site. Dalrymple and Ryder (1993) obtained chronological data on the range of Apollo 15 impact melts defined by Ryder and Spudis (1987). All but one of the datable samples gave ages around 3.86–3.88 Ga, the other one gave an age of 3.84 ± 0.02 Ga. These data suggest an age for the ejecta blanket of 3.85 ± 0.02 Ga. The Cayley Plains at Apollo 16 are less definitive but nearly all of the impact melts must pre-date Imbrium and nearly all have ages greater than about 3.86 Ga. The main exception is a significant group with a composition similar to local regolith, to be described in the next section. Thus the Apollo 16 data are consistent with that from Apollo 15 as an upper limit on the age of Imbrium.

The Apennine Bench Formation (Hackmann 1966) has the physical features of a volcanic unit. Gamma-ray orbital data (Apollo 15 and Lunar Prospector missions) show that the unit has thorium abundances identical with those of the volcanic KREEP basalts found as small fragments and a common regolith constituent at the Apollo 15 landing site (Hawke and Head 1978; Spudis 1978; Ryder 1987). These volcanic rocks have a well-defined age of 3.85 ± 0.02 Ga, indistinguishable from the upper limit for Imbrium defined by its ejecta. Thus both the upper and lower absolute age limits for Imbrium are the same, establishing the Imbrium basin as 3.85 ± 0.02 Ga (Table 5.10).

Arguments for a 3.77 ± 0.02 Ga age. This age has been derived from detailed Consortium studies of the Apollo 14 and 16 highland breccia samples (e.g., Stöffler et al. 1981, 1985, 1989; Stadermann et al. 1991 and references therein). The main arguments for the 3.77 Ga age are given in Deutsch and Stöffler (1987) and supplemented by Stadermann et al. (1991). The youngest lithic clast of the basement breccias of the Apollo 14 and 16 sites, representing the Imbrium related Fra Mauro and Cayley Formations, respectively, must provide the age of the parent basin. At both sites, there are "young crystalline impact melt rocks" ranging in age from 3.71 ± 0.03 Ga to 3.81 ± 0.01 Ga (Deutsch and Stöffler 1987). The 3.77 ± 0.02 Ga age is mainly based on the group of anorthositic-noritic melt rocks (3 Apollo 16 melt rocks clustering at 3.75 ± 0.01 Ga) and on the group of youngest Apollo 14 melt rocks, which are chemically distinct from them. The age of 3.77 ± 0.02 Ga is covered by the age uncertainties of the two groups of subophitic melt rocks (Deutsch and Stöffler 1987).

The Apollo 14 and 16 samples younger than 3.82 Ga belong to different textural and chemical groups and range in size from the cm- to the m-scale (e.g., boulder 68415/416). The subophitic samples (e.g., 14310 and 68415/416) represent clast-free, relatively coarse-grained and therefore slowly cooled impact-melt rocks, particularly critical for the arguments against a post-Imbrian origin of these rocks. According to arguments given in Deutsch and Stöffler (1987), these rocks are considered to originate from large pre-Imbrian impact crater formations (melt sheets and polymict breccia deposits); they cannot be derived from erratic clasts ejected from local or distant post-Imbrium craters.

Deutsch and Stöffler (1987) questioned both the argument that the Apennine Bench Formation is younger than Imbrium and that it is composed of the same type of KREEP basalts that occur as clasts at the Apollo 15 landing site dated at 3.85 ± 0.05 Ga (e.g., Carlson and Lugmair 1979). There is no direct geologic evidence that the volcanic "light plains" of the Apennine Bench extend to the Apollo 15 site forming the substratum of the mare basalts and covering Imbrium ejecta (Fig. 7.13 in Spudis 1993) because these assumed relationships are not exposed at the Apollo 15 site. Deutsch and Stöffler (1987) argue therefore, that the Apennine Bench Formation is pre-Imbrian in age and formed on top of an older terra unit that assumed its present position between the inner ring and the main rim of the Imbrium multiring basin as a parautochthonous megablock of the pre-impact target not completely flooded by mare basalt flows.

6.3.2. Age of the Orientale impact basin. Orientale is the youngest of the multiring basins on the Moon (Wilhelms et al. 1987; Spudis 1993), but its absolute age cannot be determined directly from measured ages because samples related to Orientale have not been identified with any certainty at any of the landing sites. This is not surprising because only ray material could be present at the landing sites and would be difficult to identify in the sample collections. Wilhelms et al. (1987) contended that Orientale must have been formed between 3.85 and 3.72 Ga, assuming that 3.85 Ga is the age of Imbrium and 3.72 Ga is a lower limit set by the oldest age of nearby exposed mare basalts of Upper Imbrian age. Based on relative crater densities of these basins, he proposed a tentative age of 3.8 Ga for Orientale. However, it could be almost as old as Imbrium, i.e., 3.84 Ga. Based on the proposed age of 3.77 ± 0.02 Ga for Imbrium (see above) and on the relative crater densities, Orientale should be equal to or younger than 3.75 Ga and could be as young as 3.72 Ga (Table 5.10).

6.4. Late Imbrian Epoch

6.4.1. Age of Apollo 17 basalt surfaces (3.70 - 3.75 Ga). Apollo 17 landed on mare plains of the same intermediate-age group of the Upper Imbrian Series that occupies northern Mare Tranquillitatis near Apollo 11 (Fig. 5.29). The site is located in a mare-flooded valley, a radial graben in the massifs that form a main topographic rim of the Serenitatis basin (Fig. 5.30). The subfloor basalt at the landing site is about 1.4 km thick. Much of the surface of massifs and mare in the area is covered with a "dark mantling material," correlated with volcanic orange glass sampled at the site. The samples collected on the valley floor near numerous fresh clustered craters are dominantly mare basalts and some regolith breccias, and dark mantle material was sampled as orange glass deposits.

The Apollo 17 mare basalt samples collected over a wide area of several kilometers are high-Ti basalt. They fall into distinct chemical groups (Table 5.8) that represent at least four distinct extrusions (Warner et al. 1979; Neal et al. 1990; Ryder 1990b). Most of the samples are group A (3.75 Ga) or the more complex group B (3.70 Ga). Group C (~3.75 Ga) samples have been identified only among the few Shorty Crater samples, and group D (~3.85 Ga ?) only by one sample from the Van Serg regolith core. Samples from boulders at the rim of the 650 m diameter Camelot Crater (Station 5) presumably represent the deepest excavated basalt, perhaps 100 m; and all belong to group A basalts which occur at all mare sampling locations. Samples of boulders 150 m from the rim of 600 m diameter Steno Crater (Station 1) presumably represent a shallower level and belong to group B basalts. Group B basalts are found throughout the mare sampling locations except Shorty Crater. These relationships suggest that group A basalts underlie group B basalts, consistent with their radiometrically determined ages (Table 5.8). Wolfe et al. (1981) suggested that Group C basalts, dominating the ejecta of the small Shorty Crater, were the youngest, but radiometric ages show that they are older than Group B and similar in age to Group A. Thus the youngest basalts, which flood at least the eastern end of the Taurus-Littrow valley, are the group B basalts (3.70 Ga). However, to the west the covering by group B basalts may be patchy leaving group C and A basalts as the topmost bedrock (3.75 Ga).

Geologic relationships at the Apollo 17 site make relating a radiometric age to a crater density or crater degradation parameter an uncertain task. These include the presence of dark mantle deposits, which appear to be correlated with the sampled orange volcanic glass, with a preferred age of ~3.5 Ga (Tera and Wasserburg 1976); the apparently patchy distribution of the lava flows; and the considerable obscuration of the older cratering history by the production of the central cluster of craters (Lucchitta and Sanchez 1975) at about 110 Ma. However, it seems likely that the mare plain at least to the immediate east of the landing site consists of lava flows with an age of 3.70 Ga, while those extending out into Mare Serenitatis and Mare Tranquillitatis might be slightly older. It is unlikely that the oldest sampled basalts, group D, form any extensive surface in the region. We infer that an age of 3.75 Ga probably best represents the crater densities measured in basalts just inside the southeast rim of Serenitatis (Table 5.10).

6.4.2. Age of Apollo 11 basalt surfaces (3.58 Ga and 3.80 Ga). The landing site, 40 km north-northeast of the nearest highlands region at the Kant Plateau (Fig. 5.30), is on intermediate-age-group basalts of the Upper Imbrian Series, the southern of two belts separated by the youngest-age group (Fig. 5.29). Three patchy units of mare basalt of different age are in the area within at least several tens of kilometers of the landing site (Grolier 1970a,b). For details of the landing site geology, see Heiken et al. (1991) and Stöffler and Ryder (2001).

The mare basalt samples collected from approximately 400 m west of a sharp-rimmed, rayed crater approximately 180 m in diameter and 30 m deep (West Crater) (Fig. 5.30) are all high-Ti varieties. Beaty and Albee (1978) suggested that most of the samples collected were ejected from West Crater. The samples have a range of compositions and ages that represent at least four separately extruded basalt types (Table 5.8). Group A (3.58 Ga), the high-K basalt, is most abundant. Group B1-B3 (3.70 Ga), a complex group, comprises most of the rest of the samples, while the two oldest groups B2 (3.80 Ga) and D (3.85 Ga) are comparatively minor. Exposure data (Geiss et al. 1977) indicate that the group A samples came from a surface exposure, and that the low-K basalts (groups B and D) came from a shielded site, most excavated in a single impact (possibly West Crater, which is only 30 m deep, but possibly from much further away).

Galileo and Clementine spectral reflectance data (Staid et al. 1996) indicate that the landing site lies in, but close to the edge of, a western unit that is both the youngest and the highest in Ti in Tranquillitatis, which they correlate with the group A basalts. A much more extensive nearby unit identified spectrally is older and extends a coherent surface as far north as the Apollo 17 landing site, consistent with this unit being the group B1-3 basalts, which are similar in both age and composition to Apollo 17 basalts. Even older basalts identified spectrally as a little lower in Ti may correspond with group B2 or D (or both). Of the crater density units referred to by Wilhelms et al. (1987) and Neukum and co-workers (e.g., Neukum and Ivanov 1994), we infer that the young one is the 3.58 Ga group A basalts, and the older one is group B2 or D, which is about 3.80 Ga. For a more detailed discussion see Stöffler and Ryder (2001).

6.4.3. Age of Luna 16 basalt surface (3.41 Ga).

Luna 16 landed on mare lavas that flood the 690 km diameter Fecunditatis basin of pre-Nectarian age, about 400 km south of highlands formed by the ejecta blanket of Crisium basin (Fig. 5.30). The mare floods are thin; probably slightly thicker than a kilometer in the center and about 300 m at the landing site (De Hon and Waskom 1976). The basalt plains are in the middle to upper part of the Upper Imbrian System (Fig. 5.29). The sampling site is on a dark unit whose spectral class indicates a higher Ti content than most of Mare Fecunditatis. Presumably this is the unit to which the crater density listed in the Table of Neukum and Ivanov (1994) corresponds.

The samples consist of 101 g of dark gray regolith, obtained by drilling to a depth of 35 cm (Vinogradov 1971). Most of the few particles >3 mm are of feldspathic mare basalt or minerals derived from them; others are glassy agglutinates and regolith breccias. A small amount of feldspathic highland material is present (e.g., Keil et al. 1972). The tiny mare basalt fragments available from the Luna 16 regolith appear to be mainly a coherent chemical group that is more aluminous than typical mare basalts and with intermediate Ti contents (4-5% TiO_2) (Grieve et al. 1972; Keil et al. 1972; Kurat et al. 1976; Ma et al. 1979); they probably represent a single flow or related flows. The basalt fragments are all fine-grained, suggesting either a thin flow or a series of similar, overlapping thin flows. A Rb-Sr isochron and a ^{40}Ar-^{39}Ar age on a single fragment are consistent with an age of 3.41 Ga for this basalt group (Table 5.8; Huneke et al. 1972; Papanastassiou and Wasserburg 1972). Two separate fragments (3.45 ± 0.06 Ga and 3.30 ± 0.15 Ga) are consistent with this age (Cadogan and Turner 1977). The regolith composition (Reid et al. 1972) indicates that the age of this group of basalts is representative of the mare surface at the Luna 16 landing site.

6.4.4. Age of Apollo 15 basalt surface (3.30 Ga).

The site is located on a mare plain of the youngest group of the Upper Imbrian Series (Fig. 5.26), about 2 km from Hadley Rille, whose walls expose a layered mare basalt sequence (Fig. 5.30). The maria flood an embayment in the Apennine front, a scarp that is the main rim crest of the Imbrium basin. Extensive lava plains occur to the west of the landing site. Sampling was in the mare plains near the Hadley rille edge and in an area called South Cluster (Fig. 5.30). Samples from the plains are mainly mare basalts and regolith breccias.

The mare basalt samples collected from the mare plains on the Apollo 15 missions are dominated by two low-Ti varieties, the olivine-normative mare basalts (3.30 Ga) and the quartz-normative mare basalts (Ryder and Schuraytz 2001), which are 3.35 Ga old (Table 5.8). The other rare mare basalt fragments are of a similar age, but were found as exotic fragments on the Apennine Front. Various types of volcanic glasses (~3.3–3.6 Ga, Spangler et al. 1984) occur only locally or dispersed in the regolith. Stratigraphically the olivine-normative mare basalts appear to be the highest and are dominant among small rock samples (compilation in Ryder 1985). This is consistent with their younger radiometric ages.

The chemical composition of the Apollo 15 mare regolith samples demonstrates the domination by the olivine-normative mare basalt (Korotev 1987), even at Dune Crater. The difference in ages of all the Apollo 15 mare basalt types (and probably the glass as well) is in any case so small and the total thickness so great that the crater density measured for this part of Palus Putredinis can be ascribed to an age of 3.30 Ga with confidence (Table 5.10).

6.4.5. Age of Luna 24 basalt surface (3.22 Ga).

Luna 24 landed on the Upper Imbrian mare plains that flood the Crisium basin (Fig. 5.30), about 40 km north of the basin rim (Butler and Morrison 1977; Florensky et al. 1977). Mare Crisium is fairly uniform but three successive main units have been mapped by Head et al. (1978a). Luna 24 landed on the upper part of the middle age group that is common in the northern part of the basin but also forms exposed patches in the south. Various lines of evidence suggest that the basalts at the site are about 1–2 km thick. The Luna 24 core (12 mm diameter) was 160 cm long and ~200 cm deep (Barsukov 1977; Florensky et al. 1977). Some larger particles, up to 10 mm in size, are basaltic rocks ranging from very-low-Ti basalt to olivine basalt. Other fragments include glasses, breccias, and agglutinates. Most of the basaltic fragments and at least a large proportion of the coarser mineral fragments from all levels of the Luna 24 regolith core represent a distinct very-low-Ti aluminous mare basalt type (Ryder and Marvin 1978; Taylor et al. 1978; Graham and Hutchison 1980). Metabasalts, impact melts, and glasses have the same composition, indicating that it is a dominant component of the regolith, although other lithic types are present.

The available ages, which appear to all be on low-Ti mare basalts and metabasalts, show a rather narrow range around 3.22 Ga (Burgess and Turner 1998). That the metabasalts have the same ages suggest that they are metamorphosed flow margins and that the basalts consist of a sequence of overlapping flows of similar composition. It is possible that a slightly younger age of 2.93 Ga for one particle should be considered more reliable. Nonetheless it would appear that the basalt particles are dominated by a single component with an age of 3.22 Ga (Table 5.8).

The bulk regolith is very similar in chemical composition for both major and minor elements to that of the very-low-Ti basalts, indicating that these basalts are the surface unit at the landing site. This is inconsistent with the remote-sensing data that show that surfaces with such low Ti do not exist within tens of kilometers of the nominal landing site (Blewett et al. 1997). This implies that Luna 24 did not land where it was reported to have landed, or that the basalts collected are representative of only a very small area surrounded by basalts with higher Ti that were not collected and thus not dated. However, according to Wilhelms et al. (1987), Mare Crisium is stratigraphically among the most uniform, and therefore we consider the 3.22 Ga age to be correlated with the typical crater density of the southern Mare Crisium (Table 5.10).

6.5. Eratosthenian Period

6.5.1. Age of Apollo 12 basalt surface (3.15 Ga). The Apollo 12 site is in a region of mare basalts of a younger age (Eratosthenian) and spectral type different from those at the Apollo 11 site (Fig. 5.30). Highland islands within about 15 km suggest that the basalts are quite thin, and the area is topographically complex. Nearly all of the sampled terrain is dominated by ejecta of several craters larger than 100 m (Fig. 5.30). The site is close to the rim of the 300 m diameter Surveyor Crater. The rock samples are mainly mare basalts, with some regolith breccias.

On the basis of chemical and isotopic characteristics, the collection of more than 40 mare basalt rocks from the Apollo 12 landing site represent three numerically subequal groups (olivine basalts, pigeonite basalts, and ilmenite basalts) and a single fragment of a fourth group (feldspathic basalt) (Neal et al. 1994). The ilmenite and pigeonite basalt groups have very similar ages (3.15–3.17 Ga), with the olivine basalts and the feldspathic basalt being perhaps slightly older (3.22 Ga; Table 5.8). This is consistent with stratigraphic relationships, where the ilmenite basalts are the only type found around the smaller craters, and the pigeonite and olivine basalts required excavation from larger craters (Surveyor, 200 m diameter and Middle Crescent, 400 m diameter). This would indicate that the ilmenite basalt is about 40 m thick (Rhodes et al. 1977). Details of the sample provenance and the relationships with the remotely-sensed data are discussed in Stöffler and Ryder (2001). They note some inconsistency in the use of crater density in plots of this site, even by the same author group. The crater densities are reported to be lower than those of the Apollo 15 site in some papers and higher in others (Neukum et al. 1975a; Neukum and Wise 1976; Neukum 1977; Wilhelms et al. 1987). The crater density data for Apollo 12 (e.g., Neukum et al. 1975a) show an unusual kink at the critical point around 1–2 km sizes, deviating from a standard calibration. Possibly some secondary craters have not been identified and the actual count is indeed lower for Apollo 12 than for Apollo 15. We suggest that the interpolated lower count correlates with the surface basalt age of 3.15 Ga (Table 5.10).

6.6. Copernican Period

6.6.1. Age of Autolycus and Aristillus (2.1 Ga). Early geologic analysis of the Apollo 15 landing site showed that a ray from either of the rayed craters Aristillus or the older but nearer Autolycus crossed the landing site and deposited exotic material. KREEP basalt fragments with an original crystallization age of ~3.84 Ga collected at the landing site were shocked and thermally heated, and in one case shock-melted, at 2.1 Ga (Ryder et al. 1991). Autolycus lies in the Apennine Bench Formation, correlated with Apollo 15 KREEP basalts (Spudis 1978) and is expected to contribute more material to the landing site than Aristillus, which would contribute mainly mare basalt fragments. Thus it is a reasonable inference that Autolycus formed at 2.1 Ga (Table 5.10). If so, and assuming that Autolycus is indeed a Copernican crater, then that Period commenced earlier than commonly assumed from the inference that Copernicus itself is less than 1 Ga (see next section). Unfortunately, and in part because Autolycus has been degraded by the later Aristillus ejecta, crater density measurements have not been made, although D_L measurements (180 ± 20) suggest that it is very close to the Copernican-Eratosthenian boundary (Wilhelms et al. 1987). Autolycus is definitely one of the most degraded Copernican craters.

Although an age of ~1.3 Ga has been suggested for the stratigraphically younger Aristillus crater on the basis of the age of a 1-m block of KREEP impact melt on the Apennine Front (sample 15405; Bernatowicz et al. 1978), this correlation is unreliable (Table 5.10). Aristillus is unlikely to have both created and ejected the melt in question to such a distance, and a delivery from Aristillus floor material by a later impact seems unlikely. It is more likely that sample 15405 is of a more local origin.

6.6.2. Age of Copernicus (0.8 Ga). The Apollo 12 mare landing site is heavily contaminated with KREEP materials, although the nearest non-mare outcrops are about 25 km away. Rays from Copernicus cross the landing site, and Meyer et al. (1971) suggested that KREEP glass in the samples was produced and ejected by the Copernicus event and thus could be used to date it. Subsequent ^{40}Ar-^{39}Ar dating of such materials suggested appreciable degassing at about 800 Ma (Eberhardt et al. 1973; Alexander et al. 1976). U,Th-Pb data also yielded an age of 850 ± 100 Ma for regolith disturbance (Silver 1971). Bogard et al. (1994) found that a granite fragment encased in KREEP glass had been almost completely degassed at 800 ± 15 Ma. These ages are all from samples 12032 and 12033, the most immature and most KREEP rich regoliths, which were probably both collected at Head Crater (Korotev et al. 2000).

The 800 ± 15 Ma age is most commonly accepted as that of Copernicus (Table 5.10). If the dated samples are from Copernicus' rays, then the age of Copernicus is well established. However, this interpretation could be wrong. Not all of the KREEP at the site can be from Copernicus, even in a concentrate in a ray, and most of it may have arrived by other means (Korotev et al. 2000, Jolliff et al. 2000a). The dated samples are all from a restricted site (Head Crater) and the KREEP glass is apparently not found elsewhere, whereas a ray as seen from orbit might be expected to more widely distribute materials. In addition, Copernicus itself does not seem to have excavated dominantly KREEP materials, although KREEP might have been an early jet phase excavating shallow material. With these caveats, then either the age of Copernicus is well-defined at 800 ± 15 Ma (Table 5.10), or it is known only to be younger than about 2 Ga.

6.6.3 Age of Tycho (0.1 Ga). The dating of the crater Tycho (98 km diameter) rests on the inference that a landslide on the slope of the South Massif (Apollo 17) was triggered by ejecta of Tycho, which is about 2200 km away. The exposure age near 0.1 Ga of landslide material then represents the age of Tycho (Wolfe et al. 1975; Arvidson et al 1976; Drozd et al. 1977; Lucchitta 1977). The "Central Cluster" craters at the Apollo 17 site also show an exposure age of about 0.1 Ga, and were interpreted as secondary craters of Tycho (Wolfe et al. 1975; Lucchitta 1977). Thus, Drozd et al. (1977) proposed an age for Tycho of 109 ± 4 Ma (Table 5.10). However, the geological evidence for the South Massif landslide and the Central Cluster craters being formed by distal ejecta from Tycho are somewhat equivocal.

6.6.4. Ages of Cone, North Ray, and South Ray Craters (25, 50, and 2 Ma). These
young Copernican craters are of prime interest among all other young craters which have been
dated on the basis of cosmic ray exposure ages because (1) samples collected from their ejecta
deposits provide the basis for the age determination of the Nectaris and the Imbrium basins
(see Sections 6.2 and 6.3) and (2) crater frequency data measured on their ejecta blankets are
available (Moore et al. 1980; Table 5.10). According to the exposure age data (Drozd et al.
1974, 1977; Stadermann et al. 1991; Eugster 1999) the ages of Cone Crater (Apollo 14 landing
site) and of North Ray and South Ray Craters (Apollo 16 landing site) are 25.1 ± 1.2 Ma, 50.3
± 0.8 Ma and 2.0 ± 0.2 Ma, respectively (Table 5.10).

6.7. Derivation of a revised, time-calibrated lunar stratigraphy

As discussed in the foregoing section, revised absolute ages can be assigned to the rock-
stratigraphic or time-stratigraphic units (Fig. 5.25) as they have been established since the early
work of Shoemaker and Hackmann (1962) and later documented in most detail by Wilhelms
et al. (1987). The scheme of Wilhelms et al. (1987) and earlier versions of it (Wilhelms 1980,
1984) have been adopted by virtually all textbooks and review articles related to the Moon
(e.g., Taylor 1982; Hartmann et al. 1984; Heiken et al. 1991; Spudis 1993). This scheme has
to be revised in terms of absolute ages only slightly as shown in Fig. 5.31. Most critical are the
absolute ages of the following stratigraphic boundaries: Pre-Nectarian-Nectarian, Nectarian-
Imbrian, and Eratosthenian-Copernican. Although we have adopted a 3.92 Ga age for the first,
a 3.85 Ga age cannot be ruled with absolute certainty. Two optional ages for the Nectarian-
Imbrian boundary have to be kept at this point (3.85 and 3.77 Ga; Stöffler and Ryder 2001).
The Eratosthenian-Copernican boundary is most variable in the literature ranging from nearly
2 Ga (Stöffler and Ryder 2001) to 1.5 Ga (Neukum and Ivanov 1994), to 1.2 Ga (Wilhelms
et al. 1987), and to less than 1.0 Ga (Stöffler and Ryder 2001). We propose for future
consideration to use the age of Copernicus itself as the boundary age: 800 Ma (Fig. 5.31). This
would of course imply that the presence of rays at impact craters is no longer a criterion for a
Copernican age. Instead, the crater density measured on the ejecta blanket of Copernicus and
calibrated by its age (800 Ma) would be the main criterion. As discussed in Section 6, there
are a number of problems with rayed craters and their lifetime as exemplified by Autolycus,
whose possible age would shift the Copernican Period very far back in time. Using Copernicus
as the boundary event would make the definition of the main stratigraphic boundaries on the
Moon more consistent since all boundaries would then be defined by major impact events (Fig.
5.31). Moreover, the age sequence 3.92, 3.89, 3.87, 3.85 or 3.77, and 3.75 Ga for Nectaris,
Crisium, Serenitatis, Imbrium, and Orientale, respectively, would be most compatible with the
differences of the crater densities for these impact basins (Wilhelms et al. 1987).

7. THE LUNAR CRATERING HISTORY

7.1. The lunar cratering rate as a function of time: The data base

The most recent determinations of the absolute ages of datable lunar surface formations
have been presented by Stöffler and Ryder (2001) and related to the available measurements of
the cumulative frequency of superimposed impact craters. The following section is essentially
based on this recent review. The age data, the crater frequencies, and the widely used
parameter of crater degradation (D_L) are summarized in Table 5.10. They are used to derive
revised calibration curves for the crater retention ages of lunar surfaces of varied age ranging
from about 4 Ga to the present (Figs. 5.32 and 5.33). The same data have also been used for
the revised calibration curve of Neukum et al. (2001). As recognized very early in the Apollo
lunar science program (e.g., Hartmann 1970; Soderblom and Lebofsky 1972), such calibration
curves are of fundamental importance for (1) determining the cratering rate in the Earth-Moon
system as a function of time, (2) establishing an absolute lunar stratigraphy, and (3) providing

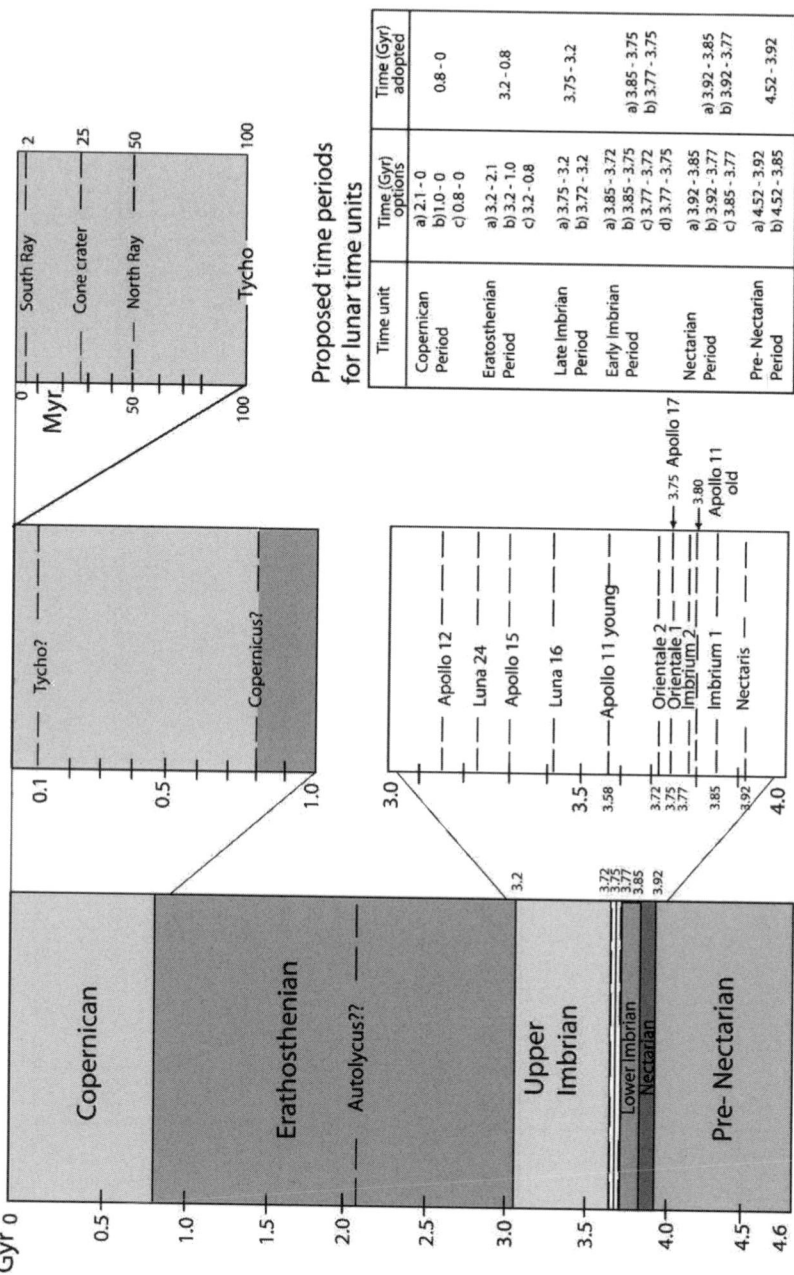

Proposed time periods for lunar time units

Time unit	Time (Gyr) options	Time (Gyr) adopted
Copernican Period	a) 2.1 - 0 b) 1.0 - 0 c) 0.8 - 0	0.8 - 0
Eratosthenian Period	a) 3.2 - 2.1 b) 3.2 - 1.0 c) 3.2 - 0.8	3.2 - 0.8
Late Imbrian Period	a) 3.75 - 3.2 b) 3.72 - 3.2	3.75 - 3.2
Early Imbrian Period	a) 3.85 - 3.72 b) 3.85 - 3.75 c) 3.77 - 3.72 d) 3.77 - 3.75	a) 3.85 - 3.75 b) 3.77 - 3.75
Nectarian Period	a) 3.92 - 3.85 b) 3.92 - 3.77 c) 3.85 - 3.77	a) 3.92 - 3.85 b) 3.92 - 3.77
Pre- Nectarian Period	a) 4.52 - 3.92 b) 4.52 - 3.85	4.52 - 3.92

Figure 5.31. Proposal for a revised time-calibrated lunar stratigraphy based on data from Stöffler and Ryder (2001) and from references therein. The proposal differs from these authors regarding the age of the Eratosthenian-Copernican boundary. The data points for the lunar landing sites refer to ages of mare basalt surfaces. The optional age of 3.85 Ga for Nectaris basin is not shown here because the 3.92 Ga age is preferred.

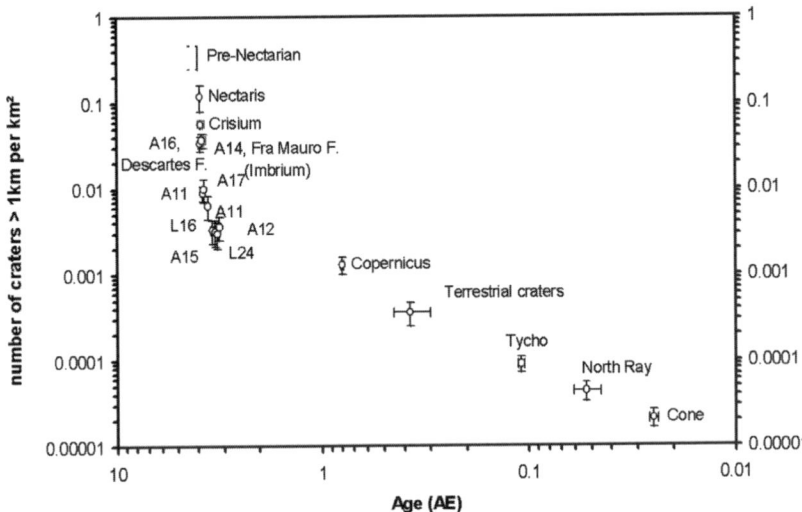

Figure 5.32. Cumulative crater frequencies for craters > 1 km per km² as a function of the age of dated lunar surfaces; data are from Table 5.10 in a log-log-plot; for some impact basins alternative ages are given according to sets a) and b) of Table 5.10 (from Stöffler and Ryder 2001).

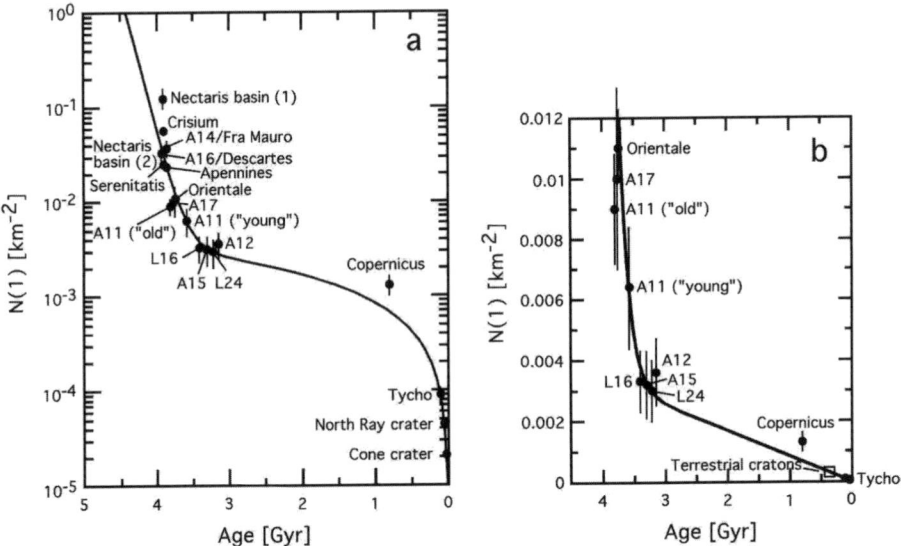

Figure 5.33. Graphical representation of Equation (5.13) (lunar cratering chronology) with data points from Table 5.10 (see also Table VI of Stöffler and Ryder 2001); N-values for Serenitatis and Apennines are from G. Neukum (pers. comm. of unpub. data) and are not contained in Table 5.10; a) logarithmic scale for N, b) part of the lunar N/T function with linear scale for N. Data for terrestrial craters are shown assuming a constant cratering rate (see Fig. 5.23) with error bars of factor of $2^{1/2}$.

a standard reference curve for stratigraphic time applicable for other planetary bodies of the inner solar system (cf. Section 3).

Calibration curves for the lunar cratering rate and the absolute crater retention ages have been presented previously by a number of authors (Hartmann 1972; Soderblom and Lebofsky 1972; Neukum et al. 1975b, Neukum and König 1976; Hartmann et al. 1981; Neukum and Ivanov 1994) which used another set of ages derived for specific surface areas from isotope ages of lunar samples. These curves were reproduced in a large number of reference books and textbooks such as Taylor (1982), Wilhelms et al. (1987), and Heiken et al. (1991), and have been widely and sometimes, as will be shown below, uncritically accepted by the planetary science community.

During the evaluation of the presently available database (Sections 4–6), it became evident that several problems exist with previously used age calibration curves of the lunar crater frequency data (relative crater retention ages) from which absolute crater retention ages have been derived for lunar surface units of unknown age. The problems relate to the following:

1. The definition of coeval surface units for a specific set of crater counts

2. Incorrect derivation of mare surface ages from ranges of mare basalt ages

3. The use of outdated or even incorrect absolute ages (including incorrect uncertainties) of surface units based on wrong interpretations of ages of lunar rocks

4. Incorrect ages of multiring basins including incorrect uncertainties

5. The unsubstantiated assignment of an absolute age of >4.3 Ga to terrains of oldest lunar crust ("ancient highland," "lunar uplands") displaying the highest values for measured crater frequencies

For some of these issues or open questions we present solutions or suggestions; some others remain open or at least disputable.

We propose new best estimates for *ages of mare surfaces* at the Apollo and Luna landing sites and new errors for these ages, which are lower than the errors used in previous calibration curves by the Basaltic Volcanism Study Project (Hartmann et al. 1981) and by Neukum and Ivanov (1994): 3.75 ± 0.01 Ga (Apollo 17), 3.80 ± 0.02 Ga (Apollo 11 older surface unit), 3.58 ± 0.01 Ga (Apollo 11 younger surface unit), 3.41 ± 0.04 Ga (Luna 16), 3.30 ± 0.02 Ga (Apollo 15), 3.22 ± 0.02 Ga (Luna 24), and 3.15 ± 0.04 Ga (Apollo 12). These data and the corresponding values used previously are given in Tables 5.8 and 5.10 and used for the Figures 5.32 and 5.33.

For the *ages of multiring basins* of the Nectarian and Imbrian Systems we are proposing two differing sets of data for which arguments are discussed in Section 6. They may be used in parallel for the calibration curve until better data become available (Table 5.10, Figs. 5.32 and 5.33). The differences in ages for Imbrium (3.85 Ga vs. 3.77 Ga) and the various ages proposed for Nectaris ranging from 3.92 Ga to 3.85 Ga as well as the somewhat differing ages for Crisium and Serenitatis do not have much influence on the shape of the calibration curves (Figs. 5.32 and 5.33). However, there is an effect on the curves due to the deletion of old, outdated ages and of unjustified errors (Hartmann et al. 1981; Neukum and Ivanov 1994) for the Nectaris basins (4.1 ± 0.1 Ga), the Descartes Formation (3.90 ± 0.1 Ga), the Imbrium basin, and the Fra Mauro Formation (3.91 ± 0.1 Ga). An additional important effect comes from discarding very old ages for the pre-Nectarian highlands such as the specific age of 4.35 ± 0.1 Ga for the "ancient highlands" (Neukum and Ivanov 1994) and the age ranges of 4.0–4.4 Ga and 4.35–4.55 Ga for the "most densely cratered province" and the "uplands" (Hartmann et al. 1981). For these old ages no firm geologic evidence combined with any clear isotope data basis exists.

For the *ages of Eratosthenian and Copernican craters* and the related ejecta blankets appreciable uncertainties still exist. Although some of the youngest ages are indisputable and perfectly constrained (Cone, North Ray and South Ray Craters), others are isotopically well constrained ages (radiometric and exposure ages) but their geological interpretation is uncertain or equivocal (2.1, 0.8, and 0.1 Ga for Autolycus, Copernicus, and Tycho, respectively), and the remaining age (1.3 Ga for Aristillus), little more than speculative.

The improved data base has been incorporated into the absolute age calibration for the lunar cratering rate as shown in Figures 5.32 and 5.33 by Stöffler and Ryder (2001) and by Neukum et al. (2001). These new calibration curves, which are also adopted here, affect all calibration curves for other terrestrial planets (see original curves in Hartmann et al. 1981, Figs. 8.6.1 to 8.6.5 as reprinted in Taylor 1982, p. 105, and in Neukum and Ivanov 1994, Fig. 16). The calibration curve published in the Lunar Sourcebook (Heiken et al. 1991, Fig. 4.15) contains large errors (e.g., the data points for L24, A14, and A16 are incorrect) and is therefore misleading by suggesting large uncertainties for the determination of absolute crater retention ages of lunar surfaces. For example, for a surface with 10^{-4} craters >4 km/km², the minimum and maximum values for its age differ by some 1.7 Ga (!) compared to about 0.9 Ga read from the revised calibration. For an area with 5×10^{-4} craters >4 km/km² the corresponding value drops from about 0.55 Ga to some 0.15 Ga. This problem holds similarly with the lunar standard curve of Hartmann et al. (1981, Fig. 8.6.1) but somewhat less with Neukum and Ivanov's (1994) curve because it contains the correct value for Luna 24 and smaller error bars for Apollo 16 and 14 than that in Heiken et al. (1991).

7.2. Interpretations of the lunar cratering rate and the ongoing debate about a possible late cataclysm

What are the major implications of the new calibration curve (Figs. 5.32 and 5.33, Table 5.10), which is based on Stöffler and Ryder (2001) and Neukum et al. (2001), for the lunar cratering rate as a function of time?

First, it is important to note that the data base is compatible with the analytical formula for the crater production rate originally proposed by Neukum (1983) and discussed in detail in Neukum and Ivanov (1994) and Neukum et al. (2001):

$$N(1) = 5.44 \times 10^{-14} \left[\exp(6.93\,T) - 1 \right] + 8.38 \times 10^{-4}\,T \qquad (5.13)$$

which relates the number of craters equal to and larger than 1 km in diameter per km² in an area with the crater accumulation time (crater retention age) T in Ga. Assuming a constant shape in time for the "Size-Frequency Distribution" (SFD) for the projectiles (see Section 3), Equation (5.13) is valid for any crater diameter because we assume a constant slope of the crater size-frequency distribution.

The data base presented above also demonstrates with a first order accuracy that (a) during the last 3 Ga the lunar impactor flux was relatively constant with possible variations by a factor of 2, which is in accordance with age data for young impact-melt rocks to be discussed below, and (b) before 3 Ga ago, the impactor flux ("early heavy bombardment flux") was much higher and rapidly decaying in time. Figure 5.33, for example, shows the graphical representation of Equation (5.13) and demonstrates that on approximately 4 Ga old surfaces 95% of all craters were formed between 3 and 4 Ga ago and only 5% of craters are younger than 3 Ga. The time-derivative of the number frequency - time relationship ($N - T$ function) gives an expression for the cratering rate, dN/dt. The results are shown in Fig. 5.34. One can see that the cratering rate 4 Ga ago was 500 times higher than the constant rate during the last 3 Ga in accord with earlier models by Hartmann (1970), Hartmann et al. (1984), and Neukum (1983).

The recommended calibration curve (Figs. 5.32 and 5.33) is better constrained, with small errors in the age range from about 4.0 to 3.0 Ga, corresponding to cumulative crater frequencies of about 1.5×10^{-1} craters >1 km/km² to about 2×10^{-3} craters >1 km/km². However, major uncertainties still exist for the pre-Nectarian Period (> ca. 4 Ga) and for the Eratosthenian and

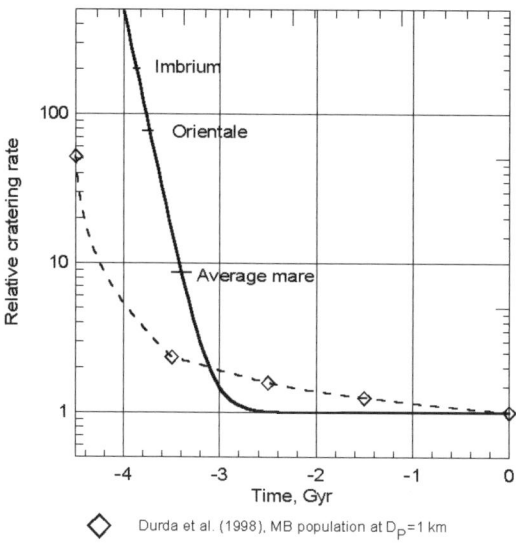

Figure 5.34. Estimate of the cratering rate as a function of time. The curve is a time derivative of Equation (5.13), normalized to the modern impact rate. Diamonds present the model by Durda et al. (1998) where the gradual decrease of the number of the Main Belt asteroids is due to the collision evolution only (no losses to planet crossing orbits).

Copernican Periods (< about 3 Ga). The steepness of the calibration curve above about 3.75 Ga, the possibility that the pre-Nectarian surfaces for which crater counts exist (see Table 5.10 and Fig. 5.26) may not be older than 4.2 Ga (Wilhelms et al. 1987), and the fact that impact melt lithologies older than 4.15 Ga are lacking indicate that the cratering rate may not increase smoothly according to the present calibration curve from 3.75 Ga up to the time of the formation of the Moon. Ryder (1990a) argued that a smooth increase would be incompatible with the accretion rates required for the size of the Moon, contrasting with opposite views (Hartmann et al. 2000). The observations at least mean that the cratering rate between 4.5 and 4.0 Ga is not known and that there is still room for speculation about a possible late lunar cataclysm (Tera et al. 1974; Ryder 1990a; Hartmann et al. 2000).

Recent age data obtained for impact-melt rock clasts of lunar meteorites (Cohen et al. 2000; Cohen 2002) have been interpreted by these authors in favor of the cataclysm hypothesis (see also Kring and Cohen 2002), referring to a peak in the age distribution around 3.9 Ga and to a lack of ages older than ~4.0 Ga. This view has been questioned by Hartmann (2003) since the ages of impact-melt rocks measured by Cohen et al. (2000) and Cohen (2000) in fact show a wide range from ~4.2 to ~0.5 Ga, in agreement with the corresponding age data from the Apollo 16 landing site (Stöffler et al. 1985; Deutsch and Stöffler 1987; see also Sections 5.2 and 6.2). Moreover, Hartmann claims that the lack of ages of lunar impact-melt rocks older than ~4.1 Ga is a consequence of the early heavy bombardment which lead to a "pulverization" of melt rocks prior to ~4.1 Ga (his model implies survival half-lives of melt rocks of <100 Ma prior to 4.1 Ga ago: "lack of old impact melts does not mean lack of old impacts"). Hartmann (2003) argued strongly against a late cataclysm ("The hypothesis of a lunar cataclysmic cratering episode between 3.8 and 3.9 Ga ago lacks proof"). He argued also against the view of Kring and Cohen (2002) that a late cataclysm affected the whole inner solar system by showing that the age data (Apollo lunar rocks, lunar meteorites, asteroidal meteorites) cannot be interpreted in this way. We believe that the review of all available data discussed in this chapter provides no solid arguments for assuming a low cratering rate early in the Moon's history and one distinct, late cataclysm (around 3.8–3.9 Ga ago), but it leaves the possibility for the existence of discrete spikes in the pre-Imbrian cratering rate, which in general appears to decline smoothly from a maximum rate existing very early in the Moon's history (see also Hartmann 2003). In fact, the impactor flux required to accrete the Earth in ~50 Ma and declining with half-lives ranging from a few Ma early in time to about 20–30 Ma later in time can be fit to the mathematical model function of Neukum et al. (2001) discussed above (Hartmann et al. 2000; Hartmann 2003).

The cratering rates for the Eratosthenian and Copernican Periods are also not sufficiently well constrained because reliable absolute ages for surfaces formed between 3 Ga and 1 Ga are conspicuously lacking. In spite of the uncertain ages of the craters Autolycus, Aristillus, and Copernicus, the tentative figures for their ages are largely compatible with a steady state and constant cratering flux since about 3 Ga, although the data for Copernicus (Figs. 5.32 and 5.33) may indicate a slightly increased flux in the past 1 Ga. The possibility of a non-constant flux has been confirmed recently by the non-uniform distribution of ^{40}Ar-^{39}Ar ages of 155 glass spherules collected from the Apollo 14 regolith (Culler et al. 2000). Culler et al. suggested that the cratering rate decreased since about 3.5 Ga by a factor of 2 to 3 to a minimum value at about 0.5–0.6 Ga and increased by a factor of 3.7 ± 1.2 in the past 0.4 Ga in accordance with data for terrestrial craters (Grieve and Shoemaker 1994) and astronomical constraints (Shoemaker et al. 1994). We argued in Section 3 that a reevaluation of the terrestrial cratering record and of the astronomical data sets does not support a recent increase of the impact flux as suggested by Culler al. (2000). The question of whether such changes in the post-Imbrian cratering flux of the Earth-Moon system are real or not (see Ryder 2000; Hörz 2000; Muller et al. 2000) has important implications for the chronostratigraphy of Mars and other terrestrial planets.

7.3. Open questions and future work

The most burning open questions concerning the time-calibrated impact rate and the absolute time scale for the lunar stratigraphy relate to the following issues:

(a) The lack of datable lunar samples from which age data could be derived for lunar highland terrains that are older than 4.0 Ga such as intensely cratered highland regions of the far side of the Moon or surface material deposited by the South Pole-Aitken impact basin.

(b) The lack of datable lunar samples from Eratosthenian and possibly Copernican mare basalt surfaces (extending to at least 1.2 Ga ago as proposed by Hiesinger et al. 2003) and from large Eratosthenian and Copernican rayed craters such as Autolycus, Copernicus, and Tycho.

(c) Continued uncertainties in the unequivocal dating of lunar multiring basins, in particular, of the Nectaris and Imbrium basins and the lack of age data for the Orientale impact basin.

(d) The existence (or not) of a late cataclysm or of spikes in the impact flux prior to 3.75 Ga as a consequence of issues (a) and (c); this problem is related to the lack of impact-melt rock samples older than about 4.2 Ga.

(e) The source and nature of projectiles impacting the Moon during the Early Heavy Bombardment and/or during discrete spikes of the impact flux, e.g., a late cataclysm.

The problem with the lack of fundamental ground truth data concerning the lunar cratering rates in the "old" pre-Nectarian and in the "younger" Eratosthenian and Copernican time periods—outlined in the issues (a) and (b)—can only be solved by new sample return missions to pre-Nectarian, Eratosthenian, and Copernican regions of the Moon. In terms of the fundamental task to improve the lunar standard reference for the cratering flux in the inner solar system, such missions should be given an equally high priority as sample return missions to Mars.

Part of the problem related to issues (c) and (d) is model dependent and part of it is due to insufficient evaluation and sometimes incorrect interpretation of lunar highland samples and of their absolute ages. This holds particularly for polymict breccias and for their geological setting and provenance. The first part of the problem concerns the lack of understanding of impact cratering mechanics of large basins, especially of the effects of ballistic sedimentation

(Oberbeck 1975) on the stratigraphy and nature of samples at the highland landing sites such as Apollo 14, 15, 16, and 17. Haskin and coworkers have reinvestigated this problem (Haskin et al. 2002, 2003) and proposed "average stratigraphies" for the relevant landing sites. However, these models do not account for important details such as the influence of pre-existing relief on the thickness distribution of the secondary mass flow, which could be much thinner at specific places than predicted by the model (see the ground truth at the Ries crater ejecta blanket, Hörz et al. 1983; Pohl et al. 1977). Such model stratigraphies are of limited use for specific local interpretations; for example, it cannot be proven that post-Imbrian craters such as North Ray Crater (Apollo 16) did not penetrate into Nectaris deposits as predicted by the model because the real, local thicknesses of the Imbrium deposits are not known. The distinct differences in the distributions of ages of clasts found in breccias at North Ray Crater rim as opposed to those in the Cayley plains (Apollo 16), and in breccias at Cone Crater rim as opposed to those of the main landing site of Apollo 14 speak to the contrary of the model assumptions (e.g., Stöffler et al. 1985; Deutsch and Stöffler 1987; Stadermann et al. 1991).

The second major problem of issue (d) is the lack of old impact melt lithologies, which has been used in favor of a late cataclysm (e.g., Ryder 1990a; Kring and Cohen 2002). The absence of old impact melts has been explained by Hartmann (2003) as an effect of the very heavy early bombardment in erasing the record of impact-melt rocks older than 4.1 Ga. We believe that multiple impact cratering cannot completely erase previously formed, old impact melt clasts because (1) resetting of ages by remelting of old melt lithologies is extremely inefficient and (2) impact brecciation typically does not lead to very fine grained material which could no longer be dated (in fact all returned highland breccias are not at all fine-grained, e.g., Heiken et al. 1991). In our view it appears more likely that the specific geological setting of the Apollo highland landing sites (14, 15, 16, and 17) lead to a very selective sampling of material dominated by Nectaris, Serenitatis, and Imbrium ejecta deposits, all being younger than 3.92 Ga, and by material from smaller craters, locally reworked by the ejecta of these three basins from craters formed shortly before those large impact events, e.g., later than 4.2 Ga ago. One important additional aspect related to the problem, which has been largely neglected so far, is that coherent melt sheets of very large craters could be differentiated similar to the terrestrial Sudbury Igneous Complex (Pye et al. 1984; Grieve et al. 1991) and be developed as clast-free impact melt lithologies with "igneous" textures that would not necessarily be recognized as impact-melt rocks. Members of the very old "Mg-suite" of plutonic rocks having ages between 4.5 and 4.2 Ga (see Table 5.6) could be old impact-melt rocks. In conclusion, data from the study of the Apollo and Luna samples cannot solve the issues of the early bombardment history of the Moon. The study of lunar meteorites is somewhat more promising but the final answers can only come from new sample-return missions.

The extremely high impact rate during the early heavy bombardment (EHB) leads to the question—outlined in issue (e)—about the nature of projectiles responsible for this effect: asteroids, comets, or left-over planetesimals? A detailed discussion of this issue is beyond the scope of this article and the reader is referred to a recent review by Hartmann et al. (2000). Clearly, the source and nature of projectiles during the EHB period and the decay of the EHB flux in time are important topics for future studies. An equally enigmatic issue relates to the source and nature of projectiles that are potentially responsible for spikes in the flux during the EHB or for a late or terminal cataclysm, if it can be proven.

8. ACKNOWLEDGMENTS

The first author is greatly indebted to the technical staff and students of the Institute of Mineralogy, Museum für Naturkunde, Berlin, for excellent assistance. In particular, thanks are due to Hwa Ja Nier, Kirsten Born, Ingo Herter, Daniel Lieger, Karin Reineck, and Heidi Wolff. The third author's (B.A.I.) participation is supported by the Humboldt Foundation, Germany.

Critical reviews by Elisabetta Pierazzo and Peter Schultz helped to improve the manuscript and are appreciated.

9. REFERENCES

Ahrens TJ, O'Keefe JD (1972) Shock melting and vaporization of lunar rocks and minerals. The Moon 4:59-94

Ahrens TJ, O'Keefe JD (1977) Equations of state and shock-wave attenuation on the Moon. *In*: Impact and Explosion Cratering. Roddy DJ, Pepin RO, Merrill RB (eds), Pergamon, p 639-656

Ahrens TJ, Takata T, O'Keefe JD, Orton GS (1994) Impact of comet Shoemaker-Levy 9 on Jupiter. Geophys Res Letters 21:1087-1090

Alexander EC Jr, Bates A, Coscio MR Jr, Dragon JC, Murthy VR, Pepin RO, Venkatesan TR (1976) K/Ar dating of lunar soils II. Proc Lunar Sci Conf 7:625-648

Anderson CE (1987) An overview of the theory of hydrocodes. Int J Impact Engng 3:33-59

Anderson JLB, Schultz PH, Heineck JT (2003) Asymmetry of ejecta flow during oblique impacts using three-dimensional particle image velocimetry. J Geophys Res 108(E8):13.1-10, doi:10.1029/2003JE002075

Artemieva NA (2001) Tektites and Martian Meteorites in Numerical Modeling of Impacts. Meteorit Planet Sci 36:A12

Artemieva NA, Shuvalov VV (2001) Motion of a fragmented meteoroid through the planetary atmosphere. J Geophys Res 106:3297-3310

Artemieva NA, Ivanov BA (2001) Numerical simulation of oblique impacts: impact melt and transient cavity size. Lunar Planet Sci XXXII:1321

Artemieva NA, Ivanov BA (2002) Ejection of Martian meteorites - can they fly? Lunar Planet Sci XXXIII:1113

Artemieva NA, Ivanov BA (2004) Launch of martian meteorites in oblique impacts. Icarus 171:84-101

Arvidson R, Drozd RJ, Guinness E, Hohenberg CM, Morgan CJ, Morrison RH, Oberbeck VR (1976) Cosmic ray exposure ages of Apollo 17 samples and the age of Tycho. Proc Lunar Sci Conf 7:2817-2832

Baldwin RB (1949) The Face of the Moon. University of Chicago Press

Baldwin RB (1981) On the tsunami theory of the origin of multi-ring basins. *In*: Multi-Ring Basins. Schultz PH, Merrill RB (eds), Pergamon, p 275-288

Barsukov VL (1977) Preliminary data for the regolith core brought to earth by automatic lunar station Luna 24. Proc Lunar Sci Conf 8:3303-3318

Bauer JF (1979) Experimental shock metamorphism of mono- and polycrystalline olivine. A comparative study. Proc Lunar Planet Sci Conf 10:2573-2596

Beaty DW, Albee AL (1978) Comparative petrology and possible genetic relations among the Apollo 11 basalts. Proc Lunar Planet Sci Conf 9:359-463

Benz W, Asphaug E (1999) Catastrophic disruptions revisited. Icarus 142:5-20

Bernatowicz TJ, Hohenberg CM, Hudson B, Kennedy BM, Podosek FA (1978) Argon ages for lunar breccias 14064 and 15405. Proc Lunar Planet Sci Conf 9:905-919

Binder A (1998) Lunar Prospector: overview. Science 281:1475-1476

Bischoff A, Stöffler D (1984) Chemical and structural changes induced by thermal annealing of shocked feldspar inclusion in impact melt rocks from Lappajärvi Crater, Finnland. Proc Lunar Planet Sci Conf 14:B645-B656

Bischoff A, Stöffler D (1992) Shock metamorphism as a fundamental process in the evolution of planetary bodies: information from meteorites. Eur J Mineral 4:707-755

Bjork RL (1961) Analysis of the formation of Meteor Crater, Arizona: a preliminary report. J Geophys Res 66:3379-3387

Bjorkman MD, Holsapple KE (1987) Velocity scaling impact melt volume. Int J Impact Eng 5:155–163

Blewett DT, Lucey PG, Hawke BR, Jolliff BL (1997) Clementine images of the lunar sample-return stations: Refinement of FeO and TiO2 mapping techniques. J Geophys Res 102:16,319-16,325

Bogard D, Hörz F, Johnson P (1987) Shock effects and argon loss in samples of the Leedey L6 chondrites experimentally shocked to 29-70 GPa pressures. Geochim Cosmochim Acta 51:2035-2044

Bogard D, Hörz F, Stöffler D (1988) Loss of radiogenic argon from shocked granitic clasts in suevite deposits from the Ries Crater. Geochim Cosmochim Acta 52:2639-2649

Bogard D, Garrison DH, Shih CY, Nyquist LE (1994) [40]Ar-[39]Ar dating of two lunar granites: The age of Copernicus. Geochim Cosmochim Acta 58:3093-3100

Bottomley RJ, York D, Grieve RAF (1990) [40]Argon-[39]Argon dating of impact craters. Proc Lunar Planet Sci Conf 20:421-431

Boyce JM, Dial AL Jr. (1975) Relative ages of flow units in Mare Imbrium and Sinus Iridum. Proc Lunar Sci Conf 6:2585-2595

Braslau D (1970) Partitioning of energy in hypervelocity impact against loose sand targets. J Geophys Res 75:3987-3999

Bridgman PW (1922) Dimensional Analysis. Yale University Press

Buckingham E (1914) On physically similar systems: Illustrations of the use of dimensional equations. Phys Rev 4:345-376

Burchell MJ, MacKay NG (1998) Crater ellipticity in hypervelocity impacts on metals. J Geophys Res 103:22,761-22,774

Burgess R, Turner G (1998) Laser ^{40}Ar-^{39}Ar age determinations of Luna 24 mare basalts. Meteorit Planet Sci 33:921-935

Butler P, Morrison DA (1977) Geology of the Luna 24 landing site. Proc Lunar Sci Conf 8:3281-3301

Cadogan PH, Turner G (1977) ^{40}Ar-^{39}Ar dating of Luna 16 and Luna 20 samples. Phil Trans Royal Soc London 284:167-177

Cameron AGW (2001) From interstellar dust to the Earth-Moon system. Meteorit Planet Sci 36:9-22

Cameron AGW, Benz W (1991) The origin of the Moon and the single impact hypothesis. Icarus 92:204-216

Campo Bagatin A, Cellino A, Davis DR, Farinella P, Paolicchi P (1994a) Wavy size distribution for collisional systems with a small-size cutoff. Planet Space Sci 42:1049-1092

Campo Bagatin A, Farinella P, Petit J-M (1994b) Fragment ejection velocities and the collisional evolution of asteroids. Planet Space Sci 42:1099-1107

Canup RM, Asphaug E (2001) Outcomes of planet-scale collisions. Lunar Planet Sci XXXII:1952

Carlson RW, Lugmair GW (1979) Sm-Nd constraints on early lunar differentiation and the evolution of KREEP. Earth Planet Sci Lett 45:123-132

Chabai AJ (1965) On scaling dimensions of craters produced by buried explosives. J Geophys Res 70:5075-5098

Chapman CR, McKinnon WB (1986) Cratering of planetary satellites. *In*: Satellites. Burns JA, Matthews MS (eds), Univ Arizona Press, p 529-533

Cintala MJ (1979) Mercurian crater rim heights and some interplanetary comparisons. Proc Lunar Planet Sci Conf 10:2635-2650

Cintala MJ (1992) Impact-induced thermal effects in the lunar and mercurian regoliths. J Geophys Res 97:947-974

Cintala MJ, Grieve RAF (1994) The effects of differential scaling of impact melt and crater dimensions on lunar and terrestrial craters: Some brief examples. *In*: Large Meteorite Impacts and Planetary Evolution. GSA Spec Paper 293:51-59

Cintala MJ, Grieve RAF (1998) Scaling impact-melt and crater dimensions: Implications for the lunar cratering record. Meteorit Planet Sci 33:889-912

Cintala MJ, Hörz F (1990) Regolith evolution in the laboratory: Scaling dissimilar comminution experiments. Meteoritics 25:27-40

Cintala MJ, Wood CA, Head JW (1977) The effects of target characteristics on fresh crater morphology: Preliminary results for the Moon and Mercury. Proc Lunar Sci Conf 8:3409-3425

Cintala MJ, Head JW, Veverka J (1979) Characteristics of the cratering process on small satellites and asteroids. Proc Lunar Planet Sci Conf 9:3803-3830

Cintala MJ, Shelfer TD, Hörz F (1999) Growth times of impact craters formed in fine-grained sand. Lunar Planet Sci XXX:1958

Cohen BA (2002) Geochemical and geochronological constraints on early lunar bombardment history. Lunar Planet Sci XXXII:1984

Cohen BA, Swindle TD, Kring DA (2000) Support for the lunar cataclysm hypothesis from lunar meteorite impact melt ages. Science 290:1754-1756

Collins GS, Melosh HJ, Morgan JV, Warner MW (2002) Hydrocode simulations of Chicxulub crater collapse and peak-ring formation. Icarus 157:24-33

Collins GS, Melosh HJ (2004) Numerical modeling of the South Pole-Aitkin impact. Lunar Planet Sci XXXV:1375

Crawford DA (1997) Comet Shoemaker-Levy 9 fragment size and mass estimates from light flux observations. Lunar Planet Sci XXVIII:267

Crawford DA, Boslough MB, Trucano TJ, Robinson AC (1994) The impact of comet Shoemaker-Levy 9 on Jupiter. Shock Waves 4:47-50

Croft SK (1978) Lunar crater volume: Interpretation by models of impact cratering and upper crustal structure. Proc Lunar Planet Sci Conf 9:3711-3733

Croft SK (1980) Cratering flow fields: Implications for the excavation and transient expansion stage of crater formation. Proc Lunar Planet Sci Conf 11:2347-2378

Croft SK (1981a) The modification stage of basin formation: Conditions of ring formation. *In*: Multi-Ring Basins. Schultz PH, Merrill RB (eds) Pergamon, p 227-257

Croft SK (1981b) The excavation stage of basin formation: A qualitative model. *In*: Multi-Ring Basins. Schultz PH, Merrill RB (eds) Pergamon, p 207-225

Croft SK (1982) A first-order estimate of shock heating and vaporization in oceanic impacts. *In*: Geological Implications of Impacts of Large Asteroids and Comets on the Earth. GSA Spec Paper 190:143-152

Croft SK (1985) The scaling of complex craters. Proc Lunar Planet Sci Conf 15, In J Geophys Res 89:C828-C842

Culler TS, Becker TA, Muller RA, Renne PR (2000) Lunar impact history from ^{40}Ar/^{39}Ar dating of glass spherules. Science 287:1785-1788

Dahl JM, Schultz PH (2001) Measurement of stress wave asymmetries in hypervelocity projectile impact experiments. Int J Impact Eng 26:145-155

Dalrymple GB (1991) The Age of the Earth. Stanford University Press

Dalrymple GB, Ryder G (1993) $^{40}Ar/^{39}Ar$ ages of Apollo 15 impact melt rocks by laser step heating and their bearing on the history of lunar basin formation. J Geophys Res (Planets) 98:13085-13095

Dalrymple GB, Ryder G (1996) $^{40}Ar/^{39}Ar$ age spectra of Apollo 17 highlands breccia samples by laser step-heating and the age of the Serenitatis basin. J Geophys Res (Planets) 101:26069-26084

Davis D, Weidenschilling SJ, Farinella P, Paolicchi P, Binzel RP (1989) Asteroid collisional history: Effects on sizes and spins. *In*: Asteroids II. Binzel RP, Gehrels T, Matthews MS (eds) Univ Arizona Press, p 805-826

De Hon RA, Waskom JD (1976) Geologic structure of the eastern mare basins. Proc Lunar Sci Conf 7:2729-2746

Dence MR, Grieve RAF, Robertson PB (1977) Terrestrial impact structures: Principal characteristics and energy considerations. *In*: Impact and Explosion Cratering. Roddy DJ, Pepin RO, Merrill RB (eds) Pergamon Press, p 247-276

Deutsch A, Schärer U (1994) Dating terrestrial impact events. Meteoritics 29:301-322

Deutsch A, Stöffler D (1987) Rb-Sr analyses of Apollo 16 melt rocks and a new age estimate for the Imbrium basin: Lunar basin chronology and the early heavy bombardment of the Moon. Geochim Cosmochim Acta 51:1951-1964

Dressler BO, Reimold WU (2001) Terrestrial impact melt rocks and glasses. Earth Sci Rev 56:205-284

Drozd RJ, Hohenberg CM, Morgan CJ, Ralston CE (1974) Cosmic-ray exposure history at the Apollo 16 and other lunar sites: Lunar surface dynamics. Geochim Cosmochim Acta 38:1625-1642

Drozd RJ, Hohenberg CM, Morgan CJ, Podosek FA, Wroge ML (1977) Cosmic-ray exposure history at Taurus-Littrow. Proc Lunar Sci Conf 8:3027-3043

Durda D, Greenberg R, Jedicke R (1998) Collisional models and scaling laws: A new interpretation of the shape of the Main-Belt asteroid distribution. Icarus 135:431-440

Eberhardt P, Geiss J, Grögler N, Stettler A (1973) How old is the crater Copernicus? The Moon 8:104-114

Eugster O (1999) Chronology of dimict breccias and the age of South Ray crater at the Apollo 16 site. Meteorit Planet Sci 34:385-391

Faure G (1986) Principles of Isotope Geology (2nd edition). Wiley

Feldman WC, Gasnault O, Maurice S, Lawrence DJ, Elphic RC, Lucey PG, Binder AB (2002) Global distribution of lunar composition: New results from Lunar Prospector. J Geophys Res 107(E3):5-1–5-14, doi:10.1029/2001JE001506

Florensky CP, Basilevsky AT, Ivanov AV, Pronin AA, Rode OD (1977) Luna 24: Geological setting of landing site and characteristics of sample core (preliminary data). Proc Lunar Sci Conf 8:3257-3279

French BM (1998) Traces of Catastrophe: A handbook of shock-metamorphic effects in terrestrial meteorite impact structures. LPI Contribution No. 954, Lunar and Planetary Institute

French BM, Short NM (1968) Shock Metamorphism of Natural Materials. Mono Book Corporation

Fritz J, Greshake A, Stöffler D (2003) Launch conditions for Martian meteorites: Plagioclase as a shock pressure barometer. Lunar Planet Science XXXIV:1335

Gault DE, Heitowit ED (1963) The partition of energy for hypervelocity impact craters formed in rock. Proc Hypervelocity Impact Symp 6:419-456

Gault DE, Wedekind JA (1978) Experimental studies of oblique impacts. Proc Lunar Planet Sci Conf 9:3843-3875

Gault DE, Quaide WL, Oberbeck VR (1968) Impact cratering mechanics and structures. *In*: Shock Metamorphism of Natural Materials. French BM, Short NM (eds) Mono Book Corporation, p 87-99

Geiss J, Eberhardt P, Grögler N, Guggisberg S, Maurer P, Settler A (1977) Absolute time scale of lunar mare formation and filling. Royal Soc London Phil Trans 285:151-158

Gilbert GK (1893) The Moons face a study of the origin of its features. Phil Soc Wash Bull 12:241-292

Graham AL, Hutchison R (1980) Mineralogy and petrology of fragments from the Luna 24 core. Royal Soc London Phil Trans A 297:15-22

Grier JA, McEwen AS, Lucey PG, Milazzo M, Strom RG (2001) Optical maturity of ejecta from large rayed lunar craters. J Geophys Res 106: 32,847-32862

Grieve RAF (1987) Terrestrial impact structure. Ann Rev Earth Planet Sci 15:245-270

Grieve RAF (1991) Terrestrial impact: the record of the rocks. Meteoritics 26:175-194

Grieve RAF, Cintala MJ (1992) An analysis of differential impact melt-crater scaling and implications for the terrestrial impact record. Meteoritics 27 526–538

Grieve RAF, Garvin JB (1984) A geometric model for excavation and modification at terrestrial simple impact craters. J Geophys Res 89:11561-11572

Grieve RAF, Shoemaker EM (1994) The record of past impacts on Earth. *In*: Hazards Due to Comets and Asteroids. Gehrels T (ed) Univ Arizona Press, p 417-462

Grieve RAF, Pilkington M (1996) The signature of terrestrial impacts. AGSO J Australian Geol Geophys 16:399-420

Grieve RAF, Therriault A (2000) Vredefort, Sudbury, Chicxulub: Three of a kind? Annual Rev Earth Planet Sci 28: 305-338

Grieve RAF, McKay GA, Weill DF (1972) Microprobe studies of three Luna 16 basalt fragments. Earth Planet Sci Lett 13:233-242

Grieve RAF, Robertson PB, Dence MR (1981) Constraints on the formation of ring impact structures based on terrestrial data. Multi-Ring Basins, Proc Lunar Planet Sci 12A:37-57

Grieve RAF, Stöffler D, Deutsch A (1991) The Sudbury Structure - Controversial or misunderstood? J Geophys Res 96: 22,753-22,764

Grieve RAF, Langenhorst F, Stöffler D (1996) Shock metamorphism of quartz in nature and experiment: II. Significance in geoscience. Meteorit Planet Sci 31:6-35

Grieve RAF, Garvin JB, Coderre JM, Rupert J (1989) Test of a geometric model for the modification stage of simple impact crater development. Meteoritics 24:83-88

Grolier MJ (1970a) Geologic map of Apollo site 2 (Apollo 11); Part of Sabine D region, southwestern Mare Tranquillitatis. USGS Map I-619 [ORB II-6 (25)], scale 125000

Grolier MJ (1970b) Geologic map of the Sabine region on the Moon, Lunar Orbiter site II P-6, southwestern Mare Tranquillitatis, including landing site 2. USGS Map I-618 [ORB II-6 (100)], scale 1100000

Hackmann RJ (1966) Geologic map of the Montes Apenninus quadrangle of the Moon. USGS Map I-463 (LAC-41), scale 11000000

Hale WS, Grieve RAF (1982) Volumetric analysis of complex lunar craters: Implications for basin ring formation. Proc Lunar Planet Sci Conf 13, In J Geophys Res 87:A65-A76

Hale WS, Head JW (1979) Central peaks in lunar craters: Morphology and morphometry. Proc Lunar Planet Sci Conf 10:2623-2633

Halliday AN (2000) Terrestrial accretion rates and the origin of the Moon. Earth Planet Sci Lett 176:17-30

Hartmann WK (1965) Terrestrial and lunar flux of meteorites in the last two billion years. Icarus 4:157-165

Hartmann WK (1966) Martian cratering. Icarus 5:565-576

Hartmann WK (1970) Lunar cratering chronology. Icarus 13:299-301

Hartmann WK (1972) Paleocratering of the Moon Review of post-Apollo data. Astrophys Space Science 17:48-64

Hartmann WK (1995) Planetary cratering I: Lunar highlands and tests of hypotheses on crater populations. Meteoritics 30:451-467

Hartmann WK (2003) Megaregolith evolution and cratering cataclysm models – Lunar cataclysm as a misconception (28 years later). Meteorit Planet Sci 38:579-593

Hartmann WK, Neukum G (2001) Crater chronology and the evolution of Mars. *In*: Chronology and Evolution of Mars. Kallenbach R, Geiss J, Hartmann WK (eds) Kluwer Dordrecht, p 165-194

Hartmann WK, Wood CA (1971) Moon: Origin and Evolution of Multi-Ring Basins. The Moon 3:3-78

Hartmann WK, Farinella P, Vokrouhlický D, Weidenschilling SJ, Morbidelli A, Marzari F, Davis DR, Ryan E (1999) Reviewing the Yarkovsky effect: New light on the delivery of stone and iron meteorites from the asteroid belt. Meteorit Planet Sci 34:161-167

Hartmann WK, Phillips RJ, Taylor GJ (1984) Origin of the Moon. Lunar & Planetary Institute

Hartmann WK, Ryder G, Dones L, Grinspoon DH (2000) The time-dependent intense bombardment of the primordial Earth-Moon system. *In*: Origin of the Earth and Moon. Righter RM, Canup R (eds) Univ Arizona Press, p 493-512

Hartmann WK, Strom RG, Weidenschilling SJ, Balsius KR, Woronow A, Dence MR, Grieve RAF, Diaz J, Chapman CR, Shoemaker EM, Jones KL (1981) Chronology of planetary volcanism by comparative studies of planetary craters. *In*: Basaltic Volcanism on the Terrestrial Planets. Pergamon Press, p 1049-1128

Haskin LA (1998) The Imbrium impact event and the thorium distribution at the lunar highlands surface. J Geophys Res 103:1679-1689

Haskin LA, Korotev RL, Gillis JJ, Jolliff BL (2002) Stratigraphies of Apollo and Luna highland landing sites and provenances of materials from the perspective of basin impact ejecta modeling. Lunar Planet Sci XXXIII:1364

Haskin LA, Moss BE, McKinnon WB (2003) On estimating contributions of basin ejecta to regolith deposits at lunar sites. Meteorit Planet Sci 38:13-33

Hawke BR, Head JW (1977) Impact melt on lunar crater rims. *In*: Impact Explosion Cratering. Roddy DJ, Pepin RO, Merrill RB (eds) Pergamon, p 815-841

Hawke BR, Head JW (1978) Lunar KREEP volcanism: Geologic evidence for history and mode of emplacement. Proc Lunar Planet Sci Conf 9:3285-3309

Hayhurst CJ, Ranson HJ, Gardner DJ, Birnman NK (1995) Modelling of microparticle hypervelocity oblique impacts on thick targets. Int J Impact Eng 17:375-386

Head JN, Melosh HJ, Ivanov BA (2002) Martian meteorite launch: High-speed ejecta from small craters. Science 298:1752-1756

Head JW (1976) The significance of substrate characteristics in determining morphology and morphometry of lunar craters. Proc Lunar Sci Conf 7:2913-2929

Head JW (1977) Origin of outer rings in lunar multi-ring basins: Evidence from morphology and ring spacing. *In*: Impact and Explosion Cratering. Roddy DJ, Pepin RO, Merrill RB (eds) Pergamon, p 567-573

Head JW, Adams JB, McCord TB, Pieters CM, Zisk SH (1978a) Regional stratigraphy and geologic history of Mare Crisium. Lunar and Planetary Institute, compiler, Mare Crisium: The view from Luna 24: Conf on Luna 24, Houston, Texas, 1977 Proceedings. Geochim Cosmochim Acta Supp 9:43-74

Heiken G, Vaniman D, French BM (eds) (1991) The Lunar Sourcebook: A User's Guide to the Moon. Lunar and Planetary Institute and Cambridge Univ Press

Herrick RR, Forsberg-Taylor NK (2003) The shape and appearance of craters formed by oblique impact on the Moon and Venus. Meteorit Planet Sci 38: 1551–1578

Hiesinger H, Head JW, Wolf U, Jaumann R, Neukum G, (2003) Ages and stratigraphy of mare basalts in Oceanus Procellarum, Mare Nubium, Mare Cognitum, and Mare Insularum. J Geophys Res 108(E7):1-1–1-27, doi: 10.1029/2002JE001985

Hildebrand AR, Pilkington M, Connors M, Ortiz-Aleman C, Chavez RE (1995) Size and structure of the Chicxulub crater revealed by horizontal gravity gradients and cenotes. Nature 376:415-417

Holsapple KA (1980) The equivalent depth of burst for impact cratering. Proc Lunar Planet Sci Conf 11:2379-2401

Holsapple KA (1987) The scaling of impact phenomena. Int J Impact Eng 5:343-355

Holsapple KA (1993) The scaling of impact processes in planetary sciences. Ann Rev Earth Planet Sci 21:333-373

Holsapple KA, Schmidt RM (1980) On the scaling of crater dimensions 1 Explosive processes. J Geophys Res 85: 7247-7256

Holsapple KA, Schmidt RM (1982) On the scaling of crater dimensions 2 Impact processes. J Geophys Res 87:1849-1870

Holsapple KA, Schmidt RM (1987) Point source solutions and coupling parameters in cratering mechanics. J Geophys Res 92:6350-6376

Hornemann U, Müller WF (1971) Shock-induced deformation twins in clinopyroxene. Neues Jb Min Mh 6: 247-256

Hörz F (2000) Time-variable cratering rates? Science 288:2095a

Hörz F, Ostertag R, Rainey DA (1983) Bunte Breccia of the Ries: Continuous deposits of large impact craters. Rev Geophys Space Phys 21:1667-1725

Housen KR, Schmidt RM, Holsapple KA (1983) Crater ejecta scaling laws: Fundamental forms from dimensional analysis. J Geophys Res 88:2485-2499

Howard KA (1974) Fresh lunar impact craters: Review of variations with size. Proc Lunar Sci Conf 5:61-69

Howard KA, Wilshire HG (1975) Flows of impact melt in lunar craters. J Res US Geological Survey 3:237-251

Hughes DW (2000) A new approach to the calculation of the cratering rate of the Earth over the last 125±20 Ma. Monthly Notices Royal Astron Soc 317:429-437.

Huneke JC, Podosek FA, Wasserburg GJ (1972) Gas retention and cosmic-ray exposure ages of a basalt fragment from Mare Fecunditatis. Earth Planet Sci Lett 13:375-383

Ivanov BA (1994) Geomechanical models of impact cratering: Puchezh-Katunki structure. *In*: Large Meteorite Impacts and Planetary Evolution. Dressler BO, Grieve RAF, Sharpton VL (eds) GSA Spec Paper **293**:81-91

Ivanov BA (2001) Mars/moon cratering rate ratio estimates. *In*: Chronology and Evolution of Mars. Kallenbach R, Geiss J, Hartmann WK (eds) Kluwer Dordrecht, p 87-104

Ivanov BA (2002) Deep drilling results and numerical modeling Puchezh-Katunki impact crater, Russia. Lunar Planetary Science XXXIII:1286

Ivanov BA, Artemieva NA (2001) Transient cavity scaling for oblique impacts. Lunar Planetary Science XXXII:1327

Ivanov BA, Artemieva NA (2002) Numerical modeling of the formation of large impact craters. *In*: Catastrophic Events and Mass Extinctions: Impact and Beyond. GSA Spec Paper 356:619-630

Ivanov BA, Basilevsky AT, Neukum G (1997) Atmospheric entry of large meteoroids: Implication to Titan. Planet Space Sci 45:993-1007

Ivanov BA, Neukum G, Wagner R (1999) Impact craters NEA and Main Belt asteroids Size-frequency distribution. Lunar Planet Sci XXX:1583

Ivanov BA, Neukum G, Wagner R (2001) Size-frequency distributions of planetary impact craters and asteroids. *In*: Collisional Processes in the Solar System. Rickman H, Marov M (eds) Kluwer, p 1-34

Ivanov BA, Neukum G, Bottke W, Hartmann WK (2003) The comparison of size-frequency distributions of impact craters and asteroids and the planetary cratering rate. *In*: Asteroids III. Bottke W, Cellino A, Paolicchi P, Binzel RP (eds) University Arizona Press, p 89-101

Ivezic Z, Tabachnik S, Rafikov R, Lupton RH, Quinn T, Hammergren M, Eyer L, Chu J, Armstrong JC, Fan X, Finlator K, Geballe TR, Gunn JE, Hennessy GS, Knapp GR, Leggett SK, Munn JA, Pier JR, Rockosi CM, Schneider DP, Strauss MA, Yanny B, Brinkmann J, Csabai I, Hindsley RB, Kent S, Lamb DQ, Margon B, McKay TA, Smith JA, Waddel P, York DG, and the SDSS Collaboration (2001) Solar system objects observed in the Sloan Digital Sky Survey Commissioning Data. Astron J 122:2749-2784

James OB (1981) Petrologic and age relations of the Apollo 16 rocks: Implications for the subsurface geology and the age of the Nectaris basin. Proc Lunar Planet Sci Conf 12:209-233

Jedicke R, Metcalfe TS (1998) The orbital absolute magnitude distributions of Main Belt asteroids. Icarus 131:245-260

Jessberger EK (1983) ^{40}Ar-^{39}Ar dating of North Ray Crater ejecta I. Lunar Planet Sci XIV:349-350

Jessberger EK, Kirsten T, Staudacher T (1977) One rock and many ages - further data on consortium breccia 73215. Proc Lunar Sci Conf 8:2567-2580

Jolliff BL, Rockow KM, Korotev RL, Haskin LA (1996) Lithologic distribution and geologic history of the Apollo 17 site: The record in soils and small rock particles from the highland massifs. Meteor Planet Sci 31:116-145

Jolliff BL, Gillis JJ, Korotev RL, Haskin LA (2000a) On the origin of nonmare materials at the Apollo 12 landing site. Lunar Planet Sci XXXII:1671

Jolliff BL, Gillis JJ, Haskin LA, Korotev RL, Wieczorek MA (2000b) Major lunar crustal terranes: Surface expressions and crust-mantle origins. J Geophys Res (Planets) 105:4197-4216

Keil K, Kurat G, Prinz M, Green JA (1972) Lithic fragments glasses and chondrules from Luna 16 fines. Earth Planet Sci Lett 13:243-256

Kettrup B, Deutsch A, Masaitis VL (2003) Homogeneous impact melts produced by a heterogeneous target? Sr-Nd isotopic evidence from Popigai crater, Russia. Geochim Cosmochim Acta 67:733-750

Kieffer SW (1977) Impact conditions required for formation of melt by jetting of silicates. *In*: Impact and Explosion Cratering. Roddy DJ, Pepin RO, Merrill RB (eds) Pergamon Press, p 751-769

Kieffer SW, Schaal RB, Gibbons RV, Hörz F, Milton DJ, Dube A (1976) Shocked Basalt from Lonar Impact Crater, Indis, and Experimental Analogues. Proc Lunar Sci Conf 7:1391-1412

Kitamura M, Goto T, Syono Y (1977) Intergrowth textures of diaplectic glass and crystal in shock loaded P-anorthite. Contr Mineral Petrol 61:299-304

Kitamura M., Tsuchiyama A, Watanabe S, Syono Y, Fukuoka K (1992) Shock recovery experiments on chondritic material. *In*: High-Pressure Research: Applications to Earth and Planetary Sciences. Syono Y, Manghnani MH (eds) Terra Sci. Publ., p 33-34

Korotev RL (1987) Mixing levels the Apennine Front soil component and compositional trends in the Apollo 15 soils. Proc Lunar Planet Sci Conf 17:E411-E431

Korotev RL, Jolliff BL, Zeigler RA (2000) The KREEP components of the Apollo 12 regolith. Lunar Planet Sci XXXI: 1363

Kring DA, Cohen BA (2002) Cataclysmic bombardment throughout the inner solar system 3.9-4.0 Ga. J Geophys Res 107(E2):4-1-4-6

Kurat G, Kracher A, Keil K, Warner R, Prinz M (1976) Composition and origin of Luna 16 aluminous mare basalts. Proc Lunar Sci Conf 7:1301-1321

Lakomy R (1990) Implications for cratering mechanics from a study of the Footwall Breccia of the Sudbury impact structure, Canada. Meteoritics 25:195-207

Landau LD, Lifshitz EM (1987) Fluid Mechanics. Pergamon Press

Lee DC, Halliday AN, Snyder GA, Taylor LA (1997) Age and origin of the Moon. Science 278:1098-1103

Love S, Ahrens TJ (1996) Catastrophic impacts on gravity dominated asteroids. Icarus 124:141-155

Lucchitta BK (1977) Crater clusters and light mantle at the Apollo 17 site: A result of secondary impact from Tycho. Icarus 30:80-96

Lucchitta BK, Sanchez AG (1975) Crater studies in the Apollo 17 region. Proc Lunar Sci Conf 6:2427-2441

Ma MS, Schmitt RA, Nielsen RL, Taylor GJ, Warner RD, Keil K (1979) Petrogenesis of Luna 16 aluminous mare basalts. Geophys Res Lett 6:909-912

Maurer P, Eberhardt P, Geiss J, Grögler N, Stettler A, Brown GM, Peckett A, Krähenbühl U (1978) Pre-Imbrian craters and basins Ages compositions and excavation depths of Apollo 16 breccias. Geochim Cosmochim Acta 42:1687-1720

Maxwell DE (1973) Cratering Flow and Crater Prediction Methods. Physics International Co

Maxwell DE (1977) Simple Z model of cratering, ejection, and overturned flap. *In*: Impact and Explosion Cratering. Roddy DJ, Pepin RO, Merrill RB (eds) Pergamon, p 1003-1008

McCormick KA, Taylor GJ, Keil K, Spudis PD, Grieve RAF, Ryder G (1989) Sources of clasts in terrestrial impact melts: Clues to the origin of LKFM. Proc Lunar Planet Conf 19:691-696

McGetchin TR, Settle M, Head JW (1973) Radial thickness variation in crater ejecta: Implications for lunar basin models. Earth Planet Sci Lett 20:226-236

Melosh HJ (1984) Impact ejection spallation and the origin of meteorites. Icarus 59:234-260

Melosh HJ, Sonett CP (1986) When worlds collide: Jetted vapor plumes and the Moon's origin. *In*: Origin of the Moon. Hartmann WK, Phillips RJ, Taylor GJ (eds) Lunar and Planetary Institute, p 621-642

Melosh HJ (1989) Impact Cratering - A Geologic Process. Oxford Univ Press

Melosh HJ, Ivanov BA (1999) Impact crater collapse. Ann Rev Earth Planetary Science 27:385-415

Melosh HJ, Ryan EV (1997) Asteroids shattered but not dispersed. Icarus 129:562-564

Meyer C Jr, Brett R, Hubbard NJ, Morrison DA, McKay DS, Aitken FK, Takeda H, Schonfeld E (1971) Mineralogy, chemistry and origin of the KREEP component in soil samples from the Ocean of Storms. Proc Lunar Sci Conf 2:393-411

Mileikowsky C, Cucinotta FA, Wilson JW, Gladman B, Horneck G, Lindegren L, Melosh HJ, Rickmann H, Valtonen M, Zheng JQ (2000) Natural transfer of viable microbes in space - 1. From Mars to Earth and Earth to Mars. Icarus 145:391-427

Morbidelli A, Petti JM, Gladman B, Chambers J (2001) A plausible cause of the late heavy bombardment. Meteorit Planet Sci 36:371-380

Moore HJ, Hodges CA, Scott DH (1974) Multi-ringed basins - Illustrated by Orientale and associated features. Proc Lunar Sci Conf 5:71-100

Moore HJ, Boyce JM, Hahn DA (1980) Small impact craters in the lunar regolith - their morphologies relative ages and rates of formation. The Moon and the Planets 23:231-252

Morgan J, Warner M (1999) Chicxulub: The third dimension of a multi-ring impact basin. Geology 27:407-410

Muehlberger WR, Hörz F, Sevier JR, Ulrich GE (1980) Mission objectives for geological exploration of the Apollo 16 landing site. Proceedings of the Lunar and Planetary Institute compiler Conf on the Lunar Highland Crust, Houston, Texas 1979. Geochim Cosmochim Acta suppl. 12:1-49

Müller WF, Hornemann U (1969) Shock-induced planar deformation structures in experimentally shock-loaded olivines and in olivines from chondritic meteorites. Earth Planet Sci Lett 7:251-264

Muller RA, Becker TA, Culler TS, Karner DB, Renne PR (2000) Time-variable cratering rates? Science 288:2095a

Neal CR, Taylor LA, Hughes SS, Schmitt RA (1990) The significance of fractional crystallization in the petrogenesis of Apollo17 type A and B high-Ti basalts. Geochim Cosmochim Acta 54:1817-1833

Neal CR, Hacker MD, Snyder GA, Taylor LA, Liu Y-G, Schmitt RA (1994) Basalt generation at the Apollo 12 site. Part 1: New data classification and re-evaluation. Meteoritics 29:334-348

Nemtchinov IV, Shuvalov VV, Kosarev IB, Artemieva NA, Svetsov VV, Ivanov BA, Loseva TV, Neukum G, Hahn G, de Niem D (1997) Assessment of comet Shoemaker-Levy 9 fragment sizes using light curves measured by Galileo spacecraft instruments. Planet Space Sci 45:311-326

Neukum G (1977) Different ages of lunar light plains. The Moon 17:383-393

Neukum G (1983) Meteoritenbombardement und Datierung planetarer Oberflächen. Habilitation Dissertation for Faculty Membership Ludwig-Maximilians-University Munich, 186 p

Neukum G, Ivanov BA (1994) Crater size distribution and impact probabilities on Earth from lunar terrestrial-planet and asteroid cratering data. *In*: Hazards Due to Comets and Asteroids. Gehrels T (ed) Univ Arizona Press, p 359-416

Neukum G, König B (1976) Dating of individual lunar craters. Proc Lunar Sci Conf 7:2867-2881

Neukum G, Wise DU (1976) Mars: A standard crater curve and possible new time scale. Science 194:1381-1387

Neukum G, König B, Arkani-Hamed J (1975a) A study of lunar impact crater size-distributions. The Moon 12:201-229

Neukum G, König B, Fechting H, Storzer D (1975b) Cratering in the Earth-moon system: Consequences for age determination by crater counting. Proc Lunar Sci Conf 6:2597-2620

Neukum G, Ivanov B, Hartmann WK (2001) Cratering records in the inner solar system. *In*: Chronology and Evolution of Mars. Kallenbach R, Geiss J, Hartmann WK (eds) Kluwer, p 55-86

Nozette S and the Clementine team (1994) The Clementine mission to the Moon: Scientific overview. Science 266:1835-1839

Nyquist LE, Shih CY (1992) The isotopic record of lunar volcanism. Geochim Cosmochim Acta 56:2213-2234

Nyquist LE, Bogard DD, Shih CY (2001) Radiometric chronology of Moon and Mars. *In:* The Century of Space Science. Kluwer Publishers, p 1325-1376

Oberbeck VR (1975) The role of ballistic erosion and sedimentation in lunar stratigraphy. Rev Geophys Space Phys 13:337-362

Oberbeck VR, Morrison RH (1976) Candidate areas for *in situ* ancient lunar materials. Proc Lunar Sci Conf 7:2983-3005

Oberbeck VR, Morrison RH, Hörz F, Quaide WL, Gault DE (1974) Smooth plains and continuous deposits of craters and basins. Proc Lunar Sci Conf 5:111-136

Oberbeck VR, Hörz F, Morrison RH, Quaide WL, Gault DE (1975) On the origin of the lunar smooth-plains. The Moon 12:19-54

O`Keefe JD, Ahrens TJ (1977) Impact-induced energy partitioning melting and vaporization on terrestrial planets. Proc Lunar Sci Conf 8:3357-3374

O`Keefe JD, Ahrens TJ (1999) Complex craters: Relationship of stratigraphy and rings to impact conditions. J Geophys Res 104:27091-27104

Öpik EJ (1971) Cratering and the Moons surface. Academic Press

Orphal DL, Borden WF, Larson SA, Schultz PH (1980) Impact melt generation and transport. Proc Lunar Planet Sci Conf 11:2309-2323

Ostertag R (1983) Shock experiments on feldspar crystals. Proc Lunar Planet Sci Conf 14, In J Geophys Res (Suppl) 88:B364-B376

Palme H, Janssens M-J, Takahashi H, Anders E, Hertogen J (1978) Meteoritic material at five large impact craters. Geochim Cosmochim Acta 42:313-323

Papanastassiou DA, Wasserburg GJ (1972) The Rb-Sr age of a crystalline rock from Apollo 16. Earth Planet Sci Lett 16:289-298

Papike JJ, Ryder G, Shearer CK (1998) Lunar samples. Rev Mineral 36:5-1–5-234

Phinney WC, Simonds CH (1977) Dynamical implications of the petrology and distribution of impact melt rocks. *In*: Impact and Explosion Cratering. Roddy DJ, Pepin RP, Merrill RB (eds) Pergamon, p 771-790

Pierazzo E, Melosh HJ (2000a) Hydrocode modeling of oblique impacts: The fate of the projectile. Meteorit Planet Sci 35:117-130

Pierazzo E, Melosh HJ (2000b) Understanding oblique impacts from experiments, observations and modeling. Ann Rev Earth Planet Sci 28: 141-167

Pierazzo E, Melosh HJ (2000c) Melt production in oblique impacts. Icarus 145:252-261

Pierazzo E, Vickery AM, Melosh HJ (1997) A re-evaluation of impact melt production. Icarus 127:408-423

Pieters CM, Head JW, Gaddis L, Jolliff B, Duke M (2001) Rock types of South Pole- Aitken Basin and extent of basaltic volcanism. J Geophys Res 106:28001-28022

Pike RJ (1974) Depth/diameter relations of fresh lunar craters: Revision from spacecraft data. Geophys Res Letters 1: 291-294

Pike RJ (1977a) Size-dependence in the shape of fresh impact craters on the Moon. *In*: Impact and Explosion Cratering. Roddy DJ, Pepin RO, Merrill RB (eds) Pergamon, p 489-510

Pike RJ (1977b) Apparent depth/apparent diameter relations for lunar craters. Proc Lunar Sci Conf 8:3427-3436

Pike RJ (1980a) Formation of complex impact structures: Evidence from Mars and other planets. Icarus 43:1-19

Pike RJ (1980b) Geometric Interpretation of Lunar Craters. US Geol Survey Prof Paper 1046-C, US Govt Printing Office

Pike RJ (1985) Some morphologic systematics of complex impact structures. Meteoritics 20:49-68.

Pike RJ (1988) Geomorphology of impact craters on Mercury. *In*: Mercury. Vilas F, Chapman CR, Matthews MS (eds) Univ. Arizona Press, p 165-273

Pike RJ, Spudis P (1987) Basin-ring spacing on the Moon, Mercury, and Mars. Earth, Moon, and Planets 39:129-194

Podosek FA, Huneke JC, Gancarz AJ, Wasserburg GJ (1973) The age and petrography of two Luna 20 fragments and inferences for widespread lunar metamorphism. Geochim Cosmochim Acta 37:887-904

Pohl J, Stöffler D, Gall H, Ernstson K (1977) The Ries impact crater. *In*: Impact and Explosion Cratering. Roddy DJ, Pepin RP, Merrill RB (eds) Pergamon, p 343-404

Pye EG, Naldrett AJ, Giblin PE (1984) The Geology and Ore Deposits of the Sudbury Structure. Ontario Geological Survey Special Volume 1

Rabinowitz D, Helin E, Lawrence K, Pravdo S (2000) A reduced estimate of the number of kilometre-sized near-Earth asteroid. Nature 403:165-166

Ravine MA, Grieve RAF (1986) An analysis of morphologic variation in simple lunar craters. Proc Lunar Planet Sci Conf 17, In J Geophys Res 91:E75-E83

Reid AM, Ridley WI, Harmon RS, Warner JL, Brett R, Jakes P, Brown RW (1972) Highly aluminous glasses in lunar soils and the nature of the lunar highlands. Geochim Cosmochim Acta 36:903-912

Reimold WU, Stöffler D (1978) Experimental shock metamorphism of dunite. Proc Lunar Planet Sci Conf 9:2805-2824

Rhodes JM, Blanchard DP, Dungan MA, Brannon JC, Rodgers KV (1977) Chemistry of Apollo 12 mare basalts: Magma types and fractionation processes. Lunar Science Conf 8:1305-1338

Robertson PB, Grieve RAF (1977) Shock attenuation at terrestrial impact structures. *In*: Impact and Explosion Cratering. Roddy DJ, Pepin RP, Merrill RB (eds) Pergamon, p 687-702

Roddy DJ, Pepin RO, Merrill RB (eds) (1977) Impact and Explosion Cratering. Pergamon Press

Russel SS, Zolensky M, Righter K, Folco L, Jones R, Connolly HC Jr, Grady MM, Grossman JN (2006) The Meteoritical Bulletin, Bo. 89, 2005 September. Meteor Planet Sci 40:A201-A263

Ryder G (1981) Distribution of rocks at the Apollo 16 landing site. *In*: Workshop on Apollo 16. LPI Tech Rpt 81-01. James OB, Hörz F (eds) Lunar and Planetary Institute, p 112-119

Ryder G (1985) Catalog of Apollo 15 rocks. JSC Publ No 20787, Curatorial Branch Publ 72, NASA Johnson Space Center

Ryder G (1987) Petrographic evidence for nonlinear cooling rates and a volcanic origin for Apollo 15 KREEP basalts. Proc Lunar Planet Sci Conf 17, In J Geophys Res 92:E331-E339

Ryder G (1990a) Lunar samples, lunar accretion and the early bombardment of the Moon. EOS Trans Amer Geophys Union 71:313-333

Ryder G (1990b) A distinct variant of high-titanium mare basalt from the Van Serg core Apollo 17 landing site. Meteoritics 25:249-258

Ryder G (1994) Coincidence in time of the Imbrium basin impact and the Apollo 15 volcanic flows: The case for impact-induced melting. Geol Soc Am Spec Paper 293:11-18

Ryder G (2000) Glass beads tell a tale of lunar bombardment. Science 287:1768-1769

Ryder G, Marvin UB (1978) On the origin of Luna 24 basalts and soils. *In*: Mare Crisium: The View from Luna 24. Merrill RB, Papike JJ (eds) Pergamon, p 339-355

Ryder G, Schuraytz BC (2001) The chemical variation of the large Apollo 15 olivine-normative mare basalt rock samples. J Geophys Res 106(E1):1435-1451

Ryder G, Spudis PD (1987) Chemical composition and origin of Apollo 15 impact melts. Proc Lunar Planet Sci Conf 17, in J Geophys Res 92:E432-E446

Ryder G, Bogard DD, Garrison D (1991) Probable age of Autolycus and calibration of lunar stratigraphy. Geology 19: 143-146

Schaal RB, Hörz F (1977) Shock metamorphism of lunar and terrestrial basalts. Proc Lunar Sci Conf 8:1697- 1729

Schaal RB, Hörz F (1980) Experimental shock metamorphism of lunar soil. Proc Lunar Planet Sci Conf 11:1679-1695

Schaal RB, Hörz F, Thompson TD, Bauer JF (1979) Shock Metamorphism of granulated lunar basalt. Proc Lunar Planet Sci Conf 10:2547-2571

Schmidt RM (1977) A centrifuge cratering experiment: Development of a gravity-scaled yield parameter. *In*: Impact and Explosion Cratering. Roddy DJ, Pepin RP, Merrill RB (eds) Pergamon, p 1261-1278

Schmidt RM (1980) Meteor crater energy of formation - implications of centrifuge scaling. Proc Lunar Planet Sci Conf 11:2099-2128

Schmidt RM, Holsapple KA (1978) A gravity-scaled energy parameter relating impact and explosive crater size. EOS, Trans Amer Geophys Union 59:1121

Schmidt RM, Holsapple KA (1982) Estimates of crater size for large-body impact: Gravity-scaling results. *In*: Geological Implications of the Impacts of Large Asteroids and Comets on the Earth. GSA Spec Paper 190:93-102

Schmidt RM, Housen KR (1987) Some recent advances in the scaling of impact and explosion cratering. Int J Impact Engin 5:543-560

Schmitt RT (2000) Shock experiments with the H6 chondrite Kernouvé: Pressure calibration of microscopic shock effects. Meteorit Planet Sci 35:545-560

Schultz PH (1976) Moon Morphology. Univ Texas Press

Schultz PH (1988) Cratering on Mercury: A relook. *In*: Mercury. Vilas F, Chapman CR, Matthews MS (eds) University of Arizona Press, p 274-335

Schultz PH (1996) Effect of impact angle on vaporization. J Geophys Res 101(E9):21117-21136

Schultz PH, Gault DE (1975) Seismic effects from major basin formations on the moon and Mercury. Moon 12, 159-177

Schultz PH, Gault DE (1985) Clustered impacts: Experiments and implications. J Geophys Res 90:3701-3732

Schultz PH, Gault DE (1990) Prolonged global catastrophes from oblique impacts. *In*: Global Catastrophes in Earth History: An Interdisciplinary Conf on Impacts Volcanism and Mass Mortality. GSA Spec Paper 247:239-261

Schultz PH, Merrill RB (eds) (1981) Multi-Ring-Basins. Proc Lunar Planet Sci Conf 12A. Pergamon

Schultz PH, Spudis PH (1983) The beginning and end of lunar mare volcanism. Nature 302:233-236

Schultz PH, Mustard JF (2004) Impact melts and glasses on Mars. J Geophys Res 109(E01001) doi:10.1029/2002JE002025

Schultz PH, Orphal D, Miller B, Borden WF, Larson SA (1981) Multi-ring basin formation: Possible clues from impact cratering calculations. Proc Lunar Planet Sci Conf (Multiring Basins) 12A:181-195

Sharpton VL, Burke K, Camargo-Zanoguera A, Hall SA, Lee DS, Marin LE, Suarez-Reynoso G, Quezada-Muneton JM, Spudis PD, Urrutia-Fucugauchi J (1993) Chicxulub multi-ring impact basin size and other characteristics derived from gravity analysis. Science 261:1564-1567

Shoemaker EM (1962) Interpretation of lunar craters. *In*: Physics and Astronomy of the Moon. Kopal Z (ed) Academic Press, p 283-359

Shoemaker EM, Hackman RJ (1962) Stratigraphic basis for a lunar time scale. *In*: The Moon. Kopal Z, Mikhailov ZK (eds) Academic Press, p 289-300

Shoemaker EM, Weissman PR, Shoemaker CS (1994) The flux of periodic comets near earth. *In*: Hazards Due to Comets and Asteroids. Gehrels T (ed) Univ Arizona Press, p 313-336

Shuvalov VV (2002) Displacement of target material due to impact. Lunar Planet Sci XXXIII:1259

Silver LT (1971) U-Th-Pb isotope systems in Apollo 11 and 12 regolithic materials and a possible age for the Copernican impact. EOS, Trans Amer Geophys Union 52:534

Smith EI, Sanchez AG (1973) Fresh lunar craters: Morphology as a function of diameter a possible criterion for crater origin. Modern Geology 4:51-59

Snee LW, Ahrens TJ (1975) Shock-induced deformation features in terrestrial peridot and lunar dunite. Proc Lunar Sci Conf 6:833-842

Snyder GA, Borg LE, Nyquist LE, Taylor LA (2000) Chronology and isotopic constraints on lunar evolution. *In:* The Origin of the Earth and the Moon. Univ Arizona Press, p 361-395

Soderblom LA (1970) A model for small-impact erosion applied to the lunar surface. J Geol Res 75:2655-2661

Soderblom LA, Lebofsky LA (1972) Technique for rapid determination of relative ages of lunar areas from orbital photography. J Geophys Res 77:279-296

Spangler RR, Warasila R, Delano LW (1984) ^{40}Ar-^{39}Ar ages for the Apollo 15 green and yellow volcanic glasses. Proc Lunar Planet Sci Conf 14:B487-B497

Spray JG, Thompson LM (1995) Friction melt distribution in terrestrial multi-ring impact basins. Nature 373:130-132

Spudis PD (1978) Composition and origin of the Apennine Bench Formation. Proc Lunar Planet Sci Conf 9:3379-3394

Spudis PD (1993) The Geology of Multi-ring Impact Basins: The Moon and Other Planets. Cambridge University Press

Spudis PD, Ryder G (1981) Apollo 17 impact melts and their relation to the Serenitatis basin Multi-ring basins. Proc Lunar Planet Sci Conf 12:133-148

Spudis PD, Hawke BR, Lucey PG (1984) Composition of Orientale Basin deposits and implications for the lunar basin-forming process. Proc Lunar Planet Sci Conf 15, In J Geophys Res 89:C197-C210

Spudis PD, Ryder G, Taylor GJ, McCormick KA, Keil K, Grieve RAF (1991) Sorces of mineral fragments in impact melts 15445 and 15455: Towards an origin of Low-K Frau Mauro basalt. Proc Lunar Planet Sci Conf 21:151-165

Stadermann FJ, Heusser E, Jessberger EK, Lingner S, Stöffler D (1991) The case for a younger Imbrium basin: New ^{40}Ar-^{39}Ar ages of Apollo 14 rocks. Geochim Cosmochim Acta 55:2339-2349

Stähle V (1972) Impact glasses from the suevite of the Nördlinger Ries. Earth Planet Sci Lett 17:275-293

Staid MI, Pieters CM (2001) Mineralogy of the last lunar basalts: Results from Clementine. J Geophys Res 106(E11): 27,887-27,900

Staid MI, Pieters CM, Head JW III (1996) Mare Tranquillitatis Basalt emplacement history and relation to lunar samples. J Geophys Res (Planets) 101:23,213-23,228

Staudacher T, Jessberger EK, Dominik B, Kirsten T, Schaeffer OA (1982) ^{40}Ar - ^{39}Ar ages of rocks and glasses from the Nördlinger Ries crater and the temperature history of impact breccias. J Geophys 51:1-11

Steiger RH, Jäger E (1977) Subcommission on Geochronology: Convention on the use of decay constants in geo- and cosmochronology. Earth Planet Sci Lett 36:359-362

Stephan T, Jessberger E (1992) Isotope systematics and shock-wave metamorphism III K-Ar in experimentally and naturally shocked rocks; the Haughton impact structure, Canada. Geochim Cosmochim Acta 56:1591-1605

Stöffler D (1972) Deformation and transformation of rock-forming minerals by natural and experimental shock processes. I. Behavior of minerals under shock compression. Fortschr Mineral 49:5477-5488

Stöffler D (1974) Deformation and transformation of rock forming minerals by natural and experimental shock processes. II. Physical properties of shocked minerals. Fortschr Mineral 51:256-289

Stöffler D (1984) Glasses formed by hypervelocity impact. Journal Non-Cryst Solids 67:465- 502

Stöffler D, Hornemann U (1972) Quartz and feldspar glasses produced by natural and experimental shock. Meteoritics (7):371-394

Stöffler D, Reimold WU (1978) Experimental shock metamorphism of dunite. Proc Lunar Planet Sci Conf 9:2805-2824

Stöffler D, Langenhorst F (1994) Shock metamorphism of quartz in nature and experiment: I. Basic observation and theory. Meteoritics 29:155-181

Stöffler D, Grieve RAF (1994) Classification and nomenclature of impact metamorphic rocks: a proposal to the IUGS Subcommission on the Systematics of Metamorphic Rocks. Lunar Planet Sci XXV:1347

Stöffler D, Grieve RAF (1996) IUGS classification and nomenclature of impact metamorphic rocks: Towards a final proposal. International Symposium on the Role of Impact Processes in the Geological and Biological Evolution of Planet Earth, Postojna, Slovenia, 27.9.-2.10.1996, Abstract

Stöffler D, Ryder G (2001) Stratigraphy and isotope ages of lunar geologic units: Chronological standard for the inner solar system. *In*: Chronology and Evolution of Mars. Kallenbach R, Geiss J, Hartmann WK (eds) Kluwer, p 9-54

Stöffler D, Grieve RAF (2006) Towards a unified nomenclature of metamorphic petrology: 11. Impactites - A proposal on behalf of the IUGS Subcommission on the Systematics of Metamorphic Rocks. IUGS Blackwell Publishers (in press)

Stöffler D, Gault DE, Wedekind J, Polkowski G (1975) Experimental hypervelocity impact into quartz sand Distribution and shock metamorphism of ejecta. J Geophys Res 80:4062-4077

Stöffler D, Knöll HD, Maerz U. (1979) Terrestrial and lunar impact breccias and the classification of the lunar highland rocks. Proc Lunar Planet Sci Conf 10:639-675

Stöffler D, Knöll HD, Marvin UB, Simonds CH, Warren PH (1980) Recommended classification and nomenclature of lunar highland rocks - a committee report. Proceedings of the Lunar and Planetary Institute compiler Conf on the Lunar Highlands Crust, Houston, Texas 1979. Geochim Cosmochim Acta suppl 12:51-70

Stöffler D, Ostertag R, Reimold WU, Borchardt R, Malley J, Rehfeldt A (1981) Distribution and provenance of lunar highland rock types at North Ray Crater Apollo 16. Proc Lunar Planet Sci Conf 12:185-207

Stöffler D, Bischoff A, Borchardt R, Burghele A, Deutsch A, Jessberger EK, Ostertag R, Palme H, Spettel B, Reimold WU, Wacker K, Wänke H (1985) Composition and evolution of the lunar crust in the Descartes Highlands Apollo 16. Proc Lunar Planet Sci Conf 15, In J Geophys Res 89:C449-C506

Stöffler D, Ostertag R, Jammes C, Pfannschmidt G, Sen Gupta PR, Simon SB, Papike JJ, Beauchamp RH (1986) Shock metamorphism and petrography of the Shergotty achondrite. Geochim Cosmochim Acta 50:889-903

Stöffler D, Bischoff L, Oskierski W, Wiest B (1988a) Structural deformation, breccia formation, and shock metamorphism in the basement of complex terrestrial impact craters: implications for the cratering process. *In*: Deep drilling in crystalline bedrock. Volume 1. Boden A, Eriksson KG (eds) Springer-Verlag, p 277-297

Stöffler D, Bischoff A, Buchwald V, Rubin AE (1988b) Shock effects in meteorites. *In*: Meteorites and the early solar system. Kerridge JF, Matthews MS (eds) University of Arizona Press, p 165-202

Stöffler D, Bobe KD, Jessberger EK, Lingner S, Palme H, Spettel B, Stadermann F, Wänke H (1989) Fra Mauro Formation Apollo 14. IV. Synopsis and synthesis of consortium studies. *In*: Workshop on Moon in Transition Apollo 14 KREEP and evolved lunar rocks. Taylor GJ, Warren PH (eds) LPI Tech Rpt 89-03:145-148

Stöffler D, Keil K, Scott RD (1991) Shock metamorphism of ordinary chondrites. Geochim Cosmochim Acta 55: 3845- 3867

Stöffler D, Deutsch A, Avermann M, Bischoff L, Brockmeyer P, Buhl D, Lakomy R, Müller-Mohr V (1994) The formation of the Sudbury structure, Canada: Toward a unified impact model. GSA Spec Pap 293:303-318

Stöffler D, Artemieva NA, Pierazzo E (2002) Modeling the Ries-Steinheim impact event and the formation of the moldavite strewn field. Meteorit Planet Sci 37:1893-1907

Strom RG, Neukum G (1988) The cratering record on Mercury and the origin of impacting objects. *In*: Mercury. Vilas F, Chapman CR, Matthews MS (eds) University of Arizona Press, p 336-373

Strom RG, Croft SK, Barlow NG (1992) The Martian impact cratering record. *In*: Mars. Kieffer HH, Jakosky BM, Snyder C, Matthews MS (eds) Univ Arizona Press, p 383-423

Stuart JS (2001) A near-Earth asteroid population estimate from the LINEAR survey. Science 294:1691-1693

Sugita S, Schultz PH (1999) Spectroscopic characterization of hypervelocity jetting: Comparison with a standard theory. J Geophys Res 104(E12):30825-30845

Swindle TD, Spudis PD, Taylor GJ, Korotev RL, Nichols RH Jr, Olinger CT (1991) Searching for Crisium basin ejecta: Chemistry and ages of Luna 20 impact melts. Proc Lunar Planet Sci Conf 21:167-181

Tagle RA, Erzinger J, Hecht L, Schmitt RT, Stöffler D, Claeys P (2004) Platinum group elements in impactites of the ICDP Chicxulub drill core Yaxcopoil-1: Are there traces of the projectile? Meteorit Planet Sci 39:1009-1016

Taylor SR (1975) Lunar Science: A post-Apollo View. Pergamon

Taylor SR (1982) Planetary Science: A Lunar Perspective. Lunar and Planetary Institute

Taylor LA, Onorato PIK, Uhlmann DR, Coish RA (1978) Subophitic basalts from Mare Crisium: Cooling rates. *In*: Mare Crisium: The View from Luna 24. Merrill RB, Papike JJ (eds) Pergamon, p 473-482

Tera F, Wasserburg GJ (1976) Lunar ball games and other sports. Lunar Planet Sci VII:858-860

Tera F, Papanastassiou DA, Wasserburg GJ (1974) Isotopic evidence for a terminal lunar cataclysm. Earth Planet Sci Lett 22:1-21

Therriault AM, Fowler AD, Grieve RAF (2002) The Sudbury Igneous Complex: A differentiated impact melt sheet. Econ Geol 97:1521-1540

Thompson SL, Lauson HS (1972) Improvements in the Chart-D radiation hydrodynamic code. III. Revised analytical equation of state. Tech Rep SC-RR-710714 Sandia Nat Labs

Tompkins S, Pieters CM (1999) Mineralogy of the lunar crust: results from Clementine. Meteorit Planet Sci 34:25-41

Turner G (1977) Potassium-argon chronology of the Moon. Phys Chem Earth 10:145-195

Ulrich GE, Hodges CA, Muehlberger WR (eds) (1981) Geology of the Apollo 16 area central lunar highlands. USGS Professional Paper 1048

Vinogradov AP (1971) Preliminary data on lunar ground brought to Earth by automatic probe "Luna 16". Proc Lunar Sci Conf 2:1-16

Vinogradov AP (1973) Preliminary data on lunar soil collected by the Luna 20 unmanned spacecraft. Geochim Cosmochim Acta 37:721-729

Wacker K, Müller N, Jessberger EK (1983) ^{40}Ar-^{39}Ar dating of North Ray Crater ejecta II. Meteoritics 18:201

Warner RD, Taylor GJ, Conrad GH, Northrup HR, Barker S, Keil K, Ma MS, Schmitt R (1979) Apollo 17 high-Ti mare basalts: New bulk compositional data, magma types and petrogenesis. Proc Lunar Planet Sci Conf 10:225-247

Wieczorek MA, Phillips RJ (1999) Lunar multiring basins and the cratering process. Icarus 139: 246–259

Wieczorek MA, Zuber MT (2001) A Serenitatis origin for the Imbrian grooves and South Pole-Aitken thorium anomaly. J Geophys Res 106:27,853-27,864

Weiss BP, Kirschvink JL, Baudenbacher FJ, Vali H, Peters NT, Macdonald FA, Wilks00 JP (2000) A low temperature transfer of ALH 84001 from Mars to Earth. Science 290:791-795

Whitehead J, Grieve RAF, Spray J (2002) Mineralogy and petrology of melt rocks from the Popigai impact structure, Russia. Meteorit Planet Sci 37:623-648

Wilhelms DE (1980) Stratigraphy of part of the lunar near side. USGS Professional Paper 1046-A A1-A71

Wilhelms DE (1984) The Moon. *In*: The Geology of the Terrestrial Planets. NASA SP-469:107-205

Wilhelms DE, McCauley JF, Trask NJ (1987) The Geologic History of the Moon. US Geol Survey Prof Paper 1348

Williams KK, Zuber MT (1998) Measurement and analysis of lunar basin depths from Clementine altimetry. Icarus 131:107-122

Wolfe EW, Lucchitta BK, Reed VS, Ulrich GE, Sanchez AG (1975) Geology of the Taurus-Littrow valley floor. Proc Lunar Sci Conf 6th, Pergamon Press, p. 2463-2482.

Wolfe EW, Bailey NG, Lucchitta BK, Muehlberger WR, Scott DH, Sutton RL, Wilshire HG (1981) The geologic investigation of the Taurus-Littrow valley Apollo 17 landing site. USGS Professional Paper 1080

Wünnemann K, Ivanov BA (2003) Numerical modelling of the impact crater depth–diameter dependence in an acoustically fluidized target. Planet Space Sci 51:831-845

Wood CA, Andersson L (1978) New morphometric data for fresh lunar craters. Proc Lunar Planet Sci Conf 9th, Pergamon Press, p 3669-3689

Xie X, Chen M, Dai C, El Goresy A, Gillet P (2001) A comparative study of naturally and experimentally shocked chondrites. Earth Planet Sci Lett 187:345-356

Zahnle K, Mac Low M-M (1994) The collision of Jupiter and Comet Shoemaker-Levy 9. Icarus 108:1-17

Zukas JA (2004) Introduction to Hydrocodes. Elsevier

Reviews in Mineralogy & Geochemistry
Vol. 60, pp. 597-656, 2006
Copyright © Mineralogical Society of America

6

Development of the Moon

Michael B. Duke

Colorado School of Mines
Golden, Colorado, 80401, U.S.A.
e-mail: mduke@mines.edu

Lisa R. Gaddis

U. S. Geological Survey, Astrogeology Team
2255 N. Gemini Dr.
Flagstaff, Arizona, 86001, U.S.A.

G. Jeffrey Taylor

Hawaii Institute of Geophysics & Planetology
University of Hawaii
2525 Correa Rd.
Honolulu, Hawai'i, 96822, U.S.A.

Harrison H. Schmitt

P.O. Box 90730
Albuquerque, New Mexico, 87199, U.S.A.

1. INTRODUCTION

This book focuses largely on new results from recent missions and on their implications for how we interpret results from older missions. The new results have also renewed awareness of the Moon as a future target for exploration and many people see development of the Moon, particularly its resources, as a key step in the future exploration of the solar system (Aldridge et al. 2004). The interpretation of the lunar data sets in the context of future exploration and development of the Moon is, therefore, parallel to new scientific interpretations. This is in a sense a forward-looking view inspired in part by New Views of the Moon perspectives. It is also timely, as the United States is currently reconsidering its space exploration program, with a greater focus on renewed exploration of the Moon.

Earth's Moon can be looked upon as an enormous Earth-orbiting Space Station, a natural satellite outside of Earth's gravity well, with raw materials that can be put to practical use as humanity expands outward into the Universe. As outlined in previous chapters, new remote-sensing data for the Moon have reinvigorated lunar science and improved understanding of the Moon's composition, the ages of its prominent formative events, and the character of the earliest lunar crust and its subsequent geologic evolution. In this chapter, we consider how we might use lunar materials for exploration, utilization, and development of the Moon. The Moon offers a nearby location from which to develop resources and capabilities to explore further in the Solar System. The natural resources of the Moon include minerals, rocks, and soils, which can be processed to produce metals, oxygen, glass, ceramics, and other useful products (McKay et al. 1992). Water ice may exist near the poles and low concentrations of volatiles deposited by solar wind (H, He, C, N) are trapped in the regolith. A low gravity field, limited atmosphere, and a location in space near the Earth provide an environment in which

1529-6466/06/0060-0006$05.00   DOI: 10.2138/rmg.2006.60.6

resource development is challenging, but possible, with products that are useful in space or, in limited cases, on Earth. Proper utilization of these resources, in the establishment of lunar habitations and as fuel for planetary exploration, will provide an invaluable stepping-stone for humanity as it grows beyond the limits of Earth and of our Solar System. Near term uses of these resources can have economic implications in space industrialization and opening up virtually unlimited sources of energy for a growing world economy. Scientific and economic considerations must be melded into plans for future lunar exploration and development, and commercial interests can be expected to join the scientific and technical objectives that have formed the basis for previous lunar exploration programs.

This chapter addresses the major topics that must be considered in developing a strategy for the exploration and development of the Moon:

- *Why go to the Moon?* The objectives of lunar exploration will be based on human society's need for more information, energy, and raw materials that constitute the principal underpinnings of economic growth. The chief themes addressed here are the expansion of humans into space, space industrialization (possibly including tourism), space transportation, the search for more energy, the Moon as a laboratory for planetary science, and astronomy and other science on the Moon.

- *Getting There and Back.* Although the Moon is near enough to Earth to be considered part of an economic system with Earth, routine and safe transportation within the Earth-Moon system will need to be developed.

- *Lunar Material Resources.* The resources of the Moon are known in a preliminary sense, but further exploration for resources and the determination of their practical economic utility will be required. The scale of resource development on the Moon may be small by terrestrial standards, but for applications in space, production of relatively small amounts of materials (e.g., propellants) from the Moon may be economically feasible because transportation costs can be much smaller from the Moon than from the Earth.

- *The Lunar Environment.* In developing the Moon, attention must be paid to the preservation of the environmental characteristics of the Moon that are unique or those that form the basis for its beneficial utilization. Large-scale economic activities on the Moon could locally modify the lunar atmosphere, possibly interfere with uses such as astronomy, or produce visible effects that some might consider undesirable.

- *Humans on the Moon.* Human outposts and settlements on the Moon appear to be feasible, but will have to operate under a variety of economic and environmental constraints, if they are to be viable in the long term.

- *A strategy for lunar exploration and development.* A strategy is advanced as a baseline from which to consider alternative approaches and prioritize investments in science and technology.

The strategic exploration and development of the Moon are greatly facilitated by previous and ongoing analyses of sample and remote sensing data from the Apollo and Luna missions, Earth-based telescopic data for the Moon, lunar meteorites, and from the more recent Galileo, Clementine and Lunar Prospector missions. It is anticipated that the gathering of more scientific information will continue to be critical to the identification of opportunities for, and constraints on, lunar development. Currently, the European Space Agency (ESA) is conducting the SMART-1 mission to the Moon. Japan, India and China are developing or planning lunar missions. Recently, NASA has been directed to develop plans for a renewed human exploration program beyond low Earth orbit, with robotic missions to the Moon starting in 2008 and human missions beginning in 2015-2020.

2. WHY GO TO THE MOON? EXPLORATION RATIONALES AND MOTIVATIONS

Past exploration of the Moon has been driven by political forces (e.g., Apollo, Luna missions) or by science and technology (e.g., Galileo, Clementine, Lunar Prospector, ESA, and Japanese missions). The impetus for missions being planned by India and China may be driven in part by a desire to demonstrate technological innovation. Future lunar exploration may be driven by economic or strategic purposes. What is our current motivation for further exploration and possible utilization of the Moon? Here we identify six themes as those that are most likely to guide the strategic exploration and utilization of the Moon in future years.

2.1. Expansion of humans into space

Humanity has evolved to occupy nearly every available niche on Earth. This evolution is a result of the inherent need to survive and prosper that is a fundamental drive of living organisms. There is no reason to believe that this drive will cease at the edge of space. Some would argue that the movement of humans to the planets and out of the Solar System is a basic survival strategy for the species, because of the potential for disastrous natural and human-caused events on a single planet (Gott 1993).

Constraints to the expansion of humans beyond Earth are primarily technological and economic, but health and safety of crewmembers also is a major concern. Space travel for humans, using existing chemical propulsion systems, is limited to the Earth-Moon system, Venus and Mars, because travel times become much longer when traveling beyond the nearest planets. Advanced propulsion systems (solar and nuclear electric, nuclear thermal, nuclear fusion) are being studied that might reduce the one-way transportation time for Mars (Carlson 2003; Frisbee 2003; Head et al. 2003). Human missions beyond Mars are unlikely for the next 30 to 50 years.

Human health and performance is a principal concern in human missions that remain away from Earth for extended periods of time. Our experience base is deficient for Mars exploration, in which a trip of several months in weightlessness is followed by a surface stay that may either be short (30 to 90 days) or long (~500 days), followed by a several month long return trip to Earth. Russian cosmonauts spent more than 1 year in weightlessness, which is approximately the time that would be spent in space in a human mission to Mars (e.g., six months of weightlessness each way). But three days is the longest period of human experience in a low gravity planetary surface environment by an Apollo astronaut. It is known that humans adapt to weightlessness in ways that are often detrimental to the human body when they return to the Earth's gravitational field. Exercise and medication are to some extent effective countermeasures. Astronauts who exercise vigorously while in space appear to readapt to the Earth's gravity field quite rapidly (D. R. Pettit, pers. comm.). A lunar outpost program must include a program to establish the limitations of the human body for long stays on the Moon and readaptation when returning to Earth. These data will be applicable to the human exploration of Mars.

Operational strategies are also important for long-duration missions. The balance between autonomy of crews on a planet's surface and the support that they receive from the Earth will change markedly with the distance of the planet from the Earth. For the Moon, communication is slightly delayed (a few seconds); however in Mars missions, the round trip light time may be 16 to 30 minutes, depending on the relative location of Mars and Earth. This increases the crew's need for autonomous operation in periods of emergency. A lunar outpost can provide a realistic environment for testing operational approaches for humans and robots supported by mission controllers on Earth, with growing autonomy over time.

It would be foolish for humans to plan to stay in space or on planetary surfaces for several years or longer without producing food. Plants are essential for food and also for long-term stable life-support systems, as they can recycle H_2O and CO_2. Cultivation of plants for food can provide a basic diet for astronauts as well as psychological benefits for people living for years in cramped quarters. The Moon can provide a laboratory for testing the viability of plants and animals in the space environment and the possibility of plant nutrients from indigenous materials (Henninger and Ming 1989).

Spacecraft systems need to be tested for time periods that are consistent with the times required for human missions to Mars. For voyages lasting many years (2.5 to 3 years for a human mission to Mars), testing over equivalent or longer periods of time is needed in relevant environments. The Moon may provide a "test bed," a facility where technologies that are intended for use on planetary surfaces may be tested. A lunar test bed could eventually grow into a sophisticated activity, including the capability to maintain, repair and improve these technologies. Technologies that might be useful for Mars exploration that could be developed and tested on the Moon are shown in Table 6.1.

The Moon is an ideal place to establish a technology test-bed facility because it is much closer to Earth than Mars, which can allow easier recovery from accidents or anomalous conditions, while providing relevant environmental conditions for the crew and hardware.

Table 6.1. Systems technology test beds at a lunar outpost.

Mars Application	*Lunar Demonstration*	*Comments*
Highly reusable EVA suits	Long term performance in representative environment: -Operational tests of agility -Long duration operations -Multiple uses -Maintenance and repair	A suit designed for lunar gravity may not be useful on Mars, due to its higher gravity. However, a suit designed to meet Mars requirements should be fully testable on the Moon
Long range tele-operated rover	Operated from Earth to simulate crew operations on Mars	Communication delay times of a few seconds may also be realistic for Mars in the case that astronauts operate the equipment from a Martian outpost.
Closed life support systems	Long term operation in representative environment, including maintenance and repair	1/6 g may cause more severe effects than on Mars. Therefore, a system designed for the Moon should be applicable for Mars.
Nuclear reactor power system	Robotic emplacement, shielding using indigenous materials; monitoring of radiation environment with robotic systems.	The design and operation of a nuclear power system should be very similar for both Moon and Mars, though the Martian atmosphere will need to be considered in terms of its effects on system design.
In-situ resource utilization	Subsystems, long-duration operations: -Electrolyzers -Liquefaction of cryogens -Fluid transfer -Storage	Detailed extraction processes will not be the same, but the components will be similar.
Human health and performance	Long-duration tests at 1/6-g with many subjects.	More severe environment than Mars; if humans can flourish on the Moon, they will be able to adapt to Mars.

It should also be possible to create a robust testing capability on the Moon with flexibility to change if new and better technologies or operational protocols are developed. A routine transportation system to the Moon would be required to support such a test-bed facility. Eckart (1999) has provided a thorough examination of the technical issues associated with establishment of lunar bases.

2.2. The search for energy alternatives

Resources on the Moon, especially solar energy or ^{3}He, could provide new sources of energy for Earth. Continued economic prosperity of the Earth is a strong function of the availability of inexpensive energy (e.g., Criswell 1994; Schmitt 1997). The current energy use of the developed nations of the world is approximately 6 kWt (kilowatts of thermal energy) of continuous power per person, but much of the world survives on far less. Most of the world's energy is provided by the combustion of fossil biomass (especially coal and oil). Global demand for energy will likely increase by a factor of at least eight by the mid-point of the 21st Century due to a combination of population increase, new energy intensive technologies, and aspirations for improved standards of living in the less-developed world (NRC 2001). If the Earth's population increases to 10 billion people and all people of Earth are provided with electricity at the current average in the developed world (2 kWe/person (kilowatts of electrical energy), 20 billion kWe (20,000 GWe) of energy production would be required. At this level of power, known and predicted sources of fossil fuel would soon be consumed. Because so much more energy is required than is now produced, the energy sources should also be cleaner. Two sources of abundant clean energy have been identified as potentially capable of filling that need—space solar power and nuclear fusion. The Moon could play a substantial role in the development of either.

2.2.1. Solar Power Satellites. The concept of Solar Power Satellite (SPS) systems located in geostationary Earth orbit (GEO) was formulated by Glaser (1968). In his concept, large arrays would be assembled in GEO and located above the cities or regions that needed power (Fig. 6.1, after Mankins 2001a). The arrays could be hundreds of square kilometers in dimension. Typical designs for these arrays have yielded masses of 50,000 metric tons (mt) of material for

Figure 6.1. Solar Power Satellite: A recent concept (after Mankins 2001a) in medium or low Earth orbit.

each 2GW of power capability. Recent advances in technology (Grey 2001; Mankins 2001a,b) have somewhat reduced the estimated mass and improved the estimated performance of these systems, but the high cost of Earth-to-orbit transportation is the overwhelming barrier to the deployment of the system. If the cost of transportation from Earth to Low Earth Orbit (LEO) can be lowered to $400/kg (a factor of 25 lower than current costs), recent studies (Mankins 1995) indicate that the cost of energy can be lowered to about $0.05/kWh, but it will be difficult for SPS to compete with energy from fossil fuels for the next 20 years or more (Greenberg 2000).

As noted by DuBose (1985), essentially all of the materials needed to construct solar energy collection, conversion, and power-beaming facilities in space or on the Moon are available from lunar materials (Table 6.2). Recent research has suggested that silicon solar photovoltaic (PV) devices can be produced from lunar materials with an investment in hardware shipped from Earth that is quite small in comparison to the amount of energy that can be produced (Ignatiev et al. 1998).

Bock (1979) modeled the deployment of a SPS system using lunar materials, and using processing approaches studied by Criswell (1978) and Miller (1978-1979). The lunar regolith could provide at least 90% of the required materials (a major part of the total mass of the SPS is structure) in the Bock study and 99% in the DuBose (1985) study. The performance of structural elements derived from the Moon would probably be lower than those brought from Earth, so the mass of the DuBose SPS was about 10% greater than required for the same system constructed entirely from terrestrial materials. Transportation will remain a principal hurdle. The Bock (1979) study envisioned using an electromagnetic launching system (mass driver) capable of very low $/kg launch costs from the Moon (Billingham and Gilbreath 1979). A mass driver efficiently converts electrical to kinetic energy using a linear motor concept. If the cost of power on the Moon were as low as that on Earth, transportation costs from the Moon could be very low. However, such a launch system would require significant engineering development and its cost is currently unknown.

In general, the economic viability of lunar resources depends on a tradeoff between the cost of developing lunar manufacturing systems and the cost of transportation to space. If lunar processing systems can be designed that produce large quantities of materials in their useful lifetime, even though the development cost is high, the cost of production spread against the amount

Table 6.2. Solar Power Satellite elements and materials (from DuBose 1985).

Functional Element	*% of system*	*Principal materials*	*% available on Moon*
Silicon photovoltaic arrays	45		
- Cell covers and substrates		Fused silica	100
- Solar cells		Silicon	>99.99
- Interconnects, support wires, connectors, tensioning system		Aluminum	100
Structure	14	Aluminum	100
Power Management, Distribution	6.5	Aluminum	100
- Electrical conductor buses		Aluminum, fused silica	90
- Switches, energy storage		(flywheel), silver, copper	
Microwave transmitters (klystron amplifiers, waveguides)	33	Aluminum, titanium, steel (with Co, V), tungsten, copper, carbon, silica, etc.	98
Miscellaneous other	1.5	Various	96

of material produced can be reasonable. The cost of transportation from the Moon to space can become far smaller than the cost of transportation from Earth, because the Moon's gravitational field is smaller and complications of the Earth's atmosphere are eliminated (see Section 4.5).

2.2.2. Lunar Power System. A Lunar Power System (LPS) is similar in principal to the SPS, but consists of large PV arrays at both limbs of the Moon beaming energy to Earth by microwaves (Fig. 6.2). This system could be built the lunar surface of materials available on the Moon (Criswell and Waldron 1990). In the Criswell and Waldron (1990) concept, solar energy is collected by photovoltaic cells deposited directly onto the lunar surface. The electricity is conducted to a microwave transmitter that beams microwaves onto a reflector aimed at Earth. A complete system includes many of these assemblages that complete a filled aperture as viewed from Earth. Although the distance to Earth is much greater than that from GEO, the focusing effect of the large lunar array can be efficient in transmitting power to Earth. The machinery required for production of solar arrays on the Moon is small in mass compared to the energy collected and transmitted to Earth. Transportation costs are minimized compared to those required for installing a SPS in Earth orbit.

The micrometeorite and radiation environments of space and the lunar surface will cause slow degradation of photovoltaic arrays. In space, if high efficiency cells are used, this

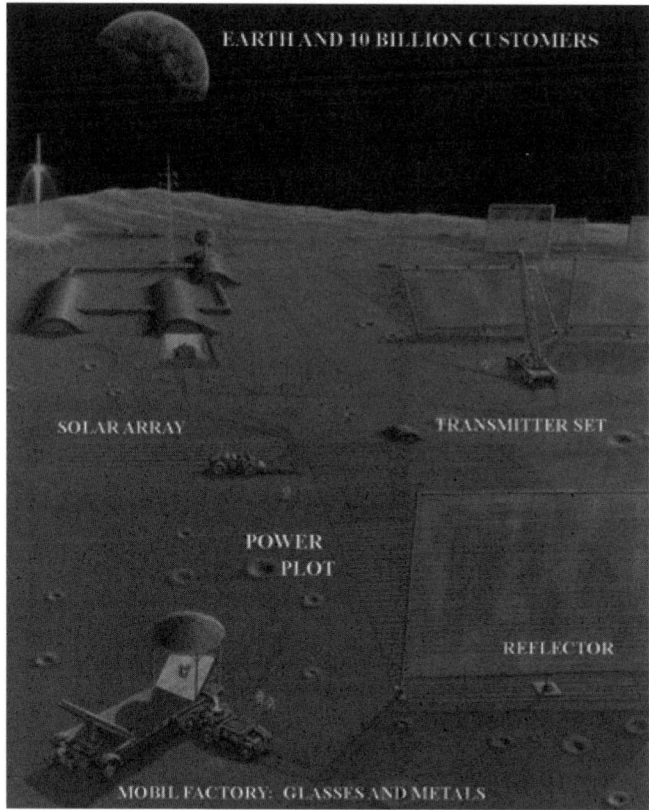

Figure 6.2. Power plots within a demonstration lunar power base (© David R. Criswell 2002). The power plots are areas of solar cells, connected to transmitters that beam microwaves toward the reflector. The full set of reflectors constitutes a filled array as observed from Earth.

degradation can be severe. As a lunar power system would use lower-efficiency solar cells, the radiation degradation is reduced and the replacement of micrometeorite-damaged solar arrays on the Moon by the same process by which they were produced should constitute a minor cost to system operations. Care will be needed to control dust generation near the lunar arrays; however, it is interesting to note that the laser retro-reflectors emplaced on the Moon by Apollo have performed without serious degradation for over 30 years and are still in use (e.g., Williams et al. 2001).

The technologies required for LPS are generally highly advanced (Criswell 1996). The requirement for Moon-to-space transportation of materials (a major factor in an SPS made of lunar material) is substantially reduced. If the machinery for production of solar arrays and beaming hardware can itself be manufactured from lunar materials, the transportation requirement from the Earth can be further reduced. Automation of production and installation equipment reduces the number of people required to operate the lunar production facilities. The transportation of materials from Earth to the Moon is more expensive than taking the same materials to GEO; however, the amount of equipment and materials brought from Earth to the Moon is greatly reduced, and the time required for construction should also be decreased. The availability of lunar propellant and propellant depots in lunar orbit or at an Earth-Moon Lagrangian point such as L-1 can reduce the transportation costs from Earth and Moon.

The principal materials that are required for constructing the LPS are similar to those listed in Table 6.2 for a SPS constructed in space from lunar materials, with the exception of the supporting structure for the silicon PV arrays, which is provided by the lunar surface in the Criswell and Waldron (1990) concept.

If an LPS can be constructed on the Moon, it follows that power on the Moon could be very inexpensive (perhaps a factor of 10 less than the cost of the beamed power on Earth) because the requirements for beaming and receiving the energy would be eliminated. Criswell and Thompson (1996) predict very low energy costs on Earth (<$.01/kWh) with a fully developed LPS. No analysis has been made of the economic implications of essentially free power on the Moon but some speculation is possible. The energy to transport a kilogram of material from the Moon to space (including to the Earth) is less than 1 kWh. If the cost of electricity on the Moon were $0.01/kWh, the energy cost of transportation from the Moon to Earth, for example by an electromagnetic launch system, could become vanishingly small. Additionally, as most resource processing is energy intensive, low cost lunar power can also lower the cost of commodities produced on the Moon.

2.2.3. ³He. The Moon also might also play a role in the development of nuclear fusion as a long-term source of power on Earth (Kulcinski et al. 1989). The Moon's surface is covered with many meters of regolith that stores low but ubiquitous concentrations of Helium-3 (^{3}He), an isotope of helium that undergoes fusion reactions with deuterium (D-^{3}He) and with itself (^{3}He-^{3}He), which may ultimately be tapped for energy. The atom ^{3}He is quite rare on Earth, but has been implanted into the surface of the particles of the regolith covering the Moon. Concentrations are very low (8-10 ppb in maria) (Taylor 1993, 1994) so about 21 km^2 of regolith would have to be mined to a depth of 3 m to extract one metric tonne (mt), if the bulk density of lunar soil is 1.6. However, the energy content of ^{3}He is very high, so very little would have to be transported from Earth in proportion to the electricity production requirements. The energy equivalent value of ^{3}He relative to $7 per barrel crude oil is $1 billion per mt; 40 mt of ^{3}He could provide current U.S. energy needs for a year (Kulcinski and Schmitt 1992). There is estimated to be about 1 million mt of ^{3}He in the upper 3m of the entire Moon (Taylor and Kulcinski 1999) and the Tranquillitatis Ti-rich lunar mare alone is estimated to contain at least 10,000 tonnes of ^{3}He (Cameron 1993). By-products of lunar ^{3}He extraction, largely H, O, N, C, and H_2O, have large potential markets in space and would add to the economic attractiveness of this business opportunity.

A concept for a lunar volatiles miner and processor was developed by Sviatoslavsky (1992). This system would mine its way through the lunar regolith on normally parallel linear tracks, avoiding craters and boulder fields that are beyond the capabilities of the bucket wheel excavation system. Coarse material would be separated from relatively fine material (about 50-60% of the regolith) and discarded or saved for later use as construction aggregate. The fine material would be heated in the processor to 700–800 °C (avoiding the decomposition of troilite, FeS) to extract 80–90% of the solar-wind volatiles. The volatiles would be stored temporarily in tanks that would be picked up and taken to a central base refinery for separation of ^{3}He, H_2 and various other by-products, including water. This miner/processor concept can be used with a "spiral mining" architecture (Schmitt 1992) that transports the extracted volatiles by pipe to a semi-mobile central habitat and refinery that would avoid eventually long distance transport to a central base. There is no doubt that extensive engineering development would be required to create an effective process for extraction of ^{3}He and transporting it to Earth, but no known technological breakthroughs are required. The most important of these developments would be in the areas of low-maintenance excavation systems, highly effective thermal-extraction systems and the large power sources that would be necessary for heating.

Environmental questions related to lunar ^{3}He mining have not been completely addressed. Vondrak (1989) concluded that human activities at a modest lunar outpost could easily add more to the lunar atmosphere than natural processes (principally from the solar wind), but the effects should be localized as gases are removed rapidly. For larger activities, current understanding of the lunar environment is insufficient to predict where and when the environment could be degraded for other activities. Maintenance of high vacuum conditions for astronomical observations is probably the most demanding requirement. Establishment of major or long-term activities that can generate atmospheric contamination should be preceded by experimental verification of the behavior of the lunar environment.

The current state of fusion energy research is an issue relative to the use of this lunar resource and parallel research in this area would be required. U.S. federal fusion resources have been concentrated on the D-T reaction and not on alternative fuels such as ^{3}He. No D-T magnetic containment fusion device that produces more energy than it consumes has yet been developed and the containment requirements for D-^{3}He fusion reaction are more severe than those for the D-T reaction. However, other containment approaches, such as inertial electrostatic confinement (IEC), are being studied. IEC devices may have other applications than large scale electrical power production, including isotope production, mobile power sources for ships and planes, nuclear waste transmutation, and several others (Kulcinski 1996). The development of lunar ^{3}He could also lead to the development of fusion rocket propulsion systems, with long-term implications for interplanetary missions in terms of reduced trip times, associated reductions in astronaut exposure to the low-gravity and radiation environment of space.

2.3. The industrialization of space

Various types of production facilities might eventually be located in space as profit-making ventures or to relieve environmental pressures on Earth. Most investigations of material processing in space to date have been aimed at producing new and unique materials by taking advantage of the micro-gravity environment. Little research has been aimed at production in space, because the cost of space transportation is so high that only products that sell at prices several times the cost/mass of space transportation can be considered. Currently, the bulk cost of transportation from Earth to orbit is approximately $10,000/kg or $10/g, so any product returned from space must carry a price far above that of gold (~$10/g) and probably several times that much (depending on manufacturing costs), unless launch costs are reduced.

The Commercial Space Transportation Study (CSTS) (Boeing et al. 1994) considered space manufacturing in the context of the current infrastructure available to experimenters.

The CSTS concluded that a space manufacturing facility could be profitable if the cost of Earth to orbit transportation fell to about $1200/kg. Even if transportation costs from Earth reached that level, there would still be few products valuable enough to merit the transportation costs of the raw materials into space.

The CSTS study considered the most likely customers for an initial space industrial park to be microprocessor and medical tissue producers, as well as other industries seeking access to microgravity or high vacuum to improve production processes. Few actual products have been identified; however, possibilities have been identified that will be tested in the International Space Station. For example, it has been discovered that certain porous and inter-metallic ceramic materials that can only be produced in microgravity may allow strong and effective bone replacements (Zhang et al. 1999) that could last the lifetime of the patient, rather than the 10 years typical of current materials. Each of the approximately 10,000 hip replacements made in the U.S. each year might use 1 kg of space-produced materials. The capability of producing the required materials on the Moon and transporting them to an orbital manufacturing facility could be feasible if the total costs were a small fraction of the cost of a hip replacement operation ($10,000). A space business park could require a number of similar applications to be viable.

The development of the infrastructure (transportation, habitats in space, power) for space manufacturing could enable other uses, such as space tourism. Tourists in space would require a large range of supplies and materials, including food, which might be brought from the Moon, if production and transportation costs were low enough. Propellant produced on the Moon would find use at the space industrial park as well as helping to lower transportation costs from the Moon. Aeroshells for spacecraft returning to Earth could be fabricated from lunar materials to lower the cost of returning manufactured products. Some of the products that a lunar outpost might provide to a space industrial park are listed in Table 6.3.

2.4. Exploration and development of the Solar System

The human exploration and development of the inner Solar System is critically dependent on low cost space transportation. Development of the Moon's resources can help to reduce the cost of space transportation. In the near term, humans in space will be limited to trips of a few months to a few years, in rockets that use chemical propulsion, particularly liquid hydrogen

Table 6.3. Potential lunar products for space industrialization and tourism.

Product	Use in Space	Source
Propellant; H_2/O_2 for life support	Space transportation, station-keeping for orbital assets; life support consumables; energy storage (e.g. fuel cell reactants)	Regolith; ilmenite; polar ice deposits
Raw materials (metals, oxides, ceramics)	Large structural platforms; specialty manufacturing; Earth entry aeroshells	Mare and highland regolith
Organic constituents (e.g. plastics)	Manufactured parts for use in space	Lunar polar volatiles; recycled space industrial wastes
Inorganic reagents (e.g. H_2SO_4)	Chemical processing	Mare regolith; lunar polar volatiles
Solar PV cells	Power systems	Lunar silicon
Food; manufactured goods	Earth orbit tourist facility	Recycled products; Regolith

and oxygen in the Earth's neighborhood, or methane and oxygen for Mars missions. For any mission to high Earth orbits (e.g., GEO) or beyond (Moon, Mars), approximately 75% of the mass that must be lifted from Earth is propellant to take the spacecraft beyond LEO. Thus, an expansion of exploration and utilization in the inner Solar System is propellant intensive. If the cost of propellant delivered to particular points in space from the Moon is less than the cost of delivering that amount of propellant from Earth, propellant production on the Moon can be economically competitive if the total demand is large enough to amortize the cost of installing the production facilities.

For a round trip to the Moon by a team of astronauts, the availability of propellant on the Moon reduces the total mass that must be lifted from Earth to LEO by more than a factor of 2 (Siegfried and Santa 1999). If a high-Earth orbit or lunar orbit refueling depot were to be supplied with propellant from the Moon, another factor of 2 reduction can be obtained, because the propellant required for landing on the Moon no longer would have to be brought from Earth. Thus, the Earth to orbit transportation vehicles could be a factor of four smaller and current space transportation systems, such as the Space Shuttle or large expendable vehicles.

Recent NASA exploration strategies (e.g., NASA 1998) have focused on the use of the Earth-Moon Lagrange (libration) point L-1 as a high-Earth orbit transportation node. The Lagrange points are points of equilibrium in the Earth-Moon and Sun-Earth systems (Fig. 6.3). The Earth–Moon L-1 point provides a convenient location in high Earth orbit with the special property that it is fixed in position with respect to the lunar surface and trips to any place on the Moon from L-1 can be carried out at any time. NASA studies have identified low-energy trajectories between Lagrange points in the Sun–Earth–Moon system (Lo and Ross 2001). Concepts exist for using the L-1 point as a location where large space telescopes could be constructed and then moved farther from the Earth to the Sun–Earth L-2 point.

With fueling both on the lunar surface and an L-1 depot, a single Space Shuttle launch could propel a fully fueled crew module on a three-day trip to the L-1 point, where the crew could transfer to a lunar descent/ascent stage for a one-day trip to the Moon. When they were finished working on the Moon, their descent/ascent stage would have been refueled on the Moon and they would return to L-1, transfer to their crew module and return to Earth. At that point (assuming that lunar propellant is inexpensive to produce and transport), the cost of going to the Moon might be only slightly larger than going to L-1 or GEO.

In a recent Mars reference mission strategy (NASA 1998), an interplanetary vehicle and its H_2/O_2 propellant would be taken from LEO to L-1 using an efficient ion propulsion system. Crews would be transported separately to L-1 in a crew transfer vehicle, which could be left at L-1 for their return to Earth. The H_2/O_2 interplanetary vehicle would be used to take the crew to Mars and return. Availability of propellant from the Moon would reduce the size of the ion

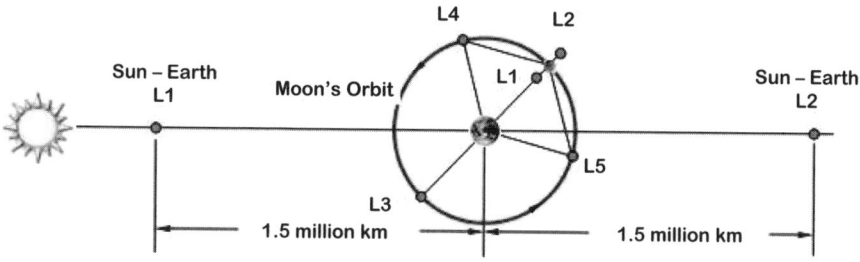

Figure 6.3. Schematic view of Earth–Moon Lagrangian points of gravitational stability in the Sun-Earth and Earth-Moon system.

propulsion system required and could influence the strategy for using the crew transfer vehicle (it might be used for lunar missions while the Mars crew were in transit).

The discovery of hydrogen enhancements on the Moon (Feldman et al. 2000, 2001), probably within permanently shadowed craters, increases the likelihood that propellants can be produced on the Moon. Water concentrations of 1 to 10%, may exist, the total quantities may be on the order of 10^{10} mt, and other useful volatile species may be present. The water could be removed by heating the regolith above ambient temperature (<100 K) and condensing the water vapor. Energy requirements would not be excessive, but development of technology for working within the permanent shadow may be difficult (Duke et al. 1998).

It is also possible to extract hydrogen or produce water from typical lunar regolith, where it exists in 50–100 ppm concentrations, by heating of the regolith to ~800 °C (Gibson and Johnson 1971). A concept for extraction was developed by Eagle Engineering (1988) and has been updated by Diaz et al. (2004). A large quantity of regolith must be mined and heated, so excavation, hauling, and extraction processes need to be very efficient in terms of their hardware mass, to avoid large transportation costs for the equipment. The economic extraction of hydrogen from the regolith depends highly on the cost of energy on the Moon, because of the energy required for heating. If energy becomes inexpensive on the Moon, it may become feasible to extract hydrogen in equatorial regions of the Moon. In addition, the carbon and nitrogen by-products of heating the regolith could be used for human support. These constituents also could be valuable byproducts of ^{3}He production (Schmitt 1997).

Abundant and relatively inexpensive water in space would find a variety of uses other than propellant, including life support and chemical processing, which might otherwise be shunned because of the cost of replacing water lost from the system during processing. Water also can provide an effective radiation shield for space habitats in which crews must spend extended times. Propellant production on the Moon would probably be the initial development step, because its availability reduces the launch masses required from Earth for many purposes. Production of a wide variety of materials that may be useful for space industrialization and exploration could follow. Materials ranging from simple uses of lunar regolith for shielding of space assets (McKay et al. 1992) to production of photovoltaic cells (Ignatiev et al. 1998) have been suggested. Thus, the development of lunar resources, first for transportation and, later, for construction, infrastructure development, and support of humans can pave the way for human expansion throughout the inner Solar System.

2.5. The Moon as a planetary science laboratory

The Moon provides a natural platform for its own study and for the study of planetary processes, particularly volcanism, crustal evolution and impact. The importance of lunar science and what the Moon can reveal about Solar System and planetary history is the subject of much of the preceding chapters in this book. This section addresses only those aspects of lunar science that would be enabled by the development of a lunar outpost and much easier access to the Moon by humans. Problems that might be addressed if a lunar outpost were established include:

- *Highland lithologies and the history of the lunar crust.* Because the impact flux prior to 4.0 Ga ago was much higher, the Moon is believed to have a "megaregolith" of fragmental material, perhaps several kilometers in thickness. In the megaregolith a wide variety of ancient crustal rocks may be found. The capability to recognize and extract the range of rock fragments it contains, describe them, and subject them to detailed analysis would benefit from an on-site ability to collect and process materials, to recognize minor constituents of the regolith and perhaps, to do much of the analysis on the Moon.

- *Mare filling history.* The near-side maria were emplaced within basins excavated by impacts prior to about 3.9 Ga ago. The history of this process probably was complex, consisting of periods of volcanic activity interspersed with periods of quiet, in which impacts modified the surface. A sequence of volcanic and regolith units may be encountered with depth. Drilling to depths of 1-2 km at various locations within the mare could provide the history of this process and an understanding of the time scale, source of material, and nature of the external environment as it existed during the basin-filling period. This is more likely to be accomplished when the Moon is readily accessible to humans.

- *Magnetic history of the Moon.* The nature and history of a lunar core might be understood if the Moon had an internally generated magnetic field, as believed by some to have been the case in the period around 4 Ga ago (e.g., Runcorn 1996). Oriented samples from successive layers of mare basalt in an older mare could both provide specimens could provide a definitive test of the existence, magnitude, and orientation of a lunar magnetic field between about 3.7 and 4.0 Ga ago. This is more likely to be accomplished if humans are present on the Moon.

- *Lunar resources.* Exploration for uncommon lunar resources may require a large number of samples to be collected over a large area. Because the regolith is highly mixed, studies might begin with the identification of a particular rock composition among regolith fragments and proceed to establishing the concentration of those fragments over an extended area. By studying large numbers of samples, the point of origin of the fragments could be defined, leading to more intensive exploration by drilling or other sampling methods. A systematic campaign of this type for economic reasons would also yield unique scientific data.

Although significant exploration objectives remain to be carried out from orbit, most future work will require detailed observations on the surface. The presence of humans working on the Moon opens the possibility for a permanent geoscience laboratory, which can contribute to making lunar exploration efficient as well as introduce new science capabilities. A geoscience laboratory on the Moon could undertake a range of investigations: exploring the features of the Moon to determine their composition, age, and role in the formative history of the Moon; and conducting experimental and observational investigations into the processes by which these features were produced. The research facilities that have been established by many nations in Antarctica can provide a useful model. The capabilities of such a laboratory could include:

- *Exploration vehicles.* Such vehicles would greatly enhance exploration and detailed study, as well as emplacement and maintenance of instruments, on a global basis.

- *An analytical laboratory.* This would be capable of conducting very sophisticated geochemical, biochemical and isotopic analysis of samples (Zumwalt 1997). This could also be an instrument development laboratory, focusing on instruments such as very high-resolution mass spectrometers, taking advantage of the vacuum environment of the Moon's surface.

- *An experimental laboratory.* This would allow properties of lunar materials to be tested under a variety of pressure and temperature regimes. This could include an active experimental capability for studying lunar atmospheric phenomena and the lunar environment.

- *Subsurface access capability.* It may be feasible to conduct drilling or tunneling operations to depths of up to 10 km in the Moon, as temperatures remain moderate. At these depths, the base of lunar maria and the subsurface structures around major impact basins could be investigated.

- *Cryogenic laboratory.* Near the lunar poles where access to permanent shadow is possible, large volume cryogenic laboratory facilities allowing the study of natural materials at temperatures associated with the outer Solar System should be feasible.

- *Curatorial facility.* Samples collected for study from all over the Moon would be stored in this facility for subsequent investigations on the Moon and for transfer to study on Earth. The facility would have the capability of storing rocks, regolith samples and cores in controlled ambient lunar conditions. Relatively large samples could be maintained in the facility, as the sample collection trips to particular locations on the Moon may be infrequent.

- *Equipment maintenance and repair facility.* This facility would be a well-equipped machine shop, capable of manufacturing replacement parts, including circuit boards, for facility equipment. This facility might double as an experimental facility for fabrication processes utilizing indigenous feedstocks.

2.6. Astronomical observatories on the Moon

The Moon may provide a particularly useful platform for large astronomical instruments (Mumma and Smith 1990). On the Moon, an optical telescope can function in the same manner as a terrestrial telescope, which would allow it to be rapidly and accurately pointed at a target in the sky. There is no atmosphere, and therefore no atmospheric disturbances at optical wavelengths. The Moon's surface is quite stable, so optical interferometers should be able to operate there (Fig. 6.4). The far side of the Moon is perpetually shielded from low-frequency noise from the Earth, so may become the place of choice for radio telescopes (Fig. 6.5) and long-wavelength infrared observations. At the lunar poles, access to low temperature cold traps could enable very large infrared telescopes to be operated at perpetual cryogenic temperature. Optical and radio telescopes serviced by a polar outpost, but located at reasonable beyond the pole as viewed from Earth could avoid electromagnetic interference from Earth.

Figure 6.4. Lunar optical interferometer. The individual telescopes send light to the central beam combination dome, where the interferogram would be constructed. Individual telescopes could be separated by 100 to 1000 m.

Although the Moon appears to offer significant advantages for astronomy, astronomers have not focused much attention on the possibility of lunar observatories, preferring to develop concepts that can be deployed in deep space. These concepts include very light-weight "gossamer" structures and space-based interferometers. The Next Generation Space Telescope (Webb Space Telescope), which is planned for the decade of 2000 to 2010, would place an observing system in the Sun-Earth L-2 Lagrange point (Goddard Space Flight Center 1999). In a subsequent generation of telescopes, instruments such as NASA's "Terrestrial Planet Finder" (Jet Propulsion Laboratory, 2000) are being considered that would have the capability of resolving earth-sized planets around nearby stars. Such instruments require light-collecting areas of 1000 m[2] or greater. The technological problems of establishing, maintaining, and operating such telescopes will be enormous. Both space and lunar telescope facilities are subject to degradation from micrometeoroids; however, degradation will be slow and the telescopes are likely to be made of segments, which could be replaced. Further examination of the tradeoffs between deep space facilities and facilities on the Moon are required.

Figure 6.5. Lunar radio telescope fills a lunar crater on the far side.

The principal arguments against siting a telescope on the Moon (van Susante 2002) appear to be that it would cost more (due to additional transportation to the lunar surface), that a significant dust problem exists on the Moon, that thermal problems due to changing solar illumination will degrade the telescope, and that an observatory farther out in the Solar System is required to avoid contamination by light scattered from interplanetary dust. In a strategy of lunar development that includes lunar propellant production, the additional cost argument is diminished. If it becomes feasible for a tourist to go to the Moon, astronomical facilities are likely to be affordable. The dust problem can be solved by appropriate protective measures during nearby operations. Thermal problems can be mitigated by locating observatories in polar regions, perhaps in permanently shadowed craters. Deep space locations of telescopes such as the Sun-Earth L1 point will be more difficult operationally if human maintenance is required, owing to the greater distance and longer transfer times from Earth and because few other activities would be planned there. On the Moon, particularly once other lunar activities are underway, access should be quick and relatively inexpensive. The opportunity to draw on infrastructure provided for a lunar outpost may reduce the cost of operating the instruments. Technicians supported largely by other programs might be available to maintain and repair instruments. Vehicles provided for geological exploration could also carry maintenance crews to visit ailing astronomical instruments. Power might be provided from central power facilities.

We do not yet know enough about the Moon and the problems of emplacement and operation of a lunar telescope to settle these issues, however, future exploration can be directed at resolving uncertainties and reducing costs. These include (a) characterizing lunar cold traps to validate their usefulness for astronomy; (b) determining the leakage of radiofrequency energy around the limbs of the Moon, to help determine the location of a low frequency radio observatory; (c) developing lunar resources for practical applications in construction of telescopes and reducing transportation costs; and (d) developing adequate power and communications infrastructures that allow operation of lunar observatories. If telescope mirrors can be fabricated on the Moon from lunar materials, more rapid expansion of lunar telescope capabilities would be feasible.

A variety of intriguing experiments in space physics, high-energy physics and cosmology are discussed by Potter and Wilson (1990). These opportunities take advantage of the lunar environment and the availability of lunar materials. Potential opportunities include studying the Earth's magnetosphere, constructing cosmic ray and neutrino telescopes, and measuring the half-life of the proton. Many of the proposed experiments could only be carried out on the Moon using natural lunar materials as important parts of the system. A proton-decay detector, for example, could require many tons of detector material (rock, water) buried 100 m below the surface, where the natural background radiation would be very low.

The use of the Moon as a platform for observatories is dependent on the preservation of the appropriate environment for the telescopes, in particular, the maintenance of pristine, high-vacuum conditions, which might be altered by large-scale resource extraction activities. This is discussed further below.

3. TRANSPORTATION BY ROCKET IN THE EARTH-MOON SYSTEM

Situated at an average distance of only 384,000 km from Earth, the Moon's location allows it to be reached in a 3–4-day trip from Earth using current technologies. The Moon provides a natural location for a human outpost or experimental station as humans test their capabilities for long-duration missions beyond low-Earth orbit. At such a station, both humans and machines can be tested in a realistic space environment. The Apollo program landed six crews of two persons each on the Moon for very short periods of time. The total amount of time spent on the lunar surface by all crewmembers was approximately 162 person hours. By contrast, trips to Mars using current rocket systems (including nuclear rockets) require either very long transit times or long stays (up to ~500 days) on the surface. Long-duration missions in an Earth-orbiting space station and long stays on the surface of the Moon are the most relevant simulations of human trips to Mars and further outward into the Solar System (Stafford 1991).

The Earth–Moon Lagrange points may provide useful transportation nodes. Trip times from the Earth for humans are reasonable (3 days), any location on the Moon can be reached at any time (about a day after leaving L-1), and a station located at L-1 can provide a safe haven to which lunar explorers can retreat in case of an unresolvable emergency on the Moon. L-1 lies outside the Earth's Van Allen radiation belts, in which the radiation environment is elevated, so it should be possible to establish a long-term outpost. It can provide a starting point for very low energy (but extended time) transfers to the Sun–Earth and other Earth–Moon Lagrange points (Lo and Ross 2001). L-1 makes a good location for transferring from one kind of space transportation system (e.g., electric propulsion) to another (e.g., chemical propulsion).

In traveling through space, the energy required for going from one orbit to another is the dominant factor that determines the size of the propulsion system. The energy required for an orbit change is related to the mass of the spacecraft and exponentially increases with the velocity change (ΔV) that must be imparted to the spacecraft. Breaking into or out of a "gravity well" associated with a planet is the most energy intensive step in reaching another planet or the Moon, as the required velocity changes are the greatest.

In going from the surface of a planet to space, chemical propulsion systems that are relatively inefficient, but can provide the high thrust needed to break away from gravitational forces, are required. Within space, chemical propulsion may be utilized, but low thrust systems, such as solar electric propulsion systems, may also play a part. The low thrust systems can have high efficiency, so they require lower amounts of fuel, but the fuels tend to be exotic (e.g., Xenon). If they are used to leave the gravitational field of a planet, the low thrust systems require very long transfer times. The SMART-1 mission, utilizing solar-electric propulsion, required about 10 months to go from Earth to Moon (Foing 2004). For quick trips in the vicinity of the Earth and Moon, chemical propulsion currently appears to be the best choice.

A variety of vehicles and systems have been studied for reusable space transportation systems. A recent example of such considerations is the OASIS study led by the Langley Space Center in 2001 (Troutman 2002). This concept utilizes a vehicle that can be propelled either by cryogenic propellants (H_2/O_2) or by electric propulsion, giving it additional flexibility for certain missions.

When utilizing chemical propellants, such as could be produced from lunar resources, the amount of propellant required can be calculated from what is known as the rocket equation (Larson and Wertz 1992).[1] This equation can be used to relate the amount of propellant required to the mass of the spacecraft and its payload and to the velocity change, ΔV. A very simplified calculation (which ignores complexities in spacecraft design) can be used to demonstrate the effectiveness of lunar propellant in the Earth-Moon system. The rocket equation takes into consideration the fact that in accelerating a rocket, some of the propellant is needed to accelerate the rest of the propellant as well as the payload. Using this equation and the characteristics of spacecraft, if the Apollo missions could have refueled on the Moon for their return to Earth, about half of the launch mass would have been required, that is, the Saturn-V launch vehicle could have been half as large. If, in addition, lunar propellant is transferred to a fuel depot in space, such as at L-1, the requirement for launching payloads from Earth is halved once more. The availability of propellant in space would also be conducive to reusable in-space transfer vehicles. If lunar propellant were used in reusable transfer vehicles going from low Earth orbit to L1 to Moon and return, the mass of the system launched from Earth can be less than one tenth of the scale of Apollo to carry the same lunar module from the Earth to Moon and return (Duke et al. 2004).

Once a productive facility is installed on the Moon, it may become economically feasible to transport other lunar products into space. Exporting material products, such as propellants, from the Moon is a more difficult economic problem than simply using products on the Moon, because about as much propellant must be produced as the payload being launched. However, if a vehicle that is reusable and fueled on the Moon from lunar resources can be developed, transporting propellant to lunar orbit, a Lagrange point, or GEO is economically favored over bringing propellant from Earth. Blair et al. (2002) examined such a case. Their scenario demonstrated a performance advantage for lunar propellant over Earth propellant, with economic returns at sufficiently large propellant demand.

The low gravity field and lack of atmosphere on the Moon could allow electromagnetic launchers (Billingham and Gilbreath 1979; O'Neill 1989) to operate at very low energy cost. Electromagnetic propulsion systems have been demonstrated in the laboratory to be able to achieve accelerations of 500-g to 1800-g (Space Studies Institute 1990). If the electromagnetic launcher can produce an acceleration of 500-g ($g = 9.8$ m/sec^2), lunar escape velocity can be reached with an accelerator 587 m in length, at 1800-g the accelerator would be 160 m in length. Once launched to space, the objects launched from the Moon must be collected. In general, the dispersions associated with minor deviations at launch lead to rather large dispersions in space, which would require propellant if all the objects were to be collected. Heppenheimer (1985) showed that certain locations on the Moon could yield orbits that converged upon a single point in space and that high accuracy could be attained for payloads launched to L-2, optimizing the capability to assemble many payloads at that point without expending large amounts of energy.

Two major strategies for transportation from the Moon can be discerned from the above comments. In the first, a reusable H$_2$/O$_2$ vehicle that could be fueled on the Moon could be utilized for round trips carrying payloads to a Lagrange-point station. If propellant were also available at the Lagrange-point station, the amount of propellant that must be transported from

---

[1] The amount of propellant required can be computed from the "rocket equation," which can be expressed as $M_i/M_f = \exp(\Delta V/g \cdot I_{sp})$, where M_i is the initial mass (rocket plus payload plus propellant) and M_f is the final mass (rocket and payload), ΔV is the change in velocity, $g = 9.8$ m/sec^2 is the acceleration of gravity at the Earth's surface and I_{sp} is specific impulse (determined at the Earth's surface). The I_{sp} of the Space Shuttle's main engines, which burn liquid oxygen and liquid hydrogen, is 460 s; a typical chemical rocket burning hydrocarbons and oxygen, 360 s; and a solid rocket, 310 s. Low thrust propulsion systems can have very high I_{sp} (1000-3000), but are not powerful enough to overcome planetary gravity fields and can be used only in free space.

Earth, and therefore the size of the launch vehicle, are substantially reduced. In the mass-driver strategy, a relatively small investment in material and access to large quantities of lunar power could enable this new transportation system. It is clear that the former approach is more feasible in the near term and, in the far term, a transportation system that does not use non-renewable lunar resources would be more desirable.

4. THE DEVELOPMENT AND USE OF LUNAR RESOURCES

The availability of lunar material resources and power is a crosscutting idea that has implications for each of the development themes listed above in Section 2. Here we examine lunar resource availability, the technology needed to access these resources, and the manufacturing capabilities that would be developed on the Moon or in space to process lunar materials.

4.1. What are the lunar resources?

The rocks of the Moon contain all of the elements found on the Earth; however, many very useful elements, such as copper, gold, chlorine, and boron, which have been concentrated by natural processes on the Earth, are in dispersed form on the Moon. Humans ultimately will use only small proportions of the available resources of the Earth and Moon, but the natural concentration mechanisms are very important economically, and civilization on Earth would have been significantly retarded if these elements did not occur in natural concentrations. Nevertheless, many useful elements exist in the rocks of the Moon in substantial concentrations. Others may be obtained as byproducts of the processing of lunar regolith and rocks for other purposes.

The regolith in both highlands and mare regions could be the primary source for many materials, particularly iron, titanium, silicon, and sodium from mare regolith and silicon,

Table 6.4. Common lunar rock chemical compositions.

	Mare Rocks				*Highland Rocks*					
	High-Ti	Low-Ti	Very-Low-Ti	Al-rich	Anortho-site	Norite	Trocto-lite	KREEP basalt	QMD	Gran-ite
SiO$_2$	39.7	45.8	46.0	46.4	45.3	51.1	42.9	50.8	56.9	74.2
TiO$_2$	11.2	2.8	1.1	2.6	<0.02	0.34	0.05	2.2	1.1	0.33
Al$_2$O$_3$	9.5	9.6	12.1	13.6	34.2	15.0	20.7	14.8	6.4	12.5
Cr$_2$O$_3$	0.37	0.56	0.27	0.4	0.004	0.38	0.11	0.31	0.16	0.002
FeO	19.0	20.2	22.1	16.8	0.50	10.7	5.0	10.6	18.6	2.32
MnO	0.25	0.27	0.28	0.26	0.008	0.17	0.07	0.16	0.28	0.02
MgO	7.8	9.7	6.0	8.5	0.21	12.9	19.1	8.2	4.7	0.07
CaO	11.2	10.2	11.6	11.2	19.8	8.8	11.4	9.7	8.3	1.3
Na$_2$O	.38	0.34	0.26	0.4	0.45	0.38	0.20	0.73	0.52	0.52
K$_2$O	0.05	0.06	0.02	0.01	0.11	0.18	0.03	0.67	2.17	8.6
P$_2$O$_5$	0.06	0.05	—	—	—	—	0.03	0.70	1.33	—
S	0.19	0.09	—	—	—	—	—	—	—	—
Total	99.7	99.7	99.7	100.2	100.6	99.9	99.6	98.9	100.5	99.9
Ref.	(1)	(2)	(3)	(4)	(5)	(6)	(7)	(8)	(9)	(10)

References: (1) Average of Apollo 11 and 17 low-K high-Ti mare basalts (Haskin and Warren 1991); (2) Average of Apollo 12 and 15 low-Ti basalts (Haskin and Warren 1991); (3) Luna 24 rock fragment 24174,7 (Taylor et al. 1991); (4) Apollo 14 aluminous mare basalt 14053, with Cr and Na from average of 14321 group 5 (Taylor et al. 1991); (5) Typical anorthosite, 60025 (Table A5.32, Papike et al. 1998); (6) Typical norite, 77075 (Table A5.33, Papike et al. 1998); (7) Troctolite, 76535 (Table A5.33, Papike et al. 1998); (8) KREEP basalt 15386 (Table A5.35, Papike et al. 1998); (9) Quartz monzodiorite (Table A5.35, Papike et al. 1998); (10) Granitic rock fragment (Table A5.36, Papike et al. 1998) calcium, and aluminum from highlands regolith. Production of metallic elements from any of these yields byproduct oxygen. KREEP basalts and rare differentiated rocks associated with them are potential sources of several minor elements, such as phosphorous, rare earth elements, and zirconium (e.g., Meyer 1977).

Essentially all of the samples so far collected on the Moon are fragments (some of them quite large) of rocks found in the regolith, associated with glasses/agglutinates produced by micrometeoroid bombardment on the lunar surface. The range of rock types found in lunar regolith, therefore, indicates the range of primary rocks that could provide useful materials and could be the targets for detailed exploration. The mare rocks sampled to date are all basalts. The highland rocks cover a great range, from anorthosites (principally consisting of feldspar) to norite and troctolite that have different major minerals (pyroxene and olivine, in addition to feldspar). The KREEP, QMD (quartz monzodiorite or *monzogabbro*) and granite are chemically evolved rocks that are related to magmatic differentiation within the Moon. The chemical compositions of the principal rock types are characterized in Table 6.4 and discussed in detail in Chapters 2 and 3.

Most lunar resources are not highly concentrated. Therefore, it is important to understand the chemical nature of their minerals, which constitute separable and concentratable portions. Table 6.5 gives the compositions of some typical minerals in lunar rocks. It is likely from these minerals that some of the less abundant elements could be extracted. Among these would be Ti from ilmenite; Mg from olivine; Na, K, Sr and Ba from plagioclase and potassium feldspar; and

Table 6.5. Major-element compositions of lunar minerals (from Papike et al. 1991).

	Pyrox	Pyrox	Pyrox	Pyrox	Oliv	Oliv	Plag	Plag	K-spar	Ilm	Sp	Apat	Whit
SiO_2	49.3	47.1	53.0	54.3	35.4	39.9	44.5	44.6	60.9	0.23		0.72	0.41
TiO_2	3.2	0.86	0.51	1.1	0.00	0.00				53.0	6.3		
Al_2O_3	2.5	0.72	0.24	0.66	0.00	0.00	35.2	35.2	22.5		11.0		
Cr_2O_3	0.45	0.06	0.09	0.64	0.11	<0.02				0.52	43.8		
FeO	11.0	41.0	22.6	4.8	33.8	12.0	0.81	0.20	0.13	45.1	33.3	0.92	1.0
MnO	0.08	0.67	0.62	0.15	0.36	0.1				0.45	0.27		
MgO	14.8	2.3	21.9	16.8	30.1	47.1	0.08	0.06	0.0	0.75	3.9	0.35	2.9
CaO	17.95	8.0	0.80	22.3	0.36	<0.02	19.2	20.0	3.9	0.10		54.4	39.3
Na_2O	0.0	0.0	0.0				0.58	0.35	3.5				0.14
K_2O							0.02	0.01	8.0				0.02
P_2O_5												39.1	42.14
BaO									1.2				
V_2O_3											0.74		
Y_2O_3												1.5	3.7
La_2O_3													1.5
Ce_2O_3													3.8
Nd_2O_3													2.3
Cl												1.8	0.04
F												1.1	0.14
Total	99.3	100.7	99.8	100.8	100.1	99.1	100.4	100.4	100.1	100.0	99.3	99.9	97.7
Ref.	(1)	(2)	(3)	(4)	(5)	(6)	(7)	(8)	(9)	(10)	(11)	(12)	(13)

References: (1) Augite in Apollo 11 high-Ti basalt 10058; (2) Sub-calcic augite in Apollo high-Ti basalt 10058; (3) Orthopyroxene in ferroan anorthosite 15415; (4) Orthopyroxene in norite 78235; (5) Olivine in low-Ti mare basalt 12035; (6) Olivine in highlands troctolite 76535; (7) Plagioclase feldspar in low-Ti mare basalt 12021; (8) Plagioclase feldspar in ferroan anorthosite 60015; (9) K-feldspar in granite 12013; (10) Ilmenite in high-Ti mare basalt 10058; (11) Spinel in low-Ti basalt 12063; (12) Apatite in breccia 14321; (13) Whitlockite (RE-merrillite) in KREEP-rich sample 12013

Mn and Cr from pyroxenes. Chromium might also be recovered from spinel, which is ubiquitous in mare basalts and is found also in anorthosites and troctolites. Chrome spinel and ilmenite concentrations, formed by crystal settling during differentiation of basalt flows or intrusions, could occur (Taylor 1990) but have not been reported. Phosphate minerals may be a useful source of rare earth elements (whitlockite) or halogens (apatite). Whitlockite (RE-merrillite) contains up to 4 wt% Ce_2O_3, the most abundant rare earth element. Apatite contains 3–4 wt% F and Cl. These have been observed as minor or trace minerals in rocks studied to date, but future exploration could search for higher concentrations and beneficiation techniques used to concentrate them. If mineral beneficiation is to be used, rocks rather than regolith may be the best source, as distinct mineral fragments in regolith are less abundant than in the source rocks.

Solar-wind atoms implanted in the grains of the lunar regolith, and grain-surface deposits produced by volcanic or impact vaporization, may constitute unique lunar resources. Regolith compositions generally mimic the bulk composition of the underlying rocks; however, the volatile elements He (and other noble gases), C, H, and N are ubiquitous in the regolith, having been implanted in the surfaces of regolith grains by the solar wind. These elements have low abundances (Table 6.6), but the solar-wind species are strongly surface-correlated (as shown in Table 6.7 by carbon variations with grain size). Most regolith grains have 1000 angstrom thick radiation-damaged rims that contain the solar-wind implanted atoms (Bibring et al 1974; Keller and McKay 1997). Consequently, the finest-grained fractions (<20 μm) of the regolith have significantly higher concentrations of He, H, C, N, and S. Solar-wind volatiles, most notably H and He, are retained preferentially by ilmenite (Cameron 1993). Other elements, volatilized and redeposited as surface coatings on regolith particles by micrometeorite impact, also may be concentrated on grain surfaces. Examples include rare volatile elements such as Cd, Hg, and Zn (Table 6.7). Grain surfaces are also enriched in vapor-deposited Fe^0 (e.g., Pieters et al. 2000;

Table 6.6. Typical concentrations (in ppm) of solar wind volatile species in lunar regolith samples (Haskin and Warren 1991).

	H	*He*	*C*	*N*	*Ne*	*Ar*
Apollo 11	20-100	20-84	96-216	45-110	2-11	1.3-12
Apollo 12	2-106	14-68	23-170	46-140	1.2-6	0.5-4.6
Apollo 14	67-105	5-16	42-225	25-130	0.14-1.6	0.4-2.2
Apollo 15	13-125	5-19	21-186	33-135	0.6-108	0.5-2.7
Apollo 16	4-146	3-36	31-280	4-209	0.4-1.2	0.6-3
Apollo 17	0.1-106	13-41	4-200	7-94	1.2-2.7	0.6-2.6

Table 6.7. Concentrations of surface elements as a function of size fraction of regolith grains in sample 72501.

	C *(ppm)*	*Cd* *(ppb)*	*Hg* *(ppb)*	*Zn* *(ppm)*
Bulk soil	71	39	3.1	17
<37 μm	236			
7-15 μm		72	6.3	33
< 2 μm		106	22	54
Ref.	(1)	(2)	(2)	(2)

References: (1) DesMarais et al. (1975); (2) Krahenbuhl (1977)

Taylor et al. 2001a). Keller and Clemett (2001) and Taylor et al. (2001a,b) recently showed that the composition of the <20 μm fraction is significantly affected by these redeposited elements.

Volcanic glasses, such as the pyroclastic[2] orange glass discovered at the Apollo 17 site, are mare basaltic in bulk composition (Table 6.8), though they tend to have even lower Al than do mare basalts (see Chapter 3). Most importantly, they also contain surface-correlated phases consisting of volatile elements such as S, Ag, Cd, Zn, and Br (e.g., Baedeker et al. 1974; Wasson et al. 1976; Delano 1986). These phases were deposited from the volatile vapor phase associated with the eruption of the particles, and are enriched in elements that are relatively rare on the Moon (Table 6.9). In most cases, bulk deposits of the orange (Apollo 17) and green (Apollo 15) glasses are enriched in volatile elements. The surfaces of the glasses are enriched compared to their interiors by factors of 3 to over 400 (Table 6.9). This suggests that chemically etching or physically abrading volcanic deposits may be an effective way to extract rare volatile elements needed for industrial processes. For a few elements, notably S, bulk concentrations are lower than in mare basalt samples, although their concentrations are still greater in the surfaces than interiors of individual glass spheres.

The lunar surface has been bombarded by meteoritic material and typical regolith contains significant concentrations (2%) of chondritic meteoritic material (Heiken et al. 1991). Metallic iron from meteorite sources may become a source on the Moon for Fe, Ni, and noble metals; however, Fe^0 in the regolith is derived from several sources, including comminution of basalts, which contain small amounts of native Fe, from nanophase Fe (e.g., Morris 1977) formed in the production of agglutinates, and as vapor-deposited nanophase Fe on the surfaces of most soil particles in a mature soil (Taylor et al. 2001a,b).

Results from the Lunar Prospector mission have shown that elevated hydrogen concentrations are likely to exist at the lunar poles, perhaps in association with shadowed craters (Feldman et al. 2000, 2001). Areas with the lowest epithermal neutrons are interpreted as areas with elevated hydrogen contents. Because water molecules released anywhere on the lunar surface have a high probability of migrating to polar cold traps (Watson et al. 1961; Arnold 1979; Butler 1997), a strong possibility is that the hydrogen is associated with water-ice deposits. The Lunar Prospector data suggest that water concentrations may be in the range of 1 to 2%, with local concentrations perhaps as high as 10% (Feldman et al. 2001). Information is not yet available with sufficient precision to designate areas where exploration for water would be most productive. Additional remote sensing should help to delineate water concentrations. Radar mapping from a polar orbiter could map the subsurface at a resolution useful for characterizing the distribution of ice (Nozette et al. 2001), if the concentrations of ice are 30% or more.

Arnold (1979) showed that there was enough water brought to the surface by micrometeoroids and comets or produced on the Moon by interactions of solar-wind hydrogen with lunar minerals to have deposited several meters of water in the cold traps over the past 2 to 3 billion years. However, there are also removal mechanisms such as sputtering (Lanzerotti et al. 1981) and activation by Lyman α radiation (Morgan and Shemansky 1991), and the calculated removal rates are close enough to the calculated deposition to leave the question in doubt. The form of the hydrogen produced by solar-wind interactions has not been determined. Crider and Vondrak (2000) have suggested that much of it may be in the form of OH⁻ ions, which would not be readily removable by sputtering or Lyman α radiation, and even thin layers of impact debris could shield ice from Lyman α radiation. If water has been deposited in the cold traps, other volatile elements may be concentrated there as well. Comets contain

---

[2] Pyroclastic refers to the process of hot-ash extrusion from volcanoes. The lunar pyroclastic glass is believed to have formed in a process similar to fire-fountaining observed in terrestrial volcanoes, where gases disperse and eject droplets of melt that quickly freeze to form glass.

Table 6.8. Compositions of some lunar volcanic glass groups (Delano 1986), listed in order of increasing TiO_2.

	(1)	*(2)*	*(3)*	*(4)*	*(5)*	*(6)*	*(7)*	*(8)*	*(9)*	*(10)*
SiO_2	48.0	44.0	42.9	40.8	40.5	39.4	38.8	37.3	35.6	33.4
TiO_2	0.26	0.91	3.5	4.6	6.9	8.6	9.3	10.0	13.8	16.4
Al_2O_3	7.7	6.9	8.3	6.2	8.0	6.2	7.6	5.7	7.2	4.6
Cr_2O_3	0.57	na	0.59	0.41	0.63	0.67	0.66	0.63	0.77	0.84
FeO	16.5	20.2	22.1	24.7	22.3	22.2	22.9	23.7	21.9	23.9
MnO	0.19	0.23	0.27	0.30	0.25	0.28	0.29	na	0.25	0.30
MgO	18.2	19.5	13.5	14.8	12.6	14.7	11.6	14.3	12.1	13.0
CaO	8.6	7.4	8.5	7.7	8.6	7.5	8.6	7.6	7.9	6.3
Na_2O	nd	0.1	0.45	0.42	0.39	0.41	0.39	0.31	0.49	0.05
K_2O	nd	nd	nd	0.10	nd	0.04	nd	nd	0.12	0.12

nd = not detected; na = not analyzed
Sample: (1) Apollo 15 green C; (2) Apollo 17 green; (3) Apollo 15 yellow; (4) Apollo 14 yellow; (5) Apollo 17 yellow; (6) Apollo 17 orange; (7) Apollo 17 orange; (8) Apollo 11 orange; (9) Apollo 15 red; (10) Apollo 12 red.

Table 6.9. Concentrations of surficial volatile elements on volcanic glasses.

	Zn (µg/g)	Ga (µg/g)	Ge (ng/g)	Cd (ng/g)	In (ng/g)	Ir (pg/g)	Au (pg/g)	F[3] (µg/g)
Apollo 17 orange glass[1]								
Interior	19	2.8	161	32	0.80	210	274	50
Surface	434	33.1	880	1150	62.4	650	860	700
Apollo 15 green glass[2]								
Interior	1.6	2.9	12.3	1.6	0.2	260	50	40
Surface	113	10.9	668	678	6.9	1200	5420	600
Other elements concentrated on surfaces of glass beads, excluding solar-wind components[4]:	B Na S Cl	K Cu Se Br	Ag Sb Te I	Hg Pb B Ta				

Note: Surface concentrations analyzed by leaching and etching; interior concentrations measured on residues. In both cases, about 10% of the glass spheres was etched away.
References: [1]Wasson et al. (1976); [2]Chou et al. (1975); [3]Goldberg et al. (1976); [4]Delano (1986)

abundant volatiles (C, N compounds), some of which might be condensed in the cold traps. Volatile metallic elements, such as Hg, may also be concentrated (Reed 1999).

Sources and sinks for potential resources cannot be resolved by modeling, but must be established by surface exploration. To prepare the way for resource utilization, it will be necessary to establish the lateral and vertical distribution of hydrogen and water in areas that are amenable to resource recovery, to determine the form of hydrogen and its compounds, and to determine the presence of other potentially valuable materials in the polar cold traps. There are also scientific investigations of significance that could be carried out in conjunction with these measurements, particularly investigation of the vertical distribution (stratigraphy) within the hydrogen deposits and the isotopic characteristics of the hydrogen-containing products as indication of their source and depositional history.

Table 6.10 provides a list of resources, their potential host material, estimated grades, and theoretically possible upper limit on grade. A discussion of how these estimates were made is given by Taylor and Martel (2003). Interesting, detailed discussions of lunar resources can be found in Haskin et al. (1993) and Fegley and Swindle (1993).

4.2. Distribution of lunar resources

The distribution of useful resources on the Moon is a result of magmatic and impact processes. Our understanding of the potential distribution of lunar resources is based on results of remote-sensing analyses, calibrated with studies of lunar samples. For a number of elements, including Fe, Ti, and Th, the global distribution is known by observation from Apollo (e.g., Metzger et al. 1973), Clementine (e.g., Lucey et al. 1998a, 2000; Giguere et al. 2000), and Lunar Prospector (e.g., Elphic et al. 1998; Lawrence et al. 1998; Prettyman et al. 2001). As the Lunar Prospector data are further reduced, it may be possible to obtain the distribution of

Table 6.10. Potential resources, possible host materials, and their grades.

Resource	Potential Host Material	Grade (average)	Grade (max. meas.)	Grade (possible)[1]
Solar wind volatiles				
H	regolith	50 µg/g	150 µg/g	150 µg/g
^{3}He	regolith	4 ng/g	30 ng/g	30 ng/g
^{4}He	regolith	14 µg/	100 µg/g	100 µg/g
C	regolith	124 µg/g	300 µg/g	300 µg/g
N	regolith	81 µg/g	150 µg/g	150 µg/g
Precious metals				
Pd	regolith/breccias	12 ng/g	28 ng/g	60 ng/g
Ir	regolith/breccias	9 ng/g	26 ng/g	50 ng/g
Au	regolith/breccias	6 ng/g	17 ng/g	35 ng/g
Iron and ferroalloy metals				
Fe	mare basalts	15 wt%	17 wt%	20 wt%
Ni	regolith	250 µg/g	730 µg/g	1500 µg/g
Co	regolith	35 µg/g	68 µg/g	140 µg/g
W	regolith	370 ng/g	1950 ng/g	3900 ng/g
Mn	mare basalts	0.2 wt%	0.3 wt%	0.3 wt%
Cr	mare basalts	0.2 wt%	1.1 wt%	2 wt%
Cr	highland norites	0.2 wt%	0.3 wt%	15 wt%
Non-ferrous metals				
Ti	High-Ti mare basalts	7 wt%	8 wt%	16 wt%
Al	anorthosite	18 wt%	—	18 wt%
	average highlands	15 wt%	—	15 wt%
Zn	volcanic glass	10 µg/g	400 µg/g	800 µg/g
Cd	Volcanic glass	16 ng/g	1150 ng/g	2200 ng/g
Non-metals				
Ce (rep. REE)	Evolved ig. rocks (KREEP)	175 µg/g	700 µg/g	3500 µg/g
Ba	Evolved ig. rocks (KREEP)	0.1 wt%	0.45 wt%	2 wt%
K	Evolved ig. rocks (KREEP)	0.8 wt%	1.8 wt%	9 wt%
P	Evolved ig. rocks (KREEP)	0.6 wt%	2.2 wt%	10 wt%
Th	Evolved ig. rocks (KREEP)	22 µg/g	50 µg/g	500 µg/g
Zr	Evolved ig. rocks (KREEP)	0.1 wt%	0.7 wt%	3.5 wt%

Notes: Data from Haskin and Warren (1991); Fegley and Swindle (1993); Papike et al. (1998). [1]"Possible Grade" estimated by extrapolation from existing data and knowledge of the processes that concentrate the elements. No data confirm these estimates.

additional elements such as Mg. For other elements, estimates can only be made by deducing the rock type indicated by the remote-sensing data and estimating elemental abundances based on their known distribution in lunar samples. The spatial resolution for most elements in the Lunar Prospector gamma ray spectrometer results is quite coarse (>25 km, see Chapter 2), as will be the X-ray spectrometry data expected from SMART-1. Thus, available data can indicate general areas where resources may be located, but not specific areas of maximum concentration. For elements present in concentrations of 1% or more, an area 100 m² mined to a depth of 1 m, could yield several hundred tons or more of product, which is larger than anticipated early uses. Thus, the 100 m resolution of the best available data from Clementine is potentially useful in locating mining operations, but locating small areas of higher ore grade will require higher resolution data sets. Such detailed studies may be best carried out with surface investigations.

The processes of formation of the lunar regolith have caused lateral mixing that reduces the sharpness of unit boundaries (e.g., Li and Mustard 2000) and the agglutinate-containing and vapor-deposited nanophase Fe limits spectral resolution. Sub-regolith rock type identifications by remote sensing are best done by analyses of materials excavated by fresh impact craters (e.g., Tompkins and Pieters 1999; Staid and Pieters 2000; Pieters et al. 2001). For small craters, the maximum depth of excavation of rocks exposed in the rims of impact craters is about half of the diameter of the crater. This material can mostly be observed within about one crater diameter of the crater rim. The resolution available in the Clementine UVVIS images (about 100 m), therefore, limits the depth resolution of crater composition mapping to >50 m from the surface. For larger craters, the depth/diameter ratio is smaller. The highest-resolution data from Lunar Prospector (e.g., thorium) may ultimately be mappable to a resolution of 15 km (A. Binder, pers. comm..), and so may be suitable for distinguishing layers that are deeper than about 10 km. Pieters et al. (2001) used Clementine UVVIS data to distinguish between anorthosite, gabbro, and olivine-bearing rocks in the South Pole-Aitken Basin. Lunar Prospector data clearly depict materials that likely to have formed at depth within the lunar crust were excavated by the Imbrium and South Pole–Aitken Basins (Pieters et al. 2001). More work of this type may be useful in locating deep layers that could contain potential resources.

Lunar pyroclastic deposits are typically observable from orbit as very dark deposits distributed across broad highland regions or as dark-halo craters within ancient floor-fractured craters (Fig. 6.6; e.g., Head 1974). Most of these deposits are Fe- and Ti-rich, but unrecognized pyroclastic deposits of other compositions may exist (e.g., Gaddis et al. 1985). The association of many pyroclastic deposits with Imbrian-aged floor-fractured craters (e.g., Head and Wilson 1979) or ancient lunar volcanic deposits (e.g., Whitford-Stark 1990; Hiesinger et al. 2000) suggest that most of these deposits are of late Imbrian age, generally 3.2 to 3.7 Ga. These ancient pyroclastic deposits have mature regolith surfaces, and they exhibit the subdued mafic absorption bands comparable to those of mature maria and highlands in the Clementine UVVIS data. The majority of these deposits consist of fragmented basalts, with substantial components of iron-bearing mafic minerals (clinopyroxene, olivine) and smaller amounts (if any) of iron-bearing volcanic glass (Gaddis et al. 2000, 2003). In addition to these mare-like deposits, two additional classes of pyroclastic deposits stand out spectrally, but are rare. These are the "black-bead" deposits similar to the devitrified orange glass found at the Apollo 17 site and the spectrally red glass-rich deposits that occur at Aristarchus Plateau. Although these deposits are areally extensive, this apparently distinctive composition is limited to a few deposits in the central nearside of the Moon. These results suggest that the pyroclastic deposits most commonly recognized as having commercial potential, such as the black-bead deposits at Taurus Littrow that have been suggested as sources of O, Fe, and Ti (e.g., Hawke et al. 1990; Allen et al. 1996), may be restricted to a few source regions on the Moon.

Remote characterization of the distribution of a valuable lunar material resources is exemplified by recent efforts to map the distribution of ³He on the Moon (e.g., Johnson et al. 1999).

It is well known that He is correlated with Ti content of the lunar regolith due to the preferential retention of solar-wind-implanted ^{3}He in the ilmenite crystal structure (e.g., Cameron 1987) (Fig. 6.7). The abundance of solar-wind implanted ^{3}He in the lunar regolith depends on surface or regolith maturity, the amount of solar-wind fluence, and Ti content. Regolith maturity can be assessed through remote-sensing data (e.g., Lucey et al. 1998b). A model of solar-wind fluence combined with maps of surface maturity and TiO_2 content derived from Clementine UVVIS data (Lucey et al. 1998a,b; 2000) can be used to map the approximate concentration of ^{3}He in the regolith (Fig. 6.8). The highest ^{3}He abundances are predicted to occur in the farside maria due to greater solar-wind fluence received (Swindle et al. 1992) and in higher TiO_2 nearside mare regions. Areas that show high-TiO_2 content and high maturity are prime locations for extraction of solar-wind volatiles, including H, C, and N.

Another example of lunar resource assessment is provided by efforts to map thorium concentrations established by Apollo and Lunar Prospector gamma ray spectrometers (Plate 2.3). It is currently believed that thorium is principally associated with the KREEP rocks of the Imbrium Basin (e.g., Haskin et al. 2000). It can be inferred from current understanding of lunar rock types that elements that are enriched in KREEP will be concentrated in areas with high Th concentrations (e.g., Jolliff et al. 2000). Among the

Figure 6.6. Clementine 750-nm images of dark pyroclastic deposits at Rima Bode (top) and in the floor of Oppenheimer crater (bottom) (after Gaddis et al. 2003).

more interesting of these are Th, U, Zr, and P, which could have significant industrial and agricultural uses. However, the resolution of the Lunar Prospector data is not sufficient to delineate the highest concentrations at scales commensurate with mining operations; surface mapping will be required.

In general, the orbital geochemical mapping tools have resolution too coarse to be used for operational resource mapping, but trends established by these tools correlated to geological features feed back through high-resolution multispectral imaging to useful resource assessment. It is possible to improve spatial resolution for X-ray and γ-ray spectrometers by using collimated instruments and by lowering the spacecraft altitude (W. Boynton, pers. comm.). The Lunar Prospector team has demonstrated that higher resolution can be obtained with more data; however, these improvements are limited by counting statistics, i.e., to improve the data by a factor of two requires four times the data. Active orbital sensing techniques, using energetic beams for probing the surface, may be feasible (Meinel et al 1990), but would require large amounts of energy if conducted from 100 km orbits.

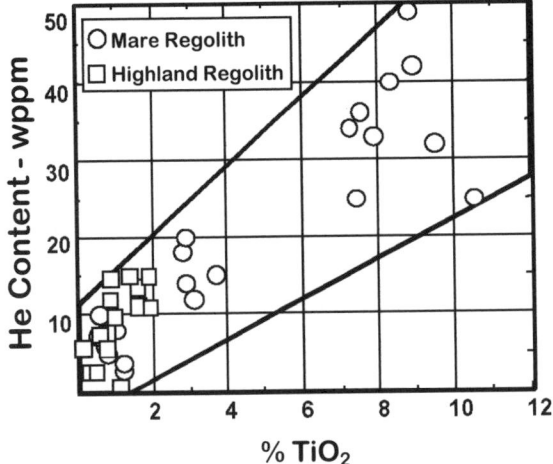

Figure 6.7. Helium content (ppm, weight) in lunar soils as a function of TiO$_2$ content (after Cameron 1987).

4.3. Prospecting for and processing lunar resources

4.3.1. Surface prospecting for lunar mineral resources. It appears that surface prospecting will be required to detect and characterize sub-surface concentrations of useful minerals. This may not be an issue for major elements, because a few percent variation in grade is not likely to cause changes in the extraction process. However, if minor minerals or subtle features are sought, it will be necessary to collect and analyze many samples to localize sub-regolith sources in mare regions. A procedure might consist of first removing agglutinates magnetically, then separating and analyzing the mineral constituents in a given size fraction of the regolith. The distribution of rock types in many samples of regolith from a local area might allow subsurface deposits of interest to be located.

An alternative approach is to use rovers equipped with analytical devices capable of measuring the abundances of target elements or elements that are diagnostic for the target elements (e.g., Th for the rare earth elements). The rovers would be sent to regions thought on the basis of remote sensing data to be rich in a specific resource. An example of how we might explore the Moon for a specific resource is the search for economic deposits of non-metals (e.g., Th, REE, Zr, P, K). This search could begin by finding the regions richest in Th by using orbital geochemical data. The spatial resolution of orbital γ-ray measurements is about 60 km. A rover could be landed inside a 60-km spatial resolution element that contained a high concentration of Th. The Th might be distributed uniformly, or it might be concentrated locally. The rover would autonomously traverse an assigned region, making measurements of a selected element or elements. For this example, a good choice would be Zr, which is relatively abundant (Table 6.8) and is correlated with Th, rare earth elements, and K. Measurements could be made every 10 or 100 meters. The data would be transmitted to orbiting satellites. The numerous analyses at known locations would allow planetary economic geologists to make contour maps of the abundance of Zr (or another element), thereby mapping out potential concentrations of the elements of interest. Regions of exceptional concentrations would be investigated further.

Another alternative is the systematic analysis of rocks excavated by recent impacts, which can be used to characterize near surface units. For example, the maria were emplaced as a sequence of flows. Some layers may have cooled slowly enough to allow differentiation and concentration of useful minerals. These might be located by systematic study of ejecta from craters. Old, buried mare regolith surfaces that might contain different concentrations of solar-wind products than the surface regolith might also be located in this manner.

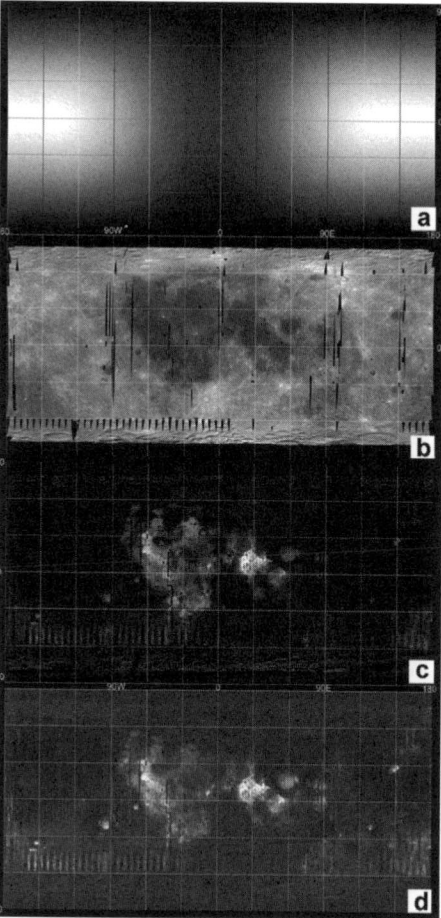

Figure 6.8. Simple cylindrical projections (left side of images corresponds to 180 W longitude; 30-degree grid shown, resolution is 0.25°/pixel) of (a) solar wind fluence model, (b) Clementine 750-nm mosaic; (c) TiO_2 abundance map from Lucey et al. (2000) displayed from 0 to 6 wt%; and (d) ^{3}He abundance map displayed from 0 to 10 ppb (from Johnson et al. 1999).

On Earth, when concentrations of ore minerals are suspected at depth, drilling is used to delineate the extent of the resource. Studying crater ejecta is an alternative to drilling, using the concept that the impact excavation reveals materials from depth. On the Moon, there will be a tradeoff between exploring for concentrations of trace minerals in subsurface rocks and beneficiation of the already-crushed materials in the regolith. It may be preferable to expend energy on beneficiation, even if it requires that much more regolith be mined and processed, than to go to the next step of processing solid rock, a subject addressed by Chambers et al. (1995). However, if a high concentration of a mineral such as chromite were found in a layered magmatic deposit, other processes, such as those used in hard rock mining on Earth, might be favored.

Surface exploration for hydrogen and water in lunar permanently shadowed craters is a special case, because useful deposits can be within 1-2 meters of the surface. A neutron spectrometer and perhaps ground-penetrating radar, active seismic, or electrical resistivity techniques could be used to characterize the content of hydrogen and possibly water ice. Thermal probes could be employed to sense the presence of condensed volatiles (it is possible that the presence of the sensing system itself, e.g., a rover, would be a sufficient energy source). Drilling and sampling to depths of 1–2 m could be used to determine the extent, thickness and grade of the deposits. All of these tools might be carried aboard a robotic rover, capable of spending many hours within the permanent shadow. The deployment of such a rover may be made feasible by Stirling dynamic isotope power systems (Thieme et al. 2000) that are being developed for outer planet and comet missions.

4.3.2. Mining. Mining is a somewhat misleading term in the context of most lunar-materials processing strategies, because most material will be mined by scooping up regolith, and the scale of activities will be rather small by terrestrial standards. This does not apply to concepts for H or ^{3}He mining, which would require the excavation of very large areas utilizing large machinery, comparable to terrestrial mining of coal and harvesting of grain. Approaches to such large-scale excavation include front-end loaders, scrapers, and continuous excavators (Chamberlain et al 1992).

Modeling of excavation of the lunar regolith (Eagle Engineering 1988; Muff et al. 2001) indicates that excavation is a low-energy process and machines should be able to excavate several times their mass each hour of operation. Most mining would be done during daylight hours at equatorial outposts. At a polar outpost, where mining in permanent shadow is required, all activities must be done in the dark. Issues associated with mining include the ability to provide energy to mobile equipment, so that mining equipment can operate flexibly, as well as maintenance and repair, which is a significant problem for terrestrial mining equipment. Henley et al. (2002) proposed laser-power beaming as a method of transmitting energy to a mobile system operating in permanent shadow. Schrunk et al. (1999) have proposed more complex lunar infrastructures based in part in transmitting power from point to point on the Moon.

4.3.3. Transportation. If the mine is separated from the processing plant by a significant distance, a surface transportation system is required. A mechanized hauler is a convenient and efficient way to provide surface transportation. Such systems, operating in the low-gravity field of the Moon, can haul more of a load per unit mass of equipment than equivalent haulers on Earth (Eagle Engineering 1988). The construction of roadways can make haulers even more effective. In special situations, for example, where the mining site is at higher elevation than the processing site, gravity can be used effectively to transport regolith to a reactor. A tramway, for example can have gondolas that are filled with material and fall in the Moon's gravitational field, simultaneously delivering regolith to the processing facility and pulling empty gondolas to the top.

If the resource to be extracted is in low abundance, such as the case for solar-wind implanted volatiles, it may not be cost-effective to haul material long distances to processing plants and a mobile processor might well be a more effective solution. However, typically, the energy required for extraction in these cases is high, so means must be found to provide power to the excavator, as proposed for the ^{3}He excavator (Sviatoslavski 1992) or the power-beaming concept of Henley et al. (2002).

4.3.4. Beneficiation. Beneficiation is the term used to designate the processing of an ore to concentrate a particularly useful mineral or element. Examples of beneficiation on the Moon could include: (1) size sorting of regolith to concentrate the fine-grained portions that preferentially retain solar-wind volatiles and native Fe; (2) concentration of ilmenite using electrostatic separation techniques (Agosto 1985), or (3) magnetic separation of Fe-bearing minerals from anorthositic regolith to concentrate anorthite or Fe (Taylor and Oder 1990), and ilmenite from regolith or crushed rock (Chambers et al. 1994). The use of beneficiation is dictated by the economics of the mineral recovery process. If the advantages of using a beneficiation process outweigh the costs and complexities of transporting and operating another piece of equipment on the Moon, the beneficiation technique will be used.

4.3.5. Thermal and chemical processing

Extraction of oxygen. Most investigations of the extraction of resources from lunar materials have focused on the extraction of oxygen (Table 6.11, from Taylor and Carrier 1992a). Table 6.12 gives characteristic feeds, plant mass, and energy requirements (Taylor and Carrier 1992a,b) for several processes. Each of the plants could produce several times its own mass in a year. Those with smaller masses, including the mass of the energy-production system, would be favored for application on the Moon, due to the high costs of transportation. In ranking processes based on consideration of technology readiness, number of major steps, process conditions, and feedstock requirements, Taylor and Carrier (1992a) selected 8 processes (bold in Table 6.11) as the "best" for further consideration.

Typically, when iron oxide in ilmenite, silicate minerals, or molten regolith is reduced, Fe^0 appears as a byproduct of oxygen production. Because the reducing agents are not abundant on the Moon, they must be recycled and conserved (see Section 4.4.7). The amount of reducing agent can be significant. For example, in the reaction $FeO + H_2 = H_2O + Fe^0$; $H_2O = H_2 + \frac{1}{2}O_2$

Table 6.11. Oxygen and volatile extraction methods (Taylor and Carrier 1992).

Solid/Gas Interaction	*Silicate/Oxide Melt*
Ilmenite reduction with Hydrogen	Molten Silicate Electrolysis
Ilmenite Reduction with C/CO	Fluxed Silicate Electrolysis
Ilmenite Reduction with Methane	Caustic Dissolution and Electrolysis
Glass Reduction with Hydrogen	Carbothermal Reduction
Reduction of FeO with Hydrogen Sulfide	Magma Partial Oxidation
Extraction of silicates with Fluorine	Li or Na Reduction of Ilmenite
Carbochlorination of silicates	
Chlorine Plasma Reduction of regolith	*Aqueous Solutions*
	HF Acid Dissolution
Pyrolysis	H_2SO_4 Acid Dissolution
Vapor Pyrolysis	
Ion Plasma Pyrolysis	*Co Product Recovery*
Plasma Reduction of Ilmenite	Hydrogen-Helium-Water from Soil

Table 6.12. Scale and energy requirements for selected oxygen plants (Taylor and Carrier 1992).

Processes	Ore (1000 tonne/yr)		Plant Mass (tonne)	Energy (MW-yr)
	Raw	Process Throughput		
Ilmenite: High-Ti Mare				
Reduction[2]	210,000	21,000[1]	200	3
CO Reduction	210,000	21,000[1]	225	3.5
CH_4 Reduction	210,000	21,000[1]	225	3.5
Mare or Highlands				
Glass reduction by H_2	160,000	80,000[2]	200	4
Molten silicate electrolysis	5000	5000[3]	70	3
Fluxed molten silicate electrolysis	5000	5000[3]	80	3.5
Vapor Pyrolysis	5000	5000[3]	40	2
Ion Plasma pyrolysis	5000	5000[3]	40	2.5

Notes: [1] Assumes feedstock with 50 wt% ilmenite from an ore with 5% available ilmenite. It is assumed that the iron oxide in 90% of the ilmenite is converted . [2] Assumes soil with 25% glass is beneficiated by a factor of two; assumes 15% FeO in glass and 75% conversion to Fe° + O_2. [3] No beneficiation required; Assumes 43 wt% O_2 in regolith with 50% recovery.

the amount of hydrogen needed is 25% of the amount of the oxygen produced per pass. In the case of $FeO + CH_4 = 4Fe^0 + CO_2 + 2H_2O$; $H_2O = H_2 + \frac{1}{2}O_2$, the amount of methane required is 50% by amount of oxygen produced. Thus, the processing must be highly effective in the recovery of the reagents or a source of reagents must be found on the Moon. Electrolytic methods using molten regolith do not require reagents, so they avoid the problem of reagent loss; however, electrodes are generally consumed during the operation of an electrolytic cell, and their replacement may be a problem of equivalent severity. Terrestrial electrolytic techniques such as the Hall process for aluminum require catalysts, but these are also consumed slowly in the production process. Pyrolysis techniques (heating to very-high temperatures to evolve metals in a vapor phase) requires a great deal of energy, but might be directly implemented on the lunar surface with concentrated energy from a solar mirror, operating in the lunar vacuum. Tradeoffs between materials and energy will have to be considered before the best processes can be selected.

The hydrogen reduction process has been demonstrated with actual lunar samples. Gibson et al. (1994) produced oxygen by reducing lunar basalt at temperatures of 900-1150 °C. Ilmenite was the primary source of oxygen, with lesser yields from olivine and pyroxene. The total oxygen release ranged from 2.9 to 4.6 wt%. Allen et al. (1994b, 1996) demonstrated oxygen release from 16 lunar soils and 3 volcanic glass samples. Hydrogen reduction at 1050 °C yielded 1.6–4.7 wt% oxygen, with yield directly correlated to each sample's total iron content. Ilmenite-rich mare soils and iron-rich glass samples yielded the most oxygen. However, it should be emphasized that these were experiments conducted in a "batch mode," and were only preliminary to demonstrate that oxygen extraction is possible.

Extraction of metals. Any process that can release oxygen from lunar materials also is producing a reduced material, e.g., metal, at the same time. For example, the hydrogen reduction of ilmenite produces metallic iron and TiO_2, and the carbothermal reduction process can produce silicon as well. Other processes, for example HF reduction and magma or molten-salt electrolysis can lead to the production of a range of metals. Although hydrogen reduction of ilmenite produces a residuum of TiO_2 and Fe^0, experimental studies of the reaction inevitably lead to Fe^0 being intimately mixed with TiO_2. In principle, a high temperature furnace at an oxygen fugacity below the ilmenite stability buffer curve could be used to heat this mixture to the melting point of iron, which could be separated from the slag. Carbon can be used to reduce silicon, as well as iron (Rosenberg et al. 1996). The iron is more readily reduced and can be extracted at lower temperature than that needed to reduce silicon. In fact, the process of producing elemental silicon is more complex, because the first product of reaction of carbon and silicon is silicon carbide. On Earth, silicon carbide in the presence of excess SiO_2 is reacted at about 1500 °C to produce elemental silicon. No complete reactor for the production of silicon by this method has been defined.

For production of elements higher on the electrochemical series, fluorine has been proposed as the reducing agent (Keller 1988; Keller and Taberaux 1991; Burt 1992; Waldron 1993). Indeed, fluorine will attack almost all constituents of lunar rocks and, by a variety of separation steps, elements or oxides could be produced. However, the process is complex, might require makeup of fluorine, which is scarce on the Moon, and could produce a great quantity of unusable products.

Magma electrolysis has also been proposed (Haskin et al. 1992). In this process, raw materials are heated to their melting temperature and metals and oxygen extracted by electrolysis. This also can in principle reduce all constituents to the elemental stage, though Fe, Si, Ti, and Al will be produced in that order. There are many technical difficulties associated with magma electrolysis, such as containment vessels and electrode stability; however, the fact that it operates at high temperature does not mean that it is overly energy intensive. The electrical energy that must be provided is the same energy that must be provided to break metal-oxygen bonds in other processes. Thermal energy can be used to heat the material to its melting point. Molten salt electrolysis, which is similar to magma electrolysis, but in which small amounts of the material to be processed are dissolved in a molten chloride or fluoride bath, may also be useful. This approach is used terrestrially for commercial production of metals such as Al and Ca, and has been proposed for production of Si (Rao 1988). A molten salt system would operate at lower temperature and would probably focus on extracting a single element, such as Si or Al from lunar anorthite. Operational problems, such as electrode stability and replacement of electrolyte, would have to be solved.

Much materials-processing research has been aimed at using regolith as the starting material. As pointed out previously, this eliminates the need for crushing or grinding of the feed material. However, the mixed chemical nature of regolith leads to the problem that either mixtures of compounds (e.g., Fe + TiO_2) that require further processing are produced, or that more material is processed than may be needed for the applications intended. One way to deal with these problems is to consider pre-processing of feedstock by chemical means. For example, lunar plagioclase might be digested in mineral acid (HCl, H_2SO_4) and relatively pure oxides (SiO_2, CaO, Al_2O_3) separated by chemical means at low temperature. The energy required for this process is principally that needed to recover the acid. Gillett (1997) has proposed a similar process in which silicates are dissolved in an acid-alcohol solution. These types of dissolutions are commonly used in preparation of silica gels for sol-gel processing in ceramic production and the intermediate products are well known. However, no research has been done to demonstrate the production and separation of the oxide phases. This type of processing would to some extent mimic the natural processes that have concentrated elements in the Earth.

If the oxides can be concentrated chemically, the chemical processing steps identified above may be more tractable. Iron oxide, separated from ilmenite, could be reduced directly with hydrogen to produce metallic iron. Silicon could be produced by carbothermal reduction of SiO_2 without the intermediate steps that involve the carbides or oxides of the other elements. Also, the oxides may be useful in their own right. Mixtures of SiO_2, CaO, and Al_2O_3 in various proportions will produce useful ceramics and cements. TiO_2 is a widely used industrial product on Earth. Chemical separation might also lead to processes for concentrating byproduct trace constituents, such as Ba or REE in plagioclase. This remains a largely uninvestigated area.

Some elements must be obtained in high purity to be useful. This is the case, for example, for Si, if it is to be used in photovoltaic devices. Silicon for these purposes must contain less than 100 ppm of contaminants. Silicon for semiconductors must be two orders of magnitude purer. On Earth, Si is purified by fractional distillation of SiF_4 in a process that requires considerable energy and is inefficient. Pre-processing of anorthite to produce SiO_2, as suggested above, may be a practical purification step in which contaminating species such as Fe and Ti can be largely eliminated.

Extraction of hydrogen and water. Hydrogen is perhaps the most important, but among the rarest, element on the Moon. It is incorporated into the regolith by solar wind implantation and exists at concentrations of 50 to 100 ppm in many regolith samples. Experiments show that the hydrogen can be extracted by heating regolith to modest temperatures (800 °C), in a process that also releases other volatile components (He, CH_4, H_2O, etc.), either from other solar wind species or from the reaction of solar wind species with oxides in the regolith (Gibson and Johnson 1971). Once released, hydrogen could be reacted with ilmenite to produce water (Allen et al. 1994b, 1996).

Eagle Engineering (1988) provided an engineering analysis of systems that would extract oxygen from lunar ilmenite by hydrogen reduction in a fluidized bed processor. Two approaches were studied. The first approach assumed that hydrogen brought from Earth would be reacted with ilmenite concentrated from the lunar regolith or from ilmenite-bearing basalt. The optimum temperature of the reaction is over 900 °C, and water has to be extracted as fast as it is formed, in order to allow the reaction to continue. This was accomplished by extracting the vapor phase through a high-temperature electrolysis cell, which produced hydrogen and oxygen. All hydrogen was returned to the reactor. In the other process studied by Eagle Engineering, a combined process whereby solar-wind hydrogen would be obtained by pyrolysis of lunar regolith and would react at the same time with regolith ilmenite. This process would produce water, from which hydrogen and oxygen would be formed by electrolysis. Figure 6.9 provides a schematic of this system, and Table 6.13 lists some of the principal characteristics of this approach, which could become economically feasible if hydrogen contents were greater than the 50 ppm assumed by Eagle Engineering.

Extraction of water from possible lunar ice deposits near the poles (e.g., Feldman et al. 2000, 2001) should be simpler and require less energy than the extraction of hydrogen and oxygen from the lunar regolith, because (a) the expected concentrations of water (~1%,

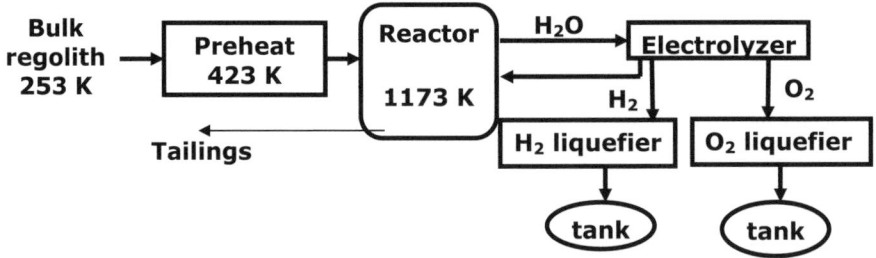

Figure 6.9. Schematic of a process to extract hydrogen from typical lunar regolith. Note the very large quantities of regolith that must be processed to extract the small amounts of H_2 and O_2. The system would have to be more complex, to deal with the other volatiles present in the lunar regolith, but these would also become useful byproducts.

Table 6.13. Characteristics of system to extract hydrogen and oxygen from lunar regolith (0.5 tonne/month LH_2; 2 tonne/month LO_2) (Eagle Engineering 1988).

System Element	Mass (tonne)	Power
Front-end loaders (3)	7.7	66 kWe
Haulers (5)	5.1	18.6 kWe
Feed and tailings bins	0.74	
H_2 extraction reactors (2)	16.6	1557kWe; 4670kWt
Other H_2 extraction eqpt.	1.1	
Electrolysis	0.1	16.8
O_2 Liquefier	0.1	1.7
H_2 Liquefier	0.1	36.3
Storage tanks	1.61	
Radiator and Thermal Control	3	48 kWt
Nuclear Power System	13	
Total	60	1611 kWe; 4670 kWt*

Note: All thermal energy was assumed to be waste heat from the nuclear reactor. kWe = kilowatts of electricity; kWt = kilowatts of thermal power.

equivalent to ~ 0.1% hydrogen) are greater than the concentration of hydrogen in the typical regolith (0.005 to 0.01% H) and (b) ice can be converted to vapor at much lower temperatures than are required for the extraction of hydrogen from the regolith. Mining in the lunar cold traps will be more difficult, perhaps, than mining elsewhere in the sunlight (Duke 1999). Extraction of water for propellant production will also lead to the production of excess oxygen, as the most energetic O/H ratio for propulsion is about 6, rather than 8 as in water. A schematic for the process is shown in Figure 6.10 and Table 6.14 estimates the characteristics of a plant to extract propellant from lunar ice. It was assumed that the ice concentration is 1% by weight and that other characteristics are scaled from the Eagle Engineering model. The principal differences are associated with the mass of regolith that needs to be processed and the lower energy to which the material must be heated. Clearly, the lunar ice extraction plant is much more productive than the one that derives hydrogen from the regolith.

Sources of chemical reactants on the Moon. One of the principal difficulties with chemical processing on the Moon is the need to provide chemical reactants. The amount of reactants required can be very large. Consider the reaction Si + 4HCl → $SiCl_4$ + $2H_2$. In

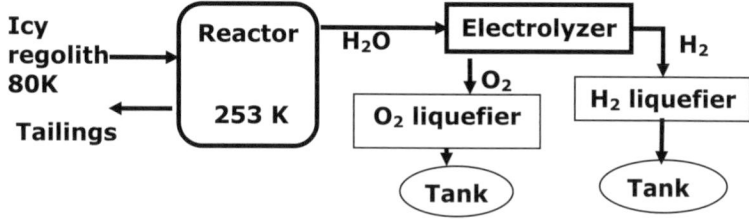

Figure 6.10. Schematic of a process to extract hydrogen from typical lunar regolith. Note the very large quantities of regolith that must be processed to extract the small amounts of H_2 and O_2. The system would have to be more complex, to deal with the other volatiles present in the lunar regolith, but these would also become useful byproducts.

Table 6.14. Conceptual plan for extracting 2.2 tonne/month of H_2 from lunar ice. For the solar power system, it is assumed that electrical power can be provided at 4kg/kW.

System Element	Mass (tonne)	Power
Front end loaders	.35	3.3
Haulers	.25	1.0
Feed and tailings bins	0.03	
H_2O extraction reactor	1.00	55.0
Other H_2O extraction eqpt.	1.10	
Electrolysis	0.10	16.8
O_2 Liquefier	0.10	1.7
H_2 Liquefier	0.10	36.3
Storage tanks	1.61	
Radiator and Thermal Control	.10	
Solar Power System	0.50	
Total	5.44	114.0

this reaction, the mass of HCl is over four times the mass of the silicon being reacted. Loss of reactants from the system is magnified with respect to production of the product. Losses can occur in several ways. Leaks of gaseous phases may occur, but systems will have to be engineered to be tight and techniques exist to isolate reacting systems from higher and lower pressure environments. The loss of reactants to unrecoverable forms may be more important. If a minor constituent reacts with HCl to form a chloride that cannot be easily recovered, this reactant will be effectively lost and will have to be replaced. This consideration drives the selection of reactions toward those for which reactants can be replaced by using indigenous lunar material. The discovery of polar hydrogen may lessen this problem; however, useful deposits of C, S, Cl and other chemical reactants remain high priorities for exploration or the development of processes to extract them from low-grade sources. The tradeoff must be made between creating a new process to extract minor constituents from the regolith and improving the design of reactors to recover the maximum amount of the reagents utilized.

4.3.6. Power for resource recovery on the Moon. Access to electrical and thermal energy is essential for inhabited facilities, scientific stations, and materials production facilities. Most of the processes required to extract basic materials on the Moon are quite energetic (e.g., Table 6.12). The requirement can be for thermal or electrical energy or both. Thermal energy can be provided by concentrated solar energy, by waste heat from nuclear sources, or by resistance heating. Photovoltaic devices, solar dynamic electrical generators, or nuclear systems can provide electrical energy.

The mass of solar photovoltaic arrays needed to produce a given power level is quite small using thin-film technologies (Tuttle et al. 2000). However, the 14-day cycle of day and night on the Moon requires that a solar-energy system include energy storage (e.g., fuel cells) for nighttime use. Fuel cells are not in themselves very massive, but the reactants needed to provide 14 days of energy supply are large. Producing these reactants on the Moon (typically H_2/O_2) would lessen the mass that must be transported. At selected near-polar sites, *nearly continuous* access to solar energy may be available to solar collectors (e.g., Bussey et al. 1999). Alternatively, nuclear reactors, which can operate continuously, will be attractive for some applications. Nuclear reactors were being designed for planetary applications in the early 1990s, but research was terminated when near term opportunities for human missions disappeared. Current NASA efforts to develop nuclear electric propulsion systems may lead also to reactors that can be used on the lunar surface. Nuclear reactors require a great deal of shielding, and will not be particularly mobile on the Moon. Lunar regolith can be used for shielding to reduce the cost of transporting radiation shielding material to the Moon.

Power at low levels has been provided in space for many years using radioisotope thermoelectric generators (RTGs) to convert heat from a radioactive source to electricity by thermoelectric conversion. Recent developments of Stirling Dynamic Isotope Power Systems (Thieme et al. 2000), which also use radioactive heat sources, have increased their efficiency of electricity production to about 25%. It may be possible to design a RTG system in which the 75% of energy generated as waste heat also can be used. These then could be competitive with other forms of energy for both fixed and mobile applications.

The production of photovoltaic cells from lunar material has been suggested by Ignatiev et al. (1998). Silicon is plentiful on the Moon. Silicon suitable for PV cells must be quite pure and techniques to purify silicon on the Moon remain to be developed. Ignatiev et al. (1998) proposed to deposit PV cells directly onto the lunar surface (as previously suggested by Criswell and Waldron 1990) by vacuum evaporation, which would eliminate the need for structural support. It might also be possible to build arrays by depositing the PV devices on lunar glass sheets, then erecting them using robotic devices. In either case, the fabrication and emplacement or erection of PV cells using lunar material could lead to very low cost power on the Moon with very little importation of material from Earth. Most needed materials exist on the Moon for producing solar concentrators, which could be used to provide thermal energy for material processing.

In cases where solar energy is tapped, energy storage can be accomplished using regenerable fuel cells, in which water is electrolyzed during the day using solar energy and the hydrogen and oxygen reacted during the night to produce electricity (Fig. 6.11). This technology is being considered for use in the DARPA "water rocket" program. In that application, water would be delivered to a spacecraft and H_2 and O_2 produced by electrolysis for use as propellant. If the water cycle were to be closed, a regenerable system would result.

Most power generation systems are more effective when centralized. Nuclear plants must be separated from other activities because of the radiation hazards that require shielding and the possibility of accidents. Solar arrays require large areas for significant power capability. Thus, it is necessary to transmit energy from central production facilities to its place of use. As on Earth, some of the energy transmission may be through the transport of chemical reactants (e.g., H_2 and O_2 pipelines). Electrical cables can be laid on the lunar surface or buried within the regolith, which is highly insulating. Over the past few years, transmission of energy by microwave or laser beams has been intensively studied in connection with the definition of solar-power satellites (e.g., Mankins 2001a). Such systems require line-of-sight

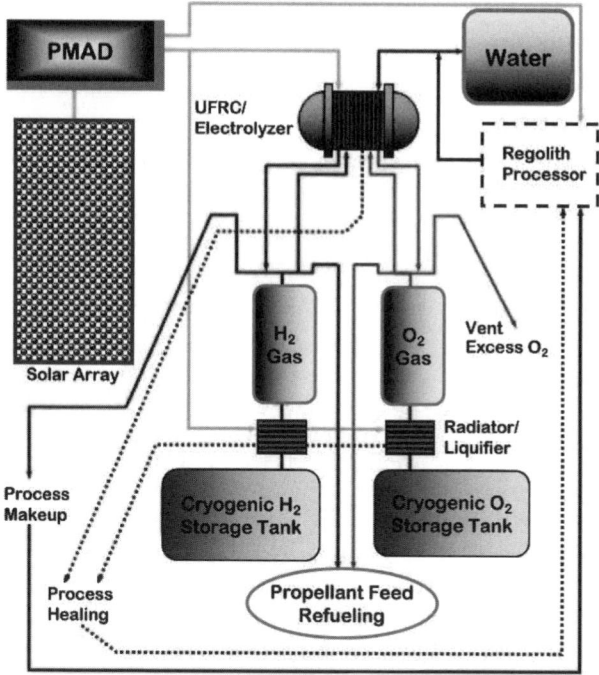

Figure 6.11. Schematic Regenerable Fuel Cell. During the day, energy from the solar array allows water (upper right) to be introduced into the electrolyzer, producing hydrogen and oxygen. At night, the process is reversed, producing energy and reconstituting the water.

communications between points on the lunar surface, but would be well suited to long-distance transmission on the Moon and perhaps particularly for work in the polar regions, where high points that are generally in sunlight lie close to and at higher altitude than points in permanent shadow (Schrunk et al. 1999), allowing creation of a lunar electrical grid.

4.3.7. Manufacturing. Propellants and H_2O and O_2 for life support are used in the form in which they are initially produced, either in liquid or gaseous form. For most other products from lunar resources that might be utilized on the Moon or in space, additional manufacturing is required. Assuming that a range of useful raw materials—metals, ceramics, glass and possibly organic materials—are available, the manufacturing techniques to be utilized will be derived from terrestrial practice. The techniques that will be favored are those that are the simplest, are tailored to the lunar environment, and can be delivered and operated at reasonable cost. The decision whether to manufacture a product on the Moon will be linked closely to the demand for the product. Products in low or intermittent demand are likely to be imported from Earth, whereas manufacturing on the Moon should be considered for products that are required consistently and in high volume.

Basic shop equipment (lathe, drill press, saws, etc., with a skilled machinist or computer controls) can be used to manufacture a wide variety of objects and even new machines. The earliest lunar outpost should include a well-equipped machine shop. In the past two decades, great advances have been made in integrating information systems with manufacturing (e.g., rapid prototyping approaches, numerically controlled machines), which increases the versatility of the manufacturing system and may be able to produce a variety of useful products with the same equipment. This approach may make it feasible to locally produce items that may be required on a more rapid timescale than can be accommodated with resupply from Earth. Consideration should be given to emphasizing the production of machine tools, e.g., drill bits, from lunar materials early in a lunar outpost program. The lunar surface offers an environment in which advanced techniques, such as laser drilling or cutting, might be applied.

Some manufacturing techniques will make beneficial use of the lunar environment. The production of silicon PV devices (Ignatiev et al. 1998) is one example, which uses the lunar vacuum and incident solar radiation in the manufacturing process. In their technique, elemental silicon (and other materials) is deposited by vacuum evaporation onto a surface of glass made by fusing lunar regolith. Thus, mechanical steps of manufacturing, such as cutting and forming are largely bypassed. Perhaps more complex vacuum evaporation techniques can be used to form three-dimensional objects.

Allen et al. (1992, 1994a) demonstrated that bricks can be manufactured by sintering regolith into compact, dense and strong forms. This process could be readily added to an extraction system that requires heating of regolith to high temperature. Mare regolith must be heated to 1000-1100 °C to sinter properly. This is somewhat above the reaction temperatures for reduction of ilmenite (900 °C) or extraction of volatiles from lunar regolith (800 °C), but it should be possible to add heat at the appropriate step in the process so that bricks can be formed. These may retain porosity if cast under surface conditions; however, pressing the bricks can densify them.

Taylor and Meek (2004, 2005) have discovered a unique property of lunar regolith/soil that presents significant potential for microwave processing. The nanophase Fe^0 present throughout the soil, associated with the abundant glass, strongly absorbs microwave radiation. Samples of lunar soil (not simulants as in previous studies) placed in a normal kitchen microwave (2.45 GHz) completely melt within a few minutes, "almost before the tea water comes to a boil." The temperatures generated by this radiation on the lunar soil is on the order of 1200–1500 °C. Microwave processing may be an effective method for releasing solar-wind implanted gases in lunar soil or heating regolith at the poles to facilitate the recovery of water. However, the energy efficiency of the process has not been investigated. Microwave sintering of regolith may be useful for paving roads (Taylor and Meek 2005), or producing bricks, structures, even antenna dishes.

Cast basalt has been used in Europe for manufacturing sewer pipes (e.g., Jakes 2000) and a wide variety of products might be produced in this manner. Molds could be constructed of mineral fragments sieved from the regolith. Anorthosite regolith sand might be a useful molding material for molten basalt. Heating could be provided by microwave, solar or waste energy from nuclear reactors.

Production of objects of specific shape and composition will be quite useful. Duke (2000) suggested that a process known as "combustion synthesis" could be used for fabricating near-net-shape objects (Moore and Feng 1995). Combustion synthesis is a process in which a mixture of solid reactants that undergo a strong exothermic reaction once ignited is formed into the desired shape. When the mixture is ignited, it can react to completion without additional energy input. Reactions are typically rapid (seconds). In cases where the reaction products are also solid, the process can be carried out in a vacuum environment. Among the products that might be made in this manner are glasses (e.g., formed by the reaction of CaO, Al_2O_3, and SiO_2 in appropriate mixture), glass-ceramics, intermetallic compounds (Fe-Ti), or ceramics (Al_2TiO_5).

Freitas and Gilbreath (1982) surveyed a wide variety of materials-processing and manufacturing techniques that might be applicable to manufacturing with lunar materials, particularly for structural materials and other solid forms. For working with metals, traditional casting, rolling, extrusion, powder metallurgy, and other techniques may be useful for manufacturing a wide variety of shapes ranging from bars to wires to pipes. In these cases, terrestrial technology may have to be adapted to the smaller throughputs expected for lunar manufacturing and the need to operate in the lunar environment. Joining techniques for use in the space environment will need to be developed that are suitable for the materials and thermal characteristics of the environment.

Many of the products now used in everyday life are based on organic materials, which are likely to be in short supply on the Moon. Carbon and nitrogen compounds will be the subjects of intense conservation on the Moon. Planning of materials to be brought from Earth should recognize the secondary value of these materials on the Moon, so that substitutions might be made on Earth where materials such as carbon-epoxy can be used in place of metallic constituents. A carbon-epoxy recycling system remains to be developed.

4.4. Uses of lunar resources

There are many ways in which lunar resources can be used in an exploration and development strategy. The principal rationale in the early phase of development would be to offset costs of transporting needed equipment, materials and supplies from the Earth to the Moon. In a second phase, the emphasis would move toward providing consumables for lunar outpost crews and exportation of goods, particularly propellant, from the Moon. In the third phase, emphasis would turn toward self-sufficiency, enhanced exports, and new services for Moon, space and Earth.

4.4.1. Propellant. With current technology, the only way in which payloads can be landed on or launched from planetary surfaces is through the use of rocket engines. Although it is possible to launch payloads from airless bodies (it has even been proposed for Earth) using electromagnetic launchers, the low accelerations that can be withstood by humans makes the establishment of an electromagnetic launcher a daunting engineering task, owing to the long acceleration distances needed. In space, more efficient means of propulsion will be developed; however, these still require propellant, only in smaller quantities with respect to payloads. For a very long time, propellant will be a valuable product in space. Propellants potentially come in many forms. Hydrogen and oxygen are the most efficient choice for rockets. Hydrogen is used in nuclear propulsion systems, though techniques for augmenting hydrogen with oxygen have been investigated (Borowski et al. 2002). Electric propulsion vehicles work best with high-mass inert gases, such as Xe, which is not likely to be produced soon outside of Earth. The most logical choice of propellants in the next few decades would be hydrogen and oxygen, for transportation in near-Earth space and for Mars missions. Hydrogen, however, requires very low temperatures for liquefaction, is low in density (requiring large tanks), and is prone to loss. Therefore, alternate fuels (e.g., methane) or forms of storage should be considered.

The demand for propellants in space is significant. Within the next few years, approximately 500 mt of propellant will be utilized annually in low Earth orbit to send communications satellites to GEO. Of course, current launch systems do not stop in LEO to refuel, so the propellant is carried on board the vehicle as it is launched from Earth. However, this use could provide a market for propellant from the Moon if the cost of lunar propellant in LEO became less than the cost of propellant transported from Earth. Blair et al. (2002) analyzed the commercial prospects for lunar propellant. To this potential market could be added military satellites, government, lunar, and Mars human exploration and development missions, and satellite servicing requirements. Nock et al (2003) described a Mars exploration architecture using cycling spacecraft between the Earth and Mars, which utilizes lunar propellant.

4.4.2. Power. Power is central to any human activities on the Moon. Presently, for a process that requires 20 kW/kg to produce (e.g., metals), the mass of the power system that must be provided is 50–80% of the total production system mass. Providing power from local resources offsets a need to import massive systems from Earth. Materials used in the collection and distribution of energy are among the most useful products that could be manufactured from lunar materials. The major constituents of photovoltaic devices are readily available on the Moon. Also, once power systems can be produced from lunar materials, the lunar outpost will be capable of expanding its own power supply, one of the principal determinants of growth rate for a lunar establishment.

If the engineering of fusion power plants burning ^{3}He is accomplished, the Moon is the closest and most likely point of supply. The process of extracting ^{3}He for use on Earth would also release immense quantities of H_2 and O_2 for use as propellant or life support consumables.

4.4.3. Life support consumables. Water, oxygen, and nitrogen (or argon) are the principal "consumables" required for people to live and work in space. The term "consumables" is in quotation marks because in space, it will be important to recycle these constituents and to lose as little as possible. The ability to produce consumables from lunar resources has two major effects. The first is to offset the need to transport them from the Earth. The second, and perhaps more important, is that they provide the opportunity to build reserve caches that would otherwise be too expensive to provide. These reserves would make a lunar outpost life support system more robust against accidental losses of consumables and could alter the way in which a human crew conducts daily life. Inert atmospheric gas to maintain breathable environments is the most difficult provision to supply from the Moon. The only known source is the solar-wind-implanted nitrogen in the regolith and small amounts of argon from solar wind and internal sources implanted in grains in the lunar regolith.

Life support systems for lunar outposts will be highly conservative of their materials. Everything possible will be recycled. Production of food at the outpost can relieve a considerable burden on the Earth-Moon transportation system and will be a key research, development, and operational activity at any outpost. If transportation costs between the Moon and LEO become much less than Earth to LEO, it may become profitable to grow food on the Moon for export to LEO, recycling organic wastes generated on Earth-orbiting stations. Addition of C, N, and H obtained on the Moon to the life support system could be used to increase the robustness of the outpost by making up losses and increasing the stored reserves. However, in terms of masses of materials required, the requirements for life support will be far smaller than requirements for propellant.

4.4.4. Construction materials. Construction materials may be utilized to build habitats and work areas. These could be in the form of metals, such as iron, or ceramics and glasses formed from silicate minerals. At an advanced stage of development with low energy costs, metallic aluminum may be produced for structural use. These same materials could be used for ducts, pipes, wires, etc. Bare metal wires may be sufficient for external use if buried in regolith. Internal wiring will be brought from Earth until practical means of producing insulating materials from lunar materials are developed.

If water is readily available, concrete made from lunar material may be useful in a variety of applications (e.g., Lin 1987). Calcium and aluminum oxides are constituents of plagioclase feldspar, which is ubiquitous on the Moon's surface, and could be residual to processes that produce Si from plagioclase. In appropriate mixtures, CaO and Al_2O_3 could be mixed with graded sand, mechanically separated from the regolith, and rock fragments to produce concrete. Concrete could have many uses, from building structures to holding tanks for water. Concrete could provide a basis for constructing pressurized habitats from lunar materials.

Structural materials might be formed by sintering of the regolith (Allen et al. 1994a) or by melting basalt (Jakes 1999). Sintered regolith might be a byproduct of high-temperature processes that extract volatiles or oxygen from lunar regolith. Aggregate derived from regolith processing will be useful in stabilizing roadways and work areas.

4.4.5. Manufactured products. A wide range of manufactured products can be conceived from lunar materials. These include items ranging from storage tanks for fluids to a wide range of other items (such as furniture or utensils). At an early lunar outpost, some of the time a crew spends will be aimed at broadening understanding of what products would be most useful. If useful manufactured items can be produced, they may be candidates for export to other locations in space. Major construction projects in space have been suggested. The largest of these would be solar-power satellite systems, each of which would require 50,000 mt of material and might be built at the rate of one per year (Mankins 2001a). If supplied from the Moon, this would also require the production of similar amounts of propellant (although electromagnetic launch is possible for inert materials). Therefore, if this type of project were to be developed, substantially larger amounts of materials would be required than the current requirements for propellants in space, and the demand could drive the development of a broad industrial capability on the Moon.

4.5. Economics of lunar resource utilization

The foregoing description outlines the availability of resources on the Moon and the approaches to their development. What is technically possible is not necessarily economically feasible. In order to determine what lunar resources should be developed and when, some level of economic justification will be required. This justification could range from saving cost (or decreasing risk) on a particular mission or program to demonstrating a new profitable industry in space. In general, the economic benefits from lunar resource utilization will be greatest on the Moon, as the lunar surface is the place with the highest transportation costs from Earth. Exporting products from the Moon to near-Earth space faces a steeper economic barrier because of added transportation costs from the Moon and smaller transportation costs from Earth. The cost associated with a lunar product will also increase with the complexity of the process to produce it. Using lunar regolith for shielding requires minimal processing, whereas producing solar cells from lunar materials requires excavation, extraction, surface transportation and manufacturing steps and will be more expensive to develop. The other side of the equation is the volume of the demand for the product. If the investment in developing the process can be distributed across a large amount of product, the economic feasibility will be much more likely than if there are only a few uses of small amounts of material. For these reasons, the production of propellant and life support consumables appears to be the most likely choice for early development of lunar resources. Propellant is required in rather large amounts (for some applications, customers already exist), the processes involved are not overly complex, and the materials are relatively easy to handle. In addition, as discussed previously, production of propellant on the Moon can reduce the transportation cost of emplacing other production equipment on the Moon.

Development of an indigenous energy supply is the second most productive early development. If means can be demonstrated to produce solar photovoltaic cells or solar thermal concentration systems from lunar materials (at competitive costs to bringing the systems from Earth), the cost associated with developing additional production systems can be reduced. Development of an indigenous energy supply on the Moon can reduce the cost of exporting products to space.

In principle, it would be possible for a largely self-sufficient settlement, capable of expanding its facilities using indigenous materials, to be established on the Moon. It would have to use intensive recycling and might be limited by supplies of volatiles, such as nitrogen, carbon and hydrogen. However, if lunar materials can be economically exported to space, the basis for a separate lunar economy could be established, supporting the possibilities for economic self-sufficiency for lunar operations.

4.5.1. A case study of the economic use of lunar resources. Blair et al. (2002) examined a scenario in which propellant is derived from icy regolith in shadowed craters near the lunar poles. The customer for the propellant is a space transportation system based at Earth-Moon L-1, which uses a space-based orbital transfer vehicle to bring communications satellites from low Earth orbit to GEO. This is a current commercial market opportunity with a target of about 30 communications satellites bound for GEO per year with an average mass of 5 mt.

The average transportation cost for these satellites is approximately $100 million per launch. Seventy percent of this cost is associated with carrying the satellites between LEO to GEO, and 30% with launching them from Earth. This means that there is a potential customer base for transporting payloads from LEO to GEO, equivalent to approximately $3.5 billion per year. If lunar propellant becomes less expensive in LEO than propellant brought from Earth, some or all of this market might be captured.

The Blair et al. (2002) space transportation system is designed to fly from L-1 to LEO, rendezvous with a communication satellite delivered by a much smaller rocket to LEO, carry the satellite to GEO, then return to L-1 for refueling and another trip. In this analysis, the production system on the Moon was designed, including the ice excavation and mining system, water extraction, electrolysis and liquefaction of propellant on the Moon. Only enough water is electrolyzed on the Moon to provide propellant for a lunar water tanker to fly to L-1 and return empty to the Moon. The tanker carries water to L-1, where another electrolyzer produces propellant for the operation of the orbital transfer vehicle. L-1 is a good location for this transportation system because it is straightforward to fly to any Earth orbit with about the same energy requirement. Aerobraking is used to decelerate the orbital transfer vehicle in the Earth's atmosphere.

The analysis considered a number of factors, including the research and development costs, the cost of producing all of the hardware, costs of operations (modeled as an annual refurbishment cost of lunar systems, the propellant depot, the lunar tanker, and the orbital transfer vehicle), the concentration of ice in the lunar deposits, and a number of economic variables such as the rate of market capture, discount rate, and financing approach. The lunar production facility was emplaced with a series of identical segments, which allows propellant produced in early phases of development to be utilized in transporting the remainder of the system to the Moon. This "bootstrapping" technique is effective for the same reasons that were defined in Section 3.0, as the burden of launching hardware from Earth is substantially diminished.

The Blair et al. (2002) analysis concluded that, while the performance of the lunar propellant production facility (measured in terms of the amount of mass that would have to be launched from Earth to LEO) was excellent, it was not clear that a commercial organization could afford to develop the capability. If the research and development costs were to be borne by a government organization (this is perhaps reasonable for basic infrastructure technology development), the scenario approached economic viability. However, to be completely viable as an investment opportunity, a number of developments would be needed, including improved technology for excavation and extraction, exploration for higher concentrations of ice, and development of technology for operating in permanently shadowed areas. These likely would have to be supported by government as part of an exploration program. It was also shown that a larger market for propellant would be beneficial to economic feasibility.

4.5.2. Strategic considerations and priorities for lunar resource development. As shown in Table 6.15, several strategies may be important in developing economically competitive lunar resources.

"Bootstrapping" is a strategy that can be applied to the development of the Moon. Bootstrapping, as defined by Criswell (1978) means using indigenous materials to produce machines, which then produce the desired products. Machines could include mining, beneficiation, extraction and manufacturing machines. It is not necessary that entire machines be produced from lunar materials. If the massive parts (e.g., iron pedestals for heavy machines or insulation for high temperature furnaces) can be made from lunar materials, significant reductions in transportation costs may be achieved. As the capability for manufacturing items from lunar materials grows, the lunar outpost can approach self-sufficiency and can begin to expand based primarily on resources from the Moon.

Table 6.15. Strategic considerations for the development of
economically competitive lunar resources.

Strategy	Approach	Utility
Produce propellants	Extraction from water at poles or from regolith or volatiles from the regolith	Reduces cost of transportation for outpost operations or expansion
Lower energy costs	Produce portions of or entire energy collection systems from lunar materials	Easier to recover lower-grade mineral deposits; expand capabilities to produce metals, other materials with large energy requirements
Beneficially use lunar water	Extract water from cold traps or from regolith	Propellant, life support; Enable new processes based on availability of water (e.g. leaching trace elements from regolith)
Utilize regolith rather than rock	Extract products from fine-grained regolith rather than from rock	Uses natural processes of impact for comminution, bypassing need for additional energy for rock crushing
Simple beneficiation	Size classification; magnetic susceptibility and electrostatic separations	Beneficiation increases concentration of useful constituents in processed materials (e.g. solar wind hydrogen concentrated in <20 μm fraction; reduce process energy by size separation; concentrate metallic iron (Taylor and Oder 1990) or ilmenite (Agosto 1985) in regolith
Throw nothing away.	Keep track of residuals in processed materials	Decomposition of anorthite as a source of SiO_2 for production of silicon. Use byproduct CaO, Al_2O_3 and Na_2O; Extraction of ^3He from lunar regolith produces ^4He, H_2O, C, N (Kulcinski et al.1989).
Recycle everything.	Design for reuse; inventory materials; recover organic wastes	Reduces imported materials and energy requirements

The development of lunar resources might be undertaken only for economic reasons; however, government programs such as the human exploration of Mars or the establishment of astronomical observatories could precede the economic development of the Moon and benefit from lunar resources. Such scenarios would be attractive to commercial investors, because governments would bear much of the cost of infrastructure development. Assuming such a government-led strategy, Table 6.16 identifies two categories: Category A includes those resource development concepts that can significantly reduce the cost and increase the robustness of a lunar outpost development strategy; Category B includes developments that can increase the robustness and safety of operations on the lunar surface and provide additional products for local use or export.

4.6. Status of U.S. and international space resource law

International space law has been viewed as a potential barrier to lunar resource development. Viikari (2003) has compared the legal regimes required by current international treaties (The Outer Space Treaty, which most countries have agreed to, and the Moon Treaty, which has not been adopted by any of the space-faring nations) and the Law of the Sea. The Outer Space Treaty contains provisions that provide for freedom of exploration, but prohibit national appropriation of an object in outer space, and calls for the avoidance of harmful

Table 6.16. Categorization of utility of use of lunar resources in a lunar outpost program.

Category A: Reductions in cost of a lunar outpost	Category B: Upgraded capabilities and products.
1. Propellant and life support—H_2 and O_2.	1. Sintered regolith materials for construction of roadways, launch and landing facilities, shielding—bulk regolith.
2. Power production and distribution systems—Si for PV devices, metallic iron for conductors and wires.	2. Machine replacement parts for mining and materials processing—iron alloys, ceramics.
3. Glass and ceramics for structural material—silicates and oxides.	3. Alloying elements for more complex manufactured products; requires additional lunar exploration

contamination. Further, nations are responsible for what they or their citizens do in outer space. These provide a starting place for the establishment of rules that could pertain to resource development on the Moon, but are as yet too general for direct application. Viikari discusses how the Law of the Sea has been developed to include mechanisms that would allow economic utilization of seabed resources. Although rather complex, these mechanisms could be adapted to the case of lunar resources. The Moon Treaty of 1979 raised more questions, particularly because it stressed the concept of the Moon's resources being the common heritage of mankind. This treaty was seen as being an impediment to lunar resource development and, consequently, the United States, Russia and other space-faring nations have not ratified it.

These observations suggest that certain international legal considerations could emerge if a project for the extraction and utilization of lunar resources were to be undertaken by a government or private group (which would have to operate under license to a government) (Schmitt 1997). Although the Space Treaty prohibits the national appropriation of the Moon or of a particular location on the Moon, the Treaty does not prohibit the extraction of materials from the Moon (Sterns and Tennen 2003). Although this would appear to exclude private ownership of resource deposits on the Moon, there are well-established mechanisms for private enterprise to extract resources in the absence of land ownership, such as grazing leases on public lands and off-shore oil platforms (Sterns and Tennen 2003). Compliance with the Treaty's guidelines by a legal corporate entity under the laws of the United States can be straightforward provided the government chooses to be enabling rather that inhibiting (Schmitt 1997).

On the other hand, if the Moon Treaty were to be ratified by major space-faring nations, a high degree of uncertainty that is antithetical to private commercial activities on the Moon would result (Schmitt 1998). The Agreement could, create a de facto moratorium on such activities. A mandated international regime would both complicate private commercial efforts and establish international political control over the permissibility, timing, and management of all commercial resource activities.

The general status of the U.S. government's regulatory involvement in space related activities has no known or foreseen road blocks to an all private lunar resource initiative. Many regulatory hurdles must be overcome, however, none have proved insurmountable for existing private entities involved in commercial endeavors involving launch, communications, remote sensing, biomedical research, and space services activities. The same conclusion can be made for current tax law, although such law certainly could be made more encouraging of all new business initiatives (Schmitt 1998).

If a private organization were to propose the development of a lunar resource and perform outside of reasonable norms, would there be any recourse by the government or the international community? First of all, the licensing nation can withdraw its launch, communications, and return licenses for cause and enforce compliance before they are re-activated. If the licensing state does not act to force compliance, the world community can bring the matter to the World Court and/or join in part or in whole to enforce sanctions against the offending entity and licensing state. Any one of these actions would have a serious impact on the private entity's ability to maintain its space operations, other than those related to safety of personnel, or its access to the capital and customer markets upon which its ability to do business depends (Schmitt 1998).

Although some uncertainty remains in the application of international law to the development of lunar resources, it appears that the international legal community has ample precedents to use in crafting a satisfactory set of regulations that would allow commercial resource development, extraction, and use.

5. THE LUNAR ENVIRONMENT

Although it is not our intention to address the topic of the lunar environment at length, it is worth a brief summary here. Preservation of the lunar environment is a cross-cutting theme that must be considered at any level of exploration or development significantly beyond that undertaken by Apollo. Many of the unique uses of the Moon are dependent on preserving its existing environment. For example, the use of optical telescopes on the Moon would be degraded if a significantly greater lunar atmosphere was created by human activities. The lunar atmosphere was modified by the Apollo landings, and effects were detectable for several months after the landings (Johnson et al. 1972). Use of the Moon's vacuum for fabrication would also be impeded if the atmosphere were to become contaminated. Volatile constituents released into the atmosphere can be condensed onto the surface or lost from the Moon. Apollo experiments measured the daytime and nighttime lunar atmosphere's concentration of natural species (e.g., Hoffman et al. 1973; see also Stern 1999). Vondrak (1989) suggested that intense human activities could lead to a permanent lunar atmosphere, but only at very high levels of effort such as those associated with ^{3}He mining. Thus, understanding of atmospheric dynamics and composition should be undertaken before human exploration becomes a continuous activity. Active experiments in which known quantities of gases are released into the atmosphere at different times of the lunar day and their fate is followed by surface instruments are required to understand the potential effects of gases released by human activities.

In addition, unique features of the Moon can be altered by incautious human contamination of the lunar environment. The unique implantation effects of the solar wind can become contaminated if additional gases with different isotopic compositions become the prevalent molecules in the lunar atmosphere. These molecules would be available for implantation in the lunar surface by ionization and acceleration in the solar wind. Thus, large-scale utilization of the Moon should be preceded by a thorough review of these effects, additional study of the implantation processes, and preservation of areas of regolith for future studies.

The lunar polar cold traps have been collecting a range of naturally released volatiles over the past several billion years. They would be subject to contamination by increased human activities. It would be prudent to study the cold trap deposits in detail before committing to intensive human exploration or development. The rates of deposition of various volatile species (including metallic species such as Na and Hg) in the cold traps should be established early in a program of lunar exploration, because they may become more difficult to establish when human activity begins.

The use of the Moon's far side for radiotelescopes has been suggested as a means of excluding intense radiofrequency contamination from Earth and other natural sources. These uses could be impeded by the presence of artificial sources in lunar orbit. The International Academy of Astronautics has commissioned a study of this issue (Maccone 2001).

Some people will be opposed to the notion of disturbing the Moon's surface for any reason, particularly economic reasons. The repair of surface disruptions will occur only at timescales of billions of years, so commitment to a surface project that alters the lunar surface should be carefully considered. Most lunar activities that can now be imagined occur at scales that cannot be observed from Earth even with large telescopes. Mining of ^{3}He is a potential exception, because very large surface areas of the Moon would be disrupted. The lunar power system concept of Criswell and Waldron (1990) would also be constructed at a scale large enough to be seen from Earth. The environmental issues associated with these very large scale project should be considered and discussed, within the same kind of framework in which environmental decisions are now made on Earth, including consideration of the economic benefits to be gained by the project.

The Outer Space Treaty prohibits adverse changes or harmful contamination of the lunar environment caused by national (or private) activities on the Moon. However, the terms "adverse changes" and "harmful contamination" are not defined (Viikari 2003) and will certainly appear differently to different parties. Thus, the basis exists for international litigation or the development of more precise language in a new treaty. Establishment of international standards probably will not occur until lunar development is imminent.

6. HUMANS ON THE MOON: ESTABLISHMENT OF LUNAR OUTPOSTS

NASA has sponsored studies of lunar outposts since the mid-1980s, and many articles have appeared in the literature (e.g., see papers in Mendell 1985, 1989; Eckart 1999) describing the functions and architecture of such facilities. Recently, NASA has been directed to develop a lunar exploration program that would lead to a human lunar outpost in 2020, as part of a long term strategy that would lead to human exploration of Mars. The major goals of the lunar program are to conduct new scientific investigations and develop technologies, including those needed to use lunar resources in support of sustainable human exploration. Thus, the major premises of the introduction to this chapter have been incorporated into NASA's planning for lunar exploration. The detailed design of the exploration and development strategy have not been defined. However, NASA, most likely in collaboration with international space agencies, will be developing concepts for the design of a multipurpose lunar outpost.

H. H. Koelle and his students have conducted long-term studies of lunar outposts (e.g., Koelle 1996) and they have categorized the functions of a lunar outpost (Table 6.17). The specific systems developed to meet these functional requirements will be the subject of lively debate for several years. The depth and breadth of the program will depend on a complex interplay of requirements, costs, and capabilities. However, there is no doubt that the facilities needed to undertake these activities could be developed within the next 15 years. There remain questions of detail, some of which have been identified in previous sections of this chapter.

6.1. Site selection for a lunar outpost

The question of where a permanent outpost on the Moon should be located is a central issue in the exploration strategy for the Moon. The selection of a site or sites for lunar outposts will be based on the intended uses of the outposts as well as operational issues such as safety of landing and return, energy availability, etc. Workshops held in 1988-1989 addressed the question of site selection for a lunar outpost, based on information available at that time (NASA 1990; Taylor and Taylor 2000). Those studies selected six candidate sites and held a mock site selection debate in which the merits of the sites were discussed by members of the science community (Table 6.18). At present, with somewhat more interest in the possible use of lunar resources, a polar site might acquire higher priority.

Table 6.17. Functions of a Lunar Outpost (after Koelle 1996).

Lunar Science and Technology

- Research laboratories
- Observatories
- Mobile research systems
- Component subsystem and system test facilities

Production of Raw Materials

- Mining and beneficiation
- Production of metals
- Production of gases, raw materials and feedstock
- Production of non-metallic products

Manufacturing of end products and services (for use on Moon and for export)

- Structural components
- Food production
- Other manufactured products
- Propellant production
- Assembly of parts and subsystems
- Collection/ conversion of energy
- Services for external customers (e.g. waste recycling)
- Tourism

Direct Support Operations

- Supervision and control
- Health and recreation
- Training and education
- Housing
- Storage
- Communication services and data management
- Space transportation services (e.g. launch services)
- Surface transportation
- Construction, maintenance, repair
- Electric and thermal power supply
- Life Support and waste recycling

Table 6.18. Criteria for selecting lunar outpost sites (adapted from NASA 1990). The site workshops conducted by NASA established criteria for selection of scientifically useful sites.

Criterion	*Rationale*
Scientific	
For astronomy, locations near (behind) the limb and near the equator	Access to both celestial hemispheres; hidden from Earth
For geology, locations near a mare-highlands boundary	Access to widest range of general problems
For resource extraction, a mare site	Greatest flexibility
For large arrays and observatories, large flat areas	Ease of construction
Avoidance of high indigenous radiation	Possible criterion for some uses
Operational	
For safety, large relatively flat areas	Minimal problem with launch and landing
For space transportation – equatorial, polar and near side mid latitudes preferred; far side mid-latitudes avoided	Minimize propellant and timing of launch and landing restrictions
Availability of long range surface mobility	Allows selection of "safe" sites and surface traverses to interesting sites

Six sites were examined in the 1988 studies. The overall rankings of sites in that study were based largely on ability to meet scientific requirements, with somewhat greater priority given to geology, reflecting a greater variety of evaluation criteria in that area. Resource availability was not given high priority. The participants in the study were focused on developing a process, not defining the best outpost site, and concluded that none of the candidate sites could be ruled out at the time of the study from scientific and resource perspectives. For a renewed program of lunar exploration, establishment of a focused set of goals and objectives of the program will lead to a new set of site-selection criteria and renewed discussions.

If polar resources are demonstrated to provide strong rationale for a lunar outpost, further work at site definition and selection will be required. The lunar poles were suggested as lunar outpost sites by Green (1978) and Burke (1989), and have received greater attention recently due to the discovery of hydrogen concentrations near the poles by Lunar Prospector, which suggests that extractable quantities of hydrogen or water exist there.

Some of the other attributes of a lunar polar outpost site could include:

- The lunar poles, particularly in conjunction with a L-1 or L-2 space depot, can be reached routinely. For sites within 5° of the poles, continuous access is possible. Spacecraft can be launched from the International Space Station every 6 to 11 days and return at similar intervals. The minimum round trip duration is 21 days, with about 10 days spent on the lunar surface. Shorter round trip times can be accomplished if Apollo-type trajectories are used (no Earth orbit rendezvous); however, these may not be consistent with a highly reusable transportation system.

- Near the poles, it may be possible to find suitable sites where nearly continuous access to sunlight is possible. "Peaks of eternal light" (Bussey et al. 1999) will provide unusual lighting conditions that require energy management, but the sun will always be low on the horizon, so midday temperatures will be moderate. (Apollo missions were carried out in the early morning to avoid extreme noon-time temperatures). The Moon's slow rotation will cause surface shadow patterns to change slowly. Thermal variation is more modest than at the equator. Because the angle of the sun is always low, regolith temperatures are also low. Radiators can be aimed away from hot surfaces for maximum effectiveness.

- Geoscience at a polar site would primarily be focused on early crustal development. There are few mare basalt units near the poles, although access to cryptomaria might be possible. As surface transportation capability improves to 500 km sorties, access to mare sites becomes feasible. Near the Earth-facing side North Pole, intensive study of Imbrium Basin KREEP ejecta would be possible. Near the front side South Pole, similar studies of South Pole–Aitken Basin ejecta could be undertaken.

- Detailed understanding of the origin, scientific significance, and economic importance of polar hydrogen concentrations would be a primary objective. Access to lunar hydrogen will be utilized to produce lunar propellant. In the early stages, propellant will be provided on the surface. Later, propellant could be delivered to the L-1 or L-2 facility to establish a refueling station there.

- The lunar polar station can have access to the far side for astronomical observations that are hampered by noise from the Earth. This was one criterion that was accepted as important in the 1988 study, although the preferred sites were equatorial limb sites. If optical or infrared astronomy were to be a program element, a lunar polar site would be hampered by having access only to one hemisphere of the sky. A southern polar location might be preferred, if only one celestial hemisphere could be observed (NASA 1990).

- The polar outpost could allow observation of the Earth's magnetosphere (Freeman 1990).

- Technology development would focus on human support for long duration stays in an environment that to some extent mimics that of the Martian surface (low surface temperatures, low atmospheric pressure, dust, possibility of ice deposits, etc.).

- The lunar South Pole lies within the South Pole–Aitken Basin, an important and enigmatic ancient lunar impact basin. Both the North and South Poles lie within lunar highlands terrain.

- Hydrogen or water deposits near the poles may be accessible, providing an economic incentive to lunar development.

Locations near the lunar poles could provide challenging environmental problems for operations. Figure 6.12 shows a flat-floored crater near the larger crater Newton on the lunar near side, at 79°S latitude. The flat deposit on the floor of the crater may consist of mare basalt flows or impact ejecta. The pole-facing crater wall appears to be in permanent shadow, and the shadow extends at least 10 km onto the flat floor of the crater. The Earth will be above the horizon, bobbing up and down throughout the month. Astronauts working on the sunlit surface would encounter surfaces that are hot (rocks with faces perpendicular to the sun will have temperatures of about 100 °C) and very cold (less than −100 °C in shadow), as were encountered on Apollo missions, which flew to within 30° of the lunar equator. Boulders on the surface will cast long shadows, and the shadows will move throughout the month, so both nighttime and eclipse temperature changes will have to be accounted for in designing surface systems. The image of Figure 6.12 was taken by Clementine during lunar winter, when daytime shadows are the longest. The permanent shadow occupies significantly less area. For an outpost located outside of the noontime shadow throughout the year, it will be located 8 to 10 km from the minimum, permanent shadow, based on the difference between solar illumination angles in winter and summer. Initial investigations of the cold trap can be carried out during South Pole summer. Lunar Prospector data have not demonstrated that there is ice in the permanent shadow of this crater, but if ice were present, the flat-floored aspect of this crater could reduce the difficulty of extractive operations.

Figure 6.12 also shows the potential location of major facilities that could be associated with a polar lunar outpost at this site near Newton crater. These include: (1) Permanently-shadowed hydrogen mine; (2) outpost and possible mare regolith processing facility; (3) highlands material processing facility; (4) farside astronomical observatory, located approximately 600 km from the outpost, on the other side of the

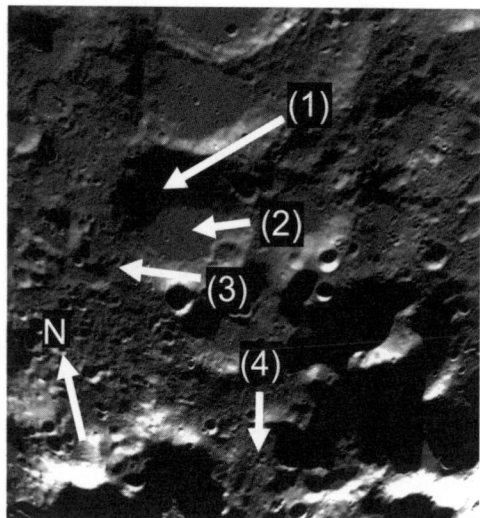

Figure 6.12. Flat-floored crater near the lunar south pole (Clementine image, UVVIS UI80S345). The north rim of the crater appears to be in permanent shadow and may be a location where ice exists. Numbers and arrows refer to potential locations of major facilities that could be associated with a polar lunar outpost at this site (see text, Section 6.1).

South Pole. This facility could also become a regional outpost to support investigations of the South Pole–Aitken basin. Field camps and shelters might be located at 200 km intervals along repeatedly used surface transportation routes.

6.2. Surface accessibility options for lunar exploration and development

Some scientific and resource development objectives require access to distributed locations on the Moon. As the specific sites currently are unknown, this is equivalent to requiring <u>global access</u>, the ability to conduct either short or long duration activities at any point on the Moon. Other objectives will be well suited to a single location on the Moon, where supporting systems can be accumulated for efficiency and robustness. These two modes can be in conflict if limited resources are available. Ten sortie missions to explore diverse geological terrain might be equivalent in expense to the same number of missions that develop a robust permanent outpost. In order to help assess the tradeoffs in exploration and development strategies, Table 6.19 compares several ways in which global access might be accomplished.

For any expected long-term utilization of the Moon for science, resource extraction or other reasons, a surface transportation system should be strongly considered. The effectiveness of a surface transportation network increases with the frequency the route is traveled. Surface transportation may be used between a central outpost and a resource mining location or a remote observatory, whereas a small lunar hopper propelled by lunar resources may be most effective for conducting sample-return missions from a large number of sites. Thus, both types of systems may find use. Both will benefit from the creation of a surface infrastructure for propellant production, maintenance, and operations.

7. A STRATEGY FOR EXPLORING AND DEVELOPING THE MOON

To make effective use of the Moon for humanity, a strategic approach to its exploration and utilization should be defined, which provides scientific and technological value at each step, but also maximizes the long-term potential of the Moon. Several themes can be identified that should be addressed in the strategy; however, the priorities will depend on a variety of considerations that cannot be specified now.

Typical planetary exploration programs have been based on an iterative strategy, in which information from one mission is utilized to define the objectives of subsequent missions. Because robotic missions typically require development times of 3-5 years, these approaches tend to be ponderous, insuring long periods of time between missions. The presence of humans can alter that strategy to the extent that on-site observations and immediate feedback into exploration plans can accelerate the rate of advancement of science and technology. In developing a strategy for the Moon, attention should be given to exploration and development strategies that consider alternative pathways based on potential discoveries and that allow for rapid incorporations of new information into the mission planning activity.

A plan for exploring and utilizing the Moon should address the key scientific questions while developing practical applications. The precise mix of robotic and human elements in the exploration strategy will be dependent upon the anticipated return from lunar exploration and development. Table 6.20 lists types of lunar activities and characteristic implementations that can be woven into a strategy.

We describe here a program that would lead to long-term human presence on the Moon. The strategy mixes robotic and human exploration in a manner that might be a template for Mars exploration. In general, robotic systems will be present whenever humans are undertaking exploration and development activities. However, in advance of human exploration missions, substantial progress can be made with purely robotic missions.

Table 6.19. Methods of achieving global access to the Moon.

Approach	Characteristics	Evaluation
Direct Launch from Earth	Expendable vehicles from Earth; low payload/launch mass	Expensive
Launch from L-1 propellant depot	Reusable landers; Opportunities for repeated missions	Requires 4-times the lander mass in propellant at L-1; Not viable without lunar propellant production
Launch from L-1 propellant depot; Propellant production on Moon both at central facility (to provide propellant to L-1) and individually (to provide propellant to individual landers.	Reusable landers; Expendable propellant production hardware; Propellant production on surface for individual missions; Transport of propellant to L-1 from central lunar production facility	Significant cost reduction after infrastructure established; Surface missions must remain in place until propellant is produced. Propellant production hardware is expendable. Propellant production opportunities differ from place to place on Moon
Reusable lunar hopper, using propellant produced on Moon to move from place to place	Lander carries propellant processing hardware.	Point to point rocket transportation has similar propellant requirements to L-1 to surface transportation; Propellant production opportunities differ from place to place on Moon
Surface transportation system	Electric vehicles with regenerable fuel cells; network of surface fuel cell regeneration systems (photovoltaic arrays).	Energy and mass efficient; Slower than rocket transportation; propellant depots tailored to local resources; Traverses of surface vehicle occupy many points instead of single point accessible to rocket vehicle; Requires high reliability vehicles; benefits from maintenance capability
Railways (Schrunk et al. 1999) or roadways	Vehicles on prepared roadways	More efficient than individual vehicles; easier maintenance; access constrained to infrastructure pathways

A "robotic outpost strategy" is proposed for the early exploration and development of the Moon. The robotic outpost is a strategy advanced by Friedman and Murray (2003) as a means of bridging the apparent gap between the capabilities of current robotic programs, such as those that have investigated the planets, asteroids and comets of the Solar System, and the human exploration missions to Mars, which tend to be two orders of magnitude larger in mass and energy requirements. In the robotic outpost concept for Mars, robotic missions are initially used to reconnoiter landing sites, then deliver instruments to the surface, and subsequently deliver payloads to the surface that prepare for human exploration (such as propellant production plants). In the case of the Moon, the robotic missions could have a more complex role, emplacing much of the infrastructure that is needed to support humans and reducing the cost of human exploration. They would be followed by the establishment of a permanent human outpost that would then grow in capability to support expanded human presence on the Moon. Robotic science missions could be conducted during all phases of the strategy.

Table 6.20. Potential lunar activities.

Lunar Activities	Characteristics of Implementation	Rationale
Scientific reconnaissance on a global basis, including emplacement of instrument networks	Robotic orbiters, landers, sample return	The absence of compelling urgency implies small per-mission investments; human global access may be too expensive
Detailed investigations of specific sites	Human sorties	Field investigations require intensive, iterative study, but not necessarily permanent facilities
Detailed investigations of hazardous sites	Robotic exploration systems with or without on-site human support	Conditions in areas such as lunar polar craters will be too severe for human access; humans nearby may be effective
Establishment of simple observatories	Robotic emplacement	Observatory systems integrated on Earth require simple site preparation and emplacement processes
Establishment of complex observatories	On-site human supervision; remote operation	Some observatories may require construction or assembly on the Moon; others may require use of local materials
Operation and maintenance of observatories	Human sorties	Observatories will perform better in the absence of humans; maintenance can be done from time to time
Experimental test bed for human mission technology development	Human tended outpost	Deployment, inspection and maintenance by humans requires repeated visits, but not necessarily permanent crew
Determination of long term effects of low gravity on humans and plants	Permanent outpost	Requires immersion of humans in environment
Establishment of materials processing facilities	Robotic emplacement	At scales of early requirements, complex systems could be emplaced and operated by tele-operated robots
Maintenance of complex materials processing facilities	Permanent outpost	Maintenance of chemical and mechanical systems is complex
Environmental studies	Robotic missions; permanent instrument networks	Understanding effects of humans on Moon requires pre-human baseline and monitor system

7.1. Robotic exploration and utilization missions

7.1.1. Reconnaissance orbiter missions. Orbital missions are needed to advance our understanding of the Moon and to help with the selection and verification of human outpost sites. Orbital missions such as Clementine and Lunar Prospector have given us a global view of the Moon. Additional orbital missions will provide improvements to the available data sets. In particular, the following items are important.

Additional global elemental information at higher spatial resolution. The Lunar Prospector (LP) Gamma Ray and Neutron Spectrometer data is limited in resolution due to the non-directional nature of the detector that limits spatial resolution on the surface to approximately the altitude of the spacecraft above the surface. The spatial resolution of LP data was improved by lowering its orbit in the last six months of its mission. Further improvements in the compositional data for the Moon can be obtained by using different detection systems and improving the signal/noise of the measured spectra. This will lead to improved spatial resolution for many elements. The SMART-1 mission includes an X-ray spectrometer with a spatial resolution of about 25 km (Marini et al. 2000). However, the SMART-1 orbit is quite high. An X-ray spectrometer carried in a 100 km orbit could provide maps with resolution of about 1 km. This may be about the highest resolution compositional information that would be useful on the Moon, due to lateral mixing by impacts.

Improved characterization of the lunar polar regions. Although Clementine obtained images of both poles (Nozette et al. 1994), the South Pole was observed during its winter, so that the extent of permanent shadow could not be distinguished from its UVVIS images. Margot et al. (1999) defined permanently shadowed area within 2.5° of the poles, based on radar observations from Earth; however, permanently shadowed areas farther from the poles and on the far side could be depicted by additional Clementine-like imaging, if it were carried out for a full year. Altimetry data for the polar regions is also important scientifically as well as to guide surface activities (Clementine obtained no altimetry southward of 80°S). The prevailing interpretation of hydrogen enrichment at the lunar poles has been that the permanently shadowed areas contain deposits of water-ice. If water-ice is present, its surface and subsurface distribution will be of interest in deciphering the processes by which it has been emplaced. However, the neutron-spectrometer data resolution is not high enough to precisely correlate hydrogen enrichments with permanently shadowed areas. Temperature maps of the lunar polar regions would be very useful. Direct observation of the topography within the shadowed craters and possibly the detection of ice might be accomplished by orbital radar sounding (Nozette et al. 2001) if enough ice is present to provide a useful response. An experiment was performed by crashing the Lunar Prospector at the end of its mission into a crater that could contain ice and looking for evidence of water vapor using terrestrial telescopes, but it detected no water vapor from the impact (Goldstein et al. 1999). Other active surface sensing techniques have been proposed (Meinel et al. 1990), but they have been disregarded in the past because of their high power requirements.

Acquisition of new high-resolution image data. Full global coverage is available at 100 to 200 m resolution from Clementine, generally at high sun angles as the mission's orbit took it over the surface close to local noontime. Some coverage was obtained at 30 m resolution. However, for the far side, there are still rather large errors (~10 km) in knowledge of the absolute location of features (R. Kirk, pers. comm..). As part of a site selection and certification program for a lunar outpost, additional high resolution imaging at 1–2 m resolution would allow detailed outpost-site planning, including positioning of key elements of the outpost.

7.1.2. Robotic surface exploration missions. Robotic surface exploration missions could continue exploration by emplacement of surface geophysical networks, exploration of regions that are difficult to access by humans, and robotic sample returns that could influence the location of a permanent outpost by verifying the existence of a valuable resource.

A global geophysical network of seismic stations. The distributed characteristic of the network and the small size of individual instruments suggest that robotic missions constitute a cost-effective means of emplacing these instruments. They would not provide direct support of lunar development.

Characterization of possible ice deposits in the polar areas. Robotic rovers, capable of conducting traverses in shadow, will be required to document the distribution of ice near the surface. This could be accomplished through traverse geophysical measurements such as ground penetrating radar and active seismic systems, or by a neutron spectrometer with an active neutron source. The three dimensional distribution of ice is a scientific issue, as an ice stratigraphy could record the history of comet impacts on the Moon. To address this issue, a drilling system capable of repeatedly drilling 2–3 m into the surface and bringing subsurface samples for analysis of water and other condensed volatiles would be appropriate.

Outpost site certification. These missions would obtain very high-resolution imaging data for the site and establish physical properties of the regolith at the proposed landing site.

Surface geological investigation and sample collection. These activities can be conducted by robots launched independently from Earth, or by teleoperated robots operated from Earth during a human exploration program. If teleoperated robots are used to collect samples, the samples could be returned to a central lunar outpost for preliminary analysis.

7.1.3. Robotic utilization and infrastructure emplacement missions. A vigorous development program for lunar resource utilization could be carried out robotically. Resource utilization missions are another step upward in complexity from those that would emplace instruments. The utilization missions would typically consist of two distinct elements, a mobile system for accessing local resources, such as an excavator/hauler, and a reaction system that produces the desired resource. Robotic systems could emplace nuclear power supplies, prepare obstacle-free landing sites, and emplace landing beacons. A robotic mission to establish a lunar photovoltaic power system has been defined by Ignatiev et al. (1998).

7.2. First permanent outpost established

The emphasis of this outpost would be on relatively local exploration, development and demonstration of resource utilization technologies, and establishment of technology test beds. Technologies that support self-reliant long-term human habitation, such as regenerable life support systems would be emphasized. Humans would be observed medically and psychologically as the first long term planetary surface missions were undertaken. This could be an essential step toward human migration beyond the Earth-Moon system and is essential if permanent human habitation of the Moon is undertaken. Robotic missions would continue, providing global surface access, sample collection and instrument placement. These systems would be refueled and maintained at the human outpost. The first economic uses of lunar resources would be made during this phase. Scientific laboratories and human-tended observatories would be established. An initial outpost might have facilities for 6–10 people, with tours of duty starting at a few months and extending to 1–2 years, potentially limited by undesirable changes in the human body in long lunar stays.

Scientific exploration by humans during this phase would be concentrated around the landing site, perhaps out to a distance of 20–30 km. Longer range surface traverses could be conducted by robotic systems, bringing back samples from distant sites and conducting reconnaissance surveys for later human expeditions. There are many experiments that could be conducted by humans within these constraints, including:

- A systematic survey and sampling of small impact craters, to establish the nature of impactors and the age distribution of recent craters.

- A deep trench to study the stratigraphy of the lunar surface in detail to a depth of several meters.

- Drilling into the mare regolith to determine the depth to bedrock as well as to penetrate bedrock, looking for contacts between volcanic units and the possibility of fossil regolith layers.

- Sampling complex rock boulders, if present in the selected area.

- Conducting resource extraction experiments to support expanded human lunar presence, economic utilization.

- Learning how to maintain complex equipment on the lunar surface.

In this phase, humans and robots would have global access for science and utilization. Surface transportation would be inexpensive and routine between several regional lunar facilities, with the potential to establish short-term field camps for human scientific and resource exploration. As economic potential develops at a given site, such a short-term field camp might grow into a regional facility. Economic utilization of lunar resources would be the driver for outpost expansion. Tourism could begin. Services provided for people on the Moon and in space would be promoted by lowered transportation costs, based on lunar resources and energy. The size of the main outpost could grow to hundreds or larger, depending on the economic productivity of the outpost, and subsidiary facilities could support groups of people as needed for scientific or commercial activities.

A number of outstanding issues are associated with this exploration strategy, including:

1. What are appropriate locations for permanently occupied outposts? This is a question that can be addressed with the robotic exploration of Phase I and includes considerations of what sites are better for science, resource extraction, and other activities that may be site-dependent.

2. What are the experiments/demonstrations that should be conducted in earlier phases to reduce the risk and mass of subsequent phases? This depends on the long-term objectives adopted for lunar development. Certainly, experiments that allow improvement in lunar operations efficiency and safety would be prominent.

3. What resources are available, and what must be found, if anything, to make the strategy viable? These can be established by early robotic missions, but as in the case of the lunar polar ice question, may require human explorers on site to make final conclusions. Lunar polar ice appears to be the highest priority resource-related issue for near term exploration.

4. What technologies should be developed that would provide the greatest advantage to development strategies? For example, is the development of technology to produce solar cells from lunar material more important than the development of a highly reusable cryogenic H_2/O_2 engine?

5. What is the proper mix of science, exploration and technology development for each phase to maintain program momentum? This is a political question, associated with the source of funding for the program. At current levels of investment in lunar missions, a significant commercial enterprise such as lunar tourism or solar power development could dwarf science and exploration efforts.

In conclusion, the discussion and vision expressed in this chapter could not have been advanced without the forgoing exploration program. In the same manner, discoveries made in carrying out the exploration strategy discussed here will form the basis for new sets of questions and conclusions about the Moon, its origin and history, and its role in the future course of human exploration and utilization of space. The newly-announced NASA emphasis on a renewed lunar program can prove to be a critical step in this progression of humanity from the Earth into the cosmos.

8. REFERENCES

Agosto WN (1985) Electrostatic concentration of lunar soil minerals. *In*: Lunar Bases and Space Activities of the 21st Century. Mendell WW (ed) Lunar and Planetary Institute, p 453-464

Aldridge EC, Fiorina CS, Jackson MP, Leshin LA, Lyles LL, Spudis PD, Tyson ND, Walker RS, Zuber MT (2004) A Journey to Inspire, Innovate and Discover, Report of the President's Commission on Implementation of United States Space Exploration Policy. U. S. Government Printing Office, Washington DC

Allen CC, Graf JC, McKay DS (1994a) Sintering bricks on the Moon. *In*: Engineering, Construction, and Operations in Space IV. Galloway RG, Lokaj S (eds) American Society of Civil Engineers, p 1220-1229

Allen CC, Graf JC, McKay DS (1994b) Experimental reduction of lunar mare soil and volcanic glass. J Geophys Res 99:23173-23185

Allen CC, Hines JA, McKay DS, Morris RV (1992) Sintering of Lunar Glass and Basalt. *In*: Engineering, Construction, and Operations in Space III. Sadeh WZ, Sture S, Miller RJ (eds) American Society of Civil Engineers, p 1209-1218

Allen, CC, Morris RV, McKay DS (1996) Oxygen extraction from lunar soils and pyroclastic glass. J Geophys Res 101:26085-26095

Arnold JR (1979) Ice at the Lunar Poles. J Geophys Res 84:5659-5668

Baedeker PA, Chou CL, Sundberg LL, Wasson JT (1974) Volatile and siderophilic trace elements in the soils and rocks of Taurus-Littrow. Proc Lunar Sci Conf 5:1625-1643

Bibring JP, Burlingame AL, Chaumont J, Langevin Y, Maurette M, Wszolek PC (1974) Simulation of lunar carbon chemistry: I. Solar wind contribution. Proc Lunar Sci Conf 5:1747-1762

Billingham J, Gilbreath W (1979) Space Resources and Space Settlements, NASA SP-428, National Aeronautics and Space Administration, Washington, DC

Blair BR, Diaz J, Duke MB, Lamassoure E, Easter R, Oderman M, Vaucher M (2002), Space Resource Economic Analysis Toolkit: The Case for Commercial Lunar Ice Mining, Final Report to the NASA Exploration Team. Colorado School of Mines, December 20, 2002

Bock E (1979) Lunar Resources Utilization for Space Construction, Contract NAS9-15560, General Dynamics – Convair Division, San Diego, CA

Boeing, General Dynamics, Lockheed, Martin Marietta, McDonnell Douglas, and Rockwell (1994) CSTS: Commercial Space Transportation Study Final Report, NASA. Washington, DC

Borowski SK, Dudzinski LA, McGuire ML (2002) Vehicle and Mission Design Options for the Human Exploration of Mars/Phobos Using "Bimodal" NTR and LANTR Propulsion, NASA TM-1998-208834/REV1, NASA Glenn Research Center, Cleveland OH

Burke, JD (1989) Merits of a Lunar Polar Base Location. *In:* The Second Conference on Lunar Bases and Space Activities of the 21st Century. Mendell WW (ed) NASA CP-3166. NASA, p 77-84

Burt D (1992) Lunar Mining of Oxygen Using Fluorine. NASA CP3166, p 423-428

Bussey DBJ, Spudis PD, Robinson MS (1999) Illumination conditions at the lunar south pole. Geophys Res Lett 26: 1187-1190

Butler B (1997) The migration of volatiles on the surfaces of Mercury and the Moon. J Geophys Res 102:19283-19291

Cameron EN (1988) Titanium in lunar regoliths and its use in selecting helium-3 mining sites, Report No. WCSAR-TR-AR3-8708. Wisconsin Center for Space Automation and Robotics, Univ. Wisconsin

Cameron EN (1993) Evaluation of the Regolith of Mare Tranquillitatis as a Source of Volatile Elements. Technical Report, WCSAR-TR-AR3-9301-1. Wisconsin Center for Space Automation and Robotics, Univ. Wisconsin

Carlson RW (2003) A brief review of infrared, visible and ultraviolet spectroscopy of Europa and recommendations for Jupiter Icy Moons Orbiter. Forum on Jupiter Icy Moons Orbiter, abstract #9042. Lunar and Planetary Institute

Chamberlain PG, Taylor LA, Podnieks ER, Miller RJ (1992), A review of possible mining applications in space. *In*: Resources of Near-Earth Space. Lewis JS, Matthews MS, Guerrieri ML (eds) University Arizona Press, p 51-68

Chambers JG, Taylor LA, Patchen A, McKay DS (1994), Mineral liberation and beneficiation of lunar high-Ti mare basalt 71055: Digital-imaging analyses. *In*: Engineering, Construction, and Operations in Space IV. Galloway RG, Lokaj S (eds) American Society of Civil Engineers, p 878-888

Chambers JG, Taylor LA, Patchen A, McKay DS (1995) Quantitative mineralogical characterization of lunar high-Ti mare basalts and soils for oxygen production. J Geophys Res 100:14391-14401

Chou CL, Boynton WV, Sundberg LL, Wasson JT (1975) Volatiles on the surface of Apollo 15 green glass and trace-element distributions among Apollo 15 soils. Proc Lunar Sci Conf 6:1701-1727.

Crider DH, Vondrak RR (2000) The solar wind as a possible source of lunar polar hydrogen deposits. J Geophys Res 105(E11):26773-26782

Criswell DR (1978) Extraterrestrial Materials Processing and Construction Final Report NSR 09-051-001, mod. 24. NASA Johnson Space Center

Criswell DR (1994) Net Growth in the Two-Planet Economy. IAA-94-IAA.8.1.704, International Academy of Astronautics

Criswell DR (1996) Lunar Solar Power System: Review of the Technology Base of an Operational LSP System, Paper IAF-96-R.2.04. American Institute of Aeronautics and Astronautics

Criswell DR, Thompson RG (1996) Data envelopment analysis of space and terrestrially-based large scale commercial power systems for Earth: a prototype analysis of their relative economic advantages. Solar Energy 56:119-131

Criswell DR, Waldron RD (1990) Lunar system to supply solar electric power to Earth. Proc of the 25th Intersociety Energy Conversion Engineering Conf 1:62-71

Delano JW (1986) Pristine lunar glasses: criteria, data, and implications. Proc Lunar Planet Sci Conf 16, in J Geophys Res 91:D201-D213

DesMarais DJ, Basu A, Hayes JM, Meinschein WG (1975) Evolution of carbon isotopes, agglutinates, and the lunar regolith. Proc Lunar Sci Conf 6:2353-2373

Diaz J, Ruiz B, Blair B, Duke MB (2004) Is Extraction of Methane, Hydrogen and Oxygen from the Lunar Regolith Economically Feasible? Space Technology and Applications International Forum -- STAIF 2004: AIP Conf Proc 699:984-991

DuBose P (1985) Solar power satellite built of lunar materials. Space Power 6, Special Issue

Duke MB (1999) Lunar Polar Regolith Mining and Materials Production. Paper IAA-99-IAA.13.02.05, International Academy of Astronautics

Duke MB (2000) The use of combustion synthesis for parts fabrication using lunar materials. *In*: Proceedings of the Fourth International Conference on Exploration and Utilisation of the Moon, ESA SP-462. Foing B, Perry M (eds) European Space Agency, p 267-270

Duke MB, Blair BR, Ruiz B, Diaz J (2004) New Space Transportation Architectures Based on the Use of Planetary Resources. Paper 04ICES-220. International Conference of Environmental Systems, SAE International

Duke MB, Gustafson RJ, Rice EE (1998) Extraction and Utilization of Lunar Polar Ice. IAA-98-IAA.1.5.09, International Astronautical Federation

Duke MB, Ignatiev A, Freundlich A, Rosenberg S, Makel D (2001) Silicon PV cell production on the Moon. J Aerospace Engineering 14:77-83

Eagle Engineering (1988) Conceptual Design of a Lunar Oxygen Pilot Plant. Contract Report EEI 88-182, Contract NAS9-17878. NASA Johnson Space Center, (available at *www.mines.edu/research/srr*)

Eckart P (1999) The Lunar Base Handbook. McGraw Hill

Elphic RC, Lawrence DJ, Feldman WC, Barraclough BL, Maurice S, Binder AB, Lucey PG (1998) Lunar Fe and Ti abundances: Comparison of Lunar Prospector and Clementine data. Science, 281:1493-1496

Fegley B, Swindle TD (1993) Lunar volatiles: implications for lunar resource utilization. *In*: Resources of Near-Earth Space. Lewis JS, Matthews MS, Guerrieri ML (eds) Univ. of Arizona Press, p 367-426

Feldman WC, Lawrence DJ, Elphic RC, Barraclough BL, Maurice S, Genetay I, Binder AB (2000) Polar hydrogen deposits on the Moon. J Geophys Res 105(E2):4175-4195

Feldman WC, Maurice S, Lawrence DJ, Little RC, Lawson SL, Gasnault O, Wiens RC, Barraclough BL, Elphic RC, Prettyman TH, Steinberg JT, Binder AB (2001) Evidence for water ice near the lunar poles. J Geophys Res 106(E10):23232-23252

Freeman JH Jr. (1990) The Moon and the magnetosphere and prospects for neutral particle imaging. *In*: Physics and Astrophysics from a Lunar Base. Potter AE, Wilson TL (eds) AIP Conf Proc 202. American Institute of Physics, p 9-16

Freitas RA Jr., Gilbreath WP (1982) Advanced Automation for Space Missions. NASA Conference Publication 2255. NASA

Friedman LD, Murray BC (2003) We can all go to Mars – The Mars outpost proposal. Geotimes, May 2003, American Geological Institute

Frisbee R (2003) Advanced propulsion for the XXIst Century, AIAA Paper #2003-2589. American Institute of Aeronautics and Astronautics

Gaddis LR, Hawke BR, Robinson MS, Coombs CR (2000) Compositional analyses of small lunar pyroclastic deposits using Clementine multispectral data. J Geophys Res 105:4245-4262

Gaddis LR, Pieters CM, Hawke BR (1985) Remote sensing of lunar pyroclastic mantling deposits. Icarus 61:461-489

Gaddis LR, Staid MI, Tyburczy JA, Hawke BR, Petro N (2003) Compositional analyses of lunar pyroclastic deposits. Icarus: 261-280

Gibson EK, Johnson SM (1971) Thermal analysis-inorganic gas release studies of lunar samples. Proc Lunar Sci Conf 2:1351-1366.

Gibson MA, Knudsen CW, Brueneman DJ, Allen CC, Kanamori H, McKay H (1994) Reduction of lunar basalt 70035 – oxygen yield and reaction product analysis. J Geophys Res 99:10887-10897

Giguere, TA, Taylor GJ, Hawke BR, Lucey PG (2000) The titanium contents of lunar mare basalts. Meteorit Planet Sci 35:193-35200

Gillett SL (1997) Toward a Silicate-Based Molecular Nanotechnology I. Background and Review. The Fifth Foresight Conference on Molecular Nanotechnology, Palo Alto California

Glaser P (1968) Power from the sun: Its future. Science 162:857-861

Goddard Space Flight Center (1999) "Next Generation Space Telescope". Goddard Space Flight Center

Goldberg RH, Tombrello TA, Burnett DS (1976) Fluorine as a constituent in lunar magmatic gases. Proc Lunar Sci Conf 7:1597-1613

Goldstein DB, Nerem RS, Barker ES, Austin1 JS, Binder AB, Feldman WC (1999) Impacting Lunar Prospector in a cold trap to detect water ice. Geophys Res Lett 26:1653-1656

Gott JR III (1993) Implications of the Copernican principle for our future prospects. Nature 363:315-319

Green J (1978) The polar lunar base. *In*: The Future United States Space Program. Proceedings of the 25th Anniversary Conference, AAS Paper 78-191. American Astronaut Soc. Univelt, p 385-425

Greenberg JS (2000) "Space Solar Power: The Economic Reality". May 2000 Aerospace America 38(4):42-46

Grey J (2001) Technical assessment of Space Solar-Power Research Program. J Aerospace Eng 14:52-58.

Haskin L, Colson RO, Lindstrom DJ, Lewis RH, Semkow KW (1992) Electrolytic smelting of lunar rock for oxygen, iron, and silicon. *In*: The Second Conference on Lunar Bases and Space Activities of the 21st Century. Mendell WW (ed) NASA Conference Publication 3166, p 411-422

Haskin L, Warren PW (1991) Lunar chemistry. *In*: Lunar Sourcebook. Heiken G, Vaniman D, French B (eds) Cambridge University Press, p 357-474

Haskin LA, Colson RO, Vaniman DT, Gillett SL (1993) A geochemical assessment of possible lunar ore formation. *In*: Resources of Near-Earth Space. Lewis JS, Matthews MS, Guerrieri ML (eds) University of Arizona Press, p 17-50

Haskin LA, Gillis JJ, Korotev RL, Jolliff BL (2000) The materials of the lunar Procellarum KREEP Terrane: A synthesis of data from geomorphological mapping, remote sensing, and sample analyses. J Geophys Res 105: 20403-20415

Hawke BR, Coombs CR, Clark B (1990) Ilmenite-rich pyroclastic deposits: An ideal lunar resource. Proc Lunar Planet Sci Conf 20:249-258

Head JW III (1974) Lunar dark-mantle deposits: Possible clues to the distribution of early mare deposits. Proc Lunar Sci Conf 5:207-222

Head JW III, Patterson GW, Collins GC, Pappalardo RT, Proctor LM (2003) Global geologic mapping of Ganymede: Outstanding questions and candidate contributions from JIMO. *In*: Forum on Jupiter Icy Moons Orbiter, abstract #9039. Lunar and Planetary Institute

Head JW III, Wilson L (1979) Alphonsus-type dark-halo craters: morphology, morphometry, and eruption conditions. Proc Lunar Planet Sci Conf 10:2861-2897

Heiken GH, Vaniman DT, French BM (eds) (1991) Lunar Sourcebook, Cambridge University Press

Henley MW, Fikes JC, Howell J, Mankins JC (2002) Space Solar Power Technology Demonstration for Lunar Polar Applications. IAC-02-R.4.04. International Astronautical Federation

Henninger D, Ming D (eds) (1989) Lunar Agriculture. American Society of Agronomy

Heppenheimer TA (1985) Achromatic Trajectories and the Industrial-Scale Transport of Lunar Resources. *In*: Lunar Bases and Space Activities of the 21st Century. Mendell WW (ed) Lunar and Planetary Institute p 142-168

Hiesinger H, Jaumann R, Neukum G, Head JW III (2000) Ages of mare basalts on the lunar nearside. J Geophys Res 105:29239-29275

Hoffman JH, Hodges RR, Johnson FS, Evans DE (1973) Lunar atmospheric composition results from Apollo 17. Proc Lunar Sci Conf 4:2865-2875

Ignatiev A, Kubricht T, Freundlich A (1998) Solar Cell Development on the Surface of the Moon. IAA-98-IAA.13.2.03. International Astronautics Federation

Jakes P (2000) Cast basalt, mineral wool and oxygen production: early industries for planetary (lunar) outposts. *In*: Workshop on Using *In-Situ* Resources for Construction of Planetary Outposts. Duke M (ed) Lunar and Planetary Institute, p 9

Jet Propulsion Laboratory (2000) Origins Science Roadmap. Jet Propulsion Laboratory, Pasadena California (see *http://origins.jpl.nasa.gov/library/scienceplan00/ index.html*)

Johnson FS, Carroll JM, Evans DE (1972) Lunar Atmosphere Measurements. Proc Lunar Sci Conf 3:2231-2242

Johnson JR, Swindle TD, Lucey PG (1999) Estimated Solar Wind-Implanted Helium-3 Distribution on the Moon. Geophys Res Lett 26 (3):385-388

Jolliff BL, Gillis JJ, Haskin LA, Korotev RL, Wieczorek MA (2000) Major lunar crustal terrains: surface expressions and crust-mantle origins. J Geophys Res 105:4197-4216

Keller L, McKay D, (1997) The nature and origin of rims on lunar soil grains, Geochim Cosmochim Acta 61:2331-2340

Keller LP, Clemett SJ (2001) Formation of nanophase iron in the lunar regolith. Lunar Planet Sci XXXII:2097

Keller R (1988) Lunar production of aluminum, silicon, and oxygen. *In* Metallurgical Processes for the Year 2000 and Beyond. Sohn HY, Geskin ES (eds) Minerals, Metals, & Materials Society, p 551-562

Keller R, Taberaux AT (1991) Electrolysis of lunar resources in molten salt. *In*: Resources in Near-Earth Space, Proc 2nd Ann. Symp. UA/NASA SERC , Tucson p 10

Koelle HH (1996) Lunar Base Facilities Development and Operation. ILR Mitt 300. Technical University

Krahenbuhl U, Grutter A, von Gunten HR, Meyer G, Wegmuller F, Wyttenbach A (1977) Volatile and non-volatile elements in grain size fractions of Apollo 17 soils 75081, 72461, and 72501. Proc Lunar Sci Conf 8:3901-3916

Kulcinski GL (1996) Near term commercial opportunities from long range fusion research. Fusion Technology 30: 411-421

Kulcinski GL, Cameron EN, Santarius JF, Sviatoslavsky IN, Wittenberg LJ, Schmitt HH (1989) Fusion Energy from the Moon for the 21st Century. *In*: Second Symposium on Lunar Bases and Space Activities of the 21st Century. Mendell WW(ed) Lunar and Planetary Institute, p 459-474

Kulcinski GL, Schmitt HH (1992) Fusion power from lunar resources. Fusion Technology 21:2221

Lanzerotti L, Brown WL, Johnson RE (1981) Ice in the polar regions of the Moon. J Geophys Res 86:3949-3950

Larson WJ, Wertz JR (eds) (1992) Space Mission Analysis and Design. Microcosm, Inc. and Kluwer Academic Publishers

Lawrence DJ, Feldman WC, Barraclough BL, Binder AB, Elphic RC, Maurice S, Thomsen DR (1998) Global elemental maps of the Moon: the Lunar Prospector gamma-ray spectrometer. Science 281:1484-1489

Li L, Mustard JF (2000) Compositional gradients across mare-highland contacts: Importance and geological implication of lateral transport. J Geophys Res 105:20431-20450

Lin TD (1987) Concrete for lunar base construction. Concr Int 9

Lo M, Ross SD (2001) The Lunar L1 Gateway: Portal to the Stars and Beyond, AIAA Paper 2001-4768. American Inst. Aeronautics and Astronautics

Lucey PG, Blewett DT, Hawke BR (1998a) Mapping the FeO and TiO_2 content of the lunar surface with multispectral imagery. J Geophys Res 103:3679-3699

Lucey PG, Blewett DT, Jolliff BL (2000) Lunar iron and titanium abundance algorithms based on final processing of Clementine ultraviolet visible images. J Geophys Res 105:20297-20305

Lucey PG, Taylor GJ, Hawke BR (1998b) Global imaging of maturity: Results from Clementine and lunar samples studies. Lunar and Planet Sci XXIX, Lunar and Planetary Institute, Houston, p 1356-1357

Maccone C (2001) The Lunar SETI Cosmic Study of IAA: Current Status and Perspectives, Paper IAA-01-IAA.9.1.05. International Academy of Astronautics, Paris

Mankins JC (1995) Space solar power: A fresh look AIAA 95-3653. American Institute of Aeronautics and Astronautics

Mankins JC (2001a) Space solar power: A major new energy option? J Aerospace Eng 14:38-45

Mankins, JC (2001b) Space solar power: An assessment of challenges and progress. J Aerospace Eng 14:46-51

Margot JL, Campbell DB, Jurgens RF, Slade MA (1999) Locations of cold traps for frozen volatiles at the lunar poles from radar topographic mapping. Science 284:1658-1660

Marini A, Racca G, Foing B (2000) SMART-1 Technology in Preparation to Future ESA Planetary Missions. In Proceedings of the Fourth International Conference on Exploration and Utilisation of the Moon. Foing B, Perry M (eds) ESA SP-462. European Space Agency, p 89-92

McKay MF, McKay DS, Duke MB (eds) (1992) Space Resources: Materials. NASA SP-509, volume 3. NASA

Meinel C, Gutheinz LM, Toepfer AJ, Vandenberg J, Reitz R (1990) Remote resource mapping of Solar System bodies. *In:* Engineering, Construction and Operations in Space II. Johnson SW, Wetzel JP (eds) ASCE, p 266-273

Mendell WW (ed) (1985) Lunar Bases and Space Activities of the 21st Century. Lunar and Planetary Institute

Mendell WW (ed) (1989) The Second Conference on Lunar Bases and Space Activities of the 21st Century, NASA CP-3166. NASA

Metzger AE, Trombka JI, Peterson LE, Reedy RC, Arnold JR (1973) Lunar radioactivity: Preliminary results of the Apollo 15 and Apollo 16 gamma-ray spectrometer experiments. Science 179:800-803

Meyer C (1977) Petrology, mineralogy, and chemistry of KREEP basalt. Phys Chem Earth 10:239-260

Miller RH (1978-1979) Extraterrestrial Processing and Manufacturing of Large Space Systems. NAS 08-32925. Massachusetts Institute of Technology

Ming DW, Henninger DL (eds) (1989) Lunar Base Agriculture: Soils for Plant Growth, ASA-CSSA-SSSA Special Publication 52

Moore JJ, Feng HJ (1995) Combustion Synthesis of Advanced Materials: Part I. Reaction Parameters. Prog Mat Sci 39:243-273

Morgan T, Shemansky D (1991) Limits to the Lunar Atmosphere. J Geophys Res 96:1351-1367

Morris RV (1977) Origins and size distribution of metallic iron particles in the lunar regolith. Proc Lunar Planet Sci Conf 11:1697-1712

Muff T, King RH, Duke MB (2001) Analysis Of A Small Robot For Martian Regolith Excavation. AIAA 2001-4616. American Institute of Aeronautics and Astronautics

Mumma MJ, Smith HJ (1990) Astrophysics from the Moon. AIP Conference Proceedings 207. American Institute of Physics

NASA (1990) A Site Selection Strategy for a Lunar Outpost. Solar System Exploration Division, Johnson Space Center

NASA (1998) Reference Mission Version 3.0 Addendum to the Human Exploration of Mars: The Reference Mission of the NASA Mars Exploration Study Team. NASA Johnson Space Center

Nock KT (2003) Cyclical Visits to Mars Via Astronaut Hotels, Global Aerospace Corporation final report to NASA Institute for Advanced Technology. *http://www.niac.usra.edu/studies/* (July 22, 2004)

Nozette S, et al. (1994) The Clementine mission to the Moon: scientific overview. Science 266:1835-1836

Nozette S, Spudis PD, Robinson M, Bussey DBJ, Lichtenberg C, Bonner R (2001) Integration of lunar polar remote-sensing data sets: evidence for ice at the lunar south pole. J Geophys Res 106 (E19):23253-23266

NRC (2001) Laying the Foundation for Space Solar Power: An Assessment of NASA's Space Solar Power Investment Strategy. National Research Council

O'Neill G (1989) The High Frontier. Space Studies Institute Press

Papike JJ, Ryder G, Shearer CK (1998) Lunar samples. Rev Mineral 36:5.1-5.234

Papike JJ, Taylor L, Simon S (1991) Lunar minerals. *In:* Lunar Sourcebook. Heiken G, Vaniman D, French B (eds) Cambridge University Press, p 121-181

Pieters CM, Head JW III, Gaddis LR, Jolliff BL, Duke M (2001) Rock Types of South Pole-Aitken Basin and Extent of Basaltic Volcanism. J Geophys Res 106(E11):28001-28022

Pieters CM, Taylor LA, Noble SK, Keller LP, Hapke B, Morris RV, Allen CC, McKay DS, Wentworth SJ (2000) Space weathering on airless bodies: Resolving a mystery with lunar samples. Meteorit Planet Sci 35:1101-1107

Potter AE, Wilson TL (1990) Physics and Astrophysics from a Lunar Base. AIP Conference Proceedings 202, American Institute of Physics

Prettyman TH, Feldman WC, Lawrence DJ, Elphic O, Gasnault O, Maurice S, Moore KR, Binder AB (2001) Distribution of iron and titanium on the lunar surface from Lunar Prospector gamma ray spectra. Lunar Planet Sci XXXII:2122

Rao GM (1988) Electrometallurgy of silicon. *In:* Metallurgical Processes for the Year 2000 and Beyond. Sohn HY, Geskin ES (eds) The Minerals, Metals and Materials Society, p 571-581

Reed GW (1999) Don't drink the water. Meteorit Planet Sci 34:809-811

Rosenberg S, Hermes P, Rice EE (1996) Carbothermal reduction of lunar materials for oxygen production on the Moon: Five metric tonne per year lunar oxygen plant design study. Paper AIAA-96-07, American Institute of Aeronautics and Astronautics

Runcorn SK (1996) The formation of the lunar core. Geochim Cosmochim Acta 60 (7):1205-1208

Schmitt HH (1992) Spiral mining for lunar volatiles. *In:* Engineering, Construction, and Operations in Space III (SPACE 92). Sadeh WZ (ed), American Society of Civil Engineering, p 1162-1170

Schmitt HH (1997) Interlune-Intermars business initiative: returning to deep space. J Aerospace Eng 10 (2):60-67

Schmitt HH (1998) Interlune-Intermars financing and management. *In:* SPACE 98. Sadeh WZ (ed) American Society of Civil Engineers, p 1-14

Schrunk DG, Sharpe BL, Cooper BL, Thangavelu M (1999) The Moon. John Wiley and Sons

Siegfried WH, Santa J (1999) Use of Propellant from the Moon in Human Exploration and Development of Space, IAA-99-IAA.13.2.02. International Academy of Astronautics

Stafford TP (1991) America at the Threshold. US Government Printing Office

Staid MI, Pieters CM (2000) Integrated spectral analysis of mare soils and craters: applications to eastern nearside basalts. Icarus 145:122-139

Stern SA (1999) The Lunar atmosphere: history, status, and current problems. Rev Geophys 37(5):453

Sterns PM, Tennen LI (2003) Privateering and profiteering on the Moon and other celestial bodies: debunking the myth of property rights in space. Adv Space Res 31:2433-2440

Sviatoslavsky IN (1992) Lunar He-3 mining: improvements on the design of the UW Mark II Lunar Miner. *In:* Engineering, Construction, and Operations in Space III (SPACE 92). Sadeh WZ (ed) American Society of Civil Engineers, p 1080-1091

Swindle TD, Burkland M K, Johnson JR, Larson SM, Morris RV, Rizk B, Singer R B (1992) Systematic Variations in Solar Wind Fluence with Lunar Location: Implications for abundances of Solar-Wind-Implanted Volatiles. Lunar Planet Sci XXIII:1395-1396

Taylor GJ, Martel LMV (2003) Lunar prospecting. Adv Space Res 31:2403-2412

Taylor GJ, Warren P, Ryder G, Delano J, Pieters C, Lofgren G (1991) Lunar rocks. *In:* Lunar Sourcebook. Heiken G, Vaniman D, French B (eds) Cambridge University Press, p 183-284

Taylor LA (1990) Rocks and minerals in the regolith of the Moon: Resources for a lunar base. *In:* Advanced Materials – Applications of Mining and Metallurgical Processing Principles. Lakshmanan VI (ed) Soc. Of Mining, Metallurgy, and Exploration, p 29-47

Taylor LA (1993) Evidence for Helium-3 on the Moon: Model assumptions and abundances. 2nd Wisc. Symp. on He-3 and Fusion Power. Kulcinski G (ed) Wisconsin Center for Space Automation and Robotics, p 49-56

Taylor LA (1994) Helium-3 on the Moon: Model assumptions and Abundances. *In:* Engineering, Construction, and Operations in Space IV. Galloway RG, Lokaj (eds) American Society of Civil Engineers, p 678-686

Taylor LA, Cahill JT, Patchen A, Pieters C, Morris RV, Keller LP, McKay DS (2001b) Mineralogical and chemical characterization of lunar highland regolith: Lessons learned from mare soils. Lunar Planet Sci XXXII:2196

Taylor LA, Carrier WD III (1992a) Oxygen production on the Moon: an overview and evaluation. *In:* Resources of Near Earth Space. Lewis JS, Matthews MS, Guerrieri ML (eds) University of Arizona Press, p 69-108

Taylor LA, Carrier WD III (1992b) Production of oxygen on the Moon: Which processes are best and why. AIAA 30: 2858-2863

Taylor LA, Kulcinski GL (1999) Helium-3 on the Moon for fusion energy: the Persian Gulf of the 21st century. Solar System Res 33:338-345

Taylor LA, Meek TT (2004) Microwave processing of Lunar soil. *In:* Proceedings. Intl Lunar Conf 2003/ILEWG5, p. 109-123. American Astronautical Society 108, Sciences & Technology Series

Taylor LA, Meek TT (2005) Microwave sintering of lunar soil: Properties, theory, and practice. J Aerospace Engr 18(3): 188-196

Taylor LA, Oder RR (1990) Magnetic beneficiation of highland and hi-Ti mare soils, rock, mineral and glass components. *In:* Engineering, Construction, and Operations in Space II. Am. Soc. Civil Eng, p 143-152

Taylor LA, Pieters C, Keller LP, Morris RA, McKay DS, Patchen A, Wentworth S (2001a) The effects of space weathering on Apollo 17 mare soils: petrographic and chemical characterization. Meteorit Planet Sci 36:285-299

Taylor LA, Taylor DH (2000) Considerations for return to the Moon and lunar base site selection workshops. J Aerospace Eng 10:68-79

Thieme LG, Qiu S, White MA (2000) Technology Development for a Stirling Radioisotope Power System for Deep Space Missions. NASA Glenn Research Center, Cleveland (see *http://www.oak.doe.gov/procure/Stirling/tp209767.pdf*)

Tompkins S, Pieters CM (1999) Mineralogy of the lunar crust: Results from Clementine. Meteorit Planet Sci 34(1):25-31

Troutman P (2002) Orbital Aggregation and Space Infrastructure Systems (OASIS), IAC-02-IAA-13.2.06. International Astronautical Federation

Tuttle JR, Szalaj A, Keane J (2000) A 15.2% Am0 / 1433 W/kg Thin-Film Cu(In,Ga)Se$_2$ Solar Cell For Space Applications. 28th IEEE Photovoltaics Specialists Conference, p 1042-1045

Van Susante PJ (2002) Design and Construction of a Lunar South Pole Infrared telescope (LSPIRT), IAF-02-Q4-3. International Astronautical Federation

Viikari L (2003) The legal regime for moon resource utilization and comparable solutions adopted for deep seabed activities. Adv Space Res 31:2427-2432

Vondrak, R. R. (1989) Lunar base activities and the lunar environment. *In:* Second Symposium on Lunar Bases and Space Activities of the 21st Century. Mendell WW (ed) Lunar and Planetary Institute, p 337-345

Waldron R (1993) Production of non-volatile materials on the Moon. *In:* Resources of Near Earth Space. Univ. of Arizona Press, p 257-295

Wasson JT, Boynton WV, Kallemeyn GW, Sundberg LL, Wai CM (1976) Volatile compounds released during lunar lava fountaining. Proc Lunar Sci Conf 7:1583-1595

Watson K, Brown H, Murray BC (1961) Behavior of volatiles on the lunar surface. J Geophys Res 66:3033

Whelan DA (2000) DARPA Orbital Express Program: Overview briefing on potential program to develop an autonomous space transfer and robotic space vehicle, Paper 4136-05. Proceedings of SPIE, volume 4136

Whitford-Stark JL (1990) The volcanotectonic evolution of Mare Frigoris. Proc Lunar Planet Sci Conf 20:175-185

Williams JG, Boggs DH, Yoder CF, Ratcliff JT and Dickey JO (2001) Lunar rotational dissipation in solid body and molten core. J Geophys Res 106:27933-27968

Zhang X, Johnson DP, Manerbino AR, Moore JJ, and Schowengerdt F (1999) Recent Microgravity Results in the Synthesis of Porous Materials. *In:* Space Technology and Applications International Forum-1999, CP458. MS El-Genk (ed) American Institute of Physics, p 88-93

Zumwalt RW (1997) Analytical instrumentation for a lunar base: performance and efficiency. *In:* A Lunar-Based Analytical Laboratory. Gehrke CW, Hobish MK, Zumwalt RW (eds) A. Deepak Publ., p 215-224

Reviews in Mineralogy & Geochemistry
Vol. 60, pp. 657-704, 2006
Copyright © Mineralogical Society of America

7

Earth-Moon System, Planetary Science, and Lessons Learned

S. Ross Taylor

Department of Geology
The Australian National University
Canberra ACT 0200, Australia
e-mail: Ross.Taylor@anu.edu.au

Carle M. Pieters

Department of Geological Sciences
Brown University
Providence, Rhode Island, 02912, U.S.A.

Glenn J. MacPherson

Department of Mineral Sciences, MRC-119
National Museum of Natural History
Smithsonian Institution
P.O. Box 37012
Washington, D.C. 20013-7012, U.S.A.

1. INTRODUCTION

The origin and evolution of the Moon is now well understood. Many bizarre theories were proposed to account for our satellite before these were swept away by the data from the Apollo missions. We now understand with the benefit of hindsight that we are looking at a very ancient object whose surface has changed little for the past 3.5 b.y.

Nevertheless the Moon remains a unique body in the solar system. The Moon formed as the result of a glancing collision with the Earth of a Mars-sized impactor late in the accretional history of the Earth when both bodies had differentiated into metallic cores and silicate mantles. The impactor is derived from the neighborhood of the Earth, as shown by similarities in O and Cr isotopes. Our satellite is mostly derived from the mantle of the impactor rather than from the mantle of the Earth. The Moon melted, perhaps entirely and a thick (50 km?) feldspathic crust, dated at 4460 ± 40 Ma floated on the ocean of magma. The closest analogue may be the crust of Mercury. The interior crystallized into cumulate zones of differing mineralogy from which the mare basalts were later derived by partial melting.

The impact history of the lunar surface has provided insights into planetary accretion, the planetesimal hypothesis and the bombardment history in the inner solar system. Debate continues whether the concentration of impact-produced basins constitutes a "cataclysm" or the tail end of accretion. A notable example and a target for future missions is the 2500 km diameter South Pole-Aitken basin that may provide insights into the nature of the deep lunar crust.

Remaining significant problems include whether the Moon has a small metallic core, while the deep interior structure is not well understood. New seismic and heat flow data are an urgent priority. The depletion of the Moon in volatile elements is well established but may reflect earlier processes in the solar nebula rather than being due to the giant impact. Likewise,

1529-6466/06/0060-0007$05.00

DOI: 10.2138/rmg.2006.60.7

the depletion of siderophile elements and the Hf-W isotopic systematics may reflect processes in precursor planetesimals. Meanwhile, the bulk lunar composition for major elements (e.g., Mg, Al, Ca, Si) is not well constrained. Neither the Earth nor Moon resemble the composition of the primordial rocky component of the solar nebula (CI) nor that of the common meteorites. This is indicative of much fractionation in the inner solar nebula occurring during formation and accretion of planetesimals.

In the following sections we discuss the advances in lunar science, both from the Apollo results and particularly from the more recent Clementine ant Lunar Orbiter Missions. We discuss the place of the Moon in the solar system, and conclude with our assessment of how to go about exploring solid planets and satellites on the basis of lessons learned from experience with the Moon.

2. ADVANCES IN LUNAR SCIENCE OVER THE PAST 10 YEARS

2.1. Planetary geochemistry: the Moon in the context of the solar system

2.1.1. Processes in the early nebula. In the current model for the early evolution of the solar nebula, (e.g., Stevenson and Lunine 1988; see review by Taylor 2001a) gas, water and elements volatile <1100K were driven out along a "snow line" at about 5 AU and so were depleted in the inner nebula. This process was most likely associated with violent solar activity (e.g., T Tauri stage) in the earliest stages of nebular evolution. The related buildup of material at the "snow line" allows for the rapid (10^5 year) runaway growth of a 10–20 Earth mass core and enables the gravitational capture of gas before the nebula is dispersed, so allowing early growth of Jupiter. The cores of Saturn, Uranus and Neptune probably formed in the same region of enhanced density and were dispersed by the prior growth of Jupiter (Thommes et al. 1999). Thus the giant planets formed at a significantly earlier stage than the inner planets that accreted later from the surviving dry refractory planetesimals after the gaseous components of the nebula had been dissipated.

The dry and refractory nature of the planetesimals is reflected by the primitive anhydrous mineralogy of chondrites (Brearley and Jones 1998) and by the differing oxygen isotopic ratios of ice and the primary meteoritic minerals (Young and Russell 1998). Ice was present far out in the asteroid belt, and its subsequent melting produced the aqueous alteration present in the CI chondrites. The water now present in the Earth and Mars (and Venus?) accreted later to the planets either by the later drift back of icy planetesimals or from comets that had formed originally in the region of Jupiter (Cyr et al. 1998, 1999; Morbidelli et al. 2000).

Timescales for the formation of planetesimals were considered by Carlson and Lugmair (2000) who concluded that "melting and internal differentiation took place on a timescale of a few million years" after T_o (4566 Ma used here; 4570 Ma is an alternative). The implication of this is that the planetesimals accreting to form the terrestrial planets were already differentiated into metallic cores and silicate mantles, with far-reaching implications for the interpretation, for example, of the Hf-W isotopic system.

These planetesimals were subsequently assembled over the succeeding 10–100 m.y. into the four inner planets, with the Moon forming during a giant collision of a body about the size of Mars late in the accretional history of the Earth (Canup and Asphaug 2001; Canup 2004).

Relative to the elemental abundances in the primitive solar nebula, that are represented by the CI carbonaceous chondrites (hereafter CI), the Moon is more strongly depleted in volatile elements than the Earth and is completely dry (late addition from cometary impacts might account for the reported presence of ice at the poles: Nozette et al. 2001, but see Campbell et al. 2003). However this depletion is similar to that observed in the parent body of the HED meteorites (4 Vesta), and in the angrites. Much of the volatile element depletion and the bone-dry state of the Moon is often attributed to the high temperature conditions during the giant impact. However the low value of the lunar initial $^{87}Sr/^{86}Sr$ ratio, LUNI argues for very early loss of volatile Rb relative to refractory Sr, consistent with the early depletion of volatile elements in the solar nebula, rather than during the impact. Moreover, the lack of fractionation in the abundance patterns of the refractory element patterns to CI are not consistent with condensation from vapor at temperatures of several thousand Kelvin that certainly existed in some regions affected by the giant impact (Cameron 2000, his Figs. 7, 8) nor indeed of much exposure to temperatures above 1100K. The K isotopes show no fractionation either between the Moon and other inner solar system bodies, including meteorites, relative to CI, indicating that the Moon did not accrete by condensation from a hot impact-induced vapor (Humayun and Clayton 1995).

Thus the question arises whether the bone-dry and volatile depleted nature of the Moon is related to the giant impact, or is a relic from the primordial dry inner solar nebula, inherited from the impactor? "Harold Urey thought of the Moon as a primitive object, so it is ironic that the Moon may after all preserve some memory of the earliest nebula during the long journey of the precursor material from primitive dust grains to the curious composition of the Moon" (Taylor 2001b).

2.1.2. Lunar interior. The lunar radius is 1738 km, lunar density is 3.3437 ± 0.0016 g/cm^3 and the moment of inertia (I/MR^2) is 0.3931 ± 0.0002 (Konopliv et al. 1998). The seismic velocity model of Nakamura (1983) modeled upper, middle and lower mantle zones with constant velocities with boundaries at 270 and 500 km. Below about 1000 km, P and S-waves become attenuated. P-waves are transmitted though this region of high attenuation, but S-waves are missing, possibly suggesting the presence of a few percent of partial melt.

Khan et al. (2000) have provided a reinterpretation of the Apollo seismic data from an inversion of the lunar free oscillation periods. However, their lower mantle seismic velocities are higher than either those of Nakamura et al. (1983), Goins et al. (1981) or Kuskov et al. (2001). Until new seismic data become available, the detailed structure of the mantle will remain uncertain and this remains one of the most important measurements to be made in the future.

2.1.3. Lunar core. Although there is a common belief that the Moon possesses a small iron core, there is in fact no decisive evidence, and the moment of inertia, magnetic sounding, seismic data, thermal evolution, and lunar rotation could equally be interpreted as indicating a Moon with a silicate core. Indeed the free oscillation study of Khan and Mosegaard (2001) are consistent with a silicate core. Core-free models are also consistent with the moment of inertia value and the allowable density range in the lower mantle for a core-free Moon is 3.49–3.52 g/cm^3 (Kuskov and Kronrod 2000; Kuskov et al. 2001).

Data from the Lunar Prospector magnetometer have refined the estimate for the size of a metallic core to one with a radius of 340 km (Hood et al. 1999). The revised moment of inertia value places narrow limits: 220–370 km diameter for an Fe core and 350–590 km diameter for an FeS core (Konopliv et al. 1998). A metallic core is consistent with the estimate of the metallic iron abundance in the Moon of $2.5 \pm 2\%$ (Dyal et al. 1977), lunar magnetism (Hood et al. 1999) and the depletion of trace siderophile elements in order of their metal-silicate partition coefficients in the lunar mantle (Righter and Drake 1996; Righter 2002). This is evidence of metal segregation somewhere, but perhaps this occurred in precursor bodies prior to the formation of the Moon and the lunar mantle siderophile element abundances, with their evidence of metal-silicate fractionation, might thus be inherited.

The two current models for the interior of the Moon (Fig. 7.1) have implications for the evolution of the Moon. It seems difficult to form a metallic core without involving whole Moon melting, consisting, in the deepest mantle, mostly of olivine and orthopyroxene cumulates from the magma ocean. These are likely to contain few heat sources, making it difficult to account for the possible presence of partial melt.

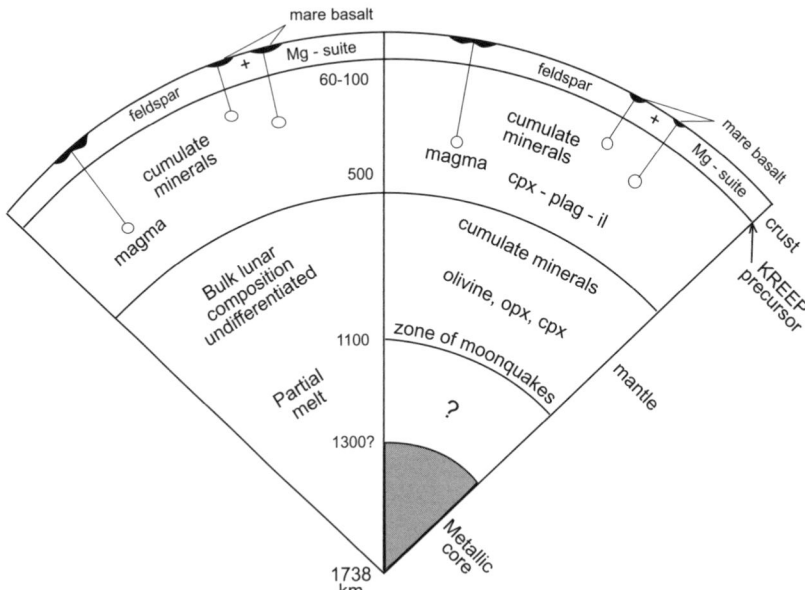

Figure 7.1. Two alternatives for the internal structure of the Moon. On the left, only half the Moon melted and differentiated and the deep interior has primitive lunar composition. Some partial melting has occurred due to the presence of K, U, and Th. This model is consistent with the lunar free oscillation periods (Amir Kahn, Univ. of Copenhagen personal communication). On the right, the Moon was totally melted and differentiated, forming a small metallic core (adapted from Taylor 2001).

In the alternative model, the absence of a core could imply half-Moon melting, with a primitive deep interior consisting of silicate. This might contain sufficient K, U and Th to generate some partial melt (1–2% is required), consistent with the geophysical interpretation. Thus our knowledge of the deep lunar interior and the question of a core must await further data. It is clear, however, that answers to these questions are crucial to understanding the origin, composition and early history of the Moon.

2.1.4. Lunar crustal structure. Geophysical mapping carried out by the Clementine and Lunar Prospector missions have resulted in greatly improved estimates of crustal thicknesses and topography (Hood and Zuber 2000) and in a better understanding of the strength of the crust. Figure 7.2 shows the mantle uplift beneath the Orientale basin. Similar mantle uplift is seen beneath lunar farside basins There is a significant negative correlation of age and mantle relief (Neumann et al. 1996, their Fig. 7b) suggesting that the mantle was somewhat weaker in earlier (pre-4.0 b.y.) times. However the relaxation is not complete and both surface topography and mantle uplift are preserved even in most of the ancient basins. It is clear that the crust has been strong and able to sustain such structures without much relaxation for over 4 b.y. This places constraints of the amount of crustal igneous activity after the solidification of the lunar crust. Figure 7.3 shows the uplift beneath the South Pole-Aitken Basin as well as that partially superimposed later uplift resulting from the excavation of the Apollo basin. The SPA structure, where most of the upper crust is missing has been preserved for over 4.1 b.y.

On the basis of the FeO and Th abundances measured by the Clementine and Lunar Prospector missions, Haskin (1998), Jolliff et al. (2000), Korotev (2000), and Wieczorek and Phillips (2000) divided the crust into three major terranes: (1) the Feldspathic Highlands Terrane (FHT) (2) the Procellarum KREEP Terrane (PKT) and (3) the South Pole-Aitken

Orientale Basin

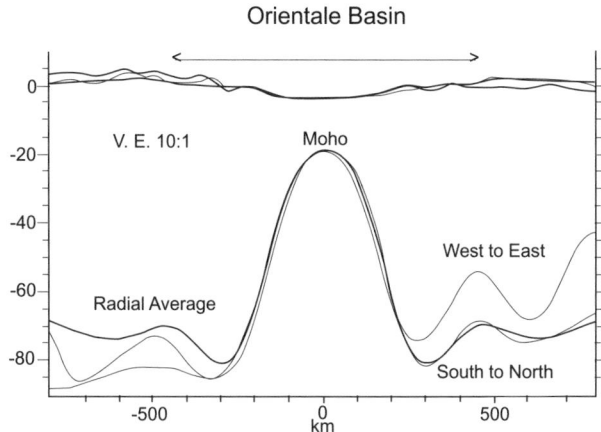

Figure 7.2. Average radial profiles of surface topography and Moho depth, assuming a 1.7 km thick mare load for Mare Orientale (adapted from Hood and Zuber 2000, Fig. 2).

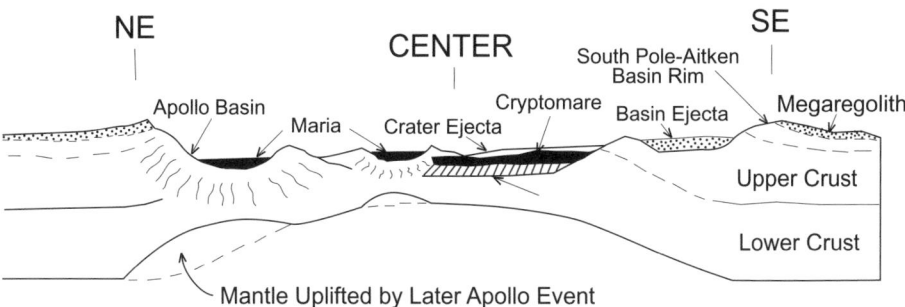

Figure 7.3. Cross-section of the South Pole-Aitken Basin (adapted from Pieters et al. 1997, Fig. 2).

(SPA) Terrane. These represent (1) the formation of the feldspathic lunar crust by accumulation from the magma ocean, (2) intrusion into the crust (or mixing into the crust and underlying mantle) of the residual KREEP liquid from the last stages of crystallization of the magma ocean and (3) the subsequent excavation of the ca. 2500 km diameter South Pole-Aitken basin, an event that stripped off most of the upper crust over that region and whose ejecta contributed significantly to the thickness of the farside anorthositic crust, north of the basin (FHT,A; Plate 1, Jolliff et al. 2000) (Figs. 7.4, 7.5).

2.1.5. Crustal thickness. Revisions of the Apollo seismic data indicate a thickness of 45 km (Khan et al. 2000) at the Apollo 12 and 14 sites rather than the earlier estimates of 60 km. The farside crust is thicker and Wieczorek (pers comm) estimates an average crustal thickness of 52 km (8.7% of lunar volume). The thickness of the lunar crust and the abundance of anorthosite in the crust are critical parameters constraining the bulk composition of the Moon. The plagioclase-rich upper crust extends to a depth of at least 30 km (from the central peak data, Wieczorek and Zuber 2001) so that decreasing the crustal thickness increases the relative proportions of the upper Al_2O_3 rich crust.

2.1.6. Siderophile element depletion. The depletion of Fe in the Moon is accompanied by a depletion of all siderophile elements (Righter and Drake 1996) that is correlated with

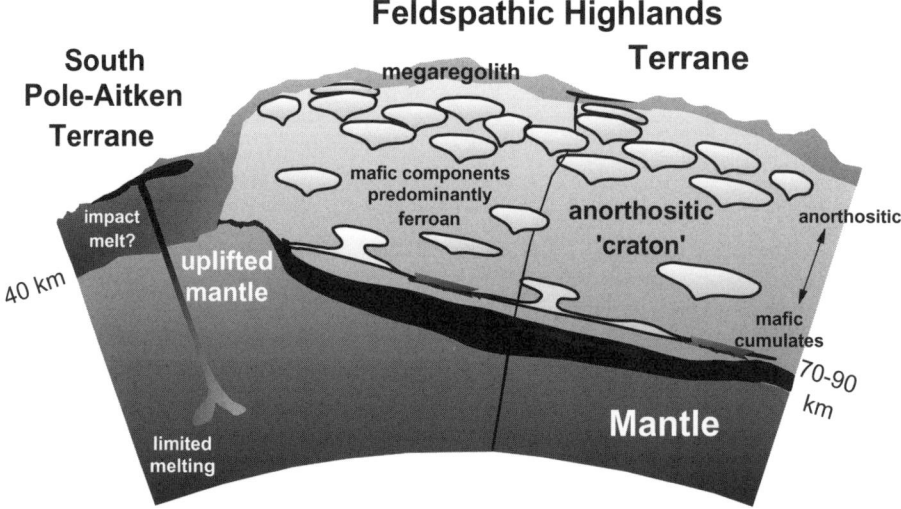

Figure 7.4. Schematic cross section of the Feldspathic Highlands Terrane (FHT)
and the South Pole-Aitken Terrane (SPA) (after Jolliff et al. 2000, Fig. 13).

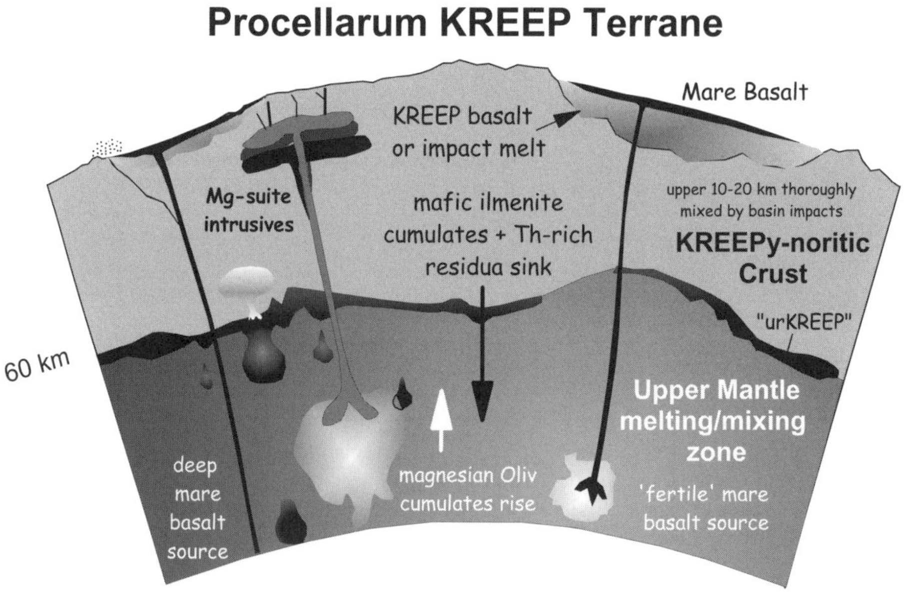

Procellarum KREEP Terrane

Figure 7.5. Schematic cross section of the Procellarum KREEP Terrane (PKT)
(after Jolliff et al. 2000, Fig. 12).

increasing metal-silicate partition coefficients (Co-W-Ni-Mo-Au-Pd-Re-Ir) (Fig. 7.6), but not with increasing volatility (Re-W-Ir-Mo-Pd-Au-Co-Ni) (Newsom 1986).

In the current model for lunar origin, the impactor core was accreted to the Earth following the collision. Various scenarios are possible: (a) the depletion of the siderophile elements occurred during core formation in the impactor or other precursor bodies and so predates the evolution of the Moon. Dates derived from W-Hf isotopic systematics may be difficult to assess in this scenario. (b) Melting of the Moon during accretion resulted in the formation of a small core which incorporated the remaining lunar budget of siderophile elements in order of their metal-silicate partition coefficients (c) Only half the Moon was melted and metal segregation under strongly reducing conditions accomplished the depletion of the remaining siderophile elements in the upper mantle, no core being formed.

2.1.7. The highly volatile elements. The Moon is dry, with less than one ppb water, except for some possible amounts trapped in permanently shadowed craters at the lunar south pole (Nozette et al. 1994; Simpson and Tyler 1999; Feldman et al. 1998, 2001, but see Campbell et al. 2003). Rare gases are indeed rare on the Moon. Those present in the lunar samples either originate from less volatile radiogenic parents or are derived from the solar wind (Ozima and Podosek 1983, Swindle et al. 1986). The absence of any indigenous noble gases, like that of indigenous H_2O, in the Moon is a first-order observation.

The highly volatile elements Bi, Tl, Cd, Br, Se, Te and In, condense below about 800 K (at 10^{-3} atm). They are depleted in the Moon by large factors relative to CI abundances and by factors of about 40 relative to their abundances in the Earth (Fig. 7.7). Many meteorites also record the depletion of the volatile elements. This depletion roughly correlates with condensation temperatures (e.g., Bi, Tl; see Wulf et al. 1995).

2.1.8. Volatile elements. There was a general depletion of volatile elements throughout the inner nebula before the formation of the inner planets (Taylor 2001a). (Fig. 7.8). Venus and the Earth have similar K/U ratios, but these are a factor of six lower than that for C1. The SNC meteorites from Mars have higher K/U ratios, indicative of a volatile element budget about 50% higher than that of the Earth, but Mars is still strongly depleted in comparison with the primordial solar-nebular values. However, the potassium isotopes show no variation in inner solar system materials thus showing no evidence of exposure to Raleigh fractionation processes involving evaporation or condensation (Fig. 7.9).

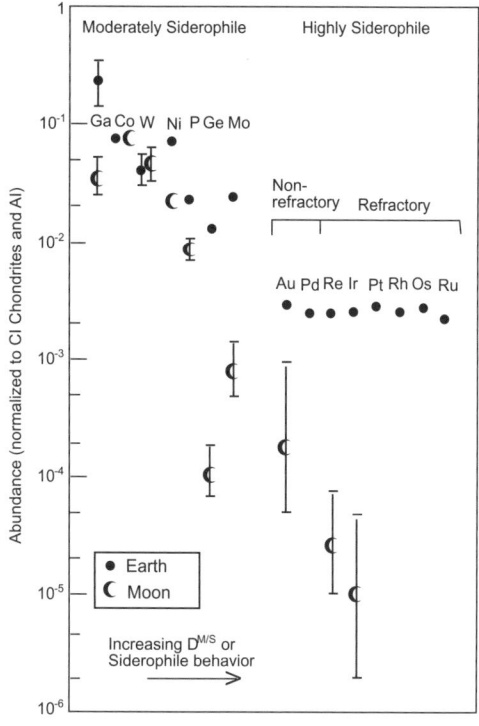

Figure 7.6. The depletion of the siderophile elements in the Earth and the Moon, relative to CI abundances (normalized to refractory Al), plotted in order of increasing metal/silicate partition coefficients. The greater depletion of Ga and Mo in the Moon is due the fact that they are also volatile elements. Courtesy H. E. Newsom, University of New Mexico (adapted from Taylor 2001).

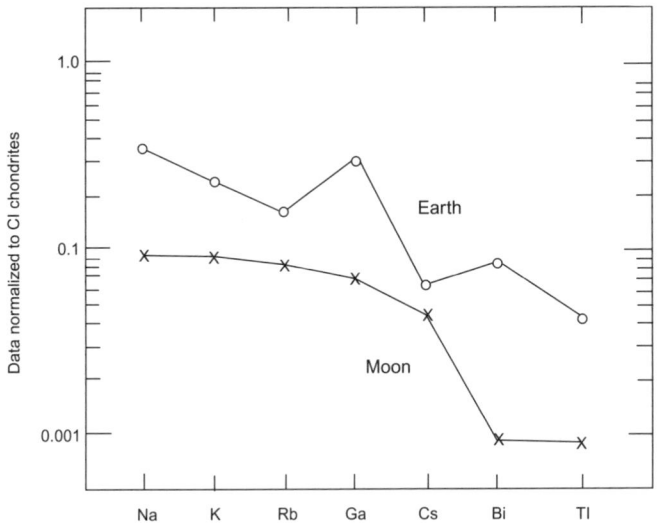

Figure 7.7. The depletion of volatile elements in the Earth and Moon, relative to the abundances in carbonaceous chondrites (CI).

Another moderately volatile element is Mn, which has the useful property that ^{53}Cr is derived in part from short-lived ^{53}Mn. The Earth and Moon have similar ^{53}Cr/^{52}Cr ratios. These contrast with higher values in meteorites (Lugmair and MacIsaac 1995) suggesting that ^{53}Mn was depleted relative to refractory Cr in the inner solar system and that there was a heliocentric gradient in Mn abundances.

Basic questions hang over the issue of volatile-element depletion in the Moon. These include where and when this depletion occurred and whether there was more than one stage of element loss. As noted earlier, it is tempting to ascribe the depletion in the Moon, relative to the Earth, as resulting from the giant Moon-forming impact; however, Venus, Earth, and Mars are all depleted, although not to the same extent as the Moon. Basaltic meteorites from Vesta, a very early object, show similar depletions of volatile and siderophile elements to lunar basalts. This, coupled with the low initial strontium isotopic values in the Moon, and the absence of evidence for high temperature (>1100 K) processing of proto-lunar material may indicate that most loss of volatiles occurred in the precursor events and that the Moon mostly inherited its dry volatile-depleted nature.

2.1.9. Vanadium, chromium and manganese. These elements have often been assigned genetic significance because of the claimed similarity between the terrestrial and lunar abundances Indeed, Cr isotopes (^{53}Cr/^{52}Cr) are similar between the Earth and Moon (Lugmair and Shukolyukov 1998), but much of this debate becomes irrelevant if most of the Moon is derived from the silicate mantle of the impactor, rather than from that of the Earth. An extensive discussion is provided by Jones and Palme (2000) that illustrates the many uncertainties inherent in trying to estimate V, Cr and Mn abundances in the Earth and Moon.

Ruzicka et al. (1998, 2000) note that "terrestrial volcanic rocks have one to two orders of magnitude lower Cr abundances than do mare basalts or eucrites." Much of this difference is probably due to the differences in redox state between the Earth and Moon with most Cr being trivalent on Earth compared with the divalent state of Cr in the Moon (Papike and Bence 1978). Ruzicka et al. (1998, 2000) have pointed out that the supposed similarity between the Earth and the Moon for V, Cr and Mn is based on model-dependent assumptions about bulk composition and conclude that "contrary to general belief, Cr data do not provide good evidence for a chemical link between Earth and Moon." (Ruzicka et al. 1997, p. 857) (Fig. 7.10). This figure shows the clear differences between the Moon and HED parent body on the one hand, and Mars and the Earth on the other (Taylor and Esat 1996). Papike (1998) has noted the increase in Mn/Fe in olivines and pyroxenes with distance from the Sun, presumably due to the volatile nature of Mn (see comments on Cr isotopes).

This apparent similarity between the composition of the Moon and that of the HED parent body (4 Vesta) raises the question whether such bodies that are depleted in volatile and siderophile elements were more common in the early solar system than previously supposed.

2.1.10. Refractory lithophile elements. There has been a longstanding controversy over the question whether the Moon is significantly enriched in refractory elements relative to the Earth. Fig. 7.11 shows one assessment (Taylor 1982) (see also section 2.1.12). The super-refractory elements Zr, Hf,

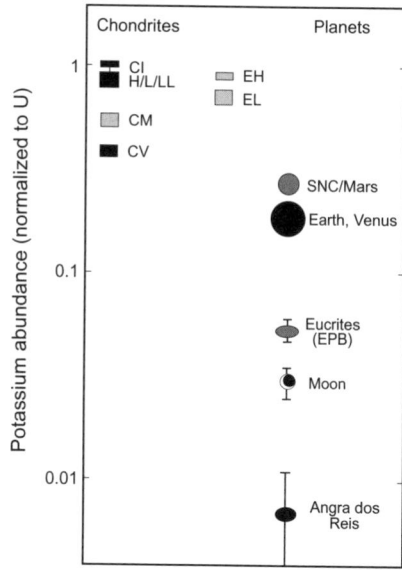

Figure 7.8. The depletion of potassium (normalized to CI and U, a refractory element) in various chondritic meteorites and solar system bodies. Courtesy Munir Humayun and Robert Clayton, University of Chicago (adapted from Taylor 2001).

Y, and Sc do not appear to be fractionated in the Moon relative to the other refractory elements, in contrast to their behavior in some meteorite minerals (e.g., hibonite) and in CAI refractory inclusions where they may be separated from the lesser refractory elements such as the REE.

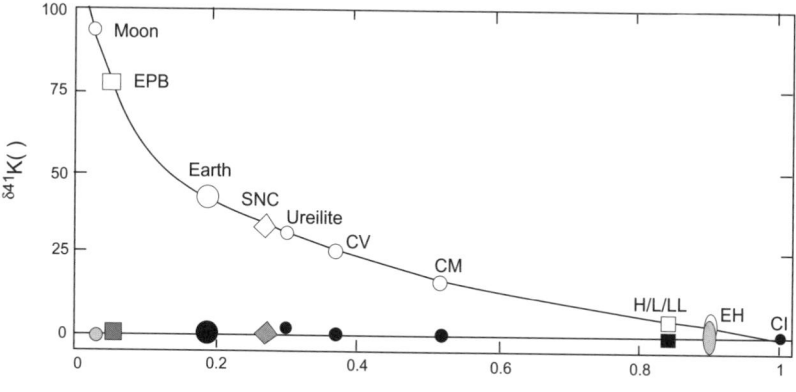

Figure 7.9. Potassium isotopic compositions for planets and meteorites. The bottom line shows the values measured. The top curve shows the calculated values that would be expected if the potassium depletion were due to loss by evaporation, beginning with a CI composition and assuming Raleigh distillation. Courtesy Munir Humayun and Robert Clayton, University of Chicago (adapted from Taylor 2001).

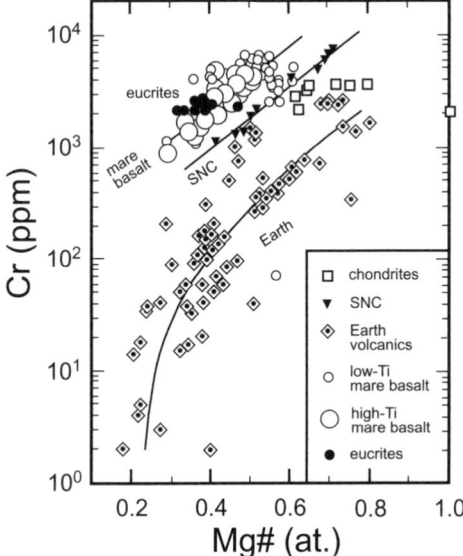

Figure 7.10. The relationship between the abundance of Cr and the Mg# (Mg/Mg + Fe) for mare basalts, eucrites and terrestrial basalts, compared to chondrites. This shows the similarity between eucrites from 4 Vesta and the Moon and their significant difference in Cr content from that of the Earth. Courtesy Alex Ruzicka, University of Tennessee, Knoxville (adapted from Taylor 2001).

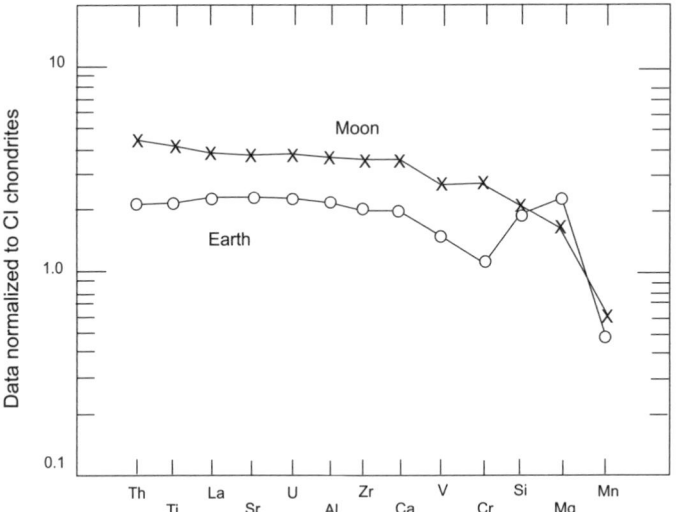

Figure 7.11. The abundances of the refractory lithophile elements in the Earth and Moon, plotted relative to the abundances in CI chondrites.

The most extreme conditions are recorded by the refractory inclusions (CAI) in meteorites. These show depletions and enrichments within the REE group based on relative volatility (Fig. 7.12). In the extreme case of the Group II inclusions, both the most volatile (Eu, Yb, Ce) and the most refractory (Sc, Y, Lu, Er) are depleted, presumably as a consequence of repeated cycles of evaporation and condensation.

In contrast, the bulk planetary REE patterns of the Earth and Moon show no depletions or enrichments (except those related to crystal-liquid fractionation) and are depleted only in

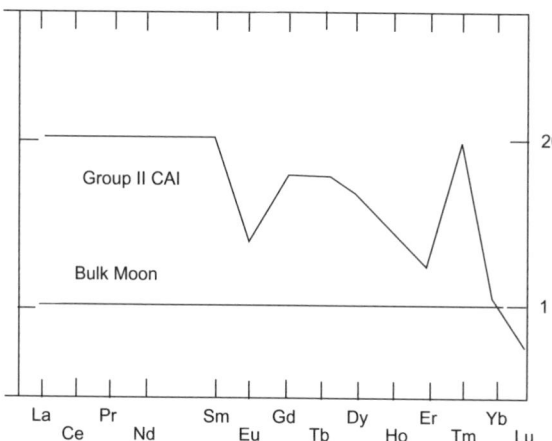

La Pr Sm Gd Dy Er Yb
Ce Nd Eu Tb Ho Tm Lu

Figure 7.12. The abundances of the REE in the Moon and in Group II CAI plotted relative to CI.

elements more volatile than the REE. Thus the REE patterns in the Moon show no sign of loss of the more volatile members of the group. These span a considerable range in volatility from moderately refractory (e.g., Eu, 1290 K) to very refractory elements (e.g., Er, 1676 K). [The temperatures given are for 50% condensation at 10^{-3} atm from Fegley (1986) and Larimer (1988)]. The lunar REE patterns are mostly sub-parallel to the CI abundances, with minor excursions attributable to crystal-liquid fractionation in igneous melts within the Moon (Fig. 7.13). The most striking deviations are the enrichments and depletions of Eu, due to its separation from the other trivalent REE. This effect occurs in the highly reducing lunar environment, where the smaller Eu^{2+} ion mimics the geochemical behavior of Sr^{2+}. We can thus attribute the behavior of relatively volatile Eu in the Moon to crystal-liquid equilibria because Yb, of about the same volatility, shows no difference from its neighbors on chondrite-normalized plots.

The most volatile of the lithophile refractory trace elements are Sr (1275K) and Ba (1227K). (The temperatures listed are for 50% condensation at 10^{-3} atm.) However, these elements are not depleted relative to the other refractory elements. It thus appears that all elements with condensation temperatures above 1100–1200 K (at 10^{-3} atm) are present in their cosmic abundance proportions. Thus the material now in the Moon was apparently not subjected to temperatures in excess of 1100–1200 K (i.e., was not vaporized) (Taylor 1983, 2001b).

2.1.11. Uranium and thorium abundances. Uranium and thorium are refractory elements and so their abundances correlate with those of Al, Ca and Ti. The terrestrial U abundance of about 20 ppb is tightly constrained by an interlocking set of isotopic and chemical abundance ratios (K/U, K/Rb, Rb/Sr, Sm/Nd).

The bulk lunar alumina value (6%) of Taylor (1982) implies a lunar uranium value of 33 ppb. The value of 6.9% alumina of Kuskov (1997) yields a bulk lunar uranium abundance of 38 ppb. Another way to get at the refractory element abundances is via the lunar K/U ratio that averages about 2500. Drake (1996) gave a minimum estimate of 27 ppb uranium based on the assumption that all the lunar potassium was concentrated in the highland crust. This value must be a minimum for two reasons. First, there is enough potassium and uranium retained in the deep interior to provide the typical values of 500 ppm potassium and 200 ppb uranium in mare basalts. Secondly, the bulk of the potassium and uranium in the Moon ends up being concentrated in the final residual melt from the crystallization of the magma ocean, that is trapped beneath the anorthositic crust.

From the Lunar Prospector and sample data, Jolliff et al. (2000, 2001) and Haskin et al. (2001) estimate that the bulk Th value for the Moon is 0.142 ppm, compared to a value of 0.125

ppm (Taylor 1982) that for Th/U = 3.8 gives a U value of 37 ppb. Although bulk lunar uranium values as low as 19 ppb (Rasmussen and Warren 1985) have been proposed, these fail to provide enough uranium (or other refractory elements) for geochemically reasonable lunar models (see detailed discussion by Taylor 1986a).

However, U, Th, K and the other incompatible elements in KREEP appear to be concentrated in the Procellarum KREEP Terrane. The implication of this strong segregation is that little information about the bulk Moon abundances for any of these elements can be derived either from returned samples or from remote sensing measurements across the lunar surface. Integration of global data and assessment of variations with depth, as can be done with crater ejecta, are required. The same comment applies to local heat flow measurements, unless a major lunar grid was established, or some other technique for establishing the global heat flux is developed. Estimates for the abundances of the trace refractory lithophile elements may be best constrained by their correlation with major refractory elements such as Al, Ca and perhaps Ti.

2.1.12. Estimates of bulk lunar compositions of the refractory elements.

As noted above, these elements are not fractionated within the group, except in the extreme cases recorded in the CAIs, an accurate determination of the bulk lunar composition of one (e.g., Al) will yield values for the others. The essential point about the bulk lunar composition is that it (a) must contain concentrations of Ca and Al that are sufficiently high to account for the thick feldspathic crust (about 8% of lunar volume) and (b) must allow plagioclase to have precipitated early enough to extract Sr and Eu from the mare basalt source region. Data from the Clementine Mission established the alumina-rich composition of the highland crust, requiring that relative to the Earth, there is an enrichment of refractory elements (e.g., Ca, Al, Ti, REE, U, Th) in the Moon (e.g., Lucey et al. 1995).

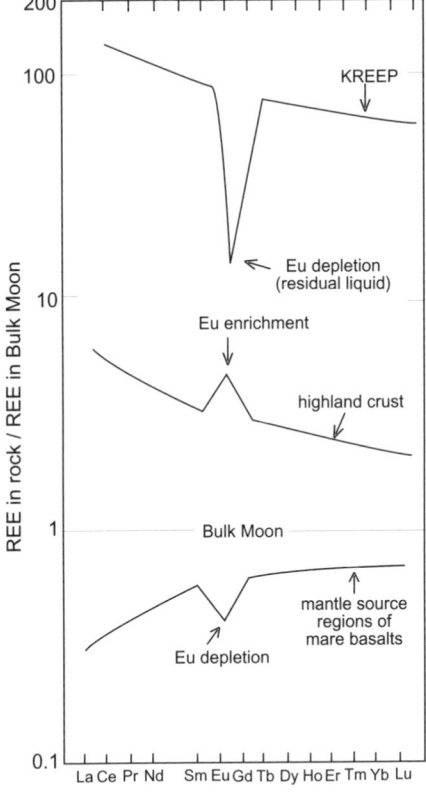

Figure 7.13. The abundances of the rare earth elements in the source regions of the mare basalts, the highland crust and KREEP, relative to the bulk Moon concentrations. These patterns result from the preferential entry of divalent europium (similar in radius to strontium) into plagioclase feldspar. This mineral floats to form the highland crust, and so depletes the interior in Eu. Mare basalts subsequently erupted from this region deep within the Moon bear the signature of this earlier depletion in Eu. KREEP is the final residue of the crystallization of the magma ocean. It is strongly depleted in Eu due to prior crystallization of plagioclase, and is enriched in the other rare earth elements, and other trace elements (e.g., K, U, Th, Ba, Rb, Cs, Zr, P) that are excluded from olivine, pyroxene and ilmenite during the crystallization of the major mineral phases of the magma ocean (adapted from Taylor 2001).

Another way to get at the problem was carried out by Hood and Jones (1987) and by Kuskov and coworkers (e.g., 1997, 1998). These workers calculated the phase relations in the lunar mantle, using the constraints of the lunar moment of inertia, density and seismic velocities, using the interpretation of the seismic velocity data by Nakamura (1983). Kuskov and Konrad (1998) then used this thermodynamic modeling to determine bulk composition, density and temperature distribution in the lunar mantle. They concluded that the bulk alumina content of the silicate portion of the Moon lies between 5.5 and 7.0% and that "the Moon is depleted in MgO and is enriched in refractory elements compared to the terrestrial mantle" and "bears no genetic relationship to the terrestrial mantle as well as to any of the known chondrites" (Kuskov and Konrad 1998, p 302).

The Moon has 50% more FeO than the Earth's mantle and in current models, is derived from the mantle of the impactor. But the terrestrial planets themselves vary in composition, the Earth having higher Al/Si and Mg/Si ratios than CI (Drake and Righter 2002). Thus the accretion of the inner planets did not sample a uniform population of planetesimals and these did not match the compositions of the common classes of meteorites. Improved lunar seismic data and heat-flow measurements are needed to resolve all these questions.

However, the question turns critically on the thickness and alumina content of the highland crust and on the depth of melting of the magma ocean and the bulk lunar composition for Al, Ca, Mg and Si remains uncertain. The issue will remain open until these questions are resolved. If the Moon turns out to be significantly enriched in refractory elements, this produces a difficult cosmochemical problem. No adequate mechanisms have yet been proposed to produce such an enrichment.

2.1.13. Oxygen isotopes. The Moon and Earth are indistinguishable on an oxygen three-isotope plot (Clayton and Mayeda 1975; Wiechert et al. 2001), but this feature is shared by the enstatite chondrites (Clayton et al. 1984) (Fig. 7.14). Both on this account, and their extremely reduced nature, enstatite chondrites are favorite candidates for forming the Earth (Javoy 1995). However, their content of volatile elements, their high K/U ratios, and low Mg/Si and Al/Si ratios, make it quite certain that neither the Earth nor the Moon is made of enstatite chondrites. Thus it is probably coincidental that these three objects share the same oxygen isotopic composition. Similarity does not imply identity, a well-known philosophical trap. The similarity between the Earth and Moon in oxygen isotopes has sometimes been used to argue for derivation of the Moon from the Earth. However the probable explanation is that this oxygen isotopic composition is characteristic of material in the primitive solar nebula during the accretion of the Earth. Thus the oxygen isotopic similarities between Earth and Moon, although essential

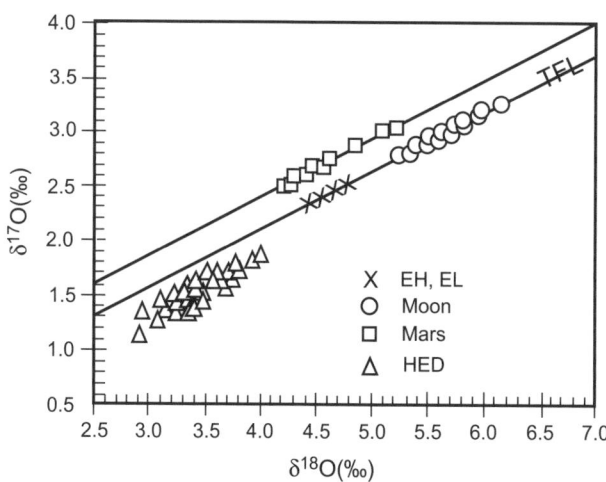

Figure 7.14. The oxygen isotope plot for the Moon, Mars HED and EH and EL meteorites. The data for the Moon plot on the terrestrial fractionation line (adapted from Wiechert et al. 2001, Fig. 2).

in providing motivation for searching for the impact-trigger hypothesis (Hartmann 1986), are by no means definitive proof of a terrestrial origin of the Moon.

2.1.14. Magma oceans. The thick anorthositic crust, the reciprocal Eu anomalies in the crust and the source regions of the mare basalts and the strong near-surface concentration of many incompatible elements, are consistent with large-scale melting of the Moon. The abundances of the incompatible elements in the highland crust require concentration from a large volume of the Moon, are consistent with enrichment due to crystal-liquid fractionation and are not related to volatility (Taylor 1973). This observation ruled out earlier suggestions that the refractory nature of the Moon was due to heterogeneous accretion (Papanastassiou and Wasserburg 1971; Gast 1972). In magma ocean models, KREEP forms the residual melt layer sandwiched between the top of the cumulate pile and the bottom of the anorthositic crust (Taylor and Jakes 1974). However KREEP, with its very high abundances of incompatible elements, including the REE, Zr, Hf, Ba, U, Th, and other refractory lithophile elements, appears to be segregated in the Procellarum KREEP Terrane on the lunar near-side (Jolliff et al. 2000).

The decisive test of the magma ocean hypothesis is the abundance of anorthosite in the lunar crust. The Clementine data show that very large areas of the crust "are composed exclusively of anorthosite as predicted by the magma ocean hypothesis" (Lucey et al. 1995). A lengthy paper on this subject by O'Hara (2000) disputes the magma ocean hypothesis. Crucial to his argument is his statement that "there is no positive europium anomaly in the average lunar highland crust" (O'Hara 2000 p.1545, see also p 1551) and this leads him to deny the existence of a magma ocean from which the crust formed by plagioclase flotation. This claim that there is no overall enrichment of Eu in the average highland crust is based on the data from the Apollo 16 site as interpreted by Korotev and Haskin (1988), but the Apollo 16 site is more mafic than much of the feldspathic highlands.

Subsequent extensive coverage of the farside highlands by the Clementine and Lunar Prospector missions has revealed that they are dominated by Ca-rich anorthosite, as discussed by Lucey et al. (1995). This highly feldspathic nature of the highlands is reinforced by studies of the lunar farside (e.g., Tompkins and Pieters 1999). Although recent data have led to new estimates of surface Al_2O_3 contents as high as 30% (Lucey et al. 1995; Wieczorek and Zuber 2001), an estimate of the integrated crustal Al_2O_3 content by Jolliff and Gillis (2002) of 25% is very similar to the earlier estimate of 24.6% (Taylor 1975).

All basalts derived from the lunar interior display a depletion in Eu, traditionally ascribed to an overall depletion of their source regions by the early crystallization of olivine and pyroxene, which exclude Eu^{2+} relative to REE^{3+}, and by the later crystallization of plagioclase, which concentrates Eu^{2+} relative to REE^{3+}, now in the highland crust (McKay 1982; McKay et al. 1990). Even the samples that come from the deepest mantle sources (probably ~400 km) have been involved in the magma ocean melting. The lunar orange and green glasses are among the most primitive lunar samples available. They possess low $^{238}U/^{204}Pb$ ratios and this is often cited as evidence that they represent samples of the primitive lunar interior. However, the glasses all possess the small but diagnostic signature of depletion in Eu that indicates that they come from a fractionated source (Papike et al. 1998). A sole exception is some USSR Luna 24 VLT basalt samples where only milligram-size amounts were available, raising serious questions about how representative were the samples that were analyzed (Neal and Taylor 1992).

The lunar highland crust that has been sampled can be interpreted as a mixture of a two component system, anorthosite and KREEP (or rocks whose incompatible element signatures are dominated by KREEP), as is apparent from the REE patterns (Fig. 7.15). If serial magmatism was the process responsible for forming the lunar crust (Walker, 1983), or if multiple igneous events occurred during the formation of the highland crust, one might expect to see the wide diversity of REE patterns that are observed in terrestrial igneous rocks, not the monotonous regularity that is observed in the REE patterns of rocks from the lunar highlands

There is thus general agreement (e.g., Warren 1985 and earlier chapters of this book) that the magma ocean concept remains the most viable hypothesis to explain the geochemical evolution of the Moon.

2.1.15. Depth of melting.

Two possible alternatives for the initial depth of the magma ocean are a depth of 500 km (about half lunar volume) or total melting of the Moon. In the first case, the deeper parts of the Moon are composed of undifferentiated material of bulk lunar composition that might contain enough K, U and Th to induce a small (1–2%) partial melting apparently seen in the seismic models; in the second case, they will consist of the earliest phases to crystallize from the magma ocean (Mg-rich olivine and orthopyroxene cumulates) and a small metallic core. These regions are likely to be barren of the heat-producing elements. The Apollo seismic velocity data do not clearly distinguish between these two models and new seismic data are required to resolve this issue.

There is very little sign of any features on the lunar surface that can be ascribed to either expansion or compression. This has been used to constrain the initial depth of melting of the Moon to about 200 km

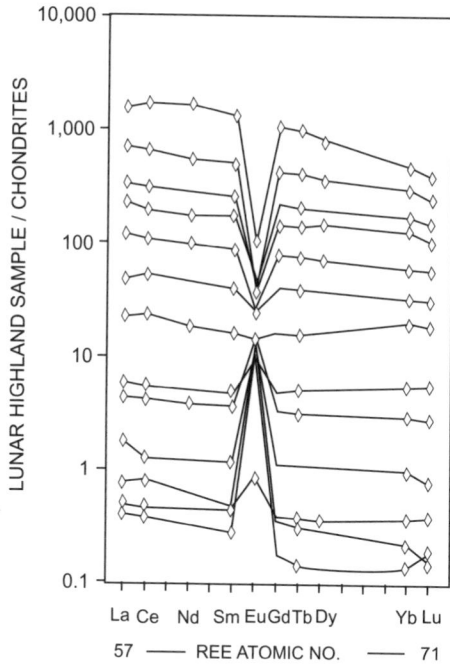

Figure 7.15. The abundances of the REE in a wide variety of highland samples, normalized to CI. There is a smooth transition from high values in KREEP with large depletions in Eu, to low concentrations in ferroan anorthosites which show strong enrichment in Eu due to its entry into plagioclase. The parallel nature of the patterns argues for a uniform petrogenesis. (adapted from Vaniman et al. 1991, Fig. 8.8).

(Solomon and Chaiken 1976). However, their limits of ±1 km change to lunar radius apply only after the end of the massive bombardment at about 3.8 Ga that is responsible for the major morphologic features of the lunar surface (DE Wilhelms, pers. comm. 1998). Another assessment of the problem reveals that "the Moon could have been initially >50% molten (with the remainder relatively close to the solidus) and yet experienced little volume change over the last 3.8 b.y." (Kirk and Stevenson 1989). In attempting to specify the initial thermal state of the Moon (molten or partially molten), Pritchard and Stevenson (2000) concluded that the widely cited restriction of Solomon and Chaiken (1976) on the depth of the magma ocean is "insufficiently precise to infer a useful constraint." The other major constraint on the depth of melting comes from the requirement in the magma-ocean hypothesis to generate the thick feldspathic crust. The revised crustal thickness models do not provide fresh constraints on the depth of melting except that deep melting of at least half the Moon is still required.

2.1.16. Mare basalts.

Although the familiar dark patches of mare basalt cover 17% of the lunar surface, their volume is trivial (Head 1976). The thin sheets of the visible maria constitute only 1% of the volume of the lunar crust or <0.1% of the Moon. Fragments of intrusive gabbros are rare in highland samples. Thus even allowing for subcrustal intrusions and cryptic maria, the lunar basalts amount at most to 1% of the Moon. Over 25 distinct varieties are known implying derivation from local regions. Accordingly the thermal requirements for the melting of the mare basalts are modest and melting must be very localized. Although there are many disputes in detail, there is broad agreement that the mare basalts were derived from zones of cumulate minerals that crystallized from the magma ocean.

Snyder et al. (2000) discuss the chronology of and isotopic constraints on the evolution of the Moon. Many complexities arise in detail but the limited sampling must engender caution. They note that "only a return to the Earth's Moon and return of samples from previously unsampled regions will allow us to answer many of the outstanding questions in lunar petrology and geochemistry." The lack of field control and the complex nature of the highland breccias make it difficult to distinguish among local (or trivial) and regional processes.

2.2. Discussion of ideas about the origin of the Earth-Moon system

2.2.1. Pre-Apollo ideas. It is difficult now, in the light of our present knowledge, to recall the state of ignorance existing before the Apollo missions cast their revealing light. Before spacecraft exploration, facts about the Moon were restricted to information about the lunar orbit, angular momentum and density. Speculations about composition and origin were unconstrained. Statements that running water carved the lunar rilles, that many of the large craters were volcanic calderas, that ice was present in a permafrost layer, that the maria were sediments, were full of dust, or were a few million years old, that the lunar highlands were composed of granite or covered with volcanic ash-flows, that a 'lunar grid' reflected tectonic stress patterns, that tektites came not only from the Moon, but from the crater Tycho, and that the Moon was essentially a primitive undifferentiated object, were all published in the serious scientific literature.

The Apollo data immediately swept away the earlier speculations. The maria were shown to be composed of basaltic lava flows, derived by partial melting from the mantle, with ages between 3–4 Ga. The postulated calderas were all due to meteoritic impact, the lunar highlands were mainly composed of anorthosite, not granite, the highland plains were ejecta sheets from large basin impacts, not siliceous volcanics, the 'lunar grid' was an artifact of the overlapping basin ejecta patterns, and tektites were terrestrial. Thus in view of our fundamental ignorance about the nature of the Moon, it is not surprising that none of the pre-Apollo theories of lunar origin survived.

The origin of the Moon had been long debated (for a review see Brush 1988, 1996). Pre-Apollo models for the origin of the Moon can be grouped into four separate categories that included (1) capture from an independent orbit, (2) fission from a rapidly rotating Earth, (3) formation as a double planet and (4) disintegration of incoming planetesimals. None of these notions survived the tests imposed of the returned lunar samples.

Hypotheses in which the Earth captured an already formed Moon have severe inherent dynamical problems and provide no obvious explanation for the exotic lunar geochemistry.

Fission hypotheses that derive the material for the Moon from the Earth's mantle encounter two basic difficulties (a) the angular momentum of the Earth-Moon system, although large, is insufficient by a factor of about 4 to allow for rotational fission (Durisen and Gingold 1986) and (b) there is no acceptable mechanism for removing this excess angular momentum following lunar formation. Despite the prediction that the chemistry of the Moon should bear some recognizable signature of the terrestrial mantle, the Moon contains, for example, 50% more iron and has distinctly different trace siderophile element signatures. It also contains lower amounts of volatile elements and probable higher concentrations of refractory elements. The hypothesis thus failed the test requiring an identifiable chemical signature of the terrestrial mantle in the Moon, despite heroic attempts by geochemists to fit the lunar compositional data to that of the silicate mantle of the Earth.

The Earth and Moon have distinctly different siderophile element patterns (Newsom 1986). Claims that the siderophile elements in the Moon show a unique terrestrial signature have been shown to be erroneous (Ruzicka et al. 1998, 2001). The similarity in V, Cr, and Mn abundances in the Earth and Moon, sometimes claimed as definitive proof of lunar origin from the terrestrial mantle (Ringwood 1986), has been shown to result from incorrect models of the bulk lunar composition (Ruzicka et al. 1998, 2001). The eucrite class of basaltic meteorites, formed close to T_0, shows similar depletions in volatile and siderophile elements to lunar basalts (Ruzicka et al. 1998, 2001).

Double-planet models that form the Moon and Earth in association possess the twin difficulties of failing to account for the angular momentum of the Earth-Moon system and of readily accounting for the density difference. They do, however, account for the similarity in oxygen isotopes between Earth and Moon. A variation of these models involved the break-up of differentiated planetesimals as they came within the Roche limit. This process was postulated to result in a circum-terrestrial ring of broken-up silicate debris, while the more coherent metallic cores of the planetesimals either accreted to the Earth or escaped. Such scenarios do not solve the angular momentum problem. Moreover, the proposed break-up of the incoming planetesimals may be difficult (Boss et al. 1990). The previous two hypotheses envisage processes that might be thought to be rather general features of planetary accretion and lead to the presence of moonlike satellites around the other terrestrial planets (e.g., Venus), where they are conspicuously absent.

2.2.2. Post-Apollo views. Any theory of the origin of the Moon must satisfy chemical constraints imposed by the unique lunar composition and by the dynamical constraints of angular momentum. Such a theory must be physically plausible as well (i.e., satisfy the rules of orbital mechanics, with boundary conditions preferably derived from conditions prevailing during the accretion of planets). Few theories proposed for the origin of the Moon can simultaneously satisfy all these conditions, and the acceptability of any theory is entirely dependent on such agreement between theory and observation.

Two first-order observations have been made on lunar composition that have significantly influenced the debate on lunar origin. Siderophile element abundances and Fe/Si depletion indicate that the Moon is composed of matter derived from the mantle of a planet. The depletion of volatile elements has generally been regarded as an important indication of the degree of thermal processing of lunar matter during formation of the Moon (Ringwood 1970, 1992; Hartmann 1986; Stevenson 1987), although Taylor (1979, 1987, 2001a,b) has argued that the volatiles were to a large extent depleted prior to lunar formation. Understanding the nature of volatile element loss from the Moon is of critical importance in determining the origin of the Moon (see discussion in Jones and Palme 2000).

Tidal calculations have often been used to assess the history of the lunar orbit, but attempts to determine whether the Moon was once very much closer to the Earth (for example, near the Roche limit at about 18,000 km), which would place significant constraints on lunar origins, produce non-unique solutions (Williams 1989, 2000; Boss and Peale 1986). During the Late Proterozoic at about 620 m.y., the mean lunar distance was 58.4 ± 1.0 Earth radii, so the Moon was only marginally closer to the Earth (The present mean Earth-Moon distance is 60.3 Earth radii). The current rate of retreat of the lunar orbit predicts that the Moon should have been close to the Roche Limit about 1.5 b.y ago. There is no evidence of such a catastrophic close approach of the Moon 2000 m.y. ago (Williams 1989, 2000; Boss and Peale 1986). Tidal deposits are known in Archean sediments about 3200 m.y. old, indicating that lunar tides, apparently similar in magnitude to those of today, were present at that ancient epoch (Eriksson 1977; Eriksson and Simpson 2000). Current theories of lunar origin form the Moon close to the Earth. If so, then the evolution of the Moon's orbit probably has varied with time, with the Moon receding from the Earth more quickly when it was closer.

2.2.3. The large impact model for the origin of the Moon. The Earth-Moon system has an anomalously high angular momentum compared to other planets and this has been the rock on which most hypotheses of lunar origin have foundered. This excess cannot arise through random small multiple impacts since these must average out. Less than 10% excess angular momentum could arise by such means. However, one very large glancing impact could account for the observed excess (Cameron 1986, 2000; Canup and Righter 2000; Canup and Asphaug 2001). The model proposes that a body larger than Mars struck the Earth and a hot disk of material was ejected to form the Moon. This "Mars-sized impactor" hypothesis thus solves the angular momentum problem by definition. This involves collision with the Earth by the next largest body in the hierarchy and assumes prior core formation in the impactor.

A primary requirement is that such bodies were common in the early solar system. It is clearly established that melting and internal differentiation of planetesimals took place within a few million years after T_0. There is no shortage of potential impactors. Kortenkamp et al. (2000) consider how to assemble planetesimals into larger bodies, with results converging towards runaway accretion forming perhaps 20 Mars-sized bodies within timescales of 10^5 years after T_0. However, it appears "that the final number, masses and spacings of planets are stochastic, and not very dependent on the intermediate embryo formation stage of accretion" (Kortenkamp et al. 2000). Random impact events capable of supplying the angular momentum of the Earth-Moon system were probably common in the early solar system.

The model provides sufficient energy to melt at least half the Moon as required by the geochemical evidence that indicates that melting occurred more or less concurrently with accretion. Indeed the energetic nature of the event provides not only the lunar magma ocean, required by the geochemical evidence, but also induces terrestrial mantle melting (Stevenson 1987). Computer simulations indicate that the material that makes up the Moon comes primarily from the silicate mantle of the impacting body, not from the Earth (Canup 2004). In the giant impact hypothesis, the impactor is assumed to have differentiated into a metallic core and a silicate mantle. The low density and low bulk Fe content of the Moon, long-standing problems, are thus resolved in the hypothesis, as in the simulations, the metallic core of the impactor accretes to the Earth following the collision. The hypothesis also resolves the geochemical dilemma that the composition of the Moon does not match that of the Earth's mantle by cutting this particular Gordian Knot.

Cameron (2000) postulated that the impact occurred when Earth was 2/3 accreted. It seems difficult to add 1/3 of the Earth subsequently without affecting the Moon. The lack of evidence of such a late addition of presumably iron-rich material to the Moon, is a serious problem for this iteration. However, in revised models by Canup and Asphaug (2001) and Canup (2004), a Mars-sized impactor (10% Earth mass) impacts at a late stage in the accretion of the Earth (Fig. 7.16) thus returning to the original Cameron models. This avoids the problem of having to park the Moon somewhere out of the way while one third of the mass of the Earth is added. The revised models form the Moon as the last major collisional event during the hierarchical accretion of the Earth.

The impacting body was clearly derived from the inner solar system as shown by similarity between the Earth and Moon for the oxygen (Wiechert et al. 2001) and chromium isotopes (Carlson and Lugmair 2000) and by the dynamical constraints of the model. Although the Hf-W isotopic system has provided new information on the separation of silicate from iron metal (Halliday et al. 2000), this is difficult to interpret. It is possible that we may never be able to disentangle all the details of the accretion of the Earth and the impactor from a series of planetesimals that had already differentiated into iron cores and silicate mantles and that had already been involved in multiple collisional episodes.

The Giant Impact model has not been without its critics. Thus Jones and Palme (2000) assess the various geochemical parameters in an attempt to decide whether the giant impact remains as a viable hypothesis. Their main concern is the problems of a lunar core and of relating

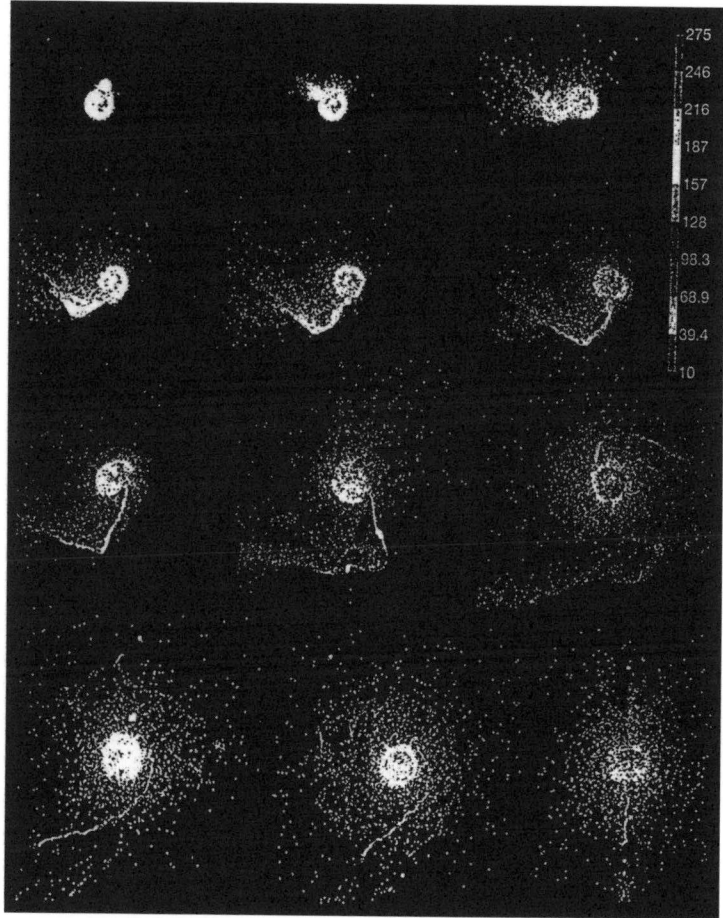

Figure 7.16. A computer simulation of the Moon-forming impact at times 0.3, 0.7, 1.4, 1.9, 3, 3.9, 5, 7.1,11.6, 17 and 23 hours after the impact. All views are in the plane of the impact except the last which is an edge on view. (courtesy Robin Canup and Erik Asphaug).

the composition of the Earth and Moon, but many of the problems disappear if the Moon is derived from the silicate mantle of the impactor rather than from the earth (Canup 2004). A small metallic lunar core might well be expected as the separation of silicate from metal during the giant impact is unlikely to be perfect. Thus not all geochemists would agree with their conclusions that the geochemical and geophysical evidence "casts doubt on the giant impact hypothesis." The stochastic nature of that event, the original depletion of volatile elements in the nebula, the unknown nature of the impactor, from which most of the Moon comes in the models, the great assemblage of precursor planetesimals and possible loss of volatiles during the impact make the task of geochemists attempting to reconstruct that distant event, model-dependent at best. It is also worth recalling that the Moon constitutes only about 5% of the mass of the impactor so perhaps is not even a uniform sample of the impactor mantle (e.g., Warren 1992).

Coupled with the very low value for the initial $^{87}Sr/^{86}Sr$ ratio (LUNI = BABI) (Nyquist 1977), the depletion of the Moon in many volatile elements might also be attributed to a volatile-depleted impactor rather than being due to the impact (Taylor 2001b).

Many complex factors need to be thought about in modeling the formation of a Moon following a giant impact. Stewart (2000) notes that the Moon "may have accreted slowly from the outer (cooler) edge of a protolunar disk, while the inner disk remained hot and molten. He points out that many of the apparent problems stem from "our primitive understanding of how the terrestrial planets formed", so that "idealized models....can also mislead the unwary" noting that "collisions between planetesimals... are inherently messy processes."

In summary, the peculiarities of the chemistry and dynamics of the Moon can be ascribed in one way or another to the extraordinary events surrounding a giant impact.

2.2.4. The nature of the impactor. A critical feature of the large-impactor hypothesis is that most of the material making the Moon comes from the silicate mantle of the impactor. Thus, the composition of the interloper is the most significant component for the geochemical constitution of the Moon.

The similarity in oxygen isotopes between the Earth and Moon indicate derivation of both the Earth and the impactor from the same region of the nebula, thus excluding models that derive the impactor from the outer reaches of the solar system. The similarity in $^{53}Cr/^{52}Cr$ ratios (^{53}Cr is derived in part from ^{53}Mn: $t_{1/2}$= 3.7 m.y.) between the Earth and the Moon and their contrast with higher meteoritic values (Lugmair and MacIsaac 1995) carries the same implication of derivation of lunar material from around 1 A.U. A third constraint is the relatively low collision velocity (Benz et al. 1989; Cameron and Benz 1991) required to produce a Moon-sized body, that once again restricts the impactor to have been formed in the neighborhood of Earth.

Current models (e.g., Canup 2004) assume that core mantle separation occurred before impact, to account for the lunar siderophile element abundances and the lunar depletion in iron (13% FeO) relative to primordial solar nebula volatile-free abundances (as shown by the CI meteorites) of 36%. The abundance of FeO in the mantle of the impactor must however have been greater than that of the terrestrial mantle (8% FeO), since the bulk Moon contains a much higher abundance (13–14.6%, Jones and Palme 2000). Such values are intermediate between the FeO abundance in the martian mantle (in which an estimate based on the SNC meteorites is 18%; Longhi et al. 1992) and that of the terrestrial mantle. Since Mars with a high mantle FeO content exists in the inner solar system, the postulate of an impactor with a high mantle FeO content is not unreasonable.

O'Neill (1991, p.1169) proposed that the impactor was an "exotic interloper.......from further out in the solar system" of CI composition, "fully oxidized and volatile rich." However, the factors noted above constrain the impactor, most likely to be volatile-depleted, to have originated in the neighborhood of the Earth. In addition, the rare gas and O isotopic evidence make CI chondrites an unsatisfactory component for building the terrestrial planets, so that there is no reason to suppose that planetesimals of CI composition were abundant in the inner nebula. The great distinction between the region now occupied by the terrestrial planets and that from which the CI chondrites are derived lies in the striking depletion of volatile elements. This depletion as shown by low K/U ratios, was common to Venus, Earth and Mars, and was thus probably due to a major event affecting the inner solar nebula (Taylor 2001a).

2.2.5. Resolution of the origin of the Moon. Of most significant immediate interest would be various geophysical data, specifically for heat flow which would resolve some geochemical questions about bulk composition, seismic data on the internal structure that will establish the nature of lunar deep structure and the presence or absence of a metallic core. These data will refine our ideas about initial thermal history and the initial depth of the magma ocean.

Another series of problems deals with events in the solar system before formation of the Moon and the history of the precursor material that is critical to the interpretation of the W-Hf isotopic systematics. Is the Moon a unique object or are there bodies (asteroids) with similar histories. What was the nature of the impactor? What was thermal history of precursor material and was the volatile-element depletion in the Moon due to the collision or inherited from the impactor, or is it a memory of events in the earliest nebula, as suggested by the low value of LUNI?

2.3. Highlights of recent discoveries

As discussed in previous chapters, a series of recent small missions to the Moon created an entirely new perspective of this intriguing planetary body, and the small amount of new data caused several paradigm shifts in our understanding of the character of the Moon. The fly-by of the Moon by the Galileo spacecraft on its way to Jupiter (Belton et al. 1992, 1994) was the first time a spacecraft had visited the Moon with modern sensors since Apollo. This was followed by the Department of Defense Clementine mission (Nozette et al. 1994), which was principally designed to test a suite of small sensors in space. NASA sponsored a science team shortly before launch to plan and implement science data. After several calibration issues (some still ongoing) the first global topography (Zuber et al. 1994) and global multispectral imagery data sets were produced for the Moon (e.g., Eliason et al. 1999). The only mission flown in this new wave of exploration whose principal objective was lunar science was the Lunar Prospector mission (Binder 1998), selected through NASA's competitive Discovery Program. This mission provided the first global assessment of elemental abundance (especially Th, K, Fe, Ti, H; see Chapter 2) (e.g., Lawrence et al. 1998, 2000, 2002; Feldman et al. 1998, 2000, 2001).

The input of data from this combined group of encounters and small missions has provided a valuable foundation upon which to reevaluate the evolution of silicate bodies. This is not to say important research had not continued via lunar sample studies and telescopic measurements of the lunar nearside during the dark decades. The origin of the Moon and the Earth/Moon system through a violent planetesimal encounter was well on the way to being understood (see Section 7.1.2). But our perception and understanding of the Moon changed with this minor amount of new data from relatively modern sensors. Three particularly important surprises are highlighted below.

2.3.1. SPA—the biggest basin in the solar system. Although several lunar scientists had long argued for a big basin on the lunar farside (e.g., Wilhelms 1987), the South Pole-Aitken basin (SPA) is missing from most commercial lunar maps produced using Apollo era data. This remarkable oversight reflects the limited character of the data available, changed, however by the first digital images from Galileo. Not only was the basin rim readily recognized, but the basin interior had a distinctively low albedo and exhibited spectral properties indicating that it was rich in mafic minerals (Belton et al. 1992). Subsequent data from Clementine illustrated the dramatic topography (up to 12 km deep!) (e.g., Zuber et al. 1994), enhanced mafic composition (e.g., Lawrence et al. 1998 2002; Lucey et al. 1998) and unusual mineralogy exposed in the interior (Pieters et al. 2001). It became apparent that SPA may be the largest documented basin in the solar system (Spudis et al. 1994; Zuber et al. 1994), and the deep-seated material of the interior represents a terrane on the Moon that is quite distinctive and unlike the lithologies documented with samples returned from the near side (Jolliff et al. 2000).

2.3.2. Thorium hot spot. The global elemental data from Lunar Prospector confirmed that the SPA interior is iron-rich (midway between feldspathic highlands and mafic mare basalt). One of the most unexpected results, however, was the very well-defined pattern of radiogenic materials, almost all of which are associated with the nearside Procellarum Basin, especially the Imbrium basin and its deposits. Since such elements (KREEP material) are expected to be concentrated in the residua of primordial lunar differentiation, it is not surprising that a large basin tapped such a concentration. The surprise is that the pattern is *only* associated with one part of the Moon and strongly associated with the Imbrium basin and circum-Imbrium deposits. Imbrium is not only smaller than SPA but also occurred near the end of the period of basin forming events. This highly localized concentration of radiogenic (incompatible) elements indicates there are enormous gaps in our understanding of the first few hundred million years of planetary evolution and differentiation.

2.3.3. Polar H (water?). The extremely low volatile content of lunar samples was one of the first characteristics noted for returned lunar samples (along with their ancient age). Thus, the contested hint of polar ice by Clementine (Nozette et al. 1996; Simpson and Tyler 1999) and the unquestioned discovery of hydrogen concentration at the lunar poles by Lunar Prospector (Feldman et al. 2000) has led to some optimism that water resources may indeed exist in the permanently shadowed regions of the Moon. However a reassessment by Campbell et al. (2003) indicates that any ice sheets in the regolith are probably only a few cm thick. This counter-intuitive concept, although reasonable in theory (Arnold 1979), is made more credible by the discovery of extensive volatile (perhaps ice) deposits in craters at Mercury's poles (Harmon et al. 1994).

2.4. Uniqueness of the Earth-Moon system

There are many peculiar features about the Moon and the Earth-Moon system that make this pair unique in the solar system. The orbit of the Moon about the Earth is neither in the equatorial plane of the Earth, nor in the plane of the ecliptic, but is inclined at 5.1° to the latter. The angular momentum of the Earth-Moon system is anomalously high compared to the other planets. The Moon has a low density relative to the terrestrial planets. The Earth-Moon mass ratio (1/81.3) is the largest in the solar system (excluding the Pluto-Charon system in the Edgeworth-Kuiper Belt).

Neither Venus, close in size to the Earth, nor Mercury have moons, while Phobos and Deimos, the two moons of Mars, may be tiny captured asteroids, or pieces of one, with what appear to be primitive compositions. The satellites of the outer planets are mostly ice-rock mixtures. Our satellite has an unusual composition. It is bone dry, depleted in elements such as K, Ga, Pb, Tl and Bi that are volatile below about 1100 K, and also in siderophile elements; and it is probably enriched in refractory elements such as Al and U relative to CI. However, in this respect the Moon is probably not unique as Vesta, at least, appears to possess similar depletions in volatile and siderophile elements (see Section 2.3 on Vesta).

The spin angular momentum of the Earth-Moon system is anomalously high compared with that of Mars, Venus or the Earth alone. Some event or process spun up the system, relative to the other terrestrial planets.

The Moon has a lower bulk density (3.34 g/cm^3) than the Earth (5.54 g/cm^3) (Fig. 7.17). However, the lunar density is close to the uncompressed density of the terrestrial mantle, implying that the Moon is composed largely of silicate minerals. Thus, it lacks the massive amounts of metallic iron that are responsible for the high densities of the inner planets.

2.5. Role of the Moon in making Earth a habitable planet

What were the implications for the Earth of the single impact origin of the Moon? It may be responsible for the 24 hour rotation period and the obliquity of the Earth, both significant parameters affecting biological evolution. An entertaining discussion by Williams and Pollard (2000) notes that the large mass and close proximity of the Moon stabilizes the obliquity of the Earth and has important implications for life and our climate, concluding that perhaps the Earth does owe its habitability and long-term sustainment thereof (and its suitability for humans) to the presence of the Moon.

According to Lascar and coworkers (1993), the obliquity of the Earth would be "chaotic with large variations reaching more than 50° in a few million years and even, in the long term, more than 85° in the absence of the Moon" and so "our satellite is a climate regulator for the Earth" (Fig. 7.18). Any primitive atmosphere was removed during the impact event. The Moon also plays a unique role by raising substantial tides in the Earth's oceans. In the context of this discussion, two questions without answers are how crucial is the presence of the Moon in producing this habitable planet and how likely are such events in other planetary systems.

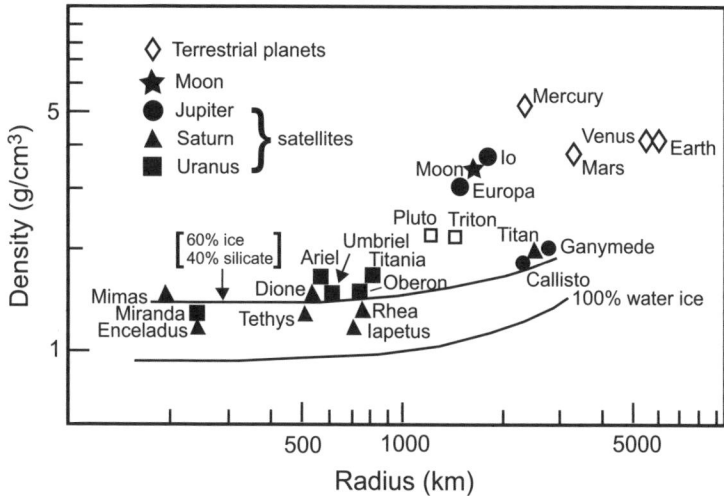

Figure 7.17. The densities (in g/cm²) of the principal satellites of the giant planets, the Moon and the terrestrial planets (adapted from Taylor 2001).

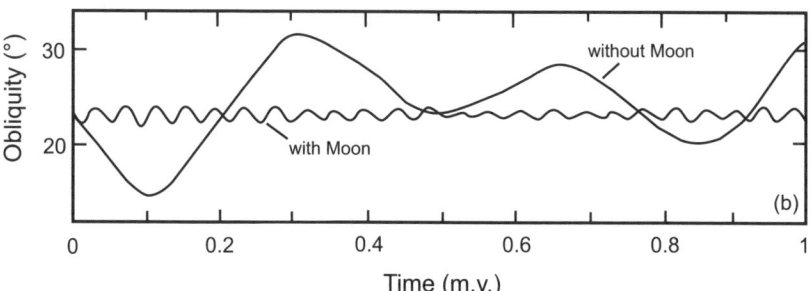

Figure 7.18. Obliquity variations for the Earth over one million years with and without a Moon (adapted from Williams and Pollard 2000, Fig 3).

3. PLACE OF THE MOON IN THE SOLAR SYSTEM

The Earth and Moon share the same part of the solar system. Since their birth they have both witnessed and experienced the same abundance of material and debris moving through our solar system. Without an atmosphere or magnetic field, however, the Moon has also constantly been subject to high-energy radiation from the sun and galactic sources as well as a constant dusting of small particles.

3.1. The Moon as a satellite

The density of the Moon is compared with that of the other satellites in the solar system in Fig. 7.17. Only Io and to a lesser degree Europa, are comparable with the Moon. The rest contain significant proportions of ice. Io owes its volcanic nature to its close proximity (six Jupiter radii) to that gas giant and so is a unique object. Ganymede, Callisto and Titan, are all larger, while Europa, Triton (and Pluto) are significantly smaller than the Moon. The similarity of Io to the Moon in density and radius has no fundamental significance, but must be due to stochastic processes. Size is not a very useful discriminator. Thus the planet Mercury is smaller than Ganymede and Titan, while seven satellites, including the Moon, outrank the largest Edgeworth-Kuiper belt object, Pluto (although larger Triton also has a case to be considered as a former resident of that belt). Most of the satellites in the solar system are mixtures of rock and water ice and have low densities. They formed in the early solar nebula, out beyond the "snow line" around 4 to 5 AU, where water-ice was stable. The captured satellites of the giant planets represent icy planetesimals left over following the formation of the giant planets, or captured bodies from the Edgeworth-Kuiper belt. The regular satellites of the giant planets, of which the Galilean satellites of Jupiter are the best known, appear to have formed from planetary subnebulae, and thus differ fundamentally in origin from the Moon.

In the inner solar system, only one rocky planet managed to grow to around the size of the Earth, but the Earth and Venus are similar only in a Jekyll-Hyde sense. Thus Venus is significant as it shows what happens when nature tried to duplicate our planet. Although Venus is close enough in mass and density to be regarded as a possible twin planet to the Earth, there are major differences between the two planets. Venus does not possess a satellite. The slow backward rotation and low obliquity of the planet are indicative of a very different collisional history.

3.2. Comparison with other airless differentiated bodies

3.1.1. Physical and compositional properties of Vesta.
Vesta was the first asteroid studied with remote spectroscopic analysis of sufficient spectral resolution to characterize its surface mineralogy (McCord et al. 1970). Based on the inferred mineralogy, a strong association was made between Vesta and the basaltic achondrite meteorites (Howardites, Eucrites, Diogenites, or HEDs). This link was strengthened as new instruments and more sensitive measurements were made, especially when it became apparent that Vesta was the only large asteroid in the main belt with properties comparable to the HEDs (e.g., Tholen 1989; Gaffey et al. 1993; Pieters and McFadden 1994).

Dynamists were skeptical of the link between the HEDs and Vesta at first because it was believed to be extremely difficult to remove rocks from Vesta's surface and get them to a part of the solar system that favored perturbations into Earth-crossing orbits. All that changed when telescopic instrumentation improved sufficiently to accurately measure more and smaller asteroids. It was discovered (Binzel and Xu 1993), that a spray of small "Vestoids" exists between Vesta and the 3:1 resonance with Jupiter, an accessible "escape hatch." The case continued to grow stronger as additional spectroscopic observations were made (Gaffey 1997; Burbine et al. 2001) and HST images showed there to be a 460 km diameter central-peak crater, 13 km deep on Vesta's surface (Thomas et al. 1997), which may account for the Vestoids.

The basaltic eucrites derived from 4 Vesta display many similar compositional features to the lunar basalts (Fig. 7.19). Vesta produced basaltic material with similar element patterns to low-Ti lunar basalts, including strong depletions in volatile elements, at about 4557 m.y. ago (Lugmair and Carlson 2000) close to T_0. The angrites and the eucrites are the only meteorites that have such low K/U ratios that are less than those of the Earth and that overlap those of the Moon. The eucrites are also depleted in siderophile elements and have very similar high Cr abundances to those of mare basalts (Ruzicka et al. 1998, 2001).

Thus, at least one asteroid went through a sequence of volatile and siderophile element depletion similar to that experienced by the Moon. But the timing of the thermal event on the HED parent body was within only a few millions of years after planetesimal formation. The canonical view is that asteroids with surfaces of achondritic composition are rare (based on astronomical surveys such as Tholen 1989), so the processes that produced the HED parent body may have been uncommon in the early solar nebula, at least in the region of the asteroid belt. However, such processes producing "igneous" planetesimals may have been more common in the inner nebula that supplied planetesimals to form the inner planets.

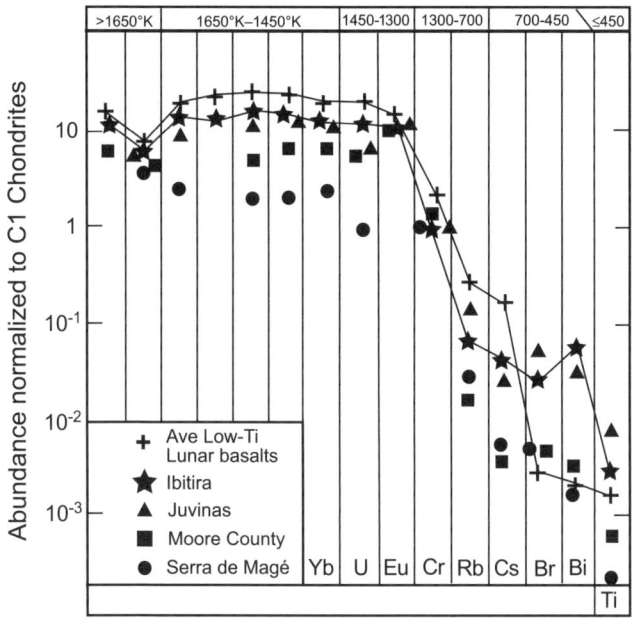

Figure 7.19. The close correspondence between the abundances of the lithophile elements in eucrites and low-Ti mare basalts (adapted from Taylor and Esat 1996, Fig. 2).

3.2.2. Physical and compositional properties of Mercury. The Moon and Mercury represent special cases even by the standards of the solar system. Mercury is unique due to its high density, with an iron-silicate ratio about twice that of the other inner planets. In contrast, the Moon is of interest because of its low density and low metal-silicate ratio. Mercury is of great interest as the scant evidence shows that the surface bears a close resemblance to that of the lunar highlands (Fig. 7.20). A comparison between the integrated spectrum for the mercurian surface and that of a laboratory spectrum for Apollo 16 lunar highlands samples shows that the mercurian data are consistent with a regolith surface. Spectral data from thermal infrared exclude both ultramafic and granitic soil compositions, and appear to be compatible with a feldspathic crust similar to that of the lunar highlands although the surface appears to be heterogeneous with feldspar compositions ranging from Ab_{50} up to anorthite (Mitchell and de Pater 1994; Sprague et al. 1997; Sprague and Roush 1998). More recent evaluations of the spectral reflectance data in the 0.4–0.65 μm region, coupled with the earlier data, indicate that the concentrations of iron and titanium appear to be near zero, while the spectral reflectance data are consistent with a surface mineralogy consisting of a 3:1 labradorite-enstatite mixture, with an average grain size a factor of two smaller than the lunar soils (Warell and Blewett 2004).

The presence of Na and K in the tenuous mercurian atmosphere is consistent with a feldspathic surface. Na and K ions are also observed up to 1200 km above the surface of the Moon. In the latter case, they are almost certainly derived by sputtering from the feldspathic surface.

Some information can be gleaned about composition from the surface morphology. There are two types of plains units on Mercury, the early intercrater plains of preTolstojan age and the later "smooth" plains of Calorian age (Spudis and Guest 1988). The origin of both sets of plains is disputed between those who equate them with the lunar Cayley-type plains (debris deposits or impact melt sheets, both produced by impacts) (e.g., Wilhelms 1976)) and those who propose that they are volcanic plains analogous to the lunar maria, and hence are formed

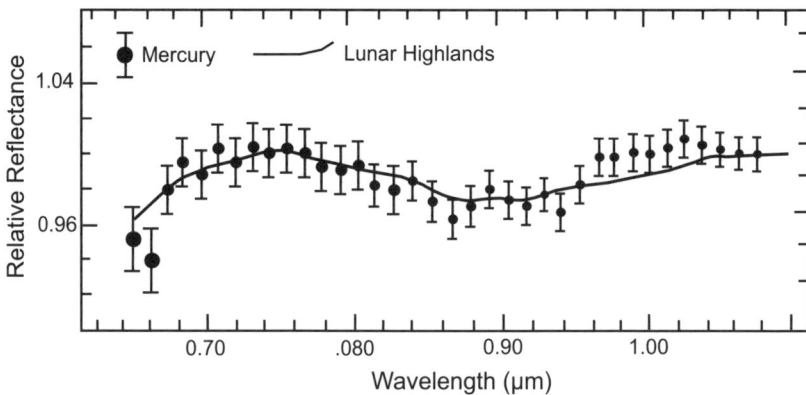

Figure 7.20. The reflectance spectrum of the surface of Mercury, covering both smooth and intercrater plains, shows a close resemblance to a laboratory spectrum of an Apollo 16 lunar highland soil sample. (adapted from Vilas et al. 1988, p 71, Fig. 6).

by rather fluid basaltic lava flows (Spudis and Guest 1988). It is interesting to note parallels with the history of the pre-Apollo interpretation of the lunar surface in this controversy.

If the intercrater plains are volcanic, the mercurian lavas must possess the same low viscosity as the lunar basalts. This assumption remains to be tested, but it would imply a similarity in mantle sources that, given the unique nature of lunar evolution, seems surprising. Their albedo is twice as high as that of the lunar maria, a fact not consistent with the presence of dark low viscosity iron-rich basaltic lavas. The wrinkle ridges on the floor of the Caloris Basin, cited in support of an analogy with lunar mare basalts, appear to be very large. Only more data will resolve this issue, but it is clear that if these features are volcanic, then their petrogenesis is distinct from that of the lunar mare basalts.

Additional light on this problem has been shed by microwave and mid-infrared data (Jeanloz et al. 1995; Mitchell and de Pater 1994; Blewett et al. 1997). The regolith of Mercury is two to three times more transparent to microwave radiation than the lunar mare regolith and 40% more transparent than that of the lunar highlands. This limits the amount of FeO and TiO_2 on the surface of Mercury to much less than that present in the lunar highlands and the concentrations of these elements may be close to zero (Warell and Blewett 2004). The apparent low content of iron on the mercurian surface is consistent with a lack of volcanic activity. It is possible that "volcanic heat piping has not played a major role in Mercury's thermal evolution" (Jeanloz et al. 1995) thus insulating the core. Perhaps the alkali and other incompatible elements were concentrated at an early stage in the crust of this one plate planet. In this scenario, the crust of Mercury may have had a similar origin (flotation of feldspar from a magma ocean) to that of the Moon. The difference between the two bodies appears to be a lack of subsequent basaltic volcanism on Mercury.

Some spectra may be compatible with that of nepheline-bearing alkali syenite (Jeanloz et al. 1995) and presence of such sodium-rich volcanic rocks could provide an additional source for the sodium ions detected in the atmosphere. Silicic volcanism, which might produce low-albedo deposits, seems an unlikely candidate on Mercury and may indeed be a mainly terrestrial phenomenon. Accordingly, magmas derived from the mercurian mantle may indeed be unique.

3.2.3. Implications of space weathering processes on airless body regolith. An understanding of regolith evolution may be key to interpreting remote observations of Mercury and other airless bodies in the Solar System. As detailed in Chapter 2, the harsh space environment alters planetary surfaces in a manner that makes remote compositional analysis difficult if regolith processes are not taken into account. Gradual physical and compositional transformations occur as a result of bombardment by micrometeorites and solar and galactic particles. Without the detailed information available from lunar soil samples, we would be ignorant about the myriad of processes that have a profound effect on observed properties of solar system surfaces. On the Moon as soil particles are exposed to the space environment, they accumulate thin (a few hundred angstroms) amorphous silicate rims that contain abundant sub-microscopic native iron, or nanophase metallic iron ($npFe^0$) (see Chapter 2 for detailed discussion). The longer a soil is exposed, the more particles develop such $npFe^0$-bearing rims and the total abundance of $npFe^0$ increases. Although $npFe^0$ is a minor component of lunar soils (<<1%), it is exceptionally optically active and absorbs radiation passing through particles in a very non-linear manner. Also, since deposits of $npFe^0$ originate as rims on individual particles, finer particles with more surface area are affected most strongly.

The effects of $npFe^0$ on optical properties depend on the size of these tiny metallic grains and their overall abundance, both of which depend on the host material and the environment in which $npFe^0$ forms. In all cases, the presence of $npFe^0$ causes the soil to darken and diagnostic mineral absorption bands to weaken. The effects are not uniform with wavelength. For lunar soils the size of $npFe^0$ ranges from about 3 to 20 nm (Keller and McKay 1993, 1997). This size of grains imparts a wavelength dependence to absorption of radiation, which is highly sensitive to the abundance of $npFe^0$ (Noble et al. 2001, 2003; Hapke 2001). For the Moon, more radiation is absorbed at shorter wavelength, creating a "red" continuum for most lunar materials. The curvature of this lunar continuum is controlled by the abundance of $npFe^0$, and to a lesser extent the size of grains (Hapke 2001; Noble et al. 2003; Noble 2004).

The origin of $npFe^0$ is linked to processes that release and mobilize iron from surface materials, the two most prominent of which are vaporization from micrometeorite impacts and solar wind sputtering (Keller and McKay 1997; Taylor et al. 2001; Hapke 2001). For the Moon most of the iron comes from FeO in regolith minerals, with minor amounts from meteoritic contaminants. Soils developed from feldspathic highland lithologies have lower amounts of FeO available to produce $npFe^0$ than soils from mare regions. Systematic differences in optical effects are thus observed as a soil matures in the space environment for mare and for highlands regoliths (Noble et al. 2001, 2003; Noble 2004). This understanding of the processes affecting space weathering allows predictions for other solar system bodies. The effects of space weathering and development of $npFe^0$ on soil particles is expected to be considerably reduced for asteroids and readily accounts for optical effects that transform ordinary chondrite parent bodies into "S-type" asteroids (Pieters et al. 2000; Hapke 2001). It should be noted that metallic grains larger than 100 nm (i.e., approaching micron scale) simply act as absorbers at all optical wavelengths (Noble et al. 2003; Noble 2004). The impact darkening observed in ordinary chondrites due to dispersed metal and sulfides is a good example of this effect (Britt and Pieters 1994). On the surface of Mercury, the effects of space weathering and development of $npFe^0$ should be enhanced (Noble and Pieters 2003), but is complicated by the extreme thermal cycling and higher flux of particles. Experimental evidence suggests the size of the $npFe^0$ is slightly larger on Mercury than on the Moon and that the total abundance of Mercurial regolith $npFe^0$ is somewhat more than typical lunar highlands (Noble et al. 2004, in prep).

4. HOW TO EXPLORE A SOLID BODY

An overview of lunar exploration and the historic sequence and context of its multiple aspects is provided in Chapter 1. Subsequent chapters provide details and depth of this incredible undertaking. It is constructive to reflect on what has been learned and some of the important lessons that might guide further detailed exploration of other solar system bodies.

4.1. What we learned (and changing perspectives)

4.1.1. From samples. In the 30 years since the last Apollo and Luna samples were brought to Earth, many review papers and books have documented the profound impact that the lunar rocks had in changing nearly all preconceived notions about the nature of the Moon (e.g., Taylor 1982; Wood 1986). Prior to Apollo, it is doubtful if anyone could have predicted some of the remarkable properties of the lunar samples. They serve still as an object lesson about the dangers of over-interpreting planetary bodies in the absence of sample-derived knowledge. Yet, if the single most important lesson to be learned from the Apollo samples is "expect the unexpected," that lesson seems not to have been universally learned. Great technological advances in robotic analytical instrumentation have led to some speculation that costly and difficult sample return might not be required to adequately explore Mars (e.g., Paige 2000). But, the fact is that there will always be major limitations to what a robotic lander or rover can do. Partly this is because of restrictions in power and mass, and partly it is because no robotic spacecraft instruments can ever conduct experiments under such controlled conditions or with sufficiently well prepared samples as is possible in an Earth laboratory. An additional very important reason is that robots cannot adapt in the way a human scientist can in response to unexpected results, bringing additional and unforeseen tests to bear.

There are of course over 37 known (as of this writing) lunar meteorites (~80 separate stones) that have been found in Antarctica and elsewhere. Even without the Apollo and Luna samples, we might by now hypothesize much about the formation and evolution of the Moon from the meteorites (albeit 10 years later; the first lunar meteorite was identified in 1981). Although their identity as such rests on comparison with the Apollo and Luna specimens, even in the absence of the latter it is reasonable to assume that a lunar origin for the meteorites would be suggested from such evidence as identity in oxygen isotope compositions with Earth rocks (recall, however that enstatite chondrites share this property), and the young ages of many mare basalts relative to most other meteorites. The meteorites are considerably less informative than mission-returned samples in the sense that the geologic context of the meteorites is completely unknown. The Apollo and Luna samples come from known localities and in some cases even from outcrops, so ages of specific sites, formations, and impact events can be determined. But the fact remains that meteorites are samples that are analyzed in terrestrial labs just as the mission-returned samples are. The only difference is the delivery mechanism and the lack of definitive information about their place of origin.

Previous reviews (e.g., Taylor 1975; Drake 1986; Wood 1986) have written in detail about what has been learned from the Apollo samples. There would be little point in restating here what has been so eloquently said elsewhere, were it not for the fact (alluded to above) that 25+ years (since Viking) of Mars exploration and exploration planning still have not resulted in a sample return mission. While admitting the high cost and technical difficulty of implementing a robotic martian sample return mission, there nonetheless remains a continuing lack of recognition of the critical importance of sample return to achieving Mars science goals. The emphasis is on building bigger and better robotic rovers with larger and more capable science payloads, in order to find "just the right place" to eventually collect samples for return to Earth. But this attitude forces sample return to be continually postponed into the almost indefinite future. Such planning either does not recognize or else ignores a single inescapable fact: The most important science goals (e.g., the existence of ancient or present life, crustal evolution, atmospheric evolution) of Mars exploration will likely not be achieved in the absence of sample return because robotic rovers will simply not be able to carry analytical instruments with the levels of precision and sensitivity of Earth-based instruments, nor will they have the flexibility to react to unknown findings with anything like the flexibility of human scientists in a lab.

Therefore, in the interest of using the lunar experience as an object lesson for the exploration of other solar system bodies—especially Mars—this section takes a somewhat different approach than previous reviews. Specifically, we highlight here a sampling (by no means exhaustive) of important lunar measurements that we could *not* have made, or make even now, with strictly robotic means; these findings derive uniquely from the study of the returned samples (including the meteorites) by humans in Earth laboratories. These measurements were necessary precursors to current concepts of lunar formation and evolution. For example, the oxygen isotopic similarity of lunar and terrestrial rocks plus the extreme depletion of volatiles in the lunar rocks and the Eu anomalies are fundamental underpinnings of the giant impact hypothesis. We do not attempt to review in detail the results themselves—that is done elsewhere in this volume—but rather to explain why laboratory analysis of returned samples plays such a critical role in understanding the evolution of the Moon—and, hence, any other planet. Not only would some of the most important lunar discoveries have been missed 25 years ago had we relied solely on telescopic/orbital/*in situ* robotic exploration, they would be missed *even today* using the much more sensitive and accurate remote sensing instruments that currently are available.

Lunar differentiation and igneous diversity. One of the most fundamental scientific properties of any rocky planet or satellite is its basic geologic structure: whether it is primitive (chondritic) or differentiated into a core, mantle, and crust. And, if it is differentiated, what are the basic crustal units and how did they evolve? Up until nearly the time of Apollo 11 some scientists believed the Moon to be not a differentiated planet but, rather, a chondritic aggregate (e.g., Urey 1965). Although remotely obtained chemical data from Surveyor missions suggested the presence of basalts in the Mare basins and hinted at feldspar-rich rocks elsewhere (e.g., Turkevich 1971), it was the preliminary examination of the Apollo 11 rocks that largely dispelled the possibility of a primitive undifferentiated Moon (LSPET 1969). Samples from the subsequent Apollo missions showed even more clearly, in their petrographic and geochemical characteristics, the strongly differentiated character of the Moon. Seismic experiments set up by the Apollo crews confirmed that the Moon has a stratified structure similar to that of Earth, with a crust, a mantle, and possibly a small core (Hood and Zuber 2000). From the samples themselves, three major igneous rocks suites were recognized—the highlands ferroan anorthosites, mare basalts, and magnesium-rich norites and gabbros. A fourth important "rock type," KREEP, is mainly a geochemical component contained to varying degrees in other suites, such as the Mg suite. Detailed geochemical and petrologic studies showed these suites to be interrelated by differentiation on a planetary scale, albeit not from a homogeneous reservoir. For example, trace-element studies showed ubiquitous negative Eu anomalies in the diverse mare basalts and complementary positive anomalies in the ferroan anorthosites. This observation suggests that the basalts originated by partial melting of a mantle source from which plagioclase of the crust was also removed, and supports the idea of formation of the anorthositic crust by plagioclase flotation and fractionation of a magma ocean.

Consider how the situation would now be different had there never been any Apollo and Luna samples, or even any lunar meteorites (or, at least, that they were not recognized as such). The Clementine and Lunar Prospector orbital data obviously reveal the chemical dichotomy between the feldspathic highlands and the mafic mare, the existence of a huge geochemical anomaly in the region of Procellarum basin (i.e., the Procellarum KREEP terrane), and the mafic South Pole-Aiken basin. What then would we learn by sending robotic landers to those regions of the Moon today for the first time, equipped with modern robotic analytical instrumentation? The Mars 2003 rovers, for example, carried an instrument package that included a rock-abrasion tool (for abrading below the surfaces of coated or weathered rocks), an alpha particle X-ray spectrometer for bulk chemical analysis, a Mössbauer spectrometer, a miniature thermal emission spectrometer, and a high-resolution imager. A similar package on the surface of the Moon could characterize the dominant mineralogy and major-element chemistry of, and thus identify, the dominant lunar igneous types such as basalts, anorthosites, and troctolites/norites. Thus we would undoubtedly recognize that the Moon is a differentiated body. But deciphering the details of that differentiation, and seeing through the devastating effects of more than 4 billion years of impact processing, requires tools that mostly are beyond the capabilities of any rovers now or in the foreseeable future. For example, recognizing and characterizing an impact melt breccia is straightforward for a petrologist in a lab using a thin section, a petrographic microscope, and an electron microprobe. The hypothetical robotic lander described above could, in fact, probably also recognize a coarse impact breccia through the presence of glass and large diverse clasts. However, barring the presence of xenoliths, recognizing a hybrid impact melt would be difficult at best for the rover. Distinguishing between a primary basalt, a fine grained basaltic breccia, and an impact-melt of basaltic composition, probably would be even more difficult (e.g., Fig. 7.21).

Petrogenetic modeling of the lunar igneous suites is critically dependent on trace elements, among which the rare earths are of course important because of how igneous processes fractionate the individual elements from one another. In terrestrial laboratories, bulk determination of REE in common igneous rocks is routinely done either by neutron activation analysis or inductively-coupled plasma mass spectrometry (ICP-MS); These elements generally are beyond the detection limit for X-ray fluorescence, and elemental peak overlaps in any case are severe. Consider that the REE abundances in lunar mare basalts range from several tens of ppm for the most abundant elements (Ce) down to less than one ppm for the less abundant (e.g., Lu), and abundances in highland anorthosites are less than one ppm for even Ce and Eu and much

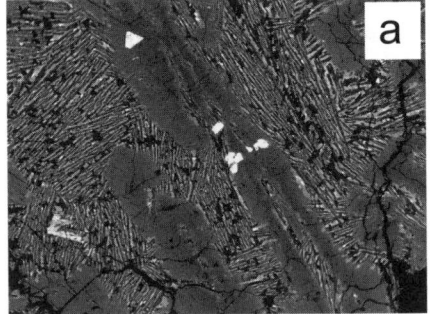

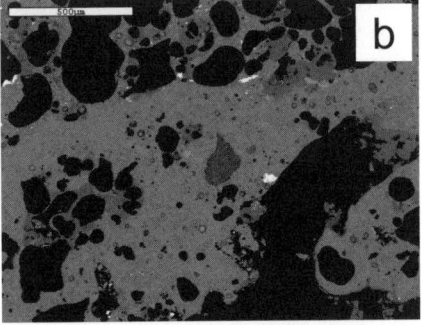

Figure 7.21. Back-scattered electron photomicrographs of three lunar basaltic rocks: (a) primary igneous basalt; (b) basaltic impact melt; (c) basaltic breccia. Photos courtesy of C. Shearer.

lower for the other REE. Even in KREEP, the abundances of Sm and heavier REE generally are several tens of ppm at most. No current generation robotic lander instrument is capable of precisely analyzing trace elements, and it is unlikely that such instruments will be capable any time soon of analyzing the REE at the levels indicated with sufficient precision to resolve intra-REE fractionation patterns. Thus without the returned Apollo and Luna samples and the lunar meteorites, there would be no accurate data for REE or other trace element abundances that are so important in constraining the petrogenesis of the dominant lunar igneous rock types and their possible relationships to one another. Without those samples, it is also unlikely that we would recognize KREEP at all beyond simply knowing that the geochemically-anomalous Procellarum region exists.

In summary, without having the Apollo/Luna samples and lunar meteorites we would not understand any of the details about lunar differentiation. This is because of the simple reason that we would not be able to bring to bear two of the most critical tools for constraining igneous petrogenesis: precise trace element determinations and classical petrographic analysis. There is no robotic lander yet that can even begin to do those things as well as a human scientist studying returned samples in a lab.

The antiquity of lunar rocks. The ages of major crustal units and the timing of global differentiation are essential facts for understanding the crustal evolution of a differentiated planet. On Earth, plate tectonic recycling and crustal reprocessing have erased most traces of the first 500 million years of Earth's history. Uplift and erosion have destroyed most traces of its early impact history as well. Therefore the Moon offers critical insight into Earth's earliest history.

Prior to Apollo, the ages of lunar surface units were estimated based on crater counts. Noting that pre-Apollo crater age estimates for the mare ranged from 3.5 Ga to a mere few million years, Taylor (1982, p 31; and refs. therein) wryly concluded that the understanding of lunar ages at that time "might best be described as chaotic." One of the fundamental discoveries from Apollo is that the lunar surface is uniformly very old. The youngest measured crystallization (igneous) age of any lunar rock is 3.1 Ga for a mare basalt, although the lunar volcanism that produced the mare basalts extended over a substantial time interval from 4.2 Ga. Ferroan anorthosite from the highlands dates back to at least 4.44 Ga (Lugmair 1987) or 4.46 Ga (Norman et al. 2003), meaning that the Moon had formed and began to solidify within about 100 my of solar system formation.

What was true 30 years ago remains equally true today: the extreme care required to make precise isotopic age determinations of ancient lunar (or any other) rocks is far beyond the capability of any robotic instrumentation and probably always will be. Epsilon-level (part per 10,000) precision in isotopic ratio measurements is achieved only through elaborate sample preparation (elimination of contamination, complete digestion of samples in ultra-pure reagents, separation of the isotopic systems of interest by ion exchange, etc.), low blank levels, the frequent measurement of known standards during unknown runs, and the use of stable, high-resolution, magnetic-sector mass spectrometers under carefully controlled and monitored conditions. Without the lunar samples, we would not have any knowledge about the absolute timing of lunar evolution. It is worth emphasizing in this regard, yet again (as other writers have done), that modern methods for using crater counts as a lunar surface dating tool have only been possible because the crater counts are calibrated against the radiometric ages of the lunar samples.

Radiometric age dating is sometimes cited as the one first-order scientific investigation that absolutely requires sample return. Thus it is very interesting that there have been recent suggestions (e.g., Swindle 2000) that it might be possible to measure radiometric ages robotically via landed rovers. This has been driven by the high cost of returning samples from other large and distant bodies, especially Mars, and the need to provide some reasonable calibration for crater counting as a means of dating planetary surfaces. It is estimated that the obtainable precision for robotic K-Ar ages might be on the order of ±10–20%. Even the most ardent proponents of robotic age dating are very careful to emphasize that such measurements are no substitute for terrestrial laboratory measurements. However, as Swindle (2000) points out, if the cratering history of some martian surfaces (for example) spans 3 Ga or more, then a robotic measurement with a precision of 10–20% to constrain those crater count ages would be useful.

Constraining rough ages of planetary surfaces is one thing, but deciphering the details of early planetary crustal evolution is another matter entirely. Current knowledge of the chronology of lunar crustal evolution required laboratory analysis of returned or meteoritic samples, and this situation will remain true until such time as we actually build manned laboratories on other worlds. The truth of this statement can be appreciated by realizing that a robotically-acquired rock age of 3.8 Ga with an uncertainty 20% could actually be anywhere from 3.0 Ga to 4.6 Ga; on Earth's Moon, that leaves its entire known igneous history unresolved. Moreover, a robotically measured age (on the Moon or anywhere else) will be very difficult to interpret correctly unless enough supplemental information is collected about the sample to reveal whether the age is meaningful, such as:

(1) Did the rock form approximately as an equilibrium system (i.e., is it igneous or metamorphic) or is it a clastic sediment?

(2) Has the rock been severely shocked or otherwise affected subsequent to initial crystallization?

(3) Is the rock an impact melt that might possibly contain (on, e.g., Mars) trapped atmospheric gas?

Answers to these questions require human scientists using detailed petrographic and electron microprobe studies of carefully prepared rock slices or thin sections; for the foreseeable future, it is difficult to envision the ability to do so via robotic means.

Identity of oxygen isotope compositions in the Earth-Moon system. The oxygen isotope compositions of rocks from the Earth and Moon lie within ±0.1–0.2‰ of the same mass-dependent fractionation line (Clayton and Mayeda 1975; Valley et al. 1995; Wiechert et al. 2001); very little else in the solar system lies on that same line (notably, enstatite chondrites and aubrites) (see Fig. 7.14). In a solar system where most solid bodies have their own characteristic oxygen isotopic fingerprints, the Earth and its Moon are alike. This observation is one of the most important constraints on the origin of the Moon, because it inextricably links together the matter from which the Earth and Moon were made. It renders unlikely whole classes of models for lunar origin; e.g., random capture of a passing body (Wood 1986). It also places severe constraints on the current best model, namely collision of a Mars-sized body with proto-Earth: either the colliding body had the same oxygen isotopic composition as Earth (again, aubrites and enstatite chondrites) or else it contributed a very small mass of material (< 3–5%; Wiechert et al. 2001) to the new moon.

Precise oxygen isotopic analysis is difficult under any circumstances. Conventional analysis of bulk material uses a highly reactive compound (BrF_5) to liberate oxygen from oxides and silicates (Clayton and Mayeda 1963). Isotopic fractionation during extraction is a significant problem unless >98% of the theoretical amount of oxygen is recovered. Microanalysis via laser ablation of small spots is done in an even more reactive environment (F_2), and because of the small amounts of gas liberated and the low abundance of ^{17}O, achieving low blanks in the extraction line are critical. Both of these techniques yield analytical precisions that generally are better than 0.2‰. Analyses of oxygen isotopes by secondary ionization mass spectrometry have (so far) significantly worse precision, generally ± ~1.0–1.5‰.

Whether twenty-five years ago or today, the prospect of designing a spacecraft to handle BrF_5 or F_2 robotically is not a concept that any spacecraft engineer would be likely to embrace enthusiastically. And setting aside such hazard issues, the blank and instrumental isotopic fractionation problems associated with bulk or laser fluorination techniques are beyond the capability of any current robotic instrumentation that we are aware of.

Without the laboratory analysis of oxygen isotopes in the Apollo, Luna, or lunar meteorite samples, either 25 years ago or today, we would not possess one of the most important existing constraints on lunar origin.

Depletion of volatiles. It was recognized almost immediately after Apollo 11 that lunar rocks are virtually anhydrous and are depleted in volatile trace elements relative to their terrestrial equivalents (e.g., LSPET 1969; Wasson 1971; Taylor 1975). Moreover, estimates of lunar and terrestrial bulk compositions suggest that the whole Moon is depleted in volatile elements relative to Earth (see review by Drake 1986). It should be noted that both Earth and the Moon are depleted in volatile elements relative to bulk solar (CI chondritic) values (Taylor and Jakes 1974; Ringwood and Kesson 1977), and the HED meteoritic basalts are equally as depleted in volatiles as the Moon. Nevertheless, lunar volatile depletion is commonly used as an argument that the lunar matter underwent high-temperature processing relative to that of the Earth even though some (e.g., Wasson 1971; Drake 1986) have pointed out that the Earth-Moon volatile differences are not entirely compatible with volatility considerations alone. For example, the Moon is less depleted in cesium relative to rubidium than the Earth is, even though cesium is the more volatile of the two elements.

Regardless, volatile elements useful for Earth-Moon comparisons are (with the exception of Na and possibly K) trace elements whose precise determination is beyond the capabilities of robotic analytical instruments even today, much less 25 years ago (e.g., Table 7.1). Even for K the analyses are difficult robotically in the case of the Moon, because K also essentially is a trace element in most lunar rocks. Even worse, Earth-Moon comparisons regarding volatile elements commonly are expressed as volatile/nonvolatile ratios such as K/U, and U in mare basalts is in the abundance range ~0.1–0.9 ppm.

Table 7.1. Typical Abundances of volatile elements in some lunar rock suites.

	Mare Basalts	Monomict Anorthosites	KREEP
Rb	0.28–6.2	0.12–<6	2.4–113
Cs	<0.1	0.015–<0.2	0.15–2.2
Zn	<2–30	0.26–93	2.6–5
Ag (ppb)	—	0.41–1.73	0.44–2.3
As	—	2–4.1 ppb	0.37
K_2O (%)	0.01–0.36	0.07–0.28	0.1–3.6

(Abundances in ppm except where indicated)
Data summarized from Lunar Sourcebook, BVSP

KREEP. Orbital gamma ray measurements by Apollo 15 and 16 demonstrated gross thorium heterogeneities in the lunar crust, with high concentrations in the Imbrium and Procellarum regions. More complete global coverage by the Lunar Prospector mission showed that the lunar surface has a profound geochemical asymmetry, with a huge geochemically-anomalous region centered on Mare Imbrium that shows up not only in Th but also in Sm, inferred from modeling neutron spectrometer data (e.g., Lawrence et al. 2000; Elphic et al. 2000). But it was laboratory studies of regolith fines and breccias brought back by Apollo 12 that first showed the presence of some mafic fragments that are highly enriched in a range of large ion lithophile elements (notably: K, REE, and P, hence the enduring acronym KREEP; Hubbard et al. 1971). It has taken 25+ years of detailed laboratory studies to reveal the nature of KREEP in ways that could never have been achieved from orbit or by *in situ* landers. Moreover, the orbital data cannot distinguish between KREEP and other lunar rocks that also have elevated abundances of some of the same trace elements (e.g., Th). In short, even though orbital measurements demonstrated the existence of a lunar geochemical anomaly, the methods required to decipher the detailed nature of the material making up that anomaly are another matter.

KREEP exists most commonly as a cryptic geochemical component in breccias and glasses. Voluminous trace-element determinations (by activation analysis, isotope dilution) showed that the characteristic KREEP "signature" is one of surprisingly constant elemental ratios despite large ranges in absolute trace element enrichments from rock to rock. Because of these constant elemental ratios, KREEP is interpreted (Taylor and Jakes 1974, Warren and Wasson 1979) to represent a late-stage residual liquid following the extensive fractional crystallization of an ultramafic liquid rather than the result of partial melting. Radiometric age determinations show that KREEP is very old. Relatively pristine KREEP basalts give ages of ~3.85–4.08 Ga (Shih et al. 1992), with formation of the KREEP source from bulk Moon being estimated at ~ 4.4 Ga (review by Nyquist and Shih 1992). For reasons already noted, none of these precise trace element and isotopic studies would be possible robotically. They were and are possible only in terrestrial laboratories and have yielded a set of KREEP properties that, taken together, suggest KREEP originated as a residue of the fractional crystallization that produced the lunar crust and mantle from a lunar magma ocean (e.g., Wood 1972; Lipin 1976; see discussion by Warren and Wasson 1979, and other references therein).

Character of the lunar regolith; effects of impacts and "space weathering." One of the great unknowns prior to Apollo was the nature of the lunar regolith: its origin, and the cause of the surficial darkening that is so apparent in comparison with ejecta near relatively young craters. The Apollo samples revealed a complexity far beyond the capability of any robotic instruments at the time to decipher: detailed studies including bulk chemistry, isotope chemistry, and electron microscopy showed that the regolith is not simply comminuted bed rock but is the result of eons of impact-induced brecciation and melting, vaporization and vapor re-deposition, and grain-surface changes induced by radiation and solar-wind-ion implantation. One early result following the return of the first Apollo samples was the recognition that the lunar regolith is not just highly comminuted lunar rocks, but has a significantly lower albedo and very unusual spectral properties (e.g., McCord and Adams 1973). This unusual regolith is produced from the uppermost lunar surface by processes that came to be loosely called "space weathering." Although it was recognized early that glass welded aggregates in regolith, or *agglutinates*, contain abundant extremely fine-grained metallic iron (e.g., Housley et al. 1973), the iron was believed to be widely dispersed and the complex agglutinate glass itself was responsible for the dominant optical properties of the lunar "soil." As laboratory spectrometers improved, it was later shown that it is actually the very fine-grained fraction of the lunar soil, not agglutinates, that dominates the optical properties and that the optically active component must occur on the surface of soil grains (Pieters et al. 1994). Recent work has taken advantage of technological advances in the field of transmission electron microscopy to show that the smallest iron metal particles in fact are concentrated within very thin rims on the surfaces of individual grains, even grains that intrinsically contain little iron (e.g., feldspar). The grain rims are generally ≤ ~200 nm thick, and the iron particles themselves are typically < 10 nm (Keller and McKay 1997).

These observations led in turn to the conclusion that the nanophase reduced iron is the result of impact vapor condensation or sputter deposition from nearby grains (Keller and McKay 1997), and contributed to the new understanding that the distinctive optical properties of the lunar regolith is in fact due mainly to the nanometer-sized iron metal on soil grains (reviews by Pieters et al. 2000; Hapke 2001). This model has now been extended to include alteration effects on S-asteroid regoliths as well (Pieters et al. 2000; Hapke 2001). The critical point, stated very clearly by Hapke (2001) in his review, was that even the transmission-electron-microscopy studies of 25 years ago showing the existence of tiny iron grains in the lunar agglutinate glass were inadequate to recognize that the iron is a ubiquitous surficial feature derived from deposition onto the grains. Only the modern analytical TEM methods (e.g., Keller and McKay 1997) sufficed to establish this fact. Studies of anticipated returned samples of martian, asteroidal, or other regoliths will require similar levels of detailed analysis to reveal their likely very different origin and evolution. For the foreseeable future, such analysis can occur only in the confines of a terrestrial laboratory.

Common themes. There are two common threads running through the science accomplishments unique to sample return. The first is the need not just for highly sensitive analytical equipment, operating without severe power and mass restrictions, but also that the analyses are able to be conducted under highly controlled conditions on samples that are ideally prepared, and with unlimited possibility for repeatability. Such conditions are necessary for valid science.

The continued importance of petrography to understanding planetary materials is manifest by the considerable effort that is currently underway to develop a robotic petrographic instrument, e.g., by using a laser Raman spectrometer with a small spot size on a rock surface (Haskin et al. 1997; Wang et al. 2003).

4.1.1. From remote sensing. Virtually all of our global information about the Moon comes from remote observations of gravity, topography, surface morphology, and composition. The only exception is the limited *in situ* seismic array emplaced during Apollo, which was able to obtain hints of the structure of the deep interior from random meteorite impacts and "moonquakes" induced by tidal stress. Even such fundamental lunar evolution concepts as the magma ocean would be meaningless without the recognition from remote data that vast areas of the exposed lunar crust consist of feldspathic dominated lithologies. The geologic context of the samples collected by Apollo and Luna have been critical to their scientific value.

That said, by today's standards of remote sensing data, exemplified by that being accomplished in the Mars Program, remote observations of the Moon are extremely primitive. Recall that the Apollo and Luna programs were accomplished in the late 1960's and early 1970's, an era without the benefit of even simple CCD detectors. With the advent of Clementine and Lunar Prospector decades later in the 1990's, the amount of geophysical and compositional data for the Moon improved significantly (any delta above zero is enormous), but the quality and detail of current lunar remotely acquired data is nevertheless embarrassingly below the standards expected of modern sensors.

This irony is mitigated by the fact that a *little* data goes a long way in a void when there is much to discover. We have learned some extremely fundamental information about the character of the Moon from this recent remotely acquired data. The data have allowed a host of additional interesting hypotheses of a more focused nature to be explored: diversity of mare basalts; impact processes; crustal evolution; local and regional geology; etc. Highlights of key lessons learned from remote sensing of the Moon of a global nature are summarized below. Much of the details, especially from Clementine and Lunar Prospector results, are discussed in Chapter 2.

Coverage is essential. If we have learned anything in the decades since Apollo, it is that ignorance is indeed bliss—and often wrong. All our models of lunar evolution were predicated on limited knowledge. Now we must rewrite aspects of early lunar evolution (and the implied consequences for the rest of the terrestrial planets) to accommodate the non-uniform distribution of heat-producing elements, the implications of the giant SPA basin, and the volatile deposits at the poles.

Diversity of approaches is fundamental to a balanced program of planetary exploration. There is no single experiment that will provide data to resolve fundamental problems like the origin of the Moon or even the origin of polar volatiles. Few interesting problems are that simple. But each step, each set of remote observations with different instruments, provides key boundary conditions to the unknown.

Capabilities (spectral resolution) cannot be compromised. Simple instruments provide simple data. We must bear in mind, however, that the worlds we explore are complex. If we measure the world in two "colors," that is all we will see. Technology advancements for most remote sensors have been astounding in expanding capabilities. Many modern developments, for example, are centered on vastly improved spectroscopic analyses. We are not providing the customer (the paying public) with the best value if we do not use the best that have been developed for remote-sensing purposes.

Spatial resolution sets the types of problems addressed. Every remote measurement made of the Moon spawns additional questions that require higher spatial resolution for resolution (the words are related). What is it, where is it, and how did it get to be that way? The cycle continues with each step.

Precision is essential. Bad data (poorly calibrated; low S/N; inadequate digitization, etc.) is usually worse than no data.

4.1.3. From combination of samples and remote sensing (1 + 1 + 1 >> 3). Our knowledge

about the Moon and its evolution has progressed through an intimately interwoven series of events that have involved both remote sensing and analysis of returned samples. Much of this progression was driven by history, politics, and the technical tools available at the time: The sequence of programs started with orbital, *in situ*, and sample return at the dawn of our first venture into space: Ranger, Surveyor, Lunar Orbiter, Apollo, Luna, …. [the big gap]……. Galileo, Clementine, Lunar Prospector. Our perspectives today are intimately colored by the progression and it is difficult, if not impossible, to identify the single source of a specific insight or breakthrough. There is nevertheless no question that we would be ignorant about the body as a whole if we had only the remote data or only the sample data with which to work. The items below are just a few examples of how the whole is *much* greater than the sum of individual parts.

Assessment of principal global properties. The returned Apollo and Luna samples showed that material from the high albedo "highlands" is feldspathic and that material from the dark "maria" is basaltic in overall character. Even the rudimentary remote gamma-ray and X-ray measurements from the Apollo Command Module showed such a pattern to be universal across the equatorial path. This first-order assessment of lunar rock types is so ingrained in our thinking that it is difficult to recognize that it is actually a full marriage between remote measurements and sample return.

Validation of magma ocean concept. The hypothesis of a magma ocean early in lunar evolution came directly from analyses of returned lunar samples. For this hypothesis to be valid, it necessarily must be consistent with global compositional data. First order assessments (feldspathic highlands, basaltic maria) have kept the magma-ocean hypothesis intact, supported by later similar global data. A detailed physical model, however, requires more complete and detailed measurements. The new remote data on asymmetric distribution of KREEP, for example, illustrates how additional data upsets the equilibrium achieved with limited data (both samples and remote sensing). Large issues such as understanding the magma ocean and early planetary evolution depend on a cyclical feedback between remote measurements and sample analyses.

Identification of the character of the lunar regolith. Before spacecraft landed on the Moon, there was definitely what could be called wild speculation about the physical and chemical properties of surface material. That is of course history now, but without the return of the lunar samples, it is likely we would still be formulating hypotheses about the unusual optical properties of lunar soils. Now that we know and understand the soil's unusual characteristics, we are able to use remote-sensing measurements with confidence.

Identification of the "dark mantling material:" a case study. The last Apollo mission was selected to be on the rim of a large basin near unusual terrain that appeared to be mantled by some mysterious "dark mantling material." This mantled area was distinct in remote measurements, but its origin was totally unknown. Even the age of this material relative to surrounding units was debatable. An unplanned discovery during Apollo 17 by geologist-astronaut Jack Schmitt led to a core sample that turned out to be the key. The core was filled with unusual particles of black and orange spheres that had the same properties as the dark mantle (Pieters et al. 1973, 1974). Subsequent detailed laboratory analyses of the samples identified them to be pyroclastic particles with volatile-enriched surfaces and a composition derived from a relatively primitive source (e.g., Heiken and McKay 1977, 1978). This direct link between the remote sensing measurements and the samples allowed a new type of lunar material to be recognized and several areas mapped across the surface.

4.2. Lessons learned

The Moon is the only extraterrestrial body in the solar system to which we have sent the two ultimate exploratory tools: humans, and landed sample return. As a result we understand more about the Moon's complex history than any body in the solar system other than Earth. Even that point can be disputed with some scientists holding the opinion that we understand the Moon better than the Earth. A reader might therefore expect that a concluding chapter in a book dedicated to our modern understanding of the Moon, especially a chapter with the title "Lessons Learned", would be a two-note symphony about the necessity of human exploration and sample return for other solar system bodies as well. In fact, the conclusion is not nearly so simple. Sample return (even robotic) and human exploration are difficult and very costly, even for the Moon, and implementing such programs for every rocky or icy body we choose to explore is simply not practical. It may be expected that the ultimate human destiny is to colonize space, but this destiny might be centuries away. In the short term we will need to rely largely on robotic exploration of our solar system. And landed robotic sample return missions (as opposed to missions such as Genesis or Stardust) will always be more difficult and costly than robotic *in situ* or orbital missions. So although human exploration coupled with sample return unquestionably results in the most profound scientific understanding of a planet, both options must be judiciously exercised and timed. But they *are* essential.

For the contributors to this volume, the "New View" of the Moon that has resulted from the entire range of exploratory tools and cross-disciplinary studies has been eye-opening. It points compellingly to a model for how future planetary exploration should be conducted, because we now know how to do it right: what kinds of missions, and (in hindsight) in what order. The lesson is timely: thirty years after the Viking landings on Mars, there still is not a firm commitment to a sample return mission. Global chemical and mineralogical mapping of the martian surface is only now beginning. And, only now is the kind of logical mission queue planning underway for Mars that the lunar experience suggests is needed.

This section, therefore, is about using the lunar experience as a model for exploration of other rocky bodies in the solar system: the "lessons learned." The first four "lessons" are programmatic ones. That is, they bear on how to explore a silicate body. The last two lessons are profound scientific ones that will apply to most future exploration of our solar system.

4.2.1. Lesson #1: Remote sensing, in situ, and sample return are ALL required. A comprehensive and balanced program for exploring a terrestrial body requires orbital remote sensing, landed in situ measurements, and return of samples to Earth-based laboratories. There is no single "silver bullet" for deciphering a planet. Orbital remote sensing in the absence of landed *in situ* exploration and laboratory analysis of returned samples can provide only the largest scale kinds of visual and spectral data, and which are difficult to interpret unambiguously without detailed knowledge acquired "on the ground." Conversely, returning samples without knowing in advance about the global chemistry and mineralogy runs the risk of those samples being systematically nonrepresentative in the way that the Apollo samples are. The Moon is certainly not unusual in this regard, considering that a landed "sample return" on Earth would miss entirely the most abundant rock on the planet—sea floor basalt. Mars' Tharsis Plateau seems to be a similar large scale anomaly on the surface of that planet. We have to assume that every other rocky body in the solar system is heterogeneous in its own way. Moreover, the need for multiple kinds of analysis holds true at every scale, from the largest (planetary) scale down to the scale of a single crystal. Consider that even in a terrestrial laboratory, no one analytical method, no matter how sensitive, and even if carried out under the most highly controlled conditions, is sufficient for understanding the genesis of a complex rock. Multiple techniques, accompanied by careful petrographic studies, are always required for meaningful laboratory analyses of geologic samples. The same holds true for orbital and landed *in situ* missions as well. The ability to conduct multiple experiments, each serving as checks on the others, is essential.

The proper model, or paradigm, for exploration of a planet or moon is field geologic mapping here on Earth. A first season to a new area consists of reconnaissance, usually based on aerial photography as well as surficial (topographic) maps, in which the main goal is to identify major distinct units and bring back representative samples of those units. Laboratory studies of the samples following the field season results in much better understanding of what the units are, their relationships to each other, and some sense of the geologic history of the field area. Work during the second field season involves detailed mapping of the principal units, determination of essential geologic structures, and detailed sampling of units to reveal any systematic variations (chemical, mineralogic, sedimentary, or otherwise). The process is iterative, and so too should be the exploration of another world. Although much can be learned by single remote sensing missions followed by a single sample return mission, the best information especially for a geologically-complex body requires multiple missions. The lunar experience also suggests that, in order to make most efficient use of sample return, global (orbital) surface chemistry/mineralogy should precede sample return. In the case of Apollo, the difficulties and danger of landing on the far side of the Moon may have resulted in doing things the way we did even if a global chemical map had been available. Nevertheless, it was not known in advance that all of the Apollo landing sites would be within or close to the chemically anomalous Procellarum KREEP Terrane. For the most part, we have done all the right things during our exploration of the Moon, but we have not necessarily done them in the most logical order. Early exploration of the Moon, up to and including Apollo, was driven as much by national political goals as by science goals. This was especially true of the imperative to land humans on the Moon by 1970.

4.2.2. Lesson #2: Expect the unexpected. From start to finish, lunar exploration provided astonishing surprises. The geologic record on the Moon turned out to be nothing like that on Earth, and many predictions about the Moon turned out to be wrong. Immediately after the first Apollo missions, analyses of returned samples revealed that the lunar chemistry, complex regolith, oxidation state, mineralogy, and petrologic diversity were in general unexpected. Later, Clementine and Lunar Prospector revealed the Moon's global dichotomy. It is likely that every planet we explore will hold unique surprises. We already know that Mars differs from both Earth and the Moon in very macroscopic ways: like the Moon it retains the scars of a long bombardment history, suggesting an absence of global crustal recycling, yet is also has an oxidizing atmosphere and did—and perhaps still does—have a hydrosphere. Thus, it seems unlikely that Mars' regolith will retain many of the small-scale signatures of impact processes the way that the lunar regolith does. But, especially considering that no known samples of martian regolith are represented in the meteorite collections, we have little idea what to expect in that regolith other than its apparently highly-oxidized state. Similarly, much remains to be learned about how the martian crust has evolved or what igneous rock types commonly occur there other than basalt and ultramafic rocks. If these rocks are derived from a mantle that has undergone fractionation, it may be very difficult to see back through such events to a primitive martian composition. We know far less about other planets, moons, and asteroids in the solar system. Exploration programs should be planned with no preconceptions. In particular, utilizing only orbital global imaging and hyperspectral studies, even in conjunction with sophisticated *in situ* landers, is likely to provide a very incomplete or even misleading picture of the body. They must be followed up with sample return.

4.2.3. Lesson #3: There is no substitute for the ultimate mobile sensor: a human. Robotic landers are becoming increasingly sophisticated: they can "see" things in ways the human eye cannot. Thus there is no doubt that robotic exploration—being both less costly and less risky then human exploration—is a practical way of exploring more bodies in the solar system and more places on each individual body. Yet if the Apollo experience taught anything, it is that the human ability to recognize interesting features quickly and then independently act to follow up on that information can lead to important discoveries. The Apollo astronauts received extensive geology training prior to their missions, as a result of which they were able to recognize features that were unusual relative to the local surroundings.

One of the most famous examples occurred during the Apollo 15 mission, when a suspiciously white rock was spotted sitting perched atop a small pedestal of soil (Fig. 7.22). The "Genesis Rock", 15415, turned out to be nearly pure sample of the kind of anorthosite that Wood et al. (1970) had predicted might compose the lunar highlands. The astronauts were on the lookout for anorthosite, and immediately recognized 15415 for what it was and collected it. It is quite possible, even likely, that a robotic rover on the surface of the Moon would have collected this sample once the human controllers on earth spotted it from panoramic images; unquestionably, 15415 stands out from it surroundings on account of its white color. However, the near-instantaneous recognition and collection of the sample by the astronauts is nothing like the laborious process of finding it with a rover and then programming the rover to travel to and collect the sample. Humans are completely autonomous and process large amounts of complex information very quickly.

Another example, which illustrates a very different advantage of human explorers, is the story of the "seatbelt basalt," 15016. During a transit by the Apollo 15 crew back to the lunar module, astronaut Scott spotted an unusually vesicular rock along the way. Judging that they would not be given permission to make an unscheduled stop for the sole purpose of picking up a sample, Scott instead stopped to "fix a seatbelt." At the end of the stop, the sample coincidentally ended up inside the LEM (Scott's pocket, actually). Outside of science fiction, robotic rovers do not fib or otherwise evade instructions in order to take advantage of a scientific find. Humans are independent and creative.

Figure 7.22. The Genesis Rock, 15415, sitting on its perch just prior to collection by the A15 crew. [NASA photo AS15-90-12227]

A third famous example is the collection of the orange glass by astronaut Schmitt on Apollo 17. In the process of investigating an interesting boulder, Schmitt noticed that his walking and scuffing of the ground had exposed orange soil underneath a centimeter-thick gray surface layer. Although orange soil subsequently was found to be exposed on the surface nearby, the lighting made it difficult to recognize; Schmitt's attention was drawn to it because of the footprint exposures. Pronounced colors are rare on the Moon, and orange in particular suggested (erroneously, as it turned out) the possibility of oxidized iron. Thus this famous discovery raised considerable excitement. It is debatable whether a robotic rover might have made the discovery at all. The clue was visual color alone, under difficult lighting conditions. A geologist's instinctive attraction to an unexpected and potentially significant color, and his ability to instantly act upon a serendipitous discovery, led to the collection of the orange glass soil.

The point of these anecdotes is not to advocate that all planetary exploration should be done by human missions—that is not realistic. Rather, it is to emphasize the kinds of discoveries that are possible when humans are present: ones that require quick decisions, astute judgment arising from intense training, and the ability and willingness to take quick advantage of serendipity (even if it means sometimes using subterfuge to do so!). Our knowledge of the Moon would be much poorer if we had not sent astronauts. There is no question that similar discoveries will be made when (hopefully) we send astronauts to Mars.

4.2.4. Lesson #4: Science requires exceptionally sensitive instrumentation. Achieving the best science requires the most precise and sensitive analytical instrumentation, whether terrestrial laboratory-based or spacecraft-based. One of the great successes of the Apollo program was NASA's far-sighted investment in building up a new generation of laboratory instrumentation in terrestrial labs prior to the return of the first samples. Given what turned out to be the very low volatile element inventory of lunar samples, the new instrumentation was essential in accurate determinations not only of elemental compositions but also for precise radiometric age determinations. Refined techniques for activation analysis also permitted trace element determinations for very small samples. Had NASA not implemented the instrumentation program ahead of time, much valuable sample and time would have been wasted. As an added bonus, the new laboratory instruments fortuitously were in place when two extremely important meteorites (Allende and Murchison) fell in the same year in which the first lunar samples came back. Because work on these chondritic meteorites entailed analyzing individual components (chondrules, calcium-aluminum-rich inclusions, matrix) rather than bulk meteorite samples, the ability to do highly sensitive and precise measurements on the minute masses of material was perhaps even more critical than for the Apollo samples. We are now at a similar critical juncture in space exploration. The first Mars surface-sample return is likely to bring back 0.5 kg or less of material that must satisfy the demands of the entire world scientific community. The only laboratories that will be able to compete successfully for sample allocations are those suitably equipped to work on very small amounts of material, and few are fortunate enough to be at the leading edge of analytical technology. For some foreseeable missions, no laboratories have suitable equipment for the simple reason that no current analytical technology is capable of making the critical measurements. For example, the Genesis (Discovery class) mission to collect implanted solar wind has as one of its most important goals the determination of the oxygen isotopic composition of the solar wind. Instruments to do this are (as of this writing) under development with NASA funding, but they do not yet exist and when they do only one or two labs will have them. The comet dust that will be collected by the Stardust mission (also Discovery class) can be analyzed by many currently available techniques, but relatively few laboratories are equipped with state-of-the-art equipment that can do so. For both of these missions, the amount of material that will be available to satisfy the world demand is tiny, and only the most capable laboratories will compete successfully for samples. A consortia approach employing the skills of several laboratories might be employed, as was successfully done with

some Apollo samples. Fortunately a new funding program to correct this deficit has been put in place by NASA, but only recently and (arguably) just barely in time. The lesson is that studying astromaterials is intensely instrument-dependent, and there must always be an ongoing program to extend the analytical state-of-the-art and make it available to many investigators.

This lesson does not apply to laboratory instruments alone, although the lesson does not actually derive from past lunar experience. For some time there has been a disconnect between the design of spacecraft instrumentation and the actual scientific requirements for what those instruments need to do to make meaningful measurements. Worse, currently there is no requirement for proposed space-flight instruments to prove their performance capabilities— precision, accuracy, and sensitivity—prior to being selected for flight. The result is that instruments can fly without necessarily being able to return highly meaningful data. Robotic *in situ* landers are going to be one of the chief means of exploring extraterrestrial bodies, and it is critical that the analytical instruments they carry be capable of providing remote sensing data that seriously addresses important problems and can be interpreted confidently by the scientists back on Earth. The counterexample is the experience with the Mars Pathfinder Sojourner rover, whose major-element chemical data remain ambiguous to this day on account of poor spectral resolution and lack of standardization prior to flight. Space-flight analytical instruments must be developed to satisfy the precision and accuracy requirements of specific scientific problems. Perhaps there should be a programmatic requirement that all such instruments undergo double-blind tests on Earth using standard materials in order to establish performance characteristics, before they are selected for flight.

Although analysis of samples in terrestrial laboratories will almost always be superior to what is possible with analysis of samples by spacecraft *in situ* instruments, it must be stressed that such spacecraft instruments can and must be made much better. For example, the laboratory state-of-the-art at the time the Apollo samples were first returned included such capabilities as major elemental and mineralogical analysis at the scale of tens of microns; these capabilities are today feasible for robotic analysis *in situ* on planetary surfaces, but such instruments are not yet flight-certified. These and other analytical capabilities must be constantly improved for flight instruments. It is imperative that the scientific community—especially those who do actual laboratory analysis of planetary materials and thus understand the real state-of-the-art— work closely with instrument developers and communicate challenging but achievable science requirements and technical goals. Together with a more coordinated and vigorous program of instrument development funding and testing than currently exists, this will help close the gap between what is available and what is possible.

Finally, two profound scientific lessons were learned from the lunar experience that are worth repeating here, even though they have been emphasized by so many others in so many places that they are understood by even the most beginning students of planetary geology:

(1) Basaltic volcanism is a widespread phenomenon and a natural consequence of melting and differentiation of a body having chondritic (solar) bulk composition. The magnificent *Basaltic Volcanism Study Project* volume (BVSP 1984) is eloquent testimony to this fact.

(2) Impacts are profoundly important in the evolution of planetary surfaces. Even for planets (e.g., Earth and Venus) where much of the long record of bombardment has been obliterated, we now understand that impacts played an enormous role. For bodies where the record is preserved, notably the Moon, understanding the effects of eons of impacts is essential to interpreting geochemical data. Identifying primary lunar lithologies can be difficult even in the lab. Doing so with robots and remote-sensing instrumentation would be more so, and failing to recognize the hybrid nature of (e.g.) martian rocks on a robotic mission would render certain kinds of measurements, such as age determinations, nearly meaningless.

4. SUMMARY

As the stepping stone, cornerstone, capstone, and keystone of planetary science, exploration of the Moon has provided the foundation, inspiration, and bridge to understanding the planets of our solar system. These concepts are highlighted with countless examples throughout various chapters of this book.

Much of our detailed geochemical understanding of this extraterrestrial neighbor comes, of course, from the samples returned during the Apollo era. The more recent remote measurements of Clementine and Lunar Prospector nevertheless jolted a complacent community. The decades-long neglect of lunar exploration after Apollo stunted what had been an enormous growth of knowledge in the years immediately following the return of the initial samples. The small pulse of new data in the early 1990's made an enormous impact. The "New View" of possible accumulation of water ice at the lunar poles, the unique character of the enormous South Pole-Aitken basin on the farside, and the asymmetric concentration of heat-producing elements are fundamental aspects of lunar science whose importance was unrecognized a decade ago. All three of these paradigm shifts in our understanding beg explanations. They remind us in no uncertain terms, that our knowledge is terribly incomplete.

Clearly, remote observations, *in situ* measurements, and sample return all play a significant role in exploration of planetary bodies. There is ample evidence that to extend exploration beyond simple reconnaissance in fact requires both *in situ* and orbital measurements as well as—and integrated with—analysis of returned samples. There are arguments (and examples) about which sequence is the most efficient. On one hand, detailed remote analyses coupled with targeted *in situ* measurements allow assessment of regional properties and selection of sites that will optimize the value of samples returned from limited targets. On the other hand, the enormous amount of information inherent in the detailed analyses performed on samples in Earth-based laboratories has repeatedly identified broad-scale issues that were unimagined with the more limited remote measurements. A useful strategy could involve two stages. Firstly a simple (and less costly) grab sample from the surface (regolith) that would enable an interpretation of orbital data. The second, more costly phase would target specific localities identified from this combination of orbital and *in situ* measurements.

Like most areas of science, discovery and progress in understanding how planets work is cyclical in nature. As valuable data are first acquired in one area, the information derived lays the foundation for all subsequent questions and the definition of new steps. The danger, of course, is that our ignorance can never tell us what we have missed. We must make the best use of lessons learned.

5. REFERENCES

Adams JB, McCord TB (1971) Alteration of lunar optical properties: Age and composition effects. Science 171:567-571

Adams JB, McCord TB (1971) Optical properties of mineral separates, glass, and anorthositic fragments from Apollo mare samples. Proc Lunar Sci Conf 2:2183-2195

Adams JB, McCord TB (1973) Vitrification in the lunar highlands and identification of Descartes material at the Apollo 16 site. Proc Lunar Sci Conf 4:163-177

Anders E (1977) Chemical compositions of the Moon Earth and eucrite parent body. Phil Trans R Soc A 285:23-40

Anderson JD, Sjogren WL, Schubert G (1996) Galileo gravity results and the internal structure of Io. Science 272: 709-712

Arnold JR (1979) Ice in the lunar polar regions. J Geophys Res 84:5659-5668

Belton MJS, Greeley R, Greenberg R, McEwen AS, Klaasen KP, Head JW III, Pieters CM, Neukum G, Geissler PE, Heffernan C, Breneman H, Anger CD, Carr MH, Davies ME, Fanale FP, Gierasch PJ, Ingersoll AP, Johnson TV, Pilcher CB, Thompson R, Veverka J, Sagan C (1994) Galileo multispectral imaging of the North Polar and Eastern Limb regions of the Moon. Science 264:1112-1115

Belton MJS, Head JW III, Pieters CM, Greeley R, McEwen AS, Neukum G, Klaasen KP, Anger CD, Carr MH, Chapman CR, Davies ME, Fanale FP, Gierasch PJ, Greenberg R, Ingersoll AP, Johnson TB, Paczkowski B, Pilcher CB, Veverka J (1992) Lunar impact basins and crustal heterogeneity: New western limb and farside data from Galileo. Science 255(5044):570-576

Binder AB (1998) Lunar Prospector: Overview. Science 281:1475-1476

Binzel RP, Xu S (1993) Chips off Asteroid 4 Vesta: Evidence for the parent body of basaltic achondrite meteorites. Science 260:186-191

Blewett DT, Hawke BR, Lucey PG, Taylor GJ, Jaumann R, Spudis PD (1995) Remote sensing and geologic studies of the Schiller-Schickard region of the Moon. J Geophys Res 100:16,959-16,997

Blewett DT, Lucey PG, Hawke BR (1997) A comparison of mercurian reflectance and spectral quantities with those of the Moon. Icarus 129:217-231

Borg L, Norman M, Nyquist L, Bogard D, Snyder G, Taylor L, Lindstrom M (1999) Isotopic studies of ferroan anorthosite 62236: A young lunar crustal rock from a light-rare-earth-element-depleted source. Geochim Cosmochim Acta 63:2679-2691

Boss AP, Cameron AGW, Benz W (1990) Tidal disruption of inviscid protoplanets. Lunar Planet Sci XXI:117-118, Lunar and Planetary Institute, Houston

Boss AP, Peale SJ (1986) Dynamical constraints on the origin of the Moon. *In*: Origin of the Moon. Hartmann WK, Phillips RJ, Taylor GJ (eds) Lunar and Planetary Institute, p 59-101

Brearley A, Jones JH (1998) Chondritic meteorites. Rev Mineral 36:3-01–3-398

Britt DT, Pieters CM (1994) Darkening in black and gas-rich ordinary chondrites: The spectral effects of opaque morphology and distribution. Geochim Cosmochim Acta 58:3905-3919

Brophy JG, Basu A (1989) Europium anomalies in mare basalts as a consequence of mafic cumulate fractionation from an initial lunar magma. Proc Lunar Planet Sci Conf 20:25-30

Burbine TH, Buchanon PC, Binzel RP, Bus SJ, Hiroi T, Hinrichs JL, Meibom A, McCoy TJ (2001) Vesta, vestoids, and the howardite, eucrite, diogenite group: Relationships and the origin of spectral differences. Meteor Planet Sci 36:761-781

Bussey DBJ and Spudis PD (1997) Compositional analysis of the Orientale basin using full resolution Clementine data: Some preliminary results. Geophys Res Lett 24:445-448

BVSP (1981) Basaltic Volcanism on the Terrestrial Planets. Pergamon

Cameron AGW (1986) The impact theory for the origin of the Moon. *In*: Origin of the Moon. Hartmann WK, Phillips RJ, Taylor GJ (eds) Lunar and Planetary Institute, p 609-616

Cameron AGW (2000) Higher-resolution simulations of the giant impact. *In*: Origin of the Earth and Moon Canup RM, Righter K (eds) Univ Arizona Press, p 133-144

Cameron AGW (2001) From interstellar gas to the Earth-Moon system. Meteorit Planet Sci 36:9-22

Campbell BA, Campbell DB, Chandler JF, Hine AA, Nolan MC, Perillat PJ (2003) Radar mapping of the lunar poles. Nature 426:137-138

Canup R, Asphaug E (2001) Origin of the Moon in a giant impact near the end of the Earth's formation. Nature 412:708-712

Canup RM (2004) Simulations of a late lunar-forming impact. Icarus 168:433-456

Canup RM, Righter K (2000) Accretion of the terrestrial planets. *In*: Origin of the Earth and Moon Canup RM, Righter K (eds) Univ Arizona Press, p 113-129

Canup RM, Righter K (eds) (2000) Origin of the Earth and Moon. Univ Arizona Press

Carlson RW, Lugmair GW (2000) Timescales of planetesimal formation and differentiation based on extinct and extant radioisotopes. *In*: Origin of the Earth and Moon Canup RM, Righter K (eds) Univ Arizona Press, p 25-44

Clayton RN, Mayeda TK (1963) The use of bromine pentafluoride in the extraction of oxygen from oxides and silicates for isotopic analysis. Geochim Cosmochim Acta 27:43-52

Clayton RN, Mayeda TK (1975) Genetic relations between the Moon and meteorites. Proc Lunar Sci Conf 6:1761-1769

Clayton RN, Mayeda TK, Rubin AE (1984) Oxygen isotopic compositions of enstatite chondrites and aubrites. J Geophys Res 8:C245-C249

Drake MJ (1986) Is lunar bulk material similar to Earth's mantle? *In*: Origin of the Moon. Hartmann WK, Phillips RJ, Taylor GJ (eds) Lunar and Planetary Institute, p 105-124

Drake MJ, Righter K (2002) Determining the composition of the Earth. Nature 416:39-43

Durisen RH, Gingold RA (1986) Numerical simulations of fission. *In*: Origin of the Moon. Hartmann WK, Phillips RJ, Taylor GJ (eds) Lunar and Planetary Institute, p 487-498

Elphic RC, Lawrence DJ, Feldman WC, Barraclough BL, Maurice S, Binder AB, Lucey PG (2000) Lunar rare earth element distribution and ramifications for FeO and TiO_2: Lunar Prospector neutron spectrometer observations. J Geophys Res 105:20,333-20,345

Eriksson KA (1977) Tidal deposits from the Archaean Moodies Group, Barberton Mountain Land, South Africa. Sediment Geol 18:257-281

Eriksson KA, Simpson EL (2000) Quantifying the oldest tidal record: The 3.2 Ga Moodies Group, Barberton Greenstone Belt, South Africa. Geology 28:831-834

Eshleman VR, Parks GA (1999) No ice on the Moon? Science 285:531

Feldman W, Maurice S, Lawrence DJ, Little RC, Lawson SL, Gasnault O, Wiens RC, Barraclough BL, Elphic RC, Prettyman TH, Steinberg JT, Binder AB (2001) Evidence for water ice near the lunar poles. J Geophys Res 106: 23,231-23,251

Feldman WC, Lawrence DJ, Elphic RC, Barraclough BL, Maurice S, Genetay I, Binder AB (2000) Polar hydrogen deposits on the Moon. J Geophys Res 105:4175-4195

Feldman WC, Maurice S, Binder AB, Barraclough BL, Elphic RC, Lawrence DJ (1998) Fluxes of Fast and Epithermal Neutrons from Lunar Prospector: Evidence for water ice at the lunar poles. Science 281:1496-1500

Gaffey MJ (1997) Surface lithologic heterogeneity of Asteroid 4 Vesta. Icarus 127:130-157

Gaffey MJ, Burbine TH, Binzel RP (1993) Asteroid spectroscopy, progress and perspectives, Meteoritics 28:161-187

Gast PW (1972) The chemical composition and structure of the Moon. Moon 5:121-148

Gonzalez G (1998) Spectroscopic analyses of the parent stars of extrasolar planetary system candidates. Astronom Astrophys 334:221-229

Halliday AN, Lee D-C, Jacobsen SB (2000) Tungsten isotopes, the timing of metal-silicate fractionation, and the origin of the Earth and Moon. *In*: Origin of the Earth and Moon. Canup RM, Righter K (eds) Univ Arizona Press, p 45-62

Hapke B (2001) Space weathering from Mercury to the asteroid belt. J Geophys Res 106:10,039-10,073

Hapke B, Cassidy W, Wells EN (1975) Effects of vapor-phase deposition processes on the optical, chemical and magnetic properties of the lunar regolith. The Moon 13:339-353

Harmon JK, Slade MA, Velez RA, Crespo A, Dryer MJ, Johnson JM (1994) Radar mapping of Mercury's polar anomalies. Nature 369:213-215

Hartmann WK, Ryder G, Dones L, Grinspoon D (2000) The time-dependent intense bombardment of the primordial Earth/Moon system. *In*: Origin of the Earth and Moon. Canup RM, Righter K (eds) Univ Arizona Press, p 493-512

Haskin LA (1998) The Imbrium impact event and the thorium distribution at the lunar highlands surface. J Geophys Res 103:1679-1689

Haskin LA, Gillis JJ, Korotev TL, Jolliff BL (2000) The materials of the lunar Procellarum KREEP Terrane: A synthesis of data from geomorphological mapping remote sensing and sample analyses. J Geophys Res 105:20403-20415

Haskin LA, Wang A, Rockow KM, Jolliff BL, Korotev RL, Viscupic KM (1997) Raman spectroscopy for mineral identification and quantification for *in situ* planetary surface analysis: A point-count method. Journal of Geophysical Research. J Geophys Res 103:19,293-19,306

Head JW (1976) Lunar volcanism in space and time. Rev Geophys Space Phys 14:265-300

Heiken G, McKay DS (1977) A model for eruption behavior of a volcanic vent in eastern Mare Serenitatis. Proc Lunar Sci Conf 8:3243-3255

Heiken G, McKay DS (1978) Petrology of a sequence of pyroclastic rocks from the valley of Taurus-Littrow (Apollo 17 landing site). Proc Lunar Sci Conf 9:1933-1943

Heiken G, Vaniman D, French BM (eds) (1991) Lunar Sourcebook, Cambridge University Press

Hood LL, Jones JH (1987) Geophysical constraints on lunar bulk composition and structure: A reassessment. J Geophys Res 92:E396-E410

Hood LL, Mitchell DL, Lin RP, Acuña MH, Binder AB (1999) Initial measurement of the lunar induced magnetic dipole moment using Lunar Prospector magnetometer data. Geophys Res Lett 26:2327

Hood LL, Zuber MT (2000) Recent refinements in geophysical constraints on lunar origin and evolution. I *In*: Origin of the Earth and Moon. Canup RM, Righter K (eds) Univ Arizona Press, p 397-412

Housley RM, Grant RW, Paton NE (1973) Origin and characteristics of excess Fe metal in lunar glass welded aggregates. Proc Lunar Sci Conf 4:2737-2749

Humayun M, Cassen P (2000) Processes determining the volatile abundances of the meteorites and terrestrial planets. *In*: Origin of the Earth and Moon. Canup RM, Righter K (eds) Univ Arizona Press, p 3-23

Humayun M, Clayton RN (1995) Potassium isotope chemistry: Implications of volatile element depletion. Geochim Cosmochim Acta 59:2131-2148

Javoy M (1995) The integral enstatite chondrite model of the Earth. Geophys Res Lett 22:2219-2222

Jeanloz R, Mitchell DL, Sprague AL, de Pater I (1995) Evidence for a basalt-free surface on Mercury and implications for internal heat. Science 268:1455-1457

Jolliff BL, Gillis JJ (2002) Lunar crustal and bulk composition. Workshop on The Moon Beyond 2002: Next Steps in Lunar Science and Exploration, #3056, Taos, New Mexico

Jolliff BL, Gillis JJ, Haskin LA (2000) Thorium mass balance for the Moon from Lunar Prospector and sample data: Implications for thermal evolution. Lunar Planet Sci 31:1763

Jolliff BL, Gillis JJ, Haskin LA, Korotev RL, Wieczorek MA (2000) Major lunar crustal terranes: Surface expressions and crust-mantle origins. J Geophys Res 105:4197-4216

Jones JH, Delano JW (1989) A three-component model for the bulk composition of the Moon. Geochim Cosmochim Acta 53:513-527

Jones JH, Palme H (2000) Geochemical constraints on the origin of the Earth and Moon. *In*: Origin of the Earth and Moon. Canup RM, Righter K (eds) Univ Arizona Press, p 197-216

Kahn A, Mosegaard K (2001) New information on the deep lunar interior from an inversion of lunar free oscillation periods. Geophys Res Lett 28:1791-1794

Keller L, McKay D (1997) The nature and origin of rims on lunar soil grains. Geochim Cosmochim Acta 61:2331-2340

Keller LP, McKay DS (1993) Discovery of vapor deposits in the lunar regolith. Science 261:1305-1307

Khan A, Mosegaard K, Rasmussen KL (2000) A new seismic velocity model for the Moon from a Monte Carlo inversion of the Apollo lunar seismic data. Geophys Res Lett 27:1591-1594

Kirk RL, Stevenson DJ (1989) The competition between thermal contraction and differentiation in the stress history of the Moon. J Geophys Res 94:12133-12144

Kokubo E, Canup RM, Ida S (2000) Lunar Accretion from an Impact-Generated Disk. *In*: Origin of the Earth and Moon. Canup RM, Righter K (eds) Univ Arizona Press, p 145-163

Konopliv AS, Binder AB, Hood LL, Kucinskas AB, Sjogren WL, Williams JG (1998) Improved gravity field of the Moon from Lunar Prospector. Science 281:1476-1480

Korotev RL (2000) The great lunar hot spot and the composition and origin of the Apollo mafic ("LKFM") impact-melt breccias. J Geophys Res 105:4317-4345.

Korotev RL, Haskin LA (1988) Europium mass balance in polymict samples and implications for plutonic rocks of the lunar crust. Geochim Cosmochim Acta 52:1795-1813

Kortenkamp SJ, Kobuko E, Weidenschilling SJ (2000) Formation of planetary embryos. *In*: Origin of the Earth and Moon. Canup RM, Righter K (eds) Univ Arizona Press, p 85-100

Kuskov OL (1997) Constitution of the Moon: 4. Composition of the mantle from seismic data. Phys Earth Planet Inter 102:239-257

Kuskov OL, Kronrod VA (1998) Constitution of the Moon: 5 Constraints on composition density temperature and radius of a core. Phys Earth Planet Inter 107:285-306

Kuskov OL, Kronrod VA (2000) Resemblance and difference between the constitution of the Moon and Io. Planet Space Sci 48:717-726

Laskar J, Joutel F, Robutel P (1993) Stabilization of the Earth's orbit by the Moon. Nature 361:615-617

Lawrence DJ, Feldman WC, Barraclough BL, Binder AB, Elphic RC, Maurice S, Thomsen DR (1998) Global elemental maps of the Moon: The Lunar Prospector gamma-ray spectrometer. Science 281:1484-1489

Lawrence DJ, Feldman WC, Barraclough BL, Binder AB, Elphic RC, Maurice S, Miller MC, Prettyman TH (2000) Thorium abundances on the lunar surface. J Geophys Res 105:20,307-20,331

Lawrence DJ, Feldman WC, Elphic RC, Little RC, Prettyman TH, Maurice S, Lucey PG, Binder AB (2002) Iron abundances on the lunar surface as measured by the Lunar Prospector gamma-ray and neutron spectrometers. J Geophys Res 107(E12):DOI:10.1029/2001JE001530

Lee D-C, Halliday AN, Snyder GA, Taylor LA (1997) Age and origin of the Moon. Science 278:1098-1103

Lipin BR (1976) The origin of Fra Mauro basalts. Lunar Sci VII:495-497

Longhi J (1992) Experimental petrology and petrogenesis of mare volcanics. Geochim Cosmochim Acta 56:2235-2251

Longhi J (2003) A new view of lunar ferroan anorthosites: Post magma ocean petrogenesis. J Geophys Res 108:5083, doi:10.1029/2002JE001941

LSPET (1969) Summary of the Apollo 11 Lunar Science Conference. Science 167:449-451

Lucey PG, Taylor GJ, Hawke BR, Spudis PD (1998) FeO and TiO$_2$ concentrations in the South Pole-Aitken basin: Implications for mantle composition and basin formation. J Geophys Res 103:3701-3708

Lucey PG, Taylor GJ, Malaret E (1995) Abundance and distribution of iron on the Moon. Science 268:1150-1153

Lugmair G (1987) The age of the lunar crust: 60025—Methuselah's legacy. Lunar Planet Sci XVIII:584-585

Lugmair GW, Shukolyukov A (1998) Early solar system timescales according to ^{53}Mn/^{53}Cr systematics. Geochim Cosmochim Acta 62:2863-2886

McCord TB, Adams JB (1973) Progress in remote optical analysis of lunar surface composition. The Moon 7:453-474

McCord TB, Adams JB, Johnson TB (1970) Asteroid Vesta: spectral reflectivity and compositional implications. Science 168:1445-1447

McKay GA (1982) Partitioning of REE between olivine, plagioclase, and synthetic melts: Implications for the origin of lunar anorthosite. Lunar Planet Sci XIII:493-494

McKay GA, Wagstaff J, Le L (1990) REE distribution coefficients for pigeonite: Constraints on the origin of the mare basalt europium anomaly. *In:* Workshop on Lunar Volcanic Glasses: Scientific and Resource Potential. Delano JW, Heiken GH (eds) LPI Tech Rpt. 90-02:48-49

Mitchell DL, de Pater I (1994) Microwave imaging of Mercury's thermal emission at wavelengths from 03-205 cm. Icarus 110:2-32

Morbidelli A, Chambers J, Lunine JI, Petit JM, Robert F, Valsecchi GB, Cyr KE (2000) Source regions and time scales for the delivery of water to the Earth. Meteor Planet Sci 35:1309-1320

Mueller S, Taylor GJ, Phillips R (1988) Lunar composition: A geophysical and petrological synthesis. J Geophys Res 93:6338-6352

Nakamura Y (1983) Seismic velocity structure of the lunar mantle. J Geophys Res 88:677-686

Neal CR (2001) Interior of the Moon: The presence of garnet in the primitive deep lunar mantle. J Geophys Res 106: 27,865-27,855

Neal CR, Taylor LA (1992) Petrogenesis of mare basalts: a record of lunar volcanism. Geochim Cosmochim Acta 56: 2177-2211

Neumann GA, Zuber MT, Smith DE, Lemoine FG (1996) The lunar crust: Global structure and signature of major basins. J Geophys Res101:16481-16863

Newsom HE (1986) Constraints on the origin of the Moon from the abundance of molybdenum and other siderophile elements. *In*: Origin of the Moon. Hartmann WK, Phillips RJ, Taylor GJ (eds) Lunar and Planet. Inst., p 203-229

Newsom HE, Taylor SR (1989) Geochemical implications of the formation of the Moon by a single giant impact. Nature 338:29-34

Noble SK (2004) Turning Rocks into Regolith: The Physical and Optical Consequences of Space Weathering in the Inner Solar System. Brown University, Providence, RI

Noble SK, Pieters CM (2003) Space Weathering on Mercury: Implications for Remote Sensing. Solar System Res 37: 31-35

Noble SK, Pieters CM, Keller LP (2003) The optical properties of nanophase iron: Investigation of a space weathering analog. Lunar Planet Sci 34:1172

Noble SK, Pieters CM, Taylor LA, Morris RV, Allen CC, McKay DS, Keller LP (2001) The optical properties of the finest fraction if lunar soil: Implications for space weathering,. Meteorit Planet Sci 36:31-42

Norman MD, Borg LE, Nyquist LE and Bogard DD (2003) Chronology, geochemistry, and petrology of a ferroan noritic anorthosite clast from Descartes breccia 67215: Clues to the age, origin, structure, and impact history of the lunar crust. Meteor Planet Sci 38:645–661

Nozette S, Lichtenberg CL, Spudis PD, Bonner R, Ort W, Malaret E, Robins M, Shoemaker EM (1996) The Clementine Bistatic Radar Experiment. Science 274:1495-1498

Nozette S, Rustan P, Pleasance LP, Horan D, Regeon MP, Shoemaker EM, Spudis PD, Acton CH, Baker DN, E. BJ, Buratti BJ, Corson MP, Davies ME, Duxbury TC, Eliason EM, Jakosky BM, Kordas JF, Lewis IT, Lichtenberg CL, Lucey PG, Malaret E, Massie MA, Resnick JH, Rollins CJ, Park HS, McEwen AS, Priest RE, Pieters CM, Reisse RA, Robinson MS, Smith DE, Sorenson TC, Vorder Breugge RW, Zuber MT (1994) The Clementine mission to the Moon: Scientific overview. Science 266:1835-1839

Nozette S, Spudis PD, Robinson MS, Bussey DBJ, Lichtenberg C, Bonner R (2001) Integration of lunar polar remote-sensing data sets: Evidence for ice at the lunar south pole. J Geophys Res 106:23,253-23266

Nyquist LE (1977) Lunar Rb-Sr chronology. Phys Chem Earth 10:103-142

Nyquist LE, Shih C-Y (1992) The isotopic record of lunar volcanism. Geochim Cosmochim Acta 56:2213-2234

O'Hara MJ (2000) Flood basalts basalt floods or topless bushvelds? Lunar petrogenesis revisited. J Pet 41:1545-1651

O'Neill HSC (1991) The origin of the Moon and the Earth-A chemical model. Geochim Cosmochim Acta 55:1135-1157

Paige DA (2000) Mars exploration strategies: Forget about sample return! *In*: Concepts and Approaches for Mars Exploration. Lunar Planet Inst Contrib 1062:243-244

Papanastassiou DP, Wasserburg GJ (1971) Lunar chronology and evolution from Rb-Sr studies of Apollo 11 and 12 samples. Earth Planet Sci Lett 11:37-62

Papike JJ (1998) Comparative planetary mineralogy: Chemistry of melt-derived pyroxene, feldspar, and olivine. Rev Mineral 36:7-1–7-11

Papike JJ, Bence AE (1978) Lunar mare basalts versus Mid-Ocean Ridge basalts: Planetary constraints. Geophys Res Lett 5:803-806

Papike JJ, Ryder G, Shearer CK (1998) Lunar Samples. Rev Mineral 36:5-1–5-234

Pieters C, Taylor L, Noble S, Keller L, Hapke B, Morris R, Allen C, McKay D, Wentworth S (2000) Space weathering on asteroids: Resolving a mystery with Lunar samples. Meteorit Planet Sci 35:1101-1107

Pieters CM, Fischer EM, Rode O, Basu A (1993) Optical effects of space weathering: the role of the finest fraction. J Geophys Res 98(E11):20,817-20,824

Pieters CM, Head JW, III, Gaddis L, Jolliff BL, Duke M (2001) Rock types of the South Pole-Aitken basin and extent of basaltic volcanism. J Geophys Res 106:28,001-28,022

Pieters CM, McCord TB, Charette MP, Adams JB (1974) Lunar surface: Identification of the mantling material in the Apollo 17 soil samples. Science 183:1191-1194

Pieters CM, McCord TB, Zisk SH, Adams JB (1973) Lunar black spots and the nature of Apollo 17 landing area. J Geophys Res 78:5867-5875

Pieters CM, McFadden LA (1994) Meteorite and asteroid reflectance spectroscopy: clues to early solar system processes. Ann Rev Earth Planet Sci 22:457-497

Pieters CM, Taylor LA (2003) Systematic global mixing and melting in lunar soil evolution. Geophys Res Lett 30:2048 doi:10.1029/2003GL018212

Pieters CM, Tompkins S (1999) Tsiolkovsky crater: A window into crustal processes on the lunar farside. J Geophys Res 104:21935-21949

Pieters CM, Tompkins S, Head JW, Hess PC (1997) Mineralogy of the mafic anomaly in the South Pole-Aitken Basin: Implications for excavation of the lunar mantle. Geophys Res Lett 24:1903-1906

Premo WR, Tatsumoto M, Misawa K, Nakamura N, Kita NI (1999) Pb-isotopic systematics of lunar highland rocks (>39 Ga): Constraints on early lunar evolution. *In*: Planetary Geology and Geochemistry. Snyder GA, Neal CR, Ernst WG (eds) Geological Society America, p 207-240

Pritchard ME, Stevenson DJ (2000) Thermal aspects of a lunar origin by giant impact. *In*: Origin of the Earth and Moon. Canup RM, Righter K (eds) Univ Arizona Press, p 179-196

Righter K (2002) Does the Moon have a metallic core? Constraints from giant-impact modeling and siderophile elements. Icarus 158:1-13

Righter K, Drake MJ (1996) Core formation in Earth's Moon, Mars and Vesta. Icarus 124:513-529

Righter K, Walker RJ, Warren PH (2000) The origin and significance of highly siderophile elements in the lunar and terrestrial mantles. *In*: Origin of the Earth and Moon. Canup RM, Righter K (eds) Univ Arizona Press, p 291-322

Ringwood AE (1986) Terrestrial origin of the Moon. Nature 322:323-328

Ringwood AE (1990) Earliest history of the Earth-Moon system. *In*: Origin of the Earth. Newsom HE, Jones JH (eds) Oxford Univ Press, p 101-134

Ringwood AE, Kesson SE (1977) Basaltic magmatism and the bulk composition of the Moon, II. Siderophile and volatile elements in Moon, Earth, and chondrites: Implications for lunar origin. The Moon 16:425-464

Ruzicka A, Snyder GA, Taylor LA (1998) Giant impact and fission hypotheses for the origin of the Moon: A critical review of some geochemical evidence. Int Geol Rev 40:851-864

Ruzicka A, Snyder GA, Taylor LA (2001) Comparative geochemistry of basalts from the Moon Earth HED asteroid and Mars: Implications for the origin of the Moon. Geochim Cosmochimica Acta 65:979-997

Ryder G (1994) Coincidence in time of the Imbrium basin impact and Apollo 15 KREEP volcanic flows: The case for impact-induced melting. Geol Soc Am Spec Paper 293:11-18

Ryder G. (1990) Lunar samples, lunar accretion and the early bombardment of the Moon. Eos, Transactions AGU 71: 313-323

Schultz PH, Spudis PD (1983) Beginning and end of lunar mare volcanism. Nature 302:233-236

Shearer CK, Floss C (2000) Evolution of the Moon's mantle and crust as reflected in trace-element microbeam studies of lunar magmatism. *In*: Origin of the Earth and Moon. Canup RM, Righter K (eds) Univ Arizona Press, p 339-359

Shearer CK, Papike JJ (1989) Is plagioclase removal responsible for the negative Eu anomaly in the source regions of mare basalts? Geochim Cosmochim Acta 53:3331-3336

Shih C-Y, Nyquist LE, Bansal BM, Wiesmann H (1992) Rb-Sr and Sm-Nd chronology of an Apollo 17 KREEP basalt. Earth Planet Sci Lett 108:203-215

Short NM (1975) Planetary Geology. Prentice Hall

Simpson RA, Tyler GL (1999) Reanalysis of Clementine bistatic radar from the lunar South Pole. J Geophys Res 104: 3845-3862

Snyder GA, Borg LE, Nyquist LE, Taylor LA (2000) Chronology and isotopic constraints on lunar evolution. *In*: Origin of the Earth and Moon. Canup RM, Righter K (eds) Univ Arizona Press, p 361-395

Solomon SC, Chaiken J (1976) Thermal expansion and thermal stress in the Moon and terrestrial planets: clues to early thermal history. Proc Lunar Sci Conf 7:3229-3243

Sprague AL, Nash DB, Witteborn FC, Cruikshank DP (1997) Mercury's feldspar connection - Mid-IR measurements suggest plagioclase. Adv Space Res 19:1507

Sprague L, Roush TL (1998) Comparison of laboratory emission spectra with Mercury telescopic data. Icarus 133: 174-183

Spudis PD, Guest JE (1988) Stratigraphy and geologic history of Mercury. *In*: Mercury. Vilas F, Chapman C, Matthews M (eds) Univ Arizona Press, p 152-164

Spudis PD, Reisse RA, Gillis JJ (1994) Ancient multi-ring basins on the Moon revealed by Clementine laser altimetry. Science 266:1848-1851

Stacey NJS et al. (1997) Arecibo radar mapping of the lunar poles: A search for ice deposits. Science 276:1527-1530

Stern SA (1999) The lunar atmosphere: History status current problems and context. Rev Geophys 37:453-491

Stevenson DJ (1987) Origin of the Moon-the collision hypothesis. Ann Rev Earth Planet Sci 15:271-315

Steward GR (2000) Outstanding questions for the giant impact hypothesis. *In*: Origin of the Earth and Moon. Canup RM, Righter K (eds) Univ Arizona Press, p 217-223

Swindle TD (2000) (2000) *In situ* noble-gas based chronology on Mars. Concepts And Approaches For Mars Exploration. Lunar and Planetary Institute Contribution 1062:294-295

Taylor LA, Pieters CM, Morris RV, Keller LP, McKay DS (2001) Lunar mare soils: Space weathering and the major effects of surface-correlated nanophase Fe. J Geophys Res 106:27,985-28,000

Taylor SR (1975) Lunar Science: A Post-Apollo View; Scientific Results and Insights from the Lunar Samples. Pergamon Press

Taylor SR (1982) Planetary Science: A Lunar Perspective. Lunar and Planetary Institute

Taylor SR (1987) The unique lunar composition and its bearing on the origin of the Moon. Geochim Cosmochim Acta 51:1297-1309

Taylor SR (1998) Destiny or Chance: Our Solar System and its Place in the Cosmos. Cambridge University Press

Taylor SR (1999) On the difficulties of forming Earth-like planets. Meteor Planet Sci 34 317-329

Taylor SR (2001a) Solar System Evolution: A New Perspective (2nd ed). Cambridge University Press

Taylor SR (2001b) Does the lunar composition retain a memory from the early solar nebula? Meteor Planet Sci 36: 1567-1569

Taylor SR (2001c) Flood basalts basalt floods or topless Bushvelds? Lunar petrogenesis revisited: A critical comment. J Pet 42:1219-1220

Taylor SR, Esat TM (1996) Geochemical constraints on the origin and evolution of the Moon. *In*: Earth Processes: Reading the Isotopic Code. Basu A, Hart SR (eds) AGU Geophysical Monograph 95:33-46

Taylor SR, Jakes P (1974) The geochemical evolution of the Moon. Proc Lunar Sci Conf 5:1287-1305

Tholen DJ (1989) Asteroid taxonomic classifications. *In*: Asteroids II. Binzel RP, Gehrels T, Matthews MD (eds) Univ. of Arizona Press, p 1139-1150

Thomas PC, Binzel RP, Gaffey MJ, Storrs AD, Wells EN, Zellner BH (1997) Impact excavation on asteroid 4 Vesta. Hubble Space Telescope Results:1492-1495

Tompkins S, Pieters CM (1999) Mineralogy of the lunar crust: Results from Clementine. Meteorit Planet Sci 34:25-41

Turkevich A (1971) Comparison of the analytical results from the Surveyor, Apollo, and Luna missions. Proc Lunar Sci Conf 2:1209-1215

Urey HC (1965) Meteorites and the Moon. Science 147:1262-1265

Vilas F, Chapman C, Matthews M (eds) (1988) Mercury. University of Arizona Press

Walker D (1983) Lunar and terrestrial crust formation. Proc Lunar Planet Sci Conf 14, in J Geophys Res 88:B17-B25

Walter MJ, Newsom HE, Ertel W, Holzheid A (2000) Siderophile elements in the Earth and Moon: Metal/silicate partitioning and implications for core formation. *In*: Origin of the Earth and Moon. Canup RM, Righter K (eds) Univ Arizona Press, p 265-289

Wang A, Haskin LA, Lane AL, Wdowiak TJ, Squyres SW, Wilson RJ, Hovland LE, Manatt KS, Raouf N and Smith CD (2003) Development of the Mars microbeam Raman Spectrometer. J Geophys Res 108(E1):5005, doi:10.1029/2002JE001902

Wänke H, Dreibus G (1986) Geochemical evidence for the formation of the Moon by impact-induced fission of the proto-Earth. *In*: Origin of the Moon. Hartmann WK, Phillips RJ, Taylor GJ (eds) Lunar and Planetary Institute, p 649-672

Ward PD, Brownlee D (2000) Rare Earth: Why Complex Life Is Uncommon in the Universe. Springer-Verlag

Warell J, Blewett DT (2004) Properties of the Hermean regolith: V. New optical reflectance spectra, comparison with lunar anorthosites, and mineralogical modelling. Icarus 168:257-276

Warren PH (1992) Inheritance of silicate differentiation during lunar origin by giant impact. Earth Planet Sci Lett 112: 101-116.

Warren, PH (2001) Early lunar crustal genesis: The ferroan anorthosite epsilon-neodymium paradox. Meteor Planet Sci 36:A219

Warren PH, Wasson JT (1979) The origin of KREEP. Rev Geophys Space Phys 17:73-88

Wasson JT (1971) Volatile elements on the Earth and Moon. Earth Planet Sci Lett 11:219-225

Wiechert U, Halliday AN, Lee D-C, Snyder GA, Taylor LA, Rumble D (2001) Oxygen isotopes and the Moon-forming giant impact. Science 294:345-348

Wieczorek MA, Phillips RJ (1997) The structure and compensation of the lunar highland crust. J Geophys Res 102: 10,933-10,943

Wieczorek MA, Phillips RJ (1999) Lunar multiring basins and the cratering process. Icarus 139:246-259

Wieczorek MA, Phillips RJ (2000) The Procellarum KREEP terrane: Implications for mare volcanism and lunar evolution. J Geophys Res 105:20,417-20,430

Wieczorek MA, Zuber MT (2001) A Serenitatis origin for the Imbrium grooves and South Pole-Aitken thorium anomaly. J Geophys Res 106:27,825-27,840

Wieczorek MA, Zuber MT (2001) The composition and origin of the lunar crust: Constraints from central peaks and crustal thickness modeling. Geophys Res Lett 28:4023-4026

Wieczorek MA, Zuber MT, Phillips RJ (2001) The role of magma buoyancy on the eruption of lunar basalts. Earth Planet Sci Lett 185:71-83

Wilhelms DE (1976) Mercurian volcanism questioned. Icarus 28:551-558

Wilhelms DE (1987) The Geologic History of the Moon. US Geol Surv Prof Paper 1348, Washington, DC

Williams DE, Pollard D (2000) Earth-Moon interactions: Implications for terrestrial climate and life. *In*: Origin of the Earth and Moon. Canup RM, Righter K (eds) Univ Arizona Press, p 513-525

Williams GE (1989) Late Precambrian tidal rhythmites in South Australia and the history of the Earth's rotation. J Geol Soc London 146:97-111

Williams GE (2000) Geological constraints on the Precambrian history of Earth's rotation and the Moon's orbit. Rev Geophys 38:37-59

Wood JA (1972) Fragments of terra rock in the Apollo 12 soil samples and a structural model of the Moon. Icarus 16: 462-501

Wood JA (1986) Moon over Mauna Loa. *In*: Origin of the Moon. Hartmann WK, Phillips RJ, Taylor GJ (eds) Lunar and Planetary Institute, p 17-55

SUBJECT INDEX

Note: in most instances where a topic is discussed in a section spanning several pages, only the first page is listed.

1529-6466/06/0060-0ind$05.00

DOI: 10.2138/rmg.2006.60.ind